ABC PROTEINS

From Bacteria to Man

ABC PROTEINS

From Bacteria to Man

Editors:
I. Barry Holland,
Susan P.C. Cole,
Karl Kuchler
and
Christopher F. Higgins

An imprint of Elsevier Science

Amsterdam Boston London New York Oxford Paris
San Diego San Francisco Singapore Sydney Tokyo

This book is printed on acid-free paper.

Academic Press
An Imprint of Elsevier Science
84 Theobald's Road, London WC1X 8RR, UK
http://www.academicpress.com

Academic Press
An Imprint of Elsevier Science
525 B Street, Suite 1900, San Diego, California 92101-4495, USA
http://www.academicpress.com

ISBN 0-12-352551-9

Library of Congress Catalog Number: 2002103782

A catalogue record for this book is available from the British Library

Cover acknowledgments
Left: Scanning electron micrograph of bacterial cells.

Centre: An *Arabidopsis* plasma membrane-localized multispecific ATP-binding cassette transporter, AtMRP4, extrudes glutathione-conjugates and confers tolerance to organic xenobiotics. Courtesy of Mingsheng Peng, Sixue Chen, Rocío Sánchez-Fernández, Honglan Gu, Shun Liang and Philip A. Rea, Plant Science Institute, Department of Biology, University of Pennsylvania.

Right: The human figure illustrates the localization of some important ABC proteins implicated in human disease (for more details see Chapter 3).
Commissioned from Barking Dog Art, Gloucestershire.

Designed by Elaine C Monaghan
(elaine.monaghan@virgin.net)

Typeset by Charon Tec Pvt. Ltd, Chennai, India
Printed and bound in Hong Kong by RDC

02 03 04 05 06 07 RD 9 8 7 6 5 4 3 2 1

Contents

Contributors

Rando Allikmets
Columbia University
Departments of Ophthalmology and Pathology
Eye Institute Research Rm 715
630 West 168th Street
New York, NY 10032
USA
Email: rla22@columbia.edu

Shlomo Almashanu
Howard Hughes Medical Institute and
The Institute of Genetic Medicine
Johns Hopkins University School of Medicine
802 PCTB, 725 N. Wolfe St
Baltimore, MD 21205
USA
Email: shlomo@jhmi.edu

Frances M. Ashcroft
University Laboratory of Physiology
University of Oxford
Parks Road
Oxford OX1 3PT
UK
Email: frances.ashcroft@physiol.ox.ac.uk

Susan E. Bates
Cancer Therapeutics Branch
National Cancer Institute
National Institutes of Health
Bethesda, MD 20892
USA
Email: sebates@helix.nih.gov

Bettina E. Bauer
Institute of Medical Biochemistry
Department of Molecular Genetics
University and BioCenter of Vienna
Dr. Bohr-Gasse 9/2, A-1030
Vienna
Austria
Email: bettina@mol.univie.ac.at

Houssain Benabdelhak
Institut de Génétique et Microbiologie
Université de Paris-Sud
91405 Orsay Cedex
France
Email: benh@igmors.u-psud.fr

Mark A. Blight
Institut de Génétique et Microbiologie
Université de Paris-Sud
91405 Orsay Cedex
France
Email: mark.blight@igmors.u-psud.fr

Piet Borst
The Netherlands Cancer Institute
Division of Molecular Biology and
Center for Biomedical Genetics
Plesmanlaan 121
1066 CX Amsterdam
The Netherlands
Email: PBorst@nki.nl

Richard Callaghan
Nuffield Department of Clinical Laboratory
Sciences
John Radcliffe Hospital
University of Oxford
Oxford OX3 9DU
UK
Email: richard.callaghan@ndcls.ox.ac.uk

Olivier Chambenoit
Centre d'Immunologie de Marseille
Luminy INSERM-CNRS-Université de la
Mediterranée, Case 906
Parc Scientifique de Luminy
13288 Marseille Cedex 09
France
Email: chambeno@ciml.univ-mrs.fr

Geoffrey A. Chang
The Scripps Research Institute
Department of Molecular Biology
10550 N. Torrey Pines Road
La Jolla, CA 92037
USA
Email: gchang@scripps.edu

Sixue Chen
Plant Science Institute
Department of Biology
University of Pennsylvania
Philadelphia, PA 19104-6018
USA
Email: sixue@sas.upenn.edu

Giovanna Chimini
Centre d'Immunologie de Marseille
Luminy INSERM-CNRS-Université de la
Mediterranée, Case 906
Parc Scientifique de Luminy
13288 Marseille Cedex 09
France
Email: chimini@ciml.univ-mrs.fr

Susan P.C. Cole
Division of Cancer Biology and Genetics
Queen's University Cancer Research Institute
Kingston, Ontario
Canada K7L 3N6
Email: coles@post.queensu.ca

Yunhai Cui
Division of Tumor Biochemistry
Deutsches Krebsforschungszentrum
D-69120 Heidelberg
Germany
Email: y.cui@dkfz-heidelberg.de

Elie Dassa
Unité de Programmation Moléculaire et
Toxicologie génétique,
CNRS URA 1444 Département des
Biotechnologies
Institut Pasteur
25 Rue du Dr Roux
75724 Paris Cedex 15
France
Email: elidassa@pasteur.fr

Michael Dean
Human Genetics Section
Laboratory of Genomic Diversity
Bld. 560, Rm 21-18
National Cancer Institute-Frederick
Frederick, MD 21702
USA
Email: dean@ncifcrf.gov

Roger G. Deeley
Division of Cancer Biology and Genetics
Queen's University Cancer Research Institute
Kingston, Ontario
Canada K7L 3N6
Email: deeleyr@post.queensu.ca

Andrea de Lima Pimenta
Institut de Génétique et Microbiologie
Université de Paris-Sud
91405 Orsay Cedex
France
Email: pimenta@igmors.u-psud.fr

Christopher Fielding
Cardiovascular Research Institute
University of California
San Francisco
CA 94143-0130
USA
Email: cfield@itsa.ucsf.edu

Markus Geisler
Laboratory for Molecular Physiology
Institute of Plant Sciences
University of Zurich
Zolikerstrasse 107
8008 Zurich
Switzerland
Email: markus.geisler@bota.unine.ch

Martina Gentzsch
Mayo Foundation and
S.C. Johnson Medical Research Center
Mayo Clinic
Scottsdale, AZ 85259
USA
Email: gentzsch.martina@mayo.edu

John W. Hanrahan
Department of Physiology
McGill University
Montréal, Québec
Canada, H3G 1Y6
Email: hanrahan@physio.mcgill.ca

Christopher F. Higgins
MRC Clinical Sciences Centre
Faculty of Medicine
Imperial College
Hammersmith Hospital Campus
Du Cane Rd
London W12 0NN
UK
Email: christopher.higgins@csc.mrc.ac.uk

I. Barry Holland
Institut de Génétique et Microbiologie
Université de Paris-Sud
91405 Orsay Cedex
France
Email: Holland@igmors.u-psud.fr

Dietrich Keppler
Division of Tumor Biochemistry
Deutsches Krebsforschungszentrum
D-69120 Heidelberg
Germany
Email: d.keppler@dkfz-heidelberg.de

Ian D. Kerr
School of Biomedical Sciences
Medical School
University of Nottingham
Nottingham NG7 2UH
UK
Email: kerrid@cancer.org.uk

Gyula Kispal
Institute of Biochemistry
Medical Faculty
University of Pecs
Szigeti ut 12
7624 Pecs
Hungary
Email: kispal@apacs.pote.hu

Markus Klein
Laboratory for Molecular Physiology
Institute of Plant Sciences
University of Zurich
Zolikerstrasse 107
8008 Zurich
Switzerland
Email: markus.klein@bota.unine.ch

Jörg König
Division of Tumor Biochemistry
Deutsches Krebsforschungszentrum
D-69120 Heidelberg
Germany
Email: j.koenig@dkfz.de

Wil N. Konings
Department of Microbiology
Groningen Biomolecular Sciences and
Biotechnology Institute
University of Groningen
NL-9751 NN Haren
The Netherlands
Email: W.N.Konings@biol.rug.nl

Karl Kuchler
Institute of Medical Biochemistry
Department of Molecular Genetics
University and BioCenter of Vienna
Dr. Bohr-Gasse 9/2
A-1030 Vienna
Austria
Email: kaku@mol.univie.ac.at

Brigitte Lankat-Buttgereit
Institut für Physiologische Chemie
Klinikum, Universität Marburg
Karl-von-Frisch-Str. 1
D-35043 Marburg
Germany
Email: IanKatbu@Mailer.uni-Marburg.de

Danielle Légaré
Centre de Recherche en Infectiologie and
Département de Biologie médicale
Université Laval
2705, boul. Laurier, Sainte-Foy, QC
Canada G1V 4G2
Email: danielle.legare@crchul.ulaval.ca

Roland Lill
Institut für Zytobiologie und
Zytopathologie der Philipps-Universität
Marburg, Robert-Koch-Str. 5
35033 Marburg
Germany
Email: Lill@mailer.uni-marburg.de

Kenneth J. Linton
MRC Clinical Sciences Centre
Faculty of Medicine
Imperial College
Hammersmith Hospital Campus
Du Cane Rd,
London W12 0NN
UK
Email: kenneth.linton@csc.mrc.ac.uk

Enrico Martinoia
Laboratory for Molecular Physiology
Institute of Plant Sciences
University of Zurich
Zolikerstrasse 107
8008 Zurich
Switzerland
Email: enrico.martinoia@botinst.unizh.ch

Michinori Matsuo
University Laboratory of Physiology
University of Oxford
Parks Road
Oxford OX1 3PT
UK
Email: michi.matsuo@physiol.ox.ac.uk

Anne T. Nies
Division of Tumor Biochemistry
Deutsches Krebsforschungszentrum
D-69120 Heidelberg
Germany
Email: a.nies@dkfz.de

Ronald Oude Elferink
Academic Medical Center
Laboratory for Experimental Hepatology
Meibergdreef 9
1105 AZ Amsterdam
The Netherlands
Email: r.p.oude-elferink@amc.uva.nl

Marc Ouellette
Centre de Recherche en Infectiologie and
Département de Biologie médicale
Université Laval
2705, boul. Laurier
Sainte-Foy, QC
Canada G1V 4G2
Email: marc.ouellette@crchul.ulaval.ca

Mingsheng Peng
Plant Science Institute
Department of Biology
University of Pennsylvania
Philadelphia, PA 19104-6018
USA
Email: pengm@sas.upenn.edu

Gerrit J. Poelarends
College of Pharmacy, PHR 4.116
University of Texas at Austin
Austin, Texas 78712
USA
Email: gjpoelarends@mail.utexas.edu

Bert Poolman
Department of Biochemistry
Groningen Biomolecular Sciences and Biotechnology Institute
University of Groningen
Nijenborgh 4
9747 AG Groningen
The Netherlands
Email: b.poolman@chem.rug.nl

Philip A. Rea
Plant Science Institute
Department of Biology,
University of Pennsylvania
Philadelphia, PA 19104-6018
USA
Email: parea@sas.upenn.edu

Glen Reid
The Netherlands Cancer Institute
Division of Molecular Biology and
Center for Biomedical Genetics
Plesmanlaan 121
1066 CX Amsterdam
The Netherlands
Email: g.reid@nki.nl

John R. Riordan
Mayo Foundation and
S.C. Johnson Medical Research Center, Mayo Clinic
Scottsdale, AZ 85259
USA
Email: riordan@mayo.edu

Mark F. Rosenberg
Department of Biomolecular Sciences
UMIST
Manchester M60 1QD
UK
Email: mark.rosenberg@umist.ac.uk

Christopher B. Roth
The Scripps Research Institute
10550 N. Torrey Pines Road
La Jolla, CA 92037
USA
Email: cbroth@scripps.edu

Jean-Marie Ruysschaert
Labaratoire de Chimie Physique des Macromolécules aux Interfaces
Université Libre de Bruxelles
B1050 Brussels
Belgium
Email: jmruyss@ulb.ac.be

Timothy Ryder
University Laboratory of Physiology
University of Oxford
Parks Road
Oxford OX1 3PT
UK
Email: Timothy.Ryder@physiol.ox.ac.uk

Andrey Rzhetsky
Department of Medical Informatics and
Columbia Genome Center
Columbia University
New York, NY 10032
USA
Email: andrey@genome2.cpmc.columbia.edu

Tohru Saeki
Laboratory of Molecular Nutrition
Department of BRC
Faculty of Agricultural Science
Kyoto Prefectural University
Nakaragi, Shimogamo
Sakyo-ku, Kyoto 606-8522
Japan
Email: tsaeki@nki.nl

Rocío Sánchez-Fernández
Plant Science Institute
Department of Biology
University of Pennsylvania
Philadelphia, PA 19104-6018
USA
Email: rocio.sanchez_fernandez@sungene.de

Balázs Sarkadi
National Institute of Haematology and Immunology
Membrane Research Group of
the Hungarian Academy of Sciences
Daróczi ut 24
1113 Budapest
Hungary
Email: sarkadi@biomembrane.hu

Lutz Schmitt
Institute Biochemistry
Johann Wolfgang Goethe University
Frankfurt, Marie-Curie Strasse 9
60439 Frankfurt
Germany
Email: lschmitt@em.uni-frankfurt.de

Erwin Schneider
Hunboldt Universitaet zu Berlin
Institut fuer Biologie/Bakterienphysiologie
Chaussee Str. 117
D-10115 Berlin
Germany
Email: Erwin.schneider@rz.hu-berlin.de

Christoph Schüller
Institute of Medical Biochemistry
Department of Molecular Genetics
University and BioCenter of Vienna
Dr. Bohr-Gasse 9/2
A-1030 Vienna
Austria
Email: christoph@mol.univie.ac.at

Frances J. Sharom
Department of Chemistry and Biochemistry
University of Guelph
Guelph, ON N1G 2W1
Canada
Email: sharom@chembio.uoguelph.ca

Robert Tampé
Institut für Biochemie, Biozentrum
Universität Frankfurt, Marie-Curie-Str. 9
D-69047 Frankfurt a.M.
Germany
Email: tampe@em.uni-frankfurt.de

Gábor E. Tusnády
Institute of Enzymology
Hungarian Academy of Sciences
Karolina ut 29
1113 Budapest
Hungary
Email: tusi@enzim.hu

Kazumitsu Ueda
Laboratory of Biochemistry
Division Applied Life Sciences
Kyoto University
Kitashirakawa, Sakyo
Kyoto 606-8502
Japan
Email: uedak@kais.kyoto-u.ac.jp

David Valle
Howard Hughes Medical Institute and
The Institute of Genetic Medicine
Johns Hopkins University School of Medicine
802 PCTB, 725 N. Wolfe St
Baltimore, MD 21205
USA
Email: dvalle@jhmi.edu

Tiemen van der Heide
Department of Biochemistry
Groningen Biomolecular Sciences and
Biotechnology Institute
University of Groningen
Nijenborgh 4
9747 AG Groningen
The Netherlands
Email: Heidet@chem.rug.nl

Gerrit van Meer
Academic Medical Center
Laboratory of Cell Biology and Histology
Meibergdreef 15
1105 AZ Amsterdam
The Netherlands
Email: G.vanMeer@AMC.UvA.NL

Hendrik W. van Veen
Department of Pharmacology
University of Cambridge
Tennis Court Road
Cambridge CB2 1QJ
UK
Email: hwv20@cam.ac.uk

András Váradi
Institute of Enzymology
Hungarian Academy of Sciences
Karolina ut 29
1113 Budapest
Hungary
Email: varadi@enzim.hu

Koen H.G. Verschueren
H-2-Netherlands Cancer Institute
Plesman Laan 121
Amsterdam 1066 CX
The Netherlands
Email: Koenv@nki.nl

Catherine Vigano
Labaratoire de Chimie Physique des
Macromolécules aux Interfaces
Université Libre de Bruxelles
B1050 Brussels
Belgium
Email: catherine.vigano@ulb.ac.be

Peter Wielinga
The Netherlands Cancer Institute
Division of Molecular Biology and
Center for Biomedical Genetics
Plesmanlaan 121
1066 CX Amsterdam
The Netherlands
Email: p.wielinga@nki.nl

Anthony J. Wilkinson
Structural Biology Laboratory
Department of Chemistry
University of York
York YO10 5DD
UK
Email: ajw@yorvic.york.ac.uk

Joanne Young
EMBL
Meyerhof Str1 Postfach 10
10.2209
Heidelberg 6900
Germany
Email: joanne.young@embl.heidelberg.de

Noam Zelcer
The Netherlands Cancer Institute
Division of Molecular Biology and
Center for Biomedical Genetics
Plesmanlaan 121
1066 CX Amsterdam
The Netherlands
Email: zelcer@nki.nl

The Editors

I. Barry Holland

Susan P.C. Cole

Karl Kuchler

Christopher F. Higgins

ABC Transporters: an Introduction and Overview

Christopher F. Higgins and Kenneth J. Linton

Introduction

ABC (ATP-binding cassette) transporters comprise one of the largest of all paralogous protein families. ABC transporters, as far as we know, are found in all cells of all species from the lowliest microbe to man. Almost 5% of the entire *Escherichia coli* genome encodes components of ABC transporters (Linton and Higgins, 1998); 28 are encoded by the yeast genome (Schüller *et al.*, Chapter 14) and around 50 by the human genome (Dean *et al.*, Chapter 3). The diverse scientific backgrounds of the authors of this volume is one illustration of how ABC transporters play central roles in every kind of physiological system.

It is intriguing that genes encoding ABC transporters played a key role in the development of genetics in the first half of the twentieth century – the *Drosophila* eye colour genes encode ABC transporters. However, ABC transporters only became 'sexy' in recent years following the realization that they are of considerable medical, industrial and economic importance. ABC transporters play a major role in resistance to antibiotics and antifungal agents (Poelarends *et al.*, Chapter 12; Bauer *et al.*, Chapter 15), the resistance of *Plasmodium* to antimalarials (Ouellette and Légaré, Chapter 16), and plants to herbicides (Rea *et al.*, Chapter 17). In man, mutations in genes encoding ABC transporters underlie diverse genetic diseases including cystic fibrosis (Hanrahan *et al.*, Chapter 29), Tangier disease (Chimini *et al.*, Chapter 23), Dubin–Johnson syndrome (König *et al.*, Chapter 20), sight disorders (Allikmets, Chapter 28) and adrenoleukodystrophy (Almashanu and Valle; Chapter 24). The Hollywood film 'Lorenzo's Oil', starring Nick Nolte, is about a child with a mutation in an ABC transporter gene. Tens of thousands of cancer deaths each year are a consequence of overexpression of ABC transporters which confer resistance to chemotherapeutic drugs (Bates, Chapter 18).

Transport across Cell Membranes

The cell membrane is not simply a passive barrier, but provides the major interface between the cytoplasm of the cell and the extracellular milieu. Without the selective permeability properties of the cell membrane, the cytoplasm could not be maintained at a different chemical composition from the external environment, and life as we know it could not exist. Membrane-bound proteins determine these selective permeability properties and around 10% of all genes encode proteins predicted to have a role in membrane permeability. Many of these proteins are ion channels and facilitators – proteins which permit the passive (energy-independent) movement of a solute across the membrane down its electrochemical gradient. However, the accumulation of solute against a concentration gradient requires the input of energy: active transport. Evolution has devised only a limited number of 'templates' for the active transport of solutes across cellular membranes, and each of these 'templates' has been adapted to many types of solute and hence diverse physiological needs. ABC transporters form the largest and most physiologically diverse of these transport families.

It is frequently forgotten that ABC transporters were first identified and characterized in bacteria, and that the majority of fundamental principles relating to their structure and function emerged from basic studies of bacterial physiology rather than with any view to medical or commercial goals, particularly work on the histidine and maltose systems from the laboratories of Giovanna Ferro-Luzzi Ames, Maurice Hofnung and Hiroshi Nikaido. It became apparent in the 1970s, from the study of nutrient uptake by *E. coli* and *Salmonella typhimurium*, that these species have multiple, independent transport systems, which fall into a small number of classes: the phosphotransferase systems; secondary, shock-insensitive transporters energized by the electrochemical gradient; and primary, shock-sensitive systems energized directly by the hydrolysis of ATP (Berger and Heppel, 1974). This latter class of transporter was characterized by a sensitivity to osmotic shock which was due to the loss of an essential substrate-binding protein from the periplasm (see Wilkinson and Verschueren, Chapter 10). Thus, these transporters were also termed periplasmic binding protein-dependent transport systems (see Holland, Chapter 8; Schneider, Chapter 9). Each transporter was found to require a different periplasmic binding protein with a defined substrate-binding specificity. Genetic and biochemical studies showed that at least one additional protein, located in the cell membrane, was required for each transporter.

In 1982 the first complete sequence of the genes encoding one of these binding protein-dependent transporters, the histidine transporter of *S. typhimurium*, was published (Higgins *et al.*, 1982). In addition to the periplasmic substrate-binding protein (HisJ), the transporter was found to have three membrane-associated components (HisQ, M and P). Shortly thereafter, the sequence of a component of the *E. coli* maltose transporter (MalK) was determined and its similarity to HisP led to the suggestion that such transporters might have evolved from a common ancestor (Gilson *et al.*, 1982). When the sequences of other homologues were derived it was noted that these highly conserved proteins included a consensus nucleotide-binding motif (Higgins *et al.*, 1985), similar to that previously identified in ATP synthase, myosin and adenylate kinase – the so-called 'Walker' A and B motifs (Walker *et al.*, 1982). This led to the suggestion that these domains couple ATP hydrolysis to the transport process – experimental evidence that these domains bind ATP (Hobson *et al.*, 1984; Higgins *et al.*, 1985) and couple ATP hydrolysis to transport (Bishop *et al.*, 1989; Mimmack *et al.*, 1989) soon followed.

Only in 1986 was it was recognized that these ATP-binding subunits define a large superfamily of transport proteins, and the now-accepted organization of four core membrane-associated domains proposed (Higgins *et al.*, 1986). Intriguingly, it was also noted that a few of these ATP-binding domains were 'orphans' which appeared to have been recruited to couple ATPase activity to other biological processes such as DNA repair. At least one of these bacterial ABC transporters, HlyB (Holland *et al.*, Chapter 11), was found to be an exporter (Felmlee *et al.*, 1985), in contrast to the previously characterized uptake systems. The identification of the first eukaryotic ABC transporter, the human multidrug resistance P-glycoprotein (Pgp), and its subsequent detailed characterization by the groups of Michael Gottesman, Phillipe Gros, Igor Roninson and Vic Ling, was a landmark in the field (Chen *et al.*, 1986; Gerlach *et al.*, 1986; Gros *et al.*, 1986). Pgp was correctly predicted to be an exporter, rather than an importer like the binding protein-dependent ABC transporters, based on its similarity to HlyB. In 1990 the name ABC (ATP-binding cassette) transporter was coined (Hyde *et al.*, 1990), cementing the general recognition of the importance of this functionally diverse family of proteins. Subsequent genome sequencing and phylogenetic analyses have confirmed that the NBDs of ABC transporters, whether of prokaryotic or eukaryotic origin and irrespective of their substrate specificity and physiological roles, have a common evolutionary origin (Dassa, Chapter 1).

SUBSTRATE SPECIFICITY AND PHYSIOLOGICAL ROLES

The number of ABC transporters differs widely between species. Bacteria that live in diverse environments and can adapt to variable external conditions have large numbers – the *E. coli* chromosome encodes around 70 ABC transporters occupying 4.8% of the genome. In contrast, bacterial species that occupy a more restricted growth niche (for example, mammalian pathogenic species) have far fewer ABC transporters. Eukaryotic cells generally have fewer ABC transporters, presumably because other more sophisticated mechanisms for moving solutes across membranes have evolved. For example, cells which can absorb nutrients by endocytosis do not require ABC transporters for the uptake of nutrients, although ABC transporters still play an important role in the uptake of solutes into organelles (Almashanu and Valle, Chapter 24; Lill and Kispal, Chapter 25). Indeed, to date, there is no example of a eukaryotic ABC transporter with a role in uptake into the cytoplasm – all are exporters. Nevertheless, ABC transporters in eukaryotic cells have adapted to serve a wide variety of specialized roles, for example in antigen presentation (Lankat-Buttgereit and Tampé, Chapter 26), leukotriene transport (Deeley and Cole, Chapter 19), detoxification (König *et al.*, Chapter 20), lipid transport (Borst *et al.*, Chapter 22), and many others (Borst *et al.*, Chapter 21).

It is a general, though not universal, rule that each ABC transporter is specific for its own particular substrate(s) or group of chemically related substrates. In itself this is not unusual – we are familiar with enzymes exhibiting such 'lock and key' specificity. What is, however, most remarkable is that despite this there is an ABC transporter for essentially every class of molecule imaginable: there are ABC transporters specific for small molecules, large molecules, highly charged molecules and highly hydrophobic molecules; for inorganic ions, sugars, amino acids, proteins, and complex polysaccharides. Furthermore, even though most ABC transporters are highly substrate specific, some are multispecific – the oligopeptide transporter handles essentially all di- and tripeptides (Tame *et al.*, 1994) – while others (the bacterial LmrA and human Pgp) have an extremely broad specificity for hydrophobic compounds (van Veen and Callaghan, Chapter 5). It is remarkable that a family of evolutionarily and presumably structurally related transport proteins can accommodate such an enormous diversity of substrates. This diversity is reflected in the diversity of physiological roles played by ABC transporters in the cell – there is hardly a biological molecule that is not moved around by an ABC transporter.

It is not yet clear what determines the directionality of the vectorial transport process (see below) or its substrate specificity (van Veen and Callaghan, Chapter 5). In consequence, sequence analysis alone is a poor guide to predicting function and many misassignments have been made based on this approach – the ABC transporter most closely related to a newly identified family member is not necessarily its functional counterpart. As an example, the closest homologue to the human multidrug transporter Pgp is the human MDR3 protein, which is about 80% identical, yet MDR3 transports phosphatidylcholine (van Helvoort *et al.*, 1996). Conversely, the bacterial LmrA protein is only about 40% identical to Pgp yet transports a similar range of hydrophobic drugs to Pgp (van Veen *et al.*, 1998) (see also Chapter 12).

The vast majority of ABC transporters that have been characterized are active transporters. However, it is important to note that there are exceptions. Several proteins which from their sequences look to all the world like a typical ABC transporter have diversified in function. Thus, the cystic fibrosis protein, CFTR, is not an active transporter but a chloride channel (Hanrahan *et al.*, Chapter 29) – to date the only ABC protein found to be a channel rather than an active transporter. Elegant studies of the structure and function of CFTR by the groups of Jack Riordan, Michael Welsh and others have provided key insights into the structure–function of the ABC family, and influenced our thinking about issues such as the difference between a channel and a transporter. Similarly, the sulfonylurea receptor (SUR) has no known transport activity but imposes nucleotide regulation on a heterologous potassium channel (Aguilar-Bryan *et al.*, 1995; Matsuo *et al.*, Chapter 27): a similar regulatory role has been ascribed to the bacterial Sap/Trk system of *S. typhimurium/E. coli* (Parra-Lopez *et al.*, 1994). ABC transporters appear to influence physiological processes by indirect mechanisms, reflecting a function additional to their primary transporter activity – the influence of Pgp on cell volume, for example (Valverde *et al.*, 1992). So, although it is often possible to anticipate the function of an ABC transporter from sequence analysis, the adaptable nature of ABC transporters ensures that exceptions will confound our expectations – until our understanding of how structure relates to function is very much more advanced, there will be no substitute for experimental verification.

Domain Organization and Structure

ABC transporters consist of four core domains, two transmembrane domains (TMDs) and two nucleotide-binding domains (NBDs). The four core domains of some bacterial ABC transporters are often encoded as separate polypeptides. In other transporters the domains are fused in one of a number of ways into multidomain polypeptides **(Figure 1)**. In cases in which one of the NBDs or TMDs appears to be absent, one of the remaining domains functions as a homodimer to maintain the full complement.

The two TMDs span the membrane multiple times via predicted α-helices. The recent crystal structure of MsbA demonstrates that, at least for this transporter, these predictions reflect reality

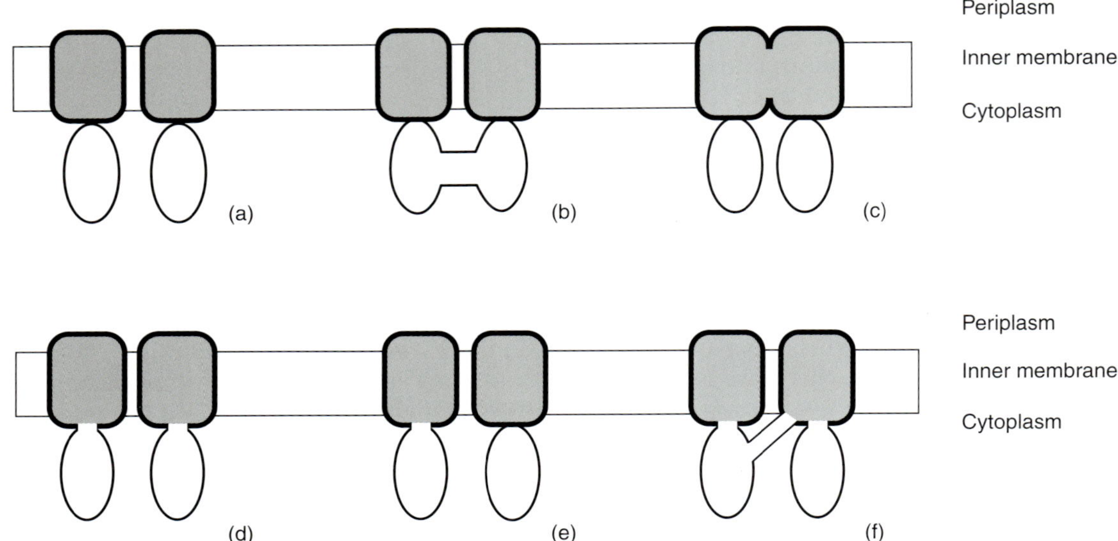

Figure 1. Domain organization of ABC transporters. The typical ABC transporter has two membrane-associated domains (TMDs) depicted by shaded squares. The two ATP-binding domains (NBDs or ABC domains) are depicted by ovals at the intracellular face of the membrane. The domains can be fused in any one of a number of ways: (a) they can be encoded as four separate polypeptides (as for the oligopeptide transporter OppBCDF); (b) fused NBDs (as for the ribose transporter RbsA); (c) fused TMDs (as for the iron-chelate transporter FhuCB); (d) one ABC fused to one NBD, with the hybrid protein functioning as a homodimer (as for LmrA); (e) one TMD fused to one NBD, with the other TMD and NBD as separate polypeptides (as for YhiGHI of E. coli*); (f) all four domains fused into a single polypeptide, often found in eukaryotic ABC transporters. Reproduced by permission of Blackwell Publishing from Linton and Higgins (1998).*

(Chang and Roth, 2001; Chapter 7). Typically, there are six membrane-spanning α-helices per domain, a total of twelve per transporter (Váradi *et al.*, Chapter 2). Some variation on this formula may reflect auxiliary functions for the α-helices, such as roles in membrane insertion or regulation of transporter activity. The 10–12 transmembrane α-helices are probably the minimum necessary to generate a pathway (pore) of an appropriate size to accommodate the range of substrates known to be handled by ABC transporters. It is clear for ABC transporters which have been purified and reconstituted, such as Pgp and CFTR, that the TMDs alone form the transmembrane pathway, although for other ABC transporters such as some of the protein translocators, additional proteins may also be involved (Holland *et al.*, Chapter 11). Suggestions that the NBDs might span the membrane and contribute to the pore or pathway through the membrane are unlikely based on recent biochemical (Blott *et al.*, 1999) and structural data (Chang and Roth, 2001; Locher *et al.*, 2002). Importantly, the TMDs are the primary determinants of substrate specificity, through specific substrate-binding sites, although auxiliary proteins such as the periplasmic binding proteins can also play a role in some transporters (Wilkinson and Verschueren, Chapter 10). For several ABC transporters, for example Pgp, there is known to be more than one substrate-binding site per transporter, although the amino acid residues involved in generating these sites, and the location of these sites on the protein, are still unclear (van Veen and Callaghan, Chapter 5).

The ATP- or nucleotide-binding domains (NBDs) are hydrophilic and interact with intracellular loops of the TMDs at the cytoplasmic face of the membrane. The NBDs consist of a core of 215 or so amino acids, the conserved ABC domain by which these transporters are defined (Higgins *et al.*, 1986). This domain includes the Walker A and B and so-called 'ABC-signature' motifs. It is important to emphasize that it is the conservation of this entire domain which is important in defining and delimiting the family; other ATP-binding proteins which possess the Walker A and Walker B motifs (e.g. the ATP-binding domains of the arsenate transporter ArsA) but do not have the other conserved sequences are therefore not

ABC transporters. The crystal structures of several isolated ABC domains have now been determined, all of which (not unexpectedly) show very similar folds (Linton *et al.*, Chapter 4). However, the structures differ significantly at the subunit interface when dimers are present. Consequently, the faces of the domains involved in such interactions are still unclear.

Although there has been much debate, it now seems clear from biochemical, genetic and structural data that the basic functional unit of an ABC transporter is a monomer of the four core domains (Linton *et al.*, Chapter 4). That is to say, the four core domains are both necessary and sufficient to mediate transmembrane translocation of solute. This does not, of course, exclude the requirement of auxiliary proteins or domains for some ABC transporters, or the possibility that two or more molecules may form part of a complex and influence each other's activities – this is the case for the sulfonylurea receptor SUR (Matsuo *et al.*, Chapter 27).

For many ABC transporters auxiliary domains have been recruited to add on specific functions. In bacterial uptake systems the periplasmic binding proteins (PBPs) bind substrate outside the cell membrane and deliver it to the membrane-associated transport complex (Wilkinson and Verschueren, Chapter 10). The PBPs have two distinct but related functions: (a) To impart high affinity and specificity – binding protein-dependent transporters often show remarkably high affinity and specificity compared with transporters which lack such a component. It is possible that the flexibility in the TMDs inherent in their transport function precludes the 'lock and key' fit required for such high affinity and specificity, such that an additional component is required. (b) To confer directionality. There is a perfect correlation between the presence of a PBP and solute uptake, and between the absence of a PBP and solute export. Although this does not prove that the PBP determines directionality, the fact that interaction of the PBP with the transporter at the outside of the cell can trigger ATP hydrolysis at the cytoplasmic face of the membrane strongly implies such a role (Davidson *et al.*, 1992).

Other ABC transporters have additional intrinsic domains which add functions to the core transporter, such as the R-domain of CFTR, which plays a role in regulation of CFTR activity by phosphorylation (Hanrahan *et al.*, Chapter 29), and the N-terminal domain of MRP1, which consists of five transmembrane α-helices but can be deleted with no loss of activity and whose function remains unclear.

STRUCTURES AND MECHANISMS OF TRANSPORT

Despite a plethora of elegant biochemical studies from many laboratories, for example that of Michael Gottesman (see Sharom, Chapter 6), which have led to insights into how ABC transporters work, structural data are required in order to understand fully the mechanisms of transport. Such data have been extremely difficult to obtain. Crystal structures of several NBDs, mainly prokaryotic, have provided insights into ATP binding and hydrolysis (see Linton *et al.*, Chapter 4). Two approaches to the structures of complete ABC transporters have recently provided intriguing insights, the X-ray structure to high resolution of the bacterial MsbA and BtuCD proteins (Chang and Roth, 2001; Locher *et al.*, 2002; Chapter 7), and electron cryomicroscopy structures of the multidrug resistance Pgp and other ABC transporters (see Linton *et al.*, Chapter 4). These different structures share many similarities but also some important differences, which have yet to be resolved. These differences may be a result of different methodologies or inappropriate dimer interfaces forming during purification and crystallography, or may be a reflection of the major changes in conformation which occur during the transport cycle (Rosenberg *et al.*, 2001).

Based on studies of both pro- and eukaryotic ABC transporters, and the not unreasonable assumption that there are mechanistic similarities, an overall picture is emerging of the mechanisms by which ABC transporters move solute across the cell membrane. With the caveat that there is almost no point on which there is complete consensus, the transport mechanism can be summarized as follows (see also Chapter 6).

Transport clearly involves major conformational changes and a conventional enzyme-like mechanism. The transport cycle is initiated by the interaction of substrate with a specific binding site(s) on the TMDs. For Pgp and at least some other ABC transporters the interaction is within the inner leaflet of the lipid bilayer. (For bacterial binding protein-dependent importers the substrate – considered as the PBP–substrate complex – interacts at the extracellular face of the membrane to trigger ATP hydrolysis. Subsequent steps are not thought to be very different for importers and exporters, but full reconciliation into a single model awaits further data.) Substrate binding induces a conformational change in the TMDs, which is transmitted to the NBDs to initiate ATP binding. There is now compelling, but not yet conclusive, evidence that it is ATP binding (rather than hydrolysis) which induces the major conformational changes responsible for altering the affinity and orientation of the substrate-binding site(s) such that substrate is released at the extracellular face of the membrane (Martin *et al.*, 2001). Subsequent ATP hydrolysis and ADP/Pi release 'resets' the transporter for another cycle (Sauna and Ambudkar, 2000). Both NBDs bind and hydrolyze ATP, and there is strong evidence in support of the 'alternating catalytic cycle' mechanism (Senior *et al.*, 1995). However, it is still unclear whether 1 or 2 ATP molecules are hydrolyzed per molecule of substrate transported; determination of this number, together with determination of the exact number of substrate-binding sites and the nature of the conformational changes involved, is crucial to complete elucidation of the transport cycle.

Conclusions

ABC transporters are now recognized as one of the most intriguing of all protein families and are of considerable medical significance. The study of ABC transporters, in all their guises, has now become a minor industry. This is a far cry from the esoteric beginnings, and provides a wonderful example of scientific serendipity – how fundamental studies on model systems with no obvious commercial or medical implications, can unexpectedly have a significant impact in unimagined arenas of biology. The diversity of this volume is a testimony to this truism.

References

Aguilar-Bryan, L., Nichols, C.G., Wechsler, S.W., Clement, J.P., Boyd, A.E., Gonzalez, G., Herrera-Sosa, H., Nguy, K., Bryan, J. and Nelson, D.A. (1995) The beta cell high affinity sulfonylurea receptor: a regulator of insulin secretion. Science 268, 423–426.

Berger, E.A. and Heppel, L.A. (1974) Different mechanisms of energy coupling for the shock-sensitive and shock-resistant amino acid permeases of *Escherichia coli*. J. Biol. Chem. 249, 7747–7755.

Bishop, L., Agbayani, R., Ambudkar, S.V., Maloney, P.C. and Ames, G.F.-L. (1989) Reconstitution of a bacterial periplasmic permease in proteoliposomes and demonstration of ATP hydrolysis concomitant with transport. Proc. Natl Acad. Sci. USA 86, 6953–6957.

Blott, E.J., Higgins, C.F. and Linton, K.J. (1999) Cysteine-scanning mutagenesis provides no evidence for the extracellular accessibility of the nucleotide-binding domains of the multidrug resistance transporter P-glycoprotein. EMBO J. 23, 6800–6808.

Chang, G. and Roth, C.B. (2001) Structure of MsbA from *E. coli*: a homolog of the multidrug resistance ATP binding cassette (ABC) transporters. Science 293, 1793–1800.

Chen, C.-J., Chin, J.E., Ueda, K., Clark, D.P., Pastan, I., *et al.* (1986) Internal duplication and homology with bacterial transport proteins in the mdr1 (P-glycoprotein) gene from multidrug-resistant human-cells. Cell 47, 381–389.

Davidson, A.L. and Nikaido, H. (1991) Purification and characterization of the membrane-associated components of the maltose transport system from *Escherichia coli*. J. Biol. Chem. 266, 8946–8951.

Davidson, A.L., Shuman, H.A. and Nikaido, H. (1992) Mechanism of maltose transport in *Escherichia coli*: transmembrane signalling by periplasmic binding proteins. Proc. Natl Acad. Sci. USA 89, 2360–2364.

Felmlee, T., Pellett, S. and Welch, R.A. (1985) Nucleotide sequence of an *Escherichia coli* chromosomal hemolysin. J. Bacteriol. 163, 93–105.

Gerlach, J.H., Endicott, J.A., Juranka, P.F., Henderson, G., Sarangi F., *et al.* (1986) Homology between P-glycoprotein and a bacterial haemolysin transport protein suggests a model for multidrug resistance. Nature 324, 485–489.

Gilson, E., Higgins, C.F., Hofnung, M., Ames, G.F.-L. and Nikaido, H. (1982) Extensive homology between membrane-associated components of histidine and maltose transport systems of *Salmonella typhimurium* and *Escherichia coli*. J. Biol. Chem. 257, 9915–9918.

Gros, P., Croop, J. and Housman, D. (1986) Mammalian multidrug resistance gene – complete cDNA sequence indicates strong homology to bacterial transport proteins. Cell 47, 371–380.

Higgins, C.F., Haag, P.D., Nikaido, K., Ardeshir, F., Garcia, G. and Ferro-Luzzi Ames, G. (1982) Complete nucleotide sequence and identification of membrane components of the histidine transport operon of *S. typhimurium*. Nature 298, 723–727.

Higgins, C.F., Hiles, I.D., Whalley, K. and Jamieson, D.J. (1985) Nucleotide-binding by membrane components of bacterial periplasmic binding protein-dependent transport systems. EMBO J. 4, 1033–1040.

Higgins, C.F., Hiles, I.D., Salmond, G.P.C., Gill, D.R., Downie, J.A., Evans, I.J., *et al.* (1986) A family of related ATP-binding subunits coupled to many distinct biological processes in bacteria. Nature 323, 448–450.

Hobson, A.C., Weatherwax, R. and Ferro-Luzzi Ames, G. (1984) ATP-binding sites in the membrane components of histidine permease, a periplasmic transport system. Proc. Natl Acad. Sci. USA 81, 7333–7337.

Hyde, S.C., Emsley, P., Hartshorn, M.J., Mimmack, M.M., Gileadi, U., Pearce, S.R., Gallagher, M.P., Gill, D.R., Hubbard, R.E. and Higgins, C.F. (1990) Structural and functional relationships of ATP-binding proteins associated with cystic fibrosis, multidrug resistance and bacterial transport. Nature 346, 362–365.

Linton, K.J. and Higgins, C.F. (1998) The ABC (ATP-binding cassette) transporters of *Escherichia coli*. Mol. Microbiol. 28, 5–13.

Locher, K.P., Lee, A.T. and Rees, D.C. (2002) The *E. coli* BtuCD structure: a framework for ABC transporter architecture and mechanism. Science 296, 1091–1098.

Martin, M., Higgins, C.F. and Callaghan, R. (2001) The vinblastine binding site adopts high and low affinity conformations during the transport cycle of P-glycoprotein. Biochemistry 40, 15733–15742.

Mimmack, M.L., Gallagher, M.P., Hyde, S.C., Pearce, S.R., Booth, I.R. and Higgins, C.F. (1989) Energy coupling to periplasmic binding protein-dependent transport systems: ATP hydrolysis during transport. Proc. Natl Acad. Sci. USA 86, 8257–8261.

Parra-Lopez, C., Lin, R., Aspedon, A. and Groisman, E.A. (1994) A salmonella protein that is required for resistance to antimicrobial peptides and transport of potassium. EMBO J. 13, 3964–3972.

Rosenberg, M.F., Velarde, G., Ford, R.G., Martin, C., Berridge, G., Kerr, I.D., Linton, K.J. and Higgins, C.F. (2001) Repacking of the transmembrane domains of P-glycoprotein, during the transport ATPase cycle. EMBO J. 20, 5615–5625.

Sauna, Z.E. and Ambudkar, S.V. (2000) Evidence for a requirement for ATP hydrolysis at two distinct steps during a single turnover of the catalytic cycle of human P-glycoprotein. Proc. Natl Acad. Sci. USA 97, 2515–2520.

Senior, A.E., Al-Shawi, M.K. and Urbatsch, I.L. (1995) The catalytic cycle of P-glycoprotein. FEBS Lett. 377, 285–289.

Tame, J.R.H., Murshudov, G.N., Dodson, E.J., Neil, T.K., Dodson, G.G., Higgins, C.F. and Wilkinson, A.J. (1994) The structural basis of sequence-independent peptide binding by OppA protein. Science 264, 1578–1581.

Valverde, M.A., Diaz, M., Sepulveda, F.V., Gill, D.R., Hyde, S.C. and Higgins, C.F. (1992) Volume-regulated chloride channel associated with the human multidrug resistance P-glycoprotein. Nature 355, 830–833.

van Helvoort, A., Smit, A.J., Sprong, H., Fritzche, I., Schinkel, A.H., Borst, P. and van Meer, G. (1996) MDR1 P-glycoprotein is a lipid translocase of broad specificity, while MDR3 P-glycoprotein specifically translocates phosphatidylcholine. Cell 87, 507–517.

van Veen, H.W., Callaghan, R., Soceneantu, L., Sardini, A., Konings, W.N. and Higgins, C.F. (1998) A bacterial antibiotic resistance gene that complements the human multidrug resistance P-glycoprotein gene. Nature 391, 291–295

Walker, J.E., Saraste, M., Runswick, M.J. and Gay, N.J. (1982) Distantly related sequences in the a- and b-subunits of ATP synthase, myosin, kinases and other ATP-requiring enzymes and a common nucleotide-binding fold. EMBO J. 1, 945–951.

Part I

Phylogeny and Evolution of ABC Transporters

CHAPTER 1

Phylogenetic and Functional Classification of ABC (ATP-binding Cassette) Systems*

Elie Dassa

This paper is dedicated to the memory of Maurice Hofnung (1942–2001), a pioneer in the study of ABC (ATP-binding cassette) systems. Two decades ago, by noticing a strong sequence similarity between HisP and MalK, the two first-described ABC proteins, he initiated the studies that led to the identification and characterization of this large superfamily.

Introduction

ATP-binding cassette (ABC) systems constitute one of the most abundant families of proteins. At the time of writing this review, we have identified more than 2000 ABC ATPase domains or proteins in translated nucleic acid sequence databases. A total of about 6000 proteins were found when the partners of ATPases were taken into account. The size of this mass of sequences is therefore similar to the coding capacity of a bacterial genome. Several properties of members of this superfamily have been reviewed in the last decade (Ames and Lecar, 1992; Ames *et al.*, 1990, 1992; Doige and Ames, 1993; Higgins, 1992; Higgins *et al.*, 1988; Holland and Blight, 1999). The most prominent characteristic of these systems is that they share a highly conserved ATPase domain, the ABC, which has been demonstrated to bind and hydrolyze ATP, thereby providing energy for a large number of biological processes. The amino acid sequence of this cassette displays three major conserved motifs, the Walker A and Walker B motifs commonly found in ATPases together with a specific signature motif, usually commencing LSGG-, and also known as the linker peptide (Schneider and Hunke, 1998). The crystal structures of some ABC proteins are presented in Chapters 4 and 7.

ABC systems are involved not only in the import or export of a wide variety of substances, but also in many cellular processes and in their regulation. Importers constitute mainly the prokaryotic transporters dependent upon a substrate-binding protein (BPD), whose function is to provide bacteria with essential nutrients even if the latter are present in submicromolar concentrations in the environment (Boos and Lucht, 1996). Exporters are found in both prokaryotes and eukaryotes and are involved in the extrusion of noxious substances, the secretion of extracellular toxins and the targeting of membrane components (Fath and Kolter, 1993). The third type of ABC system is apparently not involved in transport but rather in cellular processes such as DNA repair, translation or regulation of gene expression. Since ATP is found principally in the cytosol, we define import as the inwardly directed transport of a molecule into the cytosol. By contrast, export is

**ABSCISSE, a database of ABC systems, which includes functional, sequence and structural information, is available on the internet at the following address: www.pasteur.fr/recherche/unites/pmtg/abc/index.html.*

ABC Proteins: From Bacteria to Man
ISBN 0-12-352551-9

the translocation of a molecule out of the cytosol, even if its final location is an intracellular organelle. ABC systems of the three types can be distinguished on the basis of the design of their component parts. All the *transporters* are composed of four structural domains: two very hydrophobic membrane-spanning or integral membrane domains (IMs) and two hydrophilic cytoplasmic domains containing the ABC, peripherally associated with IM on the cytosolic side of the membrane. (a) *Importers* have in general the four domains encoded as independent polypeptides and they need for function an extracellular substrate-binding protein. (b) In most well-characterized *exporters*, the transmembrane domains are fused to the ABC domains in several ways. However, some systems with separated IM and ABC domains have been reported to act as exporters although the complete characterization of their transport mechanism awaits more studies. Prokaryote exporters also require accessory proteins and these will be discussed in the specific sections dealing with these transporters. (c) Systems involved in *cellular processes* other than transport do not have IM domains and are composed of two ABC domains fused together.

Inventory and Classification of ABC Systems

To understand the complexity and diversity of ABC systems, computer-assisted methods have been applied by several authors based on comparisons of the ABC ATPase domain, the most highly conserved element. These methods were instrumental in the early definition of the superfamily on the basis of primary sequence comparisons (Higgins *et al.*, 1986). However, in most cases, the ABC proteins of a given organism (Braibant *et al.*, 2000; Linton and Higgins, 1998; Quentin *et al.*, 1999) or ABC systems with clear functional similarity (Fath and Kolter, 1993; Hughes, 1994; Kuan *et al.*, 1995) were compared. The presence of the highly conserved ATPase domain permitted more global comparisons, for example (Paulsen *et al.*, 1998). The first general phylogenetic study specifically devoted to the ABC superfamily (Saurin *et al.*, 1999) was recently updated to include the analysis of about 600 ATPase proteins or domains (Dassa and Bouige, 2001). The sequences segregate in

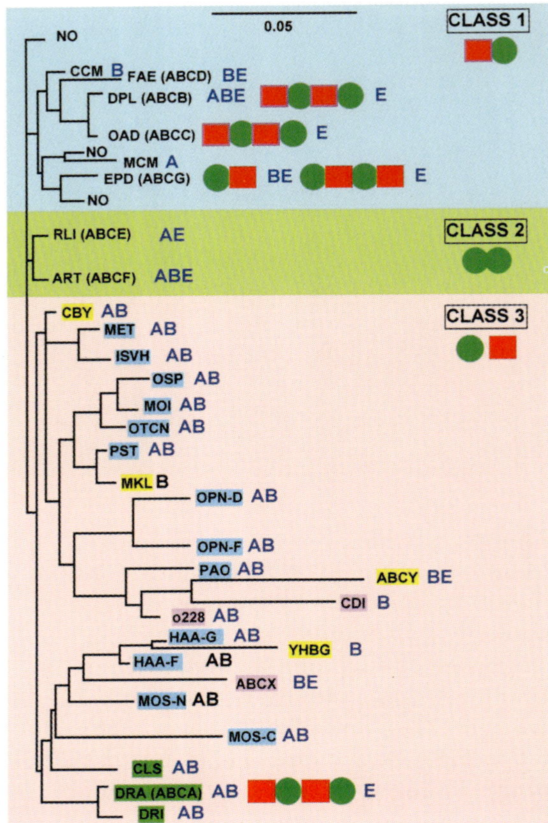

Figure 1.1. Unrooted simplified phylogenetic tree of ABC proteins and domains. For the sake of clarity, only the branches pointing to families have been drawn. The major subdivisions of the tree are indicated according to the nomenclature used in the text. Class 1: systems with fused ABC and IM domains (exporters); class 2: systems with no known transmembrane domains (antibiotic resistance, translation, etc.); class 3: systems with IM and ABC domains carried by independent polypeptide chains (BPD importers and other systems). Under the name of the class, the minimal consensus organization of ABC systems is represented by colored symbols in a linear fashion. IM proteins or domains are represented by red rectangles and ABC proteins or domains by green circles. When the organization of a system in a family does not fit exactly with the consensus, it is indicated on the same line as the system name. In class 3, BPD transporters are highlighted in blue, while systems that are not conclusively related to import are highlighted in purple; systems that could be importers are colored in yellow and systems that could be exporters in green. The sequences of UVR family proteins were omitted from this analysis (see the section on the UVR family for details). Family names are abbreviated

(continued)

33 clusters on the phylogenetic tree shown in **Figure 1.1**. Some clusters comprise obviously highly related proteins known to function together; for example, the two ATPases of oligopeptide importers were fused into a single family. The final 29 families are listed in **Table 1.1**. Since a general nomenclature for ABC systems is not yet available, **Table 1.1** provides the present nomenclature and the equivalent alternative adopted for transporters in general (Saier, 2000) or specifically for human ABC systems (see Chapter 3).

FAMILIES OF ABC SYSTEMS IN LIVING ORGANISMS

This classification was derived solely on the basis of the comparison of the sequences of the highly conserved ATPase domain. The families of systems will be described as they appear from the top to the bottom of **Table 1.1** and the names adopted here are explained in the legend of this table. The most striking finding is that ABC proteins or domains fall into three main subdivisions or classes. Class 1 comprises systems with fused ABC and IM domains, class 2 comprises systems with two duplicated, fused ABC domains and no IM domains and class 3 contains systems with IM and ABC domains carried by independent polypeptide chains (Dassa and Bouige, 2001). This disposition matches fairly well, although there are a few exceptions, with the three functional types of ABC systems mentioned in the Introduction. Class 1 **(Figure 1.2)** is composed essentially of all known exporters with fused ABC and IM domains. Class 2 contains systems involved in cellular processes other than transport and in antibiotic resistance. Class 3 contains all known BPD transporters and systems with ill-characterized function or transport mechanism, some of the latter being considered as exporters. This classification is indeed useful for predicting the putative functions of open reading frames (ORFs) of unknown function based on primary sequence similarities. This concept is justified by the fact that proteins or protein domains that participate in similar functions are found in the same phylogenetic cluster. However, within this cluster, proteins handling different substrates are clearly separated (see, for example, **Figure 1.3B** showing the different dispositions of the highly conserved but functionally different MDR1, MDR3 and BSEP proteins). The second important issue of this classification is that it does not reflect the universal classification of living organisms. The consequences of these issues will be discussed in the 'Conclusions and Perspectives' section at the end of the chapter. In the following sections, I shall discuss the known or predicted functions of the ABC systems found in each class. The organization of ABC systems will be schematized by using the IM (for integral membrane) and ABC (for the ATPase) symbols, as explained in the legends of **Figures 1.2 to 1.8**.

Figure 1.1. ***(continued)*** ***according to the conventions used in Table 1.1 and throughout the text and the nomenclature of human ABC systems is given in parentheses after the name of the family. NO represents a few sequences with unknown function and apparently unrelated to neighboring families. They are not discussed in the text. OPN-D, OPN-F; HAA-F, HAA-G and MOS-N, MOS-C correspond to the two different ABC subunits of OPN, HAA and MOS systems, respectively. The distribution of the systems in the three kingdoms of life is indicated as follows: A (archaea), B (bacteria) and E (eukaryotes). The scale at the top of the figure corresponds to 5% divergence per site between sequences.***

CLASS 1 COMPRISES ESSENTIALLY ALL KNOWN EXPORTERS WITH FUSED ABC AND IM DOMAINS

The FAE (ABCD) family putatively involved in very long chain fatty acid export

The IM and ABC domains of the proteins of this family are fused into a single polypeptide chain and their organization can be represented as IM-ABC **(Figure 1.2D)**. The properties of this medically important family are reviewed in Chapter 24. The most characterized members of this family are two homologous peroxisome-associated proteins PXA1 and PXA2. These form heterodimers and when inactivated, cause impaired growth on oleic acid and a reduced ability to oxidize oleate (Shani *et al.*, 1995). In humans, the adrenoleukodystrophy protein ALDp (ABCD1) is defective in X chromosome-linked adrenoleukodystrophy (ALD), a neurodegenerative disorder with impaired peroxisomal oxidation of very long chain fatty acids (Fanen *et al.*, 1994). Three other proteins, highly

Table 1.1. Classes, families and subfamilies of ABC systems

The three classes of ABC systems are the following. Class 1: systems with fused ABC and IM domains; class 2: systems with two duplicated fused ABC domains and no IM domains; class 3: systems with IM and ABC domains carried by independent polypeptide chains. Family names are abbreviations of the substrate or the biological process handled by systems. For families comprising systems of unknown function, an arbitrary name is given.

The number (Nbr) of systems within families and subfamilies is given, followed by a very short definition of their properties (Function).

For each family or subfamily a typical ABC protein (Model) is indicated as an example, and when available, the Swissprot ID or the PIR accession number of the protein is given. Cross-reference to the nomenclatures adopted by the Human Gene Nomenclature Committee (HGNC) http://www.gene.ucl.ac.uk/nomenclature/genefamily/abc.html and by the Transport Commission (TC) http://www-biology.ucsd.edu/~msaier/transport/classf.html is given.

Some phylogenetic families described in this table are separated by the TC into subfamilies according to substrate type.

(1) = CPSE + LPSE

(2) = PhoT + MolT + SulT + FeT + POPT + ThiT + BIT

(3) = QAT + NitT + TauT

(4) = VB12T + FeCT

The last column (Taxon) indicates the occurrence of members of a given family in the different taxa of living organisms. A: archaea; B: bacteria; E: eukaryotes.

Family	Subfamily	Nbr	Function	Model	HGNC	TC	Taxon
Class 1 systems (exporters)							
FAE		24	Very long chain fatty acid export, putative	ALD_HUMAN	ABCD	FAT	BE
DPL		272	Drug, peptides and lipid export				ABE
	LAE	24	Lantibiotic export	NIST_LACLA		Pep1E	B
	BAE	21	Bacteriocin and peptide export	MESD_LEUME		Pep2E	AB
	CYD	10	Cytochrome *bd* biogenesis	CYDC_ECOLI			B
	HMT	17	[Fe/S] cluster export	ATM1_YEAST	ABCB	HMT	BE
	CHV	4	Beta-1,2-glucan export	CHVA_AGRTU		GlucanE	B
	MDL	9	Mitochondrial peptide export	MDL1_YEAST	ABCB		E
	SID	4	Siderophore biogenesis	YBTP (T17437)			B
	LIP	18	Lipid A or glycerophospholipid export	MSBA_ECOLI		LipidE	B
	PED	12	Prokaryote drug export	LMRA_LACLA		DrugE2	AB
	LLP	19	LIP-like exporters, putative	YFIB_BACSU			AB
	ARP	9	Antibiotic resistance or production, putative	STRW (S57562)		DrugE3	B
	PRT	20	Proteases, lipases, S-layer protein export	PRTD_ERWCH		Prot2E	B
	HLY	19	RTX toxin export	HLYB_ECOLI		Prot1E	B
	TAP	19	Peptide export	TAP1_HUMAN	ABCB	TAP	E
	Pgp	65	Eukaryote multiple drug resistance and lipid export	MDR1_MOUSE	ABCB	MDR	E
OAD		65	Organic anion and conjugate drug export				E
	CFTR	13	Chloride anion channel	CFTR_HUMAN	ABCC	CFTR	E
	MRP	44	Conjugate drug exporters	MRP1_HUMAN	ABCC	CT1-2	E
	SUR	8	Potassium channel regulation	SUR1_HUMAN	ABCC		E
EPD		66	Eye pigment precursors and drugs				BE
	WHI	34	Eye pigment precursors and drugs	WHIT_DROME	ABCG	EPP	BE
	PDR	32	Pleiotropic drug resistance	PDR5_YEAST		PDR	E
CCM		13	Cytochrome *c* biogenesis	CCMA_ECOLI		HemeE	ABE
MCM		4	Unknown	ATWA (D64507)			A

(continued)

TABLE 1.1. *(continued)*

Family	Subfamily	Nbr	Function	Model	HCGN	TC	Taxon
Class 2 systems with no transmembrane domains and involved in non-transport cellular processes and antibiotic resistance							
RLI		12	RNase L inhibitor	RNASELI (S63672)	ABCE		AE
ART		66	Antibiotic resistance and translation regulation				ABE
	EF-3	7	Translation elongation	EF3_YEAST			E
	REG	39	Translation regulation	GC20_YEAST	ABCE		BE
	ARE	18	Macrolide antibiotic resistance	MSRA_STAEP		DrugRA2	B
UVR		29	DNA repair and drug resistance	UVRA_ECOLI			AB
Class 3 systems with unfused transmembrane and ATP-binding domains; binding protein-dependent importers							
MET		41	Metals	ZNUC_ECOLI		MZT	AB
ISVH		55	Iron-siderophores, vitamin B_{12} and hemin	FHUC_ECOLI		(4)	AB
OSP		98	Oligosaccharides and polyols	MALK_ECOLI		CUT1	AB
MOI		116	Mineral and organic ions	POTD_ECOLI		(2)	AB
OTCN		50	Osmoprotectants, taurine, cyanate and nitrate phosphonates	TAUB_ECOLI		(3)	AB
OPN		93	Oligopeptides and nickel	OPPD_SALTY		PepT	AB
PAO		57	Polar amino acid and opines	HISP_SALTY		PAAT	AB
HAA		23	Hydrophobic amino acids and amides	LIVG_ECOLI		HAAT	AB
MOS		54	Monosaccharides	RBSA_ECOLI		CUT2	AB
Class 3 systems of unknown function that could be importers							
CBY		34	Cobalt uptake and unknown function				
	CBU	16	Cobalt uptake, putative	CBIO_SALTY		CoT	AB
	Y179	18	CBU-like systems, unknown function	Y179_METJA			AB
MKL		14	Cell surface integrity, putative	MKL_MYCLE			BE
ABCY		10	Unknown function	ABC_ECOLI			BE
YHBG		23	Unknown function	YHBG_ECOLI			B
Class 3 systems which are not known to be importers							
o228		58	Lipoprotein release	LOLD_ECOLI			AB
ABCX		23	[Fe/S] cluster assembly, putative	ABCX_CYAPA			ABE
CDI		9	Cell division	FTSE_ECOLI			B
Class 3 systems which could be exporters							
DRA		67	Drug and antibiotic resistance				ABE
	DRR	28	Polyketide drug resistance	DRRA_STRPE		DrugE1	AB
	NOD	10	Nodulation	NODI_RHISM		LOSE	B
	NAT		Na^+ extrusion	NATA_BACSU			AB
	ABCA		Lipid trafficking	ABC1_HUMAN	ABCA	CPR	E
DRI		103	Drug resistance, bacteriocin and lantibiotic immunity				AB
	BAI	8	Bacteriocin immunity	BCRA_BACLI			B
	LAI	21	Lantibiotic immunity	SPAF (I40516)			B
	DRB	51	Drug resistance, putative	PAB1845(E75122)			AB
	NOS	15	Nitrous oxide reduction	NOSF_PSEST			ABE
CLS		41	Extracellular polysaccharide export	KST1_ECOLI		(1)	AB
NO		39	Unclassified systems				ABE

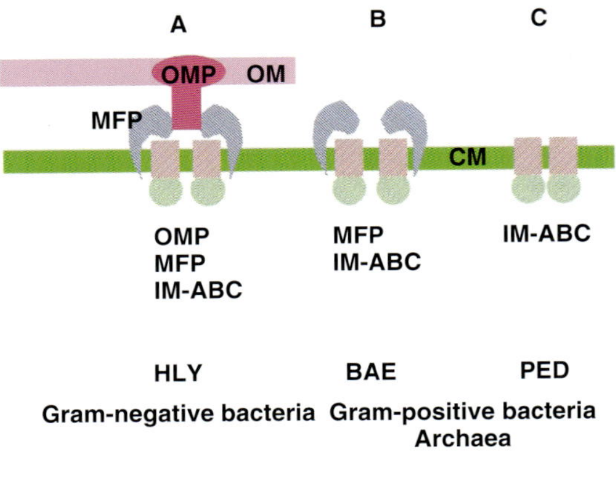

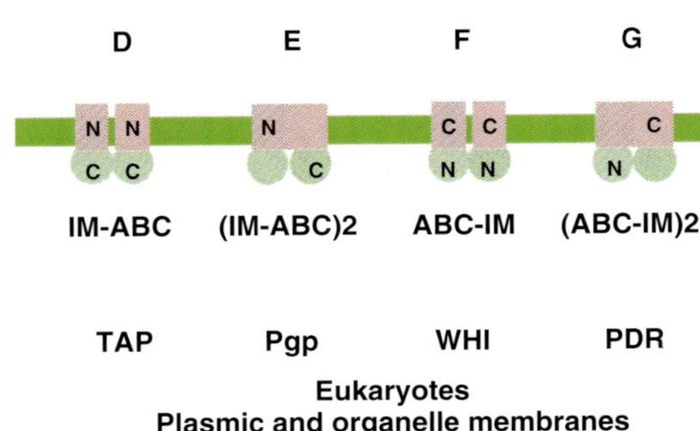

Figure 1.2. Typical organization of class 1 exporters. The membranes are represented schematically; OM: outer membrane of Gram-negative bacteria, CM: cytoplasmic membrane. The class 1 systems are characterized by the fusion of the integral membrane protein (IM) domain to the ATP-binding domain (ABC) in two different ways: the IM domain could be at either the N-terminus (IM-ABC) or the C-terminus (ABC-IM) of the protein (indicated by C or N on the domain). The functional transporter is composed of two IMs (red hatched rectangles) and two ABC subunits (green hatched circles). Different hatches in IM and ABC mean that different gene products are associated within the same system. From the top to the bottom of the figure are represented: a schematic organization of the transporters; the types of proteins encoded by the genes that determine the system; the subfamily of the system and the distribution among living organisms.
Prokaryote systems: A, The HLY subfamily systems (e.g. the hemolysin exporter of Gram-negative bacteria) comprise a TolC-like trimeric outer membrane protein (OMP), a probable trimer of a membrane fusion protein (MFP) and a homodimeric complex of an IM-ABC protein. B, In Gram-positive organisms, the OMP is lacking as shown for the lacticin M exporter (BAE subfamily) but a homologue of the MFP could be found. C, PED subfamily systems (e.g. protein LmrA) apparently lack both OMP and MFP. Eukaryote systems: No accessory proteins are known. D, The TAP1/TAP2 heterodimer involved in the transport of MHC-peptides. E, The Pgp subfamily proteins probably originate from the fusion of two TAP-like proteins. F, The white/brown heterodimer involved in eye pigment metabolite export (WHI subfamily). G, The PDR subfamily (pleiotropic drug resistance) systems originate probably from the fusion of two WHI-like proteins.

similar to ABCD1 were identified in the human genome: ALDR (ABCD2), PMP70 (ABCD3) and PMP69 (ABCD4). A mutated form of PMP70 was associated with certain manifestations of Zellweger syndrome, a group of genetically heterogeneous disorders affecting peroxisome biogenesis (Gartner *et al.*, 1992). The actual function of these transporters is unknown, but it has been proposed that they could export into peroxisomes very long chain fatty acids or the enzyme(s) responsible for their degradation (Hettema and Tabak, 2000). Interestingly, nine proteins strongly similar to ALDp over the entire sequence length were detected in bacteria, but their functions remain to be investigated.

The DPL family involved in drug, peptide and lipid export

The DPL family is composed of transporters that are significantly similar over the entire sequence length. A simplified phylogenetic tree of the ABC domains of the members of this huge family that illustrates their sequence relationships is presented in **Figure 1.3**. The typical organization of these transporters is IM-ABC for prokaryote systems and several eukaryote systems. The (IM-ABC)2 type of organization is apparently restricted to the P-glycoprotein-like systems found exclusively in eukaryotes **(Figures 1.2G and 1.3)**. This family can be subdivided into 15 subfamilies on the basis of sequence similarity. The systems with an IM-ABC organization will be described first.

The LAE subfamily involved in lantibiotic export

Lantibiotics are peptides containing post-translationally modified amino acids such as dehydrated amino acids and lanthionine residues that form intramolecular thioether rings, and are secreted by several Gram-positive

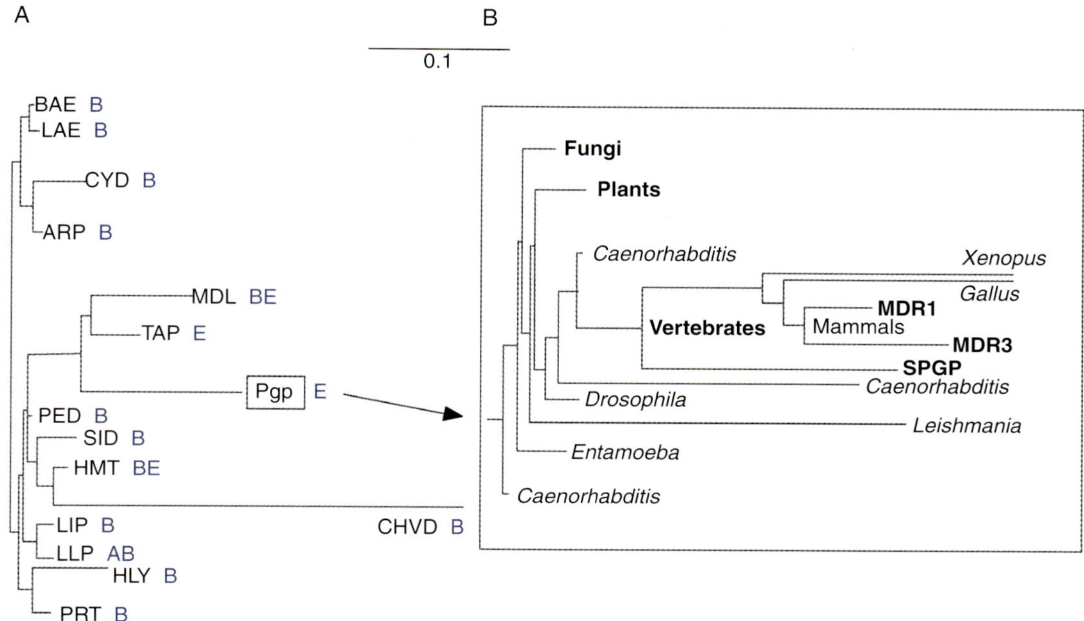

Figure 1.3. ***Simplified phylogenetic trees of the DPL family. Same conventions as in Figure 1.1. All subfamilies, with the exception of the Pgp subfamily, are composed of systems with an IM-ABC organization. A, A simplified tree of the whole DPL family. B, A simplified tree of the Pgp subfamily showing the distribution of the proteins in eukaryotes and the segregation of the three functionally different proteins MDR1, MDR2/3 and SPGP in mammals.***

bacteria (Otto and Gotz, 2001). LAE systems are involved in the processing and export of lantibiotics. These transporters carry an N-terminal cytosolic proteolytic domain that is involved in the processing of the lantibiotic precursor (Havarstein *et al.*, 1995). The operons containing these transporters contain a single IM-ABC transporter that is predicted to function as a homodimer **(Figure 1.2B)**. Although functionally very similar to bacteriocin exporters, the LAE subfamily is clearly distinguishable from the BAE subfamily in the phylogenetic trees.

The BAE subfamily involved in bacteriocin and competence peptide export

These systems are very similar to LAE systems but they are involved in the export of non-post-translationally modified peptides such as bacteriocins (O'Keeffe *et al.*, 1999) and the competence-stimulating peptides of Gram-positive bacteria (Hui and Morrison, 1991).

The CYD subfamily putatively implicated in cytochrome bd *biogenesis*

The CydC and CydD proteins are important for the formation of cytochrome *bd* terminal oxidase and for periplasmic *c*-type cytochromes. CydCD may determine a hetero-oligomeric complex important for heme export into the periplasm (Poole *et al.*, 1994) or according to another hypothesis, could be involved in the maintenance of the proper redox state of the periplasmic space (Goldman *et al.*, 1996). However, in *Bacillus subtilis*, the absence of CydCD does not affect the presence of holo-cytochrome *c* in the membrane and this observation suggests that CydCD proteins are not involved in the export of heme, at least in this organism (Winstedt *et al.*, 1998).

The HMT subfamily of mitochondrial and bacterial transporters

This subfamily (described in Chapter 25) is composed of proteins homologous to the *Saccharomyces pombe* HMT1 protein, a vacuolar phytochelatin transporter involved in heavy metal resistance by a sequestration mechanism (Ortiz *et al.*, 1995), and to the yeast ATM1 protein, essential for the transport of iron/sulfur clusters from the mitochondrial matrix to the cytosol (Lill and Kispal, 2001). Close homologues of these proteins were identified in several eukaryotes and two examples, RP205 and RP214, in the intracellular parasitic bacterium *Rickettsia prowazekii*. This observation is in line with the hypothesis suggesting that *Rickettsia* and mitochondria evolved from a common ancestor. The human orthologue of ATM1, ABCB7 (ABC7), is implicated in the X-linked inherited disease sideroblastic anemia and ataxia (Allikmets *et al.*, 1999). A second human mitochondrial transporter, ABCB6 (MTABC3), was found to be able to compensate for the defects in the yeast ATM1 mutant, as was ABCB7 (Mitsuhashi *et al.*, 2000).

The CHV family involved in beta-1,2-glucan export

This very small family comprises proteins ChvA and NdvA and a few ORFs detected in the genomes of various bacteria. ChvA is required for the attachment of *Agrobacterium tumefaciens* to plant cells, an early step in crown gall tumor formation. Strains defective in *chvA* do not secrete normal amounts of cyclic beta-1,2-glucan, although they contain three times more beta-1,2-glucan in their cytoplasm than the wild-type strain. It was concluded that ChvA is a transporter involved in the export of cyclic glucans. The NdvA protein is very probably an orthologue of ChvA in *Rhizobium meliloti*.

The MDL subfamily of mitochondrial and bacterial transporters

This distinct subfamily of mitochondrial targeted transporters comprises proteins similar to those of the TAP family (see below). The yeast Mdl1 and Mdl2 proteins belong to this family. Recently, the Mdl1 protein has been shown to be required for mitochondrial export of peptides generated by proteolysis of inner membrane proteins by the m-AAA protease in the mitochondrial matrix (Young *et al.*, 2001). Several homologues were found in eukaryotes, including two proteins in mammals M-ABC1 (ABCB8) and M-ABC2 (ABCB10) (Hogue *et al.*, 1999; Zhang *et al.*, 2000a).

The SID subfamily

This subfamily is composed of systems encoded by genes located near genes encoding peptide/polyketide synthases involved in the non-ribosomal synthesis of peptide-siderophores. A typical example is the YbtP-YbtQ system of *Yersinia pestis*, composed of two IM-ABC transporters. The *ybtP* and *ybtQ* genes are found in the operon encoding the enzymes responsible for the synthesis of the siderophore yersiniabactin. Cross-feeding experiments suggested that this system could be involved in the acquisition of iron chelated to yersiniabactin (Fetherston *et al.*, 1999).

The LIP subfamily involved in the export of the lipid A moiety of lipopolysaccharide

Thermosensitive mutations in the *msbA* gene encoding an IM-ABC transporter essential for growth cause the accumulation in the inner membrane of hexa-acylated lipid A and glycerophospholipids, which are precursors of lipopolysaccharides, at the nonpermissive temperature (Zhou *et al.*, 1998). It was proposed that MsbA encodes a lipid A or a glycerophospholipid transporter, thus delivering these precursors to the outer membrane during lipopolysaccharide biosynthesis. Most importantly, as described in Chapter 7, the first high-resolution structure of an entire ABC transporter has been obtained for the *Escherichia coli* MsbA protein (Chang and Roth, 2001). Homologues of MsbA proteins were found in several Gram-negative bacteria (Holland and Wolk, 1990; McDonald *et al.*, 1997).

The PED subfamily involved in prokaryote drug export

This subfamily is closely related to the MsbA subfamily and comprises systems involved in peptide or drug resistance. The LmrA protein of *Lactobacillus* (van Veen *et al.*, 2001), involved in the resistance to several unrelated hydrophobic drugs, is representative of this subfamily and is discussed in Chapter 12 **(Figure 1.2C)**. Putative drug exporters encoded by genes located in the vicinity of genes involved in the biosynthesis of the cyclic decapeptide antibiotic tyrocidine and of the glycolipid antibiotic vancomycin belong to this subfamily.

The LLP subfamily of LIP-like exporters of unknown function

This family is composed exclusively of pairs of ORFs detected in the completed genomes of prokaryotes and encoding putative IM-ABC transporters. They cluster near the LIP family system and it is possible that they encode heterodimeric ABC transporters.

The ARP family involved in production of or resistance to antibiotics

The genetic region determining resistance towards tetracycline in *Corynebacterium striatum* contains genes *tetA* and *tetB* encoding two ABC transporters with an IM-ABC organization. These genes were able to confer upon a sensitive strain of *Corynebacterium glutamicum* resistance to tetracycline, oxytetracycline and the structurally and functionally unrelated beta-lactam antibiotic oxacillin. It was proposed that these antibiotics would be exported by the TetAB heterodimer (Tauch *et al.*, 1999). Similar genes, *strV* and *strW*, were found in the cluster for the biosynthesis of 5′-hydroxystreptomycin in *Streptomyces glaucescens* (Beyer *et al.*, 1996). The *ramA* and *ramB* genes that belong to this family were shown to be involved in the development of aerial hyphae in *Streptomyces* species. It was suggested that the *ram* gene products are involved in the transport of a factor essential for normal development (Keijser *et al.*, 2000).

The PRT subfamily involved in export of hydrolytic enzymes and S-layer proteins

This subfamily is involved in the one-step secretion of proteases, glycanases, and S-layer proteins in Gram-negative bacteria (reviewed in Chapter 11). The vast majority of the proteins exported by this family of systems display a characteristic but variable number of glycine-rich repeats (RTX) forming a calcium-binding site. A typical system **(Figure 1.2A)** comprises an IM-ABC transporter, expected to function as a homodimer, a cytoplasmic membrane component belonging to the membrane fusion protein family and an outer membrane protein (Létoffé *et al.*, 1990). All these components are essential for export. The outer membrane proteins are very similar to TolC, a protein shown to be involved in the export of *E. coli* hemolysin A (Wandersman and Delepelaire, 1990) and to participate in several ABC-independent drug efflux systems. The recently established three-dimensional structure of TolC revealed that this trimeric protein is folded in such a way that it forms a large 'channel-tunnel', which spans both the outer membrane and periplasmic space (Koronakis *et al.*, 2000).

The HLY subfamily involved in RTX toxin export

This subfamily contains all hemolysin and toxin exporters (reviewed in Chapter 11). Such large toxins, which contribute to the virulence of bacteria, also have the RTX motifs mentioned above. The protein composition of HLY subfamily systems is identical to that of PRT subfamily systems. Despite their similarity, the ABC domains of HLY subfamily systems cluster apart from those of the PRT subfamily. Interestingly, it was found that the proteins exported by HLY systems differ significantly from those exported by PRT systems in a very short C-terminal sequence known to constitute part of the secretion signal (Young and Holland, 1999). These observations suggest either that the sequences of the IM domains, thought to contain substrate recognition sites, exert a constraint on the sequence of the ABC domain or, alternatively, that the ABC domain by itself might participate in the constitution of such a substrate recognition site.

The TAP subfamily involved in eukaryote peptide export

The transporter associated with antigen processing (TAP) in mammals is essential for peptide presentation to the major histocompatibility complex (MHC) class I molecules on the cell surface and necessary for T-cell recognition (reviewed in Chapter 26). The complete TAP system is composed of a heterodimeric complex TAP1 (ABCB2) and TAP2 (ABCB3), two ABC transporters with an IM-ABC organization **(Figure 1.2D)**, encoded by genes lying in the MHC class II region encoding a cluster of genes for antigen processing (Beck *et al.*, 1992). Peptides generated from cytosolic proteins by the proteasome are translocated to the endoplasmic reticulum by the TAP transporter, where they are bound to nascent MHCI molecules, thereby allowing their transport to the cell surface (Abele and Tampe, 1999; Karttunen *et al.*, 1999). Very recently, the crystal structure of the ABC domain of human TAP1 was published (Gaudet and Wiley, 2001). Sequences orthologous to TAP1 and TAP2 are found in vertebrates. However, sequences similar to these proteins have a larger distribution but their functions are unknown. For example, the human TAP-L protein (ABCB9) was found to be associated with lysosomes and highly expressed in testes (Yamaguchi *et al.*, 1999; Zhang *et al.*, 2000b). ORFs highly similar to TAP-L were identified in invertebrates (four in *Caenorhabditis elegans*) and in *Arabidopsis thaliana*.

The Pgp subfamily involved in eukaryote multiple drug resistance and lipid export

The MDR1 gene (ABCB9), responsible for multidrug resistance in human cells, encodes a broad specificity efflux pump P-glycoprotein or Pgp. Pgp consists of two similar halves **(Figure 1.2E)**, each half including a hydrophobic transmembrane region and a nucleotide-binding domain (IM-ABC)2. Homologues of MDR1 are found almost exclusively in eukaryotes, and the handful of examples of prokaryotic proteins with an (IM-ABC)2 configuration are probably due to sequencing errors.

A recent review has dealt with the properties of this vast and medically important subfamily of proteins (Borst *et al.*, 2000), and thus, only the evolutionary aspects will be briefly reported here. In the Pgp subfamily, proteins are clustered according to the taxonomy of eukaryotes, with clusters corresponding to parasite, fungal, insect, worm, plant and vertebrate proteins. This disposition suggests that Pgp family proteins descend from a single ancestor but that multiple Pgps in each of these taxa have arisen by independent duplication events. In mammals, three different groups of sequences are detected and correspond to MDR1-like (ABCB9) proteins, involved in multidrug resistance, MDR3-like (ABCB3) proteins and BSEP-like (ABCB11) proteins, involved in the export of phosphatidylcholine and bile salts, respectively, through the liver canalicular membrane. Mutations in MDR3 and BSEP have been found in two forms of progressive familiar intrahepatic cholestasis in humans, PFIC2 and PFIC3, respectively.

The OAD family involved in organic anion and conjugate drug export and in ion channel regulation

The OAD family is composed of systems involved in ion channel regulation, ion channel formation and the efflux of organic anions across cellular membranes. Some systems are linked to resistance to cytotoxic drugs but in contrast with DPL family systems described above, drug resistance is achieved by the efflux of drugs conjugated or associated with anionic molecules such as glutathione or glucuronide derivatives. This family is found exclusively in eukaryotes and the proteins have an (IM-ABC)2 organization. The phylogenetic tree shows three main branches corresponding to three subfamilies.

The CFTR subfamily of anion selective channels

In contrast to most other members of the ABC transporters, CFTR (ABCC7) forms an anion-selective channel involved in epithelial chloride transport (reviewed in Chapter 29). In secretory epithelia of vertebrates, it is located in the apical membrane, where it regulates transepithelial Cl^- secretion (Sheppard and Welsh, 1999). Cystic fibrosis, one of the most frequent inherited human diseases, is caused by mutations in the CFTR protein (Riordan *et al.*, 1989). CFTR displays the typical organization (IM-ABC)2 but in addition carries a characteristic hydrophilic R-domain that separates the first half of the protein from the second. This domain participates in the control of channel gating by a kinase-mediated phosphorylation mechanism (Naren *et al.*, 1999).

The MRP family involved in conjugate drug resistance

The MRP subfamily is widely distributed among eukaryotes. The biological roles of mammalian MRP family systems are quite diverse (see Chapters 18–21). In addition to the core structure (IM-ABC)2, most MRP subfamily proteins have an additional, large N-terminal hydrophobic region predicted to contain four to six transmembrane helices (Tusnady *et al.*, 1997). This N-terminal region is apparently not essential for the function or final localization of human MRP1 (ABCC1) (Bakos *et al.*, 1998). Moreover, the mammalian MRP4 (ABCC4) and MRP5 (ABCC5) proteins lack this domain. Like Pgp, the clusterings of MRP subfamily proteins follow the taxonomy of eukaryotes. MRP subfamily proteins have been identified in plants, fungi and parasites and they show a large variety of cellular functions. *A. thaliana* AtMRP2 (see Chapter 17) encodes a multispecific ABC transporter involved in the transport of both glutathione S conjugates and chlorophyll catabolites (Lu *et al.*, 1998). In yeast, the YCF1 protein is a vacuolar glutathione S conjugate pump that mediates cadmium and arsenite resistance by a vacuole sequestration mechanism (Li *et al.*, 1996) and the BAT1 (YLL048c) protein mediates ATP-dependent bile acid transport (Ortiz *et al.*, 1997). In *Leishmania*, amplification of the MRP family protein PgpA is associated with arsenite and antimonyl tartrate resistance mediated via a glutathione-coupled sequestration mechanism (Haimeur *et al.*, 2000; Legare *et al.*, 2001).

The SUR subfamily of potassium channel regulators

ATP-sensitive potassium (K-ATP) channels in pancreatic β-cells regulate insulin secretion (Ashcroft, 2000). The cloning and reconstitution of the subunits of these channels demonstrate that they are octameric hetero-oligomeric complexes of inwardly rectifying potassium channel subunits (KIR6.x) and SUR1 (ABCC8) sulfonylurea receptors with a (KIR6.x-SUR)4 stoichiometry (Aguilar-Bryan *et al.*, 1995; Clement *et al.*, 1997; Shyng and Nichols, 1997). Persistent hyperinsulinemic hypoglycemia of infancy, a rare genetic disease due to defective regulation of insulin secretion, is associated with mutations in the gene encoding SUR1. An isoform of SUR1, SUR2 (ABCC), is expressed more ubiquitously (Isomoto *et al.*, 1996). SUR proteins are strongly related to MRP proteins and also possess an N-terminal additional transmembrane domain. Their properties are discussed in more detail in Chapter 27. A SUR-like protein was found in *Drosophila melanogaster*, and when expressed in *Xenopus* oocytes, determined the appearance of a characteristic glibenclamide-sensitive potassium channel activity (Nasonkin *et al.*, 1999).

The EPD family involved in eye pigment precursor transport, lipid transport regulation and drug resistance

The EPD family systems display a unique organization with an N-terminal ABC domain fused to a C-terminal IM domain. The proteins segregate within two subfamilies, the WHI subfamily with an ABC-IM organization **(Figure 1.2F)** and the PDR subfamily with an (ABC-IM)2 organization **(Figure 1.2G)**.

The WHI subfamily

The white, brown and scarlet genes of *D. melanogaster* encode ABC transporters that are believed to transport guanine and tryptophan, which are precursors of the red and brown eye color pigments, respectively. It is thought that the white and brown proteins form a heterodimeric complex involved in guanine transport, while the white and scarlet proteins form a tryptophan transporter (Ewart *et al.*, 1994) **(Figure 1.2F)**. It has generally been assumed that these proteins are localized in the plasma membrane and are involved in the import of eye pigment precursor molecules from the hemolymph into the cells. However, a recent analysis suggests that they export a metabolic intermediate (such as 3-hydroxy kynurenine) from the cytoplasm into the pigment granules of the *Drosophila* eye cells (Mackenzie *et al.*, 2000). Close homologues of these systems have been identified in several diptera but recently the availability of a large number of complete genomes has revealed an even broader distribution of these transporters. In addition to the three genes mentioned above, the genome of *D. melanogaster* contains 13 homologues of the white gene. Homologues of these genes were found in *Saccharomyces cerevisiae* (1 gene ADP1), *C. elegans* (8 genes) and *A. thaliana* (8 genes). Bacteria such as *Mycobacterium tuberculosis* and *Synechocystis* sp. PCC3803 were found to carry WHI family systems. This indicates that the range of functions performed by this family of transporters is broader than eye pigment precursor transport. Homologues of these transporters were also identified in mammals. The human and mouse white gene homologue ABCG1 (ABC8) is highly induced in lipid-loaded macrophages, suggesting a role in cholesterol and phospholipid trafficking (Klucken *et al.*, 2000; Venkateswaran *et al.*, 2000). Another homologue, ABCG2 (MXR, BCRP), is associated with anthracyclin drug resistance when overexpressed in certain cell lines (Allikmets *et al.*, 1998; Doyle *et al.*, 1998). Recently it was found that phytositosterolemia (elevation of plasma levels of plant sterols due to enhanced intestinal absorption and reduced removal) was caused by mutations in the human ABCG5 and ABCG8 transporters (Berge *et al.*, 2000). The properties of eukaryote WHI family systems have been reviewed recently (Schmitz *et al.*, 2001) (see also Chapter 28).

The PDR subfamily involved in pleiotropic drug resistance

Systems of this subfamily were probably generated by the duplication followed by the fusion of a WHI subfamily system **(Figure 1.2G)**. PDR subfamily systems are found in fungi and in plants (8 proteins in *A. thaliana*). In fungi, they are involved in the efflux of a wide variety of noxious substances (Bauer *et al.*, 1999; Wolfger *et al.*, 2001). Their properties are reviewed in Chapter 14. In the aquatic plant *Spirodela polyrrhiza*, the expression of the TUR2 gene is induced by environmental stress treatments such as low temperature or high salt (Smart and Fleming, 1996).

The CCM family involved in bacterial cytochrome *c* biogenesis

Bacterial CcmA (ABC), CcmB (IM) and CcmC (IM) proteins are required for cytochrome *c* synthesis and are thought to constitute the subunits of an ABC transporter. Despite the fact that they are not fused to the IM domains, the ABC proteins of this family cluster within class 1. The possible hypotheses raised to understand the functions of this transporter have been discussed recently (Goldman and Kranz, 2001). One hypothesis proposes that a complex consisting of two CcmA subunits and one each of CcmA and CcmB is involved in the transport of reduced heme into the periplasm. The second hypothesis concludes that a CcmA-CcmB heterodimeric ABC transporter does not transport heme but some other substrate required for cytochrome *c* biogenesis (Schulz *et al.*, 1999). Homologues of these proteins were found in the mitochondrial genome of some protists and red algae. ORFs homologous to CcmB and CcmC were found in the mitochondrial genome of plants (Bonnard and Grienenberger, 1995; Jekabsons and Schuster, 1995; Schuster, 1994), and this suggests that the missing gene encoding the ABC subunit has moved into the chromosome. This hypothesis has been recently proven to be true in the case of *A. thaliana* (Dassa, unpublished). ORFs similar to CcmC were found otherwise only in archaea.

The MCM family

This very small family comprises four proteins found in methanogenic bacteria. They consist of two ABC modules fused together although they cluster in class 1. Only one protein, AtwA of *Methanothermobacter thermautotrophicus*, has been investigated. It was found to be essential for the *in vitro* activity of the nickel enzyme methyl-coenzyme M reductase, which catalyzes the terminal step of methane formation in methanogenic archaea (Kuhner *et al.*, 1993; Rouviere *et al.*, 1985). The gene encoding this protein was located apart from the *mcr* operon encoding the subunits of methyl-coenzyme M reductase. However, more recent purification procedures of the enzyme demonstrated that AtwA was dispensable for activity and was probably a contaminant (Ellermann *et al.*, 1989), so the actual function of this protein is not known.

CLASS 2 CONTAINS SYSTEMS WITH NO KNOWN IM DOMAINS AND INVOLVED IN ANTIBIOTIC RESISTANCE AND CELLULAR PROCESSES OTHER THAN TRANSPORT

These families are characterized by the fact that the ABC subunit is made up of duplicated, fused ABC modules (ABC2). No known transmembrane proteins or domains are associated with these proteins **(Figure 1.4)**.

The RLI family

The mammalian interferon-induced 2′,5′-oligoadenylate/RNase L system is considered as a central pathway of interferon action and could possibly play a more general physiological role, for instance, in the regulation of RNA stability in mammalian cells (Bisbal *et al.*, 2000). The activity of RNase L is modulated by an ABC

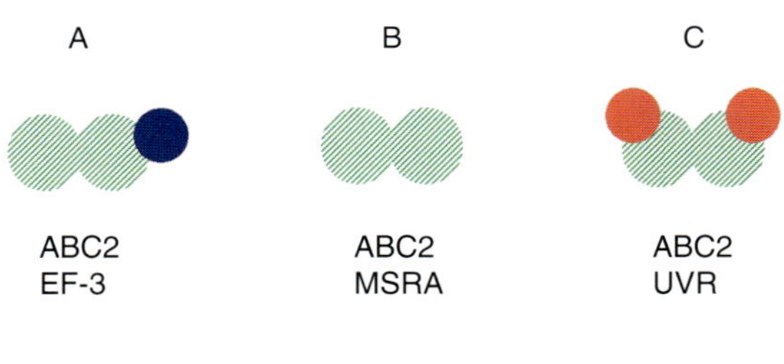

Figure 1.4. ***Typical organization of class 2 systems (non-transport processes). Same conventions as Figure 1.2 for the representation of ABC domains. A, Protein EF-3 (EF-3 subfamily) involved in the elongation of polypeptides in translation in yeast. The ribosome interaction domain is represented by a blue circle. B, Protein MsrA (ARE subfamily) involved in erythromycin resistance. C, Protein UvrA (UVR family) involved in DNA repair. The zinc finger domains that lie between the Walker motifs A and B are represented by red circles.***

protein called RNase L inhibitor or RLI (ABCE1) (Bisbal *et al.*, 1995). RLI proteins display a characteristic 90 amino acid long N-terminal domain similar to an iron–sulfur center. Proteins homologous to RLI were identified in lower eukaryotes and archaea but their function has not yet been investigated.

The ART family of systems involved in antibiotic resistance and in translation of mRNA and its regulation

On the basis of multiple alignments and phylogenetic trees, three subfamilies could be distinguished.

The EF-3 subfamily

Fungi appear to be unique in their requirement for a third soluble translation elongation factor EF-3 **(Figure 1.4A)**. This was first described in *S. cerevisiae* and has subsequently been identified in a wide range of fungal species (Chakraburtty, 2001). EF-3 stimulates binding of aminoacyl-tRNA to the ribosomal A-site by facilitating release of deacylated tRNA from the exit site (E-site). The YEF3 gene encoding EF-3 is essential for the survival of yeast. The deduced amino acid sequence of EF-3 has revealed the presence of duplicated ABC domains. The carboxy-terminus of EF-3 contains blocks of lysine boxes essential for its functional interaction with yeast ribosomes (Chakraburtty and Triana-Alonso, 1998). A homologue of EF-3 is carried by the genome of a large virus that infects eukaryotic chlorella-like green algae and is expressed during the entire infection process (Yamada *et al.*, 1993).

The REG subfamily of proteins involved in the regulation of diverse phenomena

This subfamily is comprised of eukaryote and prokaryote proteins known to participate in regulatory functions and of several prokaryote systems with unknown function. The most studied eukaryote ABC protein of this type is the yeast protein GCN20. This was shown to associate with another protein GCN1, in order to stimulate the activity of GCN2, a kinase that phosphorylates the eukaryotic translation initiation factor eIF2. This leads to increased translation of the transcriptional activator GCN4 in amino acid-starved cells (Marton *et al.*, 1997). GCN20 contains a lysine-rich N-terminal domain of about 200 residues, which is essential for binding to GCN1. A human homologue of GCN20, ABC50 (ABCF1) was recently shown to interact with eIF2 and to associate with ribosomes in an ATP-dependent manner (Tyzack *et al.*, 2000). Interestingly, several ORFs detected in bacterial complete genomes are homologous to GCN20, raising the possibility that they could be implicated in regulatory processes. Indeed, the *A. tumefaciens* ChvD protein was found to be inactivated in mutants selected for the reduced transcription of the *virA* and *virG* genes (Winans *et al.*, 1988).

The ARE subfamily of antibiotic resistance determinants in prokaryotes

The most thoroughly investigated representatives of this family are the staphylococcal MsrA and Vga proteins involved in virginiamycin and erythromycin resistance, respectively (Allignet *et al.*, 1992; Ross *et al.*, 1990) and several *Streptomyces* proteins involved in the immunity of bacteria against the antibiotics that they produce (Mendez and Salas, 2001). The mechanism of resistance is still an open question. The genes encoding these resistance determinants are located on plasmids and they are sufficient to provide antibiotic resistance (Ross *et al.*, 1996). No transmembrane protein partners for these ABC proteins have ever been detected **(Figure 1.4B)**.

The UVR family involved in DNA repair and drug resistance

Excision of damaged DNA in *E. coli* is accomplished by three proteins designated UvrA, UvrB and UvrC (Goosen and Moolenaar, 2001). The UvrA protein is composed of two fused ABC domains **(Figure 1.4C)**. In this protein, a large intervening sequence consisting of one zinc finger domain, separates the Walker motif A from the signature motif. This is the reason why the UvrA-like proteins were omitted from the multiple alignment since it was observed that their presence altered the quality of the multiple alignment used to compute the tree. However, pairwise local alignment programs using portions of UvrA deleted from the intervening sequences were used to assess the position of these proteins in class 2. UvrA is found mainly in eubacteria but an ORF probably orthologous to UvrA is present in the genome of the archaeon *M. thermautotrophicus* but not in the other archaea whose complete genome sequences are available. Interestingly, several *Streptomyces* species that produce antibiotics and drugs possess, in addition to the UvrA protein involved in DNA repair, a UvrA-like protein, which is involved in antibiotic self-immunity. This is the case with the DrrC protein of *Streptomyces peucetius*, which is able to confer daunorubicin resistance upon sensitive strains of *Streptomyces*. It was postulated that these UvrA-like proteins determined a new type of resistance mechanism, different from the drug efflux mechanism promoted by other ABC transporters (Lomovskaya *et al.*, 1996).

CLASS 3 CONTAINS SYSTEMS WITH UNFUSED IM AND ABC DOMAINS COMPRISING ALL KNOWN BPD TRANSPORTERS AND MORE

This class comprises all binding protein-dependent (BPD) systems, which are largely represented in archaea and eubacteria and which are primarily involved in scavenging solutes from the environment. BPD transporters require an extracytoplasmic substrate-binding protein (BP). The structure of BPs are discussed in detail in Chapter 10. This protein, an essential component for transport, is a periplasmic protein in Gram-negative bacteria (PBP) and a surface-anchored lipoprotein in Gram-positive bacteria and archaea. Very recently, it was shown that certain BPs of Archaea are attached to the membrane by an amino-terminal transmembrane segment (Albers *et al.*, 1999). The IMs of BPD transporters display a distinctive signature, the EAA motif, a 20 amino acid conserved sequence located at about 100 residues from the C-terminus. The motif is hydrophilic and it was found to reside in a cytoplasmic loop located between the penultimate and the antepenultimate transmembrane segment in all proteins with a known topology (Saurin *et al.*, 1994). The conservation of this motif argues for an important functional role and we found that it constitutes a site of interaction with the so-called helical domain of ABC proteins (Hunke *et al.*, 2000; Mourez *et al.*, 1997).

In addition to BPD importers, several systems of unknown function or that have been proposed to be involved in the export of drugs and polysaccharides were found in this class. These will be discussed in later sections after BPD importers for clarity, but it should be kept in mind that their ABC proteins do not cluster independently of those of ABC importers.

The MET family specific for metallic cations

This family is composed of systems involved in the uptake of various metallic cations such as iron, manganese and zinc (Claverys, 2001). Putative systems belonging to the MET family were found in the genomes of prokaryotes and in the cyanelle genome of the photoautotrophic protist *Cyanophora paradoxa*. The ATPases of these systems are strongly related to those of iron-siderophore uptake systems (ISVH family), suggesting that they arose from a common ancestor (Saurin *et al.*, 1999). Weaker but significant similarities could be detected between IM of the MET and ISVH families.

The ISVH family specific for iron-siderophores, vitamin B_{12} and hemin

The substrates handled by the ISVH family systems are quite different. Their common characteristic is to chelate iron (ferrichrome, enterobactin, achromobactin, anguibactin, citrate, exochelin, hemin, vibriobactin) or cobalt (vitamin B_{12}). All these systems are associated with high-affinity outer membrane receptors in Gram-negative bacteria, the activity of which is dependent on a transmembrane protein complex composed of TonB, ExbC and ExbD whose function is to transduce energy to the outer membrane **(Figure 1.5D)**. Once released from the outer membrane receptor, the substrate is translocated through the inner membrane thanks to an ABC BPD importer (Köster, 2001).

The OSP family specific for di- and oligosaccharides and polyols

The OSP family includes transport systems for malto-oligosaccharides, cyclodextrins, trehalose/maltose, cellobiose/cellotriose, arabinose oligomers and lactose. Members of this family also transport several polyols such as mannitol, arabitol, sorbitol (glucitol) and glycerol-3-phosphate (Schneider, 2001). Some systems can mediate the uptake of several oligosaccharides such as the raffinose/melibiose/isomaltotriose system of *Streptococcus mutans* (Russell *et al.*, 1992). Systems of this family have a highly conserved organization comprising a BP, two IMs and one ABC **(Figure 1.5B)**. In *Streptomyces reticuli*, it was demonstrated that a single ABC MsiK is involved in the energization of two different transporters specific for

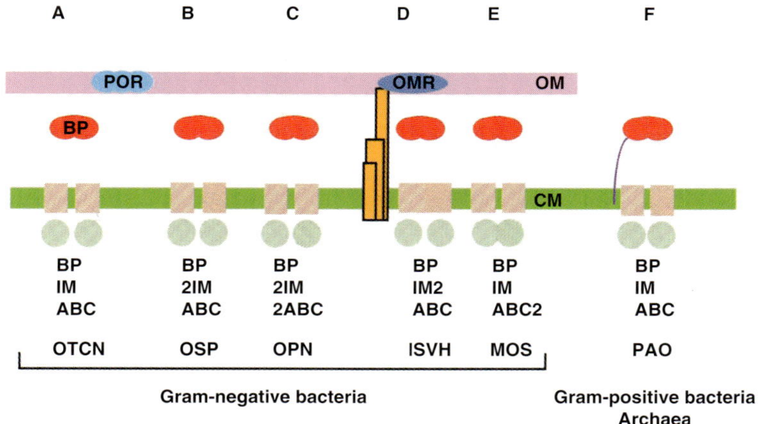

Figure 1.5. Typical organization of class 3 binding protein-dependent ABC importers. Same conventions as Figure 1.2 for the representations of membranes and for the IM and ABC domains. Gram-negative bacteria (A to E): All systems share the same organization: (i) An outer membrane channel that may be a general or a substrate-specific trimeric porin (POR) or for TonB-dependent systems (ISVH family), a high-affinity outer membrane receptor (OMR). The energy needed by the latter to translocate substrates into the periplasmic space is transduced from the cytoplasmic membrane to the outer membrane by the TonB, ExbB and ExbD complex (orange rectangles). (ii) A periplasmic solute-binding protein (BP). (iii) A cytoplasmic membrane complex. Gram-positive bacteria and Archaea: The solute-binding protein is a surface lipoprotein inserted into the membrane via a lipid anchor.

A, Glycine-betaine importer (OTCN family) composed of a homodimer of IM and a homodimer of ABC. B, Maltose importer (OSP family): a heterodimer of IM and a homodimer of ABC. C, Oligopeptide importer (OPN family): two heterodimers of IMs and ABCs. D, Ferric-hydroxamate importer (ISVH subfamily): the two IM domains are fused in a single polypeptide chain. E, Ribose importer (MOS family): the two ABC domains are fused in a single polypeptide chain. F, Glutamine importer (PAO family): a homodimer of IM and a homodimer of ABC.

maltose and cellobiose (Schlösser *et al.*, 1997). This property might be general for Gram-positive OSP transporters since several completely sequenced genomes display a large excess of IMs and BPs over ABCs (Quentin *et al.*, 1999). The best-characterized system of this family, the *E. coli* maltose/maltodextrin transporter energized by MalK, is reviewed in Chapter 9. The crystal structure of the archaeon *Thermococcus litoralis* MalK protein was reported with a resolution of 1.9 Å (Diederichs *et al.*, 2000).

The MOI family specific for mineral and organic ions

The MOI family includes transport systems for inorganic anions such as thiosulfate and sulfate (Kertesz, 2001), molybdate (Self *et al.*, 2001), and organic anions such as polyamines (Igarashi *et al.*, 2001) and thiamine. Members of this family also transport ferric iron (Köster, 2001). However, iron might be transported as a salt since crystals of the iron-binding protein of *Haemophilus influenzae* show that iron is coordinated by water and phosphate (Bruns *et al.*, 1997). The ABC component of importers specific for phosphate cluster apart from the MOI family. However, the IMs are clustered with the IMs of the MOI family. The MOI family is the largest family of BPD systems. Opines like mannopines and chrysopine are transported by MOI family systems similar to the polyamine transporters. Most systems of the MOI family have two IMs but ferric iron transporters have the two IM domains fused into a single polypeptide chain (IM2), while molybdate and thiamine transporters have only one IM.

The OTCN family involved in the uptake of osmoprotectants, taurine (alkyl sulfonates), alkyl phosphonates, phosphites, hypophosphites, cyanate and nitrate

This family comprises systems involved in the transport of apparently unrelated solutes. ABC and IM are grouped respectively in a single cluster. Analysis of BP sequences led to the identification of two non-overlapping clusters.

The first cluster groups systems involved in the transport of osmoprotectants, consisting of small modified peptides that contain an N,N,N trimethyl ammonium group like glycine-betaine, choline and carnitine (Hosie and Poole, 2001). The properties of the transporters specific for osmoprotectants were recently reviewed (Kempf and Bremer, 1998). The most characterized system is the osmoregulatory ProU system of *E. coli*, determining a glycine-betaine transporter, which consists of genes encoding ProV (ABC), ProW (IM) and ProX (BP) . They display an organization typical of BPD transporters. The OPU transporter of *Lactococcus lactis* (described in more detail in Chapter 13) constitutes a remarkable exception to this organization scheme, where an extracytoplasmic domain corresponding to the BP is fused to the C-terminus of the IM (Obis *et al.*, 1999). The second cluster is composed of systems involved in the uptake of nitrate, cyanate, N-alkylsulfonates, alkylphosphonates, phosphites and hypophosphites.

The OPN family specific for di- and oligopeptides and nickel

Oligopeptides constitute an important source of nutrients and several systems are also involved in cell–cell communication (Detmers *et al.*, 2001). Members of the OPN family have been found in all prokaryotic genera and are characterized by the fact that the two ABC subunits are encoded by different genes **(Figure 1.5C)**. Oligopeptide-like transporters have been implicated in the uptake of a class of opines such as agrocinopines, agropinic and mannopinic acids (Hayman *et al.*, 1993).

Nickel is an essential cofactor for a number of enzymatic reactions. The Nik system of *E. coli* provides Ni^{2+} ions for the anaerobic biosynthesis of hydrogenases and is similar in its composition and in the primary sequence of its components to the oligopeptide ABC transporters (Navarro *et al.*, 1993). Nik importers appear to be more restricted in their distribution than oligopeptide transporters since homologues could be identified only in the genomes of *Staphylococcus aureus*, *Bacillus halodurans* and *Deinococcus radiodurans*.

The PAO family specific for polar amino acids and opines

The PAO family includes transport systems for amino acids that have polar or charged side chains: lysine, histidine, ornithine, arginine, glutamine, glutamate, cystine and diaminopimelic acid (Hosie and Poole, 2001). Opines like octopine (*N*2-(1-carboxyethyl)-L-arginine) and nopaline (*N*2-(1,3-dicarboxypropyl)-L-arginine) are transported in agrobacteria by PAO family transporters. Typical systems have in general two IMs with the exception of the cystine- and the glutamine-specific systems, which have only one IM **(Figure 1.5A, 1.5F)**. The BPs specific to glutamine are homologous to the extracellular portion of eukaryote ionotropic glutamate receptors. Recent studies indicated that glutamate receptors share with the bacterial PAO family BPs the fundamental mechanism of amino acid recognition (Lampinen *et al.*, 1998). The best-characterized system of this family is the *Salmonella typhimurium* histidine transporter, which is energized by HisP, the first ABC protein whose crystal structure was reported with a resolution of 1.5 Å (Hung *et al.*, 1998).

The HAA family specific for hydrophobic branched-chain amino acids and amides

The HAA family comprises systems specific for the transport of the hydrophobic amino acids leucine, isoleucine and valine (Hosie and Poole, 2001). A transport system involved in the uptake of urea and short-chain aliphatic amines in *Methylophilus methylotrophus* belongs to this family (Mills *et al.*, 1998). This system is homologous to the *Synechocystis* and *Anabaena* systems for the uptake of neutral amino acids Ala, Val, Phe, Ile, and Leu (Montesinos *et al.*, 1997). It is therefore possible that the urea transporter of *M. methylotrophus* could also transport such amino acids. Systems of the HAA family have a characteristic organization made up of one or several BPs, two IMs and two ABCs. The eukaryote gamma-aminobutyric acid type B (GABA(B)) receptors and the related metabotropic glutamate receptor-like family of G-protein-coupled receptors have their extracellular domains homologous to the bacterial leucine-binding protein. Furthermore, the effect of point mutations can be explained by the Venus flytrap model, which proposes that the initial step in the activation of the receptor by the agonist results from the closure of the two lobes of the binding domain (Galvez *et al.*, 1999).

The MOS family specific for monosaccharides

The MOS family systems are involved in the uptake of monosaccharides (pentoses and hexoses) like arabinose, D-allose, galactose, ribose and xylose. The typical organization of these systems consists of one BP, one IM and one ABC **(Figure 1.5E)**. This ABC subunit is made up of two homologous halves, suggesting that a primordial gene duplication and subsequent fusion event occurred in the generation of the ancestral MOS system (Schneider, 2001). In the *B. subtilis*, *Treponema pallidum*, *Borrelia burgdorferi*, *Archeoglobus fulgidus* and *Aeropyrum pernix* sequenced genomes, several putative MOS family transporters were identified. However, the putative operons encoding these systems were apparently devoid of a typical substrate-binding protein. Rather, they were associated with secreted proteins homologous to a family of lipoproteins of unknown function, the so-called basic membrane proteins C (BMPC), which constitute potent immunogens in pathogenic bacteria. Psi-Blast analyses show that these lipoproteins display significant similarity to MOS family substrate-binding proteins. We therefore speculate that at least some BMPCs might be involved in the uptake of a yet unidentified monosaccharide.

Class 3 systems of unknown function that could be importers

The CBY family

The CBU subfamily putatively involved in cobalt uptake

This subfamily comprises systems which are found in operons encoding genes involved in cobalamin biosynthesis. Cobalamin is derived from uroporphyrinogen III, a precursor of heme, siroheme and chlorophylls, and a cobalt ion is chelated in the center of the molecule. The genes necessary for cobalamin production are organized in a single operon in *S. typhimurium* (Jeter and Roth, 1987). In addition to genes known to encode enzymes catalyzing steps of the cobalamine biosynthetic pathway, the products of *cbiQ*, *cbiO* and *cbiN* were proposed to constitute a cobalt uptake system since CbiN and CbiQ are integral membrane proteins and CbiO is an ABC ATPase **(see Figure 1.6A)** (Roth *et al.*, 1993). However, direct evidence supporting this idea is lacking. Mutations affecting cobalamin biosynthesis were never found in *cbiNOQ* genes. The CbiQ proteins are related to the ABCs of the MET–ISVH families. No substrate-binding protein has been identified in the *cbi* operons and the exact function of the *cbiNOQ* genes remains unknown. CBU systems are found in several bacteria and archaea.

The Y179 subfamily of unknown function

The genomes of several eubacteria and archaea contain ORFs homologous to the *cbiO* gene, but they lack *cbiNQ* genes. Such ORFs cluster near the ABCs of the CBU subfamily and they constitute the Y179 subfamily. A typical system is composed of two genes encoding ABC proteins followed by one gene encoding the IM. These systems are not located close to cobalamin biosynthetic genes and their function has never been investigated.

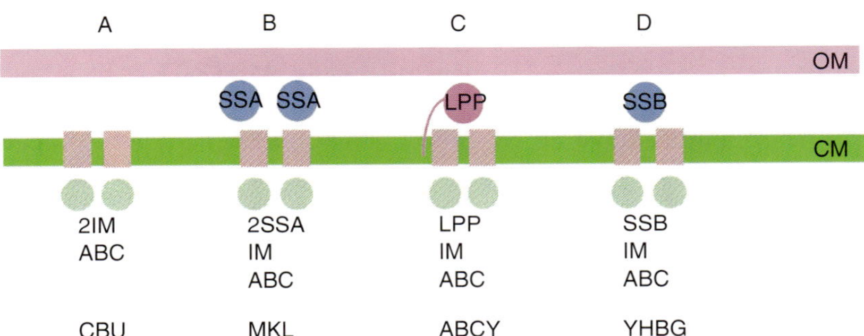

Figure 1.6. Typical organization of class 3 systems that could be importers. Same conventions as Figure 1.2 for the representations of membranes and for the IM and ABC domains. A, CBU subfamily system putatively involved in the import of cobalt. B, MKL family system of unknown function (SSA: periplasmic protein), a system contains at least two homologous genes encoding such an SSA. C, ABCY family system of unknown function (LPP: periplasmic lipoprotein). D, YHBG family system of unknown function (SSB: periplasmic protein).

The MKL family

This family is composed of systems with unknown function found mainly in Gram-negative bacteria. A typical system is composed of genes encoding one ABC, one IM and two SSA proteins with a putative signal sequence **(Figure 1.6B)**. The ABC proteins are related to those of the MOI family but no significant similarity could be detected between SSA proteins and BPs of BPD transporters. Some systems have been partially characterized. In *Pseudomonas putida*, a Tn*5* insertion within the *ttg2A* gene, encoding an MKL family ABC, renders the cells sensitive to toluene (Vermeij *et al.*, 1999). In *Shigella flexneri*, a mutation in the gene encoding the SSA protein VspC was found to inhibit intracellular spreading of bacteria but not invasion, and to promote an increase in secreted virulence proteins and in sensitivity to sodium dodecylsulfate (Hong *et al.*, 1998). In *Campylobacter jejuni*, the IamA (ABC) IamB (IM) system has been associated with the phenotype of adherence to and invasion of epithelial cells (Carvalho *et al.*, 2001). Interestingly, the IM and the SSA proteins of an MKL family system in *M. tuberculosis* have been subjected to extensive gene duplications, so that eight genes for IMs and ten genes for SSA proteins were identified in the genome. It was reported that IM Rv0169 (Mcep) protein was associated with the entry and survival of the bacterium inside cells (Arruda *et al.*, 1993). It is therefore tempting to speculate that MKL systems are implicated in the maintenance of bacterial outer cell surface integrity. MKL systems are apparently not restricted to prokaryotes since a typical system was found in the nuclear genome of *A. thaliana*.

The ABCY family

Systems of the ABCY family have an overall organization similar to that of BPD transporters with the difference that the extracytoplasmic protein is a lipoprotein (LPP), even in Gram-negative bacteria. They are composed of one ABC, one IM and the LPP **(Figure 1.6C)**. IMs display strong similarity to those of the OTCN family and have the characteristic EAA motif found exclusively in a cytoplasmic loop of IMs of the BPD import systems. ATPases of the family cluster near the ATPases of the PAO family. This unusual feature of the components of ABCY systems might indicate that they originate from the association of components from different families of BPD transporters. LPPs display a slight similarity to BPs of the OTCN family and belong to a family of surface LPPs, which includes the NlpA lipoprotein of *E. coli*. The function of ABCY systems has not been investigated. They could be involved in the import of an as yet unidentified molecule. Interestingly, in *Salmonella enteritidis*, the ABCY family SfbABC system was located in a pathogenicity islet of 4 kilobases. It is inducible by iron limitation and by acidic pH and it was found that inactivation of the *sfbA* gene encoding LPP resulted in a mutant that is avirulent and induces protective immunity in BALB/c mice (Pattery *et al.*, 1999).

The YHBG family

The ABCs of the YHBG family are related to those of the HAA family. The genes encoding these polypeptides are very often located 5′ to the *ntrA* gene encoding sigma factor 54, involved in nitrogen regulation and diverse physiological functions in bacteria. The genes for these ABCs are associated with a putative secreted protein (SSB) and with an IM **(Figure 1.6D)**. Indirect evidence suggests that *yhbG* is an essential gene for *R. meliloti*, whose transcription is not linked to that of *ntrA* (Albright *et al.*, 1989).

CLASS 3 SYSTEMS WHICH APPARENTLY ARE NOT INVOLVED IN IMPORT

The o228 family involved in release of lipoproteins from the cytoplasmic membrane

A typical system of this family comprises one ABC and two IMs **(Figure 1.7A)**. Only one system has been experimentally characterized, the LolC (IM) LolD (ABC) LolE (IM) system of *E. coli*. Consequently, it is difficult to generalize to the whole family the properties of this system. Lipoproteins directed to the outer membrane are released from the inner membrane in an ATP-dependent manner through the formation of a complex with LolA, a periplasmic chaperone. The LolCDE complex catalyzes the release of lipoproteins from the cytoplasmic membrane to LolA into the periplasmic space prior to their targeting to the outer membrane (Yakushi *et al.*, 2000). An outer membrane lipoprotein, LolB, then mediates the transfer of lipoproteins to their final location in the outer membrane (Tanaka *et al.*, 2001). Since this complex is involved in neither export nor import of any

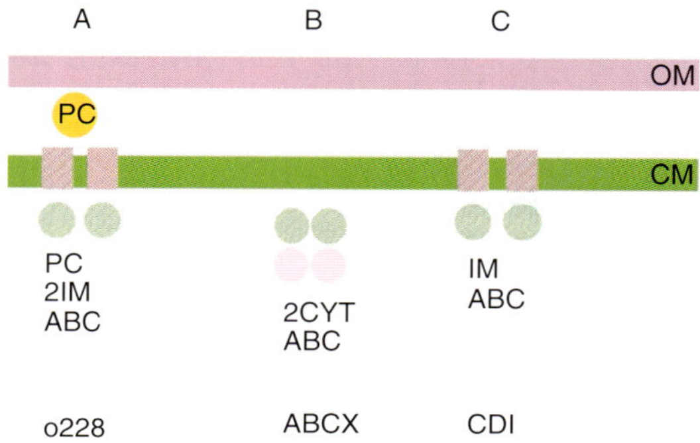

Figure 1.7. ***Typical organization of class 3 systems that are not known to be importers. Same conventions as Figure 1.2 for the representations of membranes and for the IM and ABC domains. A, o228 subfamily system involved in the release of lipoproteins from the cytoplasmic membrane (PC: periplasmic chaperone). B, ABCX subfamily system that could be involved in [Fe/S] center formation. C, CDI family system involved in cell division.***

known molecules, it may constitute a new type of ABC system. The genes encoding the complete LolABCDE system are found in Gram-negative bacteria. However, homologues of the LolCDE system have been found also in Gram-positive bacteria and archaea, but in these cases the homologues of LolA and LolB are lacking. This suggests that the members of this family might be involved in a more general lipoprotein-releasing mechanism common to all prokaryotes or in an as yet unidentified function.

The ABCX family

These systems are found in the genomes of several bacteria and archaea and on the plastid genome of red algae. They consist of one gene encoding the ABC protein almost always associated with two genes encoding two conserved cytosolic proteins (CYT) **(Figure 1.7B)**. In most eubacteria, these three genes are found in an operon containing genes encoding two ORFs displaying homology to the NifS and to the IscA proteins, respectively. NifS is a pyridoxal 5′-phosphate-dependent L-cysteine desulfurase producing alanine and elemental sulfur and it seems to play a general role in the mobilization of sulfur for iron/sulfur cluster biosynthesis (Zheng *et al.*, 1993). IscA is a protein involved in the transfer of iron/sulfur center to apoproteins (Tokumoto and Takahashi, 2001). Plastid genomes of red algae usually encode only one CYT protein and it could be speculated that the second one has moved into the nuclear genome. Plant chloroplast genomes usually lack these systems. However, three ORFs homologous to each ABC and CYT proteins were found in the nuclear genome of *A. thaliana*. This observation strengthened the hypothesis suggesting that plant ABCX systems migrated to the nuclear genome of plants. The function of ABCX family systems is unknown and it could be speculated that these ABCs do not function as transporters since no IMs have been associated with them. A genetic screen identified the *A. thaliana* CYT protein AtABC1, whose inactivation determines a long hypocotyl phenotype and the accumulation of protoporphyrin IX. It was suggested that functional atABC1 is required for the transport and correct distribution of protoporphyrin IX, which may act as a light-specific signaling factor involved in coordinating intercompartmental communication between plastids and the nucleus (Moller *et al.*, 2001). Recently, it was found that the SufC ABC protein of the plant pathogen *Erwinia chrysanthemi*, encoded within a typical ABCX operon, is essential for virulence and for the SoxR-dependent oxidative stress response. It was concluded that SufC could be a versatile ATPase that can associate either with the other Suf proteins to form a Fe–S cluster-assembling machinery or with membrane proteins encoded elsewhere in the chromosome to form an Fe–S ABC exporter (Nachin *et al.*, 2001).

The CDI family involved in cell division

CDI family systems are found only in eubacteria and comprise two proteins: the FtsE ABC and the FtsX IM expected to homodimerize in order to form a transporter **(Figure 1.7C)**. The ABCs cluster very closely with those of the PAO family. However, the sequences of the IM do not show significant similarity to those of the PAO family and they have no EAA motif. The *ftsE*(Ts) mutation of *E. coli* causes defects in cell division and cell growth. An *ftsE* null mutant showed filamentous growth and appeared viable on high-salt medium only, indicating a role for FtsE in cell division and/or salt transport (de Leeuw *et al.*, 1999). Recently, it was shown that the membrane insertion of the KdpA potassium transporter is affected in *ftsE* mutants (Ukai *et al.*, 1998). It is therefore possible that CDI systems play a role in the proper membrane targeting or insertion of some proteins essential for septum formation.

CLASS 3 SYSTEMS WHICH COULD BE INVOLVED IN EXPORT

The great majority of these systems are composed of ABC and IM domains carried by independent polypeptide chains. They are found in prokaryotes, with one exception: the ABCA subfamily discussed below. Although these systems have ABC subunits related to those of importers, indirect experimental evidence led to the idea that they could be involved in antibiotic resistance by an efflux mechanism (the DRA and DRI families) and in the export of complex polysaccharides (the NOD subfamily and the CLS family). The IM proteins do not have the EAA motif.

The DRA family

This vast family is characterized by a strong sequence conservation of ABC subunits contrasting with a wide variety of associated functions. In eubacteria and archaea, the typical system consists of one gene encoding the ABC and one or two genes encoding the IMs leading to the presumed organization shown in **Figure 1.8A**. This family may be subdivided into four subfamilies on the basis of the clustering of ABC proteins or domains. Among these, the ABCA subfamily is exceptional in the sense that it is composed of systems found exclusively in eukaryotes and having the ABC domains fused to the IM domains.

The DRR subfamily of systems involved in polyketide drug resistance

Doxorubicin, daunorubicin, oleandomycin and mythramycin are antibiotics synthesized by multifunctional polyketide synthases. *Streptomyces* species that produce these drugs are resistant to their action. The best-characterized system is the daunorubicin resistance determinant of *S. peucetius*, which consists of two proteins, DrrA (ABC) and DrrB (IM), believed to export the antibiotic out of the cell (Guilfoile and Hutchinson, 1991), although active efflux of the antibiotics has never been demonstrated (Mendez and Salas, 2001). Homologues of these proteins are found only in eubacterial and archaeal genomes.

The NOD subfamily involved in nodulation

Rhizobial lipochito-oligosaccharidic Nod factors mediate specific recognition between leguminous plants and their prokaryotic symbionts. Mutations in *nodI* (ABC) or *nodJ* (IM) induce a delayed phenotype in plant nodulation and it was suggested that NodI and NodJ proteins

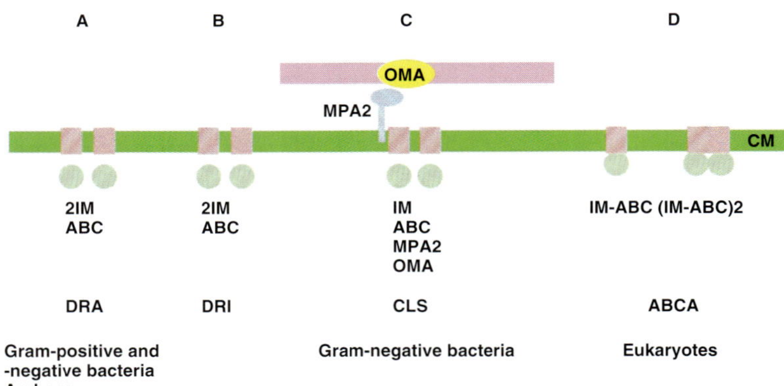

Figure 1.8. Typical organization of class 3 systems that could be exporters. Same conventions as Figure 1.2 for the representations of membranes and for the IM and ABC domains. A, Organization of DRA family systems involved in the resistance to polyketide drugs, in nodulation and sodium ion extrusion. B, Representative organization of DRI family systems involved in the resistance to peptide drugs, in bacteriocin and lantibiotic immunity. C, A CLS family system involved in capsular polysaccharide export. A typical system comprises an outer membrane protein (OMA), a periplasmic membrane protein (MPA2), a cytoplasmic membrane protein complex composed of a homodimer of an integral membrane protein (IM) and a homodimer of the ATP-binding cassette subunit (ABC). D, Two types of proteins of the ABCA subfamily. The systems with an IM-ABC organization are found in completed genomes and none has been characterized. It is not known if they determine homo- or heterodimeric transporters. The ABCA1 and the ABCR proteins display an (IM-ABC)2 organization.

play a role in the efficiency of secretion of Nod factors (Spaink *et al.*, 1995). It has been proposed that the NodIJ system is a transporter that mediates the export of Nod factors. However, strains carrying disrupted *nodIJ* genes are still able to secrete such Nod factors at a reduced rate as compared to wild-type strains (Cardenas *et al.*, 1996). Members of the NOD subfamily are exclusively found in rhizobia and sequence comparison studies revealed that NodIJ proteins are homologous to drug resistance proteins of the DRA family (Reizer *et al.*, 1992).

The NAT subfamily

A transposition mutant of *B. subtilis*, isolated on the basis of growth inhibition by Na^+ at elevated pH, was found to be deficient in energy-dependent Na^+ extrusion. The site of transposition was in an operon encoding NatA (ABC) and NatB (IM), which were proposed to constitute a sodium extrusion pump (Cheng *et al.*, 1997). Systems with proteins homologous to NatA and NatB were found in prokaryotes only.

The ABCA subfamily involved in lipid trafficking

Members of this subfamily are exclusively found in eukaryotes and display an IM-ABC or an (IM-ABC)2 organization **(Figure 1.8D)**, similar to that of the DPL family of exporters (see above). The ABC domains are highly similar to the ABC proteins of the DRR and NOD subfamilies (see the sections above), while no significant similarity could be detected between IM domains of ABCA proteins and the IM proteins of the DRR–NOD subfamilies. The properties of mammalian ABCA subfamily transporters have been recently reviewed (Broccardo *et al.*, 1999) and will be discussed in more detail in Chapter 23. The best-studied systems are the human ABC1 (ABCA1) and ABCR (ABCA4) proteins. The ABCA1 protein is involved in the inherited Tangier disease, characterized by a defect in cellular cholesterol removal, which results in the absence of high-density lipoproteins (HDL) in plasma and in massive tissue deposition of cholesteryl esters (Bodzioch *et al.*, 1999; Rust *et al.*, 1999). In fibroblasts, an ABCA1-dependent release of cholesterol was demonstrated (Orso *et al.*, 2000). ABCA1 regulates HDL levels and is considered to control the first step of cellular reverse cholesterol transport from the periphery to the liver by transferring cellular cholesterol and phospholipids to apolipoproteins. However, its direct role in promoting cholesterol efflux is still questioned (Groen *et al.*, 2001; Wang *et al.*, 2001). The ABCR protein is involved in Stargardt disease and in age-related macular degeneration (Allikmets *et al.*, 1997; Azarian and Travis, 1997) (see also Chapter 28). The ABCR protein is located in retina rod outer segment disks. Analysis of the phenotype of ABCR knockout mice suggests that the protein functions as a flippase for *N-N*-retinylidene-phosphatidyl ethanolamine that translocates this molecule to the cytosolic side of the disk membrane, where it is reduced to all-*trans*-retinol and subsequently released into the cytoplasm (Weng *et al.*, 1999). At least 11 other human genes encoding ABCA systems have been identified. These include ABCA2, which is highly expressed in brain, ABCA3, which is homologous to the *C. elegans* ced-7 gene involved in the engulfment of cell corpses during programmed cell death (Wu and Horvitz, 1998), and ABCA7, putatively involved in macrophage transmembrane lipid transport (Kaminski *et al.*, 2000). Twelve different members of this family were found in the genome of *A. thaliana*, eight in that of *C. elegans* and 14 in that of *D. melanogaster* but none in *S. cerevisiae*. Six of the *A. thaliana* proteins have an IM-ABC type organization.

The DRI family involved in drug resistance, and bacteriocin and lantibiotic immunity

This family is composed of systems having the same global organization as those of the DRA family. However, the ABCs of the two families cluster independently **(Figure 1.1)**. No significant similarity could be detected between IM proteins of these two families. DRI family systems are found in prokaryotes and some of these are involved in antibiotic resistance and in bacteriocin and lantibiotic immunity.

The BAI subfamily involved in bacteriocin immunity

Immunity towards bacteriocins such as lacticin RM, pediocin A and butyrivibriocin AR10 is conferred by transporters composed of one ABC and one IM. These bacteriocins are synthesized by peptide synthases, similar to the enzymes that produce peptide antibiotics. The CylAB system, involved in the production of hemolysin and pigment in *Streptococcus agalactiae*, belongs to this subfamily (Spellerberg *et al.*, 1999).

The LAI subfamily of lantibiotic immunity systems

The gene clusters determining the biosynthesis, modification and export of lantibiotics also contain genes involved in self immunity (Otto and Gotz, 2001). Immunity towards lantibiotics is achieved by genes encoding one ABC and two IMs. Inactivation of any of these genes resulted in the complete loss of the immunity phenotype. It was proposed that immunity towards lantibiotics is mediated by active efflux through the transporter (Otto *et al.*, 1998). However, an alternative hypothesis, where the transporter mediates the import of the lantibiotic into the cytoplasm, where it is subsequently degraded, was not excluded.

The DRB subfamily involved in peptide antibiotic resistance

The best-characterized system in this subfamily is the branched cyclic dodecyl peptide bacitracin resistance determinant BcrABC in *Bacillus licheniformis* (Podlesek *et al.*, 1995). This antibiotic is synthesized non-ribosomally by large multienzymatic polypeptide synthases. A typical system comprises one or two genes encoding IMs and one gene encoding the ABC; the latter is expected to homodimerize in the transporter, leading to the presumed organization described in **Figure 1.8B**. It is thought that such systems determine antibiotic resistance owing to an active efflux of the drug through the membrane, but this hypothesis awaits direct experimental support (Mendez and Salas, 2001). In the BcrABC system, it was shown that expression of the IMs BcrB and BcrC was sufficient to provide a significant level of bacitracin resistance, which was increased when the ABC BcrA was coexpressed (Podlesek *et al.*, 2000). *E. coli* has only a BcrC homologue, which is involved in the intrinsic bacitracin resistance of this bacterium (Harel *et al.*, 1999).

The NOS subfamily

In bacteria capable of dissimilatory nitrous oxide (NO) reduction, the genes essential for this function are organized in an operon containing the gene *nosZ* encoding the NO reductase followed by three genes encoding a periplasmic protein NosD, an ABC NosF and an IM NosY, respectively. Transposon insertion downstream of *nosZ* has a nitrous-oxide-reduction-negative phenotype and the NosZ protein is produced in an apo form, devoid of copper. It was therefore suggested that the NosDFY system encodes a copper ABC importer (Zumft *et al.*, 1990), but this has not yet been supported by direct experimental evidence. Homologues of NosDFY were found in several prokaryotes. In particular, a NOS family system was found to be essential for type IV pilus biogenesis in *Myxococcus xanthus* (Wu *et al.*, 1998).

The CLS family involved in extracellular polysaccharide export

Among the variety of membrane-linked or extracellular polysaccharides excreted by bacteria, only capsular polysaccharides, lipopolysaccharides and teichoic acids have been shown to be exported by ABC transporters (Silver *et al.*, 2001). A typical system consists of one gene encoding the ABC and one gene encoding the IM. These proteins are predicted to homodimerize in order to lead to the organization presented in **Figure 1.8C**. In addition to these proteins, capsular polysaccharide exporter systems require two 'accessory' proteins to perform their function: a periplasmic (*E. coli*) or a lipid-anchored outer membrane protein called OMA (*Neisseria meningitidis* and *H. influenzae*, for example) and a cytoplasmic membrane protein MPA2 (Paulsen *et al.*, 1997). The proteins that are common to the CLS family (IM and ABC) segregate within two distinct clusters corresponding to capsular polysaccharide exporters and lipopolysaccharide exporters, respectively. The IM and ABC of other extracellular polysaccharide (such as teichoic acids) exporters in several Gram-positive bacteria, mycobacteria and archaebacteria cluster in the lipopolysaccharide-specific group of sequences.

Lessons from Genome Comparisons

The complete nucleotide sequence of several genomes is now available and efforts have been developed to build complete inventories of ABC systems, in yeast (Decottignies and Goffeau, 1997), *E. coli* (Dassa *et al.*, 1999; Linton and Higgins, 1998), *B. subtilis* (Quentin *et al.*,

1999), and *M. tuberculosis* (Braibant *et al*., 2000). Global comparisons of the ABC protein content of several genomes have also appeared (Paulsen *et al*., 1998, 2000; Tomii and Kanehisa, 1998). In the course of the construction of ABSCISSE, our database of ABC systems (see the internet address in the footnote on the first page of this chapter), we have also analyzed the composition of 31 completely sequenced genomes **(Table 1.2)**.

When the total number of ABC systems is plotted against the genome size, the number of ABC systems is about 15, while genome size varies from 0.5 to 1.5 megabases (Mb). Most of the bacteria within this genome size range are intracellular parasites. In such bacteria able to grow inside cells, the presence of homologous host genes or the availability of a metabolite can lead to gene inessentiality and to subsequent disruption or deletion of the gene. It is

TABLE 1.2. GENOME STATISTICS OF ABC PROTEINS IN LIVING ORGANISMS

The complete genomes of representatives of three taxa of life were analyzed: eubacteria (B, 21 species), archaea (A, 6 species) and eukaryotes (E, 4 species). The number of bacterial systems may be larger than indicated since a single ATPase might energize more than one system. For each species, the genome size in megabases (Size, Mb), the gene number (Total ORFs) and the total number of proteins displaying the ABC signature (Total ABC) are given. In the next columns, the number of ABC proteins belonging to each class of ABC systems as defined in the text and in Table 1.1 is indicated. See Chapter 3 for details of human ABSs.

Genome	Taxon	Size (Mb)	Total ORFs	Total ABC	Class 1	Class 2	Class 3	NO
Mycoplasma genitalium	B	0.58	484	14	2	1	11	
Ureaplasma urealyticum	B	0.75	613	16	3	1	11	1
Mycoplasma pneumoniae	B	0.81	689	15	2	1	11	1
Chlamydia trachomatis	B	1.05	894	14	1	2	11	
Rickettsia prowazekii	B	1.11	834	15	6	2	6	1
Treponema pallidum	B	1.14	1031	17	0	2	15	
Chlamydia pneumoniae AR39	B	1.23	1110	15	1	2	12	
Borrelia burgdorferi	B	1.44	850	14		2	12	
Aquifex aeolicus	B	1.55	1522	13	2	1	10	
Campylobacter jejuni	B	1.64	1654	28	3	3	22	
Helicobacter pylori J99	B	1.64	1491	20	3	2	13	2
Helicobacter pylori	B	1.66	1553	19	3	2	12	2
Haemophilus influenzae Rd	B	1.83	1709	45	8	5	32	
Thermotoga maritima	B	1.86	1846	64	6	1	55	2
Neisseria meningitidis	B	2.27	2025	23	3	5	15	
Deinococcus radiodurans R1	B	3.28	3124	61	6	5	49	1
Synechocystis sp. PCC6803	B	3.57	3169	54	11	3	39	1
Bacillus subtilis	B	4.21	4100	84	8	5	67	4
Mycobacterium tuberculosis	B	4.41	3918	38	9	3	25	1
Escherichia coli K12	B	4.64	4289	78	7	5	63	3
Pseudomonas aeruginosa	B	6.26	5565	88	12	6	67	3
Methanococcus jannaschii	A	1.66	1715	16	0	3	13	0
Aeropyrum pernix	A	1.67	2694	41	0	1	38	2
Methanobacterium thermoautotrophicum	A	1.75	1869	16	0	4	12	0
Pyrococcus abyssi	A	1.76	1765	33	2	1	29	1
Pyrococcus horikoshii	A	1.73	2064	33	2	1	29	1
Archaeoglobus fulgidus	A	2.18	2420	40	0	1	38	1
Saccharomyces cerevisiae	E	13	6280	29	21	6	0	2
Caenorhabditis elegans	E	87.56	19256	60	48	4	8	0
Drosophila melanogaster	E	132.5	13600	55	38	4	13	0
Arabidopsis thaliana	E	115.7	25498	116	88	8	16	4

therefore possible that the ABC systems that are common in these species constitute the minimal requirement of ABC systems for life. As the size of the genome increases, the number of ABC systems apparently increases linearly, in agreement with the observation that the number of transporters of all categories (ion gradient-driven, PTS, ABC, facilitators) is approximately proportional to genome size (Paulsen *et al.*, 1998). There are, however, some exceptions. The genome of *Thermotoga maritima* has a very high content of ABC systems compared with that of species of similar genome size. This is due to the extensive amplification of operons encoding ABC systems putatively involved in the uptake of oligosaccharides (11 systems) and oligopeptides (12 systems). On the other hand, the genome of *M. tuberculosis* (4.4 Mb) has only 38 systems. This number is significantly lower than that found in *E. coli* (4.6 Mb, 78 systems) or in *B. subtilis* (4.2 Mb, 84 systems). Since it was found that the total number of transporters was fairly constant among prokaryotes (Paulsen *et al.*, 2000), this means that bacteria with low ABC system contents compensate for this deficiency by having a higher number of transporters from other functional categories. Eukaryotes display a smaller number of ABC systems with respect to genome size when compared with prokaryotes, and this is particularly evident in the case of *S. cerevisiae*, a free-living microorganism which shares with bacteria almost the same ecological niches. Indeed, high-affinity BPD importers are lacking in eukaryotes.

Class 1 ABC systems (exporters with fused ABC and IM domains) are not well represented in the genomes of bacteria and are virtually absent from the genomes of archaea. By contrast, they represent the major fraction of ABC systems in eukaryotes. Class 2 ABC systems (ABC2 organization, no IM domains) are found in all genomes, even in the smallest ones. This observation establishes the physiological importance of this class of ABC systems, which contains proteins experimentally or putatively involved in the regulation of gene expression. The number of class 2 systems by genome ranges from one to eight when the genome sizes vary from 0.58 to 132.5 Mb. Class 3 systems (mostly importers) are almost exclusively found in prokaryote genomes with one exception: the ABCA subfamily of eukaryote systems (Broccardo *et al.*, 1999). Uncompleted class 3 systems are also found in the genomes of eukaryotes and are probable remnants of BPD transporters present on the genome of the ancestor of organelles.

Very few families of ABC systems appear to be species- or kingdom-specific **(Table 1.2)**. Apart from the MCM family, which is found only in methanogenic archaea, and the PDR subfamily, which is found only in plants and fungi, most families have been identified in more than one kingdom.

CONCLUSIONS AND PERSPECTIVES

1. Each of the three classes of ABC systems contains proteins from the three kingdoms: archaea, bacteria and eukaryotes. The separation of eukaryotic from prokaryotic systems does not occur at the root of the clusters. Homologous systems from the three kingdoms are present at the tip of the branches of the tree. This suggests that ABC systems began to specialize very early, probably before the separation of the three kingdoms of living organisms (Saurin *et al.*, 1999), and that functional constraints on the ABC domain were responsible for the conservation of sequences. Another explanation would be the occurrence of horizontal gene transfer between the three kingdoms.
2. There is a quite good correlation between the sequence of the ABC ATPase and the overall function of the system to which it belongs. This is probably due to the fact that ABC domains segregate mostly according to sequence differences in the so-called helical domain that lies between the Walker motifs A and B. In the maltose system, we have demonstrated that this region is critical for the interaction between the ABC MalK and the conserved EAA loop in IMs MalF and MalG (Hunke *et al.*, 2000; Mourez *et al.*, 1997). Indeed, in the crystal structure of MsbA, a cytoplasmic loop of the transmembrane domain, which is highly conserved amongst members of the DPL family, is in close contact with the helical portion of the ABC domain (Chang and Roth, 2001, Chapter 7). The relationship with substrate specificity would reflect more general constraints imposed by the interaction of the ATPase with its partners.
3. The divergence between import, export and other systems probably occurred once in the

history of ABC systems. However, in addition to BPD importers, class 3 contains several transporters whose function is unknown or could not be conclusively related to import. Some systems of unknown function could be predicted to be importers in view of the strong similarity of their constituents with those of experimentally characterized importers. Others, like the systems of the DRA family (involved in drug and antibiotic resistance, see above) and the CLS family (involved in the biogenesis of capsular polysaccharides, lipopolysaccharides and teichoic acids, see above), have been suggested to participate in the export of such molecules. The fact that these transporters are clustered in phylogenetic analyses with the binding-protein-dependent systems may suggest either that they are not directly involved in the export of their presumed substrates or alternatively that the transport polarity of some families may change during evolution.

4. From this analysis, a hypothetical scenario on the evolution of ABC systems could be proposed (Dassa and Bouige, 2001; Saurin *et al.*, 1999). The ancestor 'progenote' cells already had all classes of ABC systems. Prokaryotes inherited all ABC classes. Eukaryotes probably acquired IM-ABC and ABC-IM (class 1) and ABC2 (class 2) systems from the symbiotic bacteria that are the putative ancestors of organelles. It is noteworthy that most eukaryote IM-ABC systems are specifically targeted to organelle membranes, which probably descended from a prokaryote ancestor. For instance, the mammalian TAP proteins, involved in the presentation of antigenic peptides to the class I major histocompatibility complex, are inserted into the endoplasmic reticulum, and the ALD proteins, putatively involved in the export of very long chain fatty acid from the cytosol into peroxisomes, are targeted to the peroxisomal membrane. From genes encoding IM-ABC or ABC-IM systems, eukaryotes developed specific systems by several independent duplication–fusion events, for instance those that led to the constitution of the proteins of the PDR (fungal pleiotropic drug resistance) family (ABC-IM)2, the P-glycoprotein-like proteins (IM-ABC)2 and the ABCA family proteins.
5. Our systematic study of primary sequence similarities between ABC domains led to a comprehensive phylogenetic and functional classification of ABC systems, including those involved in non-transport cellular processes. The sequences of more than 2000 different ABC systems are presently deposited in the Genbank database. About half of these comprise systems discovered during genome sequencing projects and their annotation is very limited. We are maintaining a database of these systems, which includes functional, sequence and structural information. This database will be helpful in accurately annotating ABC systems and in identifying the partners of the ABC ATPases. The first release is available on the Institut Pasteur Web server: www.pasteur.fr/recherche/unites/pmtg/abc/index.html.

REFERENCES

Abele, R. and Tampe, R. (1999) Function of the transport complex TAP in cellular immune recognition. Biochim. Biophys. Acta 1461, 405–419.

Aguilar-Bryan, L., Nichols, C.G., Wechsler, S.W., Clement, J.P.T., Boyd, A.E.R., Gonzalez, G., Herrera-Sosa, H., Nguy, K., Bryan, J. and Nelson, D.A. (1995) Cloning of the beta cell high-affinity sulfonylurea receptor: a regulator of insulin secretion. Science 268, 423–426.

Albers, S.V., Elferink, M.G., Charlebois, R.L., Sensen, C.W., Driessen, A.J. and Konings, W.N. (1999) Glucose transport in the extremely thermoacidophilic *Sulfolobus solfataricus* involves a high-affinity membrane-integrated binding protein. J. Bacteriol. 181, 4285–4291.

Albright, L.M., Ronson, C.W., Nixon, B.T. and Ausubel, F.M. (1989) Identification of a gene linked to *Rhizobium meliloti ntrA* whose product is homologous to a family of ATP-binding proteins. J. Bacteriol. 171, 1932–1941.

Allignet, J., Loncle, V. and El Solh, N. (1992) Sequence of a staphylococcal plasmid gene, *vga*, encoding a putative ATP-binding protein involved in resistance to virginiamycin A-like antibiotics. Gene 117, 45–51.

Allikmets, R., Shroyer, N.F., Singh, N., Seddon, J.M., Lewis, R.A., Bernstein, P.S., *et al.* (1997) Mutation of the Stargardt disease gene (ABCR) in age-related macular degeneration. Science 277, 1805–1807.

Allikmets, R., Schriml, L.M., Hutchinson, A., Romano-Spica, V. and Dean, M. (1998)

A human placenta-specific ATP-binding cassette gene (ABCP) on chromosome 4q22 that is involved in multidrug resistance. Cancer Res. 58, 5337–5339.

Allikmets, R., Raskind, W.H., Hutchinson, A., Schueck, N.D., Dean, M. and Koeller, D.M. (1999) Mutation of a putative mitochondrial iron transporter gene (ABC7) in X-linked sideroblastic anemia and ataxia (XLSA/A). Hum. Mol. Genet. 8, 743–749.

Ames, G.F. and Lecar, H. (1992) ATP-dependent bacterial transporters and cystic fibrosis: analogy between channels and transporters. FASEB J. 6, 2660–2666.

Ames, G.F.-L., Mimura, C. and Shyamala, V. (1990) Bacterial periplasmic permeases belong to a family of transport proteins operating from *Escherichia coli* to human traffic ATPases. FEMS Microbiol. Rev. 75, 429–446.

Ames, G.F., Mimura, C.S., Holbrook, S.R. and Shyamala, V. (1992) Traffic ATPases: a superfamily of transport proteins operating from *Escherichia coli* to humans. Adv. Enzymol. Relat. Areas Mol. Biol. 65, 1–47.

Arruda, S., Bomfim, G., Knights, R., Huima-Byron, T. and Riley, L.W. (1993) Cloning of an *M. tuberculosis* DNA fragment associated with entry and survival inside cells. Science 261, 1454–1457.

Ashcroft, S.J.H. (2000). The beta-cell K-ATP channel. J. Membr. Biol. 176, 187–206.

Azarian, S.M. and Travis, G.H. (1997) The photoreceptor rim protein is an ABC transporter encoded by the gene for recessive Stargardt's disease (ABCR). FEBS Lett. 409, 247–252.

Bakos, E., Evers, R., Szakacs, G., Tusnady, G.E., Welker, E., Szabo, K., *et al.* (1998) Functional multidrug resistance protein (MRP1) lacking the N-terminal transmembrane domain. J. Biol. Chem. 273, 32167–32175.

Bauer, B.E., Wolfger, H. and Kuchler, K. (1999) Inventory and function of yeast ABC proteins: about sex, stress, pleiotropic drug and heavy metal resistance. Biochim. Biophys. Acta 1461, 217–236.

Beck, S., Kelly, A., Radley, E., Khurshid, F., Alderton, R.P. and Trowsdale, J. (1992) DNA sequence analysis of 66 kb of the human MHC class II region encoding a cluster of genes for antigen processing. J. Mol. Biol. 228, 433–441.

Berge, K.E., Tian, H., Graf, G.A., Yu, L.Q., Grishin, N.V., Schultz, J., Kwiterovich, P., Shan, B., Barnes, R. and Hobbs, H.H. (2000) Accumulation of dietary cholesterol in sitosterolemia caused by mutations in adjacent ABC transporters. Science 290, 1771–1775.

Beyer, S., Distler, J. and Piepersberg, W. (1996) The *str* gene cluster for the biosynthesis of 5′-hydroxystreptomycin in *Streptomyces glaucescens* GLA.0 (ETH 22794): new operons and evidence for pathway-specific regulation by StrR. Mol. Gen. Genet. 250, 775–784.

Bisbal, C., Martinand, C., Silhol, M., Lebleu, B. and Salehzada, T. (1995) Cloning and characterization of a RNAse L inhibitor. A new component of the interferon-regulated 2-5A pathway. J. Biol. Chem. 270, 13308–13317.

Bisbal, C., Silhol, M., Laubenthal, H., Kaluza, T., Carnac, G., Milligan, L., Le Roy, F. and Salehzada, T. (2000) The 2′-5′ oligoadenylate/ RNase L/RNase L inhibitor pathway regulates both MyoD mRNA stability and muscle cell differentiation. Mol. Cell. Biol. 20, 4959–4969.

Bodzioch, M., Orso, E., Klucken, J., Langmann, T., Bottcher, A., Diederich, W., *et al.* (1999) The gene encoding ATP-binding cassette transporter 1 is mutated in Tangier disease. Nat. Genet. 22, 347–351.

Bonnard, G. and Grienenberger, J.M. (1995) A gene proposed to encode a transmembrane domain of an ABC transporter is expressed in wheat mitochondria. Mol. Gen. Genet. 246, 91–99.

Boos, W. and Lucht, J.M. (1996) Periplasmic binding protein-dependent ABC transporters. In: *Escherichia coli* and *Salmonella* Cellular and Molecular Biology, 2nd ed. (ed. F.C. Neidhardt), pp. 1175–1209. Washington: ASM Press.

Borst, P., Zelcer, N. and van Helvoort, A. (2000) ABC transporters in lipid transport. Biochim. Biophys. Acta 1486, 128–144.

Braibant, M., Gilot, P. and Content, J. (2000) The ATP binding cassette (ABC) transport systems of *Mycobacterium tuberculosis*. FEMS Microbiol. Rev. 24, 449–467.

Broccardo, C., Luciani, M. and Chimini, G. (1999) The ABCA subclass of mammalian transporters. Biochim. Biophys. Acta 1461, 395–404.

Bruns, C.M., Nowalk, A.J., Arvai, A.S., McTigue, M.A., Vaughan, K.G., Mietzner, T.A. and McRee, D.E. (1997) Structure of *Haemophilus influenzae* Fe(+3)-binding protein reveals convergent evolution within a superfamily. Nat. Struct. Biol. 4, 919–924.

Cardenas, L., Dominguez, J., Santana, O. and Quinto, C. (1996) The role of the *nodI* and *nodJ* genes in the transport of Nod metabolites in *Rhizobium etli*. Gene 173, 183–187.

Carvalho, A.C., Ruiz-Palacios, G.M., Ramos-Cervantes, P., Cervantes, L.E., Jiang, X. and Pickering, L.K. (2001) Molecular characterization of invasive and noninvasive *Campylobacter jejuni* and *Campylobacter coli* isolates. J. Clin. Microbiol. 39, 1353–1359.

Chakraburtty, K. (2001) Translational regulation by ABC systems. Res. Microbiol. 152, 391–399.

Chakraburtty, K. and Triana-Alonso, F.J. (1998) Yeast elongation factor 3: structure and function. Biol. Chem. 379, 831–840.

Chang, G. and Roth, C.B. (2001) Structure of MsbA from *E. coli*: a homolog of the multidrug resistance ATP binding cassette (ABC) transporters. Science 293, 1793–1800.

Cheng, J.B., Guffanti, A.A. and Krulwich, T.A. (1997) A two-gene ABC-type transport system that extrudes Na+ in *Bacillus subtilis* is induced by ethanol or protonophore. Mol. Microbiol. 23, 1107–1120.

Claverys, J.P. (2001) A new family of high-affinity ABC manganese and zinc permeases. Res. Microbiol. 152, 231–243.

Clement, J.P.t., Kunjilwar, K., Gonzalez, G., Schwanstecher, M., Panten, U., Aguilar-Bryan, L. and Bryan, J. (1997) Association and stoichiometry of K(ATP) channel subunits. Neuron 18, 827–838.

Dassa, E. and Bouige, P. (2001) The ABC of ABCs: a phylogenetic and functional classification of ABC systems in living organisms. Res. Microbiol. 152, 211–229.

Dassa, E., Hofnung, M., Paulsen, I.T. and Saier, M.H., Jr (1999) The *Escherichia coli* ABC transporters: an update. Mol. Microbiol. 32, 887–889.

de Leeuw, E., Graham, B., Phillips, G.J., ten Hagen-Jongman, C.M., Oudega, B. and Luirink, J. (1999) Molecular characterization of *Escherichia coli* FtsE and FtsX. Mol. Microbiol. 31, 983–993.

Decottignies, A. and Goffeau, A. (1997) Complete inventory of the yeast ABC proteins. Nat. Genet. 15, 137–145.

Detmers, F.J.M., Lanfermeijer, F.C. and Poolman, B. (2001) Peptides and ATP binding cassette peptide transporters. Res. Microbiol. 152, 245–258.

Diederichs, K., Diez, J., Greller, G., Muller, C., Breed, J., Schnell, C., Vonrhein, C., Boos, W. and Welte, W. (2000) Crystal structure of MalK, the ATPase subunit of the trehalose/maltose ABC transporter of the archaeon *Thermococcus litoralis*. EMBO J. 19, 5951–5961.

Doige, C.A. and Ames, G.F.L. (1993) ATP-dependent transport systems in bacteria and humans – relevance to cystic fibrosis and multidrug resistance. Annu. Rev. Microbiol. 47, 291–319.

Doyle, L.A., Yang, W.D., Abruzzo, L.V., Krogmann, T., Gao, Y.M., Rishi, A.K. and Ross, D.D. (1998) A multidrug resistance transporter from human MCF-7 breast cancer cells. Proc. Natl Acad. Sci. USA 95, 15665–15670.

Ellermann, J., Rospert, S., Thauer, R.K., Bokranz, M., Klein, A., Voges, M. and Berkessel, A. (1989) Methyl-coenzyme-M reductase from *Methanobacterium thermoautotrophicum* (strain Marburg). Purity, activity and novel inhibitors. Eur. J. Biochem. 184, 63–68.

Ewart, G.D., Cannell, D., Cox, G.B. and Howells, A.J. (1994) Mutational analysis of the traffic ATPase (ABC) transporters involved in uptake of eye pigment precursors in *Drosophila melanogaster* – implications for structure–function relationships. J. Biol. Chem. 269, 10370–10377.

Fanen, P., Guidoux, S., Sarde, C.O., Mandel, J.L., Goossens, M. and Aubourg, P. (1994) Identification of mutations in the putative ATP-binding domain of the adrenoleukodystrophy gene. J. Clin. Invest. 94, 516–520.

Fath, M.J. and Kolter, R. (1993) ABC transporters – bacterial exporters. Microbiol. Rev. 57, 995–1017.

Fetherston, J.D., Bertolino, V.J. and Perry, R.D. (1999) YbtP and YbtQ: two ABC transporters required for iron uptake in *Yersinia pestis*. Mol. Microbiol. 32, 289–299.

Galvez, T., Parmentier, M.L., Joly, C., Malitschek, B., Kaupmann, K., Kuhn, R., Bittiger, H., Froestl, W., Bettler, B. and Pin, J.P. (1999) Mutagenesis and modeling of the GABA(B) receptor extracellular domain support a Venus flytrap mechanism for ligand binding. J. Biol. Chem. 274, 13362–13369.

Gartner, J., Moser, H. and Valle, D. (1992) Mutations in the 70K peroxisomal membrane protein gene in Zellweger syndrome. Nat. Genet. 1, 16–23.

Gaudet, R. and Wiley, D.C. (2001) Structure of the ABC ATPase domain of human TAP1, the transporter associated with antigen processing. EMBO J. 20, 4964–4972.

Goldman, B.S. and Kranz, R.G. (2001) ABC transporters associated with cytochrome *c* biogenesis. Res. Microbiol. 152, 323–329.

Goldman, B.S., Gabbert, K.K. and Kranz, R.G. (1996) Use of heme reporters for studies of cytochrome biosynthesis and heme transport. J. Bacteriol. 178, 6338–6347.

Goosen, N. and Moolenaar, G.F. (2001) Role of ATP hydrolysis by UvrA and UvrB during nucleotide excision repair. Res. Microbiol. 152, 401–409.

Groen, A.K., Bloks, V.W., Bandsma, R.H.J., Ottenhoff, R., Chimini, G. and Kuipers, F. (2001) Hepatobiliary cholesterol transport as not impaired in Abca1-null mice lacking HDL. J. Clin. Invest. 108, 843–850.

Guilfoile, P.G. and Hutchinson, C.R. (1991) A bacterial analog of the *mdr* gene of mammalian tumor cells is present in *Streptomyces peucetius*, the producer of daunorubicin and doxorubicin. Proc. Natl Acad. Sci. USA 88, 8553–8557.

Haimeur, A., Brochu, C., Genest, P., Papadopoulou, B. and Ouellette, M. (2000) Amplification of the ABC transporter gene PGPA and increased trypanothione levels in potassium antimonyl tartrate (SbIII) resistant *Leishmania tarentolae*. Mol. Biochem. Parasitol. 108, 131–135.

Harel, Y.M., Bailone, A. and Bibi, E. (1999) Resistance to bacitracin as modulated by an *Escherichia coli* homologue of the bacitracin ABC transporter BcrC subunit from *Bacillus licheniformis*. J. Bacteriol. 181, 6176–6178.

Havarstein, L.S., Diep, D.B. and Nes, I.F. (1995) A family of bacteriocin ABC transporters carry out proteolytic processing of their substrates concomitant with export. Mol. Microbiol. 16, 229–240.

Hayman, G.T., Von Bodman, S.B., Kim, H., Jiang, P. and Farrand, S.K. (1993) Genetic analysis of the agrocinopine catabolic region of *Agrobacterium tumefaciens* Ti plasmid pTiC58, which encodes genes required for opine and agrocin 84 transport. J. Bacteriol. 175, 5575–5584.

Hettema, E.H. and Tabak, H.F. (2000) Transport of fatty acids and metabolites across the peroxisomal membrane. Biochim. Biophys. Acta 1486, 18–27.

Higgins, C.F. (1992) ABC transporters: from microorganisms to man. Annu. Rev. Cell Biol. 8, 67–113.

Higgins, C.F., Hiles, I.D., Salmond, G.P.C., Gill, D.R., Downie, J.A., Evans, I.J., *et al.* (1986) A family of related ATP-binding subunits coupled to many distinct biological processes in bacteria. Nature 323, 448–450.

Higgins, C.F., Gallagher, M.P., Mimmack, M.L. and Pearce, S.R. (1988) A family of closely related ATP-binding subunits from prokaryotic and eucaryotic cells. BioEssays 8, 111–116.

Hogue, D.L., Liu, L. and Ling, V. (1999) Identification and characterization of a mammalian mitochondrial ATP-binding cassette membrane protein. J. Mol. Biol. 285, 379–389.

Holland, D. and Wolk, C.P. (1990) Identification and characterization of *hetA*, a gene that acts early in the process of morphological differentiation of heterocysts. J. Bacteriol. 172, 3131–3137.

Holland, I.B. and Blight, M.A. (1999) ABC-ATPases, adaptable energy generators fuelling transmembrane movement of a variety of molecules in organisms from bacteria to humans. J. Mol. Biol. 293, 381–399.

Hong, M., Gleason, Y., Wyckoff, E.E. and Payne, S.M. (1998) Identification of two *Shigella flexneri* chromosomal loci involved in intercellular spreading. Infect. Immun. 66, 4700–4710.

Hosie, A.H.F. and Poole, P.S. (2001) Bacterial ABC transporters of amino acids. Res. Microbiol. 152, 259–270.

Hughes, A.L. (1994) Evolution of the ATP-binding-cassette transmembrane transporters of vertebrates. Mol. Biol. Evol. 11, 899–910.

Hui, F.M. and Morrison, D.A. (1991) Genetic transformation in *Streptococcus pneumoniae*: nucleotide sequence analysis shows *comA*, a gene required for competence induction, to be a member of the bacterial ATP-dependent transport protein family. J. Bacteriol. 173, 372–381.

Hung, L.W., Wang, I.X.Y., Nikaido, K., Liu, P.Q., Ames, G.F.L. and Kim, S.H. (1998) Crystal structure of the ATP-binding subunit of an ABC transporter. Nature 396, 703–707.

Hunke, S., Mourez, M., Jehanno, M., Dassa, E. and Schneider, E. (2000) ATP modulates subunit–subunit interactions in an ATP-binding cassette transporter (MalFGK2) determined by site-directed chemical cross-linking. J. Biol. Chem. 275, 15526–15534.

Igarashi, K., Ito, K. and Kashiwagi, K. (2001) Polyamine uptake systems in *Escherichia coli*. Res. Microbiol. 152, 271–278.

Isomoto, S., Kondo, C., Yamada, M., Matsumoto, S., Higashiguchi, O., Horio, Y., Matsuzawa, Y. and Kurachi, Y. (1996) A novel sulfonylurea receptor forms with BIR (Kir6.2) a smooth muscle type ATP-sensitive K+ channel. J. Biol. Chem. 271, 24321–24324.

Jekabsons, W. and Schuster, W. (1995) orf250 encodes a second subunit of an ABC-type heme transporter in *Oenothera* mitochondria. Mol. Gen. Genet. 246, 166–173.

Jeter, R.M. and Roth, J.R. (1987) Cobalamin (vitamin B12) biosynthetic genes of *Salmonella typhimurium*. J. Bacteriol. 169, 3189–3198.

Kaminski, W.E., Orso, E., Diederich, W., Klucken, J., Drobnik, W. and Schmitz, G. (2000) Identification of a novel human sterol-sensitive ATP-binding cassette transporter (ABCA7). Biochem. Biophys. Res. Commun. 273, 532–538.

Karttunen, J.T., Trowsdale, J. and Lehner, P.J. (1999) Antigen presentation: TAP dances with ATP. Curr. Biol. 9, R820–R824.

Keijser, B.J., van Wezel, G.P., Canters, G.W., Kieser, T. and Vijgenboom, E. (2000) The ram-dependence of *Streptomyces lividans* differentiation is bypassed by copper. J. Mol. Microbiol. Biotechnol. 2, 565–574.

Kempf, B. and Bremer, E. (1998) Uptake and synthesis of compatible solutes as microbial stress responses to high-osmolality environments. Arch. Microbiol. 170, 319–330.

Kertesz, M.A. (2001) Bacterial transporters for sulfate and organosulfur compounds. Res. Microbiol. 152, 279–290.

Klucken, J., Buchler, C., Orso, E., Kaminski, W.E., Porsch-Ozcurumez, M., Liebisch, C., *et al.* (2000) ABCG1 (ABC8), the human homolog of the *Drosophila* white gene, is a regulator of macrophage cholesterol and phospholipid transport. Proc. Natl Acad. Sci. USA 97, 817–822.

Koronakis, V., Sharff, A., Koronakis, E., Luisi, B. and Hughes, C. (2000) Crystal structure of the bacterial membrane protein TolC central to multidrug efflux and protein export. Nature 405, 914–919.

Köster, W. (2001) ABC transporter-mediated uptake of iron, siderophores, heme and vitamin B-12. Res. Microbiol. 152, 291–301.

Kuan, G., Dassa, E., Saurin, W., Hofnung, M. and Saier, M.H. (1995) Phylogenetic analyses of the ATP-binding constituents of bacterial extracytoplasmic receptor-dependent ABC-type nutrient uptake permeases. Res. Microbiol. 146, 271–278.

Kuhner, C.H., Lindenbach, B.D. and Wolfe, R.S. (1993) Component A2 of methylcoenzyme M reductase system from *Methanobacterium thermoautotrophicum* delta H: nucleotide sequence and functional expression by *Escherichia coli*. J. Bacteriol. 175, 3195–3203.

Lampinen, M., Pentikainen, O., Johnson, M.S. and Keinanen, K. (1998) AMPA receptors and bacterial periplasmic amino acid-binding proteins share the ionic mechanism of ligand recognition. EMBO J. 17, 4704–4711.

Legare, D., Richard, D., Mukhopadhyay, R., Stierhof, Y.D., Rosen, B.P., Haimeur, A., Papadopoulou, B. and Ouellette, M. (2001) The *Leishmania* ATP-binding cassette protein PGPA is an intracellular metal-thiol transporter ATPase. J. Biol. Chem. 276, 26301–26307.

Létoffé, S., Delepelaire, P. and Wandersman, C. (1990) Protease secretion by *Erwinia chrysanthemi*: the specific secretion functions are analogous to those of *Escherichia coli* alpha-haemolysin. EMBO J. 9, 1375–1382.

Li, Z.S., Szczypka, M., Lu, Y.P., Thiele, D.J. and Rea, P.A. (1996) The yeast cadmium factor protein (YCF1) is a vacuolar glutathione S-conjugate pump. J. Biol. Chem. 271, 6509–6517.

Lill, R. and Kispal, G. (2001) Mitochondrial ABC transporters. Res. Microbiol. 152, 331–340.

Linton, K.J. and Higgins, C.F. (1998) The *Escherichia coli* ATP-binding cassette (ABC) proteins. Mol. Microbiol. 28, 5–13.

Lomovskaya, N., Hong, S.K., Kim, S.U., Fonstein, L., Furuya, K. and Hutchinson, R.C. (1996) The *Streptomyces peucetius drrC* gene encodes a UvrA-like protein involved in daunorubicin resistance and production. J. Bacteriol. 178, 3238–3245.

Lu, Y.P., Li, Z.S., Drozdowicz, Y.M., Hortensteiner, S., Martinoia, E. and Rea, P.A. (1998) AtMRP2, an *Arabidopsis* ATP binding cassette transporter able to transport glutathione S-conjugates and chlorophyll catabolites – functional comparisons with atMRP1. Plant Cell 10, 267–282.

Mackenzie, S.M., Howells, A.J., Cox, G.B. and Ewart, G.D. (2000) Sub-cellular localisation of the white/scarlet ABC transporter to pigment granule membranes within the compound eye of *Drosophila melanogaster*. Genetica 108, 239–252.

Marton, M.J., Vazquez de Aldana, C.R., Qiu, H., Chakraburtty, K. and Hinnebusch, A.G. (1997) Evidence that GCN1 and GCN20, translational regulators of GCN4, function on elongating ribosomes in activation of eIF2alpha kinase GCN2. Mol. Cell. Biol. 17, 4474–4489.

McDonald, M.K., Cowley, S.C. and Nano, F.E. (1997) Temperature-sensitive lesions in the *Francisella novicida valA* gene cloned into an *Escherichia coli msbA lpxK* mutant affecting deoxycholate resistance and lipopolysaccharide assembly at the restrictive temperature. J. Bacteriol. 179, 7638–7643.

Mendez, C. and Salas, J.A. (2001) The role of ABC transporters in antibiotic-producing organisms: drug secretion and resistance mechanisms. Res. Microbiol. 152, 341–350.

Mills, J., Wyborn, N.R., Greenwood, J.A., Williams, S.G. and Jones, C.W. (1998) Characterisation of a binding-protein-dependent, active transport system for short-chain amides and urea in the methylotrophic bacterium *Methylophilus methylotrophus*. Eur. J. Biochem. 251, 45–53.

Mitsuhashi, N., Miki, T., Senbongi, H., Yokoi, N., Yano, H., Miyazaki, M., *et al.* (2000) MTABC3, a novel mitochondrial ATP-binding cassette protein involved in iron homeostasis. J. Biol. Chem. 275, 17536–17540.

Moller, S.G., Kunkel, T. and Chua, N.H. (2001) A plastidic ABC protein involved in intercompartmental communication of light signaling. Genes Dev. 15, 90–103.

Montesinos, M.L., Herrero, A. and Flores, E. (1997) Amino acid transport in taxonomically diverse cyanobacteria and identification of two genes encoding elements of a neutral amino acid permease putatively involved in recapture of leaked hydrophobic amino acids. J. Bacteriol. 179, 853–862.

Mourez, M., Hofnung, M. and Dassa, E. (1997) Subunit interactions in ABC transporters: a conserved sequence in hydrophobic membrane proteins of periplasmic permeases defines an important site of interaction with the ATPase subunits. EMBO J. 16, 2066–2077.

Nachin, L., El Hassouni, M., Loiseau, L., Expert, D. and Barras, F. (2001) SoxR-dependent response to oxidative stress and virulence of *Erwinia chrysanthemi*: the key role of SufC, an orphan ABC ATPase. Mol. Microbiol. 39, 960–972.

Naren, A.P., Cormet-Boyaka, E., Fu, J., Villain, M., Blalock, J.E., Quick, M.W. and Kirk, K.L. (1999) CFTR chloride channel regulation by an interdomain interaction. Science 286, 544–548.

Nasonkin, I., Alikasifoglu, A., Ambrose, C., Cahill, P., Cheng, M., Sarniak, A., Egan, M. and Thomas, P.M. (1999) A novel sulfonylurea receptor family member expressed in the embryonic *Drosophila* dorsal vessel and tracheal system. J. Biol. Chem. 274, 29420–29425.

Navarro, C., Wu, L.F. and Mandrand-Berthelot, M.A. (1993) The *nik* operon of *Escherichia coli* encodes a periplasmic binding protein-dependent transport system for nickel. Mol. Microbiol. 9, 1181–1191.

Obis, D., Guillot, A., Gripon, J.C., Renault, P., Bolotin, A. and Mistou, M.Y. (1999) Genetic and biochemical characterization of a high-affinity betaine uptake system (BusA) in *Lactococcus lactis* reveals a new functional organization within bacterial ABC transporters. J. Bacteriol. 181, 6238–6246.

O'Keeffe, T., Hill, C. and Ross, R.P. (1999) Characterization and heterologous expression of the genes encoding enterocin A production, immunity, and regulation in *Enterococcus faecium* DPC1146. Appl. Environ. Microbiol. 65, 1506–1515.

Orso, E., Broccardo, C., Kaminski, W.E., Bottcher, A., Liebisch, G., Drobnik, W., *et al.* (2000) Transport of lipids from Golgi to plasma membrane is defective in Tangier disease patients and ABC1-deficient mice. Nat. Genet. 24, 192–196.

Ortiz, D.F., Ruscitti, T., McCue, K.F. and Ow, D.W. (1995) Transport of metal-binding peptides by HMT1, a fission yeast ABC-type vacuolar membrane protein. J. Biol. Chem. 270, 4721–4728.

Ortiz, D.F., St Pierre, M.V., Abdulmessih, A. and Arias, I.M. (1997) A yeast ATP-binding cassette-type protein mediating ATP-dependent bile acid transport. J. Biol. Chem. 272, 15358–15365.

Otto, M. and Gotz, F. (2001) ABC transporters of staphylococci. Res. Microbiol. 152, 351–356.

Otto, M., Peschel, A. and Gotz, F. (1998) Producer self-protection against the lantibiotic epidermin by the ABC transporter EpiFEG of *Staphylococcus epidermidis* Tu3298. FEMS Microbiol. Lett. 166, 203–211.

Pattery, T., Hernalsteens, J.P. and De Greve, H. (1999) Identification and molecular characterization of a novel *Salmonella enteritidis* pathogenicity islet encoding an ABC transporter. Mol. Microbiol. 33, 791–805.

Paulsen, I.T., Beness, A.M. and Saier, M.H. (1997) Computer-based analyses of the protein constituents of transport systems catalysing export of complex carbohydrates in bacteria. Microbiology 143, 2685–2699.

Paulsen, I.T., Sliwinski, M.K. and Saier, M.H., Jr (1998) Microbial genome analyses: global comparisons of transport capabilities based on phylogenies, bioenergetics and substrate specificities. J. Mol. Biol. 277, 573–592.

Paulsen, I.T., Nguyen, L., Sliwinski, M.K., Rabus, R. and Saier, M.H. (2000) Microbial genome analyses: comparative transport capabilities in eighteen prokaryotes. J. Mol. Biol. 301, 75–100.

Podlesek, Z., Comino, A., Herzogvelikonja, B., Zgurbertok, D., Komel, R. and Grabnar, M. (1995) *Bacillus licheniformis* bacitracin-resistance ABC transporter: relationship to mammalian multidrug resistance. Mol. Microbiol. 16, 969–976.

Podlesek, Z., Comino, A., Herzog-Velikonja, B. and Grabnar, M. (2000) The role of the bacitracin ABC transporter in bacitracin resistance and collateral detergent sensitivity. FEMS Microbiol. Lett. 188, 103–106.

Poole, R.K., Gibson, F. and Wu, G.H. (1994) The *cydD* gene product, component of a heterodimeric ABC transporter, is required for assembly of periplasmic cytochrome-*c* and of cytochrome-*bd* in *Escherichia coli*. FEMS Microbiol. Lett. 117, 217–224.

Quentin, Y., Fichant, G. and Denizot, F. (1999) Inventory, assembly and analysis of *Bacillus subtilis* ABC transport systems. J. Mol. Biol. 287, 467–484.

Reizer, J., Reizer, A. and Saier, M.H. (1992) A new subfamily of bacterial ABC-type transport systems catalyzing export of drugs and carbohydrates. Protein Sci. 1, 1326–1332.

Riordan, J.R., Rommens, J.M., Kerem, B., Alon, N., Rozmahel, R., Grzelczak, Z., *et al.* (1989) Identification of the cystic fibrosis gene: cloning and characterization of complementary DNA. Science 245, 1066–1072.

Ross, J.I., Eady, E.A., Cove, J.H., Cunliffe, W.J., Baumberg, S. and Wootton, J.C. (1990) Inducible erythromycin resistance in Staphlyococci is encoded by a member of the ATP-binding transport super-gene family. Mol. Microbiol. 4, 1207–1214.

Ross, J.I., Eady, E.A., Cove, J.H. and Baumberg, S. (1996) Minimal functional system required for expression of erythromycin resistance by *msrA* in *Staphylococcus aureus* RN4220. Gene 183, 143–148.

Roth, J.R., Lawrence, J.G., Rubenfield, M., Kieffer-Higgins, S. and Church, G.M. (1993) Characterization of the cobalamin (vitamin B12) biosynthetic genes of *Salmonella typhimurium*. J. Bacteriol. 175, 3303–3316.

Rouviere, P.E., Escalante-Semerena, J.C. and Wolfe, R.S. (1985) Component A2 of the methylcoenzyme M methylreductase system from *Methanobacterium thermoautotrophicum*. J. Bacteriol. 162, 61–66.

Russell, R.R.B., Aduse-Opoku, J., Sutcliffe, I.C., Tao, L. and Ferretti, J.J. (1992) A binding protein-dependent transport system in *Streptococcus mutans* responsible for multiple sugar metabolism. J. Biol. Chem. 267, 4631–4637.

Rust, S., Rosier, M., Funke, H., Real, J., Amoura, Z., Piette, J.C., *et al.* (1999) Tangier disease is caused by mutations in the gene encoding ATP-binding cassette transporter 1. Nat. Genet. 22, 352–355.

Saier, M.H. (2000) A functional–phylogenetic classification system for transmembrane solute transporters. Microbiol. Mol. Biol. Rev. 64, 354–411.

Saurin, W., Köster, W. and Dassa, E. (1994) Bacterial binding protein-dependent permeases: characterization of distinctive signatures for functionally related integral cytoplasmic membrane proteins. Mol. Microbiol. 12, 993–1004.

Saurin, W., Hofnung, M. and Dassa, E. (1999) Getting in or out: early segregation between importers and exporters in the evolution of ATP-binding cassette (ABC) transporters. J. Mol. Evol. 48, 22–41.

Schlösser, A., Kampers, T. and Schrempf, H. (1997) The *Streptomyces* ATP-binding component MsiK assists in cellobiose and maltose transport. J. Bacteriol. 179, 2092–2095.

Schmitz, G., Langmann, T. and Heimerl, S. (2001) Role of ABCG1 and other ABCG family members in lipid metabolism. J. Lipid Res. 42, 1513–1520.

Schneider, E. (2001) ABC transporters catalyzing carbohydrate uptake. Res. Microbiol. 152, 303–310.

Schneider, E. and Hunke, S. (1998) ATP-binding-cassette (ABC) transport systems: functional and structural aspects of the ATP-hydrolyzing subunits/domains. FEMS Microbiol. Rev. 22, 1–20.

Schulz, H., Fabianek, R.A., Pellicioli, E.C., Hennecke, H. and Thöny-Meyer, L. (1999) Heme transfer to the heme chaperone CcmE during cytochrome *c* maturation requires the CcmC protein, which may function independently of the ABC-transporter CcmAB. Proc. Natl Acad. Sci. USA 96, 6462–6467.

Schuster, W. (1994) The highly edited orf206 in *Oenothera* mitochondria may encode a component of a heme transporter involved in cytochrome *c* biogenesis. Plant Mol. Biol. 25, 33–42.

Self, W.T., Grunden, A.M., Hasona, A. and Shanmugam, K.T. (2001) Molybdate transport. Res. Microbiol. 152, 311–321.

Shani, N., Watkins, P.A. and Valle, D. (1995) PXA1, a possible *Saccharomyces cerevisiae* ortholog of the human adrenoleukodystrophy gene. Proc. Natl Acad. Sci. USA 92, 6012–6016.

Sheppard, D.N. and Welsh, M.J. (1999) Structure and function of the CFTR chloride channel. Physiol. Rev. 79, S23–45.

Shyng, S. and Nichols, C.G. (1997) Octameric stoichiometry of the KATP channel complex. J. Gen. Physiol. 110, 655–664.

Silver, R.P., Prior, K., Nsahlai, C. and Wright, L.F. (2001) ABC transporters and the export of capsular polysaccharides from Gram-negative bacteria. Res. Microbiol. 152, 357–364.

Smart, C.C. and Fleming, A.J. (1996) Hormonal and environmental regulation of a plant PDR5-like ABC transporter. J. Biol. Chem. 271, 19351–19357.

Spaink, H.P., Wijfjes, A.H.M. and Lugtenberg, B.J.J. (1995) Rhizobium NodI and NodJ proteins play a role in the efficiency of secretion of lipochitin oligosaccharides. J. Bacteriol. 177, 6276–6281.

Spellerberg, B., Pohl, B., Haase, G., Martin, S., Weber-Heynemann, J. and Lutticken, R. (1999) Identification of genetic determinants for the hemolytic activity of *Streptococcus agalactiae* by ISS1 transposition. J. Bacteriol. 181, 3212–3219.

Tanaka, K., Matsuyama, S.I. and Tokuda, H. (2001) Deletion of *lolB*, encoding an outer membrane lipoprotein, is lethal for *Escherichia coli* and causes accumulation of lipoprotein localization intermediates in the periplasm. J. Bacteriol. 183, 6538–6542.

Tauch, A., Krieft, S., Puhler, A. and Kalinowski, J. (1999) The *tetAB* genes of the *Corynebacterium striatum* R-plasmid pTP10 encode an ABC transporter and confer tetracycline, oxytetracycline and oxacillin resistance in *Corynebacterium glutamicum*. FEMS Microbiol. Lett. 173, 203–209.

Tokumoto, U. and Takahashi, Y. (2001) Genetic analysis of the isc operon in *Escherichia coli* involved in the biogenesis of cellular iron-sulfur proteins. J. Biochem. (Tokyo) 130, 63–71.

Tomii, K. and Kanehisa, M. (1998) A comparative analysis of ABC transporters in complete microbial genomes. Genome Res. 8, 1048–1059.

Tusnady, G.E., Bakos, E., Varadi, A. and Sarkadi, B. (1997) Membrane topology distinguishes a subfamily of the ATP-binding cassette (ABC) transporters. FEBS Lett. 402, 1–3.

Tyzack, J.K., Wang, X.M., Belsham, G.J. and Proud, C.G. (2000) ABC50 interacts with eukaryotic initiation factor 2 and associates with the ribosome in an ATP-dependent manner. J. Biol. Chem. 275, 34131–34139.

Ukai, H., Matsuzawa, H., Ito, K., Yamada, M. and Nishimura, A. (1998) *ftsE*(Ts) affects translocation of K+-pump proteins into the cytoplasmic membrane of *Escherichia coli*. J. Bacteriol. 180, 3663–3670.

van Veen, H.W., Higgins, C.F. and Konings, W.N. (2001) Multidrug transport by ATP binding cassette transporters: a proposed two-cylinder engine mechanism. Res. Microbiol. 152, 365–374.

Venkateswaran, A., Repa, J.J., Lobaccaro, J.M.A., Bronson, A., Mangelsdorf, D.J. and Edwards, P.A. (2000) Human white/murine ABC8 mRNA levels are highly induced in lipid-loaded macrophages – a transcriptional role for specific oxysterols. J. Biol. Chem. 275, 14700–14707.

Vermeij, P., Wietek, C., Kahnert, A., Wuest, T. and Kertesz, M.A. (1999) Genetic organization of sulphur-controlled aryl desulphonation in *Pseudomonas putida* S-313. Mol. Microbiol. 32, 913–926.

Wandersman, C. and Delepelaire, P. (1990) TolC, an *Escherichia coli* outer membrane protein required for hemolysin secretion. Proc. Natl Acad. Sci. USA 87, 4776–4780.

Wang, N., Silver, D.L., Thiele, C. and Tall, A.R. (2001) ATP-binding cassette transporter A1 (ABCA1) functions as a cholesterol efflux regulatory protein. J. Biol. Chem. 276, 23742–23747.

Weng, J., Mata, N.L., Azarian, S.M., Tzekov, R.T., Birch, D.G. and Travis, G.H. (1999) Insights into the function of Rim protein in photoreceptors and etiology of Stargardt's disease from the phenotype in ABCR knock-out mice. Cell 98, 13–23.

Winans, S.C., Kerstetter, R.A. and Nester, E.W. (1988) Transcriptional regulation of the *virA* and *virG* genes of *Agrobacterium tumefaciens*. J. Bacteriol. 170, 4047–4054.

Winstedt, L., Yoshida, K., Fujita, Y. and von Wachenfeldt, C. (1998) Cytochrome *bd* biosynthesis in *Bacillus subtilis*: characterization of the *cydABCD* operon. J. Bacteriol. 180, 6571–6580.

Wolfger, H., Mamnun, Y.M. and Kuchler, K. (2001) Fungal ABC proteins: pleiotropic drug resistance, stress response and cellular detoxification. Res. Microbiol. 152, 375–389.

Wu, S.S., Wu, J., Cheng, Y.L. and Kaiser, D. (1998) The *pilH* gene encodes an ABC transporter homologue required for type IV pilus biogenesis and social gliding motility in *Myxococcus xanthus*. Mol. Microbiol. 29, 1249–1261.

Wu, Y.C. and Horvitz, H.R. (1998) The *C. elegans* cell corpse engulfment gene *ced-7* encodes a protein similar to ABC transporters. Cell 93, 951–960.

Yakushi, T., Masuda, K., Narita, S., Matsuyama, S. and Tokuda, H. (2000) A new ABC transporter mediating the detachment of lipid-modified proteins from membranes. Nat. Cell. Biol. 2, 212–218.

Yamada, T., Fukuda, T., Tamura, K., Furukawa, S. and Songsri, P. (1993) Expression of the gene encoding a translational elongation factor 3 homolog of Chlorella virus CVK2. Virology 197, 742–750.

Yamaguchi, Y., Kasano, M., Terada, T., Sato, R. and Maeda, M. (1999) An ABC transporter homologous to TAP proteins. FEBS Lett. 457, 231–236.

Young, J. and Holland, I.B. (1999) ABC transporters: bacterial exporters-revisited five years on. Biochim. Biophys. Acta 1461, 177–200.

Young, L., Leonhard, K., Tatsuta, T., Trowsdale, L. and Langer, T. (2001) Role of the ABC transporter Mdl1 in peptide export from mitochondria. Science 291, 2135–2138.

Zhang, F., Hogue, D.L., Liu, L., Fisher, C.L., Hui, D., Childs, S. and Ling, V. (2000a) M-ABC2, a new human mitochondrial ATP-binding cassette membrane protein. FEBS Lett. 478, 89–94.

Zhang, F., Zhang, W.D., Liu, L., Fisher, C.L., Hui, D., Childs, S., Dorovini-Zis, K. and Ling, V. (2000b) Characterization of ABCB9, an ATP binding cassette protein associated with lysosomes. J. Biol. Chem. 275, 23287–23294.

Zheng, L., White, R.H., Cash, V.L., Jack, R.F. and Dean, D.R. (1993) Cysteine desulfurase activity indicates a role for NifS in metallocluster biosynthesis. Proc. Natl Acad. Sci. USA 90, 2754–2758.

Zhou, Z.M., White, K.A., Polissi, A., Georgopoulos, C. and Raetz, C.R.H. (1998) Function of *Escherichia coli* MsbA, an essential ABC family transporter, in lipid A and phospholipid biosynthesis. J. Biol. Chem. 273, 12466–12475.

Zumft, W.G., Viebrock-Sambale, A. and Braun, C. (1990) Nitrous oxide reductase from denitrifying *Pseudomonas stutzeri*. Genes for copper-processing and properties of the deduced products, including a new member of the family of ATP/GTP-binding proteins. Eur. J. Biochem. 192, 591–599.

Membrane Topology of the Human ABC Transporter Proteins

CHAPTER 2

András Váradi, Gábor E. Tusnády and Balázs Sarkadi

Introduction

The most recent annotation* of the human genome sequence revealed 48 genes for ABC proteins, which were grouped into seven subclasses, from ABCA to ABCG (see: http://nutrigene.4t.com/humanabc.htm; see also Chapter 3). As detailed elsewhere in this book, ABC proteins in all organisms can be recognized by their conserved motifs within the ATP-binding domains. The majority of these proteins are membrane embedded and fulfill various membrane transport or regulatory functions (hence the designation ABC transporters), and our present review deals with such human proteins. In contrast, the human ABCE and ABCF subfamilies contain proteins with no known transmembrane domains; therefore these are outside the scope of this chapter.

It is generally accepted that the minimum functional unit requirement for an ABC transporter is the presence of two transmembrane domains (TMD; IM in Chapter 1) and two ATP-binding cassette (ABC) units. These may be present within one polypeptide chain ('full transporters'), or within a membrane-bound homo- or heterodimer of 'half transporters'.

At present there are no high-resolution structural data available for any mammalian ABC transporter; therefore computer modeling and laborious biochemical experiments are necessary to elucidate membrane topology, i.e. the position and orientation of membrane-spanning segments within the polypeptide chain. The generally applied experimental methods include epitope insertion, localization of glycosylation sites, limited proteolysis and immunochemical techniques. Several computer-assisted empirical prediction methods are available to generate the hydrophobicity profile for a putative transmembrane protein, but such an analysis may provide only a basis for developing experimental strategies for the elucidation of the actual membrane topology.

In the present review we summarize the available data on the membrane topology for various ABC transporters by examining the distinct subfamilies. We discuss predicted topology models and their experimental reinforcement or negation, and assess the relationship between phylogenetic linkages and the arrangement of membrane topology patterns. We call the reader's attention to a recent thematic review, also analyzing the problems of membrane topology models within the ABC-protein kingdom (Dean *et al.*, 2001).

Methodology

In the present study, for subfamily classification, we used the Human ABC Proteins Database

This chapter reflects the information available in June, 2001.

ABC Proteins: From Bacteria to Man
ISBN 0-12-352551-9

(http://nutrigene.4t.com/humanabc.htm, last update May 20, 2001). We included in the analysis only those members of a given subfamily whose full sequences were known. We chose the sequences of the longest and/or the most common splicing variants, in cases where alternatively spliced cDNAs were described. We considered a membrane topology to be 'established' if it was supported by independent experimental data.

Protein sequences were aligned by the ClustalW server (http://www.ebi.ac.uk/clustalw), and hydrophobicity plots were generated according to a described method (von Heijne, 1992). For computer-assisted predictions the HMMTOP (Hidden Markov Model for Topology Prediction) transmembrane topology prediction server was applied (Tusnády and Simon, 1998, 2001), which is freely available for noncommercial users (http://www. enzim.hu/hmmtop). This method is based on the principle that the topology of transmembrane proteins is determined by considering the maximal divergence in the amino acid composition of defined sequence segments.

The results of the above analyses are presented as hydrophobicity plots of the aligned sequences within a given subfamily, and the phylogenetic trees of the subfamilies are also shown together with the plots.

MEMBRANE TOPOLOGY OF THE ABC TRANSPORTERS

ABCA SUBFAMILY

The Human ABC Database lists 14 members for the ABCA subfamily, from which currently seven are represented by full cDNA sequences. We have analyzed the membrane topology of the following seven proteins: A1 = 2261 aa; A2 = 2436 aa; A3 = 1704 aa; A4 = 2273 aa; A7 = 2146 aa; A8 = 1581 aa; and A12 = 2277 aa. The ABCA subfamily, as examined here, contains 'full transporters', and it is noteworthy that these proteins have the largest molecular masses within the entire human ABC protein family.

Within this subfamily the function and membrane topology of only two members have been studied in detail. These are *ABCA1*, the gene associated with Tangier disease, and *ABCA4*, the retina-specific ABC transporter whose mutations cause severe retinopathies. Both of these proteins (see Chapters 23 and 28) are located in the plasma membrane, and they share 50% identical amino acids. All the suggested membrane topology models agree that both halves of these proteins contain one TMD and one ABC domain, and the TMDs consist of six transmembrane helices. However, there are several different models for the actual distribution of the helices.

In the case of *ABCA1*, the first model, elaborated by Luciani *et al.* (1994), predicted a large cytoplasmic domain at the extreme N-terminal part, and another large cytoplasmic domain ('regulatory domain') in the central portion of the protein. A hydrophobic segment with a hairpin orientation was predicted to be located within this cytoplasmic regulatory domain, thus anchoring it to the plasma membrane.

After the cloning of *ABCA4*, a topology model smilar to ABCA1 was first suggested (Allikmets *et al.*, 1997). However, Illing *et al.* (1997) described a different model for ABCA4, with two large extracellular domains (ECDs) between the first and second transmembrane helices in each predicted TMD. In this model the first TM helix within the N-terminal half was placed close to the N-terminus, while the first predicted TM helix in the C-terminal half was the very same hydrophobic segment suggested to form a hairpin with a membrane-associated structure in the previous models. This latter model thus predicted no large intracellular regulatory domain for ABCA4. A combination of the two different models was also published (Azarian and Travis, 1997; Sun *et al.*, 2000), in which the N-terminal half of the ABCA4 protein was represented by the Illing model, while the C-terminal half contained a large cytoplasmic (regulatory) domain, as suggested by the model of Luciani *et al.* (1994).

In the case of ABCA4, an elegant experimental proof for the Illing model has been recently published (Bungert *et al.*, 2001). Eight functional N-glycosylation sites were mapped by mutagenesis within the bovine ABCA4 sequence, four in the N-terminal half, and four within the C-terminal half. These results support the presence of a 600 amino acid extracellular domain within the N-terminal half, and a 275 amino acid extracellular domain within the C-terminal half of the protein. The authors also presented experimental data suggesting that the two extracellular domains are linked by disulfide bridge(s).

Regarding ABCA1, after the initial description of its coding region, an in-frame, upstream methionine was discovered, and translation from this codon results in a 60 amino acid extension of the originally expected polypeptide chain (Costet *et al.*, 2000; Pullinger *et al.*, 2000; Santamarina-Fojo *et al.*, 2000). Within this N-terminal 60 amino acid chain, sequence analysis predicted the presence of a transmembrane helix (between residues 22 and 44), and a potentially cleavable signal sequence (between amino acids 45 and 46). These predictions triggered experiments to test the membrane topology dictated by the presence of a TM helix (TMH) in the proximity of the N-terminus, which may serve as an anchor signal and orient a large loop in the N-terminal half of the protein extracellularly. Fitzgerald *et al.* (2001) expressed various truncated and tagged versions of human ABCA1, and demonstrated that the loop from amino acids 44 to 640 indeed has an extracellular orientation. Tanaka *et al.* (2001) obtained similar results, and they also predicted a large extracellular loop within the C-terminal half of the protein (between amino acids 1368 and 1655). However, the two publications contain contradicting data concerning whether cleavage occurs at position 45. According to the results of Tanaka *et al.* (2001), the polypeptide chain is cleaved at this position during maturation, and they suggest that ABCA1 is present in the plasma membrane with a large N-terminal segment extending to the extracellular space. Fitzgerald *et al.* (2001), using a bulkier N-terminal GFP-tag, found no cleavage of ABCA1 at this position.

In summary, the experimental data obtained for ABCA1 and ABCA4 support a similar membrane topology for the two proteins, with a domain arrangement of TMH1-ECD1-TMH(2–6)-ABC1-TMH7-ECD2-TMH(8–12)-ABC2, where H indicates helix and ECD, an extracellular domain. The primary sequence alignment and hydropathy analysis of the aligned ABCA sequences (as presented in **Figure 2.1**) are in line with this assumption. In the projection of **Figure 2.1**, the location of TM helices within ABCA1, as predicted by Tanaka *et al.*, is used as a guide to label similarities within the hydrophobicity patterns of the members in this subfamily. Indeed, a similar domain arrangement and membrane topology can be predicted for all the currently known members of the ABCA subfamily. However, the alignment also reveals that ABCA3 and ABCA8 have relatively short ECDs within their N-terminal and C-terminal halves.

ABCB subfamily

Three members of this subfamily are 'full transporters', with ABCB1 (MDR1 or Pgp), ABCB4 (MDR3) and ABCB11 (sisterPgp or BSEP), all localized in the plasma membrane, in polarized cells in the apical membrane. All the other family members are 'half transporters', including ABCB2 and ABCB3 (TAP1 and TAP2) residing in the endoplasmic reticulum. ABCB6, ABCB7, ABCB8 and ABCB10 are mitochondrial membrane proteins, while ABCB9 has a putative lysosomal localization. It is well established that ABCB2/TAP1 and ABCB3/TAP2 form a noncovalent heterodimer, and actively translocate peptides into the endoplasmic reticulum (see Chapter 26). Dimer formation for the mitochondrial half transporters has not been experimentally shown, but the presence of mitochondrial targeting signals within the N-terminal regions of both ABCB7 (Csere *et al.*, 1998), and ABCB8 (Hogue *et al.*, 1999) has been demonstrated.

For the first recognized mammalian ABC transporter, the original topology model of ABCB1 (MDR1-Pgp) predicted six TM helices in both TMDs of the protein, each followed by an ABC domain (Chen *et al.*, 1986). This membrane topology has been fully supported by later epitope insertion experiments (Kast *et al.*, 1995, 1996). ABCB4 and ABCB11 are close relatives of ABCB1, and similar membrane topology arrangements can be predicted for both of these proteins **(Figure 2.2)**.

Hydrophobicity plots of the aligned sequences reveal that the positioning of predicted helices in the ABCB family of half transporters is more closely related to the C-terminal halves (TMD2) of the full transporters than to the N-terminal halves, as presented in **Figure 2.2**. The hydrophobicity plots of the half transporters support the six TM helix model in their TMDs, although the N-terminal regions are clearly extended, and contain hydrophobic regions which may correspond to TM helices. Indeed, in the case of TAP1 and TAP2, additional four and three TM helices, respectively, were predicted (Abele and Tampe, 1999). It is worth mentioning that currently no reliable algorithms are available for such predictions, and visual inspection and heuristic adjustments

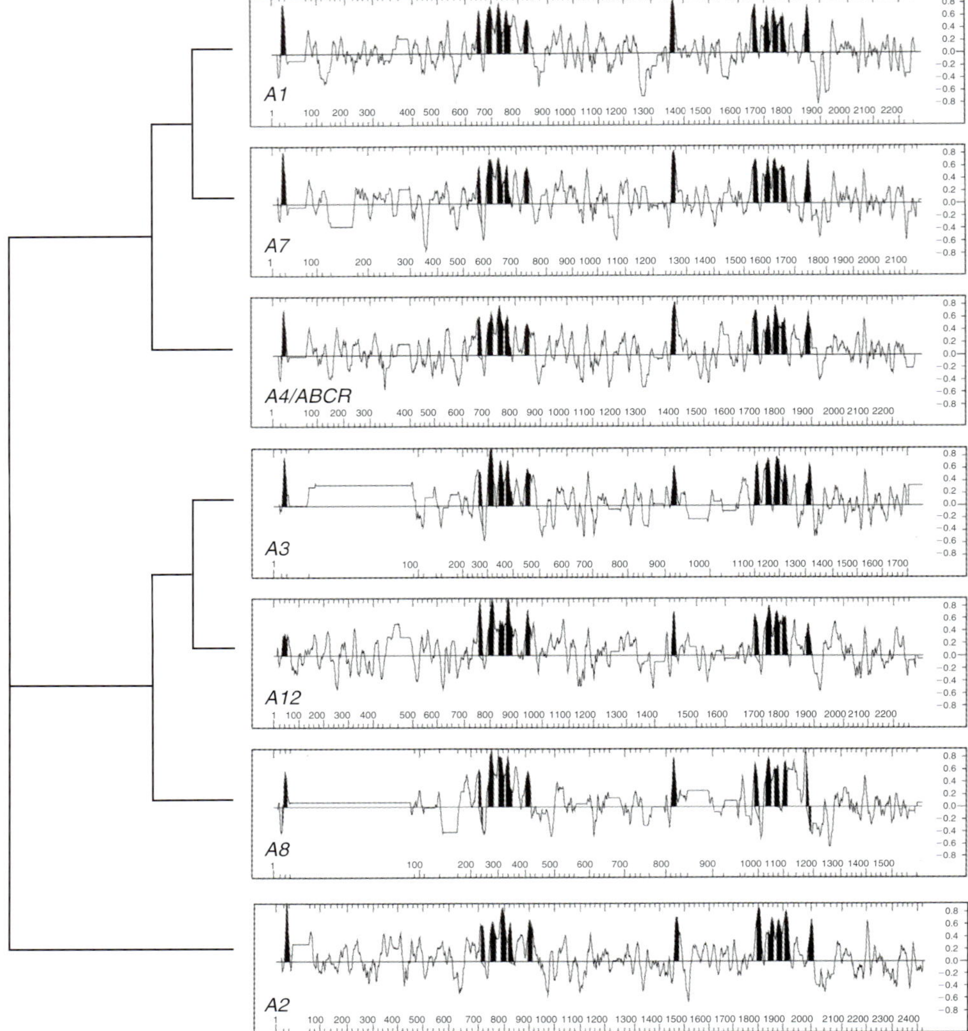

Figure 2.1. ***Relative similarity dendrogram and the hydrophobicity plots of the aligned sequences of the ABCA subfamily transporters. The shaded areas show transmembrane helices with locations supported by experimental results from studies of ABCA1 and ABCA4 (see text).***

of the sequences can be most helpful for predicting the probable location of some TM helices, such as TM2 in ABCB6, or TM6 in ABCB8.

ABCC SUBFAMILY

The ABCC subfamily consists of 11 members in the human genome and most of these (ABCC1–6 or MRP1–6) have been identified as active membrane transporters for various organic anions, several of which are discussed in other chapters in this volume. Although the majority of the characterized ABC proteins are active pumps, the cystic fibrosis transmembrane conductance regulator, ABCC7 (CFTR), is a chloride channel which may also regulate other channel proteins. The sulfonylurea receptors, ABCC8 (SUR1) and ABCC9 (SUR2), are best described as intracellular ATP sensors, regulating the permeability of specific K^+ channels (with which they form transmembrane complexes). Nothing is currently known about the function of ABCC10 and ABCC11.

The membrane topology of human CFTR/ABCC7 was originally predicted based on the MDR1 model, and supported experimentally by glycosylation site insertion mutagenesis (Chang *et al.*, 1994). This study strongly supported the original model, with a TMD1-ABC1-R-TMD2-ABC2 domain arrangement. Each TMD is predicted to consist of six TM helices, and a regulatory domain (R) is present between

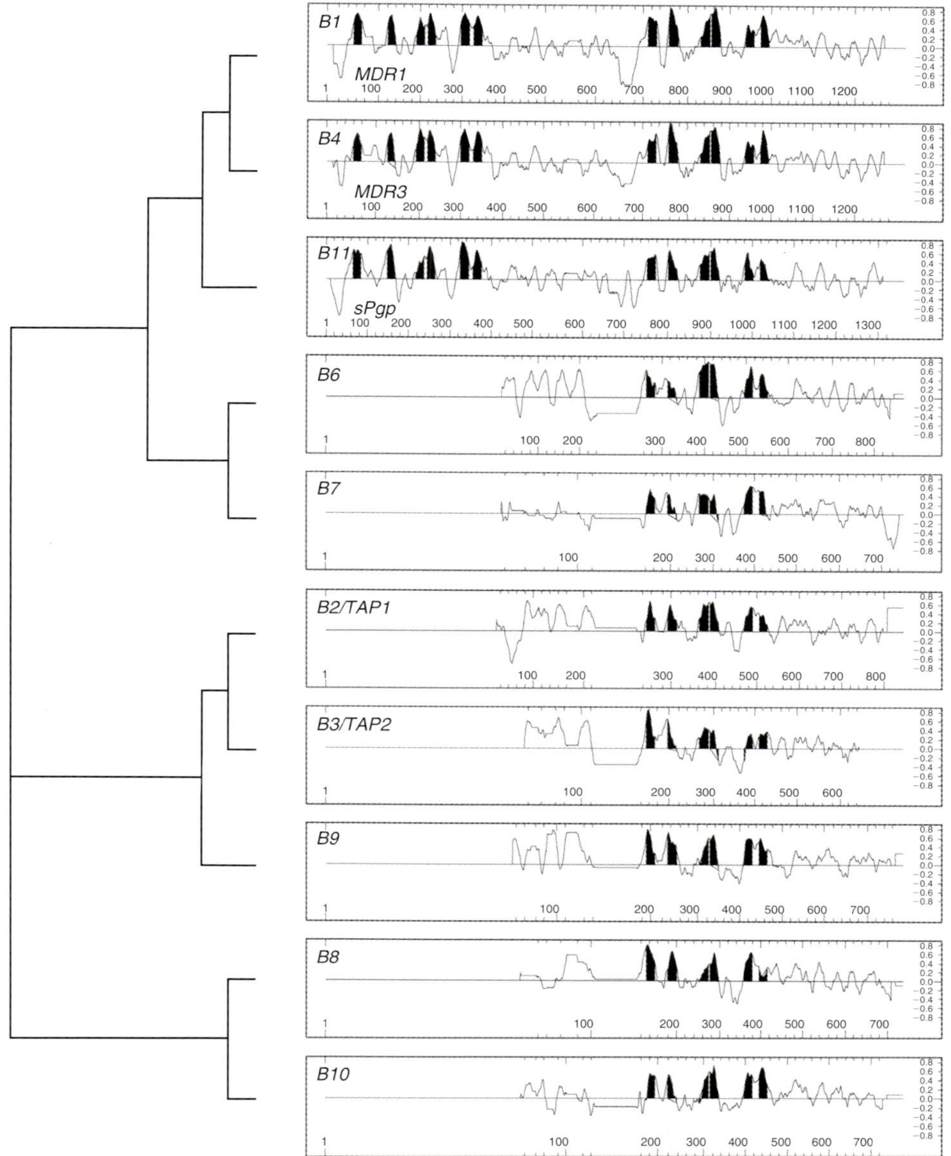

Figure 2.2. ***Relative similarity dendrogram and the hydrophobicity plots of the aligned sequences of the ABCB subfamily transporters. The shaded areas show transmembrane helices, predicted with locations as projected from experimental results based on ABCB1 (see text).***

the two halves of the protein, which seems to be unique for CFTR.

The membrane topology models for human ABCC1 have been independently formulated by two research groups (Bakos *et al.*, 1996; Stride *et al.*, 1996). They found that when the CFTR/ABCC7 and MRP1/ABCC1 sequences were aligned, the hydrophobicity analysis of the aligned sequences yielded a close matching of putative transmembrane segments, thus also suggesting a six plus six transmembrane helix topology for MRP1/ABCC1. However, ABCC1 contains an additional N-terminal segment of about 230 amino acids, which has no counterpart in CFTR/ABCC7. On the basis of the hydropathy profiles and limited proteolysis experiments, the hydrophobic N-terminal segment of ABCC1 was suggested to be membrane embedded, with four to six transmembrane helices (Bakos *et al.*, 1996; Stride *et al.*, 1996).

Subsequent investigations of the membrane topology of human ABCC1 by epitope insertion (Kast and Gros, 1997, 1998) and by mutation of glycosylation sites (Hipfner *et al.*, 1997) fully supported the above topology and revealed the presence of five TM helices in the N-terminal segment. These results indicated a domain arrangement of TMD0-L0-TMD1-ABC1-L-TMD2-ABC2, in which TMD0 represents the N-terminal five TM helix extension, while L0 and L represent intracellular linker sequences. In fact, by aligning the linear sequences and the hydrophobicity plots for all full-size ABC transporters present in the sequence database in 1997, we concluded that a common membrane topology and domain arrangement distinguishes a subfamily (MRP or ABCC subfamily) within the ABC kingdom (Tusnády *et al.*, 1997).

In the present review the amino acid sequences of 11 members of the ABCC subfamily (those with full cDNA sequences) were aligned, and the hydrophobicity plots of the aligned protein sequences were determined **(Figure 2.3)**. This comparison indicates that seven proteins in the family (ABCC1–3, ABCC6, and ABCC8–10) form a subcluster within which each member possesses the N-terminal TMD0 domain, first described for ABCC1. Thus this subgroup is characterized by the TMD0-L0-TMD1-ABC1-L-TMD2-ABC2 arrangement. The N-terminal TMD0 domain is absent from ABCC4, ABCC5 and CFTR/ABCC7.

Recent studies revealed that the TMD0 domain of ABCC1 does not play a crucial role either in the transport activity of the protein, or in the proper routing of the protein into the basolateral membrane compartment. However, the presence of the L0 region (together with the TMD1-ABC1-L-TMD2-ABC2 core) is necessary for the ABCC1 GS-conjugate transport activity, and for the proper intracellular routing of the protein (Bakos *et al.*, 1998). Thus, the ABCC1 L0 polypeptide was found to be membrane associated, and a 10 amino acid deletion within this region, encompassing a putative amphipathic helix, abolished the L0–membrane interaction and eliminated transport function, while not affecting membrane routing (Bakos *et al.*, 2000). We have concluded from these studies that the L0 region forms a distinct structural and functional domain, which interacts with both the membrane and the core region of the transporter.

In harmony with the above conclusions, the cytoplasmic amino-terminal of CFTR/ABCC7 (which corresponds to the L0 domain) was found to have a major role in the control of CFTR channel gating, via physical interaction with the regulatory (R) domain (Naren *et al.*, 1999). Of course, the actual sites of interactions still need to be explored in these different proteins.

ABCD subfamily

Four half transporters with TMD-ABC arrangements, ABCD1/ALDP, ABCD2/ALDR, ABCD3/PMP70 and ABCD4/PMP70R, are the members of this family. They are localized to the peroxisomal membrane and their mutant forms are involved in different inherited peroxisomal disorders. It has been proposed that peroxisomal transporters need to dimerize in order to exert their function. Co-immunoprecipitation experiments demonstrated the homodimerization of ABCD1, while a heterodimerization of ABCD1 with ABCD3 or ABCD2, and heterodimerization of ABCD2 with ABCD3 were found (Liu *et al.*, 1999).

The presence of six TM helices in the TMDs of the ABCD half transporters is generally predicted, but the experimental verification of this prediction has yet to be approached.

ABCG subfamily

The members of this subfamily are also half transporters, with a unique domain arrangement of ABC-TMD, i.e. the ABC domain located at the N-terminus. Four proteins have been described with full sequences in this subgroup. The best-characterized transporter is ABCG2 (MXR/BCRP), whose overexpression confers multidrug resistance. Interestingly, this protein performs an active drug extrusion from the cells and is N-glycosylated in its mature form. These data suggest that ABCG2 is localized in the plasma membrane, while several other half ABC transporters in the ABCB and ABCD subfamilies have been suggested to localize in membranes of various intracellular organelles, e.g. the TAP proteins are located in the ER. There is genetic evidence that ABCG5 and ABCG8 form heterodimers, as it was found that mutations of either of these genes cause the recessive genetic disease sitosterolemia (Berge *et al.*, 2000; Lee *et al.*, 2001). On the other hand, overexpression of the human ABCG2 multidrug resistance protein in a heterologous insect cell system generated an active, drug-stimulated ATPase activity, strongly suggesting that this protein can act as a homodimer (Özvegy *et al.*, 2001).

No experimental data are available as yet on the exact membrane topology of the ABCG transporters. However, here we combined some experimental data with empirical predictions to localize the transmembrane helices. We used the

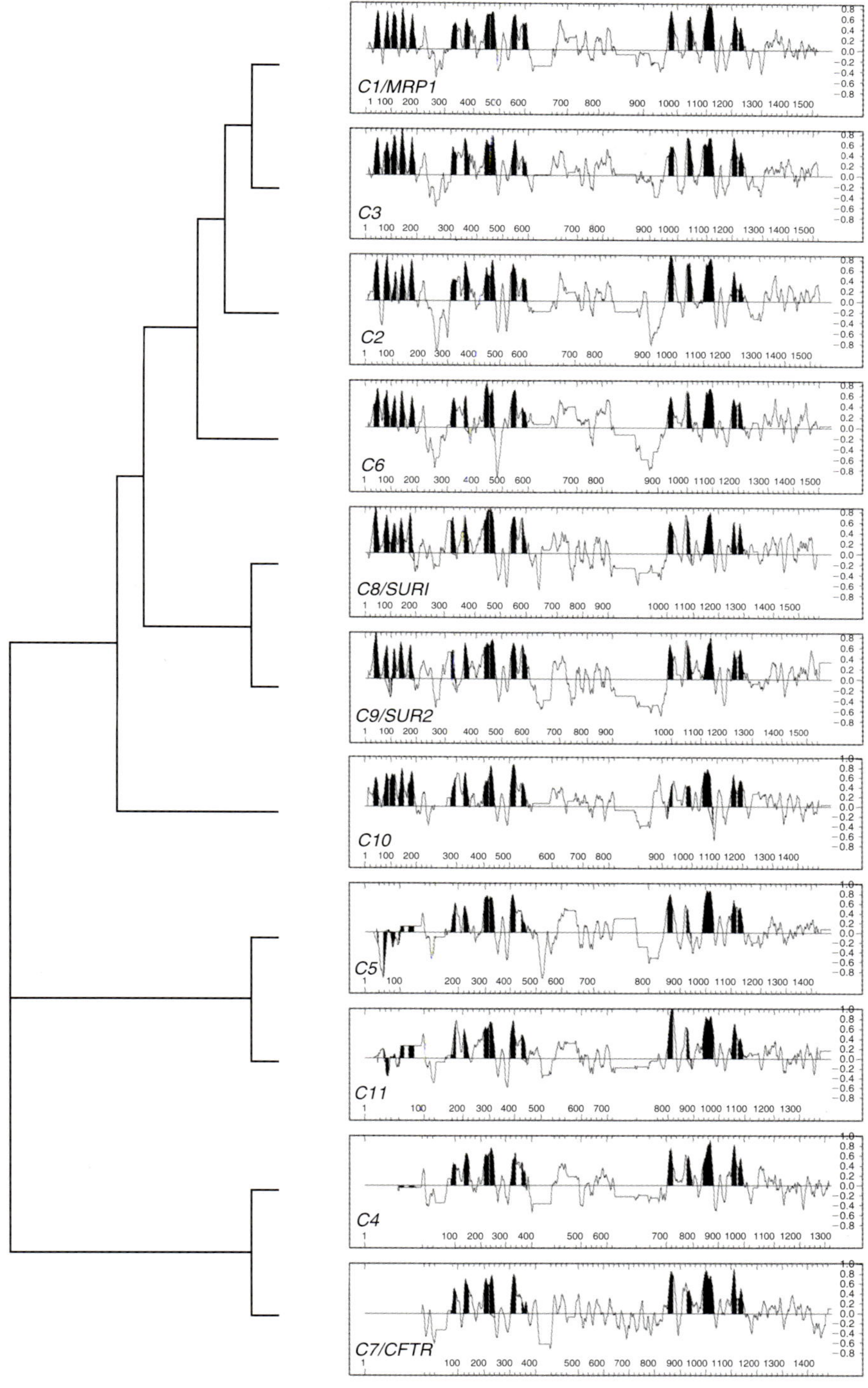

Figure 2.3. ***Relative similarity dendrogram and the hydrophobicity plots of the aligned sequences of the ABCC subfamily transporters. The shaded areas show transmembrane helices, predicted with locations as projected from experimental results based on ABCC1 (see text).***

HMMTOP transmembrane topology prediction server for this purpose in ABCG2, with the following restrictions: the N-terminal 290 amino acid chain, which harbors the ABC domain, should be intracellular, and – as the protein was shown to be N-glycosylated (Özvegy *et al.*, 2001) – the predicted extracellular loops should contain consensus N-glycosylation site(s). The model predicted six TM helices with the following localization in the linear sequence of

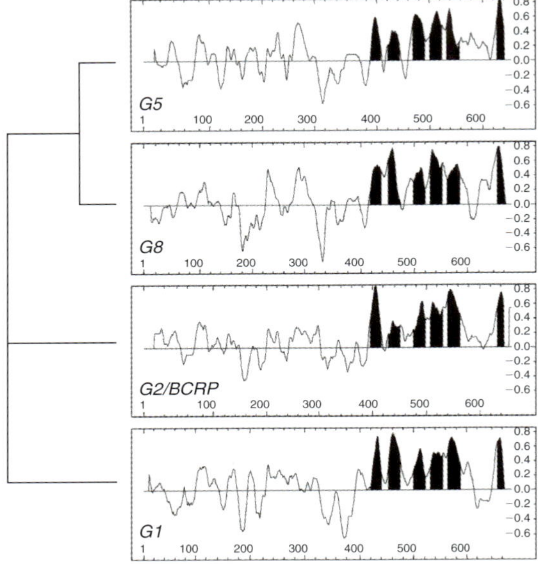

Figure 2.4. ***Relative similarity dendrogram and the hydrophobicity plots of the aligned sequences of the ABCG subfamily transporters. The shaded areas show transmembrane helices, predicted with locations as projected from the HMMTOP server data and experimental results, based on ABCG2 (see text).***

ABCG2: 394–416, 427–449, 474–497, 506–530, 539–563 and 632–651. Two of the potential N-glycosylation sites (at positions 418 and at 596) were predicted within extracellular loops 1 and 3, respectively. **Figure 2.4** shows the hydrophobicity plots of the aligned ABCG sequences, with the HMMTOP predicted TMDs superimposed. This analysis suggests a similar location of the six TM helices in the TMDs for all ABCG proteins.

CONCLUSIONS

Membrane topology models without experimental studies are just 'educated hallucinations' (to quote J. Riordan), but even at this stage they have an important stimulatory role in searching for structure–function relationships. As demonstrated in the case of CFTR or Pgp-MDR1, biochemical experiments may efficiently validate such models, while in other cases, e.g. in the ABCA subfamily, experiment-based models still remain contradictory. The various membrane topology predictions in the ABCC-MRP family have led to numerous experimental and theoretical studies, yielding important information regarding functional domains and those involved in membrane routing/targeting.

A membrane topology is still difficult to define for any given ABC transporter, but a correct prediction for domain arrangements may provide a major help in devising useful antibodies, site-directed mutants and even specific functional modulators or inhibitors. We can hardly wait for the detailed crystal-based structures of these large human membrane proteins, but until then there is still much fun to be had in working out more and more sophisticated and accurate models.

After completing this chapter, the structure of a bacterial ABC transporter, MsbA of *Escherichia coli*, determined by X-ray crystallography to a resolution of 4.5 Å (Chang and Roth, 2001, see Chapter 7) was published. MsbA is a half transporter with a TMD-ABC domain arrangement, and the functional protein is a homodimer. The published structure reveals that each MsbA subunit contains a TMD with six transmembrane helices, an ABC domain, and an 'intracellular domain' which is composed of three intracellular loops, connecting TMH2 to TMH3, TMH4 to TMH5, and TMH6 to the ABC domain. A very important result of the published MsbA structure is the first demonstration that the membrane-spanning segments of this ABC transporter are indeed α-helices. Helix organization and interactions similar to those in the MsbA protein most probably will be characteristic for many other ABC transporters.

REFERENCES

Abele, R. and Tampe, R. (1999) Function of the transport complex TAP in cellular immune recognition. Biochim. Biophys. Acta 461, 405–419.

Allikmets, R., Singh, N., Sun, H., Shroyer, N.F., Hutchinson, A., Chidambaram, A., *et al.* (1997) A photoreceptor cell-specific ATP-binding transporter gene (ABCR) is mutated in recessive Stargardt macular dystrophy. Nat. Genet. 15, 236–246.

Azarian, S.M. and Travis, G.H. (1997) The photoreceptor rim protein is an ABC transporter encoded by the gene for recessive Stargardt's disease (ABCR). FEBS Lett. 409, 247–252.

Bakos, É., Hegedûs, T., Holló, Z., Welker, E., Tusnády, G.E., Zaman, G.J., Flens, M.J., Váradi, A. and Sarkadi, B. (1996) Membrane topology and glycosylation of the human multidrug resistance-associated protein. J. Biol. Chem. 271, 12322–12326.

Bakos, É., Evers, R., Szakács, G., Tusnády, G.E., Welker, E., Szabó, K., *et al.* (1998) Functional multidrug resistance protein (MRP1) lacking the N-terminal transmembrane domain. J. Biol. Chem. 273, 32167–32173.

Bakos, É., Evers, R., Calenda, G., Tusnády, G., Szakács, G., Váradi, A. and Sarkadi, B. (2000) Characterization of the amino-terminal regions in the human multidrug resistance protein (MRP1). J. Cell Sci. 113, 4451–4461.

Berge, K.E., Tian, H., Graf, G.A., Yu, L., Grishin, N.V., Schultz, J., Kwiterovich, P., Shan, B., Barnes, R. and Hobbs, H.H. (2000) Accumulation of dietary cholesterol in sitosterolemia caused by mutations in adjacent ABC transporters. Science 290, 1771–1775.

Bungert, S., Molday, L.L. and Molday, R.S. (2001) Membrane topology of the ATP binding cassette transporter ABCR and its relationship to ABC1 and related ABCA transporters. Identification of N-linked glycosylation sites. J. Biol. Chem. 276, 23539–23546.

Chang, G. and Roth, C.B. (2001) Structure of MsbA from *E. coli*: a homolog of the multidrug resistance ATP binding cassette (ABC) transporters. Science 293, 1793–1800.

Chang, X.B., Hou, Y.X., Jensen, T.J. and Riordan, J.R. (1994) Mapping of cystic fibrosis transmembrane conductance regulator membrane topology by glycosylation site insertion. J. Biol. Chem. 269, 18572–18575.

Chen, C.J., Chin, J.E., Ueda, K., Clark, D.P., Pastan, I., Gottesman, M.M. and Roninson, I.B. (1986) Internal duplication and homology with bacterial transport proteins in the mdr1 (P-glycoprotein) gene from multidrug-resistant human cells. Cell 47, 381–389.

Costet, P., Luo, Y., Wang, N. and Tall, A.R. (2000) Sterol-dependent transactivation of the ABC1 promoter by the liver X receptor/retinoid X receptor. J. Biol. Chem. 275, 28240–28245.

Csere, P., Lill, R. and Kispal, G. (1998) Identification of a human mitochondrial ABC transporter, the functional orthologue of yeast Atm1p. FEBS Lett. 441, 266–270.

Dean, M., Hamon, Y. and Chimini (2001) The human ATP-binding cassette (ABC) transporter superfamily. J. Lipid Res. 42, 1007–1017.

Fitzgerald, M.L., Mendez, A.J., Moore, K.J., Andersson, L.P., Panjeton, H.A. and Freeman, M.W. (2001) ATP-binding cassette transporter A1 contains an NH2-terminal signal anchor sequence translocates the protein's first hydrophilic domain to the exoplasmic space. J. Biol. Chem. 276, 15137–15145.

Hipfner, D.R., Almquist, K.C., Leslie, E.M., Gerlach, J.H., Grant, C.E., Deeley, R.G. and Cole, S.P. (1997) Membrane topology of the multidrug resistance protein (MRP). J. Biol. Chem. 272, 23623–23630.

Hogue, D.L., Liu, L. and Ling, V. (1999) Identification and characterization of a mammalian mitochondrial ATP-binding cassette membrane protein. J. Mol. Biol. 285, 379–389.

Illing, M., Molday, L.L. and Molday, R.S. (1997) The 220-kDa rim protein of retinal rod outer segments is a member of the ABC transporter superfamily. J. Biol. Chem. 272, 10303–10310.

Kast, C. and Gros, P. (1997) Topology mapping of the amino-terminal half of multidrug resistance-associated protein by epitope insertion and immunofluorescence. J. Biol. Chem. 272, 26479–26487.

Kast, C. and Gros, P. (1998) Epitope insertion favors a six transmembrane domain model for the carboxy-terminal portion of the multidrug resistance-associated protein. Biochemistry 37, 2305–2313.

Kast, C., Canfield, V., Levenson, R. and Gros, P. (1995) Membrane topology of P-glycoprotein as determined by epitope insertion. Biochemistry 34, 4402–4411.

Kast, C., Canfield, V., Levenson, R. and Gros, P. (1996) Transmembrane organization of mouse P-glycoprotein determined by epitope insertion and immunofluorescence. J. Biol. Chem. 271, 9240–9248.

Lee, M.H., Lu, K., Hazard, S., Yu, H., Shulenin, S., Hidaka, H., *et al.* (2001) Identification of a gene, ABCG5, important in the regulation of dietary cholesterol absorption. Nat. Genet. 27, 79–83.

Liu, L.X., Janvier, K., Berteaux-Lecellier, V., Cartier, N., Benarous, R. and Aubourg, P. (1999) Homo- and heterodimerization of peroxisomal ATP-binding cassette half-transporters. J. Biol. Chem. 274, 32738–32743.

Luciani, M.F., Denizot, F., Savary, S., Mattei, M.G. and Chimini, G. (1994) Cloning of two novel ABC transporters mapping on human chromosome 9. Genomics 21, 150–159.

Naren, A.P., Cormet-Boyaka, E., Fu, J., Villain, M., Blalock, J.E., Quick, M.W. and Kirk, K.L. (1999) CFTR chloride channel regulation by an interdomain interaction. Science 286, 544–548.

Özvegy, C., Litman, T., Szakács, G., Nagy, Z., Bates, S., Váradi, A. and Sarkadi, B. (2001) Functional characterization of the human multidrug transporter, ABCG2, expressed in insect cells. Biochem. Biophys. Res. Commun. 285, 111–117.

Pullinger, C.R., Hakamata, H., Duchateau, P.N., Eng, C., Aouizerat, B.E., Cho, M.H., Fielding, C.J. and Kane, J.P. (2000) Analysis of hABC1 gene 5′ end: additional peptide sequence, promoter region, and four polymorphisms. Biochem. Biophys. Res. Commun. 271, 451–455.

Santamarina-Fojo, S., Peterson, K., Knapper, C., Qiu, Y., Freeman, L., Cheng, J.F., *et al.* (2000) Complete genomic sequence of the human ABCA1 gene: analysis of the human and mouse ATP-binding cassette A promoter. Proc. Natl Acad. Sci. USA 97, 7987–7992.

Stride, B.D., Valdimarsson, G., Gerlach, J.H., Wilson, G.M., Cole, S.P. and Deeley, R.G. (1996) Structure and expression of the messenger RNA encoding the murine multidrug resistance protein, an ATP-binding cassette transporter. Mol. Pharmacol. 49, 962–971.

Sun, H., Smallwood, P.M. and Nathans, J. (2000) Biochemical defect in ABCR protein variants associated with human retinopathies. Nat. Genet. 26, 242–246.

Tanaka, A.R., Ikeda, Y., Abe-Dohmae, S., Arakawa, R., Sadanami, K., Kidera, A., *et al.* (2001) Human ABCA1 contains a large amino-terminal extracellular domain homologous to an epitope of Sjogren's Syndrome. Biochem. Biophys. Res. Commun. 283, 1019–1025.

Tusnády, G.E. and Simon, I. (1998) Principles governing amino acid composition of integral membrane proteins: application to topology prediction. J. Mol. Biol. 283, 489–506.

Tusnády, G.E. and Simon, I. (2001) The HMMTOP transmembrane topology prediction server. Bioinformatics 17, 849–850.

Tusnády, G.E., Bakos, É., Váradi, A. and Sarkadi, B. (1997) Membrane topology distinguishes a subfamily of the ATP-binding cassette (ABC) transporters. FEBS Lett. 402, 1–3.

von Heijne, G. (1992) Membrane protein structure prediction. Hydrophobicity analysis and the positive-inside rule. J. Mol. Biol. 225, 487–494.

CHAPTER 3

Human and *Drosophila* ABC Proteins

Michael Dean, Andrey Rzhetsky and Rando Allikmets

Introduction

From the sequencing of the human (Lander *et al.*, 2001; Venter *et al.*, 2001), *Drosophila* (Myers *et al.*, 2000), and *Caenorhabditis elegans* genomes (Consortium, 1998) the full complement of ABC genes in each of these species has been characterized. **Figure 3.1** is an attempt to portray the major locations of some of the protein products of these genes in the human body. The eukaryotic ABC genes are organized either as full transporters containing two sets of transmembrane domains (TMDs) and two nucleotide-binding domains (NBDs), or as half transporters containing one TMD and one NBD (Hyde *et al.*, 1990). Half transporters must form

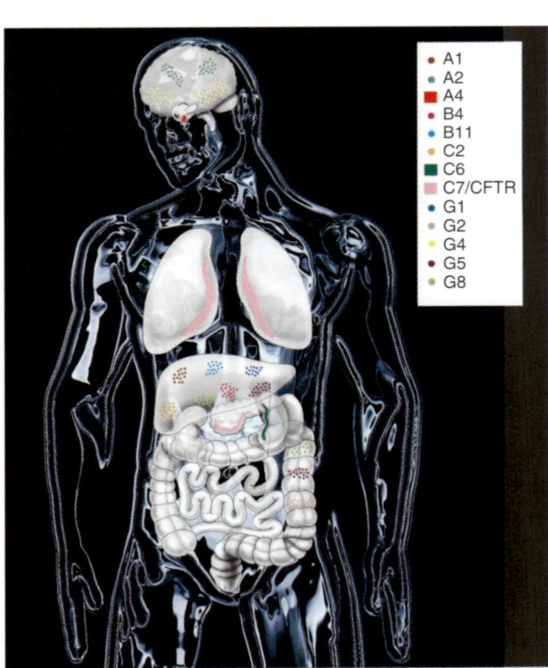

Figure 3.1. *Anatomy of human ABC proteins. A diagram of the human body is shown with the location of selected ABC transporters. ABC genes with a clearly defined tissue expression and/or disease association are shown. ABCA2 and ABCG4 are highly expressed in the brain, ABCA4 is exclusively expressed in the retina and mutations cause several retinal disorders. CFTR is expressed in the lung and pancreas and CF patients display pathologies in these organs as well as in the intestine and the vas deferens (not shown). ABCC6 is expressed in the kidney (and also liver, not shown), but leads to pathologies in the skin, eyes, and arteries (not shown). ABCG5 and ABCG8 are expressed in the liver and intestine, and mutations in these genes lead to aberrant sterol transport in these organs. ABCB4, ABCB11, ABCC2 and ABCG1 are expressed in the liver and play a role in the transport of bile components. ABCA1 is expressed in peripheral cells and liver and regulates cholesterol transport. ABCG2 is expressed in the placenta (not shown) and the intestine and probably serves to transport xenobiotics and toxic cell metabolites. (This figure was prepared by Barking Dog Art, Gloucestershire.)*

ABC Proteins: From Bacteria to Man
ISBN 0-12-352551-9

either homodimers or heterodimers in order to produce a functional transporter. ABC genes are abundant in all vertebrate and invertebrate eukaryotic genomes, indicating that most of these genes have existed since the beginning of eukaryotic evolution. The genes can be divided into subfamilies based on similarity in gene structure (half versus full transporters), order of the domains and sequence homology in the NBDs and TMDs. There are seven mammalian ABC gene subfamilies, five of which are also found in the *Saccharomyces cerevisiae* genome.

PHYLOGENETIC ANALYSIS OF HUMAN ABC GENES

The identification of the complete set of 48 human ABC genes **(Table 3.1)** (Dean *et al.*, 2001) has allowed a comprehensive phylogenetic analysis of the superfamily. **Figure 3.2** shows a neighbor-joining tree displaying the relationships of all human ABC genes. The nomenclature of ABC transporters is in

TABLE 3.1. LIST OF HUMAN ABC GENES, CHROMOSOMAL LOCATION AND FUNCTION

Family	Symbol	Alias	Location	Expression	Function
ABCA	ABCA1	ABC1	9q31.1	Ubiquitous	Cholesterol efflux onto HDL
	ABCA2	ABC2	9q34.4	Brain	Drug resistance
	ABCA3	ABC3, ABCC	16p13.3	Lung	Surfactant production
	ABCA4	ABCR	1p21.3	Rod photoreceptors	*N*-retinylidiene-PE efflux
	ABCA5		17q24.3	Muscle, heart, testes	
	ABCA6		17q24.3	Liver	
	ABCA7		19p13.3	Spleen, thymus	
	ABCA8		17q24.3	Ovary	
	ABCA9		17q24.3	Heart	
	ABCA10		17q24.3	Muscle, heart	
	ABCA12		2q34	Stomach	
	ABCA13		7p12.3	Low in all tissues	
ABCB	ABCB1	PGY1, MDR	7q21.12	Adrenal, kidney, brain	Multidrug resistance
	ABCB2	TAP1	6p21	All cells	Peptide transport
	ABCB3	TAP2	6p21	All cells	Peptide transport
	ABCB4	PGY3	7q21.12	Liver	Phosphatidylcholine transport
	ABCB5		7p21.1	Ubiquitous	
	ABCB6	MTABC3	2q35	Mitochondria	Iron transport
	ABCB7	ABC7	Xq21-q22	Mitochondria	Fe/S cluster transport
	ABCB8	MABC1	7q36.1	Mitochondria	
	ABCB9		12q24.31	Heart, brain	
	ABCB10	MTABC2	1q42.13	Mitochondria	
	ABCB11	SPGP	2q24.3	Liver	Bile salt transport
ABCC	ABCC1	MRP1	16p13.12	Lung, testes, PBMC	Drug resistance
	ABCC2	MRP2	10q24.2	Liver	Organic anion efflux
	ABCC3	MRP3	17q21.33	Lung, intestine, liver	Drug resistance
	ABCC4	MRP4	13q32.1	Prostate	Nucleoside transport
	ABCC5	MRP5	3q27.1	Ubiquitous	Nucleoside transport
	ABCC6	MRP6	16p13.12	Kidney, liver	
	CFTR	ABCC7	7q31.31	Exocrine tissues	Chloride ion channel
	ABCC8	SUR	11p15.1	Pancreas	Sulfonylurea receptor
	ABCC9	SUR2	12p12.1	Heart, muscle	

(continued)

TABLE 3.1. *(continued)*

Family	Symbol	Alias	Location	Expression	Function
	ABCC10	MRP7	6p21.1	Low in all tissues	
	ABCC11	MRP8	16q12.1	Low in all tissues	
	ABCC12	MRP9	16q12.1	Low in all tissues	
ABCD	ABCD1	ALD	Xq28	Peroxisomes	VLCFA transport regulation
	ABCD2	ALDL1, ALDR	12q11	Peroxisomes	
	ABCD3	PXMP1,PMP70	1p22.1	Peroxisomes	
	ABCD4	PMP69, P70R	14q24.3	Peroxisomes	
ABCE	ABCE1	OABP, RNS4I	4q31.31	Ovary, testes, spleen	Oligoadenylate-binding protein
ABCF	ABCF1	ABC50	6p21.1	Ubiquitous	
	ABCF2		7q36.1	Ubiquitous	
	ABCF3		3q27.1	Ubiquitous	
ABCG	ABCG1	ABC8, White	21q22.3	Ubiquitous	Cholesterol transport?
	ABCG2	ABCP, MXR, BCRP	4q22	Placenta, intestine	Toxin efflux, drug resistance
	ABCG4	White2	11q23	Liver	
	ABCG5	White3	2p21	Liver, intestine	Sterol transport
	ABCG8		2p21	Liver, intestine	Sterol transport

HDL, high density lipoprotein; VLCFA, very long chain fatty acid.

excellent agreement with the phylogenetic groups obtained. In particular, all major ABC transporter families are represented in the human tree by stable clusters with high bootstrap values.

This analysis provides evidence for frequent domain duplication of ATP-binding domains in ABC transporters. In nearly all cases, both ATP-binding domains encoded within a gene are more closely related to each other than to ATP-binding domains from ABC transporter genes of other subfamilies. This is unlikely to represent a concerted evolution of domains within the same gene, as the two domains within each gene are usually substantially diverged. A far more likely scenario suggests several independent duplication events rather than a single ancestral duplication.

DROSOPHILA ABC GENES

Analysis of the *Drosophila* genome sequence identified 56 ABC genes (Dean *et al.*, 2001) with at least one representative of each of the known mammalian subfamilies **(Table 3.2)**. To confirm the subfamily groupings the ATP-binding domain amino acid sequences were used to perform phylogenetic analyses (full transporters are represented with two ATP-binding domains each; **Figure 3.3**). Genes from the same subfamily cluster together and confirm the initial assignments made by inspection.

Both the human and *Drosophila* ABC genes are largely dispersed in the genome **(Figures 3.4 and 3.5)**. In the human genome there are five clusters of two genes and one cluster of five genes. For *Drosophila* ABC genes there are four clusters of two genes and one cluster of four genes. One of these *Drosophila* clusters (on chromosome 2L, band 37B9) is composed of an ABCB and an ABCC gene, indicating that this is a chance grouping of genes. The remaining clusters are composed of genes from the same subfamily and arranged in a head-to-tail fashion consistent with gene duplication. The one exception is the human ABCG5 and ABCG8 genes, which are arranged head-to-head (Berge *et al.*, 2000). Since the clusters themselves are dispersed and involve different subfamilies they presumably represent independent gene duplication events.

There are 15 ABCG genes in the *Drosophila* genome, making this the most abundant ABC subfamily. This is in sharp contrast to the only 5 and 6 known ABCG genes in the human and mouse genomes, respectively. The *Drosophila* ABCG genes are highly dispersed in the genome with only two pairs of linked genes. In addition, they are quite divergent phylogenetically, suggesting that there were many

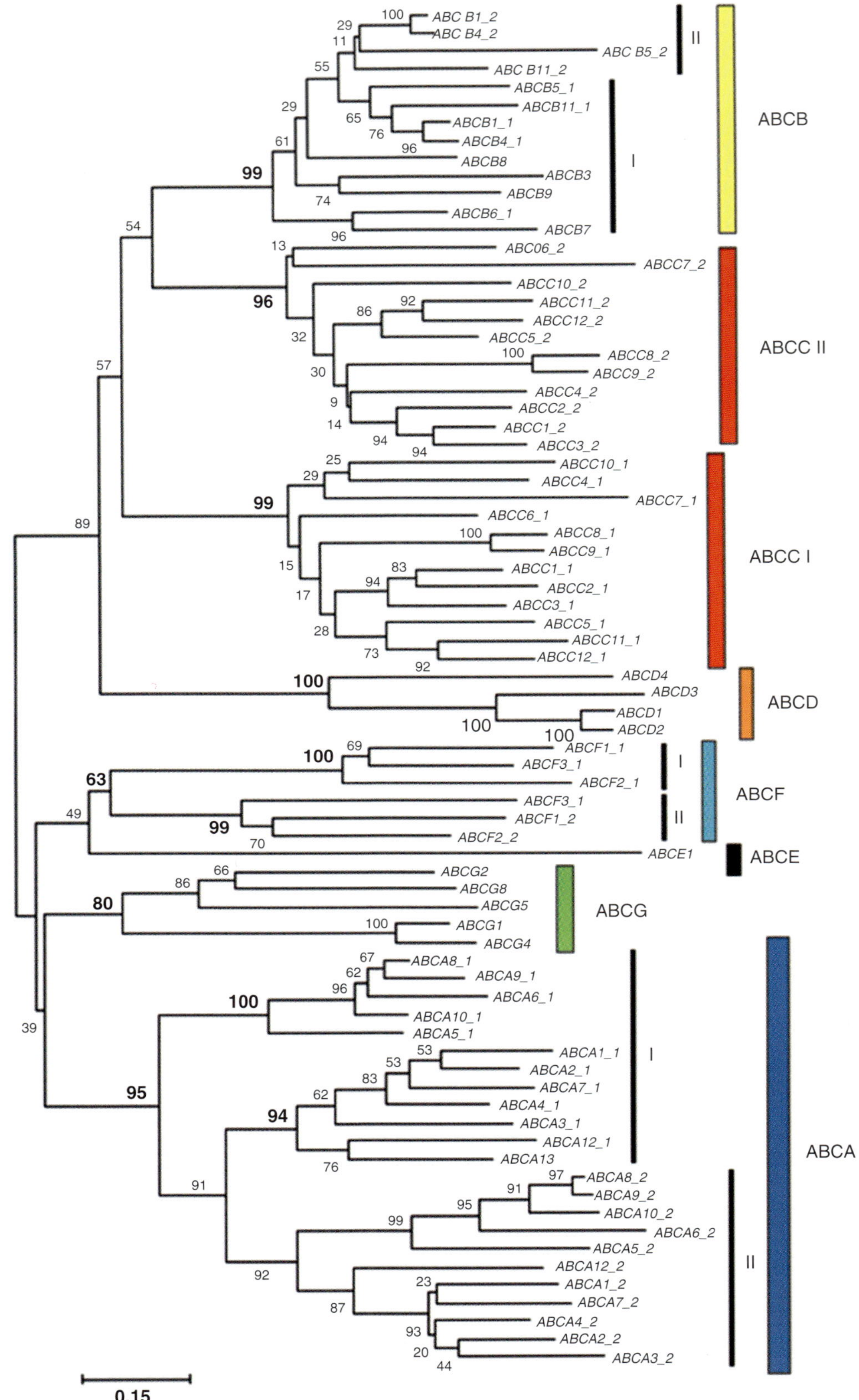

Figure 3.2. ***Phylogenetic tree of the human ABC genes. Amino acid sequences containing ATP-binding-domain proteins were identified with the model ABC_tran (accession PF00005) of the pfam database (Bateman* et al.*, 1999) as described earlier (Dean* et al.*, 2001).***

TABLE 3.2. *DROSOPHILA* ABC GENES

Gene	Alias	Size (aa)	Family	Location (Chr. Nuc)	Cyto. Loc.
CG3156			B	X 252038-254671	1B4
CG2759	w	696	G	X 2545753-2539884	3B4
CG1703		901	E	X 11393813-11396731	10C10
CG1824		761	B	X 12363742-12360802	11B16
CG9281		611	E	X 15454374-15450765	13E14
CG8473		2556	A	X 15513659-15523896	13E18–F1
CG12703		618	D	X 19494615-19497465	18F1–F2
CG1819		1500	A	X 20757531-20763638	19F1
CG1718		1713	A	X 20909795-20902146	19F2
CG1801		1511	A	X 20924492-20917580	19F2
CG1494		1197	A	X 20896205-20901578	19F2
CG3164		620	G	2L 123902-117541	21B
CG4822		643	G	2L 112000-116000	21B
CG17646		627	G	2L 1720498-1727693	22B3
CG9892		615	G	2L 2649300-2658596	23A6
CG9664		609	G	2L 3211844-3209624	23E4–23E5
CG9663		812	G	2L 3214000-3220000	23E4–23E5
CG3327		729	G	2L 3257267-325948	23F
CG2969	Atet	832	G	2L 4251813-4262480	24F8
CG11147		705	H	2L 5656028-5653232	26A1
CG7806		1487	C	2L 8212839-8218079	29A3-A4
CG7627		1327	C	2L 8262316-8256791	29B1
CG5853		689	G	2L 9854119-9847658	30E1–30E3
CG5772	Sur	2250	C	2L 10105357-10089272	31A2
CG6214		1896	C	2L 12619174-12641593	33F2
CG7491		324	A	2L 13675599-13676775	34D1
CG17338		1275	B	2L 18829742-18834099	37B9
CG10441		1307	B	2L 18835157-18839979	37B9
CG9270		1014	C	2L 20741821-20738317	39A2
CG8799		1344	C	2R 4426560-4431236	45D1
CG3879	Mdr49	1279	B	2R 7940090-7934079	49E1
CG8523	Mdr50	1313	B	2R 9235904-9241222	50F1
CG8908		1382	A	2R 15203694-15208725	56F11
CG10505		1283	C	2R 16226805-16222698	57D2
CG17632	bw	755	G	2R 18476505-18465883	59E3
CG7955		606	B	3L1597621-1602155	62B1
CG10226		1320	B	3L 6180561-6175400	65A14
Mdr65		1302	B	3L 6186691-6181468	65A14
CG5651		611	E	3L 8895129-8892720	66E3–E4
CG7346		597	G	3L 11555624-11559309	68C10–C11
CG4314	st	666	G	3L 16398050-16400715	73A3
CG5944		1463	A	3L 17695681-17689489	74E3–E4
CG6052		1660	A	3L 17627439-17622025	74E3–E4
CG9330		708	E	3L 1971540-1947231	76B6
CG14709		1307	C	3R 7362645-7369141	86F1
CG4225		866	B	3R 11615803-11612420	89A11–A12
CG4562		1348	C	3R 15626899-15619809	92B9
CG4794		711	A	3R 15725586-15728807	92C1
CG5789		1239	C	3R 29281221-20277309	96A7
CG18633		702	G	3R 29625526-29622829	96B5
CG11069		602	G	3R 20635134-20637920	96B6
CG6162		535	H	3R 22087630-22088417	97B1
CG9990		808	H	3R 24409613-24429503	98F1
CG11898		1302	C	3R 24887241-24892598	99A
CG11897		1346	C	3R 24881629-24885998	99A
CG2316		730	D	4 154260-145146	101F

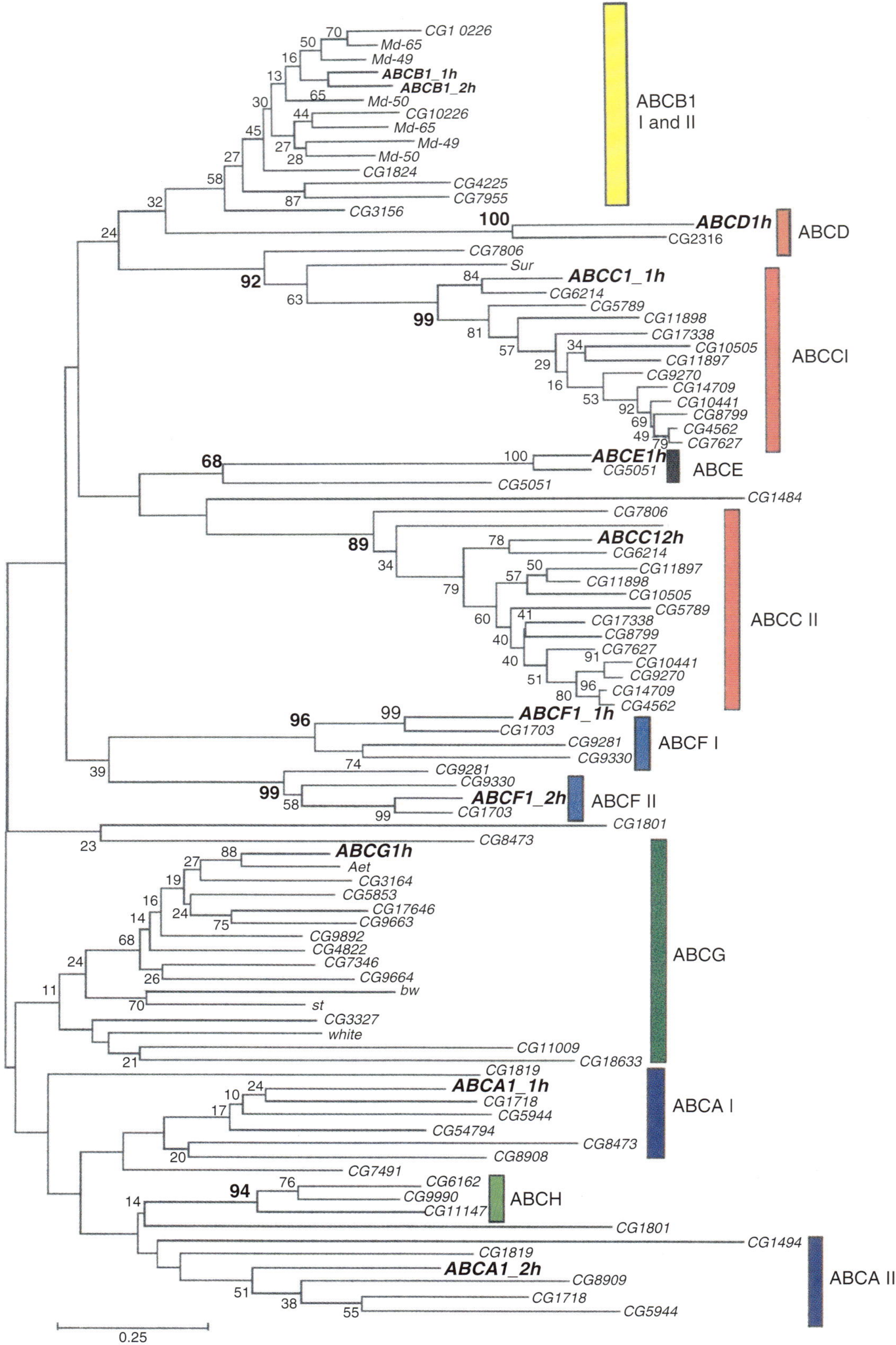

Figure 3.3. Phylogenetic tree of the **Drosophila** *ABC genes. Analysis (as described for Figure 3.2) was performed with all extracted* **Drosophila** *predicted protein sequences and a representative of each human subfamily. N- and C-terminal ATP-binding domains of full transporters are included as separate units.*

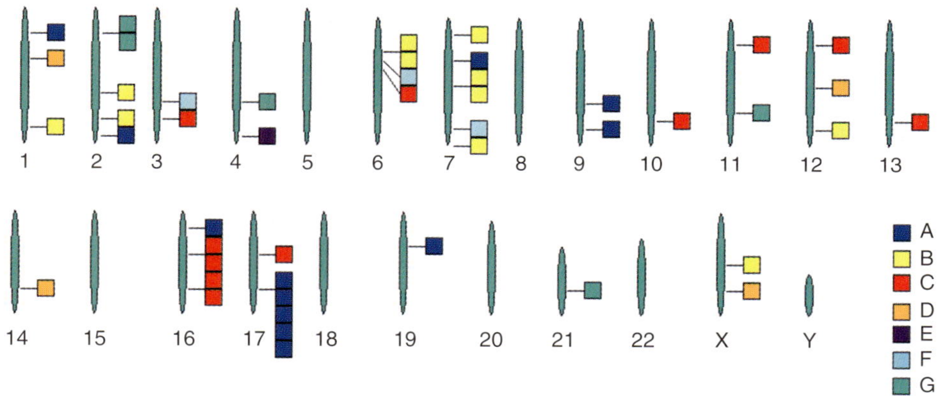

Figure 3.4. Map of human ABC genes. A schematic map is shown for each human chromosome, with the approximate location of all ABC genes. Clustered genes have a single line connecting to the chromosome. The key indicates the subfamily for each gene (A = ABCA, etc.).

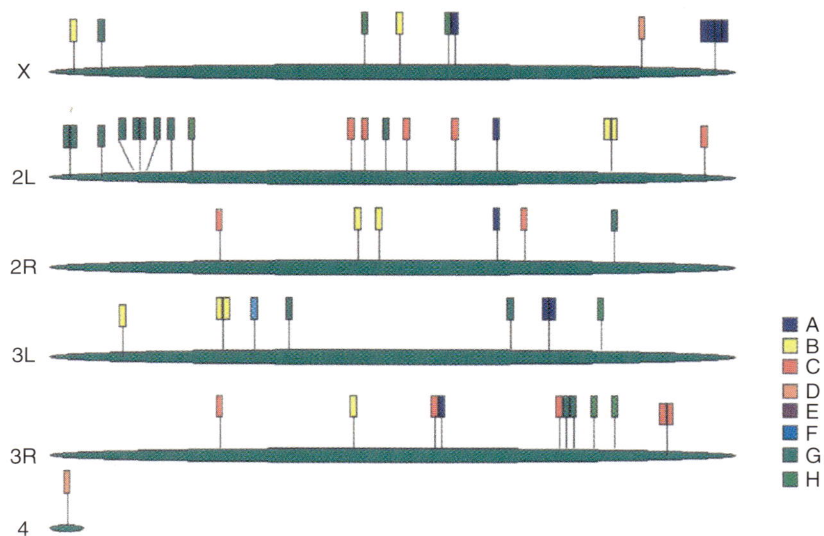

Figure 3.5. Map of **Drosophila** *ABC genes. A map is shown for each* **Drosophila** *chromosome with the approximate location of all identified ABC genes. Clustered genes have a single line connecting to the chromosome. The key indicates the subfamily for each gene (A = ABCA, etc.).*

independent and ancient gene duplication events.

Several *Drosophila* ABCB genes, *Mdr49*, *Mdr50* and *Mdr65*, have been well characterized. A fourth member of this group, *CG10226*, was identified as clustered with *Mdr65* **(Figure 3.5)**. All these genes are closely related to the human and mouse P-glycoproteins (ABCB1, ABCB4) and disruption of Mdr49 results in sensitivity to colchicines (Wu *et al.*, 1991).

Three genes, *CG9990*, *CG6162* and *CG11147*, were identified that do not fit into any of the known subfamilies and, in fact, are most closely related to ABC genes from bacteria (i.e. *Rhizobium* NodI and *E. coli* YhiH (subfamily NOD and DRI, respectively; see also Chapter 1). There are no close homologues to these genes in any other eukaryotic genome, including worms and plants. The three genes are within large sequence contigs and have introns, therefore excluding the possibility of contamination from bacterial sequences. In addition, this group forms a distinct cluster on the *Drosophila* tree. Apart from the eye pigment precursor transporters white, scarlet and brown, very few *Drosophila* genes are associated with known functions. Knockout technology will have to be employed to begin to elucidate the functions of these genes. In addition, very few *Drosophila* genes have clear orthologues in the human genome, suggesting consistent duplication and loss of ABC genes during the evolution of eukaryotic ABC genes.

Human ABC Gene Subfamilies

ABCA (ABC1)

This subfamily comprises 12 full transporters **(Table 3.1)**, which are further divided into two subgroups based on phylogenetic analysis and intron structure (Arnould *et al.*, 2001; Broccardo *et al.*, 1999). The first group includes seven genes dispersed on six different chromosomes (*ABCA1–A4, A7, A12, A13*), whereas the second group contains five genes (*ABCA5–A6, A8–A10*) arranged in a cluster on chromosome 17q24 (Arnould *et al.*, 2001). The ABCA subfamily contains some of the largest ABC genes, several of which encode over 2100 amino acids. Representative examples of the major human ABC transporters, including ABCA4, are described in detail in several chapters in this volume.

The *ABCA4* gene is expressed exclusively in photoreceptors, where it transports retinol (vitamin A) derivatives from the photoreceptor outer segment disks into the cytoplasm (Allikmets *et al.*, 1997). The chromophore of a visual pigment rhodopsin, retinal, or its conjugates with phospholipids are the likely substrates for ABCA4, as they stimulate the ATP hydrolysis of the intact protein (Sun *et al.*, 1999). Mice lacking *Abca4* show increased levels of all-*trans*-retinaldehyde (all-*trans*-RAL) following light exposure, elevated phosphatidylethanolamine (PE) in outer segments, accumulation of the protonated Schiff base complex of all-*trans*-RAL and PE (*N*-retinylidene-PE), and striking deposition of a major lipofuscin fluorophore (A2-E) in retinal pigment epithelium (RPE) (Weng *et al.*, 1999). These data suggest that ABCR is an outwardly directed flippase for *N*-retinylidene-PE.

Mutations in the *ABCA4* gene have been associated with multiple eye disorders (Allikmets, 2000). A complete loss of ABCA4 function leads to retinitis pigmentosa whereas patients with at least one missense allele have Startgardt disease (STGD). STGD is characterized by juvenile to early adult onset of macular dystrophy with loss of central vision. Carriers of the *ABCA4* mutation also occur at increased frequency in age-related macular degeneration (AMD) patients. AMD patients display a variety of phenotypic features, including the loss of central vision, after the age of 60. The causes of this complex trait are poorly understood, but a combination of genetic and environmental factors plays a role. The abnormal accumulation of retinoids, due to ABCA4 deficiency, has been postulated to be one mechanism by which this process could be initiated. Consistent with this hypothesis, mice heterozygous for *Abca4* mutations accumulate lipofuscin-containing particles in their RPE cells (Mata *et al.*, 2001).

Tangier disease is characterized by deficient efflux of lipids from peripheral cells, such as macrophages, and a very low level of high-density lipoproteins (HDL). The disease is caused by alterations in the *ABCA1* gene, implicating this protein in the pathway of removal of cholesterol and phospholipids onto HDL particles (Young and Fielding, 1999). Patients with hypolipidemia have also been described who are heterozygous for *ABCA1* mutations, suggesting that ABCA1 variations may play a role in regulating the level of HDLs in the blood (Marcil *et al.*, 1999). *ABCA1* gene expression is regulated by sterols (Langmann *et al.*, 1999) and current models for ABCA1 function place it at the plasma membrane mediating the transfer of phospholipid and cholesterol onto lipid-poor apolipoproteins to form nascent HDL particles. The ABCA1-mediated efflux of cholesterol is regulated by nuclear hormone receptors, such as oxysterol receptors (LXRs) and the bile acid receptor (FXR), which form heterodimers with retinoid X receptors (RXRs) (Repa *et al.*, 2000).

ABCB (MDR/TAP)

The ABCB subfamily is unique in that it contains both full transporters and half transporters. Four full transporters and seven half transporters have been currently described as members of this subfamily. *ABCB1* (*MDR/PGY1*) was the first human ABC transporter cloned and characterized through its ability to confer a multidrug resistance phenotype to cancer cells (Juliano and Ling, 1976). ABCB1 was demonstrated to transport several hydrophobic substrates including drugs such as colchicine, VP16, adriamycin and vinblastine as well as lipids, steroids, xenobiotics and peptides (reviewed in Ambudkar and Gottesman, 1998). The gene is thought to play an important role in removing toxic metabolites from cells, and is also expressed in cells at the blood–brain barrier, where it plays a role in transporting into the brain compounds such as ivermectin and cortisol that cannot be delivered by diffusion. ABCB1 also affects the pharmacology of drugs that are substrates, and a common polymorphism in the gene influences digoxin uptake (Hoffmeyer *et al.*, 2000).

Several ABC transporters are specifically expressed in the liver. These play a role in the secretion of components of the bile, and are responsible for several forms of progressive familial intrahepatic cholestasis (PFIC), through intracellular accumulation of bile salts. PFICs are a heterogeneous group of autosomal recessive liver disorders, characterized by early onset of cholestasis, which leads to liver cirrhosis and failure (Alonso *et al.*, 1994). The *ABCB4* (PGY3) gene transports phosphatidylcholine across the canalicular membrane of hepatocytes (van Helvoort *et al.*, 1996). Mutations in this gene cause PFIC3, which results in a defect in the transport of phosphatidylcholine across the canalicular membrane of the hepatocyte (Deleuze *et al.*, 1996; de Vree *et al.*, 1998). PFIC3 is also associated with intrahepatic cholestasis of pregnancy (Dixon *et al.*, 2000). The *ABCB11* gene was originally identified based on homology to *ABCB1* (Childs *et al.*, 1995). *ABCB11* is highly expressed on the liver canalicular membrane and has been demonstrated to be the major bile salt export pump. Mutations in *ABCB11* are found in patients with PFIC2, a disease associated with very low secretion of biliary bile salts (Strautnieks *et al.*, 1998).

The *ABCB2* and *ABCB3* (TAP) genes are half transporters that form heterodimers to transport into the ER peptides that are presented as antigens by the Class I HLA molecules. The closest homologue of the TAPs, the ABCB9 half transporter, has been localized to lysosomes. Several half transporters of the MDR/TAP subfamily have been localized to the inner membrane of the mitochondria. The yeast orthologue of ABCB7, Atm1, has been implicated in mitochondrial iron homeostasis, as a transporter in the biogenesis of cytosolic Fe/S proteins (Kispal *et al.*, 1997). Two distinct missense mutations in *ABCB7* are associated with the X-linked anemia and ataxia (muscle non-coordination) (XLSA/A) phenotype (Allikmets *et al.*, 1999). Three more half transporters from this subfamily, *ABCB6*, *ABCB8* and *ABCB10*, have also been localized to mitochondria **(Table 3.1)**.

ABCC (CFTR/MRP)

The ABCC subfamily contains 12 full transporters with diverse functional spectra including toxin secretion activities, ion transport, and regulation of a cell surface receptor. The *ABCC1* gene was identified in the small cell lung carcinoma cell line NCI-H69, a multidrug resistant cell that did not overexpress *ABCB1* (Cole *et al.*, 1992). The ABCC1 pump confers resistance to doxorubicin, daunorubicin, vincristine, colchicines and several other compounds, a very similar profile to that of ABCB1. However, unlike ABCB1, ABCC1 transports drugs that are conjugated to glutathione (Borst *et al.*, 2000). ABCC1 can also transport leukotrienes such as leukotriene C4 (LTC4). LTC4 is an important signaling molecule for the migration of dendritic cells. Migration of dendritic cells from the epidermis to lymphatic vessels is defective in *Abcc1* −/− mice (Robbiani *et al.*, 2000). ABCC2 and C3 also transport drugs conjugated to glutathione and other organic anions. The ABCC4, C5, C11 and C12 proteins are smaller than the other MRP1-like genes and lack a proximal domain near the N-terminus (Borst *et al.*, 2000), which is not essential for transport function (Bakos *et al.*, 2000). The ABCC4 and C5 proteins confer resistance to nucleosides including the drug 9-(2-phosphonylmethoxyethyl)adenine (PMEA) and purine analogues.

The rat *Abcc2* gene was found to have a frameshift mutation in the strain defective in canalicular multispecific organic anion transport, the TR^- rat (Paulusma *et al.*, 1996). The TR^- rat is an animal model of Dubin–Johnson syndrome and mutations in *ABCC2* have been identified in Dubin–Johnson syndrome patients (Wada *et al.*, 1998). The ABCC2 protein is expressed on the canalicular side of the hepatocyte and mediates organic anion transport, important for conjugation to and detoxification of many endogenous and xenobiotic lipophilic compounds in the liver. Patients with Dubin–Johnson syndrome display hyperbilirubinemia, deposition of melanin-like pigment in liver cells, and in some cases, hepatomegaly and abdominal pain.

The CFTR/ABCC7 protein is a chloride ion channel that plays a role in all exocrine secretions, and mutations in CFTR cause cystic fibrosis (Quinton, 1999). Cystic fibrosis is the most common fatal childhood disease in Caucasian populations, reaching frequencies ranging from 1/900 to 1/2500. The most common allele is a deletion of three base pairs (ΔF508). This allele is found in 85% of CF chromosomes in some populations, particularly northern Europeans.

At least two populations have a high frequency of other CF alleles. The W1282X allele constitutes 51% of the CF alleles in the Ashkenazi Jewish population and the 1677delTA allele has been found at a high frequency in Georgians and is also present at an elevated level in Turkish and Bulgarian populations. This has led several groups to hypothesize that these alleles arose through selection of an advantageous phenotype in the heterozygotes. It is through CFTR as a surface receptor that some bacterial pathogens such as cholera and *E. coli* cause increased fluid flow by release of toxins in the intestine, and resulting diarrhea. Therefore several researchers have proposed that the CF mutations have been selected for in response to these disease(s). This hypothesis is supported by studies showing that indeed CF homozygotes fail to secrete chloride ions in response to a variety of stimulants, and a study in mice in which heterozygous null animals showed reduced intestinal fluid secretion in response to cholera toxin (Gabriel *et al.*, 1993). CFTR is also the receptor for *Salmonella typhimurium* and is implicated in the innate immunity to *Pseudomonas aeruginosa* (Pier *et al.*, 1998).

Patients with two severe CFTR alleles such as ΔF508 typically display severe disease with inadequate secretion of pancreatic enzymes (Quinton, 1999), leading to nutritional deficiencies, bacterial infections of the lung, and obstruction of the vas deferens, leading to male infertility. Patients with at least one partially functional allele display enough residual pancreatic function to avoid the major nutritional and intestinal deficiencies (Dean *et al.*, 1990), and subjects with very mild alleles display only congenital absence of the vas deferens with none of the other symptoms of CF. Recently, heterozygotes of CF mutations have been found to have an increased frequency of pancreatitis (Cohn *et al.*, 1998) and bronchiectasis (Pignatti *et al.*, 1995). Thus, there is a spectrum of severity in the phenotypes caused by this gene that is inversely related to the level of CFTR activity. Clearly other modifying genes and the environment also affect disease severity, particularly the pulmonary phenotypes.

The *ABCC8* gene is a high-affinity receptor for the drug sulfonylurea. Sulfonylureas are a class of drugs widely used to increase insulin secretion in patients with non-insulin-dependent diabetes. These drugs bind to the ABCC8 protein and inhibit an associated potassium channel, under the control of ABCC8. Familial persistent hyperinsulinemic hypoglycemia of infancy (PHHI) is an autosomal recessive disorder in which subjects display unregulated insulin secretion. The disease was mapped to 11p15-p14 by linkage analysis, and mutations in the ABCC8 gene are found in PHHI families (Thomas *et al.*, 1995). The *ABCC8* gene has also been implicated in the insulin response in Mexican-American subjects (Goksel *et al.*, 1998) and in type 2 diabetes in French Canadians (Reis *et al.*, 2000) but not in a Scandinavian cohort (Altshuler *et al.*, 2000).

ABCD (ALD)

The ABCD subfamily contains four genes in the human genome and two each in the *Drosophila* and yeast genomes. The yeast PXA1 and PXA2 products dimerize to form a functional transporter involved in very long chain fatty acid oxidation in the peroxisome (Shani and Valle, 1998). All yeast and human ABCD genes encode half transporters that are located in the peroxisome, where they function as homo- and/or heterodimers in the regulation of very long chain fatty acid transport.

Adrenoleukodystrophy (ALD) is an X-linked recessive disorder characterized by neurodegenerative phenotypes with onset typically in late childhood and caused by mutations in the *ABCD1* gene (Mosser *et al.*, 1993). Adrenal deficiency commonly occurs and the presentation of ALD is highly variable. Adrenomyeloneuropathy (AMN), childhood ALD and adult-onset forms are recognized, but there is no apparent correlation to *ABCD1* alleles. ALD patients have an accumulation of unbranched saturated fatty acids with a chain length of 24 to 30 carbons, in the cholesterol esters of the brain and in adrenal cortex. The ALD protein, like its yeast homologue, is located in the peroxisome, where it is believed to be involved in the transport of very long chain fatty acids.

ABCE (OABP) and ABCF (GCN20)

The ABCE and ABCF subfamilies contain genes that have ATP-binding domains that are clearly derived from ABC transporters but they have no TMD and are not known to be involved in any membrane transport functions. The ABCE subfamily is composed solely of the oligoadenylate-binding protein, a molecule that recognizes oligoadenylate produced in response to infection by certain viruses. This gene is found in multicellular eukaryotes but not in yeast, suggesting it is part of innate immunity. Each ABCF gene contains a pair of NBDs. The best-characterized member, the *S. cerevisiae* GCN20 gene, mediates the activation of the eIF-2 alpha kinase (Marton *et al.*, 1997) and a human homologue, ABCF1, is associated with the ribosome and appears to play a similar role (Tyzack *et al.*, 2000).

ABCG (WHITE)

The human ABCG subfamily is composed of six half transporters that have an NBD at the N-terminus and a TMD at the C-terminus. The most intensively studied ABCG gene is the *white* locus of *Drosophila*. The white protein, together with brown and scarlet, transport precursors of eye pigments (guanine and tryptophan) in the eye cells of the fly (Chen *et al.*, 1996). The mammalian *ABCG1* gene is involved in cholesterol transport regulation (Klucken *et al.*, 2000).

Two half-transporter genes, *ABCG5* and *ABCG8,* were identified (Berge *et al.*, 2000; Lee *et al.*, 2001; Shulenin *et al.*, 2001), located head-to-head on the human chromosome 2p15-p16, and regulated by the same promoter. These genes are both mutated in families with sitosterolemia, a disorder characterized by defective transport of plant and fish sterols and cholesterol. Most likely, the two half transporters form a functional heterodimer. However, since *ABCG5* is more frequently mutated in Asian and *ABCG8* in Caucasian populations, they may also act as homodimers. The *ABCG1* gene is also regulated by cholesterol (Klucken *et al.*, 2000) and *ABCG3* is highly expressed in the liver, suggesting that these two genes may also be involved in cholesterol transport **(Table 3.1)**.

Analysis of cell lines resistant to mitoxantrone that do not overexpress ABCB1 or ABCC1 led several laboratories to identify the *ABCG2* (*ABCP, MXR1, BCRP*) gene as a drug transporter (Allikmets *et al.*, 1998; Doyle *et al.*, 1998; Miyake *et al.*, 1999). *ABCG2* confers resistance to anthracycline anticancer drugs and is amplified or involved in chromosomal translocations in cell lines selected with topotecan, mitoxantrone or doxorubicin treatment. It is suspected that ABCG2 functions as a homodimer, because transfection of the gene into cells is sufficient to confer resistance to chemotherapeutic drugs. ABCG2 can also transport several dyes such as Rhodamine and Hoechst 33462 and the gene is highly expressed in a subpopulation of hematopoietic stem cells (side population) that stain poorly for these dyes (Zhou *et al.*, 2001). However, the normal function of the gene in these cells is unknown. *ABCG2* is highly expressed in the trophoblast cells of the placenta. This suggests that the pump is responsible either for transporting compounds into the fetal blood supply, or for removing toxic metabolites. The gene is also expressed in the intestine and inhibitors could be useful in making substrates orally available.

Other ABCG genes include *ABCG3*, to date exclusively found in rodents (Mickley *et al.*, 2000), and the *ABCG4* gene, which is expressed predominantly in human brain. The functions of these two genes are unknown.

CONCLUSIONS AND PERSPECTIVES

The complete identification of all the ABC genes in several eukaryotic genomes allows a comprehensive picture of the evolution of these genes to be ascertained. Surprisingly, there are more ABC genes in both the *Drosophila* and *C. elegans* genomes than there are in the human genome **(Table 3.3)**. Although all of these species have the same seven ABC gene subfamilies,

TABLE 3.3. ABC GENE SUBFAMILIES IN CHARACTERIZED EUKARYOTES

Subfamily	Human	*Drosophila*	*C. elegans*	Yeast
A	12	10	7	0
B	11	10	23	4
C	12	12	8	7
D	4	2	5	2
E	1	1	1	0
F	3	3	3	5
G	5	15	11	10
H	0	3	0	0
Other	0	0	0	3[a]
Total	48	56	58	31

[a]Includes YDRO91c, YFL028c, YDR061w.

and a fairly similar number of genes in each subfamily, it seems that most of these genes arose by unique gene duplication and deletion events. Therefore, very few ABC genes have retained conserved functions throughout eukaryotic evolution. The exceptions appear to be the organelle specific genes such as the ABCD genes found in the peroxisome and the ABCB genes in the mitochondria, all of which represent half transporters.

Identifying the function for ABC genes is a challenging task. There is very little correlation between the gene sequence and the specificity of substrates identified to date. For instance, ABCB1, ABCC1 and ABCG2 all transport overlapping sets of hydrophobic drugs, yet derive from separate subfamilies and are highly divergent. The lack of evolutionary conservation and, therefore, direct orthologues of ABC genes in eukaryotic model organisms means that the worm and fly ABC genes cannot be used to determine the function of mammalian genes. Thus, laborious genetic and biochemical experiments will be required to reveal the function of most ABC genes. A recent breakthrough in the three-dimensional structure of a prokaryotic ABC gene (Chang and Roth, 2001; see Chapter 7) promises to provide a better understanding of these proteins. Hopefully, similar approaches will prove successful with eukaryotic ABC proteins.

REFERENCES

Allikmets, R. (2000) Simple and complex ABCR: genetic predisposition to retinal disease. Am. J. Hum. Genet. 67, 793–799.

Allikmets, R., Singh, N., Sun, H., Shroyer, N.F., Hutchinson, A., Chidambaram, A., *et al.* (1997) A photoreceptor cell-specific ATP-binding transporter gene (*ABCR*) is mutated in recessive Stargardt macular dystrophy. Nat. Genet. 15, 236–246.

Allikmets, R., Schriml, L.M., Hutchinson, A., Romano-Spica, V. and Dean, M. (1998) A human placenta-specific ATP-binding cassette gene (ABCP) on chromosome 4q22 that is involved in multidrug resistance. Cancer Res. 58, 5337–5339.

Allikmets, R., Raskind, W.H., Hutchinson, A., Schueck, N.D., Dean, M. and Koeller, D.M. (1999) Mutation of a putative mitochondrial iron transporter gene (ABC7) in X-linked sideroblastic anemia and ataxia (XLSA/A). Hum. Mol. Genet. 8, 743–749.

Alonso, E.M., Snover, D.C., Montag, A., Freese, D.K. and Whitington, P.F. (1994) Histologic pathology of the liver in progressive familial intrahepatic cholestasis. J. Pediatr. Gastroenterol. Nutr. 18, 128–133.

Altshuler, D., Hirschhorn, J.N., Klannemark, M., Lindgren, C.M., Vohl, M.C., Nemesh, J., *et al.* (2000) The common PPARgamma Pro12Ala polymorphism is associated with decreased risk of type 2 diabetes. Nat. Genet. 26, 76–80.

Ambudkar, S.V. and Gottesman, M.M. (1998) ABC transporters: biochemical, cellular, and molecular aspects. Methods Enzymol. 292, 1–853.

Arnould, I., Schriml, L., Prades, C., Lachtermacher-Triunfol, M., Schneider, T., Maintoux, C., *et al.* (2001) Identification and characterization of a cluster of five new ATP-binding cassette transporter genes on human chromosome 17q24: a new sub-group within the ABCA sub-family. GeneScreen 1, 157–164.

Bakos, E., Evers, R., Calenda, G., Tusnady, G.E., Szakacs, G., Varadi, A. and Sarkadi, B. (2000) Characterization of the amino-terminal regions in the human multidrug resistance protein (MRP1). J. Cell Sci. 113, 4451–4461.

Bateman, A., Birney, E., Durbin, R., Eddy, S.R., Finn, R.D. and Sonnhammer, E.L. (1999) Pfam 3.1: 1313 multiple alignments and profile HMMs match the majority of proteins. Nucleic Acids Res. 27, 260–262.

Berge, K.E., Tian, H., Graf, G.A., Yu, L., Grishin, N.V., Schultz, J., *et al.* (2000) Accumulation of dietary cholesterol in sitosterolemia caused by mutations in adjacent ABC transporters. Science 290, 1771–1775.

Borst, P., Evers, R., Kool, M. and Wijnholds, J. (2000) A family of drug transporters: the multidrug resistance-associated proteins. J. Natl Cancer Inst. 92, 1295–1302.

Broccardo, C., Luciani, M. and Chimini, G. (1999) The ABCA subclass of mammalian transporters. Biochim. Biophys. Acta 1461, 395–404.

Chang, G. and Roth, C.B. (2001) Structure of MsbA from *E. coli*: a homolog of the multidrug resistance ATP binding cassette (ABC) transporters. Science 293, 1793–1800.

Chen, H., Rossier, C., Lalioti, M.D., Lynn, A., Chakravarti, A., Perrin, G. and Antonarakis, S.E. (1996) Cloning of the cDNA for a human homologue of the *Drosophila* white gene and mapping to chromosome 21q22.3. Am. J. Hum. Genet. 59, 66–75.

Childs, S., Yeh, R.L., Georges, E. and Ling, V. (1995) Identification of a sister gene to P-glycoprotein. Cancer Res. 55, 2029–2034.

Cohn, J.A., Friedman, K.J., Noone, P.G., Knowles, M.R., Silverman, L.M. and Jowell, P.S. (1998) Relation between mutations of the cystic fibrosis gene and idiopathic pancreatitis. N. Engl. J. Med. 339, 653–658.

Cole, S.P.C., Bhardwaj, G., Gerlach, J.H., Mackie, J.E., Grant, C.E., Almquist, K.C., Stewart, A.J., *et al.* (1992) Overexpression of a transporter gene in a multidrug-resistant human lung cancer cell line. Science (Washington, DC) 258, 1650–1654.

Consortium, The *C. elegans* Sequencing (1998) Genome sequence of the nematode *C. elegans*: a platform for investigating biology. The *C. elegans* Sequencing Consortium. Science 282, 2012–2018.

de Vree, J.M., Jacquemin, E., Sturm, E., Cresteil, D., Bosma, P.J., Aten, J., Deleuze, J.F., *et al.* (1998) Mutations in the MDR3 gene cause progressive familial intrahepatic cholestasis. Proc. Natl Acad. Sci. USA 95, 282–287.

Dean, M., White, M.B., Amos, J., Gerrard, B., Stewart, C., Khaw, K.T. and Leppert, M. (1990) Multiple mutations in highly conserved residues are found in mildly affected cystic fibrosis patients. Cell 61, 863–870.

Dean, M., Rzhetsky, A. and Allikmets, R. (2001) The human ATP-binding cassette (ABC) transporter superfamily. Genome Res. 11, 1156–1166.

Deleuze, J.F., Jacquemin, E., Dubuisson, C., Cresteil, D., Dumont, M., Erlinger, S., *et al.* (1996) Defect of multidrug-resistance 3 gene expression in a subtype of progressive familial intrahepatic cholestasis. Hepatology 23, 904–908.

Dixon, P.H., Weerasekera, N., Linton, K.J., Donaldson, O., Chambers, J., Egginton, E., *et al.* (2000) Heterozygous MDR3 missense mutation associated with intrahepatic cholestasis of pregnancy: evidence for a defect in protein trafficking. Hum. Mol. Genet. 9, 1209–1217.

Doyle, L.A., Yang, W., Abruzzo, L.V., Krogmann, T., Gao, Y., Rishi, A.K. and Ross, D.D. (1998) A multidrug resistance transporter from human MCF-7 breast cancer cells. Proc. Natl Acad. Sci. USA 95, 15665–15670.

Gabriel, S.E., Clarke, L.L., Boucher, R.C. and Stutts, M.J. (1993) CFTR and outward rectifying chloride channels are distinct proteins with a regulatory relationship. Nature 363, 263–268.

Goksel, D.L., Fischbach, K., Duggirala, R., Mitchell, B.D., Aguilar-Bryan, L., Blangero, J., *et al.* (1998) Variant in sulfonylurea receptor-1 gene is associated with high insulin concentrations in non-diabetic Mexican Americans: SUR-1 gene variant and hyperinsulinemia. Hum. Genet. 103, 280–285.

Hoffmeyer, S., Burk, O., von Richter, O., Arnold, H.P., Brockmoller, J., Johne, A., *et al.* (2000) Functional polymorphisms of the human multidrug-resistance gene: multiple sequence variations and correlation of one allele with P-glycoprotein expression and activity in vivo. Proc. Natl Acad. Sci. USA 97, 3473–3478.

Hyde, S.C., Emsley, P., Hartshorn, M.J., Mimmack, M.M., Gileadi, U., Pearce, S.R., *et al.* (1990) Structural model of ATP-binding proteins associated with cystic fibrosis, multidrug resistance and bacterial transport. Nature 346, 362–365.

Juliano, R.L. and Ling, V.A. (1976) A surface glycoprotein modulating drug permeability in Chinese hamster ovary cell mutants. Biochim. Biophys. Acta 455, 152–162.

Kispal, G., Csere, P., Guiard, B. and Lill, R. (1997) The ABC transporter Atm1p is required for mitochondrial iron homeostasis. FEBS Lett. 418, 346–350.

Klucken, J., Buchler, C., Orso, E., Kaminski, W.E., Porsch-Ozcurumez, M., Liebisch, G., *et al.* (2000) ABCG1 (ABC8), the human homolog of the *Drosophila* white gene, is a regulator of macrophage cholesterol and phospholipid transport. Proc. Natl Acad. Sci. USA 97, 817–822.

Lander, E.S., Linton, L.M., Birren, B., Nusbaum, C., Zody, M.C., Baldwin, J., *et al.* (2001) Initial sequencing and analysis of the human genome. International Human Genome Sequencing Consortium. Nature 409, 860–921.

Langmann, T., Klucken, J., Reil, M., Liebisch, G., Luciani, M.F., Chimini, G., *et al.* (1999) Molecular cloning of the human ATP-binding cassette transporter 1 (hABC1): evidence for sterol-dependent regulation in macrophages. Biochem. Biophys. Res. Commun. 257, 29–33.

Lee, M.H., Lu, K., Hazard, S., Yu, H., Shulenin, S., Hidaka, H., *et al.* (2001) Identification of a gene, ABCG5, important in the regulation of dietary cholesterol absorption. Nat. Genet. 27, 79–83.

Marcil, M., Brooks-Wilson, A., Clee, S.M., Roomp, K., Zhang, L. H., Yu, L., *et al.* (1999) Mutations in the ABC1 gene in familial HDL deficiency with defective cholesterol efflux. Lancet 354, 1341–1346.

Marton, M.J., Vazquez de Aldana, C.R., Qiu, H., Chakraburtty, K. and Hinnebusch, A.G. (1997) Evidence that GCN1 and GCN20, translational regulators of GCN4, function on elongating ribosomes in activation of eIF2alpha kinase GCN2. Mol. Cell Biol. 17, 4474–4489.

Mata, N.L., Tzekov, R.T., Liu, X., Weng, J., Birch, D.G. and Travis, G.H. (2001) Delayed dark-adaptation and lipofuscin accumulation in abcr +/− mice: implications for involvement of ABCR in age-related macular degeneration. Invest. Ophthalmol. Vis. Sci. 42, 1685–1690.

Mickley, L., Jain, P., Miyake, K., Schriml, L.M., Rao, K., Fojo, T., *et al.* (2000) An ATP-binding cassette gene (ABCG3) closely related to the multidrug transporter ABCG2 (MXR/ABCP) has an unusual ATP-binding domain. Mammalian Genome 12, 86–88.

Miyake, K., Mickley, L., Litman, T., Zhan, Z., Robey, R., Cristensen, B., *et al.* (1999) Molecular cloning of cDNAs which are highly overexpressed in mitoxantrone-resistant cells: demonstration of homology to ABC transport genes. Cancer Res. 59, 8–13.

Mosser, J., Douar, A.M., Sarde, C.O., Kioschis, P., Feil, R., Moser, H., *et al.* (1993) Putative X-linked adrenoleukodystrophy gene shares unexpected homology with ABC transporters. Nature 361, 726–730.

Myers, E.W., Sutton, G.G., Delcher, A.L., Dew, I.M., Fasulo, D.P., Flanigan, M.J., *et al.* (2000) A whole-genome assembly of *Drosophila*. Science 287, 2196–2204.

Paulusma, C.C., Bosma, P.J., Zaman, G.J.R., Bakker, C.T.M., Otter, M., Scheffer, G.L., *et al.* (1996) Congenital jaundice in rats with a mutation in a multidrug resistance-associated protein gene. Science 271, 1126–1128.

Pier, G.B., Grout, M., Zaidi, T., Meluleni, G., Mueschenborn, S.S., Banting, G., *et al.* (1998) *Salmonella typhi* uses CFTR to enter intestinal epithelial cells. Nature 393, 79–82.

Pignatti, P.F., Bombieri, C., Marigo, C., Benetazzo, M. and Luisetti, M. (1995) Increased incidence of cystic fibrosis gene mutations in adults with disseminated bronchiectasis. Hum. Mol. Genet. 4, 635–639.

Quinton, P.M. (1999) Physiological basis of cystic fibrosis: a historical perspective. Physiol. Rev. 79, S3–S22.

Reis, A.F., Ye, W.Z., Dubois-Laforgue, D., Bellanne-Chantelot, C., Timsit, J. and Velho, G. (2000) Association of a variant in exon 31 of the sulfonylurea receptor 1 (SUR1) gene with type 2 diabetes mellitus in French Caucasians. Hum. Genet. 107, 138–144.

Repa, J.J., Liang, G., Ou, J., Bashmakov, Y., Lobaccaro, J.M., Shimomura, I., *et al.* (2000) Regulation of mouse sterol regulatory element-binding protein-1c gene (SREBP-1c) by oxysterol receptors, LXRalpha and LXRbeta. Genes Dev. 14, 2819–2830.

Robbiani, D.F., Finch, R.A., Jager, D., Muller, W.A., Sartorelli, A.C. and Randolph, G.J. (2000) The leukotriene C(4) transporter MRP1 regulates CCL19 (MIP-3beta, ELC)-dependent mobilization of dendritic cells to lymph nodes. Cell 103, 757–768.

Shani, N. and Valle, D. (1998) Peroxisomal ABC transporters. Methods Enzymol. 292, 753–776.

Shulenin, S., Schriml, L.M., Remaley, A.T., Fojo, S., Brewer, B., Allikmets, R. and Dean, M. (2001) An ATP-binding cassette gene (ABCG5) from the ABCG (White) gene subfamily maps to human chromosome 2p21 in the region of the Sitosterolemia locus. Cytogenet. Cell Genet. 92, 204–208.

Strautnieks, S., Bull, L.N., Knisely, A.S., Kocoshis, S.A., Dahl, N., Arnell, H., *et al.* (1998) A gene encoding a liver-specific ABC transporter is mutated in progressive familial intrahepatic cholestasis. Nat. Genet. 20, 233–238.

Sun, H., Molday, R.S. and Nathans, J. (1999) Retinal stimulates ATP hydrolysis by purified and reconstituted ABCR, the photoreceptor-specific ATP-binding cassette transporter responsible for Stargardt disease. J. Biol. Chem. 274, 8269–8281.

Thomas, P.M., Cote, G.J., Wohllk, N., Haddad, B., Mathew, P.M., Rabl, W., *et al.* (1995) Mutations in the sulfonylurea receptor gene in familial persistent hyperinsulinemic hypoglycemia of infancy. Science 268, 426–429.

Tyzack, J.K., Wang, X., Belsham, G.J. and Proud, C.G. (2000) ABC50 interacts with eukaryotic initiation factor 2 and associates with the ribosome in an ATP-dependent manner. J. Biol. Chem. 275, 34131-34139.

van Helvoort, A., Smith, A.J., Sprong, H., Fritzsche, I., Schinkel, A.H., Borst, P. and

van Meer, G. (1996) MDR1 P-glycoprotein is a lipid translocase of broad specificity, while MDR3 P-glycoprotein specifically translocates phosphatidylcholine. Cell 87, 507–517.

Venter, J.C., Adams, M.D., Myers, E.W., Li, P.W., Mural, R.J., Sutton, G.G., *et al.* (2001) The sequence of the human genome. Science 291, 1304–1351.

Wada, M., Toh, S., Taniguchi, K., Nakamura, T., Uchiumi, T., Kohno, K., *et al.* (1998) Mutations in the canilicular multispecific organic anion transporter (cMOAT) gene, a novel ABC transporter, in patients with hyperbilirubinemia II/Dubin–Johnson syndrome. Hum. Mol. Genet. 7, 203–207.

Weng, J., Mata, N.L., Azarian, S.M., Tzekov, R.T., Birch, D.G. and Travis, G.H. (1999) Insights into the function of Rim protein in photoreceptors and etiology of Stargardt's disease from the phenotype in abcr knockout mice. Cell 98, 13–23.

Wu, C.T., Budding, M., Griffin, M.S. and Croop, J.M. (1991) Isolation and characterization of *Drosophila* multidrug resistance gene homologs. Mol. Cell Biol. 11, 3940–3948.

Young, S.G. and Fielding, C.J. (1999) The ABCs of cholesterol efflux. Nat. Genet. 22, 316–318.

Zhou, S., Schuetz, J.D., Bunting, K.D., Colapietro, A.M., Sampath, J., Morris, J.J., *et al.* (2001) The ABC transporter Bcrp1/ABCG2 is expressed in a wide variety of stem cells and is a molecular determinant of the side-population phenotype. Nat. Med. 7, 1028–1034.

Part II

Fundamental Aspects of the Mechanism of Action of ABC Transporters

CHAPTER 4

Structure of ABC Transporters

Kenneth J. Linton, Mark F. Rosenberg, Ian D. Kerr and Christopher F. Higgins

Introduction

In order to understand the mechanism by which ABC transporters translocate solute across cellular membranes, structural data are essential. Such data have been hard won. This is primarily because of the difficulty inherent in overexpressing and purifying these proteins in an active form. As for many membrane proteins, ABC transporters are often toxic to the cell or misfold when overproduced, and their vectorial active transport function is disrupted as they are purified. Perhaps most importantly, the activity of many ABC transporters is influenced by their lipid membrane environment so ensuring that any purified protein is fully active, and therefore properly folded, is non-trivial. There is also increasing evidence that the transmembrane domains (TMDs) of ABC transporters are highly flexible, a characteristic not conducive to ready crystallization.

Structural data have gradually emerged from a variety of approaches. Many bacterial ABC transporters are multi-protein complexes with each of the four core domains encoded as a separate polypeptide (see introductory chapter to this volume). Several of the relatively hydrophilic nucleotide-binding domains (NBDs) have been overexpressed (including one of eukaryotic origin), purified and characterized at high resolution by X-ray crystallography **(Table 4.1)**. Although such data tell us much about interactions with ATP, they have had little impact on our understanding of the mechanism of transport. This is because binding of the translocated substrate is a property of the TMDs, and transport requires the interaction of all four domains. Structural data for a complete transporter came

Table 4.1. High-resolution structures of NBDs

NBD	Organism	Function	Resolution (Å)	Reference
RbsA	*E. coli*	Ribose uptake	2.5	Armstrong *et al.* (1998)
HisP	*S. typhimurium*	Histidine uptake	1.5	Hung *et al.* (1998)
MalK	*T. litoralis*	Maltose uptake	1.9	Diederichs *et al.* (2000)
MJ0796	*M. jannaschii*	Unknown	2.7	Yuan *et al.* (2001)
MJ1267	*M. jannaschii*	Amino acid transport	1.6	Karpowich *et al.* (2001)
TAP1	*H. sapiens*	Antigen presentation	2.4	Gaudet and Wiley (2001)
Rad50	*P. furiosus*	Double-strand DNA repair	2.5	Hopfner *et al.* (2000)
SMC	*T. maritima*	Double-strand DNA repair	3.1	Lowe *et al.* (2001)

ABC Proteins: From Bacteria to Man
ISBN 0-12-352551-9

initially from single-particle analyses and electron crystallography, and low- to medium-resolution structures are now available for three mammalian ABC transporters: the multidrug resistance P-glycoprotein (Pgp) to 25 Å by transmission electron microscopy (TEM) of single particles and to approximately 10 Å by electron cryomicroscopy (ECM) of two-dimensional (2-D) crystals (Rosenberg *et al.*, 1997, 2001); MRP1 to 22 Å resolution by TEM of single particles and 2-D crystals (Rosenberg *et al.*, 2001); and TAP to approximately 35 Å resolution (Velarde *et al.*, 2001). The structure of YvcC from *Bacillus subtilis* has also been resolved to 25 Å by ECM. The only X-ray crystallographic data for any complete ABC transporter came from a *tour de force* approach for the lipid A transporter (MsbA) from *Escherichia coli* (Chang and Roth, 2001; Chapter 7). In that study, over twenty *E. coli* transporters were expressed and then tested under 96 000 crystallization conditions to yield an example that crystallized and diffracted (to a resolution of 4.5 Å).

STRUCTURES OF NBDs

Although not the first NBD structure to be obtained (Stauffacher and colleagues solved the structure of the N-terminal NBD of *E. coli* RbsA earlier; Armstrong *et al.*, 1998), the HisP structure (Hung *et al.*, 1998) was the first published description of an NBD at atomic resolution (1.5 Å). HisP, the NBD of the *Salmonella* histidine uptake system, is a single polypeptide domain of which two copies associate with the TMDs (HisQ and HisM) in the intact transporter (Kerppolla *et al.*, 1991). The structure of HisP is shown in **Figure 4.1A**. The most convenient description is that it comprises two 'arms' oriented approximately perpendicular to one another. Arm-I contains an ABC-specific β-sheet subdomain (Karpowich *et al.*, 2001), with the Walker A motif in a typical phosphate-binding loop conformation together with the Walker B motif (Walker *et al.*, 1982). The perpendicular Arm-II, an α-helical subdomain, contains the 'ABC signature' motif. The two other conserved motifs of ABC transporter NBDs, namely the Q-loop (Diederichs *et al.*, 2000) and the H-loop (Linton and Higgins, 1998), are located at the interface of the two arms.

HisP was crystallized in the presence of ATP (Hung *et al.*, 1998). The environment of the ATP molecule is depicted in **Figure 4.1B**. A number of side-chains interact with nucleotide, stabilizing its binding. Among these are the expected interactions with the consecutive acidic amino acids at the C-terminus of the Walker B motif (D178 and E179 in HisP), and with the polar side-chains of the Walker A motif (K45, S46 and T47). The conserved polar amino acids of the Q-loop and the H-loop also make contact with ATP through water molecules bound within the ATP pocket. One additional interaction is noteworthy: the adenine ring

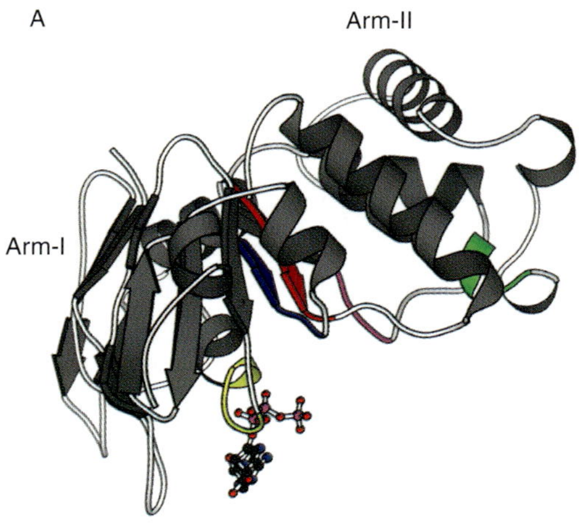

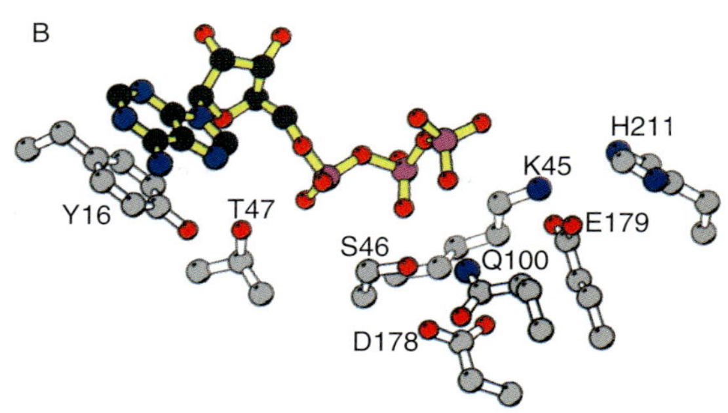

Figure 4.1. The structure of HisP and the environment of nucleotide within the ATP-binding pocket. A, the two perpendicular arms of HisP are displayed in cartoon format with ribbons denoting α-helices and arrows representing β-sheets. The bound ATP molecule is displayed in ball-and-stick format. The colours represent: yellow, Walker A motif; red, Walker B motif; blue, H-loop; magenta, Q-loop; green, ABC signature. B, The ATP molecule and side-chains of residues with which it interacts are displayed in ball-and-stick format. Amino acid positions are indicated. To ensure clarity, the backbone of the Walker A motif has been removed. These and other structural diagrams were produced using the program Molscript (Kraulis, 1991). Carbon atoms are in gray (darker in the ATP molecule), oxygen atoms are red, nitrogen blue and phosphorus pink. Reproduced with permission from Kerr (2002).

stacks against the side-chain of a well-conserved aromatic residue (Y16 in HisP). However, there are relatively few contacts on one side of the ATP molecule, and compared to other ATPases, the nucleotide appears somewhat exposed (Dreusicke *et al.*, 1988); **(Figure 4.1)**.

CONFORMATIONAL CHANGES IN THE NBDs

There are now structural data for NBDs in different conformational states (i.e. nucleotide-free, or complexed with ADP or ATP). The most pertinent comparisons are made between different conformations of the same NBD. Otherwise, to assess the conformational changes invoked by ATP binding, or by ATP hydrolysis followed by loss of Pi, NBDs from different ABC transporters must be compared. This limits any analysis to regions conserved between NBDs.

Conformational change associated with ATP binding

Rad50 provides the best model for comparison of nucleotide-free and ATP-bound conformations of NBDs. This bacterial protein is involved in the repair of double-strand breaks in DNA and thus is an unusual paradigm for the interaction of NBDs in ABC transporters. However, Rad50 contains sequences that unequivocally identify an ABC-transporter-like NBD, and the structure displays the characteristic L-shaped domain (Hopfner *et al.*, 2000). Analysis of the ATP-bound and ATP-free forms of Rad50 indicates that there is a pronounced ordering of the Walker A, Walker B and Q-loop motifs upon interaction with nucleotide. This suggests an induced fit of ATP. This conclusion must be tempered with some caution as it is possible that the ATP-dependent dimerization of Rad50 might be responsible for this effect. The implications of this dimer are discussed below. However, analysis of the ADP-bound and ADP-free forms of another NBD protein (MJ1267) supports the hypothesis of induced nucleotide fit (see below). A second change observed in the ATP-bound form as compared with the ATP-free form of RAD50 is rotation of the α-helical subdomain relative to the β-subdomain. Of course, it is difficult to assess whether rotation of the α-helical subdomain is a consequence of ATP binding or NBD:NBD dimerization. However, evidence from other NBDs suggests that within these domains there is a degree of conformational flexibility between the subdomain motifs.

Conformational change associated with ATP hydrolysis and release of Pi

The HisP structure (complexed with ATP) and the MJ1267 structure (complexed with Mg.ADP) have enabled comparison of NBDs in a pre-hydrolytic and post-hydrolytic state (Karpowich *et al.*, 2001). These changes are illustrated in **Figure 4.2**. The ABC-specific subdomain (Arm-II) undergoes a 12–15° inward rotation, bringing it closer to Arm-I in the ATP-bound structure compared with the ADP-bound structure. In another allosteric membrane protein (the nicotinic acetylcholine receptor) a relatively small rotation at the presumed ligand-binding sites results in a substantial rotational movement in the channel-lining region (Unwin, 1993, 1995). Thus, although this 12–15 Å rotation may appear slight, its effects on the TMDs could be highly significant. The rotation observed in MJ1267 appears to be centred on the Walker B motif, which is located close to the hinge between the two 'arms'. The most dramatic change in orientation of residues involves the conserved glutamine of the Q-loop (Karpowich *et al.*, 2001). In the ATP-bound conformation of HisP this amino acid is located within 5 Å of the β-phosphate of ATP and interacts with it through a bound water molecule. In stark contrast, the ADP-bound conformation sees this amino acid withdrawn from the nucleotide such that its closest approach is now >12 Å from the β-phosphate (Karpowich

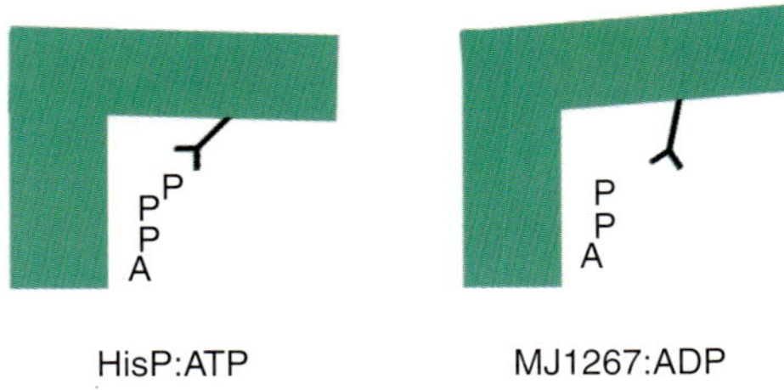

Figure 4.2. The conformational changes associated with ATP hydrolysis. Arm-II of the NBD undergoes a 12–15° outward rotation following hydrolysis of ATP and loss of phosphate. The Q-loop (shown in black) is moved 7 Å further from the bound nucleotide such that it can no longer interact with the β-phosphate. Reproduced with permission from Kerr (2002).

et al., 2001). Thus, the Q-loop is a strong candidate for transmitting the conformational change associated with ATP hydrolysis to other domains of an ABC transporter. In addition, the ABC signature motif (which is located in Arm-II) is displaced by about 6 Å as a result of rotation of the α-helical subdomain and thus may also be involved in transmission of conformational change (Karpowich *et al.*, 2001). Similar hinge-motions have been hypothesized in analyses of the structures of HisP (Hung *et al.*, 1998), TAP1 (Gaudet and Wiley, 2001) and another *Methanococcus* NBD (MJ0796; Yuan *et al.*, 2001). Whether these changes are the result of ATP hydrolysis or the release of phosphate from the post-hydrolytic intermediate remains unanswered. Pharmacological studies indicate that conformational changes in the drug-binding site (TMDs) of Pgp occur upon phosphate release, rather than upon ATP hydrolysis (Martin *et al.*, 2001; Rosenberg *et al.*, 2001). The determination of NBD structures complexed with ADP.vanadate may confirm at which step the changes in NBD conformation occur.

Conformational change associated with release of ADP

Conformational differences between the ADP-bound and nucleotide-free structures of the NBD of the branched-chain amino acid transporter of *Methanococcus jannaschii* (MJ1267; Karpowich *et al.*, 2001) may be seen as a model for the release of ADP from a post-hydrolytic NBD. Studies of other ATPases suggest that the release of dinucleotide may often be accompanied by substantial conformational change (Scheirlinckx *et al.*, 2001; Zhou and Adams, 1997). The two predominant changes are (i) a general destabilization of the NBD in the absence of nucleotide and (ii) alteration in the conformation of the H-loop. The destabilization is reflected in higher crystallographic B-factors in several regions that provide interactions with the nucleotide. The B-factor is a measure of the degree of flexibility in a region of a structure. Thus, the Walker A, Walker B, H-loop and the β-strand containing the adenine ring-interacting aromatic residue (Tyr-16 in HisP) all exhibit higher B-factors in the absence of nucleotide, suggesting that release of ADP from the post-hydrolytic NBD is accompanied by a relaxation of the domain. Put another way, it suggests an induced fit of nucleotide with NBD (Karpowich *et al.*, 2001). The second structural effect is the change in conformation of the backbone of the conserved H-loop, which displaces the side-chains of the H-loop by as much as 12 Å. This suggests that the H-loop may be involved in transmitting post-hydrolytic conformational changes in an intact transporter. No other changes of comparable magnitude are observed elsewhere (except in loops that are not conserved across the ABC transporter family).

As previously stated, caution must be applied when comparing structural data on NBDs from different ABC transporters as crystal-packing forces may contribute to the conformational changes described. However, it is particularly interesting that molecular dynamics simulations of NBDs in the presence or absence of nucleotide demonstrate both nucleotide-dependent rotation of the α-helical subdomain and withdrawal of the Q-loop (Campbell and Sansom, personal communication).

Interaction between NBDs

The structural data described above have been obtained for isolated, monomeric NBDs (in the case of RbsA, which contains two NBDs in a single polypeptide, only the first 259 amino acids corresponding to the N-terminal NBD were crystallized; Armstrong *et al.*, 1998). Clearly, our understanding of the function and dynamics of ABC transporters would be greatly enhanced by a description of the structural and conformational interactions between domains. Both NBDs in an ABC transporter are required for function (Azzaria *et al.*, 1989; Gill *et al.*, 1992). In the alternating catalytic cycle model only one ATP molecule is hydrolyzed at a time, with the ATPase activity alternating between the two NBDs (Hrycyna *et al.*, 1999; Senior and Gadsby, 1997). Although we should not overlook the possibility that the two NBDs influence each other indirectly through their cognate TMDs, the simplest explanation is that they interact directly with each other. In this respect, it is interesting that several of the monomeric NBDs have formed a crystallographic dimer (Kerr, 2002). These associations can be considered as hypothetical models for the interaction of NBDs in an intact ABC transporter.

Four alternative models have been presented for NBD:NBD association and are represented schematically in **Figure 4.3**. HisP forms a back-to-back crystallographic dimer in which

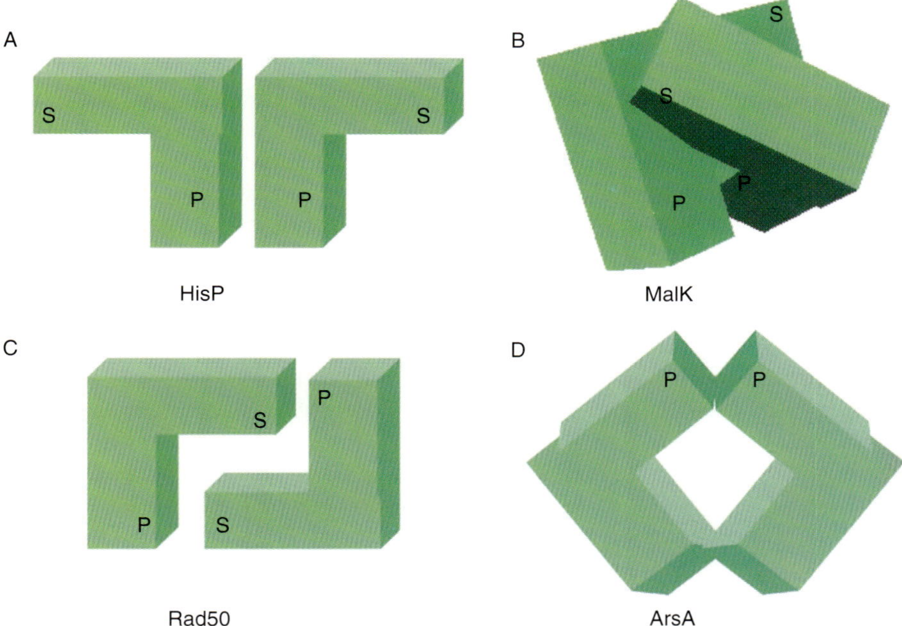

Figure 4.3. ***Models for NBD association. In each case 'P' refers to the location of the phosphate-binding loop (Walker A motif), while 'S' represents the signature motif (absent from ArsA). A, HisP; B, MalK; C, Rad50; D, ArsA. Reproduced with permission from Kerr (2002).***

the domains interact with each other through the exposed β-strands that constitute the ABC-specific β-sheet subdomain in Arm-I **(Figure 4.3A)**. The perpendicular Arm-II and the ABC signature motif are proposed to interact with the TMDs, HisQ and HisM (Hung *et al.*, 1998). The big drawback of this interface is the very small surface area buried by dimer formation (about 1000 Å^2). Detailed comparison of dimer interfaces in proteins suggests such a small buried surface area may be the result of crystal-packing forces, rather than a physiologically relevant dimer (Kerr, 2002). The NBDs of the thermophilic maltose transporter (MalK) interlock in the crystal structure as shown in **Figure 4.3B**, with close contacts between the hinge regions of the two NBDs. In particular, the Q-loops are in close contact across the dimer interface (4 Å), consistent with a role in transmission of inter-domain conformational change. In this dimer, the TMDs would be in close apposition to the ABC signature motifs (Diederichs *et al.*, 2000). For Rad50, the crystallographic dimer **(Figure 4.3C)** shows the two monomers interact in a head-to-tail fashion. Interestingly, nucleotide binds at the NBD:NBD interface in Rad50 and is coordinated by interactions with the Walker A and B motifs of one NBD, and the ABC signature motif of the other NBD. This orientation provides extra stability for the nucleotide and provides a possible explanation for domain:domain interaction upon ATP hydrolysis through the ABC signature motifs (Hopfner *et al.*, 2000). A further potential dimer-interface model is derived from the structure of ArsA **(Figure 4.3D)**, the ATP-hydrolytic domain of the bacterial arsenic transporter (Zhou *et al.*, 2000). Although ArsA is not a member of the ABC transporter family (as it does not possess the characteristic ABC signature motif), it may be informative since the ArsA dimer is formed from a single polypeptide with two ATP-binding sites. Thus, the association of the two domains cannot be an artifact of crystallization conditions. In this structure the two ATP-binding pockets are considerably closer together (about 15 Å) than in the other three models (Zhou *et al.*, 2000).

In attempting to assess the validity of the four dimer interface models, a number of considerations are necessary. First, what biochemical evidence supports the association state? Second, are there theoretical considerations that may have an impact? Third, does the interface between the NBD and the TMD, as suggested crystallographically for MsbA (Chang and Roth, 2001), rule out any of the proposed models? The biochemical data are diverse and consist of attempts to measure *direct* interactions (e.g. by crosslinking of cysteine residues), as well as *indirect* interactions (e.g. by examining the effects that mutations in NBDs have on the transport complex). Data for MalK suggest that mutation of residues proximal to the Q-loop to cysteine can result in dimerization of MalK, consistent with the close apposition of these loops in the MalK dimer model (Hunke *et al.*, 2000). Cysteine crosslinking data for Pgp suggest that the two Walker A motifs may be as close as 15–20 Å apart (Loo and Clarke, 2000; Urbatsch *et al.*, 2001). This appears to be consistent only with the ArsA structure, as these motifs are more than 25 Å apart in the Rad50, HisP and MalK dimer models. However, fluorescence resonance experiments on Pgp suggest that the distance between Walker A motifs might be 30–35 Å, which would be satisfied by models other than ArsA (Qu and Sharom, 2001).

A considerable amount of data has been obtained for bacterial ABC transporters that is rather more indirect, in that it pertains more to the interaction of NBDs with the TMDs. However, the nature of this interaction could clearly provide a considerable constraint on potential NBD:NBD interactions. For the histidine and maltose transporters, the interaction of the TMDs with the NBDs has been assessed by mutagenesis and co-purification studies (Liu *et al.*, 1999; Petronilli and Ames, 1991). Mutations on the face of HisP that is proposed to interact with the TMDs (i.e. the upper surface of Arm-II in **Figure 4.1A**) diminished co-purification with the TMDs, suggesting that the association of the domains is disrupted. Similarly, for the maltose transporter, mutations in the conserved region linking TM α-helices 4 and 5 of MalF and MalG (containing the so-called 'EAA' motif) that disrupted function could be rescued by 'suppressor' mutations in the α-helical ABC-specific domain (Arm-II; Mourez *et al.*, 1997). Some residues within the EAA motif could also be crosslinked to three residues in the same α-helix of the α-helical subdomain of MalK (Hunke *et al.*, 2000). Again this helix is on the upper face of Arm-II as viewed in **Figure 4.1A**. The main caveat is that the results can also be explained by allostery, i.e. that the mutations leading to a loosening of the $HisQMP_2$ or $MalGFK_2$ complexes are not necessarily at the domain:domain interfaces but are downstream effects of these mutations.

The interaction of nucleotide with NBD proteins has been cited as support for the Rad50 dimer model (Hopfner *et al.*, 2000; Karpowich *et al.*, 2001). Rad50 in the absence of nucleotide is a monomer in solution and in the crystal. However, ATP (or indeed the non-hydrolyzable analogue AMP-PNP) leads to dimerization of Rad50 in solution and in the crystal. Mutations in the ABC signature motif disrupt ATP-dependent dimerization in Rad50 (Hopfner *et al.*, 2000). However, several studies have indicated that ATP does not promote the dimerization of other NBDs (see Kerr, 2002). Furthermore, SMC, another DNA-interacting protein with an ABC-transporter-like NBD, does not show the interlocking-L arrangement that characterizes Rad50. Instead, SMC adopts a hexameric association in the crystal irrespective of the presence of nucleotide (Lowe *et al.*, 2001).

Recently, a structure has been obtained at 4.5 Å resolution for the entire prokaryotic lipid A transporter MsbA (Chang and Roth, 2001). This protein is a 'half transporter', containing a single NBD and a single TMD within the same polypeptide. Although a considerable proportion of the NBD is not resolved in the crystal structure, the structural data do illustrate the interface between the TMD and the NBD, and so constrain putative NBD dimer interfaces (see above). The most noticeable feature of the MsbA structure (**Figure 4.4**; a complete description of the structure is given in Chapter 7 of this volume) is the recognition of intracellular subdomains (ICDs) linking the TMD and NBD. In the orientation shown **(Figure 4.4)** the sequences in the NBD that interact with the ICDs are clearly visible (purple). Three conserved motifs of the NBDs which pack against the ICDs are the Walker B β-strand, the Q-loop and the first α-helix of the α-helical subdomain. All three can be mapped onto the *lower* surface of Arm-II in HisP **(Figure 4.1A)**. The implication of this admittedly incomplete structure is that the MalK dimer interface (in which the Q-loops are in very close apposition; Diederichs *et al.*, 2000) is not a feasible model for an NBD:NBD dimer interface in intact ABC transporters. How the crosslinking data for MalK (see above), which suggest that the *upper* face of the α-helical subdomain is in contact with the TMDs, can be reconciled is also unclear.

In conclusion, the current data seem to favour the orientation of the two ATP-hydrolytic domains of Rad50 (and possibly ArsA) as a model for NBD:NBD interactions within an intact ABC transporter. It is worth pointing out that the published MsbA structure consists of two molecules tilted together at the extracellular face and splayed apart at the intracellular side, separating their NBDs such that they do not share an interface (Chang and Roth, 2001). Thus, perhaps the NBDs do not interact or

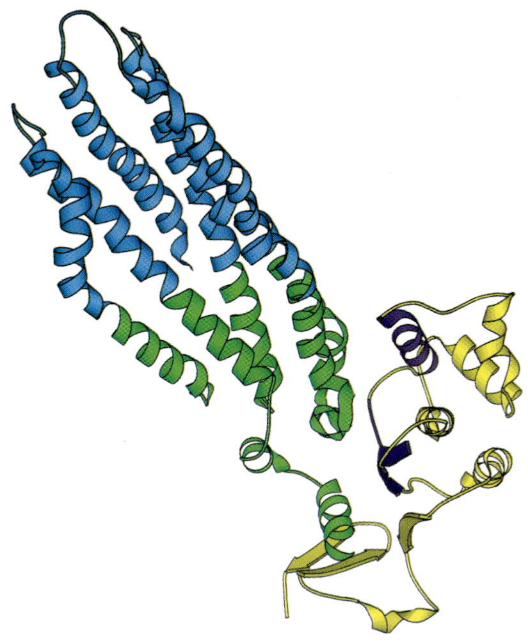

Figure 4.4. ***Domain:domain interactions in MsbA. The structure of a single MsbA molecule is displayed in a cartoon format. The α-helices of the TMDs are displayed in blue, while their intracellular extensions which comprise the ICD are colored green. The NBD is predominantly colored yellow, but the three regions which interact with the ICDs are colored purple. These comprise the Q-loop and Walker B motif and the first α-helix of Arm-II.***

interact only transiently during the transport cycle. Progress towards a higher-resolution structure of mammalian ABC transporters will be required to resolve this issue.

STRUCTURES OF INTACT ABC TRANSPORTERS

Pgp is the best-characterized mammalian ABC transporter and this section is focused on that protein. Data obtained for other mammalian ABC transporters are broadly consistent and, where appropriate, are compared and contrasted. The MsbA structure is described in detail elsewhere in this volume (Chapter 7): a comparison with Pgp illustrates unresolved questions.

Structure determination requires not only pure protein, but a reasonable degree of confidence that the purified protein has retained its native fold. This poses a significant problem when working with large molecular weight membrane proteins. Pgp activity is dependent on the lipid environment of the membrane (Callaghan *et al.*, 1997), while the first step in purification requires disruption of the lipid bilayer and solubilization of the protein components using detergent. Not surprisingly, this destroys measurable activity of Pgp. To demonstrate that the solubilized and purified protein has retained, or can regain, the native protein fold, the detergent must be replaced by lipid so that activity can, once again, be measured. The choice of detergent is therefore crucial to the success of the purification process. The detergent must solubilize the membrane protein without irreversible denaturation and it must be possible to replace the solubilizing detergent with lipids. The non-ionic detergent, dodecyl-β-D-maltoside has the requisite characteristics and has proved invaluable for solubilization of Pgp from multidrug resistant Chinese hamster ovary cells (Callaghan *et al.*, 1997).

A LOW-RESOLUTION STRUCTURE FOR PGP

A low-resolution (25 Å) structure of active Pgp (shown by drug-binding and drug-stimulated ATPase activity) was determined by TEM of single particles. In this technique, multiple images of single particles with a similar orientation were aligned and averaged to produce an image of higher signal:noise ratio (Rosenberg *et al.*, 1997). Single particle analysis was initially carried out on reconstituted Pgp because, compared with solubilized protein, the lipid bilayer confines all the particles in the *z* axis (i.e. in the plane of the membrane). Thus, the molecules only exhibit rotational freedom in the *x*, *y* dimensions, limiting the potential orientations that they can adopt. Furthermore, when reconstituted into a lipid environment, Pgp was shown to exhibit drug binding and ATPase activity similar to that of the protein in its native membranes. The reconstituted protein was examined under negative-stain, such that only the stain-accessible surface of the molecule was observed. More than 70% of the protein reconstituted into the lipid bilayer adopted a single orientation. TEM of this material **(Figure 4.5A)** projected an electron-dense ring of protein 12 nm in diameter with both twofold and sixfold symmetry surrounding a central chamber of approximately 5 nm diameter. The twofold symmetry in this structure was consistent with

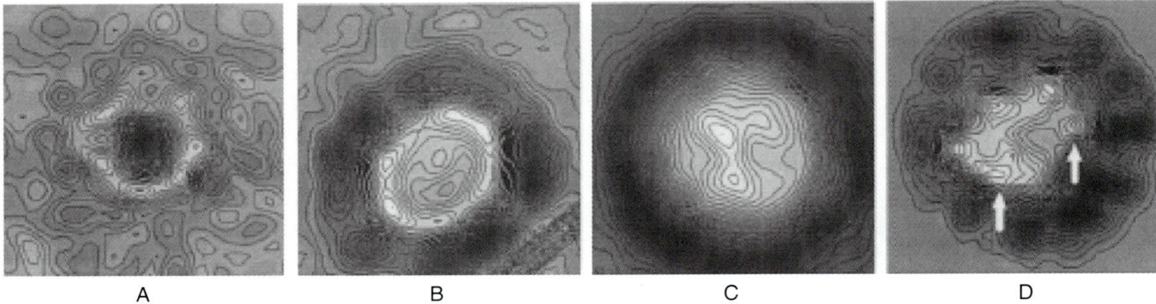

Figure 4.5. ***Projection maps of Pgp. Images derived from single particle alignment and averaging of negatively stained Pgp particles, using contours and shading to delineate stain*** (**black**) ***and protein*** (**white**) ***boundaries. A, averaged projection map of Pgp reconstituted into proteoliposomes. B, C and D, averaged projection maps of solubilized Pgp particles. Three classes of particle were observed, which differ in their orientations with respect to the electron beam. B, face-on projection of the extracellular face of Pgp. C, face-on projection of the cytoplasmic face of Pgp. D, side-on view of Pgp; the two NBDs are indicated. Reproduced with permission from Rosenberg*** **et al.** ***(1997).***

the presence of two homologous TMDs and the sixfold symmetry is consistent with six pairs of transmembrane α-helices, each pair linked by an extracellular loop.

To obtain images of different surfaces, the particles were examined as detergent solubilized material. This provided several different views of the particles **(Figure 4.5B, C and D)**. One projection **(Figure 4.5B)** closely resembled that of the reconstituted protein **(Figure 4.5A)**. Preferential labeling of this surface of solubilized Pgp particles by lectin-gold shows it to be glycosylated and therefore consistent with a view of the extracellular surface of the TMDs of Pgp. Because the chamber accumulated the uranyl acetate stain it is likely to be aqueous and thus open to the extracellular milieu. Futhermore, at least for this surface, the protein has a similar fold in the lipid bilayer and when solubilized in detergent. A second projection **(Figure 4.5C)** was also circular and 10–12 nm in diameter. However, this surface, with no central chamber and two 3 nm diameter lobes, was distinct from the extracellular view. These lobes are an appropriate size for the 200 amino acid NBDs. The third projection **(Figure 4.5D)** was asymmetric in shape, with three small lobes on one half of the particle and two larger lobes on the other half. This projection probably represents a side view, in which the two larger lobes correspond to the two NBDs and the three small lobes correspond to the TMDs (the hexagonal symmetry of the TMDs when viewed from above would be expected to project three electron-dense lobes when viewed from the side).

Thus, the shape of the Pgp particle approximates a short and fat cylinder, 10 nm in diameter and about 8 nm high. The lipid bilayer is about 4 nm in thickness, suggesting that about one-half of the molecule resides within the membrane. The TMDs form a chamber in the membrane, open at the extracellular face. The chamber is closed at the cytoplasmic face of the membrane and the two 3 nm lobes probably correspond to the NBDs.

MEDIUM-RESOLUTION STRUCTURE OF P-GLYCOPROTEIN

Single particle analysis and negative stain limit the resolution of the data. To obtain higher resolution data 2-D crystals of Pgp were obtained and imaged using low-dose electron cryomicroscopy (ECM). Precipitant-induced, large, well-ordered 2-D crystals of Pgp can be grown reproducibly at the air/water interface of a droplet. Projection images of frozen-hydrated, 2-D crystals displayed reflections to approximately 8 Å resolution (Rosenberg *et al.*, 2001). Importantly, because crystallized protein cannot be analyzed for function, the unit cell of the crystals was very similar to the size and shape of the single particles, making it likely that the native protein fold had been preserved. Tantalizingly, the resolution of the processed image remains around 10 Å, just outside that required for recognition of secondary structural features.

ECM of the holoenzyme, unlike negative stain, produced a 2-D projection of all electron densities in the 3-D protein **(Figure 4.6)**. The

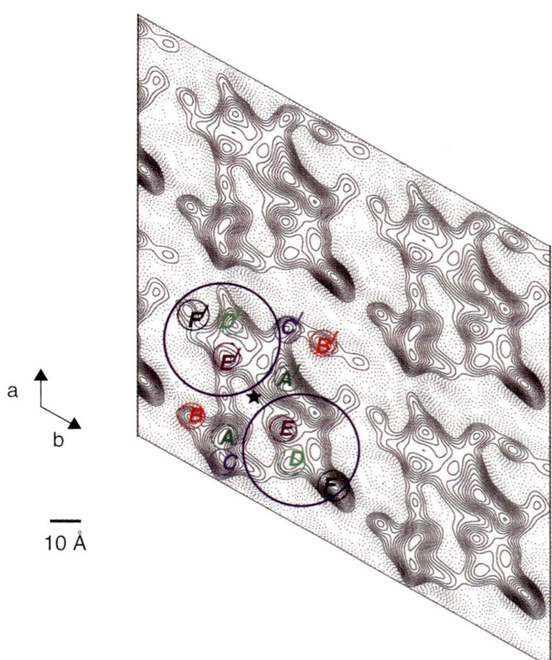

Figure 4.6. ***Projection map of Pgp determined by electron cryomicroscopy at 10 Å resolution. Solid lines indicate density above the mean. Twelve major densities (A–F with their pseudosymmetric densities A′–F′) are related by a pseudo-twofold symmetry axis centered at the star. A region of low protein density corresponds to an aqueous chamber within the membrane. The areas circled in blue at opposite ends of the molecule probably include densities corresponding to the NBDs; the three-dimensional reconstruction shows they are at the cytoplasmic face of the membrane. Reproduced with permission from Rosenberg* et al. *(2001).***

projection structure approximates to an elliptical ring of 91 Å × 60 Å with a slightly asymmetric, central low-density region entirely consistent with the surface views obtained by analysis of single particles in negative stain. As expected, the protein has distinct pseudo-twofold symmetry with several pairs of clearly related density peaks **(Figure 4.6)**.

The consistency with the low-resolution structure is more easily recognized in a 3-D map of the protein, generated from negatively stained 2-D crystals by imaging the crystal lattice from different angles. Although at lower resolution, this analysis provided information about the spatial organization of the four domains. Sections through the 3-D map at approximately 10 Å intervals through the plane of the membrane are shown in **Figure 4.7**. Working up from what is thought to represent the intracellular surface of the transporter, two large densities (see filled arrows in **Figure 4.7A**) are related by twofold pseudosymmetry. These domains presumably reflect the NBDs and are of an appropriate size to each accommodate HisP (see open arrow, **Figure 4.7A**). In the next section through the transporter **(Figure 4.7B)** the NBDs are still apparent (indicated by arrows) but there are now two extra electron densities either from a second lobe of each of the NBDs, or from intracellular loops of the TMDs (or, possibly, a combination of both). The four lobes now surround a distinct centre of low electron density indicating that the central chamber of the transporter extends deep into the plane of the membrane. At this resolution it is not possible to ascertain whether the specific residues which block the chamber come from the cytoplasmic loops of the TMDs or from the NBDs (although no evidence for such a role for the NBDs could be found by *in vivo* labeling studies; Blott *et al.*, 1999). Towards the midpoint of the membrane, two arcuate domains form almost a complete ring of protein around the central chamber. Each domain has three higher electron densities, which presumably correspond to pairwise clustering of the six putative transmembrane α-helices of each TMD (labeled 1–3 and 4–6 in **Figure 4.7D**). There are noticeable 'gaps' between the two TMDs within the plane of the membrane, potentially permitting side-access to the chamber from the lipid phase (arrows in **Figure 4.7D**). At the extracellular surface of the transporter **(Figure 4.7E)** the electron densities probably include contributions from both the TM α-helices and the extracellular loops and a pronounced gap in the protein ring is evident.

Comparison of Pgp with low-resolution structures of MRP1 and TAP

Structures for MRP1 and TAP have been resolved to 22 Å resolution (Rosenberg *et al.*, 2001) and around 35 Å (Velarde *et al.*, 2001), respectively. MRP1, TAP and Pgp are members of different subfamilies of ABC transporters. MRP1 has an extra TMD (TMD0) in addition to the four core domains, and TAP is a heterodimer of two 'half ABC transporters', TAP1 and TAP2. The primary sequence of the TMDs of TAP is particularly different from those of Pgp and MRP1, and probably reflects the specialized nature of TAP for antigen presentation to MHC class I molecules. Despite these differences, single particle images of MRP1 and TAP

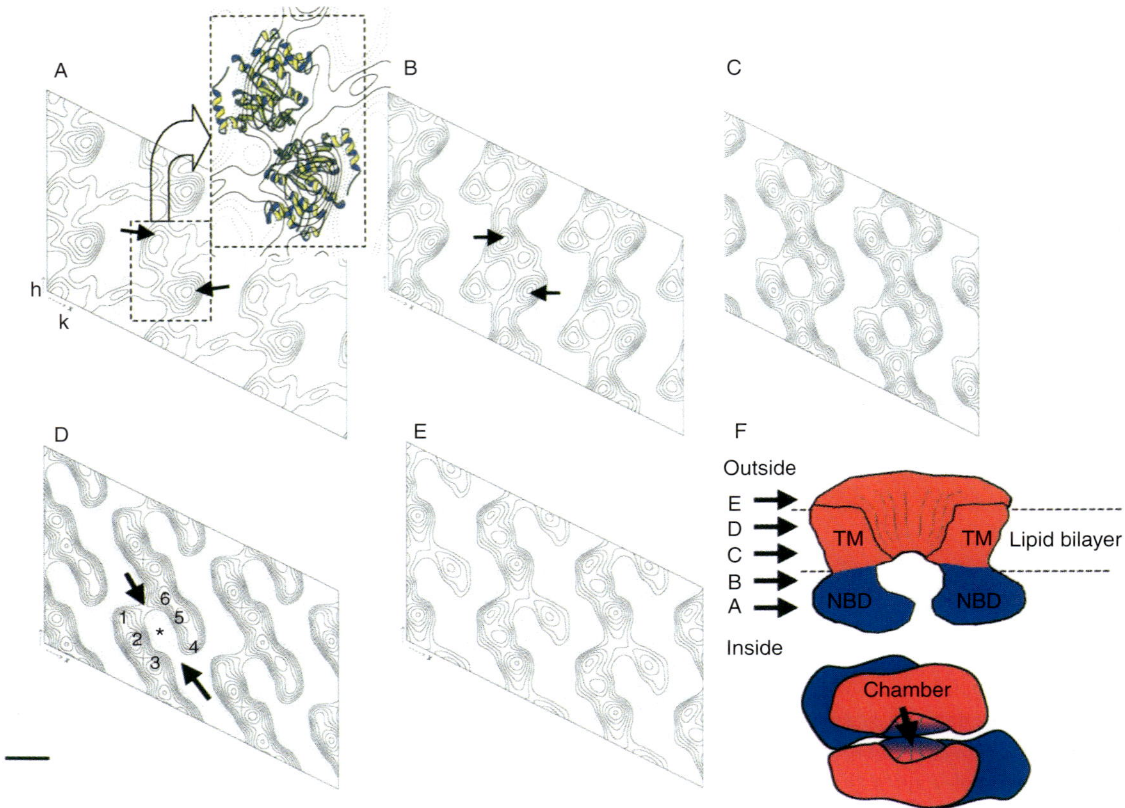

Figure 4.7. 3-D map of Pgp. Panels A to E represent slices in the plane of the crystal (x, y) (parallel to the plane of the membrane), with each slice having an approximate thickness of 10 Å and progressively moving from the intracellular (A) to the extracellular (E) side of the membrane. The scale bar = 28 nm. In the inset of panel A, two HisP monomers (Hung et al.*, 1998) to the same scale are modeled onto the structure as ribbon diagrams. Panel F shows two sketched views of Pgp with the TMDs in red and the NBDs in blue. The upper panel shows a cross-sectional sketch of the Pgp molecule indicating the approximate plane of sections. The lower panel shows a view of Pgp in the plane of the membrane as viewed from the extracellular face. Reproduced with permission from Rosenberg* et al. *(2001).*

are similar in size and shape to those for Pgp. Each structure consists of a ring of protein surrounding a large central hydrophilic chamber, which is open at one side of the membrane and closed at the other with two large protein lobes. In the MRP1 and TAP structures it is not possible to say whether the 'open' side is equivalent to the intra- or extracellular face of the membrane, although it is expected to be extracellular by virtue of comparison with Pgp and because the two large lobes are probably the intracellular NBDs. Interestingly, analysis of TAP2 by itself suggests that the TMD of TAP2 can form an arcuate structure independently of TAP1, consistent with the interpretation that each TMD contributes half of the chamber in the membrane (as also seen for the MsbA X-ray structure; see below). In single particles of MRP1, a particularly large protein density at the outer side of the ring may represent the additional TMD0. Although MRP1 crystallizes as a dimer, at current resolution it is not possible to demonstrate that the crystallization dimer is 'double barreled' with a separate chamber per molecule, although this would seem likely given the structure of single particles of MRP1.

Comparison of Pgp with the structures of MsbA and YvcC

MsbA is a 'half transporter' with one NBD and one TMD. It is expected to function as a homodimer to transport lipid A across the inner membrane of *E. coli*. The higher-resolution structure (4.5 Å) for MsbA represents an important advance (Chang and Roth, 2001; Chapter 7). For the first time it is definitively demonstrated that

TMDs of ABC transporters cross the membrane via α-helices (six in the case of MsbA). The interdigitation of intracellular loops of the TMD and structural elements of the NBD define this interdomain interface and suggest a mechanism for signal and energy transduction between domains. In MsbA the two TMDs form a block of α-helices angled towards each other at approximately 45° to the perceived plane of the membrane. Only towards the outer leaflet of the membrane do the two TMDs of the homodimer interact, forming a structure that resembles a pitched roof with the apex towards the extracellular surface of the membrane and the chamber below. Extending this analogy, the gable ends are missing from the structure, leaving large gaps between the TMDs towards the inner leaflet of the membrane. (Similar, but much less pronounced gaps are evident in the Pgp structure and may represent a route for substrate to access the center of the chamber directly from the lipid phase.) The MsbA chamber is therefore closed at the extracellular surface and open at the intracellular surface, in stark contrast to the mammalian examples. At the cytoplasmic face of the membrane, the NBDs of MsbA are distinct from each other and do not appear to interact.

YvcC is also a bacterial 'half transporter', homologous with MsbA, LmrA and both halves of Pgp. YvcC purifies as a monomer in detergent micelles and forms higher-order structures when the detergent is removed. These higher-order structures have been analyzed by ECM and resolved to 25 Å. The multimers form rather beautiful ring shapes, 40 nm in diameter, comprising 48 monomers of YvcC. These are arranged in two layers of 12 dimers. Of particular interest here are the three dimer interfaces within the particle. Two of these interfaces resemble the arrangement found in the MsbA dimer and form a chamber, closed at the extracellular surface. The third arrangement appears to share a large interface between the NBDs and would form a chamber open at the extracellular surface.

In comparison with the lower-resolution structures of mammalian ABC transporters the MsbA structure is both similar and different. From all structures, it is clear that a chamber is formed in the membrane from two 'half transporters' and that gaps between the two TMDs could provide access to the lipid phase within the plane of the membrane. The major difference between the structures is inherent in the angle at which the transmembrane α-helices cross the membrane and this dictates the opening of the chamber either to the cytoplasmic side of the membrane (as in MsbA) or to the extracellular side of the membrane (as in Pgp). How to reconcile this difference? YvcC does not help as different dimer interfaces within the particles are consistent with either structure. In the absence of data demonstrating function of the purified MsbA and YvcC, it is difficult to be confident that this material represents the native protein fold or that the physiological dimer interface has survived the purification protocol (YvcC solubilized from the membrane is monomeric). It is worth remembering that crystals grown from isolated NBDs have reported a number of different dimer interfaces, not all of which can be correct (see above). Furthermore, the arangement of the transmembrane α-helices in MsbA is inconsistent with *in vivo* crosslinking data for the TMDs of Pgp (DRS, KJL and CFH, unpublished data; Loo and Clarke, 1996b), although crosslinking data can be misleading. Alternatively, flexibility is a prerequisite for vectorial transport and Pgp is demonstrably very flexible (see below) and therefore the apparent differences could reflect different conformational states. Additional data are required before it will be possible to fully interpret structural and biochemical data obtained to date.

Comparison of the structure of ABC transporters with other transporters

The overall structure of ABC transporters appears different from that of the P-type ion-transporting ATPases. Principally, the large chamber in the membrane formed by the TMDs of Pgp contrasts with the relatively tight packing of the transmembrane α-helices of the ion-translocating ATPases (Toyoshima *et al.*, 2000). The structure of NhaA, a secondary transporter mediating H^+/Na^+ antiport, has been determined by similar methods (Williams *et al.*, 1999), as has the bacterial multidrug transporter EmrE (Tate *et al.*, 2001). Again, in contrast to Pgp the membrane-spanning segments of these proteins are packed into a relatively asymmetric tight bundle, which lacks a large central pore. The reason for these differences is unclear, but may be related to the nature of the substrates transported. Individual ABC transporters are often very specific for a given substrate (Pgp is unusual in this respect), but when different members of the superfamily are considered it seems clear that a similar domain architecture is capable of handling a wide diversity of substrates. Cations, amino acids, large polypeptides and phospholipids are all substrates of ABC transporters and a chamber in the membrane may provide a universal architecture which is more adaptable to different-sized substrates than closely packed α-helices. Interestingly, in the ER membrane the translocon for general protein export also has a central chamber of similar dimensions to that of Pgp (Hamman *et al.*, 1997).

Monomer or dimer?

The quaternary organization of Pgp and other ABC transporters has long been a subject of debate. There are biochemical data in support of both monomeric (Loo and Clarke, 1996a; Taylor *et al.*, 2001) and multimeric (Boscoboinik *et al.*, 1990; Naito and Tsuruo, 1992; Poruchynsky and Ling, 1994) Pgp. Structural studies strongly suggest the Pgp particle and unit cell is a monomer (Rosenberg *et al.*, 2001). Lectin-gold labeling also suggests that the unit cell of the crystal lattice and the single particles are composed of monomeric Pgp (Rosenberg *et al.*, 1997). Lectin recognizes the sugar residues added to the first extracellular loop of Pgp. When single particles of Pgp were labeled with lectin-gold and analyzed by TEM only one gold particle was observed per Pgp particle, consistent with only one copy of the first extracellular loop per Pgp particle (or unit cell). Thus, we can be reasonably certain that the structural unit of Pgp is a monomer with 12 membrane-spanning α-helices forming the pathway through the membrane. Is this the functional unit? Evidence that monomeric Pgp, observed by single particle analysis, binds drugs and has a drug-stimulated ATPase activity not substantially different from native Pgp (Callaghan *et al.*, 1997; Rosenberg *et al.*, 1997) strongly supports this view. However, perhaps the clearest evidence comes from genetic analysis of LmrA, a bacterial homologue of Pgp (van Veen *et al.*, 2000). LmrA is a half molecule of Pgp with one NBD and one TMD. Dominant-negative mutant and wild-type LmrA molecules were purified, mixed in different ratios, reconstituted into proteoliposomes, and tested for transport activity. The level of activity varied with the ratio of wild-type to mutant forms in a manner consistent only with a model in which two monomers of LmrA form a functional homodimer, equivalent in domain structure to a monomer of Pgp.

In contrast to Pgp, MRP1 crystallizes as a homodimer, although the functional relevance of this in biological membranes is unknown and single particles of the protein are monomeric (see above). Recent studies on CFTR suggest that dimerization mediated by a cytoplasmic accessory protein (CAP70) can lead to the formation of a double-barreled channel complex, which can alter the regulation of the channel (Wang *et al.*, 2000) but is not necessary for function (Ramjeesingh *et al.*, 2001). Thus, although the activity of Pgp and other ABC transporters may be regulated by dimerization (transient or otherwise), the available evidence suggests that the minimal functional unit is a monomer, and that dimers would have two aqueous chambers each formed by 10 to 12 TM α-helices.

CONFORMATIONAL CHANGES IN THE TMDS OF PGP DURING TRANSPORT

It is clear from biochemical studies of Pgp that the protein is very flexible within the membrane and that the transport cycle involves changes in conformation (Chapter 6). Such changes have now been seen directly (Rosenberg *et al.*, 2001). Pgp function can be blocked at different stages of the ATP catalytic cycle by the non-hydrolyzable ATP analogue AMP-PNP, or trapping ADP in the NBDs using vanadate (Senior and Gadsby, 1997). Projection maps of the extracellular surface of the protein in negative stain showed that the TMDs of Pgp adopt significantly different conformations depending on the occupancy of the NBDs **(Figure 4.8)**. Such conformational changes were predicted by biochemical data (Chapter 6) but their magnitude is unexpected. In the presence of AMP-PNP, which mimics the ATP-bound form of Pgp, the overall shape was roughly triangular with three distinct protein densities surrounding the central pore. In the presence of ADP and vanadate, Pgp was trapped in an intermediate catalytic conformation.

Thus, three distinct conformations of Pgp correspond to the nucleotide-free, nucleotide-bound, and post-hydrolytic (vanadate-trapped) steps of the catalytic cycle. These can be combined with biochemical data to generate a model for Pgp transport **(Figure 4.8)**. For clarity, the ATPase cycle for only one NBD is shown, although both NBDs are known to be required for function. ATP binding to Pgp generates the first major conformational change. ATP binding has also been shown to reduce significantly the affinity of Pgp for vinblastine (Martin *et al.*,

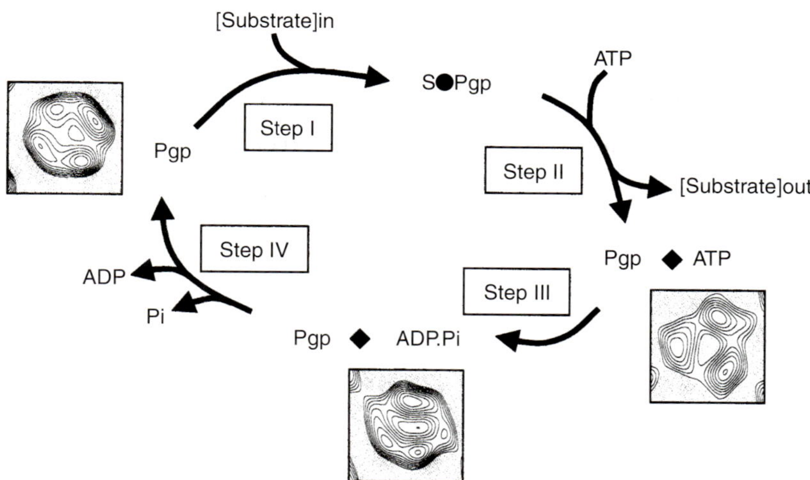

Figure 4.8. ***Conformational changes during the transport cycle of Pgp. The model is based on data presented here and published elsewhere (Senior and Gadsby, 1997). For clarity the ATPase cycle for one NBD is shown, although in Pgp both NBDs are required and operate in an alternating catalytic cycle. Pgp in its native state (the absence of bound nucleotide) binds drug substrate from the inner leaflet of the lipid bilayer at the intracellular face of the membrane (step I). Subsequently, ATP is bound by the NBD(s) (step II), inducing a conformational change which results in a reduction in the affinity of the drug-binding site and its reorientation such that it is exposed to the extracellular milieu (probably the aqueous phase in the chamber formed by the TMDs of Pgp). ATP hydrolysis via at least one intermediate (steps III and IV) returns the protein to its starting configuration. Reproduced with permission from Rosenberg et al. (2001).***

2001; Rosenberg *et al.*, 2001). Thus, vinblastine binding to its high-affinity site must precede ATP binding in the transport cycle (**Figure 4.8**, step I). Upon binding ATP, a major conformational change is induced in the TMDs, which reduces the affinity of the vinblastine-binding site, and hence permits release of drug (**Figure 4.8**, step II). Presumably, this conformational change also reorientates the drug-binding site so that it is exposed extracellularly. This most probably involves release of drug into the aqueous chamber; Pgp and its bacterial homologue LmrA are known to bind drug directly from the inner leaflet of the membrane (Bolhuis *et al.*, 1996; Shapiro and Ling, 1997) and for maximal efficiency they would be expected to release drug directly to the extracellular environment. Thus, any drug re-entering the cell would have to diffuse across the phosphate head groups of the outer leaflet and flip to the lipid chains of the inner leaflet, where it would again be accessible to Pgp. Consistent with this idea, it has been shown for the bacterial drug transporter LmrA that ATP binding results in reorientation of the drug-binding site such that it is exposed to the extracellular rather than intracellular face of the membrane (van Veen *et al.*, 2000). It is therefore likely that the major conformational change following ATP binding both reduces affinity and reorientates the drug-binding site. Thus, ATP binding, rather than hydrolysis, appears to provide the energy for drug translocation, although this is still a controversial issue. Following ATP hydrolysis, but prior to release of ADP or Pi (the vanadate-trapped state), there is a further conformational change (**Figure 4.8**, step III), although the drug-binding site remains low affinity. Only on release of ADP and/or Pi (**Figure 4.8**, step IV) does the protein return to the original configuration with high-affinity drug binding, and so is 'reset' to enter another transport cycle (Sauna and Ambudkar, 2001).

Conclusions

We can be confident of the two-arm, L-shaped structure of the NBDs of ABC transporters. The recent MsbA structure has provided important information as to how the NBD interacts with its cognate TMD through so-called ICDs consisting of intracellular loops of the TMD. However, to fully understand transport, it is necessary to understand domain–domain interactions: the availability of structural data for full-length ABC transporters is still limited. Nevertheless, data now appearing provide intriguing insights, guide the design of new experiments, and demonstrate that technical problems can be overcome. Transport is, by definition, a vectorial process, which is likely to require conformational changes, and thus single structures will never tell the whole story. This is evidenced by the major rearrangement of transmembrane domains seen in P-glycoprotein during the transport cycle. Only a combination of structures obtained under a variety of conditions, for example with and without substrate and/or ATP, together with biochemical data will give a complete insight into precisely how transmembrane transport is achieved.

Recently the structure of another bacterial ABC transporter, BtuCD, has been published (Locher *et al.*, 2002). The structure differs substantially from the MsbA structure in a number of respects. The NBDs (BtuD) form a dimer interface similar to that of Rad50, with the signature motif of one NBD in a position to co-ordinate with ATP bound to the other NBD. The TMDs (BtuC) interact to form a small chamber in the membrane open to the extracellular melieu, which the authors plausibly suggest could accommodate the substrate, vitamin B12. There are no real equivalents of the ISDs of MsbA. The TMDs, therefore, form a structure much more similar to that seen at lower resolution for Pgp and other mammalian ABC transporters, and very different from the unexpected arrangement reported for MsbA. Although the BtuCD structure was obtained by a similar approach to that of MsbA, and is subject to the same caveats, it appears more plausible. We must either conclude that the TMDs of different ABC transporters adopt very different architectures, which also impose different NBD:NBD interactions, or that one or other of the structures reflects at least in part crystallographic artefacts which do not reflect the physiological structure. Much more effort, both structural and biochemical, is required to resolve these issues.

ACKNOWLEDGMENTS

We are grateful to Cancer Research UK, Wellcome Trust and Medical Research Council UK for support for our research.

REFERENCES

Armstrong, S., Tabernero, L., Zhang, H., Hermodsen, M. and Stauffacher, C. (1998) Powering the ABC transporters: the 2.5 Å crystal structure of the ABC domain of RbsA. Pediatr. Pulmonol. 17, 91–92.

Azzaria, M., Schurr, E. and Gros, P. (1989) Discrete mutations introduced in the predicted nucleotide-binding sites of the mdr1 gene abolish its ability to confer multidrug resistance. Mol. Cell. Biol. 9, 5289–5297.

Blott, E.J., Higgins, C.F. and Linton, K.J. (1999) Organization of the nucleotide-binding domains of the multidrug resistance transporter P-glycoprotein. EMBO J. 18, 6800–6808.

Bolhuis, H., van Veen, H.W., Molenaar, D., Poolman, B., Driessen, A.J. and Konings, W.N. (1996) Multidrug resistance in *Lactococcus lactis*: evidence for ATP-dependent drug extrusion from the inner leaflet of the cytoplasmic membrane. EMBO J. 15, 4239–4245.

Boscoboinik, D., Debanne, M.T., Stafford, A.R., Jung, C.Y., Gupta, R.S. and Epand, R.M. (1990) Dimerization of the P-glycoprotein in membranes. Biochim. Biophys. Acta 1027, 225–228.

Callaghan, R., Berridge, G., Ferry, D.R. and Higgins, C.F. (1997) The functional purification of P-glycoprotein is dependent on maintenance of a lipid–protein interface. Biochim. Biophys. Acta 1328, 109–124.

Chang, G. and Roth, C.B. (2001) Structure of MsbA from *E. coli*: a homolog of the multidrug resistance ATP binding cassette (ABC) transporters. Science 293, 1793–1800.

Diederichs, K., Diez, J., Greller, G., Muller, C., Breed, J., Schnell, C., Vonrhein, C., Boos, W. and Welte, W. (2000) Crystal structure of MalK, the ATPase subunit of the trehalose/maltose ABC transporter of the archaeon *Thermococcus litoralis*. EMBO J. 19, 5951–5961.

Dreusicke, D., Karplus, P.A. and Schulz, G.E. (1988) Refined structure of porcine cytosolic adenylate kinase at 2.1 Å resolution. J. Mol. Biol. 199, 359–371.

Gaudet, R. and Wiley, D.C. (2001) Structure of the ABC ATPase domain of human TAP1, the transporter associated with antigen processing. EMBO J. 20, 4964–4972.

Gill, D.R., Hyde, S.C., Higgins, C.F., Valverde, M.A., Mintenig, G.M. and Sepulveda, F.V. (1992) Separation of drug transport and chloride channel functions of the human multidrug resistance P-glycoprotein. Cell 71, 23–32.

Hamman, B.D., Chen, J.-C., Johnson, E.E. and Johnson, A.E. (1997) The aqueous pore through the translocon has a diameter of

40–60 Å during cotranslational protein translocation at the ER membrane. Cell 89, 535–544.

Hopfner, K.P., Karcher, A., Shin, D.S., Craig, L., Arthur, L.M., Carney, J.P. and Tainer, J.A. (2000) Structural biochemistry and interaction architecture of the DNA double-strand break repair Mre11 nuclease and Rad50-ATPase. Cell 101, 789–800.

Hrycyna, C.A., Ramachandra, M., Germann, U.A., Cheng, P.W., Pastan, I. and Gottesman, M.M. (1999) Both ATP sites of human P-glycoprotein are essential but not symmetric. Biochemistry 38, 13887–13899.

Hung, L.W., Wang, I.X., Nikaido, K., Liu, P.Q., Ames, G.F.-L. and Kim, S.H. (1998) Crystal structure of the ATP-binding subunit of an ABC transporter. Nature 396, 703–707.

Hunke, S., Mourez, M., Jehanno, M., Dassa, E. and Schneider, E. (2000) ATP modulates subunit–subunit interactions in an ATP-binding cassette transporter (MalFGK2) determined by site-directed chemical cross-linking. J. Biol. Chem. 275, 15526–15534.

Karpowich, N., Martsinkevich, O., Millen, L., Yuan, Y.R., Dai, P.L., MacVey, K., Thomas, P.J. and Hunt, J.F. (2001) Crystal structures of the MJ1267 ATP binding cassette reveal an induced-fit effect at the ATPase active site of an ABC transporter. Structure 9, 571–586.

Kerppola, R.E., Shymala, V.K. Klebba, P. and Ames, G.F.-L. (1991) The membrane-bound proteins of periplasmic permeases form a complex. Identification of the histidine permease HisQMP complex. J. Biol. Chem. 266, 9857–9865.

Kerr, I.D. (2002) Structure and association of ATP-binding cassette transporter nucleotide-binding domains. Biochim. Biophys. Acta. In press.

Kraulis, P.J. (1991) MOLSCRIPT: a program to produce both detailed and schematic plots of protein structures. J. Appl. Crystallogr. 24, 946–950.

Linton, K.J. and Higgins, C.F.H. (1998) The *Escherichia coli* ATP-binding cassette (ABC) proteins. Mol. Microbiol. 28, 5–13.

Liu, P.Q., Liu, C.E. and Ames, G.F. (1999) Modulation of ATPase activity by physical disengagement of the ATP-binding domains of an ABC transporter, the histidine permease. J. Biol. Chem. 274, 18310–18318.

Locher, K.P., Lee, A.T. and Rees, D.C. (2002) The *E. coli* BtuCD structure: a framework for ABC transporter architecture and mechanism. Science 296, 1091–1098.

Loo, T.W. and Clarke, D.M. (1996a) The minimum functional unit of human P-glycoprotein appears to be a monomer. J. Biol. Chem. 271, 27488–27492.

Loo, T.W. and Clarke, D.M. (1996b) Inhibition of oxidative cross-linking between engineered cysteine residues at positions 332 in predicted transmembrane segments (TM) 6 and 975 in predicted TM12 of human P-glycoprotein by drug substrates. J. Biol. Chem. 271, 27482–27487.

Loo, T.W. and Clarke, D.M. (2000) Drug-stimulated ATPase activity of human P-glycoprotein is blocked by disulfide cross-linking between the nucleotide-binding sites. J. Biol. Chem. 275, 19435–19438.

Lowe, J., Cordell, S.C. and van den Ent, F. (2001) Crystal structure of the SMC head domain: an ABC ATPase with 900 residues antiparallel coiled-coil inserted. J. Mol. Biol. 306, 25–35.

Martin, C., Higgins, C.F. and Callaghan, R. (2001) Drug binding sites on P-glycoprotein are altered by ATP binding prior to nucleotide hydrolysis. Biochemistry 40, 15733–15742.

Mourez, M., Hofnung, M. and Dassa, E. (1997) Subunit interactions in ABC transporters: a conserved sequence in hydrophobic membrane proteins of periplasmic permeases defines an important site of interaction with the ATPase subunits. EMBO J. 16, 3066–3077.

Naito, M. and Tsuruo, T. (1992) Functionally active homodimer of P-glycoprotein in multidrug-resistant tumor cells. Biochem. Biophys. Res. Commun. 185, 284–290.

Petronilli, V. and Ames, G.F. (1991) Binding protein-independent histidine permease mutants. Uncoupling of ATP hydrolysis from transmembrane signaling. J. Biol. Chem. 266, 16293–16296.

Poruchynsky, M.S. and Ling, V. (1994) Detection of oligomeric and monomeric forms of P-glycoprotein in multidrug resistant cells. Biochemistry 33, 4163–4174.

Qu, Q. and Sharom, F.J. (2001) FRET analysis indicates that the two ATPase active sites of the P-glycoprotein multidrug transporter are closely associated. Biochemistry 40, 1413–1422.

Ramjeesingh M., Li C., Kogan I., Wang Y., Huan L.J. and Bear C.E. (2001) A monomer is the minimum functional unit required for channel and ATPase activity of the cystic fibrosis transmembrane conductance regulator. Biochemistry 40, 10700–10706.

Rosenberg, M.F., Callaghan, R., Ford, R.C. and Higgins, C.F. (1997) Structure of the multidrug resistance P-glycoprotein to 2.5 nm resolution determined by electron microscopy and image analysis. J. Biol. Chem. 272, 10685–10694.

Rosenberg, M.F., Velarde, G., Ford, R.C., Martin, C., Berridge, G., Kerr, I.D., *et al.* (2001) Repacking of the transmembrane domains of P-glycoprotein during the transport ATPase cycle. EMBO J. 20, 5615–5625.

Sauna, Z.E. and Ambudkar, S.V. (2001) Characterization of the catalytic cycle of ATP hydrolysis by human P-glycoprotein. The two ATP hydrolysis events in a single catalytic cycle are kinetically similar but affect different functional outcomes. J. Biol. Chem. 276, 11653–11661.

Scheirlinckx, F., Buchet, R., Ruysschaert, J.M. and Goormaghtigh, E. (2001) Monitoring of secondary and tertiary structure changes in the gastric H+/K+ -ATPase by infrared spectroscopy. Eur. J. Biochem. 268, 3644–3653.

Senior, A.E. and Gadsby, D.C. (1997) ATP hydolysis and mechanisms in P-glycoprotein and CFTR. Sem. Cancer Biol. 8, 143–150.

Shapiro, A.B. and Ling, V. (1997) Extraction of Hoechst 33342 from the cytoplasmic leaflet of the plasma membrane by P-glycoprotein. Eur. J. Biochem. 250, 122–129.

Tate, C.G., Kunji, E.R., Lebendiker, M. and Schuldiner, S. (2001) The projection structure of EmrE, a proton-linked multidrug transporter from *Escherichia coli*, at 7 Å resolution. EMBO J. 20, 77–81.

Taylor, J.C., Horvath, A.R., Higgins, C.F. and Begley, G.S. (2001) The multidrug resistance P-glycoprotein. Oligomeric state and intramolecular interactions. J. Biol. Chem. 276, 36075–36078.

Toyoshima, C., Nakasako, M., Nomoura, H. and Ogawa, H. (2000) Crystal structure of the calcium pump of sarcoplasmic reticulum at 2.6 Å resolution. Nature 405, 647–655.

Unwin, N. (1993) Nicotinic acetylcholine receptor at 9 Å resolution. J. Mol. Biol. 229, 1101–1124.

Unwin, N. (1995) Acetylcholine receptor channel imaged in the open state. Nature 373, 37–43.

Urbatsch, I.L., Gimi, K., Wilke-Mounts, S., Lerner-Marmarosh, N., Rousseau, M.E., Gros, P. and Senior, A.E. (2001) Cysteines 431 and 1074 are responsible for inhibitory disulfide cross-linking between the two nucleotide-binding sites in human P-glycoprotein. J. Biol. Chem. 276, 26980–26987.

van Veen, H.W., Callaghan, R., Soceneantu, L., Sardini, A., Konings, W.N. and Higgins, C.F. (1998) A bacterial antibiotic resistance gene that complements the human multidrug resistance P-glycoprotein gene. Nature 391, 291–295.

van Veen, H.W., Margolles, A., Muller, M., Higgins, C.F. and Konings, W.N. (2000) The homodimeric ATP-binding cassette transporters LmrA and P-glycoprotein mediate multidrug transport by an alternating two-site (two-cylinder engine) mechanism. EMBO J. 19, 2503–2514.

Velarde, G., Ford, R.C., Rosenberg, M.F. and Powis, S.J. (2001) Three-dimensional structure of transporter associated with antigen processing (TAP) obtained by single particle image analysis. J. Biol. Chem. 276, 46054–46063.

Walker, J.E., Saraste, M., Runswick, M.J. and Gay, N.J. (1982) Distantly related sequences in the alpha- and beta-subunits of ATP synthase, myosin, kinases and other ATP-requiring enzymes and a common nucleotide binding fold. EMBO J. 1, 945–951.

Wang, S., Yue, H., Derin, R.B., Guggino, W.B. and Li, M. (2000) Accessory protein facilitated CFTR–CFTR interaction, a molecular mechanism to potentiate the chloride channel activity. Cell 103, 169–179.

Williams, K.A., Kaufer, U.G., Padan, E., Schuldiner, S. and Kuhlbrandt, W. (1999) Projection structure of NhaA, a secondary transporter from *Escherichia coli*, at 4.0 Å resolution. EMBO J. 18, 3558–3563.

Yuan, Y.R., Blecker, S., Martsinkevich, O., Millen, L., Thomas, P.J. and Hunt, J.F. (2001) The crystal structure of the MJ0796 ATP-binding cassette. Implications for the structural consequences of ATP hydrolysis in the active site of an ABC transporter. J. Biol. Chem. 276, 32313–32321.

Zhou, J. and Adams, J.A. (1997) Participation of ADP dissociation in the rate-determining step in cAMP-dependent protein kinase. Biochemistry 36, 15733–15738.

Zhou, T., Radaev, S., Rosen, B.P. and Gatti, D.L. (2000) Structure of the ArsA ATPase: the catalytic subunit of a heavy metal resistance pump. EMBO J. 19, 4838–4845.

CHAPTER 5

SUBSTRATE-BINDING SITES IN ABC TRANSPORTERS

HENDRIK W. VAN VEEN AND RICHARD CALLAGHAN

GENERAL INTRODUCTION

As is evident from this volume, the ABC family of membrane transporters comprises a fascinatingly diverse range of proteins, which mediate a variety of different types of transport processes. A particularly distinguishing feature of this family is that the substrates recognized by ABC proteins appear to know no chemical, physical or functional boundaries. Perhaps this is not surprising given that gaining membership to this family is purely based on structural features rather than the nature of substrate translocated. With a few apparent exceptions (e.g. CFTR, SUR), all ABC proteins are active transporters that move substrates against their concentration gradients. The 'engine room' of all ABC proteins comprises the two nucleotide-binding domains (NBDs), whose catalytic activity drives the transport process, and the NBDs share a high level of sequence similarity and common overall mechanism. Does this mean that the substrate interactions and translocation processes in these proteins, primarily involving the transmembrane domains, also display common themes, irrespective of the array of physiological processes in which ABC transporters are involved?

Another issue addressed in this chapter concerns the number of substrate-binding sites present in ABC transporters. Many ABC transporters interact with a single substrate (or class of substrates) and this is particularly evident with the bacterial import pumps, which are often associated with dedicated substrate 'capture and delivery' proteins. These transporters provide a marked contrast to the so-called *multidrug* pumps, which interact with a myriad of compounds. ABC transporters may contain a single all-encompassing substrate-binding site, or multiple sites with highly selective substrate specificity. There is a paucity of information on where substrate-binding sites are located on ABC proteins. If ABC proteins have more than one site, do these sites interact and, if so, what is the nature of the communication between them?

The aim of this chapter will be to summarize our current insights into the physicochemical aspects of substrate interactions with the different types of ABC proteins mentioned above. The information available will be used to speculate on possible common molecular mechanisms of substrate translocation amongst ABC transporters.

ABC Proteins: From Bacteria to Man
ISBN 0-12-352551-9

PROPERTIES OF SUBSTRATES RECOGNIZED BY ABC TRANSPORTERS

ABC PROTEINS MEDIATE A VARIETY OF DIFFERENT TRANSPORT PROCESSES

The variety of substrates handled by different ABC transporters is enormous. As shown in **Figure 5.1**, the substrates transported include the majority of organic and inorganic chemical classes found in cells: amino acids, sugars, inorganic ions, lipids, polysaccharides, peptides and even proteins, in addition to compounds that are foreign to the organism itself. Hence, the ABC transporter family is very different from other transport protein families, which are characterized by, or even named after, the compound(s) translocated (van Winkle, 1999).

Whereas most eukaryotic ABC transporters appear to mediate substrate efflux only, prokaryotic members are divided into import and export proteins (Higgins, 1992). The bacterial importers usually interact with accessory proteins (e.g. periplasmic binding proteins) that bind and deliver substrate to the translocation machinery. The high-resolution structures for periplasmic binding proteins shown to interact with oligopeptides (OppA), histidine (HisJ) and ribose (RBP) demonstrate the presence of binding sites with high specificity for the transported substrate (Mowbray and Cole, 1992; Tame *et al.*, 1994; Yao *et al.*, 1994). On the other hand, most prokaryotic export proteins (e.g. those involved in antibiotic extrusion) mediate highly selective transport that is independent of substrate-binding proteins. Furthermore, several prokaryotic and eukaryotic ABC transporters display an apparent 'broad selectivity' for substrates and are known as multidrug pumps. Well-known examples are P-glycoprotein (Pgp) and MRP1, overexpression of which are major causes of resistance of human tumors to chemotherapy (Cole and Deeley, 1998; Gottesman *et al.*, 1995; Lum *et al.*, 1993), and LmrA, a bacterial homologue of Pgp which mediates transport of amphiphilic and toxic compounds, and of clinically relevant antibiotics (Margolles *et al.*, 1999; Putman *et al.*, 2000; van Veen *et al.*, 1996, 1998). MRP1, which has a broad specificity for conjugated drugs, is the only ABC protein demonstrated thus far to mediate symport, by virtue of its ability to co-transport drugs and glutathione (Cole and Deeley, 1998).

The heterogeneity of ABC transporters has hindered the elucidation of the molecular basis of substrate recognition by these proteins, and the subsequent steps involved in the translocation process. Many ABC transporters

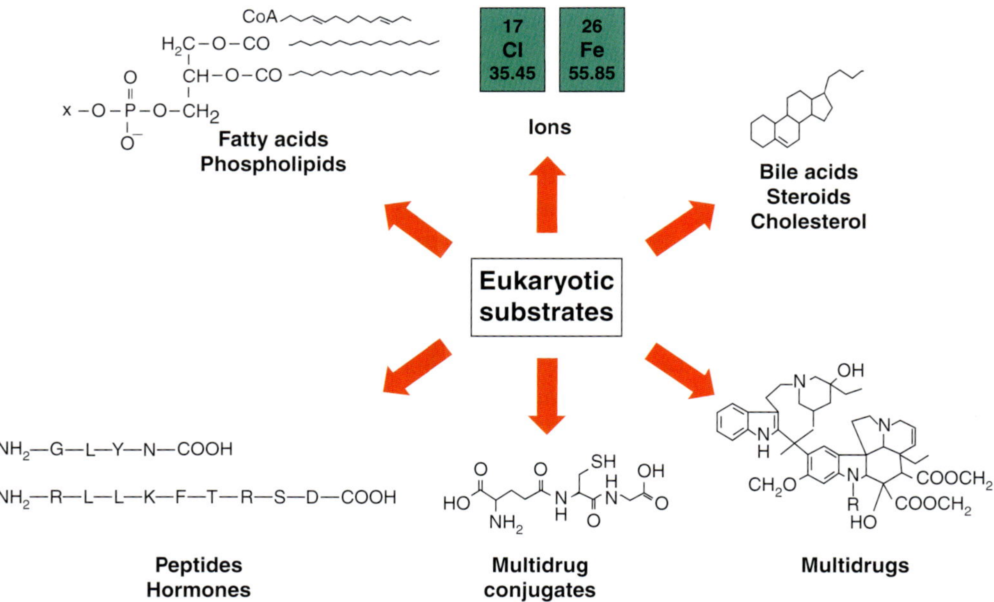

Figure 5.1. Illustration of the various general chemical species that are recognized and transported by eukaryotic ABC proteins.

are intimately involved in disease states and the elucidation of the molecular details of their substrate-binding site(s) would prove invaluable in designing drugs to target specific proteins in the clinical setting.

SPECIFIC ABC PROTEIN–SUBSTRATE INTERACTIONS

Specific characteristics required of transport substrates for recognition by ABC transporters have been delineated most thoroughly for the eukaryotic TAP transporters. These proteins mediate the transport of peptides across the endoplasmic reticulum to facilitate their loading on the MHC class I complex (Lankat-Buttgereit and Tampe, 1999; Uebel and Tampe, 1999). Peptides containing 9–16 amino acids are the preferred substrates, although lengths of up to 40 residues are possible, with reduced efficiency (Koopmann *et al.*, 1996). The involvement of TAP transporters in cellular antigen processing and the corresponding variability in peptide substrates suggests a low-specificity binding site. However, an elegant investigation using a combinatorial peptide library has revealed marked selectivity (Uebel *et al.*, 1997). The transporter exhibits preference for hydrophobic and positively charged amino acids on the C-terminus of peptide substrates, whilst aspartate or glycine residues are not tolerated (Uebel *et al.*, 1997). Positions 1 and 3 at the N-terminus of peptides also greatly affect binding to TAP, although a strict pharmacophoric preference is not obvious. The interaction of positions 1, 3 and 9 of a nonapeptide substrate with the TAP-binding site appears to involve contributions from the peptide backbone and the side-chains. In contrast, positions 4–8 provide almost no determinants for substrate–protein interaction. Together these physicochemical characteristics of peptide–TAP interaction suggest that the peptide 'docks' at two sites by virtue of its N- and C-terminal residues, whilst the central amino acids span a cavity within the transporter structure.

THE CONCEPT OF MULTIDRUG PUMPS

Unfortunately, no such directed investigations have been possible for the ABC proteins able to transport a variety of compounds that share no discernible structural similarities. Initially, these 'multidrug transporters' were thought to contravene a central dogma of substrate recognition by proteins; namely the 'lock–key hypothesis' postulated by Emil Fischer in 1894. This hypothesis, or its adaptation to an 'induced fit model' (Koshland, 1987), adequately describes the interactions of enzymes or transporters with hydrophilic substrates. Such systems are characterized by highly specific interactions between protein side-chains and the substrate. It is inconceivable that *all* the compounds recognized by multidrug pumps are able to invoke such specific interactions with a single protein. Most of the compounds recognized by multidrug pumps are hydrophobic or amphiphilic organic molecules. Their interaction with these pumps is perhaps governed by a different set of 'rules' than those observed for hydrophilic agents. Consequently, it was suggested that the recognition site in multidrug pumps might be a simple non-specific hydrophobic core or pocket (Gottesman and Pastan, 1993). This notion may now be discounted owing to the quite large observed differences in the relative affinities of compounds for interaction with multidrug ABC transporters such as Pgp (Martin *et al.*, 1999; Sharom *et al.*, 1999), LmrA (van Veen *et al.*, 1998), and PDR5 (Rogers *et al.*, 2001). Therefore, the multidrug pumps should not be considered as non-specific transporters, but rather as transporters displaying polyspecific recognition of substrates, not necessarily different from TAP (see above) or the OppA oligopeptide-binding protein, where three-dimensional (3-D) structure has been solved (Tame *et al.*, 1994). Unfortunately, multidrug pumps, by virtue of this characteristic, often provide a general pathway for mediating drug resistance, which is not restricted to a single drug, in a variety of clinical settings (Lum *et al.*, 1993; Nikaido, 1994).

SPECIFIC CHARACTERISTICS OF SUBSTRATES TRANSPORTED BY MULTIDRUG TRANSPORTERS

The promiscuity with which Pgp recognizes substrates has sparked many investigations into the identification of potent blockers of the transport process in tumor cells. The reader is directed to two reviews that provide a guide to the many different clinically relevant compounds identified with Pgp inhibitory actions (Lum *et al.*, 1993; Sikic, 1997). However, these studies have not provided significant information on the overall chemical features that characterize Pgp substrates. To provide such information, several groups embarked on manipulation of the physicochemical properties of known Pgp substrates in an effort to elucidate key molecular constituents for recognition by the protein (Chiba *et al.*, 1996; Ford *et al.*, 1990; Horton *et al.*, 1993; Lawrence *et al.*, 2001; Pearce *et al.*, 1989; Tang-Wai *et al.*, 1993; Toffoli *et al.*, 1995). **Figure 5.2** shows the general structures of the compounds used. Unfortunately, these investigations failed to elucidate precise and conserved pharmacophoric elements for Pgp substrates. However, they did highlight some key physical requirements. Hydrophobicity is a key element and planar aromatic groups contribute significantly to this property. A basic nitrogen atom is frequently observed and a tertiary amino moiety is associated with the ability of compounds to display high-affinity interaction with the protein. Compounds for which the nitrogen is located within non-aromatic rings display the greatest potency to interact with, and bind to, Pgp.

Hydrogen bonds play major roles in the interactions of many biological molecules and may impart a high specificity by virtue of their requirement for directionality (Fersht, 1998). This prompted a screen of compounds that had previously been examined for their interaction with Pgp, to determine the percentage of constituent groups capable of mediating hydrogen bonds (Seelig, 1998). This exhaustive screen established that compounds known to interact with Pgp display noticeable clustering of electron donor groups that are vital to create hydrogen bonds. Moreover, these clusters display characteristic relative spatial arrangements of their electron donor groups. On this basis, the author suggested that two distinct spatial patterns of electron donor groups are required for a compound to interact with Pgp. Electron donor groups with spatial separations of 2.5 ± 0.3 Å are classified as type I units. Type II units on the other hand have a separation of 4.6 ± 0.6 Å between two donor groups, or the outer two groups in a series of three. Interestingly, mutation of residues 939 and 941 in Pgp to non-hydrogen-bonding side-chain groups impairs interaction of the protein with substrates (Kajiji *et al.*, 1993). These residues are located in transmembrane segment (TM) 11, which has been

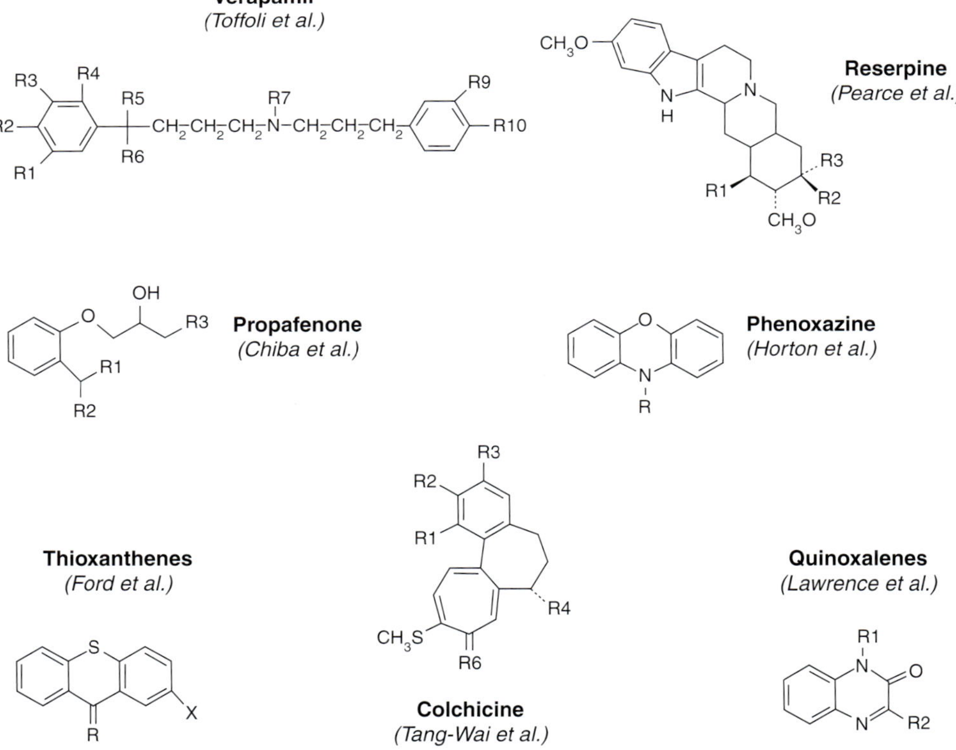

Figure 5.2. Chemical structures of template compounds used to ascertain structure–affinity relationships for pharmacologic agents known to inhibit Pgp. Positions indicated by Rn denote areas of chemical substitutions.

strongly implicated in drug binding (Loo and Clarke, 1999a). More recently, the screen of hydrogen-bonding groups was used to identify or correlate to the type of drug interaction with Pgp (Seelig and Landwojtowicz, 2000). There was a positive correlation between the propensities of a drug to form hydrogen bonds and *inhibition* of Pgp function. Compounds such as cyclosporin A, which may form an extensive network of hydrogen bonds, are potent and long-lasting inhibitors owing to the resultant low dissociation rate from the protein (Seelig and Landwojtowicz, 2000). It is thought that such compounds will be poorly transported by Pgp. MRP1 substrates share chemical properties with Pgp substrates, and often contain at least one electrically neutral type I unit together with one negatively charged type I unit or two electrically neutral type I units (Seelig *et al.*, 2000). Compounds with cationic type I units, which are good substrates for Pgp, are not transported by MRP1.

In summary, the large number of exhaustive studies employing chemical modifications of substrates or correlations with biophysical properties have given guidelines for physicochemical properties required for interaction of molecules with Pgp. Unfortunately, however, they have not produced a clear idea of what constitutes a substrate of a multidrug pump such as Pgp.

CAN A DISTANT STRANGER PROVIDE THE CLUE TO UNRAVELING REQUISITE FEATURES OF SUBSTRATE INTERACTION?

The concept of multidrug pumps is by no means restricted to ABC proteins. Secondary transporter families such as the major facilitator superfamily (MFS), the small multidrug resistance family (SMR), the resistance-nodulation-cell division (RND) family, and the multidrug and toxic compounds extrusion (MATE) family all contain multidrug pumps (for reviews see Higgins, 1992; Marger and Saier, 1993; Paulsen *et al.*, 1996; Putman *et al.*, 2000; Saier *et al.*, 1994). There are over a hundred known multidrug pumps and their distribution is widespread, with examples in mammalian cells, lower eukaryotes, eubacteria and archaea. Interestingly, there is a small selection of compounds that appear to be 'universal' substrates for unrelated multidrug pumps, such as human Pgp, *Lactococcus lactis* LmrA, *Escherichia coli* MdfA and *Bacillus subtilis* Bmr **(Figure 5.3)**.These

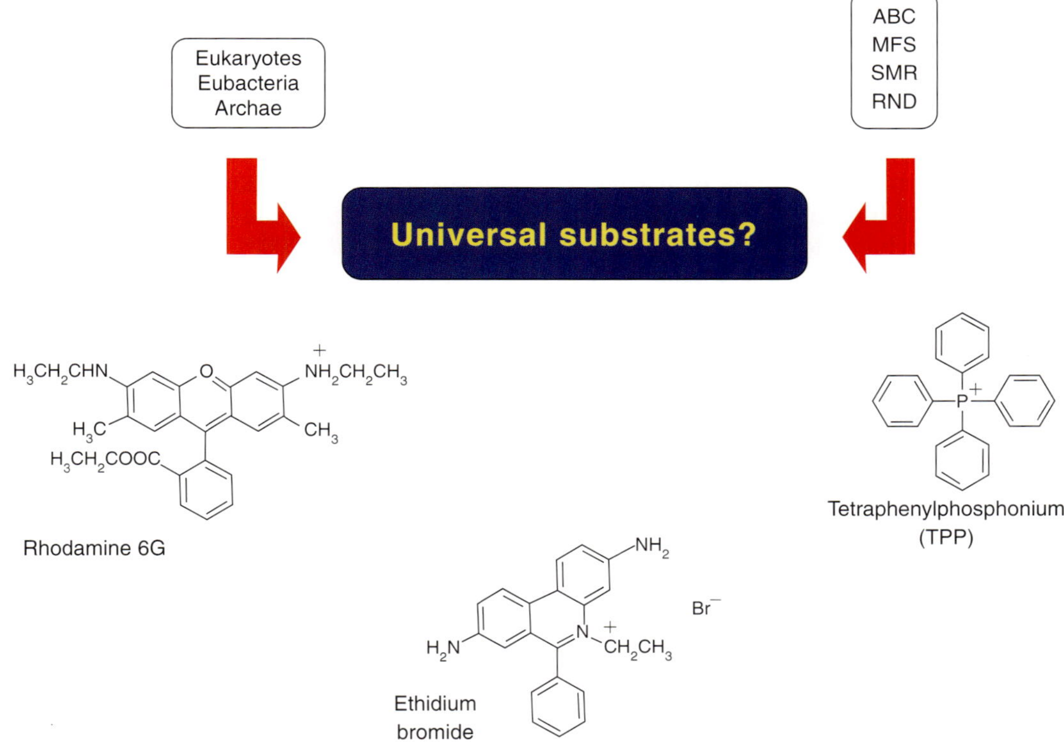

Figure 5.3. ***Rhodamine 6G, ethidium bromide and tetraphenylphosphonium are transported by multidrug pumps from a wide distribution of organisms and belonging to many different transporter families.***

compounds were included in a screen of potential Pgp substrates (Seelig, 1998), and display the characteristic physicochemical features and spatial arrangements of electron donor groups necessary for interaction.

Investigations into drug resistance in *B. subtilis* have provided a new avenue to understanding substrate interaction with multidrug pumps in the ABC family. The expression of the Bmr transporter is regulated by the BmrR transcription factor, which is activated by binding of aromatic cationic substrates for the Bmr transporter (Markham *et al.*, 1997). Elucidating the crystal structure of this transcriptional activator to 2.8 Å resolution in the presence of substrate has provided considerable insight into the molecular basis of multidrug recognition (Zheleznova *et al.*, 1999). The authors argue that two main factors are involved in this polyspecific interaction. Firstly, the binding of substrates has a minimal hydrogen-bonding component, but relies instead on contributions from Van der Waals forces, stacking interactions, electrostatic forces and the hydrophobic effect to reduce water contacts (Zheleznova *et al.*, 1999). Secondly, the drug-binding pocket is capable of producing structural rearrangements to accommodate the substrate in a manner analogous to the 'induced-fit' model. The processes underlying drug–protein interactions deviate from the rigid molecular specificity associated with dedicated unisubstrate transporters, and may contribute to the ability of multidrug ABC transporters such as Pgp, MRP, LmrA and others to confound chemotherapy in a variety of clinical settings. Clearly, it is vital that the location of binding sites within multidrug pumps are elucidated in order that we may inhibit the actions of these proteins.

LOCATION(S) AND NUMBER OF SUBSTRATE-BINDING SITES IN ABC TRANSPORTERS

LOCATION(S) OF SUBSTRATE-BINDING SITES

In order to mediate the translocation of substrates across biological membranes, ABC transporters must contain domains and regions that interact with the transported substrate. The periplasmic binding proteins are essential in determining substrate specificity in solute uptake systems of Gram-negative bacteria. However, mutant bacterial strains without substrate-binding proteins still exhibit specific uptake of maltose and histidine via their respective ABC transporters (Petronilli and Ames, 1991; Treptow and Shuman, 1985). Furthermore, mutations that alter the selectivity of the histidine transporter in *Salmonella typhimurium* from L-histidine to L-histidinol were found to localize as amino acid deletions in the membrane-bound HisM protein (Payne *et al.*, 1985). The ability of the maltose transporter in *E. coli* to transport *p*-nitrophenyl-α-maltoside was shown to be dependent on mutations in the transmembrane domains (TMDs) of the transporter (Reyes *et al.*, 1986). These early investigations of ABC transport systems clearly demonstrate that TMDs are involved in substrate recognition and translocation, even for those transporters with a periplasmic binding protein.

For several ABC transporters that do not have an extracellular substrate-binding protein, the TMDs have also been demonstrated to mediate substrate recognition. For example, the replacement of charged residues lining the transmembrane pore of the chloride channel CFTR changes its ion selectivity profile (Anderson *et al.*, 1991). The SUR1 and SUR2 receptor proteins confer different responsiveness of their associated Kir6.1 protein to the channel opener glibenclamide. The differential effects of the sulfonylurea receptor proteins SUR1 and SUR2 have also been related to differences in the primary structures of the TMDs in these proteins (Babenko *et al.*, 2000; Morbach *et al.*, 1993). Investigations using chimeric TAP1/2 transporters in which the TAP1 TMD was replaced by the TAP2 TMD, and vice versa, have revealed that both TMDs are essential for high-affinity peptide binding. In the absence of either, substrate binding to the TAP complex is impaired, indicating specific roles for each TMD in substrate recognition (Arora *et al.*, 2001). A more detailed analysis of residues in TAP2 controlling peptide binding/recognition utilized transporters composed of the 'a' and 'u' allele-encoded rat TAP2 proteins. These TAP2 proteins display different substrate specificities. Chimeras of the respective proteins demonstrated that critical residues in determining substrate specificity are located in putative cytoplasmic loops in the TMD, close to the plasma membrane (Momburg *et al.*, 1996).

The role of TMDs in substrate specificity has been most extensively investigated for Pgp. Evidence is accumulating that the drug–protein interactions, which determine the binding specificity of Pgp, are organized within the TMDs of the proteins. Independent photoaffinity labeling and epitope mapping studies of Pgp involving the 1,4-dihydropyridine derivative azidopine (Bruggemann *et al.*, 1992), iodoaryl-azidoprazosin (Greenberger, 1993; Isenberg *et al.*, 2001), iodoaryl-azidoforskolin (Busche *et al.*, 1989), the 3′- and 7′-benzophenone analogues of taxol (Wu *et al.*, 1998), and the daunomycin derivative iodomycin (Demmer *et al.*, 1997) have identified the same two major photobinding regions within the TMDs. The sites encompass transmembrane segments (TMS) 5 and 6 in the N-terminal half and TMS 11 and 12 in the C-terminal half. Moreover, a deletion mutant of Pgp consisting of the TMDs in the absence of the NBDs retained the ability to interact with drugs (Loo and Clarke, 1999b).

Mutational analyses have also proved useful in attempting to pinpoint specific regions within the TMDs involved in substrate binding. An informative approach was to generate a chimera of Pgp (encoded by the MDR gene) with the human MDR3 gene product. MDR3-Pgp is a phosphatidylcholine transporter with a low affinity for multiple multidrug substrates and is present in the canalicular membrane of hepatocytes. Pgp (MDR1) and MDR3 share about 80% sequence identity at the amino acid sequence level. In chimeras, the replacements limited to TMS 12 severely impaired Pgp (MDR1)-mediated transport of actinomycin D, vincristine and doxorubicin, but not colchicine, suggesting the importance of TMS 12 in the specificity to certain drugs (Zhang *et al.*, 1995b). Mutating residues that are not conserved amongst the MDR1- and MDR3-Pgp isoforms from different species provided evidence of further involvement of TMS 12 in conferring specificity. The results indicated that non-conserved residues within the amino-terminal half of TMS 12 determine the relative rates of transport for a variety of different substrates (Hafkemeyer *et al.*, 1998).

The literature is replete with investigations that have mutated residues throughout the protein in an attempt to localize the drug-binding sites, and these are well summarized in a published review (Ambudkar *et al.*, 1999). Drug specificity appears to be particularly sensitive to mutations in TMS 5, 6, 11 and 12, but many other mutations that affect substrate specificity are scattered throughout the polypeptide. The investigations have usually determined the effect of mutations on (i) the ability of Pgp to confer drug resistance to whole cells, (ii) alterations in steady-state cellular accumulation of Pgp substrates, or (iii) modifications in drug-stimulated ATP hydrolysis by Pgp. These studies are difficult to interpret as simple changes in drug–Pgp interaction, since altered activity may be manifest by a number of possible factors including drug binding, communication between TMDs and NBDs, or conformational changes involved in the translocation step. Furthermore, altered drug recognition and binding may be due to changes in protein–drug interactions within binding sites, or due to conformational changes in binding sites related to long-range perturbations in the global structure of the protein. Recently, however, cysteine scanning mutagenesis of all the predicted TMSs of Pgp (1 through 6, and 7 through 12), combined with thiol modification using the thiol-reactive substrates dibromobimane and methanethiosulfonateverapamil, demonstrated that residues in TMS 4, 5, 6, 10, 11 and 12 directly interact with drug molecules (Loo and Clarke, 1999b, 2000, 2001). A model was proposed wherein all of these segments contribute to a large domain consisting of multiple recognition elements for transported substrates.

Although the NBDs in Ppg are known to interact with non-transported modulators that compete with nucleotides for binding (e.g. flavenoids) (Conseil *et al.*, 1998), at present there is no evidence that the NBDs in ABC transporters play a direct role in determining substrate specificity. Experiments with chimeric multidrug resistance genes argue against such a role (Buschman and Gros, 1991). Mouse Mdr1 confers resistance to drugs whereas mouse Mdr2 acts as a phosphatidylcholine transporter. A chimeric protein in which the NBDs of Mdr2 replaced those in Mdr1 still transported drugs, while the replacement of TMDs of Mdr1 by those in Mdr2 abolished drug transport.

In summary, the TMDs of ABC transporters clearly mediate substrate-binding events (although in bacterial uptake systems with a periplasmic binding protein, the initial event is binding by the PBP) in both uptake and efflux pumps, and TMSs involved in this process have been identified for a number of transporters. However, we still require significant effort to elucidate the precise molecular components of drug-binding sites.

Number of substrate-binding sites

Two regions of Pgp photolabeled by drugs may contribute to a single binding surface, which is able to interact with different drugs (Bruggemann *et al.*, 1992). Alternatively, the labeled regions may each represent separate sites, giving two distinct drug-binding sites per Pgp monomer (Dey *et al.*, 1997). A number of observations suggest that Pgp and related ABC multidrug transporters possess two or more drug-binding sites:

(a) Kinetic functional assays: Using steady-state drug accumulation assays in whole cells, competitive and non-competitive and cooperative interactions have been detected between different transport substrates (Ayesh *et al.*, 1996). Non-competitive interactions demonstrate the presence of multiple transport-competent drug-binding sites in Pgp and cooperative effects suggest that these sites may interact with more than one ligand. Rhodamine 123 and Hoechst 33342 (Shapiro and Ling, 1997), colchicine and synthetic hydrophobic peptides (Sharom *et al.*, 1996), and colchicine and tetramethylrhosamine (Lu *et al.*, 2001) stimulated transport of each other using isolated membranes or purified Pgp preparations. Shapiro and Ling suggested that this effect in Pgp may arise from positively cooperative interactions between two transport-competent drug-binding sites, denoted the R site and H site. The non-competitive interactions between drugs in their ability to stimulate ATP hydrolysis also provides strong evidence for multiple binding sites on the protein (Orlowski *et al.*, 1996; Pascaud *et al.*, 1998). Positively cooperative interactions between transported substrates have also been observed for other ABC transporters. For bacterial LmrA, vinblastine and Hoechst 33342 stimulated the transport of each other, pointing to the presence of two transport-competent drug-binding sites in the transporter with overlapping drug specificities (van Veen *et al.*, 2000). The stimulation of the transport of non-conjugated drugs by glutathione, and vice versa, in MRP1 and MRP2 (Evers *et al.*, 2000; Loe *et al.*, 2000a, 2000b) also suggests the presence of at least two positively cooperative transport-competent substrate-binding sites in the proteins, one for glutathione and the other for unconjugated drugs (Evers *et al.*, 2000).

(b) Conformational change assays: The quenching by vinblastine and verapamil of the fluorescence of intrinsic tryptophan residues and 2-[4′-maleimidyl-anilino]naphthalene 6-sulfonic acid (MIANS)-labeled Pgp has been shown to exhibit a biphasic profile (Liu *et al.*, 2000; Sharom *et al.*, 1999). This was suggested to provide evidence of multiple sites of interaction for vinblastine. However, for the MIANS-labeled protein, this observation may also be explained by a differential sensitivity of the two labeled NBDs to allosteric effects caused by the drug. Biphasic displacement of [^{3}H]-verapamil binding by vinblastine to Pgp was also observed by using a radioligand-binding assay (Doppenschmitt *et al.*, 1999). Studies on the Pgp conformation-sensitive monoclonal antibody UIC2 suggested that the stoichiometric binding of two vinblastine molecules per Pgp was required to conformationally increase the accessibility of the UIC2 epitope (Druley *et al.*, 2001).

(c) Photoaffinity labeling approaches: The modulator *cis*(*Z*)-flupentixol increased the affinity of iodoarylazidoprazosin for the C-terminal half of Pgp (C site) without changing the affinity for the N-terminal half (N site) (Dey *et al.*, 1997). In addition, iodoarylprasozin binding to these sites was differentially inhibited by both vinblastine and cyclosporin A.

(d) Direct measurements of binding: Equilibrium or kinetic radioligand-binding assays provide a *direct* insight into drug–protein interaction. The increased dissociation rate of [^{3}H]-vinblastine from Pgp and bacterial LmrA by various modulators provided the first solid pharmacological proof for the existence of multiple drug-binding sites on the proteins (Ferry *et al.*, 1992, 2000; Malkhandi *et al.*, 1994; Martin *et al.*, 1997, 1999; van Veen *et al.*, 1998). More recently, the kinetic data for Pgp were combined with Schild analyses of drug–drug interactions at equilibrium to demonstrate that Pgp contains at least four distinct sites involved in drug binding (Martin *et al.*, 2000a). For LmrA, vinblastine equilibrium binding experiments provide evidence for the presence of two vinblastine-binding sites in the homodimeric transporter (van Veen *et al.*, 2000).

Collectively, these data show that ABC multidrug transporters must contain more than one drug-interaction site. The drug-interaction sites could represent two (or more) physically and spatially distinct binding sites or, alternatively, be present in a single flexible binding region within the transporters.

STRUCTURAL AND FUNCTIONAL PROPERTIES OF SUBSTRATE-BINDING SITES

STRUCTURAL PROPERTIES OF SUBSTRATE-BINDING SITES

We still know very little about the structural elements in multidrug transport proteins that dictate drug specificity. However, crystal structures at 2.7 Å resolution, in the absence and presence of drug, obtained for the polyspecific transcription regulator BmrR from *B. subtilis* provide some insight (Zheleznova *et al.*, 1999). One of the universal multidrug pump ligands, tetraphenylphosphonium **(Figure 5.3)**, appears to penetrate into the hydrophobic core of BmrR, where it forms van der Waals and stacking interactions with hydrophobic and aromatic residues, and makes an ion pair interaction with a buried glutamic residue Glu134 (Vazquez-Laslop *et al.*, 1999; Zheleznova *et al.*, 1999). The Bmr transporter also binds and translocates tetraphenylphosphonium (Neyfakh *et al.*, 1991). Similar drug–protein interactions are likely to occur in the transcription regulator and the Bmr transporter and, by inference, other multidrug transporters (Zheleznova *et al.*, 2000). Indeed, the presence of acidic residues within TMSs of secondary multidrug transporters MdfA and EmrE in *E. coli* (Edgar and Bibi, 1999; Muth and Schuldiner, 2000; Yerushalmi and Schuldiner, 2000), QacA in *Staphylococcus aureus* (Paulsen *et al.*, 1996) and LmrP in *L. lactis* (van Veen, 2001) was shown to be related to the cation specificity of these transporters. More recently, similar observations have been made for mammalian ABC multidrug transporters. A glutamate residue in human MRP (Glu1089) in predicted TMS 14 appears to be critical for the ability of the protein to confer resistance to cationic drugs, such as anthracyclines, whereas this residue is not critical for its ability to transport endogenous glutathione conjugates and glucuronides (e.g., leukotriene C4 and 17-β-estradiol 17-β-D-glucuronide) (Zhang *et al.*, 2001).

Similar to the role of acidic residues in determining the specificity for cationic drugs, basic residues can play a role in the specificity for anionic drugs. For example, the specificity of human MRP2 for glutathione conjugates (leukotriene C4 and 2,4-dinitrophenyl-*S*-glutathione) is related to the presence of a lysine at position 325 and an arginine at position 586, in TMS 6 and 11, respectively (Ito *et al.*, 2001). For the breast cancer resistance protein (BCRP), different versions of the BCRP cDNA have been obtained from cell lines that had been selected for resistance by drug exposure and display distinctly different MDR phenotypes. These versions of BCRP contain different amino acid substitutions at position 482 (arginine, glycine or threonine). Surprisingly, BCRP protein containing an arginine at position 482 is unable to transport (cationic) rhodamine 123, whereas BCRP with a neutral residue at this position (glycine or threonine) does transport this substrate. These data support a role for charged residues in determining the drug specificity of BCRP (Litman *et al.*, 2001).

Although acidic and basic residues in TMSs of ABC multidrug transporters appear to play a role in the selectivity towards charged drugs, Pgp, LmrA and others do not possess charged residues in their TMDs, and yet, do transport cationic amphipathic drugs (see above). Hence, alternative mechanism(s) for cation selectivity must exist. Interestingly, it has been shown that cations can bind to the π face of the aromatic ring structures of tyrosine, phenylalanine and tryptophan residues (Dougherty, 1996). Since, in the hydrophobic environment of the phospholipid bilayer, this binding can be as strong as the electrostatic interactions between ion pairs, aromatic residues may determine cation selectivity in ABC multidrug transporters. This notion is supported by the observation that site-directed substitution of aromatic residues in the protein does affect the specificity or potency of drug interaction with Pgp (Kwan *et al.*, 2000; Ueda *et al.*, 1997). Analysis of the transmembrane α-helices in Pgp and LmrA has revealed that aromatic residues and polar amino acid residues with hydrogen donor side-chains are often clustered together on one side of a helix, with amino acid residues with non-hydrogen-bonding side-chains on the other side (Seelig and Landwojtowicz, 2000; van Veen, 2001). Hence, the TMSs could be oriented with their non-interactive residues facing the hydrophobic phospholipid bilayer, and their interactive residues facing a translocation pore. Within this pore, π electrons may enable cation binding and may even provide a 'slide guide' system for cationic drugs, whereas hydrogen bonds and stacking interactions facilitate the interaction with electroneutral moieties within the amphipathic drug molecules (Seelig, 1998).

How do substrates access the binding sites?

The binding sites for drugs on multidrug ABC transporters appear to exist in two conformational states that display high- or low-affinity binding for their specific ligand. These binding sites exist in an equilibrium that, in the case of Pgp, may be switched between affinities by allosteric communication from either (i) binding of drug at an alternate site (Martin *et al.*, 2000a), or (ii) events during the hydrolytic cycle in the NBDs (Martin *et al.*, 2000b; Sauna and Ambudkar, 2001). The orientation and location of the high- and low-affinity conformations of drug-binding sites remain elusive to date for any multidrug transporter. The conventional view of substrates gaining access to multidrug ABC transporters via the aqueous phase seems a bit naive given the highly hydrophobic nature of many of the compounds. In the case of Pgp and LmrA (Chapter 12), several lines of evidence support the proposal that drugs access their binding sites via the lipid bilayer. For example, fluorescent compounds such as doxorubicin enable the membrane-localized probe 5-[^{125}I]-iodonapthalene-1-azide to label Pgp via direct energy transfer between the compounds. This labeling can only occur over a short range and within the bilayer (Raviv *et al.*, 1990). Further proof has been obtained from investigations into the transport of the acetoxy-methyl ester of calcein (calcein-AM) and BCECF (BCECF-AM) in cells. These compounds are rapidly metabolized to the highly fluorescent fluorescein derivatives by cytoplasmic esterases. However, only the parent non-fluorescent acetoxy-methyl esters are substrates for efflux by Pgp and LmrA. Cells expressing Pgp or LmrA exhibit measurable fluorescence only following inhibition of these multidrug pumps, indicating that these proteins actively extrude the acetoxy-methyl esters before they can reach the cytoplasm (i.e. from within the bilayer) (Bolhuis *et al.*, 1996; Homolya *et al.*, 1993). The transport of substrates from the lipid bilayer may also be relevant for other ABC transporters with hydrophobic substrates. The human *MDR3*-encoded Pgp transports phosphatidylcholine from the cytoplasmic leaflet of the bile canalicular membrane of hepatocytes into the bile (Ruetz and Gros, 1994; Smit *et al.*, 1993). In addition, the *E. coli* α-hemolysin transporter HlyB appears to bind the signal sequence of α-hemolysin when the signal sequence forms an amphiphilic helix that binds to the cytoplasmic leaflet of the plasma membrane (Sheps *et al.*, 1995; Zhang *et al.*, 1995a) (but see also Chapter 11).

Whilst our understanding of binding-site properties is growing, we still have little knowledge regarding the molecular consequences of substrate interaction on the local protein structure of ABC transporters.

Coupling between substrate-binding sites and nucleotide-binding domains

General principles of transport and coupling

The movement of molecules against their concentration gradients by ABC proteins requires a series of coordinated events as outlined in **Figure 5.4**. The 'driven substrate' (e.g. drug) and the 'driving substrate' (ATP) will interact with the transporter in a mutually dependent fashion to prevent ATP hydrolysis in the absence of translocation (Jencks, 1980; Krupka, 1993). Understanding how these two processes are coupled is fundamental to elucidating the mechanism by which ABC proteins translocate substrates. The mechanism of translocation is composed of many discrete stages leading to two major events: (i) reorientation of a substrate-binding site across the membrane and (ii) an alteration in the binding site from high to low affinity for transported substrate. Transport of compounds against a concentration gradient will be driven by the Gibbs energy change of the overall process. The individual steps play vital roles to ensure a reasonable rate of turnover that is free of 'bottlenecks' (Jencks, 1980). Energy produced by the binding and dissociation of transported molecules, ATP and its metabolites, and the energy produced by the hydrolysis of ATP will be used to ensure adequate turnover rates and efficient coupling (Jencks, 1980; Krupka, 1993).

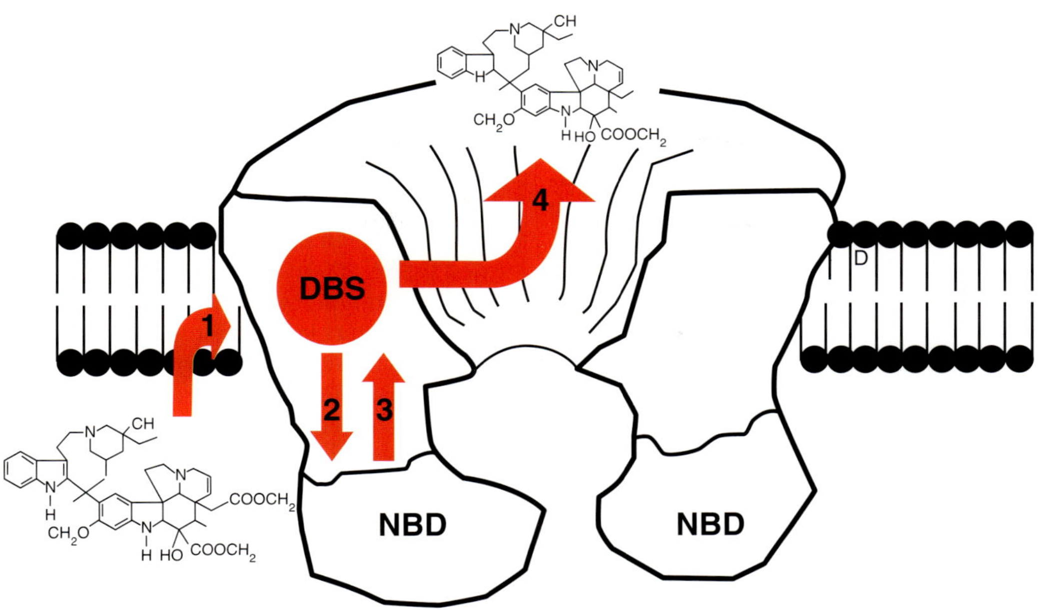

Figure 5.4. ***Schematic presentation of the steps involved in the translocation of drug across a plasma membrane by Pgp. Step 1 is the initial interaction of drug with the drug-binding site (DBS) in the TMD via the lipid bilayer. Step 2 represents the signal to stimulate ATP hydrolysis in the nucleotide-binding domain (NBD). Step 3 is the signal to initiate conformational changes in the TMD during a catalytic cycle. Step 4 is the translocation and release of drug across the membrane.***

Substrate binding and nucleotide hydrolysis events are spatially distinct in ABC proteins and therefore a series of communication pathways will be required to ensure efficient coupling.

ABC PROTEINS: COUPLED OR NOT?

Despite intensive research efforts, the mechanism of coupling in ABC transporters remains elusive. An obvious and seemingly straightforward question is whether the NBDs require a stimulus to hydrolyze ATP. When associated with their compatriot membrane proteins, the NBDs of all ABC transporters are capable of hydrolyzing ATP. When isolated from their membrane-bound domains the situation is not as clear. The NBDs of the maltose transporter (MalK) and the histidine permease (HisP) display high levels of ATP hydrolysis (0.5–1.0 μmol ATP min^{-1} mg^{-1}) when expressed separately from the membrane domains of the whole transporter (Morbach *et al.*, 1993; Nikaido *et al.*, 1997). When HisP and MalK are associated with the membrane-bound subunits of their respective transporters, the ATPase activity is inhibited and only reaches the high levels observed for isolated domains when the transported substrate is present (Davidson and Nikaido, 1991; Liu and Ames, 1998). In contrast with these bacterial importers, isolated NBDs of the eukaryotic proteins Pgp (Dayan *et al.*, 1996), MRP (Kern *et al.*, 2000) and CFTR (Ko and Pedersen, 1995) only display low activities of approximately 0.05 μmol min^{-1} mg^{-1}. The activities of these domains within full-length human isoforms of Pgp (Loo and Clarke, 1995; Ramachandra *et al.*, 1998), MRP (Chang *et al.*, 1997) and TAP (Gorbulev *et al.*, 2001) have been reported to be greater than 1 μmol min^{-1} mg^{-1}. These disparities between the behavior of isolated NBDs and that found when they are associated with the TMDs indicate a significant degree of functional interaction between the two types of domains. The opposing influence of TMDs on the activity of NBDs observed between the prokaryotic and eukaryotic members suggests that distinct mechanisms of coupling may occur.

However, the role of substrate binding and dissociation in regulating ATP hydrolysis at NBDs is of universal importance within the ABC transporter family. Does this provide any insight into a possible coupling mechanism? **Table 5.1** shows the degree to which several transported agents are able to stimulate ATP hydrolysis by their target ABC transporters. The maltose, histidine and TAP transporters, which display transport with high specificity, do not display appreciable basal or intrinsic ATPase activity in the absence of substrate (Davidson *et al.*, 1992; Gorbulev *et al.*, 2001; P.-Q. Liu *et al.*, 1999). For these 'dedicated' transporters it appears that the process of ATP hydrolysis is tightly coupled to substrate binding, and therefore transport. Interestingly, the polyspecific or multidrug efflux pumps Pgp, LmrA and MRP display a high basal ATPase activity with only modest amounts of stimulation by substrate (Callaghan *et al.*, 1997; Chang *et al.*, 1997; Ramachandra *et al.*, 1998; Shapiro and Ling, 1994). The compounds used to stimulate Pgp, LmrA and MRP ATPase activities **(see Table 5.1)** vary markedly in their affinities to bind to the proteins (50 nM to 50 μM), yet the degree to which they stimulate ATP hydrolysis is similar (1.5–4 fold). This lack of correlation between maximal ATPase activity (reflecting transport rate) and substrate affinity is a characteristic feature of passive transport processes. Clearly, Pgp, LmrA and MRP are active transporters and therefore this characteristic has been used to argue for a partially uncoupled transport mechanism (Krupka, 1999). An alternative explanation for the high basal ATPase activity of Pgp and other multidrug transporters may be that, even in the absence of added drugs, these systems encounter endogenous substrates (e.g. lipids) in the biological membranes in which they are embedded (Ferte, 2000).

The catalytic cycle and drug-binding events must be intertwined to some degree in order to produce the vectorial transport by single substrate specific and multidrug ABC proteins. The subsequent sections explore how these two processes interact and which events during the transport cycle are critical to the coupling.

TABLE 5.1. STIMULATION OF ATP HYDROLYSIS BY SUBSTRATES FOR TRANSPORT IN PROKARYOTIC AND EUKARYOTIC ABC TRANSPORTERS

Transporter and stimulating agent	Fold stimulation
Human Pgp (Ramachandra *et al.*, 1998)	
Verapamil	4.2
Nicardipine	3.6
Vinblastine	2.0
TPP 1	3.8
Colchicine	3.1
Human MRP (Chang *et al.*, 1997; Mao *et al.*, 1999)	
LTC4	1.3–1.5
GSSG	1.5
Estradiol-glucuronide	1.5
Doxorubicin	1.3
Human TAP (Gorbulev *et al.*, 2001)	
RRYQKSTEL	2.8
Bacterial LmrA (van Veen *et al.*, 1998)	
Verapamil	2.4
Bacterial histidine transporter (P.-Q. Liu *et al.*, 1999)	
Liganded HisJ	23
Bacterial maltose transporter (Davidson *et al.*, 1992)	
Liganded MalE	>300

WHAT DRIVES TRANSLOCATION: SUBSTRATE BINDING OR ATP HYDROLYSIS?

Investigations with the histidine transporter of *S. typhimurium* have provided significant insight into the interaction between substrate binding and ATP hydrolysis. As indicated above, the membrane segments of the transporter (HisQM) regulate the intrinsic ATP hydrolytic capability of HisP (Liu *et al.*, 1997) and a significant increase in ATPase activity is observed in the presence of liganded substrate-binding protein (HisJ). How is this signal transmitted? The HisJ protein undergoes significant and, importantly, substrate-dependent conformational changes upon binding histidine (Wolf *et al.*, 1996). It was therefore concluded that the conformation of HisJ produced by ligand binding provides the driving force to stimulate ATP hydrolysis and initiate transport through HisQM. However, a subsequent investigation from the same laboratory demonstrated (i) no direct correlation between the affinities of different carbohydrates to bind to HisJ and their translocation rates and (ii) a poor correlation between translocation rates and substrate-induced stimulation of ATPase activity (C.E. Liu *et al.*, 1999). These findings at first appeared difficult to reconcile with a coupled vectorial ATP-dependent translocation process (Jencks, 1980). A more recent publication has demonstrated that binding of ATP to the HisQMP2 complex, *prior* to association of liganded HisJ, initiates quaternary structural changes within the complex (P.Q. Liu *et al.*, 1999). These changes in association of subunits in turn *facilitate* the ability of liganded HisJ to stimulate further ATP hydrolysis, which is essential to 'open' the translocation pathway.

Can this model of coupling in the histidine transporter act as a template for all ABC proteins? None of the eukaryotic efflux pumps associate with a ligand-binding protein such as HisJ, which suggests an alternative coupling mechanism. The heterodimeric TAP1/2 complex provides a useful model to examine coupling in a transporter that shows strictly defined substrate specificity and without the need for an accessory binding protein (Uebel and Tampe, 1999). The TAP1/2 transporter exhibits negligible ATPase activity in the absence of substrate (Gorbulev *et al.*, 2001) and has an absolute requirement for two functional NBDs to sustain this activity (Neefjes *et al.*, 1993). It has also recently been suggested that the NBDs of the TAP1 and 2 monomers mediate distinct, yet cooperative roles within a single catalytic cycle (Saveanu *et al.*, 2001). Taken together, these observations point to a highly coupled process underlying peptide translocation. The process of translocation begins with rapid binding of peptide, which leads to slow conformational changes within the protein (Neumann and Tampe, 1999). These conformational changes produce closer contact between the two NBDs, which is thought to affect the regulation and rate of ATP hydrolysis. In turn, nucleotide binding initiates and stabilizes specific conformations of the TAP1/2 during the transport cycle as demonstrated by antibody accessibility studies (van Endert, 1999). Interestingly, the maximal rate of ATP hydrolysis, and therefore translocation, is independent of the substrate-binding affinity at the membrane domains (Gorbulev *et al.*, 2001), indicating that conformational changes produced by ATP hydrolysis, but not peptide binding, determine the overall rate of translocation. Peptide binding itself may provide a trigger that initiates rather than assists the formation of intermediary states of TAP1/2 during a transport cycle. Similar mechanisms may be relevant for ABC multidrug transporters. The existence of interactions between membrane domains and NBDs is strongly supported by the observations that: (i) substrates for Pgp-mediated transport stimulate, whilst some modulators inhibit, ATPase activity (Pascaud *et al.*, 1998), (ii) binding of transported substrates produces spectral changes in fluorescently labeled NBDs (Liu and Sharom, 1996), and (iii) nucleotide binding causes altered binding of transported substrates (Martin *et al.*, 2000b; van Veen *et al.*, 2000).

In many tightly coupled active transport systems, free energy changes produced by substrate binding play major roles in the movement of proteins between intermediate states of a transport cycle (Jencks, 1980; Krupka, 1993). Does this occur for the promiscuous ABC multidrug pumps or, alternatively, is substrate binding a passive event, being driven by conformational changes induced by nucleotide hydrolysis? This question has been addressed for Pgp and other multidrug transporters, with several different approaches employed to measure conformational changes elicited by drugs and nucleotides. Neither drug nor nucleotide was able to produce significant global alteration of secondary structural elements in Pgp, LmrA and MRP1 as measured by ATR- FITR (Grimard *et al.*, 2001; Manciu *et al.*, 2001; Sonveaux *et al.*, 1996; Vigano *et al.*, 2000). However, by measuring the kinetics of $^{2}H/H$ exchange it was demonstrated that both MgATP binding and hydrolysis produce major changes in the tertiary structure. Similarly, investigations of Pgp using proteolytic enzyme accessibility, or differential immunoreactivity of the UIC2 antibody, demonstrated that the protein undergoes distinct conformational transitions during the various stages of a catalytic cycle (Mechetner *et al.*, 1997; Wang *et al.*, 1998). It is becoming apparent from these unrelated techniques that many transported substrates, which in some cases cause stimulation of ATP hydrolysis, produce small or negligible overall conformational changes in Pgp (Grimard *et al.*, 2001; Mechetner *et al.*, 1997; Sonveaux *et al.*, 1996, 1999; Wang *et al.*, 1998). The effects of drugs are, however, considerably greater in magnitude when used in conjunction with nucleotides. This suggests that major tertiary conformational transitions in Pgp and other systems are primarily driven by the catalytic cycle, whilst substrate binding appears to attenuate or modulate these changes. The large shifts in binding affinities essential to a transport cycle are most likely instigated by nucleotide-induced effects and the role of substrates is somewhat passive. It may be likened to the early proposal that the Pgp-mediated transport cycle resembles a slow-moving waterwheel, driven by ATP hydrolysis, with substrates 'hitching' a ride though the membrane. In the case of multidrug pumps the waterwheel is constantly turning, but may be speeded up by substrates. Transporters with more limited specificity, as described above, only switch on the waterwheel following substrate binding.

WHERE DOES THE COMMUNICATION BETWEEN TMDs AND THE NBDs TAKE PLACE?

The three-dimensional organization of ABC proteins will dictate how signals at the TMD trigger events at the NBD and vice versa. Presumably it is the residues located at the interface between these domains that will play a major role. However, in the absence of any high-resolution structural data, it has not been possible to identify such regions in ABC proteins to date. Sequence analyses have been instrumental in identifying many potentially important functional domains within ABC proteins. Analysis of 61 different prokaryotic ABC proteins constituting the membrane domains of uptake systems may have provided an important clue to identify a region of TMD–NBD contact (Saurin *et al.*, 1994). The short sequence (EAA---G---------I-LP), known as the EAA-loop, is highly conserved between these proteins and located 100 residues from the C-terminus, lying between TMS 4 and 5. Mutations of this loop have been demonstrated to produce dramatic reductions in the transport activity of the maltose (Dassa, 1993), iron (Koster and Bohm, 1992), and phosphate uptake systems (Webb *et al.*, 1992). The precise molecular consequences of mutations within the EAA-loop have not been elucidated yet. However, genetic analyses suggest a role in mediating contact between the TMDs and the NBDs (Mourez *et al.*, 1997). The peroxisomal ABC transporter Pxa1p, which is the yeast orthologue of human ADLP, is the only eukaryotic protein for which an EAA-loop has been assigned (Shani *et al.*, 1995, 1996). Mutations in the EAA-loop of ADLP have been observed in patients with adrenoleukodystrophy (Ligtenberg *et al.*, 1995), whilst their introduction in Pxa1p abrogates transport (Shani *et al.*, 1996). This highlights the role of the EAA-loop in many different ABC proteins.

However, the vast majority of eukaryotic ABC proteins do not have a characteristic EAA-loop. Perhaps this is born out of the tendency to adopt structures comprising a single polypeptide chain compared with the oligomeric organization of prokaryotic transporters (Holland and Blight, 1999). Deletion of the intracellular loop between TMS 4 and 5, which is the equivalent location of the EAA-loop, caused destabilization of the fully open-conductance state of the CFTR protein (Xie *et al.*, 1995). The movement of CFTR between conductance states has been shown to depend on nucleotide binding/hydrolysis (Gadsby and Nairn, 1994) and disruption of this movement may indicate a role in the interaction between the TMDs and NBDs. The precedent for a role of intracellular loops in mediating the communication between TMDs and NBDs is also well established for Pgp. This fact was recognized in 1988, when specific cell lines exposed to high concentrations of colchicine developed an altered spectrum of resistance to cytotoxic drugs than was previously described (Choi *et al.*, 1988). The high selection pressure with colchicine caused a glycine to valine mutation at position 185, which is located in the first intracellular loop of Pgp (Choi *et al.*, 1988; Kioka *et al.*, 1989). The altered resistance spectrum of the G185V mutant Pgp was mirrored by alterations in the drug potencies to inhibit photoaffinity labeling (Safa *et al.*, 1990) and stimulate ATP hydrolysis (Muller *et al.*, 1996). Mutation of endogenous glycine residues to valine in other intracellular loops also affected the ability of Pgp to confer cellular resistance against many cytotoxic drugs (Loo and Clarke, 1994). Another investigation localized many more sites within the first intracellular loop, mutations of which affected the degree and potency of drugs to stimulate ATP hydrolysis by Pgp (Kwan and Gros, 1998). Whilst all of these investigations strongly suggest that intracellular loops provide the link that couples drug binding to ATP hydrolysis, further investigations are required to uncover the molecular process underlying this critical domain–domain interaction.

WHAT FUNCTIONAL EFFECTS DO THE NBDs HAVE ON DRUG-BINDING SITES?

The major approach used to examine the cross talk between membrane domains and NBDs in ABC transporters has been to measure substrate binding to different conformationally stabilized transition states in the catalytic cycle. The technique of vanadate trapping has been used to 'lock' Pgp immediately post hydrolysis of ATP. Photoaffinity (Dey *et al.*, 1997; Sauna and Ambudkar, 2000) and equilibrium binding (Martin *et al.*, 2000b) approaches have shown that the vinblastine, iodo-arylazidoprazosin and XR9576 binding sites change from high to low affinity in the ADP/vanadate trapped species. In vanadate-trapped LmrA, the low-affinity site for vinblastine and the photoreactive drug *N*-(4′,4′-azo-*n*-pentyl)-21-deoxy-[^{3}H]ajmalinium(APDA) was show to be localized on the outside surface of the transporter (van Veen *et al.*, 2000). Thus it was assumed that Pgp and LmrA utilized the process of ATP hydrolysis to drive the reorientation of drug-binding sites during transport.

Interestingly, binding of the non-hydrolyzable analogues AMP-PNP or ATP-γ-S produced a switch in the vinblastine-binding site to a low-affinity conformation in Pgp (Martin *et al.*, 2000b). This finding was in agreement with earlier preliminary evidence that AMP-PNP reduced the photolabeling of Pgp by azidopine, another transported compound (Urbatsch and Senior, 1995). These results suggest that the binding energy produced by nucleotide interaction with Pgp is sufficient to trigger reorientation of the drug-binding sites on Pgp. Such an effect of AMP-PNP binding is consistent with the conformational changes outlined in biophysical studies described earlier. However, the binding sites for the non-transported modulator XR9576 and the prazosin analogue IAAP were unaffected by binding of non-hydrolyzable nucleotides (Martin *et al.*, 2000b; Sauna and Ambudkar, 2000). As described above, it appears that the site to which XR9576 binds can indeed undergo conversion to a low-affinity conformation, but it cannot do so if this non-transported compound is bound to it. This may suggest that the binding site for XR9576 is modulatory rather than transport competent. However, the transported dye Hoechst 33342 binds at the identical site to XR9576 on Pgp (Martin *et al.*, 2000a). This indicates that it is the nature of substrate bound that is critical in dictating the response of specific TMSs or binding sites to signals emanating from the NBDs.

How is the low-affinity drug-binding site 'reset' to its original high-affinity conformation? Two possibilities are that (i) a second ATP hydrolysis step is required or (ii) the release of inorganic phosphate from the NBD after hydrolysis of ATP provides energy to reconfigure the binding site. The first possibility is supported by a recent publication using the prazosin analogue IAAP (Kerr *et al.*, 2001). However, the observation that photolabeling of Pgp by IAAP had not recovered following dissociation of ADP.Vi could also be complicated by the fact that only a single ligand concentration (at 1/150th of the K_d) was used to measure recovery of binding. In another study, full dose–response analysis revealed that complete recovery of high-affinity vinblastine binding was possible in the absence of ATP or a second hydrolytic event (Martin *et al.*, 2000b). Furthermore, recent data for Pgp suggest that the release of phosphate during the catalytic cycle provides the trigger to 'reset' the vinblastine drug-binding site (Martin *et al.*, 2001). Release of phosphate has previously been implicated as the stage of the catalytic cycle corresponding to the release of free energy from ATP hydrolysis (Urbatsch *et al.*, 1995). Clearly further investigations are required to resolve the differences in response of the IAAP and vinblastine drug-binding sites to stages of the catalytic cycle of Pgp.

THE TRANSPORT CYCLE OF ABC TRANSPORTERS

TRANSPORT MODELS

To integrate the observations mentioned in the previous sections into more universal transport models for ABC transporters, it is useful to describe the general features of a transport cycle. Data relating to the transport characteristics of several ion pumps (e.g. Na^+/K^+-ATPase, SERCA, H^+-ATPase) indicate four essential aspects of protein function are required in an active translocation process (Jencks, 1980; Krupka, 1993; Tanford, 1983a, 1983b):

1. The transporter must be capable of assuming at least two *conformational states* that allow access of binding sites for transported substrates on either side of the membrane.
2. The *binding-site affinities* are different on opposite sides of the membrane to facilitate capture (high affinity) and release (low affinity) of the transported species by the protein.
3. The transporter must adopt structures that allow the binding site to *move or reorient* between sides of the membrane.
4. The transporter must be *reset* to its original state at the end of each reaction cycle.

In primary active transport systems the changes in binding-site orientation and/or affinities are facilitated when coupled to the generation of energy by nucleotide hydrolysis. The scheme in **Figure 5.5** outlines a simple efflux system involving two conformations of the transporter with the substrate-binding site exposed to the cytoplasm (E1) or the extracellular space (E2). The E1 conformation has high affinity for substrate and is converted to the outward facing, low-affinity E2 conformation at a specific stage of the ATP hydrolytic cycle.

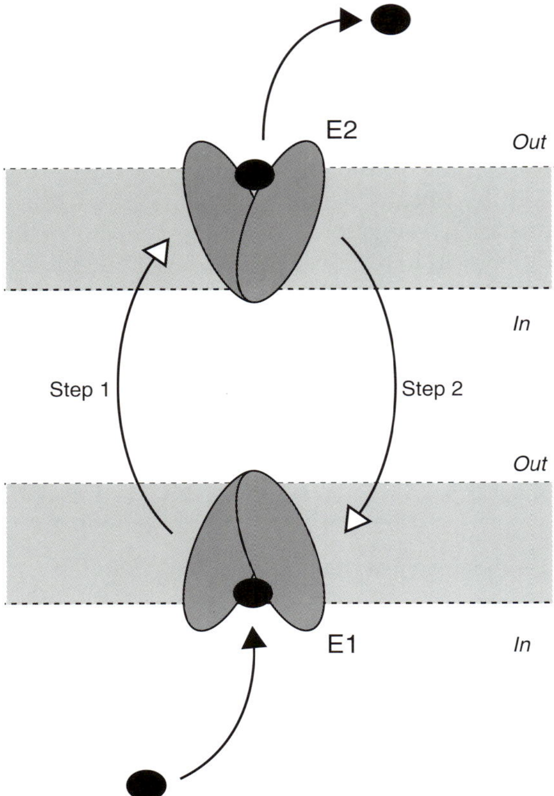

Figure 5.5. ***General scheme for the transport cycle of a solute export system. E1 refers to the conformation in which the substrate binds with high affinity to a substrate-binding site at the inside surface the membrane. E2 refers to the conformation in which substrate dissociates from a low-affinity substrate-binding site at the outside surface of the membrane. During the operation of the transport cycle, the E1 and E2 conformations are interconverted. For a solute uptake system, the substrate-binding affinities of the E1 and E2 conformations are reversed.***

Dissociation of the transported substrate and/or another stage of the ATP hydrolytic cycle convert the E2 state to the original E1 conformation. Conversely, an import protein will display high-affinity binding of substrate in the E2 conformation. The transport parameters outlined above remain elusive for the majority of ABC proteins and therefore no precise cycle has been elucidated (see also Chapters 6 and 9).

Based on these fundamental principles, three transport models have been proposed for the multidrug pumps Pgp and LmrA. Each of the models integrates observations for the individual transporters with two general observations that (i) the NBDs operate in an alternating cycle to produce ATP hydrolysis (Senior, 1998; Senior and Bhagat, 1998) and (ii) the cycle of ATP binding and hydrolysis elicits conformational changes that alter affinity of drug-binding sites (see above). The appearance of three different models has arisen to account for the observations, as discussed in previous sections, that Pgp, LmrA and MRP proteins contain multiple drug-interaction sites. The *alternating access/single-site model* (Bruggemann *et al.*, 1992; Martin *et al.*, 2000a; Shapiro *et al.*, 1999) proposes that these drug-interaction sites may be organized in the transporters within a single drug-binding region that is alternately exposed to the inner or outer membrane surface of the phospholipid bilayer. The molecular interactions between drugs and the transporter that occur within this drug-binding region may be organized in a similar way as has been described for the soluble transcription regulator BmrR in *B. subtilis* (Zheleznova *et al.*, 1999). The *fixed two-site model* proposes the presence of two static drug-binding regions: an 'on-site' with high binding affinity in the C-terminal half of Pgp and an 'off-site' in the N-terminal half (Dey *et al.*, 1997). Upon ATP hydrolysis at one NBD, a conformational change decreases the affinity of the 'on-site' for drugs, and as a result, the drugs move from the 'on-site' to the 'off-site' through a pore-like structure. Subsequently, ATP-hydrolysis at the second NBD is required either to drive drug translocation from the 'off-site' to the external medium or to reset the high-affinity 'on' site (Sauna and Ambudkar, 2001). Finally, the *alternating two-site (two-cylinder engine) model* combines some properties of these two models. It proposes the presence of two drug-binding regions, as suggested in the fixed two-site model, with alternate exposure to the inner and outer membrane surface of the phospholipid bilayer, as suggested in the single-site model. The binding and/or hydrolysis of ATP results in the movement of both sites from the inner leaflet of the membrane to the outer leaflet of the membrane, simultaneously, or in a sequential or alternating fashion, with a concomitant change to low affinity (van Veen *et al.*, 2000).

THE LINK BETWEEN SINGLE-SITE AND TWO-SITE MODELS

The transport scheme presented in **Figure 5.6** depicts the alternating access of the binding site(s) for transport species according to the single- and two-site models described above.

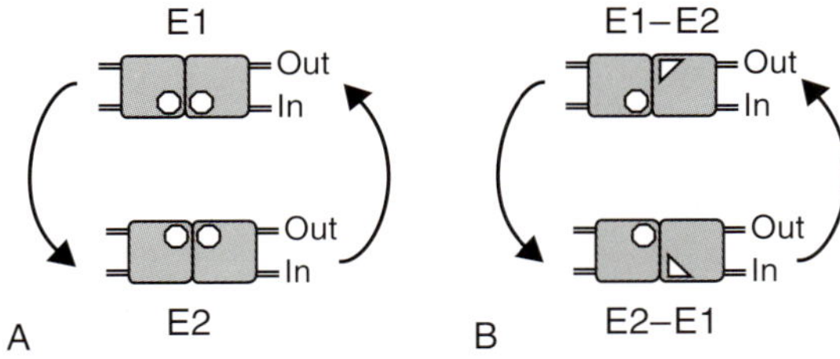

Figure 5.6. Link between single-site (A) and two-site models (B). The ABC proteins are represented by two squares each corresponding to a TMD and an NBD. In the single-site model (A), each half transporter may contribute a partial drug-binding site (represented by a circle) to a general drug-binding region that contains multiple drug-recognition sites at the interface between the half transporters. Both half transporters cycle together between the E1 and E2 conformations, and hence, both partial drug-binding sites move simultaneously from one membrane surface to the other. In the two-site model (B), the partial drug-binding sites move in a sequential or alternating fashion from one membrane surface to the other. Hence, both half transporters cycle separately between the E1 and E2 conformations.

The binding site (or region which may recognize multiple transported species) cycles between E1 and E2 conformations in the single-site model. The two-site model is described by a mixed E1–E2 to E2–E1 shift in the respective substrate-binding sites in each half of the transporter. Both models conform to the four fundamental principles of any transport mechanism and were born out of investigations into the characteristics of substrate binding to ABC transporters described in previous sections. Although, at first sight, the alternating single- and two-site models may seem different, they are not necessarily mutually exclusive. Most ABC transporters have a common domain organization with two half transporters, each consisting of a membrane domain and an NBD. If ABC transporters contained a single, general drug-binding region at the interface between both half transporters, each half transporter might contribute a partial drug-binding site to this drug-binding region. For example, it is conceivable that in MRP1 one half transporter contains the glutathione-binding site whereas the other half transporter contains the binding site for hydrophobic drugs. The presence of the two binding sites in MRP1 in close proximity to each other at the interface between the two half transporters would also allow the binding of glutathione–drug conjugates. In the case of TAP, peptides may bind with their N-terminal end to a peptide-binding site in one half transporter, and with their C-terminal end to the peptide-binding site in the other half transporter, giving rise to single-site binding kinetics (Neumann and Tampe, 1999). However, in other ABC transporters such as LmrA and the binding protein-dependent transporters BusAB (Obis *et al.*, 1999) and OpuA (van der Heide and Poolman, 2000), the two partial substrate-binding sites may not be simultaneously accessible, e.g. owing the presence of these sites on opposite sides of the membrane (Jones and George, 2000; van Veen, 2001). Hence, the differences between single- and two-site models for ABC transporters may relate to differences in the localization of the substrate-binding sites in these proteins. Clearly, our understanding of transport mechanisms requires molecular details of the regions involved in the interaction with transported species.

Concluding Remarks

Although many details of substrate-binding sites in ABC transporters have been discovered in the past years, much remains to be learned about the molecular basis of substrate specificity and transport by these proteins. ABC transporters may operate by a common mechanism or they may have different mechanisms that relate to the nature of the transported substrate, the direction of transport, and the phylogenetic origin of the transporters. It is not so easy to predict which of these two hypotheses will eventually prevail.

Acknowledgments

Work in the authors' laboratories is funded by Cancer Research UK (formely Cancer Research Campaign). In addition, the van Veen laboratory receives funding from the Association for International Cancer Research (AICR), the Royal Society, Biotechnology and Biological Sciences Research Council (BBSRC), Medical Research Council and Molecular Devices Ltd. We would also like to thank Catherine Martin for helpful discussions and critical reading of the manuscript.

REFERENCES

Ambudkar, S.V., Dey, S., Hrycyna, C.A., Ramachandra, M., Pastan, I. and Gottesman, M.M. (1999) Biochemical, cellular, and pharmacological aspects of the multidrug transporter. Annu. Rev. Pharmacol. Toxicol. 39, 361–398.

Anderson, M.P., Gregory, R.J., Thompson, S., Souza, D.W., Paul, S., Mulligan, R.C., Smith, A.E. and Welsh, M.J. (1991) Demonstration that CFTR is a chloride channel by alteration of its anion selectivity. Science 253, 202–205.

Arora, S., Lapinski, P.E. and Raghavan, M. (2001) Use of chimeric proteins to investigate the role of transporter associated with antigen processing (TAP) structural domains in peptide binding and translocation. Proc. Natl Acad. Sci. USA 98, 7241–7246.

Ayesh, S., Shao, Y.-M. and Stein, W.D. (1996) Co-operative, competitive and non-competitive interactions between modulators of P-glycoprotein. Biochim. Biophys. Acta 1316, 8–18.

Babenko, A.P., Gonzalez, G. and Bryan, J. (2000) Pharmaco-topology of sulfonylurea receptors. Separate domains of the regulatory subunits of K(ATP) channel isoforms are required for selective interaction with K(+) channel openers. J. Biol. Chem. 275, 717–720.

Bolhuis, H., van Veen, H.W., Molenaar, D., Poolman, B., Driessen, A.J. and Konings, W.N. (1996) Multidrug resistance in *Lactococcus lactis*: evidence for ATP-dependent drug extrusion from the inner leaflet of the cytoplasmic membrane. EMBO J. 15, 4239–4245.

Bruggemann, E.P., Currier, S.J., Gottesman, M.M. and Pastan, I. (1992) Characterization of the azidopine and vinblastine binding site of P-glycoprotein. J. Biol. Chem. 267, 21020–21026.

Busche, R., Tummler, B., Cano-Gauci, D.F. and Riordan, J.R. (1989) Equilibrium, kinetic and photoaffinity labelling studies of daunomycin binding to P-glycoprotein containing membranes of multidrug resistant Chinese hamster ovary cells. Eur. J. Biochem. 183, 189–197.

Buschman, E. and Gros, P. (1991) Functional analysis of chimeric genes obtained by exchanging homologous domains of the mouse mdr1 and mdr2 genes. Mol. Cell. Biol. 11, 595–603.

Callaghan, R., Berridge, G., Ferry, D.R. and Higgins, C.F. (1997) The functional purification of P-glycoprotein is dependent on maintenance of a lipid–protein interface. Biochim. Biophys. Acta 1328, 109–124.

Chang, X.-B., Hou, Y.-X. and Riordan, J.R. (1997) ATPase activity of purified multidrug resistance-associated protein. J. Biol. Chem. 272, 30962–30968.

Chiba, P., Ecker, G., Schmid, D., Drach, J., Tell, B., Goldenberg, S. and Gekeler, V. (1996) Structural requirements for activity of propafenone-type modulators in P-glycoprotein mediated multidrug resistance. Mol. Pharmacol. 49, 1122–1130.

Choi, K., Chen, C.-J., Kriegler, M. and Roninson, I.B. (1988) An altered pattern of cross-resistance in multidrug resistant human cells results from spontaneous mutations in the *mdr*1 (P-glycoprotein) gene. Cell 53, 519–529.

Cole, S.P.C. and Deeley, R.G. (1998) Multidrug resistance mediated by the ATP-binding cassette transporter protein MRP. BioEssays 20, 931–940.

Conseil, G., Baubichon-Cortay, H., Dayan, G., Jault, J.M., Barron, D. and Di Pietro, A. (1998) Flavonoids: a class of modulators with bifunctional interactions at vicinal ATP- and steroid-binding sites on mouse P-glycoprotein. Proc. Natl Acad. Sci. USA 95, 9831–9836.

Dassa, E. (1993) Sequence–function relationships in MalG, an inner membrane protein from the maltose transport system in *Escherichia coli*. Mol. Microbiol. 7, 39–47.

Davidson, A.L. and Nikaido, H. (1991) Purification and characterization of the membrane-associated components of the maltose transport system from *Escherichia coli*. J. Biol. Chem. 266, 8946–8951.

Davidson, A.L., Shuman, H.A. and Nikaido, H. (1992) Mechanism of maltose transport in *Escherichia coli*: transmembrane signaling by periplasmic binding proteins. Proc. Natl Acad. Sci. USA 89, 2360–2364.

Dayan, G., Baubichon-Cortay, H., Jault, J.-M., Cortay, J.-C., Deleage, G. and Di Pietro, A. (1996) Recombinant N-terminal nucleotide-binding domain from mouse P-glycoprotein. J. Biol. Chem. 271, 11652–11658.

Demmer, A., Thole, H., Kubesch, P., Brandt, T., Raida, M., Fislage, R. and Tummler, B. (1997) Localization of the iodomycin binding site in hamster P-glycoprotein. J. Biol. Chem. 272, 20913–20919.

Dey, S., Ramachandra, M., Pastan, I., Gottesman, M.M. and Ambudkar, S.V. (1997) Evidence for two nonidentical drug-interaction sites in the human P-glycoprotein. Proc. Natl Acad. Sci. USA 94, 10594–10599.

Doppenschmitt, S., Langguth, P., Regardh, C.G., Andersson, T.B., Hilgendorf, C. and Spahn-Langguth, H. (1999) Characterization of binding properties to human P-glycoprotein: development of a [3H]verapamil radioligand-binding assay. J. Pharmacol. Exp. Ther. 288, 348–357.

Dougherty, D.A. (1996) Cation-pi interactions in chemistry and biology: a new view of benzene, Phe, Tyr, and Trp. Science 271, 163–168.

Druley, T.E., Stein, W.D. and Roninson, I.B. (2001) Analysis of MDR1 P-glycoprotein conformational changes in permeabilized cells using differential immunoreactivity. Biochemistry 40, 4312–4322.

Edgar, R. and Bibi, E. (1999) A single membrane-embedded negative charge is critical for recognizing positively charged drugs by the *Escherichia coli* multidrug resistance protein MdfA. EMBO J. 18, 822–832.

Evers, R., de Haas, M., Sparidans, R., Beijnen, J., Wielinga, P.R., Lankelma, J. and Borst, P. (2000) Vinblastine and sulfinpyrazone export by the multidrug resistance protein MRP2 is associated with glutathione export. Br. J. Cancer 83, 375–383.

Ferry, D.R., Russell, M.A. and Cullen, M.H. (1992) P-glycoprotein possesses a 1,4-dihydropyridine selective drug acceptor site which is allosterically coupled to a vinca alkaloid selective binding site. Biochem. Biophys. Res. Commun. 188, 440–445.

Ferry, D., Boer, R., Callaghan, R. and Ulrich, W.R. (2000) Localization of the 1,4-dihydropyridine drug acceptor of P-glycoprotein to a cytoplasmic domain using a permanently charged derivative N-methyl dexniguldipine. Int. J. Clin. Pharmacol. Ther. 38, 130–140.

Fersht, A. (1998) Structure and Mechanism in Protein Science: A Guide to Enzyme Catalysis and Protein Folding. New York: W.H. Freeman.

Ferte, J. (2000) Analysis of the tangled relationships between P-glycoprotein-mediated multidrug resistance and the lipid phase of the cell membrane. Eur. J. Biochem. 267, 277–294.

Ford, J.M., Bruggemann, E.P., Pastan, I., Gottesman, M.M. and Hait, W.N. (1990) Cellular and biochemical characterization of thioxanthenes for reversal of multidrug resistance in human and murine cell lines. Cancer Res. 50, 1748–1756.

Gadsby, D.C. and Nairn, A.C. (1994) Regulation of CFTR channel gating. Trends Biochem. Sci. 19, 513–518.

Gorbulev, S., Abele, R. and Tampe, R. (2001) Allosteric crosstalk between peptide-binding, transport, and ATP hydrolysis of the ABC transporter TAP. Proc. Natl Acad. Sci. 98, 3732–3737.

Gottesman, M.M. and Pastan, I. (1993) Biochemistry of multidrug resistance by the multidrug transporter. Annu. Rev. Biochem. 62, 385–427.

Gottesman, M.M., Hrycyna, C.A., Schoenlein, P.V., Germann, U.A. and Pastan, I. (1995) Genetic analysis of the multidrug transporter. Annu. Rev. Genet. 29, 607–649.

Greenberger, L.M. (1993) Major photoaffinity labeling sites for iodoaryl azidoprazosin in P-glycoprotein are within or immediately C-terminal to transmembrane domains 6 and 12. J. Biol. Chem. 268, 11417–11425.

Grimard, V., Vigano, C., Margolles, A., Wattiez, R., van Veen, H.W., Konings, W.N., Ruysschaert, J.M. and Goormaghtigh, E. (2001) Structure and dynamics of the membrane-embedded domain of LmrA investigated by coupling polarized ATR-FTIR spectroscopy and (1)H/(2)H exchange. Biochemistry 40, 11876–11886.

Hafkemeyer, P., Dey, S., Ambudkar, S.V., Hrycyna, C.A., Pastan, I. and Gottesman, M.M. (1998) Contribution to substrate specificity and transport of nonconserved residues in transmembrane domain 12 of human P-glycoprotein. Biochemistry 37, 16400–16409.

Higgins, C.F. (1992) ABC transporters: from microorganisms to man. Annu. Rev. Cell Biol. 8, 67–113.

Holland, I.B. and Blight, M.A. (1999) ABC-ATPases, adaptable energy generators fuelling transmembrane movement of a variety of molecules in organisms from bacteria to humans. J. Mol. Biol. 293, 381–399.

Homolya, L., Hollo, Z., Germann, U.A., Pastan, I., Gottesman, M.M. and Sarkadi, B. (1993) Fluorescent cellular indicators are extruded by the multidrug resistance protein. J. Biol. Chem. 268, 21493–21496.

Horton, J.K., Thimmaiah, K.N., Haewood, F.C., Kuttesch, J.F. and Houghton, P.J. (1993)

Pharmacological characterization of N-substituted phenoxazines directed toward reversing vinca alkaloid resistance in multidrug resistant cancer cells. Mol. Pharmacol. 44, 552–559.

Isenberg, B., Thole, H., Tummler, B. and Demmer, A. (2001) Identification and localization of three photobinding sites of iodoarylazidoprazosin in hamster P-glycoprotein. Eur. J. Biochem. 268, 2629–2634.

Ito, K., Suzuki, H. and Sugiyama, Y. (2001) Charged amino acids in the transmembrane domains are involved in the determination of the substrate specificity of rat Mrp2. Mol. Pharmacol. 59, 1077–1085.

Jencks, W.P. (1980) The utilization of binding energy in coupled vectorial processes. Adv. Enzymol. 51, 75–106.

Jones, P.M. and George, A.M. (2000) Symmetry and structure in P-glycoprotein and ABC transporters what goes around comes around. Eur. J. Biochem. 267, 5298–5305.

Kajiji, S., Talbot, F., Grizzuti, K., Van Dyke-Phillips, V., Agresti, M., Safa, A.R. and Gros, P. (1993) Functional analysis of P-glycoprotein mutants identifies predicted transmembrane domain 11 as a putative drug binding site. Biochemistry 32, 4185–4194.

Kern, A., Felfoldi, F., Sarkadi, B. and Varadi, A. (2000) Expression and characterization of the N- and C-terminal ATP-binding domains of MRP1. Biochem. Biophys. Res. Commun. 273, 913–919.

Kerr, K.M., Sauna, Z.E. and Ambudkar, S.V. (2001) Correlation between steady-state ATP hydrolysis and vanadate-induced ADP trapping in human P-glycoprotein. J. Biol. Chem. 276, 8657–8664.

Kioka, N., Tsubota, J., Kakehi, Y., Komano, T., Gottesman, M.M., Pastan, I. and Ueka, K. (1989) P-glycoprotein gene (MDR1) cDNA from human adrenal: normal P-glycoprotein carries Gly185 with an altered pattern of multidrug resistance. Biochem. Biophys. Res. Commun. 162, 224–231.

Ko, Y.H. and Pedersen, P.L. (1995) The first nucleotide binding fold of the cystic fibrosis transmembrane conductance regulator can function as an active ATPase. J. Biol. Chem. 270, 22093–22096.

Koopmann, J.O., Post, M., Neefjes, J.J., Hammerling, G.J. and Momburg, F. (1996) Translocation of long peptides by transporters associated with antigen processing (TAP). Eur. J. Immunol. 26, 1720–1728.

Koshland, D.E. (1987) Evolution of catalytic function. Cold Spring Harbor Symp. Quant. Biol. 52, 1–7.

Koster, W. and Bohm, B. (1992) Point mutations in 2 conserved glycine residues within the integral membrane protein FhuB affect iron(III) hydroxamate transport. Mol. Gen. Genet. 232, 399–407.

Krupka, R.M. (1993) Coupling mechanisms in ATP-driven pumps. Biochim. Biophys. Acta 1183, 114–122.

Krupka, R.M. (1999) Uncoupled active transport mechanisms accounting for low selectivity in multidrug carriers: P-glycoprotein and SMR transporters. J. Membr. Biol. 172, 129–143.

Kwan, T. and Gros, P. (1998) Mutational analysis of the P-glycoprotein first intracellular loop and flanking transmembrane domains. Biochemistry 37, 3337–3350.

Kwan, T., Loughrey, H., Brault, M., Gruenheid, S., Urbatsch, I.L., Senior, A.E. and Gros, P. (2000) Functional analysis of a tryptophan-less P-glycoprotein: a tool for tryptophan insertion and fluorescence spectroscopy. Mol. Pharmacol. 58, 37–47.

Lankat-Buttgereit, B. and Tampe, R. (1999) The transporter associated with antigen processing TAP: structure and function. FEBS Lett. 464, 108–112.

Lawrence, D.S., Copper, J.E. and Smith, C.D. (2001) Structure–activity studies of substituted quinoxalinones as multiple-drug-resistance antagonists. J. Med. Chem. 44, 594–601.

Ligtenberg, M.J., Kemp, S., Sarde, C.O., van Geel, B.M., Kleijer, W.J., Barth, P.G., Mandel, J.L., van Oost, B.A. and Bolhuis, P.A. (1995) Spectrum of mutations in the gene encoding the adrenoleukodystrophy protein. Am. J. Hum. Genet. 56, 44–50.

Litman, T., Druley, T.E., Stein, W.D. and Bates, S.E. (2001) From MDR to MXR: new understandings of multidrug resistance systems, their properties and clinical significance. Cell. Mol. Life Sci. 58, 1–29.

Liu, C.E., Liu, P.-Q. and Ames, G.F.-L. (1997) Characterization of the adenosine triphosphatase activity of the periplasmic histidine permease, a traffic ATPase (ABC transporter). J. Biol. Chem. 272, 21883–21891.

Liu, C.E., Liu, P.-Q., Wolf, A., Lin, E. and Ames, G.F.-L. (1999) Both lobes of the soluble receptor of the histidine permease, an ABC transporter (traffic ATPase), interact with the membrane-bound complex. J. Biol. Chem. 274, 739–747.

Liu, P.-Q. and Ames, G.F.-L. (1998) *In vitro* disassembly and reassembly of an ABC transporter, the histidine permease. Proc. Natl Acad. Sci. USA 95, 3495–3500.

Liu, P.-Q., Liu, C.E. and Ames, G.F.-L. (1999) Modulation of ATPase activity by physical disengagement of the ATP-binding domains of an ABC transporter, the histidine permease. J. Biol. Chem. 274, 18310–18318.

Liu, R. and Sharom, F.J. (1996) Site-directed fluorescence labeling of P-glycoprotein on cysteine residues in the nucleotide binding domains. Biochemistry 35, 11865–11873.

Liu, R., Siemiarczuk, A. and Sharom, F.J. (2000) Intrinsic fluorescence of the P-glycoprotein multidrug transporter: sensitivity of tryptophan residues to binding of drugs and nucleotides. Biochemistry 39, 14927–14938.

Loe, D.W., Deeley, R.G. and Cole, S.P. (2000a) Verapamil stimulates glutathione transport by the 190-kDa multidrug resistance protein 1 (MRP1). J. Pharmacol. Exp. Ther. 293, 530–538.

Loe, D.W., Oleschuk, C.J., Deeley, R.G. and Cole, S.P. (2000b) Structure–activity studies of verapamil analogs that modulate transport of leukotriene C(4) and reduced glutathione by multidrug resistance protein MRP1. Biochem. Biophys. Res. Commun. 275, 795–803.

Loo, T.W. and Clarke, D.M. (1994) Functional consequences of glycine mutations in the predicted cytoplasmic loops of P-glycoprotein. J. Biol. Chem. 269, 7243–7248.

Loo, T.W. and Clarke, D.M. (1995) Rapid purification of human P-glycoprotein mutants expressed transiently in HEK 293 cells by nickel-chelate chromatography and characterization of their drug stimulated ATPase activities. J. Biol. Chem. 270, 21449–21452.

Loo, T.W. and Clarke, D.M. (1999a) Identification of residues in the drug binding domain of human P-glycoprotein. J. Biol. Chem. 274, 35388–35392.

Loo, T.W. and Clarke, D.M. (1999b) The transmembrane domains of the human multidrug resistance P-glycoprotein are sufficient to mediate drug binding and trafficking to the cell surface. J. Biol. Chem. 274, 24759–24765.

Loo, T.W. and Clarke, D.M. (2000) Identification of residues within the drug-binding domain of the human multidrug resistance P-glycoprotein by cysteine-scanning mutagenesis and reaction with dibromobimane. J. Biol. Chem. 275, 39272–39278.

Loo, T.W. and Clarke, D.M. (2001) Defining the drug-binding site in the human multidrug resistance P-glycoprotein using a methanethiosulfonate analog of verapamil, MTS-verapamil. J. Biol. Chem. 276, 14972–14979.

Lu, P., Liu, R. and Sharom, F.J. (2001) Drug transport by reconstituted P-glycoprotein in proteoliposomes. Effect of substrates and modulators, and dependence on bilayer phase state. Eur. J. Biochem. 268, 1687–1697.

Lum, B., Gosland, M.P., Kaubisch, S. and Sikic, B.I. (1993) Molecular targets in oncology: implications of the multidrug resistance gene. Pharmacotherapy 13, 88–109.

Malkhandi, J., Ferry, D.R., Boer, R., Gekeler, V., Ise, W. and Kerr, D.J. (1994) Dexniguldipine-HCl is a potent allosteric inhibitor [3H]vinblastine binding to P-glycoprotein of CCRF ADR5000 cells. Eur. J. Pharmacol. 288, 105–114.

Manciu, L., Chang, X.-B., Riordan, J.R., Buyse, F. and Ruysscheart, J.M. (2001) Nucleotide-induced changes in the human multidrug resistance protein MRP1 are related to the capacity of chemotherapeutic drugs to accumulate or not in resistant cells. FEBS Lett. 493, 31–35.

Mao, Q., Leslie, E.M., Deeley, R.G. and Cole, S.P.C. (1999) ATPase activity of purified and reconstituted multidrug resistance protein MRP1 from drug-selected H69AR cells. Biochim. Biophys. Acta 1461, 69–82.

Marger, M.D. and Saier, M.H. (1993) A major superfamily of transmembrane facilitators that catalyse uniport, symport and antiport. Trends Biochem. Sci. 18, 13–20.

Margolles, A., Putman, M., van Veen, H.W. and Konings, W.N. (1999) The purified and functionally reconstituted multidrug transporter LmrA of *Lactococcus lactis* mediates the transbilayer movement of specific fluorescent phospholipids. Biochemistry 38, 16298–16306.

Markham, P.N., LaGuidice, J. and Neyfakh, A.A. (1997) Broad ligand specificity of the transcriptional regulator of the *Bacillus subtilis* multidrug transporter BmrR. Biochem. Biophys. Res. Commun. 239, 269–272.

Martin, C., Berridge, G., Higgins, C.F. and Callaghan, R. (1997) The multi-drug resistance reversal agent SR33557 and modulation of vinca alkaloid binding to P-glycoprotein by an allosteric interaction. Br. J. Pharmacol. 122, 765–771.

Martin, C., Berridge, G., Mistry, P., Higgins, C., Charlton, P. and Callaghan, R. (1999)

The molecular interaction of the high affinity reversal agent XR9576 with P-glycoprotein. Br. J. Pharmacol. 128, 403–411.

Martin, C., Berridge, G., Higgins, C.F., Mistry, P., Charlton, P. and Callaghan, R. (2000a) Communication between multiple drug binding sites on P-glycoprotein. Mol. Pharmacol. 58, 624–632.

Martin, C., Berridge, G., Mistry, P., Higgins, C., Charlton, P. and Callaghan, R. (2000b) Drug binding sites on P-glycoprotein are altered by ATP binding prior to nucleotide hydrolysis. Biochemistry 39, 11901–11906.

Martin, C., Higgins, C.F. and Callaghan, R. (2001) The vinblastine site adopts high- and low-affinity conformations during a transport cycle of P-glycoprotein. Biochemistry 40, 15733–15742.

Mechetner, E.B., Schott, B., Morse, B.S., Stein, W.D., Druley, T., Davis, K.A., Tsuruo, T. and Roninson, I.B. (1997) P-glycoprotein function involves conformational transitions detectable by differential immunoreactivity. Proc. Natl Acad. Sci. USA 94, 12908–12913.

Momburg, F., Armandola, E.A., Post, M. and Hammerling, G.J. (1996) Residues in TAP2 peptide transporters controlling substrate specificity. J. Immunol. 156, 1756–1763.

Morbach, S., Tebbe, S. and Schneider, E. (1993) The ATP-binding cassette (ABC) transporter for maltose/maltodextrins of *Salmonella typhimurium*. J. Biol. Chem. 268, 18617–18621.

Mourez, M., Hofnung, M. and Dassa, E. (1997) Subunit interactions in ABC transporters: a conserved sequence of periplasmic permeases defines an important site of interaction with the ATPase subunits. EMBO J. 16, 3066–3077.

Mowbray, S.L. and Cole, L.B. (1992) 1.7A X-ray structure of the periplasmic ribose receptor from *Escherichia coli*. J. Mol. Biol. 225, 155–175.

Muller, M., Bakos, E., Welker, E., Varadi, A., Germann, U.A., Gottesman, M.M., Morse, B.S., Roninson, I.B. and Sarkadi, B. (1996) Altered drug-stimulated ATPase activity in mutants of the human multidrug resistance protein. J. Biol. Chem. 271, 1877–1883.

Muth, T.R. and Schuldiner, S. (2000) A membrane-embedded glutamate is required for ligand binding to the multidrug transporter EmrE. EMBO J. 19, 234–240.

Neefjes, J.J., Momburg, F. and Hammerling, G.J. (1993) Selective and ATP-dependent translocation of peptides by the MHC-encoded transporter. Science 261, 769–771.

Neumann, L. and Tampe, R. (1999) Kinetic analysis of peptide binding to the TAP transport complex: evidence for structural rearrangements induced by substrate binding. J. Mol. Biol. 294, 1203–1213.

Neyfakh, A.A., Bidnenko, V.E. and Chen, L.B. (1991) Efflux-mediated multidrug resistance in *Bacillus subtilis*: similarities and dissimilarities with the mammalian system. Proc. Natl Acad. Sci. USA 88, 4781–4785.

Nikaido, H. (1994) Prevention of drug access to bacterial targets: permeability barriers and active efflux. Science 264, 382–388.

Nikaido, K., Liu, P.-Q. and Ames, G.F.-L. (1997) Purification and characterization of HisP, the ATP-binding subunit of a traffic ATPase (ABC transporter), the histidine permease of *Salmonella typhimurium*. J. Biol. Chem. 272, 27745–27752.

Obis, D., Guillot, A., Gripon, J.C., Renault, P., Bolotin, A. and Mistou, M.Y. (1999) Genetic and biochemical characterization of a high-affinity betaine uptake system (BusA) in *Lactococcus lactis* reveals a new functional organization within bacterial ABC transporters. J. Bacteriol. 181, 6238–6246.

Orlowski, S., Mir, L.M., Belehradek, J. and Garrigos, M. (1996) Effects of steroids and verapamil on P-glycoprotein ATPase activity: progesterone, desoxycorticosterone and verapamil are mutually non-exclusive modulators. Biochem. J. 317, 515–522.

Pascaud, C., Garrigos, M. and Orlowski, S. (1998) Multidrug resistance transporter P-glycoprotein has distinct but interacting binding sites for cytotoxic drugs and reversing agents. Biochem. J. 333, 351–358.

Paulsen, I.T., Skurray, R.A., Tam, R., Saier, M.H., Turner, R.J., Weiner, J.H., Goldberg, E.B. and Grinius, L.L. (1996) The SMR family. A novel family of multidrug efflux proteins involved with the efflux of lipophilic drugs. Mol. Microbiol. 19, 1167–1175.

Payne, G.M., Spudich, E.N. and Ames, G.F. (1985) A mutational hot-spot in the hisM gene of the histidine transport operon in *Salmonella typhimurium* is due to deletion of repeated sequences and results in an altered specificity of transport. Mol. Gen. Genet. 200, 493–496.

Pearce, H.L., Safa, A.R., Bach, N.J., Winter, M.A., Cirtain, M.C. and Beck, W.T. (1989) Essential features of the P-glycoprotein pharmacophore as defined by a series of reserpine analogs that modulate multidrug resistance. Proc. Natl Acad. Sci. USA 86, 5128–5132.

Petronilli, V. and Ames, G.F. (1991) Binding protein-independent histidine permease mutants. Uncoupling of ATP hydrolysis from transmembrane signaling. J. Biol. Chem. 266, 16293–16296.

Putman, M., van Veen, H.W. and Konings, W.N. (2000) Molecular properties of bacterial multidrug transporters. Microbiol. Mol. Biol. Rev. 64, 672–693.

Ramachandra, M., Ambudkar, S.V., Chen, D., Hrycyna, C.A., Dey, S., Gottesman, M.M. and Pastan, I. (1998) Human P-glycoprotein exhibits reduced affinity for substrates during a catalytic transition state. Biochemistry 37, 5010–5019.

Raviv, Y., Pollard, H.B., Bruggemann, E.P., Pastan, I. and Gottesman, M.M. (1990) Photosensitized labeling of a functional multidrug transporter in living drug-resistant tumor cells. J. Biol. Chem. 265, 3975–3980.

Reyes, M., Treptow, N.A. and Shuman, H.A. (1986) Transport of p-nitrophenyl-alpha-maltoside by the maltose transport system of *Escherichia coli* and its subsequent hydrolysis by a cytoplasmic alpha-maltosidase. J. Bacteriol. 165, 918–922.

Rogers, B., Decottignies, A., Kolaczkowski, M., Carvajal, E., Balzi, E. and Goffeau, A. (2001) The pleitropic drug ABC transporters from *Saccharomyces cerevisiae*. J. Mol. Microbiol. Biotechnol. 3, 207–214.

Ruetz, S. and Gros, P. (1994) Phosphatidylcholine translocase: a physiological role for the mdr2 gene. Cell 77, 1071–1081.

Safa, A.R., Stern, R.K., Choi, K., Agresti, M., Tamai, I., Mehta, N.D. and Roninson, I.B. (1990) Molecular basis of preferential resistance to colchicine in multidrug resistant human cells conferred by Gly185–Val185 substitution in P-glycoprotein. Proc. Natl Acad. Sci. USA 87, 7225–7229.

Saier, M.H., Tam, R., Reizer, A. and Reizer, J. (1994) Two novel families of bacterial membrane proteins concerned with nodulation, cell division and transport. Mol. Microbiol. 11, 841–847.

Sauna, Z.E. and Ambudkar, S.V. (2000) Evidence for a requirement for ATP hydrolysis at two distinct steps during a single turnover of the catalytic cycle of human P-glycoprotein. Proc. Natl Acad. Sci. USA 97, 2515–2520.

Sauna, Z.E. and Ambudkar, S.V. (2001) Characterization of the catalytic cycle of ATP hydrolysis by human P-glycoprotein. The two ATP hydrolysis events in a single catalytic cycle are kinetically similar but affect different functional outcomes. J. Biol. Chem. 276, 11653–11661.

Saurin, W., Koster, W. and Dassa, E. (1994) Bacterial binding protein-dependent permeases: characterization of distinctive signatures for functionally related cytoplasmic membrane proteins. Mol. Microbiol. 12, 993–1004.

Saveanu, L., Daniel, S. and van Endert, P.M. (2001) Distinct functions of the ATP-binding cassettes of the transporters associated with antigen processing: a mutational analysis of Walker A and B sequences. J. Biol. Chem. 276, 22107–22113.

Seelig, A. (1998) A general pattern for substrate recognition by P-glycoprotein. Eur. J. Biochem. 251, 252–261.

Seelig, A. and Landwojtowicz, E. (2000) Structure–activity relationship of P-glycoprotein substrates and modifiers. Eur. J. Pharm. Sci. 12, 31–40.

Seelig, A., Blatter, X.L. and Wohnsland, F. (2000) Substrate recognition by P-glycoprotein and the multidrug resistance-associated protein MRP1: a comparison. Int. J. Clin. Pharmacol. Ther. 38, 111–121.

Senior, A.E. (1998) Catalytic mechanism of P-glycoprotein. Acta. Physiol. Scand. Suppl. 643, 213–218.

Senior, A.E. and Bhagat, S. (1998) P-glycoprotein shows strong catalytic cooperativity between the two nucleotide sites. Biochemistry 37, 831–836.

Shani, N., Watkins, P.A. and Valle, D. (1995) PXA1, a possible *Saccharomyces cerevisiae* ortholog of the human adrenoleukodystrophy gene. Proc. Natl Acad. Sci. USA 92, 6012–6016.

Shani, N., Sapag, A. and Valle, D. (1996) Characterization and analysis of conserved motifs in a peroxisomal ATP-binding cassette transporter. J. Biol. Chem. 271, 8725–8730.

Shapiro, A.B. and Ling, V. (1994) ATPase activity of purified and reconstituted P-glycoprotein from Chinese Hamster Ovary cells. J. Biol. Chem. 269, 3745–3754.

Shapiro, A.B. and Ling, V. (1997) Positively cooperative sites for drug transport by P-glycoprotein with distinct drug specificities. Eur. J. Biochem. 250, 130–137.

Shapiro, A.B., Fox, K., Lam, P. and Ling, V. (1999) Stimulation of P-glycoprotein-mediated drug transport by prazosin and progesterone. Evidence for a third drug-binding site. Eur. J. Biochem. 259, 841–850.

Sharom, F.J., Yu, X., DiDiodato, G. and Chu, J.W. (1996) Synthetic hydrophobic peptides are substrates for P-glycoprotein and stimulate drug transport. Biochem. J. 320, 421–428.

Sharom, F.J., Yu, X., Lu, P., Liu, R., Chu, J.W., Szabo, K., *et al.* (1999) Interaction of the P-glycoprotein multidrug transporter (MDR1) with high affinity peptide chemosensitizers in isolated membranes, reconstituted systems, and intact cells. Biochem. Pharmacol. 58, 571–586.

Sheps, J.A., Cheung, I. and Ling, V. (1995) Hemolysin transport in *Escherichia coli*. Point mutants in HlyB compensate for a deletion in the predicted amphiphilic helix region of the HlyA signal. J. Biol. Chem. 270, 14829–14834.

Sikic, B. (1997) Pharmacologic approaches to reversing multidrug resistance. Sem. Hematol. 34, 40–47.

Smit, J.J.M., Schinkel, A.H., Oude Elferink, R.P.J., Groen, A.K., Wagenaar, E., van Deemter, L., *et al.* (1993) Homozygous disruption of the murine MDR2 P-glycoprotein gene leads to a complete absence of phospholipid from bile and to liver disease. Cell 75, 451–462.

Sonveaux, N., Shapiro, A.B., Goormaghtigh, E., Ling, V. and Ruysschaert, J.M. (1996) Secondary and tertiary structure changes of reconstituted P-glycoprotein. A Fourier transform attenuated total reflection infrared spectroscopy analysis. J. Biol. Chem. 271, 24617–24624.

Sonveaux, N., Vigano, C., Shapiro, A.B., Ling, V. and Ruysschaert, J.M. (1999) Ligand-mediated tertiary structure changes of reconstituted P-glycoprotein. J. Biol. Chem. 274, 17649–17654.

Tame, R.H., Murshudov, G.N., Dodson, E.J., Neil, T.K., Dodson, G.G., Higgins, C.F. and Wilkinson, A.J. (1994) The structural basis of sequence-independent peptide binding by the OppA protein. Science 264, 1578–1581.

Tanford, C. (1983a) Mechanism of free energy coupling in active transport. Annu. Rev. Biochem. 52, 379–409.

Tanford, C. (1983b) Translocation pathway in the catalysis of active transport. Proc. Natl Acad. Sci. USA 80, 3701–3705.

Tang-Wai, D.F., Brossi, A., Arnold, L.D. and Gros, P. (1993) The nitrogen of the acetamido group of colchicine modulates P-glycoprotein-mediated multidrug resistance. Biochemistry 32, 6470–6476.

Toffoli, G., Simone, F., Corona, G., Rashack, M., Cappelletto, B., Gigante, M. and Boiocchi, M. (1995) Structure–activity relationship of verapamil analogs and reversal of multidrug resistance. Biochem. Pharmacol. 50, 1245–1255.

Treptow, N.A. and Shuman, H.A. (1985) Genetic evidence for substrate and periplasmic-binding-protein recognition by the MalF and MalG proteins, cytoplasmic membrane components of the *Escherichia coli* maltose transport system. J. Bacteriol. 163, 654–660.

Uebel, S. and Tampe, R. (1999) Specificity of the proteasome and the TAP transporter. Curr. Opin. Immunol. 11, 203–208.

Uebel, S., Kraas, W., Kienle, S., Weismuller, K.-H., Jung, G. and Tampe, R. (1997) Recognition principle of the TAP transporter disclosed by combinatorial peptide libraries. Proc. Natl Acad. Sci. USA 94, 8976–8981.

Ueda, K., Taguchi, Y. and Morishima, M. (1997) How does P-glycoprotein recognize its substrates? Semin. Cancer Biol. 8, 151–159.

Urbatsch, I.L. and Senior, A.E. (1995) Effects of lipids on ATPase activity of purified Chinese Hamster P-glycoprotein. Arch. Biochem. Biophys. 316, 135–140.

Urbatsch, I.L., Sankaran, B., Weber, J. and Senior, A.E. (1995) P-glycoprotein is stably inhibited by vanadate induced trapping of nucleotide at a single catalytic site. J. Biol. Chem. 270, 19383–19390.

van der Heide, T. and Poolman, B. (2000) Osmoregulated ABC-transport system of *Lactococcus lactis* senses water stress via changes in the physical state of the membrane. Proc. Natl Acad. Sci. USA 97, 7102–7106.

van Endert, P.M. (1999) Role of nucleotides and peptide substrate for stability and functional state of the human ABC family transporters associated with antigen presentation. J. Biol. Chem. 274, 14632–14638.

van Veen, H.W. (2001) Towards the molecular mechanism of prokaryotic and eukaryotic multidrug transporters. Semin. Cell Dev. Biol. 12, 239–245.

van Veen, H.W., Venema, K., Bolhuis, H., Oussenko, I., Kok, J., Poolman, B., Driessen, A.J. and Konings, W.N. (1996) Multidrug resistance mediated by a bacterial homolog of the human multidrug transporter MDR1. Proc. Natl Acad. Sci. USA 93, 10668–10672.

van Veen, H.W., Callaghan, R., Soceneantu, L., Sardini, A., Konings, W.N. and Higgins, C.F. (1998) A bacterial antibiotic-resistance gene that complements the human multidrug resistance P-glycoprotein gene. Nature 391, 291–295.

van Veen, H.W., Margolles, A., Muller, M., Higgins, C.F. and Konings, W.N. (2000) The homodimeric ATP-binding cassette transporter LmrA mediates multidrug transport by an alternating (two-cylinder engine) mechanism. EMBO J. 19, 2503–2514.

Van Winkle, L.J. (1999) Biomembrane Transport. San Diego: Academic Press.

Vazquez-Laslop, N., Markham, P.N. and Neyfakh, A.A. (1999) Mechanism of ligand recognition by BmrR, the multidrug-responding transcriptional regulator: mutational analysis of the ligand-binding site. Biochemistry 38, 16925–16931.

Vigano, C., Margolles, A., van Veen, H.W., Konings, W.N. and Ruysscheart, J.M. (2000) Secondary and tertiary structure changes of reconstituted LmrA induced by nucleotide binding or hydrolysis. J. Biol. Chem. 275, 10962–10967.

Wang, G., Pincheira, R. and Zhang, J.-T. (1998) Dissection of drug-binding-induced conformational changes in P-glycoprotein. Eur. J. Biochem. 255, 383–390.

Webb, D.C., Rosenberg, H. and Cox, G.B. (1992) Mutational analysis of the *Escherichia coli* phosphate-specific transport system, a member of the traffic ATPase (or ABC) family of membrane transporters – a role for proline residues in transmembrane helices. J. Biol. Chem. 267, 24661–24668.

Wolf, A., Lee, K.C., Kirsch, J.F. and Ames, G.F.-L. (1996) Ligand-dependent conformational plasticity of the periplasmic histidine-binding protein HisJ. J. Biol. Chem. 271, 21243–21250.

Wu, Q., Bounaud, P.Y., Kuduk, S.D., Yang, C.P., Ojima, I., Horwitz, S.B. and Orr, G.A. (1998) Identification of the domains of photoincorporation of the 3′- and 7-benzophenone analogues of taxol in the carboxyl-terminal half of murine mdr1b P-glycoprotein. Biochemistry 37, 11272–11279.

Xie, J., Drumm, M.L., Ma, J. and Davis, P.B. (1995) Intracellular loop between transmembrane segments IV and V of cystic fibrosis transmembrane conductance regulator is involved in regulation of chloride conductance state. J. Biol. Chem. 270, 28084–28091.

Yao, N., Trakhanov, S. and Quiocho, F.A. (1994) Refined 1.89-A structure of the histidine-binding protein complexed with histidine and its relationship with many other active transport/chemosensory proteins. Biochemistry 33, 4769–4779.

Yerushalmi, H. and Schuldiner, S. (2000) An essential glutamyl residue in EmrE, a multidrug antiporter from *Escherichia coli*. J. Biol. Chem. 275, 5264–5269.

Zhang, D.W., Cole, S.P. and Deeley, R.G. (2001) Identification of a non-conserved amino acid residue in multidrug resistance protein (MRP) 1 important for substrate specificity: evidence for functional interaction between transmembrane helices 14 and 17. J. Biol. Chem. 27, 27.

Zhang, F., Yin, Y., Arrowsmith, C.H. and Ling, V. (1995a) Secretion and circular dichroism analysis of the C-terminal signal peptides of HlyA and LktA. Biochemistry 34, 4193–4201.

Zhang, X., Collins, K.I. and Greenberger, L.M. (1995b) Functional evidence that transmembrane 12 and the loop between transmembrane 11 and 12 form part of the drug-binding domain in P-glycoprotein encoded by MDR1. J. Biol. Chem. 270, 5441–5448.

Zheleznova, E.E., Markham, P.N., Neyfakh, A.A. and Brennan, R.G. (1999) Structural basis of multidrug recognition by BmrR, a transcription activator of a multidrug transporter. Cell 96, 353–362.

Zheleznova, E.E., Markham, P., Edgar, R., Bibi, E., Neyfakh, A.A. and Brennan, R.G. (2000) A structure-based mechanism for drug binding by multidrug transporters. Trends Biochem. Sci. 25, 39–43.

CHAPTER 6

Probing of Conformational Changes, Catalytic Cycle and ABC Transporter Function

Frances J. Sharom

Introduction

In this chapter, the various biochemical and biophysical methods that have been employed to monitor conformational changes in ABC proteins, and the way in which these changes may be related to substrate and nucleotide binding, the proposed catalytic cycle, and the mechanism of substrate transport are described. The ABC family member that has been studied most intensively from this point of view is the MDR1 P-glycoprotein (Pgp), and for that reason, much of the work described will focus on this protein. Information on other ABC proteins is included where it is available.

The human Pgp gene family comprises two genes encoding closely related proteins that share ~75% sequence identity. The MDR1 Pgp is a multidrug transporter responsible for exporting a wide variety of structurally unrelated hydrophobic drugs, natural products and peptides from cells (for a more complete list of Pgp substrates, see Sharom, 1997). Drug efflux takes place via active transport, driven by the energy of ATP hydrolysis. Pgp can generate a drug concentration gradient across the membrane of about 5- to 20-fold, depending on the substrate, presumably maintaining the cytosolic drug concentration low enough to allow cell survival, and hence drug resistance. The MDR3 gene product, on the other hand, is expressed at the apical surface of the liver canalicular cells, where it exports phosphatidylcholine (PC) into the bile (Ruetz and Gros, 1994). MDR3 appears to be able to transport drugs at a low rate, and (as discussed below) the MDR1 Pgp can export short-chain fluorescent lipid derivatives. Although these two ABC proteins appear to have each evolved to efficiently transport a different group of substrates, they may share many aspects of their structure and mechanism of action.

One unique feature of the MDR1 Pgp is the existence of a second group of compounds, known as modulators or chemosensitizers, which are able to greatly reduce multidrug resistance in intact cells by blocking its action. Like drug substrates, modulators appear to interact directly with Pgp, and compete with the drug-binding site(s) on the protein (see Chapter 5). Two widely used modulators, verapamil and cyclosporin A, are transported by Pgp (for a more extensive list of modulators, see Sharom, 1997). Several modulator drugs have already been used clinically in conjunction with anti-cancer agents in the treatment of human tumors, with some initial success, and more effective and less toxic third-generation compounds are currently under development by the pharmaceutical industry. The molecular mechanism by which modulators reverse drug resistance may be intimately connected to the relationship between the Pgp transporter, drugs, and the lipid bilayer component of the membrane. A better understanding of the factors involved may lead to the rational design of more effective modulators.

ABC Proteins: From Bacteria to Man
ISBN 0-12-352551-9

THE CATALYTIC CYCLE OF PGP AND OTHER ABC TRANSPORTERS – AN OVERVIEW

The catalytic cycle of Pgp and other ABC transporters is believed to involve the following steps: nucleotide and drug binding, hydrolysis of ATP, release of Pi, and release of ADP. Movement of the drug substrate to the other face of the membrane, and its subsequent release, take place during the ATP hydrolysis cycle, at a point which remains to be established. Vanadate trapping of ADP results in the formation of the complex ADP-Vi-M^{2+} (where M can be Mg^{2+}, Mn^{2+} or Co^{2+}), which is thought to resemble the transition state of the transporter. This has been a very useful tool employed to access intermediate steps of the catalytic cycle.

Any proposed mechanism for the transport by Pgp must account for the spatial aspects of the various processes taking place during the catalytic cycle. Access of drug substrates to the binding site(s) of the transporter probably takes place within the membrane bilayer itself, with access from the cytoplasmic leaflet (see also Chapter 12). Drug may be released either to the aqueous phase on the opposite side of the membrane, or, potentially in the case of lipid-like substrates, into the extracellular leaflet of the membrane. The exact spatial relationship between the sites where substrates bind and are released, and how they are interconnected to the transport process, remain to be determined (reviewed in Chapter 5).

VACUUM CLEANER MODEL FOR ABC MULTIDRUG TRANSPORTERS

DRUG SUBSTRATES MAY GAIN ACCESS TO PGP FROM THE MEMBRANE

Most of the transport substrates for Pgp are hydrophobic, and would thus be expected to show greater solubility in lipid bilayers than in water. Binding of substrates and modulators to

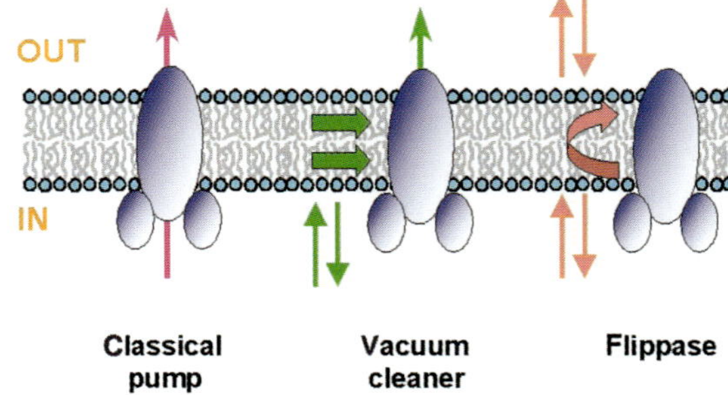

Figure 6.1. The classical pump, vacuum cleaner, and flippase models for drug transport by Pgp. In the pump model, drug molecules in the aqueous phase at the cytosolic side of the plasma membrane interact with Pgp, are pumped across the membrane and released into the aqueous phase on the extracellular side. Drugs move through a transport channel within the protein, but do not contact the lipid bilayer phase of the membrane. In the vacuum cleaner model, hydrophobic drugs partition into the lipid bilayer and subsequently interact with Pgp, which then expels them into the aqueous phase on the extracellular side. Drug builds up to a higher concentration extracellularly relative to the cytosol, thus establishing a gradient across the membrane. In the flippase model, drugs partition into the lipid bilayer, interact with a region of Pgp within the cytoplasmic membrane leaflet, and are then translocated, or flipped, into the outer leaflet, where they build up to a higher concentration. Re-partitioning of drug into the aqueous extracellular medium will result in a higher external drug concentration, again giving rise to a concentration gradient.

Pgp thus appears to take place within the membrane itself. Higgins and Gottesman (1992) first suggested that, rather than pumping drugs from one aqueous compartment to another, Pgp may remove them directly from the bilayer, thus functioning as a 'hydrophobic vacuum cleaner' or 'flippase' **(Figure 6.1)**. A two-step recognition process was proposed, consisting first of partitioning of drug into the lipid bilayer, followed by interaction with a relatively nonselective substrate-binding site within the protein.

Over the years, substantial evidence has accumulated supporting the proposal that substrates gain access to Pgp from the membrane. The fluorescence emission maximum of rhodamine 123, a Pgp transport substrate, was indicative of a molecule in a hydrophobic environment in drug-sensitive cells, or in multidrug-resistant (MDR) cells treated with a modulator, indicating that the drug may be primarily located within the membrane. In contrast, the fluorescence spectrum of the dye in MDR cells was characteristic of a molecule in a hydrophilic aqueous environment, suggesting that it had been expelled from the bilayer (Kessel, 1989). Photoactivation of INA (5-iodonaphthalene-1-azide), a labile lipid-soluble probe, can be achieved by fluorescence resonance energy transfer (FRET) from drugs such as Rhodamine 123 or doxorubicin. Activation resulted in non-specific labeling of many different membrane proteins in drug-sensitive cells, whereas in MDR cells, Pgp was the only protein labeled by INA, suggesting that a specific interaction takes place within the membrane (Raviv *et al.*, 1990).

It seems likely that Pgp intercepts drugs at the plasma membrane, before they have the opportunity to enter the cytosol. Hydrophobic acetoxymethyl esters of several fluorescent indicator dyes (e.g. calcein-AM) are readily transported by Pgp. When the non-fluorescent acetoxymethyl derivative reaches the cytosol, it is rapidly cleaved by esterase enzymes to give the highly fluorescent free acid form of the dye. Since the free dye is not a Pgp substrate, it is trapped in the cytosol at this point, and an increase in cellular fluorescence is observed (Homolya *et al.*, 1993). However, in MDR cells, the rate of fluorescence increase due to accumulation of the free dye is negligible compared to that seen in their drug-sensitive counterparts, implying that the acetoxymethyl ester is effluxed from the membrane by Pgp, and in effect never reaches the cytosol. Shapiro and Ling (1997, 1998b) showed that purified Pgp reconstituted into lipid bilayers pumped the fluorescent dyes Hoechst 33342 and LDS-751 out of the bilayer environment, where their fluorescence emission is greatly enhanced due to the hydrophobic milieu, into the aqueous phase, where their fluorescence is highly quenched. Based on additional FRET experiments using lipid fluorophores, they also proposed that these two dyes were removed from the cytoplasmic leaflet (Shapiro and Ling, 1997, 1998b), which is consistent with the idea that access to the drug-binding site of the transporter is from the cytoplasmic side of the plasma membrane. Strong support for the membrane bilayer being the source of substrate for Pgp comes from experiments indicating that it can translocate fluorescent lipid derivatives from the cytoplasmic to the extracellular leaflet of intact cells (van Helvoort *et al.*, 1996), or in reconstituted proteoliposomes (Romsicki and Sharom, 2001). This lipid 'flippase' activity (Higgins and Gottesman, 1992) appears to be closely related to the drug-binding properties of Pgp, indicating that drugs and lipids are probably transported via the same path within the protein (Romsicki and Sharom, 2001). Thus Pgp may be a drug flippase, moving its substrates from the cytoplasmic to the extracellular leaflet of the membrane **(Figure 6.1)**. At present, it is not known whether membrane access is an absolute requirement for binding and transport of all drugs, or whether the binding sites within Pgp are also accessible from the aqueous phase in the case of more hydrophilic water-soluble substrates.

Other ABC transporters with lipophilic substrates may also operate by a vacuum cleaner-type mechanism. As reviewed in detail in Chapter 12, the bacterial multidrug transporter LmrA shares a high degree of sequence similarity with mammalian Pgp, and can functionally complement human Pgp in intact cells (van Veen *et al.*, 2000b). It also transports hydrophobic substrates and, in fact, LmrA recognizes many of the same drugs as Pgp. LmrA reconstituted into proteoliposomes transported Hoechst 33342 out of the bilayer, and was also capable of translocating fluorescence phospholipid derivatives, which suggests that it interacts with its substrates within the membrane (Margolles *et al.*, 1999). MRP1 displays overlapping substrate specificity with Pgp, and it has also been shown to flip a variety of fluorescent phospholipid and sphingolipid derivatives into the extracellular leaflet of the membrane (Dekkers *et al.*, 1998; Kamp and Haest, 1998; Raggers *et al.*, 1999; van Helvoort *et al.*, 1996). Thus, a common characteristic of ABC proteins with hydrophobic substrates appears to be their ability to interact with substrates within the bilayer, probably at the cytoplasmic leaflet, and either expel them from the membrane, or translocate them to the extracellular leaflet.

PARTITIONING OF PGP DRUGS AND MODULATORS INTO MEMBRANES

Since the majority of drugs that interact with Pgp are relatively hydrophobic, they would be expected to partition into the lipid bilayer, and thus be concentrated within the membrane relative to the aqueous phase. According to the original 'flippase' proposal (Higgins and Gottesman, 1992) the most important factor determining the selectivity of drug binding would actually be the lipid–water partition coefficient, P_{lip}. Recent work has demonstrated that the apparent affinity for many drugs and modulators, as measured by fluorescence quenching, covers a range of over 1000-fold (Sharom *et al.*, 1998a, 1999). This suggests that Pgp is in fact capable of discriminating effectively between many different compounds based on their binding affinity. Crude estimates of the extent of membrane partitioning may be obtained from the value of the octanol–water partition coefficient, P_{ow}, for a particular compound. However, phospholipid bilayers are ordered structures with charged polar headgroup regions, and differ in this respect from a homogeneous solvent. Positively charged compounds can interact electrostatically with the phosphate moieties of the lipid headgroups, while the hydrophobic portion inserts into the nonpolar region of the membrane interior, resulting in interfacial partitioning of the drug. This phenomenon probably accounts for the much higher than expected membrane partitioning of these types of drugs, based on P_{ow} values (Austin *et al.*, 1995; Krämer *et al.*, 1998; Zeng *et al.*, 1999). Favorable interactions of this type will be important for drugs with protonated amino groups, such as daunorubicin, vinblastine and verapamil, all of which are Pgp substrates. For this reason, a direct approach involving experimental measurement of P_{lip} seems warranted if relationships involving drug partitioning are to be examined. Measurements of P_{lip} have been made for partitioning of various compounds (Rodrigues *et al.*, 2001; Rogers and Davis, 1980; Zeng *et al.*, 1999), and some Pgp drugs and modulators (see, for example, Romsicki and Sharom, 1999) into PC liposomes.

IMPLICATIONS FOR PGP TRANSPORT FUNCTION AND THE CATALYTIC CYCLE

Because of their intrinsic hydrophobicity, many Pgp substrates are expected to show strong partitioning into lipid bilayers. This has been confirmed by experimental measurements. For example, P_{lip} for liposomes composed of egg PC was 267 for vinblastine, 507 for verapamil, and 425 for daunorubicin (Romsicki and Sharom, 1999), confirming that the bulk of these drugs will be located within the membrane, where they will reach relatively high concentrations (a 10 μM solution of daunorubicin will reach a lipid concentration of 4 mM). Thus the true affinity of Pgp for drugs and modulators may be quite low. The binding process is favored because these compounds are concentrated in the membrane before they interact with the protein **(Figure 6.2)**.

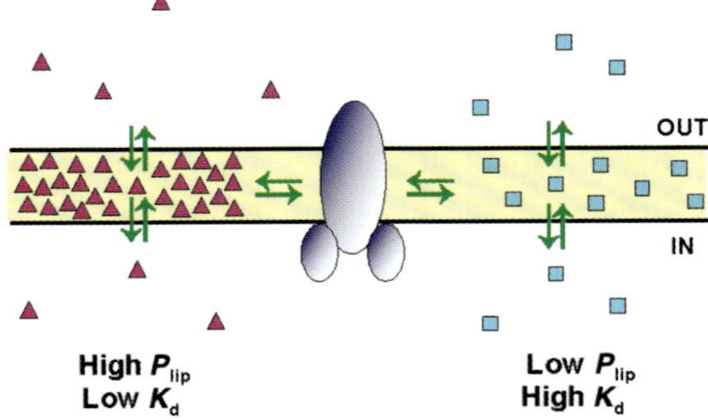

Figure 6.2. The effect of membrane partitioning on drug binding by Pgp. The measured affinity of binding of a drug to Pgp may be related to the lipid:water partition coefficient, P_{lip}*. A substrate with a high value of* P_{lip} *(left side of the figure) will accumulate to a relatively high concentration within the membrane relative to a substrate with a low value of* P_{lip} *(right side of the figure). A higher membrane concentration of drug will push the equilibrium for binding to Pgp in the forward direction, leading to the observation of a low apparent* K_d *value (high apparent binding affinity). A drug with a low* P_{lip} *will have a lower membrane concentration, and will thus appear to have a high* K_d *(low apparent binding affinity).*

The membrane-bound binding sites on Pgp for drugs are probably located within the cytoplasmic leaflet of the membrane. A recent FRET study indicated that the drug substrate Hoechst 33342 was indeed bound to Pgp on the cytoplasmic side of the membrane (Qu and Sharom, 2002). Compounds that cross membranes slowly, or not at all, will not be able to interact with Pgp in intact cells, and we would predict that they would appear to be non-substrates. It has in fact been observed that a positively charged derivative of the high-affinity MDR modulator dexniguldipine (Ferry *et al.*, 2000) and several hydrophobic peptides (Sharom *et al.*, 1998b) interact well with Pgp in plasma membrane systems, where a substantial fraction of the vesicles are inside-out with their cytoplasmic face directly accessible to the drug, but these same compounds are ineffective in reversing MDR in intact cells. Any model for the catalytic cycle and mechanism of drug transport of Pgp should take into account the possible intrinsic low binding affinity, and the location of the drug-binding site within the cytoplasmic leaflet of the membrane.

The intimate association of both Pgp and its drug substrates with the lipid bilayer would be expected to result in functional modulation of

Pgp by the membrane. This has indeed proved to be the case; both the apparent affinity of binding of substrates and modulators to Pgp (Romsicki and Sharom, 1999) and the rate of drug transport (Lu *et al.*, 2001) are influenced by the properties of the membrane, probably mediated via changes in drug partitioning. Callaghan *et al.* (1993) added various lipid-like molecules to intact MDR cells, and noted that changes in the physical properties of the membrane affected drug accumulation. Also, collateral sensitivity of Pgp-expressing cells to narcotics appeared to correlate with changes in the physical properties and fluidity of the membrane (Callaghan and Riordan, 1995).

The mode of action of modulators may be related to their ability to cross lipid bilayers. Eytan and co-workers noted that the rate of movement of various compounds across lipid bilayer membranes correlated with their classification as either substrates or modulators (Eytan *et al.*, 1996b). Substrates tended to cross membrane bilayers relatively slowly, thus allowing Pgp to build up a concentration gradient across the membrane, resulting in drug resistance. Modulators, on the other hand, crossed membranes very rapidly, so that they would be expected to re-partition into the membrane after extrusion by Pgp, move rapidly to the inner leaflet, and interact with the transporter once more. Thus, Pgp-mediated efflux of the drug will be unable to keep pace with re-entry, and no drug gradient will be established. The transporter will essentially operate in a futile cycle in the presence of modulator drugs, with a high turnover rate for transport and ATP hydrolysis. No net transport of modulator will be observed, even though the compound is being translocated by Pgp.

These initial observations were supported by later work showing that the rate of transmembrane movement was the major factor determining the efficacy of the Pgp-mediated efflux of a series of rhodamine dyes from MDR cells (Eytan *et al.*, 1997). Pgp did not effectively exclude compounds with a rapid rate of transmembrane movement, whereas dyes that crossed the membrane slowly were effectively kept out of the cells. This suggests that highly effective modulators should display two important characteristics; high-affinity binding to Pgp, and also a rapid rate of transbilayer diffusion. Thus, both these criteria need to be considered in strategies for the development of effective new Pgp modulators for clinical application. At present, there is no obvious way to predict the rate of transbilayer movement of a particular chemical, and some studies designed to address this issue would clearly be useful.

Compounds that affect membrane fluidity may act as 'nonspecific' modulators, without interacting with Pgp, by increasing the rate of transbilayer diffusion of drugs so that Pgp-mediated extrusion cannot keep pace with re-entry to the cytoplasmic leaflet of the membrane. Various detergents that are able to greatly increase the flip-flop rate of membrane phospholipids (Pantaler *et al.*, 2000) may also increase the rate of transbilayer movement of other hydrophobic compounds present within the lipid bilayer. Thus, we might predict the existence of a class of 'nonspecific' modulators, comprising detergents, surfactants and fluidizers, that do not themselves interact with Pgp. In this respect, MDR can be reversed effectively by various membrane-active surfactants, such as Cremophor EL and Solutol HS15 (Kessel *et al.*, 1995; Woodcock *et al.*, 1992) and membrane fluidizers (Sinicrope *et al.*, 1992). Compounds such as these may be useful clinically in combination with modulator drugs that interact specifically with Pgp.

Probing Conformational Changes of Pgp during the Catalytic Cycle

Several different approaches, both biochemical and biophysical, have been used to monitor changes in the conformation of Pgp during the catalytic cycle. Such changes may be associated with binding of nucleotide or drugs, as well as the steps that follow binding, such as ATP hydrolysis and release of Pi and ADP. In some cases, it has been possible to infer changes in the affinity of the protein for substrates at different stages of the transport process, which can in turn provide clues as to the mechanism of transport.

Protease susceptibility

Susceptibility to protease digestion is a sensitive technique that can be used to detect the conformational changes induced in a protein

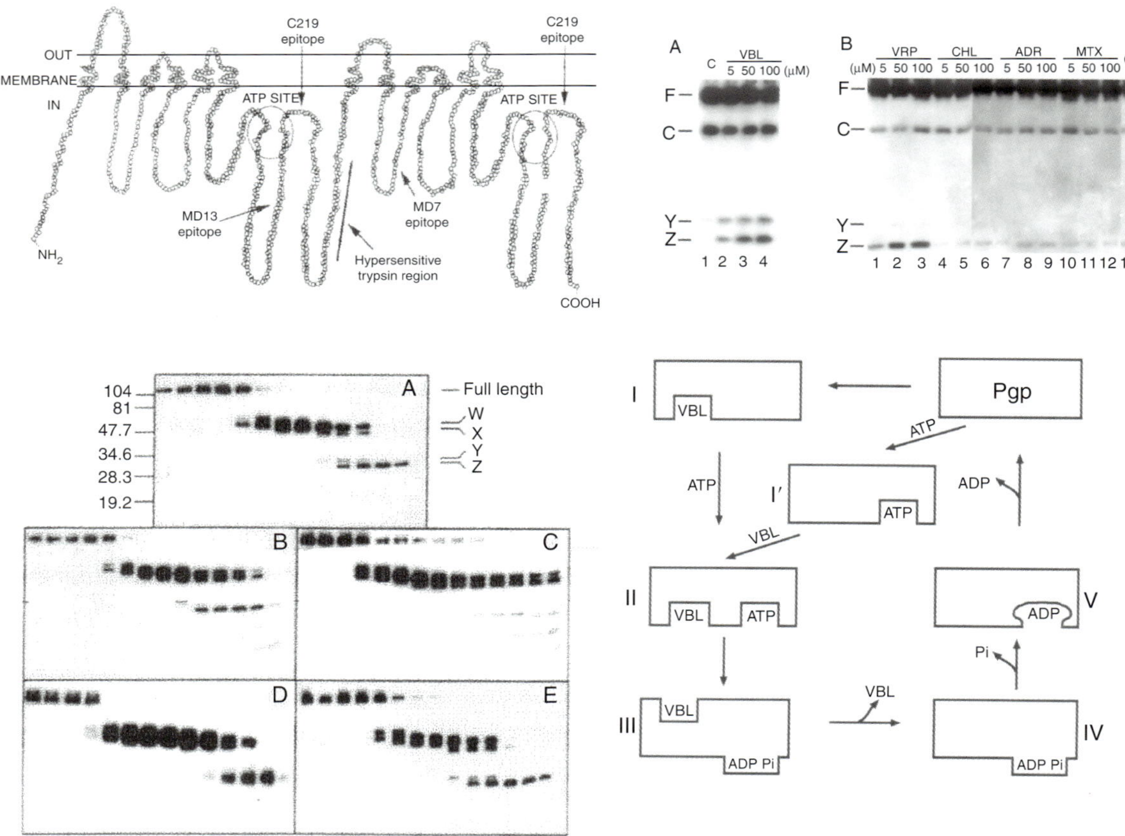

Figure 6.3. Conformational changes in Pgp induced by binding of nucleotides and drugs as assessed by sensitivity to proteolysis. Top left panel: Topology of Pgp and location of the epitopes for the mAbs C219, MD7 and MD13. Bottom left panel: Conformational changes taking place on nucleotide binding. Trypsin digestion profiles of murine Mdr3 Pgp in the presence of (A) Mg^{2+} alone, (B) MgAMP-PNP, (C) MgATP+ vanadate (trapped transition state), (D) MgADP and (E) MgATP. Purified protein was incubated for 15 min with the various nucleotides, as indicated, followed by digestion with trypsin for 15 min at 37°C, at various ratios of trypsin:protein. After stopping the reaction, peptide fragments were visualized by SDS–PAGE, followed by Western blotting with the mAb C219. *Left panel figures were reprinted from Julien and Gros (2000) with permission. Top right panel: Conformational changes taking place on drug binding. Trypsin digestion profiles of Pgp after treatment with increasing concentrations of (A) vinblastine (VBL) (C = control, no drug treatment), and (B) verapamil (VRP), colchicine (CHL), adriamycin (ADR) and methotrexate (MTX). Plasma membrane vesicles were treated with trypsin for 2 h at 37°C, stopped with inhibitors, and collected by centrifugation. Peptide fragments were visualized by SDS–PAGE, followed by Western blotting with the* mAb MD7. *Bottom right panel: A schematic model (Wang* et al.*, 1998) depicting the conformational changes proposed to take place during the catalytic and transport cycle of Pgp. Binding of the drug vinblastine to Pgp generates conformation I, binding of MgATP to Pgp generates conformation I′, while binding of both drug and nucleotide results in conformation II. ATP hydrolysis drives a change in conformation to III, which results in movement of the drug to the other side of the membrane. Release of drug generates conformation IV, and dissociation of Pi leads to conformation V. The starting conformation is then regenerated by loss of ADP. Right panel figures were reprinted from Wang* et al. *(1998) with permission of Blackwell Science Ltd.*

by ligand binding, or alterations arising as a result of point mutations. To examine conformational changes taking place following nucleotide binding to Pgp, Zhang and co-workers used trypsin to digest the protein in isolated inside-out membrane vesicles from MDR cells (Wang *et al.*, 1997). The tryptic fragment pattern was visualized using SDS–PAGE followed by Western blotting with the monoclonal antibody **(mAb) MD7**, which was generated against an epitope in the loop between transmembrane segments TM8 and TM9 (**Figure 6.3**, top left panel). The peptide profile showed two major fragments, both derived from the C-terminal half of the protein. Addition of MgATP or MgADP led to the appearance of a third peptide fragment, indicating that a conformational change had taken place on nucleotide binding. The concentration dependence of these changes in the tryptic peptide profile was consistent with the K_m of Pgp for ATP, and was readily reversible on removal of nucleotide. The conformational change induced by MgATP was probably a result of subsequent hydrolysis to MgADP, since it was prevented by treatment with *N*-ethylmaleimide, which abolishes ATPase activity by reacting covalently with the Cys residues of the Walker A motif. Non-hydrolyzable ATP analogues such as adenosine-(βγ-imido)-5′-triphosphate (AMP-PNP) altered the tryptic digestion pattern in a different way, suggesting that they stabilize another Pgp conformation. Trapping of ADP and vanadate in one of the nucleotide-binding domains (NBDs) led to yet another change in the peptide profile, intermediate between that seen for the ATP-bound state (AMP-PNP) and the ADP-bound state. Based on these trypsin sensitivity experiments, the authors proposed the existence of four different Pgp conformations: the unbound state, the ATP-bound state, the transition state with ADP · Pi bound that is stabilized by vanadate trapping, and the ADP-bound state formed after ATP hydrolysis and Pi dissociation. Later work showed that the nucleotide-bound states were in a conformation less sensitive to further degradation by trypsin than the unbound state (Wang *et al.*, 1998).

This group went on to explore the effects of drug binding on Pgp conformation, again using trypsin digestion and **MD7** antibody detection of the fragments as a tool (Wang *et al.*, 1998). Addition of vinblastine and verapamil resulted in an altered proteolysis pattern (which was different from that seen following nucleotide binding), whereas several other drugs produced no change at all, although they could compete for the vinblastine-induced change (**Figure 6.3**, top right panel). This suggested that vinblastine and verapamil bind to the same site on Pgp, and induce the same conformational change, whereas the other drugs bind to a different site, and do not give rise to a change detectable with trypsin. When both drugs and MgATP were added, the resulting peptide pattern was different from that seen for either ligand alone, and did not have the characteristics seen for Pgp with bound drug. This was interpreted as showing that drug is no longer bound to Pgp following hydrolysis of ATP to ADP and Pi. Proteolysis of Pgp in the transition state conformation (with trapped ADP and vanadate) together with bound drug was also different from that for the individual ligands. Based on this proteolysis approach, five different Pgp conformations were proposed at different points around the catalytic cycle (see **Figure 6.3**, bottom right panel).

Julien and Gros (2000) also used trypsin sensitivity to examine the effect of nucleotide binding to wild-type murine Pgp, and proteins carrying site-directed mutations in the NBDs. Expressed Pgp was purified and reconstituted into lipid bilayer vesicles, so all cleavage sites are likely to be fully accessible to the protease, which was tested over a wide concentration range. Tryptic peptides were detected by SDS–PAGE followed by Western blotting with **mAb C219**, which recognizes a conserved sequence in the NBD of both halves of the protein (**Figure 6.3**, top left panel). Four well-defined stable peptide products were observed (**Figure 6.3**, bottom left panel). Thus, different proteolysis profiles following binding to wild-type Pgp of MgADP, MgATP, and MgAMP-PNP were obtained, indicative of a conformation more resistant to protease cleavage. In the case of Pgp with trapped ADP and vanadate, a dramatic change in trypsin sensitivity resulted, giving rise to a very different digestion pattern **(see Figure 6.3)**. The catalytic transition state appeared to have a unique conformation with greatly enhanced stability and resistance to trypsin. Julien and Gros (2000) also examined Pgps with single and double mutations in the Walker A and B motifs that abolish ATPase activity and vanadate trapping, but not ATP binding. These mutant proteins were slightly more trypsin-sensitive than the wild-type, but none of them showed the change seen for wild-type protein following vanadate treatment, indicating that they cannot adopt the transition state conformation of the wild-type protein.

FLUORESCENCE SPECTROSCOPY

Fluorescence studies of Pgp have provided substantial evidence for conformational changes taking place following binding of drugs and nucleotides. Liu and Sharom (1996) carried out site-directed labeling of purified Pgp on two conserved Cys residues, one in each of the Walker A motifs of the NBDs, using the sulfhydryl-specific fluorophore MIANS (2-(4′-maleimidylanilino)-naphthalene-6-sulfonic

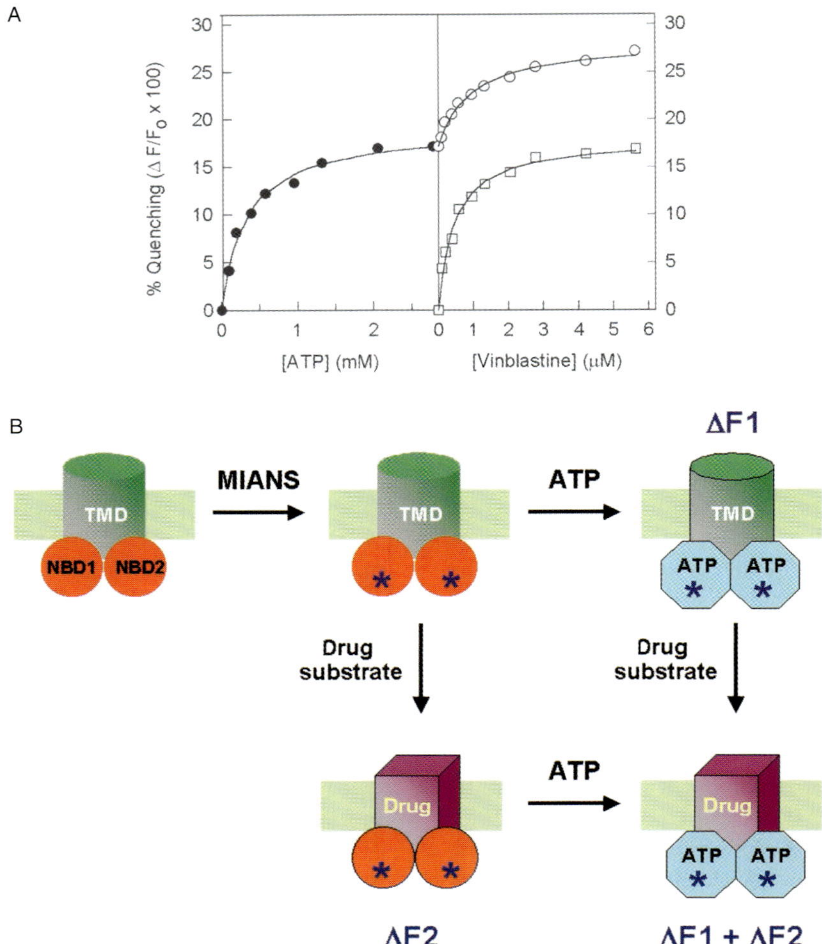

Figure 6.4. Effect of binding of drugs and ATP on the conformation of the various domains of Pgp, as determined by fluorescence quenching. A, Additive conformational changes in Pgp following drug and nucleotide binding. Purified Pgp was labeled with MIANS at two Cys residues, one within each of the Walker A motifs of the NBDs. MIANS-Pgp was titrated in the presence of the phospholipid asolectin with increasing concentrations of ATP (left panel, ●) followed by vinblastine (right panel, ○) or titrated with vinblastine alone (right panel, □). The percent quenching of the fluorescence was calculated relative to the fluorescence of MIANS-labeled Pgp in the absence of drug and ATP. The continuous lines represent computer fitting of the data to an equation describing binding to a single type of binding site. Binding of ATP and drugs appears to be independent and additive. Reproduced from Liu and Sharom (1996) with permission. B, Conformational changes in Pgp following binding of nucleotides and drugs as determined by fluorescence quenching studies. Pgp is labeled with MIANS (indicated by asterisks). Accessibility of the MIANS group to the dynamic quenchers acrylamide and I^- changes on ATP binding, suggesting that a conformational change takes place (Liu and Sharom, 1997). Binding of ATP leads to quenching of the MIANS fluorescence (ΔF1), probably via a direct effect on the quantum yield of the fluorophore. Binding of drug substrates also results in a conformational change in the NBDs which causes MIANS quenching (ΔF2). Quenching of MIANS-Pgp induced by ATP and drugs appears to be independent and additive, suggesting that Pgp does not require ordered addition of nucleotide and the transported drug.

acid). MIANS-modified Pgp lost its catalytic activity, but was still able to bind both nucleotides and drugs with unchanged affinity, providing a means to dissect the conformational changes occurring as a result of substrate binding from those involved in ATP hydrolysis and transport. The covalently linked MIANS group proved to be sensitive to binding of nucleotides. Binding of ATP, ADP and non-hydrolyzable analogues like AMP-PNP resulted in saturable quenching of MIANS fluorescence **(Figure 6.4A)**. The results were fitted to an equation describing binding to a single type of site, resulting in an estimate of the affinity of nucleotide binding, K_d. Quenching of the MIANS fluorescence probably arises via a direct effect on the local environment of the fluorophore, since it is located close to the site of ATP binding within the active site (Liu and Sharom, 1996). Binding of drugs and modulators to the substrate-binding sites, which are believed to be located within the membrane-bound regions of the protein, also led to saturable concentration-dependent quenching of the MIANS fluorescence **(Figure 6.4A)**. The fact that quenching takes place suggests that there is 'crosstalk' between the drug-binding sites within the membrane and the ATPase active sites of the protein. In other words, drug binding elicits a 'signal' which results in a conformational change within the active site of the NBDs. Such a conformational change presumably results in the observed stimulation of the ATPase activity of Pgp by drugs and modulators, and is probably part of the mechanism by which drug transport is coupled to ATP hydrolysis. ATP and drug binding each led to an independent, additive change in fluorescence quenching **(Figure 6.4A)**, suggesting that each causes separate changes in conformation that are not dependent on prior binding of the other **(Figure 6.4B)**. Therefore, it was proposed that nucleotide and drug bind to Pgp in a random order (Liu and Sharom, 1996).

MIANS-Pgp fluorescence is also quenched by collisional quenching agents such as acrylamide or iodide ions, which provide information on the solvent accessibility of the region surrounding the bound fluorophore. A change in quenching efficiency in the presence of a ligand is a good indicator of conformational change induced by binding. Liu and Sharom (1997) used three collisional quenchers differing in charge (acrylamide, iodide ions and cesium ions) to probe the solvent accessibility of the MIANS groups within the active site of Pgp. Low quenching efficiency (as indicated by the value of the Stern–Volmer quenching constant, K_{SV}) indicated that the MIANS group is buried in a relatively inaccessible region of the protein. When ATP was added to MIANS-Pgp, the value of K_{SV} changed for all three quenchers, providing evidence for a conformational change in the NBD as a result of nucleotide binding **(Figure 6.4B)**. Reduced quenching by acrylamide following ATP binding suggested that the change in conformation leads to reduced solvent accessibility of the active site. However, this change was not large (K_{SV} was reduced by only ~10%), suggesting that the conformational change induced by nucleotide binding is small.

Intrinsic tryptophan fluorescence studies of Pgp have also demonstrated the existence of conformational changes associated with binding of nucleotides and drugs. Trp residues are highly conserved across the Pgp family, and may be involved in substrate recognition and binding within the membrane-bound regions of the protein, via stacking of aromatic rings (found in many substrates) with Trp side-chains (Pawagi *et al.*, 1994). Three Trp residues are located within the transmembrane (TM) regions of the protein, and they appear to be responsible for most of the intrinsic fluorescence emission of purified Pgp (Liu *et al.*, 2000).

Sonveaux *et al.* (1999) used acrylamide quenching of the Trp fluorescence of purified reconstituted Pgp to examine the changes in aqueous accessibility induced by binding of substrates and nucleotides. They reported a large increase in the K_{SV} for acrylamide quenching following binding of nucleotides and various anthracycline derivatives. Their experiments indicated that Pgp adopts a different tertiary structure, with slightly increased solvent accessibility, following binding of nucleotides, with ATP giving a much larger change than the non-hydrolyzable analogue adenosine-5′-*O*-(3-(thio) triphosphate) (ATPγS). In contrast, another study of the intrinsic fluorescence of Pgp found that there were only small changes in the Stern–Volmer quenching constants following binding of nucleotide and drugs, suggesting that changes in Trp accessibility as a result of substrate binding are also small (Liu *et al.*, 2000). These results argue against major changes in protein conformation that alter the environment of the Trp residues following nucleotide and drug binding. The reasons for the discrepancy between these two reports is not clear. However, the first study examined only a narrow range of acrylamide concentrations (0–0.08 M) compared with that used in the later study (0–0.5 M) and the observed changes in fluorescence were very small (4–10%), even at the highest acrylamide concentration. In addition, in some experiments, no quenching at all was noted (K_{SV} was essentially zero), suggesting that the data should be interpreted with caution.

Infrared spectroscopy

Attenuated total reflection Fourier transform infrared spectroscopy (ATR-FTIR) has proved to be a very useful technique in the study of membrane protein structure (Goormaghtigh *et al.*, 1999). Of particular interest is the amide I band at 1700–1600 cm^{-1}, which is assigned to the ν (C=O) of the peptide bond, and is sensitive to the secondary structure of the protein. Thus, changes in this region of the ATR-FTIR spectrum can be indicative of alterations in protein conformation. The rate of exchange of the amide hydrogens of a protein with D_2O is also a measure of the solvent accessibility of the NH group of the peptide bonds, and can be sensitively measured by the loss of the amide II band intensity at 1500–1570 cm^{-1}, and a corresponding increase in the ^2H-exchanged amide II region at 1450 cm^{-1}. The $^2H/^1H$ exchange process can be conveniently (although somewhat arbitrarily) fitted to several exponential functions, representing values of the period, T_i, for three classes of protons, which exchange very rapidly, at an intermediate rate, or very slowly. A kinetic study of $^2H/^1H$-exchange can thus be a useful indicator of global changes in tertiary structure.

Sonveaux and co-workers were the first to use ATR-FTIR to examine the secondary and tertiary structure changes taking place following binding of nucleotides and drugs to purified Pgp reconstituted into lipid bilayers (Sonveaux *et al.*, 1996). No changes in the overall content of α-helix, β-sheet, β-turn, or random coil were observed following addition to Pgp of MgATP, MgADP, or MgATP + verapamil, indicating that (as might be expected) no gross changes in protein secondary structure take place on binding of nucleotides or drugs. Examination of the deuteration exchange kinetics showed that a large fraction of the amino acids within Pgp are poorly accessible to the aqueous medium. These regions probably represent the transmembrane helices of the transporter, which are protected by the membrane bilayer, as well as other folded domains outside the membrane. Binding of MgATP (but not MgADP) led to an increase in the fraction of solvent-accessible amino acids within the protein, from ~50% to 56%, suggesting that a conformational change had taken place. Addition of verapamil alone had no effect on the exchange kinetics. On the other hand, addition of both MgATP and verapamil led to a substantial decrease in the fraction of exchangeable amino acids, from ~50% to 46%, reflecting a conformation different from that of the ATP-bound protein. The change in solvent accessibility was interpreted in terms of some amino acids moving from the rapidly exchanging pool to the slow and intermediate pools. The conformational change arising from simultaneous binding of MgATP and drug may represent a tightly coupled conformation that buries some exposed residues. Thus, although ATR-FTIR does not give precise details of the conformational changes taking place, some useful information can be obtained at the molecular level.

PHOTOAFFINITY LABELING

Drug binding to Pgp has frequently been assessed by labeling of the protein with photoactive substrate analogues, such as azidopine and iodoarylazidoprazosine (IAAP), which are usually used in radiolabeled form. Early photoaffinity labeling experiments identified two regions of Pgp that were able to interact with drugs, one in each half of the protein, and later studies demonstrated that these two regions probably represent two non-identical drug-binding sites (Dey *et al.*, 1997). Ambudkar and co-workers studied Pgp following purification and reconstitution into lipid bilayer vesicles (Ramachandra *et al.*, 1998), where it retained the ability to bind both drugs and ATP, and displayed drug-stimulated ATPase activity. Purified Pgp was strongly photolabeled by [^{125}I]IAAP, and neither ADP nor ATP binding had any significant effect on the intensity of labeling. However, the vanadate-trapped state of Pgp showed very low levels of photolabeling with IAAP, suggesting that the transition state of the protein has a greatly reduced (>30-fold) affinity for binding drug substrates (**Figure 6.5**, top panel) (Ramachandra *et al.*, 1998; Sauna and Ambudkar, 2000). Vanadate inhibition of photoaffinity labeling required ATP hydrolysis (it did not occur in Pgp with an inactivating point mutation in the NBD), and was also observed for the drug azidopine. In any model for drug transport mediated by Pgp, ATP hydrolysis must be linked to changes in 'sidedness' of the drug-binding site, to allow for translocation of the substrate from one side of the membrane to the other, and also changes in binding affinity. Presumably, drug must bind tightly to the drug-binding site when it faces the cytosolic side of the membrane (or the inner membrane leaflet), and must bind very weakly to the binding site when it faces the extracellular side of the membrane, so that it can be released outside the cell (or into the outer membrane leaflet). Studies with the myosin ATPase, which can also be trapped in a transition-like state using ATP and vanadate, have shown that the release of Pi from the post-hydrolysis complex is responsible for generating a profound conformational change (Spudich, 1994). Since Pi binds to Pgp only very weakly, it seems likely that the dissociation of Pi is accompanied by a large release of free energy, which may be harnessed to power transport of drug across the membrane.

Further work by Sauna and Ambudkar (2000) suggested that nucleotide binding alone was insufficient to generate the Pgp conformation with low drug-binding affinity. Thus, ATP hydrolysis is a requirement for the conformational change at the substrate-binding site to take place. Following ATP hydrolysis, the high-affinity drug-binding conformation of Pgp is regenerated. However, after the initial formation of the vanadate-trapped transition state species, which requires one turnover of ATP hydrolysis, high-affinity drug binding was not restored in the presence of the non-hydrolyzable ATP analogue AMP-PNP (Sauna and Ambudkar, 2000). Thus, it appeared that drug binding only recovers if an additional round of ATP hydrolysis takes place. This observation led to the proposal that two molecules of ATP are used per transport cycle (**Figure 6.5**, bottom panel). Hydrolysis of the first molecule of ATP generates the transition state conformation, which has low drug binding affinity, and presumably leads to substrate release as part of the transport mechanism. Presumably, the substrate is moved from a high affinity 'on' site to a low affinity 'off' site. The other molecule of ATP is subsequently hydrolyzed to 're-set' the protein back to the starting conformation (**Figure 6.5**, bottom panel). Such a mechanism predicts that two molecules of ATP should be hydrolyzed per molecule of drug transported, and stoichiometries in this range have indeed been reported (Ambudkar *et al.*, 1997; Eytan *et al.*, 1996a). However, it should be pointed out that reliable determination of the stoichiometry of ATP hydrolysis is difficult, because of the high constitutive ATPase activity of Pgp, and some estimates cover a range between 1 and 2 (see, for example, Shapiro and Ling, 1998a). Further exploration of the outcome of the two proposed ATP hydrolysis events indicated that the Pgp conformation formed after the first round of ATP hydrolysis also has a drastically (>30-fold) reduced affinity for nucleotide binding (Sauna and Ambudkar, 2001), which is coincident with the reduction in drug-binding affinity noted earlier. The proposal was made that while one catalytic site is in the transition state conformation, the other site cannot bind nucleotide. Only after release of occluded ADP from the first site is the second site able to bind and hydrolyze ATP, thus resulting in alternating site catalysis. ADP release is the most likely rate-limiting step in the catalytic cycle of Pgp (Kerr *et al.*, 2001; Senior *et al.*, 1995), and it is this step (following the first round of ATP hydrolysis) that leads to recovery of high drug-binding affinity.

Antibody reactivity

Several mAbs have been developed that recognize external and internal epitopes of Pgp (e.g. MRK16, C219, UIC2, MM12.10). Roninson and co-workers were the first to show that the binding affinity and number of binding sites for UIC2 in Pgp are sensitive to the conformational changes associated with drug transport (Mechetner *et al.*, 1997). In MDR1-expressing cells, UIC2 binds to only a small fraction of the available Pgp molecules. UIC2 labeling of intact cells was increased 2- to 5-fold following addition of several substrates and modulators, as detected by flow cytometry, indicating the existence of different Pgp conformations, one of which binds UIC2. If cellular ATP was depleted, or if the NBDs of Pgp were rendered nonfunctional by mutation, enhanced UIC binding was also observed, and it was suggested that the conformational transition detected by the antibody might result from either stimulation of ATP hydrolysis, or dissociation of ATP from the NBDs. UIC2 binding results in inhibition of Pgp transport function, and it was suggested that it probably trapped the protein in a conformation part-way along the catalytic cycle.

UIC2 immunoreactivity analysis was recently extended to permeabilized MDR cells, which permitted a more quantitative analysis of Pgp interactions with drugs and nucleotides at various stages of the catalytic cycle. This approach allowed Roninson and co-workers to elucidate the origin of the change in UIC2 reactivity of Pgp (Druley *et al.*, 2001). Using intact cells permeabilized with *Staphylococcus aureus* toxin, thus removing most of the intracellular ATP, they were able to test the effect of different nucleotides and nucleotide analogues on UIC2 reactivity. They found that binding of nucleotide, whether ATP, ADP or non-hydrolyzable analogues such as ATPγS and AMP-PNP, resulted in a decrease in UIC2 reactivity, suggesting that nucleotide binding, and not hydrolysis, is responsible for the change in antibody reactivity. The previously noted ability of vinblastine, and presumably

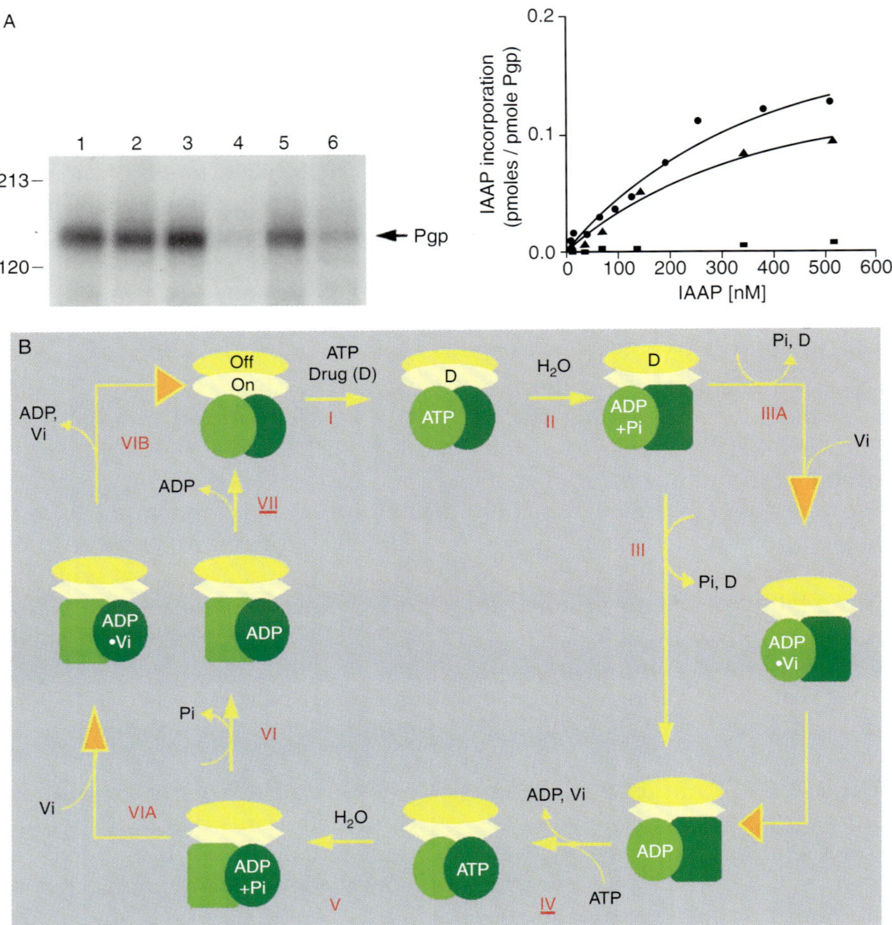

Figure 6.5. Photoaffinity labeling suggests that Pgp exists in a conformation with low drug-binding affinity following trapping of ADP and vanadate at one of the NBDs. A, Left panel: Membranes containing Pgp were labelled with [^{125}I]IAAP after pre-treatment at 37°C with ATP or 8-azido-ATP (which is also hydrolysed by Pgp), in the absence or presence of vanadate. In the presence of both ATP or 8-azido-ATP and vanadate, nucleotide hydrolysis and trapping of vanadate in one of the NBDs will take place. Samples were visualized by SDS–PAGE and autoradiography. Untreated Pgp (Lane 1); Pgp pre-treated with ATP (Lane 2), vanadate alone (Lane 3), ATP and vanadate (Lane 4), 8-azido-ATP (Lane 5), and 8-azido-ATP and vanadate (Lane 6). Right panel: Incorporation of [^{125}I]IAAP into normal and transition state Pgp in photolabelling experiments. Membrane samples were untreated (●), or treated with non-hydrolysable AMP-PNP (▲) or ATP (■) in the presence of V_i then labelled with IAAP. Reproduced from Sauna and Ambudkar (2000) with permission. B, Proposed reaction scheme for the catalytic cycle of Pgp. Shown are the high-affinity (ON) and low-affinity (OFF) substrate-binding sites, and the two NBDs (a green circle indicates a conformation that binds ATP, an empty square with rounded edges indicates a conformation that has reduced affinity for ATP). **Step I:** *Drug substrate binds to the ON site, ATP binds to both NBDs.* **Step II:** *ATP is hydrolyzed and drug moves to the low-affinity OFF site.* **Step III:** *Pi is released, and drug is released on the opposite side of the membrane.* **Step IIIA:** *Vi replaces Pi forming the vanadate-trapped complex, which has a reduced affinity for both drug and nucleotide.* **Step IV:** *ADP and Vi dissociate from the complex, producing a conformation which still has low drug-binding affinity, but has regained its affinity for nucleotide, so that another molecule of ATP binds to the empty NBD.* **Step V:** *the second molecule of ATP is hydrolyzed.* **Step VIA:** *When Vi is present, the vanadate-trapped complex is again formed, which has a low affinity for drug.* **Step VIB and Step VII:** *Dissociation of ADP restores Pgp to its original conformation, with high-affinity drug binding. The observed low affinity for nucleotides at the second NBD when the first NBD is in the catalytic transition state provides a basis for the alternating site catalytic model* *(see also Chapter 4)**. ADP release* (at steps IV and VII, *underlined) appears to be the rate-limiting step in the catalytic cycle. Reproduced from Sauna and Ambudkar (2001) with permission.*

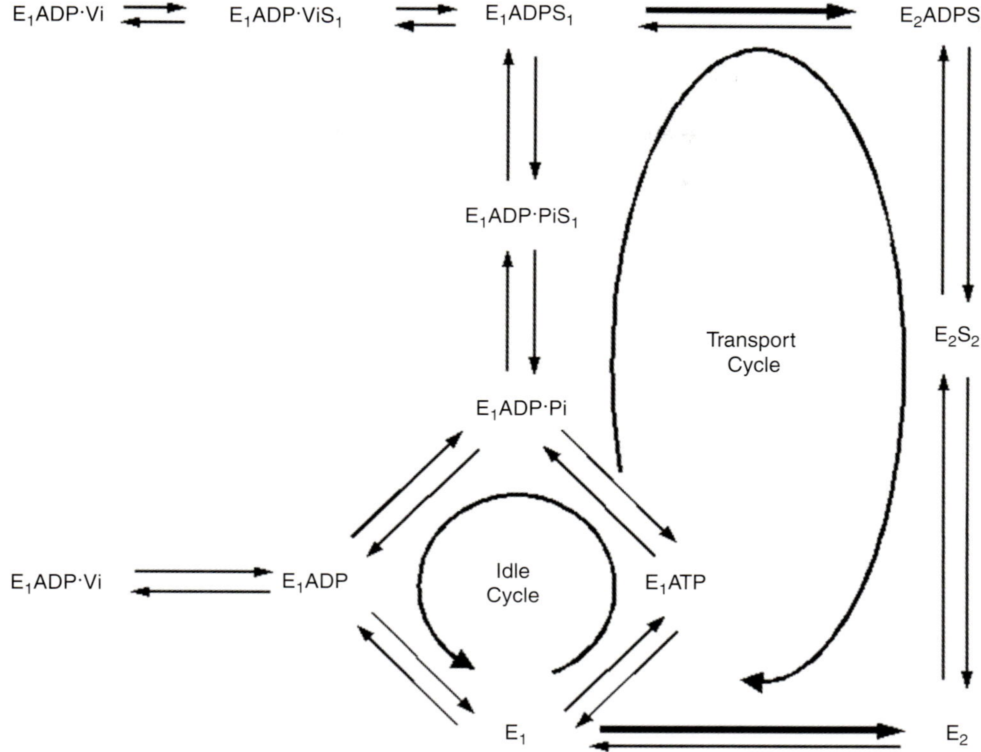

Figure 6.6. ***Proposed cycle of Pgp function based on conformational changes detected by binding of the mAb UIC2.*** ***E_1 is defined as the Pgp conformational state in which the drug-binding site is in an intracellular location, and has low reactivity with the mAb UIC2. E_2 is defined as the Pgp conformational state in which the drug-binding site is in an extracellular location, and has high UIC2 reactivity. Conformational transitions between E_1 and E_2, and between $E_1 \cong ADP \cong S_1$ and $E_2 \cong ADP \cong S_2$, can be detected by UIC2. The vanadate-trapped transition state has a low UIC2 reactivity. The idle cycle refers to the constitutive ATP hydrolysis displayed by Pgp in the absence of drugs, while the presence of drug substrate S commits Pgp to the transport cycle that releases S at the extracellular side of the membrane. Reproduced from Druley*** **et al.** ***(2001) with permission.***

other substrates, to increase the UIC2 reactivity of Pgp arose from the finding that binding of this drug greatly reduces the affinity of nucleotides for binding to Pgp, i.e. drug binding stimulates dissociation of nucleotides from the NBDs. Thus the UIC2 antibody appears to be able to distinguish between two different conformations of Pgp; one bound to nucleotides, and one with empty catalytic sites.

Based on the results of this study, a detailed transport and ATPase catalytic cycle was proposed for Pgp (Druley *et al.*, 2001). The transporter was suggested to exist in one of two different conformations, E_1 and E_2 **(Figure 6.6)**. The E_1 conformation was proposed to have a low reactivity with UIC2, with its drug-binding site(s) available at the cytoplasmic face of the membrane, whereas the E_2 conformation was proposed to have high UIC2 reactivity, with its drug-binding site(s) available at the extracellular face of the membrane. Drug would bind to E_1 and be released by E_2. Binding of nucleotide shifts the protein into the E_1 state, resulting in low UIC2 reactivity. In contrast, binding of the drug vinblastine promotes nucleotide dissociation, and shifts Pgp into the highly UIC2-reactive E_2 state.

Nagy *et al.* (2001) recently investigated competition between UIC2 and another mAb which appears to share some of its recognition epitopes, likely to be located in extracellular loops 4 and 6. They found that drugs and modulators differed in their ability to influence the competition process. Verapamil and Tween 80 had little or no effect on antibody competition, whereas when cyclosporin A, vinblastine, and valinomycin interacted with Pgp, they altered the conformation so that binding of the first antibody abolished subsequent binding of the other. These observations suggest that substrates and modulators may fall into two distinct classes, which can be distinguished by their effects on the conformation of the transporter.

Cysteine crosslinking

Transmembrane segments TM6 and TM12 are directly connected to the ATP-binding domain in each homologous half of Pgp, and these regions are implicated in making up the drug-binding site(s) within the protein. Loo and Clarke (1997) introduced pairs of Cys residues at different positions in TM6 and TM12, and investigated which residues could be oxidatively crosslinked. The observed pattern of crosslinking indicated that these two helices probably interact along their length, in a left-handed coiled-coil arrangement. Pgp mutant proteins with pairs of Cys residues located at the putative interactive faces of the two helices were examined for the functional consequences of crosslinking these residues (Loo and Clarke, 1997). The results showed that crosslinking between P350C and S993C inhibited verapamil-stimulated ATPase activity by about 75%. Activity was fully restored when the disulfide crosslink was broken by dithiothreitol. Thus, movement between these two helices appears to be essential for drug-stimulated ATPase activity, and is likely to be a part of the conformational change required for catalysis and/or transport **(Figure 6.7)**.

More recent work by the same group focused on crosslinking of pairs of Cys residues within the two NBDs (Loo and Clarke, 2000). The double Cys replacement mutant L439C(NBD1)/Cys-1074(NBD2) had similar drug-stimulated ATPase activity relative to Cys-less Pgp. However, oxidative crosslinking of the two Cys residues led to almost complete inhibition of drug-stimulated ATPase activity, which could be largely recovered after breaking the –S–S-link with dithiothreitol. Thus, the two NBDs appear to be close to each other, and crosslinking them inhibits ATP hydrolysis. A conformational change in this region of the protein, possibly the movement of one NBD relative to the other, may be a necessary step in the catalytic cycle.

Direct visualization of conformational changes

Recently (Rosenberg *et al.*, 2001) have provided direct evidence for conformational changes during the transport cycle. Using cryo-electron

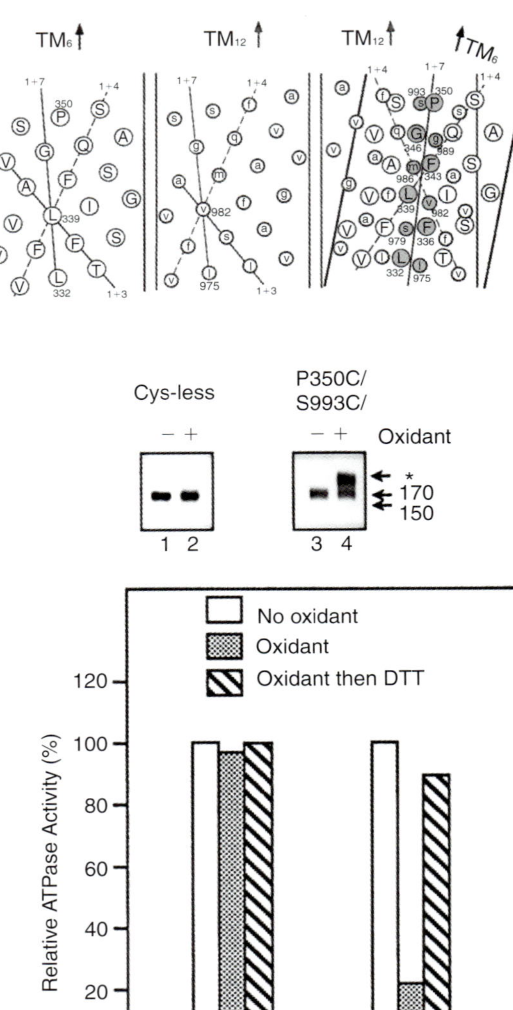

Figure 6.7. ***Model of proposed conformational changes between TM6 and TM12 during drug transport. Top panel; residues in TM6 (large circles) and TM12 (small circles) arranged as α-helical nets were superimposed in a left-handed coiled coil. The i + 3 axis of TM6 is superimposed on the i + 4 axis of TM12. The residues from each helix that face each other are shown along the i + 7 axis (shaded circles), with L332 (TM6) and L975 (TM12) as the starting point. The arrows point towards the cytoplasmic face of the membrane. The paired residues F336/S979, L339/V982, F343/M986, G346/G989, and P350/S993 are predicted to be close to each other. Middle panel: Immunoblot analysis of His-tagged Pgp purified from HEK 293 cells expressing Cys-less and P350C/S993C mutants that were untreated (−) or treated (+) with the oxidant Cu^{2+} (phenathroline)$_3$ to crosslink the Cys residues prior to nickel column chromatography. A protein***

(continued)

microscopy of Pgp trapped at different stages of the catalytic cycle, they were able to demonstrate that the transmembrane domains undergo very significant reorganization upon binding ATP, and following ATP hydrolysis. Importantly, the major conformational change was associated with ATP binding and this correlated with a change in drug binding from high to low affinity (Rosenberg *et al.*, 2001), suggesting that ATP binding is the crucial step in translocation and that ATP hydrolysis 'resets' the transporter. However, as discussed above, not all data are consistent with the model and further study is necessary to fully elucidate the transport cycle.

CONFORMATIONAL CHANGES IN OTHER ABC TRANSPORTERS

MRP1

Like Pgp, MRP1 acts as an energy-dependent efflux pump for cytotoxic drugs, and is also responsible for MDR in tumor cells (Hipfner *et al.*, 1999). MRP1 confers resistance to a similar, but not identical, spectrum of drugs. MRP1 transports organic anionic species, including glutathione conjugates, glucuronides and sulfates, and it can also co-transport some drugs, such as vincristine and daunorubicin, together with free glutathione (Loe *et al.*, 1998). MRP1 has a similar core structure to Pgp, but also contains a third membrane-bound domain at the N-terminus, which probably consists of five membrane-spanning segments. The two NBDs of Pgp appear to be very similar, or identical, both structurally and functionally. However, the sequences of the two NBDs of MRP1 are considerably more divergent than those of Pgp, and, indeed, they appear to be non-equivalent. Thus, in MRP1 nucleotide binding appears to take place exclusively in the N-terminal NBD, and vanadate trapping appears to occur mainly in the C-terminal NBD (Hou *et al.*, 2000; Nagata *et al.*, 2000). In addition, complex allosteric interactions were noted between the two catalytic sites (Hou *et al.*, 2000).

ATR-FTIR spectroscopy was used to detect conformational changes in MRP1 induced by nucleotide binding, in a purified reconstituted system where it retained both its ATPase and drug transport activity (Manciu *et al.*, 2000). Binding of various nucleotides, including MgATP, MgATPγS, MgADP and MgADP + Pi, did not alter the secondary structure of MRP1, which comprised ~46% α-helix, ~26% β-sheet, ~11% β-turn and ~17% random coil. The rate of ^1H–^2H exchange was used as an indicator of nucleotide-induced conformational changes in the protein **(Figure 6.8)**. About 39% of the amide hydrogen atoms of MRP1 exchange very slowly with the aqueous medium; this group of residues is probably made up of those shielded by the membrane, plus additional folded domains external to the membrane. Addition of MgATP, MgATPγS or MgADP + Pi (but not ADP alone) resulted in a substantial increase in the rapidly exchanging amino acid population, indicating that a conformational change had increased the exposure of a large number of residues to the aqueous phase. The vanadate-trapped state of MRP1 showed a similar set of changes. Thus, unlike the situation observed for Pgp, ATR-FTIR studies indicate the existence of only two conformational states during the catalytic cycle, although it is possible that one or more intermediate states may have evaded identification because of the more rapid exchange rate for MRP1 relative to Pgp. MRP1 has 30 Trp residues, 10 of which are predicted to be located in the membrane-bound regions of the protein. A preliminary Trp quenching experiment was also carried out (Manciu *et al.*, 2000), and confirmed that the solvent accessibility of MRP1 to acrylamide was enhanced by binding of MgATP, MgATPγS and MgADP + Pi, but not MgADP.

More recently, another study examined the intrinsic Trp fluorescence of MRP1 in more depth (Manciu *et al.*, 2001). This study employed several structurally related anthracycline derivatives, some of which did not gain access to the

Figure 6.7. ***(continued)***
product with lower mobility is observed (*), indicative of crosslinking between this pair of Cys residues (*), which are thus located close together. Bottom panel: Drug-stimulated ATPase activity of Cys-less Pgp and the P350C/S993C mutant without oxidation (no crosslinking), after treatment with oxidant to crosslink the Cys residues, and after treatment with oxidant and subsequent treatment with dithiothreitol (DTT) to break the crosslinks. Pgps were mixed with lipid and treated with the modulator verapamil to stimulate the ATPase activity. Reproduced from Loo and Clarke (1997) with permission.

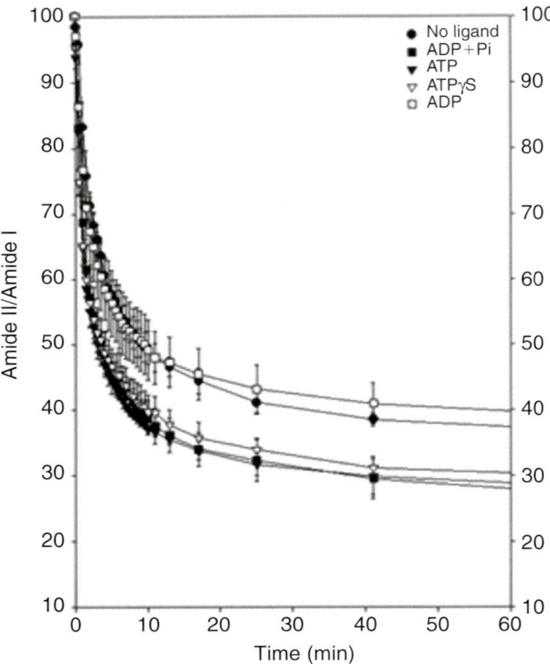

Figure 6.8. ***Conformational changes in MRP1 induced by nucleotide binding as assessed by ATR-FTIR spectroscopy. The aqueous accessibility of the amide bonds of purified MRP1 reconstituted into liposomes of asolectin was monitored by deuterium exchange. H–D exchange kinetics were followed by monitoring the relative decrease in the area of the amide II band as a function of time to exposure to D_2O. MRP1 was examined in the absence of bound ligands (●), and after addition of ADP + Pi (■), ATP (▼), ATPγS (▽), or ADP (□), at a molar ratio of nucleotide:protein of 6:1. Each exchange curve displays multi-exponential decay, which can be fitted to three different half-times, T_1 = 1 min (rapidly exchanging), T_2 = 9 min (medium exchange rate), and T_3 = 666 min (slowly exchanging). Changes in the proportion of amino acid residues in these three classes following ligand binding is an indication that a conformational change has taken place. Reproduced from Manciu*** **et al.** ***(2000) with permission.***

interior of MRP1-expressing cells, i.e. a concentration gradient was built up across the plasma membrane, since they are presumably effluxed via MRP1. The other drugs used were able to gain access to the interior of cells (i.e. a concentration gradient is not built up across the plasma membrane), but they were clearly MRP1 substrates, since they stimulated the ATPase activity of the protein. It was suggested that these compounds interact with the high-affinity drug-binding site of MRP1 but are not transported, so that the catalytic cycle is blocked. Alternatively, if these compounds have a high rate of passive diffusion across the plasma membrane, they may rapidly re-enter the cell and cause MRP1 to undergo futile cycling, in which case it would not be possible to build up a drug concentration gradient. Purified reconstituted MRP1 was treated with various combinations of anthracycline drugs, glutathione, and MgATP or MgATPγS, and acrylamide quenching of Trp residues was used as an indicator of the exposure of these residues to the aqueous environment. Binding of MgATP caused a conformational change in MRP1, but a difference in the pattern of changes was noted in the presence of drugs which accumulated in MRP1-expressing cells, relative to those that did not, suggesting that the coupling between the catalytic site and the drug-binding site is different in each case (Manciu *et al.*, 2001).

TAP1/TAP2

TAP1 and TAP2 are two 'half transporter' members of the ABC superfamily (described in detail in Chapter 26) that play a vital role in cell surface presentation of intracellular peptides (for example, peptides of viral origin) to cytotoxic T-cells. The TAP1 and TAP2 proteins form a functional heterodimeric complex in the membrane of the endoplasmic reticulum that is responsible for ATP-dependent translocation of short peptides (8–16 residues) from the cytosol into the lumen of the endoplasmic reticulum (Reits *et al.*, 2000b). Once delivered to this location, the peptides can bind to MHC class I proteins, which present them at the cell surface. The TAP complex was recently shown to have peptide-stimulated ATPase activity after reconstitution into proteoliposomes (Gorbulev *et al.*, 2001), and it is believed to function similarly to Pgp, except that its peptide substrates are hydrophilic, and are expected to be located primarily in the aqueous phase, prior to transport.

Radiolabeled peptides bind to the TAP complex with high efficiency at low temperature (4°C), but are not transported, allowing dissection of the binding and transport processes. Peptide binding in microsomes or permeabilized cells does not appear to depend on the prior addition of nucleotides. Studies with partially purified, reconstituted TAP confirmed that ATP is not necessary for peptide binding (Gorbulev *et al.*, 2001). The TAP complex binds peptides inefficiently at 37°C, and appeared to

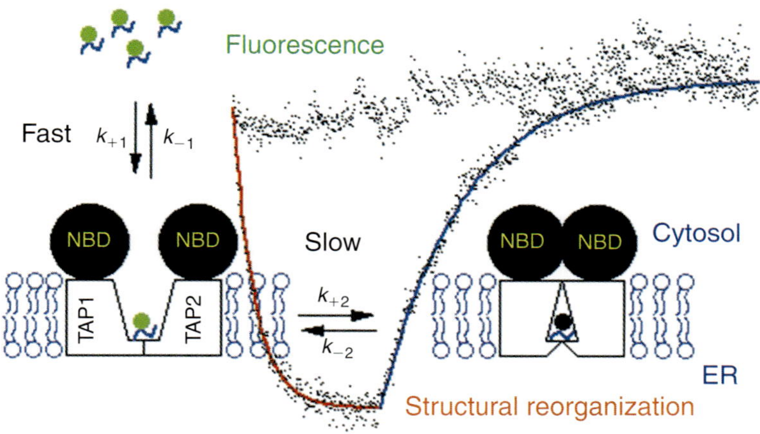

Figure 6.9. A two-step model proposed for peptide binding to the TAP1/TAP2 complex. Binding of the fluorescein-labeled peptide substrate RRY KSTEL (where ϕ = cysteine-acetamido-fluorescein) to the TAP complex in microsomes is composed of a fast bimolecular association step (k_{+1}/k_{-1}), followed by a slow isomerization reaction (k_{+2}/k_{-2}), which causes quenching of the peptide fluorescence in the substrate-binding pocket of TAP. The bottom fluorescence trace shows the time-dependent quenching that takes place on peptide binding to TAP, followed by the recovery of fluorescence when the bound peptide is displaced by an excess of competing unlabeled peptide. Microsomes that do not contain the TAP complex have no effect on the fluorescence of the peptide (top trace). Reproduced from Neumann and Tampé (1999) with permission.

be unstable, undergoing rapid inactivation, which could be prevented by di- or tri-nucleotides (Van Endert, 1999). Binding of high-affinity peptide substrates to TAP1/TAP2 also protected the protein complex from inactivation, although they were much less effective than ATP (Van Endert, 1999). These observations suggest that nucleotide binding to the NBDs (and to a lesser extent, peptide binding to the substrate-binding sites) stabilizes the TAP1/TAP2 dimer in a functional conformation. This conformation may be related to a step along the peptide transport pathway. Non-functional TAP complexes appear to adopt a conformation that is no longer recognized by several TAP-specific antibodies, and also does not participate in peptide binding, suggesting that the conformational changes associated with nucleotide binding are transmitted to other domains of the protein complex.

Kinetic analysis of peptide binding has also provided evidence for structural rearrangements induced by substrate binding. Neumann and Tampé (1999) employed fluorescein-labeled peptide substrates, which were bound and transported by the TAP complex. They observed a striking decrease in the fluorescence emission intensity of the peptides upon binding to the transporter **(Figure 6.9)**. This quenching effect was due only to binding of the peptide, and was ATP-independent. The association and dissociation of the fluorescent peptide could be followed in real time, which allowed determination of the kinetics of the TAP–peptide interaction. Both association and dissociation displayed mono-exponential kinetics, in accordance with the expected 1:1 binding model. However, the association rate, which was expected to show a linear dependence on the peptide concentration, instead showed saturation kinetics, reaching a constant value at high peptide concentrations. This type of effect is indicative of a two-step process: a fast step corresponding to association of peptide with the TAP complex, followed by a slow unimolecular isomerization reaction. This second step appeared to be responsible for quenching of the peptide fluorescence, and was attributed to a conformational change in the transporter complex **(Figure 6.9)**. This conformational change appeared to trigger the movement of a proton-donating group into the vicinity of the N-terminus of the bound peptide, resulting in the protonation of its carboxyl group, and consequent quenching. This altered conformation may represent an intermediate in the peptide transport cycle, which would indicate successful loading of peptide into the substrate-binding site. The structural reorganization on peptide binding could bring about communication between the substrate-binding site and the NBDs, initiating peptide translocation.

More recently, fluorescence recovery after photobleaching (FRAP) experiments have provided biophysical evidence for conformational changes taking place in the TAP complex during the process of peptide transport (Reits *et al.*, 2000a). Measurement of the lateral mobility of a green fluorescent protein (GFP)-TAP1/TAP2 complex in living cells indicated a diffusion coefficient of $\sim 4 \times 10^{-10} cm^2\ s^{-1}$, similar to that of many integral proteins. Microinjection of high-affinity peptide substrate into the cells resulted in a saturable decrease in TAP complex mobility of ~20%. In contrast, depletion of the cells of endogenous peptide substrates, or depletion of cellular ATP, led to an increase in lateral mobility of ~14%. Both of these conditions should lead to TAP being in a transport-incompetent state. It appears, therefore, that when the TAP complex is actively transporting peptides it has a substantially reduced lateral mobility. Microinjection in the presence of ATP of a large peptide that binds to TAP1/TAP2, but inhibits the transporter, led to an even larger decrease in the lateral diffusion coefficient. These results suggest that the change in lateral mobility is caused by a peptide- and ATP-dependent conformational change in the TAP complex. The TAP complex thus moves more rapidly when it is unoccupied by peptide substrate and ATP, and more slowly when it is engaged in peptide translocation. Conformational changes could result in an alteration in the dimensions of the complex, or rotation/tilting of the membrane-spanning segments within the bilayer.

BACTERIAL ABC TRANSPORTERS

Several bacterial transporters are members of the ABC superfamily (Higgins *et al.*, 1986; Schneider and Hunke, 1998). Two of the best-characterized bacterial ABC proteins are components of the maltose and histidine permeases, which are involved in the import of maltose and histidine, respectively, from the periplasmic space into the cytosol (reviewed in Chapter 9). The various domains typical of the ABC protein family are most often present as separate subunits in this group of proteins. The maltose permease comprises two membrane-bound subunits, MalF and MalG, and two copies of the NB subunit, MalF, which associate to give the complex $MalFGK_2$. The histidine permease is similarly made up of the membrane-bound subunits HisQ and HisM, and two copies of the NBD HisP, to form the complex $HisQMP_2$. Each transport complex is also associated with a soluble periplasmic binding protein (MalE or HisJ), which binds the substrate with high affinity, and delivers it to the transporter complex in the cytoplasmic membrane. Unlike the situation with Pgp, the hydrolysis of ATP is strictly coupled to substrate transport in the intact $MalFGK_2$ or $HisQMP_2$ complexes. The isolated NB subunits have (lower) constitutive ATPase activity when they are studied in isolation. The interaction of liganded binding protein with the periplasmic face of the membrane-bound complex is thought to transmit a signal to the NB subunits on the cytoplasmic side of the membrane, thus activating them. Hydrolysis of ATP would then trigger additional conformational changes leading to translocation of the substrate molecule into the cytosol. Over the past few years, considerable progress has been made in understanding the mechanism of action of these permeases, and the existence of conformational changes has been demonstrated by both biochemical and biophysical approaches.

Schneider and co-workers (1994) were the first to show that the binding to MalK of ATP and GTP (but not ADP or non-hydrolyzable ATP analogues) induces a change in its conformation, as assessed by quenching of intrinsic Trp fluorescence. The helical domain (Arm-II of the HisP structure (Hung *et al.*, 1998)) of the protein was proposed to be involved in the structural change induced by ATP, since it was substantially more resistant to trypsin proteolysis in the presence of nucleotide. These findings were supported by a recent site-directed chemical crosslinking study, which identified protein–protein contacts within the protein complex, and found that the relative positions of the subunits of the transporter complex were changed by ATP binding (Hunke *et al.*, 2000). One of the changes may involve insertion and de-insertion of the MalK subunit into the membrane-bound portion of the complex, as has been proposed for HisP (see below).

Davidson and co-workers explored the mechanism of maltose transport in detail, using purified protein components that can be assembled *in vitro* and reconstituted into proteoliposomes (Chen *et al.*, 2001). Vanadate inhibits the ATPase and transport activity of the maltose permease in a similar fashion to Pgp, by stable trapping of $Mg^{2+} \cdot ADP \cdot Vi$ in the active site of the NB subunit after ATP hydrolysis and release of Pi (Sharma and Davidson, 2000). As in the case of

Pgp, the vanadate-trapped state is believed to represent a conformation close to that of the transition state. The maltose-binding protein, MalE, is required to form the vanadate-inhibited species, and was found to be tightly bound to the membrane-bound inhibited complex under these conditions (Chen *et al.*, 2001). However, it had lost its bound maltose, suggesting that transport of the sugar had already taken place before formation of the vanadate-inhibited complex. This observation is comparable to similar features noted for Pgp, whereby formation of the vanadate-trapped state is associated with loss of high-affinity drug binding (see above). These results were interpreted in terms of a mechanism whereby concerted conformational changes take place within the maltose permease complex (Chen *et al.*, 2001).

Certain mutations in MalF and MalG have been isolated that allow ATP hydrolysis and maltose transport in the absence of MalE. These binding protein-independent mutants have proved useful in investigation of the conformational changes taking place within the complex during maltose transport. The fluorophore MIANS was covalently linked to a reactive Cys residue in the Walker A motif of the active site of the MalK subunit, to act as a reporter group (Mannering *et al.*, 2001).

When comparing wild-type and binding protein-independent mutant complexes labeled with MIANS, it was noted that the environment around the fluorophore was more hydrophobic in the mutants. Quenching experiments indicated that the nucleotide-binding pocket around the bound MIANS was (as expected) poorly accessible to the aqueous medium. However, this effect was more pronounced for the mutant protein complexes, suggesting a conformational difference between them. This difference disappeared when the wild-type complex was labeled with MIANS after vanadate trapping, in the presence of the binding protein. Thus the conformation of the mutants appears to resemble that of the vanadate-trapped transition state. This was confirmed by the finding that the mutants bound the binding protein tightly, and had a high affinity for ATP, both hallmarks of the transition state. A model was proposed in which the MalK subunits are normally positioned apart in the absence of the binding protein, thus preventing them from hydrolyzing ATP (Mannering *et al.*, 2001). The binding protein would bring the two NB subunits together, activating ATP hydrolysis, and making the active sites less accessible to solvent **(Figure 6.10)**. In the binding protein-independent mutants, the two MalK subunits would already be positioned close together (explaining their reduced accessibility to solvent), so that ATP hydrolysis can take place without the need for the binding protein.

The histidine permease can also be assembled from purified subunits *in vitro* (Liu and Ames, 1998), which has allowed detailed study of the conformational changes taking place during histidine transport. Although the HisP subunits have been reported to form a dimer both in solution and in the membrane-bound complex, a study with inactivated mutant HisP proteins indicated that histidine transport can be powered by only one of the two subunits, although at half the rate (Nikaido and Ames, 1999). The HisP protein has the properties of both a peripheral and an integral membrane protein, and appears to physically disengage from the membrane-bound complex in a cycle of association/dissociation (Liu *et al.*, 1999). Constitutive binding protein-independent mutant HisP proteins show a 'looser' structure, indicating that they already exist in a disengaged conformation.

The conformational changes taking place during the various stages of the histidine transport process were investigated using Cys-specific crosslinking, circular dichroism (CD) spectroscopy, and intrinsic Trp fluorescence measurements (Kreimer *et al.*, 2000). The $HisQMP_2$ complex contains a total of six Cys residues (two in the HisP subunit), all of which could be modified by reaction with the sulfhydryl reagent, monobromobimane (mBBr). The rates of reaction of the two Cys residues in HisP were different, probably because Cys-51 is buried, as indicated by the X-ray crystal structure (Hung *et al.*, 1998). However, two other positively charged sulfhydryl reagents labeled only half of the available HisP, suggesting the existence of differences in reactivity between the two subunits, which thus do not appear to behave identically. The rate of labeling of HisP with the sulfhydryl-reactive fluorophore, MIANS, was reduced by the presence of MgATP in a saturable manner with a $K_{0.5}$ of ~1 mM, similar to the measured affinity for ATP. These results indicate that a conformational change takes place upon nucleotide binding to HisP, resulting in a change in accessibility to MIANS. Additional experiments using far-UV CD spectroscopy showed an α-helical content of ~43% for the $HisQMP_2$ complex, consistent with the HisP crystal structure and sequence analysis of HisQ and HisM.

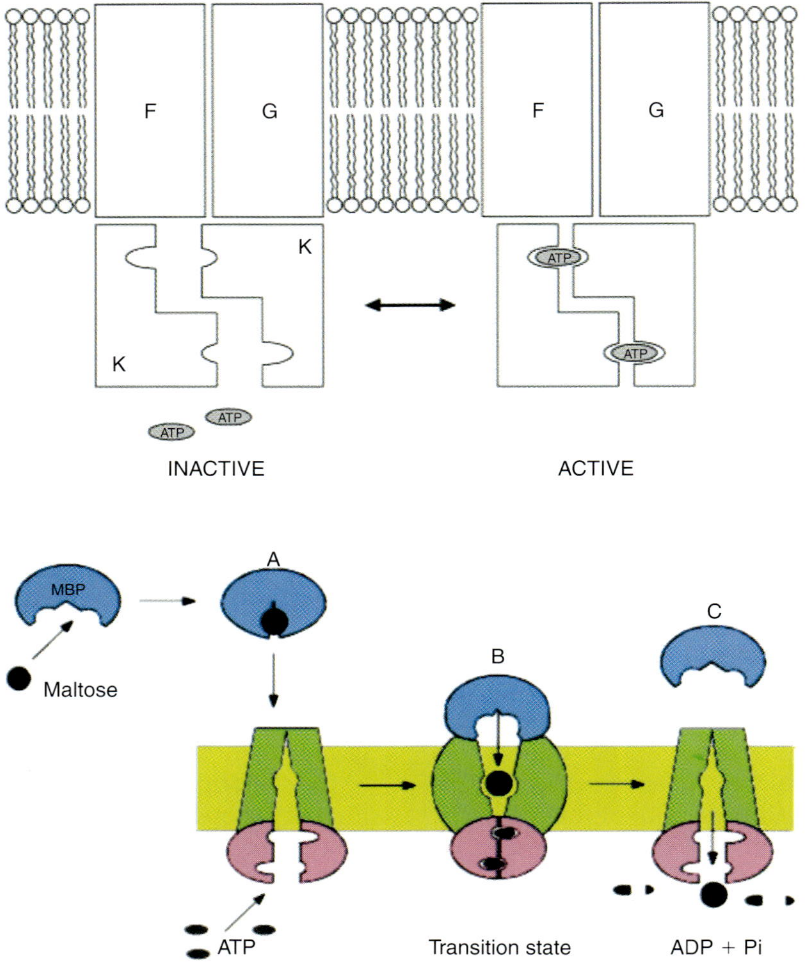

Figure 6.10. *Conformational changes taking place in the maltose permease during ATP hydrolysis and maltose transport, as proposed by Davidson and co-workers (Chen* et al.*, 2001). Top panel: Activation of the ATPase activity of the MalFGK$_2$ complex by dimerization of the MalK subunit. Based on the crystal structure of the ABC protein Rad50cd (Hopfner* et al.*, 2000), the two MalK subunits are aligned in a head-to-tail orientation. The ATP-binding sites are located along the dimer interface, and both MalK subunits contribute residues to the nucleotide-binding sites, providing a mechanism for regulating ATPase activity via MalK subunit association and dissociation within the complex. It was suggested that the MalK subunits are positioned apart from each other in the wild-type permease complex, preventing ATP hydrolysis in the absence of the MalE maltose-binding protein. In this conformation, the Walker A motif of the MalK active site is relatively more accessible to solvent. Association of MalE brings the two MalK subunits closer together, activating the ATPase activity, and making the binding sites less accessible to the aqueous solution. In binding protein-independent mutants, the MalK subunits are proposed to be positioned close together, thus allowing ATP hydrolysis to take place in the absence of the binding protein. Reproduced from Mannering* et al. *(2001) with permission. Bottom panel: Conformational changes taking place during maltose transport. A, The maltose-binding protein (MBP) binds maltose and changes from an open to a closed conformation, generating a high-affinity sugar-binding site. The liganded MBP binds to the MalFGK$_2$ complex to initiate ATP hydrolysis and maltose transport. B, In the transition state, MBP becomes tightly bound to the membrane-bound complex, and both proteins open their binding sites to each other. The maltose is released from MBP and is transferred to a low-affinity sugar-binding site within the MalFGK$_2$ complex. C, Maltose is released at the other side of the membrane and MBP is released. The MalK subunits are shown as undergoing an ATP-induced dimerization and activation step, brought about by binding of liganded MBP. Reproduced from Chen* et al. *(2001) with permission.*

Addition of the binding protein HisJ to the membrane-bound complex led to a reorganization of the α-helical secondary structure, indicative of a conformational change. It was speculated that this change might involve the helical domain (Arm-II) of HisP. This was corroborated by intrinsic Trp fluorescence experiments using a Trp-less HisJ, where the fluorescence of the membrane-bound complex was monitored. Binding of ATP to the complex in the absence of HisJ, or addition of HisJ alone (unliganded HisJ) or HisJ bound to histidine (liganded HisJ) in the absence of ATP, led to a small amount of quenching of $HisQMP_2$. Addition of liganded and unliganded HisJ in the presence of ATP gave rise to higher quenching (although different for each case), suggesting the existence of larger conformational changes when a 'signal' is provided by the binding protein. Dynamic quenching experiments with iodide ions confirmed a significant change in Trp accessibility to the aqueous medium when ATP and liganded HisJ are added to the membrane-bound complex. Since the transport complex is only fully active in the presence of both ATP and liganded HisJ, such changes represent those that would occur during the translocation process.

LmrA

Many bacterial multidrug transporters are drug/H^+ antiporters, rather than ABC proteins. LmrA was the first bacterial ABC multidrug transporter to be identified. The protein appears to be a 'half transporter' (each half comprises six putative TM helices and one NBD) and probably operates as a homodimer in the membrane (van Veen *et al.*, 1999). Based on equilibrium binding, photoaffinity labeling and drug transport experiments, it was proposed that LmrA transports drugs by an alternating two-site mechanism (van Veen *et al.*, 2000a). In the homodimeric state, each LmrA monomer is proposed to possess two drug-binding sites: a high-affinity site on the cytoplasmic face of the membrane, which accepts drug, and another low-affinity site on the extracellular face of the membrane, which releases drug to the exterior. Hydrolysis of ATP is proposed to mediate the interconversion of these two sites, via an intermediate transition state in which the drug transport site is inaccessible. Each 'half' of the homodimer was envisaged as being a drug transport unit, with the two halves operating in tandem.

Ruysschaert and co-workers carried out ATR-FTIR and fluorescence quenching studies of purified LmrA reconstituted into extruded liposomes of *Escherichia coli* lipid and egg PC (Vigano *et al.*, 2000), with the objective of examining the conformational changes taking place in the transporter following binding and hydrolysis of nucleotides. Binding of ATP, ATPγS, or ADP + Pi was associated with a 10% increase in β-sheet secondary structure, coupled with a corresponding decrease in the % β-turn. The α-helical content of the protein (~30%), which represents both the membrane-spanning helices and highly structured domains outside the membrane, remained unchanged after nucleotide binding. Hydrolysis of ATP and release of Pi from the protein were necessary to regain the original secondary structure. It was suggested that this increase in secondary structure, which is not seen for either Pgp (Sonveaux *et al.*, 1996) or MRP1 (Manciu *et al.*, 2000), might be related to ATP-mediated reorganization of LmrA into a dimeric form, which would then dissociate following ATP hydrolysis. Fluorescence experiments indicated that ATP hydrolysis induced the formation of a conformation in which the Trp residues of LmrA have an increased accessibility to the aqueous quencher acrylamide. This conformation is not promoted by binding of ADP + Pi or ATPγS, despite the observed increase in secondary structure, suggesting that these two changes are not linked.

DOMAIN INTERACTIONS IN ABC TRANSPORTERS

The energy obtained from ATP hydrolysis at the catalytic sites in the NBDs of ABC proteins is used to drive translocation of substrates across the membrane. Communication between the different functional domains of ABC proteins is therefore essential for the completion of the catalytic cycle. Such domain communication must come about via conformational changes taking place within the protein. To date, there is considerable experimental evidence supporting the existence of such domain communication, most of it obtained for the MDR1 Pgp.

INTERACTIONS BETWEEN THE TWO NBDs AND THE TMDs OF PGP

Loo and Clarke explored the possible interactions between the membrane-bound domains and the NBDs in the two halves of Pgp by co-expressing each as a separate polypeptide, and testing for associations by co-immunoprecipitation (Loo and Clarke, 1995). When two complete 'half-molecules' (TMD–NBD) were co-expressed, physical association between them was observed, which had previously been noted to restore drug-stimulated ATPase activity (Loo and Clarke, 1994b). Similarly, each membrane-bound domain was found to be associated with the other half-molecule when they were co-expressed, presumably via interactions between the two TM regions in each half. Each of the NBDs could be recovered in association with the membrane-bound domain from the same half of the protein, but not with that from the other half, suggesting that each NBD specifically associates with the appropriate transmembrane domain within each half of Pgp. Similarly, the two NBDs could also be co-immunoprecipitated by antibodies directed towards tags on either the N-terminal or the C-terminal domain, suggesting that they may dimerize.

These results clearly indicate that there are specific non-covalent interactions between the four domains of Pgp, which presumably play an important role in various aspects of its function. Based on structural and biochemical information, it is widely believed that the two NBDs of many ABC transporters function as dimers (Diederichs *et al.*, 2000; Hopfner *et al.*, 2000; Hung *et al.*, 1998; Jones and George, 1999), and the physical association of the two NBDs may be necessary to achieve this functional coupling. Dimerization of the two NBDs may thus be required for ATPase catalytic activity. The association between the NBD and the membrane-bound domain in each half may be important in coupling drug transport to the conformational changes induced by ATP hydrolysis. Such interactions could be mediated by the large cytoplasmic loops that connect the transmembrane segments, mutations in which are known to alter substrate specificity (Loo and Clarke, 1994a). Similar co-expression and co-immunoprecipitation approaches have been used to establish domain interactions in the peptide transporter complex TAP1/TAP1 (Lapinski *et al.*, 2000). In this case, the substrate-binding sites are thought to reside at the boundary of the membrane-bound and cytosolic regions of one TM helix of each protein, and may include some cytosolic loops. These regions also appear to be important for formation of the TAP1/TAP2 complex (Lapinski *et al.*, 2000). The yeast ABC protein STE6 can be functionally reconstituted from separately expressed half-molecules, or a quarter molecule and a three-quarter molecule, and, again, a physical association between the two protein fragments was demonstrated by co-immunoprecipitation (Berkower *et al.*, 1996).

POTENTIAL ROLE OF THE LINKER DOMAIN OF PGP

The two homologous halves of Pgp are connected by a linker region of approximately 75 amino acid residues, which is highly divergent in MDR1 transporters across many species (Germann, 1996). Deletion of a substantial region in the central core of this linker (Δ34) led to expression of a nonfunctional Pgp molecule that could not confer drug resistance, whereas insertion of a flexible 17-residue segment had no effect (Hrycyna *et al.*, 1998). The deletion mutant was able to bind drug and nucleotide, but showed no drug-stimulated ATP hydrolysis or drug transport activity, and did not bind the conformationally sensitive antibody UIC2 under conditions where the wild-type protein was able to do so, suggesting that the conformation was different from that of wild-type Pgp. The linker region was previously implicated in the formation of Pgp dimers (Juvvadi *et al.*, 1997). Expression of the N-terminal 48 amino acids of the linker region as a peptide was sufficient to induce the formation of stable dimers. However, later work by Hrycyna *et al.* (1998) indicated that the integrity of the linker region does not appear to be essential for Pgp function. Normal function of the Δ34 mutant discussed above could be fully restored by replacement of the deleted region with a flexible 17-residue peptide. Overall, it appears that a region of flexible secondary structure is probably all that is required for Pgp to attain a functional conformation, possibly by allowing the NBDs in the two halves of the protein to interact with each other in the correct manner.

REFERENCES

Ambudkar, S.V., Cardarelli, C.O., Pashinsky, I. and Stein, W.D. (1997) Relation between the turnover number for vinblastine transport and for vinblastine-stimulated ATP hydrolysis by human P-glycoprotein. J. Biol. Chem. 272, 21160–21166.

Austin, R.P., Davis, A.M. and Manners, C.N. (1995) Partitioning of ionizing molecules between aqueous buffers and phospholipid vesicles. J. Pharm. Sci. 84, 1180–1183.

Berkower, C., Taglicht, D. and Michaelis, S. (1996) Functional and physical interactions between partial molecules of STE6, a yeast ATP-binding cassette protein. J. Biol. Chem. 271, 22983–22989.

Callaghan, R. and Riordan, J.R. (1995) Collateral sensitivity of multidrug resistant cells to narcotic analgesics is due to effects on the plasma membrane. Biochim. Biophys. Acta 1236, 155–162.

Callaghan, R., Stafford, A. and Epand, R.M. (1993) Increased accumulation of drugs in a multidrug resistant cell line by alteration of membrane biophysical properties. Biochim. Biophys. Acta 1175, 277–282.

Chen, J., Sharma, S., Quiocho, F.A. and Davidson, A.L. (2001) Trapping the transition state of an ATP-binding cassette transporter: evidence for a concerted mechanism of maltose transport. Proc. Natl Acad. Sci. USA 98, 1525–1530.

Dekkers, D.W.C., Comfurius, P., Schroit, A.J., Bevers, E.M. and Zwaal, R.F.A. (1998) Transbilayer movement of NBD-labeled phospholipids in red blood cell membranes: outward-directed transport by the multidrug resistance protein 1 (MRP1). Biochemistry 37, 14833–14837.

Dey, S., Ramachandra, M., Pastan, I., Gottesman, M.M. and Ambudkar, S.V. (1997) Evidence for two nonidentical drug-interaction sites in the human P-glycoprotein. Proc. Natl Acad. Sci. USA 94, 10594–10599.

Diederichs, K., Diez, J., Greller, G., Muller, C., Breed, J., Schnell, C., Vonrhein, C., Boos, W. and Welte, W. (2000) Crystal structure of MalK, the ATPase subunit of the trehalose/maltose ABC transporter of the archaeon *Thermococcus litoralis*. EMBO J. 19, 5951–5961.

Druley, T.E., Stein, W.D. and Roninson, I.B. (2001) Analysis of MDR1 P-glycoprotein conformational changes in permeabilized cells using differential immunoreactivity. Biochemistry 40, 4312–4322.

Eytan, G.D., Regev, R. and Assaraf, Y.G. (1996a) Functional reconstitution of P-glycoprotein reveals an apparent near stoichiometric drug transport to ATP hydrolysis. J. Biol. Chem. 271, 3172–3178.

Eytan, G.D., Regev, R., Oren, G. and Assaraf, Y.G. (1996b) The role of passive transbilayer drug movement in multidrug resistance and its modulation. J. Biol. Chem. 271, 12897–12902.

Eytan, G.D., Regev, R., Oren, G., Hurwitz, C.D. and Assaraf, Y.G. (1997) Efficiency of P-glycoprotein-mediated exclusion of rhodamine dyes from multidrug-resistant cells is determined by their passive transmembrane movement rate. Eur. J. Biochem. 248, 104–112.

Ferry, D., Boer, R., Callaghan, R. and Ulrich, W.R. (2000) Localization of the 1,4-dihydropyridine drug acceptor of P-glycoprotein to a cytoplasmic domain using a permanently charged derivative N-methyl dexniguldipine. Int. J. Clin. Pharmacol. Ther. 38, 130–140.

Germann, U.A. (1996) P-glycoprotein – a mediator of multidrug resistance in tumour cells. Eur. J. Cancer 32A, 927–944.

Goormaghtigh, E., Raussens, V. and Ruysschaert, J.M. (1999) Attenuated total reflection infrared spectroscopy of proteins and lipids in biological membranes. Biochim. Biophys. Acta 1422, 105–185.

Gorbulev, S., Abele, R. and Tampé, R. (2001) Allosteric crosstalk between peptide-binding, transport, and ATP hydrolysis of the ABC transporter TAP. Proc. Natl Acad. Sci. USA 98, 3732–3737.

Higgins, C.F. and Gottesman, M.M. (1992) Is the multidrug transporter a flippase? Trends Biochem. Sci. 17, 18–21.

Higgins, C.F., Hiles, I.D., Salmond, G.P.C., Gill, D.R., Downie, J.A., Evans, I.J., *et al.* (1986) A family of related ATP-binding subunits coupled to many distinct biological processes in bacteria. Nature 323, 448–450.

Hipfner, D.R., Deeley, R.G. and Cole, S.P.C. (1999) Structural, mechanistic and clinical aspects of MRP1. Biochim. Biophys. Acta 1461, 359–376.

Holland, I.B. and Blight, M.A. (1999) ABC-ATPases, adaptable energy generators fuelling transmembrane movement of a variety of molecules in organisms from bacteria to humans. J. Mol. Biol. 293, 381–399.

Homolya, L., Hollo, Z., Germann, U.A., Pastan, I., Gottesman, M.M. and Sarkadi, B. (1993) Fluorescent cellular indicators are extruded by the multidrug resistance protein. J. Biol. Chem. 268, 21493–21496.

Hopfner, K.P., Karcher, A., Shin, D.S., Craig, L., Arthur, L.M., Carney, J.P. and Tainer, J.A. (2000) Structural biology of Rad50 ATPase: ATP-driven conformational control in DNA double-strand break repair and the ABC-ATPase superfamily. Cell 101, 789–800.

Hou, Y., Cui, L., Riordan, J.R. and Chang, X. (2000) Allosteric interactions between the two non-equivalent nucleotide binding domains of multidrug resistance protein MRP1. J. Biol. Chem. 275, 20280–20287.

Hrycyna, C.A., Airan, L.E., Germann, U.A., Ambudkar, S.V., Pastan, I. and Gottesman, M.M. (1998) Structural flexibility of the linker region of human P-glycoprotein permits ATP hydrolysis and drug transport. Biochemistry 37, 13660–13673.

Hung, L.W., Wang, I.X., Nikaido, K., Liu, P.Q., Ames, G.F. and Kim, S.H. (1998) Crystal structure of the ATP-binding subunit of an ABC transporter. Nature 396, 703–707.

Hunke, S., Mourez, M., Jéhanno, M., Dassa, E. and Schneider, E. (2000) ATP modulates subunit–subunit interactions in an ATP-binding cassette transporter ($MalFGK_2$) determined by site-directed chemical cross-linking. J. Biol. Chem. 275, 15526–15534.

Jones, P.M. and George, A.M. (1999) Subunit interactions in ABC transporters: towards a functional architecture. FEMS Microbiol. Lett. 179, 187–202.

Julien, M. and Gros, P. (2000) Nucleotide-induced conformational changes in P-glycoprotein and in nucleotide binding site mutants monitored by trypsin sensitivity. Biochemistry 39, 4559–4568.

Juvvadi, S.R., Glavy, J.S., Horwitz, S.B. and Orr, G.A. (1997) Domain organization of murine mdr1b P-glycoprotein: the cytoplasmic linker region is a potential dimerization domain. Biochem. Biophys. Res. Commun. 230, 442–447.

Kamp, D. and Haest, C.W. (1998) Evidence for a role of the multidrug resistance protein (MRP) in the outward translocation of NBD-phospholipids in the erythrocyte membrane. Biochim. Biophys. Acta 1372, 91–101.

Kerr, K.M., Sauna, Z.E. and Ambudkar, S.V. (2001) Correlation between steady-state ATP hydrolysis and vanadate-induced ADP trapping in human P-glycoprotein – evidence for ADP release as the rate-limiting step in the catalytic cycle and its modulation by substrates. J. Biol. Chem. 276, 8657–8664.

Kessel, D. (1989) Exploring multidrug resistance using rhodamine 123. Cancer Commun. 1, 145–149.

Kessel, D., Woodburn, K., Decker, D. and Sykes, E. (1995) Fractionation of Cremophor EL delineates components responsible for plasma lipoprotein alterations and multidrug resistance reversal. Oncol. Res. 7, 207–212.

Krämer, S.D., Braun, A., Jakits-Deiser, C. and Wunderli-Allenspach, H. (1998). Towards the predictability of drug-lipid membrane interactions: the pH-dependent affinity of propranolol to phosphatidylinositol containing liposomes. Pharm. Res. 15, 739–744.

Kreimer, D.I., Chai, K.P. and Ames, G.F.L. (2000) Nonequivalence of the nucleotide-binding subunits of an ABC transporter, the histidine permease, and conformational changes in the membrane complex. Biochemistry 39, 14183–14195.

Lapinski, P.E., Miller, G.G., Tampé, R. and Raghavan, M. (2000) Pairing of the nucleotide binding domains of the transporter associated with antigen processing. J. Biol. Chem. 275, 6831–6840.

Liu, P.Q. and Ames, G.F. (1998) In vitro disassembly and reassembly of an ABC transporter, the histidine permease. Proc. Natl Acad. Sci. USA 95, 3495–3500.

Liu, P.Q., Liu, C.E. and Ames, G.F.L. (1999) Modulation of ATPase activity by physical disengagement of the ATP-binding domains of an ABC transporter, the histidine permease. J. Biol. Chem. 274, 18310–18318.

Liu, R. and Sharom, F.J. (1996) Site-directed fluorescence labeling of P-glycoprotein on cysteine residues in the nucleotide binding domains. Biochemistry 35, 11865–11873.

Liu, R. and Sharom, F.J. (1997) Fluorescence studies on the nucleotide binding domains of the P-glycoprotein multidrug transporter. Biochemistry 36, 2836–2843.

Liu, R., Siemiarczuk, A. and Sharom, F.J. (2000) Intrinsic fluorescence of the P-glycoprotein multidrug transporter: sensitivity of tryptophan residues to binding of drugs and nucleotides. Biochemistry 39, 14927–14938.

Loe, D.W., Deeley, R.G. and Cole, S.P. (1998) Characterization of vincristine transport by the M(r) 190,000 multidrug resistance protein (MRP): evidence for cotransport with reduced glutathione. Cancer Res. 58, 5130–5136.

Loo, T.W. and Clarke, D.M. (1994a) Functional consequences of glycine mutations in the predicted cytoplasmic loops of P-glycoprotein. J. Biol. Chem. 269, 7243–7248.

Loo, T.W. and Clarke, D.M. (1994b) Reconstitution of drug-stimulated ATPase activity following co-expression of each half of human P-glycoprotein as separate polypeptides. J. Biol. Chem. 269, 7750–7755.

Loo, T.W. and Clarke, D.M. (1995) P-glycoprotein. Associations between domains and between domains and molecular chaperones. J. Biol. Chem. 270, 21839–21844.

Loo, T.W. and Clarke, D.M. (1997) Drug-stimulated ATPase activity of human P-glycoprotein requires movement between transmembrane segments 6 and 12. J. Biol. Chem. 272, 20986–20989.

Loo, T.W. and Clarke, D.M. (2000) Drug-stimulated ATPase activity of human P-glycoprotein is blocked by disulfide cross-linking between the nucleotide-binding sites. J. Biol. Chem. 275, 19435–19438.

Lu, P., Liu, R. and Sharom, F.J. (2001) Drug transport by reconstituted P-glycoprotein in proteoliposomes – effect of substrates and modulators, and dependence on bilayer phase state. Eur. J. Biochem. 268, 1687–1697.

Manciu, L., Chang, X.B., Riordan, J.R. and Ruysschaert, J.M. (2000) Multidrug resistance protein MRP1 reconstituted into lipid vesicles: secondary structure and nucleotide-induced tertiary structure changes. Biochemistry 39, 13026–13033.

Manciu, L., Chang, X., Riordan, J.R., Buyse, F. and Ruysschaert, J.M. (2001) Nucleotide-induced conformational changes in the human multidrug resistance protein MRP1 are related to the capacity of chemotherapeutic drugs to accumulate or not in resistant cells. FEBS Lett. 493, 31–35.

Mannering, D.E., Sharma, S. and Davidson, A.L. (2001) Demonstration of conformational changes associated with activation of the maltose transport complex. J. Biol. Chem. 276, 12362–12368.

Margolles, A., Putman, M., van Veen, H.W. and Konings, W.N. (1999) The purified and functionally reconstituted multidrug transporter LmrA of *Lactococcus lactis* mediates the transbilayer movement of specific fluorescent phospholipids. Biochemistry 38, 16298–16306.

Mechetner, E.B., Schott, B., Morse, B.S., Stein, W.D., Druley, T., Davis, K.A., Tsuruo, T. and Roninson, I.B. (1997) P-glycoprotein function involves conformational transitions detectable by differential immunoreactivity. Proc. Natl Acad. Sci. USA 94, 12908–12913.

Nagata, K., Nishitani, M., Matsuo, M., Kioka, N., Amachi, T. and Ueda, K. (2000) Nonequivalent nucleotide trapping in the two nucleotide binding folds of the human multidrug resistance protein MRP1. J. Biol. Chem. 275, 17626–17630.

Nagy, H., Goda, K., Arceci, R., Cianfriglia, M., Mechetner, E. and Szabó, G., Jr (2001) P-glycoprotein conformational changes detected by antibody competition. Eur. J. Biochem. 268, 2416–2420.

Neumann, L. and Tampé, R. (1999) Kinetic analysis of peptide binding to the TAP transport complex: evidence for structural rearrangements induced by substrate binding. J. Mol. Biol. 294, 1203–1213.

Nikaido, K. and Ames, G.F.L. (1999) One intact ATP-binding subunit is sufficient to support ATP hydrolysis and translocation in an ABC transporter, the histidine permease. J. Biol. Chem. 274, 26727–26735.

Pantaler, E., Kamp, D. and Haest, C.W.M. (2000) Acceleration of phospholipid flip-flop in the erythrocyte membrane by detergents differing in polar head group and alkyl chain length. Biochim. Biophys. Acta 1509, 397–408.

Pawagi, A.B., Wang, J., Silverman, M., Reithmeier, R.A. and Deber, C.M. (1994) Transmembrane aromatic amino acid distribution in P-glycoprotein. A functional role in broad substrate specificity. J. Mol. Biol. 235, 554–564.

Qu, Q. and Sharom, F.J. (2002) Proximity of bound Hoechst 33342 to the ATPase catalytic sites places the drug binding site of P-glycoprotein within the cytoplasmic membrane leaflet. Biochemistry 41, 4744–4752.

Raggers, R.J., van Helvoort, A., Evers, R. and van Meer, G. (1999) The human multidrug resistance protein MRP1 translocates sphingolipid analogs across the plasma membrane. J. Cell Sci. 112, 415–422.

Ramachandra, M., Ambudkar, S.V., Chen, D., Hrycyna, C.A., Dey, S., Gottesman, M.M. and Pastan, I. (1998) Human P-glycoprotein exhibits reduced affinity for substrates during a catalytic transition state. Biochemistry 37, 5010–5019.

Raviv, Y., Pollard, H.B., Bruggemann, E.P., Pastan, I. and Gottesman, M.M. (1990) Photosensitized labeling of a functional multidrug transporter in living drug-resistant tumor cells. J. Biol. Chem. 265, 3975–3980.

Reits, E.A., Vos, J.C., Gromme, M. and Neefjes, J. (2000a) The major substrates for TAP in vivo are derived from newly synthesized proteins. Nature 404, 774–778.

Reits, E.A.J., Griekspoor, A.C. and Neefjes, J. (2000b) How does TAP pump peptides?

Insights from DNA repair and traffic ATPases. Immunol. Today 21, 598–600.

Rodrigues, C., Gameiro, P., Reis, S., Lima, J.L.F.C. and De Castro, B. (2001) Spectrophotometric determination of drug partition coefficients in dimyristoyl-L-a-phosphatidylcholine/water: a comparative study using phase separation and liposome suspensions. Anal. Chim. Acta 428, 103–109.

Rogers, J.A. and Davis, S.S. (1980) Functional group contributions to the partitioning of phenols between liposomes and water. Biochim. Biophys. Acta 598, 392–404.

Romsicki, Y. and Sharom, F.J. (1999) The membrane lipid environment modulates drug interactions with the P-glycoprotein multidrug transporter. Biochemistry 38, 6887–6896.

Romsicki, Y. and Sharom, F.J. (2001) Phospholipid flippase activity of the reconstituted P-glycoprotein multidrug transporter. Biochemistry 40, 6937–6947.

Rosenberg, M.F., Velarde, G., Ford, R.C., Martin, C., Berridge, G., Kerr, I.D., *et al.* (2001) Repacking of the transmembrane domains of P-glycoprotein during the transport ATPase cycle. EMBO J. 20, 5615–5625.

Ruetz, S. and Gros, P. (1994) Phosphatidylcholine translocase: a physiological role for the mdr2 gene. Cell 77, 1071–1081.

Sauna, Z.E. and Ambudkar, S.V. (2000) Evidence for a requirement for ATP hydrolysis at two distinct steps during a single turnover of the catalytic cycle of human P-glycoprotein. Proc. Natl Acad. Sci. USA 97, 2515–2520.

Sauna, Z.E. and Ambudkar, S.V. (2001) Characterization of the catalytic cycle of ATP hydrolysis by human P-glycoprotein – the two ATP hydrolysis events in a single catalytic cycle are kinetically similar but affect different functional outcomes. J. Biol. Chem. 276, 11653–11661.

Schneider, E. and Hunke, S. (1998) ATP-binding-cassette (ABC) transport systems: functional and structural aspects of the ATP-hydrolyzing subunits/domains. FEMS Microbiol. Rev. 22, 1–20.

Schneider, E., Wilken, S. and Schmid, R. (1994) Nucleotide-induced conformational changes of MalK, a bacterial ATP binding cassette transporter protein. J. Biol. Chem. 269, 20456–20461.

Senior, A.E., al-Shawi, M.K. and Urbatsch, I.L. (1995) The catalytic cycle of P-glycoprotein. FEBS Lett. 377, 285–289.

Shapiro, A.B. and Ling, V. (1997) Extraction of Hoechst 33342 from the cytoplasmic leaflet of the plasma membrane by P-glycoprotein. Eur. J. Biochem. 250, 122–129.

Shapiro, A.B. and Ling, V. (1998a) Stoichiometry of coupling of rhodamine 123 transport to ATP hydrolysis by P-glycoprotein. Eur. J. Biochem. 254, 189–193.

Shapiro, A.B. and Ling, V. (1998b) Transport of LDS-751 from the cytoplasmic leaflet of the plasma membrane by the rhodamine-123-selective site of P-glycoprotein. Eur. J. Biochem. 254, 181–188.

Sharma, S. and Davidson, A.L. (2000) Vanadate-induced trapping of nucleotides by purified maltose transport complex requires ATP hydrolysis. J. Bacteriol. 182, 6570–6576.

Sharom, F.J. (1997) The P-glycoprotein efflux pump: how does it transport drugs? J. Membr. Biol. 160, 161–175.

Sharom, F.J., Liu, R. and Romsicki, Y. (1998a) Spectroscopic and biophysical approaches for studying the structure and function of the P-glycoprotein multidrug transporter. Biochem. Cell Biol. 76, 695–708.

Sharom, F.J., Lu, P., Liu, R. and Yu, X. (1998b) Linear and cyclic peptides as substrates and modulators of P-glycoprotein: peptide binding and effects on drug transport and accumulation. Biochem. J. 333, 621–630.

Sharom, F.J., Liu, R., Romsicki, Y. and Lu, P. (1999) Insights into the structure and substrate interactions of the P-glycoprotein multidrug transporter from spectroscopic studies. Biochim. Biophys. Acta 1461, 327–345.

Sinicrope, F.A., Dudeja, P.K., Bissonnette, B.M., Safa, A.R. and Brasitus, T.A. (1992) Modulation of P-glycoprotein-mediated drug transport by alterations in lipid fluidity of rat liver canalicular membrane vesicles. J. Biol. Chem. 267, 24995–25002.

Sonveaux, N., Shapiro, A.B., Goormaghtigh, E., Ling, V. and Ruysschaert, J.M. (1996) Secondary and tertiary structure changes of reconstituted P-glycoprotein. A Fourier transform attenuated total reflection infrared spectroscopy analysis. J. Biol. Chem. 271, 24617–24624.

Sonveaux, N., Vigano, C., Shapiro, A.B., Ling, V. and Ruysschaert, J.M. (1999) Ligand-mediated tertiary structure changes of reconstituted P-glycoprotein. A tryptophan fluorescence quenching analysis. J. Biol. Chem. 274, 17649–17654.

Spudich, J.A. (1994) How molecular motors work. Nature 372, 515–518.

Van Endert, P.M. (1999) Role of nucleotides and peptide substrate for stability and functional state of the human ABC family transporters associated with antigen processing. J. Biol. Chem. 274, 14632–14638.

van Helvoort, A., Smith, A.J., Sprong, H., Fritzsche, I., Schinkel, A.H., Borst, P. and van Meer, G. (1996) MDR1 P-glycoprotein is a lipid translocase of broad specificity, while MDR3 P-glycoprotein specifically translocates phosphatidylcholine. Cell 87, 507–517.

van Veen, H.W., Putman, M., Margolles, A., Sakamoto, K. and Konings, W.N. (1999) Structure-function analysis of multidrug transporters in *Lactococcus lactis*. Biochim. Biophys. Acta 1461, 201–206.

van Veen, H.W., Margolles, A., Müller, M., Higgins, C.F. and Konings, W.N. (2000a) The homodimeric ATP-binding cassette transporter LmrA mediates multidrug transport by an alternating two-site (two-cylinder engine) mechanism. EMBO J. 19, 2503–2514.

van Veen, H.W., Putman, M., Margolles, A., Sakamoto, K. and Konings, W.N. (2000b) Molecular pharmacological characterization of two multidrug transporters in *Lactococcus lactis*. Pharmacol. Ther. 85, 245–249.

Vigano, C., Margolles, A., van Veen, H.W., Konings, W.N. and Ruysschaert, J.M. (2000) Secondary and tertiary structure changes of reconstituted LmrA induced by nucleotide binding or hydrolysis. A Fourier transform attenuated total reflection infrared spectroscopy and tryptophan fluorescence quenching analysis. J. Biol. Chem. 275, 10962–10967.

Wang, G., Pincheira, R., Zhang, M. and Zhang, J.T. (1997) Conformational changes of P-glycoprotein by nucleotide binding. Biochem. J. 328, 897–904.

Wang, G., Pincheira, R. and Zhang, J.T. (1998) Dissection of drug-binding-induced conformational changes in P-glycoprotein. Eur. J. Biochem. 255, 383–390.

Woodcock, D.M., Linsenmeyer, M.E., Chojnowski, G., Kriegler, A.B., Nink, V., Webster, L.K. and Sawyer, W.H. (1992) Reversal of multidrug resistance by surfactants. Br. J. Cancer 66, 62–68.

Zeng, Y.C., Han, X.L. and Gross, R.W. (1999) Phospholipid-subclass-specific partitioning of lipophilic ions in membrane-water systems. Biochem. J. 338, 651–658.

CHAPTER

7

X-ray Structure of an Intact ABC Transporter, MsbA

Christopher B. Roth and Geoffrey A. Chang

Introduction: The Multidrug Resistance (MDR) ABC Transporters

ABC exporters transport a diverse array of substrates, including peptides, toxins, lipids and hydrophobic drug molecules, from the cytoplasmic side of the cell membrane to either the outer membrane leaflet or the outside of the cell. Overexpression of a subset of these exporters is the most frequent cause of resistance to cytotoxic agents including antibiotics and anticancer drugs. These ABC transporters have been likened to 'hydrophobic vacuum cleaners' because of their ability to remove drugs from the inner membrane leaflet (see Chapter 12). Many of these transporters are believed to be 'flippases', transporting or 'flipping' drugs and/or lipids from the inner to the outer membrane leaflet. Some of the best-studied MDR-ABC transporters are the human P-glycoprotein (Pgp) and other related drug transporters. Human Pgp is located in the plasma membrane of numerous cell types and transports a remarkably broad array of hydrophobic compounds (Ambudkar *et al.*, 1999). Overexpression of human Pgp in the plasma membrane can confer resistance to chemotherapeutic drugs by intercepting them in the membrane and pumping them to the extracellular medium or the outer membrane leaflet. Human Pgp is a 170 kDa polypeptide that has the four domains of a typical ABC transporter. The two transmembrane domains each consist of six predicted membrane-spanning α-helices separated by hydrophilic loops. These transmembrane domains (TMDs) are thought to recognize substrates and form a pathway by which hydrophobic compounds are transported across the cell membrane. The nucleotide-binding domains (NBDs), are located at the cytoplasmic face of the membrane and couple ATP hydrolysis to substrate transport.

Recent advances in microbial genome sequencing projects have revealed prokaryotic ABC exporters that are similar in protein sequence to human Pgp. Protein sequence alignment using the TMDs of selected ABC exporters using the program ClustalW reveals a distinct phylogenetic grouping **(Figure 7.1)**. One of these, LmrA, from the bacterium *Lactococcus lactis* is a close functional homologue of human Pgp (van Veen *et al.*, 1998). Whereas human Pgp consists of four fused domains, LmrA is organized as a homodimer of two polypeptides (van Veen *et al.*, 2001). Hence, LmrA and most bacterial mdr-ABC transporters are called half

**Parts of this chapter are reprinted with permission from the American Association for the Advancement of Science. Please see Acknowledgments section for details.*

ABC Proteins: From Bacteria to Man
ISBN 0-12-352551-9

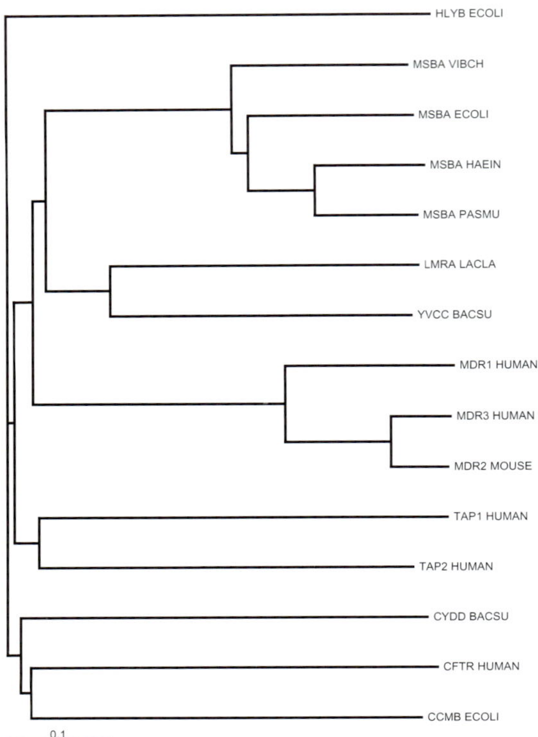

Figure 7.1. ***Phylogenetic tree of the transmembrane domain portion of various ABC efflux pump genes (ClustalW with the PHYLIP output option). The GENBANK notations are used for gene and species designation.***

transporters with each polypeptide containing a TMD and an NBD. LmrA and the human Pgp extrude a remarkably similar spectrum of amphiphilic cationic compounds.

By protein sequence comparison using scores derived from the BLAST algorithm, the lipid flippase MsbA from *Escherichia coli* (Eco-MsbA) is one of the closest bacterial ABC transporters to human Pgp **(Figure 7.1)**. The complete Eco-MsbA transporter is predicted to be a functional homodimer with a total molecular mass of 129.2 kDa. MsbA transports lipid A, a major component of the bacterial outer cell membrane, and is essential for cell viability (Karow and Georgopoulos, 1993). Loss of MsbA expression in the cell membrane, or a disruption of transport by mutation, results in a lethal accumulation of lipid A in the cytoplasmic leaflet (Doerrler *et al.*, 2001; Zhou *et al.*, 1998). Numerous bacterial homologues of MsbA have been reported in more than 30 divergent prokaryotic species (McDonald *et al.*, 1997).

STRUCTURAL STUDIES OF MDR-ABC TRANSPORTERS

Although a significant effort has been focused on understanding the biochemical mechanism of substrate translocation of ABC transporters, the evidence gathered to date has lacked a structural foundation on which to build a complete functional model. Some of the key unresolved issues include the structural basis of coupling ATP hydrolysis to the movement of substrates and the achievement of specificity by the TMD. One of the early pioneering attempts to generate a three-dimensional (3-D) model of an intact MDR-ABC transporter used cryo-electron microscopy to study both single particles and 2-D crystalline arrays of human Pgp (Rosenberg *et al.*, 1997). In this groundbreaking work, a 25 Å EM structure revealed a significant chamber with no evidence of close contact between the NBDs. Cryo-EM studies of Pgp trapped in distinct catalytic states revealed dramatic rearrangements of the TMD during the transport cycle. Although the precise boundaries of the transmembrane and NBD elements cannot be unambiguously determined, there is a substantial opening in the plane of the cell membrane that was clearly resolved and appears to form a chamber that could accept substrates directly from the lipid bilayer. The existence of a chamber within the bilayer is consistent with biochemical data supporting a 'flippase' model for human Pgp, and potentially other members of the MDR-ABC transporter family (Higgins and Gottesman, 1992).

The recent X-ray structure of the lipid flippase MsbA from *E. coli* at 4.5 Å in resolution establishes the overall structural architecture of an ABC transporter and suggests a model for the structural basis of the flipping mechanism that moves hydrophobic substrates from the inner to the outer membrane leaflet of the cell membrane (Chang and Roth, 2001). The close protein sequence homology to other MDR-ABC exporters strongly suggests a common mechanism that is general for this family. As Eco-MsbA is the first member of the ABC transporter family to be elucidated by X-ray crystallography, we will overview some of the techniques and methods used for membrane protein structure determination.

PROTEIN EXPRESSION

Membrane protein X-ray structure determination of transporters and ion channels presents new challenges in several areas including: the relatively lower natural abundance of these molecules in the cell membrane; difficulties in protein purification due to the presence of detergent; and disorder caused by inherent movements of the transmembrane α-helices. Because of these difficulties, we adopted a general strategy of rapidly exploring crystallization space by cloning, overexpressing, and purifying more than 20 full-length bacterial ABC transporters from 12 bacterial species. Our anticipation was that one or more of these naturally occurring proteins would be optimal for protein expression, purification and crystal formation. Plasmids expressing full-length MsbA were generated by excision of the target genes amplified by polymerase chain reaction (PCR). The PCR products were cloned into Novagen pET 19b expression vector, which contains a leader sequence containing an amino-terminal decahistidine tag. The ligated DNA product vector was then transformed into the XL-1 (Strategene) *E. coli* strain. Testing of protein expression was performed by growing a single clone overnight in Terrific Broth (TB; 12 g tryptone, 24 g yeast) containing carbenicillin (50 μg ml^{-1}) and then transferring 5 ml of this culture to TB for trial protein expression. Cultures were induced for expression during mid-log phase ($OD_{600} \approx$ 0.6–0.8) using isopropyl-β-D-thiogalactoside (IPTG).

Large-scale expression of MsbA was accomplished by batch fermentation in an 80-liter BioPilot (New Brunswick Scientific). A small overnight culture from a single bacterial clone was grown to $OD_{600} \approx 0.6$–0.8 before transferring into the fermentor containing TB + carbenicillin (100 μg/μl^{-1}) and 3% glycerol as a supplemental carbon source. Oxygen saturation was maintained above 40% during fermentation by adjusting the agitation and air flow. Protein expression was induced for 3 hours with 3 mM IPTG when the cell density approached an $OD_{600} \approx 7$–8. The entire fermented culture was harvested in under an hour by centrifugation (6000 × *g*). Cell pellets (0.6–1.2 kg) were flash frozen in liquid nitrogen and then stored at −80°C to facilitate lysing and subsequent purification.

PROTEIN PURIFICATION

The general purification strategy involved the (1) preparation of cell membranes, (2) solubilization in one or more detergents, and (3) protein purification by metal affinity, ion exchange, and size exclusion chromatography. In the case of MsbA from *E. coli*, the alpha isomer of dodecylmaltoside (α-DDM) was used to solubilize and extract MsbA directly from the processed membranes as an intact homodimer, as suggested by gel filtration. The bacterial cell wall was removed from the crude cell membranes by freeze-thawing and stirring the cell paste for 2 hours at 4°C in cracking buffer (20 mM Tris pH 7.5, 100 mM NaCl, 20 mM imidazole pH 8.0) containing lysozyme (0.5 mg l^{-1}). Following centrifugation at 6000 × *g* for 40 minutes, the cells were resuspended in lysing buffer (20 mM Tris pH 7.5, 20 mM NaC1) containing DNase I (0.1 mg l^{-1}) and stirred at 4°C for 2 hours. Cell membranes were then harvested by centrifugation at 6000 × *g* for 30 minutes and resuspended in an equal volume of solubilization buffer (20 mM Tris pH 7.5, 100 mM NaCl, 30 mM imidazole pH 8.0) containing 1% dodecyl-α-D-maltoside (α-DDM, Anatrace, Maumee, WI) for 3 hours. The insoluble fraction was separated from the crude detergent-solubilized extract at 29 000 × *g* for 1 hour.

The solubilized MsbA protein was purified first by metal affinity chromatography using a column loaded with nickel-NTA superflow resin (Qiagen) or nickel-NTA fast-flow sepharose (Pharmacia Biotech) that was in a low-salt buffer. Following an extensive wash, nonspecific binding was reduced with high salt buffer. The column was then re-equilibrated with low-salt buffer and another stringency step was introduced using 60 mM imidazole buffer. The MsbA protein, which was still bound to the resin, was eluted using 300 mM imidazole buffer and the resultant fractions combined and diluted fourfold. The protein was then loaded on an anion exchange column containing Source 30Q or Source 15Q resin (Amersham-Pharmacia). Following a brief wash, the protein was eluted by a single high-salt step. At this stage, a few minor contaminants were observed and eliminated using a preparative gel filtration chromatography using Superdex 26/60 columns. Eco-MsbA runs as a dimer as confirmed by running against protein molecular weight markers by gel filtration chromatography.

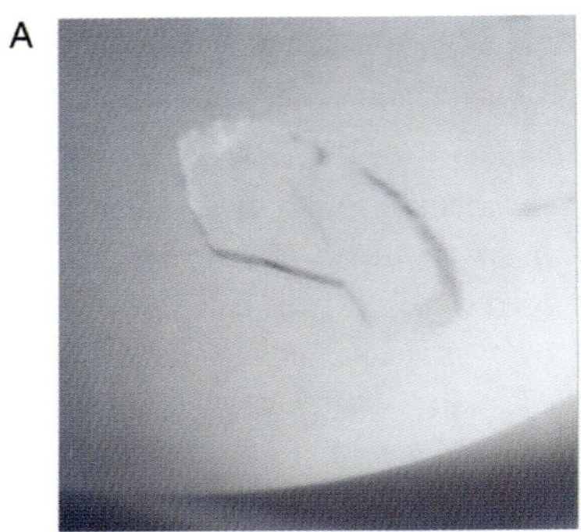

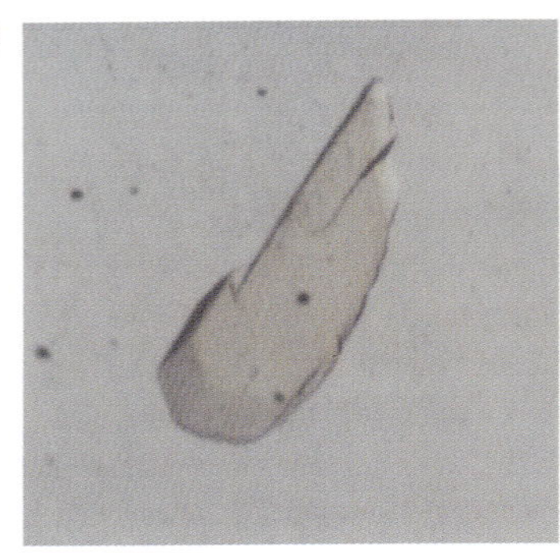

Figure 7.2. ***Crystals of Eco-MsbA grown in 0.05% α-DDM. A, Two fused crystals, which were split to use in data collection as a native and derivative. B, Color change of Eco-MsbA crystal upon uptake of $OsCl_3$, which significantly improved the diffraction quality and resolution.***

The purified MsbA transporter protein was concentrated by a smaller Source 30Q column and then desalted using a HiPrep 26/10 desalting column (Amersham-Pharmacia) into a stabilizing buffer (20 mM Tris pH 7.5, 20 mM NaCl, 0.05% α-DDM) and modestly concentrated using a Centricon 100 filtration concentrator (Amicon) prior to freezing aliquots of the protein at −80°C. Final yields of purified MsbA transporter of about 0.5 mg g^{-1} of cell were typical. Protein purity was assayed by sodium dodecylsulfate (SDS) gel electrophoresis with Coomassie staining, Western blotting by anti-histidine tag antibodies, and monitored using MALDI-TOF (matrix-assisted laser desorption/ionization – time of flight) mass spectroscopy at the Scripps Center for Mass Spectrometry. Samples of MsbA were exceptionally clean with molecular weights within 0.05% of expected.

PROTEIN CRYSTALLIZATION

The crystallization of integral membrane proteins is still a highly empirical process involving a large number of important parameters, which include precipitating reagents, temperature, pH, salt, protein concentration, detergent concentration and detergent type. As such, we adopted an approach of screening several detergents and crystallization conditions in parallel using experiments derived from the X-ray structure determination of the mechanosensitive ion channel from *Mycobacterium tuberculosis* (Chang *et al.*, 1998). After screening and refining approximately 96 000 crystallization conditions for 12 MDR bacterial ABC transporters and detergents, we have grown and analyzed nearly 35 distinct crystal forms. At present, the structure of MsbA from *E. coli* (Eco-MsbA) in α-DDM has been successfully determined. Crystals of Eco-msbA were grown at 5°C by the sitting-drop method using protein at approximately 10–15 mg ml^{-1} and a final detergent concentration of 0.05% α-DDM **(Figure 7.2A)**. Crystallization trials were performed using a multivariate crystallization matrix of temperatures, detergents and several precipitants. The protein was mixed in a ratio of 1:1 or 3:1 with reservoir solution containing 100–200 mM citrate buffer (pH 4.8 to 5.4), 15 to 20% PEG 300, 80–120 mM Li_2SO_4 or $(NH_4)_2SO_4$ and 0.05% α-DDM. Crystals appeared within 3 weeks and continued to grow for 2 months to a full size (0.4 mm × 0.8 mm × 0.3 mm). Gel electrophoresis using SDS–PAGE, Western blotting using anti-histidine tag antibodies, and mass spectroscopy clearly indicated that the crystals were composed of Eco-MsbA. To further verify the identity, crystals were washed and dissolved, and the NH_2-terminal amino acid sequence was determined to five residues.

CRYSTALLOGRAPHIC ANALYSIS

Crystals of Eco-MsbA using α-DDM grew in space group P1 (a = 107.8 Å, b = 126.1 Å, c = 206.6 Å, α = 83.5°, β = 76.3°, γ = 84.1°). The native crystals diffracted to a resolution of ~6.2 Å using synchrotron radiation at the Stanford Synchrotron Radiation Laboratory (SSRL) but the data was fairly anisotropic. In an effort to improve protein lattice contacts and decrease the disorder within the crystals, we applied a crystal refinement strategy that was also used for the structure determination of MscL (Chang *et al.*, 1998), which included the screening of an extensive matrix of detergent types, detergent concentrations, salts, temperatures, organics, additives, deuterium oxide, and heavy metals. One compound, $OsCl_3$, greatly improved the diffraction quality to a limiting resolution of 4.5 Å with spots observed to 3.8 Å. The binding of $OsCl_3$ made the crystals turn yellowish-brown in color **(Figure. 7.2B)**. This heavy atom compound was later found to bind at crystal lattice contacts between the NBDs of two transporters in the unit cell. In view of the greatly improved diffraction quality, data collected from this soaked crystal was used as the 'native' data set for X-ray structure determination. Protein phases were determined by the method of single isomorphous replacement (SIR/SAS) using two evenly split fragments from a single crystal and also by phase combination with a two-wavelength multiple anomalous dispersion (MAD) using the computer package PHASES (Furey and Swaminathan, 1997). Initial electron density maps clearly revealed that the asymmetric unit contained four complete Eco-MsbA transporters (eight monomers) in a pseudo-222 arrangement consistent with the operators of the self-rotation function. The Matthews coefficient is 5.0 $Å^3$ per dalton, which is a reasonable value for most membrane protein crystals and corresponds to a solvent/detergent content of ~75%. Electron density correlations on experimentally derived maps indicated that there were some differences between the transporters that did and did not bind osmium heavy atoms. Transporters that did not bind $OsCl_3$ did not adhere to a perfect twofold relating the monomers within a dimer. These differences were accommodated in the averaging masks and subsequent density modification procedures. Iterative eightfold non-crystallographic symmetry averaging, solvent flattening/flipping, phase extension, and amplitude sharpening using in-house programs yielded electron density maps of excellent quality for tracing the polypeptide chain. A chemical model was built using the program CHAIN (Sacks, 1988), and the protein sequence registration was established for bulky aromatic groups in the transmembrane helices from the electron density maps. Regions of the histidine ABC domain HisP (Hung *et al.*, 1998) served as a useful guide for modeling the NBD of Eco-MsbA (Hung *et al.*, 1998). Numerous rounds of vector refinement were performed using the program XPLOR to best fit the model into the sharpened electron density maps. Using this preliminary model, anisotropic corrections to the diffraction intensities were applied using the program XPLOR.

The crystal structure refinement of Eco-MsbA was complicated by a rapid decrease in the intensity of the diffraction pattern as a function of resolution, corresponding to an overall temperature factor of ~150 $Å^2$. Similar temperature factors were reported for the original K^+ ion channel from *Streptomyces lividans* (KcsA) as well as for MscL (Chang *et al.*, 1998; Doyle *et al.*, 1998). The experimentally phased electron density maps, however, appeared to be of much higher quality than one would expect for this temperature factor. This suggested that there were predominant orientations for the transporters in the crystal that were probably stabilized by heavy-atom binding with additional orientations that introduced a degree of positional disorder. As a consequence, whereas the diffraction intensities remained relatively strong at lower resolution, the scattering contributions from this ensemble of transporters and associated detergent interfered at higher resolution, resulting in a rapid decrease in diffraction intensities at a higher resolution. Standard crystallographic refinement protocols are generally inadequate for modeling this type of positional disorder and, as a result, we were unable to refine a single model of Eco-msbA to values of R and R_{free} below ~38 and ~45%, respectively. In an effort to better simulate the data, we used a multicopy refinement procedure (mc refinement) using 16 copies of the asymmetric unit (eight molecules per asymmetric unit) against the $OsCl_3$ data with very strict eightfold non-crystallographic harmonic constraints (2000 kcal mol^{-1}) between the monomers using a modified version of the program XPLOR **(Figure 7.3)** (Brunger *et al.*, 1987; Gros *et al.*, 1990; Kuriyan *et al.*, 1991; Pellegrini *et al.*, 1997). After molecular dynamics refinement, an ensemble of very similar models (average root mean square deviation of C_a atoms among models of 1.4 Å) was achieved with a crystallographic R value of 27% and an R_{free} of 38%. Multicopy refinements were also done using 5, 8 and 10 copies in the asymmetric unit and these mc refinements yielded similar crystallographic R factors and

A

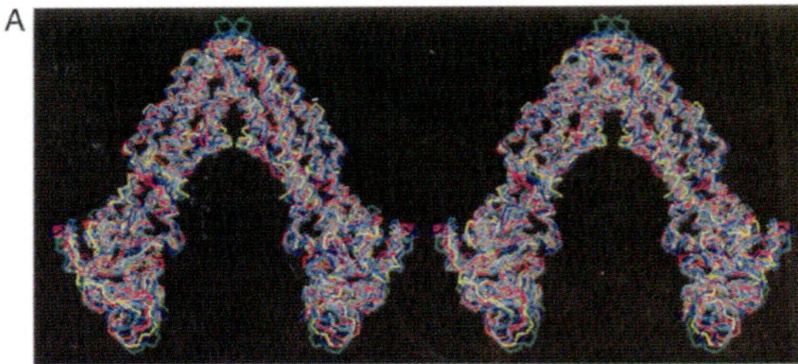

B

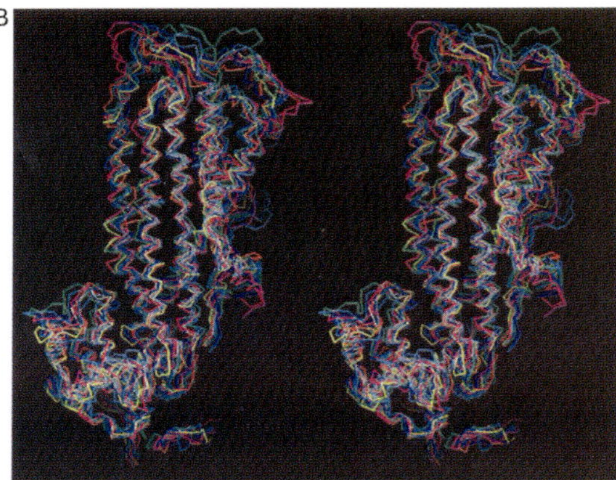

*Figure 7.3. **Stereoviews of 16 superimposed models from multicopy (mc) refinement of Eco-msbA. The models are shown in different colors. A, View of the dimer looking into the chamber opening. B, View of the dimer looking from the lipid bilayer at the external (embedded) surface of the chamber opening. Helical regions are more similar while loop regions are generally less ordered. B-factors for all atoms were fixed at 90 Å^2 during the mc refinement. An averaged model was computed and used for analysis. Reprinted with permission from G. Chang and C.B. Roth, Science 293, 1793 (2001). Copyright 2001. American Association for the Advancement of Science.***

R_{free} values. Most importantly, the significant drop in the R_{free} values during the multicopy refinements and the similarity of the independently averaged models suggested a proper fit of the data. The choice of using 16 copies of the asymmetric unit provided an opportunity to observe their spatial distribution more finely and was the maximum limit of the computing resources at the time. Residue positions in helical regions were very well defined in these models, whereas the loop regions were less ordered, as expected. An averaged model with good stereochemistry was computed and used for structural analysis. B factors for all atoms were fixed at 90 Å^2 during the multicopy refinement.

STRUCTURE DESCRIPTION

The X-ray crystal structure of Eco-MsbA transporter is consistent with the molecule being a homodimer and each subunit is composed of two domains **(Figure 7.4)**. We have identified a third domain bridging the TMD and the NBD. Eco-MsbA is approximately 120 Å in length, with the TMD, including the membrane-spanning region, accounting for approximately 52 Å. All the transmembrane α-helices are tilted between 30° and 40° from the normal for the membrane, forming a cone-shaped structure with two substantial openings on either side facing the lipid bilayer. These openings are approximately 25 Å wide in the longest dimension and lead into a large cone-shaped chamber in the interior of the molecule's TMD. The leaflet half of the membrane domain in the outer membrane leaflet domain forms the inter-molecular contacts holding the two monomers of the transporter together. The dimer interface, which is mostly contributed by the second and fifth membrane-spanning α-helices, buries approximately 850 Å^2 of solvent accessible surface area. The base of the chamber facing the cytoplasm is approximately 45 Å in the widest dimension, and the volume of the chamber can easily accommodate lipid A molecules. The resolved regions of the NBDs share no intermolecular contact and are separated by approximately 50 Å in the closest dimension. Approximately 200 residues are not resolved in this particular crystal form of the molecule, which is free of nucleotide and ligand. We assume that this part of the structure is disordered in the absence of either ligand or nucleotide and probably plays an important role in the mechanism of substrate translocation.

TRANSMEMBRANE DOMAIN STRUCTURE

Eco-MsbA begins with the NH_2-terminus on the cytoplasmic side as a helix (residue 10–21) that is parallel with the lipid bilayer **(Figure 7.4)**. Residues Trp^{10}, Phe^{13} and Trp^{17} would intercalate into the inner leaflet side of the cell membrane. The polypeptide chain continues into the first transmembrane helix (TM1, residues 22–52) and leads into the first extracellular loop (EC1, residues 53–64), which crosses

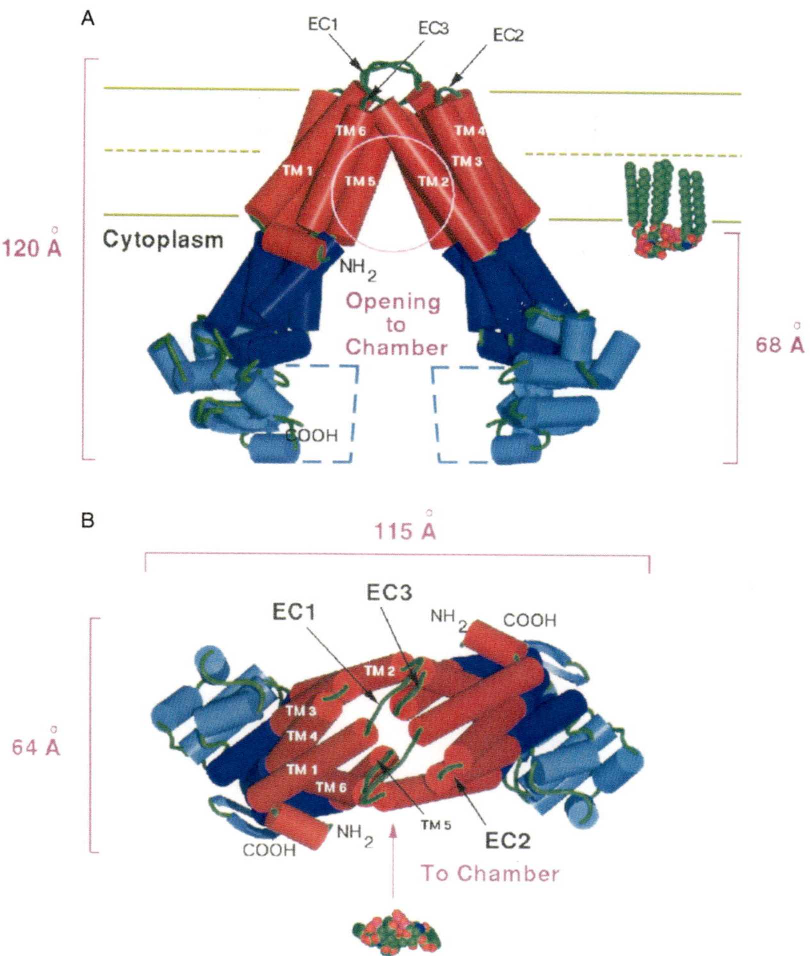

Figure 7.4. ***Structure of Eco-MsbA. A, View of the dimer looking into the chamber opening. The transmembrane domain, nucleotide-binding domain and intracellular domain are colored red, cyan and dark blue, respectively. Transmembrane α-helices are marked and the connecting loops are shown in green. A model of lipid A (not in the crystal structure) is shown to the right embedded in the lower bilayer leaflet. Solid and dotted green lines represent the boundaries of the membrane bilayer leaflets. Dotted cyan lines indicate the approximate location of the disordered region in the NBD. B, View of Eco-MsbA from the extracellular side, perpendicular to the membrane with a model of lipid A. Transporter dimensions are labeled and images were rendered using BOBSCRIPT and RASTER 3D. Reprinted with permission from G. Chang and C.B. Roth, Science 293, 1793 (2001). Copyright 2001. American Association for the Advancement of Science.***

over to the far side of the transporter. The electron densities for residues 58–63 of EC1 are diffuse. The second transmembrane helix (TM2, residues 65–96) forms part of the opening of the flippase chamber and appears mobile when comparing TM2 α-helices from other copies of Eco-MsbA in the unit cell. The third and fourth transmembrane helices (TM3, residues 140–164 and TM4, residues 168–192) are connected by a very short extracellular loop (EC2, residues 165–167), which has a polar character. Strong electron density is present for several bulky residues, which include Phe^{161}, Tyr^{162} and Tyr^{163} in TM3 and Trp^{165} in EC2. The fifth transmembrane helix (TM5, residues 253–272) begins with a kink caused by Pro^{253} and forms half of the opening facing the bilayer. TM5 is connected to the sixth transmembrane helix (TM6, residues 281–301) by an extracellular loop (EC3, residues 273–280). Although the absolute orientation of Eco-MsbA in the membrane is not known, the putative third extracellular loop (EC3) in human Pgp aligns by sequence homology to Eco-MsbA EC3 and has been shown to be located on the cell surface by *in vivo* topology studies using the antibody UIC2 (Zhou *et al.*, 1999). TM2 and TM5 from opposing monomers within the dimer form the major dimerization contact and are well positioned to serve as a hinge allowing the transporter to undergo significant structural rearrangements.

NBD STRUCTURE

The NBD (colored cyan in **Figure 7.4**) is the most conserved feature of the ABC transporter family and contains the Walker A/B motif along with the ABC signature motif. In the absence of ATP or nucleotide analogue in this crystal form, residues 341–418, which include the Walker A motif, are disordered in our electron density maps. The crystal packing, however, suggests sufficient volume to accommodate the mass of the Walker A region. The remaining portion of the NBD is well resolved in our structure and includes an α-helix (residues 331–340) and residues 418–561, which contain the ABC signature motif (colored pink in **Figure 7.5**) and the Walker B region (colored gray in **Figure 7.5**). The NBD of Eco-MsbA has significant similarity in sequence and structure to the corresponding regions found in histidine ABC domain (HisP) with a root mean square (rms) deviation of ~1.5 Å (C_α) for residues 445–528. NBD residues that are in direct contact with the intracellular domain (residues 420–448, 500–508 and 531–556) are among some of the most highly conserved regions among all members of the multidrug resistance transporters. The COOH-terminus of Eco-MsbA (residues 565–582) is not resolved.

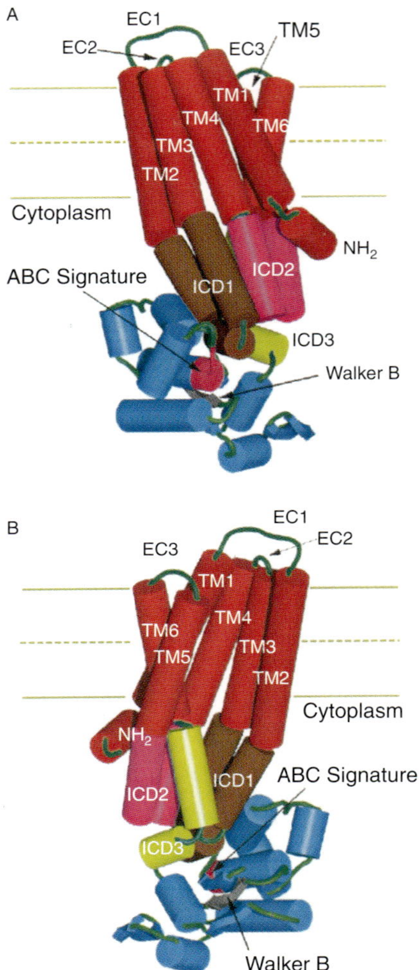

Figure 7.5. View of Eco-MsbA looking (A) from the lipid bilayer at the external (embedded) surface of the chamber opening and (B) looking at the interior of the chamber. The transmembrane domain is colored red and the transmembrane α-helices are marked. The NBD is colored cyan with the Walker B motif and ABC signature motif highlighted in gray and pink respectively. ICD1, ICD2 and ICD3 are colored brown, violet and yellow respectively for clarity. The estimated cell membrane (~35 Å) and the boundary between the bilayer leaflets are illustrated as solid and dotted yellow lines. Figure was rendered using BOBSCRIPT and RASTER 3D. Reprinted with permission from G. Chang and C.B. Roth, Science 293, 1793 (2001). Copyright 2001. American Association for the Advancement of Science.

INTRACELLULAR DOMAIN (ICD) STRUCTURE

A distinctive feature of the MsbA is two extensive intracellular regions, which we called ICD1 (residues 97–193) and ICD2 (residues 193–252). We have identified a third intracellular subdomain, which connects TM6 with the NBD and which we label ICD3 (residues 302–327). We collectively designate the region located between the TMD and the NBD as the intracellular domain (colored dark blue in **Figure 7.4**). Most of the residues that correspond to structural elements in ICD1 that are in contact with the NBD are highly conserved throughout multidrug resistance ABC transporters. ICD1 (colored brown in **Figure 7.5**), which is sandwiched between ICD2 (colored violet in **Figure 7.5**) and the NBD, is composed of three α-helices connected by short loops to form a 'U'-like structure. The second α-helix of ICD1 (residues 111–121) is highly conserved and is nestled against residues 420–430 of the NBD. The well-ordered portions of ICD2 in our electron density maps are mostly α-helical (residues

A

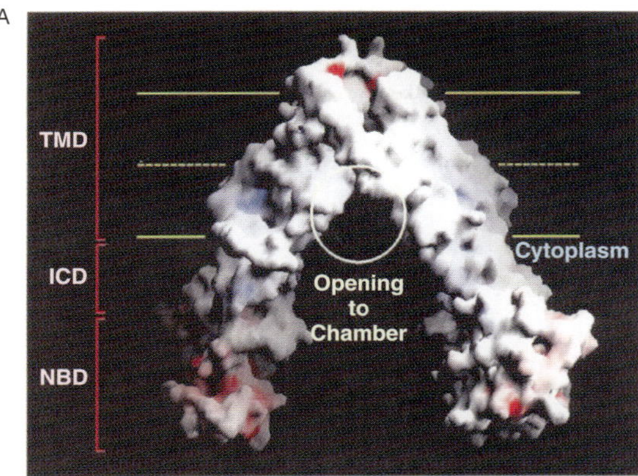

B

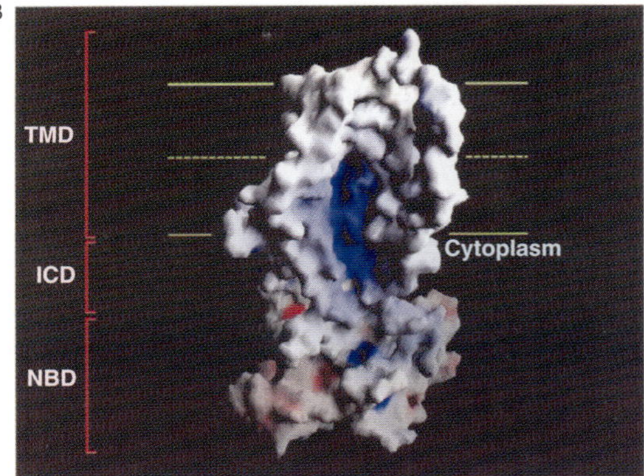

Figure 7.6. Molecular surface rendering of Eco-MsbA with electrostatic potentials. This figure was generated with the program GRASP (54) assuming an ionic strength of 100 mM NaCl and dielectric constant of 2 and 80 for protein and solvent, respectively. The surface potential varies continuously from blue (positive) to red (negative). The cell membrane, which is ~35 Å wide, is represented with yellow lines. The transmembrane domain (TMD), intracellular domain (ICD) and NBD are indicated. A, Side view of dimer looking into the chamber. The opening of the chamber spans the lower bilayer leaflet. B, View of Eco-MsbA monomer looking at the inner surface of the chamber. The surface potential within the lower half of the chamber is highly positive due to a clustering of lysine and arginine residues. Reprinted with permission from G. Chang and C.B. Roth, Science 293, 1793 (2001). Copyright 2001. American Association for the Advancement of Science.

193–207 and 237–252). The electron densities for residues 208–236, however, are diffuse. ICD3 links TM6 and the NBD and forms two α-helices connected by short loops (colored yellow in **Figure 7.5B**). The α-helix just preceding the NBD (residues 318–329) is conserved and is in direct contact with both ICD1 and ICD2.

Chamber structure

The chamber is a symmetric structure that is formed from two Eco-MsbA transmembrane domains and extends along the pseudo-twofold axis perpendicular to the cell membrane **(Figure 7.6)**. The chamber has an opening (~25 Å) on either side facing the bilayer, providing free access of substrate from the cytoplasmic leaflet of the lipid bilayer while excluding molecules from the outer leaflet. The openings to the chamber are defined by intermolecular interactions between TM2 of one monomer and TM5 of another **(Figure 7.6A)**. Residues 268–273 of TM2 lining the opening of the chamber are partially shielded from the bilayer by TM5. The residues lining the chamber are contributed by all 12 transmembrane α-helices and could be highly solvated. The inner membrane leaflet side of the chamber contains a cluster of positively charged residues (Arg^{148}, Arg^{183}, Lys^{187}, Arg^{190}, Lys^{194} and Arg^{296}) **(Figure 7.6B)**, which contrasts the significantly less charged and more hydrophobic environment within the outer membrane leaflet side.

Possible Flipping Mechanism

The structure of MsbA suggests a mechanism for hydrophobic substrate translocation **(Figure 7.7)**. Studies of human Pgp function indicate that residues lining the proposed chamber opening (TM2, TM5 and TM6) play an important role in substrate recognition (Ambudkar *et al.*, 1999). There is significant evidence indicating a cooperative interaction between the two opposing NBDs (Senior and Bhagat, 1998). The intracellular domains (colored purple in **Figure 7.7**) bridge the TMD and the NBD serving to couple ATP hydrolysis by the NBD to tertiary arrangements of the transmembrane α-helices. Upon binding of a lipid A molecule, conformational changes relayed from the transmembrane domain to the intracellular domain stimulate nucleotide binding by the NBD (step 1 in **Figure 7.7**). Movement of TM2, TM5, TM6 and the two NBDs serves to recruit the substrate and close the chamber (step 2 in **Figure 7.7**). Binding of ATP by the NBDs may result in a conformational shift that promotes the interaction of the adjacent NBDs. The cluster of charges lining the chamber on the inner membrane leaflet side

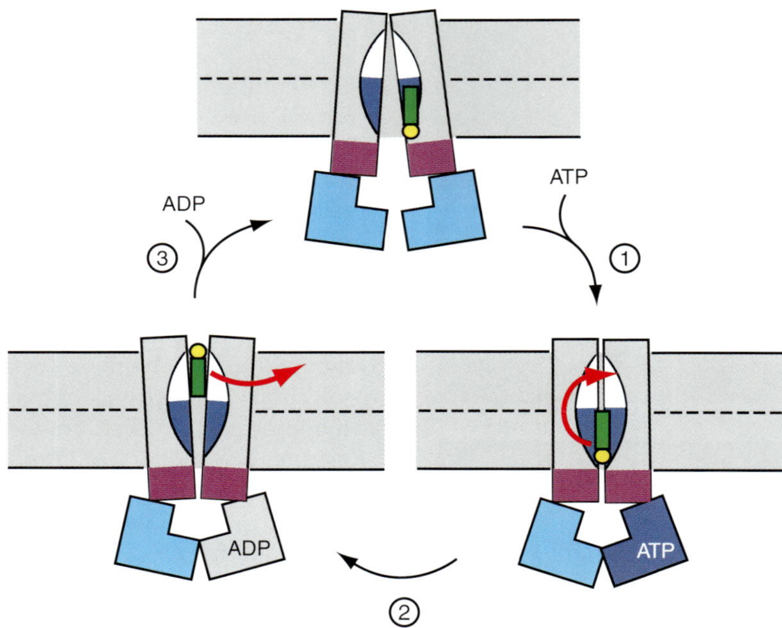

Figure 7.7. Model for lipid A transport by Eco-MsbA. Stages 1–3 begin at right and proceed clockwise. See text for details. (1) Lipid A binding, triggering of ATP hydrolysis, and recruitment of substrate to chamber. (2) Closure of the chamber and translocation of lipid A. Interaction between the two NBDs is possible. (3) Opening of the chamber, movement of TM2/TM5, release of lipid A to the outer bilayer leaflet, and nucleotide exchange. A small yellow circle and a green rectangle denote the hydrophobic tails and sugar headgroups of lipid A, respectively. The cell membrane is represented as a set of two horizontal lines separated by a dash to indicate the separation of bilayer leaflets. Blue regions indicate positive charge lining the chamber, and purple regions represent the intracellular domain. The gray region on the outer membrane side of the chamber is hydrophobic. Red and black arrows show the movement of substrate and the structural changes of MsbA, respectively.

creates an energetically unfavorable micro-environment for a hydrophobic substrate. The asymmetric charge distribution in the interior of the chamber is consistent with the vectored transport of substrate from the inner to the outer membrane leaflet. Faced with both the charge and the highly polar contribution of potentially bound solvent, the lipid A molecule 'flips' into an energetically more favorable position within the outer membrane leaflet side of the chamber (colored gray in **Figure 7.7**), where it can form hydrophobic interactions. The substrate is now properly orientated to enter the outer bilayer leaflet. The flipping of substrate initiates nucleotide hydrolysis by the NBD, causing the chamber to undergo structural rearrangements that enable the expulsion of the substrate to the outer membrane leaflet (step 3 in **Figure 7.7**). Alternatively, it has been proposed that some drug transporters can extrude their substrates directly into the extracellular medium from the inner leaflet of the bilayer (Bolhuis *et al.*, 1996). This entropically driven 'flip-flop' mechanism could account for the unusually broad range of hydrophobic drugs and lipids transported by members of this transporter family. Several other variations of this mechanism are plausible and will be refined using biochemical and structural information derived from additional MsbA studies.

Although we believe that MsbA may not be representative of the ABC transporters that import hydrophilic substrates, there is evidence that suggests human Pgp and other multidrug transporters share transport mechanisms and structural components similar to Eco-MsbA (Higgins and Gottesman, 1992). First, human Pgp and LmrA are believed to be lipid flippases like Eco-MsbA and have been shown to transport (van Helvoort *et al.*, 1996) phosphatidylcholine (Smith *et al.*, 1994) and phosphatidylethanolamine (PE) (Margolles *et al.*, 1999), respectively. This suggests a common functional ancestry between the eukaryotic multidrug transporter group and the bacterial lipid flippases. Secondly, the size and shape of the chamber of Eco-MsbA could accommodate a wide variety of amphipathic molecules, which could explain how human Pgp can extrude such an unusually broad range of substrates. The crystal structure of MsbA reveals that it is not a pore through the cell membrane, but a molecular machine scanning the lower bilayer leaflet for substrates, accepting them laterally, and flipping them to the outer membrane leaflet. The recent 25 Å cryo-EM reconstruction of the MsbA homologue YvcC from *Bacillus subtilis* reveals a structural architecture that is consistent with Eco-MsbA solved by X-ray crystallography (Chami *et al.*, 2002). The MsbA model when docked into the EM electron density map suggests that the putative YvcC transmembrane domains form a chamber within the membrane. Although they cannot be resolved at this resolution, the 30–40° tilt of all the transmembrane helices is clearly evident. Strong electron densities for both the intracellular domain and the NBDs allow the crystal structure of MsbA to be docked into the cryo-EM density with confidence, providing the first evidence of a conserved architecture among the bacterial exporters. The X-ray crystal structure of Eco-MsbA provides a first insight into the detailed structure of an ABC transporter, and a foundation for understanding the transport mechanisms of the ABC transporter family.

Adding a second chapter to the ABC transporter story is the recent 3.5 Å structure of the *E. coli* vitamin B_{12} importer BtuCD. Although BtuCD shares little homology to the MDR class exporters that translocate hydrophobic substrates, the structure reveals a new snapshot of an ABC transporter with contact between the ABC cassettes in a manner much like the Rad50 dimer. This finding is significant, for contact between the nucleotide binding domains has been thought necessary to explain the cooperative kinetics of ATP hydrolysis during the transport cycle. Interestingly, the ABC cassettes remain in contact in the absence of bound nucleotide, which is not the case with MsbA. Similar to MsbA, however, contact between the ABC cassette and the membrane-spanning domain is maintained by the ICD, through a conserved set of residues that the authors name the L-loop. This observation reinforces the idea that the ICD domain serves as the bridge coupling ATP hydrolysis by the ABC cassette to rearrangements within TMD. The mechanistic model presented by Rees *et al.* proposes that the two ABC cassettes of BtuCD maintain contact throughout the transport cycle, while converting hydrolytic energy to leverage on TMD in a 'power stroke'. The BtuCD structure presents yet another snapshot of the transport cycle, and suggests a mechanism that may be applied to other ABC importers.

ACKNOWLEDGMENTS

Excerpts are reprinted with permission from Chang and Roth, X-ray structure of an intact ABC-transporter MsbA, Science 293, 1793. Copyright 2001, American Association for the Advancement of Science.

REFERENCES

Ambudkar, S.V., Dey, S., Hrycyna, C.A., Ramachandra, M., Pastan, I. and Gottesman, M.M. (1999) Biochemical, cellular, and pharmacological aspects of the multidrug transporter. Annu. Rev. Pharmacol. Toxicol. 39, 361–398.

Bolhuis, H., van Veen, H.W., Molenaar, D., Poolman, B., Driessen, A.J. and Konings, W.N. (1996) Multidrug resistance in *Lactococcus lactis*: evidence for ATP-dependent drug extrusion from the inner leaflet of the cytoplasmic membrane. EMBO J. 15, 4239–4245.

Brunger, A.T., Kuriyan, J. and Karplus, M. (1987) Crystallographic R-factor refinement by molecular dynamics. Science 235, 458–460.

Cano-Gauci, D.F., Seibert, F.S., Safa, A.R. and Riordan, J.R. (1995) Selection and characterization of verapamil-resistant multidrug resistant cells. Biochem. Biophys. Res. Commun. 209, 497–505.

Chami, M., Steinfels, E., Orelle, C., Jault, J.M., Di Pietro, A., Rigaud, J.L. and Marco, S. (2002) Three-dimensional structure by cryo-electron microscopy of YvcC, an homodimeric ATP-binding cassette transporter from *Bacillus subtilis*. J. Mol. Biol. 315, 1075–1085.

Chang, G. and Roth, C.B. (2001) Structure of MsbA from *E. coli*: a homolog of the multidrug resistance ATP binding cassette (ABC) transporters. Science 293, 1793–1800.

Chang, G., Spencer, R.H., Lee, A.T., Barclay, M.T. and Rees, D.C. (1998) Structure of the MscL homolog from *Mycobacterium tuberculosis*: a gated mechanosensitive ion channel. Science 282, 2220–2226.

Dean, M., Hamon, Y. and Chimini, G. (2001) The human ATP-binding cassette (ABC) transporter superfamily. J. Lipid Res. 42, 1007–1017.

Doerrler, W.T., Reedy, M.C. and Raetz, C.R. (2001) An *Escherichia coli* mutant defective in lipid export. J. Biol. Chem. 276, 11461–11464.

Doyle, D.A., Morais Cabral, J., Pfuetzner, R.A., Kuo, A., Gulbis, J.M., Cohen, S.L., Chait, B.T. and MacKinnon, R. (1998) The structure of the potassium channel: molecular basis of K^+ conduction and selectivity. Science 280, 69–77.

Furey, W. and Swaminathan, S. (1997) A program package for processing and analyzing diffraction data from macromolecules. Methods Enzymol. 277, 590–620.

Gros, P., van Gunsteren, W.F. and Hol, W.G. (1990) Inclusion of thermal motion in crystallographic structures by restrained molecular dynamics. Science 249, 1149–1152.

Higgins, C.F. and Gottesman, M.M. (1992) Is the multidrug transporter a flippase? TIBS 17, 18–21.

Horio, M., Gottesman, M.M. and Pastan, I. (1988) ATP-dependent transport of vinblastine in vesicles from human multidrug-resistant cells. Proc. Natl Acad. Sci. USA 85, 3580–3584.

Hung, L.W., Wang, I.X., Nikaido, K., Liu, P.Q., Ames, G.F. and Kim, S.H. (1998) Crystal structure of the ATP-binding subunit of an ABC transporter. Nature 396, 703–707.

Karow, M. and Georgopoulos, C. (1993) The essential *Escherichia coli* msbA gene, a multicopy suppressor of null mutations in the htrB gene, is related to the universally conserved family of ATP-dependent translocators. Mol. Microbiol. 7, 69–79.

Kuriyan, J., Osapay, K., Burley, S.K., Brunger, A.T., Hendrickson, W.A. and Karplus, M. (1991) Exploration of disorder in protein structures by X-ray restrained molecular dynamics. Proteins 10, 340–358.

Margolles, A., Putman, M., van Veen, H.W. and Konings, W.K. (1999) The purified and functionally reconstituted multidrug transporter LmrA of *Lactococcus lactis* mediates the transbilayer movement of specific fluorescent phospholipids. Biochemistry 38, 16298–16306.

McDonald, M.K., Cowley, S.C. and Nano, F.E. (1997) Temperature-sensitive lesions in the *Francisella novicida* valA gene cloned into an *Escherichia coli* msbA lpxK mutant affecting deoxycholate resistance and lipopolysaccharide assembly at the restrictive temperature. J. Bacteriol. 179, 7638–7643.

Pellegrini, M., Gronbech-Jensen, N., Kelly, J.A., Pfluegl, G.M. and Yeates, T.O. (1997) Highly constrained multiple-copy refinement of protein crystal structures. Proteins 29, 426–432.

Rosenberg, M.F., Callaghan, R., Ford, R.C. and Higgins, C.F. (1997) Structure of the multidrug resistance P-glycoprotein to 2.5 nm resolution determined by electron microscopy and image analysis. J. Biol. Chem. 272, 10685–10694.

Sack, J.S. (1988) A crystallographic modeling program. J. Mol. Graphics 6, 224–225.

Senior, A.E. and Bhagat, S. (1998) P-glycoprotein shows strong catalytic cooperativity between the two nucleotide sites. Biochemistry 37, 831–836.

Smith, A.J., Timmermans-Hereijgers, J.L., Roelofsen, B., Wirtz, K.W., van Blitterswijk, W.J., Smit, J.J., Schinkel, A.H. and Borst, P. (1994) The human MDR3 P-glycoprotein promotes translocation of phosphatidylcholine through the plasma membrane of fibroblasts from transgenic mice. FEBS Lett. 354, 263–266.

van Helvoort, A., Smith, A.J., Sprong H., Fritzsche, I., Schinkel, A.H., Borst, P. and van Meer, G. (1996) MDR1 P-glycoprotein is a lipid translocase of broad specificity, while MDR3 P-glycoprotein specifically translocates phosphatidylcholine. Cell 87, 507–517.

van Veen, H.W., Callaghan, R., Soceneantu, L., Sardini, A., Konings, W.N. and Higgins, C.F. (1998) A bacterial antibiotic-resistance gene that complements the human multidrug-resistance P-glycoprotein gene. Nature 391, 291–295.

van Veen, H.W., Higgins, C.F. and Konings, W.N. (2001) Molecular basis of multidrug transport by ATP-binding cassette transporters: a proposed two-cylinder engine model. J. Mol. Microbiol. Biotechnol. 3, 185–192.

Zhou, Y., Gottesman, M.M. and Pastan, I. (1999) The extracellular loop between TM5 and TM6 of P-glycoprotein is required for reactivity with monoclonal antibody UIC2. Arch. Biochem. Biophys. 367, 74–80.

Zhou, Z., White, K.A., Polissi, A., Georgopoulos, C. and Raetz, C.R. (1998) Function of *Escherichia coli* MsbA, an essential ABC family transporter, in lipid A and phospholipids biosynthesis. J. Biol. Chem. 273, 12466–12475.

Part III

Prokaryote ABC Transporters

CHAPTER 8

INTRODUCTION TO BACTERIAL ABC PROTEINS

I. BARRY HOLLAND

In Chapter 1 in this volume, E. Dassa has reviewed the classification of ABC proteins, including prokaryote representatives and their transport substrates in the many cases where these have been identified. Previous general reviews have also discussed the ABC proteins in *Escherichia coli* (Linton and Higgins, 1998), *Bacillus subtilis* (Quentin *et al.*, 1999) and *Mycobacterium tuberculosis* (Braibant *et al.*, 2000) and more specifically concerning bacterial ABC exporters in *E. coli* (Fath and Kolter, 1993; Young and Holland, 1999). The purpose of this introductory chapter is therefore briefly to highlight some of the major characteristics of bacterial ABC systems and the breadth of their functions.

NATURE AND COMPOSITION OF THE ABC TRANSPORTER

Prokaryote ABC-dependent transport systems, whether exporters or importers, all adhere to the usual formula of a basic four-unit structure, two membrane components and two units of ABC-ATPases. The membrane components and the ABCs may be identical or non-identical and can be fused pairwise in different combinations as shown in Chapter 1, although unlike those commonly found in eukaryotes no examples of all four subunits fused together have been identified in prokaryotes. In describing ABC-dependent transport systems, it is important to emphasize that the term ABC (ATP-binding cassette, Hyde *et al.*, 1990) is synonymous with ABC-ATPase, whether present as a subdomain or an independent polypeptide. The term ABC transporter, on the other hand, describes the ABC-ATPase (also called a traffic ATPase; Ames and Lecar, 1992) plus its associated integral membrane domains, whether fused to the ABC or separately encoded. This core transporter or translocation complex may be further supplemented with essential accessory or auxiliary subunits (usually encoded separately): the external ligand-binding protein in the case of ABC importers, or the MFP (membrane fusion protein) and the OMP-F (outer membrane protein/factor) or OMA (outer membrane auxiliary) integral to the inner membrane and outer membrane, respectively. In the case of ABC transporters, the whole complex may sometimes be referred to as the translocon, whilst for the importers, the term permease is also used to describe the entire complex.

Whilst ATP is *the* substrate for the ABC-ATPase, the molecule or ion being transported by the ABC transporter is variously described as a substrate or a transport substrate or an allocrite. Since in our view, in the vast majority of cases, the component being transported remains unmodified by the process, the term 'substrate' is inappropriate, and we prefer allocrite, a term we coined, loosely derived from the Greek meaning a substance transported or

ABC Proteins: From Bacteria to Man
ISBN 0-12-352551-9

exported (Blight and Holland, 1990; Young and Holland, 1999).

An Abundance of Different Types of ABC Proteins in Prokaryotes

Probably the earliest detailed studies of ABC proteins were carried out in bacteria in the 1970s and 1980s, concerning the mechanism of uptake of solutes such as histidine and maltose, mediated by the ABC proteins, HisP (Ames and Nikaido, 1978) and MalK (Bavoil *et al.*, 1980) in *Salmonella typhimurium* and *E. coli,* respectively. These proteins were initially recognized as binding ATP and subsequently as energy generators for transport (Hobson *et al.*, 1984; Shuman and Silhavy, 1981), through the hydrolysis of ATP. As seen in Chapter 9, although we still have much to learn concerning the mechanism of transport driven by, for example, HisP and MalK, structural and genetic studies of the importing ABCs continue to be the most advanced.

ABC-ATPases are now recognized as one of the major superfamilies of proteins, represented in all three kingdoms of life and found in all organisms so far analyzed. ABC proteins are particularly abundant in prokaryotes, with genes constituting from close to 1% and up to more than 3% amongst the 19 eubacteria and 6 archaea, respectively, surveyed in Chapter 1. The very recent sequence of the *Agrobacterium tumefaciens* genome describes the highest number recorded so far, 153, excluding orphan ABCs (with no discernible membrane domain associates). Thus, ABC transporters constitute 60% of all transporters and about 3% of all predicted polypeptides in the *A. tumefaciens* genome (Goodner *et al.*, 2001; Wood *et al.*, 2001). All the known bacterial genomes, with one exception, *Treponema pallidum* (only 1.14 Mb), encode all the three main categories of ABC protein discussed below, the exporters, orphans and importers. Curiously, *T. pallidum* and four out of the six archaeal genomes listed in Chapter 1, apparently do not encode any exporters.

In Chapter 1, based on cluster or phylogeny analysis of sequences constituting the ABC polypeptides, over 600 examples out of the more than 2000 entries in the current databases, 33 distinct clusters were identified. These are assigned so far to three major classes, all strongly represented in bacteria. Class 1 contains the large family of exporters. Class 2 is a small family of orphans, with no known membrane protein associates and, at least in some cases, with no connection to membrane transport processes, for example the bacterial UvrA protein essential for specific DNA repair processes. Class 3 is functionally probably a more heterogeneous family, since it probably contains both importers and exporters. This heterogeneity may necessitate a future separation into at least two distinct classes.

An Additional Class of Bacterial ABCs Involved in DNA Recombination and Repair

Importantly, an additional important group of ABC proteins present in both bacteria and eukaryotes, which are not involved in transport but concerned with DNA repair or recombination, have yet to be classified as class 1, 2 or 3 and may well constitute a completely new class. Such an example, the ABC domain of Rad50 from *Pyrococcus furiosus*, involved in homologous recombination, has recently been crystallized and the structure determined (Hopfner *et al.*, 2000). The ABC domain contains the two characteristic lobes or arms found in HisP (Hung *et al.*, 1998). This contains all the expected, highly conserved motifs, the Walker A, Q-loop, Walker B and the downstream histidine (Linton and Higgins, 1998), present in Arm-I, the RecA-like, catalytic domain (Geourjon *et al.*, 2001). Similarly, Rad50 contains the signature motif in the smaller Arm-II, sometimes referred to as the helical (Ames and Lecar, 1992) or signaling/regulatory domain (Holland and Blight, 1999). In reality, in the intact Rad50 molecule, the helical or signaling domain is interrupted by the insertion into helix 3 of 600 residues forming a long coiled coil region, thereby separating the Walker A from the Walker B domain. Interestingly, as discussed in Chapter 11, structural studies so far indicate that functionally different types of ABC protein display the greatest variation in structural organization in the helical domain, frequently affecting helix 3.

The extensive coiled coil region of Rad50, facilitating dimerization of these large molecules, restoring the close proximity of the Walker A and B motifs for nucleotide binding, is in fact diagnostic of a large family of bacterial and eukaryote SMC (structural maintenance of chromosome) proteins (Melby *et al.*, 1998; Soppa, 2001), many of which are involved in condensation of DNA, including the SMC protein in *B. subtilis* required for chromosomal segregation (Graumann *et al.*, 1998). Notably, whereas Rad50 has a relatively well-conserved LSGG motif compared with the 'classical' ABC proteins, other SMCs have a more 'degenerate' version of this signature motif. Finally, perhaps the most distant relatives, but still considered as ABC proteins (Aravind *et al.*, 1999), are the DNA repair enzymes such as the bacterial MutS. These proteins contain minimal Walker A and B motifs and have the same overall fold for the catalytic domain as HisP (Lamers *et al.*, 2000), but the signature motif is significantly diverged from that of HisP, and indeed much of the region equivalent to the helical domain of HisP is absent (Geourjon *et al.*, 2001).

EXPORTERS

Class 1 ABC-ATPases (fused to a membrane domain), and apparently some class 3 proteins (encoded independently from the membrane domain), constitute at least eight distinct families, all concerned with the export of a wide range of compounds. These include extremely large polypeptides, greater than 400 kDa in some cases (Chapter 11), polysaccharides, a wide variety of antibiotics, many drugs (Chapter 12), and certain lipids (Chapter 7). A fascinating adaptation of the modular structure of an ABC protein is shown in the ABC component of the translocators for non-lantibiotics secreted by Gram-positive bacteria. In these cases the N-terminal domain of the ABC transporter carries a cytoplasmic extension to the membrane domain (Havarstein *et al.*, 1995), which constitutes a cysteine protease, necessary for processing the antibiotic peptide as it exits from the cell (see Chapter 11). Some evidence suggests that class 1 ABCs are also involved in exporting fatty acids and Na^+ ions as transport substrates or allocrites. As reviewed in Chapter 1, however, firm evidence for the identity of allocrites in many cases is still lacking. Importantly, whilst inferences regarding potential allocrites for class 1 transporters can be drawn from cluster analysis through guilt by association with well-characterized transporters, this approach is not necessarily reliable. One of the largest exporter families, DPL (see Chapter 1), contains at least 11 subfamilies of bacterial ABCs, which are involved in the export of allocrites as diverse as lipids, large polypeptides, or a wide range of drugs. Of course, we cannot rule out the possibility that some of these transporters export in reality more than one type of compound, as has been demonstrated for Pgp (Johnstone *et al.*, 2000; Raymond *et al.*, 1992). As a further complication, the ABC transporters in the Prt and Hly clusters in the heterogeneous DPL family require additional, specific auxiliary membrane proteins in order to complete, if not provide, the actual translocation pathway (Chapter 11).

Interestingly, from knowledge that is available so far, the bacterial exporters appear to fulfill a variety of important cellular functions, for example the secretion of factors required for dominating other bacterial species in the environment, for colonization of plant, insect or animal hosts leading to pathogenic infection or symbiosis, for the removal of toxic compounds and for the biogenesis of several constituents of the organism's own cellular envelope. Many of the latter are essential for respiratory functions, the integrity of the bilayer, simple surface protection and even movement of the bacteria. Moreover, some ABC exporters have been implicated in various developmental and differentiation programs, although their precise roles and allocrites transported in these cases are mostly obscure. For further information and literature sources on several of these aspects, see other chapters in Parts I and II in this volume.

CLASS 2, ORPHAN ABCS

The class 2 group of ABC proteins are present in all organisms but are curious exceptions to the rule that the ABC proteins are always involved in transport processes across membranes. The functions of these proteins as a group are quite diverse and surprising, being involved in translation of polypeptides, drug and antibiotic resistance, and in DNA repair, although only the latter two have been documented in bacteria so far (Chapter 1). It is intriguing to know what common principles might govern the action of a highly conserved ABC domain involved in processes as different as membrane transport, DNA repair and protein synthesis. Interestingly, in the bacterial UvrA protein, a tandemly duplicated ABC, there is an insertion of a DNA binding motif, a zinc finger, between the Walker A and the signature motif in each ABC domain (see, for example, Husain *et al.*, 1986; Yamamoto *et al.*, 1996). This insertion occurs in a position close to the equivalent of the interface between the two lobes in the HisP structure, which presumably must affect the regulation of UvrA function. As a further curiosity, if not a mystery, ABCs in this group of class 2 orphans include proteins, also with duplicated ABC domains, from, for example, *Staphylococcus aureus* and *Streptomyces antibioticus* (Mendez and Salas, 2001; Ross *et al.*, 1995), responsible for resistance (and immunity in some cases) to certain drugs and antibiotics. The simplest explanation would be that these ABCs do work in conjunction with some membrane protein to export the drugs but, despite intensive efforts, such proteins have not yet been identified.

IMPORTERS

The class 3 ABC transporters in bacteria constitute an enormous family of import systems for small molecules. The transport complex is composed of two molecules of an independently encoded ABC protein(s), a hetero- or homodimer of integral membrane proteins constituting the translocation pathway, and an external ligand-binding protein, amongst which the most characterized, the periplasmic binding proteins in Gram-negative bacteria are considered in Chapter 10. The class 3 importers have been assigned to at least nine major families in the phylogeny analysis in Chapter 1. The allocrites transported cover a wide range of essential and non-essential molecules, including several metal ions, iron chelates, vitamin B_{12}, mono-, di- and oligosaccharides, polyols, polyamines, inorganic anions such as sulfate, nitrate and phosphate, phosphonates, peptide osmoprotectants and ssother di- and oligopeptides. From these examples, the import systems for histidine and maltose will be considered in some detail in Chapter 9, and for uptake of osmoprotectants in Chapter 13. As already indicated, despite the range of allocrites transported in this very large family, the nature of the different translocators is surprisingly uniform: an external ligand-binding protein, free in the periplasm in Gram-negative bacteria whilst it may be anchored to the membrane surface in Gram-positive bacteria; two membrane proteins for transport, carrying the EAA interaction motif; and a highly conserved ABC protein on the cytoplasmic side of the inner membrane. Since evidence of exchangeability of one ABC component for another in these otherwise very similar systems has been rarely indicated in the literature, we must assume that each ABC is tailormade for contact and intramolecular signaling with its cognate membrane domains.

Recent studies of two ABC-dependent solute uptake systems responsible for transport of general amino acids and branched amino acids in *Rhizobium leguminosarum* have revealed the surprising finding that such systems can apparently also *export* these amino acids. Moreover, the same phenomenon was demonstrated with histidine transport in *S. typhimurium* (Hosie *et al.*, 2001). This reverse transport or bidirectional capacity of these ABC transporters raises some complex questions concerning the solute pathway in the two different directions. In addition, it is not yet clear whether ATPase activity is required for the efflux process (P. Poole, personal comunication).

MEMBRANE DOMAINS OF THE BACTERIAL TRANSPORTERS ARE POORLY UNDERSTOOD

Whereas great progress has been made in the comparative, phylogenic analysis of the ABC domains, leading to prediction of possible function in the absence of other evidence in many cases, the cluster analysis of membrane domains has lagged far behind. This clearly hampers insights into the mechanistic role of these domains as potential translocation pathways and these are poorly understood. Nevertheless, as discussed in Chapter 9, the early recognition (Dassa and Hofnung, 1985) of the EAA motif, apparently completely conserved without exception within a cytoplasmic loop of the membrane components of all the bacterial ABC importers, has ultimately led to the identification of this as a specific point of contact with a region of the helical domain of the ABC-ATPase.

This is presumably also a critical point in the intramolecular signaling pathway, coordinating transport and energy generation.

Importantly, the EAA motif is not present in any of the exporters, indicating that during evolution ABC-ATPases, in bacteria at least, have associated with more than one type of membrane domain. Furthermore, the failure so far to detect any kind of conserved motif in the membrane domains of ABC exporters perhaps emphasizes, in contrast to the importers, the wide variation in both the mechanism and the pathway of molecular signaling between the membrane and ABC components of the exporters. As indicated below and discussed in Chapter 7, the elucidation of the structure of the membrane domain of the *E. coli* MsbA protein will now enormously stimulate this aspect of ABC studies.

STRUCTURE AND FUNCTION OF THE ABC TRANSPORTERS

Notably, some of the most advanced structural studies of ABC transporters have come from bacterial import and, more recently, bacterial export systems. Thus, we now have high-resolution structures for ABC importers, HisP (Hung *et al.*, 1998), a MalK from *Thermococcus litoralis* (Diederichs *et al.*, 2000), one ABC in the family of branched-chain amino acid transporters and one of unknown function (Karpowich *et al.*, 2001; Yuan *et al.*, 2001). In this laboratory, we have recently obtained the high-resolution structure of the ABC domain of HlyB (Schmitt *et al.*, in preparation), a member of the large DPL family, which includes the mammalian TAP and Pgp (Mdr1) proteins. The implications of all these structural advances will be considered in other chapters. As discussed in Chapter 7, a very major and exciting advance in the field was made by the presentation of the first structural data at 4.5 Å for the intact bacterial exporter MsbA from *E. coli* (Chang and Roth, 2001). This provides the first sign of the nature of the membrane domain, and, in particular, that of the membrane-spanning domains. These are finally shown to be helices, settling some previous controversies. Most crucially, of course, this overall structure of MsbA has profound implications for at least a global understanding of how the action of the membrane and ABC domains may be coordinated. Chang and Roth (see also Higgins and Linton, 2001) on the basis of this structure have already proposed an exciting solution to a long-standing puzzle – how close are the ABC domains in the transporter? – that most likely they are interfaced at some point in the catalytic cycle (see also Chapter 6), but under the influence of the membrane domains they are well separated in the absence of any transport substrate. Unfortunately, mechanistic studies of the nature of the catalytic cycle of ABC proteins in bacteria, and its relationship to the transport function, have lagged relatively far behind those for some of the mammalian proteins. However, recent advances in purifying and reconstituting proteins of the maltose and histidine uptake systems (see Chapter 9), combined with the power of microbial genetics, promise much for the future.

Excitingly, as this volume goes to press the high-resolution structure of the bacterial ABC import system for vitamin B12, BtuCD, is reported (Locher *et al.*, Science 296, 1091–1098), providing many new insights into the mechanism of ABC-dependent transport.

REFERENCES

Ames, G.F.-L. and Lecar, H. (1992) ATP-dependent bacterial transporters and cystic fibrosis: analogy between channels and transporters. FASEB J. 6, 2660–2666.

Ames, G.F. and Nikaido, K. (1978) Identification of a membrane protein as a histidine transport component in *Salmonella typhimurium*. Proc. Natl Acad. Sci. USA 75, 5447–5451.

Aravind, L., Walker, D.R. and Koonin, E.V. (1999) Conserved domains in DNA repair proteins and evolution of repair systems. Nucleic Acids Res. 27, 1223–1242.

Bavoil, P., Hofnung, M. and Nikaido, H. (1980) Identification of a cytoplasmic membrane-associated component of the maltose transport system of *Escherichia coli*. J. Biol. Chem. 255, 8366–8369.

Blight, M.A. and Holland, I.B. (1990) Structure and function of haemolysin B, P-glycoprotein and other members of a novel family of membrane translocators. Mol. Microbiol. 4, 873–880.

Braibant, M., Gilot, P. and Content, J. (2000) The ATP binding cassette (ABC) transport systems of *Mycobacterium tuberculosis*. FEMS Microbiol. Rev. 24, 449–467.

Chang, G. and Roth, C.B. (2001) Structure of MsbA from *E. coli*: a homolog of the multidrug resistance ATP binding cassette (ABC) transporters. Science 293, 1793–1800.

Dassa, E. and Hofnung, M. (1985) Sequence of gene *malG* in *E. coli* K12: homologies between integral membrane components from binding protein-dependent transport systems. EMBO J. 4, 2287–2293.

Diederichs, K., Diez, J., Greller, G., Müller, C., Breed, J., Schnell, C., Vonrhein, C., Boos, W. and Welte, W. (2000) Crystal structure of MalK, the ATPase subunit of the trehalose/maltose ABC transporter of the archaeon *Thermococcus litoralis*. EMBO J. 19, 5951–5961.

Fath, M.J. and Kolter, R. (1993) ABC transporters: bacterial exporters. Microbiol. Rev. 57, 995–1017.

Geourjon, C., Orelle, C., Steinfels, E., Blanchet, C., Deleage, G., Di Pietro, A. and Jault J.M. (2001) A common mechanism for ATP hydrolysis in ABC transporter and helicase superfamilies. Trends Biochem. Sci. 26, 539–544.

Goodner, B., Hinkle, G., Gattung, S., Miller, N., Blanchard, M., Qurollo, B., *et al.* (2001) Genome sequence of the plant pathogen and biotechnology agent *Agrobacterium tumefaciens* C58. Science 294, 2323–2328.

Graumann, P.L., Losick, R. and Strunnikov, A.V. (1998) Subcellular localization of *Bacillus subtilis* SMC, a protein involved in chromosome condensation and segregation. J. Bacteriol. 180, 5749–5755.

Havarstein, L.S., Diep, D.B. and Nes, I.F. (1995) A family of bacteriocin ABC transporters carry out proteolytic processing of their substrates concomitant with export. Mol. Microbiol. 16, 229–240.

Higgins, C.F. and Linton, K.J. (2001) The xyz of ABC transporters. Science 293, 1782–1784.

Hobson, A.C., Weatherwas, R. and Ames, G.F. (1984) ATP-binding sites in the membrane components of histidine permease, a periplasmic transport system. Proc. Natl Acad. Sci. USA 81, 7333–7337.

Holland, I.A. and Blight, M.A. (1999) ABC-ATPases, adaptable energy generators fuelling transmembrane movement of a variety of molecules in organisms from bacteria to humans. J. Mol. Biol. 293, 381–399.

Hopfner, K.-P., Karcher, A., Shin, D.S., Craig, L., Arthur, L.M., Carney, J.P. and Tainer, J.A. (2000) Structural biology of Rad50 ATPase: ATP-driven conformational control in DNA double-strand break repair and the ABC-ATPase superfamily. Cell 101, 789–800.

Hosie, A.H.F., Allaway, D., Jones, M.A., Walshaw, D.L., Johnston, A.W.B. and Poole, P.S. (2001) Solute-binding protein-dependent ABC transporters are responsible for solute efflux in addition to solute uptake. Mol. Microbiol. 40, 1449–1459.

Hung, L.-W., Wang, I.X., Nikaido, K., Liu, P.-Q., Ames, G.F.-L. and Kim, S.-H. (1998) Crystal structure of the ATP-binding subunit of an ABC transporter. Nature 396, 703–707.

Husain, I., van Houten, B., Thomas, D.C. and Sancar, A. (1986) Sequences of the *uvrA* gene and protein reveal two potential ATP binding sites. J. Biol. Chem. 261, 4895–4901.

Hyde, S.C., Emsley, P., Hartshorn, M.J., Mimmack, M.M., Gileadi, U., Pearce, S.R., Gallagher, M.P., Gill, D.R., Hubbard, R.E. and Higgins, C.F. (1990) Structural model of ATP-binding proteins associated with cystic fibrosis, multidrug resistance and bacterial transport. Nature 346, 362–365.

Johnstone, R.W., Ruefli, A.A. and Smyth, M.J. (2000) Multiple physiological functions for multidrug transporter P-glycoprotein? Trends Biochem. Sci. 25, 1–6.

Karpowich, N., Martsinkevich, O., Millen, L., Yuan, Y.-R., Dai, P.L., MacVey, K., Thomas, P.J. and Hunt, J.F. (2001) Crystal structures of the MJ1267 ATP binding cassette reveal an induced-fit effect at the ATPase active site of an ABC transporter. Structure 9, 571–586.

Lamers, M.H., Perrakis, A., Enzlin, J.H., Winterwerp, H.H.K., de Wind, N. and Sixma T.K. (2000) The crystal structure of DNA mismatch repair protein MutS binding to a G.T mismatch. Nature 407, 711–717.

Linton, K.J. and Higgins, C.F. (1998) The *Escherichia coli* ATP-binding cassette (ABC) proteins. Mol. Microbiol. 28, 5–13.

Melby, T.E., Ciampaglio, C.N., Briscoe, E. and Erickson, H.P. (1998) The symmetrical structure of structural maintenance of chromosomes (SMC) and MukB proteins: long, antiparallel coiled coils, folded at a flexible hinge. J. Cell Biol. 142, 1595–1604.

Mendez, C. and Salas, J.A. (2001) The role of ABC transporters in antibiotic-producing organisms: drug secretion and resistance mechanisms. Res. Microbiol. 152, 341–350.

Quentin, Y., Fichant, G. and Denizot, F. (1999) Inventory, assembly and analysis of *Bacillus subtilis* ABC transport systems. J. Mol. Biol. 287, 467–484.

Raymond, M., Gros, P., Whiteway, M. and Thomas, D.Y. (1992) Functional complementation of yeast *ste6* by a mammalian multidrug resistance *mdr* gene. Science 256, 232–234.

Ross, J.I., Eady, E.A., Cove, J.H. and Baumberg, S. (1995) Identification of a chromosomally encoded ABC-transport system with which the staphylococcal erythromycin exporter MsrA may interact. Gene 153, 93–98.

Shuman, H.A. and Silhavy, T.J. (1981) Identification of the malK gene product. A peripheral membrane component of the *Escherichia coli* maltose transport system. J. Biol. Chem. 256, 560–562.

Soppa, J. (2001) Prokaryotic structural maintenance of chromosomes (SMC) proteins: distribution, phylogeny, and comparison with MukBs and additional prokaryotic and eukaryotic coiled-coil proteins. Gene 278, 253–264.

Wood, D.W., Setubal, J.C., Kaul, R., Monks, D.E., Kitajima, J.P., Okura V.K., *et al.* (2001) The genome of the natural genetic engineer *Agrobacterium tumefaciens* C58. Science 294, 2317–2323.

Yamamoto, N., Kato, R. and Kuramitsu, S. (1996) Cloning, sequencing and expression of the *uvrA* gene from an extremely thermophilic bacterium, *Thermus thermophilus* HB8. Gene 171, 103–106.

Young, J. and Holland, I.B. (1999) ABC transporters: bacterial exporters-revisited five years on. Biochim. Biophys. Acta 1461, 177–200.

Yuan, Y.R., Blecker, S., Martsinkevich, O., Millen, L., Thomas, P.J. and Hunt, J.F. (2001) The crystal structure of the MJO796 ATP-binding cassette. Implications for the structural consequences of ATP hydrolysis in the active site of an ABC transporter. J. Biol. Chem. 276, 32313–32321.

CHAPTER 9

Import of Solutes by ABC Transporters – the Maltose and Other Systems

Erwin Schneider

This article is dedicated to Professor Dr Karlheinz Altendorf on the occasion of his 60th birthday.

Introduction and General Overview

Starch is one of the major sources of carbon and energy available to heterotrophic bacteria and archaea. For example, microorganisms living in soil and aquatic environments readily gain access to starch derived from decomposing plant material, while those that colonize the gastrointestinal tract of humans can feed on starch that escaped digestion in the small bowel. The latter is estimated to lie in the range of 10% of intake in subjects on Western diets (Cummings and Macfarlane, 1991). Since polysaccharides cannot penetrate the cell membrane, a wide variety of microorganisms secrete amylases that produce maltose and maltodextrins (oligosaccharides of two or more – up to seven – α-1,4 linked glucose units) as major degradation products of starch. The uptake of the latter is usually mediated by an ABC transport system that belongs to a subclass of ABC importers recently designated as the CUT1 (*c*arbohydrate *u*ptake *t*ransporter) or OSP (*o*ligosaccharides and *p*olyols) family by Saier (2000; http://www-biology.ucsd.edu/~msaier/transport/3_A_1.html) and Dassa and Bouige (2001; http://www.pasteur.fr/recherche/unites/pmtg/abc/index.html), respectively (see also Chapter 1).

Members of the family transport a variety of di- and oligosaccharides, glycerol-phosphate and polyols **(Table 9.1)**[1] and are composed of the extracellular substrate-binding protein, which mainly determines the specificity of the transporter, two integral membrane proteins, each usually spanning the membrane six times, and two copies of an ATPase subunit (also referred to as ABC protein/domain from here on) (reviewed in Schneider, 2001). The hydrophobic subunits contain the cytoplasmic 'EAA' sequence motif (consensus: EAA-X_3-G-X_9-I-X-LP) typically shared by all membrane-spanning subunits of prokaryotic ABC importers. The ATPase subunit is recognized by the characteristic set of Walker A and B boxes and by the ABC signature sequence ('LSGGQ' motif) (reviewed in Schneider and Hunke, 1998). However, this differs from a classical consensus ABC domain, having a carboxy-terminal extension of approximately 120 to 150 amino acid residues. In the *Escherichia coli* and *Salmonella typhimurium* maltose transporter, the C-terminal domain is involved in regulatory activities (reviewed in Boos and

[1] *It should be noted that in case of the archaeon* Sulfolobus solfataricus*, the ABC importer for maltose shows sequence homology to the subfamily of oligo/dipeptide transporters rather than to the CUT1/OSP cluster (Elferink* et al.*, 2001). Thus, functional classification of ABC transporters solely based on computer-aided analysis should be taken with caution.*

ABC Proteins: From Bacteria to Man
ISBN 0-12-352551-9

Shuman, 1998; see also **Box 9.1**). Several short sequence motifs and conserved amino acid residues within this peptide fragment can serve as signatures, together with a conserved sequence motif in the binding protein (Tam and Saier, 1993), to identify new members of the CUT1 family **(Figure. 9.1)**.

Some ABC domains of the CUT1 family are functionally exchangeable, thereby strengthening the above classification. For example, UgpC of *E. coli* and LacK of *Agrobacterium radiobacter* were both demonstrated to substitute for MalK in maltose transport in *E. coli* (Hekstra and Tommassen, 1993; Wilken *et al.*, 1996).

TABLE 9.1. REPRESENTATIVE MEMBERS OF CUT1/OSP FAMILY OF ABC IMPORTERS

Substrate(s) transported	Protein components	Representative organism(s)
Maltose/maltodextrins	MalEFG**K**	*E. coli, S. typhimurium*
Maltose/trehalose	MalEFG**K**	*Thermococcus litoralis*
Lactose	LacEFG**K**	*Agrobacterium radiobacter*
Melibiose, raffinose, sucrose	MsmEFG**K**	*Streptococcus mutans*
Glycerol-phosphate	UgpAEB**C**	*E. coli*
Polyols	SmoEFG**K**	*Rhodobacter sphaeroides*
Cyclodextrins	CymEFG**D**	*Klebsiella oxytoca*
Cellobiose/cellotriose	CebEFG**MsiK**	*Streptomyces reticuli*
Maltose/sucrose/trehalose	AglEFG**K**	*Sinorhizobium meliloti*
Alginate	Alg**S**M1M2Q1(?)Q2(?)	*Sphingomonas* sp.

Binding proteins are underlined and bold characters denote ABC proteins. Only those systems for which all components were clearly identified by sequence alignment and/or biochemical evidence are considered. For data bank accession numbers, see legend to Figure 9.1. Modified from Schneider (2001).

BOX 9.1. REGULATORY ACTIVITIES OF THE MALTOSE TRANSPORTER

The maltose transporter of *E. coli/S. typhimurium* is directly involved in transcriptional regulation of the maltose regulon, most probably by interaction of the MalK subunits with the positive regulator protein MalT. MalT–MalK interaction has been demonstrated *in vitro* (Panagiotidis *et al.*, 1998). Activation of MalT is achieved by binding of ATP and maltotriose, resulting in a conformational change and subsequent oligomerization of the protein, a prerequisite for the interaction with its DNA binding sites (Danot, 2001; Schreiber and Richet, 1999). Binding of MalT to assembled MalK interferes with this process, thereby repressing maltose-regulated gene expression (Boos and Böhm, 2000). Mutations in MalK that diminish or abolish its inhibitory effect on MalT action, W267G and G346S, map in the C-terminal extension of the protein (Kühnau *et al.*, 1991). In the case of W267G, the mutation did not affect binding to MalT *in vitro* (Panagiotidis *et al.*, 1998), indicating that mere physical interaction is insufficient to antagonize MalT activity. Interestingly, MalK variants carrying mutations in the ABC signature motif that cause loss of ATPase activity but still allow binding of ATP (G137A/V/T, Q140K/N/L) act as super-repressors (Kühnau *et al.*, 1991; Panagiotidis *et al.*, 1998; Schmees *et al.*, 1999b). Possibly, in this case local conformational changes in the ATPase domain of the mutant proteins affect the affinity of the C-terminal domain for its target, MalT. These findings led to the notion that substrate availability is sensed through the transporter, which, in the idling mode, binds MalT and thereby represses *mal* gene transcription. In the presence of substrate, however, transport activity is switched on, i.e. ATP is hydrolyzed at the MalK subunits, thus causing release of MalT and subsequent induction of maltose-regulated gene expression (Boos and Böhm, 2000).

The maltose transporter is also involved in a second regulatory process called 'inducer exclusion', which is part of the global carbon regulation in enteric bacteria. Here, in the presence of the preferred carbon source, glucose, the transport of inducer molecules for alternative metabolic pathways is prevented. This is achieved by inhibition of the respective transport systems via a component of the glucose transporter, the dephosphorylated enzyme IIA^{Glc} of the phosphoenolpyruvate phosphotransferase system (PTS) (Postma *et al.*, 1996). In the case of the maltose transporter, enzyme IIA^{Glc} binds to the MalK subunits, thereby inhibiting ATP hydrolysis (Dean *et al.*, 1990; Landmesser *et al.*, 2002). Again, mutations that render MalK insensitive to inhibition by enzyme IIA^{Glc} predominantly affect residues in the C-terminal domain (Dean *et al.*, 1990; Kühnau *et al.*, 1991) **(Table 9.2)**.

```
                                                       A motif
LACK_AR         -MAEVRLTDIRKSYGSL--EVIKGVNLEVSSGEFVVFVGPSGCGKSTLLRMIAGLEDISS 57
MALK_ST         -MASVQLRNVTKAWGDV--VVSKDINLDIHDGEFVVFVGPSGCGKSTLLRMIAGLETITS 57
SMOK_RS         -MGKITLRNVQKRFGEA--VVIPSLDLDIEDGEFVVFVGPSGCGKSTLLRLIAGLEDVSD 57
AGLK_SIM        -MTGLLLKDIRKSYGAV--DVIHGIDLDIKEGEFVVFVGPSGCGKSTLLRMIAGLEEITG 57
CYMD_KO         -MATVSLRKIEKRYENG-FKAVHGIDLEIHDGEFMVFVGPSGCAKSTTLRMIAGLEDISG 58
MSMK_SM         -MVELNLNHIYKKYPNSSHYSVEDFDLDIKNKEFIVFVGPSGCGKSTTLRMVAGLEDITK 59
AGLS_SSP        MVASVSIQNVVKRYDKT--TVVHGVSLDIEPGEFVVLVGPSGCGKSTTLRMVAGLEEISG 58
MSIK_SC         -MATVTFDKATRVYPGSTKPAVDGLDIDIADGEFLVLVGPSGCGKSTSLRMLAGLEDVNG 59
UGPC_EC         -MAGLKLQAVTKSWDGKTQ-VIKPLTLDVADGEFIVMVGPSGCGKSTLLRMVAGLERVTE 58
MALK_TL         -MAGVRLVDVWKVFGEVTA--VRELSLEVKDGEFMILLGPSGCGKTTTLRMIAGLEEPSR 57
                 :  : :    : :          . :::   **:::*****.*:*  **::****  .
                                              Lid
LACK_AR         GELTIGGTVMND------VDPS-KRGIAMVFQTYALYPHMTVRENMGFALRFAGMAKDEI 110
MALK_ST         GDLFIGETRMND------IPPA-ERGVGMVFQSYALYPHLSVAENMSFGLKLAGAKKEVM 110
SMOK_RS         GQIMIDGRDATE------MPCHRERGLAMVFQSYALYPHMTVKKNIAFPLRMAKMEPQEI 111
AGLK_SIM        GDMFIDGERVND------VPPS-KRGIAMVFQSYALYPHMTVYDNMAFGMRIARESKEEI 110
CYMD_KO         GEIYIGNRKVND------LPPK-DRGIAMVFQNYALYPHKTVFDNMAFGLKMQKRPKDEI 111
MSMK_SM         GELKIDGEVVND------KAPK-DRDIAMVFQNYALYPHMSVYDNMAFGLKLRHYSKEAI 112
AGLS_SSP        GTIRIDGRVIND------LAPK-DRDVAMVFQNYALYPHLNVRDNISFGLRLKRTKKSVI 111
MSIK_SC         GAIRIGDRDVTH------LPPK-DRDIAMVFQNYALYPHMSVADNMGFALKIAGVNKAEI 112
UGPC_EC         GDIWINDQRVTE------MEPK-DRGIAMVFQNYALYPHMSVEENMAWGLKIRGMGKQQI 111
MALK_TL         GQIYIGDRLVADPEKGIFVPPK-DRDIAMVFQSYALYPHMTVYDNIAFPLKLRKVPRQEI 116
                * : *.     .           .*.:.****.****** .* .*:.: :::       :
                                    ABC signature      B motif  D-loop
LACK_AR         ERRVNAAAKILELDALMDRKPKALSGGQRQRVAIGRAIVRQPDVFLFDEPLSNLDAELRV 170
MALK_ST         NQRVNQVAEVLQLAHLLERKPKALSGGQRQRVAIGRTLVAEPRVFLLDEPLSNLDAALRV 170
SMOK_RS         ERRVSNAAKILNLTNYLDRRPGQLSGGQRQRVAIGRAIVREPAAFLFDEPLSNLDAALRV 171
AGLK_SIM        DRRVRGAADMLQLTPYLDRLPKALSGGQRQRVAIGRAICRNPKVFLFDEPLSNLDAALRV 170
CYMD_KO         KRRVEDAAEKLEITELLYRKPKEMSGGQRQRVAVGRAIVRKPDVFLFDEPLSNLDAKLRV 171
MSMK_SM         DKRVKEAAQILGLTEFLERKPADLSGGQRQRVAMGRAIVRDAKVFLMDEPLSNLDAKLRV 172
AGLS_SSP        DAAVKTAADILGLQPLLERKPSDLSGGQRQRVAMGRAIVRDPKVFLFDEPLSNLDAKLRT 171
MSIK_SC         RQKVEEAAKILDLTEYLDRKPKALSGGQRQRVAMGRAIVREPQVFLMDEPLSNLDAKLRV 172
UGPC_EC         AERVKEAARILELDGLLKRRPRELSGGQRQRVAMGRAIVRDPAVFLFDEPLSNLDAKLRV 171
MALK_TL         DQRVREVAELLGLTELLNRKPRELSGGQRQRVALGRAIVRKPQVFLMDEPLSNLDAKLRV 176
                   *  .*  * :   : * *  :*********:**::  .. .**:********* **.
                                         Switch
LACK_AR         HMRVEIARLHKELN-----ATIVYVTHDQVEAMTLADKIVVMR----------GGIVEQV 215
MALK_ST         QMRIEISRLHKRLG-----RTMIYVTHDQVEAMTLADKIVVLD----------AGRVAQV 215
SMOK_RS         NMRLEITELHQSLE-----TTMIYVTHDQVEAMTMADKIVVLN----------AGRIEQV 216
AGLK_SIM        ATAIENRQAQRAMSN----TTMIYVTHDQVEAMTLADRIVVFSR---------RSTFEQF 217
CYMD_KO         SMRMKLAQLHRSLKEEGHPATMIYVTHDQTEALTLGDRICVLN----------HGNIMQV 221
MSMK_SM         SMRAEIAKIHRRIG-----ATTIYVTHDQTEAMTLADRIVIMSSTKNEDGSGTIGRVEQV 227
AGLS_SSP        QMRAEIKRLHQRLG-----TTVIYVTHDQVEAMTLADRIVVMR----------DGLIEQI 216
MSIK_SC         STRTQIASLQRRLG-----ITTVYVTHDQVEAMTMGDRVAVLK----------DGLLQQV 217
UGPC_EC         QMRLELQQLHRRLK-----TTSLYVTHDQVEAMTLAQRVMVMN----------GGVAEQI 216
MALK_TL         RMRAELKKLQRQLG-----VTTIYVTHDQVEAMTMGDRIAVMN----------RGVLQQV 221
                    :    :: :       * :******.**:*:.::: ::            .   *.

LACK_AR         GAPLALYDDPDNMFVAGFIGSPRMNFLPAVVIGQAEGGQVTVA--LKARPDTQLTVACAT 273
MALK_ST         GKPLELYHYPADRFVAGFIGSPKMNFLPVKVTATAIE-QVQVE--LPNRQQIWLPVESRG 272
SMOK_RS         GSPLTLYRNPANLFVAGFIGSPKMNLI----------------------------EGPE 247
AGLK_SIM        GAPFELYERPANFFVARFIGSPPMNVIPATITATG---QQTAV--SLAGGKSVTLDVPTN 272
CYMD_KO         DTPTDLYNYPNNKFVASFIGSPSINLIDTAIRKNNERLYVEIAPGVEILIPHSKQVLLEG 281
MSMK_SM         GTPQELYNRPANKFVAGFIGSPAMNFFDVTIKDG----HLVSKDGLTIAVTEGQLKMLES 283
AGLS_SSP        GKPMDLFLHPANTFVASFIGSPPMNLMPARIAVDS-----TQHVELNGGNRISLLPRAGT 271
MSIK_SC         DSPRNMYDKPANLFVAGFIGSPAMNLVEVPITDGG----VKFGNS---VVPVNRDALKAA 270
UGPC_EC         GTPVEVYEKPASLFVASFIGSPAMNLLTGRVNNEG----THFELDGGIELPLNGGYRQYA 272
MALK_TL         GSPDEVYDKPANTFVAGFIGSPPMNFLDAIVTEDGF---VDFGEFRLKLLPDQFEVLGEL 278
                . *  ::  * . *** ***** :*..

LACK_AR         PPQGGDAVTVGVRPEHFL--------------PAGSGDTQLTAHVDVVEHLGNTSYVYAH 319
MALK_ST         -VQVGANMSLGIRPEHLL--------------PSDIADVTLEGEVQVVEQLGHETQIHIQ 317
SMOK_RS         AAKHG-ATTIGIRPEHID--------------LSREAG-AWEGEVGVSEHLGSDTFLHVH 291
AGLK_SIM        ASENGKTASFGVRPEDLR--------------VTEADDFLFEGTVSIVEALGEVTLLYIE 318
CYMD_KO         --YINKPVCFGIRPEHIS---LA---------SDDDDLNTFEGVLTVVENMGSEKFLYFI 327
MSMK_SM         KGFKNKNLIFGIRPEDISSSLLV---------QETYPDATVDAEVVVSELLGSETMLY-K 333
AGLS_SSP        HLAPGQEVVFGIRPEDVTLDG-----------VEGSERAQIKATVDIVEPLGSESILHAT 320
MSIK_SC         SDKGDRTVTVGVRPEHFDVVELNGGAAKTLSKDSADAPAGLAVSVNVVEETGADGYIYGT 330
UGPC_EC         G----RKMTLGIRPEHIAL--------------SSQAEGGVPMVMDTLEILGADNLAHG- 313
MALK_TL         GYVG-REVIFGIRPEDLYDAMFAR--------VRVPGENLVRAVVEIVENLGSERIVHLR 329
                         .*:***..                           :   *  *     :

LACK_AR         TVPG--EQIIIEQEERRHGGRYGDEIAVGISAKTSFLFDASG---RRIR----- 363
MALK_ST         IPAI--RQNLVYRQNDVVLVEEGATFAIGLPPERCHLFREDGSACRRLHQEPGV 369
SMOK_RS         VAG---MPTLTVRTGGEFGVHHGDRVWLTPQADKIHRFGADG---KALPRE--- 336
AGLK_SIM        GLVE--NEPIIAKMPGIARVGRGDKVRFTADKAKLHLFDTNG---QSYRA---- 363
CYMD_KO         VGGK--ELIARVDTQDINPFHIGKTLRFNLNTAFCHVFDFYNE--NNLTNVRV- 376
MSMK_SM         LGQT--EFAARVDARDFH--EPGEKVSLTFNVAKGHFFDAETE--AAIR----- 376
AGLS_SSP        VGD----HSLVVKVGGLNEVHPGDPVTLHVDLTRVHLFDAQSQ--ASIY----- 363
MSIK_SC         VEVGGETKDLVVRVSSRAVPEKGATVHVVPRPGEIHVFSSSTG--ERLTD---- 378
UGPC_EC         -------RWVVVRLAHQERPTAGSTLWLHLAENQLHLFDGETG--QR------- 351
MALK_TL         VG----GVTFVGAFRSESRVREGVEVDVVFDMKKIHIFDKTTG--KAIF----- 372
                      *  . .       . *
```

Figure 9.1. Sequence alignment of ABC proteins of the CUT1/OSP family. The proteins considered are: MALK_ST (**Salmonella typhimurium**; *acc.no. spP19566), LACK_AR* (**Agrobacterium radiobacter**; *acc. no. spQ01937), SMOK_RS* (**Rhodobacter sphaeroides**; *acc. no. spP54933), AGLK_SIM* (**Sinorhizobium meliloti**; *spQ9Z3R8), MSMK_SM* (**Streptococcus mutans**; *acc.no. spQ00752), CYMD_KO* (**Klebsiella oxytoca**; *spQ48394), ALGS_SSP* (**Sphingomonas sp.**; *acc. no. gbABO11415), UGPC_EC* (**Escherichia coli**; *acc. no. spP10907), MSIK_SC* (**Streptomyces coelicolor**; *acc. no. gbAL160331), MALK_TL* (**Thermococcus litoralis**; *acc. no. gbAF121946). Conserved sequence motifs and amino acid residues are boxed. Those that are conserved throughout the ABC superfamily are highlighted in yellow, while motifs and single residues confined to CUT1/OSP subfamily members are shown in pink. See also text for details.*

Table 9.2. Mutations analyzed in the MalK proteins of *E. coli*/*S. typhimurium*

Mutation	Transport activity *in vivo*	ATP binding[a]	ATPase activity		Other properties	References
			In soluble variant	In transport complex		
delS3V4	–	nd	+	nd		Walter *et al.* (1992b)
Walker A box						
G36 + R	–	nd	nd	nd		Kühnau *et al.* (1991)
delG36 P37A	–	nd	nd	nd		Kühnau *et al.* (1991)
C40S	+	nd	+	+		Hunke and Schneider (1999)
C40G	+	+	nd	nd		Panagiotidis *et al.* (1993)
K42I,Q,E	–	±	nd	nd		Panagiotidis *et al.* (1993)
K42N	–	±	–	–		Panagiotidis *et al.* (1993) Wilken (1997) Davidson and Sharma (1997)
K42R	–	+	–	nd		Schneider *et al.* (1994)
S43T	+	nd	nd	nd		Walter and Schneider, unpubl.
E74G	+	nd	nd	nd		Stein *et al.* (1997)
Lid						
M79insert	–	nd	nd	nd		Lippincott and Traxler (1997)
Q82K	±	nd	±	nd		Walter *et al.* (1992b)
Q82E	±	nd	+	nd		Walter *et al.* (1992b)
A85C	±	nd	nd	nd		Hunke *et al.* (2000b)
A85M	+	nd	nd	nd	Suppressor of EAA loop mutations in MalFG	Mourez *et al.* (1997a)
L86F	–	+	+	–		Hunke *et al.* (2000a)
H89insert	–	nd	nd	nd		Lippincott and Traxler (1997)
E94Q, V	+	nd	nd	nd		Stein *et al.* (1997)
F98L, Y	+	+	nd	nd		Panagiotidis *et al.* (1993)
delF98	+	+	nd	nd		Panagiotidis *et al.* (1993)
K106C	+	nd	nd	nd		Hunke *et al.* (2000b)
V114C	+	nd	nd	nd		Hunke *et al.* (2000b)
V114M	±	nd	nd	nd	Affects interaction with MalFG	Scheffel and Schneider, unpublished
V117C	+	nd	nd	nd		Hunke *et al.* (2000b)
V117M	+	nd	nd	nd	Suppressor of EAA loop mutations in MalFG	Mourez *et al.* (1997a)
E119K	+	nd	nd	nd	Abolishes inducer exclusion	Kühnau *et al.* (1991)
L123F	±	nd	nd	nd	Affects interaction with MalFG	Scheffel and Schneider, unpublished
A124T	+	nd	nd	nd	Abolishes inducer exclusion	Dean *et al.* (1990)
ABC signature						
G137A,V,T	–	±	–	–	Super-repressors of *mal* gene regulation	Schmees *et al.* (1999b) Kühnau *et al.* (1991) Panagiotidis *et al.* (1993)
G137insert	–	nd	nd	nd		Lippincott and Traxler (1997)

(continued)

TABLE 9.2. *(continued)*

Mutation	Transport activity *in vivo*	ATP binding[a]	ATPase activity: In soluble variant	ATPase activity: In transport complex	Other properties	References
delQR140–141	–	±	nd	nd		Kühnau *et al.* (1991) Panagiotidis *et al.* (1993)
Q140L	–	±	–	–	Super-repressor of *mal* gene regulation	Schmees *et al.* (1999b)
Q140K,N	–	+	+	–	Super-repressor of *mal* gene expression	Schmees *et al.* (1999b)
G145S	–	nd	nd	nd	Affects interaction with MalFG	Brinkmann and Schneider, unpublished
T147insert	–	nd	nd	nd		Lippincott and Traxler (1997)
V149M,I	+	nd	nd	nd	Suppresses EAA loop mutations in MalG	Mourez *et al.* (1997a)
P152L,Q	+	nd	±	nd		Walter *et al.* (1992b)
V154I	+	nd	nd	nd	Suppresses EAA loop mutations in MalFG	Mourez *et al.* (1997a)
Walker B box						
D158N	–	±	nd	nd		Kühnau *et al.* (1991) Panagiotidis *et al.* (1993)
E159G	–	nd	nd	nd		Stein and Schneider, unpubl.
P160L	–	±	–	–		Hunke *et al.* (2000a)
D165N	–	±	–	nd		Hunke *et al.* (2000a)
A167insert	–	nd	nd	nd		Lippincott and Traxler (1997)
L179R	±	nd	+	nd		Walter *et al.* (1992b)
L172Q	+	nd	+	nd		Walter *et al.* (1992b)
M187I	+	nd	nd	nd	Suppresses EAA loop mutations in MalFG	Mourez et al. (1997a)
Switch						
H192R,L	–	nd	–	–		Davidson and Sharma (1997) Walter *et al.* (1992b) Landmesser *et al.* (2002)
R211insert	+	nd	nd	nd	Abolishes inducer exclusion	Lippincott and Traxler (1997)
R228C	+	nd	nd	nd	Abolishes inducer exclusion	Kühnau *et al.* (1991)
F241I	+	nd	nd	nd	Abolishes inducer exclusion	Dean *et al.* (1990)
W267G	+	nd	nd	nd	Eliminates *mal* gene repression	Kühnau *et al.* (1991)
V275insert	+	nd	nd	nd	Eliminates *mal* gene repression	Lippincott and Traxler (1997)
G278P	+	nd	nd	nd	Abolishes inducer exclusion	Dean *et al.* (1990)

(continued)

TABLE 9.2. (continued)

Mutation	Transport activity *in vivo*	ATP binding[a]	ATPase activity		Other properties	References
			In soluble variant	In transport complex		
S282L	+	nd	nd	nd	Abolishes inducer exclusion	Kühnau *et al.* (1991)
L291insert	+	nd	nd	nd	Eliminates *mal* gene repression	Lippincott and Traxler (1997)
G302D	+	nd	nd	nd	Abolishes inducer exclusion	Kühnau *et al.* (1991)
E306K (St)	−	nd	±	−		Hunke *et al.* (2000a)
S322F	+	nd	nd	nd	Abolishes inducer exclusion	Kühnau *et al.* (1991)
G346S	+	nd	nd	nd	Eliminates *mal* gene repression	Kühnau *et al.* (1991)
G346insert	+	nd	nd	nd	Eliminates *mal* gene repression	Lippincott and Traxler (1997)
C350S (St)	+	nd	nd	nd		Hunke and Schneider (1999)
C360S (St)	+	nd	nd	nd		Hunke and Schneider (1999)
R364insert	+	nd	nd	nd	Eliminates *mal* gene repression, abolishes inducer exclusion	Lippincott and Traxler (1997)

[a] Analyzed by photo-crosslinking with 8-azido-ATP in membrane vesicles or with purified soluble variants; del, deletion; insert, insertion of peptide linkers; St, numbering according to *S. typhimurium* MalK; +, indicates activities between 80 and 100% of control; ±, indicates activities <80 and >20% of control; −, indicates activities <20% of control.

Biochemical and genetic evidence, as well as computational analysis of complete microbial genomes that became available within recent years, revealed that ABC uptake systems, specific for maltose and/or maltodextrins, are widespread among Gram-negative and Gram-positive bacteria, including pathogens such as *S. typhimurium* (Schneider *et al.*, 1989), *Yersinia enterocolitica* (Brzostek *et al.*, 1993), *Streptococcus pneumoniae* (Puyet and Espinosa, 1993), *Vibrio cholerae* (Heidelberg *et al.*, 2000), *Aeromonas hydrophila* (Höner zu Bentrup *et al.*, 1994), *Mycobacterium tuberculosis* and *Mycobacterium leprae* (Borich *et al.*, 2000), to name just a few. Homologous transporters were also identified in archaea, such as *Thermococcus litoralis* (Horlacher *et al.*, 1998), *Pyrococcus furiosus* (DiRuggiero *et al.*, 2000) and *Sulfolobus solfataricus* (Elferink *et al.*, 2001).

The maltose transporter is composed of the extracellular (periplasmic) receptor, the maltose-binding protein (MBP or MalE), and the membrane-bound complex comprising the hydrophobic subunits, MalF and MalG, and two copies of the ATPase (ABC) subunit, MalK (Davidson and Nikaido, 1991) **(Figure 9.2)**. Interaction of the substrate-loaded binding protein triggers conformational changes that result in ATP hydrolysis at the MalK subunits and eventually in substrate translocation (Davidson *et al.*, 1992). In Gram-negative bacteria, an additional protein component, maltoporin or LamB, is required in the outer membrane to facilitate the diffusion of maltose (at low concentrations) and maltodextrins into the periplasm (Boos and Shuman, 1998; see also **Box 9.2**). In Gram-positive bacteria, which lack a periplasmic space, and in some archaea, maltose-binding proteins are lipoproteins that are anchored to the cytoplasmic membrane via fatty acids covalently coupled to an N-terminal cysteine residue (Horlacher *et al.*, 1998; Sutcliffe and Russel, 1995). In other archaea, attachment to the external side of the membrane is achieved by a carboxy-terminal transmembrane segment (Elferink *et al.*, 2001).

The genes encoding the transport components are usually clustered in one or two closely linked operons (Boos and Shuman, 1998; Heidelberg *et al.*, 2000). These, however, as

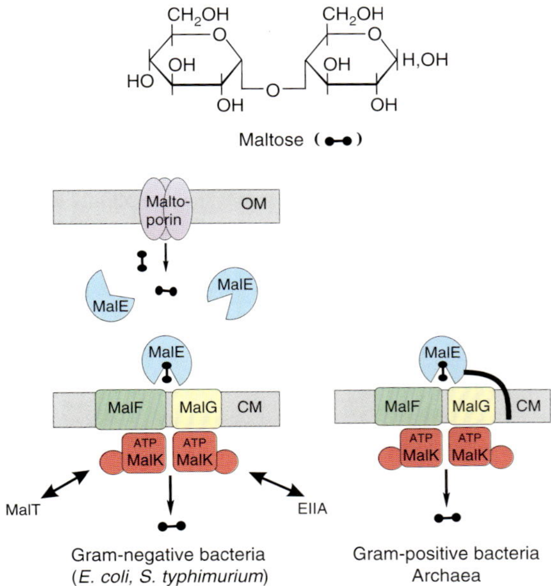

Figure 9.2. Schematic organization of components involved in maltose transport. See text for details. MalE, extracellular maltose-binding protein; MalF, MalG, hydrophobic, membrane integral subunits, presumably forming the translocation pore; MalK, ATP-hydrolyzing subunit, ABC domain. MalE can reside in an open and closed conformation. The latter is stabilized by substrate binding. In Gram-negative bacteria, the binding protein is located freely in the periplasmic space between outer and inner membrane. In Gram-positives and in some archaea, MalE is attached to the cytoplasmic membrane via an N-terminal lipid anchor. In other archaea, a transmembrane segment of the protein is used instead. In E. coli/S. typhimurium *and probably other closely related bacteria, the maltose transporter is engaged in regulatory processes that involve interactions of the MalK subunits with the positive transcriptional regulator of the mal regulon, MalT, and the dephosphorylated form of enzyme IIA of the glucose transporter (PTS). Whether similar activities exist in other Gram-negative bacteria is unknown.*

often found in Gram-positive bacteria and archaea, may lack the gene encoding the ABC protein (Greller *et al.*, 1999; Hülsmann *et al.*, 2000; Puyet and Espinosa, 1993; Quentin *et al.*, 1999; van Wezel *et al.*, 1997). This finding gave rise to the notion that a single ATPase protein could serve several transporters. Evidence in favor of this view was recently presented in the case of *Streptomyces*. Here, the ABC protein MsiK assists in the uptake of maltose and cellobiose, which is mediated by two different transporters (Schlösser *et al.*, 1997).

The ABC importer for maltose/maltodextrins of *E. coli* and *S. typhimurium* (Boos and Shuman, 1998) is by far the best-studied member of the CUT1 family. This, together with the histidine transport system of *S. typhimurium* (Doige and Ames, 1993; Liu *et al.*, 1997; P.-Q. Liu and Ames, 1998; Nikaido and Ames, 1999; Nikaido *et al.*, 1997), can serve as a model for ABC transporters in general. This chapter summarizes the current knowledge on this system, including relevant data for other members of the CUT1 family. Where appropriate, a comparative analysis with the properties of the histidine transporter is also provided. The latter is composed of the soluble substrate-binding protein HisJ and the membrane-bound complex, comprising two membrane-spanning subunits, HisQ and HisM, and two copies of the ABC subunit HisP (Kerppola *et al.*, 1991).

THE MALTOSE/ MALTODEXTRIN TRANSPORT SYSTEM OF *E. COLI* AND *S. TYPHIMURIUM*

The proteins constituting the ABC transporter for maltose in *E. coli* and *S. typhimurium* share >90% identical amino acid residues. Moreover, the components have been demonstrated to be fully exchangeable (Hunke *et al.*, 2000b). Consequently, the data summarized below will not in each case be specified with respect to the original organism of the transporter for which they have been obtained.

GENETIC ORGANIZATION AND REGULATION

The genes encoding the transport proteins for maltose are organized in two divergently transcribed operons at 91.4 min in the *malB* region of the chromosome: *malE malF malG*, and *malK lamB malM*.[2] They are part of a regulatory

[2]*The function of the product of the* malM *gene is currently unknown but it is dispensible for maltose/maltodextrin transport under all conditions tested so far (see Boos and Shuman, 1998).*

Box 9.2. Structural and functional aspects of maltoporin (LamB)

In Gram-negative bacteria, passage of maltose at low concentrations (≤10 μM), and of maltodextrins to the periplasm by facilitated diffusion, requires the presence of large amounts (40 000 copies per cell) of maltoporin in the outer membrane. (In *E. coli*, the protein serves as the receptor for bacteriophage lambda, giving rise to the alternative name, LamB.) Under these conditions, diffusion of the substrate through the outer membrane determines the overall rate of transport (Tralau *et al.*, 2000).

Maltoporin is organized as a homotrimer (molecular mass of the monomer: 47 kDa), with each monomer providing a distinct maltodextrin-binding site, which is crucial for the facilitated diffusion process (Luckey and Nikaido, 1980). The crystal structures of maltoporin from both *E. coli* (Schirmer *et al.*, 1995) and *S. typhimurium* (Meyer *et al.*, 1997) in the presence of different malto-oligosaccharides revealed that each subunit contains a channel that is formed by an 18-stranded, antiparallel β-barrel. Within a single channel, a constriction is formed by three peptide loops. The substrates are in contact with a 'greasy slide' of aromatic residues, which provides a path for translocation. There are well-defined binding sites for three consecutive glucosyl residues in the middle of the channel and one additional subsite at the extracellular end of the greasy slide (Dutzler *et al.*, 1996).

network, the 'maltose regulon', that encompasses a total of 11 genes (for review, see Boos and Shuman, 1998). Transcription of maltose-regulated genes is governed by the action of a positive regulator protein, MalT, that requires maltotriose and ATP for activity, and is affected by the functional status of the transporter (reviewed in Boos and Böhm, 2000) **(see also Box 9.1)**. In addition, the maltose regulon itself is subject to global carbon regulation of the cell (catabolite repression). Consequently, productive binding of MalT to specific nucleotide sequences upstream of the respective promoters ('MalT boxes') is brought about only in the presence of the cAMP/CAP complex (Boos and Shuman, 1998).

The subunits

In the following paragraphs, the properties of the individual components of the ABC transporter will be summarized. As maltoporin is confined to Gram-negative bacteria only and is not essential for the transport process, the interested reader is referred to **Box 9.2** for a short description of its structure and function.

Maltose-binding protein MalE

The soluble receptor MalE (molecular mass 40 kDa) binds maltose and maltodextrins with high affinity (K_D~1 μM) and is present in high concentration in the cell (~1 mM) following induction (Boos and Shuman, 1998). Whilst being crucial to the transport process, maltose-binding protein is also involved in the chemotactic response of the bacteria towards maltose by presenting the substrate to the chemoreceptor Tar (Gardina *et al.*, 1997).

MalE has been crystallized both in the absence of ligand (Sharff *et al.*, 1992) and in the presence of maltose (Spurlino *et al.*, 1991) or longer maltodextrins (Quiocho *et al.*, 1997). As found for other substrate-binding proteins, MalE consists of two nearly symmetrical lobes, between which the binding site is formed (for details, see Chapter 10). In the substrate-free form, these lobes are open and the substrate-binding site is accessible to the medium. Upon binding of ligand the two lobes move towards each other, thereby trapping the substrate inside the binding cleft. The crystallographic data further suggested that maltose may first bind to the N-terminal domain by contacting glutamate-111 at the base of the binding cleft. Subsequent ligand-induced movement of E111 may trigger the conformational change of the C-terminal lobe that eventually results in its participation in substrate binding and closing of the cleft (Sharff *et al.*, 1992).

The crystal structures of a maltose/trehalose and a maltose/maltodextrin binding protein of the hyperthermophilic archaea *T. litoralis* (Diez *et al.*, 2001) and *P. furiosus* (Evdokimov *et al.*, 2001), respectively, have recently been solved. Both are structurally related to MalE of *E. coli* despite the moderate level of sequence identity between these proteins and MalE-Ec.

The transport complex in the cytoplasmic membrane recognizes its substrate only when bound to MalE. Thus, only interaction of substrate-loaded MalE with the transport components can initiate the transport process. In fact, mathematical treatment of experimental data gave rise to the notion that the open nonliganded form of MalE can also bind to the membrane components. However, the affinity of the $MalFGK_2$ complex is five times greater for the loaded than for the unloaded form of MalE (Merino *et al.*, 1995).

Analysis of allele-specific suppressors and of dominant negative mutants has defined glycine-13 and aspartate-14 of MalE as sites of interaction with MalG, while tyrosine-210 was identified as being in contact with MalF. Thus, the N- and C-terminal lobes of MalE may interact with MalG and MalF, respectively (Hor and Shuman, 1993). In the C-terminal lobe, residues in α-helix 7 were shown by mutational analysis to play an important role in this interaction (Szmelcman *et al.*, 1997).

The binding protein of the histidine transporter of *S. typhimurium*, HisJ, is very similar in overall structure to MalE and also to other periplasmic receptors (Oh *et al.*, 1994). In addition, another soluble receptor, the lysine-arginine-ornithine binding protein (LAO), which is closely related both in primary and tertiary structure to HisJ, also delivers its substrates to the $HisQMP_2$ complex (Kang *et al.*, 1991). As in the case of MalE, both proteins move the two globular lobes close to each other upon binding of their respective ligands, thereby restoring the conformation that productively interacts with the membrane components (Wolf *et al.*, 1994). Both lobes participate in this interaction (Liu *et al.*, 1999). Strikingly, however, and in contrast to the maltose system, liganded and nonliganded HisJ have equal affinity for the membrane-bound complex (Ames *et al.*, 1996; Merino *et al.*, 1995).

The ABC protein MalK

Enzymatic properties

The MalK protein (molecular mass 40 kDa), when overproduced in the absence of the membrane-integral subunits MalF and MalG, can be purified to near homogeneity by either conventional methods (Mourez *et al.*, 1998; Schneider *et al.*, 1995a; Sharma and Davidson, 2000; Walter *et al.*, 1992a) or as an N-terminal His_6-fusion protein by Ni-NTA affinity chromatography (Hunke *et al.*, 2000a; Reich-Slotky *et al.*, 2000). Purified MalK exhibits a spontaneous ATPase activity with an apparent K_m around 0.1 mM and V_{max} values between 0.2 and 1.3 μmol min^{-1} mg^{-1} (Morbach *et al.*, 1993; Mourez *et al.*, 1998; Reich-Slotky *et al.*, 2000; Schmees *et al.*, 1999b; Schneider *et al.*, 1995a). GTP and CTP are also accepted as substrates and Mg^{2+} ions are absolutely essential for activity (Morbach *et al.*, 1993). In contrast to that of the assembled transport complex (see below), the enzymatic activity of the free protein is surprisingly insensitive to vanadate (Hunke *et al.*, 1995; Morbach *et al.*, 1993; Sharma and Davidson, 2000). Inhibition by N-ethylmaleimide was demonstrated to be due to modification of cysteine-40 within the Walker A motif thereby interfering with ATP binding (Hunke and Schneider, 1999; Morbach *et al.*, 1993). Limited proteolysis with trypsin revealed a specific conformational change upon binding of MgATP. Except GTP, other nucleotides proved to be ineffective (Mourez *et al.*, 1998; Schneider *et al.*, 1994).

When analyzed as a function of MalK concentration, ATP hydrolysis increases in a linear mode (Landmesser and Schneider, unpublished). This finding indicates that MalK is either enzymatically active as monomer or, alternatively, a putative MalK dimer (multimer) is already formed at very low (micromolar) concentrations. The latter possibility would be consistent with results of Kennedy and Traxler (1999), who found MalK dimers *in vivo* and in cell extracts. Further support for MalK being active as a dimer was provided by the observation that mixing wild-type MalK with a catalytically inactive MalK variant (H192R) resulted in an increase in ATPase activity as compared to wild type alone, thus suggesting that heterodimers were formed (Landmesser and Schneider, unpublished) (see also below). If so, the affinity of the monomers towards each other must be low since in gel filtration experiments purified MalK of *S. typhimurium* (MalK-St) eluted at the molecular mass of a monomer (Tebbe and Schneider, unpublished observation). The same result was reported for a close homologue, the MalK protein of the hyperthermophilic archaeon *T. litoralis* (Greller *et al.*, 1999).

In contrast, the ATPase activity of HisP, the ABC subunit of the histidine transporter, was observed to be non-linearly dependent on protein concentration, suggesting already from these data the formation of dimers. When applied to a molecular sieve column, only a small fraction of HisP eluted at the position of a dimer, while the bulk of HisP was found at the position of a monomer. This was taken as further evidence for the above notion but also suggested to the authors that both forms are in rapid equilibrium with each other (Nikaido *et al.*, 1997).

Other properties of the purified HisP protein were observed to be similar to those determined for MalK, including insensitivity to vanadate (Nikaido *et al.*, 1997).

Tertiary structural model

Crystals of MalK-St were obtained that diffract to about 3 Å, but the structure has not yet been solved (Schmees *et al.*, 1999a). However, the tertiary structure of a MalK homologue, isolated from the hyperthermophilic archaeon *T. litoralis* (MalK-Tl), presumably involved in maltose/trehalose transport, has recently been determined (Diederichs *et al.*, 2000). The protein was demonstrated to exhibit similar biochemical properties to those of the *S. typhimurium* MalK protein, with an optimal ATPase activity at 80°C (Greller *et al.*, 1999). Since both proteins share >50% identical amino acid residues **(Figure 9.1)** it appears safe to conclude that their crystal structures are likely to be very similar if not identical.

The crystal structure of MalK-Tl

MalK-Tl was crystallized in the presence of ADP and its tertiary structure could be solved with a resolution of 1.9 Å (Diederichs *et al.*, 2000). Two molecules are present per asymmetric unit that contact each other through the ATPase domains with the (regulatory) C-terminal domains attached at opposite poles **(Figure 9.3)**. Deviation from twofold symmetry is observed at the interface of the dimer and in regions corresponding to residues that are deduced to be in close contact to the membrane-integral subunits (see section on subunit–subunit interactions, below). In the nucleotide-binding sites, only a pyrophosphate molecule could be identified, while a density for the adenine ring of ADP was missing **(Figure 9.4)**. Although the overall fold of the ATPase domain is almost identical to that of HisP, with equivalent catalytic (ArmI) and helical (ArmII) subdomains, the structure of their dimers clearly differs. In the HisP dimer, where the crystal structure was

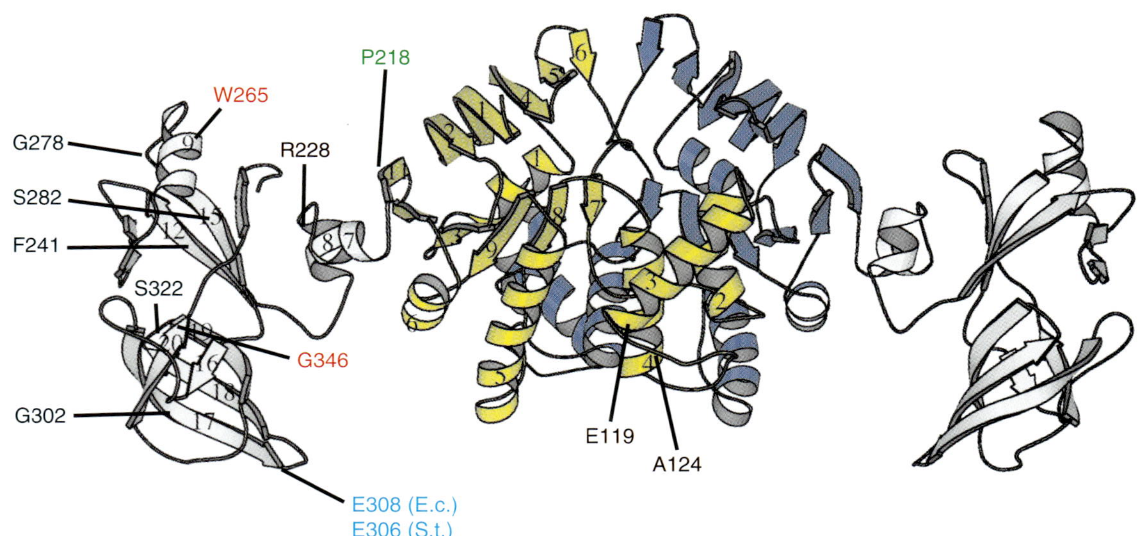

Figure 9.3. Ribbon representation of the MalK-Tl dimer. The ATPase core domains of each monomer are colored yellow and blue, respectively. The C-terminal (transcript regulatory) domains are colored gray. Labels indicate the numbers of helices and strands. The relative positions of residues discussed in the text are indicated. Numbering of the residues is according to MalK-Ec except for E308/306, where the corresponding numbering of MalK-St is also given (please note that residues $M_{260}P_{261}$ are deleted in MalK-St, resulting in a total number of 369 compared to 371 residues in MalK-Ec). Color code: black, residues when mutated that render the transporter insensitive to inducer exclusion; red, residues, when mutated that affect the repressing activity of MalK; blue, mutation to lysine reduces ATPase activity; green, residue depicted for construction of a truncated MalK variant by genetic engineering (Schmees and Schneider, 1998; see text for details). Reproduced from Diederichs et al. *(2000) with permission and modified.*

obtained in the presence of ATP, the monomers associate via antiparallel beta sheets (Hung *et al.*, 1998) that, in the MalK-Tl structure, are located at the top of the dimer **(Figures 9.3 and 9.4)**. As a consequence, in HisP the nucleotide-binding sites are located opposite to each other at the outside of the monomers, while in MalK-Tl both sites are facing each other in the center part of the core structure **(Figure 9.4)** (see Chapters 4 and 7 for a detailed description of other crystal structures of ABC proteins/domains).

Although the C-terminal regulatory domain is clearly separated from the ATPase (core) domain in MalK-Tl, mutational analysis of MalK-St (Hunke *et al.*, 2000a) (see below) and a study using truncated MalK proteins and chimeras suggested that both N- and C-terminal parts of the protein are required for its structural integrity (Schmees and Schneider, 1998). In the latter investigation, it was demonstrated that when similar sized N- and C-terminal half-molecules of MalK-St (split at L179) are expressed they assemble into a transport complex *in vivo* which is still active. On the other hand, when the site of splitting was shifted towards the C-terminal domain, transport was abolished. In particular, expression of fragments that correspond exactly to one or both of the ATPase and C-terminal domains of MalK-Tl (split at P218, see **Figure 9.3**) did not result in an active transporter, most probably due to misfolding of the peptides (Schmees and Schneider, 1998). This notion is supported by the finding that transport function was retained in chimeras composed of similar N- or C-terminal fragments of MalK, with complementing fragments of HisP (Schneider and Walter, 1991) or LacK, a close homologue of the lactose ABC transporter from *A. radiobacter* (Schmees and Schneider, 1998; Wilken *et al.*, 1996). These studies also indicated that a minimum portion necessary transcription regulation by MalK would encompass residues Q263 to V369 (Schmees and Schneider, 1998).

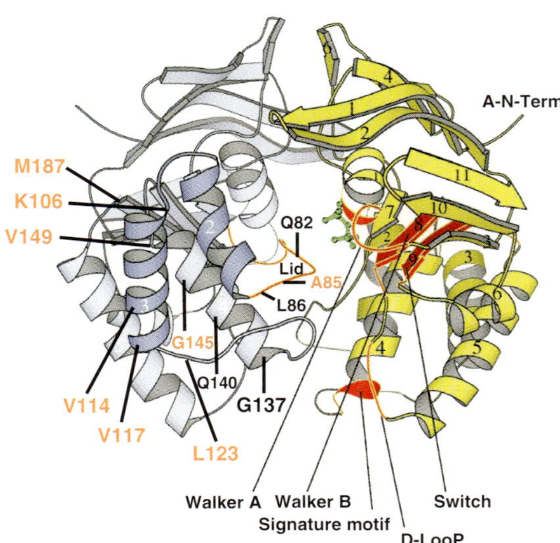

Figure 9.4. Core region of the MalK-Tl dimer. The molecule is viewed along the interface perpendicular to the pseudosymmetry axis. The relative locations of conserved motifs are indicated in the monomer colored yellow, while single residues discussed in the text are indicated in black in the monomer colored blue. The bound pyrophosphate is shown in green. Residues written in red in the lower helical (ArmII) domain are thought to interact with the membrane-integral subunits. Reproduced from Diederichs et al. *(2000) with permission and modified.*

Functional amino acid residues

The *malK* gene has been the subject of extensive mutational analyses resulting in the identification of functionally important amino acid residues and peptide fragments **(Table 9.2)**. From these studies a domain structure of the protein was postulated, with an N-terminal core (ABC) carrying the nucleotide-binding sites and residues involved in the interaction with the membrane components, together with a C-terminal domain devoted to the transcript regulatory activities of the protein (Kühnau *et al.*, 1991; Schmees and Schneider, 1998; Wilken, 1997). This view was largely confirmed by the crystal structure of MalK-Tl. Nonetheless, both domains (ABC and regulatory) are not autonomous entities but talk to each other, since mutations in both have been identified that alter the activities of the other (Hunke *et al.*, 2000a; Kühnau *et al.*, 1991; Schmees *et al.*, 1999b). The following section focuses on mutations affecting the transport activities of MalK only. (For a description of mutations that eliminate the regulatory properties of MalK, see **Box 9.1** and **Table 9.2**.)

Mutations affecting ATPase activity

As shown in **Figure 9.1**, **Table 9.2** mutations in the ATPase domain, especially those affecting the invariant lysine (K42) and aspartate (D158) residues, respectively, in the nucleotide-binding motifs A and B, usually abolish ATPase activity.

However, depending on the chemical nature of the substituting amino acid, such mutant proteins may retain the capability to bind ATP (Hunke *et al.*, 2000a; Kühnau *et al.*, 1991; Panagiotidis *et al.* 1993, Schneider *et al.*, 1994). Replacement of cysteine-40 by serine, on the other hand, is without functional consequences (Hunke and Schneider, 1999). Other mutations (P160L, D165N at the C-terminus of the B motif, also called the D-loop, see **Figure 9.1**), although remote from the ATP-binding site according to the MalK-Tl structure **(Figure 9.4)**, nonetheless reduced the ATPase activity of the soluble variants to less than 20% of the control (Hunke *et al.*, 2000a).

Amino acid substitutions in the ABC signature ('LSGGQ') motif have contrasting consequences for function. G137 cannot be replaced by other residues without complete loss of ATPase activity (Panagiotidis *et al.*, 1993; Schmees *et al.*, 1999b). However, substituting asparagine or lysine for glutamine-140 resulted in an enzymatically active MalK variant when analyzed separately but substantially reduced MalE-maltose-dependent ATPase activity in the assembled transport complex (Schmees *et al.*, 1999b). Thus, Q140 might be involved in the activation of ATPase activity upon substrate binding. From these genetic findings it was concluded that the ABC signature sequence could sense an incoming signal through its C-terminal half, while residues in the N-terminal part of the motif may assist in the catalytic reaction. This idea does not seem to be supported by the MalK-Tl dimer structure, in which the signature motif is located at the bottom of the helical domain layer **(Figure 9.4)** and thus, is distant both in *cis* and in *trans* from the nucleotide-binding sites. However, this may be different in the assembled transport complex. In fact, the first tertiary structure of a complete ABC transporter, which became available only recently, lends support to this notion. In MsbA, a protein mediating the export of the outer membrane component lipid A in *E. coli*, the signature sequence appears to be located rather closely to the Walker B motif (Chang and Roth, 2001). Moreover, by comparative analysis of the ATP-bound form of HisP with the MgADP-bound form of MJ0796, an ABC protein of the thermophilic archaeon *Methanococcus jannaschii* (Yuan *et al.*, 2001), suggested that the helical domain may rotate outward from the nucleotide-binding site upon hydrolysis, resulting in a substantial movement of the LSGGQ motif. Although attractive, a note of caution seems opportune as data from two different proteins were compared. Furthermore, in Rad50, an ABC-like protein that is not involved in transport but is a soluble DNA repair enzyme, the ABC signature motif contacts the ATPase active site in the opposing monomer (Hopfner *et al.*, 2000). Whether the structure of the Rad50 dimer can serve as a model for ABC proteins devoted to transport processes is a matter of current controversy (see also Chapter 4 for a detailed discussion). Again, one has to keep in mind that the structure of the MalK dimer in solution and of the other ABC transporter subunits for which structural data are available might differ from that in the assembled transport complex. This aspect will be discussed further below.

The conserved sequence motif around glutamine-82 (termed 'lid', see **Figure 9.4**) was found in the MalK-Tl structure near the nucleotide-binding site (Diederichs *et al.*, 2000). Substitution of lysine or glutamate for Q82 in MalK-St reduced but did not abolish transport activity *in vivo* (Walter *et al.*, 1992b). Thus, the absolute necessity of a glutamine residue at this position can be excluded but the chemical nature of the substitutes, K or E, does not rule out a role in polarizing the water molecule that attacks the γ-phosphate of ATP during catalysis, as suggested from the HisP structure (Hung *et al.*, 1998). However, such a role is not supported by the MalK-Tl structure as the corresponding Q residue is too far away from the pyrophosphate **(Figure 9.4)**. Other candidates for polarizing the water molecule (E64, E94), as suggested by sequence comparison (Yoshida and Amano, 1995), were eliminated by mutational analysis (Stein *et al.*, 1997).

Another highly conserved residue from the 'lid' region, L86, when mutated to phenylalanine, was shown to cause the same phenotype as the Q140K/N mutations described above. Thus, the purified variant exhibits ATPase activity comparable to wild type in the reconstituted transport complex, and ATP hydrolysis is abolished (Hunke *et al.*, 2000a). These results suggest that L86 may be involved in activating the enzymatic activity of MalK upon binding of substrated-loaded MalE to the complex and thus, be in close contact to MalF/MalG. Consistent with this notion is the finding that in MsbA, the region encompassing the corresponding residue (L428) is in direct contact with an intracellular domain that connects the membrane-spanning helices to the ABC domain (Chang and Roth, 2001).

Residues within the so-called 'switch' region of the ABC domain **(see Figures 9.1 and 9.4)**, located carboxy-terminal to the Walker B site, are believed to propagate conformational changes triggered by ATP hydrolysis, in analogy to evidence provided by the crystal structure of the *E. coli* recA protein (Story and Steitz, 1992; Yoshida and Amano, 1995). In line with this notion are results from mutational analysis of the highly conserved histidine-192 in the switch motif. Replacement by arginine was shown to cause defective transport *in vivo* (Walter *et al.*, 1992b) and *in vitro* (Davidson and Sharma, 1997), and loss of ATPase activity of the purified variant (Landmesser and Schneider, unpublished). (The previously reported retention of ATPase activity by this mutant (Walter *et al.*, 1992b) could not subsequently be confirmed when using an optimized purification protocol.) How this residue might make contact with the nucleotide-binding site is not obvious from the MalK-Tl structure (Diederichs *et al.*, 2000). However, the authors proposed that other conformations of the protein may exist that, by analogy with the situation seen in the HisP structure, could promote contact through an interspersed water molecule.

So far, one residue located in the very C-terminal, regulatory domain of MalK-St was shown to affect the ATPase activity of the protein. E306 (E308 in *E. coli* MalK), when mutated to lysine, resulted in the loss of transport activity *in vivo* and the purified MalK variant exhibited strongly reduced ATPase activity (Hunke *et al.*, 2000a). Although the crystal structure of MalK-Tl does not provide any clue for a possible function of this residue, E306 is highly conserved among members of the 'MalK' subfamily **(Figure 9.1)**. Its function, like that of the other conserved residues in the C-terminal domain of these proteins, remains to be elucidated.

Mutations affecting interactions with the membrane components

Mutant analyses and biochemical evidence have identified residues in MalK that are involved in the functional and/or structural interaction with the membrane integral subunits. Wilken *et al.* (1996) isolated variants of the homologous LacK protein (V114M, L123F, G145S) that partially or fully replace MalK in maltose transport. Consequently, when introduced into MalK, the same mutations reduced or abolished transport activity (Scheffel, Brinkmann and Schneider, unpublished). Mourez *et al.* (1997a) screened for MalK mutants that could restore transport in *E. coli* strains carrying mutations in the conserved 'EAA' loops of MalF and/or MalG (A85M, V117M, V149M/I, M187I). With the exception of A85 (part of the 'lid') and M187 (part of the 'switch'), all are located in the largely α-helical peptide connecting the Walker A and B sites **(Figure 9.4)**. In addition, limited proteolysis of MalK in the presence and absence of MalFG-containing membrane vesicles suggested that K106 at the end of helix 3 is also in close contact with the membrane integral subunits (Mourez *et al.*, 1998). (This theme will be continued in a later section with evidence from crosslinking experiments.)

The membrane-integral subunits MalG and MalF

MalG (molecular mass 32 kDa), as shown by extensive topological analysis using PhoA fusions, very probably spans the membrane six times, although two slightly differing models have been proposed (Boyd *et al.*, 1993; Dassa and Muir, 1993). Thus, the protein represents a typical hydrophobic domain of an ABC transporter. Linker insertion mutagenesis defined regions in MalG that are crucial for transport, assembly and protein stability, respectively (Dassa, 1993; Nelson and Traxler, 1998). Accordingly, most of transmembrane helix or segment 1 (TMS1) and parts of the first and second periplasmic loop are tolerant to variations in the primary structure and thus may be dispensable for function. Interestingly, mutation of isoleucine-154 in the second periplasmic loop to a serine renders the transport complex independent of the binding protein (see topology, **Figure 9.7**). Binding protein-independent mutants exhibit a much lower affinity for maltose (K_m of 2 mM compared to 1 μM for wild type) and have lost the ability to transport maltodextrins (Treptow and Shuman, 1985). Thus, residue 154 may be part of a substrate-binding site (Covitz *et al.*, 1994). Linker insertions close to the conserved 'EAA' motif in the third cytoplasmic loop (here: EAAALDG), shown in **Figure 9.7**, are deficient in assembly into the transport complex and thus abolish function *in vivo*. Moreover, single but radical mutations of the third (A3D) and seventh (G7P) residue of the motif in MalG eliminate transport and result in a dislocation of MalK from the membrane. In contrast, moderate changes in the chemical nature of the side-chain of the same residues (A3S, G7A) or replacement of the conserved glutamate at position 1 had no significant effect on function (Mourez *et al.*, 1997a). Mutations in the second half of the first periplasmic loop and those in TMSs 2, 4 and 5 were shown to affect protein stability (Nelson and Traxler, 1998). A region essential for substrate specificity was identified near the C-terminus as insertion mutations resulted in the loss of maltodextrin but not of maltose utilization (Dassa, 1993).

MalF (molecular mass 57 kDa) is somewhat unusual among ABC membrane-spanning domains as it is predicted to contain eight transmembrane helices (see also HlyB, Chapter 11). However, this topology seems to be confined to the enterobacterial MalF proteins and a few other examples (Ehrmann *et al.*, 1998), as most MalF proteins lack TM helices 1 and 2. In fact, in *E. coli* MalF, the first membrane-spanning helix is dispensable for function (Ehrmann and Beckwith, 1991). In addition, the enterobacterial MalF proteins contain a large periplasmic peptide loop connecting the third and fourth transmembrane helices, which is also not conserved in evolution (see **Figure 9.7**). Nevertheless, mutations in this region are mostly not tolerated with respect to transport function because they affect MalK localization to the membrane (Tapia *et al.*, 1999). Interestingly enough, overexpression of this periplasmic loop caused the induction of a protein involved in the extracytoplasmic stress response of *E. coli* (Mourez *et al.*, 1997b). Again, linker mutagenesis identified regions in cytoplasmic loops 2 and 5, in periplasmic loop 4 and in TM helix 8 as being involved in the transport mechanism, while mutations in cytoplasmic loop 3 and periplasmic loop 2 affected assembly (Tapia *et al.*, 1999). Strikingly, single mutations in the EAA loop of MalF (here: EASAMDG) differ in their phenotypic consequences from those affecting the homologous positions in MalG. For example, in contrast to the results described above, replacing the conserved glycine at position 7 by proline had no effect on transport *in vivo* (Mourez *et al.*, 1997a). As in the case of MalG, substitution of different residues for glutamate-1 also had no major effect on function. However, when the glutamate residues in both EAA loops were replaced by either lysine or leucine, transport was completely abolished (Mourez *et al.*, 1997a). These results clearly indicate an asymmetric but nonetheless crucial function of the motif in both subunits, probably involving contact with the MalK subunits (see above and below).

Based on a detailed mutational analysis, Ehrmann and collaborators assigned putative functions to the TM helices 3–8 of MalF (Ehrle *et al.*, 1996; Steinke *et al.*, 2001). Most mutations in TM5 and those in TMs 3, 4 and 7 interfered with MalF assembly. The defects of two of the mutants in TM7 could be cured by second-site mutations in TM helices 6 or 8 (Ehrle *et al.*, 1996), indicating close physical contact between these helices. Mutations affecting substrate specificity, that is resulting in a loss of maltodextrin utilization while maltose uptake is retained, clustered in TM6 and TM8 and were also found in TM helix 5 (L323Q). The L323Q mutation is close to L334, which when mutated to tryptophan, caused the transporter to accept lactose as a substrate (Merino and Shuman, 1997). The very same mutation also renders the system binding protein independent, when combined with a second mutation in either MalF or MalG (Covitz *et al.*, 1994). Together, these data support the notion that residues in TM5 facing the periplasmic side of the membrane contribute to a substrate-binding site. TMs 6, 7 and 8, in which other mutations resulting in a binding protein-independent transporter were identified, are also likely to participate in substrate binding. Based on the above data and additional evidence from other systems as well as on computational analysis of transmembrane domains of other ABC transporters, Ehrmann *et al.* (1998) have proposed a hypothetical model for the arrangement of MalF and MalG in the membrane **(Figure 9.5)**. According to this proposal, helices that form a channel for substrate translocation include TMs 2, 3, 4 and 5 of MalG and TMs 4, 5, 6 and 7 of MalF. The implications for a possible transport mechanism are discussed below.

THE MALFGK$_2$ COMPLEX

Enzymatic properties

The maltose transport complex (MalFGK$_2$) can be purified from overproducing strains either by conventional methods (Davidson and Nikaido, 1991) or by affinity-tag technology (Davidson and Sharma, 1997; Landmesser *et al.*, 2002; Reich-Slotky *et al.*, 2000; Schmees *et al.*, 1999b) (see **Box 9.3** for details). In detergent solution, most preparations exhibit a low basal ATPase activity (0.04 μmol min^{-1} mg^{-1},

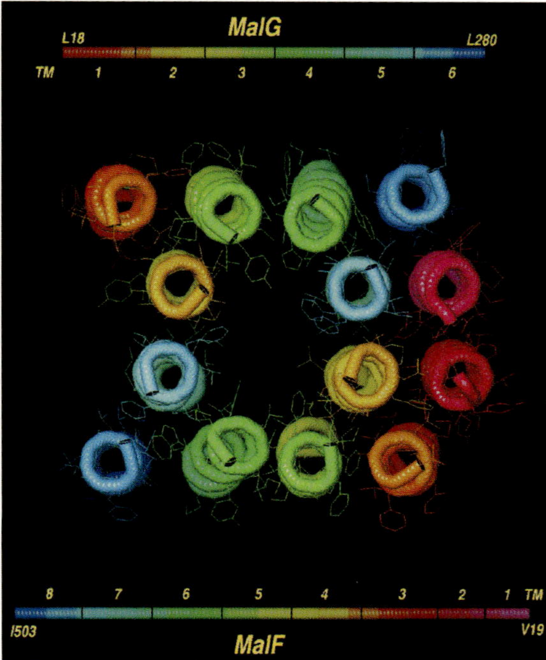

Figure 9.5. ***Model of transmembrane domains of MalF and MalG. A view of the transmembrane helices from the extracellular (periplasmic) side of the membrane is presented. Individual helices are color coded as indicated by horizontal bars at the top (MalG) and bottom (MalF). MalG TM1 is on the upper left side, MalG TM6 on the upper right side. MalF TM1 and 2 are in purple, MalF TM3 on the lower right side, MalF TM8 on the lower left side. Homologous TMs of both subunits are shown in the same color. Reproduced from Ehrmann* et al. *(1998), with permission.***

Landmesser *et al.*, 2002), which is enhanced five- to sixfold in the presence of maltose-binding protein and maltose (0.2 μmol min^{-1} mg^{-1}, Chen *et al.*, 2001; Landmesser *et al.*, 2002). To obtain rates of ATP hydrolysis that are coupled to ligand translocation the complex must be incorporated into liposomes (see **Boxes 9.3 and 9.4** for technical details). Under these conditions, V_{max} values of MalE-maltose-dependent ATPase activity were obtained in the range of 4–5 μmol min^{-1} mg^{-1} with Michaelis constants of 0.1 to 0.2 mM (Chen *et al.*, 2001; Landmesser *et al.*, 2002; Reich-Slotky *et al.*, 2000). The K_m values are in good agreement with those reported for the soluble MalK protein (see above). In proteoliposomes, ATP is hydrolyzed cooperatively (Davidson *et al.*, 1996) and two intact copies of the MalK subunit are required for function (Davidson and Sharma, 1997). Maltose transport activity as a function of ATP hydrolysis was also demonstrated with proteoliposomes, yielding widely varying rates of uptake between 1.2 (Chen *et al.*, 2001; Davidson and Sharma, 1997) and 61 nmol min^{-1} mg^{-1} (Landmesser *et al.*, 2002).

The preparation of an uncoupled $MalFGK_2$ complex that exhibits high ATPase activity in detergent solution, even in the absence of the substrate-loaded binding protein was recently reported. Addition of MalE/maltose resulted in a marked stimulation of the catalytic activity. When incorporated into liposomes, the complex returned to being dependent on the binding protein. Whether this unusual finding is due to the location of the affinity tag that, unlike in other preparations, is fused to the C-terminal end of MalG, remains unclear (Reich-Slotky *et al.*, 2000).

In contrast to soluble MalK, the ATPase activity of the transport complex is sensitive to micromolar concentrations of vanadate (Hunke *et al.*, 1995). Inhibition is caused by the trapping of ADP in the binding pocket after hydrolysis of the γ-phosphate of ATP (Sharma and Davidson, 2000). Under these conditions, that is, when the transporter is locked in the transition state, a tight association of the unloaded substrate-binding protein with the transport complex is observed (Chen *et al.*, 2001).

Binding protein-independent transport complexes carrying certain mutations in MalF and/or MalG exhibit a constitutive ATPase activity (Covitz *et al.*, 1994) and can be purified by standard protocols (Davidson *et al.*, 1992; Sharma and Davidson, 2000). The mutations do not significantly alter affinity, cooperativity, vanadate sensitivity or substrate specificity of the ATPase catalytic site (Davidson *et al.*, 1996). However, differences in fluorescence after binding a fluorophore to the MalK subunits suggested different conformations of wild type and these mutant forms of MalFG in the transporter. Moreover, the binding protein-independent complexes containing these MalFG mutant proteins seem to resemble the transition (ADP.Pi) state of the wild-type transporter (Mannering *et al.*, 2001).

THE HisQMP$_2$ COMPLEX

The other well-characterized ABC importer, the purified histidine transporter ($HisQMP_2$), exhibits essentially the same enzymatic properties as the maltose transporter (Ames *et al.*, 2001; Liu *et al.*, 1997). In the reconstituted

BOX 9.3. PRACTICAL ASPECTS. I: PURIFICATION OF THE *S. TYPHIMURIUM* MALTOSE TRANSPORTER

Principle:

The first purification of the *E. coli* maltose transporter from an overproducing strain was achieved by Davidson and Nikaido (1991) using conventional biochemical techniques. Today, purification at high protein yield of the maltose transporter and of ABC transporters in general is largely facilitated by applying affinity-tag technology. To this end, one of the subunits, usually the ABC protein, is synthesized as a fusion protein carrying a peptide or protein sequence at either the N- or C-terminal end that is specifically recognized by a corresponding chromatography matrix. The most popular approach is a fusion to six consecutive histidine residues that allows binding of the transport complex to Ni-nitrilotriacetic acid, immoblized to agarose beads. After removal of unbound material, the complex is readily eluted by imidazole. The following protocol combines the benefits of a newly constructed expression plasmid (Landmesser *et al.*, 2002) with a purification procedure basically devised by Davidson and Sharma (1997).

Procedure:

Cells of *E. coli* strain JM109 harboring plasmid pBB1 (his$_6$-*malK*, *malF*, *malG* on expression vector pQE9 under the control of the T5 promoter) are grown in rich medium to an $OD_{650} = 0.25$. Expression of *malK*, *malF*, *malG* is induced by the addition of 0.5 mM isopropyl β-D-thiogalactopyranoside (IPTG) and growth continues to $OD_{650} = 4$. Cells are harvested by centrifugation for 10 min at 9000 × *g*, resuspended in 150 ml of buffer 1 (50 mM Tris-HCl, pH 8, 5 mM $MgCl_2$, 20% glycerol, 0.1 mM phenylmethylsulfonylfluoride) and disruptured by one passage through a French press at 18 000 psi. Following a low speed spin for 15 min at 10 000 × *g*, membrane vesicles are recovered by centrifugation for 1 h at 200 000 × *g*, resuspended in buffer 1 and stored at −80°C until use. Solubilization of the transport complex is achieved by adding *n*-dodecyl-β-D-maltoside (DM, final concentration: 1.1%) to membrane vesicles at 5 mg ml^{-1} in buffer 1. After incubation for 1 h on ice under constant stirring, solubilized proteins are separated from the remaining membranes by ultracentrifugation for 1 h at 200 000 × *g*. Subsequently, the supernatant is mixed with Ni-NTA agarose (1 ml of slurry per 9 ml of supernatant), equilibrated with buffer 1 containing 0.01% DM (= buffer 2) and incubated for 1 h on a shaking device in a cold room. The mixture is then poured into a disposable column and the matrix is washed with 15 bed volumes of buffer 2, followed by 15 bed volumes of buffer 2 containing 20 mM imidazole. The transport complex is finally eluted with 15 bed volumes of buffer 2 supplemented with 50 mM imidazole. Peak fractions are combined, concentrated by ultrafiltration through Amicon filter YM30 and dialyzed against 500 volumes of buffer 2 to remove imidazole. Finally, the protein is shock-frozen in liquid nitrogen and stored at −80°C. Typically, 4–5 mg of highly purified complex are obtained from 1 liter of culture.

BOX 9.4. PRACTICAL ASPECTS. II: RECONSTITUTION OF SUBSTRATE-STIMULATED ATPASE ACTIVITY AND ATP-INDUCED SUBSTRATE TRANSPORT IN PROTEOLIPOSOMES

Principles:

The function of purified ABC importers, such as the $MalFGK_2$ complex or the $HisQMP_2$ complex, can be analyzed by studying the binding protein/substrate-dependent ATPase activity and/or by monitoring ATP-dependent translocation of radiolabeled substrate across a phospholipid bilayer. To this end, incorporation of the protein complexes into liposomes is a prerequisite. The procedures used currently were introduced by Davidson and Nikaido (1991) (see also Hall *et al.*, 1998), based on the fundamental work by E. Racker and collegues (1979). To analyze MalE/maltose-dependent ATPase activity, proteoliposomes containing the $MalFGK_2$ complex are formed by detergent dilution in the presence of maltose-binding protein and maltose. As the orientation of the complexes in the proteoliposomes cannot be controlled, hydrolysis of added ATP is only due to the activity of those complexes that expose the MalK subunits to the medium **(see Figure 9.6A)**. Ames and collegues found that incubating proteoliposomes with high concentrations of Mg^{2+} induces some leakage, thereby allowing substrate molecules and binding proteins to diffuse into the vesicles (Liu *et al.*, 1997). Thus, under these conditions, the ATPase activity of transport complexes of both orientations can be monitored.

To measure the ATP-dependent uptake of radiolabeled maltose into the lumen of the proteoliposomes, the latter are preloaded with ATP and the reaction is initiated by adding MalE and maltose to the medium. In this case, only the transport complexes that orient their MalK subunits to the interior of the vesicles add to the activity (Hall *et al.*, 1998) **(Figure 9.6B)**. Alternatively, preformed proteoliposomes can be loaded with ATP by several cycles of freezing and thawing, followed by passage through a filter to regenerate unilamellar vesicles (Chen *et al.*, 2001; Liu and Ames, 1997).

Procedures:

Preparation of proteoliposomes for analyzing ATPase activity (according to Hall *et al.*, 1998)
The reconstitution mixture (300 μl) contains 50 μg purified $MalFGK_2$ complex, 120 μg purified maltose-binding protein (MalE) (prepared according to Dean *et al.*, 1992), 60 mM maltose, 1% (w/v) octylglucoside and 2.5 mg liposomes. The liposomes are preformed from soy bean or *E. coli* phospholipids by ultrasonication in 20 mM Tris-HCl, pH 8, 1 mM dithiothreitol. After incubation on ice for 30 min under gentle stirring, 25 ml of 20 mM Tris-HCl, pH 8, 1 mM dithiothreitol, are added and the mixture is centrifuged for 1 h at 200 000 × *g*. The resulting proteoliposomes are resuspended in 100 μl of 20 mM Tris-HCl, pH 8, 1 mM $MgCl_2$, 10 μM maltose and stored on ice until use.

ATP hydrolysis assay (according to Nikaido *et al.*, 1997)
100 μl of proteoliposomes diluted in 20 mM Tris-HCl, pH 8, 1 mM $MgCl_2$, 10 μM maltose to a final concentration of $MalFGK_2$ complex of 40 μg/ml are equilibrated at 37°C for 5 min and the reaction is initiated by the addition of 10 mM $MgCl_2$ and 2 mM ATP (final concentrations). Samples (25 μl) are taken at 1 min intervals and placed into microtiter plate wells containing 25 μl of 12% SDS. The amount of Pi liberated is determined by a colorimetric assay (Chifflet *et al.*, 1988) using Na_2HPO_4 as a standard.To this end, 50 μl of a solution containing 30 mg ml^{-1} ascorbic acid in 1 N HCl and 0.5% ammonium molybdate are added to each well and the mixture is incubated for 5 min. The reaction is terminated by adding 75 μl of a solution containing 2% each of sodium citrate, sodium arsenate and acetic acid. After incubation at room temperature for 20 min, absorption is measured in a microtiter plate reader at 750 nm.

Preparation of proteoliposomes for analyzing maltose transport (according to Hall *et al.*, 1998)
Proteoliposomes are essentially prepared as described above with the following modifications: in the mixture, maltose-binding protein and maltose are replaced by 5 mM ATP and the resulting proteoliposomes are resuspended in 100 μl 20 mM Tris-HCl, pH 8, 3 mM $MgCl_2$.

Maltose transport assay (according to Hall *et al.*, 1998)
Proteoliposomes (30–60 μl) are diluted with 20 mM Tris-HCl, pH 8, 3 mM $MgCl_2$, to a final volume of 135 μl and the reaction is initiated by adding 15 μl of a solution containing maltose-binding protein (final concentration: 1 μM) and ^{14}C-maltose (final concentration: 10 μM; 0.86 μCi). Samples (25 μl) are taken at 10 second intervals, diluted in 225 μl 20 mM Tris-HCl, pH 8, 3 mM $MgCl_2$ and filtered through a Millipore filter (0.22 μm GSFT). The filters are quickly washed with 5 ml ice-cold 50 ml LiCl, air-dried, and counted in a liquid scintillation counter.

system, ATP is hydrolyzed with a V_{max} value of 2.1 μmol min^{-1} mg^{-1} of protein in the presence of liganded HisJ, while the maximal intrinsic activity of the soluble protein is about 15-fold lower. ATPase activity is vanadate-sensitive and displays positive cooperativity. A transport rate for L-histidine of 8 nmol $min^{-1} mg^{-1}$ was reported (Liu *et al.*, 1997).

Mutations that render the histidine transporter independent of the binding protein were mapped exclusively in the *hisP* gene (Speiser and Ames, 1991). The isolated mutant transport complexes display a constitutive ATPase activity, very similar to the values obtained with the HisJ-stimulated wild-type complex, although no cooperativity for ATP was observed (Liu *et al.*, 1999). However, and in marked contrast to mutations that lead to binding protein-independent maltose transport complexes (Covitz *et al.*, 1994; Davidson *et al.*, 1992), ATP hydrolysis in the HisP constitutive mutants is only poorly coupled to ligand transport unless HisJ is present (Liu *et al.*, 1999). Thus, depending on the subunit that is mutated, the phenotype of constitutive ATPase activity can reflect different steps in the transport process.

SUBUNIT–SUBUNIT INTERACTIONS

Suppressor analyses (Mourez *et al.*, 1997a; Wilken *et al.*, 1996) and biochemical evidence (Mourez *et al.*, 1998) suggested that residues in ArmII, the helical domain (mostly on α3 and connecting loops, see **Figure 9.4**) of MalK, make contact with the EAA loops of MalF and MalG, respectively (see above). This notion was confirmed by site-directed chemical crosslinking in membrane vesicles containing monocysteine variants of the respective subunits (Hunke *et al.*,

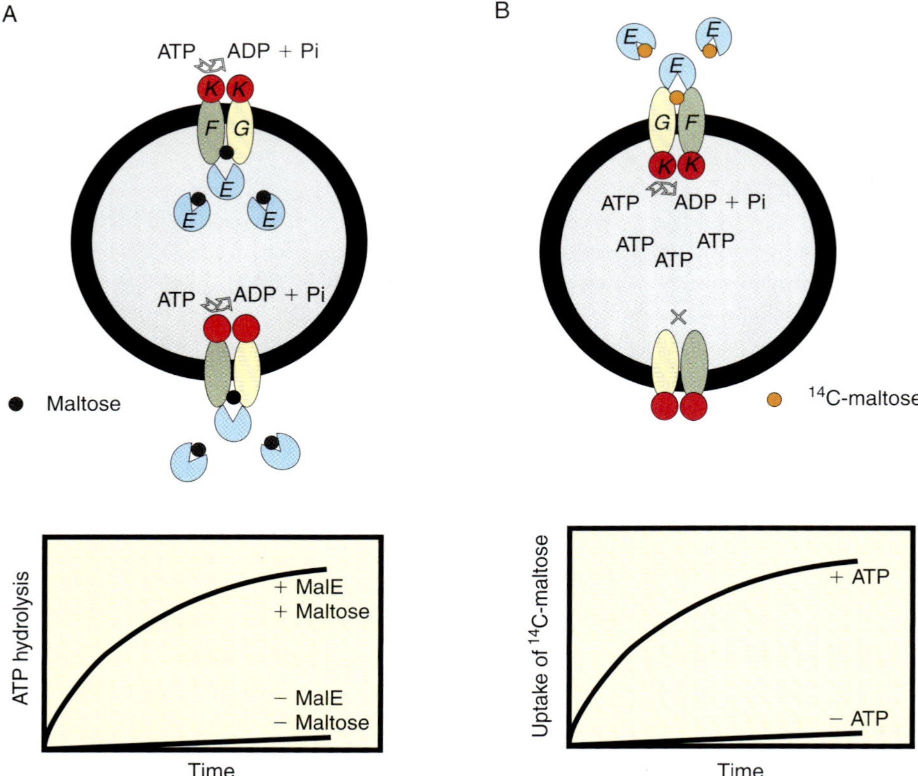

Figure 9.6. ***Experimental protocol for assaying transporter activities in proteoliposomes. See Box 9.4 for details. A, MalE/maltose-stimulated ATP hydrolysis catalyzed by the MalFGK$_2$ complex is monitored by assaying the release of inorganic phosphate. B, ATP-dependent transport activity of the MalFGK$_2$ complex is monitored by assaying the accumulation of radiolabeled maltose in the lumen of the proteoliposomes.***

2000b) **(Figure 9.7)**. According to this study, MalK-K106, MalK-V117 and, to a lesser extent, MalK-A85 contact the serine residue at position 3 in the EAA loop of MalF. In the conserved EAA loop of MalG, alanine-3 and glycine-7 are in close proximity to MalK-A85, while a looser association was observed between G7 and MalK-K106. Moreover, as revealed by crosslinking, both MalK monomers contact each other via K106 **(Figure 9.7A)**. These interactions were altered in the presence of ATP **(Figure 9.7B)**. In these conditions, MalK dimers also formed intermolecular contacts at A85 that simultaneously came within crosslinking distance of S3 in MalF. In addition, MalK-V114 also contacts MalF-S3. In MalG, a loose contact of A3 to MalK-K106 was induced by the presence of ATP. These data not only confirmed the notion that the MalK subunits interact asymmetrically with MalF and MalG (Mourez *et al.*, 1997a) but also provided additional evidence for an ATP-induced conformational change of MalK (Mourez *et al.*, 1998; Schneider *et al.*, 1994). A similar conclusion was recently drawn for the histidine transport complex from the analysis of sulfhydryl modification by thiol-specific reagents, CD spectroscopy and intrinsic fluorescence measurements (Kreimer *et al.*, 2000).

The observed crosslink between both MalK monomers at alanine-85 is consistent with the relative positions of these residues in the crystal structure of MalK-Tl dimer **(Figure 9.4)**. In contrast, the observed contact between the MalK subunits at K106 seems less likely as the residue is located in a loop connecting helices 2 and 3 that, in the MalK-Tl dimer, are positioned at opposite ends. However, this might be taken as evidence that the orientation of the MalK monomers towards each other is different in the assembled transport complex from that seen in the crystal.

Together, both the genetic and biochemical evidence in favor of helix 3 and connecting loops of MalK to be crucial for interaction with MalFG was beautifully confirmed by the

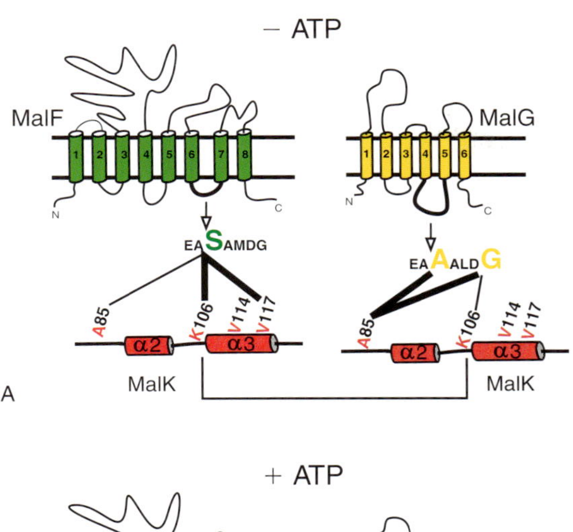

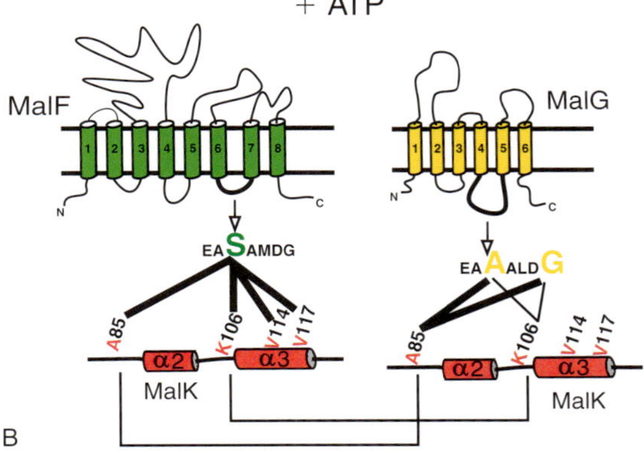

Figure 9.7. ***ATP-modulated subunit–subunit interactions in the MalFGK$_2$ transporter. Results from site-directed crosslinking experiments as described in Hunke* et al. *(2000b) are summarized. Residues in MalK (A85, K106, V114, V117) and in the 'EAA' loops of MalF (S3) and MalG (A3, G7), respectively, that were substituted for by cysteines and subjected to crosslinking are depicted in their relative positions in the topological models of MalF and MalG and in the secondary structural elements of MalK, respectively. Thick lines between MalK and MalF/MalG denote strong crosslinks induced preferentially by Cu^{2+}-phenanthroline, thin lines represent flexible crosslinks observed with Cu^{2+}-phenanthroline and/or chemical linkers.***

crystal structure of MsbA. Here, the corresponding region of the ABC domain was found to make most intimate contact to an intracellular domain (ICD1) connecting TM2 and TM3 (Chang and Roth, 2001).

Nature and assembly of the transporter complex

Assembly of the MalFGK$_2$ complex *in vivo* apparently requires the initial formation of a MalK dimer that subsequently interacts with membrane-associated MalFG (Kennedy and Traxler, 1999). MalF and MalG incorporate spontaneously and independently into the membrane. Upon interaction with MalG and MalK, MalF apparently changes its conformation as suggested by limited proteolysis (Traxler and Beckwith, 1992). The MalK dimer is formed also in the absence of MalFG, both *in vivo* and *in vitro* (Kennedy and Traxler, 1999). Furthermore, binding of purified MalK to MalFG-containing membrane vesicles that were isolated from cells lacking the *malK* gene is favored in the presence of ATP (Mourez *et al.*, 1998). The data of this study also suggested that binding of MalK occurs cooperatively and not linearly. Possibly, the formation of the MalK dimer is induced by ATP as in the case of the Rad50 dimer (Hopfner *et al.*, 2000), although no direct experimental proof for this is available. However, as the experiments by Kennedy and Traxler (1999) were performed with intact cells and cell extracts that usually contain millimolar concentrations of ATP, the MalK proteins were likely to be in the ATP-bound form. Also, based on the crosslinking data a role for alanine-85 in ATP-dependent dimer formation would be attractive but needs to be elucidated.

These data seem to contradict the reported failure to detect dimers of purified MalK by size exclusion chromatography (see above). However, as already mentioned, it cannot be excluded that the stability of the dimer is low and thus, dissociation is favored under the experimental conditions used.

In contrast to the above findings, functional rebinding of MalK to MalFG-containing proteoliposomes that were previously depleted for endogenous MalK by urea occurs independently of ATP (Landmesser *et al.*, 2002) (see also below). Thus, the conformational changes observed with purified MalK upon binding of ATP (Schneider *et al.*, 1995b) seem not to affect the site(s) of interaction with the membrane-integral subunits. Apparently, assembly of a pair of 'naive' MalF and MalG subunits with MalK (Mourez *et al.*, 1998) has different requirements than reassembly of a previously dissociated intact complex.

The study by Mourez *et al.* (1998) also identified lysine-106 as being protected against proteolytic attack by MalFG-containing vesicles, thereby adding to the notion that interaction with the membrane components is mediated in particular by helix 3 in the helical domain. Moreover, partial insertion of this peptide fragment into a pore formed by MalFG might explain the tight association of the subunits, as indicated by the high (molar) concentrations of chaotropic reagents, such as urea, that are required for dissociation of MalK (Landmesser *et al.*, 2002). This may relate to previous findings, demonstrating an accessibility of MalK to protease from the periplasmic side of the membrane (Schneider *et al.*, 1995b). Similar experiments performed with HisP (Baichwal *et al.*, 1993), KpsT (Bliss and Silver, 1997) and the isolated ABC domain of the mammalian CFTR protein (Gruis and Price, 1997) also suggested a transmembrane orientation of ABC subunits/ domains. However, as discussed by Blott *et al.* (1999), who failed to detect accessibility of Mdr1 from the external side of the membrane, the physiological significance of these findings is controversial. Also, the crystal structure of the MsbA protein does not support the above notion (Chang and Roth, 2001). Thus, to clarify unequivocally this matter, we shall have to await other crystal structures to be solved, especially that of an ABC importer.

In the histidine system, assembly of the subunits was studied after dissociation of HisP from membrane vesicles containing the wild-type complex (P.-Q. Liu and Ames, 1998). Treatment with 7.3 M urea resulted in only about 40% dissociation of HisP from the membrane, while in the presence of relatively high concentrations of Mg^{2+} ions and ATP (15 mM each), 6.6 M urea were sufficient to obtain vesicles completely depleted of HisP. The authors concluded that binding of ATP results in disengagement of the HisP subunits from HisQM (P.-Q. Liu *et al.*, 1999). Similar experiments with the maltose transporter from *S. typhimurium* showed no effect of ATP on dissociation of MalK (Landmesser *et al.*, 2002).

Analysis of the reassembly process of HisP with HisQM was then found to occur independently of ATP, as in the case of MalK when rebound to depleted MalFG-containing proteoliposomes (Landmesser *et al.*, 2002). Moreover, and in contrast to the findings discussed above for the maltose transporter, both copies of HisP are apparently recruited separately per HisQM.

Together, the available data suggest that the maltose transporter assembles from a membrane-associated MalFG subcomplex that interacts with a MalK homodimer. The latter is likely to be in an ATP-bound state. However, with respect to the results reported for the histidine transporter, ABC importers belonging to other subfamilies may require different assembly pathways, owing to variations in subunit structure, e.g. the lack of a C-terminal extension.

CURRENT TRANSPORT MODELS

Recent models describing putative individual steps in the translocation of maltose through $MalFGK_2$ are essentially based on two lines of experimental evidence:

- the ATPase activity of the purified transport complex, incorporated into liposomes, is substantially stimulated by liganded MalE (Davidson and Nikaido, 1991);
- binding protein-independent transport complexes exhibit a spontaneous ATPase activity (Covitz *et al.*, 1994; Davidson *et al.*, 1992).

These findings suggested a series of signaling events initiated by interaction of substrate-loaded binding protein with the transport complex at the extracellular side of the membrane. Subsequent conformational changes would then result in coupling the hydrolysis of ATP to the opening of a pore, which eventually leads to translocation of the substrate molecule to the cytoplasm (Davidson *et al.*, 1992). Recent findings by Davidson and collaborators using vanadate to lock the transporter in the transition state (Chen *et al.*, 2001) changed this view in that upon association of liganded MalE with the membrane-bound complex, ATP hydrolysis and release of maltose from the binding protein occur rather simultaneously. In the following scenario **(Figure 9.8)** it is intended to combine these and other data summarized above into a tentative model that also takes into account alternative views, especially that put forward by Ames and co-workers for histidine transport (Nikaido and Ames, 1999).

In the absence of substrate, the transport complex is envisaged to reside in the ground state with the MalK subunits partially inserted into a pore formed by MalFG (1 in **Figure 9.8**). Lysine-106 and helix 3 of MalK are postulated to be involved in this interaction (Hunke *et al.*, 2000b; Mourez *et al.*, 1998; Wilken, 1997). (In **Figure 9.8**, the orientation of the MalK monomers towards each other is totally arbitrary, although it resembles the structure of the HisP dimer in solution. However, with respect to the variations in dimer crystal structures

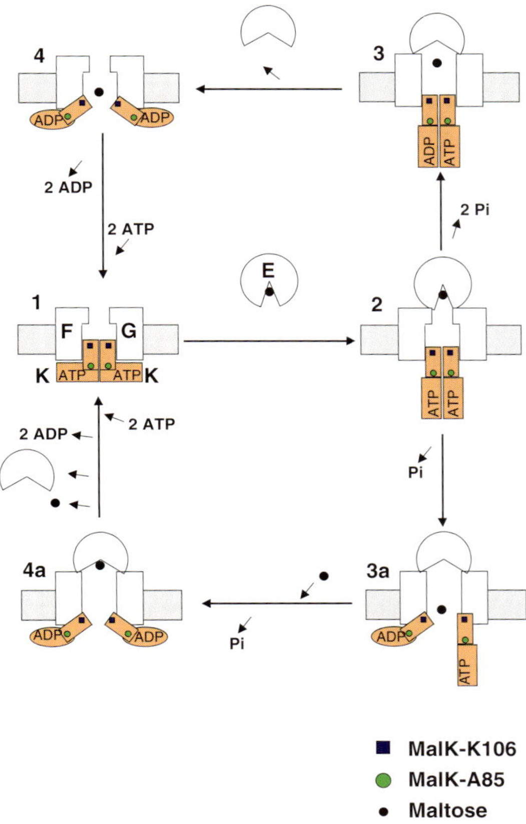

Figure 9.8. Tentative models of binding protein-dependent transport. See text for details.

observed so far and with no experimental evidence that one of these holds true for the assembled complex, no preference is given to any of the known tertiary structures.) Both MalK subunits are depicted in the ATP-bound form, as in the *E. coli* cell the ATP concentration is in the millimolar range and thus above the K_d (ATP) determined for the transport complex. As a consequence, based on crosslinking experiments (Hunke *et al.*, 2000b), alanine-85 of both monomers are in close contact with each other.

In the histidine transport model, the HisP subunits are proposed to be rather deeply embedded in the HisQM core but disengage in the ATP-bound state (Nikaido and Ames, 1999). This notion is based on the observations that (i) ATP facilitates dissociation of HisP by urea (P.-Q. Liu and Ames, 1998) (already mentioned above) and (ii) mutant HisP subunits in binding protein-independent transport complexes are disengaged in the absence of ATP (P.-Q. Liu *et al.*, 1999).

When maltose becomes available, substrate-loaded MalE specifically interacts with extracellularly peptide loops of MalFG (Hor and Shuman, 1993) (2 in **Figure 9.8**), thereby initiating conformational changes by which the ATPase activity of MalK becomes activated. Since fluorescence measurements indicated that residues in the nucleotide-binding site are less accessible to solvent in a vanadate-trapped complex than in the ground state, activation was suggested to occur by moving both catalytic sites closer together (Mannering *et al.*, 2001). Mutations in MalK that substantially reduce ATPase activity in the complex but allow ATP hydrolysis in the purified subunit (L86F, Hunke *et al.*, 2000a; Q140/N/K, Schmees *et al.*, 1999b) may thus interfere with a correct orientation of both subunits towards each other. The fact that hydrolysis of the soluble variants remains unaffected adds to the notion that the structure of the complex-associated dimer may differ from that in solution. Whether the tight association of the MalK subunits upon binding of liganded MalE is best viewed by assuming a localization of the nucleotide-binding sites at the dimer interface as in the Rad50 dimer, which is also not excluded by the MalK-Tl structure, is open for discussion and will not be considered further. However, it is intruiging that in the Rad50 dimer the ABC signature motif of one subunit interacts with the ribose and triphosphate moieties of the nucleotide in the opposite subunit (Hopfner *et al.*, 2000). If so in MalK, this could provide a possible explanation for the role of Q140 (helix 4, **Figure 9.4**) in the activation process (Schmees *et al.*, 1999b).

Concomitantly with ATP hydrolysis, MalE is thought to release maltose by lowering its affinity for the substrate through switching into the open conformation (3 in **Figure 9.8**). This idea is based on the finding that radiolabeled maltose was not associated with the stable complex formed between MalE and $MalFGK_2$ upon vanadate trapping (Chen *et al.*, 2001). Moreover, the unliganded form of MalE displays a five times lower affinity for the membrane-bound complex than maltose-loaded MalE (Merino *et al.*, 1995). As a consequence of initial binding of liganded MalE and/or ATP hydrolysis, a translocation pathway is opened to allow passage of the released ligand. According to a model by Ehrmann *et al.* (1998) (see also **Figure 9.5**), substrate binding initially occurs at hydrophilic residues in the transmembrane helices 4 and 5 of MalG. However, further translocation is blocked by hydrophobic residues in helix 5. The conformational changes induced by ATP hydrolysis at one site and possibly transmitted via changes in the position of the helical domain relative to the EAA loop in MalG (Hunke *et al.*, 2000b) would remove this hindrance. As a result, the substrate is transferred to a binding site on transmembrane helix 4 of MalF. Release of maltose to the cytoplasm would then occur via helix 6 of MalF, provided hydrophobic residues of helix 5 are moved out of the pathway by another conformational change. This might be accomplished by ATP hydrolysis at the second copy of MalK (4 in **Figure 9.8**). As a consequence of this step, the MalK subunits also move apart in the 'lid' region, as suggested from the failure to form a disulfide bond between alanine-85 in both monomers in the absence of ATP (Hunke *et al.*, 2000b). The change in position of the residue corresponding to glutamine-82 observed in MJ0796 compared to the ATP-bound form of HisP (Yuan *et al.*, 2001) may also be taken as evidence in favor of this notion. Moreover, crosslinking studies indicated less intimate contacts between K106 and residues in helix 3 to the EAA loops in MalFG in the ADP-bound state (Hunke *et al.*, 2000b). In a final step, the free energy of ATP binding powers the return of the complex to the ground state. Taking into account the observed positive cooperativity of ATP hydrolysis (Davidson *et al.*, 1996), ATP may first bind to one MalK subunit, thereby increasing the affinity of the second ATP-binding site, which then would also bind ATP.

This scenario, illustrated in steps 3,4 in **Figure 9.8**, requires the hydrolysis of two molecules of ATP per substrate molecule transported. Alternatively (lower part of **Figure 9.8**), hydrolysis at one site might be sufficient to remove both channel-blocking transmembrane helices from the pathway. Hydrolysis at the other MalK subunit (after dissociation of Pi and/or ADP from the first site?) could then allow translocation of a second substrate molecule (3a/4a in **Figure 9.8**). This would imply that either a second liganded binding protein enters the cycle or that a bound receptor may sequester a second substrate molecule. An apparent stoichiometry of one to two molecules of ATP hydrolyzed per molecule of substrate transported was found *in vivo* (Mimmack *et al.*, 1989), thus providing no clue in favor of one of the above alternatives. However, the histidine transport complex, for which the latter model has been proposed, was demonstrated to display equal affinity for both forms of its binding protein, HisJ (Ames *et al.*, 1996), indicating that reloading of a bound receptor is in the range of possibility. In addition, this model is essentially based on a study involving a HisP variant that carries a mutation in the highly conserved histidine residue in the 'switch' region (H211R) (see also **Figure 9.1**). The mutant protein by itself is catalytically inactive in solution but apparently forms heterodimers with wild-type HisP that display substantial ATPase activity. Moreover, when the heterodimers were reassembled with HisQM-containing membranes previously depleted of endogenous HisP, about half of the ATPase and transport activity of the wild type were obtained. Thus, the authors concluded that in the histidine transporter only one intact HisP monomer is required for function (Nikaido and Ames, 1999).

It should be noted, however, that in a similar study no transport activity was observed with a maltose transport complex of *E. coli* carrying the same mutation (H192R) in one of the MalK subunits (Davidson and Sharma, 1997). On the other hand, partially active heterodimer formation of wild-type and mutant MalK variants in solution were also observed (Landmesser and Schneider, unpublished). The reason for this discrepancy is currently unknown but is probably due to different experimental protocols. It is obvious that a comparative study with both transport complexes under identical experimental conditions would help to unravel this problem.

Conclusions and Perspectives

The huge body of experimental evidence that has been accumulated on the transport systems for maltose by numerous groups and for L-histidine by Ames and collaborators, respectively, has contributed considerably to our current understanding of the mechanism by which ABC importers exert their functions. Both systems are extensively characterized by various means at the levels of intact cells, membrane vesicles and, in recent years, proteoliposomes containing the purified transport complexes. Furthermore, the determination of the crystal structures of their ABC subunits, MalK-Tl and HisP-St, the discovery of a tight association between the membrane-bound transporter and the soluble substrate-binding protein at a particular stage of the translocation process, and the identification of amino acid residues involved in subunit–subunit interactions have provided important details on structural and functional aspects of the system.

Nonetheless, most of the events during a translocation cycle, including the energy consuming step, still remain to be elucidated at the molecular level. Moreover, the very same data have created new questions, for example which of the contrasting structures of the MalK-Tl and HisP-St dimers more likely reflects the situation in the assembled complex. A matter that becomes even more complicated when other currently available crystal structures of ABC proteins are also considered. Thus, attempts to crystallize an intact ABC importer and to solve its structure are among the most obvious efforts for the upcoming years. This holds true even though the first tertiary structure of an ABC transporter recently became available. However, the MsbA protein, delivering a hydrophobic substrate (lipid A) to the exterior of the *E. coli* cell, is unlikely to be a close structural representative of transporters designed to translocate hydrophilic compounds to the cytoplasm. Both the maltose and histidine transporters of *E. coli/ S. typhimurium,* for which sufficient amounts of highly purified preparations are at hand, are among the best candidates to achieve this goal. However, the use of transport complexes from thermophilic microorganisms may prove to be advantageous in this respect, as the successful crystallizations of MalK-Tl, two ABC subunits from *M. jannaschii* and the Rad50 protein from *P. furiosus* have taught us. Unfortunately, and regardless of recent progress, crystallization of membrane proteins is still an empirical venture, which makes estimates of when a first structure will become available highly unpredictable. Thus, to gain further insights into the architecture of an ABC importer in the absence of a tertiary structure we shall in addition have to rely on other approaches. These will include two-dimensional crystallization and single particle image analysis to obtain low-resolution structures, which has successfully been used in the case of the mammalian P-glycoprotein P (Rosenberg *et al.*, 2001) and the TAP1/TAP2 transporter (Velarde *et al.*, 2001). Moreover, a combination of well-established genetic and biochemical means will provide further details of protein–protein interactions in the assembled complex, combined with the analysis of their relevance to structural integrity and function.

Nevertheless, even with a tertiary structure at our disposal, unraveling the dynamics of the transport process, preferably on the level of proteoliposomes, will require the increased use of biophysical approaches, such as fluorescence energy transfer measurements. Again, because of their modular organization, and with mutant transport complexes at hand that allow site-directed modifications of any subunit with fluorophores at will, both the maltose and histidine transporters are most suited to further serve as model systems for the investigation of ABC transporters in general.

At the time of proof-reading this manuscript the first structure of an ABC importer, mediating the uptake of vitamin B_{12} in *E. coli,* was published (Locher, K.P., Lee, A.T. and Rees, D.C. (2002) The *E. coli* BtuCD structure: a framework for ABC transporter architecture and mechanism. Science 296, 1091–1098). This report provides further support for the role of the EAA loop in contacting the ABC domains, as well as for the LSGGQ motif being part of the nucleotide-binding site as in Rad50 (Hopfner *et al.*, 2000).

Acknowledgments

I thank Wolfram Welte (University of Konstanz) and Michael Ehrmann (University of Cardiff) for providing computer files of Figures 9.3, 9.4 and 9.5. Work from the author's laboratory was supported by the Deutsche Forschungsgemeinschaft (SCHN274/6-4/7-2) and by the Fonds der Chemischen Industrie.

References

Ames, G.F.-L., Liu, C.E., Joshi, A.K. and Nikaido, K. (1996) Liganded and unliganded receptors interact with equal affinity with the membrane complex of periplasmic permeases, a subfamily of traffic ATPases. J. Biol. Chem. 271, 14264–14270.

Ames, G.F.-L., Nikaido, K., Wang, I.X., Liu, P.-Q., Liu, C.E. and Hu, C. (2001) Purification and characterization of the membrane-bound complex of an ABC transporter, the histidine permease. J. Bioenerg. Biomembr. 33, 79–92.

Baichwal, V., Liu, D. and Ames, G.F.-L. (1993) The ATP-binding component of a prokaryotic traffic ATPase is exposed to the periplasmic (external) surface. Proc. Natl Acad. Sci. USA 90, 620–624.

Bliss, J.M. and Silver, R.P. (1997) Evidence that KpsT, the ATP-binding component of an ATP-binding cassette transporter, is exposed to the periplasm and associates with polymer during translocation of polysialic acid capsule of *Escherichia coli* K1. J. Bacteriol. 179, 1400–1403.

Blott, E.J., Higgins, C.F. and Linton, K.J. (1999) Cysteine-scanning mutagenesis provides no evidence for the extracellular accessibility of the nucleotide-binding domains of the multidrug resistance transporter P-glycoprotein. EMBO J. 23, 6800–6808.

Boos, W. and Böhm, A. (2000) Learning new tricks from an old dog. MalT of the *Escherichia coli* maltose system is part of a complex regulatory network. Trends Genet. 16, 404–409.

Boos, W. and Shuman, H. (1998) Maltose/maltodextrin system of *Escherichia coli*: transport, metabolism, and regulation. Microbiol. Mol. Biol. Rev. 62, 204–229.

Borich, S.M., Murray, A. and Gormley, E. (2000) Genomic arrangement of a putative operon involved in maltose transport in the *Mycobacterium tuberculosis* complex and *Mycobacterium leprae*. Microbios 102, 7–15.

Boyd, D., Traxler, B. and Beckwith, J. (1993) Analysis of the topology of a membrane protein by using a minimum number of alkaline phosphatase fusions. J. Bacteriol. 175, 553–556.

Brzostek, K., Heleszko, H. and Hrebenda, J. (1993) Maltoporins and maltose-binding protein of *Yersinia enterocolitica*. J. Gen. Microbiol. 139, 195–201.

Chang, G. and Roth, C.B. (2001) Structure of MsbA from *E. coli*: a homolog of the multidrug resistance ATP binding casette (ABC) transporters. Science 293, 1793–1800.

Chen, J., Sharma, S., Quiocho, F.A. and Davidson, A.L. (2001) Trapping the transition state of an ATP-binding cassette transporter: evidence for a concerted mechanism of maltose transport. Proc. Natl Acad. Sci. USA 98, 1525–1530.

Chifflet, S., Torriglia, A., Chiesa, R. and Tolosa, S. (1988) A method for the determination of inorganic phosphate in the presence of labile organic phosphate and high concentrations of protein: application to lens ATPases. Anal. Biochem. 168, 1–4.

Covitz, K.-M.Y., Panagiotidis, C.H., Hor, L.-I., Reyes, M., Treptow, N.A. and Shuman, H.A. (1994) Mutations that alter the transmembrane signalling pathway in an ATP binding cassette (ABC) transporter. EMBO J. 13, 1752–1759.

Cummings, J.H. and Macfarlane, G.T. (1991) The control and consequences of bacterial fermentation in the human colon. J. Appl. Bacteriol. 70, 443–459.

Danot, O. (2001) A complex signaling module governs the activity of MalT, the prototype of an emerging transactivator family. Proc. Natl Acad. Sci. USA 98, 435–440.

Dassa, E. (1993) Structure–function relationships in MalG, an inner membrane protein from the maltose transport system of *Escherichia coli*. Mol. Microbiol. 7, 39–47.

Dassa, E. and Bouige, P. (2001) The ABC of ABCs: a phylogenetic and functional classification of ABC systems in living organisms. Res. Microbiol. 152, 211–229.

Dassa, E. and Muir, S. (1993) Membrane topology of MalG, an inner membrane protein from the maltose transport system of *Escherichia coli*. Mol. Microbiol. 7, 29–38.

Davidson, A.L. and Nikaido, H. (1991) Purification and characterization of the membrane-associated components of the maltose transport system from *Escherichia coli*. J. Biol. Chem. 266, 8946–8951.

Davidson, A.L. and Sharma, S. (1997) Mutation of a single MalK subunit severely impairs maltose transport activity in *Escherichia coli*. J. Bacteriol. 179, 5458–5464.

Davidson, A.L., Shuman, H.A. and Nikaido, H. (1992) Mechanism of maltose transport in *Escherichia coli*: transmembrane signaling by periplasmic binding proteins. Proc. Natl Acad. Sci. USA 89, 2360–2364.

Davidson, A.L., Laghaeian, S.S. and Mannering, D.E. (1996) The maltose transport system of *Escherichia coli* displays positive cooperativity in ATP hydrolysis. J. Biol. Chem. 271, 4858–4863.

Dean, D.A., Reizer, J., Nikaido, H. and Saier, M.H. (1990) Regulation of the maltose transport system of *Escherichia coli* by the glucose-specific enzyme III of the phosphoenolpyruvate-sugar phosphotransferase system. Characterization of inducer exclusion-resistant mutants and reconstitution of inducer exclusion in proteoliposomes. J. Biol. Chem. 265, 21005–21010.

Dean, D.A., Hor, L.I., Shuman, H.A. and Nikaido, H. (1992) Interaction between maltose-binding protein and the membrane-associated maltose transporter complex in *Escherichia coli*. Mol. Microbiol. 6, 2033–2040.

Diederichs, K., Diez, J., Greller, G., Müller, C., Breed, J., Schnell, C., Vonrhein, C., Boos, W. and Welte, W. (2000) Crystal structure of MalK, the ATPase subunit of the trehalose/maltose ABC transporter of the archaeon *Thermococcus litoralis.* EMBO J. 19, 5951–5961.

Diez, J., Diederichs, K., Greller, G., Horlacher, R., Boos, W. and Welte, W. (2001) The crystal structure of a liganded trehalose/maltose-binding protein from the hyperthermophilic archaeon *Thermococcus litoralis* at 1.85 Å. J. Mol. Biol. 305, 905–915.

DiRuggiero, J., Dunn, D., Maeder, D.L., Holley-Shanks, R., Chatard, J., Horlacher, R., Robb, F.T., Boos, W. and Weiss, R.B. (2000) Evidence of recent lateral gene transfer among hyperthermophilic archaea. Mol. Microbiol. 38, 684–693.

Doige, C.A. and Ames, G.F.-L. (1993) ATP-dependent transport systems in bacteria and humans: relevance to cystic fibrosis and multidrug resistance. Annu. Rev. Microbiol. 47, 291–319.

Dutzler, R., Wang, Y.-F., Rizkallah, P.J., Rosenbusch, J.P. and Schirmer, T. (1996) Crystal structures of various maltooligosaccharides bound to maltoporin reveal a specific sugar translocation pathway. Structure 4, 127–134.

Ehrle, R., Pick, C., Ulrich, R., Hofmann, E. and Ehrmann, M. (1996) Characterization of transmembrane domains 6, 7, and 8 of MalF by mutational analysis. J. Bacteriol. 178, 2255–2262.

Ehrmann, M. and Beckwith, J. (1991) Proper insertion of a complex membrane protein in the absence of its amino-terminal export signal. J. Biol. Chem. 266, 16530–16533.

Ehrmann, M., Ehrle, R., Hofmann, E., Boos, W. and Schlösser, A. (1998) The ABC maltose transporter. Mol. Microbiol. 29, 685–694.

Elferink, M.G.L., Albers, S.-V., Konings, W.N. and Driessen, A.J.M. (2001) Sugar transport in *Sulfolobus solfataricus* is mediated by two families of binding protein-dependent ABC transporters. Mol. Microbiol. 39, 1494–1503.

Evdokimov, A.G., Anderson, D.E., Routzahn, K.M. and Waugh, D.S. (2001) Structural basis for oligosaccharide recognition by *Pyrococcus furiosus* maltodextrin-binding protein. J. Mol. Biol. 305, 891–904.

Gardina, P.J., Bormans, A.F., Hawkins, M.A., Meeker, J.W. and Manson, M.D. (1997) Maltose-binding protein interacts simultaneously and asymmetrically with both subunits of the Tar chemoreceptor. Mol. Microbiol. 23, 1181–1191.

Greller, G., Horlacher, R., DiRuggiero, J. and Boos, W. (1999) Molecular and biochemical analysis of MalK, the ATP-hydrolyzing subunit of the trehalose/maltose transport system of the hyperthermophilic archeon *Thermococcus litoralis*. J. Biol. Chem. 274, 20259–20264.

Gruis, D.B. and Price, E.M. (1997) The nucleotide binding folds of the cystic fibrosis transmembrane conductance regulator are extracellularly accessible. Biochemistry 36, 7739–7745.

Hall, J.A., Davidson, A.L. and Nikaido, H. (1998) Preparation and reconstitution of the membrane-associated maltose transporter complex of *Escherichia coli*. Methods Enzymol. 292, 20–29.

Heidelberg, J.F., Eisen, J.A., Nelson, W.C., Clayton, R.A., Gwinn, M.L., Dodson, R.J., *et al.* (2000) DNA sequence of both chromosomes of the cholera pathogen *Vibrio cholerae*. Nature 406, 477–483.

Hekstra, D. and Tommassen, J. (1993) Functional exchangeability of the ABC proteins of the periplasmic binding protein-dependent transport systems Ugp and Mal of *Escherichia coli*. J. Bacteriol. 175, 6546–6552.

Higgins, C.F. (2001) ABC transporters: physiology, structure and mechanism – an overview. Res. Microbiol. 152, 205–210.

Höner zu Bentrup, K., Schmid, R. and Schneider, E. (1994) Maltose transport in *Aeromonas hydrophila:* purification, biochemical characterization and partial protein sequence analysis of a periplasmic maltose-binding protein. Microbiology 140, 945–951.

Hopfner, K.-P., Karcher, A., Shin, D.S., Craig, L., Arthur, L.M., Carney, J.P. and Tainer, J.A. (2000) Structural biology of Rad50 ATPase: ATP-driven conformational control in DNA double-strand break repair and the ABC-ATPase superfamily. Cell 101, 789–800.

Hor, L.-I. and Shuman, H.A. (1993) Genetic analysis of periplasmic binding protein dependent transport in *Escherichia coli*. Each lobe of maltose-binding protein interacts with a different subunit of the $MalFGK_2$ membrane transport complex. J. Mol. Biol. 233, 659–670.

Horlacher, R., Xavier, K.B., Santos, H., DiRuggiero, J., Kossmann, M. and Boos, W. (1998) Archaeal binding protein-dependent ABC transporter: molecular and biochemical analysis of the trehalose/maltose transport system of the hyperthermophilic archaeon *Thermococcus litoralis*. J. Bacteriol. 180, 680–689.

Hülsmann, A., Lurz, R., Scheffel, F. and Schneider, E. (2000) Maltose and maltodextrin transport in the thermoacidophilic gram-positive bacterium *Alicyclobacillus acidocaldarius* is mediated by a high-affinity transport system that includes a maltose-binding protein tolerant to low pH. J. Bacteriol. 182, 6292–6301.

Hung, L.-W., Wang, I.X., Nikaido, K., Liu, P.-Q., Ames, G.F.-L. and Kim, S.-H. (1998) Crystal structure of the ATP-binding subunit of an ABC transporter. Nature 396, 703–707.

Hunke, S. and Schneider, E. (1999) A cys-less variant of the bacterial ATP binding cassette protein MalK is functional in maltose transport and regulation. FEBS Lett. 448, 131–134.

Hunke, S., Dröse, S. and Schneider, E. (1995) Vanadate and bafilomycin A1 are potent inhibitors of the ATPase activity of the reconstituted bacterial ATP-binding cassette transporter for maltose ($MalFGK_2$). Biochem. Biophys. Res. Commun. 216, 589–594.

Hunke, S., Landmesser, H. and Schneider, E. (2000a) Novel missense mutations that affect the transport function of MalK, the ATP-binding-cassette subunit of the *Salmonella enterica* serovar typhimurium maltose transport system. J. Bacteriol. 182, 1432–1436.

Hunke, S., Mourez, M., Jéhanno, M., Dassa, E. and Schneider, E. (2000b) ATP modulates subunit–subunit interactions in an ATP-binding-cassette transporter ($MalFGK_2$) determined by site-directed chemical cross-linking. J. Biol. Chem. 275, 15526–15534.

Kang, C.-H., Shin, W.-C., Yamagata, Y., Gokcen, S., Ames, G.F.-L. and Kim, S.-H. (1991) Crystal structure of the lysine-, arginine-, ornithine-binding protein (LAO) from *Salmonella typhimurium* at 2.7 Å resolution. J. Biol. Chem. 266, 23893–23899.

Kennedy, K.A. and Traxler, B. (1999) MalK forms a dimer independent of its assembly into the $MalFGK_2$ ATP-binding cassette transporter of *Escherichia coli*. J. Biol. Chem. 274, 6259–6264.

Kerppola, R.E., Shyamala, V.K., Klebba, P. and Ames, G.F.-L. (1991) The membrane-bound proteins of periplasmic permeases form a complex. Identification of the histidine permease HisQMP complex. J. Biol. Chem. 266, 9857–9865.

Kreimer, D.I., Chai, K.P. and Ames, G.F.-L. (2000) Nonequivalence of the nucleotide-binding subunits of an ABC transporter, the histidine permease, and conformational changes in the membrane complex. Biochemistry 39, 14183–14195.

Kühnau, S., Reyes, M., Sievertsen, A., Shuman, H.A. and Boos, W. (1991) The activities of the *Escherichia coli* MalK protein in maltose transport, regulation, and inducer exclusion can be separated by mutations. J. Bacteriol. 173, 2180–2186.

Landmesser, H., Stein, A., Blüschke, B., Brinkmann, M., Hunke, S. and Schneider, E. (2002) Large-scale purification, dissociation and functional reassembly of the maltose ATP-binding-cassette transporter ($MalFGK_2$) of *Salmonella typhimurium*. Biochim. Biophys. Acta, in press.

Lippincott, J. and Traxler, B. (1997) MalFGK complex assembly and transport and regulatory characteristics of MalK insertion mutants. J. Bacteriol. 179, 1337–1343.

Liu, C.E. and Ames, G.F.-L. (1997) Characterization of transport through the periplasmic histidine permease using proteoliposomes reconstituted by dialysis. J. Biol. Chem. 272, 859–866.

Liu, C.E., Liu, P.-Q. and Ames, G.F.-L. (1997) Characterization of the adenosine triphosphatase activity of the periplasmic histidine permase, a traffic ATPase (ABC transporter). J. Biol. Chem. 272, 21833–21891.

Liu, C.E., Liu, P.-Q., Wolf, A., Lin, E. and Ames, G.F.-L. (1999) Both lobes of the soluble receptor of the periplasmic histidine permease, an ABC transporter (traffic ATPase), interact with the membrane-bound complex. J. Biol. Chem. 274, 739–747.

Liu, P.-Q. and Ames, G.F.-L. (1998) *In vitro* disassembly and reassembly of an ABC transporter, the histidine permase. Proc. Natl Acad. Sci. USA 95, 3495–3500.

Liu, P.-Q., Liu, C.E. and Ames, G.F.-L. (1999) Modulation of ATPase activity by physical disengagement of the ATP-binding domains of an ABC transporter, the histidine permease. J. Biol. Chem. 274, 18310–18318.

Luckey, M. and Nikaido, H. (1980) Specificity of diffusion channels produced by λ-phage receptor protein of *Escherichia coli*. Proc. Natl Acad. Sci. USA 77, 165–171.

Mannering, D.E., Sharma, S. and Davidson, A.L. (2001) Demonstration of conformational changes associated with activation of the maltose transport complex. J. Biol. Chem. 276, 12362–12368.

Merino, G. and Shuman, H.A. (1997) Unliganded maltose-binding protein triggers lactose transport in an *Escherichia coli* mutant with an alteration in the maltose transport complex. J. Bacteriol. 179, 7687–7694.

Merino, G., Boos, W., Shuman, H.A. and Bohl, E. (1995) The inhibition of maltose transport by the unliganded form of the maltose-binding protein of *Escherichia coli*: experimental findings and mathematical treatment. J. Theor. Biol. 177, 171–179.

Meyer, J.E.W., Hofnung, M. and Schulz, G.E. (1997) Structure of maltoporin from *Salmonella typhimurium* ligated with nitrophenyl-maltotrioside. J. Mol. Biol. 266, 761–775.

Mimmack, M.L., Gallagher, M.P., Pearce, S.R., Hyde, S.C., Booth, I.R. and Higgins, C.F. (1989) Energy coupling to periplasmic binding protein-dependent transport systems: Stoichiometry of ATP hydrolysis during transport *in vivo*. Proc. Natl Acad. Sci. USA 86, 8257–8261.

Morbach, S., Tebbe, S. and Schneider, E. (1993) The ATP-binding cassette (ABC) transporter for maltose/maltodextrins of *Salmonella typhimurium*. Characterization of the ATPase activity associated with the purified MalK subunit. J. Biol. Chem. 268, 18617–18621.

Mourez, M., Hofnung, M. and Dassa, E. (1997a) Subunit interaction in ABC transporters. A conserved sequence in hydrophobic membrane proteins of periplasmic permeases define sites of interaction with the helical domain of ABC subunits. EMBO J. 16, 3066–3077.

Mourez, M., Skouloubris, S., Betton, J.M. and Dassa, E. (1997b) Heat shock induction by a misassembled cytoplasmic membrane protein complex in *Escherichia coli*. Mol. Microbiol. 26, 821–831.

Mourez, M., Jéhanno, M., Schneider, E. and Dassa, E. (1998) *In vitro* interaction between components of the inner membrane complex of the maltose ABC transporter of *Escherichia coli*: modulation by ATP. Mol. Microbiol. 30, 353–363.

Nelson, B.D. and Traxler, B. (1998) Exploring the role of integral membrane proteins in ATP-binding cassette transporters: analysis of a collection of MalG insertion mutants. J. Bacteriol. 180, 2507–2514.

Nikaido, K. and Ames, G.F.-L. (1999) One intact ATP-binding subunit is sufficient to support ATP hydrolysis and translocation in an ABC transporter, the histidine permease: J. Biol. Chem. 274, 26727–26735.

Nikaido, K., Liu, P.-Q. and Ames, G.F.-L. (1997) Purification and characterization of HisP, the ATP-binding subunit of a traffic ATPase (ABC transporter), the histidine permease of *Salmonella typhimurium*. J. Biol. Chem. 272, 27745–27752.

Oh, B.-H., Kang, C.-H., De Bond, H., Kim, S.-H., Nikaido, K., Joshi, A.K. and Ames, G.F.-L. (1994) The bacterial periplasmic histidine-binding protein. Structure/function analysis of the ligand-binding site and comparison with related proteins. J. Biol. Chem. 269, 4135–4143.

Panagiotidis, C.H., Reyes, M., Sievertsen, A., Boos, W. and Shuman, H.A. (1993) Characterization of the structural requirements for assembly and nucleotide binding of an ATP-binding cassette transporter. The maltose transport system of *Escherichia coli*. J. Biol. Chem. 268, 23685–23696.

Panagiotidis, C.H., Boos, W. and Shuman, H.A. (1998) The ATP-binding cassette subunit of the maltose transporter MalK antagonizes MalT, the activator of the *Escherichia coli mal* regulon. Mol. Microbiol. 30, 535–546.

Postma, P.W., Lengeler, J.W. and Jacobson, G.R. (1996) Phosphoenolpyruvate:carbohydrate phosphotransferase system. In: *Escherichia coli* and *Salmonella*: Cellular and Molecular Biology (ed. F.C. Neidhardt, R. Curtiss III, M. Riley, M. Schaechter and H.E. Umbarger), pp. 1149–1174. Washington, DC: American Society for Microbiology.

Puyet, A. and Espinosa, M. (1993) Structure of the maltodextrin-uptake locus of *Streptococcus pneumoniae*. Correlation to the *Escherichia coli* maltose regulon. J. Mol. Biol. 230, 800–811.

Quiocho, F.A., Spurlino, J.C. and Rodseth, L.E. (1997) Extensive features of tight oligosaccharide binding revealed in high-resolution structures of the maltodextrin transport/chemosensory receptor. Structure 5, 997–1015.

Quentin, Y., Fichant, G. and Denizot, F. (1999) Inventory, assembly and analysis of *Bacillus subtilis* ABC transport systems. J. Mol. Biol. 287, 467–484.

Racker, E., Violand, B., O'Neal, S., Alfonzo, M. and Telford, J. (1979) Reconstitution, a way of biochemical research; some new approaches to membrane-bound enzymes. Arch. Biochem. Biophys. 198, 470–477.

Reich-Slotky, R., Panagiotidis, C., Reyes, M. and Shuman, H.A. (2000) The detergent-soluble maltose transporter is activated by maltose binding protein and verapamil. J. Bacteriol. 182, 993–1000.

Rosenberg, M.F., Velarde, G., Ford, R.C., Martin, C., Berridge, G., Kerr, I.D., *et al.* (2001) Repacking of the transmembrane domains of P-glycoprotein during the transport ATPase cycle. EMBO J. 20, 5615–5625.

Saier, M.H. Jr. (2000) Families of transmembrane sugar transport proteins. Mol. Microbiol. 35, 699–710.

Schirmer, T., Keller, T.A., Wang, Y.F. and Rosenbusch, J.P. (1995) Structural basis of sugar translocation through maltoporin channels at 3.1 Å resolution. Science 267, 512–514.

Schlösser, A., Kampers, T. and Schrempf, H. (1997) The *Streptomyces* ATP-binding component MsiK assists in cellobiose and maltose transport. J. Bacteriol. 179, 2092–2095.

Schmees, G. and Schneider, E. (1998) Domain structure of the ATP-binding-cassette (ABC) protein MalK of *Salmonella typhimurium* as assessed by coexpressed half molecules and LacK'-'MalK chimeras. J. Bacteriol. 180, 5299–5305.

Schmees, G., Höner zu Bentrup, K., Schneider, E., Vinzenz, D. and Ermler, U. (1999a) Crystallization and preliminary X-ray analysis of the bacterial ATP-binding-cassette (ABC)-protein MalK. Acta Crystallogr. D-Biol. Cryst. 55, 285–286.

Schmees, G., Stein, A., Hunke, S., Landmesser, H. and Schneider, E. (1999b) Functional consequences of mutations in the conserved 'signature sequence' of the ATP-binding-cassette protein MalK. Eur. J. Biochem. 266, 420–430.

Schneider, E. (2001) ABC transporters catalyzing carbohydrate uptake. Res. Microbiol. 152, 303–310.

Schneider, E. and Hunke, S. (1998) ATP-binding-cassette (ABC) transport systems: functional and structural aspects of the ATP-hydrolyzing subunits/domains. FEMS Microbiol. Rev. 22, 1–20.

Schneider, E. and Walter, C. (1991) A chimeric nucleotide-binding protein, encoded by a *hisP-malK* hybrid gene, is functional in maltose transport in *Salmonella typhimurium*. Mol. Microbiol. 5, 1375–1383.

Schneider, E., Bishop, L., Schneider, E., Alfandary, V. and Ames, G.-F. (1989) Fine-structure genetic map of the maltose transport operon of *Salmonella typhimurium*. J. Bacteriol. 171, 5860–5865.

Schneider, E., Wilken, S. and Schmid, R. (1994) Nucleotide-induced conformational changes of MalK, a bacterial ATP binding cassette transporter protein. J. Biol. Chem. 269, 20456–20461.

Schneider, E., Linde, M. and Tebbe, S. (1995a) Functional purification of a bacterial ATP-binding cassette transporter protein (MalK) from the cytoplasmic fraction of an overproducing strain. Protein Expression Purif. 6, 10–14.

Schneider, E., Hunke, S. and Tebbe, S. (1995b) The MalK protein of the ATP-binding cassette transporter for maltose of *Escherichia coli* is accessible to protease digestion from the periplasmic side of the membrane. J. Bacteriol. 177, 5364–5367.

Schreiber, V. and Richet, E. (1999) Self-association of the *Escherichia coli* transcription activator MalT in the presence of maltotriose and ATP. J. Biol. Chem. 274, 33220–33226.

Sharff, A.J., Rodseth, L.E., Spurlino, J.C. and Quiocho, F.A. (1992) Crystallographic evidence of a large ligand-induced hinge-twist motion between the two domains of the maltodextrin binding protein involved in active transport and chemotaxis. Biochemistry 31, 10657–10663.

Sharma, S. and Davidson, A.L. (2000) Vanadate-induced trapping of nucleotides by purified maltose transport complex requires ATP hydrolysis. J. Bacteriol. 182, 6570–6576.

Speiser, D.M. and Ames, G.F.-L. (1991) *Salmonella typhimurium* histidine periplasmic permease mutations that allow transport in the absence of histidine-binding protein. J. Bacteriol. 173, 1444–1451.

Spurlino, J.C., Lu, G.-Y. and Quiocho, F.A. (1991) The 2.3-Å resolution structure of the maltose- or maltodextrin-binding protein, a primary receptor of bacterial active transport and chemotaxis. J. Biol. Chem. 266, 5202–5219.

Stein, A., Hunke, S. and Schneider, E. (1997) Mutational analysis eliminates Glu64 and Glu94 as candidates for 'catalytic carboxylate' in the bacterial ATP-binding-cassette protein MalK. FEBS Lett. 413, 211–214.

Steinke, A., Grau, S., Davidson, A., Hofmann, E. and Ehrmann, M. (2001) Characterization of transmembrane segments 3, 4, and 5 of MalF by mutational analysis. J. Bacteriol. 183, 375–381.

Story, R.M. and Steitz, T.A. (1992) Structure of the recA protein-ADP complex. Nature 355, 374–376.

Sutcliffe, I.C. and Russel, R.B. (1995) Lipoproteins of gram-positive bacteria. J. Bacteriol. 177, 1123–1128.

Szmelcman, S., Sassoon, N. and Hofnung, M. (1997) Residues in the α helix 7 of the bacterial maltose binding protein which are important in interactions with the Mal FGK_2 complex. Protein Sci. 6, 628–636.

Tam, R. and Saier, M.H. Jr. (1993) Structural, functional, and evolutionary relationships among extracellular solute-binding receptors of bacteria. Microbiol. Rev. 57, 320–346.

Tapia, M.I., Mourez, M., Hofnung, M. and Dassa, E. (1999) Structure–function study of MalF protein by random mutagenesis. J. Bacteriol. 181, 2267–2272.

Tralau, C., Greller, G., Pajatsch, M., Boos, W. and Bohl, E. (2000) Mathematical treatment of transport data of bacterial transport systems to estimate limitation in diffusion through the outer membrane. J. Theor. Biol. 207, 1–14.

Traxler, B. and Beckwith, J. (1992) Assembly of a hetero-oligomeric membrane protein complex. Proc. Natl Acad. Sci. USA 89, 10852–10856.

Treptow, N.A. and Shuman, H.A. (1985) Genetic evidence for substrate and periplasmic-binding-protein recognition by the MalF and MalG proteins, cytoplasmic membrane components of the *Escherichia coli* maltose transport system. J. Bacteriol. 163, 654–660.

van Wezel, G.P., White, J., Bibb, M.J. and Postma, P.W. (1997) The malEFG gene cluster of *Streptomyces coelicolor* A3 (2): characterization, disruption and transcriptional analysis. Mol. Gen. Genet. 254, 604–608.

Velarde, G., Ford, R.C., Rosenberg, M.F. and Powis, S.J. (2001) 3D-structure associated with antigen processing (TAP) obtained by single particle image analysis. J. Biol. Chem. 276, 46054–46063.

Walter, C., Höner zu Bentrup, K. and Schneider, E. (1992a) Large-scale purification, nucleotide binding properties and ATPase activity of the MalK subunit of *S. typhimurium* maltose transport complex. J. Biol. Chem. 267, 8863–8869.

Walter, C., Wilken, S. and Schneider, E. (1992b) Characterization of site-directed mutations in conserved domains of MalK, a bacterial member of the ATP-binding cassette (ABC) family. FEBS Lett. 303, 41–44.

Wilken, S. (1997) Funktionelle Domänen der MalK-Untereinheit des ABC-Transportsystems für Maltose bei *S. typhimurium*. Ph.D. Thesis, University of Osnabrueck, Germany.

Wilken, S., Schmees, G. and Schneider, E. (1996) A putative helical domain in the MalK subunit of the ATP-binding-cassette transport system for maltose of *Salmonella typhimurium* ($MalFGK_2$) is crucial for interaction with MalF and MalG. A study using the LacK protein of *Agrobacterium radiobacter* as a tool. Mol. Microbiol. 22, 655–666.

Wolf, A., Shaw, E.W., Nikaido, K. and Ames, G.F.-L. (1994) The histidine-binding protein undergoes conformational changes in the absence of ligand as analyzed with conformation-specific monoclonal antibodies. J. Biol. Chem. 269, 23051–23058.

Yoshida, M. and Amano, T. (1995) A common topology of proteins catalyzing ATP-triggered reactions. FEBS Lett. 359, 1–5.

Yuan, Y.-R., Blecker, S., Martsinkevich, O., Millen, L., Thomas, P.J. and Hunt, J.F. (2001) The crystal structure of the MJ0796 ATP-binding cassette: implications for the structural consequences of ATP hydrolysis in the active site of an ABC-transporter. J. Biol. Chem. 276, 32313–32321.

Crystal Structures of Periplasmic Solute-binding Proteins in ABC Transport Complexes Illuminate Their Function

CHAPTER 10

Anthony J. Wilkinson and Koen H.G. Verschueren

Introduction

The periplasmic binding protein-dependent permeases constitute a large and important class of active transport systems for the uptake of nutrients by Gram-negative bacteria (Ames, 1986; Furlong, 1987; Higgins, 1992). These ATP-binding cassette (ABC) transporters have a common organization consisting of five core functional units, these being (i) a pair of integral membrane protein domains, each of which probably spans the cytoplasmic membrane at least six times and which together form a channel through which the substrate passes, (ii) a pair of ATPase domains associated with the cytoplasmic surface of the membrane, which couple ATP hydrolysis to solute translocation and (iii) an abundant receptor protein, which resides in the periplasmic space (Higgins *et al.*, 1982). The periplasmic solute-binding proteins confer specificity on the transport system, capturing extracellular substrates and delivering them to the cognate membrane assembly for transport. Analogous transporters exist in Gram-positive bacteria. However, in these organisms there is no outer membrane and the receptor protein is anchored to the cell surface through a lipid group attached at its N-terminus (Gilson *et al.*, 1988). The ABC transporters of eukaryotic cells, invariably exporters, which have a similar arrangement of the membrane and ATPase components, function in the absence of an accessory solute-binding protein.

The periplasmic binding protein (PBP) components of bacterial ABC transporters are absolutely required for solute uptake. They capture their ligands with large association rate constants ($1–10 \times 10^7\ M^{-1}\,s^{-1}$) enabling rapid responses to the presence of the solute (Miller *et al.*, 1980). PBP concentrations in the periplasmic space (in the range 0.1–1 mM) greatly exceed those of the transporter components in the membrane and, under most circumstances, those of the cognate substrates in the extracellular environment ($\leq 1\ \mu M$). As the binding proteins exhibit high ligand affinities (in the range of 0.01–10 μM), the concentration of the liganded PBP will be in the millimolar range. This is likely to facilitate efficient uptake of solute against a net uphill concentration gradient. Uptake of nutrients can be accomplished in the absence of a functional PBP in strains that harbor compensating mutations in the other membrane components. However, transport in these mutant strains is much slower (the value of K_m is 1000-fold higher) and it is inefficient in terms of the number of ATP molecules hydrolyzed per solute molecule taken up (Davidson *et al.*, 1992).

Analysis of the *Escherichia coli* genome suggests there are between 40 and 50 periplasmic binding proteins. The PBPs are monomers ranging in size from M_r 25 000 to 60 000, serving transport systems which handle a range of substrates including oxyanions, amino acids, sugars, peptides, polyamines, vitamins and metal

ABC Proteins: From Bacteria to Man
ISBN 0-12-352551-9

ions or their chelates. The sizes of the ligands accommodated by PBPs range from single ions such as Zn^{2+} (bound by TroA) with a volume of 1.5 $Å^3$ to pentapeptides with volumes up to 1000 $Å^3$ (bound in OppA). A subset of PBPs, including those associated with galactose/glucose, ribose, maltose and dipeptide transport, has a second function serving as receptors for chemotactic signals. The solubility of the PBPs and their relative abundance in the periplasm has facilitated their overexpression and purification. In general PBPs are amenable to crystallization and once they appear, their crystals more often than not diffract X-rays to high resolution. As a result, accurate structures of a large number of PBPs have been determined and this has provided the basis for the development of detailed insights into their specificity and evolution.

STRUCTURE AND SUBSTRATE BINDING

GENERAL CHARACTERISTICS

The protein data bank has over a hundred files containing coordinates of 'periplasmic binding proteins' representing the structures of the 21 different members listed in **Table 10.1**. A panel of PBP structures is shown in **Figure 10.1**, from which common and distinct features can be deduced. Despite the absence of significant sequence similarity across the set and the diversity of the ligands bound, most of the PBPs have a common overall organization, which has been termed a periplasmic ligand-binding protein fold. This comprises two globular domains of similar topology, each of which contains a central β-pleated sheet flanked by sets of α-helices **(Figure 10.1)**. The two domains are connected by two and sometimes three segments of the polypeptide, which are usually in extended conformation. As a result each domain is made up of non-contiguous segments of the polypeptide. The β-sheets from the respective domains are oriented towards one another, giving the molecule an elongated ellipsoid form. The substrates are bound in a cleft between the domains usually in a manner that sequesters them completely from the solvent. In the larger periplasmic binding proteins such as maltose-binding protein, MBP, the dipeptide-binding protein, DppA, and the oligopeptide-binding protein, OppA, extra subdomains and even domains **(Figure 10.1f)** are present. Surprisingly, a calcium-binding site was discovered in the crystal structure of galactose/glucose-binding protein (**Figure 10.1i**; Vyas *et al.*, 1987). It is remote from the sugar-binding pocket and the putative chemotaxis receptor-binding surface. It is likely that metal cation binding has a structural rather than a regulatory role, since affinity measurements suggest that the site will be fully occupied at physiological Ca^{2+} concentrations (Vyas *et al.*, 1989).

Ligand binding in the PBPs is accompanied by large relative movements of the two domains that close around the substrate according to a mechanism which is often likened to the action of a Venus fly-trap. This notion is supported by data from small angle X-ray scattering experiments (Newcomer *et al.*, 1981; Shilton *et al.*, 1996). For a number of PBPs including the receptors for maltose (Sharff *et al.*, 1992), ribose (Björkman and Mowbray, 1998), glutamine (Hsaio *et al.*, 1996), lysine–arginine–ornithine (Oh *et al.*, 1993), dipeptide (Dunten and Mowbray, 1995; Nickitenko *et al.*, 1995), and oligopeptide (Sleigh *et al.*, 1997) transport, crystal structures have been determined of both the unliganded and liganded proteins, revealing 'open' and 'closed' conformations, respectively **(Figures 10.1 and 10.4)**. The three-dimensional structure of the individual domains does not alter significantly between the open and closed forms but the relative orientation of the two domains does. The opening and closing of the structure is the result of changes in the main-chain torsion angles of just a handful of residues located in the segments connecting the two domains which serve as a hinge. In the majority of the liganded structures of the PBPs, the ligand is sequestered from the solvent in a cavity framed by both lobes of the protein as well as the hinge segments that connect them. Protein crystallography has also revealed structures of a closed unliganded form of galactose–glucose-binding protein (**Figure 10.1c**; Flocco and Mowbray, 1994), and an open liganded form of leucine–isoleucine–valine-binding protein, the latter obtained by soaking leucine into crystals of the unliganded protein (Sack *et al.*, 1989a). There is a second crystal structure of an open liganded PBP, that of maltose-binding protein (MBP) in complex with the cyclic heptasaccharide β-cyclodextrin (Sharff *et al.*, 1993). β-Cyclodextrin is not transported by the maltose permease, probably because although the cyclic sugar binds MBP with high affinity, the domains are unable to close over the ligand **(Figure 10.1d)**.

Unliganded PBPs are viewed as an ensemble of structures in which the relative orientation of the domains varies. The angle of opening observed among the crystal structures of these proteins ranges from 26° to 64°, though the extent of opening in each case is likely to be influenced by crystal packing constraints. For ribose-binding protein three different open-form crystal structures, in which the angle of opening ranges from 43° to 64°, have been determined (Björkman and Mowbray, 1998). These open forms can be related to each other and to the liganded protein by rotations about a similar set of bonds. It seems therefore that domain opening can be described as a 'fairly pure hinge motion' (Björkman and

TABLE 10.1. THE PERIPLASMIC SOLUTE BINDING PROTEINS WHOSE STRUCTURES ARE KNOWN

The proteins are arranged according to their structural organization. Where there are a number of entries for a particular PBP (structures to different resolutions, or in complex with different ligands), the details are given for the structure which has been solved to the highest resolution.

	Ligand	No. of entries	Highest resolution (Å)	PDB entry	Organism	Citation (highest res.)
Family 1: *L-arabinose binding protein-like*						
D-Ribose-binding protein	β-D-Ribose	6	1.60	2DRI	*E. coli*	Björkman *et al.* (1994)
L-Arabinose-binding protein	D-Galactose	9	1.49	8ABP	*E. coli*	Vermersch *et al.* (1991)
D-Allose-binding protein	D-Allose	1	1.80	1RPJ	*E. coli*	Chaudhuri *et al.* (1999)
Galactose/glucose-binding protein	Galactose	5	1.70	1GCA	*S. typh.*	Zou *et al.* (1993)
Leucine/isoleucine/ valine-binding protein	No ligand	1	2.40	2LIV	*E. coli*	Sack *et al.* (1989a)
Leucine-binding protein	No ligand	1	2.40	2LBP	*E. coli*	Sack *et al.* (1989b)
Family 2: *Phosphate binding protein-like*						
Oligopeptide-binding protein (OppA)	KAK	32	1.20	1JET	*S. typh.*	Tame *et al.* (1996)
Lysine/arginine/ornithine-binding protein (LAO)	Lysine	5	1.80	1LST	*S. typh.*	Oh *et al.* (1993)
Sulfate-binding protein	Sulfate	1	1.70	1SBP	*S. typh.*	Pflugrath and Quiocho (1988)
Phosphate-binding protein	Phosphate	13	0.98	1IXG	*E. coli*	Wang *et al.* (1997)
D-Maltodextrin-binding protein	Maltose	14	1.67	1ANF	*E. coli*	Quiocho *et al.* (1997)
Ferric-binding protein	Fe(3+)	3	1.60	1MRP	*Haem. infl.*	Bruns *et al.* (1997)
Dipeptide-binding protein	No ligand	2	2.00	1DPE	*E. coli*	Nickitenko *et al.* (1995)
Histidine-binding protein	Histidine	2	1.89	1HSL	*E. coli*	Yao *et al.* (1994)
Spermidine/putrescine-binding protein (PotD)	Spermidine	2	1.80	1POT	*E. coli*	Sugiyama *et al.* (1996)
Glutamine-binding protein	Glutamine	2	1.94	1WDN	*E. coli*	Sun *et al.* (1998)
Molybdate-binding protein	Tungstate	3	1.20	1ATG	*Azot. vinel.*	Lawson *et al.* (1997)
Putrescine receptor (PotF)	1,4-diaminobutane	1	2.20	1A99	*E. coli*	Vassylyev *et al.* (1998)
Others						
Surface antigen PsaA	Zn(2+)	1	2.00	1PSZ	*Strept. pneu.*	Lawrence *et al.* (1998)
Zinc-binding protein TroA	Zn(2+)	1	1.80	1TOA	*Trep. pal.*	Lee *et al.* (1999)
Ferric siderophore-binding protein (FhuD)	Gallichrome	1	1.90	1EFD	*E. coli*	Clarke *et al.* (2000)

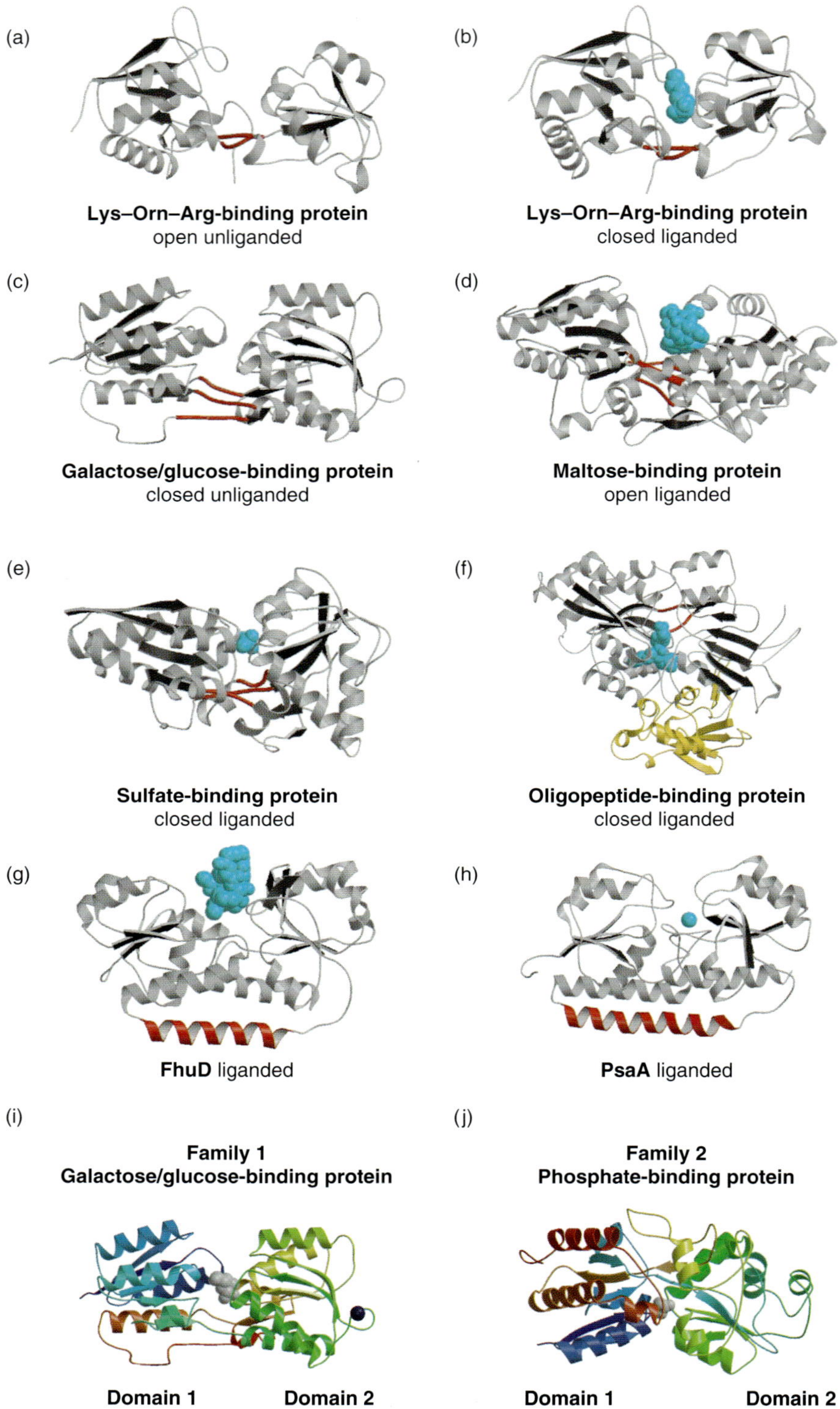

Figure 10.1. Ribbon diagrams of the structures of selected periplasmic binding proteins. The ligands, where present, are in space-filling representation. They are lysine in lysine/arginine/ornithine-binding protein (b), β-cyclodextrin in maltose-binding protein (d), sulfate in sulfate-binding protein (e), the tetrapeptide Lys–Lys–Lys–Ala in OppA (f), gallichrome in FhuD (g), Zn^{2+} in PsaA (h), glucose in galactose–glucose-binding protein (i) and phosphate in phosphate-binding protein (j). In (a) to (h) the segments connecting the two ligand-binding domains are colored red. The extra domain in the

(continued)

Mowbray, 1998; http://alpha2.bmc.uu.se/usf/pics/rbp_closure_anim.gif; http://bioinfo.mbb.yale. edu/Mol MovDB/). These open forms are in an equilibrium with one another and with the unliganded closed form (Flocco and Mowbray, 1994; Wolf *et al.*, 1994).

The closed unliganded form has been observed in crystals of galactose–glucose-binding protein (**Figure 10.1c**; Flocco and Mowbray, 1994). Evidence for its existence in solution is provided by studies of the interactions of histidine-binding protein (HisJ) with conformation-specific monoclonal antibodies (Wolf *et al.*, 1994). These antibodies, which have epitopes formed by residues on both lobes of the protein, trap HisJ in the closed unliganded form in the absence of histidine. These observations are consistent with the idea that unliganded PBPs are in equilibrium between open and closed forms in solution and that the mAbs bind to and sequester the closed form.

It is anticipated that the ligand combines with the open form of the protein, initially interacting with just one of the two domains, since in most liganded PBP structures one of the domains contributes a significantly greater proportion of the ligand-binding surface than the other domain **(Figure 10.2)**. The domains subsequently come together and the ligand becomes buried within the protein. As the substrate now makes interactions with both domains of the protein, ligand binding will shift the equilibrium towards the closed form.

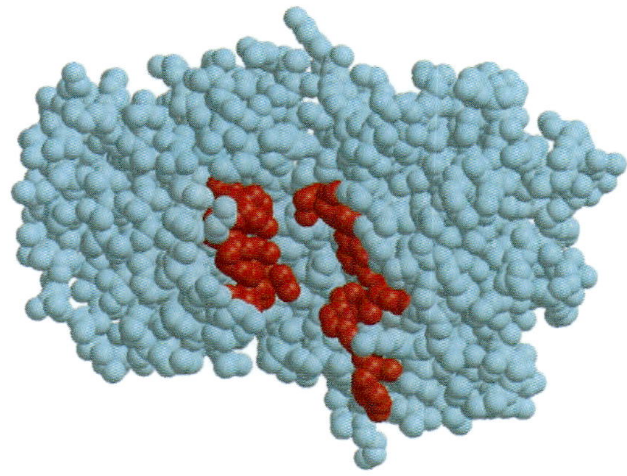

Figure 10.2. The ligand-binding residues in maltose-binding protein (MBP). The open unliganded form of the protein is shown in space-filling representation. Residues that possess atoms that are within 4.0 Å of the bound ligand in the MBP complex with maltotriotol have been colored in red (Quiocho **et al.**, *1997). The figure illustrates that the ligand-binding residues are exposed in the open unliganded form and that surfaces on both domains contribute to the binding site.*

Metal cation receptors

The recently determined structures of receptors for metal cation transporters show that the characteristic hinge peptide segments that mediate conformational change in other PBPs are missing. The structures of PsaA, a putative receptor for an ABC transporter of Mn^{2+}/Zn^{2+} in *Streptococcus pneumoniae* (Lawrence *et al.*, 1998), TroA, the periplasmic zinc-binding protein of *Treponema pallidum* (Lee *et al.*, 1999) and FhuD, the receptor for ferric siderophore transport in *E. coli* (Clarke *et al.*, 2000), retain the two α/β domain organization. However, the domains are connected by just a single segment of the polypeptide, which takes the form of an α-helix that spans the length of the molecule **(Figure 10.1g and h)**. As a result, each domain is formed from a contiguous segment of polypeptide and may be viewed as an independently folding entity. In the structure of PsaA, a Zn^{2+} ion is enclosed in the protein, tetrahedrally coordinated to a pair of imidazole groups emanating from one domain and a pair of carboxylate groups supplied by the other domain. A five coordinate Zn^{2+} species is observed bound in a similar manner in TroA. In the crystal structure of FhuD, the gallichrome (a Ga^{3+} chelate) ligand is bound in a shallow groove between the protein domains so that only 45% of its surface area is buried in the complex **(Figure 10.1g)**.

The domain-spanning α-helix packs onto secondary structure elements in each lobe of the molecule and contributes to the close packed

Figure 10.1. (continued)
oligopeptide-binding protein is colored gold. In (i) and (j) the chains are color-ramped from the N-terminus (blue) to the C-terminus (red) to emphasize the chain topology in the class I and class II PBPs. Figure (c) is a type I PBP while (a), (d), (e) and (f) are type II PBPs. In the galactose–glucose-binding protein, the bound calcium ion is shown as a dark blue sphere.

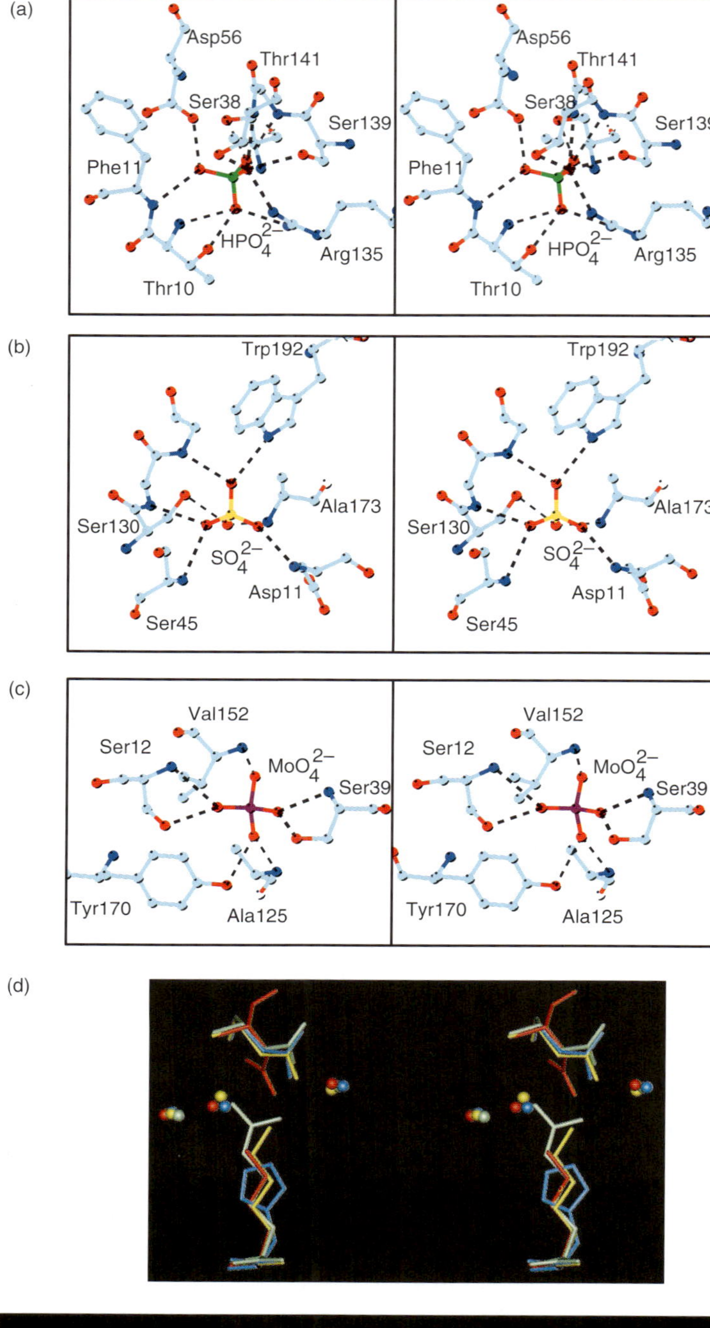
(a)
Asp56
Thr141
Ser38
Ser139
Phe11
HPO_4^{2-}
Arg135
Thr10
(b)
Trp192
Ala173
Ser130
SO_4^{2-}
Asp11
Ser45
(c)
Val152
Ser12
MoO_4^{2-}
Ser39
Tyr170
Ala125
(d)

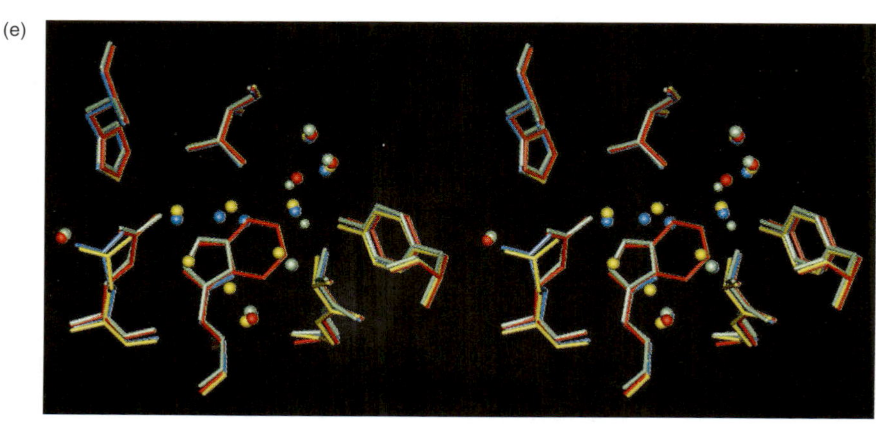
(e)

structure **(Figure 10.1g and h)**. Although this helix must be the location of hinge bending motion in TroA and PsaA required for ligand entry and exit, it is clearly a much less flexible entity than the inter-domain β-strand linkages prevalent in the other PBPs. This suggests that the angle of opening will not be as large in the cation receptors. It has been pointed out that a divalent zinc ion with a volume of 1.5 Å^3 is much smaller than a sulfate ion (67 Å^3), which is the smallest of the ligands observed in crystal structures of the conventional PBPs. Moreover, the Zn^{2+} in TroA is bound noticeably further from its hinge than the SO_4^{2-} species in sulfate-binding protein **(Figure 10.1e and h)**. These considerations together suggest that a hinge-bending angle as small as 9° would allow ligand entry and exit (Lee *et al.*, 1999). Inter-domain rotation may also be restricted in FhuD, where in contrast to other PBPs, the domain interface is hydrophobic (Clarke *et al.*, 2000). In this siderophore transport system, the Fe^{3+} ligand is initially recognized, captured and enclosed by a low molecular weight organic chelator that may be regarded in some sense, as a co-receptor. The periplasmic binding protein FhuD binds the siderophore only after it has chelated the metal. Perhaps because the metal ion is already enclosed through binding to the siderophore, there is no further enclosure mediated by domain motions in the PBP. A fuller understanding of the conformational changes accompanying ligand binding in these metal cation transporters requires the determination of crystal structures of the unliganded forms.

LIGAND BINDING

A recurring observation in the crystal structures of the PBPs is that the substrates reside in an enclosed pocket formed by surfaces from both lobes of the protein **(Figures 10.1 and 10.2)**. Enclosure in this way is inevitably associated with numerous interactions between the protein and the ligand. This accounts for the high affinity of PBPs for their cognate ligands with K_ds in the range 0.01–10 μM, as well as the impressive selectivity achieved by this class of protein.

EXQUISITE OXYANION SELECTIVITY

An example of the sharp discrimination achieved by the PBPs, and one where protein crystallography at high resolution has revealed the structural basis of specificity, is that among oxyanions. Sulfate and phosphate are transported into bacterial cells by different transporters with their own binding proteins. Although sulfate-binding protein and phosphate-binding protein share 30% sequence identity over 50 or so residues, the conserved residues do not encompass the binding pockets, which are quite different in the two proteins **(Figure 10.3a and b)**. Sulfate permease handles sulfate but not phosphate, while the phosphate permease handles phosphate and ignores sulfate. The selectivity is impressive when we consider that the respective anions are similar in (i) their size, (ii) their shape (tetrahedral) and (iii) their net charge at pH 7, which is −2. They differ, however, in one respect,

Figure 10.3. Stereoviews of ligand binding to selected PBPs. The binding of (a) phosphate to phosphate-binding protein (Wang et al., *1997), (b) sulfate to sulfate-binding protein (Pflugrath and Quiocho, 1988) and (c) molybdate to ModA (Hu* et al., *1997). Hydrogen bonding/electrostatic interactions between the bound anion and the surrounding protein are indicated by dashed lines. Atoms are colored according to type; carbon (cyan), oxygen (red), nitrogen (blue), phosphorus (green), sulfur (yellow) and molybdenum (purple). d, Comparison of the binding of basic amino acids to lysine/arginine/ornithine-binding protein (Oh* et al., *1994a). The structures of complexes of LAOBP with lysine (yellow), arginine (cyan), ornithine (red) and histidine (blue) were overlapped by least squares methods applied to protein C_α atoms. The side-chain of Asp 11 (above) adjusts its conformation according to the nature of the amino acid ligand (below). The side-chain of the ligand in the LAOBP–arginine complex displaces a water molecule present in the other complexes. e, Comparison of the binding of the tripeptides Lys–X–Lys where X = Gly (yellow), Asp (blue), His (cyan) and Trp (red) to OppA (Sleigh* et al., *1999). The structures were superimposed by least squares matching of the positions of protein C_α atoms and displayed in the region of the second side-chain binding pocket, which is circumscribed in clockwise orientation by residues Glu32, His405, Thr438, Tyr274, Ala414 and Gly415. A variable number of water molecules are displaced by the second side-chain of the ligand according to its size. Panels a–c were drawn in BOBSCRIPT (Esnouf, 1997), panels d and e were produced with the program QUANTA.*

which is in their pK_a, so that at pH 7 or thereabouts, sulfate exists as SO_4^{2-} while phosphate exists as HPO_4^{2-}. This difference in protonation state is exploited in the respective binding proteins. In the structures, the cognate ions are stripped of solvent water molecules in a buried pocket **(Figure 10.3a and b)**. The anion-binding pocket in phosphate-binding protein presents 12 hydrogen-bonding groups to the bound phosphate ligand. Eleven of these groups serve as hydrogen bond donors to the phosphate species, which has abundant capacity as a hydrogen bond acceptor (Luecke and Quiocho, 1990). The twelfth group, the carboxylate of Asp56, plays the decisive role in discrimination. The proximity of this carboxylate, which will harbor a negative charge at neutral pH, to the ligand determines that the phosphate binds so that its –OH group is oriented towards Asp56. This has two consequences; firstly it allows a favorable charge–dipole interaction to be formed with protonated anions such as HPO_4^{2-} **(Figure 10.3a)**. Secondly, it prevents a fully ionized species such as SO_4^{2-} from binding, because binding would closely appose 'like' charges. In contrast the substrate-binding site in sulfate-binding protein presents the sulfate ion exclusively with hydrogen bond donor groups, which favors the binding of the unprotonated anion (**Figure 10.3b**; Pflugrath and Quiocho, 1985). These considerations also explain how the poisons selenate (SeO_4^{2-}) and chromate (CrO_4^{2-}) enter cells via the sulfate permease, and how arsenate ($HAsO_4^{2-}$) can sneak in via the phosphate permease.

The crystal structure of a third anion-binding protein, the periplasmic receptor for molybdate transport, ModA, suggests that discrimination may also be determined by size (Hu *et al.*, 1997; Lawson *et al.*, 1997). Molybdate binds to ModA as tetrahedral MoO_4^{2-}. ModA has a closely similar tertiary structure to sulfate-binding protein, though their sequences and ligand-binding sites are not alike. The two anion-binding pockets are notable for the absence of either water molecules or positively charged residues in the vicinity of the bound ligands **(Figure 10.3b and c)**. ModA binds molybdate and a non-physiological ligand, tungstate (WO_4^{2-}), with similar affinity in the μM range and 1000-fold more tightly than it binds phosphate or sulfate. Molybdate is a significantly larger anion than sulfate (the mean Mo–O and S–O bond lengths are 1.77 Å and 1.47 Å respectively) and this is manifested in (i) a lengthening of the mean distance between the central Mo/S atom of the anion and the protein atoms donating hydrogen bonds to the molybdate/sulfate oxygens and (ii) a 25% expansion in the volume of the binding pocket in ModA (Hu *et al.*, 1997). The extensive use of main-chain groups, rather than side-chain groups, in hydrogen bonding the anion may confer rigidity on the binding sites so that they cannot easily expand and contract to accommodate one another's ligands **(Figure 10.3b and c)**.

LIMITED TOLERANCE IN AMINO ACID, POLYAMINE AND SUGAR TRANSPORTERS

There are a number of instances where a single PBP is used as a receptor for the transport of a small set of closely structurally related ligands. Thus, lysine–arginine–ornithine-binding protein (LAOBP) binds the three basic amino acids which give the protein its name with similar high affinity ($K_d \sim 0.02$ μM) and a fourth basic amino acid, histidine, somewhat less tightly ($K_d = 0.5$ μM) (Nikaido and Ames, 1992). These four amino acids differ in the shape and size of their side-chains though each is positively charged. Crystal structures of LAOBP from *Salmonella typhimurium* in complex with all four substrates have been determined at 1.8–2.1 Å resolution (**Figure 10.3d**; Oh *et al.*, 1994a). The overall conformation of the protein is closely similar in all four liganded forms. LAOBP has a binding pocket large enough to accommodate the bulkiest of its substrates, arginine, whose guanidinium group forms multiple polar contacts with the protein, some of which are mediated via a buried water molecule. There is a direct salt-bridge to the carboxylate of Asp11, whose side-chain is flexible and able to form similar ionic interactions with the charged amino groups of lysine and ornithine, but not histidine **(Figure 10.3d)**. An extra water molecule is retained in the protein when the three smaller ligands are bound and this water molecule mediates further polar contacts with the protein.

There is a second receptor for the transport of basic amino acids, HisJ. HisJ has highest affinity for its preferred substrate histidine ($K_d = 0.04$ μM), but it also binds arginine ($K_d = 0.7$ μM), lysine and ornithine. HisJ and LAOBP are 70% identical in their sequences and both proteins use the same set of membrane components ($HisQMP_2$) to effect translocation of their substrates. The structures of HisJ from *S. typhimurium* and *E. coli* in complex with histidine have been determined to 2.5 and 1.9 Å spacing, respectively (Oh *et al.*, 1994b; Yao *et al.*, 1994). As expected, the overall structure and the ligand-binding pocket are closely similar to those of LAOBP. The only difference in the residues lining the binding site is the replacement of Phe52 in LAOBP by Leu in HisJ. The leucine-52 side-chain forms a hydrophobic interaction with the imidazole ring of the histidine ligand, though it is not readily apparent why the residue-52 substitution should change the relative ligand affinities.

In polyamine transport, the receptor for the spermidine (NH_3^+–$(CH_2)_3$–NH_2^+–$(CH_2)_4$–NH_3^+) transporter, PotD, will also accommodate putrescine (NH_3^+–$(CH_2)_4$–NH_3^+), albeit with 30-fold lower affinity. However, the putrescine receptor PotF does not bind spermidine. The two proteins share 35% homology in their amino acid sequences and as expected the crystal structures of PotD and PotF, solved in complex with their cognate ligands, are closely superimposable, with the root mean squared deviation of all C_α positions between the two proteins being 1.5 Å (Sugiyama *et al.*, 1996; Vassylyev *et al.*, 1998). In both structures each of the positively charged amine groups of the substrate makes one or more ionic interactions with protein carboxylates, while apolar side-chains, and in particular tryptophan residues, pack against the aliphatic portions of the polyamine. The smaller putrescine will fit perforce into the spermidine-binding pocket of PotD; lower-affinity binding is presumably the result of the less extensive interactions achievable by the smaller ligand. The structural basis for PotF's exclusion of spermidine would be expected to be steric hindrance. Comparative analysis of PotD and PotF suggests that this is not achieved by simply occluding the extra aminopropyl moiety of spermidine by closing off the binding pocket in PotF with a bulky protein side-chain. Instead modeling studies imply that tight anchoring of the polyamine's N1 atom prevents spermidine from achieving a conformation that fits the shape of the PotF binding pocket (Vassylyev *et al.*, 1998).

The receptor for arabinose transport binds both the α and the β anomers of the sugar with similar affinities and rates (Miller *et al.*, 1983). In the crystal structure of L-arabinose-binding protein, the sugar is bound in the pyranose form in a chair conformation (Quiocho and Vyas, 1984). Refinement of the structure against high-resolution data (1.7 Å spacing) revealed that both arabinose anomers are present in the crystal in approximately equal proportions (Quiocho and Vyas, 1984). The alternative stereochemistry at the sugar's C1 atom is accommodated by the strategic positioning of the carboxylate of Asp90 so that it can form an ion pair with the C1–OH in either the α or the β configuration. A very similar strategy is used in dual substrate recognition by the receptor for the galactose–glucose transporter. Galactose–glucose-binding protein (GGBP) binds both D-galactose (K_d = 0.4 μM) and D-glucose (K_d = 0.2 μM) tightly, and crystal structures of GGBP in complex with both sugars have been determined to 2.0 Å spacing (Vyas *et al.*, 1994). D-galactose and D-glucose are epimers that differ in the stereochemistry at the C4 position. The two sugars are accommodated identically in the binding site and the hydroxyls they share form similar hydrogen bonding interactions with the protein. Recognition of the two epimers is mediated by the carboxylate of Asp14, whose $O_\delta1$ atom is used to form a charge dipole interaction with the equatorial C4 hydroxyl of glucose and whose $O_\delta2$ is used to make a similar interaction with the axially positioned C4–OH of bound galactose.

Accommodating Diversity in Peptide Transport

Peptide transport in bacteria is mediated by a set of transporters with overlapping specificities. In *E. coli* and *S. typhimurium*, two of the peptide permeases are periplasmic binding protein-dependent transporters, these being the dipeptide permease (Dpp) and the oligopeptide permease (Opp) (Abouhamed *et al.*, 1991; Hiles *et al.*, 1987). Dpp handles mainly dipeptides, with a lesser affinity for tripeptides. Opp is the most versatile of the PBP-dependent transport systems handling peptides 2–5 amino acid residues in length essentially regardless of their sequence. As a result, the potential substrates of Dpp and Opp, and by inference the number of ligands bound by the dipeptide-binding protein DppA and the oligopeptide-binding protein OppA, number in the thousands and millions, respectively.

The structures of DppA and OppA are known. DppA from *E. coli* has been solved in the open unliganded state to 1.7 Å resolution (Nickitenko *et al.*, 1995) and in the closed form with the dipeptide Gly–Leu bound to 3.2 Å spacing (Dunten and Mowbray, 1995). Crystal structures of OppA from *S. typhimurium* have been solved in two open unliganded forms and in the closed form in complex with a series of di- tri- and tetrapeptide ligands (Davies *et al.*, 1999; Sleigh *et al.*, 1997, 1999; Tame *et al.*, 1994, 1995, 1996; L. Wright, unpublished observations). As might be expected, DppA and OppA exploit the features common to all peptides, namely the main-chain, in achieving high-affinity binding. An aspartic acid carboxylate forms an ion pair with the main-chain α-amino group of the bound Gly–Leu in DppA, while the peptide's α-carboxylate group forms an ion-pairing interaction with the side-chain of Arg355. Main-chain hydrogen-bonding groups provided by each of the two domains form interactions with the peptide main-chain.

In the structures of di-, tri- and tetrapeptide complexes of OppA, the ligand's α-amino group is anchored via a salt-bridge to Asp419 and the peptide is bound in an extended conformation. In consequence the α-carboxylate group is situated a variable distance along the binding pocket according to the peptide's length. In the case of tripeptide and tetrapeptide ligands, the interactions of the peptide main-chain with OppA are reminiscent of a β-sheet. Positively charged side-chains feature in the binding of the peptide's α-carboxylate group in OppA as in DppA. However, the residues involved and the nature of the interactions are variable. In a remarkable feat of versatility, a series of positively charged side-chains are positioned along the ligand-binding pocket poised to counter the negative charge on the peptide's α-carboxylate group. The carboxylate groups of dipeptide ligands form water-mediated interactions with both Arg403 and Arg413, while those of tripeptide and tetrapeptide ligands form direct ion-pairs with Arg413 and His371, respectively. The mode of binding of pentapeptides is as yet unknown as no crystal structures are available. However, an interaction of the pentapeptide carboxylate with Lys307 is indicated by the occasional presence of acetate ions, originating from the crystallization mother liquor, close to the $\varepsilon\text{–NH}_3^+$ of this residue in some of the OppA-tripeptide complexes.

The manner in which OppA accommodates peptide side-chains that can vary in size, polarity and stereochemistry has been examined by a combination of X-ray crystallography and isothermal titration microcalorimetry experiments using Lys–Lys–Lys as a reference tripeptide. In the crystal structure of OppA-trilysine, the peptide side-chains project into distinct pockets, which can be described as voluminous and hydrated. It is notable that there are few or no direct hydrogen bonding/electrostatic interactions between the protein and the ligand side-chains. Were such interactions to form they would presumably lead to discrimination, since favoring a ligand side-chain with one polarity would tend to exclude a ligand with a side-chain of the opposite polarity. The remarkable observation from the thermodynamic analysis of 20 peptides of the sequence Lys–X–Lys, where X varies across the series of commonly occurring amino acids, is that K_d varies over only a 150-fold range even though X ranges from Ala to Trp, from Arg to Glu, and from Gly to Pro (Sleigh *et al.*, 1999). As shown in **Figure 10.3e**, a variable number of generally well-ordered water molecules are found in the second side-chain binding pocket according to the side-chain present. These water molecules act in some sense as molecular cushions, (i) their small size enables them to fill the voids that would otherwise be left around the smaller ligand side-chains, (ii) their capacity to act as hydrogen bond donors and acceptors provides flexibility in the hydrogen bonding arrangements in the side-chain binding cavity according to the ligand's polarity and (iii) the polarizability of water makes it capable of dissipating charges harbored by acidic and basic side-chains, minimizing unfavorable Coulombic interactions (Sleigh *et al.*, 1999; Tame *et al.*, 1996). The thermodynamics of peptide binding are characterized by marked enthalpy–entropy compensation whereby the ΔH and $T\Delta S$ terms associated with binding vary by 35 kJ mol^{-1} across the Lys–X–Lys series, while the free energy of binding ΔG varies over only 8 kJ mol^{-1}.

INTERACTIONS OF PBPs WITH MEMBRANE COMPONENTS

Having captured its target ligand, the next step is for the PBP to deliver the substrate to the cognate set of membrane components for transport. In the case of the galactose–glucose-, ribose-, maltose- and dipeptide-binding proteins, which also serve as receptors for chemotactic signals, the ligand-binding event may also be transduced into an intracellular signal via the Tar receptor to activate the flagellar motor and cell motility. While high-resolution crystal structures have given lucid explanations for the structural basis of ligand specificity in PBPs, relatively little is known in structural terms about the interactions between the PBPs and partner proteins in the membrane. As the PBPs are present in large excess over the other transporter components, efficient translocation of solutes to the cytoplasm requires that the membrane complexes discriminate clearly between the liganded and unliganded forms of the PBP. There is no requirement for the membrane components to increase the rates of ligand dissociation from the PBPs because the intrinsic rate constants for ligand dissociation are in the range 1–100 s^{-1} and large enough to account for the observed rates of ligand transport *in vivo* and *in vitro* (Miller *et al.*, 1983).

Genetic screens have identified mutants in histidine-, maltose- and ribose-binding proteins in which transport of the ligand is impaired even though binding is not. In other studies allele-specific suppressors of mutations in the membrane-spanning domains that lead to transport defects have been identified in the PBPs (Treptow and Shuman, 1988). The sites of these mutations can be mapped onto the structures of the PBPs to build up a picture of the surfaces important for transport. The location of sites of mutation which impair transport and/or chemotaxis without affecting sugar binding are illustrated for ribose-binding protein (RBP) in **Figure 10.4** (Binnie *et al.*, 1992; Björkman and Mowbray, 1998; Eym *et al.*, 1996). In RBP, as in other PBPs which have been similarly examined (Hor and Shuman, 1993; Oh *et al.*, 1993; Prossnitz *et al.*, 1988), mutations map to one face of the protein close to the edge of the ligand-binding pocket and they are distributed across the two domains. As these residues are situated on the opposite side of the molecule from the hinge, the surface they form is altered dramatically between the closed and open forms of the protein as ligands are bound and released **(Figure 10.4)**. The surface encompassing these receptor-binding residues is contiguous only in the closed, predominantly liganded form of the protein. In the open forms, which will generally not contain ligand, this surface is disrupted. The available evidence indicates that each lobe of the binding protein interacts with a different subunit of the transport/ chemotaxis complex (Hor and Shuman, 1993; Zhang *et al.*, 1992).

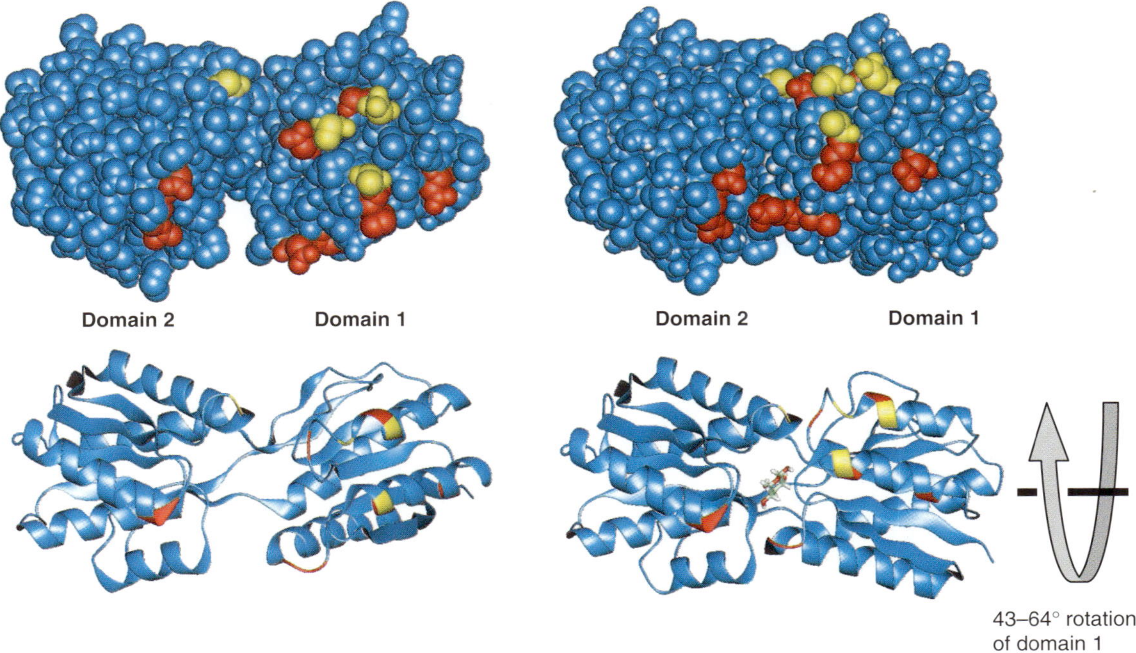

Figure 10.4. Open unliganded (left) and closed liganded (right) structures of ribose-binding protein in space-filling (top) and ribbon (bottom) representation. The ribose ligand is in ball-and-stick in the bottom right-hand panel and is completely buried in the protein. Residues whose mutation causes defects in ribose transport are colored red, while those whose mutation affects both chemotaxis and transport are colored yellow. The domain opening angle in ribose-binding protein varies from 43° to 64° in crystal structures (Björkman and Mowbray, 1998). The figure was made with the program QUANTA.

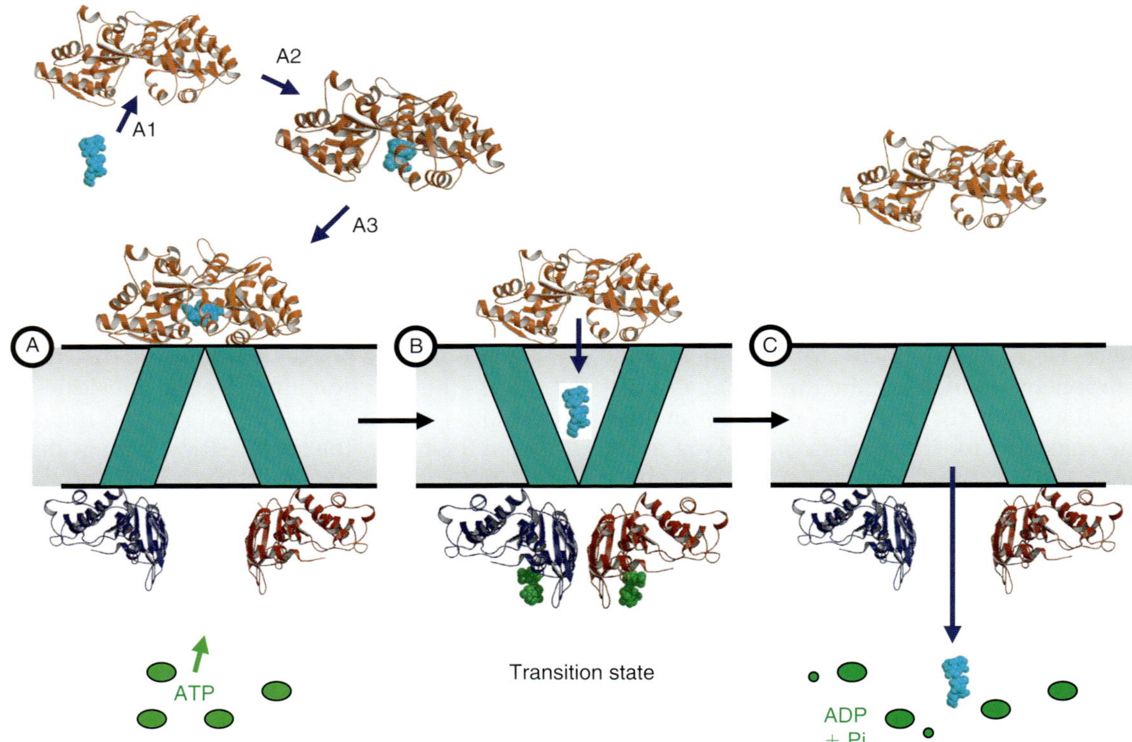

Figure 10.5. Possible scheme for membrane transport by periplasmic binding protein-dependent transporters. The model is similar to that presented by Chen **et al.**, *2001. In A, the PBP is observed capturing its ligand and in doing so, undergoing a conformational change from an open (A1) to a closed (A2) conformation. This is illustrated for maltose (blue space-filling) binding to MBP. The liganded PBP then binds to the membrane-spanning components that are represented as aquamarine bars (A3). This interaction triggers a conformational change in the latter (B), which results in ATP (green) binding and hydrolysis by the distally positioned nucleotide-binding subunits. This is shown as ATP binding to a pair of HisP subunits (red and blue ribbons). This step is also associated with opening of the PBP and release of the substrate. Release of ADP and phosphate from the ATPase subunits (C) results in a loosening of their interaction, passage of the solute into the cytoplasm and dissociation of the PBP from the membrane complex.*

The interactions of PBPs with transmembrane components in the ABC transporters have been studied most extensively in the maltose and histidine permeases. Analysis of mutant *E. coli* strains that can grow on maltose in the absence of maltose-binding protein revealed that active transport of maltose was still being accomplished albeit with a measured K_m 1000-fold higher than that for the wild-type transporter. This implies that the membrane components themselves possess a maltose-binding site (Shuman, 1982). The sites of these MBP by-pass mutations map to the two membrane-spanning components *malF* and *malG* (Treptow and Shuman, 1986). Maltose transport in the mutated transporter complexes has been examined biochemically following their reconstitution in proteoliposomes (Davidson *et al.*, 1992). Whereas rapid ATP hydrolysis in the wild-type complex takes place only in the presence of both MBP and maltose, in the mutant complexes ATP hydrolysis is constitutive. These studies point to an important function of MBP in transmitting an extracellular signal, relayed through the membrane-spanning MalF and MalG proteins, which stimulates ATP hydrolysis by the MalK protein situated on the intracellular surface of the cytoplasmic membrane (Davidson *et al.*, 1992).

The nature of the interactions among the components of reconstituted maltose transport complexes has been further illuminated in a recent study of the mechanism of inhibition of maltose transport by vanadate (Chen *et al.*, 2001). Vanadate traps ADP in one of the two nucleotide-binding sites of MalK immediately following ATP hydrolysis, presumably because the ADP.VO_4^{3-} species acts as a transition state mimic of the γ-phosphate of ATP during hydrolysis. In the presence of vanadate, maltose-binding protein becomes tightly associated with the membrane components, concomitantly losing its high affinity for maltose. These data reinforce the notion that MBP and MalK communicate with one another via the membrane components. Moreover, they point to a concerted mechanism of ATP-driven ligand transport, in which the PBP serves to stabilize the transition state in ATP hydrolysis by the nucleotide-binding components (**Figure 10.5**; Chen *et al.*, 2001).

In the histidine permease, mutations that suppress a defective histidine-binding protein (HisJ), which binds substrate normally but interacts poorly with the membrane components, have been studied (Petronilli and Ames, 1991). Histidine uptake in these suppressor strains can take place in the complete absence of HisJ (Speiser and Ames, 1991). As in the maltose system, these by-pass mutants support constitutive ATP hydrolysis – that is, ATP hydrolysis that is uncoupled from substrate translocation. Unexpectedly the suppressor mutations map to the nucleotide-binding subunit HisP. Proteolysis and chemical modification experiments suggest that HisP is accessible from both sides of the membrane. There is no evidence, however, for direct protein:protein contacts between HisJ and HisP; instead crosslinking experiments both *in vitro* and *in vivo* suggest close contacts between HisJ and the membrane-spanning component HisQ (Prossnitz *et al.*, 1988).

The crystal structure of HisP from *S. typhimurium* has provided the first visualization of an ABC transporter ATPase (Hung *et al.*, 1998). The more recent crystal structure of MalK from *Thermococcus litoralis* reveals a similar fold (Diederichs *et al.*, 2000). HisP is an L-shaped molecule with two arms, one of which (Arm-I) contains the ATP-binding site and mediates dimer formation **(Figure 10.5b)**. Mutations leading to constitutive ATP hydrolysis by HisP map to Arm-II, suggesting that the latter mediates contacts with the membrane components HisQ and HisM. Analysis of transport complexes containing protomers of HisP harboring these mutations reveals that they have lost the capacity to bind ATP cooperatively. Moreover they associate more loosely with the membrane-spanning components (Liu *et al.*, 1999). This has led to the proposal of a general model for ABC transporter function in which the ATPase components disengage and reengage the membrane components as part of the transport cycle.

Evolution of Periplasmic Binding Proteins

The large body of structural information on PBPs and related proteins presents the opportunity to examine the evolution of a family of proteins where sequence identity of most pairs of PBPs is too low to permit inferences to be drawn with confidence. A comparative analysis of the structures of the periplasmic binding proteins reveals two types of topological arrangement of the central β-sheets within the domains (Murzin *et al.*, 1995). In the first group (type I) the core structure of each domain is a parallel five-stranded β-pleated sheet with the strands in the order $\beta_B\beta_A\beta_C\beta_D\beta_E$ and the polypeptide crossing over from one domain to the other after β_E. In the type II proteins there is again a five-stranded β-structure as the core of the molecule but in this instance the strand order is $\beta_B\beta_A\beta_C\beta_n\beta_D$, where strand β_n occurs just after the first crossover from the N-domain to the C-domain and vice versa. Strand β_n runs antiparallel to the other four strands. The various members within each grouping differ in the number of helices connecting the strands of the β-sheet and in the extent of additional elements of structure appended at the C-terminus of the protein.

A genealogical chart of three-dimensional structure in the PBP family compiled on the basis of detailed structural and sequence comparisons has been presented by Fukami-Kobayashi *et al.* (1999). In their scheme, galactose–glucose-binding protein is situated at the root of the tree as the progenitor of the type I binding proteins. Their analysis suggests that at some time in evolution, but on only one occasion, a domain dislocation took place whereby strands β_E from each domain changed their residence, becoming integrated between strands β_C and β_D in the opposing domain. This gave rise to a hypothetical type II PBP precursor, from which the rest of the subfamily members were elaborated.

Whence did the type I PBP progenitor emerge? The $(\beta\alpha)_5$ fold of each of the two domains in galactose–glucose-binding protein is identical in chain topology to the phosphorylation domains of proteins of the response regulator family, whose archetypal member is CheY (Stock *et al.*, 1989). Response regulators are the downstream elements in the two

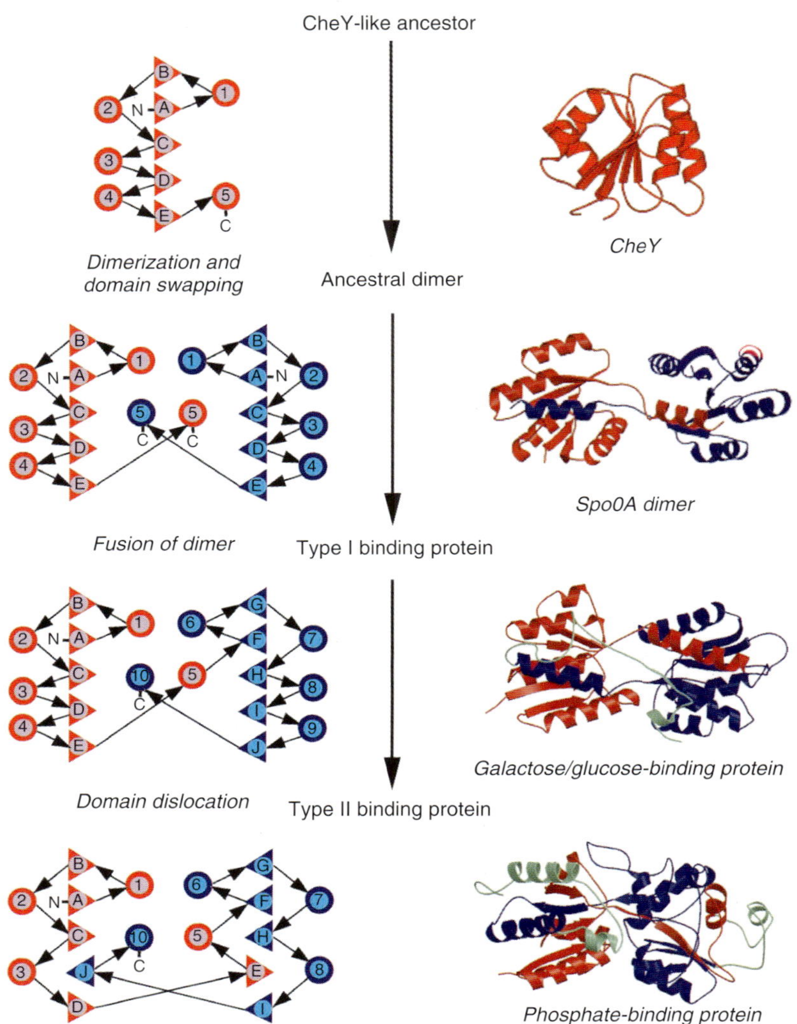

Figure 10.6. The evolution of the type I and type II periplasmic binding proteins from a CheY-like ancestor and a domain-swapped response regulator protein dimer (adapted from Fukami-Kobayashi et al.*, 1999). The right-hand panels show ribbon depictions of proteins drawn in the program BOBSCRIPT (Esnouf, 1997); the left-hand panels are a set of corresponding topology diagrams in which α-helices are shown as circles and denoted by numbers and β-strands are shown as triangles denoted by letters. The same/opposite directions of the triangles indicate parallel/anti-parallel β-strands, respectively. The secondary structural elements of the two chains of the ancestral dimer and their descendents are color-coded. Segments of structure depicted in green are elaborations on the basic scaffold, which are generally specific to each protein.*

component signal transduction systems widespread in bacteria, fungi and plants (Hoch and Silhavy, 1995). Environmental signals are transduced through phosphorylation of the response regulator components on a conserved aspartic acid residue. This naturally led to the suggestion that the progenitor type I PBP could have arisen via the duplication and fusion of a response regulator coding sequence. Fukami-Kobayashi *et al.* (1999) postulated that this process was likely to involve a CheY-like dimer intermediate; moreover they speculated that the C-terminal helices (α5) from each domain might be exchanged in a helix-swapping step, illustrated in **Figure 10.6**. Such a helix-swapping step, they argued, was necessary to form the hinging segments that connect the domains that mediate the ligand binding-associated conformational changes in the functioning PBPs.

The topology described in this 'ancestral dimer' has subsequently been observed in crystals of the regulatory domain of the response regulator Spo0A grown at pH 4 (Lewis *et al.*, 2000). In this structure, helices α5 from each monomer project away from the protomer to which they belong and pack on to the β-sheet of the partner molecule in the dimer **(Figure 10.6)**. The packing interactions formed by helix α5 are identical to those observed in the monomeric protein except for the fact that they are intermolecular. As a result of this helix exchange, which is an example of a wider phenomenon of '3-D domain swapping' (Schlunegger *et al.*, 1997), the only part of the polypeptide whose conformation changes substantially is the loop connecting β_E and α_5, where there is a *cis*-to-*trans* isomerization of a Lys–Pro peptide bond. N-Spo0A dimer formation by domain swapping is almost certainly of no significance for its physiological function, as high protein concentrations and low pH have been observed in other systems to promote domain swapping. Instead this structure is important in demonstrating that helix α5 in a response regulator protein is susceptible to domain swapping in the manner hypothesized by Fukami-Kobayashi *et al.* (1999). It is noticeable in N-Spo0A that the carboxyl-terminus of α5 of one subunit is in close proximity to the amino-terminus of the other subunit in the dimer so that a very short linker peptide would be sufficient to connect these ends in forming the two-domain monomer (**Figure 10.6**; Lewis *et al.*, 2000).

A further interesting aspect of the N-Spo0A dimer is that the active sites in the respective monomers are oriented towards one another, formed as they are, by residues at the C-termini of the β strands and the loops that connect them to the amino-termini of the following α-helices. If the hypothesis presented in **Figure 10.6** is correct, these are the residues that evolution has shaped into the ligand-binding pockets of the PBPs.

The more recently described structures of the periplasmic receptors for cation import by ABC transporters clearly place these proteins in a separate class from the other PBPs **(Table 10.1)**. PsaA and TroA are clearly closely related, each having a pair of symmetrical domains with four-stranded parallel β-sheet topology in which the strand order is $\beta_B\beta_A\beta_C\beta_D$. In FhuD, there is a five-stranded β-sheet in each domain with strand order $\beta_C\beta_B\beta_A\beta_D\beta_E$. However, whereas the sheet is a parallel one in the amino-terminal domain, in the carboxy-terminal domain, β_B runs in an anti-parallel sense to the other strands.

Proteins with Related Folds to the PBPs

Domain closure as exhibited by the PBPs transforms a ligand-binding event into a change in macromolecular conformation and not surprisingly many sensor and signaling systems in prokaryotes and eukaryotes have exploited the PBP fold. The binding cleft between the two domains of PBP-like proteins also serves as a scaffold on which chemistry can be developed, as in the active sites of enzymes such as porphobilinogen deaminase from *E. coli* and thiaminase of *Paenibacillus thiaminolyticus* (Campobasso *et al.*, 1998; Louie *et al.*, 1992).

Sequence comparisons led to the early prediction that the cofactor-binding domains of *lac* repressor-type transcriptional regulators would have similar folds to PBPs and this has been confirmed by protein crystallography (Friedman *et al.*, 1995; Hars *et al.*, 1998; Lewis *et al.*, 1996; Muller-Hill, 1983; Schumacher *et al.*, 1994). In LacI and PurR, evolution has grafted a helix-turn-helix containing DNA-binding head-piece onto the N-termini of a pair of PBP structures, which then forms a dimer **(Figure 10.7A)**. In LacI, lactose analogues serve as transcriptional inducers, while in PurR, hypoxanthine is a co-repressor. In each case the ligand is buried between the lobes of the cofactor-binding unit by a domain rotation and closure event. The inter-domain opening angles in the crystal structures of unliganded LacI and PurR are 6° and 20°, respectively, much smaller than the 45–65° openings observed in the most closely related PBP, which is ribose-binding protein (Bell and Lewis, 2000; Lewis *et al.*, 1996; Mowbray and Björkman, 1999; Schumacher *et al.*, 1995). Larger hinge motions in the repressors are precluded by the need to maintain the dimeric state, which greatly constrains the extent of inter-domain movement. Nevertheless these modest rotations are quite sufficient to allow ligand entry and exit. Ligand binding alters the affinity of the respective proteins for operator sequences on the DNA but in opposing ways: whereas hypoxanthine stimulates PurR binding to DNA, IPTG binding to LacI inhibits DNA binding.

It has subsequently been shown that the cofactor-binding domains of the LysR-type transcriptional regulators (LTTRs), CysB and OxyR also have a PBP-type fold (**Figure 10.7B**; Choi *et al.*, 2001; Tyrrell *et al.*, 1997). The LTTRs are gene activators as well as repressors, usually

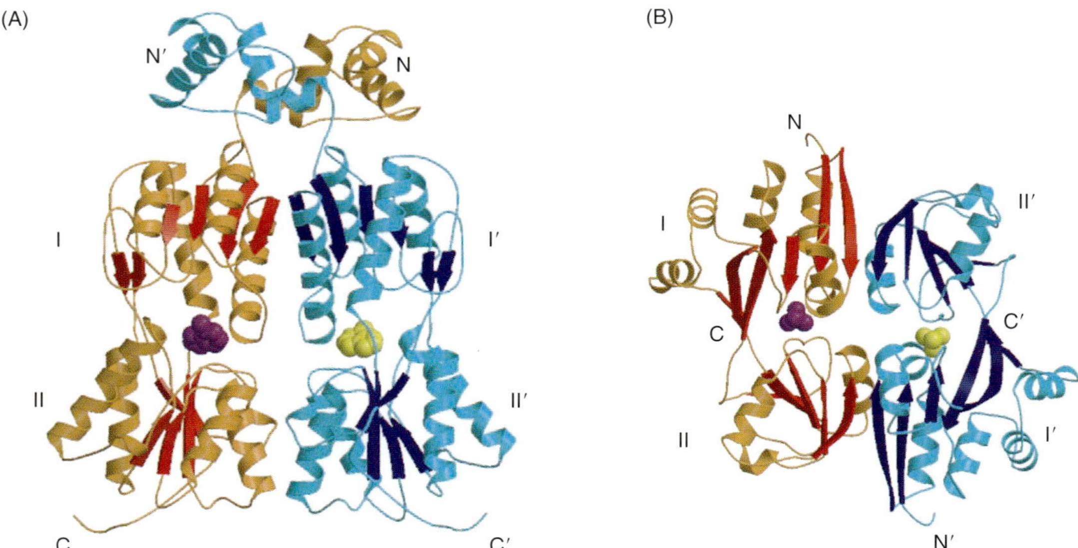

Figure 10.7. ***Ribbon diagrams of (A) PurR with bound hypoxanthine and (B) the cofactor-binding domain of CysB bound to sulfate. A dimer is shown in each case with one monomer colored in red and orange, the other is in blue and light blue. The N- and C-termini of the proteins are labeled and the domains are labeled with Roman numerals. The DNA-binding head-pieces in PurR are above the cofactor-binding domains. The DNA-binding domains are not present in the structure of CysB but these will be attached at the N-termini, which are at the top of the left-hand monomer and at the bottom of the right-hand monomer. The figure was made with the program BOBSCRIPT (Esnouf, 1997).***

regulated by cofactor binding. OxyR is an interesting exception in that its active and inactive states are inter-converted through reversible disulfide bond formation. In the LTTRs, DNA-binding domains are again attached at the N-terminus and again the cofactor-binding domains mediate dimer formation. However, the symmetry in the LTTR dimers is different from that in the LacI family dimers. As shown in **Figure 10.7**, in the PurR dimers, domain I forms the majority of its interactions with domain I of its partner in the dimer, and likewise pairs of domains II interact. In CysB, domain I interacts predominantly with domain II of its partner in the dimer and vice versa. As a result, the juxtaposition of the DNA-binding domains with respect to each other will be quite different. Interestingly, whereas PurR and LacI each belong to the type I PBP subfamily, CysB and OxyR are in the type II subfamily. If, as argued earlier, domain dislocation between the type I and type II subfamilies happened only once in evolution, then this must mean that the LacI and LysR families are the result of independent gene splicing events in which sequences encoding a DNA-binding domain became attached to a sequence encoding a type I or a type II periplasmic substrate-binding protein. A type I PBP fold is observed in another intracellular regulator, the AmiC protein of *Pseudomonas aeruginosa* (Pearl *et al.*, 1994) Acetamide binding to the central cleft of AmiC controls complex formation with the transcription anti-terminator AmiR, which in turn regulates the amidase operon (O'Hara *et al.*, 1999).

The ligand-binding domains of the ionotropic (iGluR) and metabotropic (mGluR) glutamate receptors of eukaryotes which mediate excitatory synaptic transmission also exhibit PBP folds, and crystal structures demonstrate that agonist/antagonist binding to the inter-domain cleft results in enclosure of the ligand by domain rotations (Armstrong *et al.*, 1998; Armstrong and Gouaux, 2000; Kunishima *et al.*, 2000). Even though these receptors have related functions and both bind glutamate, mGluR has a type I PBP fold whereas iGluR has a type II PBP fold. Dimers are observed in the structures of mGluR and the hormone-binding domain of the atrial natriuretic peptide receptor, which also has a PBP-like fold (Kunishima *et al.*, 2000; Van Den Akker *et al.*, 2000).

Perspectives

Studies of PBPs have taught us much about molecular recognition and protein evolution. Complete genome sequencing and the downstream activities of functional genomics will soon define for us the number of PBPs present in each organism and we can anticipate that crystal structures of orthologues of all of the PBPs possessed, for example, by *E. coli* will be known in the next few years. There will undoubtedly be new insights into chemistry and evolution particularly as it would be surprising if PBP folds were not revealed in yet more diverse areas of molecular biology. For the PBPs themselves, the future challenge is to gain crystallographic insights into how the cognate ABC transporter's membrane components are recognized and how ligand binding activates ATP-coupled solute translocation.

Acknowledgment

We are grateful to past and present members of the Structural Biology Group at York who have contributed to aspects of the work described herein and to the BBSRC and the Wellcome Trust for their support.

References

Abouhamed, W.N., Manson, M., Gibson, M.M. and Higgins, C.F. (1991) Peptide transport and chemotaxis in *Escherichia coli* and *Salmonella typhimurium*: characterization of the dipeptide permease (Dpp) and the dipeptide binding protein. Mol. Microbiol. 5, 1035–1047.

Ames, G.F.L. (1986) Bacterial periplasmic transport-systems – structure, mechanism, and evolution. Annu. Rev. Biochem. 55, 397–425.

Armstrong, N. and Gouaux, E. (2000) Mechanisms for activation and antagonism of an AMPA-Sensitive glutamate receptor: crystal structures of the GluR2 ligand binding core. Neuron 28, 165–181.

Armstrong, N., Sun, Y., Chen, G.-Q. and Gouaux, E. (1998) Structure of a glutamate-receptor ligand-binding core in complex with kainate. Nature 395, 913–917.

Bell, C.E. and Lewis, M. (2000) A closer view of the conformation of the Lac repressor bound to operator. Nat. Struct. Biol. 7, 209–214.

Binnie, R.A., Zhang, H., Mowbray, S. and Hermodson, M.A. (1992) Functional mapping of the surface of *Escherichia coli* ribose-binding protein: mutations that affect chemotaxis and transport. Protein Sci. 1, 1642–1651.

Björkman, A.J. and Mowbray, S.L. (1998) Multiple open forms of ribose-binding protein trace the path of its conformational change. J. Mol. Biol. 279, 651–664.

Björkman, A.J., Binnie, R.A., Zhang, H., Cole, L.B., Hermodson, M.A. and Mowbray, S.L. (1994) Probing protein–protein interactions. The ribose-binding protein in bacterial transport and chemotaxis. J. Biol. Chem. 269, 30206–30211.

Bruns, C.M., Nowalk, A.J., Arvai, A.S., McTigue, M.A., Vaughan, K.G., Mietzner, T.A. and McRee, D.E. (1997) Structure of *Haemophilus influenzae* Fe(+3)-binding protein reveals convergent evolution within a superfamily. Nat. Struct. Biol. 4, 919–924.

Campobasso, N., Costello, C.A., Kinsland, C., Begley, T.P. and Ealick, S.E. (1998) Crystal structure of thiaminase-I from *Bacillus thiaminolyticus* at 2.0 Å resolution. Biochemistry 37, 15981–15989.

Chaudhuri, B.N., Ko, J., Park, C., Jones, T.A. and Mowbray, S.L. (1999) Structure of D-allose binding protein from *Escherichia coli* bound to D-allose at 1.8 Å resolution. J. Mol. Biol. 286, 1519–1531.

Chen, J., Sharma, S., Quiocho, F.A. and Davidson, A.L. (2001) Trapping the transition state of an ATP-binding cassette transporter: evidence for a concerted mechanism of maltose transport. Proc. Natl Acad. Sci. 98, 1525–1530.

Choi, H.J., Kim, S.J., Mukhopadhyay, P., Cho, S., Woo, J.R., Storz, G. and Ryu, S.E. (2001) Structural basis of the redox switch in the OxyR transcription factor. Cell 105, 103–113.

Clarke, T.E., Ku, S.-Y., Dougan, D.R., Vogel, H.J. and Tari, L.W. (2000) The structure of the ferric siderophore binding protein FhuD complexed with gallichrome. Nat. Struct. Biol. 7, 287–291.

Davidson, A.L., Shuman, H.A. and Nikaido, H. (1992) Mechanism of maltose transport in *Escherichia coli*: trans-membrane signaling by periplasmic binding proteins. Proc. Natl Acad. Sci. USA 89, 2360–2364.

Davies, T.G., Hubbard, R.E. and Tame, J.R.H. (1999) Relating structure to thermodynamics: The crystal structures and binding affinity of eight OppA-peptide complexes. Protein Sci. 8, 1432–1444.

Diederichs, K., Diez, J., Greller, G., Muller, C., Breed, J., Schnell, C., Vonrhein, C., Boos, W. and Welte, W. (2000) Crystal structure of MalK, the ATPase subunit of the trehalose/maltose ABC transporter of the archaeon *Thermococcus litoralis*. EMBO J. 19, 5951–5961.

Dunten, P. and Mowbray, S.L. (1995) Crystal structure of the dipeptide-binding protein from *Escherichia coli* involved in active transport and chemotaxis. Protein Sci. 4, 2327–2334.

Esnouf, R.M. (1997) An extensively modified version of Molscript that includes greatly enhanced colouring capabilities. J. Mol. Graph. 15, 132–134.

Eym, Y., Park, Y. and Park, C. (1996) Genetically probing the regions of ribose-binding protein involved in permease interaction. Mol. Microbiol. 21, 695–702.

Flocco, M.M. and Mowbray, S.L. (1994) The 1.9 Ångstrom X-ray structure of a closed unliganded form of the periplasmic glucose/galactose receptor from *Salmonella typhimurium*. J. Biol. Chem. 269, 8931–8936.

Friedman, A.M., Fischmann, T.O. and Steitz, T.A. (1995) Crystal structure of lac repressor core tetramer and its implications for DNA looping. Science 268, 1721–1727.

Fukami-Kobayashi, K., Tateno, Y. and Nishikawa, K. (1999) Domain dislocation: a change of core structure in periplasmic binding proteins in their evolutionary history. J. Mol. Biol. 286, 279–290.

Furlong, C.E. (1987) Osmotic shock-sensitive transport systems. In: Cellular and Molecular Biology (ed. F.C. Neidhardt) pp. 768–796. Washington, DC: American Society for Microbiology.

Gilson, E., Alloing, G., Schmidt, T., Claverys, J.P., Dudler, R. and Hofnung, M. (1988) Evidence for high affinity binding-protein-dependent transport systems in Gram positive bacteria and in *Mycoplasma*. EMBO J. 7, 3971–3974.

Hars, U., Horlacher, R., Boos, W., Welte, W. and Diederichs, K. (1998) Crystal structure of the effector binding domain of the trehalose repressor of *Escherichia coli*, a member of the LacI family in its complex with inducer trehalose-6-phosphate and non-inducer trehalose. Protein Sci. 7, 2511–2521.

Higgins, C.F. (1992) ABC transporters: from microorganisms to man. Annu. Rev. Cell Biol. 8, 67–113.

Higgins, C.F., Haag, P.D., Nikaido, K., Ardeshir, F., Garcia, G. and Ames, G.F.-L. (1982) Complete nucleotide sequence and identification of membrane components of the histidine transport operon of *S. typhimurium*. Nature 298, 723–727.

Hiles, I.D., Gallagher, M.P., Jamieson, D.J. and Higgins, C.F. (1987) Molecular characterization of the oligopeptide permease of *Salmonella typhimurium*. J. Mol. Biol. 195, 125–142.

Hoch, J.A. and Silhavy, T.J. (1995) Two-component Signal Transduction. Washington, DC: American Society for Microbiology.

Hor, L.-I. and Shuman, H.A. (1993) Genetic analysis of periplasmic binding protein dependent transport in *Escherichia coli*: each lobe of maltose binding protein interacts with a different subunit of MalFKG(2) membrane-transport complex. J. Mol. Biol. 233, 659–670.

Hsiao, C.D., Sun, Y. J., Rose, J. and Wang, B.C. (1996) The crystal structure of glutamine-binding protein from *Escherichia coli*. J. Mol. Biol. 262, 225–242.

Hu, Y., Rech, S., Gunsalus, R.P. and Rees, D.C. (1997) Crystal structure of the molybdate binding protein ModA. Nat. Struct. Biol. 4, 703–707.

Hung, L.-W., Wang, I.X., Nikaido, K., Liu, P.-Q., Ames, G.F.-L. and Kim, S.-H. (1998) Crystal structure of the ATP-binding subunit of an ABC transporter. Nature 396, 703–707.

Kunishima, N., Shimada, Y., Tsuji, Y., Sato, T., Yamamoto, M., Kumasaka, T., Nakanishi, S., Jingami, H. and Morikawa, K. (2000) Structural basis of glutamate recognition by a dimeric metabotropic glutamate receptor. Nature 407, 971–977.

Lawrence, M., Pilling, P.A., Epa, V.C., Berry, A.M., Ogunniyi, A.D. and Paton, J.C. (1998) The crystal structure of pneumococcal surface antigen PsaA reveals a metal-binding site and a novel structure for a putative ABC-Type binding protein. Structure 6, 1553–1561.

Lawson, D.M., Williams, C.E.M., White, D.J., Choay, A.P., Mitchenall, L.A. and Pau, R.N. (1997) Protein ligands for molybdate. Specificity and charge stabilisation at the anion-binding sites of periplasmic and intracellular molybdate-binding proteins of *Azotobacter*. J. Chem. Soc. Dalton Transactions 21, 3981–3984.

Lee, Y.-H., Deka, R.K., Norgard, M.V., Radolf, J.D. and Hasemann, C.A. (1999) *Treponema pallidum* TroA is a periplasmic zinc-binding protein with a helical backbone. Nat. Struct. Biol. 6, 628–633.

Lewis, M., Chang, G., Horton, M.C., Kercher, M.A., Pace, H.C., Schumacher, M.A., Brennan, R.G. and Lu, P.Z. (1996) Crystal structure of the lactose operon repressor and its complex with DNA and inducer. Science 271, 1247–1254.

Lewis, R.J., Muchová, K., Brannigan, J.A., Barák, I., Leonard, G. and Wilkinson, A.J. (2000) Domain-swapping in the sporulation response regulator Spo0A. J. Mol. Biol. 297, 757–770.

Liu, P.-Q., Liu, C.E. and Ames, G.F.-L. (1999) Modulation of ATPase activity by physical disengagement of the ATP-binding domains of an ABC-transporter, the histidine permease. J. Biol. Chem. 274, 18310–18318.

Louie, G.V., Brownlie, P.D., Lambert, R., Cooper, J.B., Blundell, T.L., Wood, S.P., Warren, M.J., Woodcock, S.C. and Jordan, P.M. (1992) Structure of porphobilinogen deaminase reveals a flexible multidomain polymerase with a single catalytic site. Nature 359, 33–39.

Luecke, H. and Quiocho, F.A. (1990) High specificity of a phosphate-transport protein determined by hydrogen-bonds. Nature 347, 402–406.

Miller, D.M., Olson, J.S. and Quiocho, F.A. (1980) The mechanism of sugar binding to the periplasmic receptor for galactose chemotaxis and transport in *Escherichia coli*. J. Biol. Chem. 255, 2465–2471.

Miller, D.M., Olson, J.S., Pflugrath, J.W. and Quiocho, F.A. (1983) Rates of ligand binding to periplasmic proteins involved in bacterial transport and chemotaxis. J. Biol. Chem. 258, 13665–13672.

Mowbray, S.L. and Björkman, A.J. (1999) Conformational changes of ribose-binding protein and two related repressors are tailored to fit the functional need. J. Mol. Biol. 294, 487–499.

Muller-Hill, B. (1983) Sequence homology between Lac and Gal repressors and sugar binding periplasmic proteins. Nature, 302, 163–164.

Murzin, A.G., Brenner, S.E., Hubbard, T. and Chothia, C. (1995) SCOP: a structural classification of proteins database for the investigation of sequences and structures. J. Mol. Biol. 247, 536–540.

Newcomer, M.E., Lewis, B.A. and Quiocho, F.A. (1981) The radius of gyration of L-arabinose-binding protein decreases upon binding of ligand. J. Biol. Chem. 256, 3218–3222.

Nickitenko, A.V., Trakhanov, S. and Quiocho, F.A. (1995) 2 Å resolution structure of DppA, a periplasmic dipeptide transport/chemosensory receptor. Biochemistry 34, 16585–16595.

Nikaido, K. and Ames, G.F.-L. (1992) Purification and characterisation of the periplasmic lysine-binding, arginine-binding, ornithine-binding protein (LAO) from *Salmonella typhimurium*. J. Biol. Chem. 267, 20706–20712.

Oh, B.H., Pandit, J., Kang, C.H., Nikaido, K., Gokcen, S., Ames, G.F.-L. and Kim, S.H. (1993) Three-dimensional structures of the periplasmic lysine/arginine/ornithine-binding protein with and without a ligand. J. Biol. Chem. 268, 11348–11355.

Oh, B.H., Ames, G.F-L. and Kim, S.H. (1994a) Structural basis for multiple ligand specificity of the periplasmic lysine-, arginine-, ornithine-binding protein. J. Biol. Chem. 269, 26323–26330.

Oh, B.H., Kang, C.H., De Bondt, H., Kim, S.H., Nikaido, K., Joshi, A.K. and Ames, G.F. (1994b) The bacterial periplasmic histidine-binding protein. Structure/function analysis of the ligand-binding site and comparison with related proteins. J. Biol. Chem. 269, 4135–4143.

O'Hara, B.P., Norman, R.A., Wan, P.T.C., Roe, S.M., Barrett, T.E., Drew, R.E. and Pearl, L.H. (1999) Crystal structure and induction mechanism of AmiC-AmiR: a ligand regulated transcription anti-termination complex. EMBO J. 18, 5175–5186.

Pearl, L., O'Hara, B., Drew, R. and Wilson, S. (1994) Crystal structure of AmiC: the controller of transcription antitermination in the amidase operon of *Pseudomonas aeruginosa*. EMBO J. 13, 5810–5817.

Petronilli, P. and Ames, G.F.-L. (1991) Binding protein-dependent histidine permease mutants: uncoupling of ATP hydrolysis from transmembrane signalling. J. Biol. Chem. 266, 16293–16296.

Pflugrath, J.W. and Quiocho, F.A. (1985) Sulphate sequestered in the sulphate-binding protein of *Salmonella typhimurium* is bound solely by hydrogen bonds. Nature 314, 257–260.

Pflugrath, J.W. and Quiocho, F.A. (1988) The 2 Å resolution structure of the sulphate-binding protein involved in active-transport in *Salmonella typhimurium*. J. Mol. Biol. 200, 163–180.

Prossnitz, E., Nikaido, K., Ulbrich, S.J. and Ames, G.F.-L. (1988) Formaldehyde and photoactivatable cross-linking of the periplasmic binding protein to a membrane component of the histidine transport system of *Salmonella typhimurium*. J. Biol. Chem. 263, 17917–17920.

Quiocho, F.A. and Vyas, N.K. (1984) Novel stereospecificity of the L-arabinose-binding protein. Nature 310, 381–386.

Quiocho, F.A., Spurlino, J.C. and Rodseth, L.E. (1997) Extensive features of tight oligosaccharide binding revealed in high resolution structures of the maltodextrin transport chemosensory receptor. Structure 5, 997–1015.

Sack, J.S., Saper, M.A. and Quiocho, F.A. (1989a) Periplasmic binding protein structure and function. Refined X-ray structures of the leucine/isoleucine/valine-binding protein and its complex with leucine. J. Mol. Biol. 206, 171–191.

Sack, J.S., Trakhanov, S.D., Tsigannik, I.H. and Quiocho, F.A. (1989b) Structure of the L-leucine-binding protein refined at 2.4 Å resolution and comparison with the Leu/Ile/Val-binding protein structure. J. Mol. Biol. 206, 193–207.

Schlunegger, M.P., Bennett, M.J. and Eisenberg, D. (1997) Oligomer formation by 3D domain swapping: A model for protein assembly and misassembly. Adv. Protein Chem. 50, 61–122.

Schumacher, M.A., Choi, K.Y., Zalkin, H. and Brennan, R.G. (1994) Crystal structure of LacI member, PurR, bound to DNA: minor groove binding by alpha helices. Science 266, 763–770.

Schumacher, M.A., Choi, K.Y., Lu, F., Zalkin, H. and Brennan, R.G. (1995) Mechanism of corepressor-mediated specific DNA binding by the purine repressor. Cell 83, 147–155.

Sharff, A.J., Rodseth, L.E., Spurlino, J.C. and Quiocho, F.A. (1992) Crystallographic evidence of a large ligand-induced hinge-twist motion between the 2 domains of the maltodextrin binding-protein involved in active-transport and chemotaxis. Biochemistry 31, 10657–10663.

Sharff, A.J., Rodseth, L.E. and Quiocho, F.A. (1993) Refined 1.8 Ångstrom structure reveals the mode of binding of β-cyclodextrin to the maltodextrin binding protein. Biochemistry 32, 10553–10559.

Shilton, B.H., Flocco, M.M., Nilsson, M. and Mowbray, S.L. (1996) Conformational changes of three periplasmic receptors for bacterial chemotaxis and transport: the maltose-, glucose/galactose- and ribose-binding proteins. J. Mol. Biol. 264, 350–363.

Shuman, H.A. (1982) Active transport of maltose in *Escherichia coli* K12 – role of the periplasmic maltose binding protein and evidence for a substrate recognition site in the cytoplasmic membrane. J. Biol. Chem. 257, 5455–5461.

Sleigh, S.H., Tame, J.R.H., Dodson, E.J. and Wilkinson, A.J. (1997) Peptide binding in OppA, the crystal structures of the periplasmic oligopeptide binding protein in the unliganded form and in complex with lysyllysine. Biochemistry 36, 9747–9758.

Sleigh, S.H., Seavers, P.R., Wilkinson, A.J., Ladbury, J.E. and Tame, J.R.H. (1999) Crystallographic and calorimetric analysis of peptide binding to OppA protein. J. Mol. Biol. 291, 393–415.

Speiser, D.M. and Ames, G.F.-L. (1991) *Salmonella typhimurium* histidine periplasmic permease mutations that allow transport in the absence of histidine-binding protein. J. Bacteriol. 173, 1444–1451.

Stock, A.M., Mottonen, J.M., Stock, J.B. and Schutt, C.E. (1989) Three dimensional structure of CheY, the response regulator of bacterial chemotaxis. Nature 337, 745–749.

Sugiyama, S., Vassylyev, D.G., Matsushima, M., Kashiwagi, K., Igarashi, K. and Morikawa, K. (1996) Crystal structure of PotD, the primary receptor of the polyamine transport system in *Escherichia coli*. J. Biol. Chem. 271, 9519–9525.

Sun, Y.J., Rose, J., Wang, B.C. and Hsiao, C.D. (1998) The structure of glutamine-binding protein complexed with glutamine at 1.94 Å resolution: comparisons with other amino acid binding proteins. J. Mol. Biol. 278, 219–229.

Tame, J.R.H., Murshudov, G.N., Dodson, E.J., Neil, T.K., Dodson, G.G., Higgins, C.F. and Wilkinson, A.J. (1994) The structural basis of sequence-independent peptide binding by OppA protein. Science 264, 1578–1581.

Tame, J.R.H., Dodson, E.J., Murshudov, G.N., Higgins, C.F. and Wilkinson, A.J. (1995) The crystal structures of the oligopeptide-binding protein OppA complexed with tripeptide and tetrapeptide ligands. Structure 3, 1395–1406.

Tame, J.R.H., Sleigh, S.H., Wilkinson, A.J. and Ladbury, J.E. (1996) The role of water in sequence-independent ligand binding by an oligopeptide transporter protein. Nat. Struct. Biol. 3, 998–1001.

Treptow, N.A. and Shuman, H.A. (1986) Genetic evidence for substrate and periplasmic binding protein recognition by the MalF and MalG proteins, cytoplasmic membrane components of the *Escherichia coli* maltose transport system. J. Bacteriol. 163, 654–660.

Treptow, N.A. and Shuman, H.A. (1988) Allele-specific *malE* mutations that restore interactions between maltose binding protein and the inner membrane components of the maltose transport system. J. Mol. Biol. 202, 808–822.

Tyrrell, R., Verschueren, K.H.G., Dodson, E.J., Murshudov, G.N., Addy, C. and Wilkinson, A.J. (1997) The structure of the cofactor-binding fragment of the LysR family member, CysB: a familiar fold with a surprising subunit arrangement. Structure 5, 1017–1022.

Van Den Akker, F., Zhang, X., Miyagi, M., Huo, X., Misono, K.S. and Yee, V.C. (2000) Structure of the dimerized hormone-binding domain of a guanylyl-cyclase-coupled receptor. Nature 406, 101–104.

Vassylyev, D.G., Tomitori, H., Kashiwagi, K., Morikawa, K. and Igarashi, K. (1998) Crystal structure and mutational analysis of the *Escherichia coli* putrescine receptor. Structural basis for substrate specificity. J. Biol. Chem. 273, 17604–17609.

Vermersch, P.S., Lemon, D.D., Tesmer, J.J. and Quiocho, F.A. (1991) Sugar-binding and crystallographic studies of an arabinose-binding protein mutant (Met108Leu) that exhibits enhanced affinity and altered specificity. Biochemistry 30, 6861–6866.

Vyas, N.K., Vyas, M.N. and Quiocho, F.A. (1987) A novel calcium binding site in the galactose-binding protein of bacterial transport and chemotaxis. Nature 327, 635–638.

Vyas, N.K., Jacobsen, B.L. and Quiocho, F.A. (1989) The calcium binding site in the galactose chemoreceptor protein. J. Biol. Chem. 264, 20817–20821.

Vyas, M.N., Vyas, N.K. and Quiocho, F.A. (1994) Crystallographic analysis of the epimeric and anomeric specificity of the periplasmic transport/chemosensory protein receptor for D-glucose and D-galactose. Biochemistry 33, 4762–4768.

Wang, Z., Luecke, H., Yao, N. and Quiocho, F.A. (1997) A low energy short hydrogen bond in very high resolution structures of protein receptor phosphate complexes. Nat. Struct. Biol. 4, 519–522.

Wolf, A., Shaw, E.W., Nikaido, K. and Ames, G.F.-L. (1994) The histidine-binding protein undergoes conformational changes in the absence of ligand as analysed with conformation-specific monoclonal antibodies. J. Biol. Chem. 269, 23051–23058.

Yao, N., Trakhanov, S. and Quiocho, F.A. (1994) Refined 1.89Å structure of the histidine-binding protein complexed with histidine and its relationship with many other active transport/chemosensory proteins.

Zhang, Y., Conway, C., Rosato, M., Suh, Y. and Manson, M.D. (1992) Maltose chemotaxis involves residues in the N-terminal and C-terminal domains on the same face of maltose-binding protein. J. Biol. Chem. 267, 22813–22820.

Zou, J.Y., Flocco, M.M. and Mowbray, S.L. (1993) The 1.7 Å refined X-ray structure of the periplasmic glucose/galactose receptor from *Salmonella typhimurium*. J. Mol. Biol. 233, 739–752.

Bacterial ABC Transporters Involved in Protein Translocation

CHAPTER 11

I. Barry Holland, Houssain Benabdelhak, Joanne Young, Andrea de Lima Pimenta, Lutz Schmitt and Mark A. Blight

Introduction

Many ABC transporters have now been identified, as illustrated in **Table 11.1**, which secrete high molecular weight polypeptides. These include both pore-forming toxins and hydrolytic enzymes, important determinants for virulence in humans, plants and animals. Examples include in humans, toxins secreted from uropathogenic *Escherichia coli* (Hacker *et al.*, 1983; Welch *et al.*, 1981) and the adenyl cyclase toxin from *Bordetella pertussis* (Glaser *et al.*, 1988), and in plants, colonization and infection by *Erwinia* and other species (involving secretion of proteases, lipases, cellulases (Zhang *et al.*, 1999)). ABC transporters are also involved in secretion of several proteins required for formation of nitrogen-fixing nodules in Leguminosa (Economou *et al.*, 1990; Finnie *et al.*, 1997; York and Walker, 1997), for the formation of heterocysts in *Anabaena* spp. (Fiedler *et al.*, 1998), or for development in *Myxococcus xanthus* (Ward *et al.*, 1998). Proteins forming surface layers in some bacteria, which provide protection (Awram and Smit, 1998) or even movement (i.e. gliding (Hoiczyk and Baumeister, 1997)), are also secreted by the ABC-dependent pathway.

However, many ABC transporters, composed of appropriate membrane and ABC components, are concerned with import or export of relatively small molecules. Many of these encounter the ABC protein via the membrane bilayer or, in the case of bacterial importers, only *after* the transport substrate has largely crossed the bilayer (see Chapter 9). In contrast, ABC transporters in bacteria required for secretion of RTX toxins and related proteins have been the exception, seemingly embracing a number of different concepts in order to account for translocation of protein substrates, in some cases with sizes over 400 kDa. In all probability, such substrates, secreted by the so-called type 1 pathway, directly access the interior of the transporter from the cytoplasm, by-passing the bilayer. In this chapter we shall try to reconcile the implications of such *mammoth transport substrates*, or our preferred term, *allocrite* (Blight and Holland, 1990), with a transport mechanism which still probably shares many of the same features fundamental to other ABC proteins. Notwithstanding this, as we shall see, such transporters require at least one additional accessory or auxiliary protein to facilitate movement of the protein allocrite across the cytoplasmic membrane. In this review we shall concentrate on the best-studied examples of the type 1 system, which are in Gram-negative bacteria, where two membranes have to be negotiated. In this case, at least one further auxiliary protein in the outer membrane is required to provide the

ABC Proteins: From Bacteria to Man
ISBN 0-12-352551-9

TABLE 11.1 EXAMPLES OF RTX PROTEINS AND OTHER POLYPEPTIDES SECRETED BY THE ABC-DEPENDENT PATHWAY

Organism	Protein type	Example	Function	Reference
Escherichia coli	RTX toxins	HlyA	Cytotoxic Ca^{2+}/K^{+} pore; uropathogenic infections and pyleonephritis	1
	Microcins	ColV*	Peptide antibacterial pore forming, active against other *E. coli*	2
Serratia marcescens	Proteases	PrtA	Colonization/infection in plants	3
	Lipase	LipA	Pathogenicity?	4
	Heme binding	HasA[a]	Iron-scavenging protein	5
	S-Layer	SlaA	Possible defence against host antibacterial systems	6
Pseudomonas fluorescens	Protease	AprA	?	4
	Lipase	TliA	Pathogenicity	7
	Heme binding	HasA[a]	Iron scavenging	4
Pseudomonas aeruginosa	Protease	AprA	Pathogenicity factor?	8
	Heme binding	HasA[a]	Iron scavenging	9
Erwinia chrysanthemi	Proteases	PrtB	Colonization and infection of plant tissue	10
Cyanobacterium	Surface fibrils	Oscillin	Calcium-binding protein essential for gliding movement of filaments	11
Rhizobium leguminosarum	Nodulation protein	NodO	Calcium-binding protein implicated in infection of legumes	12
	Glycanase	EglA[a]	Symbiosis nodulation; exopolysaccharide processing	13
Caulobacter crescentus	S-layer	RsaA	?	14
Bordetella pertussis	RTX toxin	CyaA	Adenyl cylase toxin-pathogenicity factor	15
Actinobacillus pleuropneumoniae	RTX toxin	Hly	Pore-forming toxin associated with swine fever	16
Vibrio cholerae	RTX toxin	RtxA	Targets G-actin to alter cellular morphology	17
Neisseria meningitidis	RTX protein	FrpA	RTX protein with role in pathogenicity?	18
Lactococcus lactis	Lantibiotic (peptide)	NisinA[a]	Antibacterial compounds	19

*Non-RTX-type, N-terminal signal cleaved.
[a]Non-RTX.
(1) O'Hanley *et al.*, 1991; (2) Gilson *et al.*, 1990; (3) Hines *et al.*, 1988; (4) Omori *et al.*, 2001; (5) Letoffe *et al.*, 1994; (6) Kawai *et al.*, 1998; (7) Ahn *et al.*, 1999; (8) Guzzo *et al.*, 1991; (9) Idei *et al.*, 1999; (10) Delepelaire and Wandersman, 1990; (11) Hoiczyk and Baumeister, 1997; (12) Economou *et al.*, 1990; (13) Geelen *et al.*, 1995; (14) Awram and Smit, 1998; (15) Glaser *et al.*, 1988; (16) Frey *et al.*, 1993; (17) Fullner and Mekalanos, 2000; (18) Thompson and Sparling, 1993; (19) van der Meer *et al.*, 1994.

exit to the external medium. The organization of the complete translocator as we understand it at the moment is illustrated in its simplest form in **Figure 11.1**.

We shall consider in particular the three major examples of this kind of ABC transporter which have been studied in the most detail, HlyB (HlyA toxin transport), PrtD (protease transport) and HasD (transport of a heme-binding protein, HasA). All these transporters are required for the type 1 or ABC-dependent secretion pathway in Gram-negative bacteria. By definition, as shown in **Figure 11.1**, this secretion system depends upon an *ABC transporter*, an *MFP* (*membrane fusion protein*) anchored in the inner membrane and connecting the ABC protein across the periplasm to its partner in the outer membrane, and the final component of the translocator, an *OMF* (*outer membrane factor*) such as TolC (*E. coli*).

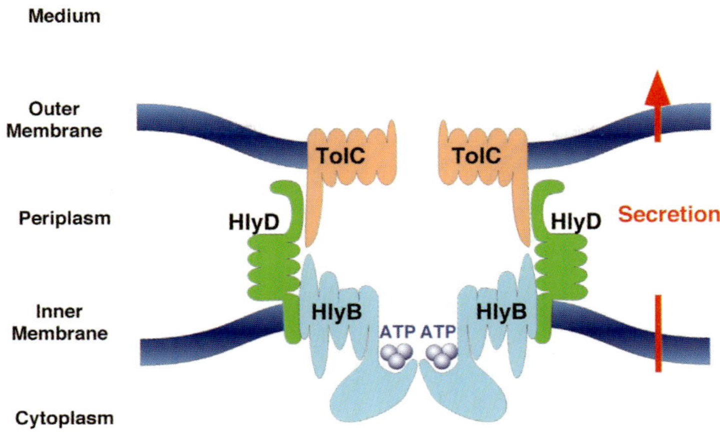

Figure 11.1. Model of the type 1, ABC-dependent translocator for protein secretion. The model is illustrated by the example of the Hly complex for secretion of the hemolysin, HlyA, from E. coli. *For simplicity, HlyD (MFP) and TolC (OMF), which in reality are at least trimers, are represented as dimers. The interactions represented between all three proteins have been demonstrated experimentally but the positioning of HlyB as the core of the translocator, rather than HlyD, with HlyB occupying the outside position is completely arbitrary.*

PHYLOGENY OR CLUSTER ANALYSIS OF THE ABC-ATPASE INVOLVED IN TYPE 1 SECRETION

As described in Chapter 1, this class of ABC transporter separates from the import group, such as HisP and MalK, and belongs to the class 1, export branch of ABCs, specifically the DPL subfamily. This includes important eukaryote ABC transporters (ABCB group; see Chapter 2), both single unit M-ABC (membrane domain plus ABC) and tandemly duplicated M_1-ABC_1-M_2-ABC_2 forms. Surprisingly, some of the closest relatives of HlyB found in the DPL subfamily are the ATPase domains of human Mdr1, whose major substrates/allocrites appear to be relatively hydrophobic antitumor drugs or lipids. Other close relatives of HlyB are, however, the ATPases of the TAP1 and TAP2 transporters (also in the group ABCB), whose physiological substrates are 'foreign' peptides (see Chapter 26) generated in the cytoplasm by proteolysis of infecting agents.

Figure 11.2 shows a similarity plot comparing the sequences of HlyB with TAP1, 2 and Mdr1 (Pgp). In addition to the high level of conservation within the ABC domain including the Walker A and B and signature motifs, there are, however, lower but significant levels of similarity between these proteins extending well into the distal region of the membrane domain (Holland and Blight, 1996). This we have suggested implies conservation of some aspect of the transport mechanism, involving coordinated action between this distal region of the membrane domain and the ABC-ATPase. However, this remains to be established.

GENETIC BASIS OF THE TYPE 1 SECRETION SYSTEM

The organization of genes required for secretion of hemolysin (HlyA) from *E. coli*, metalloprotease PrtA from *Erwinia chrysanthemi,* and HasA from *Serratia marcescens* is compared in **Figure 11.3**. The figure indicates that the ABC protein and the inner membrane, MFP, which spans the periplasm, are invariably encoded by adjacent genes, immediately downstream of that for the transport substrate itself. MFPs form a group of proteins of similar size and structural organization with sequence homology confined to a few discrete regions (Saier *et al.*, 1994). Originally thought to be present only in Gram-negative bacteria, several examples of similar proteins have now been detected in Gram-positive bacteria including *Bacillus subtilis*

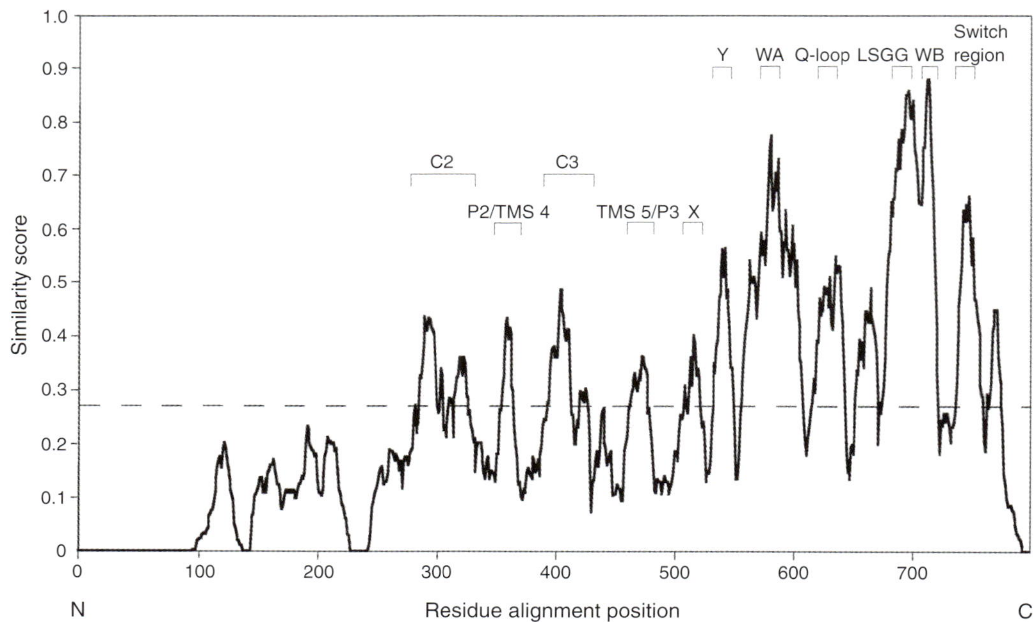

Figure 11.2. *Scanning for regions of similarity in the HlyB, TAP and Mdr1 (Pgp) molecules. Similarity plot comparing regions of homology between HlyB and close relatives (with respect to the ABC domains) TAP1, 2 and Mdr1 (Pgp). Some of the regions displaying highest levels of similarity in both the N-terminal membrane domain (approximately residues 1–550 on this scale) and the ABC domain are indicated (and see text). WA, WB, Walker motifs for nucleotide binding; LSGG-, the C- or signature motif; Switch region, containing the highly conserved histidine residue; regions immediately dowstream of TMS 6 in the membrane domain also showing significant similarity are X, containing (numbers according to HlyB sequence) S_{440}, L_{444}, L_{448}, N_{449}, P_{451}, and Y, containing G_{466}, F_{470}, F_{475}, L_{485}. C2, C3 and P2, P3 are cytoplasmic and periplasmic 'loops', respectively; the positions of these and TMS 4, 5 are indicated in Figure 11.6.*

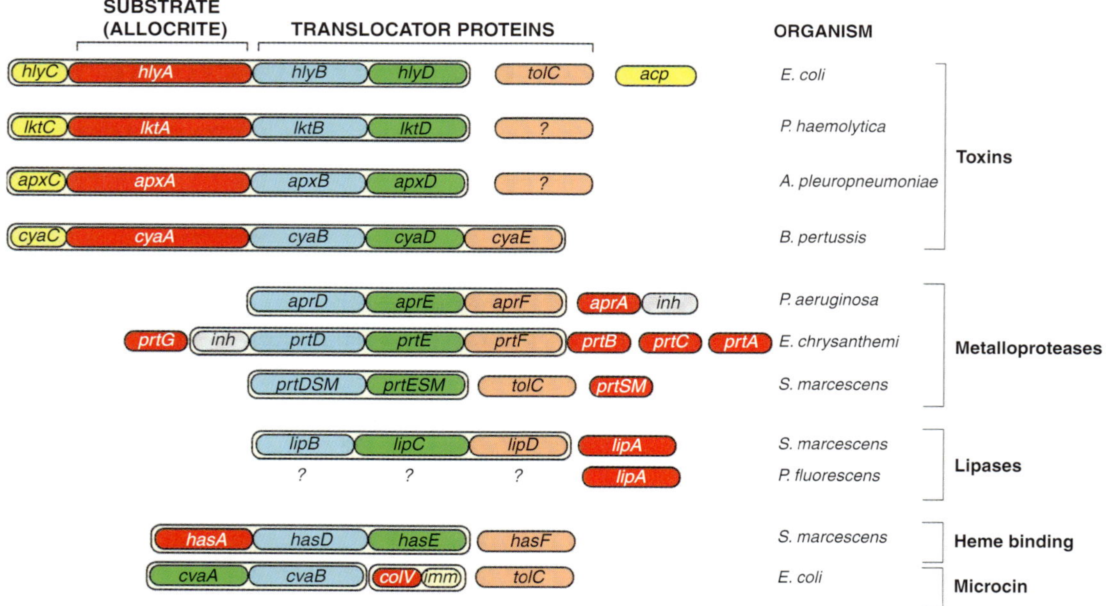

Figure 11.3. *Schematic representation of the genetic organization of the determinants for ABC-dependent secretion. Red = allocrite; three well-conserved components of the secretion apparatus, blue = ABC transporters, green = membrane fusion protein (MFP), salmon = outer membrane component (OMF); yellow = toxin activator and Acp (acyl carrier protein); gray = inhibitor of protease.*

(Johnson and Church, 1999). The precise role of the MFP, bridging the periplasm to connect the OMF directly with the inner membrane ABC transporter or to bring together the two membranes, is still unclear. These roles would not be mutually exclusive and evidence for a membrane fusion activity by a distant member of the MFP family has recently been obtained (Zgurskaya and Nikaido, 2000). Some genetic and biochemical evidence indicates that HlyD forms a specific part of the transport pathway (see later). The outer membrane component (OMF) of the translocator, which provides the final exit to the medium, may also be encoded in the same gene cluster, but may, as in *E. coli,* be encoded by the unlinked *tolC* gene. Upstream of the allocrite gene are often found genes encoding proteins which modify the activity of the substrate in some way. This may be by direct covalent fatty acid modification, required for activity of the toxin (Issartel *et al.*, 1991), in the case of HlyA, or a specific inhibitor of proteases in the Prt system (Letoffe *et al.*, 1989).

Another gene shown in **Figure 11.3** is *acp*, encoding the acyl carrier protein essential for fatty acid biosynthesis, which functions, together with HlyC, to activate HlyA by a specific acylation reaction (Issartel *et al.*, 1991). In addition, but not indicated in the figure, SecB is involved in chaperoning some early stage in the secretion of HasA, a heme-binding protein (Delepelaire and Wandersman, 1998), and GroEL, but not SecB or GroES, is implicated in HlyA secretion (Whitehead, 1993).

Many genetic studies have shown that the MFP, OMF and the ABC protein are absolutely required for secretion of allocrites to the medium. The inactivation of any of these proteins, however, leads to accumulation of the allocrite in the cytoplasm and no periplasmic intermediates have ever been reported (Felmlee and Welch, 1988; Gray *et al.*, 1986, 1989; Koronakis *et al.*, 1989). Deletion of the modifying gene encoding HlyC for activating HlyA, on the contrary, has no effect upon secretion (Nicaud *et al.*, 1985).

PROMISCUITY OF THE ABC SECRETION SYSTEM

Several studies have demonstrated that the C-terminal region of HlyA, containing the secretion signal, can promote the HlyBD-dependent secretion of a vast array of peptides and polypeptides, fused N-terminal to the signal (Gentschev *et al.*, 1996; Kenny *et al.*, 1991; Tzschaschel *et al.*, 1996). The secretion signals of PrtB and the S-protein of *Caulobacter crescentus* in targeting fusion proteins to the homologous ABC translocator, appear to be equally promiscuous (Bingle *et al.*, 2000; Delepelaire and Wandersman, 1990; Letoffe and Wandersman, 1992). The size of the allocrite appears not to be limiting since, for example, a β-galactosidase fusion of over 200 kDa is secreted efficiently, although in this particular case the great majority of secreted molecules remain attached to the cell surface, accessible to exogenous trypsin (unpublished, this laboratory). This may reflect a limiting step in the secretion mechanism, the efficient folding of the secreted passenger domain of the fusion (see later section on the form of type 1 proteins during translocation). Indeed, some evidence indicates that the RTX-repetitive, glycine-rich motifs which bind Ca^{2+} upstream of the secretion signal may be required for efficient secretion (Gentschev *et al.*, 1996; Létoffé and Wandersman, 1992). As discussed later, this may be linked to the efficiency of folding of the secreted molecules in a Ca^{2+}-dependent step, following or during late stages in secretion.

In our hands the only consistent failures to secrete a passenger protein fused to the C-terminal of HlyA, via the HlyBD translocator, concerns polypeptides which naturally form dimers or higher multimers, for example glutathione S-transferase (GST) (unpublished data). In some way this form of allocrite is incompatible with the translocator. On the other hand, the position of the secretion signal in the fusion protein appears to be crucial and secretion is blocked when the targeting signal is placed N-terminal to the passenger (Kenny, 1990). Moreover, the presence of a short peptide or even a single amino acid added to the C-terminal can block secretion of different RTX proteins (unpublished, this laboratory; Ghigo and Wandersman, 1994).

In seeking to understand the role of the ABC transporter HlyB in type 1 secretion, it is therefore necessary to take account of this broad range of transport substrates essentially any kind of monomeric polypeptide, which can be secreted provided the specific HlyA secretion signal is present at the C-terminus. We presume that this signal peptide must in some way be capable of docking with the translocator complex of HlyB and HlyD (MFP), in order to initiate translocation across the cytoplasmic membrane, and then the outer membrane, to the external medium. In subsequent sections we shall first consider the nature of the secretion signal itself; we shall then discuss in particular the topology of the membrane domain of HlyB, the structure and function of HlyB from genetic and biochemical analysis, recent progress towards the determination of the structure of the ABC domain, and finally the overall mechanism of secretion of RTX proteins, including the role of the ABC transporter and the auxiliary components of the translocator.

Allocrites for Type 1 Secretion Systems

The majority of *transport substrates* (*allocrites;* defined in Blight and Holland, 1990) for the ABC transporter-dependent export systems vary from polypeptides or peptides to, in some cases, the transport of lipids, or polysaccharides of the β-1,2-glucan type (Young and Holland, 1999). Cluster analysis of the ABC-ATPase domains (see Saurin *et al.*, 1999; Chapter 1, this volume) nevertheless separates the ABC transporters of large bacterial polypeptides from the rest. Concerning the secreted proteins themselves, although otherwise quite different in sequence, virtually all share a characteristic, highly conserved, glycine-rich, 9-residue motif, repeated many times in the region between the C-terminal secretion signal and the upstream biologically active domain. This repeat was first identified in toxins of the HlyA type (Felmlee *et al.*, 1985; Welch *et al.*, 1992), giving rise to the group name of RTX proteins (repeat in toxins) for proteins secreted by the type 1 pathway. The RTX repeats constitute high-affinity Ca^{2+}-binding sites (Baumann *et al.*, 1993), whose deletion may affect the efficiency of secretion. RTX protein is now something of a misnomer since many proteins carrying these repeats are not toxins but, for example, proteases, lipases or cellulases. However, this term, for lack of a better one, will continue to be employed in this review when referring to proteins carrying the specific nona peptide repeats with the consensus sequence GGXGXD(L/I/F)X.

Some important exceptions of proteins lacking the specific RTX repeats include HasA from *S. marcescens* (Letoffe *et al.*, 1994) and cell-associated exopolysaccharide-processing enzymes from, for example, *Rhizobium meliloti* (Geelen *et al.*, 1995). Interestingly, the latter groups nevertheless carry novel repeat motifs also implicated, at least in some cases, in Ca^{2+} binding.

Amongst the largest natural substrates for type 1 transport are the RtxA protein from *Vibrio cholerae* of more than 450 kDa (Fullner and Mekalanos, 2000) and adenyl cyclase toxin from *B. pertussis*, close to 180 kDa (Glaser *et al.*, 1988). β-Galactosidase fused to the C-terminal part of the hemolysin toxin, combined molecular weight 200 kDa, is also secreted efficiently by the HlyB, HlyD translocator on to the external surface of cells (this laboratory, unpublished), although only small amounts are released to the medium (Kenny *et al.*, 1991). At the other extreme are HasA (188 residues; Letoffe *et al.*, 1994), colicin V (a preprotein of 103 residues; Gilson *et al.*, 1990) and short peptides, termed lantibiotics and non-lantibiotics, from Gram-positive bacteria, which will be discussed in later sections.

Type 1 Secretion of Large Polypeptides Involves a C-Terminal Targeting Signal

More than 40 bacterial species secreting RTX proteins have now been identified (see Kuhnert *et al.*, 1997), with in several cases evidence for a specific C-terminal secretion signal also established. Initial deletion studies first identified a novel secretion signal at the C-terminal of HlyA, which included the last 27 residues of the toxin (Gray *et al.*, 1986; Holland *et al.*, 1990; Mackman *et al.*, 1987; Nicaud *et al.*, 1986). This was shown to be essential for secretion of HlyA by the HlyB, ABC transporter. This secretion signal is not, however, removed by cleavage during transport and may indeed be important for folding the secreted protein. Subsequent studies localized the *HlyA secretion signal* to the *C-terminal 50–60 amino acids*, based on mutagenesis studies (see below), deletion analysis and autonomous secretion of C-terminal peptides (Jarchau *et al.*, 1994; Koronakis *et al.*, 1989). Moreover, the presence of such a specific targeting signal for type 1 secretion was confirmed by fusion of variable lengths of the HlyA C-terminus to otherwise non-secretable polypeptides (Kenny *et al.*, 1991; Mackman *et al.*, 1987). Similar studies have subsequently identified C-terminal secretion signals in, for example, the *E. chrysanthemi* PrtG protease (Ghigo and Wandersman, 1994), *Pseudomonas fluorescens* lipase and $HasA_{PF}$ (Omori *et al.*, 2001), and the adenyl cyclase toxin (Sebo and Ladant, 1993). In the case of HasA, unusually cleavage of the C-terminal by extracellular proteases does take place but can occur at several sites. However, there is no evidence that proteolytic cleavage at any of these sites is related to the secretion process (Izadi-Pruneyre *et al.*, 1999) and this phenomenon cannot therefore be used to identify the precise proximal boundary of the secretion signal.

GENETIC ANALYSIS OF TYPE 1 C-TERMINAL TARGETING SIGNALS

As we showed previously (Blight *et al.*, 1994a) comparison of the sequence of the last 60 residues at the C-terminals of several RTX proteins secreted by type 1 pathways identified two major subfamilies (HlyA-like toxins and protease, respectively). A phylogenetic analysis of the terminal domains covering the secretion signal and the RTX repeats of 16 proteins indeed confirmed this separation into two distinct subfamilies (Kuhnert *et al.*, 1997) and this is illustrated in **Figure 11.4**. First, the figure shows that the C-terminal secretion signal of RTX proteins, unlike an N-terminal signal sequence, is not particularly hydrophobic. In addition, the C-terminal region of these groups of allocrites is clearly not conserved at the level of primary sequence. On the other hand, within the HlyA subgroup of very closely related proteins, a few dispersed residues may be conserved, whilst many residues are conserved in the small Prt subgroup. All proteins in the HlyA family can be secreted by HlyB with high efficiency when expressed in *E. coli*. Moreover, despite the even greater divergence in the primary sequence between the two subfamilies, low levels of secretion of the proteases by the Hly-transporter have also been detected, indicating that the HlyB,D transporter can be recognized by the targeting signals of the Prt subfamily (see below).

These two subfamilies of RTX proteins, the HlyA-like and PrtB-like, can also be distinguished by the relative conservation of a particular short, 4–5 residue, motif at the extreme C-terminus. In the case of the HlyA subfamily, this C-terminal contains a preponderance of

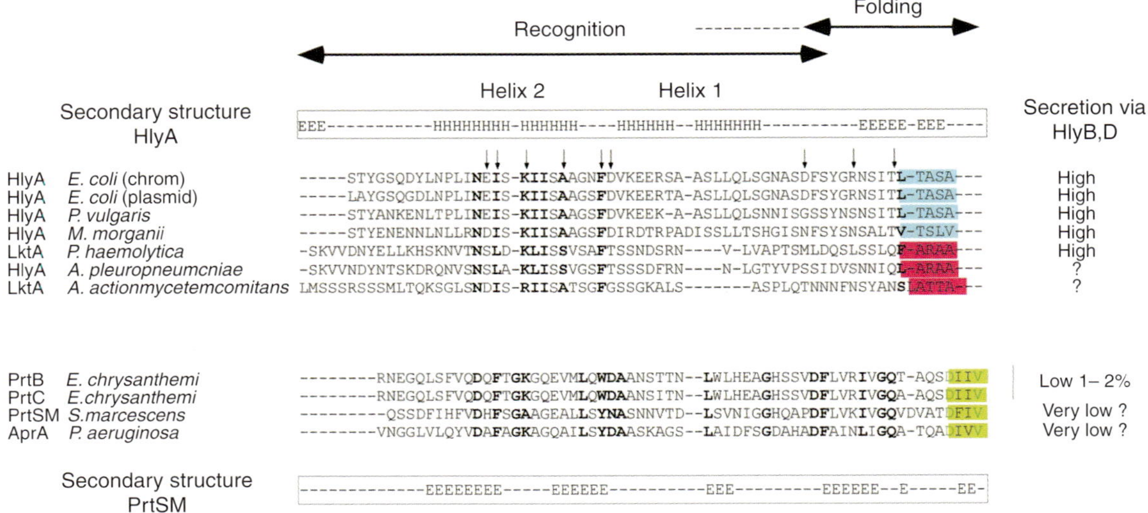

Figure 11.4. Alignment of C-terminal secretion signal regions of two major families of RTX proteins, the Hly toxin and Prt protease families. Strongly conserved residues in bold and the extreme C-terminals are highlighted in color (see text for more details). The secondary structure is predicted for HlyA; that for PrtSM is based on the structure of the secreted proteases (Baumann, 1994). The division of the signal region into recognition and folding functions is based on genetic analysis discussed in the text. Downward arrows indicate the sites of point mutations which can reduce secretion levels of HlyA substantially. To the right is indicated the level of secretion of these proteins transported by the heterologous HlyB, D, TolC translocator (see text for other details). H, helix; E, β-strand.

hydroxylated residues (Ser, Thr). Alanine is almost invariably the terminal residue, although we have shown that this can be replaced by proline in HlyA without effect on secretion (Chervaux and Holland, 1996). In contrast to the HlyA type, as illustrated in **Figure 11.4**, the proteases such as PrtB (*E. chrysanthemi*) contain at the C-terminus, three hydrophobic residues preceded by an aspartate. Whilst such C-termini may be characteristic of particular subfamilies, as shown in **Figure 11.5**, when a broad spectrum of proteins, representing most of the different types of large and small (peptides) molecules secreted by the type 1 system, are compared with respect to the C-terminal, it is clear that *no primary sequence motifs* of any kind are detectably conserved. Indeed the figure indicates the remarkable lack of conservation overall.

Returning to the HlyA and PrtD subfamilies, as shown in **Figure 11.4** these also differ markedly in terms of secondary structure, in that the C-terminal 60-residue peptide of the HlyA group is predicted to be largely helical, whilst that of the PrtB group is largely β-strand. Crystal structures of a number of the latter group of RTX proteases have confirmed this β-strand structure in the *mature, folded protein* (Baumann, 1994; Baumann *et al.*, 1993, 1995). Nevertheless, the actual structure of the secretion signal for type 1 substrates as it presents itself to the translocator *in vivo*, before the polypeptide folds, remains to be determined, although some *in vitro* evidence, as described in the next section, indicates that this may be largely devoid of secondary structure.

The most detailed genetic analysis of the function of the type 1 secretion signal has concerned HlyA, the hemolysin toxin, secreted by uropathogenic strains of *E. coli*. Many point mutations in the C-terminal 60 amino acids have been isolated by saturation mutagenesis, with the majority having little or no effect on the detectable level of secretion of the toxin (Chervaux and Holland, 1996; Kenny *et al.*, 1992, 1994; Stanley *et al.*, 1991). Moreover, large deletions into either the proximal or distal regions of the C-terminal 50–60 residues, although substantially reducing secretion, still permit detectable levels of transport of allocrites such as HlyA (Koronakis *et al.*, 1989; Zhang *et al.*, 1993). On the other hand, a few point mutations were found to reduce secretion levels by 50–70%, including replacement of F^{989}, which is completely conserved in all members of the HlyA subfamily of very closely related toxins, by several different residues (Blight *et al.*, 1994a; Chervaux and Holland, 1996). By combining three mutations, $E^{978}K$, $F^{989}L$ and $D^{1009}R$, Kenny *et al.* (1994) (see **Figure 11.4**) were able to reduce secretion

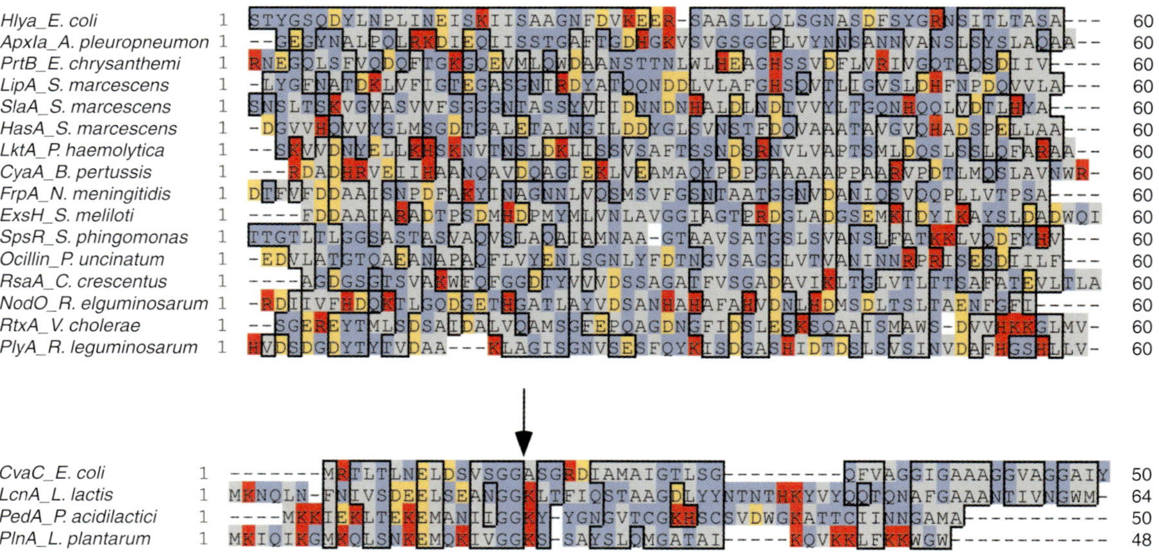

Figure 11.5. Comparison of C- and N-terminal secretion signals for the type 1 pathway. C-terminal targeting signal regions of a wide range of proteins (upper blocks) and the N-terminal signal region of small antibacterial peptides (lowers blocks), terminated by the GG, cleavage motif, also secreted by an ABC transporter complex. Acid and basic residues in yellow and red respectively, hydrophobic residues in gray, others in blue.

levels of HlyA to less than 1% of wild type. The additive effect of these point mutations in reducing secretion levels provided the best evidence that the minimum secretion signal covers at least 32 residues. Such mutations were also individually incorporated into the C-terminal signal region of a LacZ–HlyA fusion (containing the 23 kDa C-terminal of HlyA) and co-expressed in cells in competition with wild-type HlyA toxin. This competition experiment showed that all three mutations were recessive, since the LacZ fusion carrying them failed to affect secretion of the wild-type toxin. We concluded that these three residues were specifically required for *docking* with the *translocator* (Kenny *et al.*, 1994).

Stanley *et al.* (1991), in an alternative view, formulated a much more complex model for the function of the HlyA secretion signal. This was based on predictions of a single large amphipathic helix between residues −49 and −23 (now in fact accepted as a helix-turn-helix, see below), and secretion levels of mutated HlyA, with primarily multiple mutations and several frameshift mutations (generating novel sequences of varying lengths from position −20 or later).

This model essentially envisaged an interaction with HlyB restricted to the C-terminal eight residues. On the other hand, the model visualized the proposed amphipathic helix targeting the bilayer, looping first through the inner membrane, then the outer membrane, triggering fusion of the membranes and ensuring in some way direct extrusion of the rest of HlyA to the exterior. First of all, in our view, the use of such complex mutants, combined with a relatively insensitive secretion assay, makes interpretation of the results of such an analysis difficult, if not impossible. In addition, subsequent genetic and structural studies of the termini of different RTX proteins have not confirmed the presence of a conserved amphipathic helix, which might conceivably play such a role. Therefore, in line with the generally agreed sequence redundancy, the lack of hydrophobicity and lack of any obviously *conserved* primary or secondary structure in the type 1 C-terminals, we would continue to argue that docking with the translocator, involving a few residues at key positions in the secretion signal of perhaps about 50 residues, is the most likely basis for initial recognition of the translocon, the triggering of the activation of the ABC-ATPase and entry of the allocrite into the transport pathway.

The secretion signal is a largely unstructured peptide

Figure 11.4 shows the now generally accepted view that the C-terminal of HlyA itself is predicted to contain a helix (helix 2) with potential amphipathic properties, separated by a short turn from a second helical region (helix 1). In our saturation mutagenesis studies, mutations in the region of *helix 1* had little effect on secretion (see also Stanley *et al.*, 1991), whilst *helix 2* and the adjacent linker region, containing the essential F^{989}, appeared to constitute a *hot spot* for residues required for secretion. However, the analysis of the nature of the mutations and their effects on secretion did not appear to correlate with potential amphipathic properties of helix 2 (Chervaux and Holland, 1996; Kenny *et al.*, 1992). A more extensive study, using a combinatorial approach to vary the sequence of the HlyA targeting signal in the region of the two predicted helices in HlyA, (Hui *et al.*, 2000), confirmed the importance of the most proximal helix 2 and the adjacent linker for efficient sercretion, whilst changes to the distal helix 1 had little effect. This study did provide some support for the importance of the amphipathic nature of helix 2. Nevertheless, it is important to emphasize that a role for specific secondary structures in the recognition of the translocon by the allocrite has not generally been supported so far by structural studies. Thus, CD and NMR analysis of isolated RTX secretion signal peptides have indicated an unstructured peptide under aqueous conditions (Izadi-Pruneyre *et al.*, 1999; Wolff *et al.*, 1994, 1997; Zhang *et al.*, 1995; this laboratory, unpublished). In addition, the absence of overall conservation of type 1 secretion signals at the level of secondary structure, combined with the fact that several examples of the secretion of non-cognate allocrites by heterologous translocators have been reported, albeit at lowered efficiency, supports the idea that a specific secondary structure is not essential for docking with the MFP/ABC translocator. We therefore envisage a secretion signal *in vivo* that is relatively unstructured, with docking with the translocator dependent, as proposed previously, upon the side-chains of a few specific amino acids. This would provide a mechanism reminiscent of class I peptide antigen docking with the MHC-complex (see Chapter 26) in the endoplasmic reticulum.

From the foregoing discussion it is clear that final resolution of the structural (primary or secondary) determinants of the secretion signal, and in particular those that interact with the translocator, will require co-crystallization of the C-terminal of an RTX protein with the relevant portions of the ABC translocator (both HlyB and HlyD) involved in initial recognition (see below).

MUTATIONS CAN ALTER THE SPECIFICITY OF AN ALLOCRITE FOR DIFFERENT TRANSLOCATORS

The evidence discussed above clearly emphasizes the lack of detectable structural features essential for functioning of type 1 secretion signals. This is further underlined by several examples of the secretion of allocrites, albeit at reduced efficiency, by heterologous transporters (Duong *et al.*, 1994, 1996; Fath *et al.*, 1991; Guzzo *et al.*, 1991). Examples of such promiscuity include low levels of crossover between the putatively β-strand and helical structured signals of the Prt and Hly families, respectively. Moreover, substitution of the HlyA C-terminal for that of a leucotoxin from *Pasteurella haemolytica*, having a completely different primary sequence, permits almost wild-type levels of secretion of the haemolysin hybrid by the HlyB,D system (Zhang *et al.*, 1993). On the other hand, an interesting example of a specificity determinant was revealed by a study of a variety of quite different allocrites secreted naturally by a single ABC-dependent translocon, LipBCD, in *S. marcescens.*

In this case a particular triplet motif with an invariant N-terminal valine, located approximately 19 residues from the C-terminus, is essential for efficient secretion through the Lip translocator. Moreover, insertion of the motif VAL converts HasA from *S. marcescens*, normally not secreted by Lip, into an efficient allocrite for secretion. Finally, it was demonstrated in competition experiments that this motif was required for recognition of the cognate translocator (Omori *et al.*, 2001). Nevertheless, this study did not exclude the existence of other important motifs within the C-terminal 50 residues (except the extreme five residues, which appeared dispensible) necessary for secretion via LipB. The results moreover are not in conflict with the idea that a few key residues, dispersed throughout the signal region, play a key role in recognition of the translocator as proposed for HlyA (see below).

THE RTX SECRETION SIGNAL HAS A DUAL FUNCTION

In contrast to the recessive mutations described above which reduced HlyA secretion, another mutation, *hlyA99* (Chervaux and Holland, 1996), containing four substitutions in the final six C-terminal residues, which produced greatly reduced halo sizes on blood agar plates, was dominant in competition experiments. Thus, the LacZ fusion carrying the HlyA99 mutation at the C-terminal inhibited secretion of wild-type HlyA coexpressed in the same cells (Chervaux, 1995). This suggested that this mutant can still recognize and enter the translocator but is defective at a late stage in secretion. In fact, subsequent studies showed that HlyA99 is defective in hemolytic activity due to *incorrect folding* of the protein, rather than in secretion (this laboratory, unpublished). Moreover, the passenger protein β-lactamase, fused to the C-terminal of wild-type HlyA, has β-lactamase activity in the culture supernatant but the enzyme is inactive when fused to the HlyA secretion signal carrying the HlyA99 mutation (C. Chervaux and I.B. Holland, unpublished).

As indicated above, we concluded from the genetic analysis of HlyA that the three amino acids E^{978}, F^{989} and D^{1009} encompass a region extending at least from residues −15 to −46, with respect to the C-terminus, which is essential for recognition and docking with the translocator. The results with HlyA99, in contrast, indicated that at least the most distal 5–6 C-terminal residues of HlyA may be involved in a second function promoting the folding of the secreted polypeptide. On the other hand, deletion of the terminal six residues of HlyA (Stanley *et al.*, 1991) was reported to reduce secretion of the polypeptide by about 70% (activity was not tested). Ghigo and Wandersman (1994) also reported that deletion of the four C-terminal residues of PrtG, DFLV (representing a conserved motif, D,hb,hb,hb, restricted to the proteases of the PrtA subfamily) completely blocked secretion. These results may indicate a true secretion function for the residues at the extreme C-terminal of RTX proteins, whilst not ruling out an additional role in folding the secreted protein. However, it should be noted that failure to detect a protein in the culture supernatant may be an insufficient indication of a defect in secretion of type 1 allocrites. In the case of the LacZ–HlyA fusion, large amounts of secreted molecules remain tightly bound to the external cell surface after secretion by the HlyA type 1 system (perhaps especially if incorrectly folded) and are therefore not detected in the medium (unpublished this laboratory). Interestingly, Omori *et al.* (2001) have analyzed the signal of an allocrite secreted by LipB in *S. marcescens*, with the sequence ELLAA at the C-terminus, and found that elimination of the glutamate or its re-positioning in all possible positions in the downstream sequence had no detectable effect upon the secretion of the lipase polypeptide. These authors concluded that this C-terminal motif was not therefore involved in secretion. Unfortunately, Omori *et al.* did not report whether the mutations had any effect on the activity/folding of the secreted lipase.

In our view, therefore, it remains a possibility that the C-terminal 40–50 residues of the type 1 polypeptides may include *overlapping functions*, a targeting *signal* as well as an important element in promoting *folding* of the secreted polypeptide. In fact, the specificity required for an interaction of the signal sequence of type 1 proteins in *trans* with the cognate translocator, and in particular for an interaction in *cis* with its own N-terminal domain, which is required for final folding, might be expected to produce marked sequence divergence in the C-terminal during evolution.

Several Antibacterial Peptides Secreted via the Type 1 System Employ an N-Terminal Targeting Signal

In previous sections we have discussed the evidence that many polypeptides, including the so-called RTX proteins, carry a non-processed, C-terminal signal, targeting these allocrites to the ABC transporter complex. Placing such a signal at the N-terminus, or even short extensions to the signal at its normal C-terminal position, blocks its function (this laboratory unpublished; Sebo and Ladant, 1993). Nevertheless, more recently it has become clear that Gram-positive non-lantibiotics, plus some lantibiotics and related antibacterial peptides in Gram-negative bacteria, are secreted through an ABC pathway, dependent on an N-terminal targeting signal. These compounds, previously designated bacteriocins, are now more correctly defined as microcins, owing to their small size. In distinction to non-lantibiotics, lantibiotics are characterized by major modifications to a number of amino acids. Non-lantibiotics and a few lantibiotics carry a specific hydrophilic leader peptide of 15–30 residues. This includes some conserved residues and is terminated by two glycines (Havarstein *et al.*, 1995; see **Figure 11.5**). This leader is cleaved apparently during transport by a cysteine protease which, remarkably, constitutes the N-terminal (cytoplasmic) extension of the ABC transporter itself. This cleavage occurs immediately following the two glycines, and, interestingly, mutations which abolish the cleavage site in colicin V also block secretion (Gilson *et al.*, 1990), suggesting that this region is involved directly in targeting or that cleavage is a prerequisite for subsequent docking with the translocator. Evidence that the leader peptide of the double glycine type does constitute a secretion signal was provided by the demonstration that colicin V, leucocin A and lactococcin A leader peptides, fused to the N-terminal of a bacteriocin normally secreted by a different pathway, promoted its secretion now by the ABC pathway (van Belkum *et al.*, 1997). Evidently, in these cases the allocrite signal is recognized by the N-terminal protease domain of the ABC protein. However, no other information is available in relation to other possible docking sites and these are not excluded. Interestingly, secretion of the microcins with the double glycine leader peptide from Gram-positive bacteria also requires an *MFP homologue* as an *essential accessory* (see, for example, Franke *et al.*, 1996), even though the outer membrane, with which the MFP interacts in *E. coli*, is absent in Gram-positive bacteria.

Members of a second group of lantibiotics are also secreted via an ABC transporter but without a requirement for an MFP accessory. In this system secretion is dependent on a hydrophilic but distinctive leader peptide, which is eventually also removed by cleavage. However, in this case cleavage takes place *after* secretion of the pre-peptide to the medium by a specific, independently encoded serine protease, in a reaction which is not apparently linked to the secretion process (van der Meer *et al.*, 1994). It should be emphasized that *direct* evidence that the N-terminal of this second group of lantiobiotics constitutes a specific secretion or targeting signal is apparently still lacking.

Composition of the Type 1 Secretion Translocator

A variety of studies have now provided evidence that the three presumed components of the translocator, the ABC, MFP and OMF (see **Figure 11.1**), do indeed interact, although this complex has not yet been purified and reconstituted in an *in vitro* system. Remarkably, all three proteins, together with the allocrite itself, HasA, PrtC or HlyA, can be co-purified from membranes, using an affinity tag (Letoffe *et al.*, 1996) or following *in vivo* crosslinking (Thanabalu *et al.*, 1998). The choice of method may depend upon the procedure for solubilization of membrane proteins; inclusion of urea, for example, may prevent recovery of the complex unless previously crosslinked. Detailed studies using either procedure have demonstrated clearly that HlyB and HlyD interact, even in the absence of TolC or HlyA (Thanabalu *et al.*, 1998; Young, 1999). Simultaneous co-purification of all three components of the translocator, the ABC (HlyB), MFP (HlyD) and OMF (TolC), however, was shown to require the presence of the allocrite. This finding was used to develop an interesting model which proposes that the incorporation of the OMF into the translocator in a crosslinkable form only occurs when required, and is presumably triggered by the allocrite after initial interaction with the ABC/MFP complex (Balakrishnan *et al.*, 2001; Thanabalu *et al.*, 1998).

Other studies concerning the assembly of the complex have, however, been more contradictory, in some cases suggesting that TolC, HlyD and HlyB may interact in some way even in the absence of the allocrite. Thus, whilst the ABC and MFP have been shown to mutually stabilize each other (Hwang *et al.*, 1997; Pimenta *et al.*, 1999), further suggesting an interaction between these proteins, the stability of HlyD, involved in hemolysin secretion, also requires TolC. HlyD becomes extremely labile in the absence of TolC when HlyB (ABC) is also present, suggesting an HlyB:HlyD interaction which affects the structure of the latter, including the promotion of its oligomerization (see below). Notably, these effects on the stability of HlyD are observed in the absence of the allocrite (Pimenta *et al.*, 1999). This may indicate, in contrast to the studies of Thananbalu *et al.* described above, that HlyD and TolC do indeed interact to produce some form of complex in the absence of the allocrite. Other more indirect evidence supports this view. Strains expressing the Hly genes and TolC are hypersensitive to vancomycin, an antibiotic normally too large to penetrate the outer membrane effectively. An analysis of this effect suggested that the antibiotic can use a TolC channel, dependent on both HlyB and D, to cross the outer membrane (Wandersman and Letoffe, 1993). Attempts to determine which components of the translocator or the allocrite itself are required for vancomycin uptake have unfortunately given conflicting results. Schlor *et al.* (1997) demonstrated that vancomycin sensitivity, apparently dependent upon TolC and HlyD, did not require HlyA, suggesting that a TolC, HlyD interaction occurred independently of active secretion of the allocrite. Similarly, Wandersman and Letoffe (1993) concluded that HlyA was not required for vancomycin sensitivity. In contrast, Blight and co-workers (Blight *et al.*, 1994b; Pimenta *et al.*, 1999) demonstrated that only cells expressing and actively secreting HlyA were hypersensitive to vancomycin, a result more in favor of the idea that the allocrite is required for the recruitment of TolC to form a fully functional *trans*-envelope channel. The disagreement between these studies concerning the role of HlyA in recruiting TolC remains unresolved.

Regarding the stoichiometry of the Hly translocon, TolC itself has been shown to form trimers (Koronakis *et al.*, 1997), whilst HlyB is likely to form dimers. In detailed studies in this laboratory, we have shown that both TolC and HlyB (but not HlyA) are required for the detection of HlyD dimers, trimers and possibly tetramers following DSP crosslinking. Moreover, several *hlyD* or *hlyB* point mutations were shown to abolish this HlyD multimerization (Young, 1999). Thananbalu *et al.* (1998) also reported the formation of HlyD trimers, employing the crosslinker DSG, which has a shorter fixed arm spacer than DSP. However, these authors found trimer formation to be independent not only of HlyA, but also of TolC and HlyB. This discrepancy is puzzling but could be reconciled if HlyD trimers simply become more compacted and easier to crosslink with DSP in the presence of HlyB, consistent with other evidence that interaction with HlyB induces some structural change in HlyD rendering it more labile to endogenous proteases in the absence of TolC (Pimenta *et al.*, 1999).

As a result of all these studies, it seems likely that the *stoichiometry* of the type 1 translocator, ABC:MFP:OMP, is 2:3:3, although the presence of more HlyD subunits is not excluded. Whilst we may presume, as discussed below, that the ABC component provides energy and is possibly an integral component of the transport pathway, different studies of the type 1 translocator indicate that HlyB also specifically interacts in this system with the accessory MFP component, with, at least in our hands, a resultant change in the latter's structural and oligomerization state.

Both MFP and ABC Components of the Type 1 Translocator may be Involved in Recognition of the Secretion Signal of the Transport Substrate

Several mix and match experiments to investigate the *in vivo* function of various combinations of the ABC, MFP and OMF proteins from two different type 1 secretion systems indicated that the ABC protein played a particularly important role in the secretion of the homologous allocrite (Binet and Wandersman, 1995). This was taken to indicate the recognition of the signal peptide by the ABC protein, whilst not ruling out a complementary role for the MFP in initial recognition. Indeed, it is well established that deletion of either the MFP or the ABC transporter components of the translocator causes accumulation of the corresponding allocrite in the cytoplasm, suggesting that both components could be involved in initial recognition of the secretion signal. With respect to the ABC component, Letoffe *et al.* (1996) have provided some biochemical evidence for the binding of the protease PrtC to the ABC protein. On the other hand, certain point mutations in the periplasmic domain of HlyD apparently block secretion of HlyA at an early step in the secretion process (Pimenta, 1995). Moreover, in this laboratory we have shown that deletion of the first 40 N-terminal residues of HlyD block secretion (Pimenta *et al.*, 1999; Young, 1999). In addition, other studies, most recently by Balakrishnan *et al.* (2001), have clearly provided evidence for a role for HlyD in an early step in secretion of hemolysin. Thus, HlyD even in the absence of HlyB recruited HlyA into a complex that could be crosslinked *in vivo*. In addition Balakrishnan *et al.* identified a cytoplasmic region of HlyD (residues 1–45) necessary for this interaction. Moreover, in the absence of this region, the HlyB,D_{del} in the presence of HlyA fails to recruit TolC into a crosslinkable complex. The authors proposed the exciting idea that the N-terminal of HlyD is implicated in transduction of a signal (generated by HlyA binding) across the cytoplasmic membrane to the periplasmic domain of HlyD in order to effect recruitment of TolC into a functional *trans*-envelope channel or in our interpretation, the *stabilization* or activation of a pre-existing HlyD–TolC channel. Finally, the results of that study also indicated that the detection of HlyB,D complexes did not require the N-terminal 45 residues of HlyD, indicating that the region of interaction between these proteins was in the membrane or periplasmic regions.

General Organization of ABC Transporters for Type 1 Secretion: Some Have a Specific Protease Domain

ABC transporters involved in the type 1 secretion pathway are invariably so-called half transporter polypeptides. These are single polypeptides composed of an N-terminal domain of approximately 320 residues, apparently containing six transmembrane segments, extended in some cases by an N-terminal region of 130–150 residues, which might contain additional transmembrane segments, fused to a highly conserved ABC-ATPase domain of approximately 260 residues. HlyB, for example (see **Figure 11.6**), contains the extended N-terminal region of approximately 130 residues compared with PrtD (and Pgp/Mdr1, for example). There is no clearly defined function for this region and as described later, there is contradictory genetic evidence concerning any specific role for this region in the secretion of the hemolysin toxin. Interestingly, the *E. coli* colicin V ABC transporter and ABC transporters in Gram-positive bacteria, required for secretion of certain antibacterial peptides, also contain an extended N-terminal region (see **Figure 11.6**). In fact, this region has been shown to constitute a conserved serine protease domain, required for the intracellular cleavage of a specific leader peptide, apparently essential for ultimate secretion of these peptides (Havarstein *et al.*, 1995; Zhong *et al.*, 1996). The HlyB N-terminal domain shows little similarity with these protease domains. Moreover, it lacks the highly conserved cysteine residue essential for activity in this protease family (Havarstein *et al.*, 1995). We can therefore discount the possibility that this HlyB domain is a cysteine protease.

TOPOLOGY OF THE MEMBRANE DOMAIN OF HLYB

In considering the topology of HlyB it is important to emphasize that ABC transporters in bacteria engaged in export, including the type 1 secretion systems, in contrast to ABC-dependent importers (see Chapter 9), do not contain any conserved EAA motif in the membrane domain. Any corresponding motif, implicated in signaling between the membrane domain and the ABC-ATPase as demonstrated for HisP and MalF, has yet to be recognized in the export family.

Reported detailed topology studies of ABC transporters for type 1 secretion mechanisms have been largely restricted to the HlyB protein. As discussed previously (Holland and Blight, 1996), hydropathy plots of many ABC transporters, certainly including HlyB and some of its close relatives, do not give clear-cut indications of the position and number of membrane-spanning regions. Alignments using the most recent algorithms confirm this ambiguity in comparison with membrane proteins of known structure such as bacteriorhodopsin (T. Molina and I.B. Holland, unpublished). This difficulty may reflect the probable presence of large intermembranous loops in ABC transporters (see Chapter 2). On the other hand, it has been proposed that the transmembrane segments (TMSs) may contain significant amounts of β-strand structure rather than the conventional helices (Jones and George, 1998), although as discussed in Chapter 12, for the multidrug transporter LmrA in *Lactococcus lactis*, analyzed by ATR-FTIR spectroscopic techniques, this does not seem to be the case (Grimard *et al.*, 2001). Moreover, the recent exciting appearance of the first crystal structure to include the membrane domain (see Chapter 7) has shown the presence of six α-helices, spanning the bilayer in the lipid A transporter MsbA from *E. coli* (Chang and Roth, 2001).

We have previously sought to determine the topological organization of HlyB using β-*lactamase fusions* targeted to 29 positions throughout the predicted membrane domain of HlyB (Wang *et al.*, 1991). The results indicated the positioning of the ABC-ATPase domain in the cytoplasm and the presence of six (numbers 1–6) distal TMSs. Four of these were in relatively good agreement with those predicted by simple hydropathy analysis, whilst the positions of TMS 2 and 4 were clearly not. In addition, the results indicated two more TMSs (TM_{x1}, TM_{x2}), close to the N-terminus, which were not all predicted from simple hydropathy profiles, although they are predicted by some algorithms (Holland and Blight, 1996). A subsequent study by Gentschev and Goebel (1992), using alkaline phosphatase and β-galactosidase fusions, nevertheless also detected eight possible TMSs in HlyB, positioned in most cases in reasonable agreement with the β-lactamase data. In **Figure 11.7**, we have combined all these results to produce the composite, best-fit figure for all the published experimental data, with the added assumption that most TMSs will be composed of 25 amino acids. This model differs from the original proposal of Wang *et al.* (1991), in that TMS 2 and 4 are positioned more towards the C-terminal with consequent reduction in the size of the external domain P1 and an increase in the cytoplasmic domains C2 and C3. This topology indicates the presence of two similar-sized, relatively small, external loops, Px and P1, and two very short, 3- and 8-residue loops, P2 and P3. Loops P1 to P3 in particular are candidates for interactions with HlyD to form the continuation of the translocator through the periplasm and outer membrane. The cytoplasmic domains

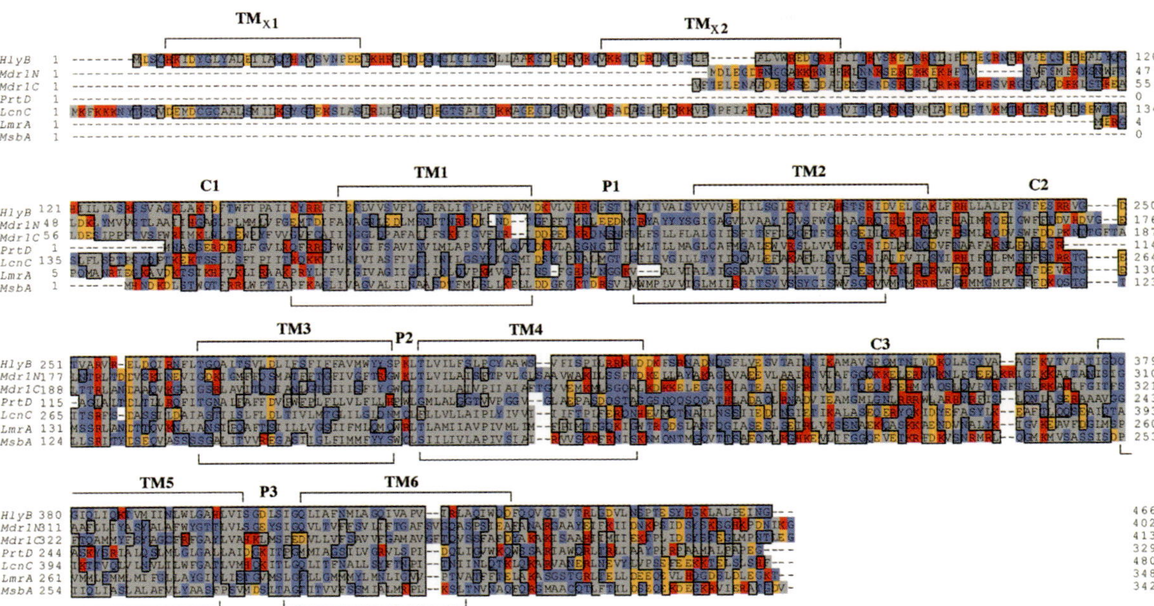

Figure 11.6. Alignment of the membrane domains of several important ABC transporters. HlyB (secretion of large toxin), Mdr1 (multidrugs, mammals), PrtD (protease), LcnC (non-lantibiotic), LmrA (multidrugs, bacterial), MsbA (Lipid A). Similar residues are boxed. The figure shows the extended but unrelated N-terminal regions of HlyB (function unknown) and LcnC (protease for release of the N-terminal secretion signal of the non-lantibiotic transported). Above blocks, positions experimentally determined for TMSs (and periplasmic loops, P, cytoplasmic loops, C) of HlyB, with below the blocks, the position of the TMSs for MsbA from the crystal structure (Chang and Roth, 2001). Color code as in Figure 11.5. Note the remarkable homology between HlyB and Mdr1 (N), extending from mid-TM5 into TM6.

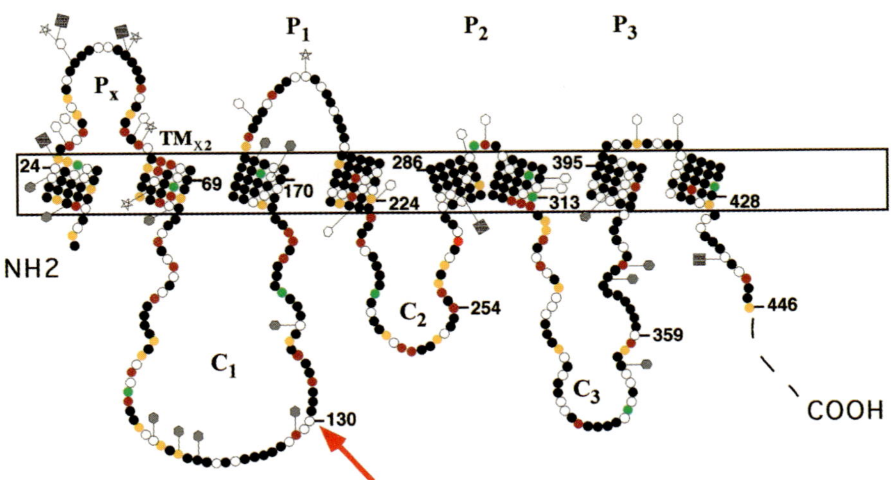

Figure 11.7. Topological organization of the HlyB membrane domain calculated from fusion analysis. The position of β-lactamase (βla) fusions giving rise to resistance (external βla) to ampicillin is indicated by open hexagons; fusions associated with ampicillin sensitivity (internal β1a) by solid hexagons (Wang et al., *1991); active phoA fusions (external) by stars; lacZ fusions (internal) by solid squares (Gentschev and Goebel, 1992). Assumption is made that all TMSs (TM$_{x1,x2}$, 1–6, left to right) except TMS 2 (30) are 25 residues. Residues in yellow, Asp and Glu; red, Arg and Lys; green, proline; black, hydrophobic; open circles, polar residues. Note that the region in HlyB, N-terminal to residue 130 (arrow), is absent from MsbA and many other ABC transporters; in this region TM$_{x1}$ and TM$_{x2}$ are poorly predicted by most algorithms.*

of HlyB are predicted to include a large 86-residue 'loop' (CI) two similar-sized 'loops' (C2 and C3) and the ABC-ATPase domain, commencing at approximately residue 460.

The experimentally determined TMS 1, 3, 5 and 6 in HlyB are positioned in line with most predictive algorithms based on hydropathy. TMS 2 and 4 are still shifted significantly downstream of those predicted. In fact, TMS 2 and 4 in the model are predicted to contain a number of charged residues, possibly raising questions about their reality. In addition the model in **Figure 11.7** predicts the presence of proline residues in TMS 1, 4 and 6, which is unusual. However, in this respect it is interesting to note that the crystal structure of MsbA indeed includes transmembrane helices containing up to four charged residues, and proline residues are present in TMS 2, 4 and 6 as shown in **Figure 11.6**. Nevertheless, alignment of MsbA with Mdr1, HlyB and other ABC transporters like LmrA, for example, indicates that these proline and charged residues are not conserved in the predicted TMSs. Therefore, it is not possible to make any generalizations regarding a special requirement for charged residues in membrane-spanning domains of ABC transporters. Remarkably, as clearly also shown in **Figure 11.6**, the experimentally determined TMSs for HlyB in the model line up very closely with those revealed by the MsbA structure, giving some confidence that these may be correct.

GENETIC ANALYSIS OF HLYB

Genetic analysis of HlyB could ultimately provide an informative basis for the dissection of its function. This is expected to include a possible interaction with the MFP HlyD, docking with HlyA involving one or both domains of HlyB, and the coupling of the energy of ATP hydrolysis to translocation of HlyA, signaled perhaps by direct interaction between the membrane and ABC domains of HlyB. In the complete absence of HlyB, hemolysin secretion is abolished and the HlyA polypeptide accumulates in the cytoplasm. However, as described below, the analysis of HlyB mutations restoring (suppressing) the secretion of HlyA with a defective targeting signal has so far failed to identify a specific docking site. Other studies have mostly involved random mutagenesis in attempts to identify regions of HlyB implicated in the secretion process.

From the topology studies described above, the HlyB molecule can be subdivided into approximately four regions: the first 100–150 residues of the N-terminal region, predicted from some experimental data and by some algorithms to contain two TMSs; the following membrane domain (approximately residues 150–440) encompassing six calculated TMSs, including some conserved residues, in particular in the cytoplasmic loops, C2, C3, found in other ABC-transporters (see **Figures 11.2, 11.6**; data not shown); a linker region (440–467); and the C-terminal (cytoplasmic) 27 kDa, ABC-ATPase (residues approximately 467–707).

MUTATIONS AFFECTING THE CONSERVED RESIDUES OF THE ABC DOMAIN

Several mutations in the ABC domain affecting the signature motif (LSGG-), the Walker A and B motifs and the highly conserved His^{662} in the switch region (see **Figure 11.2**), all block secretion *in vivo* and ATPase activity *in vitro* (Koronakis *et al.*, 1995; this laboratory, unpublished data), demonstrating that fixation or hyrolysis of ATP is essential for secretion of HlyA. Interestingly, $P^{624}L$ (Blight *et al.*, 1994b) is an identical substitution to that described by Ames and co-workers in HisP (Petronilli and Ames, 1991; Shyamala *et al.*, 1991). This mutation renders the ATPase activity of HisP constitutive *in vitro* and histidine import independent of HisJ, the periplasmic binding protein, *in vivo*. This Pro residue is conserved in many ABC domains and is present in a loop (we designate this the *Pro-loop*) joining the Walker B motif in the catalytic domain to the signature motif in the regulatory domain (see **Figure 11.11**), i.e. a position perhaps critical for intramolecular signaling.

MUTATIONS AFFECTING THE MEMBRANE DOMAIN

The most N-terminal region of HlyB (approximately the first 130 residues) is absent from most other ABC transporters, including some of its close relatives involved in secretion of polypeptides by the type 1 secretion pathway. This suggests that this region may not have a fundamental role in the secretion mechanism. Indeed, we showed that replacement of the first 25 residues of HlyB by the first 21 residues of

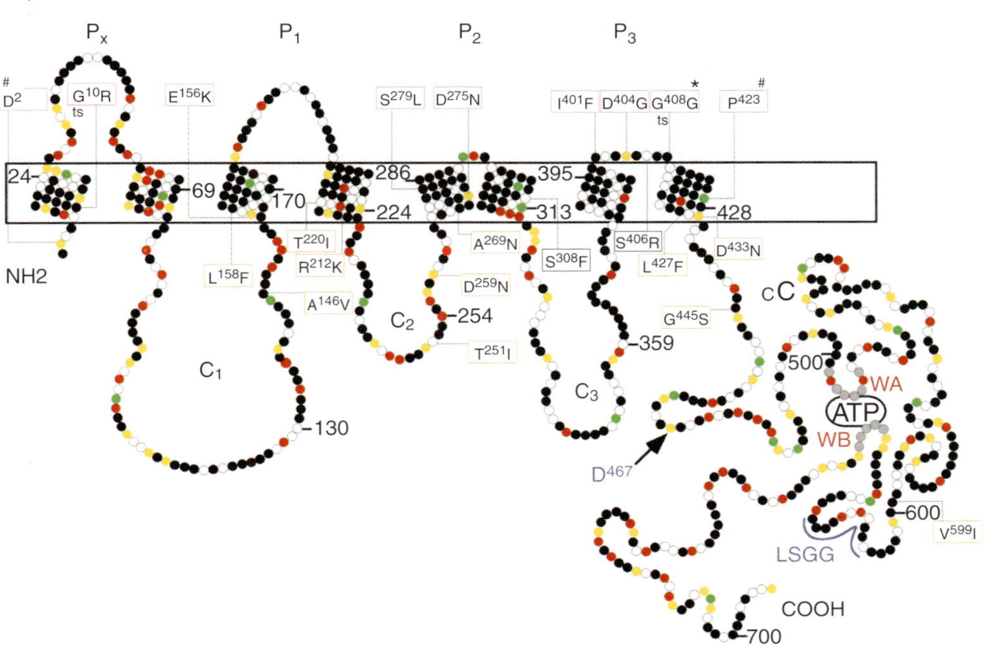

*Figure 11.8. Genetic analysis of HlyB. The positions of mutations localized to the membrane domain are shown, with mutations affecting the conserved motifs in the NBD omitted for clarity (but see text). Mutated residues (pink boxes) presented above 'bilayer' containing the TMD confer a secretion defect, whilst those presented below do not reduce secretion. Replacement of the first 21 residues of HlyB by the λ-Cro sequence does not affect secretion. ts, mutants which are defective in secretion at 42°C; *, defective in secretion at 37°C and cells fail to grow at 42°C; #, linker insertions; mutations suppressing HlyA signal mutations are boxed in yellow; mutations $E^{156}K$, $S^{279}L$ affect the oligomerization of HlyD. The arrow marks the position, D^{467}, of the N-terminal of the isolated ABC domain described in the section on properties of the purified ABC domain. For other details see text and summary of genetic analysis in Holland and Blight (1996).*

the λ-Cro protein – presumably removing the first putative TMS – had no effect on secretion **(see Figure 11.8)** activity (Blight *et al.*, 1994b). Similarly, fusion of GST to position 3 at the N-terminus of HlyB allows targeting to the membrane and apparently normal activity in the secretion of HlyA *in vivo* (Young, 1999). Conversely, insertion of the C494 epitope at the N-terminal rendered the HlyB molecule unstable and therefore defective in secretion (Juranka *et al.*, 1992). Somewhat paradoxically, a G-R mutation (temperature sensitive) at position 10 blocked secretion of HlyA at 42°C (although other evidence indicated that the protein was still assembled into the membrane), which may indicate that the N-terminal region does have a secretory function under some conditions, although other interpretations are possible (Blight *et al.*, 1994b). As described above, this N-terminal region of HlyB resembles in size the protease domain of some microcin transporters. However, as also indicated before, this region in HlyB lacks the critical active site residues of the serine proteases and it seems unlikely that this functions as a protease.

As summarized in **Figure 11.8**, mutations involving the membrane domain were also isolated by random or directed mutagenesis, with the majority of these giving a major reduction in secretion of HlyA (Blight *et al.*, 1994b). Juranka *et al.* (1992) also investigated mutations in *hlyB* by the creation of 6–7 amino acid residue (KpnI site) insertions, one in TMS 6 and six others in the ABC domain, and all abolished secretion. The initial random mutagenesis (Blight *et al.*, 1994b) involved a further screen for temperature-sensitive secretion and in addition to $G^{10}R$ and $P^{624}L$ described above, $G^{408}D$ was isolated. The G^{408} mutant was in fact conditionally lethal for the bacteria (at 42°C). This mutation is presumed to be located in the external loop 3. Blight *et al.* (1994b) also site directed changes to the region, P3, and identified completely *secretion defective mutants*, $I^{401}T$, $S^{402}P$ or $D^{404}K$ or G. Therefore, together with the $G^{408}D$ mutation, four out of eight residues predicted to be located in P3 were shown to be essential for function, either assembly of HlyB or actual translocation of HlyA. In a more recent random mutagenesis study, two additional mutations of interest, reducing but not abolishing secretion, were identified. These mutations, $S^{279}L$ (TMS 3) and to a lesser extent $E^{156}K$ (TMS 1), also abolished the oligomerization of HlyD, which is dependent on HlyB (see above). These are the first mutations in HlyB describing a possible interaction with the MFP protein (H. Benabdelhak and J. Young, unpublished).

SEARCHING FOR SITES IN HLYB IMPLICATED IN RECOGNITION OF THE HLYA SIGNAL SEQUENCE

In ABC transporters, the membrane domain provides the transport pathway and presumably the specificity for a given transporter, whilst the ABC-ATPase provides an important energizing step. From the discussion above it seems most likely that in the case of the type 1 secretion pathway, both the auxiliary MFP protein (for example, HlyD) and the ABC component of the translocator participate in the early stages of the recognition of the allocrite and the initiation of its translocation across the membrane. In consequence, it is generally assumed that recognition of the allocrite is manifest by the membrane domain of the ABC transporter, rather than the highly conserved ATPase component. Of necessity this implies allocrite-dependent intramolecular signaling from the membrane domain to the ABC-ATPase in order to activate the latter.

As described above, most point mutations in the HlyA signal peptide have no effect or only reduce secretion levels by a small percentage. Even relatively strong mutations such as $F^{989}L$ (Kenny *et al.*, 1994), reducing secretion of HlyA by 70%, are not sufficient to reduce halo sizes significantly on blood agar plates. Consequently, screening for suppressor mutations on blood plates after random mutagenesis of *hlyB* is impossible. However, in the triple mutant, $E^{978}K$, $F^{989}L$, $D^{1009}R$, described earlier in the section on genetic analysis of type 1 C-terminal targeting signals, in which secretion levels of toxin are reduced by more than 99%, the halo size is greatly reduced. Nevertheless, using this as a screen for suppressor mutations after random hydroxylamine mutagenesis of both *hlyD* and *hlyB*, we failed to identify any extragenic suppressors (Chervaux, 1995). In another study, Ling and colleagues employed a mutant of HlyA deleted for the terminal 29 amino acid residues, and an internally deleted mutant lacking almost the entire proximal half of the secretion signal. Both of these mutants gave very small colony haloes on blood agar, and in this case it was possible to identify 11 such suppressor mutations, out of more than 6×10^5 clones tested. Interestingly, all these suppressors can be mapped to the cytoplasmic regions in the model of the HlyB molecule, with 10 out of 11 localized to the membrane domain **(Figure 11.8)**. Nevertheless, since these latter mutations are distributed through several regions of this domain rather than clustering (Sheps *et al.*, 1995; Zhang *et al.*, 1995), interpreting their significance is difficult. Unfortunately, the value of this study is further limited since the mutations were selected in the absence of a large portion of the secretion signal, the presumed docking sequence. By definition this precludes determination of the allele specificity of the mutations and consequently no information on specific interpeptide contacts can be deduced. Moreover, the suppressed mutants still only secreted a few percent of the wild-type levels of HlyA. Therefore, whether these mutations define binding regions in HlyB for the allocrite remains a moot point and the possibility remains that the effect of these suppressor mutations is more indirect.

Surprisingly, we now have more recent evidence which indicates that the C-terminal region of HlyA does in fact interact directly with the ABC-ATPase domain *in vitro*, as revealed by surface plasmon resonance studies (Schmitt *et al.*, in preparation). These results will be considered again in the final section.

ATTEMPTS TO PURIFY THE INTACT ABC TRANSPORTER ASSOCIATED WITH TYPE 1 SECRETION

Few attempts have been described so far to purify the intact ABC transporter involved in type 1 secretion systems. One exception, the PrtD protein from *E. chrysanthemi*, was successfully overexpressed in *E. coli* and partially purified after detergent solubilization (Delepelaire,

1994). The protein displayed a vanadate-sensitive, although extremely low, ATPase activity (V_{max} about 25 nmol min^{-1} mg^{-1} or 1 molecule of ATP/10 s). Surprisingly, when PrtD was incubated with submicromolar amounts of the C-terminal 55-residue peptide of the allocrite PrtB, purified after secretion from culture supernatants, almost all the ATPase activity was lost. This might reflect a specific interaction between the ABC protein and its transport substrate, since another secretion signal peptide, HasA, which is not secreted *in vivo* by the Prt system, did not inhibit the PrtD activity. A small reduction in ATPase activity was also observed when the C-terminal (signal) domain of HlyA was incubated with the ABC domain of HlyB fused C-terminal to GST (Koronakis *et al.*, 1993). These results are nevertheless difficult to interpret since the effect of the allocrite on activity was inhibitory rather than the anticipated stimulation, which has been observed with a number of mammalian ABC transporters. Unfortunately, further *in vitro* studies especially with reconstituted PrtD have not been reported.

In this laboratory many attempts have been made to overexpress the intact HlyB protein in *E. coli* but without success, despite placing the *hlyB* gene under the control of a wide variety of promotors in several different plasmids (Blight, 1990). In our hands the HlyB protein in fact forms SDS-resistant aggregates in conventional electrophoresis buffers, preventing entry into the separating gel. However, this can be overcome by incorporating the zwitterionic-detergent LDAO into the electrophoresis buffer (Young, 1999). Using this procedure, we have confirmed that the protein cannot normally be overproduced to significant levels in *E. coli*. This is not due to toxicity, but rather appears to reflect at least in part post-transcriptional regulation, since high levels of *hlyB* mRNA can be detected under inducing conditions. In addition, when *hlyB* is fused downstream of the normal *lacZ* gene to produce a hybrid mRNA, very large amounts of the fusion protein can indeed be detected (Blight *et al.*, 1995). Recently, we have also found that high levels of the intact HlyB protein can be overexpressed in the heterologous host *L. lactis* (J. Kuhn and M. Blight, unpublished), suggesting that in *E. coli* some factor normally inhibits translation. Blight *et al.* (1995) in fact demonstrated that in contrast to the intact protein, the C-terminal ATPase domain, expressed from a subclone, in the absence of the membrane domain, can be overexpressed in milligram quantities.

Other extensive attempts have been made to establish an *in vitro* system with membrane vesicles for translocation of HlyA synthesized *de novo*, based on the well-established protocols for Sec-dependent transport, but without success (this laboratory, unpublished results). This may simply reflect the fragility of the inner–outer membrane fusion in such *in vitro* systems, since a continuous HlyBD-TolC structure may be essential for even the earliest stages of translocation across the cytoplasmic membrane (Balakrishnan *et al.*, 2001).

PROPERTIES OF THE PURIFIED ABC DOMAIN

Most recently, conditions have been established for the purification of the C-terminal ABC domain of HlyB, commencing at residue D^{467}, (see **Figure 11.8**) in soluble form, tagged with N-terminal histidines. The V_{max}, 0.3 μmol min^{-1} mg^{-1}, for this polypeptide is in line with other ABC-ATPases (Benabdelhak *et al.*, 2002a). On the other hand, the K_m, close to 1 mM, is much larger than that for HisP and MalK. In addition, in complete contrast to purified MalK and HisP proteins (see Chapter 9), which curiously are vandadate resistant, the activity of the HlyB-ABC domain is *sensitive to vanadate*, with a K_i of 10 μM. In fact, the behavior of HisP and MalK is especially curious since the activity of these proteins when present in a functional complex with their corresponding membrane proteins is inhibited by vanadate. In such reconstituted complexes, MalK, at least, hydrolyzes ATP with positive cooperative kinetics. The purified HlyB-ABC domain also displays cooperative kinetics in enzyme activity assays, again unlike the purified HisP and MalK proteins (Benabdelhak *et al.*, 2002b).

CD analysis of the purified HlyB, ABC domain indicates secondary structure equivalent to 37% helix and 15% β-strand, similar to HisP. CD analysis also indicates an ATP-induced conformational change (Benabdelhak *et al.*, 2002a). Because of the difficulties in obtaining the ABC domain of HlyB in soluble form, previous studies were restricted to purification of this domain fused to GST (Koronakis *et al.*, 1993). With this construct it was nevertheless possible to demonstrate that changes to several amino acid residues, essential for secretion of HlyA toxin *in vivo*, led to loss of ATPase activity of the GST–HlyB fusion *in vitro* and corresponding abolition of *in vivo* secretion activity (Koronakis *et al.*, 1995).

The activity of the purified, His-tagged HlyB, ABC-ATPase domain is reversibly inhibited at physiological salt concentrations, accompanied by a conformational change, detected by an increase in intrinsic fluorescence, and loss of ATP binding. This effect, which is not associated with a change from dimers to monomers, for example, may reflect a control switch *in vivo* preventing binding of ATP until the enzyme is activated when required (Benabdelhak *et al.*, 2002b).

PRELIMINARY HIGH-RESOLUTION STRUCTURE OF THE HLYB-ABC

The ABC domain of HlyB (from residue D^{467}) has been crystallized (Kránitz *et al.*, 2002). As this chapter goes to press we now have the preliminary data for the crystal structure of this domain at 2.55 Å. This shows the expected two domain structure, with the RecA-like nucleotide-binding domain and the smaller 'helical' domain (regulatory domain) containing the LSGG-signature sequence overlapping helix 6 in **Figure 11.11**. As indicated briefly in the final section, the regulatory domain appears to show major organizational differences, with regard to the size of helices 3 and 4, and the presence of extended loops, when compared with HisP.

HLYB-ATPASE ACTIVITY AND THE POSSIBLE ROLE OF DIMERS

Recent studies in this laboratory have shown that the *ATPase activity* of the isolated ABC domain as described above displays *positive cooperativity*, consistent with the presence of dimers (Benabdelhak *et al.*, 2002b). However, a detailed study of the oligomerization state of the ABC domain indicated a monomer–dimer equilibrium with an active monomer apparently the most prevalent species under most conditions (Benabdelhak *et al.*, 2002b). Notably, the activity of the purified ABC domain appears to be independent of dimerization since contrasting conditions, which favor monomers (high salt) on the one hand or dimers (low salt or cysteine crosslinks) on the other, have little detectable effect on the specific activity. In our view these results argue against models of dimers of the ArsA type, where it is clear that the catalytic site is only completed upon dimer formation. Rather the results support the idea that *dimer formation* is necessary for *regulation*, for example by controlling the alternating activity of the monomers through crosstalk. We conclude therefore that dimerization is not required for ATPase activity of HlyB-ABC per se. In contrast to the purified ABC domain, although not characterized in detail, dimers of the intact HlyB molecule appear relatively more stable **(Figure 11.9)** and this, together with the properties of the ABC domain *in vitro*, suggests that the membrane domains may be the main driving force for dimerization *in vivo*.

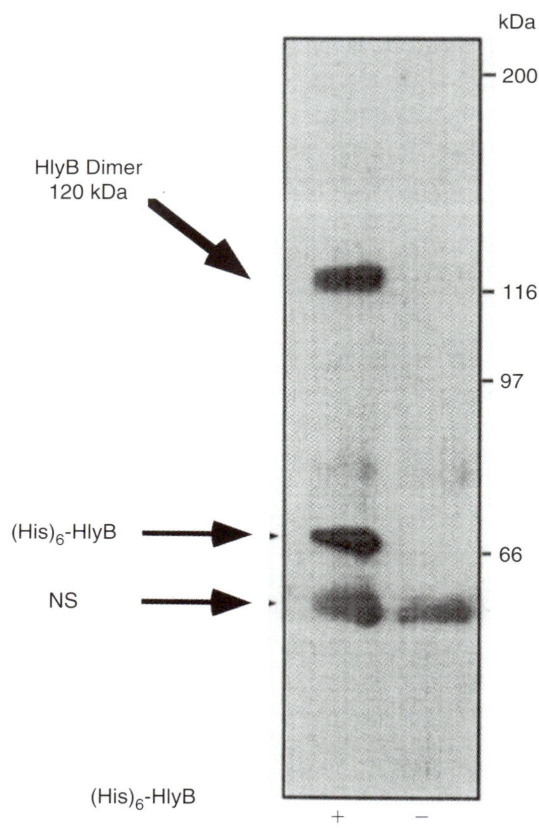

Figure 11.9. Identification of an SDS-resistant complex of 6(His)-HlyB. Strain SE5000, 6(His)-HlyB, overproducing cells (+) or non-producing SE5000 cells (−) were mixed with SDS/LDAO sample buffer at 37°C, subjected to SDS–PAGE (7% acrylamide) and transferred to nitrocellulose. The blot was probed with anti-HlyB antibodies. NS refers to a nonspecific band recognized by the anti-HlyB serum. The HlyB dimers are stable in the presence of mercaptoethanol and stable up to 70°C but are disrupted by boiling.

THE FORM OF TYPE 1 PROTEINS DURING TRANSLOCATION: FOLDED OR UNFOLDED?

Translocation of polypeptides across the cytoplasmic membrane by the classical Sec-dependent pathway apparently involves movement of an unfolded or incompletely folded molecule through a translocation 'pore' in the membrane. This is formed by a SecY–SecEG, multimeric complex (van Wely *et al.*, 2001). On the other hand, translocation of certain large redox proteins, containing a metal cofactor, to the periplasm of Gram-negative bacteria, via the Tat pathway (Chanal *et al.*, 1998), seems in contrast to involve transfer of a fully folded or largely folded molecule across the cytoplasmic membrane, via a novel protein translocase (Robinson and Bolhuis, 2001; Weiner *et al.*, 1998).

A ROLE FOR CYTOPLASMIC CHAPERONES?

By definition, polypeptides with C-terminal secretion signals must be fully synthesized before engaging with the translocator, giving opportunities for prior folding and hence a possible role for protein chaperones at some level. For HlyA neither SecB nor GroES are involved. However, using pulse-chase methods to follow secretion in wild-type and *groEL* mutants, a role for this chaperonin was indicated (Whitehead, 1993; J. Whitehead and J.M. Pratt, personal communication). These findings have unfortunately not been further pursued. The secretion of metalloproteases via the Prt-ABC pathway was also shown to be independent of SecB. However, in contrast, the SecB chaperone is involved in the secretion of the 188-residue HasA protein (which lacks RTX repeats) (Delepelaire and Wandersman, 1998). Thus, secretion but not synthesis of HasA in *E. coli* (or the natural host) was substantially reduced in a *secB* mutant, or when SecB levels were effectively depleted. This requirement, however, is lost when HasA is deleted for 10 amino acids at the N-terminal, whilst the overall level of secretion is reduced approximately twofold when this N-terminal is absent (Sapriel *et al.*, 2001). The authors concluded that SecB normally may bind – co-translationally – to the N-terminal of HasA, maintaining the protein in a state competent for translocation, thus facilitating subsequent docking of its C-terminal secretion signal with the translocator. In fact, in another study it was shown that HasA, in the absence of the translocator, accumulates in *E. coli* in a form capable of binding heme, suggesting that the tertiary structure has been acquired. Notably, this form cannot be secreted if the secretion functions are expressed subsequently. It was concluded therefore that synthesis and secretion are normally tightly coupled (Debarbieux and Wandersman, 2001). In summary, these studies indicate that HasA at least may be secreted in an 'unfolded' form. We might also speculate that SecB is required in this case, because HasA lacks the RTX repeats. These we suppose might normally fail to fold in the cytoplasm, owing to insufficient levels of free Ca^{2+} necessary to trigger folding and stabilization of the glycine-rich repeats (see below), thereby acting to limit folding of the entire RTX protein, prior to their passage across the cytoplasmic membrane.

TOLC AND HLYD MAY FORM A *TRANS*-ENVELOPE FOLDING CHAMBER FOR HLYA

The recent remarkable high-resolution structural study of TolC homotrimers (Andersen *et al.*, 2000; Koronakis *et al.*, 2000) revealed astonishingly, as shown in **Figure 11.10**, a β-strand pore structure for anchoring in the

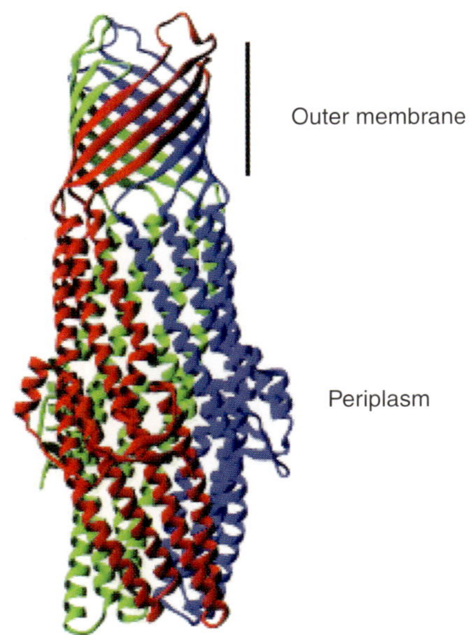

Figure 11.10. Structure of the TolC trimer at 1.2 Å. Reproduced from Koronakis et al., *2000 with permission.*

outer membrane, extending into a long tunnel-like α-helical structure which (presumably) crosses the periplasm. This structure provides some informative structural limits on the size and shape of molecules that might pass through this translocation path. As we shall briefly discuss later, the TolC structure presumably interacts with HlyD in order to form a continuous pathway across the cell envelope, connecting the ABC transporter in the cytoplasmic membrane to the outside world. The TolC tunnel, formed at the center of the TolC trimers, is 140 Å long and has an inner diameter of 30 Å in the upper half, narrowing to nearly closed at the bottom. Koronakis *et al.* (2000) have proposed an ingenious mechanism, involving realignment of a pair of inner helices of each monomer, in an 'iris'-like movement, opening up this latter entrance to 30 Å, in reponse to the presence of the transport substrate such as HlyA.

Importantly, the size of this TolC chamber through the periplasm to the outer membrane compares closely with dimensions for the chaperonin GroEL of 145 Å × 45 Å and GroES, 20–30 Å × 30 Å. Such a wide passage or chamber formed by TolC trimers could in principle provide for the transit (if not some folding, see later) of large polypeptides up to ~60 kDa, partially if not fully folded, at least in this part of the translocation complex. As noted above, however, some ABC-dependent secreted proteins exceed 400 kDa and therefore, clearly, complete folding cannot occur in this chamber. Either the true oligomerization state of the HlyD-TolC chamber has so far been underestimated or, perhaps more likely, folding of the secreted protein is completed on the surface of the bacterium.

HlyD is inserted in the inner membrane by a single TMS with an N-terminal of approximately 58 residues extending into the cytoplasm and a 40 kDa external domain, capable probably of spanning the periplasm. This domain contains an N-proximal, 20 kDa region, predicted to be largely helical, including a coiled coil motif of 41 residues. This is followed by a largely β-strand region of 20 kDa at the C-terminus (Pimenta *et al.*, 1996; Schulein *et al.*, 1992; Wang *et al.*, 1991). Crosslinking experiments *in vivo* have indicated that HlyD forms trimers or possibly larger oligomers (Thanabalu *et al.*, 1998; Young, 1999), and HlyD forms complexes with both HlyB and TolC (see section on the composition of the type 1 translocator, above). It seems likely therefore that HlyD also forms an elongated, multimeric structure across the periplasm, overlapping or interlacing with the TolC structure in a functional complex connecting the inner membrane to the exterior.

Genetic analysis of HlyD and TolC and the structure of HlyA now also provides further if indirect evidence that translocating HlyA molecules might be at least partially unfolded. Thus, *mutations* in the periplasmic *HlyD* domain have been characterized which clearly affect the *folding* rather than the secretion of HlyA (Pimenta, 1995; A. Pimenta and K. Racher, this laboratory, unpublished). Under these conditions the secreted HlyA molecules are hypersensitive to trypsin and have apparently reduced activity but can be re-folded *in vitro* to the active form. Recently, mutations in TolC have been shown to confer the same property (S. Misra, personal communication). The results in both cases show that alterations to that part of the translocator composed of HlyD and TolC affect the final folding of the haemolysin. This would be consistent with the HlyA molecules already beginning to fold after crossing the cytoplasmic membrane.

The crystal structure of proteases carrying the RTX repeats have shown that these form a specific β-strand jelly-roll structure with a Ca^{2+} ion linked to each repeat strand, thereby providing a high degree of structural stability (Baumann *et al.*, 1993). However, since the concentration of free Ca^{2+} in the cytoplasm in *E. coli* is extremely low at 0.1 to 0.2 μM, equivalent to 100 or so ions per cell, irrespective of the external Ca^{2+} concentration (Gangola and Rosen, 1987; Jones *et al.*, 1999), it seems highly unlikely that the mature form of HlyA, dependent upon substantial levels of free Ca^{2+}, would be formed intracellularly. In contrast, the periplasm can rapidly adopt the same or, apparently under some conditions, an even higher concentration of free Ca^{2+} compared with the external medium (Jones *et al.*, 2002). We conclude therefore that Ca^{2+}, essential for folding HlyA molecules, would permeate the trans-periplasmic HlyD-TolC channel and be available for folding HlyA molecules in transit to the medium – with the folding up of the repeats thus providing an auto-chaperone-like function. From studies of the crystal structure of the metalloproteases, Baumann has independently proposed that Ca^{2+} could fulfill such a catalytic folding role, following translocation of an 'unstructured' RTX protein to the exterior (Baumann *et al.*, 1993). In view of the properties of the *hlyA*99 mutation described above, which apparently affects the ability of HlyA to fold correctly, it is possible to speculate that the C-terminal of HlyA also plays an important role in the folding process.

In summary, all these studies would be consistent with unfolded HlyA molecules entering the translocation pathway, and then at least commencing folding during passage through an HlyD-TolC chamber, perhaps completing folding on the cell surface – where more than 60% of HlyA molecules in active form remain following secretion (A. Pimenta, this laboratory, unpublished data).

WHAT ROLE FOR HLYB IN THE TRANSLOCATION OF HLYA?

We may now ask the question what specific role does HlyB play in the translocation process, in particular in moving the allocrite across the cytoplasmic membrane? As discussed above, evidence for an interaction between both HlyB and HlyD and for HlyB with HlyA (see following section) has been obtained. In fact, detailed studies in this laboratory have revealed that one important function of HlyB is that it is required for both the stability of HlyD and its multimerization into at least trimers, as detected by crosslinking (Pimenta, 1995; Young 1999; this laboratory, in preparation). Certain mutations in HlyB indeed abolish multimerization of HlyD, although not a Walker A mutation (Young, 1999; this laboratory, unpublished). In addition, mutations in HlyB, including those in the Walker and LSGG motifs (Koronakis *et al.*, 1995), abolish secretion of HlyA. Unfortunately, no studies are available which might throw light on the precise nature of the role of the ABC protein in this type 1 secretion process. Whether ATP fixation and/or hydroysis is required for docking of the secretion signal, initial movement of the polypeptide through the inner membrane or for subsequent translocation through the MFP-OMF 'chamber' and the folding of the secreted protein remains a mystery. However, no mutations in HlyB were found so far to affect the activity of HlyA, and so there is no evidence that HlyB is involved in folding of the allocrite, HlyA. It is also important to remember that a previous study by Koronakis *et al.* (1991), using uncouplers, indicated a requirement for the proton motive force (PMF) for the secretion of hemolysin A from *E. coli*, which the authors equated with a role in the early steps in the secretion process reminescent of that played in protein export via the Sec pathway (van Wely *et al.*, 2001).

As discussed above, and given the fact that RTX proteins can be more than 400 kDa in size and that 'foreign' polypeptides at least as large as β-galactosidase can be secreted efficiently by the ABC transporter, it appears highly unlikely that folded proteins cross the cytoplasmic membrane through 'channels' formed by an ABC translocator, even in the form of a dimer. We may speculate therefore that, as in the case of the Sec translocator, an unfolded RTX protein is the 'substrate' for the ABC transporter, with passage across the membrane involving extrusion through the interior of the membrane domain of an HlyB dimer. An alternative possibility that merits serious consideration is that it is actually the MFP (HlyD) that forms a continuous trimeric or larger channel through the inner membrane to the cytoplasm. This would then provide the pathway for an unfolded HlyA to cross both the cytoplasmic membrane and the periplasm – with HlyB contributing simply the gating energy necessary to control entry to an HlyD translocon.

ANALYSIS OF THE INTERACTION BETWEEN HLYA AND THE HLYB-ABC DOMAIN *IN VITRO*

Direct analysis of possible interactions between intact HlyB and the allocrite HlyA has so far not been feasible, since HlyB cannot be purified in sufficient quantities. For this reason and since the nature of any possible interaction between HlyB and HlyA was completely unknown, we examined the effect of the C-terminal 25 kDa fragment of HlyA on the ATPase activity of the purified ABC domain, searching for a possible stimulation of activity under different conditions. No such stimulation was obtained, rather a slight inhibition, reminiscent of that reported for PrtD by the C-terminal of the allocrite PrtB (Delepelaire, 1994) and for the GST-HlyB-ABC (Koronakis *et al.*, 1993). We have examined this effect in great detail by surface plasmon resonance using the BiaCore system. As will be published elsewhere (Schmitt *et al.*, in preparation), a specific interaction between the HlyB NBD and HlyA was established, with an affinity constant close to 4 μM. Importantly, this interaction was abolished when the terminal 57 residues of the 25 kDa C-terminal of HlyA, encompassing the secretion signal, were deleted. Moreover, the HlyA::HlyB-ABC complex was rapidly dissociated in the presence of ATP. We have proposed, therefore, that in this case the allocrite specifically interacts with the ABC-ATPase domain via the secretion signal and that in the presence of ATP this may trigger the initiation of translocation *in vivo*, with transfer of the HlyA molecule into the channel of the translocator.

These findings will necessitate a major reassessment of the nature of signaling between the HlyA allocrite and the ABC transporter in relation to established views. However, it is worth bearing in mind that the allocrite in type 1 secretion systems approaches the translocator from the cytosol and is far removed both in size and properties from the lipid or lipophilic drug molecules transported by other ABC proteins. The role of the accessory MFP component in secretion of polypetides in this system, apparently implicated in initiation of translocation, may also be an important determinant in the precise nature of the initial docking of these transport substrates with the translocator. Consequently, in view of these perhaps necessarily radical adaptations of the ABC system to cope with large polypeptide transport substrates, some fundamental changes to the docking and intramolecular signaling mechanisms which regulate the ATPase activity in response to the presence of the allocrite should not be unexpected.

STRUCTURE OF THE ABC DOMAIN OF HLYB AND OTHER ABC PROTEINS: SEARCHING FOR REGIONS RELATING TO SPECIFICITY/IDENTITY

Although highly conserved at the sequence level, ABC domains nevertheless appear to display great specificity or *identity*, i.e. they appear to function, at least in prokaryotes, only with their homologous partners and cognate allocrites. For most ABC transporters this most probably reflects a specific interaction between the ABC domain and the cognate membrane domain(s). This in turn follows from the concept of intramolecular signaling (as discussed in other chapters) in order to couple activation of the ATPase to the eventual transport of a specific allocrite (or its modification in cases like Rad50). Some indication of features in the ATPase domain anticipated to play a role in 'identity' and intramolecular signaling are now beginning to emerge.

The high-resolution structure of the HisP monomer is shown in **Figure 11.11**. As described in Chapter 9 for MalK and HisP, regions encompassing *helix* 3 and *helix* 4 in the helical domain appear to interact directly and functionally with the region containing the *EAA-loop* of the cognate membrane domains. In the crystal structure of TAP1 (Gaudet and Wiley, 2001), a major difference shows helix 3 to be significantly truncated compared to HisP, whilst in contrast the loop connecting helix 4 to the signature motif at the beginning of helix 6 is substantially elongated, with concomitant loss of the small helix 5 found at the equivalent position in HisP. Interestingly, helix 3 in the structure of Rad50 (which functions in DNA repair, *not* transport) is completely disrupted by insertion of an extremely long helical domain (Hopfner *et al.*, 2000). In the crystal structure of MutS, another ABC protein involved in DNA repair (Junop *et al.*, 2001), the C-terminal is extended in comparison with HisP, consistent perhaps with another adaptation of the ABC structure in order to coordinate its interaction with, in this case, DNA. On the other hand, the region corresponding to *helix* 2 of HisP is absent (i.e. deleted) in the MalK structure, indicating that this is not essential for function in these two closely related import pathways.

From the preliminary analysis of the structure of the HlyB, ABC domain at 2.55 Å (Schmitt *et al.*, in preparation), it is clear that there are in particular substantial differences in the regulatory domain between HisP and HlyB. Interestingly, once again these include changes involving helices 3 and 4, in this case truncation or disruption and their apparent prolongation into loops. In consequence, this NBD appears less densely packed, compared with HisP. It is not yet clear whether these features of HlyB reflect requirements for a novel form of interaction between the regulatory and membrane domains of HlyB, or whether this is related to a particular point in the catalytic cycle present in the respective crystal forms. The

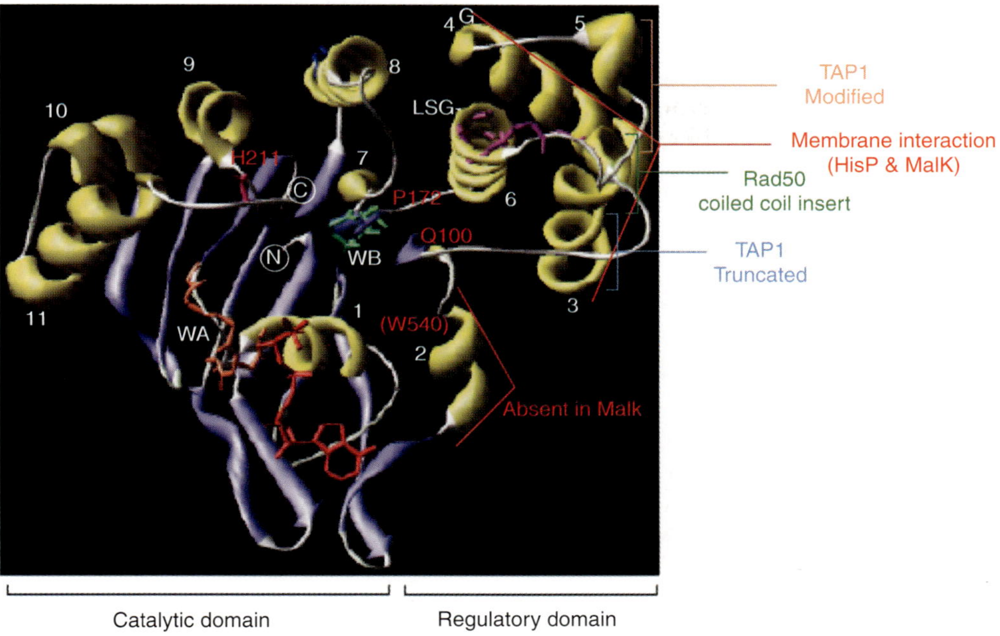

Figure 11.11. Identifying changes in the helical, signaling or regulatory domain (Arm-II, Hung **et al.**, *1998) in different ABC domains in comparison with HisP. The figure presents the structure of HisP at 1.5 Å (Hung* **et al.**, *1998) with the helices, Walker A and B, signature motif (LSGG-) and the N- and C-termini indicated in white lettering. An ATP molecule is present in the catalytic domain. The WA motif is shown in orange, the WB in green, and the conserved (switch) histidine 211 in rose; the LSGG- in pink; the positions of the conserved P^{172} and Q^{100} are presented in the Pro- and Q-loops respectively. Residue (W^{540}) indicates the equivalent position occupied by the single Trp residue in HlyB; the equivalent position of the single Cys (C^{652}) in the HlyB is also indicated in blue in helix 8. The core structure of the catalytic domain in HlyB is similar to that in HisP, but the regulatory domain shows several differences (Schmitt* **et al.**, *in preparation; see text). To the right are indicated some obvious differences between HisP and other ABC domains for which the structure is now available. In TAP1, the region between helix 4 and the signature motif (helix 6) lacks helix 5, and consequently this connecting loop is substantially elongated. The region (helix 3, 4) of HisP (and MalK) that interacts with the 'EAA' loop in the membrane domain of the import complex containing these proteins is also indicated. Other details in the text.*

crystal structure of the HlyB-NBD also identifies helix 2 (conserved in HisP, **Figure 11.11**) as the location of the single tryptophan. This residue is surface exposed, and located in this shortened helix (compared with HisP) between extended loops in the HlyB structure. We have shown that the intrinsic fluorescence of the HlyB-NBD increases markedly with the reduction in nucleotide binding observed at high salt concentrations (Benabdelhak *et al.*, 2002b). This may reflect a major structural rearrangement in this region, associated with a mechanism for controlling the ATPase activity. This will be a focus of future study. Despite the recent advances in obtaining structures of some ABC domains, much structural work remains to be done, with identification of all the conformations associated with each step in the catalytic cycle for different ABC proteins being a priority. However, the comparative results, summarized in **Figure 11.11**, encourage the view that important information concerning the mechanism of action of the ABC domain, tailored to the requirements of the cognate membrane domain, will be revealed from such structural studies.

MODEL FOR SECRETION OF TYPE 1 ALLOCRITES

The type 1 translocator illustrated by the Hly system apparently contains only three proteins, the ABC, MFP and OMF, forming a single transport pathway (tunnel or channel) across the two membranes of the Gram-negative envelope to the exterior. Some uncertainties still exist concerning the precise stoichiometry *in vivo* of the

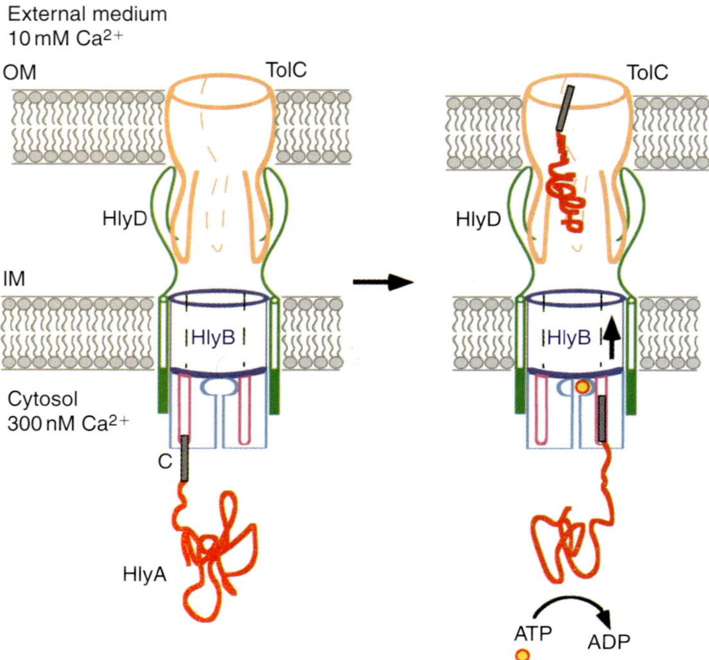

Figure 11.12. Model for the Hly translocon. This shows the interaction of the protein subunits to form the active trans-*envelope translocation chamber energized by the HlyB, ABC component. HlyD, which forms a trimer or possible higher oligomers, is presented as a dimer for simplicity. HlyB is presented as the central core of the transport pathway of the translocon but this is arbitrary and alternatively, in contrast, the N-terminal and membrane domain of HlyD, encased in HlyB, could serve this function, with the ABC protein simply controlling the opening of the pathway. The two ABC domains of HlyB are presented head to head but there is no evidence as yet for this. We envisage a binding site in the ABC domain of HlyB for the targeting signal at the C-terminus of the transport substrate, HlyA. The solid green block identifies the region in the HlyD shown to interact with HlyA and necessary for recruitment of TolC into a functional complex with HlyD (Balakrishnan* et al.*, 2001). We envisage that 'unfolded' HlyA (in complex with chaperones?) docks with the ABC domain, trigging fixation of ATP and then displacement of HlyA into the interior of the translocon and/or contact with the N-terminal of HlyD. Transport then ensues rapidly, facilitated at some point by the PMF, with the polypeptide commencing folding in the large chamber provided by TolC and HlyD, finally completing folding on the cell surface and eventual release to the medium. As some point, completely undefined so far, ATP hydrolysis takes place, perhaps to re-set the system and permit translocation of the next HlyA molecule from the other ABC domain. For other details see text.*

three proteins. In addition, with respect to the model structure depicted in **Figure 11.12**, it is not at all certain whether HlyB forms the internal 'core' of the true transport pathway across the inner membrane or whether this role is fulfilled by HlyD, with the HlyB membrane domain simply regulating the access of HlyA to the translocator HlyD-TolC. Energy for translocation probably includes both the PMF and ATP-hydrolysis by HlyB. Nevertheless, the precise role (i.e. which step in transport) for the HlyB ABC-ATPase remains unclear. The possibilities probably include important but distinctive roles for ATP fixation as well as ATP hydrolysis. Both HlyD (N-terminal) and HlyB (apparently the NBD at least) are probably involved in initiation of transport by interacting with the specific secretion signal (approximately 50 residues) at the C-terminus of HlyA. This targeting peptide in our view is relatively unstructured with a few dispersed key residues specifying recognition or docking with the translocator. The presenting type 1 polypeptide (up to 450 kDa in some cases) in all probability is maintained in the cytoplasm in an 'unfolded' state by cellular chaperones (though this remains to be verified), prior to docking and secretion. In the case of RTX proteins the low cytoplasmic Ca^{2+} concentration is probably also a major limiting factor in normal folding.

From our *in vitro* studies with the ABC domain of HlyB, we may speculate that binding of ATP is normally blocked *in vivo* in the absence of the allocrite HlyA and that initiation of translocation involves specific binding of HlyA to the ABC domain, followed by ATP-dependent displacement of HlyA into the interior of the translocator, perhaps involving the N-terminal of HlyD (Benabdelhak *et al.*, 2002b; Schmitt *et al.*, in preparation). The activity of HlyB is likely to be regulated at some step by intramolecular signaling, which we speculate involves the helix 3, 4 region of the helical domain of the ABC, immediately adjacent to the signature motif (Schmitt *et al.*, in preparation). An additional role for HlyB, although the details are still controversial, involves a specific interaction with HlyD, inducing a conformational or organizational change in the latter which facilitates its close packing into oligomeric structures. Importantly, following initial docking of HlyA with the translocator, an HlyD-dependent signal across the inner membrane results in an altered structural change in HlyD, leading to the formation of a functional complex with TolC (Balakrishnan *et al.*, 2001).

Following initiation of translocation of the unfolded polypeptide into the periplasmic outer membrane chamber, possibly formed by interlacing the extended helices of HlyD and TolC, folding of HlyA is at least initiated, dependent upon the RTX repeats, together with the high concentration of periplasmic Ca^{2+} (Jones *et al.*, 2002), the C-terminal of HlyA, and also facilitated perhaps by a chaperone-like function for HlyD and TolC. The entire HlyA protein may then complete its folding on the surface of the bacterium.

ACKNOWLEDGMENTS

We are indebted to CNRS and the Université Paris-Sud for continuing support. We are especially grateful to ABCF (Association de lutte contre la Mucoviscdose) for their generous support. HB wishes to acknowledge FRM (Fondation pour la Recherche Médicale) and Société de Secours des Amis des Science for bursary support and LS to the Deutsche Forschungsgemeinschaft (Emmy Noether program; grant number Schm1279/2-1). IBH wishes gratefully to acknowledge in addition to the present authors, the contribution of all former colleagues but in particular (it is not possible to include all in this limited space), Nigel Mackman, Jean-Marc Nicaud, Lindsay Gray, Karen Baker, Brendan Kenny, Rongchen Wang, Christian Chervaux and Kathleen Racher for their stimulating contributions, their determination and enthusiasm, which has ensured that study of the Hly system has always been an enjoyable and rewarding experience.

REFERENCES

Ahn, J.H., Pan, J.G. and Rhee, J.S. (1999) Identification of the tliDEF ABC transporter specific for lipase in *Pseudomonas fluorescens* SIK W1. J. Bacteriol. 181, 1847–1852.

Andersen, C., Hughes, C. and Koronakis, V. (2000) Channel vision. Export and efflux through bacterial channel-tunnels. EMBO Rep. 1, 313–318.

Awram, P. and Smit, J. (1998) The *Caulobacter crescentus* paracrystalline S-layer protein is secreted by an ABC transporter (type I) secretion apparatus. J. Bacteriol. 180, 3062–3069.

Balakrishnan, L., Hughes, C. and Koronakis, V. (2001) Substrate-triggered recruitment of the TolC channel-tunnel during Type I export of hemolysin by *Escherichia coli*. J. Mol. Biol. 313, 501–510.

Baumann, U. (1994) Crystal structure of the 50 kDa metallo protease from *Serratia marcescens*. J. Mol. Biol. 242, 244–251.

Baumann, U., Wu, S., Flaherty, K.M. and McKay, D.B. (1993) Three-dimensional structure of the alkaline protease of *Pseudomonas aeruginosa*: a two-domain protein with a calcium binding parallel beta roll motif. EMBO J. 12, 3357–3364.

Baumann, U., Bauer, M., Letoffe, S., Delepelaire, P. and Wandersman, C. (1995) Crystal structure of a complex between *Serratia marcescens* metallo-protease and an inhibitor from *Erwinia chrysanthemi*. J. Mol. Biol. 248, 653–661.

Benabdelhak, H., Blight, A.M., Stoven, V., Vallois, D., Boucher, J. and Holland, I.B. (2002a) Over expression, purification and initial structural and biochemical characterization of the ABC-ATPase domain of HlyB required for secretion of haemolysin from *E. coli*. Submitted to *Eur. J. Biochem.*

Benabdelhak, H., Schmitt, L., Horn, C., Jumel, K., Blight, A.M. and Holland, I.B. (2002b) The ABC-ATPase domain of HlyB *in vitro* displays activity with positive co-operativity, but dimers appear highly unstable. Submitted.

Binet, R. and Wandersman, C. (1995) Protein secretion by hybrid bacterial ABC-transporters: specific functions of the membrane ATPase and the membrane fusion protein. EMBO J. 14, 2298–2306.

Bingle, W.H., Nomellini, J.F. and Smit, J. (2000) Secretion of the *Caulobacter crescentus* S-layer protein: further localization of the C-terminal secretion signal and its use for secretion of recombinant proteins. J. Bacteriol. 182, 3298–3301.

Blight, M. (1990) Analysis of the membrane localized transport protein haemolysin B. Ph.D. Thesis, University of Leicester.

Blight, M.A. and Holland, I.B. (1990) Structure and function of haemolysin B,P-glycoprotein and other members of a novel family of membrane translocators. Mol. Microbiol. 4, 873–880.

Blight, M.A., Chervaux, C. and Holland, I.B. (1994a) Protein secretion pathway in *Escherichia coli*. Curr. Opin. Biotechnol. 5, 468–474.

Blight, M.A., Pimenta, A.L., Lazzaroni, J.C., Dando, C., Kotelevets, L., Seror, S.J. and Holland, I.B. (1994b) Identification and preliminary characterization of temperature-sensitive mutations affecting HlyB, the translocator required for the secretion of haemolysin (HlyA) from *Escherichia coli*. Mol. Gen. Genet. 245, 431–440.

Blight, M.A., Menichi, B. and Holland, I.B. (1995) Evidence for post-transcriptional regulation of the synthesis of the *Escherichia coli* HlyB haemolysin translocator and production of polyclonal anti-HlyB antibody. Mol. Gen. Genet. 247, 73–85.

Chanal, A., Santini, C. and Wu, L. (1998) Potential receptor function of three homologous components, TatA, TatB and TatE, of the twin-arginine signal sequence-dependent metalloenzyme translocation pathway in *Escherichia coli*. Mol. Microbiol. 30, 674–676.

Chang, G. and Roth, C.B. (2001) Structure of MsbA from *E. coli*: a homolog of the multidrug resistance ATP binding cassette (ABC) transporters. Science 293, 1793–1800.

Chervaux, C. (1995) Structure-function analysis of the secretion signal of *E. coli* haemolysin. Ph.D. Thesis, University of Paris XI.

Chervaux, C. and Holland, I.B. (1996) Random and directed mutagenesis to elucidate the functional importance of helix II and F-989 in the C-terminal secretion signal of *Escherichia coli* hemolysin. J. Bacteriol. 178, 1232–1236.

Debarbieux, L. and Wandersman, C. (2001) Folded HasA inhibits its own secretion through its ABC exporter. EMBO J. 20, 4657–4663.

Delepelaire, P. (1994) PrtD, the integral membrane ATP-binding cassette component of the *Erwinia chrysanthemi* metalloprotease secretion system, exhibits a secretion signal-regulated ATPase activity. J. Biol. Chem. 269, 27952–27957.

Delepelaire, P. and Wandersman, C. (1990) Protein secretion in gram-negative bacteria. The extracellular metalloprotease B from *Erwinia chrysanthemi* contains a C-terminal secretion signal analogous to that of *Escherichia coli* alpha-hemolysin. J. Biol. Chem. 265, 17118–17125.

Delepelaire, P. and Wandersman, C. (1998) The SecB chaperone is involved in the secretion of the *Serratia marcescens* HasA protein through an ABC transporter. EMBO J. 17, 936–944.

Duong, F., Soscia, C., Lazdunski, A. and Murgier, M. (1994) The *Pseudomonas fluorescens* lipase has a C-terminal secretion signal and is secreted by a three-component bacterial ABC-exporter system. Mol. Microbiol. 11, 1117–1126.

Duong, F., Lazdunski, A. and Murgier, M. (1996) Protein secretion by heterologous bacterial ABC-transporters: the C-terminus secretion signal of the secreted protein confers high recognition specificity. Mol. Microbiol. 21, 459–470.

Economou, A., Hamilton, W.D., Johnston, A.W. and Downie, J.A. (1990) The *Rhizobium* nodulation gene nodO encodes a Ca2(+)-binding protein that is exported without N-terminal cleavage and is homologous to haemolysin and related proteins. EMBO J. 9, 349–354.

Fath, M.J., Skvirsky, R.C. and Kolter, R. (1991) Functional complementation between bacterial MDR-like export systems: colicin V, alpha-hemolysin, and *Erwinia* protease. J. Bacteriol. 173, 7549–7556.

Felmlee, T. and Welch, R.A. (1988) Alterations of amino acid repeats in the *Escherichia coli* hemolysin affect cytolytic activity and secretion. Proc. Natl Acad. Sci. USA 85, 5269–5273.

Felmlee, T., Pellett, S., Lee, E.Y. and Welch, R.A. (1985) *Escherichia coli* hemolysin is released extracellularly without cleavage of a signal peptide. J. Bacteriol. 163, 88–93.

Fiedler, G., Arnold, M., Hannus, S. and Maldener, I. (1998) The DevBCA exporter is essential for envelope formation in heterocysts of the cyanobacterium *Anabaena* sp. strain PCC 7120. Mol. Microbiol. 27, 1193–1202.

Finnie, C., Hartley, N.G., Findlay, K.C. and Downie, J.A. (1997) The *Rhizobium leguminosarum* PrsDE genes are required for secretion of several proteins, some of which influence nodulation, symbiotic nitrogen fixation and exopolysaccharide modification. Mol. Microbiol. 25, 135–146.

Franke, C.M., Leenhouts, K.J., Haandrikman, A.J., Kok, J., Venema, G. and Venema, K. (1996) Topology of LcnD, a protein implicated in the transport of bacteriocins from *Lactococcus lactis*. J. Bacteriol. 178, 1766–1769.

Frey, J., Beck, M., Stucki, U. and Nicolet, J. (1993) Analysis of hemolysin operons in *Actinobacillus pleuropneumoniae*. Gene 123, 51–58.

Fullner, K.J. and Mekalanos, J.J. (2000) *In vivo* covalent cross-linking of cellular actin by the *Vibrio cholerae* RTX toxin. EMBO J. 19, 5315–5323.

Gangola, P. and Rosen, B.P. (1987) Maintenance of intracellular calcium in *Escherichia coli*. J. Biol. Chem. 262, 12570–12574.

Gaudet, R. and Wiley, D.C. (2001) Structure of the ABC ATPase domain of human TAP1, the transporter associated with antigen processing. EMBO J. 20, 4964–4972.

Geelen, D., van Montagu, M. and Holsters, M. (1995) Cloning of an *Azorhizobium caulinodans* endoglucanase gene and analysis of its role in symbiosis. Appl. Environ. Microbiol. 61, 3304–3310.

Gentschev, I. and Goebel, W. (1992) Topological and functional studies on HlyB of *Escherichia coli*. Mol. Gen. Genet. 232, 40–48.

Gentschev, I., Mollenkopf, H., Sokolovic, Z., Hess, J., Kaufmann, S.H. and Goebel, W. (1996) Development of antigen-delivery systems, based on the *Escherichia coli* hemolysin secretion pathway. Gene 179, 133–140.

Ghigo, J.M. and Wandersman, C. (1994) A carboxyl-terminal four-amino acid motif is required for secretion of the metalloprotease PrtG through the *Erwinia chrysanthemi* protease secretion pathway. J. Biol. Chem. 269, 8979–8985.

Gilson, L., Mahanty, H.K. and Kolter, R. (1990) Genetic analysis of an MDR-like export system: the secretion of colicin V. EMBO J. 9, 3875–3894.

Glaser, P., Ladant, D., Sezer, O., Pichot, F., Ullmann, A. and Danchin, A. (1988) The calmodulin-sensitive adenylate cyclase of *Bordetella pertussis*: cloning and expression in *Escherichia coli*. Mol. Microbiol. 2, 19–30.

Gray, L., Mackman, N., Nicaud, J.M. and Holland, I.B. (1986) The carboxy-terminal region of haemolysin 2001 is required for secretion of the toxin from *Escherichia coli*. Mol. Gen. Genet. 205, 127–133.

Gray, L., Baker, K., Kenny, B., Mackman, N., Haigh, R. and Holland, I.B. (1989) A novel C-terminal signal sequence targets *Escherichia coli* haemolysin directly to the medium. J. Cell Sci. Suppl. 11, 45–57.

Grimard, V., Vigano, C., Margolles, A., Wattiez, R., van Veen, H.W., Konings, W.N., Ruysschaert, J.M. and Goormaghtigh, E. (2001) Structure and dynamics of the membrane-embedded domain of LmrA investigated by coupling polarized ATR-FTIR spectroscopy and (1)H/(2)H exchange. Biochemistry 40, 11876–11886.

Guzzo, J., Duong, F., Wandersman, C., Murgier, M. and Lazdunski, A. (1991) The secretion genes of *Pseudomonas aeruginosa* alkaline protease are functionally related to those of *Erwinia chrysanthemi* proteases and *Escherichia coli* alpha-haemolysin. Mol. Microbiol. 5, 447–453.

Hacker, J., Hughes, C., Hof, H. and Goebel, W. (1983) Cloned hemolysin genes from *Escherichia coli* that cause urinary tract infection determine different levels of toxicity in mice. Infect. Immun. 42, 57–63.

Havarstein, L.S., Diep, D.B. and Nes, I.F. (1995) A family of bacteriocin ABC transporters carry out proteolytic processing of their substrates concomitant with export. Mol. Microbiol. 16, 229–240.

Hines, D.A., Saurugger, P.N., Ihler, G.M. and Benedik, M.J. (1988) Genetic analysis of extracellular proteins of *Serratia marcescens*. J. Bacteriol. 170, 4141–4146.

Hoiczyk, E. and Baumeister, W. (1997) Oscillin, an extracellular, Ca^{2+}-binding glycoprotein essential for the gliding motility of cyanobacteria. Mol. Microbiol. 26, 699–708.

Holland, I.B. and Blight, M.A. (1996) Structure and function of HlyB, the ABC-transporter essential for haemolysin secretion from *Escherichia coli*. In: Handbook Biological Physics (ed. W.N. Konings, H.R. Kaback and J.S. Lolkema). Vol. II, pp. 111–135. Amsterdam: North Holland.

Holland, I.B., Kenny, B., Steipe, B. and Pluckthun, A. (1990) Secretion of heterologous proteins in *Escherichia coli*. Methods Enzymol. 182, 132–143.

Hopfner, K.P., Karcher, A., Shin, D.S., Craig, L., Arthur, L.M., Carney, J.P. and Tainer, J.A.

(2000) Structural biology of Rad50 ATPase: ATP-driven conformational control in DNA double-strand break repair and the ABC-ATPase superfamily. Cell 101, 789–800.

Hui, D., Morden, C., Zhang, F. and Ling, V. (2000) Combinatorial analysis of the structural requirements of the *Escherichia coli* hemolysin signal sequence. J. Biol. Chem. 275, 2713–2720.

Hung, L.W., Wang, I.X., Nikaidó, K., Liu, P.-Q., Ames, G.F.L. and Kim, S.H. (1998) Crystal structure of the ATP-binding subunit of an ABC transporter. Nature 396, 703–707.

Hwang, J., Zhong, X. and Tai, P.C. (1997) Interactions of dedicated export membrane proteins of the colicin V secretion system: CvaA, a member of the membrane fusion protein family, interacts with CvaB and TolC. J. Bacteriol. 179, 6264–6270.

Idei, A., Kawai, E., Akatsuka, H. and Omori, K. (1999) Cloning and characterization of the *Pseudomonas fluorescens* ATP-binding cassette exporter, HasDEF, for the heme acquisition protein HasA. J. Bacteriol. 181, 7545–7551.

Issartel, J.P., Koronakis, V. and Hughes, C. (1991) Activation of *Escherichia coli* prohaemolysin to the mature toxin by acyl carrier protein-dependent fatty acylation. Nature 351, 759–761.

Izadi-Pruneyre, N., Wolff, N., Redeker, V., Wandersman, C., Delepierre, M. and Lecroisey, A. (1999) NMR studies of the C-terminal secretion signal of the haem-binding protein, HasA. Eur. J. Biochem. 261, 562–568.

Jarchau, T., Chakraborty, T., Garcia, F. and Goebel, W. (1994) Selection for transport competence of C-terminal polypeptides derived from *Escherichia coli* hemolysin: the shortest peptide capable of autonomous HlyB/HlyD-dependent secretion comprises the C-terminal 62 amino acids of HlyA. Mol. Gen. Genet. 245, 53–60.

Johnson, J.M. and Church, G.M. (1999) Alignment and structure prediction of divergent protein families: periplasmic and outer membrane proteins of bacterial efflux pumps. J. Mol. Biol. 287, 695–715.

Jones, H.E., Holland, I.B., Baker, H.L. and Campbell, A.K. (1999) Slow changes in cytosolic free Ca^{2+} in *Escherichia coli* highlight two putative influx mechanisms in response to changes in extracellular calcium. Cell Calcium 25, 265–274.

Jones, H.E., Holland, I.B. and Campbell, A.K. (2002) Direct measurement of the free Ca^{2+} shows different regulation of Ca^{2+} between the periplasm and the cytosol of *Escherichia coli*. Cell Calcium, in press.

Jones, P.M. and George, A.M. (1998) A new structural model for P-glycoprotein. J. Membr. Biol. 166, 133–147.

Junop, M.S., Obmolova, G., Rausch, K., Hsieh, P. and Yang, W. (2001) Composite active site of an ABC ATPase: MutS uses ATP to verify mismatch recognition and authorize DNA repair. Mol. Cell 7, 1–12.

Juranka, P., Zhang, F., Kulpa, J., Endicott, J., Blight, M., Holland, I.B. and Ling, V. (1992) Characterization of the hemolysin transporter, HlyB, using an epitope insertion. J. Biol. Chem. 267, 3764–3770.

Kawai, E., Akatsuka, H., Idei, A., Shibatani,T. and Omori, K. (1998) *Serratia marcescens* S-layer protein is secreted extracellularly via an ATP-binding cassette exporter, the Lip system. Mol. Microbiol. 27, 941–952.

Kenny, B. (1990) Analysis of the novel C-terminal secretion of the *E. coli* haemolysin protein. PhD Thesis, University of Leicester.

Kenny, B., Haigh, R. and Holland, I.B. (1991) Analysis of the haemolysin transport process through the secretion from *Escherichia coli* of PCM, CAT or beta-galactosidase fused to the Hly C-terminal signal domain. Mol. Microbiol. 5, 2557–2568.

Kenny, B., Taylor, S. and Holland, I.B. (1992) Identification of individual amino acids required for secretion within the haemolysin (HlyA) C-terminal targeting region. Mol. Microbiol. 6, 1477–1489.

Kenny, B., Chervaux, C. and Holland, I.B. (1994) Evidence that residues -15 to -46 of the haemolysin secretion signal are involved in early steps in secretion, leading to recognition of the translocator. Mol. Microbiol. 11, 99–109.

Koronakis, V., Koronakis, E. and Hughes, C. (1989) Isolation and analysis of the C-terminal signal directing export of *Escherichia coli* hemolysin protein across both bacterial membranes. EMBO J. 8, 595–605.

Koronakis, V., Hughes, C. and Koronakis, E. (1991) Energetically distinct early and late stages of HlyB/HlyD-dependent secretion across both *Escherichia coli* membranes. EMBO J. 10, 3263–3272.

Koronakis, V., Hughes, C. and Koronakis, E. (1993) ATPase activity and ATP/ADP-induced conformational change in the soluble domain of the bacterial protein translocator HlyB. Mol. Microbiol. 8, 1163–1175.

Koronakis, E., Hughes, C., Milisav, I. and Koronakis, V. (1995) Protein exporter function and in vitro ATPase activity are correlated in ABC-domain mutants of HlyB. Mol. Microbiol. 16, 87–96.

Koronakis, V., Li, J., Koronakis, E. and Stauffer, K. (1997) Structure of TolC, the outer membrane component of the bacterial type I efflux system, derived from two-dimensional crystals. Mol. Microbiol. 23, 617–626.

Koronakis, V., Sharff, A., Koronakis, E., Luisi, B. and Hughes, C. (2000) Crystal structure of the bacterial membrane protein TolC central to multidrug efflux and protein export. Nature 405, 914–919.

Kránitz, L., Benabdelhak, H., Horn, C., Blight, A.M., Holland, I.B. and Schmitt, L. (2002) Crystallization and preliminary X-ray analysis of the ATP-binding domain of the ABC-transporter haemolysin B from *E.coli*. Acta Crystallogr. D. 58, 539–541.

Kuhnert, P., Heyberger-Meyer, B., Burnens, A.P., Nicolet, J. and Frey, J. (1997) Detection of RTX toxin genes in gram-negative bacteria with a set of specific probes. Appl. Environ. Microbiol. 63, 2258–2265.

Letoffe, S. and Wandersman, C. (1992) Secretion of CyaA-PrtB and HlyA-PrtB fusion proteins in *Escherichia coli*: involvement of the glycine-rich repeat domain of *Erwinia chrysanthemi* protease B. J. Bacteriol. 174, 4920–4927.

Letoffe, S., Delepelaire, P. and Wandersman, C. (1989) Characterization of a protein inhibitor of extracellular proteases produced by *Erwinia chrysanthemi*. Mol. Microbiol. 3, 79–86.

Letoffe, S., Ghigo, J.M. and Wandersman, C. (1994) Iron acquisition from heme and hemoglobin by a *Serratia marcescens* extracellular protein. Proc. Natl Acad. Sci. USA 91, 9876–9880.

Letoffe, S., Delepelaire, P. and Wandersman, C. (1996) Protein secretion in gram-negative bacteria: assembly of the three components of ABC protein-mediated exporters is ordered and promoted by substrate binding. EMBO J. 15, 5804–5811.

Mackman, N., Baker, K., Gray, L., Haigh, R., Nicaud, J.M. and Holland, I.B. (1987) Release of a chimeric protein into the medium from *Escherichia coli* using the C-terminal secretion signal of haemolysin. EMBO J. 6, 2835–2841.

Nicaud, J.M., Mackman, N., Gray, L. and Holland, I.B. (1985) Characterisation of HlyC and mechanism of activation and secretion of haemolysin from *E. coli* 2001. FEBS Lett. 187, 339–344.

Nicaud, J.M., Mackman, N., Gray, L. and Holland, I.B. (1986) The C-terminal, 23 kDa peptide of *E. coli* haemolysin 2001 contains all the information necessary for its secretion by the haemolysin (Hly) export machinery. FEBS Lett. 204, 331–335.

O'Hanley, P., Lalonde, G. and Ji, G. (1991) Alpha-hemolysin contributes to the pathogenicity of piliated digalactoside-binding *Escherichia coli* in the kidney: efficacy of an alpha-hemolysin vaccine in preventing renal injury in the BALB/c mouse model of pyelonephritis. Infect. Immun. 59, 1153–1161.

Omori, K., Idei, A. and Akatsuka, H. (2001) *Serratia* ATP-binding cassette protein exporter, Lip, recognizes a protein region upstream of the C terminus for specific secretion. J. Biol. Chem. 276, 27111–27119.

Petronilli, V. and Ames, G.F. (1991) Binding protein-independent histidine permease mutants. Uncoupling of ATP hydrolysis from transmembrane signaling. J. Biol. Chem. 266, 16293–16296.

Pimenta, A. (1995) Structure-function analysis of HlyD, implicated in the secretion of haemolysin from *E. coli*. PhD Thesis, University Paris XI.

Pimenta, A., Blight, M., Clarke, D. and Holland, I.B. (1996) The gram-negative cell envelope 'springs' to life: coiled-coil trans-envelope proteins. Mol. Microbiol. 19, 643–645.

Pimenta, A.L., Young, J., Holland, I.B. and Blight, M.A. (1999) Antibody analysis of the localisation, expression and stability of HlyD, the MFP component of the *E. coli* haemolysin translocator. Mol. Gen. Genet. 261, 122–132.

Robinson, C. and Bolhuis, A. (2001) Protein targeting by the twin-arginine translocation pathway. Nat. Rev. Mol. Cell Biol. 2, 350–356.

Saier, M.H., Jr, Tam, R., Reizer, A. and Reizer, J. (1994) Two novel families of bacterial membrane proteins concerned with nodulation, cell division and transport. Mol. Microbiol. 11, 841–847.

Sapriel, G., Wandersman, C. and Delepelaire, P. (2001) The N-terminus of the HasA protein and the SecB chaperone cooperate in the efficient targeting and secretion of HasA via the ABC transporter. J. Biol. Chem. 6, 6.

Saurin, W., Hofnung, M. and Dassa, E. (1999) Getting in or out: early segregation between importers and exporters in the evolution of ATP-binding cassette (ABC) transporters. J. Mol. Evol. 48, 22–41.

Schlor, S., Schmidt, A., Maier, E., Benz, R., Goebel, W. and Gentschev, I. (1997) *In vivo* and *in vitro* studies on interactions between the components of the hemolysin (HlyA) secretion machinery of *Escherichia coli*. Mol. Gen. Genet. 256, 306–319.

Schulein, R., Gentschev, I., Mollenkopf, H.J. and Goebel, W. (1992) A topological model for the haemolysin translocator protein HlyD. Mol. Gen. Genet. 234, 155–163.

Sebo, P. and Ladant, D. (1993) Repeat sequences in the *Bordetella pertussis* adenylate cyclase toxin can be recognized as alternative carboxy-proximal secretion signals by the *Escherichia coli* alpha-haemolysin translocator. Mol. Microbiol. 9, 999–1009.

Sheps, J.A., Cheung, I. and Ling, V. (1995) Hemolysin transport in *Escherichia coli*. Point mutants in HlyB compensate for a deletion in the predicted amphiphilic helix region of the HlyA signal. J. Biol. Chem. 270, 14829–14834.

Shyamala, V., Baichwal, V., Beall, E. and Ames, G.F. (1991) Structure-function analysis of the histidine permease and comparison with cystic fibrosis mutations. J. Biol. Chem. 266, 18714–18719.

Stanley, P., Koronakis, V. and Hughes, C. (1991) Mutational analysis supports a role for multiple structural features in the C-terminal secretion signal of *Escherichia coli* haemolysin. Mol. Microbiol. 5, 2391–2403.

Thanabalu, T., Koronakis, E., Hughes, C. and Koronakis, V. (1998) Substrate-induced assembly of a contiguous channel for protein export from *E. coli*: reversible bridging of an inner-membrane translocase to an outer membrane exit pore. EMBO J. 17, 6487–6496.

Thompson, S.A. and Sparling, P.F. (1993) The RTX cytotoxin-related FrpA protein of *Neisseria meningitidis* is secreted extracellularly by meningococci and by HlyBD + *Escherichia coli*. Infect. Immun. 61, 2906–2911.

Tzschaschel, B.D., Guzman, C.A., Timmis, K.N. and de Lorenzo, V. (1996) An *Escherichia coli* hemolysin transport system-based vector for the export of polypeptides: export of Shiga-like toxin IIeB subunit by *Salmonella typhimurium* aroA. Nat. Biotechnol. 14, 765–769.

van Belkum, M.J., Worobo, R.W. and Stiles, M.E. (1997) Double-glycine-type leader peptides direct secretion of bacteriocins by ABC transporters: colicin V secretion in *Lactococcus lactis*. Mol. Microbiol. 23, 1293–1301.

van der Meer, J.R., Rollema, H.S., Siezen, R.J., Beerthuyzen, M.M., Kuipers, O.P. and de Vos, W.M. (1994) Influence of amino acid substitutions in the nisin leader peptide on biosynthesis and secretion of nisin by *Lactococcus lactis*. J. Biol. Chem. 269, 3555–3562.

van Wely, K.H., Swaving, J., Freudl, R. and Driessen, A.J. (2001) Translocation of proteins across the cell envelope of Gram-positive bacteria. FEMS Microbiol. Rev. 25, 437–454.

Wandersman, C. and Letoffe, S. (1993) Involvement of lipopolysaccharide in the secretion of *Escherichia coli* alpha-haemolysin and *Erwinia chrysanthemi* proteases. Mol. Microbiol. 7, 141–150.

Wang, R.C., Seror, S.J., Blight, M., Pratt, J.M., Broome-Smith, J.K. and Holland, I.B. (1991) Analysis of the membrane organization of an *Escherichia coli* protein translocator, HlyB, a member of a large family of prokaryote and eukaryote surface transport proteins. J. Mol. Biol. 217, 441–454.

Ward, M.J., Mok, K.C., Astling, D.P., Lew, H. and Zusman, D.R. (1998) An ABC transporter plays a developmental aggregation role in *Myxococcus xanthus*. J. Bacteriol. 180, 5697–5703.

Weiner, J.H., Bilous, P.T., Shaw, G.M., Lubitz, S.P., Frost, L., Thomas, G.H., Cole, J.A. and Turner, R.J. (1998) A novel and ubiquitous system for membrane targeting and secretion of cofactor-containing proteins. Cell 93, 93–101.

Welch, R.A., Dellinger, E.P., Minshew, B. and Falkow, S. (1981) Haemolysin contributes to virulence of extra-intestinal *E. coli* infections. Nature 294, 665–667.

Welch, R.A., Forestier, C., Lobo, A., Pellett, S., Thomas, W., Jr and Rowe, G. (1992) The synthesis and function of the *Escherichia coli* hemolysin and related RTX exotoxins. FEMS Microbiol. Immunol. 5, 29–36.

Whitehead, J. (1993) The role of molecule chaperones in protein export and secretion in *E. coli*. PhD Thesis, University of Liverpool.

Wolff, N., Ghigo, J.M., Delepelaire, P., Wandersman, C. and Delepierre, M. (1994) C-terminal secretion signal of an *Erwinia chrysanthemi* protease secreted by a signal peptide-independent pathway: proton NMR and CD conformational studies in membrane-mimetic environments. Biochemistry 33, 6792–6801.

Wolff, N., Delepelaire, P., Ghigo, J.M. and Delepierre, M. (1997) Spectroscopic studies of the C-terminal secretion signal of the *Serratia marcescens* haem acquisition protein (HasA) in various membrane- mimetic environments. Eur. J. Biochem. 243, 400–407.

York, G.M. and Walker, G.C. (1997) The *Rhizobium meliloti* exoK gene and prsD/prsE/exsH genes are components of independent degradative pathways which contribute to production of low-molecular-weight succinoglycan. Mol. Microbiol. 25, 117–134.

Young, J. (1999) Biochemical analysis of the translocator proteins of the haemolysin system and their interactions, *in vivo*. PhD Thesis, University Paris XI.

Young, J. and Holland, I.B. (1999) ABC transporters: bacterial exporters-revisited five years on. Biochim. Biophys. Acta 1461, 177–200.

Zgurskaya, H.I. and Nikaido, H. (2000) Multidrug resistance mechanisms: drug efflux across two membranes. Mol. Microbiol. 37, 219–225.

Zhang, F., Greig, D.I. and Ling, V. (1993) Functional replacement of the hemolysin A transport signal by a different primary sequence. Proc. Natl Acad. Sci. USA 90, 4211–4215.

Zhang, F., Yin, Y., Arrowsmith, C.H. and Ling, V. (1995) Secretion and circular dichroism analysis of the C-terminal signal peptides of HlyA and LktA. Biochemistry 34, 4193–4201.

Zhang, Y., Bak, D.D., Heid, H. and Geider, K. (1999) Molecular characterization of a protease secreted by *Erwinia amylovora*. J. Mol. Biol. 289, 1239–1251.

Zhong, X., Kolter, R. and Tai, P.C. (1996) Processing of colicin V-1, a secretable marker protein of a bacterial ATP binding cassette export system, requires membrane integrity, energy, and cytosolic factors. J. Biol. Chem. 271, 28057–28063.

Bacterial Multidrug Resistance Mediated by ABC Transporters

CHAPTER 12

Gerrit J. Poelarends, Catherine Vigano, Jean-Marie Ruysschaert and Wil N. Konings

Introduction

Microorganisms are confronted daily with numerous environmental toxins. The spectrum of these toxins ranges from naturally produced compounds (e.g. plant alkaloids), peptides (e.g. bacteriocins) and noxious metabolic products (e.g. bile salts and fatty acids in the case of enteric bacteria) to industrially produced chemicals such as organic solvents and antibiotics. Microorganisms have developed various mechanisms to resist the toxic effects of antimicrobial agents, and drug-resistant pathogens are on the rise (Cohen, 1992; Culliton, 1992; Hayes and Wolf, 1990; Nikaido, 1994). One of the resistance mechanisms involves the active extrusion of antimicrobials from the cell by drug transport systems. Some transporters, such as the tetracycline efflux proteins (Roberts, 1996), are dedicated systems which mediate the extrusion of a given drug or class of drugs. In contrast to these *s*pecific *d*rug *r*esistance (SDR) transporters, the so-called *m*ulti*d*rug *r*esistance (MDR) transporters can handle a wide variety of structurally unrelated compounds. On the basis of bioenergetic and structural criteria, multidrug transporters can be divided into two major classes: (i) secondary transporters, which are driven by a proton or sodium motive force, and (ii) ATP-binding cassette (ABC) primary transporters, which use the hydrolysis of ATP to fuel transport (for a recent review, see Putman *et al.*, 2000b).

Most bacterial multidrug transporters known to date are secondary antiport systems that remove drugs from the cell in a coupled exchange with protons or sodium ions. On the basis of size and similarities in secondary structure, these transporters are classified into four major groups: the major facilitator superfamily (MFS), the small multidrug resistance (SMR) family, the resistance nodulation cell division (RND) family, and the multidrug and toxic compound extrusion (MATE) family (Putman *et al.*, 2000b). Besides these secondary multidrug transporters, a number of ATP-dependent primary drug transporters have also been identified (e.g. Barrasa *et al.*, 1995; Linton *et al.*, 1994; Olano *et al.*, 1995; Podlesek *et al.*, 1995; Rodríguez *et al.*, 1993; Ross *et al.*, 1990). These primary drug transporters all belong to the ABC transporter superfamily, and most of them are SDR transporters. A well-known example is DrrAB, an SDR transporter of *Streptomyces peucetius*, which confers self-resistance to its secondary metabolites daunorubicin and doxorubicin (Guilfoile and Hutchinson, 1991).

In the Gram-positive bacterium *Lactococcus lactis*, an organism used in food manufacturing **(Figure 12.1)**, two distinct MDR transporters mediate resistance to toxic hydrophobic cations and antibiotics. One system, designated LmrP, is a proton/drug antiport system **(Figure 12.2)**. It belongs to the major facilitator superfamily, and is inhibited by ionophores that dissipate

ABC Proteins: From Bacteria to Man
ISBN 0-12-352551-9

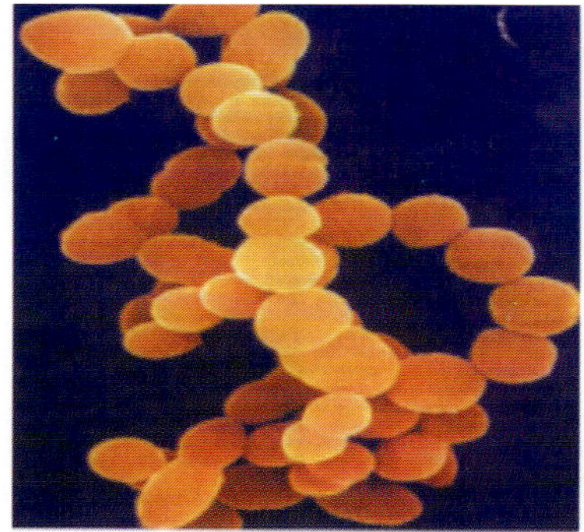

Figure 12.1. The Gram-positive lactic acid bacterium **Lactococcus lactis** *(left picture) is used in starter cultures for cheese production.*

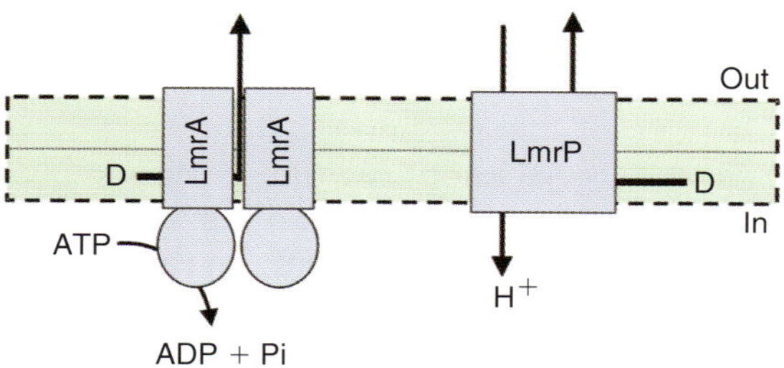

Figure 12.2. Schematic representation of two multidrug transporters found in **Lactococcus lactis.** *The ABC-type primary multidrug transporter LmrA and the secondary multidrug transporter LmrP exemplify the two major classes of multidrug transporters found in bacteria. Rectangles represent the transmembrane domains of LmrA and LmrP. Circles represent the nucleotide-binding domains of LmrA.*

the proton motive force (Bolhuis *et al.*, 1995). The other MDR system is an ATP-dependent primary transporter, designated LmrA **(Figure 12.2)** (van Veen *et al.*, 1996). The role of this chromosomally encoded primary efflux pump in multidrug resistance was first observed in an ethidium-resistant mutant of *L. lactis* subsp. *lactis* MG1363. Ethidium efflux in this mutant was inhibited by *ortho*-vanadate, an inhibitor of ABC transporters and P-type ATPases, but not upon dissipation of the proton motive force (Bolhuis *et al.*, 1994). Isolated membrane vesicles and proteoliposomes, in which purified LmrA was reconstituted, were employed to prove that transport of multiple drugs was LmrA- and ATP-dependent (Margolles *et al.*, 1999; van Veen *et al.*, 1996). Interestingly, this lactococcal LmrA protein was the first ABC-type multidrug transporter identified in bacteria.

Another ABC-type multidrug resistance pump (HorA) was discovered in *Lactobacillus brevis*, a major contaminant of spoiled beer (Sami *et al.*, 1997, 1998). This Gram-positive lactic acid bacterium can grow in beer in spite of the presence of antibacterial compounds (iso-α-acids) derived from the flowers of the hop plant *Humulus lupulus* L. The hop resistance of *Lb. brevis* is, at least in part, dependent on the expression of the *horA* gene, which is located on a 15 kb plasmid termed pRH45 (Sami *et al.*, 1997). The role of HorA in hop resistance was first suggested by a spontaneous mutant lacking the pRH45 plasmid, which displayed sensitivity to the presence of hop compounds. Reintroduction of pRH45 into this segregation mutant restored hop resistance (Sami *et al.*, 1998). These complementation studies, as well as the heterologous expression of the *horA* gene in *L. lactis*, demonstrated that HorA is involved in resistance to hop compounds. Moreover, almost all lactobacilli isolated as beer-spoilage strains possess *horA* homologues (Sami *et al.*, 1997). In addition to conferring hop resistance, HorA confers resistance to the structurally unrelated drugs novobiocin and ethidium bromide (Sami *et al.*, 1997). Drug transport studies in *L. lactis* cells and membrane vesicles and in proteoliposomes in which purified HorA was reconstituted identified this protein as a new member of the ABC family of multidrug transporters (Sakamoto *et al.*, 2001).

Here we summarize the existing data on the two bacterial ABC-type multidrug transporters LmrA and HorA, and analyze structural and mechanistic aspects of multidrug recognition and transport. In addition, the chapter will describe how attenuated total reflection Fourier transform infrared (ATR-FTIR) spectroscopy

Figure 12.3. Topology model for LmrA. The LmrA protein is predicted to contain a transmembrane domain (TMD) with six transmembrane α-helices, and a nucleotide-binding domain (NBD) with the ABC signature and Walker A/B sequences. A similar model is envisaged for HorA of **Lb. brevis.**

has provided important information about LmrA structure and the dynamic changes occurring during its catalytic cycle.

PROPERTIES OF LMRA AND HORA

STRUCTURAL ASPECTS

All ABC transporters described so far show a four-domain organization, which consists of two transmembrane domains (TMDs), which are thought to perform the transport function, and two nucleotide-binding domains (NBDs), which provide the energy for the transport process (Higgins, 1992). The four domains may be organized either in a multifunctional, single polypeptide or as separate proteins. For example, in human P-glycoprotein (MDR1), like many eukaryotic ABC transporters, the four domains are found in one single polypeptide chain arranged as TMD1-NBD1-TMD2-NBD2. As derived from the DNA sequences, bacterial LmrA is composed of 590 amino acids (calculated molecular mass of 64.6 kDa) and HorA of 583 amino acids (calculated molecular mass of 64.2 kDa). Hydropathy analysis, as shown in **Figure 12.3**, suggests a putative topology for both proteins of six membrane-spanning regions (putative α-helices) in the amino-terminal hydrophobic domain, followed by a large hydrophilic domain containing the ATP-binding site (Sami *et al.*, 1997; van Veen *et al.*, 1996). There is now experimental evidence that the membrane-spanning regions of LmrA are indeed α-helices (Grimard *et al.*, 2001). Based on the topology predictions, both the amino-terminal end and the large carboxy-terminal half are located in the cytoplasm. In addition to the NBD, there are two putative large cytoplasmic loops **(Figure 12.3)** (see also Chapter 11, HlyB). The predicted membrane topologies of LmrA and HorA still await experimental confirmation. The NBDs of both these bacterial transporters contain features diagnostic of an

ABC-type ATPase, such as the ABC signature sequence, and the Walker A and B motifs **(Figure 12.4)**. The sequence homology between full-length LmrA and HorA is around 53%. Sequence comparisons with other ABC transporters revealed that these bacterial proteins share significant overall sequence similarity with members of the subfamily of multidrug resistance P-glycoproteins, most notably the human P-glycoprotein (MDR1) (Sami *et al.*, 1997; van Veen *et al.*, 1996). For example, LmrA and each half of MDR1 share 34% identical residues and an additional 16% of conservative substitutions **(Figure 12.4)**. The ABC domain of LmrA and the ABC1 and ABC2 domains of MDR1 are 48% and 43% identical, respectively, whereas the identity between the TMD of LmrA and the amino- and carboxy-terminal TMDs of MDR1 is 23% and 27%, respectively. The sequence conservation in the TMD of LmrA includes particular regions (e.g. the region comprising transmembrane helices 5 and 6), which have been implicated as being involved in drug binding by MDR1 (Loo and Clarke, 2000). Functionally important residues in this region of LmrA are now being identified. Interestingly, LmrA shares 28% overall sequence identity with the lipid flippase MsbA from *Escherichia coli* **(Figure 12.4)**, the structure of which was recently determined by X-ray crystallography to a resolution of 4.5 Å (Chang and Roth, 2001). The overall sequence similarity between LmrA and bacterial members of other subfamilies of the ABC transporter superfamily is less than 28% and is mostly confined to the hydrophilic ABC domains.

In view of the general organization of ABC transporters, LmrA and HorA are considered to be half transporters (with the two domains arranged in TMD-NBD manner) that have to form homodimers in order to function as full four-domain transporters. Recent studies on LmrA provided evidence that this is indeed the case. First, two covalently linked wild-type LmrA monomers expressed from an engineered gene yields a functional transporter, whereas the covalent linkage of a wild-type monomer and an inactive mutant monomer (harboring the K388M mutation in the Walker A region) yields an inactive transporter (van Veen *et al.*, 2000). The latter covalently linked dimer had also lost all ATPase activity, demonstrating that both catalytic sites must be functional to allow ATP hydrolysis and drug transport. Second, LmrA solubilized from membrane vesicles prepared from LmrA-overproducing cells behaves like a dimer on native gels (our unpublished data). Third, electron microscopy analysis of purified and reconstituted LmrA revealed small, uniform particles with a diameter of 8.5 by 5 nm, similar to those previously observed for monomeric P-glycoprotein (S. Scheuring, A. Margolles, H.W. van Veen, W.N. Konings and A. Engel, unpublished data). Probably, the most convincing evidence for the dimeric nature of LmrA comes from co-reconstitution experiments into proteoliposomes of the cysteine-less wild-type LmrA and a mutant form of LmrA in which the *N*-ethylmaleimide (NEM)-reactive glycine to cysteine mutation (G386C) was introduced (van Veen *et al.*, 2000). The G386C mutant displays wild-type transport activity but is completely inactivated upon incubation with NEM, whereas wild-type LmrA activity is not affected by NEM. The transport inhibition patterns obtained with proteoliposomes, containing different ratios of wild-type and mutant proteins, upon reaction with NEM suggest strongly that the functional unit of LmrA is a dimer and not a monomer, trimer or tetramer. Taking all these data together, it is clear that the dimeric state of LmrA is a prerequisite for function, and that functional crosstalk between two monomers is essential for transport.

SUBSTRATE SPECIFICITY

The notion that inactivation of the secondary multidrug transporter LmrP increases drug extrusion mediated by the primary transporter LmrA points to the physiological importance of these multidrug transporters in *L. lactis* (Bolhuis *et al.*, 1995). However, except for the observation that LmrA might act as a lipid translocase (Margolles *et al.*, 1999), its cellular function is still under debate. The natural substrates of LmrA might be found amongst the hydrophobic compounds excreted by plants, the natural habitat of lactococci. Indeed, LmrA can extrude a wide variety of amphiphilic toxic compounds, and its classification as a multidrug transporter is evident from its currently known spectrum of substrates. LmrA substrates include anticancer drugs such as vinca alkaloids (vinblastine, vincristine) and anthracyclines (daunomycin, doxorubicin), or cytotoxic agents such as antimicrotubule drugs (colchicine) and DNA intercalators (ethidium bromide), or toxic peptides (valinomycin, nigericin), fluorescent membrane probes (Hoechst 33342, diphenylhexatriene), and fluorescent dyes such as

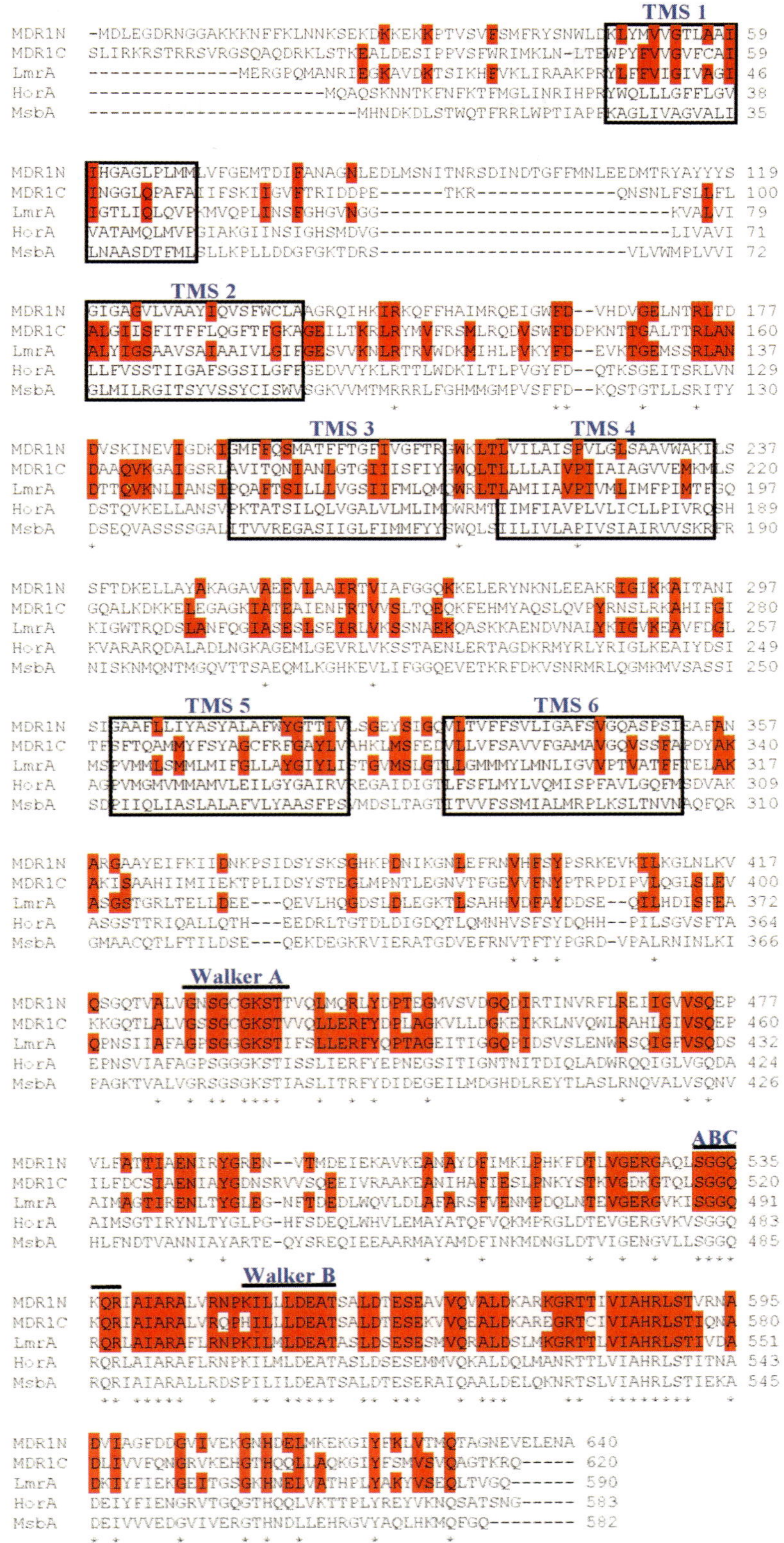

Figure 12.4. Amino acid sequence alignment. The amino acid sequence of LmrA is shown with HorA from Lb. brevis, *MsbA from* E. coli, *and the amino- and carboxy-terminal halves of human MDR1. Residues conserved throughout all sequences are indicated by an asterisk. Residues conserved between LmrA and MDR1 are shaded red. Dashes represent residues absent in other sequences. Putative transmembrane regions are boxed. The Walker A/B motifs and the ABC signature motif regions are labeled by Walker A, Walker B and ABC, respectively.*

rhodamine 6G and rhodamine 123 (Margolles *et al.*, 1999; van Veen *et al.*, 1996, 1998; see more detailed discussion in Chapter 5). LmrA modulators (i.e. compounds that reverse LmrA-mediated multidrug resistance) are also structurally unrelated to each other and include the calcium channel blockers verapamil and CP100-356 (analogue of verapamil), 1,4-dihydropyridines such as nicardipine, indolizine sulfones such as SR33557, antimalarials such as quinine and quinidine, immunosuppressants such as cyclosporin A, and the *Rauwolfia* alkaloid reserpine (van Veen *et al.*, 1999). This broad drug and modulator specificity is not only confined to LmrA. A similar range of compounds was previously found to interact with other ABC transporters, including yeast Pdr5p (Bauer *et al.*, 2000; Kolaczkowski *et al.*, 1996) and human P-glycoprotein (Ueda *et al.*, 1997).

The overlapping substrate and modulator specificities of bacterial LmrA and human P-glycoprotein reveal a functional similarity between both proteins. Expression studies of LmrA in insect and human lung fibroblast cells demonstrated that LmrA was indeed able to functionally complement P-glycoprotein (van Veen *et al.*, 1998). Surprisingly, LmrA was targeted to the plasma membrane and conferred typical multidrug resistance on the human cells. The pharmacological characteristics of LmrA and P-glycoprotein expressed in lung fibroblast cells were very similar, and reversal agents of P-glycoprotein-mediated multidrug resistance also blocked multidrug resistance mediated by LmrA. Furthermore, the affinities of both proteins for vinblastine and ATP were indistinguishable. Finally, kinetic analysis of drug dissociation from LmrA expressed in plasma membranes of insect cells revealed the presence of two allosterically coupled drug-binding sites, indistinguishable from those of P-glycoprotein (van Veen *et al.*, 1998; Chapter 5). This remarkable conservation of function between these two ABC-type multidrug transporters implies a common overall structure and transport mechanism.

L. lactis is a GRAS (generally regarded as safe) organism, that is, an organism considered to be non-pathogenic and safe to use in starter cultures for cheese production **(Figure 12.1)** (Gasser, 1994). In view of this, it is important to know whether the substrate spectrum of LmrA also includes clinically relevant antibiotics. The antibiotic specificity of LmrA was studied in cytotoxicity assays, in which the antibiotic susceptibilities of *E. coli* CS1562 cells overexpressing the transporter are compared with those of control CS1562 cells not expressing LmrA. Strain CS1562 (*tolC6::Tn10*) was used in these assays because it is hypersensitive to drugs owing to a deficiency in the TolC protein, resulting in an impaired barrier function of the outer membrane (Austin *et al.*, 1990). LmrA expression in CS1562 cells resulted in an increased resistance to 17 out of 21 clinically most used antibiotics, including broad-spectrum antibiotics belonging to the classes of aminoglycosides, lincosamides, macrolides, quinolones, streptogramins and tetracyclines **(Table 12.1)**

TABLE 12.1. EFFECT OF LMRA EXPRESSION IN *E. COLI* CS1562 ON THE RELATIVE RESISTANCE TO ANTIBIOTICS

Class	Antibiotic	Relative resistance[a] (fold)
Aminoglycosides	Gentamicin	2
	Kanamycin	3
β-Lactams	Ampicillin	2
	Ceftazidime	3
	Meropenem	1
	Penicillin	4
Glycopeptides	Vancomycin	1
Lincosamides	Clindamycin	14
Macrolides	Azithromycin	33
	Clarithromycin	23
	Dirithromycin	264
	Erythromycin	53
	Roxithromycin	35
	Spiramycin	35
Quinolones	Ciprofloxacin	2
	Ofloxacin	4
Streptogramins	Dalfopristin	163
	Quinupristin	31
	RP59500	55
Tetracyclines	Chlortetracycline	28
	Demeclocycline	12
	Minocycline	138
	Oxytetracycline	8
	Tetracycline	14
Others	Chloramphenicol	11
	Trimethoprim	1

[a]Relative resistances were determined by dividing the IC_{50} (the antibiotic concentration required to inhibit the growth rate by 50%) for cells harboring pGKLmrA by the IC_{50} for control cells harboring pGK13. For example, the latter IC_{50} values varied between 0.3 and 2 μM for kanamycin, ampicillin, erythromycin, ofloxacin, dalfopristin, and minocycline. Data obtained from Putman *et al.* (2000a) with permission.

(Putman *et al.*, 2000a). The secondary multidrug transporter LmrP also confers resistance to antibiotics, although its substrate range is smaller than that of LmrA (Putman *et al.*, 2001). The antibiotic specificity of HorA is currently being analyzed. The exceptionally broad antibiotic specificity of LmrA, the possible transfer of the *lmrA* gene to other bacteria in food or the digestive tract, and the presence of *lmrA* homologues in pathogenic microorganisms (van Veen and Konings, 1998) provide a serious threat to the efficacy of valuable antibiotics. It will be interesting to find out whether P-glycoprotein is involved in antibiotic export in human cells.

Using a fluorescence quenching technique, it has recently been demonstrated that purified and reconstituted LmrA can also transport phospholipids (Margolles *et al.*, 1999). In this study, extrusion of fluorescent (C_6-NBD-labeled) phosphatidylethanolamine from the outer leaflet of proteoliposomes by inward-facing LmrA molecules (nucleotide-binding domain exposed to the external surface) was detected in the presence of ATP, with non-hydrolyzable ATP analogues being ineffective. Phospholipid extrusion from the membrane was inhibited by vinblastine, a high-affinity substrate of LmrA. The specificity of LmrA with respect to lipid headgroup and acyl chain is now being studied, possibly leading to the identification of potential physiological lipid substrates. Several other ABC multidrug transporters have also been found to possess lipid translocation activity, including P-glycoprotein (for a recent review, see Borst *et al.*, 2000 and Chapter 22 of this volume).

SUBSTRATE RECOGNITION AND TRANSPORT MODELS

Aqueous pore versus hydrophobic vacuum cleaner and flippase models

Despite the remarkable conservation of functional properties between ABC-type multidrug transporters, there is still a considerable controversy about the mechanisms by which these proteins pump drugs from the interior of the cell to the external medium. Several transport models have been postulated for P-glycoprotein pump function **(Figure 12.5)**. These include (i) the conventional aqueous pore model, in which substrate is transported from the cytoplasm to the exterior (Altenberg *et al.*, 1994), (ii) the hydrophobic vacuum cleaner model, in which

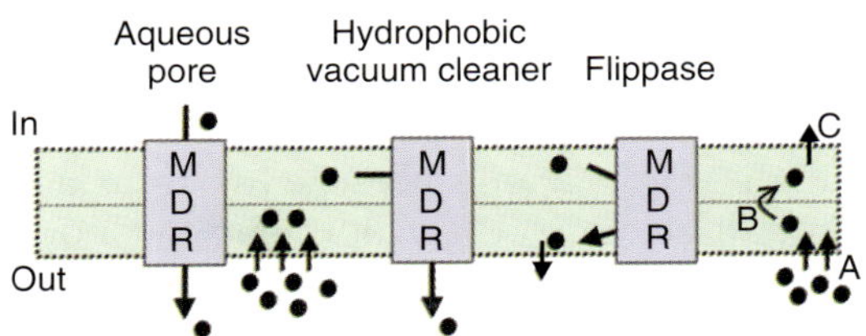

Figure 12.5. ***Possible mechanisms of drug transport across the cytoplasmic membrane. Drugs may be expelled from the cell by extrusion from the internal water phase to the external water phase (aqueous pore model) or by extrusion from the membrane to the exterior (hydrophobic vacuum cleaner and flippase models). Importantly, the hydrophobic vacuum cleaner model predicts that hydrophobic compounds are translocated by the MDR pump from the inner leaflet of the membrane to the external water phase, whereas the flippase model predicts extrusion from the inner leaflet to the outer leaflet of the membrane. A, Drug molecules reaching the cell rapidly insert into the outer leaflet of the plasma membrane. B, Flipping of the drug to the inner leaflet of the membrane is relatively slow and the rate-limiting step in entry. C, Membrane release of drug molecules.***

substrate is transported from the lipid bilayer to the exterior (Raviv *et al.*, 1990), and (iii) the flippase model, a variation on the hydrophobic vacuum cleaner model, in which substrate is transported from the inner leaflet to the outer leaflet of the lipid bilayer, after which the substrate molecules will diffuse into the external medium (Higgins and Gottesman, 1992). The latter two models take into account that most drugs that interact with multidrug transporters such as P-glycoprotein and LmrA readily intercalate into the lipid bilayer due to their high hydrophobicity and amphiphilic nature. Drug extrusion from the membrane is supported by substantial experimental evidence, including the following important observations. First, the non-fluorescent compound BCECF-AM (an acetoxymethyl ester derivative of 2′,7′-bis-(2-carboxyethyl)-5-(and-6-)-carboxyfluorescein) is excreted from P-glycoprotein- and LmrA-producing cells prior to hydrolysis into the fluorescent cellular indicator BCECF by intracellular esterases (Bolhuis *et al.*, 1996; Homolya *et al.*, 1993). Thus, LmrA and P-glycoprotein prevent the accumulation of the fluorescent indicator BCECF in the cytosol, despite the fact that BCECF-AM is rapidly cleaved by intracellular esterases and the resulting BCECF is not a substrate for LmrA and P-glycoprotein.

These observations strongly suggest that BCECF-AM is extruded from the membrane. Second, photoaffinity analogues of substrates of P-glycoprotein only label the two transmembrane domains of P-glycoprotein and not its hydrophilic ABC domains (e.g. Greenberger, 1993). Third, the affinity of binding of drugs to purified and reconstituted P-glycoprotein is modulated by their ability to intercalate into the membrane (Romsicki and Sharom, 1999). Fourth, cysteine-scanning mutagenesis, in combination with reaction with the thiol-reactive substrate dibromobimane, of all predicted transmembrane segments of P-glycoprotein indicates that the drug-binding domain of P-glycoprotein consists of residues in transmembrane segments 4, 5, 6, 10, 11 and 12 (Loo and Clarke, 2000). Taking these data together, it seems likely that these transporters recognize most, if not all, of their substrates within the membrane (hydrophobic vacuum cleaner and flippase models) and not from the cytoplasm (aqueous pore model). However, these observations do not discriminate between the vacuum cleaner model and the flippase model.

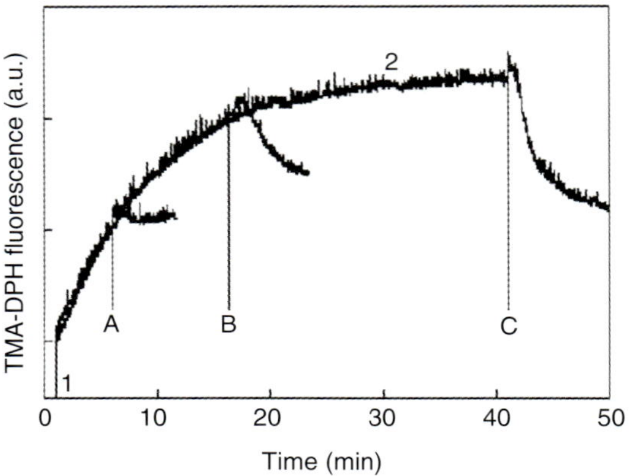

Figure 12.6. ***Time course of the rate of energy-dependent TMA-DPH extrusion. A washed cell suspension of* L. lactis *strain MG1363 (Eth^R), a mutant strain in which extrusion of TMA-DPH is LmrA-dependent, was energized with 25 mM of glucose, at 5 (A), 15 (B), and 40 min (C) after the addition of 100 nM of TMA-DPH. Data obtained from Bolhuis* et al. *(1996).***

Evidence for drug efflux from the inner leaflet of the lipid bilayer to the exterior

The most convincing evidence for drug efflux from the membrane to the aqueous phase is provided by the kinetics of ATP-dependent transport of TMA-DPH by LmrA (Bolhuis *et al.*, 1996) and of Hoechst 33342 by P-glycoprotein (Shapiro and Ling, 1997a). The amphiphilic character and the high lipid–water partition coefficients result in partitioning of these compounds into the lipid bilayer. Conveniently, these hydrophobic probes are strongly fluorescent when partitioned into the membrane but essentially non-fluorescent in an aqueous environment. Since, therefore, the fluorescence detected reflects the concentration of probe in the membrane, these properties make it possible to follow fluorimetrically the partitioning of these compounds into the lipid bilayer. The increase in fluorescence intensity due to the partitioning of TMA-DPH into the phospholipid bilayer was found to be a biphasic process **(Figure 12.6)** (Bolhuis *et al.*, 1996). This biphasic behavior reflects the fast entry (1–2 seconds) of TMA-DPH into the outer leaflet of the phospholipid bilayer (phase 1 in **Figure 12.6**), followed by a slower (several minutes) transbilayer movement from the outer to the inner leaflet of the membrane (phase 2 in **Figure 12.6**). When LmrA was energized in intact cells by the addition of glucose, it was observed that the initial rate of extrusion of TMA-DPH, monitored as a decrease in fluorescence over time, increased with an increasing concentration of TMA-DPH in the inner leaflet of the membrane **(Figure 12.6)** (Bolhuis *et al.*, 1996). The extent of extrusion never exceeded the amount of TMA-DPH present in the inner leaflet **(Figure 12.6)**, indicating that the probe cannot be extruded from the outer leaflet of the cytoplasmic membrane. When similar experiments were done with inside-out membrane vesicles with the inner leaflet now immediately accessible to drug molecules, the situation was significantly different. Upon addition of TMA-DPH to the membrane vesicle suspension, TMA-DPH rapidly intercalates into the exposed leaflet of the membrane, resulting in a maximum concentration of TMA-DPH in this leaflet. Upon energization of LmrA by the addition of ATP (the NBD of LmrA is exposed to the exterior of these vesicles), maximal rates of TMA-DPH extrusion were observed at any moment after addition of TMA-DPH and the extent of extrusion, in contrast to intact cells, now exceeded the amount of TMA-DPH present in the internal leaflet of inside-out vesicles. These observations strongly indicate that TMA-DPH is recognized as a substrate only after partitioning into the normal inner leaflet of the cellular membrane, and is directly transported to the aqueous environment as observed by the decrease in fluorescence. A similar relationship between the initial rate of transport and the concentration of substrate in the inner leaflet of the cellular membrane was observed for other multidrug transporters, including secondary transporters (Putman *et al.*, 2000b). Thus, secondary and ABC multidrug transporters appear to use the same mechanism of transport for hydrophobic drugs.

Hydrophobic vacuum cleaner model versus flippase model

It is important to note that the results presented strongly favor a vacuum cleaner mechanism of transport by the MDR pumps and are inconsistent with a flippase mechanism as proposed by Higgins and Gottesman (1992). According to the flippase mechanism the hydrophobic compounds are translocated by the MDR pump from the inner leaflet of the membrane to the outer leaflet followed by diffusion into the external medium **(Figure 12.5)**. The observation that the fluorescence of TMA-DPH or Hoechst 33342 falls rapidly upon energization of LmrA or P-glycoprotein indicates that these compounds do not stay in the lipid bilayer but are moved into the water phase.

Physiological implications of drug transport from the inner leaflet of the membrane

This mechanism of transport of hydrophobic drugs from the inner leaflet of the phospholipid bilayer to the exterior, as illustrated in **Figure 12.7**, may have several physiological implications. First, transport from the cytoplasmic leaflet of the membrane appears to be the most efficient way in which MDR transporters can prevent toxic compounds from entering the cytoplasm. As already pointed out, drug molecules reaching the cell rapidly insert into the outer leaflet of the membrane, but flipping of the drug molecules from the outer to inner leaflet is slow and the rate-limiting step in entry **(Figure 12.7)**. LmrA is able to transport drug molecules from the inner leaflet back into the external medium, counteracting the rate-limiting step in drug entry. If the transporter were to transport drugs from the outer leaflet of the membrane, it would probably not be able to compete with the high rate at which drug molecules enter this leaflet. Drug molecules would

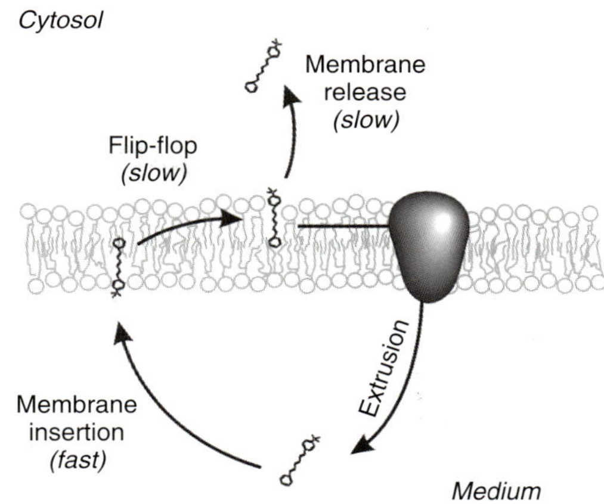

Figure 12.7. ***Proposed mechanism of LmrA-mediated TMA-DPH extrusion from a cell. The initial rate of LmrA-dependent TMA-DPH transport correlates with the amount of TMA-DPH in the inner membrane leaflet of whole cells, and with the amount of TMA-DPH in the outer leaflet of inside-out vesicles. Since the outer membrane leaflet of inside-out vesicles corresponds to the inner membrane leaflet of whole cells, both observations rule out the external leaflet of whole cells as a possible site of drug binding to LmrA.***

'escape' into the inner leaflet and subsequently enter the cytoplasm. Second, extrusion from the membrane may partially explain the lack of structural specificity and the consequent broad substrate range of multidrug transporters. The transmembrane domains of multidrug transporters are thought to form a pathway across the membrane through which solutes move, a prediction supported by structural data of P-glycoprotein (Rosenberg *et al.*, 2001) and MsbA (Chang and Roth, 2001). Assuming that the translocation pathway is only accessible from the membrane, but not from the aqueous phase, a drug molecule must be able to intercalate into the membrane in order to be recognized by the transporter. Thus, the ability to intercalate into the membrane may pre-select compounds to be transported from those which should not be transported (e.g. hydrophilic cytoplasmic components). The subsequent interaction between the intercalated substrate and the transporter would be a second determinant of specificity. This would allow the transporter to have (a) relatively non-selective substrate-binding site(s).

NUMBER OF SUBSTRATE-BINDING SITES

Studies on the kinetics of drug dissociation have revealed the presence of two distinct, but

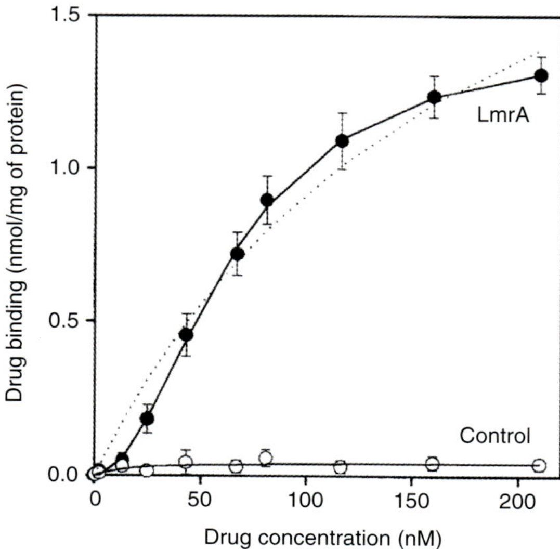

Figure 12.8. ***Vinblastine equilibrium binding to LmrA. Specific binding of [^{3}H]vinblastine to inside-out membrane vesicles without LmrA (control) or with LmrA, as a function of the free vinblastine concentration. Superimposed on the data are the best-fit curves obtained for a single-site binding model (hyperbolic, dotted curve) and the cooperative two-site binding model (sigmoidal, solid curve). Data obtained from van Veen* et al. *(2000) with permission from Oxford University Press.***

allosterically linked, drug-binding sites in the LmrA transporter (van Veen *et al.*, 1998, 2000; Chapter 5). The presence of these two drug-binding sites in LmrA is strongly supported by the finding that vinblastine equilibrium binding can best be fitted by a model in which two vinblastine-binding sites in the LmrA transporter interact cooperatively **(Figure 12.8)** (van Veen *et al.*, 2000). In this model, an initial vinblastine-binding event with low affinity initiates a second vinblastine-binding event with high affinity. Based on the model, the dissociation constants for the two vinblastine-binding sites were estimated to be approximately 150 and 30 nM vinblastine, respectively. Moreover, a direct determination of the LmrA/vinblastine stoichiometry revealed that each homodimer of LmrA binds two vinblastine molecules (van Veen *et al.*, 2000), providing convincing evidence for the presence of two sites. Importantly, these drug-binding sites seem to be directly related to drug transport as shown by the reciprocal stimulation of LmrA-mediated vinblastine and Hoechst 33342 transport at low drug concentrations, and reciprocal inhibition at high drug concentrations (van Veen *et al.*, 2000). Most probably, one of the drug-binding sites interacts preferentially with vinblastine and the other preferentially with Hoechst 33342. At lower concentrations, vinblastine binds primarily to one site and enhances transport of Hoechst 33342 bound at the other site. At higher concentrations, vinblastine is able to compete with Hoechst 33342 for binding to the same site, and therefore inhibits Hoechst 33342 transport, or vice versa. Taken together, the results strongly suggest that LmrA contains at least two distinct drug-binding sites, presumably located in the TMD, with different but overlapping specificities which interact in drug transport in a positively cooperative manner. Support for the presence of at least two positively cooperative sites for drug transport in P-glycoprotein has also been presented (e.g. Shapiro and Ling, 1997b; Sharom *et al.*, 1996). Thus, it appears that the transport process of ABC-type multidrug transporters such as LmrA and P-glycoprotein involves two general transport-competent drug-binding sites, which may be composed of multiple drug interaction sites to account for the wide range of compounds that are transported. In addition to the transport-competent drug-binding sites, LmrA and P-glycoprotein contain regulatory sites, which may reside outside of the transport-competent drug-binding sites, to which allosteric modulators bind, but are not transported (Martin *et al.*, 1997; van Veen *et al.*, 1998).

STRUCTURAL CHANGES INDUCED BY NUCLEOTIDE BINDING OR HYDROLYSIS DETECTED BY ATR-FTIR SPECTROSCOPY

Although the topology of LmrA in the lipid membrane has been deduced from its primary structure **(Figure 12.3)**, its secondary and tertiary structures are unknown since a high-resolution structure of LmrA has not yet been obtained. Such a structure would supply extremely valuable information about the overall domain organization and the interacting sites. However, such a structure would also be inherently static and would not necessarily

BOX 12.1. ATR-FTIR SPECTROSCOPY: EXPERIMENTAL PROCEDURE AND SAMPLE PREPARATION

Infrared spectroscopy is based on the absorption of electromagnetic radiation by matter owing to different vibrational modes of the chemical bonds. The infrared beam is directed into a high refractive index medium (*i*nternal *r*eflection *e*lement or IRE), which is transparent for the IR radiation of interest. The most usual design of an IRE is the trapezoidal plate, which allows the orientation of protein secondary structures to be determined by means of linear dichroism. Below a critical angle θ_c, which depends on the refractive index of the IRE (n_1) and of the external medium (n_2) according to

$$\theta_c = \sin^{-1} n_{21} \tag{1}$$

where $n_{21} = n_2/n_1$, the light beam is completely reflected when it impinges on the surface of the IRE. Several internal total reflections occur within the IRE until the beam reaches the end. Interestingly, an electromagnetic disturbance exists in the rarer medium beyond the reflecting interface. This so-called evanescent wave is characterized by its amplitude, which falls off exponentially with the distance from the interface according to

$$E = E_0 \cdot e^{-z/d_p} \tag{2}$$

where E_0 is the time-averaged electric field intensity at the interface, E is the time-averaged field intensity at a distance z from the interface in the rarer medium, and d_p is the penetration depth of the evanescent field. It is the presence of the evanescent field which makes the interaction possible between infrared light and the sample present on the surface of the IRE, within approximately the penetration depth of the field. Indeed, samples which are deposited on the IRE absorb electromagnetic radiation of the evanescent wave, and thereby reduce the intensity of the reflected light. Hence, the technique is referred to as 'attenuated total reflection-IR spectroscopy'. Since the sample has to be in close contact with the IRE, films of proteins or membranes have to be used.

The simplest sample preparation for ATR-FTIR spectroscopy has been used, in which a drop of the sample, typically 20 μl of proteoliposomes containing ~20 μg of reconstituted LmrA, is spread on the IRE surface. The solvent is slowly evaporated under a gentle N_2 flow, while a Teflon bar or pipette tip is used to spread the liquid over the useful surface of the IRE in order to make the film as uniform as possible. While evaporating, capillary forces flatten the membranes, which spontaneously form oriented multibilayer arrangements. During film preparation, many water molecules remain associated with the proteins and lipids. Consequently, protein structure is not affected by the low hydration state of the newly formed film. Moreover, the ATPase activity of reconstituted LmrA was measured after resuspension of the film. No loss of ATPase activity was observed, confirming that film preparation does not alter LmrA conformation and activity. Thus, the technique is very convenient for studying proteins inserted into lipid membranes since common reconstituted vesicles (e.g. liposomes) can be used.

represent the structure of LmrA in its native lipid environment. In view of the difficulties in obtaining high-resolution structures of membrane proteins, lower-resolution techniques such as IR (*i*nfra*r*ed) and ATR-FTIR spectroscopy can be employed to obtain global, structural information of membrane proteins. ATR-FTIR spectroscopy is particularly useful since this permits the monitoring of structural changes of membrane proteins in their native lipid environment in response to physiological modifications **(Box 12.1)**.

ANALYSIS OF THE SECONDARY STRUCTURE OF LMRA IN THE ABSENCE AND PRESENCE OF NUCLEOTIDES

In order to mediate drug transport, LmrA must couple ATP hydrolysis to conformational changes, which alter drug-binding affinity and/or accessibility of the transport-competent drug-binding sites. To investigate the nature of the conformational changes induced during the transport cycle, ATR-FTIR spectra of LmrA, reconstituted into liposomes, were recorded in the absence and presence of different nucleotides. A typical spectrum of LmrA before and after deuteration is shown in **Figures 12.9A and 12.9B**, respectively. The bands at ~3300 cm^{-1} and ~2500 cm^{-1} correspond to the O–H and O–D stretching of H_2O and D_2O, respectively. In the 1800–1700 cm^{-1} region, the band corresponding to the C=O stretching of the lipids is detected in both cases. Most importantly for the study of LmrA are the amide I (1700–1600 cm^{-1} region) and the amide II (1570–1500 cm^{-1} region) bands. The amide I band is assigned to the ν(C=O) of the peptide

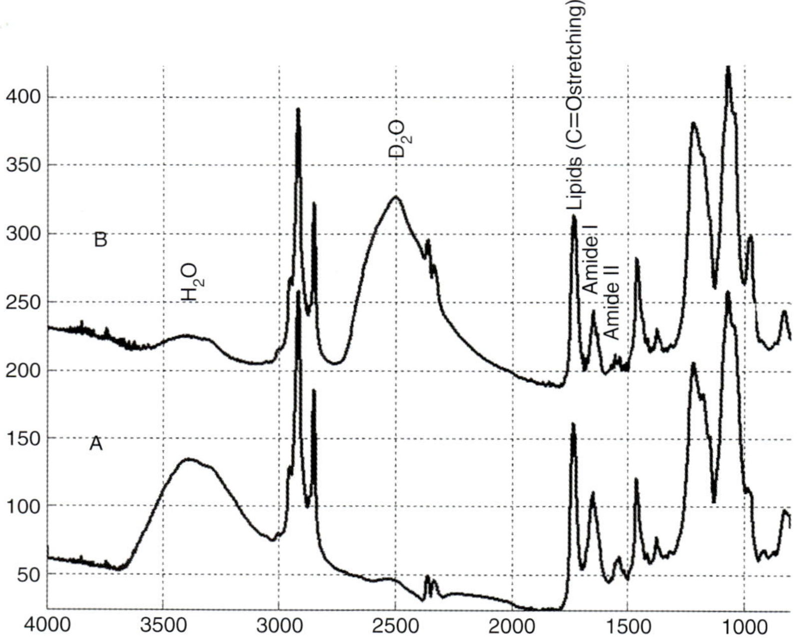

Figure 12.9. ATR-FTIR spectrum in the 4000–400 cm^{-1} region of LmrA actively reconstituted into lipids before (A) and after (B) deuteration. Thin films were obtained by slowly evaporating a sample containing 20 μg of LmrA on a Ge-attenuated total reflection element. The film was then rehydrated under a D_2O-saturated N_2 flow. Data obtained from Vigano et al. *(2000) with permission from the American Society for Biochemistry and Molecular Biology.*

bond, while the amide II band is characteristic of the δ(N–H). The amide I band is by far the most sensitive indicator of the secondary structure (Fringeli and Günthard, 1981) and is located in a region of the spectrum which is often free of other bands and is composed of 80% pure C=O vibration. The secondary structure of LmrA was determined by Fourier deconvolution and a curve-fitting analysis on the amide I region of a deuterated sample (Vigano *et al.*, 2000). H/D exchange permits differentiation of the α-helical secondary structure from random secondary structure, whose absorption band shifts from about 1655 cm^{-1} to about 1642 cm^{-1} (Goormaghtigh *et al.*, 1994).

The percentages of α-helix, β-sheet, β-turn and random coil structures of LmrA, in the presence or absence of nucleotides, are presented in **Table 12.2**. MgATPγS, a non-hydrolyzable analogue of MgATP, was used to discriminate between the influence of nucleotide binding and nucleotide hydrolysis on the structure of LmrA. In the presence of MgADP/Pi and MgADP, the structure represents the situation after ATP hydrolysis. Significant differences detected in the amide I region demonstrate that LmrA adopts two different conformations depending on the nature of the nucleotide bound to the protein **(Figure 12.10)**. First, LmrA alone or in the presence of MgADP contains 35% α-helix, 24% β-sheet, 28% β-turn and 13% random coil **(Table 12.2)**. The proportion of α-helices is higher than the proportion expected when only

TABLE 12.2. SECONDARY STRUCTURE OF LMRA DETERMINED IN THE ABSENCE AND PRESENCE OF NUCLEOTIDES[a]

Substrates[b]	Secondary structure			
	α-Helix (%)	β-Sheet (%)	β-Turn (%)	Random (%)
None	35 ± 2	24 ± 2	28 ± 1	13 ± 1
MgADP	35 ± 1	24 ± 1	31 ± 1	10 ± 1
MgATP	34 ± 2	36 ± 2	16 ± 2	14 ± 2
MgATPγS	35 ± 2	33 ± 1	18 ± 1	14 ± 1
MgADP/Pi	34 ± 1	35 ± 1	17 ± 1	14 ± 1

[a]Data obtained from Vigano *et al.* (2000) with permission from the American Society for Biochemistry and Molecular Biology.
[b]LmrA/nucleotide molar ratio = 1/5.

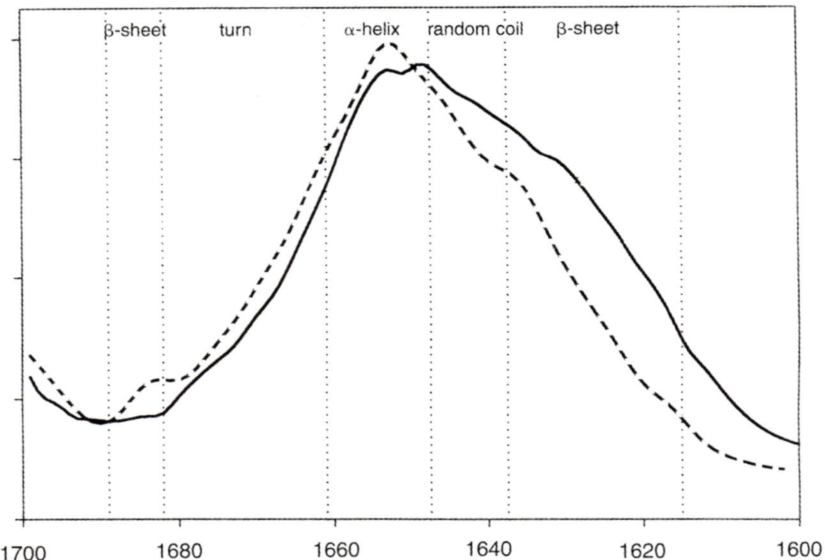

Figure 12.10. ATR-FTIR spectra between 1700 and 1600 cm^{-1} of deuterated LmrA actively reconstituted into lipids. Dotted line, in the absence of nucleotide or in the presence of MgADP (LmrA/MgADP molar ratio = 1/5). Solid line, in the presence of MgATP, MgADP/Pi or MgATPγS (LmrA/nucleotide molar ratio = 1/5). Data obtained from Vigano et al. *(2000) with permission from the American Society for Biochemistry and Molecular Biology. The results clearly show a shift indicating more **β**-sheet in the presence of non-hydrolyzable ATP or ADP/Pi compared with the ground state with no nucleotide or just ADP.*

six transmembrane segments are in an α-helical conformation (20%) (Grimard *et al.*, 2001). Therefore, α-helices, external to the membrane, have to be present, as confirmed in the recently reported crystal structure of the homologous transporter MsbA (Chang and Roth, 2001). In the presence of MgATP, MgATPγS or MgADP/Pi, the structure becomes enriched in β-sheet (35% α-helix, 34% β-sheet, 17% β-turn and 14% random coil) concomitantly with a loss of β-turn structure. Therefore, LmrA is clearly stabilized in a different secondary structure after ATP (i.e. ATPγS) binding. However, the protein returns to its initial secondary structure after Pi release, when ADP is still bound to LmrA.

The drug-binding sites of LmrA are predicted to reside within the membrane domain, which is composed of transmembrane α-helices (Grimard *et al.*, 2001). Since the α-helical content of LmrA does not change in the presence of nucleotides **(Table 12.2)**, it seems that the ATP binding-induced change of secondary structure is not related to a reorientation of the transport-competent drug-binding sites. However, it has recently been proposed for P-glycoprotein that ATP binding, not hydrolysis, drives the major conformational change associated with drug translocation, and that the reorientation of the drug-binding sites may depend on rotation and/or 'tilting' of transmembrane α-helices within the membrane (Higgins and Linton, 2001; Rosenberg *et al.*, 2001; Chapter 4). Although it is not known whether ATP binding to LmrA also results in loss of drug-binding affinity, we can not exclude the possibility that similar reorganizations, which obviously do not affect the α-helical content, occur in LmrA upon binding of ATP.

Interestingly, an ATP binding-induced enrichment in β-sheet, as observed for LmrA, was not observed for other ABC-type multidrug transporters (P-glycoprotein and MRP1) studied so far by ATR-FTIR (Manciu *et al.*, 2000; Sonveaux *et al.*, 1996). Since the ATPase and transport activities of P-glycoprotein and LmrA are very similar, it seems that the secondary structure change observed for LmrA after ATP binding is related to a behavior unique to this protein. Since LmrA must form a homodimer to be active (see earlier section on structural aspects), this raises the interesting possibility that ATP binding could mediate the secondary structural change accompanying the assembly of LmrA into its homodimeric form. After ATP hydrolysis and Pi release, the protein recovers its initial secondary structure and possibly its monomeric form.

AMIDE HYDROGEN/DEUTERIUM EXCHANGE KINETICS OF LMRA

To further investigate the effect of ATP binding and hydrolysis on the structure of LmrA, the kinetics of deuteration of reconstituted protein was monitored in the presence and absence of different nucleotides (MgATP, MgATPγS, MgADP, MgADP/Pi) (Vigano *et al.*, 2000). The rate of amide hydrogen exchange by deuterium is related to the solvent accessibility of the NH amide groups. Amide hydrogen exchange was followed by using ATR-FTIR spectroscopy to monitor the amide II absorption peak as a function of the time of exposure to D_2O-saturated N_2. The decrease of the amide II band is proportional to the number of hydrogens that have been exchanged by deuterium and provides a sensitive measure of LmrA structure and conformational changes. In the absence of ligands, approximately 31% of the amide hydrogen exchanged for deuterium within 10 s of exposure to D_2O, and an additional 15% exchanged after 2 min. The remaining 54% did not experience any exchange within 8 h of exposure to D_2O, and these protons represent the very inaccessible regions of LmrA. In the presence of MgATP and MgADP/Pi, the fraction of slowly exchanging amide protons decreases concomitant with an increase of intermediate exchanging amide protons. In the presence of MgATPγS and MgADP, the fraction of slowly exchanging amide protons is almost identical to that observed for LmrA alone. However, approximately 31% of the amide hydrogens are exchanged within 10 s in the protein alone, whereas it takes 2 min to exchange 31% of the amide hydrogens in the presence of MgADP or MgATPγS.

These H/D exchange measurements provide evidence that a large portion (54%) of LmrA is poorly accessible to the aqueous medium. The presence of a large amide population characterized by a very low exchange rate could be due, in part, to the shielding effect of the membrane on a large number of residues. To investigate this possibility, Grimard *et al.* (2001) developed a new approach (monitoring infrared linear dichroism spectra in the course of H/D exchange), which enables the recording of exchange rates of the membrane-embedded region of the protein only. This approach revealed that after 20 min 60% of the transmembrane-oriented helix amide groups have been exchanged, i.e. an unexpectedly fast exchange for the transmembrane region. In contrast, only 37% seem almost inaccessible to solvent and do not experience any exchange within 33 h of exposure to D_2O. Since the predicted transmembrane domains of LmrA account for only 20% of the total amino acids, these results demonstrate that a significant proportion of the slowly exchanging amino acids must be located outside the membrane, where they likely form highly structured domains. The kinetics of deuteration of P-glycoprotein also showed a large inaccessible fraction (53%) of the protein, where similarly only 20% of the total amino acids of the protein are predicted to be located inside the membrane (Sonveaux *et al.*, 1996).

The β-turn, β-sheet secondary structural change, observed in the presence of MgATP, MgATPγS and MgADP/Pi **(Table 12.2)**, is not responsible for the change in the global accessibility of LmrA towards the external medium, as detected by the exchange of amide protons. Indeed, LmrA in the presence of MgATPγS shows no variation in the level of inaccessible amino acids, when compared with the situation in the absence of ligand. The main changes in the accessibility of LmrA take place in the presence of MgATP, i.e. when normal hydrolysis is permitted. The inaccessible amino acids decrease from 318 to 260 in the presence of MgATP. In the presence of MgADP/Pi, the inaccessible amino acids only decrease to 289, while no changes are observed with MgADP or MgATPγS. Since the changes in accessibility observed in the presence of MgATP and MgADP/Pi are not related to the change of secondary structure, they could be correlated to different tertiary structures of LmrA.

In summary, ATR-FTIR studies reveal that LmrA undergoes a secondary structure change and passes through three different conformational states during its catalytic cycle. After binding of ATP, the protein structure becomes enriched in β-sheet, unlike that of P-glycoprotein. When ATP hydrolysis takes place, perhaps the tertiary structure of the protein changes and the protein adopts a more accessible conformation. A third conformation is reached after ATP hydrolysis, when Pi is still associated to the protein. In this conformation, the accessibility of LmrA is intermediate to the closed conformation, observed in the absence of nucleotides, and the opened conformation, observed in the presence of ATP. After Pi release, the protein recovers its initial secondary and tertiary structures. These observed conformational changes might reflect the conformational coupling of ATP hydrolysis to drug transport, as described in the following section.

COUPLING OF ATP HYDROLYSIS TO DRUG TRANSPORT

Understanding the mechanism by which ABC-type multidrug transporters couple the hydrolysis of ATP to the translocation of drugs across the membrane is a major goal. Translocation apparently involves alternate action of the two ATP-binding sites, and both catalytic sites must be functional to allow sustained drug transport (Senior *et al.*, 1995). As already pointed out, LmrA contains a low-affinity drug-binding site that is allosterically coupled to a high-affinity drug-binding site, and these two sites appear to be directly related to drug transport (see also Chapter 5). Persuasive data showing the obligatory link between the drug-binding and catalytic cycles has come from vanadate-trapping experiments (van Veen *et al.*, 2000). First, as shown in **Figure 12.11**, heterologous displacement by the drug CP100-356 of vinblastine from the vanadate-trapped LmrA transporter suggested the presence of only a single population of vinblastine-binding sites, with binding properties similar to those of the low-affinity vinblastine-binding site in the non-trapped transporter **(compare Figures 12.11A and 12.11B)**. Second, direct determination of the vinblastine/transporter stoichiometry in the vanadate-trapped transporter revealed a stoichiometry of close to one, in contrast to the value of two determined for the non-trapped transporter. These experiments demonstrated that of the two vinblastine-binding sites accessible in the LmrA transporter, only the low-affinity vinblastine-binding site is accessible in the vanadate-trapped transition state conformation of LmrA. Finally, specific photoaffinity labeling of the vanadate-trapped LmrA transporter with [^{3}H]-APDA, a drug that can be transported by LmrA, was obtained in right-side-out membrane vesicles, but not in inside-out membrane vesicles, demonstrating that the low-affinity drug-binding site is exposed to the outside (extracellular) surface of the cell membrane. The vanadate-trapped conformation of LmrA, with a single low-affinity drug-binding site exposed to the extracellular surface, is consistent with the hypothesis that an ATP hydrolysis-induced conformational change moves a high-affinity drug-binding site from the inside of the membrane to the outside with a concomitant change to a low-affinity site (van Veen *et al.*, 2000). Indeed, conformational changes in LmrA upon hydrolysis of ATP have been detected by ATR-FTIR spectroscopy (see section above).

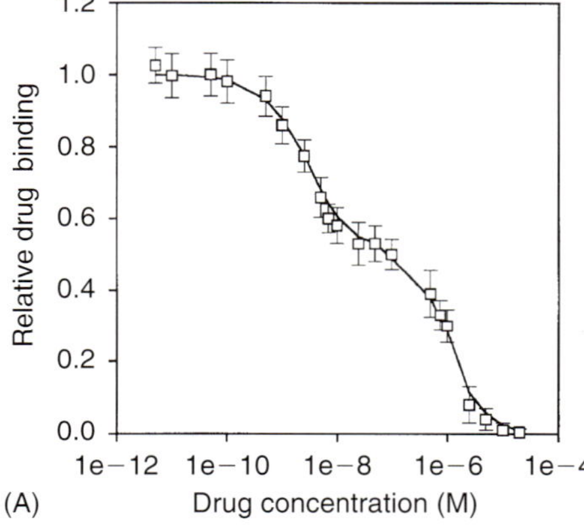

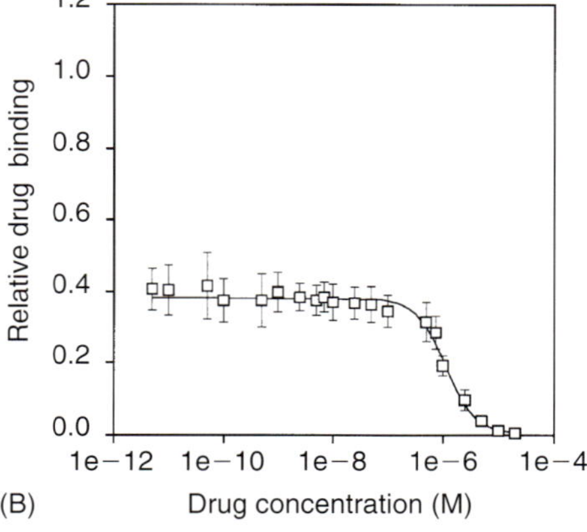

Figure 12.11. Heterologous displacement of vinblastine from LmrA by CP100-356. Non-trapped LmrA (A) and vanadate-trapped LmrA (B) were saturated with [^{3}H]vinblastine, and vinblastine displacement from LmrA by CP100-356 was measured at increasing concentrations of CP100-356. For the non-trapped transporter, the data were fitted best by a cooperative two-site drug-binding model, assuming direct competition by CP100-356 for binding to each of the two vinblastine-binding sites in the LmrA transporter. In contrast, in the vanadate-trapped transporter, the data suggest the presence of a single vinblastine-binding site with binding characteristics similar to those of the low-affinity site in the non-trapped protein. Data obtained from van Veen et al. *(2000) with permission from Oxford University Press.*

The substrate-binding data obtained with (vanadate-trapped) LmrA, when combined with the *alternating catalytic site model*, in which the ABC domains of P-glycoprotein act alternately to hydrolyze ATP (Senior *et al.*, 1995), led to the proposition of an *alternating two-site transport model* (van Veen *et al.*, 2000). In this transport model (**Figure 12.12**; also discussed in more detail in Chapter 5), the hydrolysis of ATP by the ABC domain of one monomer of the LmrA transporter is coupled to drug efflux via its TMD. This is achieved through the movement of a liganded, inside-facing, high-affinity drug-binding site (which binds a drug molecule in the inner leaflet of the membrane) to the outside of the membrane, with a concomitant change to low affinity. This last step results in release of the drug molecule into the extracellular medium. The whole process occurs via a catalytic transition state intermediate (which can be trapped with vanadate), in which the transport-competent drug-binding site is inaccessible. Importantly, ATP hydrolysis by the ABC domain of one half of the transporter is not only coupled to drug efflux via its TMD, but also must facilitate the return of an unliganded, outside-facing low-affinity site, at the membrane domain of the other half-molecule, to an inside-facing high-affinity site. The latter process should not be confused with an ATP hydrolysis-induced resetting step of a single drug-binding site in the dimeric LmrA transporter, which alternates between high- and low-affinity conformations exposed at the inner and outer membrane surfaces (recently reviewed by van Veen *et al.*, 2001). Thus, in a complete drug transport cycle, each monomer of the LmrA dimer alternates its drug-binding site from high affinity to occluded state to low affinity and back to high affinity. The affinities of the binding sites in the monomers alternate: when the binding site in one monomer is in the high-affinity state the binding site in the other monomer is in a low-affinity state and vice versa. Hence, this process is called *an alternating two-site mechanism*. Such a scenario implies that both halves of the apparently symmetric LmrA transporter are able to act asymmetrically. However, it is presently not clear whether the binding sites are present in separate transmembrane domains or at the interface between transmembrane domains.

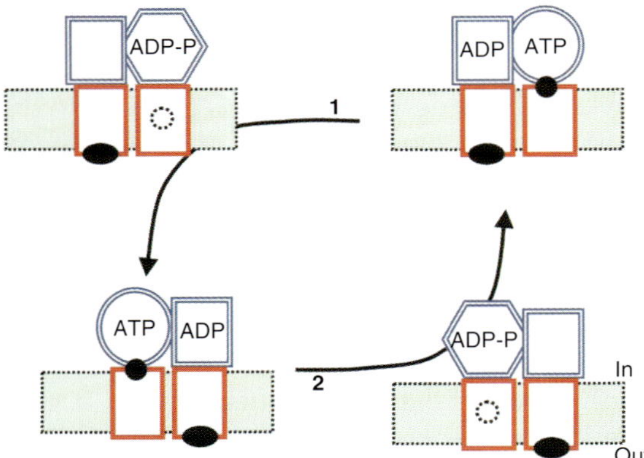

Figure 12.12. ***Alternating two-site transport model. Rectangles represent the transmembrane domains of LmrA. Circles, squares and hexagons represent different conformations of the nucleotide-binding domains. Although it is not known yet whether ATP binding, rather than hydrolysis, results in a change in drug-binding affinity, it is assumed that the ATP-bound (circle) state is associated with a high-affinity drug-binding site on the inside of the transporter. The ADP-bound (square) state is associated with a low-affinity drug-binding site on the outside of the transporter. The ADP-Pi (hexagonal) state is associated with an occluded drug-binding site, and represents the ADP/vanadate-trapped form of the ABC domain. According to the model, the transporter oscillates between two configurations, each containing a high-affinity, inside-facing, transport-competent drug-binding site, and a low-affinity, outside-facing drug-release site. The ATP-dependent interconversion of one configuration into the other proceeds via a catalytic transition state conformation in which the transport-competent site is occluded. The model is adapted from van Veen*** **et al.** ***(2000) with permission from Oxford University Press.***

CONCLUSIONS AND PERSPECTIVES

Despite the large diversity of substrates for ABC transporters, the specificity of each system is relatively high and only a few members belonging to the ABC transporter superfamily mediate multidrug resistance. Most of them are of eukaryotic origin, such as the P-glycoproteins, and only two characterized systems, LmrA and HorA, are of bacterial origin. Studies on bacterial multidrug efflux pumps are relevant because in the last few years it has been shown that pumping activities are involved in the ongoing emergence of antibiotic resistance in pathogenic bacteria. In addition, LmrA is able to complement the human multidrug resistance P-glycoprotein, supporting the clinical and academic value of studying these bacterial proteins.

The increasing availability of different microbial genomes has revealed the presence of putative ABC-type multidrug transporters, often structurally similar to LmrA and HorA, in all pathogenic microorganisms analyzed so far (see list of prokaryotic genomes on the TIGR database: http://www.tigr.org/). The physiological functions and substrate specificities of these pathogenic multidrug transporters are unknown, as are the conditions in which they are expressed. However, if these efflux pumps are expressed in clinical isolates, due to induction by antibiotic exposure or by up regulatory mutations, this may result in a multidrug resistance phenotype and further selection of such strains by antibiotic pressure. If this is indeed the case, broad-spectrum multidrug transporters are a serious threat to antibiotic therapy. Insight into the incidence of (over) expression of multidrug resistance genes in clinical strains of bacterial pathogens, the substrate selectivities of the putative efflux pumps, and the regulation of their expression in response to different antibiotics is therefore urgently needed.

The progress that has been achieved in recent years to understand the functional properties of ABC-type multidrug efflux pumps is impressive. One of the challenges that lies ahead is to understand the structural basis of how these fascinating and important proteins recognize and transport such a wide range of structurally diverse compounds. Current structures of ABC multidrug efflux pumps are of low resolution. For a detailed understanding of the mechanism of multiple drug binding and translocation, high-resolution structures of intact ABC-type multidrug transporters, both in the presence and absence of drug and nucleotide ligands, are required.

Acknowledgments

The authors thank the present and previous members of the Department of Microbiology and of the Labaratoire de Chimie Physique des Macromolécules aux Interfaces for their valuable contribution to the research presented in this chapter. We thank H. Bolhuis and J. Kok for kindly providing some of the figures.

References

Altenberg, G.A., Vanoye, C.G., Horton, J.K. and Reuss, L. (1994) Unidirectional fluxes of rhodamine 123 in multidrug resistant cells: evidence against direct extrusion from the plasma membrane. Proc. Natl Acad. Sci. USA 91, 4654–4657.

Austin, E.A., Graves, J.F., Hite, L.A., Parker, C.T. and Schnaitman, C.A. (1990) Genetic analysis of lipopolysaccharide core biosynthesis by *Escherichia coli* K12: insertion mutagenesis of the *rfa* locus. J. Bacteriol. 172, 5312–5325.

Barrasa, M.I., Tercero, J.A., Lacalle, R.A. and Jimenez, A. (1995) The *ard1* gene from *Streptomyces capreolus* encodes a polypeptide of the ABC-transporters superfamily which confers resistance to the aminonucleoside antibiotic A201A. Eur. J. Biochem. 228, 562–569.

Bauer, B.E., Wolfger, H. and Kuchler, K. (2000) Inventory and function of yeast ABC proteins: about sex, stress, pleiotropic drug and heavy metal resistance. Biochim. Biophys. Acta 1461, 217–236.

Bolhuis, H., Molenaar, D., Poelarends, G., van Veen, H.W., Poolman, B., Driessen, A.J.M. and Konings, W.N. (1994) Proton motive force-driven and ATP-dependent drug extrusion systems in multidrug-resistant *Lactococcus lactis*. J. Bacteriol. 176, 6957–6964.

Bolhuis, H., Poelarends, G., van Veen, H.W., Poolman, B., Driessen, A.J.M. and Konings, W.N. (1995) The lactococcal *lmrP* gene encodes a proton motive force-dependent drug transporter. J. Biol. Chem. 270, 26092–26098.

Bolhuis, H., van Veen, H.W., Molenaar, D., Poolman, B., Driessen, A.J.M. and Konings, W.N. (1996) Multidrug resistance in *Lactococcus lactis*: Evidence for ATP-dependent drug extrusion from the inner leaflet of the cytoplasmic membrane. EMBO J. 15, 4239–4245.

Borst, P., Zelcer, N. and van Helvoort, A. (2000) ABC transporters in lipid transport. Biochim. Biophys. Acta 1486, 128–144.

Chang, G. and Roth, C.B. (2001) Structure of MsbA from *E. coli*: a homolog of the multidrug resistance ATP binding cassette (ABC) transporters. Science 293, 1793–1800.

Cohen, M.L. (1992) Epidemiology of drug resistance: implications for a postantimicrobial era. Science 257, 1050–1055.

Culliton, B.J. (1992) Drug-resistant TB may bring epidemic. Nature 356, 473.

Fringeli, U. and Günthard, H. (1981) In: Membrane Spectroscopy (ed. E. Grell), pp. 270–332. Berlin, Heidelberg, New York: Springer.

Gasser, F. (1994) Safety of lactic acid bacteria and their occurrence in human clinical infections. Bull. Inst. Pasteur 92, 45–67.

Goormaghtigh, E., Cabiaux, V. and Ruysschaert, J.M. (1994) Determination of soluble and membrane protein structure by Fourier transform infrared spectroscopy. III. Secondary structures. Subcell. Biochem. 23, 405–450.

Greenberger, L.M. (1993) Major photoaffinity drug labelling sites for iodoaryl azidoprazosin in P-glycoprotein are within, or immediately C-terminal to, transmembrane domains 6 and 12. J. Biol. Chem. 268, 11417–11425.

Grimard, V., Vigano, C., Margolles, A., Wattiez, R., van Veen, H.W., Konings, W.N., Ruysschaert, J.-M. and Goormaghtigh, E. (2001) Structure and dynamics of the membrane-embedded domain of LmrA investigated by coupling polarized ATR-FTIR spectroscopy and $^{1}H/^{2}H$ exchange. Biochemistry 40, 11876–11886.

Guilfoile, P.G. and Hutchinson, C.R. (1991) A bacterial analog of the *mdr* gene of mammalian tumor cells is present in *Streptomyces peucetius*, the producer of daunorubicin and doxorubicin. Proc. Natl Acad. Sci. USA 88, 8553–8557.

Hayes, J.D. and Wolf, C.R. (1990) Molecular mechanisms of drug resistance. Biochem. J. 272, 281–295.

Higgins, C.F. (1992) ABC transporters: from microorganisms to man. Annu. Rev. Cell Biol. 8, 67–113.

Higgins, C.F. and Gottesman, M.M. (1992) Is the multidrug transporter a flippase? Trends Biochem. Sci. 17, 18–21.

Higgins, C.F. and Linton, K.J. (2001) The xyz of ABC transporters. Science 293, 1782–1784.

Homolya, L., Hollô, Z., Germann, U.A., Pastan, I., Gottesman, M.M. and Sarkadi, B. (1993) Fluorescent cellular indicators are extruded by the multidrug resistance protein. J. Biol. Chem. 268, 21493–21496.

Kolaczkowski, M., van der Rest, M., Cybularz-Kolaczkowski, A., Soumillion, J.-P., Konings, W.N. and Goffeau, A. (1996) Anticancer drugs, ionophoric peptides, and steroids as substrates of the yeast multidrug transporter Pdr5p. J. Biol. Chem. 271, 31543–31548.

Linton, K.J., Cooper, H.N., Hunter, I.S. and Leadleay, P.F. (1994) An ABC-transporter from *Streptomyces longisporoflavus* confers resistance to the polyether-ionophore antibiotic tetronasin. Mol. Microbiol. 11, 777–785.

Loo, T.W. and Clarke, D.M. (2000) Identification of residues within the drug-binding domain of the human multidrug resistance P-glycoprotein by cysteine-scanning mutagenesis and reaction with dibromobimane. J. Biol. Chem. 275, 39272–39278.

Manciu, L., Chang, X.B., Riordan, J.R. and Ruysschaert, J.M. (2000) Multidrug resistance protein MRP1 reconstituted into lipid vesicles: secondary structure and nucleotide-induced tertiary structure changes. Biochemistry 39, 13026–13033.

Margolles, A., Putman, M., van Veen, H.W. and Konings, W.N. (1999) The purified and functionally reconstituted multidrug transporter LmrA of *Lactococcus lactis* mediates the transbilayer movement of specific fluorescent phospholipids. Biochemistry 38, 16298–16306.

Martin, C., Berridge, G., Higgins, C.F. and Callaghan, R. (1997) The multidrug resistance reversal agent SR33557 and modulation of vinca alkaloid binding to P-glycoprotein by an allosteric interaction. Br. J. Pharmacol. 122, 765–771.

Nikaido, H. (1994) Prevention of drug access to bacterial targets: permeability barriers and active efflux. Science 264, 382–388.

Olano, C., Rodríguez, A.M., Méndez, C. and Salas, J.A. (1995) A second ABC transporter is involved in oleandomycin resistance and its secretion by *Streptomyces antibioticus*. Mol. Microbiol. 16, 333–343.

Podlesek, Z., Comino, A., Herzog-Velikonja, B., Žgur-Bertok, D., Komel, R. and Grabnar, M. (1995) *Bacillus licheniformis* bacitracin-resistance ABC transporter: relationship to mammalian multidrug resistance. Mol. Microbiol. 16, 969–976.

Putman, M., van Veen, H.W., Degener, J.E. and Konings, W.N. (2000a) Antibiotic resistance: era of the multidrug pump. Mol. Microbiol. 36, 772–774.

Putman, M., van Veen, H.W. and Konings, W.N. (2000b) Molecular properties of bacterial multidrug transporters. Microbiol. Mol. Biol. Rev. 64, 672–693.

Putman, M., van Veen, H.W., Degener, J.E. and Konings, W.N. (2001) The lactococcal secondary multidrug transporter LmrP confers resistance to lincosamides, macrolides, streptogramins and tetracyclines. Microbiology 147, 2873–2880.

Raviv, Y., Pollard, H.B., Bruggeman, E.P., Pastan, I. and Gottesman, M.M. (1990) Photosensitized labeling of a functional multidrug transporter in living drug-resistant tumor cells. J. Biol. Chem. 265, 3975–3980.

Roberts, M.C. (1996) Tetracycline resistance determinants: mechanisms of action, regulation of expression, genetic mobility, and distribution. FEMS Microbiol. Rev. 19, 1–24.

Rodríguez, A.M., Olano, C., Vilches, C., Méndez, C. and Salas, J.A. (1993) *Streptomyces antibioticus* contains at least three oleandomycin-resistance determinants, one of which shows similarity with proteins of the ABC-transporter superfamily. Mol. Microbiol. 8, 571–582.

Romsicki, Y. and Sharom, F.J. (1999) The membrane lipid environment modulates drug interactions with the P-glycoprotein multidrug transporter. Biochemistry 38, 6887–6896.

Rosenberg, M.F., Velarde, G., Ford, R.C., Martin, C., Berridge, G., Kerr, I.D., *et al.* (2001) Repacking of the transmembrane domains of P-glycoprotein during the transport ATPase cycle. EMBO J. 20, 5615–5625.

Ross, J.I., Eady, E.A., Cove, J.H., Cunliffe, W.J., Baumberg, S. and Wootton, J.C. (1990) Inducible erythromycin resistance in staphylococci is encoded by a member of the ATP-binding transport super-gene family. Mol. Microbiol. 4, 1207–1214.

Sakamoto, K., Margolles, A., van Veen, H.W. and Konings, W.N. (2001) Hop resistance in the beer spoilage bacterium *Lactobacillus brevis* is mediated by the ATP-binding cassette multidrug transporter HorA. J. Bacteriol. 183, 5371–5375.

Sami, M., Yamahita, H., Hirono, T., Kadokura, H., Kitamoto, K., Yoda, K. and Yamasaki, M. (1997) Hop-resistant *Lactobacillus brevis* contains a novel plasmid harboring a multidrug resistance-like gene. J. Ferment. Bioeng. 84, 1–6.

Sami, M., Suzuki, K., Sakamoto, K., Kadokura, H., Kitamoto, K. and Yoda, K. (1998) A plasmid pRH45 of *Lactobacillus brevis* confers hop resistance. J. Gen. Appl. Microbiol. 44, 361–363.

Senior, A.E., Al-Shawi, M.K. and Urbatsch, I.L. (1995) The catalytic cycle of P-glycoprotein. FEBS Lett. 377, 285–289.

Shapiro, A.B. and Ling, V. (1997a) Extraction of Hoechst 33342 from the cytoplasmic leaflet of the plasma membrane by P-glycoprotein. Eur. J. Biochem. 250, 122–129.

Shapiro, A.B. and Ling, V. (1997b) Positively cooperative sites for drug transport by P-glycoprotein with distinct drug specificities. Eur. J. Biochem. 250, 130–137.

Sharom, F.J., Yu, X., Didiodato, G. and Chu, J.W.K. (1996) Synthetic hydrophobic peptides are substrates for P-glycoprotein and stimulate drug transport. Biochem. J. 320, 421–428.

Sonveaux, N., Shapiro, A.B., Goormaghtigh, E., Ling, V. and Ruysschaert, J.M. (1996) Secondary and tertiary structure changes of reconstituted P-glycoprotein. A Fourier transform attenuated total reflection infrared spectroscopy analysis. J. Biol. Chem. 271, 24617–24624.

Ueda, K., Taguchi, Y. and Morishima, M. (1997) How does P-glycoprotein recognize its substrates? Semin. Cancer Biol. 8, 151–159.

van Veen, H.W. and Konings, W.N. (1998) The ABC family of multidrug transporters in microorganisms. Biochim. Biophys. Acta 1365, 31–36.

van Veen, H.W., Venema, K., Bolhuis, H., Oussenko, I., Kok, J., Poolman, B., Driessen, A.J.M. and Konings, W.N. (1996) Multidrug resistance mediated by a bacterial homolog of the human drug transporter MDR1. Proc. Natl Acad. Sci. USA 93, 10668–10672.

van Veen, H.W., Callaghan, R., Soceneantu, L., Sardini, A., Konings, W.N. and Higgins, C.F. (1998) A bacterial antibiotic-resistance gene that complements the human multidrug-resistance P-glycoprotein gene. Nature 391, 291–295.

van Veen, H.W., Putman, M., Margolles, A., Sakamoto, K. and Konings, W.N. (1999) Structure-function analysis of multidrug transporters in *Lactococcus lactis*. Biochim. Biophys. Acta 1461, 201–206.

van Veen, H.W., Margolles, A., Müller, M., Higgins, C.F. and Konings, W.N. (2000) The homodimeric ATP-binding cassette transporter LmrA mediates multidrug transport by an alternating two-site (two-cylinder engine) mechanism. EMBO J. 19, 2503–2514.

van Veen, H.W., Higgins, C.F. and Konings, W.N. (2001) Multidrug transport by ATP binding cassette transporters: a proposed two-cylinder engine mechanism. Res. Microbiol. 152, 365–374.

Vigano, C., Margolles, A., van Veen, H.W., Konings, W.N. and Ruysschaert, J.M. (2000) Secondary and tertiary structure changes of reconstituted LmrA induced by nucleotide binding or hydrolysis. J. Biol. Chem. 275, 10962–10967.

CHAPTER 13

Structure and Function of Osmoregulated ABC Transporters

Bert Poolman and Tiemen van der Heide

Introduction

General

In their natural habitats, microorganisms are often exposed to osmolality changes in the environment. For instance, soil bacteria such as *Bacillus subtilis* are alternately exposed to periods of drought and rain, to which they have to adapt. Since the cytoplasmic membrane of bacteria is highly permeable to water but forms an effective barrier for most solutes present in the medium and metabolites present in the cytoplasm, water will flow out of the cell when the outside osmolality increases ('osmotic upshift'). As a consequence of an osmotic upshift, the turgor pressure will decrease and ultimately the cell may plasmolyze. Upon osmotic downshift, water will flow into the cell and thereby increase the turgor pressure. Maintenance of cell turgor is a prerequisite for almost any form of life, as it provides a mechanical force for the expansion of the cell envelope and regulates cell growth. Generally, (micro)organisms respond to an osmotic upshift by rapidly accumulating compatible solutes to prevent the loss of water and loss of turgor pressure. Upon osmotic downshift, the cells need to rapidly export the solutes to prevent the turgor pressure becoming too high, which, ultimately, may lead to breakage of the cells. Since changes in protein expression (biosynthesis) are relatively slow, it is evident that the primary response to osmotic stress needs to be one in which transport systems or channel proteins already present in the cytoplasmic membrane are activated in order to accumulate or release solutes (Glaasker *et al.*, 1996a, 1996b). This implies that osmotic stress must be sensed by these systems and converted into a change in the appropriate activity such that the osmotic imbalance is rapidly restored.

Most osmotically controlled uptake systems are regulated at both the genetic (induction of gene expression) and the enzymatic level (direct 'activation' of the transport protein). The degree of induction can vary considerably (from 2- up to 500-fold), whereas *in vivo* activation of the transport protein by an osmotic upshift is usually in the range of 5- to 35-fold. The transport systems known to be activated by osmotic upshift are either ATP-binding cassette (ABC) transporters or so-called secondary transporters, that is, transport proteins driven by an electrochemical ion gradient (Wood, 1999). The majority of both types of these energy coupling mechanisms have the quaternary ammonium compound glycine betaine as the preferred (high affinity) substrate. Well-characterized osmotic downshift activated systems are the mechanosensitive channels, most notably the MscL protein, which open and thereby release osmolytes when the turgor pressure becomes too high. This chapter only summarizes our knowledge of the ABC-type osmoregulated transporters.

ABC Proteins: From Bacteria to Man
ISBN 0-12-352551-9

BOX 13.1. DEFINITIONS

The term hyperosmotic stress is often used to indicate an increased osmolality of the external medium. This is somewhat confusing as the cytoplasm of growing cells in complete osmotic equilibrium is generally hyperosmotic relative to the outside, even after an osmotic upshift. Therefore, we prefer to use the terms osmotic upshift and downshift for conditions that are often referred to as hyper- and hypo-osmotic stress. Turgor pressure (ΔP) is the hydrostatic pressure difference that balances the difference in internal and external osmolyte concentration.

$$\Delta P = (\mathrm{RT}/V_w)\ln(a_o/a_i) \approx \mathrm{RT}(c_i - c_o)$$

in which V_w is the partial molal volume of water, a is the water activity, c is the total osmolyte concentration, and the subscripts i and o refer inside and outside, respectively. A cell plasmolyzes when ΔP becomes negative.

Osmolality is the osmotic pressure of a solution at a particular temperature, expressed as moles of solute per kilogram of solvent (osmol kg^{-1} or osmolal). The often used term osmolarity is an approximation for osmolality and is expressed as osmol l^{-1} or osmolar (for a more extensive overview of definitions, see Wood, 1999).

BOX 13.2. COMPATIBLE SOLUTES

Compatible solutes are molecules that can be accumulated to high cytoplasmic (often molar) concentrations without affecting vital cellular processes. In many organisms, the compatible solutes glycine betaine and carnitine (and other quaternary ammonium compounds) offer the highest protection against osmotic stress.

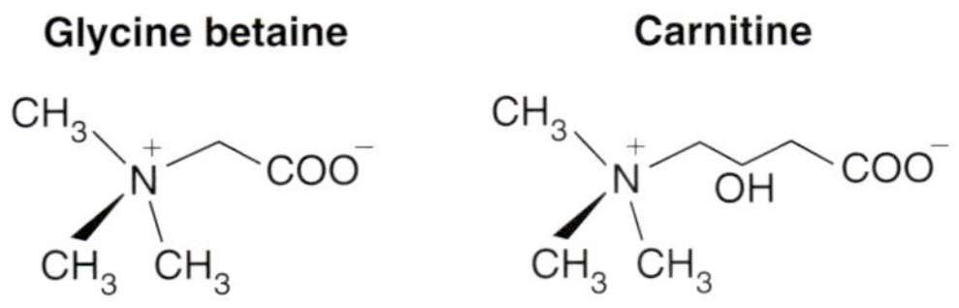

Glycine betaine

CH_3, $\overset{+}{N}$, COO^-, CH_3, CH_3

Carnitine

CH_3, $\overset{+}{N}$, OH, COO^-, CH_3, CH_3

RATIONALE FOR THE ACCUMULATION OF COMPATIBLE SOLUTES

After an osmotic upshift, cells benefit from the accumulation of compatible solutes by balancing the osmotic disturbance via an increase in cytoplasmic osmolality. High intracellular concentrations of compatible solutes are common under high osmolality conditions, but it is difficult to derive a quantitative relationship between the internal concentrations of compatible solutes and the external osmolality as multiple (macro)molecules are involved. The compatible solutes to be accumulated to high intracellular levels are restricted to a few categories (for reviews see Csonka, 1989; Csonka and Hanson, 1991; Galinski and Trüper, 1994), and they include (i) potassium ions, (ii) amino acids (glutamate, alanine, proline), (iii) amino acid analogues (taurine, *N*-acetylglutaminylglutamine amide), (iv) methyl-amines and related compounds (glycine betaine, carnitine, ectoines), and (v) polyols and sugars (glycerol, sucrose, lactose). In general, compatible solutes should not interact specifically with the (mostly negatively charged) cellular macromolecules, nor should they perturb cytoplasmic solutes via (de)hydration, precipitation, or any other (charge) interaction. Therefore, under *steady state* conditions most compatible solutes that are present in large amounts in the cytoplasm have no net charge. The accumulation of potassium ions in response to an osmotic upshift in enteric bacteria is usually only transient. Following the accumulation of potassium ions, other compatible solutes are synthesized (e.g. trehalose), and the uptake systems for glycine betaine and proline are induced. The accumulation of these solutes then eventually replaces potassium (see also Poolman and Glaasker, 1998). An exception to the rule of preferring neutral compatible solutes is found in thermophilic organisms (Bacteria and Archaea), where negatively charged compounds seem to be accumulated in response to osmotic upshifts. These compounds include mannosylglycerate, glutamate, cyclic-2,3-bisphosphoglycerate, 1,3,4,6-tetracarboxyhexane, and myo-inositolphosphate derivates (Martins and Santos, 1995; Martins *et al.*, 1996).

STRUCTURAL ANALYSIS OF OSMOREGULATED ABC TRANSPORTERS

Typical ABC-type binding protein-dependent transporters are composed of five protein(s) (domains), that is, an extracellular binding protein (receptor), two ATP-binding subunits and two integral membrane subunits. The integral membrane subunits provide the translocation

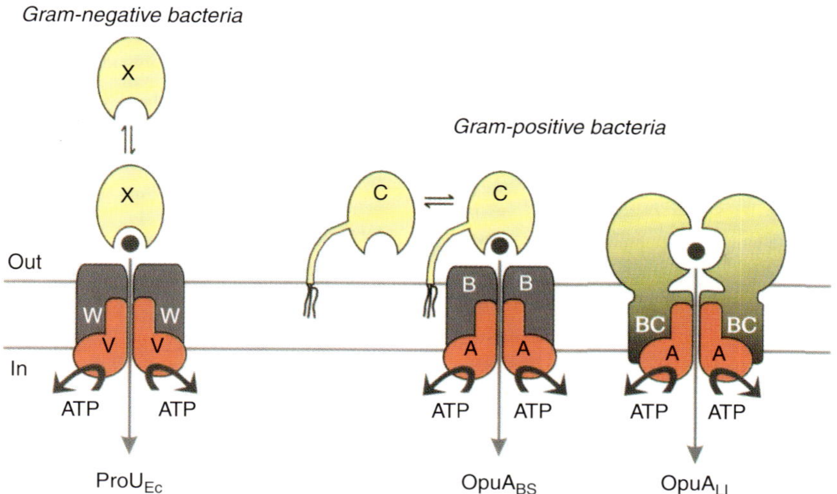

Figure 13.1. ***Structural organization of osmoregulated ABC transporters. Depicted are the ProU system from*** **E. coli,** ***$OpuA_{Bs}$ from*** **B. subtilis** ***and $OpuA_{Ll}$ from*** **L. lactis.** ***In Gram-negative bacteria, the binding protein is present in the periplasm, whereas in Gram-postive bacteria, the protein is anchored to the outer surface of the membrane via a lipid moiety (OpuAC) or fused to the translocator domain (OpuABC). For ProX and OpuAC, the equilibrium between unliganded and liganded binding protein is depicted.***

pathway. Except for the substrate-binding protein, the other subunits can be present as distinct polypeptides or fused to one another but always each entity is present twice. The osmoregulated ABC transporters belong to the OTCN family (Dassa and Bouige, 2001), of which the members transport substrates as diverse as quaternary ammonium compounds (glycine betaine, carnitine), proline, alkylsulfonates and -phosphonates, phosphites, cyanate, and nitrate.

Although biochemical evidence is not available, the osmoregulated ABC transporters most probably have two identical copies of the ATP-binding subunit. This suggestion is based on the finding that, in contrast to other families of ABC transporters, the operons specifying osmoregulated transporters only have a single gene for an ATP-binding subunit. The integral membrane components are either present as two identical copies or two homologous proteins, but they are not fused to each other or to the ATP-binding subunit **(Figure 13.1)**. The ligand-binding subunit is present as a free protein in the periplasm of Gram-negative bacteria, whereas in Gram-positive bacteria the protein can be anchored to the membrane through a fatty acid modification of the amino-terminal cysteine (Kempf *et al.*, 1997; Sankaran and Wu, 1994). Recently, it has been shown that a subset of ABC transporters has the ligand-binding protein fused to the integral membrane subunit **(Figure 13.1)**. OpuA[1] of *Lactococcus lactis* is the only well-studied representative of this subset of binding protein-dependent ABC transporters, but database searches indicate that similar systems are present in *Streptomyces coelicolor, Streptococcus pneumoniae, Streptococcus pyogenes, Chlamydia pneumoniae, Helicobacter pylori* and *Staphylococcus aureus*. By analogy with other ABC transporters, functional OpuA will most likely be composed of two integral membrane subunits and two ATP-binding subunits. Since the translocator subunit is fused to the substrate-binding protein (OpuABC subunit), the oligomeric structure implies that two receptor domains are present per functional complex. This raises questions about the observations that only a single substrate-binding protein interacts with the dimeric membrane complex of an ABC transporter, and that two lobes of a single substrate-binding protein interact with different integral membrane protein(s) (domains) (Ehrmann *et al.*, 1998; Liu *et al.*, 1999).

[1]*OpuA of* L. lactis *is composed of two different types of subunits, that is, the ATPase subunit OpuAA and the chimeric ligand-binding/translocator protein OpuABC. OpuA of* B. subtilis *has three different subunits, OpuAA, OpuAB and OpuAC, and here the translocator (OpuAB) and binding protein (OpuAC) are separate polypeptides* **(see also Figure 13.1)**. *To discriminate the* L. lactis *system from OpuA of* B. subtilis, *we use the subscripts Ll and Bs, respectively.*

THE ABC COMPONENT

The ABC or ATPase component of known osmoregulated ABC transporters has a molecular mass of about 45 kDa. Similar to the ATPase subunit (MalK) of the maltose transporters of *Escherichia coli* and *Thermococcus litoralis* (Diederichs *et al.*, 2000), the ABC component of most osmoregulated systems consists of an amino-terminal α/β type ATPase domain (~27 kDa) and a carboxyl-terminal regulatory domain (~18 kDa). The carboxyl-terminal domain in MalK of *E. coli* provides the system with a means to control the transport activity as well as the expression of the *mal* genes through interaction of MalK with regulatory proteins (Chapter 9). Whether the regulatory domain of the ATPase of the osmoregulated ABC transporters has such a role is not known. Database searches indicate that some homologues of the family of osmoregulated ABC transporters have an ATPase component with a truncated regulatory domain. These systems, e.g. ChoQ of *L. lactis*, have a molecular mass of about 35 kDa.

THE INTEGRAL MEMBRANE COMPONENT

Hydropathy profiling of the sequences of the integral membrane components of the osmoregulated ABC transporters shows that the systems fall in four subfamilies **(Figure 13.2)**. Each of the transporters has the conserved EAA motif in the equivalent cytoplasmic loop (loop V–VI in subfamilies I and III and loop III–IV in subfamilies II and IV, see **Figure 13.2**). Members of subfamily I, represented by ProW of the *E. coli* ProU system and OpuAB of *B. subtilis* $OpuA_{Bs}$ **(Figure 13.1)**, are predicted to have seven transmembrane α-helical segments with the N-terminus facing the external surface of the membrane and the C-terminus facing the cytoplasm. This membrane topology is supported by *phoA* and *lacZ* fusions, albeit a limited set of data (Haardt and Bremer, 1996). ProW differs from $OpuAB_{Bs}$ (and other subfamily I homologues) by having an unusually long amino-terminus of about 100 residues (depicted as dotted circle in **Figure 13.2**), which protrudes into the periplasmic space of *E. coli*. The members of subfamily II, represented here by OpuCB and OpuCD of the OpuC system of *B. subtilis*, have five transmembrane α-helical segments with the N- and C-termini at the external and internal side of the membrane, respectively. Thus, compared to the members of subfamily I, the first two transmembrane segments are missing.

Subfamily III, exemplified by OpuABC of the OpuA system of *L. lactis*, is similar to subfamily I except that the member proteins have the ligand-binding domain fused to the integral membrane component **(Figure 13.1)**. In this case,

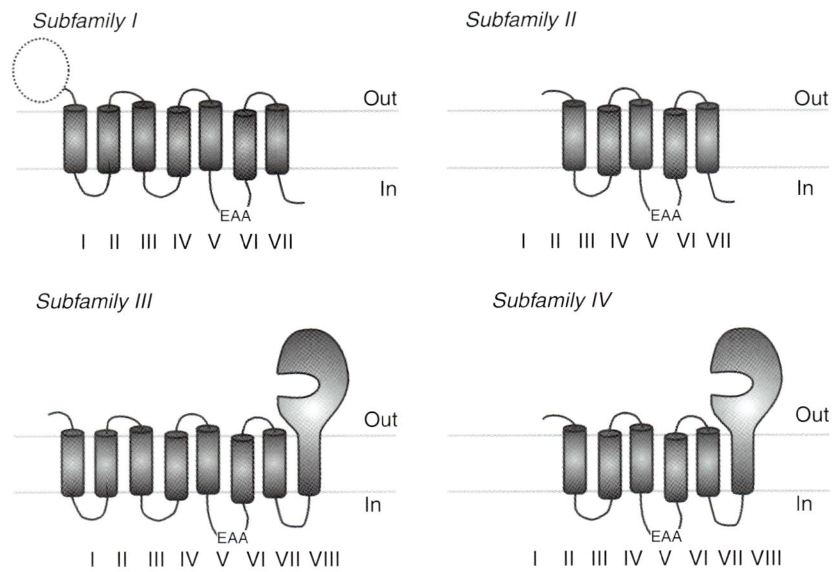

Figure 13.2. Topology models of the members of the four subfamilies of osmoregulated ABC transporters. The additional periplasmic domain in some members of subfamily I is indicated as a dotted circle. The Pacman-like structure connected to the last transmembrane segment of the members of subfamilies III and IV represents the ligand-binding domain. The EAA motif in the cytoplasmic loop is also depicted.

an eight transmembrane segment allows the binding domain to face the exterior of the cell. It is worth noting that the majority of multidomain membrane transport proteins have the individual domains fused to another via so-called flexible linker regions (Sutrina *et al.*, 1990). In the case of OpuA of *L. lactis*, the predicted end of the eighth transmembrane segment is followed by a sequence that over its entire length is highly similar (>50% identity) to the ligand-binding protein ProX of the *E. coli* ProU system. On the assumption that the amino-terminus of ProX forms an intrinsic part of the binding protein, a flexible linker will not be present between the transmembrane and ligand-binding domains of the OpuABC polypeptide of the OpuA system. Finally, members of subfamily IV, exemplified here by ProWX of *H. pylori*, and the membrane component of the ChoQ complex of *L. lactis*, are similar to those of subfamily II, except that also in this case the ligand-binding protein is fused to the integral membrane component. This gives rise to a membrane protein with six predicted transmembrane α-helical segments and an amino- and carboxyl-terminus that are facing the outer surface of the membrane. At present, it is unclear whether or not the smaller sizes of the proteins of subfamilies III and IV when compared to those of I and II are related to a different functioning of the systems.

THE SUBSTRATE-BINDING PROTEIN

Members of subfamily I and II have a soluble substrate-binding protein that in Gram-negative bacteria resides in the periplasm. In Gram-positive bacteria, the 'soluble' binding protein of subfamily I and II type transporters is anchored to the outer surface of the cytoplasmic membrane via an amino-terminal lipid moiety. Recently, ProX of *E. coli* has been crystallized (Breed *et al.*, 2001) and the structure, although not published, has been presented at scientific meetings. As expected, ProX has an overall tertiary fold typical of ligand-binding proteins belonging to ABC transporters (Quiocho and Ledvina, 1996; see Chapter 10), which is indicative of a Venus fly-trap mechanism for substrate binding. Interestingly, the substrate, glycine betaine, is bound to Trp residues via cation-π interactions similar to the observed binding of quaternary ammonium compounds to acetylcholine esterase of human brain (Bartolucci *et al.*, 2001; Bremer and Welte, unpublished). The π-electrons interact with the quaternary ammonium group of glycine betaine. One of the three Trp residues that interact with the substrate is highly conserved in the glycine betaine-binding proteins for which the primary sequences are available to date.

SPECIFICITY OF OSMOREGULATED ABC TRANSPORTERS

The substrate specificity of osmoregulated (ABC) transporters has been investigated most extensively in *B. subtilis* (Kempf and Bremer, 1998). Some systems seem very specific, e.g. OpuB only selects choline, whereas others such as $OpuA_{Bs}$ accept a wide range of substrates. It should be stressed, however, that in neither case has the specificity been determined directly, that is, as dissociation constants (K_d values) of ligand binding to the receptor. The specificity of bacterial osmoregulated ABC transporters has been determined either as apparent affinity constants of transport (K_m) or it has been inferred from the ability of a compound to offer protection during growth under hyperosmotic conditions. Given our experience with this type of analysis for the oligopeptide-binding protein of the Opp ABC transporter (Detmers *et al.*, 2000; Lanfermeijer *et al.*, 1999, 2000), we expect that differences may be more subtle than suggested by the data presented in the literature. Nevertheless, in terms of cell physiology, it is important to know if a compound does or does not offer protection against osmotic stress when taken up via a particular system. The differences in narrow versus broad specificity are not readily revealed in the primary sequences of the ligand-binding proteins/domains when analyzed by multiple sequence alignments.

For *Listeria monocytogenes* and *Lactobacillus plantarum*, it has been shown that pre-accumulated (*trans*) substrate inhibits the corresponding osmoregulated transport systems (Glaasker *et al.*, 1998; Verheul *et al.*, 1997). Upon raising the medium osmolality, the systems are rapidly activated through a diminished inhibition by *trans* substrate. Once turgor pressure has been restored, the cells are in osmotic equilibrium again and the transporters need to be inactivated or switched off. The so-called *trans*-inhibition may serve as a control mechanism to prevent the accumulation of these compatible solutes to too high levels and thereby the turgor pressure from becoming too high.

In the case of *L. monocytogenes*, carnitine is taken up via an ABC transporter that is specific for this substrate but is inhibited by high intracellular concentrations of both carnitine and glycine betaine (and perhaps other compatible solutes) depending on the osmotic status of the cells. In kinetic terms, the osmotic activation of the system results in an increase in apparent inhibition constant (K_I) for glycine betaine and carnitine at the inner surface of the membrane. Apparently, as a consequence of the water efflux following an osmotic upshift, the internal binding site for glycine betaine and carnitine, of a system specific for carnitine in the uptake reaction, is altered. Binding of compatible solutes to an internal site thus seems to represent a key step in the activation–inactivation mechanism. There is yet no molecular data on the proposed regulatory binding site in the transporters, and it is not therefore clear whether this is in the translocator protein or the ATPase component.

OSMOTIC REGULATION OF EXPRESSION OF GENES ENCODING ABC TRANSPORTERS

KINETICS OF OSMOTIC REGULATION AND OSMOTIC SIGNALS

The ProU system of *E. coli* is the best-studied ABC transporter in terms of osmotic regulation of gene expression, but despite intensive efforts, a full understanding of the molecular mechanism(s) underlying *proU* expression has not yet been achieved. In a coupled transcription–translation system, the kinetics of induction of the *proU* operon involves a lag phase of 15 to 20 min, followed by a rapid increase in expression, and, subsequently, a slow decay in the expression rate (Jovanovich *et al.*, 1989). This genetic regulation is slow in comparison to the activation of existing ProU (see below), which occurs within seconds following a change in osmolality. Therefore, the increased expression of *proU* could be mediated by signaling molecules (second messengers) that have to build up in the cytoplasm, rather than by activation directly by signal transduction pathways. Consistent with a role for specific second messenger molecules are observations that potassium glutamate is (at least partially) responsible for the induction of *proU* (Booth, 1992; Ramirez *et al.*, 1989), but other studies have rejected these claims (Csonka *et al.*, 1994; Jovanovich *et al.*, 1989). It is now thought that the stimulation of transcription of *proU in vitro* by potassium glutamate is a manifestation of the generally favorable effect of these osmolytes on macromolecular function (e.g. RNA polymerase–promoter interaction) and is not unique to osmotic regulation of the *proU* promoter (Csonka *et al.*, 1994). Taken together, the data are consistent with changes in intracellular osmolality as the signal for *proU* transcription, which would lead to maximal levels of ProU at a time that the turgor has already (largely) been restored. Such a signaling mechanism is in line with the observation that *E. coli* responds to osmotic upshift by rapidly accumulating potassium glutamate and subsequent replacement of these ionic osmolytes for neutral ones such as the substrate of the ProU system.

TRANSCRIPTION FACTORS AND OSMOREGULATED PROMOTERS

Contrary to what one would expect for a system that is tightly regulated by the osmolality of the medium (>500-fold induction), specific transcription factors do not seem to be involved or at least they have not been discovered. Genetic searches for such proteins have led to the isolation of mutants with defects in general DNA-binding proteins such as TopA, GyrAB, IHF, HU and H-NS (Kempf and Bremer, 1998). Mutations in these proteins have pleitropic effects on gene expression, for instance through alterations in DNA supercoiling, and it is unlikely that the tight osmotic control of the *proU* operon is solely mediated by these proteins. On the basis of transcriptional analysis of the P1 promoter of the *proP* gene, which encodes an osmoregulated secondary proline transporter, the suggestion has been made that the cAMP receptor protein (CRP) could function as a general osmoregulator of transcription in *E. coli* (Landis *et al.*, 1999). Binding of CRP to a site within the *proP P1* and some other promoters is destabilized after an osmotic upshift. These studies imply that CRP could have a general osmoregulatory role in addition to its function in catabolite control.

Transcription of *proU* is effected via the promoters *P1* (sigma factor σ^S) and *P2* (sigma factor σ^{70}). During exponential growth, transcription from *P2* contributes most to the expression by employing σ^{70}. Sigma factor σ^{S} generally contributes to the expression of genes in the stationary phase of growth, but the transcription of *proU* is not significantly increased under these conditions. The presence of potassium glutamate enhances transcription via σ^{S} and σ^{70}, and increases the selectivity of σ^{S} for *P1 in vitro* (Rajkumari *et al.*, 1996).

In organisms other than *E. coli*, e.g. *B. subtilis* and *L. lactis*, the expression of the genes specifying osmoregulated ABC transporters is also under osmotic control (Kempf and Bremer, 1998; van der Heide and Poolman, 2000a), but little is known about the signals and transcription factors that regulate expression. In *B. subtilis*, a general stress regulon is present, whose expression depends on the alternative sigma factor SigB (σ^{B}). In addition to salt, heat, oxidative and pH stresses also affect the expression of the σ^{B} regulon (Wood *et al.*, 2001). The induction of the σ^{B} regulon by osmotic upshift is only transient and σ^{B}-controlled proteins cannot adequately protect cells against prolonged high osmolality stress. The structural genes for the glycine betaine (OpuD) and proline (OpuE) secondary transport proteins are members of the σ^{B} regulon, but there is no evidence that the osmoregulated ABC transporters ($OpuA_{Bs}$, OpuB and OpuC) of *B. subtilis* are under the control of σ^{B}. The choline-specific transporter OpuB is under the control of the GbsR repressor, but this transcription factor is a choline sensor rather than an osmosensor/regulator (Bremer, 2002). Finally, it should be stressed that maximal rates of uptake via $OpuA_{Bs}$ and OpuC of *B. subtilis* increase only 1.5- to 3-fold when 0.4 M NaCl is added to the growth medium (Kappes *et al.*, 1996), indicating that the corresponding genes are not under tight osmotic control, as for example in the case of *proU* of *E. coli*. In *L. lactis* IL1403, the glycine betaine uptake capacity increases more than 10-fold when the cells are grown in the presence of 0.5 M KCl (van der Heide and Poolman, 2000a). The induction by osmotic stress, however, is only observed in chemically defined media and not in complex broth, suggesting that factors other than osmolality of the medium tune the expression levels to the needs of the cell. A preliminary report on the regulation of expression of the ABC transporter $OpuA_{Ll}$ suggests that a transcriptional regulator of the GnrR family acts as a repressor of the *opuA* operon (Obis *et al.*, 2000), but the signal sensed by the protein is not known.

OSMOTIC REGULATION OF ACTIVITY OF ABC TRANSPORTERS

KINETICS OF OSMOTIC (IN)ACTIVATION

In order to cope effectively with osmotic stress, cells need osmotically controllable systems in the membrane at all times as *de novo* synthesis takes too long and is not adequate for a quick response. In terms of an osmosensing/regulation mechanism, the only well-characterized ABC transport protein is OpuA of *L. lactis* (van der Heide and Poolman, 2000b; van der Heide *et al.*, 2001). The complete protein complex (OpuAA and OpuABC) has been purified and studied in proteoliposomes. By including ATP plus an ATP-regenerating system inside the vesicle lumen **(Figure 13.3)**, uptake of glycine betaine has been followed in response to osmotic shifts, as a function of membrane lipid

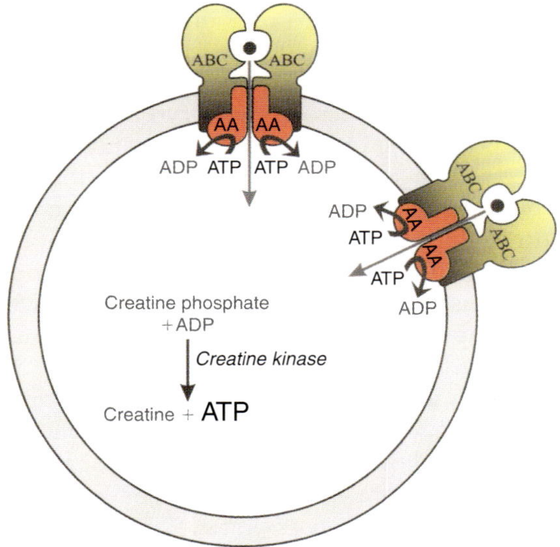

Figure 13.3. ***Proteoliposomal system to measure the activity of OpuA from*** **L. lactis*****. The purified protein complex was inserted into preformed liposomes, after which excess detergent was removed by adsorption onto polystyrene beads. The ATP-regenerating system (ATP, creatine kinase plus creatine phosphate) was included in the vesicle lumen by freeze-thawing, followed by sizing of the proteoliposomes by extrusion through polycarbonate filters with 200 nm pores (for details see van der Heide and Poolman, 2000b; van der Heide*** **et al.*****, 2001).***

composition and various water stress related parameters. It has been shown that $OpuA_{Ll}$ is activated instantaneously upon raising the osmolality of the medium, that is, when the external medium is made hyperosmotic relative to the inside. Activation has been measured as an increase in either the rate of ATP-driven substrate uptake or the rate of substrate-dependent ATP hydrolysis; for the latter measurements, ATP rather than the ATP-regenerating system was included in the proteoliposomes. Activation is elicited by ionic and non-ionic osmolytes, provided the molecules do not equilibrate across the membrane on the time scale of the transport measurements (van der Heide and Poolman, 2000b). Since (proteo)liposomes act as osmometers, that is, water diffuses across the membrane in response to the osmotic difference between the inner compartment and the outside medium, proteoliposomes are expected to decrease their volume to surface ratio when the outside osmolality is increased. The changes in membrane structure and lumen contents (osmolyte concentration) in osmotically stressed (proteo)liposomes can be compared with those in cells that are in a state of plasmolysis. Proteoliposomes with an average diameter of 200 nm change their shape from spherical to sickle-shaped as shown by cryo-electron microscopy (cryo-EM). These morphological changes occurred within milliseconds, i.e. on a time scale much shorter than the interval over which transport was measured. Upon lowering the outside osmolality to the initial value, yielding iso-osmotic conditions again, the vesicles regained their spherical shape and the transporter was deactivated (van der Heide *et al.*, 2001). Thus, osmotic activation and inactivation of $OpuA_{Ll}$ is entirely reversible, occurs on a time scale of seconds or less, and follows the shape and volume changes of the liposomes (see below, for proposed mechanism of osmosensing).

In contrast to the OpuA system from *L. lactis*, the hyperosmotic activation of ProU from *E. coli* takes several minutes (Faatz *et al.*, 1988), which makes it unlikely that decreased external water activity or reduced turgor pressure triggers the activation. A closer look at the data suggests that activation of ProU is also dependent on the presence of the transport substrate (glycine betaine), as the transport activity increases up to 3 min after an osmotic upshift in the absence but not in the presence of glycine betaine (Faatz *et al.*, 1988). Overall, the regulation of ProU transport activity seems different from that of OpuA of *L. lactis* and resembles more that of the transcription of *proU*, i.e. both are delayed upon an upshift. It should be stressed here that $OpuA_{Ll}$ has been studied in a well-defined proteoliposomal system whereas ProU has been analyzed in intact cells, which complicates a direct comparison of the data.

Mechanism(s) of osmosensing

Osmotic activation of membrane transporters may be triggered through a change in the hydration state of the transport protein complex, resulting from the altered water activity (a_w), or the signal may be transmitted to the protein via a specific signaling molecule or a change in the physicochemical properties of the surrounding membrane. Since OpuA of *L. lactis* can be activated not only by osmotic upshift but also by the insertion of cationic amphipaths in the membrane, it seems plausible that the membrane transduces the activation (van der Heide and Poolman, 2000b). The observation that an osmotic upshift leads to sickle-shaped vesicle structures already indicates that at the macroscopic level major changes take place at the membrane. These morphological changes, however, do not occur upon insertion of amphipaths into the membrane, indicating that the macroscopic alterations in membrane folding are not intrinsic to the osmosensing/regulation mechanism of $OpuA_{Ll}$. At the molecular level, physical properties such as membrane fluidity, bilayer thickness, hydration state of lipid headgroups, interfacial polarity and charge, and/or lateral pressure may vary with changes in the osmolality of the medium. To define the osmotic signaling process in more detail, $OpuA_{Ll}$ has been incorporated into liposomes of different lipid composition (van der Heide *et al.*, 2001). It was found that the fraction of anionic (charged) lipids is of major importance for the osmosensing mechanism, whereas variations in acyl chain length, degree of fatty acid saturation, position of the *cis/trans* double bond, and the fraction of non-bilayer lipids have relatively minor effects. By varying the fraction of anionic lipids (phosphatidylglycerol (PG) or phosphatidylserine (PS)) from 6 to 13 to 25%, $OpuA_{Ll}$ is converted from an inactive or tense (T) to an intermediate (I) to an osmotically controllable (R) state **(Figure 13.4)**. Moreover, at a given mol% of PG, $OpuA_{Ll}$ can be converted from R to I by adding cationic amphipaths and from I to R with anionic amphipaths. This suggests that the overall charge of the headgroup

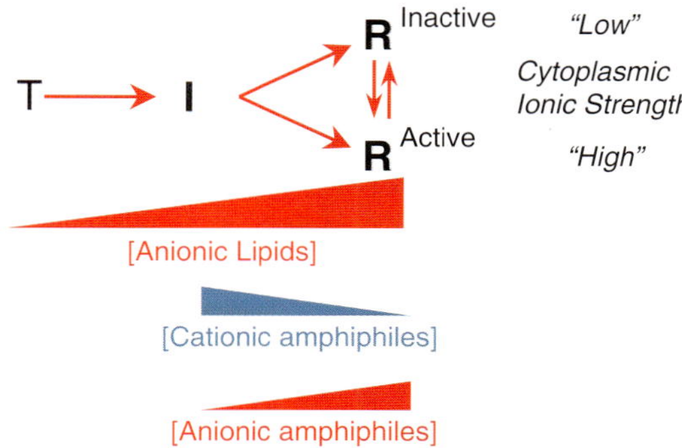

Figure 13.4. Schematic presentation of the factors affecting the osmotic activation of the membrane-embedded OpuA complex. A low fraction of anionic lipids converts OpuA from an inactive (T) to an intermediate state (C). A further increase of the anionic lipid fraction leads to an osmotically controlled (R) state. At the level of the membrane, the effect of anionic lipids can be mimicked by the addition of anionic or cationic membrane-active compounds. The effects of cytoplasmic ionic osmolytes, shifting the system from an inactive to an active state, is also depicted.

region of the membrane lipids determines the activity (kinetic state) of the transporter (van der Heide *et al.*, 2001).

The change in volume/surface ratio of the proteoliposomes upon osmotic upshift leads to an increase in the concentration of intravesicular osmolytes. By varying the intravesicular composition, it was observed that ionic osmolytes rather than internal osmolality switch the system from an inactive to an active state. In other words, the system can be activated at iso-osmotic conditions simply by an increase in the internal ionic strength. Because the effects of the ions vary with the fraction of anionic lipids in the membrane, and the threshold for osmotic activation is lowered by cationic and raised by anionic amphipaths, these experiments support the notion that osmotic signaling most probably occurs via the bilayer in which the protein complex is embedded (van der Heide *et al.*, 2001).

Nature of the Activating Signal

Why would the cell use ionic strength rather than intracellular osmolality (affecting protein hydration) or a specific signaling molecule (allosteric regulatory site on the protein)? When the osmolality of the medium is raised, the initial change in cytoplasmic water activity depends on the elasticity of the cell wall. Contrary to what is often thought, the cell wall of bacteria is not rigid but actually quite elastic (Csonka and Hanson, 1991; Doyle and Marquis, 1994). Consequently, even at turgor pressures above zero, the cytoplasmic volume decreases with increasing external osmolality, and the ion (osmolyte) concentrations increase accordingly. The increase in ionic strength accompanying the volume decrease is undesirable as too high concentrations of electrolytes interfere with macromolecular functioning in eubacteria as well as in higher organisms (Yancey *et al.*, 1982). As best documented for *E. coli* (Higgins *et al.*, 1987; Record *et al.*, 1998), eubacteria expel ionic compounds if the electrolyte concentration becomes too high and replace these molecules with neutral osmolytes such as glycine betaine to balance the cellular osmolality. The increase in electrolyte concentration (or ionic strength) upon a modest decrease in turgor pressure would thus represent an excellent trigger ('osmotic signal') for the activation of any osmoregulated transporter for neutral compatible solutes such as $OpuA_{Ll}$. Actually, it would prevent the osmotic stress from turning into 'electrolyte stress'. Why is the increase in intracellular osmolality less suitable as osmotic signal? In order to maintain a relatively constant turgor pressure at different external osmolalities, the cell will have to switch on $OpuA_{Ll}$ and take up glycine betaine with maximal activity at different internal osmolalities. In other words, the ability of (the majority of) microorganisms to grow at maximal rate over a wide range of osmolalities of the medium implies that cellular processes function optimally over wide range of intracellular osmolalities. Finally, the cell could use the osmotic upshift-dependent change in concentration of a specific molecule as signal, but ionic strength seems a more general signal for osmoresponsive systems, including signal transduction pathways to control the expression of osmoregulated genes.

Although further work is needed to elucidate all the intricacies of the osmosensing mechanism of the ABC transporter for glycine betaine in *L. lactis*, the *in vitro* studies with the purified protein complex indicate that all the regulatory properties observed *in vivo* are present in the two polypeptides comprising $OpuA_{Ll}$. The fact that osmotic activation of $OpuA_{Ll}$ in proteoliposomes mimics the regulation *in vivo* may seem surprising as the vesicles do not withstand turgor, whereas the turgor pressure in cells is several atmospheres. However, there is mounting evidence that in bacteria turgor pressure exists only across the cell wall (and outer membrane in the case of Gram-negative bacteria) and not across the cytoplasmic membrane (Cayley *et al.*, 2000). This implies that osmosensing devices present in the cytoplasmic membrane cannot repond to changes in turgor pressure; rather they must sense the consequences of the water influx or efflux, e.g. changes in ionic strength. The consequences of osmotic shifts, experienced by proteins in membrane model systems such as proteoliposomses, may thus be similar to those in the *in vivo* situation. Osmotic regulation of $OpuA_{Ll}$ requires the protein complex to be embedded in a membrane bilayer with the appropriate lipid composition (~25% anionic and some non-bilayer lipids) and the appropriate concentration of small inorganic electrolytes in the vesicle lumen. We cannot entirely rule out the possibility that additional proteinaceous factors influence the osmotic activation of $OpuA_{Ll}$ *in vivo*, as has been suggested for the proton motive force-driven proline transporter ProP (Kunte *et al.*, 1999), but we do not have any indications or need for such a factor to relate the *in vitro* and *in vivo* data.

CHALLENGES AND PERSPECTIVES

From the studies presented in the previous section, it is evident that osmosensing and regulation, be it a transporter, channel or signaling pathway, can only be fully understood if the system can be analyzed not only in the intact cell but also in artificial membrane systems with defined components. The ABC transporter for glycine betaine has some unique properties that make the protein complex ideally suitable for *in vitro* analyses. Firstly, the system has the ligand-binding domain covalently linked to the translocator moiety, which simplifies purification of the system and results in a high efficiency of the transport reaction without having to use a large excess of (soluble) ligand-binding protein. Secondly, the subunits of the OpuA complex (OpuAA and OpuABC) remain associated in the detergent-solubilized state, which greatly facilitates reconstitution in proteoliposmes. Thirdly, the system has a high catalytic efficiency, which enables us to measure accurately activities even below 5% of V_{max}.

The research on OpuA from *L. lactis* work indicates that a change in intracellular ionic strength serves as primary signal of osmotic stress for this ABC transporter. We propose that this signal is not sensed by the protein directly, i.e. via changes in surface hydration of the protein or direct effects of the ions on the protein (allosteric site); rather the membrane in which the protein is embedded serves as a mediator. Changes in cellular ionic strength are likely to alter specific interactions between (ionic) lipids and the protein, thereby affecting the transport activity. The ATPase and translocation activity of the OpuA transporter are strictly coupled, and, at present, it is unclear whether the ATPase (OpuAA) or the membrane-embedded part (translocator) of OpuABC is the actual sensor.

A challenge will now be to assign specific regions or residues in the $OpuA_{Ll}$ protein complex as sites that actually sense the changes in membrane structure (e.g. electrostatic interactions between protein and phospholipid molecules), and to translate the *in vitro* observations into proposals that can be experimentally tested *in vivo*. Moreover, one may wish to study osmotic signaling pathways that affect the transcription of genes in a manner analogous to the studies described heretofore for the ABC transporter $OpuA_{Ll}$. Past research has been restricted to mutant selection and isolation, and construction of allelic strains defective in one or more transcription factors. The *proU* regulatory circuit is clearly too complicated for a complete understanding of its osmoregulatory mechanism at this moment, but it is possible to devise protocols for the *in vitro* analysis of simpler pathways, e.g. two-component regulatory systems. One would also like to know how the cell responds to osmotic stress not only at the level of activation of transcription and transporter activity but also at the level of lipid synthesis. The membrane bilayer composition is intrinsic to the osmosensing mechanism of $OpuA_{Ll}$ and most likely other transporters as well (Rübenhagen *et al.*, 2000; Wood *et al.*, 2001), and changes therein will on the longer time scales be important for volume control of the cell. Preliminary studies have been reported on the changes in fatty acid composition of the membrane of *L. lactis* in relation to osmotic stress (Guillot *et al.*, 2000). Unfortunately, no information is available on variations in headgroup composition, e.g. fractions of anionic lipids, as these factors seem far more important for the osmoregulation of $OpuA_{Ll}$ than acyl chain length, degree of saturation, etc. Finally, the relationship between osmotic and other stresses needs to be evaluated more thoroughly at the level of transporter activity (*in vitro* and *in vivo*) and cellular glycine betaine accumulation levels.

Note added in proof: Recent analysis of genomic databases has indicated that some homologues of OpuA of *L. lactis* have two substrate-binding domains fused in tandem to the translocator moiety of the ABC transporter. These systems thus have four extracellular substrate-binding sites per functional complex. For a full description of the newly discovered chimeric substrate-binding/translocator proteins, one is referred to van der Heide and Poolman (2002) EMBO reports, in press.

Acknowledgments

This work was supported by grants from the Netherlands Organization for Scientific Research (NWO) under auspices of the Netherlands Foundation of Life Sciences. We thank Erhard Bremer for sharing information prior to publication.

References

Bartolucci, C., Perola, E., Pilger, C., Fels, G. and Lamba, D. (2001) Three-dimensional structure of a complex of galanthamine with acetylcholinesterase from *Torpedo californica*: implications for the design of new anti-Alzheimer drugs. Proteins 42, 182–191.

Booth, I.R. (1992) Regulation of gene expression during osmoregulation: the role of potassium glutamate as a secondary signal of osmotic stress. In: Alkali Cation Transport Systems in Prokaryotes (ed. E.P. Bakker), pp. 205–224, Boca Raton, FL: CRC Press.

Bourot, S., Sire, O., Trautwetter, A., Touze, T., Wu, L.F., Blanco, C. and Bernard, T. (2000) Glycine betaine-assisted protein folding in a *lysA* mutant of *Escherichia coli*. J. Biol. Chem. 275, 1050–1056.

Breed, J., Kneip, S., Gade, J., Welte, W. and Bremer, E. (2001) Purification, crystallization and preliminary crystallographic analysis of the periplasmic binding protein ProX from *Escherichia coli*. Acta Cryst. D57, 448–450.

Bremer, E. (2002) Adaptation to changing osmolarity. In: *Bacillus subtilis* and its Closest Relatives: From Genes to Cells (ed. A.L. Sonenshein *et al.*), Washington, DC: ASM Press.

Cayley, D.S., Guttman, H.J. and Record, M.T. Jr. (2000) Biophysical characterization of changes in amounts and activity of *Escherichia coli* cell and compartment water and turgor pressure in response to osmotic stress. Biophys. J. 78, 1748–1764.

Csonka, L.N. (1989) Physiological and genetic responses of bacteria to osmotic stress. Microbiol. Rev. 53, 121–147.

Csonka, L.N. and Hanson, A.D. (1991) Prokaryotic osmoregulation: genetics and physiology. Annu. Rev. Microbiol. 45, 569–606.

Csonka, L.N., Ikeda, T.P., Fletcher, S.A. and Kustu, S. (1994) The accumulation of glutamate is necessary for optimal growth of *Salmonella typhimurium* in media of high osmolarity but not induction of the *proU* operon. J. Bacteriol. 176, 6324–6333.

Dassa, E. and Bouige, P. (2001) The ABC of ABCs: a phylogenetic and functional classification of ABC systems in living organisms. Res. Microbiol. 152, 211–229.

Detmers, F.J.M., Lanfermeijer, F.C., Abele, R., Tampé, R., Konings, W.N. and Poolman, B. (2000) Combinatorial peptide libraries reveal the ligand binding properties of the oligopeptide receptor OppA of *Lactococcus lactis*. Proc. Natl Acad. Sci. USA, 97, 12487–12492.

Diederichs, K., Diez, J., Greller, G., Müller, C., Breed, J., Schnell, C., Vonrhein, C., Boos, W. and Welte, W. (2000) Crystal structure of MalK, the ATPase subunit of the trehalose/maltose ABC transporter of the archaeon *Thermococcus litoralis*. EMBO J. 19, 5951–5961.

Doyle, R.J. and Marquis, R.E. (1994) Elastic, flexible peptidoglycan and bacterial cell wall properties. Trends Microbiol. 2, 57–60.

Ehrmann, M., Ehrle, R., Hofmann, E., Boos, W. and Schlösser, A. (1998) The ABC maltose transporter. Mol. Microbiol. 29, 685–694.

Faatz, E., Middendorf, A. and Bremer, E. (1988) Cloned structural genes for the osmotically regulated binding-protein-dependent glycine betaine transport system (ProU) of *Escherichia coli* K-12. Mol. Microbiol. 2, 265–279.

Galinski, E.A. and Trüper, H.G. (1994) Microbial behaviour in salt-stressed ecosystems. FEMS Microbiol. Rev. 15, 95–108.

Glaasker, E., Konings, W.N. and Poolman, B. (1996a) Osmotic regulation of intracellular solute pools in *Lactobacillus plantarum*. J. Bacteriol. 178, 575–582.

Glaasker, E., Konings, W.N. and Poolman, B. (1996b) Glycine-betaine fluxes in *Lactobacillus plantarum* during osmostasis and hyper- and hypoosmotic shock. J. Biol. Chem. 271, 10060–10065.

Glaasker, E., Heuberger, E.H.M.L., Konings, W.N. and Poolman, B. (1998) Mechanism of osmotic activation of the quaternary ammonium compound transporter (QacT) of *Lactobacillus plantarum*. J. Bacteriol. 180, 5540–5546.

Guillot, A., Obis, D. and Mistou, M.-Y. (2000) Fatty acid membrane composition and activation of glycine betaine transport in *Lactococcus lactis* subjected to osmotic stress. Int. J. Food Microbiol. 55, 47–51.

Haardt, M. and Bremer, E. (1996) Use of *phoA* and *lacZ* fusions to study the membrane topology of ProW, a component of the osmoregulated ProU transport system of *Escherichia coli*. J. Bacteriol. 178, 5370–5381.

Higgins, C.F., Cairney, J., Stirling, D.A., Sutherland, L. and Booth, I.R. (1987) Osmotic regulation of gene expression: ionic strength as an intracellular signal? Trends Biochem. Sci. 12, 339–344.

Jovanovich, S.B., Record, M.T. and Burgess, R.R. (1989) In an *Escherichia coli* coupled transcription-translation system, expression of the osmoregulated gene *proU* is stimulated at elevated potassium concentrations and by an extract from cells grown at high osmolality. J. Biol. Chem. 264, 7821–7825.

Kappes, R.M., Kempf, B. and Bremer, E. (1996) Three transport systems for the osmoprotectant glycine betaine operate in *Bacillus subtilis*: characterization of OpuD. J. Bacteriol. 178, 5071–5079.

Kempf, B. and Bremer, E. (1995) OpuA, an osmotically regulated binding protein-dependent transport system for the osmoprotectant glycine betaine in *Bacillus subtilis*. J. Biol. Chem. 270, 16701–16713.

Kempf, B. and Bremer, E. (1998) Uptake and synthesis of compatible solutes as microbial stress responses to high-osmolarity environments. Arch. Microbiol. 170, 319–330.

Kempf, B., Gade, J. and Bremer, E. (1997) Lipoprotein from the osmoregulated ABC transport system OpuA of *Bacillus subtilis*: purification of the glycine betaine binding protein and characterization of a functional lipidless mutant. J. Bacteriol. 179, 6213–6220.

Kunte, H.J., Crane, R.A., Culham, D.E., Richmond, D. and Wood, J.M. (1999) Protein ProQ influences osmotic activation of compatible solute transporter ProP in *Escherichia coli* K-12. J. Bacteriol. 181, 1537–1543.

Lamark, T., Kaasen, I., Eshoo, M.W., Falkenberg, P., McDougall, J. and Ström, A.R. (1991) DNA sequences and analysis of the *bet* genes encoding the osmoregulatory choline-glycine betaine pathway of *Escherichia coli*. Mol. Microbiol. 5, 1049–1064.

Landis, L., Xu, J. and Johnson, R.C. (1999) The cAMP receptor protein CRP can function as an osmoregulator of transcription in *Escherichia coli*. Genes Dev. 13, 3081–3091.

Lanfermeijer, F., Picon, A., Konings, W.N. and Poolman, B. (1999) Kinetics and consequences of binding of nona- and dodecapeptides to the oligopeptide binding protein, OppA, of *Lactococcus lactis*. Biochemistry 38, 14440–14450.

Lanfermeijer, F.J., Detmers, F.J.M., Konings, W.N. and Poolman, B. (2000) On the binding mechanism of the peptide receptor of the oligopeptide transport system of *Lactococcus lactis*. EMBO J. 19, 3649–3656.

Liu, C.E., Liu, P.-Q., Wolf, A., Lin, E. and Ames, G.F.-L. (1999) Both lobes of the soluble receptor of the periplasmic histidine permease, an ABC transporter (traffic ATPase), interact with the membrane bound complex. J. Biol. Chem. 274, 739–747.

Lucht, J.M. and Bremer, E. (1994) Adaptation of *Escherichia coli* to high osmolarity environments: osmoregulation of the high-affinity glycine betaine transport system ProU. FEMS Microbiol. Rev. 14, 3–20.

Martins, L.O. and Santos, H. (1995) Accumulation of mannosylglycerate and di-*myo*-inositol-phosphate by *Pyrococcus furiosus* in response to salinity and temperature. Appl. Environ. Microbiol. 61, 3299–3303.

Martins, L.O., Carreto, L.S., Da Costa, M.S. and Santos, H. (1996) New compatible solutes related to di-*myo*-inositol-phosphate in members of the order *Thermotogales*. J. Bacteriol. 178, 5644–5651.

McLaggan, D., Naprstek, J., Buurman, E.T. and Epstein, W. (1994) Interdependence of K^+ and glutamate accumulation during osmotic adaptation of *Escherichia coli*. J. Biol. Chem. 269, 1911–1917.

Obis, D., Guillot, A. and Mistou, M.-Y. (2000) Abstract. Comp. Biochem. Physiol. A 126, S105.

Poolman, B. and Glaasker, E. (1998) Regulation of compatible solute accumulation in bacteria. Mol. Microbiol. 29, 397–407.

Quiocho, F.A. and Ledvina, P.S. (1996) Atomic structure and specificity of bacterial periplasmic receptors for active transport and chemotaxis: variation of common themes. Mol. Microbiol. 20, 17–25.

Rajkumari, K., Kusano, S., Ishihama, A., Mizuno, T. and Gowrishankar, J. (1996) Effect of H-NS and potassium glutamate on σ^S- and σ^{70}-directed transcription *in vitro* from osmotically regulated P1 and P2 promoters of *proU* in *Escherichia coli*. J. Bacteriol. 178, 4176–4181.

Ramirez, R.M., Prince, W.S., Bremer, E. and Villarjo, M. (1989) *In vitro* reconstitution of osmoregulated expression of *proU* of *Escherichia coli*. Proc. Natl Acad. Sci. USA 86, 1153–1157.

Record, M.T. Jr, Courtenay, E.S., Cayley, D.S. and Guttman, H.J. (1998) Biophysical compensation mechanisms buffering *E. coli* protein-nucleic acid interactions against changing environments. Trends Biochem. Sci. 23, 190–194.

Rübenhagen, R., Rönsch, H., Jung, H., Krämer, R. and Morbach, S. (2000) Osmosensor and osmoregulator properties of the betaine carrier BetP from *Corynebacterium glutamicum* in proteoliposomes. J. Biol. Chem. 275, 735–741.

Sankaran, K. and Wu, H.C. (1994) Lipid modification of bacterial prolipoprotein: transfer of diacylglyceryl moiety from phosphatidylglycerol. J. Biol. Chem. 269, 19701–19706.

Sukharev, S.I., Blount, P., Martinac, B. and Kung, C. (1997) Mechanosensitive channels of *Escherichia coli*: the MscL gene, protein, and activities. Annu. Rev. Physiol. 59, 633–657.

Sutrina, S.L., Reddy, P., Saier, M.H. and Reizer, J. (1990) The glucose permease of *Bacillus subtilis* is a single polypeptide chain that functions to energize the sucrose permease. J. Biol. Chem. 265, 18581–18589.

van der Heide, T. and Poolman, B. (2000a) Glycine betaine transport in *Lactococcus lactis* is osmotically regulated at the level of expression and translocation. J. Bacteriol. 182, 203–206.

van der Heide, T. and Poolman, B. (2000b) Osmoregulated ABC-transport system of *Lactococcus lactis* senses water stress via changes in the physical state of the membrane. Proc. Natl Acad. Sci. USA 97, 7102–7106.

van der Heide, T., Stuart, M.C.A. and Poolman, B. (2001) On the osmotic signal and osmosensing mechanism of an ABC transport system for glycine betaine. EMBO J. 20, 7022–7032.

Verheul, A., Glaasker, E., Poolman, B. and Abee, T. (1997) Betaine and L-carnitine transport in response to osmotic signals in *Listeria monocytogenes* Scott A. J. Bacteriol. 179, 6979–6985.

Wood, J.M. (1999) Osmosensing by bacteria: signals and membrane-based sensors. Microbiol. Mol. Biol. Rev. 63, 230–262.

Wood, J.M., Bremer, E., Csonka, L.N., Krämer, R., Poolman, B., van der Heide, T. and Smith, L. (2001) Osmosensing and osmoregulatory compatible solute accumulation by bacteria. Comp. Biochem. Physiol. A. Mol. Integr. Physiol. 130, 437–460.

Yancey, P.H., Clark, M.E., Hand, S.C., Bowlus, R.D. and Somero, G.N. (1982) Living with water stress: evolution of osmolyte systems. Science 217, 1214–1222.

Part IV

Non-mammalian Eukaryote ABC Transporters

CHAPTER 14

Inventory and Evolution of Fungal ABC Protein Genes

Christoph Schüller, Bettina E. Bauer and Karl Kuchler

Introduction

The baker's yeast *Saccharomyces cerevisiae* was the first eukaryotic organism to have its complete genome sequence determined, revealing 30 distinct genes encoding ATP-binding cassette (ABC) proteins (Bauer *et al.*, 1999; Decottignies and Goffeau, 1997; Taglicht and Michaelis, 1998). ABC proteins are ubiquitous and form one of the largest gene families known with more than 2000 distinct ABC genes present in various current databases, e.g. Interpro (www.ebi.ac.at/interpro/) or Prosite (www.expasy.ch/Prosite). All known ABC proteins share a common hallmark domain, the highly conserved ABC domain, also known as the nucleotide-binding domain (NBD). The NBD contains signature motifs found in all ABC proteins operating from bacteria to man (Higgins, 1992). Membrane-bound ABC proteins also contain variable numbers of membrane-spanning domains arranged in certain membrane architectures. Many ABC proteins transport a variety of compounds across cellular membranes by an active process that is coupled to ATP hydrolysis. These ABC proteins are therefore referred to as ABC transporters or pumps. While some pumps seem to transport various xenobiotics, others exhibit a rather narrow substrate spectrum. Notably, for many ABC proteins no defined substrates or even physiological roles are known. Interestingly, ABC proteins not only function as simple membrane translocators for molecules, they can also act as receptors, sensors, proteases, channels, channel regulators and even signal-ing components (Higgins, 1995). The question of how the highly conserved molecular architecture of ABC proteins entertains such a functional diversity remains elusive. Hence, the functions of many ABC proteins may hold surprises and many important issues remain to be discovered. In this chapter, we will discuss the structure, function and properties of fungal ABC proteins, focusing on the inventory of ABC genes in *S. cerevisiae.* Because the functional annotation of the yeast genome is fairly advanced, we will also compare the yeast ABC inventory to those from fungal pathogens (*Candida albicans* and *Aspergillus fumigatus*) whose genomes have been sequenced or are close to being sequenced.

The Inventory and Molecular Architecture of Fungal ABC Proteins

Based on their molecular architecture, one can distinguish two types of yeast ABC proteins. The first type contains at least one transmembrane domain (TMD), while the second type lacks any obvious MSD (**Figure 14.1**). The

ABC Proteins: From Bacteria to Man
ISBN 0-12-352551-9

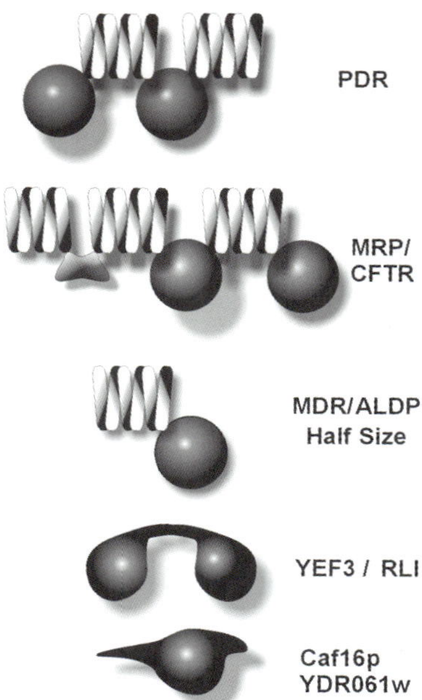

Figure 14.1. ***Molecular architecture and predicted membrane topology of yeast ABC proteins. The cartoon depicts the predicted membrane topologies and architecture present in distinct subfamilies of yeast ABC proteins.***

architecture of yeast ABC proteins also includes one or two highly conserved ABC domains or NBDs, encompassing roughly 200 amino acids. The most conserved features found in any given NBD are the Walker A and B motifs [AG]-X(4)-G-K-[ST] and [RK]-X(3)-G-X(3)-L-hydrophobic(4)-D, which are present in all ATP-binding proteins (Walker *et al.*, 1982), and the ABC signature motif [LIVMFYC]-[SA]-[SAPGLVFYKQH]-G-[DENQMW]-[KRQASP-CLIMFW]-[KRNQSTAVM]-[KRACLVM]-[LIV MFYPAN]-{PHY}-[LIVMFW]-[SAGCLIVP]-{FYWHP}-{KRHP}-[LIVMFYWSTA] – (Prosite PS00211). Moreover, two additional regions provide diagnostic sequences for ABC proteins – the center motif located between Walker A and B, and the sequences found downstream of Walker B (Michaelis and Berkower, 1995). The molecular architecture of eukaryotic ABC proteins arranges NBDs with TMDs in two possible ways. Yeast ABC proteins come in a duplicated TMD1-NBD1-TMD2-NBD2 forward or a mirror image NBD1-TMD1-NBD2-TMD2 reverse topology. The reverse architecture of such full-size transporters is found mainly in the PDR subfamily **(Table 14.1)**, while the forward orientation is similar to the one present in mammalian P-glycoproteins (Gros *et al.*, 1986). However, so-called half-size transporters of both the TMD-NBD and NBD-TMD topologies are also known **(Figure 14.1)**. Half-size ABC transporters are believed to dimerize to form functional transporter molecules. The recent elucidation of a high-resolution 3-D crystal structure of the *Escherichia coli* MsbA protein nicely illustrates this interaction (Chang and Roth, 2001). In bacteria, each domain of a given ABC protein is encoded by a single gene, although many variations on this theme also exist in prokaryotes (Young and Holland, 1999; Chapter 8).

Each TMD usually contains six predicted α-helical transmembrane-spanning segments (TMSs), although in some cases four to eight predicted TMSs per TMD are also known. In sharp contrast to NBDs encompassing the hallmark domains, only limited homology can be found within the TMDs of different ABC proteins (Decottignies and Goffeau, 1997; Michaelis and Berkower, 1995). The NBDs serve to bind and hydrolyze ATP or other NTPs, thereby fueling transport processes. However, numerous studies and genetic analyses have shown that NBDs not only serve as the fueling domains, but they appear intimately linked to the function and/or structure of individual ABC proteins. Importantly, the functions of N-terminal and C-terminal NBDs are not necessarily equivalent and thus each NBD of a eukaryotic ABC protein is indispensable. The analysis of the evolutionary sequence relationships between individual NBDs of yeast ABC proteins revealed five distinct clusters of homology. Hence, the yeast ABC gene inventory comprises 30 genes subdivided into the PDR, MDR, ALDP, MRP/CFTR, and YEF3/RLi families (Bauer *et al.*, 1999; Decottignies and Goffeau, 1997; Michaelis and Berkower, 1995).

THE PLEIOTROPIC DRUG RESISTANCE (PDR) SUBFAMILY

This subfamily includes the Pdr5p, Pdr10p, Pdr15p, Pdr11p, Pdr12p, Snq2p, Ynr070p, Adp1p and Aus1p/YOR011w ABC proteins. Their function might be linked to cellular detoxification, although in several cases no substrates have been identified. The overexpression of Pdr5p, Snq2p and Yor1p confers pleiotropic drug resistance (PDR) phenotypes. These genes confer resistance to hundreds of chemically unrelated

TABLE 14.1. THE INVENTORY OF ABC PROTEINS IN *SACCHAROMYCES CEREVISIAE*

ABC pump	Substrates	Length	Topology	Localization
MDR family				
Ste6p	**a**-factor pheromone	1290	$(TMS_6\text{-}ABC)_2$	PM, GV, ESM
Atm1p	Fe/S proteins	694	$TMS_6\text{-}ABC$	Mito IM
Mdl1p	?	696	$TMS_6\text{-}ABC$	Mito IM
Mdl2p	?	820	$TMS_6\text{-}ABC$	Mito IM
PDR family				
Pdr5p	Drugs, steroids, antifungals, PL	1511	$(ABC\text{-}TMS_6)_2$	PM
Pdr10p	?	1564	$(ABC\text{-}TMS_6)_2$	PM
Pdr15p	?	1529	$(ABC\text{-}TMS_6)_2$	PM
Snq2p	Mutagens, drugs	1501	$(ABC\text{-}TMS_6)_2$	PM
Pdr12p	Weak organic acids	1511	$(ABC\text{-}TMS_6)_2$	PM
Pdr11p	?	1411	$(ABC\text{-}TMS_6)_2$	PM
Aus1p/YOR011c	?	1394	$(ABC\text{-}TMS_6)_2$	?
Adp1p	?	1049	$TMS_2\text{-}ABC\text{-}TMS_7$	?
YNR070w	?	1333	$(ABC\text{-}TMS_6)_2$	?
YOL075c	?	1095	$(ABC\text{-}TMS_6)_2$	?
MRP/CFTR family				
Yor1p	Oligo, revero, PL	1477	$TMD0(TMS_6\text{-}ABC)_2$	PM
Ycf1p	GS-conjugates, Cd^{2+}, UCB, BA	1515	$TMD0(TMS_6\text{-}R\text{-}ABC)_2$	Vacuole
Ybt1p	BA	1661	$TMD0(TMS_6\text{-}ABC)_2$	Vacuole
Bpt1p	UCB	1559	$TMD0(TMS_6\text{-}ABC)_2$	Vacuole, ERM?
YHL035c	?	1592	$TMD0(TMS_6\text{-}ABC)_2$	?
YKR103w/YKR104c		1524	$(TMS_6\text{-}ABC)_2$	?
ALDp family				
Pxa1p	LCFA	870	$TMS_6\text{-}ABC$	Peroxisomes
Pxa2p	LCFA	853	$TMS_6\text{-}ABC$	Peroxisomes
YEF3/RLI family				
Yef3p	Hygromycin, paro	1044	ABC_2	Ribo?, Cyt?
Gcn20p		752	ABC_2	Polysomes
Hef3p		1044	ABC_2	Cytosol?
New1p/YPL226w		1196	$TMS_3\text{-}ABC_2$	?
Kre30p/YER036c		610	ABC_2	?
Rli1p/YDR091c		608	ABC_2	?
Non-classified				
YDR061w		539	ABC	?
Caf16p/YFL028c		289	ABC	?

ABC, ATP-binding cassette; TMD0, transmembrane domain; TMS, transmembrane segment; GS, glutathione S; UCB, unconjugated bilirubin; BA, bile acids; PL, phospholipids; oligo, oligomycin; revero, reveromycin A; paro, paromomycin; LCFA, long chain fatty acids; PM, plasma membrane; ERM, endoplasmic reticulum membrane; ESM, endosomal membranes; Cyt, cytoplasm; Ribo, ribosome; Mito IM, mitochondrial inner membrane.

drugs, including agricultural fungicides, benzimidazoles, dithiocarbamates, azoles, mycotoxins, herbicides, cycloheximide, sulfometuron, nigericin and anticancer drugs (Balzi *et al.*, 1987; Bissinger and Kuchler, 1994; Cui *et al.*, 1996; Hirata *et al.*, 1994; Katzmann *et al.*, 1995; Kralli *et al.*, 1995; Servos *et al.*, 1993). These ABC genes and their regulation are described in great detail in Chapter 15.

The Pdr12p pump seems to have a distinct physiological role, as it does not transport hydrophobic drugs, but confers resistance to weak organic acids. Pdr12p mediates the energy-dependent extrusion of carboxylate anions (Piper *et al.*, 1998), such as those used as food preservatives, including benzoate, sorbate and propionate, as well as C1–C7 weak organic acids, some of which are produced during normal cellular metabolism. Notably, *PDR12* mRNA synthesis is dramatically induced by sorbic acid stress and by exposure of yeast cells to low pH stress (Piper *et al.*, 1998), demonstrating that Pdr12 in fact represents a stress response gene.

Aus1p (YOR011w) is closely related to Pdr11p, sharing more than 65% sequence identity. Non-essential Aus1p appears to be involved in the uptake of sterols, as a Δ*aus*1 deletion mutant exhibits a reduced accumulation of cholesterol, while no obvious phenotypes are discernible under standard growth conditions (SGD http://genomewww.stanford.edu/Saccharomyces/). The function of other members of the PDR subfamily such as Pdr11p, Pdr10p, Pdr15p, Adp1p and Ynr070p remains unknown and no data are currently available regarding their substrates or physiological roles. Because Pdr10p and Pdr15p are tightly regulated by adverse conditions such as high osmolarity and heat shock, respectively, their functions might also be linked to a cellular response (Wolfger *et al.*, in preparation).

With the exception of Adp1p, all members of this group display a predicted NBD1-TMD1-NBD2-TMD2 structure with usually 12 predicted TMSs. Adp1p exhibits a slightly different architecture, replacing the first NBD with a large soluble domain, followed by a TMD1-NBD2-TMD2 topology **(Figure 14.1)**. It is noteworthy that all PDR members localize to the plasma membrane as shown in **Figure 14.2**. This cell surface localization further supports their purported function in cellular detoxification and cellular stress responses, although their precise roles and cellular substrates remain an enigma.

Finally, a substantial number of PDR subfamily members have been identified in other fungal species, including fungal pathogens. All PDR homologues linked to multidrug resistance are extensively discussed in Chapter 15. These ABC proteins currently total almost 50 fungal PDR genes. For many related PDR family members, a cellular function has not been established beyond the one known for the corresponding counterpart in baker's yeast. Examples for members of this impressive and growing group of fungal detoxification proteins are *Candida krusei* Abc1p, *Schizosaccharomyces pombe bfr1+*/Hba2p (Turi and Rose, 1995), *Candida glabrata* Cgr1p/Pdh1p (Miyazaki *et al.*, 1998), *Penicillium digitatum* Pmr1p (Nakaune *et al.*, 1998), *Emericella nidulans* AtrAp/ANPGP1p (Del Sorbo *et al.*, 1997) and AtrBp/ANPGP2p (Andrade *et al.*, 2000), *A. fumigatus* AtrFp, *C. albicans* Cdr1p (Prasad *et al.*, 1995), Cdr2p (Sanglard *et al.*, 1997), Cdr3p (Balan

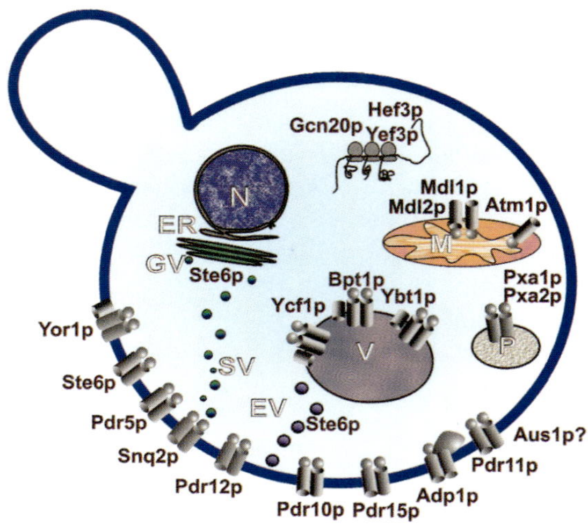

Figure 14.2. ***Subcellular localization of yeast ABC proteins. The cartoon depicts the subcellular localization of yeast ABC proteins in various cellular membranes or compartments. For a list of ABC proteins and further details see text and*** **Table 14.1.**

et al., 1997) and Cdr4p (Franz *et al.*, 1998), *Botrytis cinerea* BCPGP1p, *Cryptococcus neoformans* eCdr1p and *Magnaporte grisea* Abc1p (Urban *et al.*, 1999). This incomplete list illustrates the diversity of this ABC transporter family and hence underscores its importance, with more members surfacing at a rapid pace. The interested reader is referred to publicly accessible databases such as Swissprot (www.expasy.ch) or Interpro (www.ebi.ac.uk/Interpro) to obtain continuously updated information.

THE MULTIDRUG RESISTANCE-RELATED PROTEIN MRP/CFTR SUBFAMILY

Members of this class exhibit a membrane topology such as TMD0-TMD1-NBD1-TMD2-NBD2 (Tusnady *et al.*, 1997). The C-terminal TMD comprises 11 predicted TMSs, interrupted by a small cytoplasmic domain. Yeast MRP/CFTR-like pumps include Yor1p, Ycf1p, Bpt1p, Ybt1p, YHL035w and YKR103/YKR104w. The YKR103/YKR104w open reading frames (ORFs) include a stop codon between MSD2 and NBD2 and thus represent perhaps a pseudogene or a sequencing error.

Yor1p is probably among the best-studied members of the MRP family. The gene was initially isolated in a genetic screen for genes conferring resistance to oligomycin (Katzmann *et al.*, 1995). Yor1p is localized to the plasma membrane and has overlapping functions with PDR pumps such as Pdr5p, Snq2p and even Pdr12p, although Yor1p exhibits quite unique substrate specificities **(Table 14.1)**. The Δ*yor1* null mutant is viable, but displays increased sensitivity to a variety of compounds, including azoles, antibiotics such as tetracycline, erythromycin and oligomycin, as well as anticancer drugs like daunorubicin and doxorubicin, carboxylic acids such as acetic, propionic and benzoic acids, and heavy metals such as cadmium (Cui *et al.*, 1996). In contrast, Yor1p overproduction confers resistance to many of these compounds (Ogawa *et al.*, 1998). The function of Yor1p and its regulation is also extensively discussed in Chapter 15.

In contrast to Yor1p, Ycf1p is localized to the vacuolar membrane **(Figure 14.2)**. Nevertheless, like Yor1p, Ycf1p confers resistance to cadmium (Szczypka *et al.*, 1994). Besides vacuolar Cd^{2+} sequestration, Ycf1p is also involved in vacuolar transport of reduced glutathione and glutathione S-conjugates such as glutathione-conjugated arsenite. A homologue of Ycf1p, Bpt1p, mediates transport of unconjugated bilirubin into the vacuole. A Δ*ycf1* Δ*bpt1* double mutant is blocked for vacuolar transport of unconjugated bilirubin. Ycf1p is related to the human multidrug resistance proteins MRP1 and MRP2, and has 45% overall similarity to human CFTR (cystic fibrosis transmembrane conductance regulator) based on a ClustalW 1.4 alignment. It is interesting to note that yeast sequesters heavy metals to the vacuole, rather than extruding them. Such a 'social' behavior of a unicellular organism might be explained by a beneficial effect on immediate neighbors. Finally, Ybt1p, the yeast bile transporter (formerly Bat1p) mediates vacuolar uptake of bile acids such as taurocholate (Ortiz *et al.*, 1997). Another close homologue of Ybt1p, the YHL035w gene product, has not been studied and its physiological cargo and cellular localization has not been elucidated as yet.

ABC proteins of the MRP/CFTR family have also been identified in other fungi. However, in contrast to the large PDR family, substantially less information is available on fungal genes of this family. In *S. pombe*, YAWB (also SPAC3F10.11C) and ABC1 (Christensen *et al.*, 1997b) have been identified as MRP/CFTR family members, as well as a gene from *Neurospora crassa* (B7A16.190) and a Yor1p homologue in *C. albicans* (Ogawa *et al.*, 1998).

The ALDP Adrenoleukodystrophy Protein Subfamily

This small subfamily contains only two half-size transporters, Pxa1p and Pxa2p, displaying a TMD-NBD membrane topology. Pxa1p/ Pxa2p are yeast orthologues of human Pmp70/ABCD3/ PXMP1, ALD/ALDR/ABCD2 and ABCD4/ PXMP1L/PMP69 peroxisomal disease genes associated with neurodegenerative diseases such as adrenoleukodystrophy and Zellweger syndrome (Gartner and Valle, 1993; Holzinger *et al.*, 1997, 1999; Kamijo *et al.*, 1992). Indeed, both Pxa1p and Pxa2p localize to the peroxisomal membrane and might function as heterodimers (Shani *et al.*, 1996; Swartzman, *et al.*, 1996). They are thought to mediate peroxisomal uptake of very long chain fatty acids to undergo degradation through β-oxidation (Watkins *et al.*, 2000), which is consistent with the presence of a fatty acid-binding domain in Pxa1p/Pxa2p. The null mutants fail to grow on fatty acids such as palmitate or oleate as the sole carbon source. Although the Pxa1p/Pxa2p complex is required for peroxisome function, it is dispensable for peroxisome biogenesis or for import of peroxisomal matrix proteins. While the *PXA1* gene is only expressed when cells grow on oleate, the *PXA1* and *PXA2* promoters lack any consensus oleate-response elements, yet *PXA1*, but not *PXA2*, is oleate-induced and transcription is Oaf1p/Pip2p-dependent (Bossier *et al.*, 1994; Swartzman *et al.*, 1996). The regulators Oaf1p and Pip2p represent the two key transcription factors for peroxisome biogenesis in yeast. In contrast to the situation with the PDR family, only a few ALDP homologues have been described in other fungi, mostly from genomic sequencing approaches of other fungal genomes **(Figure 14.3 A, B)**.

The MDR Subfamily

This subfamily contains the ABC proteins Mdl1p, Mdl2p, Atm1p and Ste6p. The Ste6p

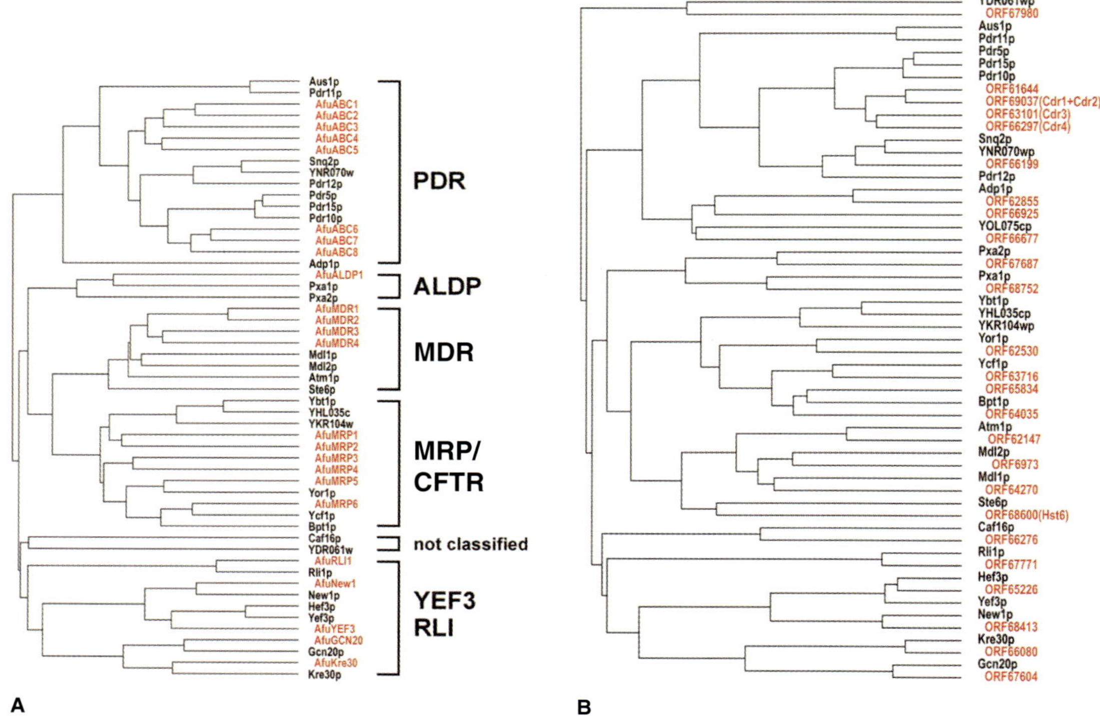

Figure 14.3. Similarity relationships of fungal ABC proteins. A, A dendrogram in which the entire yeast inventory is compared with sequences from the **Aspergillus** *genome project, including the apparent classification into yeast subfamilies. Preliminary sequence data was obtained from The Institute for Genomic Research website at http://www.tigr.org. B, The same dendrogram for the* **Candida albicans** *genome. Sequence data for* **C. albicans** *was obtained from the Stanford Genome Technology Center website at http://www-sequence.stanford.edu/group/candida. Sequencing of* **C. albicans** *was accomplished with the support of the NIDR and the Burroughs Welcome Fund.*

a-factor pheromone transporter is a full-size transporter displaying the duplicated (TMD-NBD)$_2$ topology. Ste6p is localized in Golgi vesicles, the plasma membrane and perhaps endocytic vesicles (Kölling and Hollenberg, 1994; Kuchler, 1993; Michaelis, 1993). Ste6p is a haploid-specific transporter required for the export of farnesylated **a**-factor, a pheromone absolutely required for mating in yeast. Ste6p was the first ABC transporter identified in yeast (Kuchler *et al.*, 1989; McGrath and Varshavsky, 1989), closing an evolutionary gap between the *E. coli* hemolysin transport system (Wang *et al.*, 1991) and mammalian Mdr1p P-glycoprotein mediating multidrug resistance (Chen *et al.*, 1986; Gros *et al.*, 1986). Interestingly, the steady state concentration of Ste6p was found to be highest in the Golgi vesicles, although its function is clearly required in the plasma membrane (Berkower *et al.*, 1994; Kuchler *et al.*, 1993). Because Ste6p travels through all exo- and endocytic compartments, it serves as a useful model membrane protein for intracellular trafficking, proteolytic degradation, endocytosis, and even vacuolar sorting studies (Berkower *et al.*, 1994; Kölling and Hollenberg, 1994; Kuchler, 1993; Kuchler *et al.*, 1989). Moreover, Ste6p has been subjected to extensive molecular studies to unravel the molecular mechanisms of ABC transporter-mediated peptide transport. Ste6p function can be easily tested through convenient assays such as mating (Kuchler and Egner, 1997). Notably Ste6p, although an MDR family member, does not confer typical multidrug resistance phenotypes. Extracellular **a**-factor pheromone is essential for the sexual reproduction cycle of haploid yeast cells. Ste6p functions at the plasma membrane, providing the rate-limiting step in **a**-factor export. After pheromone extrusion, Ste6p is rapidly removed from the cell surface through ubiquitin-mediated endocytosis, and delivered to the vacuole for terminal degradation (Egner *et al.*, 1995; Kölling and Hollenberg, 1994). Pheromone export occurs through a non-classical route, bypassing the vesicular secretory pathway (Kuchler, 1993).

Interestingly, severing experiments demonstrate that both Ste6p halves, when coexpressed as individual half-size transporters, mediate pheromone export (Berkower *et al.*, 1996). This indicates that both Ste6p halves are required for function and that they can interact *in vivo* to form a functional **a**-factor transporter. The transport substrate, **a**-factor, is extremely hydrophobic due to its C-terminal lipid modification and carboxy-methylation. While mutations in the structural gene encoding **a**-factor do not dramatically affect its secretion, a lack of **a**-factor farnesylation or methylation debilitates its release (Sapperstein *et al.*, 1994). Hence, the lipid moiety or its hydrophobicity may represent an essential recognition determinant for Ste6p. As with many other eukaryotic ABC transporters, Ste6p is powered by ATP hydrolysis, because many NBD mutations destroy function (Browne *et al.*, 1996), and because Ste6p binds photoactivatable ATP analogues (Kuchler *et al.*, 1993). Interestingly, Ste6p might also play a role in cell fusion, since *ste6* mutants were isolated that still mediate **a**-factor export, but fail to complete fusion of haploid mating partners (Elia and Marsh, 1996). Taken together, the precise mechanism by which the Ste6p ABC transporter mediates the actual pheromone translocation across the plasma membrane is somewhat mysterious, but it appears as if intracellular **a**-factor precursor processing and translocation across the plasma membrane are tightly coupled (Kuchler and Egner, 1997; Michaelis, 1993).

The half-size molecules Mdl1p, Mdl2p and Atm1p display a similar TMD-NBD topology and localize to the mitochondrial inner membrane **(Figure 14.2)**. Mdl1p is related to mammalian P-glycoproteins and to a greater extent to the mammalian peptide transporter of antigen presentation, TAP (Dean and Allikmets, 1995). It is required for efficient mitochondrial export of rather long peptides of 2100–2600 Da molecular mass. These peptides are proteolytic degradation products of inner membrane proteins generated by mAAA proteases Afg3p and Yta12p. However, Mdl1p fails to transport short peptides or free methionine (Young *et al.*, 2001). Notably, Mdl2p seems to play a different role in mitochondrial function, since it has not been implicated in peptide transport processes. It is therefore likely that Mdl1p and Mdl2p may form functional homodimers, which contrasts with the situation of peroxisomal Pxa1p and Pxa2p. Furthermore, Mdl1p and Mdl2p co-purify at molecular masses of approximately 200 kDa and 300 kDa, respectively, suggesting that they are part of distinct oligomeric protein complexes (Young *et al.*, 2001).

The third member of the yeast MDR group, Atm1p, is related to the human ABCB7/ABC7 protein, which is implicated in the mitochondrial X-linked sideroblastic anemia and ataxia (Allikmets *et al.*, 1999). Atm1p is required for mitochondrial DNA maintenance or stability, but this function might be an indirect phenotypic effect observed in the Δ*atm1* mutant. The *atm1-1* mutant displays a high level of damage and even loss of mitochondrial DNA during growth on rich medium. Interestingly, the *ATM1* mRNA localizes in close proximity to mitochondria in living cells, as demonstrated using a GFP fusion protein that binds to a heterologous sequence in a reporter *ATM1* mRNA (Corral-Debrinski *et al.*, 2000). Atm1p is also required for the assembly of iron–sulfur clusters of cytoplasmic iron–sulfur-containing proteins, and thus may be involved in the export of mitochondrial heme required for cluster assembly (Pelzer *et al.*, 2000).

ABC proteins of the MDR family have also been identified in other fungal species. For example *S. pombe* Mam1p is similar in length and domain structure to Ste6p and shares about 30% sequence identity, thus representing the Ste6p orthologue in fission yeast (Christensen *et al.*, 1997a). The *C. albicans* Hst6p transporter can also functionally complement a Δ*ste6* mutant (Raymond *et al.*, 1998). Further, MDR family homologues have been identified in *A. fumigatus* (Mdr2p) (Tobin *et al.*, 1997), *C. albicans* Mdl1p (Swissprot ID: P97998), *S. pombe* (YFX9 C9B6.09c) and *Rhizomucor racemosus* (Trembl ID: Q9C163/Pgy1p).

THE NON-TRANSPORTER YEF3/RLI SUBFAMILIES AND NON-CLASSIFIED ABC PROTEINS

This *S. cerevisiae* subfamily includes Yef3p, Hef3p, Rli1p, Gcn20p, Kre30p, Caf16p and New1p. Except for New1p, these ABC proteins lack any predicted TMSs normally present in other ABC transporters. Surprisingly, three ABC proteins from this class, Yef3p, Kre30p and Rli1p, are essential for viability under standard growth conditions. These proteins are involved in cellular functions that appear unrelated to transport events, and the functional role of the NBDs is in most cases not clear.

Yef3p perhaps localizes to the cytoplasm or even to ribosomes. It is also known as translation elongation factor EF-3A, which has a function in tRNA binding and dissociation from the ribosome (Chakraburtty, 1999). Yef3p displays basal ATPase activity, which is stimulated by the presence of ribosomes by two orders of magnitude, suggesting that Yef3p might at least interact with ribosomes or in fact localize to ribosomes (Gontarek *et al.*, 1998). A ribosome-binding site and a putative tRNA-binding domain is located near the C-terminus of Yef3p. ATP hydrolysis facilitates EF-3 dissociation from the ribosome. In eukaryotes only fungal homologues are known, suggesting that Yef3p is a unique fungal translation elongation factor (Sarthy *et al.*, 1998). Whole-genome transcriptome profiling of a conditional null-mutant indicates a gene expression pattern that resembles that of wild-type cells treated with cycloheximide, suggesting a role for Yef3p in blocking ribosomes *in vivo* (Hughes *et al.*, 2000). The *YEF3* mRNA levels are modulated by a variety of conditions. It is repressed by rapamycin and peroxide or heat shock stress conditions (Causton *et al.*, 2001), while hyper-expressed in high density cultures and during diauxic shift. Notably, overexpression of Yef3p renders cells hypersensitive to paromomycin and hygromycin B, two translational inhibitors (Sandbaken *et al.*, 1990).

Hef3p (also known as Yef3Bp) shares 84% overall identity with Yef3p, implying a similar or overlapping function. Indeed, Hef3p can rescue a *yef3* null mutant when expressed from the *YEF3* or *ADH1* promoter (Sarthy *et al.*, 1998). In striking contrast to loss of Yef3p, however, a Δ*hef3* null mutant has no obvious growth defect. This might be explained by the fact that Hef3p is not expressed under normal culture conditions and its promoter is therefore inactive (Maurice *et al.*, 1998). Interestingly, *HEF3* mRNA levels are highly upregulated by limiting zinc concentration in the growth medium (Yuan, 2000). The *HEF3* mRNA abundance increases during nitrogen starvation and during stationary phase, but is repressed by a shift to high osmolarity (Causton *et al.*, 2001). It will be interesting to uncover the role of Hef3p under these conditions.

Like Yef3p, and perhaps Hef3p, Gcn20p has a functional role in translation. Gcn20p is a component of a protein complex required for the response to amino acid starvation, glucose limitation and osmotic stress (Marton *et al.*, 1997). Together with Gcn1p, Gcn20p is probably involved in detection of uncharged tRNA and transmission of this signal to Gcn2p, a protein kinase which phosphorylates eIF2alpha. Gcn1p, Gcn2p and Gcn20p form a complex and the apparent role of Gcn20p is to activate Gcn2p, through the stabilization of the interaction between Gcn1p and Gcn2p (Garcia-Barrio *et al.*, 2000). The *gcn20* mutant phenotype is similar to a *gcn1* mutant, in that the null mutant is viable under normal conditions and inviable under starvation conditions (Vazquez de Aldana *et al.*, 1995). The C-terminal region of Gcn20p containing the ABC domain is dispensable for complex formation with Gcn1p and for the stimulation of Gcn2p kinase activity (Marton *et al.*, 1997), and the role of the Gcn20p NBD remains obscure.

The physiological roles of the following non-transporter ABC proteins are largely unknown and they may therefore provide some surprises in the future. Kre30p is required for viability and was initially isolated in a genetic screen for Killer toxin-resistant mutants. The cellular function of Kre30p is not known, but it seems to interact with other proteins as determined by two-hybrid assays. Interactions with several proteins, including Sma1p (spore membrane assembly) and Cbk1p (an S/T kinase required for sporulation) were discovered, but the physiological relevance of these interactions, if any, remains to be established.

The N-terminal domain of New1p, which is especially rich in glutamine and asparagine residues, is able to support prion inheritance when fused to *SUP35*. Sup35p is a translational release factor, eRF3, which interacts with Sup45p (eRF1) to form a translational release factor complex. Moreover, Sup35p is also a prion-like molecule responsible for the [PSI+] determinant (Tuite *et al.*, 1981). Although the cellular function of New1p remains elusive, it may behave as an epigenetic switch (Santoso *et al.*, 2000). The New1p sequence also includes three predicted TMSs, although they are not clustered within a classical TMD. Finally, *NEW1* mRNA levels are repressed under stress conditions such as changes in temperature, oxidation, nutrients, pH and osmolarity (Causton *et al.*, 2001; Jelinsky *et al.*, 2000).

The Caf16p and Ydr061w ABC proteins contain only a single NBD. While nothing is known about YDR061w, Caf16p seems to have a role in PoIII-dependent transcription of some, but not all, promoters. Caf16p forms a dimer and interacts with the RNA polymerase II holoenzyme components Srb9p, Ssn3p, and Ssn8p. Finally, Rli1p is similar to the human RNase L inhibitor (RLI). Its precise function has not been established, although a Δ*rli1* null mutation in yeast is lethal. Human RLI is probably a regulator of the 2′,5′-oligoadenylate-dependent RNase L, which is involved in the antiviral activity of interferons. Some viruses developed strategies to bypass the antiviral activity of RNase L by virus-induced expression of RLI (Martinand *et al.*, 1999). Interestingly, the C-terminal tail domain of yeast Ire1p displays sequence similarity to mammalian RNase L (Sidrauski and Walter, 1997). Ire1p is a regulator of the unfolded protein response pathway (UPR), which signals from the ER to the nucleus (Cox *et al.*, 1993). A direct role for Rli1p in the UPR is possible but untested as yet.

EVOLUTIONARY RELATIONSHIPS OF ABC GENES IN FUNGI

Numerous homologues of yeast ABC genes have also been identified in other fungal species through functional complementation approaches. More importantly, genome sequencing of fungal pathogens such as *C. albicans* and *A. fumigatus* provided complete sequence datasets from their genomes and the data are publicly available (TIGR: http://www.tigr.org/; Stanford Genome Technology Center: http://sequence-www.stanford.edu/). Although functional annotation of ABC genes in these fungal pathogens has been a difficult task, the comparison of various fungal ABC inventories has become possible. Because a global picture of the evolutionary relationships of ABC genes from various fungi has not been reported, we have compared the inventory of baker's yeast ABC genes to various fungal genomes. Yeast ABC genes guided a search to detect and identify homologous sequences in other fungal species, including *C. albicans* and *A. fumigatus* **(Figure 14.3 A, B)**. Previous work demonstrated that all *S. cerevisiae* NBDs generate clusters of five subfamilies **(Table 14.1)**. In a first round of comparison, NBDs were identified using a translated pattern search against the nucleotide sequence databases. The patterns were generated by alignment of the respective subgroups of *S. cerevisiae* NBD sequences. The comparison of amino acid sequence patterns with a translated nucleotide sequence minimizes the effect of sequencing errors causing truncations or frameshift mutations. In addition, the sequences were searched with the Prosite patterns for ABC proteins. In the next step, the regions surrounding hits were analyzed in detail by extracting putative NBDs. In cases where truncations due to frameshift mutations had occurred, ORFs were appropriately edited to allow for the generation of meaningful dendrograms. Next we generated an alignment using the entire set of NBDs including *S. cerevisiae* NBDs. The *A. fumigatus* candidate genes were first identified through a tblast at TIGR (http://www.tigr.org/). The dendrograms shown in **Figure 14.3** represent a graphical display of the sequence homologies as detected through the alignment, although it should be noted that this is not a phylogenetic tree. Furthermore, we intended to include a dendrogram showing the relationships to *C. neoformans* ABC genes, the sequence data of which can be publicly accessed at http://www-sequence.stanford.edu/group/C.neoformans/. However, because of the confidentiality policies of the sequencing consortium, we were prohibited from doing so.

As shown in **Figure 14.3**, each subfamily from baker's yeast has an equivalent family in other fungi. Thus, ABC proteins from other fungi form similar evolutionary relationships, and can thus be classified into similar subfamilies. The PDR subfamily contains five *Candida* homologues (Cdr1p, Cdr2p, Cdr3p, Cdr4p and Cdr99p) of Pdr5p, all of which are more similar to each other than to other yeast members of the Pdr5p-family **(Figure 14.3A)**. As in yeast, not all *CDR* genes are implicated in drug resistance. While Cdr1p and Cdr2p mediate clinical antifungal resistance, the function of Cdr3p and Cdr4p has not been linked to drug efflux. Homozygous deletion of *CDR4* did not confer hypersensitivity to fluconazole (Franz *et al.*, 1998). Interestingly, the *CDR3* gene is regulated in a cell-type-specific manner, as it appears important in morphology switching, and it is not expressed in the standard laboratory strain CAI4. However, in a WO-1 genetic strain background that switches between two morphological states, *white* and *opaque*, the *CDR3* mRNA is only present in the *opaque* form. Here, overexpression of Cdr3p did not result in increased resistance to known drug substrates of the PDR family (Balan *et al.*, 1997). Substantially less information is available on the PDR family homologues in *A. fumigatus* **(Figure 14.3)**, although in general, a clear species-specific clustering becomes immediately apparent in this family.

As expected, the yeast ALDP subfamily has equivalent orthologues in all other fungal pathogens. In *C. albicans* a homologue to both Pxa1p and Pxa2p is detectable, while in *A. fumigatus* only one close match could be identified. Concerning the MDR subfamily, single nearest matches to each Atm1p, Mdl1p, Mdl2p and Ste6p were found in *C. albicans*. The situation seems to be somewhat more complicated in *A. fumigatus*. The *A. fumigatus* MDR members cluster together and do not allow even a tentative assignment. The *C. albicans* orthologue of Ste6p has been previously described as Hst6p (Raymond *et al.*, 1992). Surprisingly, despite the diploid nature of *C. albicans*, Hst6p is able to functionally complement a *ste6* null mutant for **a**-factor transport in *S. cerevisiae*. The MRP subfamily indicates some differences between *S. cerevisiae* and *C. albicans*. No close homologue to Ste6 was identified in *A. fumigatus*. Furthermore, we do not find a close neighbor of Ybt1p and YHL035c, a fact that could also be the consequence of incomplete databases. While single orthologues to Ycf1p and Yor1p are present, two *Candida* ORFs are similar to Bpt1p. Thus, further experimental evidence will be necessary to establish the roles of the two *Candida* Bpt1ps, whether or not one of them represents a functional homologue of Ybt1p. In the *A. fumigatus* alignments we find a Yor1p and Ycf1p homologue but several candidates for Ybt1p remain. Finally, the non-transporter ABC genes from the YEF3/RLI subfamilies, as well as non-classified ABC genes, all have corresponding genes in other fungal species. For instance, both Yef3p and Hef3p cluster with the *C. albicans* homologue Tef3p. The *Candida* Eif3p, however, appears more similar to New1p than to Hef3p. Both Gcn20p and Kre30p also have close homologues in *Candida*, and Caf16p and Rli1p also have a single counterpart in *Candida*. Taken together, the inventory of ABC proteins from fungal pathogens is quite similar to the one present in baker's yeast, with similar subfamilies of close evolutionary relationships.

TRANSCRIPTOMES AND YEAST ABC GENE mRNA PROFILES

The completion of the entire yeast genome, and the availability of genomic tools such as whole-genome DNA microarrays, permitted the transcriptional profiling of many metabolic pathways. It is therefore not surprising that expression regulation of yeast ABC genes was observed in numerous studies that investigated genome-wide expression of yeast genes. For example, *PDR5, SNQ2, YOR1, PDR10, PDR11* and *PDR15* share common transcriptional regulators, such as the zinc-finger proteins Pdr1p, Pdr3p or Yrr1p (Del Sorbo *et al.*, 1997). These regulators, also instrumental for PDR development, control a number of genes of both the ABC family and non-ABC genes (DeRisi *et al.*, 2000; Wolfger *et al.*, 2001). A detailed transcriptome analysis revealed the identification of numerous potential Pdr1p/Pdr3p target genes (DeRisi *et al.*, 2000). Moreover, PDR target genes were also identified simply by the presence of potential PDRE *cis*-acting motifs in yeast gene promoters. However, a Pdr1p/Pdr3p-dependent regulation has only been experimentally verified for certain ABC genes and two MFS permeases (Wolfger *et al.*, 1997). It should be emphasized that the molecular signals, including the transduction pathways affecting transcriptional activities of Pdr1p, Pdr3p or Yrr1p, remain elusive. A specific activation of these factors by drugs has not been reported. It is tempting to speculate that PDR could evolve through increased mutation rate upon drug challenge or other adverse conditions. Apart from other regulatory influences, the mRNA levels of several ABC genes show dependencies on carbon and/or nitrogen source, stress regulation as well as cell cycle-dependent fluctuations. A closer inspection of the available literature on yeast ABC gene expression leads to the conclusion that individual mRNAs display a distinctive expression pattern. Even closely related proteins such as the PDR group display striking differences under various conditions. In many cases, the transcription factors involved remain unknown but a functional link between stress response and drug resistance is evident.

Whole genome transcriptome analysis suggested that Snq2p is induced by heat shock, H_2O_2 and rapamycin, whereas *PDR5* mRNA is only upregulated during cold shock, but not by heat shock (Causton *et al.*, 2001; Gasch *et al.*, 2000). Further, Pdr12p protein levels are specifically induced by weak organic acids (Piper *et al.*, 1998), by an as yet unknown stress response pathway. Notably, Pdr15p is upregulated in mitochondrial DNA mutants and it appears to be under general stress control through Msn2p and Msn4p (Wolfger *et al.*, in preparation). Likewise, fluctuations in *PDR12* and *YOR1* mRNAs during the cell cycle, with a peak in early G1 phase, remain unexplained, as well as the observation that *PDR5* mRNA and those of several other membrane proteins, most of them involved in nutrient metabolism, peak in the G2/M phase. It is thus not clear what the common regulatory principle affecting these genes is, but relevant hints might emerge once more physiological roles of ABC genes are established. The following chapter is devoted to comprehensive and detailed discussions on fungal ABC proteins and regulators implicated in pleiotropic or multidrug resistance phenomena. To come full circle, additional chapters in this part of the book will address the functions of plant ABC proteins, as well as ABC proteins from parasitic organisms.

ACKNOWLEDGMENTS

We are indebted to our colleagues Agnés Delahodde, Christophe D'énfert, Bertrand Favre, André Goffeau, Scott Moye-Rowley, Peter Piper, Elisabeth Presterl, Neil Ryder, Dominique Sanglard, Julius Subik, Friederike Turnowsky, Marten de Waard and Birgit Willinger for sharing unpublished information, materials, and strains as well as for many stimulating discussions. Special thanks to the Vienna EMBnet manager Martin Grabner for help with database searches. Thanks to all group members for critical comments on the manuscript. Our research is supported by grants from the 'Fonds zur Förderung der wissenschaftlichen Forschung' (FWF, P12661-BIO), by funds from the Austrian National Bank (OeNB #7421), grants from Novartis Pharma Inc., DSM Bakery Ingredients, the 'Hygiene-Fonds' of the Medical Faculty of the University of Vienna and the 'Herzfelder Foundation'. Sequence data for *Candida albicans* was obtained from the Stanford Genome Technology Center website at http://www-sequence.stanford.edu/group/candida. Sequencing of *Candida albicans* was accomplished with the support of the NIDR and the Burroughs Welcome Fund. *Aspergillus fumigatus* preliminary sequence data was retrieved from The Institute for Genomic Research website at http://www. tigr.org. Sequencing of *Aspergillus fumigatus* was funded by the National Institute of Allergy and Infectious Disease U01 AI 48830 to David Denning and William Nierman.

REFERENCES

Allikmets, R., Raskind, W.H., Hutchinson, A., Schueck, N.D., Dean, M. and Koeller, D.M. (1999) Mutation of a putative mitochondrial iron transporter gene (ABC7) in X-linked sideroblastic anemia and ataxia (XLSA/A). Hum. Mol. Genet. 8, 743–749.

Andrade, A.C., Del Sorbo, G., Van Nistelrooy, J.G. and Waard, M.A. (2000) The ABC transporter *atrB* from *Aspergillus nidulans* mediates resistance to all major classes of fungicides and some natural toxic compounds. Microbiology 146, 1987–1997.

Balan, I., Alarco, A.M. and Raymond, M. (1997) The *Candida albicans CDR3* gene codes for an opaque-phase ABC transporter. J. Bacteriol. 179, 7210–7218.

Balzi, E., Chen, W., Ulaszewski, S., Capieaux, E. and Goffeau, A. (1987) The multidrug resistance gene *PDR1* from *Saccharomyces cerevisiae*. J. Biol. Chem. 262, 16871–16879.

Bauer, B.E., Wolfger, H. and Kuchler, K. (1999) Inventory and function of yeast ABC proteins: about sex, stress, pleiotropic drug and heavy metal resistance. Biochim. Biophys. Acta 1461, 217–236.

Berkower, C., Loayza, D. and Michaelis, S. (1994) Metabolic instability and constitutive endocytosis of *STE6*, the **a-** factor transporter of *Saccharomyces cerevisiae.* Mol. Biol. Cell 5, 1185–1198.

Berkower, C., Taglicht, D. and Michaelis, S. (1996) Functional and physical interactions between partial molecules of *STE6*, a yeast ATP-binding cassette protein. J. Biol. Chem. 271, 22983–22989.

Bissinger, P.H. and Kuchler, K. (1994) Molecular cloning and expression of the *Saccharomyces cerevisiae STS1* gene product. A yeast ABC transporter conferring mycotoxin resistance. J. Biol. Chem. 269, 4180–4186.

Bossier, P., Fernandes, L., Vilela, C. and Rodrigues-Pousada, C. (1994) The yeast *YKL741* gene situated on the left arm of chromosome XI codes for a homologue of the human ALD protein. Yeast 10, 681–686.

Browne, B.L., McClendon, V. and Bedwell, D.M. (1996) Mutations within the first LSGGQ motif of Ste6p cause defects in **a**-factor transport and mating in *Saccharomyces cerevisiae*. J. Bacteriol. 178, 1712–1719.

Causton, H.C., Ren, B., Koh, S.S., Harbison, C.T., Kanin, E., Jennings, E.G., Lee, T.I., True, H.L., Lander, E.S. and Young, R.A. (2001) Remodeling of yeast genome expression in response to environmental changes. Mol. Biol. Cell 12, 323–337.

Chakraburtty, K. (1999) Functional interaction of yeast elongation factor 3 with yeast ribosomes. Int. J. Biochem. Cell Biol. 31, 163–173.

Chang, G. and Roth, C.B. (2001) Structure of MsbA from *E. coli*: a homolog of the multidrug resistance ATP-binding cassette (ABC) transporters. Science 293, 1793–1800.

Chen, C.J., Chin, J.E., Ueda, K., Clark, D.P., Pastan, I., Gottesman, M.M. and Roninson, I.B. (1986) Internal duplication and homology with bacterial transport proteins in the *mdr1* (P-glycoprotein) gene from multidrug-resistant human cells. Cell 47, 381–389.

Christensen, P.U., Davey, J. and Nielsen, O. (1997a) The *Schizosaccharomyces pombe mam1* gene encodes an ABC transporter mediating secretion of M-factor. Mol. Gen. Genet. 255, 226–236.

Christensen, P.U., Davis, K., Nielsen, O. and Davey, J. (1997b) *Abc1*: a new ABC transporter from the fission yeast *Schizosaccharomyces pombe*. FEMS Microbiol. Lett. 147, 97–102.

Corral-Debrinski, M., Blugeon, C. and Jacq, C. (2000) In yeast, the 3′ untranslated region or the presequence of *ATM1* is required for the exclusive localization of its mRNA to the vicinity of mitochondria. Mol. Cell Biol. 20, 7881–7892.

Cox, J.S., Shamu, C.E. and Walter, P. (1993) Transcriptional induction of genes encoding endoplasmic reticulum resident proteins requires a transmembrane protein kinase. Cell 73, 1197–1206.

Cui, Z., Hirata, D., Tsuchiya, E., Osada, H. and Miyakawa, T. (1996) The multidrug resistance-associated protein (MRP) subfamily (Yrs1/Yor1) of *Saccharomyces cerevisiae* is important for the tolerance to a broad range of organic anions. J. Biol. Chem. 271, 14712–14716.

Dean, M. and Allikmets, R. (1995) Evolution of ATP-binding cassette transporter genes. Curr. Opin. Genet. Dev. 5, 779–785.

Decottignies, A. and Goffeau, A. (1997) Complete inventory of the yeast ABC proteins. Nat. Genet. 15, 137–145.

Del Sorbo, G., Andrade, A.C., Van Nistelrooy, J.G., Van Kan, J.A., Balzi, E. and De Waard, M.A. (1997) Multidrug resistance in *Aspergillus nidulans* involves novel ATP-binding cassette transporters. Mol. Gen. Genet. 254, 417–426.

DeRisi, J., van den Hazel, B., Marc, P., Balzi, E., Brown, P., Jacq, C. and Goffeau, A. (2000) Genome microarray analysis of transcriptional activation in multidrug resistance yeast mutants. FEBS Lett. 470, 156–160.

Egner, R., Mahé, Y., Pandjaitan, R., Huter, V., Lamprecht, A. and Kuchler, K. (1995) In: Membrane Protein Transport 2, 57–96.

Elia, L. and Marsh, L. (1996) Role of the ABC transporter Ste6 in cell fusion during yeast conjugation. J. Cell Biol. 135, 741–751.

Franz, R., Michel, S. and Morschhauser, J. (1998) A fourth gene from the *Candida albicans* CDR family of ABC transporters. Gene 220, 91–98.

Garcia-Barrio, M., Dong, J., Ufano, S. and Hinnebusch, A.G. (2000) Association of *GCN1-GCN20* regulatory complex with the N-terminus of eIF2alpha kinase *GCN2* is required for *GCN2* activation. EMBO J. 19, 1887–1899.

Gartner, J. and Valle, D. (1993) The 70 kDa peroxisomal membrane protein: an ATP-binding cassette transporter protein involved in peroxisome biogenesis. Semin. Cell Biol. 4, 45–52.

Gasch, A.P., Spellman, P.T., Kao, C.M., Carmel-Harel, O., Eisen, M.B., Storz, G., Botstein, D. and Brown, P.O. (2000) Genomic expression programs in the response of yeast cells to environmental changes. Mol. Biol. Cell 11, 4241–4257.

Gontarek, R.R., Li, H., Nurse, K. and Prescott, C.D. (1998) The N terminus of eukaryotic translation elongation factor 3 interacts with 18 S rRNA and 80 S ribosomes. J. Biol. Chem. 273, 10249–10252.

Gros, P., Croop, J. and Housman, D. (1986) Mammalian multidrug resistance gene: complete cDNA sequence indicates strong homology to bacterial transport proteins. Cell 47, 371–380.

Higgins, C.F. (1992) ABC transporters: from microorganisms to man. Annu. Rev. Cell Biol. 8, 67–113.

Higgins, C.F. (1995) The ABC of channel regulation. Cell 82, 693–696.

Hirata, D., Yano, K., Miyahara, K. and Miyakawa, T. (1994) *Saccharomyces cerevisiae YDR1*, which encodes a member of the ATP-binding cassette (ABC) superfamily, is required for multidrug resistance. Curr. Genet. 26, 285–294.

Holzinger, A., Kammerer, S. and Roscher, A.A. (1997) Primary structure of human PMP69, a putative peroxisomal ABC-transporter. Biochem. Biophys. Res. Commun. 237, 152–157.

Holzinger, A., Mayerhofer, P., Berger, J., Lichtner, P., Kammerer, S. and Roscher, A.A. (1999) Full length cDNA cloning, promoter sequence, and genomic organization of the human adrenoleukodystrophy related (ALDR) gene functionally redundant to the gene responsible for X-linked adrenoleukodystrophy. Biochem. Biophys. Res. Commun. 258, 436–442.

Hughes, T.R., Marton, M.J., Jones, A.R., Roberts, C.J., Stoughton, R., Armour, C.D., *et al.* (2000) Functional discovery via a compendium of expression profiles. Cell 102, 109–126.

Jelinsky, S.A., Estep, P., Church, G.M. and Samson, L.D. (2000) Regulatory networks revealed by transcriptional profiling of damaged *Saccharomyces cerevisiae* cells: Rpn4 links base excision repair with proteasomes. Mol. Cell. Biol. 20, 8157–8167.

Kamijo, K., Kamijo, T., Ueno, I., Osumi, T. and Hashimoto, T. (1992) Nucleotide sequence of the human 70 kDa peroxisomal membrane protein: a member of ATP-binding cassette transporters. Biochim. Biophys. Acta 1129, 323–327.

Katzmann, D.J., Hallström, T.C., Voet, M., Wysock, W., Golin, J., Volckaert, G. and Moye-Rowley, W.S. (1995) Expression of an ATP-binding cassette transporter-encoding gene (*YOR1*) is required for oligomycin resistance in *Saccharomyces cerevisiae*. Mol. Cell. Biol. 15, 6875–6883.

Kölling, R. and Hollenberg, C.P. (1994) The ABC-transporter Ste6 accumulates in the plasma membrane in a ubiquitinated form in endocytosis mutants. EMBO J. 13, 3261–3271.

Kralli, A., Bohen, S.P. and Yamamoto, K.R. (1995) *LEM1*, an ATP-binding-cassette transporter, selectively modulates the biological potency of steroid hormones. Proc. Natl Acad. Sci. USA 92, 4701–4705.

Kuchler, K. (1993) Unusual routes of protein secretion: the easy way out. Trends Cell Biol. 3, 421–426.

Kuchler, K. and Egner, R. (1997) Unusual protein secretion and translocation pathways in yeast: implication of ABC transporters. In: Unusual Secretory Pathways: From Bacteria to Man. (ed. K. Kuchler, A. Rubartelli, and B. Holland), pp. 49–85. Austin, TX: Landes Bioscience.

Kuchler, K., Sterne, R.E. and Thorner, J. (1989) *Saccharomyces cerevisiae STE6* gene product: a novel pathway for protein export in eukaryotic cells. EMBO J. 8, 3973–3984.

Kuchler, K., Dohlman, H.G. and Thorner, J. (1993) The **a**-factor transporter (*STE6* gene product) and cell polarity in the yeast *Saccharomyces cerevisiae*. J. Cell Biol. 120, 1203–1215.

Martinand, C., Montavon, C., Salehzada, T., Silhol, M., Lebleu, B. and Bisbal, C. (1999) RNase L inhibitor is induced during human immunodeficiency virus type 1 infection and down regulates the 2-5A/RNase L pathway in human T cells. J. Virol. 73, 290–296.

Marton, M.J., Vazquez de Aldana, C.R., Qiu, H., Chakraburtty, K. and Hinnebusch, A.G. (1997) Evidence that *GCN1* and *GCN20*, translational regulators of *GCN4*, function on elongating ribosomes in activation of elF2alpha kinase *GCN2*. Mol. Cell. Biol. 17, 4474–4489.

Maurice, T.C., Mazzucco, C.E., Ramanathan, C.S., Ryan, B.M., Warr, G.A. and Puziss, J.W. (1998) A highly conserved intraspecies homolog of the *Saccharomyces cerevisiae* elongation factor-3 encoded by the *HEF3* gene. Yeast 14, 1105–1113.

McGrath, J.P. and Varshavsky, A. (1989) The yeast *STE6* gene encodes a homologue of the mammalian multidrug resistance P-glycoprotein. Nature 340, 400–404.

Michaelis, S. (1993) *STE6*, the yeast **a**-factor transporter. Semin. Cell Biol. 4, 17–27.

Michaelis, S. and Berkower, C. (1995) Sequence comparison of yeast ATP-binding cassette proteins. Cold Spring Harb. Symp. Quant. Biol. 60, 291–307.

Miyazaki, H., Miyazaki, Y., Geber, A., Parkinson, T., Hitchcock, C., Falconer, D.J., Ward, D.J., Marsden, K. and Bennett, J.E. (1998) Fluconazole resistance associated with drug efflux and increased transcription of a drug transporter gene, *PDH1*, in *Candida glabrata*. Antimicrob. Agents Chemother. 42, 1695–1701.

Nakaune, R., Adachi, K., Nawata, O., Tomiyama, M., Akutsu, K. and Hibi, T.

(1998) A novel ATP-binding cassette transporter involved in multidrug resistance in the phytopathogenic fungus *Penicillium digitatum*. Appl. Environ. Microbiol. 64, 3983–3988.

Ogawa, A., Hashida-Okado, T., Endo, M., Yoshioka, H., Tsuruo, T., Takesako, K. and Kato, I. (1998) Role of ABC transporters in aureobasidin A resistance. Antimicrob. Agents Chemother. 42, 755–761.

Ortiz, D.F., St Pierre, M.V., Abdulmessih, A. and Arias, I.M. (1997) A yeast ATP-binding cassette-type protein mediating ATP-dependent bile acid transport. J. Biol. Chem. 272, 15358–15365.

Pelzer, W., Muhlenhoff, U., Diekert, K., Siegmund, K., Kispal, G. and Lill, R. (2000) Mitochondrial Isa2p plays a crucial role in the maturation of cellular iron-sulfur proteins. FEBS Lett. 476, 134–139.

Piper, P., Mahé, Y., Thompson, S., Pandjaitan, R., Holyoak, C., Egner, R., Mühlbauer, M., Coote, P. and Kuchler, K. (1998) The Pdr12 ABC transporter is required for the development of weak organic acid resistance in yeast. EMBO J. 17, 4257–4265.

Prasad, R., De Wergifosse, P., Goffeau, A. and Balzi, E. (1995) Molecular cloning and characterization of a novel gene of *Candida albicans*, *CDR1*, conferring multiple resistance to drugs and antifungals. Curr. Genet. 27, 320–329.

Raymond, M., Gros, P., Whiteway, M. and Thomas, D.Y. (1992) Functional complementation of yeast *ste6* by a mammalian multidrug resistance mdr gene. Science 256, 232–234.

Raymond, M., Dignard, D., Alarco, A.M., Mainville, N., Magee, B.B. and Thomas, D.Y. (1998) A Ste6p/P-glycoprotein homologue from the asexual yeast *Candida albicans* transports the **a**-factor mating pheromone in *Saccharomyces cerevisiae*. Mol. Microbiol. 27, 587–598.

Sandbaken, M.G., Lupisella, J.A., DiDomenico, B. and Chakraburtty, K. (1990) Protein synthesis in yeast. Structural and functional analysis of the gene encoding elongation factor 3. J. Biol. Chem. 265, 15838–15844.

Sanglard, D., Ischer, F., Monod, M. and Bille, J. (1997) Cloning of *Candida albicans* genes conferring resistance to azole antifungal agents: characterization of *CDR2*, a new multidrug ABC transporter gene. Microbiology 143, 405–416.

Santoso, A., Chien, P., Osherovich, L.Z. and Weissman, J.S. (2000) Molecular basis of a yeast prion species barrier. Cell 100, 277–288.

Sapperstein, S., Berkower, C. and Michaelis, S. (1994) Nucleotide sequence of the yeast *STE14* gene, which encodes farnesylcysteine carboxyl methyltransferase, and demonstrates its essential role in **a**-factor export. Mol. Cell Biol. 14, 143814–143849.

Sarthy, A.V., McGonigal, T., Capobianco, J.O., Schmidt, M., Green, S.R., Moehle, C.M. and Goldman, R.C. (1998) Identification and kinetic analysis of a functional homolog of elongation factor 3, *YEF3* in *Saccharomyces cerevisiae*. Yeast 14, 239–253.

Servos, J., Haase, E. and Brendel, M. (1993) Gene *SNQ2* of *Saccharomyces cerevisiae*, which confers resistance to 4-nitroquinoline-N-oxide and other chemicals, encodes a 169 kDa protein homologous to ATP-dependent permeases. Mol. Gen. Genet. 236, 214–218.

Shani, N., Sapag, A., Watkins, P.A. and Valle, D. (1996) An *S. cerevisiae* peroxisomal transporter, orthologous to the human adrenoleukodystrophy protein, appears to be a heterodimer of two half ABC transporters: Pxa1p and Pxa2p. Ann. NY Acad. Sci. 804, 770–772.

Sidrauski, C. and Walter, P. (1997) The transmembrane kinase Ire1p is a site-specific endonuclease that initiates mRNA splicing in the unfolded protein response. Cell 90, 1031–1039.

Swartzman, E.E., Viswanathan, M.N. and Thorner, J. (1996) The *PAL1* gene product is a peroxisomal ATP-binding cassette transporter in the yeast *Saccharomyces cerevisiae*. J. Cell Biol. 132, 549–563.

Szczypka, M.S., Wemmie, J.A., Moye-Rowley, W.S. and Thiele, D.J. (1994) A yeast metal resistance protein similar to human cystic fibrosis transmembrane conductance regulator (CFTR) and multidrug resistance-associated protein. J. Biol. Chem. 269, 22853–22857.

Taglicht, D. and Michaelis, S. (1998) *Saccharomyces cerevisiae* ABC proteins and their relevance to human health and disease. Methods Enzymol. 292, 130–162.

Tobin, M.B., Peery, R.B. and Skatrud, P.L. (1997) Genes encoding multiple drug resistance-like proteins in *Aspergillus fumigatus* and *Aspergillus flavus*. Gene 200, 11–23.

Tuite, M.F., Mundy, C.R. and Cox, B.S. (1981) Agents that cause a high frequency of genetic change from [psi+] to [psi−] in *Saccharomyces cerevisiae*. Genetics 98, 691–711.

Turi, T.G. and Rose, J.K. (1995) Characterization of a novel *Schizosaccharomyces pombe* multidrug resistance transporter conferring brefeldin A resistance. Biochem. Biophys. Res. Commun. 213, 410–418.

Tusnady, G.E., Bakos, E., Varadi, A. and Sarkadi, B. (1997) Membrane topology distinguishes a subfamily of the ATP-binding cassette (ABC) transporters. FEBS Lett. 402, 1–3.

Urban, M., Bhargava, T. and Hamer, J.E. (1999) An ATP-driven efflux pump is a novel pathogenicity factor in rice blast disease. EMBO J. 18, 512–521.

Vazquez de Aldana, C.R., Marton, M.J. and Hinnebusch, A.G. (1995) *GCN20*, a novel ATP binding cassette protein, and *GCN1* reside in a complex that mediates activation of the elF-2 alpha kinase *GCN2* in amino acid-starved cells. EMBO J. 14, 3184–3199.

Walker, J.E., Saraste, M., Runswick, M.J. and Gay, N.J. (1982) Distantly related sequences in the alpha- and beta-subunits of ATP synthase, myosin, kinases and other ATP-requiring enzymes and a common nucleotide binding fold. EMBO J. 1, 945–951.

Wang, R.C., Seror, S.J., Blight, M., Pratt, J.M., Broome-Smith, J.K. and Holland, I.B. (1991) Analysis of the membrane organization of an *Escherichia coli* protein translocator, HlyB, a member of a large family of prokaryote and eukaryote surface transport proteins. J. Mol. Biol. 217(3), 441–454.

Watkins, P.A., Lu, J.F., Braiterman, L.T., Steinberg, S.J. and Smith, K.D. (2000) Disruption of a yeast very-long-chain acyl-CoA synthetase gene simulates the cellular phenotype of X-linked adrenoleukodystrophy. Cell Biochem. Biophys. 32, 333–337.

Wolfger, H., Mahé, Y., Parle-McDermott, A., Delahodde, A. and Kuchler, K. (1997) The yeast ATP-binding cassette (ABC) protein genes *PDR10* and *PDR15* are novel targets for the Pdr1 and Pdr3 transcriptional regulators. FEBS Lett. 418, 269–274.

Wolfger, H., Mamnun, Y.M. and Kuchler, K. (2001) Fungal ABC proteins: pleiotropic drug resistance, stress response and cellular detoxification. Res. Microbiol. 152, 375–389.

Young, J. and Holland, I.B. (1999) ABC transporters: bacterial exporters-revisited five years on. Biochim. Biophys. Acta 1461, 177–200.

Young, L., Leonhard, K., Tatsuta, T., Trowsdale, J. and Langer, T. (2001) Role of the ABC transporter Mdl1 in peptide export from mitochondria. Science 291, 2135–2138.

Yuan, D.S. (2000) Zinc-regulated genes in *Saccharomyces cerevisiae* revealed by transposon tagging. Genetics 156, 45–58.

CHAPTER 15

Fungal ABC Proteins in Clinical Drug Resistance and Cellular Detoxification

Bettina E. Bauer, Christoph Schüller and Karl Kuchler

Introduction

The genome of baker's yeast *Saccharomyces cerevisiae* contains 30 distinct genes encoding ATP-binding cassette (ABC) proteins (Bauer *et al.*, 1999; Decottignies and Goffeau, 1997; Taglicht and Michaelis, 1998). Expression of several yeast ABC proteins is linked to, or causes, pleiotropic drug resistance (PDR) phenomena (Wolfger *et al.*, 2001) and certain ABC genes represent orthologues of mammalian disease genes. *S. cerevisiae* is thus considered an important model organism to study the function of evolutionary conserved genes, including mammalian ABC proteins of medical importance. The PDR phenomenon is phenotypically quite analogous to multidrug resistance (MDR) as it develops in mammalian cells (Litman *et al.*, 2001), parasites, fungal pathogens or even in bacteria. MDR can be described as an initial resistance to a single drug, followed by cross-resistance to many structurally and functionally unrelated compounds (Kane, 1996; Litman *et al.*, 2001). Baker's yeast was therefore exploited to dissect the molecular mechanisms of PDR/MDR mediated by ABC transporters. For instance, cross-complementation studies yielded insights into the function of mammalian MDR transporters of the P-glycoprotein (Pgp) family (Kuchler and Thorner, 1992; Ueda *et al.*, 1993), as well as the MRP (multidrug resistance-related protein) family (Raymond *et al.*, 1992; Ruetz *et al.*, 1993; Tommasini *et al.*, 1996; Volkman *et al.*, 1995). Importantly, yeast strains lacking endogenous ABC pumps have been used to identify and clone resistance genes from fungal pathogens such as *Candida* and *Aspergillus* species. For example, the *Candida* genes *CDR1* and *CDR2*, implicated in clinical azole resistance, were initially identified by virtue of their ability to rescue the drug-hypersensitive phenotype of a mutant *S. cerevisiae* strain (Prasad *et al.*, 1995; Sanglard *et al.*, 1995, 1997). This chapter is devoted to a comprehensive discussion of ABC protein-mediated drug resistance phenomena as they have been described in model systems like *S. cerevisiae* as well as in fungal pathogens.

Pleiotropic Drug Resistance ABC Transporters in Fungi

The inventory of *S. cerevisiae* ABC proteins has been classified into five distinct subfamilies (see also Chapter 14). Several genes of the PDR and MRP/CFTR subfamilies of yeast ABC proteins **(Table 15.1)** mediate PDR, as their expression is tightly linked to compound drug resistance phenotypes. These genes are part of the PDR network **(Figure 15.1)**, which comprises several ABC transporters, as well as dedicated regulators controlling the expression of ABC target genes (Bauer *et al.*, 1999; DeRisi

ABC Proteins: From Bacteria to Man
ISBN 0-12-352551-9

TABLE 15.1. FUNGAL ABC TRANSPORTERS AND SOME RELEVANT SUBSTRATES

ABC pump	Substrates	Length	Topology	Localization
Non-pathogenic fungi				
Saccharomyces cerevisiae				
Pdr5p	Drugs, steroids, antifungals, phospholipids	1511	$(\text{ABC-TMS}_6)_2$	Plasma membrane
Snq2p	Drugs, steroids, mutagens	1501	$(\text{ABC-TMS}_6)_2$	Plasma membrane
Yor1p	Oligomycin, reveromycin A, phospholipids	1477	$(\text{TMS}_6\text{-ABC})_2$[a]	Plasma membrane
Ycf1p	GS-conjugates, Cd^{2+}, UCB, diazaborine, bile acids	1515	$(\text{TMS}_6\text{-R-ABC})_2$[a]	Vacuole
Schizosaccharomyces pombe				
Hba2p	Brefeldin A	1530	$(\text{ABC-TMS}_6)_2$	?
Pmd1p	Drugs	1362	$(\text{TMS}_6\text{-ABC})_2$	?
Hmt1p	Phytochelatin/Cd^{++}	830	$\text{TMS}_6\text{-ABC}$	Vacuole
Aspergillus nidulans				
AtrBp	Drugs	1426	$(\text{ABC-TMS}_6)_2$	?
AtrDp	Drugs, antibiotics	1348	$(\text{TMS}_6\text{-ABC})_2$	?
Pathogenic fungi				
Candida albicans				
Cdr1p	Antifungal azoles, rhodamine, drugs, dyes	1501	$(\text{ABC-TMS}_6)_2$	Plasma membrane
Cdr2p	Antifungal azoles, rhodamine, drugs, dyes	1499	$(\text{ABC-TMS}_6)_2$	?
Candida glabrata				
Pdh1p	Drugs	1542	$(\text{ABC-TMS}_6)_2$	?
CgCdr1p	Azole antifungals	1499	$(\text{ABC-TMS}_6)_2$	?
Aspergillus fumigatus				
AfuMdr1p	Drugs, cilofungin	1349	$(\text{TMS}_6\text{-ABC})_2$	?

ABC, ATP-binding cassette; TMS, transmembrane segment; PM, plasma membrane; Vac, vacuole; GS, glutathione S; UCB, unconjugated bilirubin.

[a]Since Ycf1p and Yor1p belong to the MRP/CFTR family, their membrane topology might be different, displaying an additional N-terminal transmembrane domain, but this has not been established (Tusnady *et al.*, 1997).

et al., 2000; Wolfger *et al.*, 2001). Moreover, this network contains at least two permeases of the major facilitator family (Nourani *et al.*, 1997b), and several other yeast genes (DeRisi *et al.*, 2000; Kolaczkowska, 1999). We have not included these in **Figure 15.1**, since they represent non-ABC genes.

The major *S. cerevisiae* drug efflux pumps are Pdr5p, Snq2p and Yor1p, all of which localize to the cell surface (see Chapter 14). These transporters recognize an amazingly broad spectrum of xenobiotics and hydrophobic drugs and extrude hundreds of compounds to the extracellular space (Egner *et al.*, 1998; Kolaczkowski *et al.*, 1998; Mahé *et al.*, 1996a). Thus, PDR arises from expression or induced overexpression of ABC pumps mediating cellular efflux of a great variety of different drugs or cytotoxic compounds. Although drug resistance can also be due to reduced drug uptake, target alteration and vacuolar sequestration **(Figure 15.2)**, increased efflux through membrane ABC transporters represents a major cause of acquired drug resistance phenotypes. Other closely related members of the PDR family include Pdr10p and Pdr15p, sharing about 70% identity with Pdr5p. However, no drug substrates have been identified and their expression and function appears connected to a cellular stress response (Wolfger *et al.*, in preparation). Likewise, the function of the Pdr12p efflux pump is linked to a stress response, but in this case weak organic acids rather than hydrophobic drugs were identified

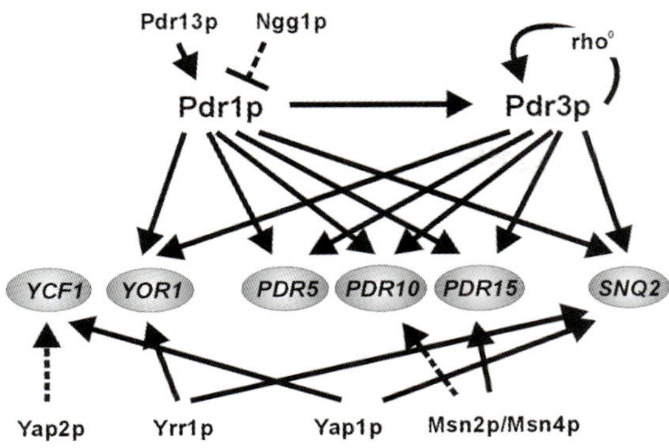

Figure 15.1. The pleiotropic drug resistance (PDR) network. The genes in the center line represent target genes of dedicated transcriptional regulators depicted above and below. Note, the cartoon only includes functional drug resistance genes of the ABC gene family. The yeast PDR network also contains non-ABC genes whose function is not always established (see text for details).

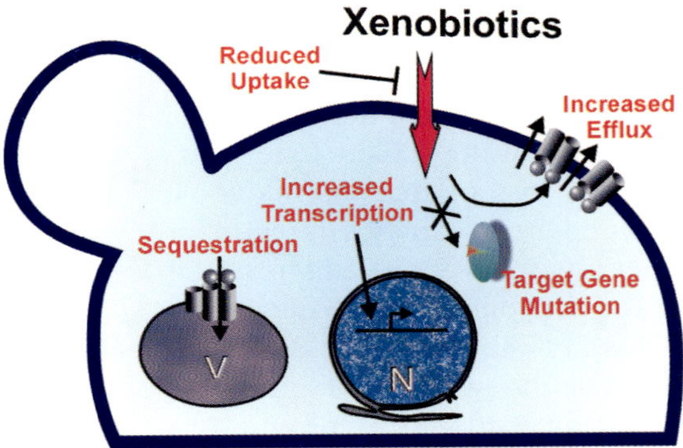

Figure 15.2. Principal mechanisms of drug resistance. Drug resistance phenotypes can arise based on several molecular principles. Pleiotropic or multidrug resistance, which displays cross-resistance to many structurally and functionally unrelated drugs, often results from the induced overexpression of cell surface ABC efflux pumps causing increased efflux of xenobiotics. Each mechanism on its own or in combination with another one can cause a drug resistance phenotype in fungal cells. N, nucleus; V, vacuole.

as physiological substrates (Holyoak *et al.*, 1999; Piper *et al.*, 1998).

A second important mechanism of PDR in yeast involves sequestration into the vacuole **(Figure 15.2)**. Vacuolar ABC pumps such as Ycf1p, Ybt1p and Bpt1p, like the plasma membrane Yor1p, belong to the MRP/CFTR subfamily, as they are more closely related to mammalian MRP, and at least to some extent to human CFTR. Xenobiotics or toxic metabolites can be sequestered into the vacuole, thereby leading to drug or even heavy metal tolerance. For example, the yeast cadmium factor (Ycf1p) is responsible for vacuolar detoxification of heavy metals as well as glutathione S-conjugates (GSH conjugates) (Li *et al.*, 1996; Szczypka *et al.*, 1994). ABC transporter genes with similar functions were also discovered in the fission yeast *Schizosaccharomyces pombe*. For instance, expression of *pmd1* and *hba2/bfr1* mediates drug resistance (Nagao *et al.*, 1995; Nishi *et al.*, 1992; Turi and Rose, 1995), while Hmt1p is involved in vacuolar sequestration of heavy metals (Ortiz *et al.*, 1992, 1995).

Because of their medical importance, ABC proteins from fungal pathogens, including *Candida* and *Aspergillus* species, have received considerable attention in recent years, particularly concerning their possible contribution to clinical antifungal resistance **(Table 15.1)**. To date, four *Candida* ABC transporters implicated in clinical drug resistance have been identified. The *CDR1* and *CDR2* genes from *Candida albicans* (Prasad *et al.*, 1995; Sanglard *et al.*, 1995, 1997), as well as *PDH1* (Miyazaki *et al.*, 1998) and *CgCDR1* (Sanglard *et al.*, 1999) from *Candida glabrata* mediate antifungal resistance both in clinical isolates and in the model system *S. cerevisiae*. For *Candida dubliniensis*, ABC transporters were also speculated to mediate clinical fluconazole resistance, and the existence of *CDR1* and *CDR2* homologues has at least been demonstrated by polymerase chain reaction (PCR) (Moran *et al.*, 1998). Several ABC transporters exist in *Aspergillus*, three of which confer drug resistance upon overexpression. *Aspergillus fumigatus AfuMDR1*, when overexpressed in a drug-sensitive *S. cerevisiae* strain, enhances resistance to the antifungal lipopeptide cilofungin, although no hyper-resistance to other compounds is observed (Tobin *et al.*, 1997). The expression of the *Aspergillus nidulans atrB* and *atrD* genes is induced by numerous drugs, suggesting a role in drug resistance. Indeed, deletion of *atrD* increases drug sensitivity (Andrade *et al.*, 2000b), and overexpression of *atrB* in a hypersensitive Δ*pdr5* yeast strain confers resistance to various compounds (Del Sorbo *et al.*, 1997). Unlike for baker's yeast, however, the literature contains only a limited amount of information as to the functional mechanisms, the regulation or even cellular localization of ABC pumps from fungal pathogens.

Genetic Analysis and Phenotypic Characterization

To study the function of ABC pumps, deletion and overexpression phenotypes should be analyzed in the individual cases. Thus, chromosomal deletions or disruptions of fungal pump genes have been generated. Remarkably, none of the yeast drug ABC transporters **(Table 15.1)** appears to be essential for viability. Hence, the physiological function of these proteins must be dispensable in cells growing under normal conditions. However, in the presence of xenobiotics, including antifungals and anticancer drugs, cells lacking Pdr5p, Snq2p or Yor1p display marked drug hypersensitivity phenotypes (Wolfger *et al.*, 2001). Such a hypersensitivity phenotype was exploited for the cloning of ABC transporters from other fungal species through functional complementation (see below). Notably, in cases like Yor1p or Pdr5p, pump deletion caused hypersensitivity for some drugs but hyperresistance for others. Such phenomena are difficult to explain at the moment, but relate to altered drug uptake, surface permeability changes due to pump deletion, the presence of intracellular drug targets or altered sequestration mechanisms **(Figure 15.2)**. Apart from hypersensitivity to mutagens like 4-nitroquinoline-*N*-oxide (4-NQO), deletion of *SNQ2* increases sensitivity to cations such as Na^+, Li^+ and Mn^{2+} (Miyahara *et al.*, 1996). Notably, deletion of *PDR5* in addition to a Δ*snq2* deletion aggravates the effect on intracellular metal ion accumulation and metal sensitivity, suggesting some functional overlap (Miyahara *et al.*, 1996). Furthermore, a deletion of *PDR5* and *SNQ2* strongly increases pregnenolone and progesterone toxicity to yeast cells (Cauet *et al.*, 1999), suggesting an intracellular target for these steroids. It has also been reported that disruption of *SNQ2* enhances the lag phase, while a Δ*pdr5* Δ*snq2* double disruption influences both lag and log phases, resulting in slower growth rates (Decottignies *et al.*, 1995). Deletion of the *YCF1* gene renders cells hypersensitive to cadmium and completely abolishes vacuolar uptake of $As(GS)_3$ (Ghosh *et al.*, 1999; Szczypka *et al.*, 1994). Finally, a loss of Yor1p causes hypersensitivity to reveromycin A, oligomycin, as well as various organic anions. Moreover, Δ*yor1* cells display cadmium hypersensitivity, indicating a functional overlap of Yor1p and Ycf1p (Cui *et al.*, 1996; Katzmann *et al.*, 1995). As expected, disruption of the two fission yeast drug transporters, *pmd1* and *hba2/bfr1*, led to a drug hypersensitivity phenotype (Nagao *et al.*, 1995; Nishi *et al.*, 1992; Turi and Rose, 1995).

Likewise, deletion analysis has been performed for the *C. albicans* transporters Cdr1p and Cdr2p (Sanglard *et al.*, 1996, 1997). While deletion of *CDR1* causes hypersensitivity to azoles, terbinafine, amorolfine and various other metabolic inhibitors, disruption of *CDR2* does not cause obvious hypersusceptibility to these compounds. However, a double disrupted Δ*cdr1* Δ*cdr2* strain displays increased sensitivity when compared to a Δ*cdr1* strain, implying that Cdr2p does play a role in drug resistance. Interestingly, spontaneous revertants of a Δ*cdr1* strain become resistant by expressing the second transporter gene *CDR2*, which is normally not overexpressed (Sanglard *et al.*, 1997). Disruption of the *C. glabrata CgCDR1* gene in a resistant clinical isolate clearly reduced azole resistance, supporting the idea that *CgCdr1p* is the drug pump mediating resistance in this isolate (Sanglard *et al.*, 1999). While a loss of the *Aspergillus* ABC proteins *atrB* and *atrD* increases susceptibility to drugs, deletion of *atrC* did not result in any drug sensitivity phenotype (Andrade *et al.*, 2000a, 2000b). Notably, deletion of *atrD* also seems to decrease the secretion of antibiotic compounds (Andrade *et al.*, 2000b), providing a case example for an ABC transporter that effluxes both physiological and non-physiological substrates.

Substrate Specificity and Mechanisms of Drug Recognition by Fungal ABC Pumps

Fungal ABC pumps and some of their relevant drug substrates are listed in **Table 15.1**. *SNQ2*, which was originally cloned as a gene conferring resistance to mutagens such as 4-nitroquinoline-*N*-oxide and triaziquone, was the first multidrug resistance ABC transporter identified in *S. cerevisiae* (Servos *et al.*, 1993). Interestingly, Snq2p also seems to modulate resistance to cations such as Na^+, Li^+ and Mn^{2+} (Miyahara *et al.*, 1996). Shortly afterwards,

PDR5 was independently isolated by several groups through its ability to mediate cycloheximide resistance (Balzi *et al.*, 1994), resistance to mycotoxins (Bissinger and Kuchler, 1994), cross-resistance to cerulenin and cycloheximide (Hirata *et al.*, 1994), as well as the transport of glucocorticoids (Kralli *et al.*, 1995). Finally, genetic screens for oligomycin and reveromycin A-resistant yeast cells led to the discovery of Yor1p, the third plasma membrane drug pump of *S. cerevisiae* (Cui *et al.*, 1996; Katzmann *et al.*, 1995).

Extensive studies on the determinants of substrate specificity revealed an extremely broad substrate specificity of fungal PDR transporters with distinct but considerably overlapping drug resistance profiles (Egner *et al.*, 1998; Kolaczkowski *et al.*, 1998; Mahé *et al.*, 1996a; Reid *et al.*, 1997; Servos *et al.*, 1993). The PDR pumps mediate extrusion of hundreds of structurally and functionally unrelated compounds, including ions, heavy metals, ionophores, antifungals, GSH-conjugates, bile acids, anticancer drugs, antibiotics, detergents, lipids, fluorescent dyes, steroids and even peptides as well as many others. Notably, Pdr5p and Yor1p may also transport phospholipids, as demonstrated by fluorescent phosphatidylethanolamine accumulation *in vivo* (Decottignies *et al.*, 1998). A similar role in phosphatidylethanolamine transport has been speculated for *C. albicans* Cdr1p (Dogra *et al.*, 1999). The leptomycin B resistance gene *pmd1* from *S. pombe* also confers cross-resistance to cycloheximide, valinomycin and staurosporine (Nishi *et al.*, 1992). The second fission yeast drug pump, Bfr1p/Hba2p, mediates MDR, with resistance to brefeldin A, cerulenin and several antibiotics (Nagao *et al.*, 1995; Turi and Rose, 1995). In contrast, Ycf1p and Hmt1p are not involved in drug efflux at the cell surface, but mediate vacuolar sequestration of heavy metals and other toxic compounds (Ortiz *et al.*, 1992, 1995; Szczypka *et al.*, 1994). Finally, the *Candida* and *Aspergillus* drug pumps were characterized mainly on the basis of their ability to cause resistance to antifungal agents such as azoles.

How such a wide variety of xenobiotics can be translocated by one transporter molecule is still not understood. The best-studied exporters in this respect are perhaps the drug-transporting mammalian Pgps, which are extensively discussed in other chapters of this book. Photo-affinity labeling studies and genetic analysis indicate that both nucleotide-binding domains (NBDs) and membrane-spanning domains (TMDs) somehow contribute to substrate recognition and transport in mammalian drug pumps (Gottesman *et al.*, 1995; Zhang *et al.*, 1995). Transport inhibition studies, mutational analyses and genetic studies identified amino acid residues required for substrate recognition and binding by Pdr5p and Cdr1p (Egner *et al.*, 1998, 2000; Kolaczkowski *et al.*, 1996; Krishnamurthy *et al.*, 1998). The possibility of genetically separating drug transport from inhibitor susceptibility indicates the existence of at least two distinct drug-binding sites in Pdr5p (Egner *et al.*, 1998, 2000), and perhaps in related transporters such as Cdr1p. In addition, the inhibition of Pdr5p-mediated rhodamine 6G fluorescence quenching supports the notion of more than one drug-binding site in fungal ABC pumps (Kolaczkowski *et al.*, 1996). At any rate, the actual drug transport mechanism and how it is linked to ATP consumption, the so-called catalytic cycle of ABC proteins originally proposed by Alan Senior (Senior *et al.*, 1995), has not been established for fungal pumps. However, it seems plausible that fungal ABC pumps may achieve substrate transport through a mechanism similar to the one described by the catalytic cycle or the alternating two-cylinder two-piston engine model for human Pgp and bacterial LmrA, respectively (Senior *et al.*, 1995; van Veen *et al.*, 2000). Extrusion of substrates might be mediated by efflux from the cytoplasm to the outside or, alternatively, they might be recognized and extruded (or flipped) from the inner leaflet of the plasma membrane to the outside through a 'molecular vacuum-cleaner' mechanism originally proposed for the human P-glycoprotein Mdr1p (Higgins and Gottesman, 1992). Given the broad substrate specificity, and the possible existence of more than one drug-binding site, one might speculate that the actual transport mechanism depends on the substrate to be transported, and that a single fungal ABC pump can actually function through several mechanisms.

While the transport mechanism has not been elucidated, the ATP dependence of drug transport is established beyond any doubt. Pdr5p and Snq2p, albeit highly homologous, display different pH optima regarding their ATPase activity and, interestingly, distinct nucleotide triphosphate (NTP) preferences. The Snq2p ATPase activity shows a sharp pH optimum at 6.0–6.5, while Pdr5p activity remains unchanged over a broad pH range from 6.0 to 9.0 (Decottignies *et al.*, 1995). As for the NTP substrates, Snq2p is more selective with a preference for ATP, whereas Pdr5p also hydrolyzes UTP and CTP, and to a lesser extent GTP and ITP (Decottignies *et al.*, 1995). UTP hydrolysis by Pdr5p and Snq2p is sensitive to vanadate and Triton X-100 inhibition. By contrast, oligomycin affects only Pdr5p UTPase activity (Decottignies *et al.*, 1995). Like mammalian P-glycoprotein and CFTR, Pdr5p and Yor1p can be photolabeled with the fluorescent ATP analogue TNP-8-azido-ATP (Decottignies *et al.*, 1998). Regarding pumps of fungal pathogens, an NTPase activity has only been shown for the *Candida* transporter Cdr1p, which exhibits both vanadate-sensitive ATPase and UTPase activities (Krishnamurthy *et al.*, 1998). Likewise, using *in vitro* uptake assays in the presence and absence of ATP (Li *et al.*, 1997; Ortiz *et al.*, 1995), the ATP dependence of the vacuolar uptake of heavy metals and glutathione conjugates via Hmt1p or Ycf1p, respectively, has also been demonstrated.

Some answers to tantalizing questions concerning the molecular mechanisms and catalytic cycles of fungal ABC pumps might emerge once 3-D crystal structures become available. An important step towards this direction is the recent elucidation of the *Escherichia coli* MsbA high-resolution crystal structure (Chang and Roth, 2001). MsbA acts as a homodimer, each subunit consisting of six transmembrane-spanning α-helices, a bridging domain and an NBD. However, despite this fascinating work, even in this case many mechanistic questions remain open or lead to ambiguous interpretations and answers (Higgins and Linton, 2001). Thus, more structures may have to be solved to obtain a physiologically relevant model of drug transport by ABC pumps. So far, only low-resolution structures are available for the eukaryotic ABC proteins MRP1, Pgp and TAP (Rosenberg *et al.*, 2001a, 2001b; Velarde *et al.*, 2001; Chapter 4), but attempts to obtain better and refined structures are well on their way in several laboratories.

MUTATIONAL ANALYSIS OF YEAST ABC PUMPS AND STRUCTURE–FUNCTION RELATIONSHIPS

To better understand the molecular basis of ABC pump function, genetic and mutational analysis is necessary. A detailed mutational analysis of Pdr5p permitted the identification of amino acid residues important for proper folding, drug substrate specificity and inhibitor susceptibility (Egner *et al.*, 1998). Non-functional mutant proteins were either the consequence of NBD mutations or caused by misfolding in the endoplasmic reticulum (ER). For instance, a C1427Y–Pdr5p exchange in the last predicted extracellular loop 6 between TMS11 and TMS12 causes Pdr5p misfolding and its efficient ER retention, followed by rapid polyubiquitination and degradation by the cytoplasmic proteasome (Plemper *et al.*, 1998). The instability of C1427Y–Pdr5p is perhaps due to a lack of disulfide bond formation between cysteines in lumenal loops, which appears as a prerequisite for correct folding and exit from the ER (Bauer *et al.*, unpublished data).

The structure–function analysis of Pdr5p also produced additional mutant transporters with altered drug substrate specificity. The S1360F exchange in the predicted TMS10 of Pdr5p is the most remarkable one. This mutation causes a highly restricted substrate specificity for the antifungal agent ketoconazole, with poor resistance to itraconazole and cycloheximide. At the same time, ketoconazole resistance is no longer reversed by the immunosuppressive drug FK506 in S1360F-Pdr5p, while azole transport of wild-type Pdr5p is completely blocked by FK506. However, when the same residue, S1360, is substituted by alanine instead of phenylalanine, the resulting S1360A-Pdr5p transporter suddenly becomes hypersensitive to FK506 inhibition (Egner *et al.*, 2000). These studies indicate that TMS10 is a major determinant of Pdr5p substrate specificity and inhibitor susceptibility. In addition, these studies allowed the genetic separation of drug transport from pump inhibitor susceptibility, again suggesting the existence of more than one drug-binding site in certain fungal pumps.

While the structure–function relationship of Snq2p has not been addressed, the MRP/CFTR family members Ycf1p and Yor1p have been subjected to detailed mutational studies. Mutations in *YCF1*, analogous to the most prominent mutations in the human CFTR protein were thus constructed. Deletions of F713 in Ycf1p and F670 in Yor1p, which are the equivalents of the ΔF508-CFTR deletion associated with cystic fibrosis, were generated and analyzed. Similar to the intracellular trafficking defect of ΔF508-CFTR in human cells, ΔF713-Ycf1p leads to ER retention, together with loss of cadmium resistance (Wemmie and Moye-Rowley, 1997). Mutations in NBDs, as well as in the regulatory (R) domain, produced two classes of mutants. First, those defective in Ycf1p biogenesis and, second, transporters causing impaired cadmium tolerance and glutathione S-conjugated leukotriene C_4 (LTC_4) transport. Interestingly, certain mutations in the R-domain and in the cytoplasmic loop 4 genetically separate cadmium resistance from LTC_4 transport (Falcon-Perez *et al.*, 1999). Likewise, a ΔF670-Yor1p mutant protein was retained in the ER and thus was unable to confer oligomycin resistance. The same effect, namely an ER retention and loss of resistance, was achieved by insertion of an alanine residue at position 652 in NBD1. Notably, replacement of a basic residue downstream of the LSGGQ motif (K715M or K715Q), despite a proper plasma membrane localization of the mutant proteins, resulted in reduced oligomycin resistance (Katzmann *et al.*, 1999).

CELLULAR DISTRIBUTION, TRAFFICKING, MEMBRANE LOCALIZATION AND PROTEOLYTIC TURNOVER

A plasma membrane localization has only been unequivocally demonstrated for Pdr5p, Snq2p and Yor1p (Decottignies *et al.*, 1995; Egner and Kuchler, 1996; Egner *et al.*, 1995; Katzmann *et al.*, 1999; Mahé *et al.*, 1996b), as well as *Candida* Cdr1p (Hernaez *et al.*, 1998). Thus, it is reasonable to assume that the majority of fungal drug transporters are active at the plasma membrane, mediating extrusion of toxic compounds from within the cell across the plasma membrane. In contrast, transporters responsible for heavy metal detoxification, such as Ycf1p and Hmt1p, reside in the vacuolar membrane (Li *et al.*, 1997; Ortiz *et al.*, 1992). This delimits the main catabolic compartment for deleterious substances, degradation products or toxic metabolites.

The yeast ABC pumps Pdr5p, Snq2p and Yor1p are rather short-lived proteins with a half-life ranging from 60 to 90 minutes (Egner *et al.*, 1995; Katzmann *et al.*, 1999; Mahé *et al.*, unpublished data). Trafficking studies revealed that cell surface proteins such as these transporters have to reach the vacuole to undergo proteolytic turnover. Yeast mutants defective in the exocytic and endocytic pathways accumulate newly synthesized Pdr5p, indicating trafficking by the normal exocytic secretion machinery (Egner *et al.*, 1995). Using strains carrying mutations in either one of the major proteolytic systems represented by the vacuole and the cytoplasmic proteasome, Pdr5p has been shown to undergo constitutive endocytosis and delivery to the vacuole for terminal degradation (Egner *et al.*, 1995). Interestingly, Pdr5p (Egner and Kuchler, 1996), Yor1p (Katzmann *et al.*, 1999) and the related Ste6p mating pheromone transporter (Kölling and Losko, 1997; Loayza and Michaelis, 1998), Snq2p (Mahé *et al.*, unpublished data), as well as several other yeast membrane proteins (Hicke, 1997), are ubiquitinated prior to endocytosis. However, this ubiquitin attachment does not target the proteins for degradation by the cytoplasmic proteasome. Instead, the ubiquitin modification, which occurs only at the cell surface (Egner and Kuchler, 1996; Kölling and Hollenberg, 1994) and is limited to a single ubiquitin, acts as an endocytosis signal (Hicke, 1997; Laney and Hochstrasser, 1999). A Pdr5p phosphorylation by Yck1p (yeast casein kinase I) might play a role in Pdr5p trafficking and turnover (Decottignies *et al.*, 1999), but any other impact of Pdr5p phosphorylation on the PDR phenotype remains unknown.

As outlined above, the physiological Pdr5p turnover requires vacuolar proteolysis but not the cytoplasmic proteasome. However, misfolded Pdr5p, which may arise from improper folding in the ER during its biogenesis, requires the proteasomal degradation system. An extensive mutational and genetic analysis of Pdr5p led to the identification of the C1427Y mutation in the last predicted extracellular loop. This mutation causes the efficient ER retention and rapid degradation of a misfolded Pdr5*p pump (Egner *et al.*, 1998) by the ER quality control system. The ER-associated degradation (ERAD) system (Fewell *et al.*, 2001) is devoted to a rapid removal of secretory membrane proteins immediately after or even during their synthesis should misfolding occur. Misfolded Pdr5*p is rapidly extracted from the ER membrane through a Sec61p-dependent retrograde pathway, becomes polyubiquitinated and subsequently degraded by the cytoplasmic proteasome (Plemper *et al.*, 1998). Similar results have been obtained for Yor1p and Ycf1p, the vacuolar heavy metal resistance transporter, as well as the **a**-factor mating pheromone transporter Ste6p (Loayza *et al.*, 1998). Since Yor1p and Ycf1p are related to human CFTR, mutations analogous to the most frequent cystic fibrosis mutation, ΔF508, were constructed in these yeast pumps (see above). Interestingly, a deletion of F713 in Ycf1p or F670 in Yor1p yields pump variants which are efficiently retained in the ER and rapidly degraded (Katzmann *et al.*, 1999; Wemmie and Moye-Rowley, 1997) by the ER quality control machinery (Plemper and Wolf, 1999). These data indicate that the basic principle of functional folding of ABC proteins is conserved in mammals and yeast, emphasizing the importance of yeast as a model system to study the biology of heterologous ABC proteins of medical importance.

FUNCTIONAL ASSAYS FOR YEAST ABC PUMPS MEDIATING DRUG RESISTANCE

A number of functional assays to study the function and substrate transport of fungal ABC proteins have been established. These assays, which are described below, include standard resistance assays, photoaffinity labeling and crosslinking studies, transport studies *in vivo* and *in vitro* using vesicles or proteoliposomes, and substrate accumulation in whole cells.

Perhaps the simplest and most widely used tests for drug resistance genes are growth inhibition assays on agar plates (Bissinger and Kuchler, 1994). Susceptibilities of various yeast strains can be tested qualitatively and semiquantitatively by spotting serial dilutions of yeast cultures on either agar plates containing various drugs at different concentrations or continuous drug-gradient plates. If both a toxic substrate and a pump inhibitor are present in the same plate, even transport inhibition or drug resistance reversal can be directly visualized by inhibition of cell growth (Egner *et al.*, 1998). Gradient plates are easy to prepare and even allow for a semi-quantitative determination of inhibitory substrate concentrations (Koch, 1999). An alternative to plate assays are halo assays, in which filter disks soaked with drug solutions are placed onto lawns of tester cells, similar to the classical antibiotic agar diffusion assay. The resulting zone of inhibition surrounding the filter disk is a direct quantitative measure of toxicity (Nakamura *et al.*, 2001). However, these assays may lead to artifacts, particularly when hydrophobic drugs with limited solubility and diffusibility in agar plates are used. Drug susceptibility profiles of filamentous fungi such as *Aspergillus* species can also be tested using a similar type of assay. Mycelial plugs from confluent plates are placed with the mycelial side down on drug plates and the radial growth is monitored after certain time periods (Andrade *et al.*, 2000b).

An excellent tool to monitor ABC transporter function *in vivo* includes the measurement of drug efflux or the cellular accumulation of radiolabeled substrates or fluorescent dyes such as rhodamine. The mitochondria-staining dyes rhodamine 6G (R6G) and rhodamine 123 (Johnson *et al.*, 1980) have thus been utilized to study both efflux and energy dependence. Dye efflux is determined either indirectly by fluorescence dequenching or directly by measuring the fluorescence of extruded rhodamine in the incubation buffer. To examine the binding of inhibitors or substrates to multidrug resistance proteins such as Pdr5p, energy-dependent rhodamine 6G fluorescence quenching has been applied (Conseil *et al.*, 2001; Kolaczkowski *et al.*, 1996). This method takes advantage of the fact that rhodamine 6G fluorescence is quenched upon dye-binding to the transporter molecule. Therefore, in the presence of a competitor, which could act as an inhibitor or any other substrate, quenching is reduced and thus fluorescence increases. The quenching assay also provides information, whether the pump inhibition is competitive and involves the same binding site, or is non-competitive due to different drug-binding sites. This approach showed that protein kinase C effectors such as staurosporine analogues are capable of inhibiting the interaction of rhodamine 6G with Pdr5p (Conseil *et al.*, 2001). Alternatively, dye accumulation within yeast cells can be monitored using a fluorescence-activated cell sorter (FACS). Such transport measurements were employed to determine the activity of Pdr5p variants (Egner *et al.*, 1998), to screen compounds for inhibitors of Pdr5p-mediated transport

(Egner *et al.*, 1998; Kolaczkowski *et al.*, 1996) or as a means to identify overexpressing *CDR1* pathogenic *Candida* strains (Maesaki *et al.*, 1999). Similarly, the non-fluorescent, membrane-permeable compound monochlorobimane can be used to monitor transport of glutathione S-conjugates, since the glutathione transfer reaction on monochlorobimane results in a highly fluorescent yet membrane-impermeable conjugate. Addition of monochlorobimane to yeast cultures and monitoring the subcellular localization of the fluorescent S-conjugate proved that Ycf1p is a major factor in the vacuolar accumulation of monochlorobimane-GS (Li *et al.*, 1996). Because certain yeast pumps such as Yor1p and possibly Pdr5p may mediate membrane flipping of phospholipids, functional assays can be used in which the movement of fluorescent phospholipid analogues such as C6-NBD-phosphatidylethanolamine (Kean *et al.*, 1997) is directly followed by time-lapse fluorescence spectroscopy (Decottignies *et al.*, 1998).

Another method to study ABC transporter activity is the use of radiolabeled substrates. For instance, a whole cell *in vivo* estradiol accumulation assay was developed to demonstrate that steroid substrates are translocated by Pdr5p and Snq2p (Mahé *et al.*, 1996a). Since overexpression of *PDR5* and *SNQ2* decreases intracellular estradiol, this approach identified steroids as new substrates of fungal pumps. These *in vivo* uptake assays can also be coupled to steroid/glucocorticoid receptor or steroid/glucocorticoid response element (ERE/GRE)-driven reporter systems (Mahé *et al.*, 1996a). Accumulation of pump substrates such as mycotoxins and environmental toxins are thus easy to measure, as these compounds display a high degree of estrogen activity (Kralli *et al.*, 1995; Mitterbauer *et al.*, 2000). Moreover, such systems also elegantly allow for the selection of mutant transporters and genetic analysis of ABC-driven substrate transport (Kralli *et al.*, 1995; Kralli and Yamamoto, 1996; Mahé *et al.*, 1996a; Tran *et al.*, 1997). Similarly, the measurement of intracellular [^{3}H]-fluconazole has been used to directly show that antifungal azoles are extruded from *Candida* cells by Cdr1p (Sanglard *et al.*, 1996). In the case of *A. nidulans*, the accumulation of the fungicide [^{14}C]-fenarimol was measured to indicate a role for *atrC* and *atrD* in drug resistance (Andrade *et al.*, 2000b).

To prove that Ycf1p mediates vacuolar sequestration of organic compounds after their conjugation to cellular glutathione, *in vitro* uptake into vacuolar membrane vesicles has been measured (Li *et al.*, 1997; Rebbeor *et al.*, 1998). For these experiments, vacuolar membrane vesicles are incubated with various radiolabeled substrate complexes, and accumulation of substrates within the vesicles is monitored by the amount of sequestered radioactivity. This assay revealed that Cd-GS_2, but not Cd-GS, transport into the vacuole requires Ycf1p. This type of assay also allows for the investigation of transport inhibition or competition by other substrates.

Finally, since ABC transporters are ATP-driven membrane translocators, following their ATP dependence and measuring ATP hydrolysis is of course an important assay. For the *S. cerevisiae* transporters Pdr5p, Snq2p and Yor1p, ATPase activity has been demonstrated. Inhibition by vanadate and oligomycin has also been reported (Decottignies *et al.*, 1994, 1995, 1998). ATP-binding by Pdr5p and Yor1p was confirmed by photolabeling of these proteins with TNP-8-azido-ATP (Decottignies *et al.*, 1998). The vanadate-sensitive (Rebbeor *et al.*, 1998) ATP consumption of Ycf1p has been shown by performing uptake assays into vacuolar membrane vesicles in the presence and absence of MgATP (Li *et al.*, 1997). However, in contrast to mammalian ABC pumps, little is known about the binding properties of individual yeast NBDs with respect to their interaction with NTPs/NDPs or the catalytic cycle of yeast drug pumps.

Heterologous Expression of Eukaryotic ABC Pumps and Functional Complementation in Yeast

S. cerevisiae has always been a valuable model organism to investigate the function of evolutionary conserved genes, including ABC proteins of medical importance and drug resistance pumps. To study functional conservation and to clone multidrug transporters, several eukaryotic drug pumps have been functionally expressed in yeast. For example, the human Pgp Mdr1p was successfully expressed in an *S. pombe* strain lacking *pmd1* (Ueda *et al.*, 1993) as well as in baker's yeast (Kuchler and Thorner, 1992). Although not properly glycosylated, the human protein was partially functional and able to confer resistance to valinomycin and actinomycin D to an otherwise sensitive yeast strain (Kuchler and Thorner, 1992; Ueda *et al.*, 1993). To learn more about the mechanism of action of Ycf1p, its human homologue MRP, sharing 63% amino acid similarity with Ycf1p, was expressed in a Δ*ycf1* strain (Tommasini *et al.*, 1996). Human MRP restores cadmium resistance to wild-type levels and facilitates transport of *S*-(2,4-dinitrobenzene)-glutathione (DNB-GS) into yeast microsomal vesicles. This was one of the first indirect indications that Ycf1p, like MRP, is a glutathione S-conjugate pump. In another approach, overexpression of the *A. fumigatus MDR1* gene yielded *S. cerevisiae* cells with increased resistance to the antifungal cilofungin (Tobin *et al.*, 1997). Certain yeast ABC pumps have also been successfully expressed in heterologous systems such as plants. For example, expression of the *PDR5* gene in tobacco confers increased resistance to the trichotecene toxin deoxynivalenol (Mitterbauer *et al.*, 2000).

The observation that MDR/PDR arises from overexpression of certain ABC transporters suggested a gene dosage strategy to clone new drug efflux genes. Genomic libraries were screened for genes which in increased dosage can confer resistance to various compounds. Taking advantage of the drug hypersensitivity phenotype of a Δ*pdr5* strain, *Candida* ABC transporters have thus been cloned by functional complementation in baker's yeast (Prasad *et al.*, 1995; Sanglard *et al.*, 1995, 1997, 1999). A fluconazole and cycloheximide supersensitive Δ*pdr5* strain was transformed with genomic *Candida* libraries and transformants resistant to the azole or the antibiotic, respectively, were selected. This approach led to the discovery of the two major *C. albicans* ABC genes *CDR1* and *CDR2*, as well as *CgCDR1* from *C. glabrata*. In addition, a gene for a transporter of the major facilitator class, *BEN*^{r}, was identified through its ability to confer benomyl resistance in *S. cerevisiae* (Sanglard *et al.*, 1995). Likewise, functional complementation studies verified that the *Aspergillus* transporter *atrB* is the orthologue of yeast Pdr5p (Del Sorbo *et al.*, 1997). Similar approaches allowed for the identification of the *S. pombe* genes *bfr1/hba2* and *pmd1* (Nagao *et al.*, 1995; Nishi *et al.*, 1992; Turi and Rose, 1995) as typical fungal MDR genes.

Clinical Relevance of ABC Pumps from Fungal Pathogens and Therapeutic Strategies

With increasing numbers of immunocompromised patients suffering from human immunodeficiency virus (HIV) infections, patients undergoing cancer chemotherapy or bone marrow and organ transplantations, the frequency of fungal infections is steadily rising (reviewed in Bastert *et al.*, 2001; White *et al.*, 1998). The increasing use of antifungal agents in prophylaxis and therapy caused resistance to emerge, and drug resistance has become a significant problem in health care during the past decade. Several classes of antifungal agents acting either fungistatically or fungicidally are in clinical use to treat local as well as systemic infections (Bastert *et al.*, 2001). Polyenes such as the fungicidal amphotericin B and nystatin interfere with ergosterol function in the plasma membrane, leading to pore formation and leakage of cellular components (Vanden Bossche *et al.*, 1994). Flucytosine is metabolized into 5-fluorouracil, which is incorporated into RNA causing disruption of protein synthesis. As shown in **Figure 15.3**, other antimycotics also act via inhibition of the ergosterol biosynthesis, the bulk sterol in the fungal plasma membrane. The Erg1p squalene epoxidase is blocked by allylamines such as terbinafine and naftifine, as well as by thiocarbamates such as tolnaftate. Morpholines such as amorolfine inhibit both the Erg24p C-14 sterol reductase and the Erg2p C-8 sterol isomerase. The fungistatic azoles, with the imidazoles ketoconazole and miconazole, and triazoles such as fluconazole, itraconazole and the newly developed voriconazole, comprise the most widely used class of ergosterol synthesis inhibitors. These azoles inhibit the Erg11p lanosterol C-14-demethylase, a cytochrome P-450 enzyme and the Erg5p C-22-desaturase. Because of their good safety profile and relatively high bioavailability, azoles are widely used to treat fungal infections (White *et al.*, 1998).

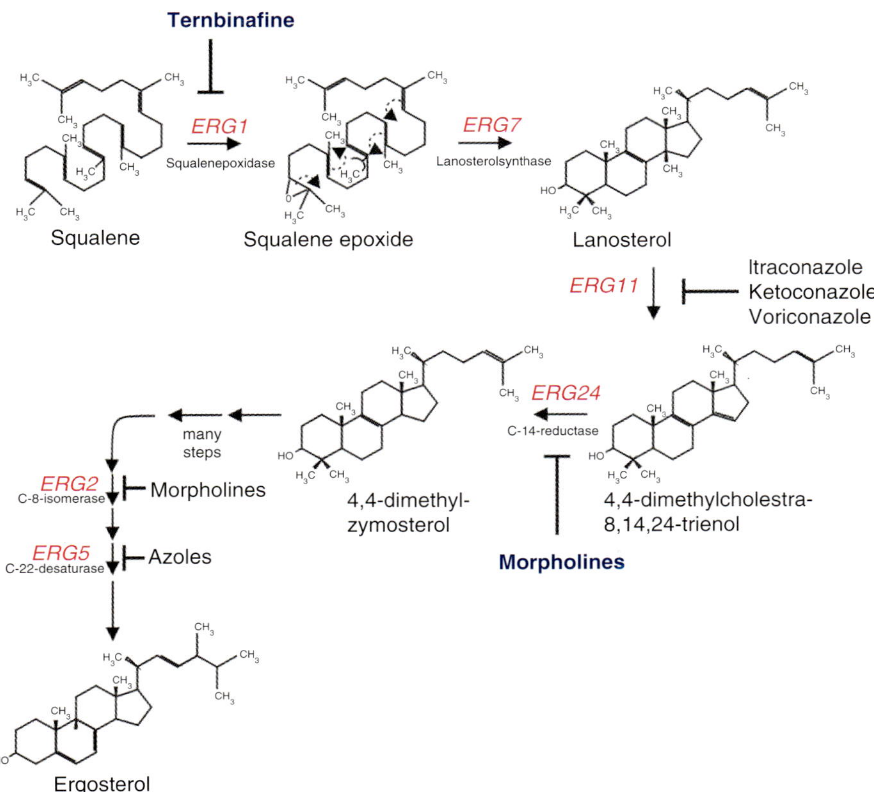

Figure 15.3. The yeast ergosterol biosynthetic pathway. This cartoon depicts the biosynthetic pathway leading to ergosterol synthesis. Only the relevant enzymatic steps are shown. Antifungal agents that act via inhibition of some of these enzymes are given, with the relevant targets indicated.

Clinical resistance to these antifungals can develop through different molecular mechanisms (reviewed in White *et al.*, 1998). These basic resistance mechanisms, depicted in **Figure 15.2**, include reduced drug uptake into the cell, alterations of the target genes by mutation or induced overexpression, changes in the ergosterol biosynthetic pathway, as well as increased drug efflux or facilitated drug diffusion from the cell. Next to target alteration, the induced overexpression of ABC efflux pumps in clinical strains represents a prime cause of clinical antifungal resistance. A number of strategies exist through which clinical drug resistance can be circumvented. As for existing drugs such as azoles, resistance reversal can be achieved by combination therapy (Ryder and Leitner, 2001). For new antifungal drugs under development, one should consider developing those that are not substrates of ABC pumps like Cdr1p or Cdr2p. Azole resistance may be manageable by reducing prophylactic treatment or by the use of specific efflux pump inhibitors in an attempt to reverse antifungal resistance. Nevertheless, the frequency of life-threatening fungal infections in immunocompromised patients is still increasing, with *Candida* and *Aspergillus* species representing the major fungal pathogens (Bastert *et al.*, 2001). Fungal organisms are becoming less susceptible to antifungal drugs, and a shift to intrinsically more resistant fungal pathogens has been observed (Bastert *et al.*, 2001; White *et al.*, 1998). This scenario clearly illustrates that there is a need to better understand the molecular basis of antifungal drug resistance and to develop improved strategies for the treatment of fungal infections.

REGULATION OF DRUG RESISTANCE GENES WITHIN THE YEAST PDR NETWORK

Certain yeast PDR genes, as well as non-PDR genes, are regulated through common transcriptional circuits, involving several dedicated transcription factors. In fact, the first yeast genes known to mediate PDR were transcription factors rather than ABC pumps. Genetic screens for drug-resistant yeast strains led to the identification of hyperactive alleles of genes encoding transcription factors such as Pdr1p (Balzi *et al.*, 1987) and Pdr3p (Delaveau *et al.*, 1994). In addition to Pdr1p and Pdr3p, another member of the Zn(II)$_2$-Cys$_6$ class of transcriptional regulators, namely Yrr1p, plays a role in the regulation of ABC pumps (Cui *et al.*, 1998; Zhang *et al.*, 2001). In all, these genes form the PDR network depicted in **Figure 15.1**. Hyperactive alleles of *PDR1*, *PDR3* or *YRR1* lead to a PDR phenotype, mainly through the induced overexpression of the target drug pumps (Carvajal *et al.*, 1997; DeRisi *et al.*, 2000; Hallström and Moye-Rowley, 2000a; Katzmann *et al.*, 1994; Mahé *et al.*, 1996b; Meyers *et al.*, 1992). The regulatory mechanisms are highly complex, since stress response regulators also contribute to the PDR regulation in concert with other yet unknown factors under both physiological and adverse growth conditions. Fungal transcription factors implicated in ABC gene regulation are listed in **Table 15.2**.

TABLE 15.2. REGULATORS OF FUNGAL PLEIOTROPIC DRUG RESISTANCE GENES

Protein	Structure	Function
Saccharomyces cerevisiae		
Pdr1p	Zn(II)$_2$Cys$_6$ TF	Regulation of PDR
Pdr3p	Zn(II)$_2$Cys$_6$ TF	Regulation of PDR
Ngg1p	TF	Inhibition of Pdr1p activity
Yrr1p	Zn(II)$_2$Cys$_6$ TF	Regulation of PDR
Yap1p	bZip TF	Oxidative stress response, Cd^{2+} and diazaborine resistance
Yap2p	bZip TF	Cd^{2+} resistance
Yap8p	bZip TF	Regulation of arsenite and arsenate resistance
Pdr13p	Hsp70 homologue	Regulation of Pdr1p
Yck1p	Casein kinase I	Modulation of azole resistance
Schizosaccharomyces pombe		
Pap1p	bZip TF	Oxidative stress response
Candida albicans		
Cap1p	bZip TF	Oxidative stress response
Fcr1p	Zn(II)$_2$Cys$_6$ TF	Deletion confers azole resistance

PDR, pleiotropic drug resistance; TF, transcription factor.

The related Pdr1p and Pdr3p regulators, like Yrr1p, belong to the family of Gal4p-like Zn(II)$_2$-Cys$_6$ transcription factors. The N-terminal zinc cluster mediates DNA binding, while the activation domains are located at the C-termini (Carvajal *et al.*, 1997; Nourani *et al.*, 1997a). However, unlike Pdr1p, Pdr3p contains an additional activation domain near its zinc finger (Delaveau *et al.*, 1994). Deletion mapping identified a serine/tyrosine-rich nuclear localization signal (NLS) that mediates Pse1/Kap121p-dependent nuclear localization of Pdr1p (Delahodde *et al.*, 2001). Both transcription factors share predicted inhibitory domains in their central region (Kolaczkowska, 1999; Nourani *et al.*, 1997a). Therefore, it is not surprising that many gain-of-function mutations map within the inhibitory motifs and the C-terminal activation domains (Delaveau *et al.*, 1994; Kolaczkowska, 1999; Nourani *et al.*, 1997a; Simonics *et al.*, 2000). Both Pdr1p and Pdr3p are phosphoproteins, localizing to the nucleus without an apparent shuttling between the nucleus and the cytoplasm (Pandjaitan *et al.*, unpublished). Moreover, as many other Zn(II)$_2$-Cys$_6$ regulators, Pdr1p and Pdr3p form both homo- and heterodimers (Pandjaitan *et al.*, unpublished) before recognizing the cognate *cis*-acting motifs in PDR target genes. These *cis*-acting elements, known as PDREs (pleiotropic drug resistance elements), have the consensus motif 5′-TCCGCGGA-3′ (Delahodde *et al.*, 1995; Katzmann *et al.*, 1994) containing everted CGG repeats also recognized by other Gal4p family members such as Leu3p (Hellauer *et al.*, 1996). Although a single PDRE is necessary and sufficient to confer regulatory control by Pdr1p/Pdr3p (Katzmann *et al.*, 1996), PDREs are found in different numbers and with a certain degree of degeneration in the promoters of Pdr1p/Pdr3p target genes (Wolfger *et al.*, 1997). Whether or not these quantitative and qualitative differences in PDREs are important for the regulation by Pdr1p/Pdr3p remains to be elucidated. *In vitro* studies suggest that recombinant Pdr1p and Pdr3p recognize and bind both perfect and degenerated PDREs (Katzmann *et al.*, 1995, 1996; Mahé *et al.*, 1996b; Nourani *et al.*, 1997b; Wolfger *et al.*, 1997).

The pool of Pdr1p/Pdr3p target genes comprises ABC transporters such as *YOR1*, *SNQ2*, *PDR5*, *PDR10* and *PDR15* (Bauer *et al.*, 1999), several stress response genes (DeRisi *et al.*, 2000), members of the major facilitator family (Nourani *et al.*, 1997b), and several genes of unknown function (DeRisi *et al.*, 2000). Interestingly, Pdr1p/Pdr3p may even play a role in membrane lipid biosynthesis as they appear to regulate the inositol phosphotransferase (*IPT1*) gene (Hallström *et al.*, 2001). The presence of PDREs in the promoters of *PDR3* and *YRR1* suggests autoregulatory loops in the control of expression (Cui *et al.*, 1998; Delahodde *et al.*, 1995; Zhang *et al.*, 2001). Interestingly, Pdr1p and Pdr3p can both positively and negatively regulate the expression of target genes (Wolfger *et al.*, 1997). Thus, it seems reasonable that Pdr1p/Pdr3p require additional factors that regulate their activity or function. This could be achieved at either the level of physical protein–protein interaction or the binding to PDRE motifs. One such candidate is Ngg1p, which, when in complex with Ada2p and Gcn5p, is involved in the regulation of Gal4p (Brandl *et al.*, 1996). Ngg1p interacts with the C-terminal activation domains of Pdr1p and reduces its regulatory activity (Martens *et al.*, 1996; Saleh *et al.*, 1997). A gene dosage screen for oligomycin hyperresistance led to the identification of another Pdr1p interaction partner. The cytoplasmic Hsp70 analogue Pdr13p acts as a positive regulator of Pdr1p function at the post-translational level (Hallström *et al.*, 1998; Hallström and Moye-Rowley, 2000a). Interestingly, the activity of Pdr3p, but not of Pdr1p, is upregulated by mitochondrial dysfunction (Hallström and Moye-Rowley, 2000b), although the signal transduction mechanisms involved remain obscure. One might speculate that Yrr1p somehow amplifies activation signals coming from Pdr1p/Pdr3p (Cui *et al.*, 1998; Zhang *et al.*, 2001), but the molecular details and precise regulatory signals, as well as signal transduction pathways within the PDR network, remain ill-defined.

THE CROSSTALK BETWEEN PDR AND STRESS RESPONSES

The complexity of regulatory circuits within the yeast PDR network is very high, as several regulators, pumps and even MFS permeases constitute this network **(Figure 15.1)**. The most striking feature is the apparent connection of PDR and cellular stress responses. For example, at least three members of the Yap family of bZip transcription factors play a role in heavy metal resistance. Overexpression of *YAP1*, a well-characterized stress regulator with an established function in response to stress (Gounalaki and Thireos, 1994; Schnell and Entian, 1991; Wendler *et al.*, 1997; Wu *et al.*, 1993), causes a PDR phenotype, although a Δ*yap1* deletion only causes hypersensitivity to heavy metals (Wemmie *et al.*, 1994). This Yap1p-mediated Cd^{2+} resistance is dependent on Ycf1p, suggesting a regulation of *YCF1* by Yap1p (Wemmie *et al.*, 1994). Deletion of two other Yap family members, *YAP2* and *YAP8*, causes sensitivity to cadmium (Wu *et al.*, 1993) as well as arsenite and arsenate, respectively (Bobrowicz *et al.*, 1997).

While the physiological substrates and function of the Pdr15p ABC pump remain to be discovered, recent data suggest *PDR15* to be subject to the control of the stress response regulators Msn2p and Msn4p (Wolfger *et al.*, unpublished). Likewise, expression of the PDR family members Pdr10p and Pdr12p is strongly influenced by adverse conditions such as high osmolarity and weak organic acid stress, respectively. The Pdr12p pump is required for adaptation to weak organic acid stress and its induced synthesis requires the function of novel stress regulators through distinct and yet unidentified signal transduction pathways (Piper *et al.*, 1998). However, Pdr12p induction requires neither known stress response regulators nor Pdr1p/Pdr3p. Nevertheless, Pdr1p/Pdr3p provide input for Pdr10p and Pdr15p regulation, since the *PDR10* and *PDR15* expression levels under normal growth conditions are strongly affected by the absence of these regulators (Wolfger *et al.*, 1997). The physiological functions or substrates of ABC genes such as *PDR15* and *PDR10* might therefore be linked to cellular stress responses but their nature remains obscure.

Still very little is known about the regulation of ABC genes in fission yeast and pathogenic fungi. Two regulators with a possible role in PDR have been identified in *C. albicans.* Cap1p regulates expression of *CaYCF1*, since its overexpression causes enhanced fluconazole resistance and increased cadmium and oxidative stress resistance (Alarco and Raymond, 1999). The absence of the zinc finger protein Fcr1p produces fluconazole hyperresistance, pointing to a negative regulation of drug resistance by Fcr1p (Talibi and Raymond, 1999). This is an example where loss-of-function in a regulatory gene causes drug resistance, whereas in other described cases only gain-of-function or overexpression of a regulator causes resistance. A recent report addressed the question of how the increased expression of *C. albicans* drug efflux pumps is achieved (Lyons and White, 2000). The study indicated that gene amplification, unlike in mammalian tumor cells (Gottesman *et al.*, 1995) or parasites (Grondin *et al.*, 1998; Ouellette and Borst, 1991), is not the cause of elevated *CDR1* and *CDR2* mRNA levels. Instead, higher transcription rates are typical causes for clinical resistance. Therefore, *trans*-acting factors most probably play a role in resistance development through transporter overexpression. However, altered gene dosage has been demonstrated as a possible mechanism for azole resistance. Duplication of a chromosome carrying the gene encoding a target enzyme of azole antifungals was detected in a resistant *C. glabrata* clinical isolate (Marichal *et al.*, 1997). In *S. pombe*, one transcription factor, Pap1p, has been described. Like its *S. cerevisiae* homologue Yap1p, Pap1p undergoes cytoplasmic-nuclear shuttling in response to oxidative stress. Acting downstream of the stress-activated kinase Sty1p, Pap1p induces transcription of the transporter genes *pmd1* and *bfr1/hba2*. Consistent with its regulatory function, *pap1* deletion results in sensitivity to the oxidizing agent diamide and the heavy metals cadmium and arsenic (Toone *et al.*, 1998).

The Physiological Functions of Yeast ABC Pumps

Although *S. cerevisiae* Pdr5p and Snq2p transport hundreds of different compounds, their normal cellular substrates or physiological roles remain obscure. ABC efflux pumps of human pathogenic fungi such as *Candida* or *Aspergillus* species have mainly been isolated and characterized by their ability to confer antifungal resistance. A function in cellular detoxification remains a feasible one, particularly considering the environment yeast cells have to cope with in nature. It has also been proposed that certain yeast ABC pumps extrude toxic catabolites that accumulate when cells enter the stationary growth phase (Egner and Kuchler, 1996). For example, Pdr12p, the closest homologue of Snq2p, extrudes both physiological substrates, such as acid metabolites, and the non-physiological weak organic acids. The plasma membrane protein Pdr12p is essential for adaptation to growth in the presence of weak organic acids such as sorbate, benzoate and propionate, which are commonly used as food preservatives. Moreover, acetate, pentanoic and hexanoic acid, toxic products of normal cellular metabolism, are also substrates of Pdr12p (Piper *et al.*, 1998). Another hypothetical function of ABC pumps might be the maintenance of the asymmetric distribution of phospholipids in the lipid bilayer of the cell surface. Certain fungal ABC pumps such as Pdr5p and Cdr1p indeed have some potential to translocate certain lipid molecules from the inner to the outer leaflet (Dogra *et al.*, 1999), but conclusive experimental evidence is not available or at least is controversial.

There are examples of yeast ABC transporters for which the physiological substrates have been identified (see also Chapter 14 for details). The *S. cerevisiae* and *S. pombe* Ste6p and Mam1p transporters secrete the peptide mating pheromones **a**-factor and M-factor, respectively. Further, the mitochondrial half-size transporter Atm1p translocates Fe/S-proteins into the cytosol and Pxa1p and Pxa2p mediate long-chain fatty acid import into the peroxisome (Bauer *et al.*, 1999). In addition, two ABC transporters appear to be key players in the vacuolar sequestration of toxic compounds. Hmt1p of *S. pombe* mediates ATP-dependent transport of phytochelatins, which act as chelators in heavy metal detoxification (Rauser, 1990), and phytochelatin-Cd^{2+} complexes into the vacuole (Ortiz *et al.*, 1995). In baker's yeast, Ycf1p is responsible for the vacuolar sequestration of heavy metals and GSH conjugates (Li *et al.*, 1996; Rebbeor *et al.*, 1998; Tommasini *et al.*, 1996). The vacuole represents the prime organelle for detoxification in both fungi and plants, and several plant ABC transporters are also implicated in the vacuolar sequestration of catabolites and environmental toxins. Hence, plant orthologues of yeast proteins might even play a more general role in vacuolar detoxification. Finally, the *atrD* gene from *A. nidulans* could be involved in the release of antibiotics, implying that ABC transporters in other filamentous fungi might also play a role in secretion of antibiotics from fungal cells (Andrade *et al.*, 2000b).

There is emerging evidence from certain plants for a physiological role for ABC transporters in host–pathogen defense mechanisms. On the one hand, ABC pumps could help a given pathogen to survive plant defense agents and antifungals. In turn, ABC pumps themselves could mediate invasion by secreting pathogenicity factors. For instance, Abc1p of the rice blast fungus *Magnaporthe grisea* is essential for invasive growth and pathogenicity (Urban *et al.*, 1999). Although the mechanism has not yet been clearly defined, Abc1p seems to provide a defense function against antimicrobial compounds produced by the host plant (Urban *et al.*, 1999). Likewise, a protective role against plant defense mechanisms has been suggested for MgAtr1p and MgAtr2p from the wheat pathogen *Mycosphaerella graminicola* (Zwiers and De Waard, 2000).

Conclusions and Perspectives

Important contributions to a better understanding of many diverse cellular roles of ABC proteins have been made during the past few years. However, despite intensive research efforts in many different laboratories, we still fall short of understanding the molecular mechanisms of a single eukaryotic ABC protein. Cures or even efficient therapies for many ABC protein-mediated diseases are still out of reach. The medical importance of many human ABC genes that are either directly or indirectly implicated in important genetic diseases illustrates the importance of understanding their biology. This understanding will certainly enter a new era once crystal structures of ABC proteins become available. Another quantum leap can be expected with methodologies that allow for genome-wide proteomic analysis of ABC proteins and how they are interlinked into cellular metabolism in all living cells. Looking back at the past decade, yeast provided important discoveries on the biology of endogenous and heterologous ABC proteins. Yeast was the first eukaryotic organism whose genome was sequenced, and genome-wide transcription analysis has become laboratory routine. Nevertheless, even the functions of many yeast ABC genes have escaped discovery as yet. Because many fungal pathogens are refractory to genetic analysis or owing to a lack of experimental tools, yeast will continue to be an important test tube for the functional characterization of eukaryotic ABC genes. In the years to come, we expect important discoveries concerning ABC genes present in other fungal pathogens, and perhaps yeast will contribute to a better understanding of other medically relevant ABC proteins which exist in mammalian or parasitic genomes.

Acknowledgments

We are indebted to our colleagues Agnés Delahodde, Christophe D'énfert, Bertrand Favre, André Goffeau, Scott Moye-Rowley, Peter Piper, Elisabeth Presterl, Neil Ryder, Dominique Sanglard, Julius Subik, Friederike Turnowsky, Marten de Waard and Birgit Willinger for sharing unpublished information, materials and strains, as well as for many stimulating discussions. Thanks to all group members for critical comments on the manuscript. Our research is supported by grants from the 'Fonds zur Förderung der wissenschaftlichen Forschung' (FWF, P12661-BIO), by funds from the Austrian National Bank (OeNB #7421), grants from Novartis Pharma Inc., DSM Bakery Ingredients, the 'Hygiene-Fonds' of the Medical Faculty of the University of Vienna and the 'Herzfelder Foundation'.

References

Alarco, A.M. and Raymond, M. (1999) The bZip transcription factor Cap1p is involved in multidrug resistance and oxidative stress response in *Candida albicans*. J. Bacteriol. 181, 700–708.

Andrade, A.C., Del Sorbo, G., Van Nistelrooy, J.G. and Waard, M.A. (2000a) The ABC transporter *atrB* from *Aspergillus nidulans* mediates resistance to all major classes of fungicides and some natural toxic compounds. Microbiology 146, 1987–1997.

Andrade, A.C., Van Nistelrooy, J.G., Peery, R.B., Skatrud, P.L. and De Waard, M.A. (2000b) The role of ABC transporters from *Aspergillus nidulans* in protection against cytotoxic agents and in antibiotic production. Mol. Gen. Genet. 263, 966–977.

Balzi, E., Chen, W., Ulaszewski, S., Capieaux, E. and Goffeau, A. (1987) The multidrug resistance gene *PDR1* from *Saccharomyces cerevisiae*. J. Biol. Chem. 262, 16871–16879.

Balzi, E., Wang, M., Leterme, S., Van Dyck, L. and Goffeau, A. (1994) *PDR5*, a novel yeast multidrug resistance conferring transporter controlled by the transcription regulator *PDR1*. J. Biol. Chem. 269, 2206–2214.

Bastert, J., Schaller, M., Korting, H.C. and Evans, E.G. (2001) Current and future approaches to antimycotic treatment in the era of resistant fungi and immunocompromised hosts. Int. J. Antimicrob. Agents 17, 81–91.

Bauer, B.E., Wolfger, H. and Kuchler, K. (1999) Inventory and function of yeast ABC proteins: about sex, stress, pleiotropic drug and heavy metal resistance. Biochim. Biophys. Acta 1461, 217–236.

Bissinger, P.H. and Kuchler, K. (1994) Molecular cloning and expression of the *Saccharomyces cerevisiae STS1* gene product. A yeast ABC transporter conferring mycotoxin resistance. J. Biol. Chem. 269, 4180–4186.

Bobrowicz, P., Wysocki, R., Owsianik, G., Goffeau, A. and Ulaszewski, S. (1997) Isolation of three contiguous genes, *ACR1*, *ACR2* and *ACR3*, involved in resistance to arsenic compounds in the yeast *Saccharomyces cerevisiae*. Yeast 13, 819–828.

Brandl, C.J., Martens, J.A., Margaliot, A., Stenning, D., Furlanetto, A.M., Saleh, A., Hamilton, K.S. and Genereaux, J. (1996) Structure/functional properties of the yeast dual regulator protein *NGG1* that are required for glucose repression. J. Biol. Chem. 271, 9298–9306.

Carvajal, E., van den Hazel, H.B., Cybularz-Kolaczkowska, A., Balzi, E. and Goffeau, A. (1997) Molecular and phenotypic characterization of yeast *PDR1* mutants that show hyperactive transcription of various ABC multidrug transporter genes. Mol. Gen. Genet. 256, 406–415.

Cauet, G., Degryse, E., Ledoux, C., Spagnoli, R. and Achstetter, T. (1999) Pregnenolone esterification in *Saccharomyces cerevisiae*. A potential detoxification mechanism. Eur. J. Biochem. 261, 317–324.

Chang, G. and Roth, C.B. (2001) Structure of MsbA from *E. coli*: a homolog of the multidrug resistance ATP binding cassette (ABC) transporters. Science 293, 1793–1800.

Conseil, G., Perez-Victoria, J.M., Jault, J.M., Gamarro, F., Goffeau, A., Hofmann, J. and Di Pietro, A. (2001) Protein kinase C effectors bind to multidrug ABC transporters and inhibit their activity. Biochemistry 40, 2564–2571.

Cui, Z., Hirata, D., Tsuchiya, E., Osada, H. and Miyakawa, T. (1996) The multidrug resistance-associated protein (MRP) subfamily (Yrs1/Yor1) of *Saccharomyces cerevisiae* is important for the tolerance to a broad range of organic anions. J. Biol. Chem. 271, 14712–14716.

Cui, Z., Shiraki, T., Hirata, D. and Miyakawa, T. (1998) Yeast gene *YRR1*, which is required for resistance to 4-nitroquinoline N-oxide, mediates transcriptional activation of the multidrug resistance transporter gene *SNQ2*. Mol. Microbiol. 29, 1307–1315.

Decottignies, A. and Goffeau, A. (1997) Complete inventory of the yeast ABC proteins. Nat. Genet. 15, 137–145.

Decottignies, A., Kolaczkowski, M., Balzi, E. and Goffeau, A. (1994) Solubilization and characterization of the overexpressed *PDR5* multidrug resistance nucleotide triphosphatase of yeast. J. Biol. Chem. 269, 12797–12803.

Decottignies, A., Lambert, L., Catty, P., Degand, H., Epping, E.A., Moye-Rowley, W.S., Balzi, E. and Goffeau, A. (1995) Identification and characterization of *SNQ2*, a new multidrug ATP binding cassette transporter of the yeast plasma membrane. J. Biol. Chem. 270, 18150–18157.

Decottignies, A., Grant, A.M., Nichols, J.W., de Wet, H., McIntosh, D.B. and Goffeau, A. (1998) ATPase and multidrug transport activities of the overexpressed yeast ABC protein Yor1p. J. Biol. Chem. 273, 12612–12622.

Decottignies, A., Owsianik, G. and Ghislain, M. (1999) Casein kinase I-dependent phosphorylation and stability of the yeast multidrug transporter Pdr5p. J. Biol. Chem. 274, 37139–37146.

Del Sorbo, G., Andrade, A.C., Van Nistelrooy, J.G., Van Kan, J.A., Balzi, E. and De Waard, M.A. (1997) Multidrug resistance in *Aspergillus nidulans* involves novel ATP-binding cassette transporters. Mol. Gen. Genet. 254, 417–426.

Delahodde, A., Delaveau, T. and Jacq, C. (1995) Positive autoregulation of the yeast transcription factor Pdr3p, which is involved in control of drug resistance. Mol. Cell. Biol. 15, 4043–4051.

Delahodde, A., Pandjaitan, R., Corral-Debrinski, M. and Jacq, C. (2001) Pse1/Kap121-dependent nuclear localization of the major yeast multidrug resistance (MDR) transcription factor Pdr1. Mol. Microbiol. 39, 304–312.

Delaveau, T., Delahodde, A., Carvajal, E., Subik, J. and Jacq, C. (1994) *PDR3*, a new yeast regulatory gene, is homologous to *PDR1* and controls the multidrug resistance phenomenon. Mol. Gen. Genet. 244, 501–511.

DeRisi, J., van den Hazel, B., Marc, P., Balzi, E., Brown, P., Jacq, C. and Goffeau, A. (2000) Genome microarray analysis of transcriptional activation in multidrug resistance yeast mutants. FEBS Lett. 470, 156–160.

Dogra, S., Krishnamurthy, S., Gupta, V., Dixit, B.L., Gupta, C.M., Sanglard, D. and Prasad, R. (1999) Asymmetric distribution of phosphatidylethanolamine in *C. albicans*: possible mediation by *CDR1*, a multidrug transporter belonging to ATP-binding cassette (ABC) superfamily. Yeast 15, 111–121.

Egner, R. and Kuchler, K. (1996) The yeast multidrug transporter Pdr5 of the plasma membrane is ubiquitinated prior to endocytosis and degradation in the vacuole. FEBS Lett. 378, 177–181.

Egner, R., Mahé, Y., Pandjaitan, R. and Kuchler, K. (1995) Endocytosis and vacuolar degradation of the plasma membrane-localized Pdr5 ATP-binding cassette multidrug transporter in *Saccharomyces cerevisiae*. Mol. Cell. Biol. 15, 5879–5887.

Egner, R., Rosenthal, F.E., Kralli, A., Sanglard, D. and Kuchler, K. (1998) Genetic separation of FK506 susceptibility and drug transport in the yeast Pdr5 ATP-binding cassette multidrug resistance transporter. Mol. Biol. Cell 9, 523–543.

Egner, R., Bauer, B.E. and Kuchler, K. (2000) The transmembrane domain 10 of the yeast Pdr5p ABC antifungal efflux pump determines both substrate specificity and inhibitor susceptibility. Mol. Microbiol. 35, 1255–1263.

Falcon-Perez, J.M., Mazon, M.J., Molano, J. and Eraso, P. (1999) Functional domain analysis of the yeast ABC transporter Ycf1p by site-directed mutagenesis. J. Biol. Chem. 274, 23584–23590.

Fewell, S.W., Travers, K.J., Weissman, J.S. and Brodsky, J.L. (2001) The action of molecular chaperones in the early secretory pathway. Annu. Rev. Genet. 35, 149–191.

Ghosh, M., Shen, J. and Rosen, B.P. (1999) Pathways of As(III) detoxification in *Saccharomyces cerevisiae*. Proc. Natl Acad. Sci. USA 96, 5001–5006.

Gottesman, M.M., Hrycyna, C.A., Schoenlein, P.V., Germann, U.A. and Pastan, I. (1995) Genetic analysis of the multidrug transporter. Annu. Rev. Genet. 29, 607–649.

Gounalaki, N. and Thireos, G. (1994) Yap1p, a yeast transcriptional activator that mediates multidrug resistance, regulates the metabolic stress response. EMBO J. 13, 4036–4041.

Grondin, K., Kundig, C., Roy, G. and Ouellette, M. (1998) Linear amplicons as precursors of amplified circles in methotrexate-resistant *Leishmania tarentolae*. Nucleic Acids Res. 26, 3372–3378.

Hallström, T.C. and Moye-Rowley, W.S. (2000a) Hyperactive forms of the Pdr1p transcription factor fail to respond to positive regulation by the Hsp70 protein Pdr13p. Mol. Microbiol. 36, 402–413.

Hallström, T.C. and Moye-Rowley, W.S. (2000b) Multiple signals from dysfunctional mitochondria activate the pleiotropic drug resistance pathway in *Saccharomyces cerevisiae*. J. Biol. Chem. 275, 37347–37356.

Hallström, T.C., Katzmann, D.J., Torres, R.J., Sharp, W.J. and Moye-Rowley, W.S. (1998) Regulation of transcription factor Pdr1p function by an Hsp70 protein in *Saccharomyces cerevisiae*. Mol. Cell. Biol. 18, 1147–1155.

Hallström, T.C., Lambert, L., Schorling, S., Balzi, E., Goffeau, A. and Moye-Rowley, W.S. (2001) Coordinate control of sphingolipid biosynthesis and multidrug resistance in *Saccharomyces cerevisiae*. J. Biol. Chem. 25, 25.

Hellauer, K., Rochon, M.H. and Turcotte, B. (1996) A novel DNA binding motif for yeast zinc cluster proteins: the Leu3p and Pdr3p transcriptional activators recognize everted repeats. Mol. Cell. Biol. 16, 6096–6102.

Hernaez, M.L., Gil, C., Pla, J. and Nombela, C. (1998) Induced expression of the *Candida albicans* multidrug resistance gene *CDR1* in response to fluconazole and other antifungals. Yeast 14, 517–526.

Hicke, L. (1997) Ubiquitin-dependent internalization and down-regulation of plasma membrane proteins. FASEB J. 11, 1215–1226.

Higgins, C.F. and Gottesman, M.M. (1992) Is the multidrug transporter a flippase? Trends Biochem. Sci. 17, 18–21.

Higgins, C.F. and Linton, K.J. (2001) Structural biology. The xyz of ABC transporters. Science 293, 1782–1784.

Hirata, D., Yano, K., Miyahara, K. and Miyakawa, T. (1994) *Saccharomyces cerevisiae YDR1*, which encodes a member of the ATP-binding cassette (ABC) superfamily, is required for multidrug resistance. Curr. Genet. 26, 285–294.

Holyoak, C.D., Bracey, D., Piper, P.W., Kuchler, K. and Coote, P.J. (1999) The *Saccharomyces cerevisiae* weak-acid-inducible ABC transporter Pdr12 transports fluorescein and preservative anions from the cytosol by an energy-dependent mechanism. J. Bacteriol. 181, 4644–4652.

Johnson, L.V., Walsh, M.L. and Chen, L.B. (1980) Localization of mitochondria in living cells with rhodamine 123. Proc. Natl Acad. Sci. USA 77, 990–994.

Kane, S.E. (1996) In: Multidrug Resistance of Cancer Cells. San Diego, CA: Academic Press.

Katzmann, D.J., Burnett, P.E., Golin, J., Mahé, Y. and Moye-Rowley, W.S. (1994) Transcriptional control of the yeast *PDR5* gene by the *PDR3* gene product. Mol. Cell. Biol. 14, 4653–4661.

Katzmann, D.J., Hallström, T.C., Voet, M., Wysock, W., Golin, J., Volckaert, G. and Moye-Rowley, W.S. (1995) Expression of an ATP-binding cassette transporter-encoding gene (*YOR1*) is required for oligomycin resistance in *Saccharomyces cerevisiae*. Mol. Cell. Biol. 15, 6875–6883.

Katzmann, D.J., Hallström, T.C., Mahé, Y. and Moye-Rowley, W.S. (1996) Multiple Pdr1p/Pdr3p binding sites are essential for normal expression of the ATP-binding cassette transporter protein-encoding gene *PDR5*. J. Biol. Chem. 271, 23049–23054.

Katzmann, D.J., Epping, E.A. and Moye-Rowley, W.S. (1999) Mutational disruption of plasma membrane trafficking of *Saccharomyces cerevisiae* Yor1p, a homologue of mammalian multidrug resistance protein. Mol. Cell. Biol. 19, 2998–3009.

Kean, L.S., Grant, A.M., Angeletti, C., Mahé, Y., Kuchler, K., Fuller, R.S. and Nichols, J.W. (1997) Plasma membrane translocation of fluorescent-labeled phosphatidylethanolamine is controlled by transcription regulators, *PDR1* and *PDR3*. J. Cell Biol. 138, 255–270.

Koch, A.L. (1999) Diffusion through agar blocks of finite dimensions: a theoretical analysis of three systems of practical significance in microbiology. Microbiology 145, 643–654.

Kolaczkowska, G.A. (1999) Regulation of pleiotropic drug resistance in yeast. Drug Resistance Updates 2, 403–414.

Kolaczkowski, M., van der Rest, M., Cybularz-Kolaczkowska, A., Soumillion, J.P., Konings, W.N. and Goffeau, A. (1996) Anticancer drugs, ionophoric peptides, and steroids as substrates of the yeast multidrug transporter Pdr5p. J. Biol. Chem. 271, 31543–31548.

Kolaczkowski, M., Kolaczowska, A., Luczynski, J., Witek, S. and Goffeau, A. (1998) *In vivo* characterization of the drug resistance profile of the major ABC transporters and other components of the yeast pleiotropic drug resistance network. Microb. Drug Resist. 4, 143–158.

Kölling, R. and Hollenberg, C.P. (1994) The ABC transporter Ste6 accumulates in the plasma membrane in a ubiquitinated form in endocytosis mutants. EMBO J. 13, 3261–3271.

Kölling, R. and Losko, S. (1997) The linker region of the ABC transporter Ste6 mediates ubiquitination and fast turnover of the protein. EMBO J. 16, 2251–2261.

Kralli, A. and Yamamoto, K.R. (1996) An FK506-sensitive transporter selectively decreases intracellular levels and potency of steroid hormones. J. Biol. Chem. 271, 17152–17156.

Kralli, A., Bohen, S.P. and Yamamoto, K.R. (1995) *LEM1*, an ATP-binding-cassette transporter, selectively modulates the biological potency of steroid hormones. Proc. Natl Acad. Sci. USA 92, 4701–4705.

Krishnamurthy, S., Chatterjee, U., Gupta, V., Prasad, R., Das, P., Snehlata, P. and Hasnain, S.E. (1998) Deletion of transmembrane domain 12 of *CDR1*, a multidrug transporter from *Candida albicans*, leads to altered drug specificity: expression of a yeast multidrug transporter in baculovirus expression system. Yeast 14, 535–550.

Kuchler, K. and Thorner, J. (1992) Functional expression of human mdr1 in the yeast *Saccharomyces cerevisiae*. Proc. Natl Acad. Sci. USA 89, 2302–2306.

Laney, J.D. and Hochstrasser, M. (1999) Substrate targeting in the ubiquitin system. Cell 97, 427–430.

Li, Z.S., Szczypka, M., Lu, Y.P., Thiele, D.J. and Rea, P.A. (1996) The yeast cadmium factor protein (*YCF1*) is a vacuolar glutathione S-conjugate pump. J. Biol. Chem. 271, 6509–6517.

Li, Z.S., Lu, Y.P., Zhen, R.G., Szczypka, M., Thiele, D.J. and Rea, P.A. (1997) A new pathway for vacuolar cadmium sequestration in *Saccharomyces cerevisiae*: *YCF1*-catalyzed transport of bis(glutathionato) cadmium. Proc. Natl Acad. Sci. USA 94, 42–47.

Litman, T., Druley, T.E., Stein, W.D. and Bates, S.E. (2001) From MDR to MXR: new understanding of multidrug resistance systems, their properties and clinical significance. Cell. Mol. Life Sci. 58, 931–959.

Loayza, D. and Michaelis, S. (1998) Role for the ubiquitin-proteasome system in the vacuolar degradation of Ste6p, the a-factor transporter in *Saccharomyces cerevisiae*. Mol. Cell. Biol. 18, 779–789.

Loayza, D., Tam, A., Schmidt, W.K. and Michaelis, S. (1998) Ste6p mutants defective in exit from the endoplasmic reticulum (ER) reveal aspects of an ER quality control pathway in *Saccharomyces cerevisiae*. Mol. Biol. Cell. 9, 2767–2784.

Lyons, C.N. and White, T.C. (2000) Transcriptional analyses of antifungal drug resistance in *Candida albicans*. Antimicrob. Agents Chemother. 44, 2296–2303.

Maesaki, S., Marichal, P., Vanden Bossche, H., Sanglard, D. and Kohno, S. (1999) Rhodamine 6G efflux for the detection of *CDR1*-overexpressing azole-resistant *Candida albicans* strains. J. Antimicrob. Chemother. 44, 27–31.

Mahé, Y., Lemoine, Y. and Kuchler, K. (1996a) The ATP binding cassette transporters Pdr5 and Snq2 of *Saccharomyces cerevisiae* can mediate transport of steroids *in vivo*. J. Biol. Chem. 271, 25167–25172.

Mahé, Y., Parle-McDermott, A., Nourani, A., Delahodde, A., Lamprecht, A. and Kuchler, K. (1996b) The ATP-binding cassette multidrug transporter Snq2 of *Saccharomyces cerevisiae*: a novel target for the transcription factors Pdr1 and Pdr3. Mol. Microbiol. 20, 109–117.

Marichal, P., Vanden Bossche, H., Odds, F.C., Nobels, G., Warnock, D.W., Timmerman, V., Van Broeckhoven, C., Fay, S. and Mose-Larsen, P. (1997) Molecular biological characterization of an azole-resistant *Candida glabrata* isolate. Antimicrob. Agents Chemother. 41, 2229–2237.

Martens, J.A., Genereaux, J., Saleh, A. and Brandl, C.J. (1996) Transcriptional activation by yeast Pdr1p is inhibited by its association with *NGG1/ADA3*. J. Biol. Chem. 271, 15884–15890.

Meyers, S., Schauer, W., Balzi, E., Wagner, M., Goffeau, A. and Golin, J. (1992) Interaction of the yeast pleiotropic drug resistance genes *PDR1* and *PDR5*. Curr. Genet. 21, 431–436.

Mitterbauer, R., Karl, T., Lemmens, M., Kuchler, K., and Adam G. (2000) Resistance to mycotoxins: a role for ABC transporter proteins in plant–pathogen interactions. In: Biology of Plant–Microbe Interactions (ed. P.J.G.M. De Wit, T. Bisseling and W.J. Stiekema), Vol. 2, pp. 352–355. IS-MPMI: St. Paul.

Miyahara, K., Mizunuma, M., Hirata, D., Tsuchiya, E. and Miyakawa, T. (1996) The involvement of the *Saccharomyces cerevisiae* multidrug resistance transporters Pdr5p and Snq2p in cation resistance. FEBS Lett. 399, 317–320.

Miyazaki, H., Miyazaki, Y., Geber, A., Parkinson, T., Hitchcock, C., Falconer, D.J., Ward, D.J., Marsden, K. and Bennett, J.E. (1998) Fluconazole resistance associated with drug efflux and increased transcription of a drug transporter gene, *PDH1*, in *Candida glabrata*. Antimicrob. Agents Chemother. 42, 1695–1701.

Moran, G.P., Sanglard, D., Donnelly, S.M., Shanley, D.B., Sullivan, D.J. and Coleman, D.C. (1998) Identification and expression of multidrug transporters responsible for fluconazole resistance in *Candida dubliniensis*. Antimicrob. Agents Chemother. 42, 1819–1830.

Nagao, K., Taguchi, Y., Arioka, M., Kadokura, H., Takatsuki, A., Yoda, K. and Yamasaki, M. (1995) *bfr1+*, a novel gene of *Schizosaccharomyces pombe* which confers brefeldin A resistance, is structurally related to the ATP-binding cassette superfamily. J. Bacteriol. 177, 1536–1543.

Nakamura, K., Niimi, M., Niimi, K., Holmes, A.R., Yates, J.E., Decottignies, A., Monk, B.C., Goffeau, A. and Cannon, R.D. (2001) Functional expression of *Candida albicans* drug efflux pump Cdr1p in a *Saccharomyces cerevisiae* strain deficient in membrane transporters. Antimicrob. Agents Chemother. 45, 3366–3374.

Nishi, K., Yoshida, M., Nishimura, M., Nishikawa, M., Nishiyama, M., Horinouchi, S. and Beppu, T. (1992) A leptomycin B resistance gene of *Schizosaccharomyces pombe* encodes a protein similar to the mammalian P-glycoproteins. Mol. Microbiol. 6, 761–769.

Nourani, A., Papajova, D., Delahodde, A., Jacq, C. and Subik, J. (1997a) Clustered amino acid substitutions in the yeast transcription regulator Pdr3p increase pleiotropic drug resistance and identify a new central regulatory domain. Mol. Gen. Genet. 256, 397–405.

Nourani, A., Wesolowski-Louvel, M., Delaveau, T., Jacq, C. and Delahodde, A. (1997b) Multiple-drug-resistance phenomenon in the yeast *Saccharomyces cerevisiae*: involvement of two hexose transporters. Mol. Cell. Biol. 17, 5453–5460.

Ortiz, D.F., Kreppel, L., Speiser, D.M., Scheel, G., McDonald, G. and Ow, D.W. (1992) Heavy metal tolerance in the fission yeast requires an ATP-binding cassette-type vacuolar membrane transporter. EMBO J. 11, 3491–3499.

Ortiz, D.F., Ruscitti, T., McCue, K.F. and Ow, D.W. (1995) Transport of metal-binding peptides by *HMT1*, a fission yeast ABC-type vacuolar membrane protein. J. Biol. Chem. 270, 4721–4728.

Ouellette, M. and Borst, P. (1991) Drug resistance and P-glycoprotein gene amplification in the protozoan parasite *Leishmania*. Res. Microbiol. 142, 737–746.

Piper, P., Mahé, Y., Thompson, S., Pandjaitan, R., Holyoak, C., Egner, R., Mühlbauer, M., Coote, P. and Kuchler, K. (1998) The Pdr12 ABC transporter is required for the development of weak organic acid resistance in yeast. EMBO J. 17, 4257–4265.

Plemper, R.K. and Wolf, D.H. (1999) Endoplasmic reticulum degradation. Reverse protein transport and its end in the proteasome. Mol. Biol. Rep. 26, 125–130.

Plemper, R.K., Egner, R., Kuchler, K. and Wolf, D.H. (1998) Endoplasmic reticulum degradation of a mutated ATP-binding cassette transporter Pdr5 proceeds in a concerted action of Sec61 and the proteasome. J. Biol. Chem. 273, 32848–32856.

Prasad, R., De Wergifosse, P., Goffeau, A. and Balzi, E. (1995) Molecular cloning and characterization of a novel gene of *Candida albicans, CDR1*, conferring multiple resistance to drugs and antifungals. Curr. Genet. 27, 320–329.

Rauser, W.E. (1990) Phytochelatins. Annu. Rev. Biochem. 59, 61–86.

Raymond, M., Gros, P., Whiteway, M. and Thomas, D.Y. (1992) Functional complementation of yeast *ste6* by a mammalian multidrug resistance mdr gene. Science 256, 232–234.

Rebbeor, J.F., Connolly, G.C., Dumont, M.E. and Ballatori, N. (1998) ATP-dependent transport of reduced glutathione on *YCF1*, the yeast orthologue of mammalian multidrug resistance associated proteins. J. Biol. Chem. 273, 33449–33454.

Reid, R.J., Kauh, E.A. and Bjornsti, M.A. (1997) Camptothecin sensitivity is mediated by the pleiotropic drug resistance network in yeast. J. Biol. Chem. 272, 12091–12099.

Rosenberg, M.F., Mao, Q., Holzenburg, A., Ford, R.C., Deeley, R.G. and Cole, S.P. (2001a) The structure of the multidrug resistance protein 1 (MRP1/ABCC1) crystallization and single-particle analysis. J. Biol. Chem. 276, 16076–16082.

Rosenberg, M.F., Velarde, G., Ford, R.C., Martin, C., Berridge, G., Kerr, I.D., *et al.* (2001b) Repacking of the transmembrane domains of P-glycoprotein during the transport ATPase cycle. EMBO J. 20, 5615–5625.

Ruetz, S., Raymond, M. and Gros, P. (1993) Functional expression of P-glycoprotein encoded by the mouse *mdr3* gene in yeast cells. Proc. Natl Acad. Sci. USA 90, 11588–11592.

Ryder, N.S. and Leitner, I. (2001) Synergistic interaction of terbinafine with triazoles or amphotericin B against *Aspergillus* species. Med. Mycol. 39, 91–95.

Saleh, A., Lang, V., Cook, R. and Brandl, C.J. (1997) Identification of native complexes containing the yeast coactivator/repressor proteins *NGG1/ADA3* and *ADA2*. J. Biol. Chem. 272, 5571–5578.

Sanglard, D., Kuchler, K., Ischer, F., Pagani, J.L., Monod, M. and Bille, J. (1995) Mechanisms of resistance to azole antifungal agents in *Candida albicans* isolates from AIDS patients involve specific multidrug transporters. Antimicrob. Agents Chemother. 39, 2378–2386.

Sanglard, D., Ischer, F., Monod, M. and Bille, J. (1996) Susceptibilities of *Candida albicans* multidrug transporter mutants to various antifungal agents and other metabolic inhibitors. Antimicrob. Agents Chemother. 40, 2300–2305.

Sanglard, D., Ischer, F., Monod, M. and Bille, J. (1997) Cloning of *Candida albicans* genes conferring resistance to azole antifungal agents: characterization of *CDR2*, a new multidrug ABC transporter gene. Microbiology 143, 405–416.

Sanglard, D., Ischer, F., Calabrese, D., Majcherczyk, P.A. and Bille, J. (1999) The ATP-binding cassette transporter gene *CgCDR1* from *Candida glabrata* is involved in the resistance of clinical isolates to azole antifungal agents. Antimicrob. Agents Chemother. 43, 2753–2765.

Schnell, N. and Entian, K.D. (1991) Identification and characterization of a *Saccharomyces cerevisiae* gene (*PAR1*) conferring resistance to iron chelators. Eur. J. Biochem. 200, 487–493.

Senior, A.E., al-Shawi, M.K. and Urbatsch, I.L. (1995) The catalytic cycle of P-glycoprotein. FEBS Lett. 377, 285–289.

Servos, J., Haase, E. and Brendel, M. (1993) Gene *SNQ2* of *Saccharomyces cerevisiae*, which confers resistance to 4-nitroquinoline-N-oxide and other chemicals, encodes a 169 kDa protein homologous to ATP-dependent permeases. Mol. Gen. Genet. 236, 214–218.

Simonics, T., Kozovska, Z., Michalkova-Papajova, D., Delahodde, A., Jacq, C. and Subik, J. (2000) Isolation and molecular characterization of the carboxy-terminal *pdr3* mutants in *Saccharomyces cerevisiae.* Curr. Genet. 38, 248–255.

Szczypka, M.S., Wemmie, J.A., Moye-Rowley, W.S. and Thiele, D.J. (1994) A yeast metal resistance protein similar to human cystic fibrosis transmembrane conductance regulator (CFTR) and multidrug resistance-associated protein. J. Biol. Chem. 269, 22853–22857.

Taglicht, D. and Michaelis, S. (1998) *Saccharomyces cerevisiae* ABC proteins and their relevance to human health and disease. Methods Enzymol. 292, 130–162.

Talibi, D. and Raymond, M. (1999) Isolation of a putative *Candida albicans* transcriptional regulator involved in pleiotropic drug resistance by functional complementation of a *pdr1 pdr3* mutation in *Saccharomyces cerevisiae*. J. Bacteriol. 181, 231–240.

Tobin, M.B., Peery, R.B. and Skatrud, P.L. (1997) Genes encoding multiple drug resistance-like proteins in *Aspergillus fumigatus* and *Aspergillus flavus*. Gene 200, 11–23.

Tommasini, R., Evers, R., Vogt, E., Mornet, C., Zaman, G.J., Schinkel, A.H., Borst, P. and Martinoia, E. (1996) The human multidrug resistance-associated protein functionally complements the yeast cadmium resistance factor 1. Proc. Natl Acad. Sci. USA 93, 6743–6748.

Toone, W.M., Kuge, S., Samuels, M., Morgan, B.A., Toda, T. and Jones, N. (1998) Regulation of the fission yeast transcription factor Pap1 by oxidative stress: requirement for the nuclear export factor *Crm1* (Exportin) and the stress-activated MAP kinase *Sty1/Spc1*. Genes Dev. 12, 1453–1463.

Tran, D.Q., Jin, L., Chen, J., McLachlan, J.A. and Arnold, S.F. (1997) Evaluation of clinical and environmental anti-estrogens with human estrogen receptor expressed in *Saccharomyces cerevisiae*: a novel role for ABC-cassette transporters in mediating anti-estrogenic activity. Biochem. Biophys. Res. Commun. 235, 669–674.

Turi, T.G. and Rose, J.K. (1995) Characterization of a novel *Schizosaccharomyces pombe* multidrug resistance transporter conferring brefeldin A resistance. Biochem. Biophys. Res. Commun. 213, 410–418.

Tusnady, G.E., Bakos, E., Varadi, A. and Sarkadi, B. (1997) Membrane topology distinguishes a subfamily of the ATP-binding cassette (ABC) transporters. FEBS Lett. 402, 1–3.

Ueda, K., Shimabuku, A.M., Konishi, H., Fujii, Y., Takebe, S., Nishi, K., Yoshida, M., Beppu, T. and Komano, T. (1993) Functional expression of human P-glycoprotein in *Schizosaccharomyces pombe.* FEBS Lett. 330, 279–282.

Urban, M., Bhargava, T. and Hamer, J.E. (1999) An ATP-driven efflux pump is a novel pathogenicity factor in rice blast disease. EMBO J. 18, 512–521.

van Veen, H.W., Margolles, A., Muller, M., Higgins, C.F. and Konings, W.N. (2000) The homodimeric ATP-binding cassette transporter LmrA mediates multidrug transport by an alternating two-site (two-cylinder engine) mechanism. EMBO J. 19, 2503–2514.

Vanden Bossche, H., Marichal, P. and Odds, F.C. (1994) Molecular mechanisms of drug resistance in fungi. Trends Microbiol. 2, 393–400.

Velarde, G., Ford, R.C., Rosenberg, M.F. and Powis, S.J. (2001) Three-dimensional structure of transporter associated with antigen processing (TAP) obtained by single particle image analysis. J. Biol. Chem. 276, 46054–46063.

Volkman, S.K., Cowman, A.F. and Wirth, D.F. (1995) Functional complementation of the *ste6* gene of *Saccharomyces cerevisiae* with the *pfmdr1* gene of *Plasmodium falciparum.* Proc. Natl Acad. Sci. USA 92, 8921–8925.

Wemmie, J.A. and Moye-Rowley, W.S. (1997) Mutational analysis of the *Saccharomyces cerevisiae* ATP-binding cassette transporter protein Ycf1p. Mol. Microbiol. 25, 683–694.

Wemmie, J.A., Wu, A.L., Harshman, K.D., Parker, C.S. and Moye-Rowley, W.S. (1994)

Transcriptional activation mediated by the yeast AP-1 protein is required for normal cadmium tolerance. J. Biol. Chem. 269, 14690–14697.

Wendler, F., Bergler, H., Prutej, K., Jungwirth, H., Zisser, G., Kuchler, K. and Högenauer, G. (1997) Diazaborine resistance in the yeast *Saccharomyces cerevisiae* reveals a link between *YAP1* and the pleiotropic drug resistance genes *PDR1* and *PDR3*. J. Biol. Chem. 272, 27091–27098.

White, T.C., Marr, K.A. and Bowden, R.A. (1998) Clinical, cellular, and molecular factors that contribute to antifungal drug resistance. Clin. Microbiol. Rev. 11, 382–402.

Wolfger, H., Mahé, Y., Parle-McDermott, A., Delahodde, A. and Kuchler, K. (1997) The yeast ATP-binding cassette (ABC) protein genes *PDR10* and *PDR15* are novel targets for the Pdr1 and Pdr3 transcriptional regulators. FEBS Lett. 418, 269–274.

Wolfger, H., Mamnun, Y.M. and Kuchler, K. (2001) Fungal ABC proteins: pleiotropic drug resistance, stress response and cellular detoxification. Res. Microbiol. 152, 375–389.

Wu, A., Wemmie, J.A., Edgington, N.P., Goebl, M., Guevara, J.L. and Moye-Rowley, W.S. (1993) Yeast bZip proteins mediate pleiotropic drug and metal resistance. J. Biol. Chem. 268, 18850–18858.

Zhang, X., Collins, K.I. and Greenberger, L.M. (1995) Functional evidence that transmembrane 12 and the loop between transmembrane 11 and 12 form part of the drug-binding domain in P-glycoprotein encoded by MDR1. J. Biol. Chem. 270, 5441–5448.

Zhang, X., Cui, Z., Miyakawa, T. and Moye-Rowley, W.S. (2001) Cross-talk between transcriptional regulators of multidrug resistance in *Saccharomyces cerevisiae.* J. Biol. Chem. 276, 8812–8819.

Zwiers, L.H. and De Waard, M.A. (2000) Characterization of the ABC transporter genes MgAtr1 and MgAtr2 from the wheat pathogen *Mycosphaerella graminicola*. Fungal Genet. Biol. 30, 115–125.

CHAPTER

16

Drug Resistance Mediated by ABC Transporters in Parasites of Humans

Marc Ouellette and Danielle Légaré

Introduction

Parasitic protozoa are responsible for some of the most devastating and prevalent human diseases and threaten the lives of more than a third of the worldwide population. Despite intensive attempts, there are no effective vaccines for the prevention of parasitic diseases. When simple prevention measures such as impregnated bednets fail or prove impractical, drugs are required for the treatment of infections that are otherwise often fatal. However, the available arsenal of antiprotozoal drugs is limited and often relies on antiquated drugs such as arsenicals for the treatment of African trypanosomiasis or antimonials for the treatment of leishmaniasis. The treatment of parasitic diseases is further complicated by the emergence of drug resistance, and several parasitic diseases including malaria and leishmaniasis were included in the World Health Organization's infamous list of the top guns of antimicrobial resistance (www.who.int/infectious-disease-report/2000/ch4.htm). With effective vaccines not yet in sight and the development of new drugs proceeding slowly, the emergence of drug resistance in parasitic protozoa is becoming a public health problem. Several mechanisms of resistance have been described in protozoa including transport-related mechanisms (reduced uptake, increased efflux, or sequestration) (see Ouellette (2001) for a recent review).

The genome sequencing of at least 10 protozoan parasites is underway (www.ebi.ac.uk/parasites; www.tigr.org) and numerous ABC transporters are being revealed. Extensive work has been carried out for only a small subset of these ABC transporters and convincing evidence is available linking some of these ABC transporters to drug resistance in parasitic protozoa. In this chapter we first describe the distribution and structural properties of the known ABC transporters in parasitic protozoa, and then provide an up-to-date summary of the known function of ABC proteins with respect to antiparasitic resistance and its clinical relevance. Where data are available the physiological function of parasite ABC proteins will be discussed.

Occurrence of ABC Transporters in Parasites

ABC transporters are ubiquitous in all organisms sequenced ranging from 11 in *Mycoplasma genitalium* to 78 in *Bacillus subtilis* (Quentin *et al.*, 1999). The increasing number of ABC transporters – more than 2000 ABC ATPase domains are now in public data banks (Dassa and Bouige, 2001) – has led to their classification according to structure and function (Dassa and Bouige, 2001; Quentin and Fichant, 2000) and several useful websites are available (e.g. http://ir2lcb.cnrs-mrs.fr/

ABC Proteins: From Bacteria to Man
ISBN 0-12-352551-9

ABCdb/presentation.html; http://nutrigene.4t.com/humanabc.htm; http://www.pasteur.fr/recherche/unites/pmtg/abc/database.html) describing the phylogeny of ABC transporters. The number of ABC transporters in parasites is increasing in parallel with their genomes being sequenced and we have attempted to provide a complete overview of the presently known parasitic ABC transporters. Since no parasite genome is complete, it is difficult to give a precise estimate of the frequency of the occurrence of ABC genes in parasite genomes. However, from parasites with close to 50% of their genome sequenced, we can extrapolate the number of ABC transporters to between 15 and 35 per genome, a number similar to that found in yeast. To date the parasites with the most full-length sequenced ABC genes are *Leishmania* with nine different genes and *Plasmodium* with five **(Table 16.1)**. The same gene may have been sequenced in several species of the same genus. The properties of these transporters were studied using prediction algorithms and ABC transporters were found with different topologies and belonging to several of the major families of ABC transporters **(Table 16.1)**. The genome sequence survey of several parasites clearly showed that several parasite ABC proteins are still to be discovered and fully characterized. In the next section we will discuss the various

TABLE 16.1. ABC TRANSPORTERS IN PROTOZOAN PARASITES

	Transporter	Accession no.	Species	No. of aa	Topology[d]	Family[f]	Subfamily[f]
Leishmania							
	MDR1	L08091	*L. enriettii*	1280	(TMD-ABC)2	DPL	Pgp
	MDR1	U63320	*L. tropica*	1341	(TMD-ABC)2	DPL	Pgp
	MDR1	L01572	*L. donovani*	1341	(TMD-ABC)2	DPL	Pgp
	MDR1	AB003329	*L. amazonensis*	1341	(TMD-ABC)2	DPL	Pgp
	ABC1	AF200948	*L. tropica*	1843	(TMD-ABC)2	DPL	Pgp
	L4468.01	AL121864[a]	*L. major*	1241	ABC-TMD	EPD	White
	ABCTP1	AC005766[b]	*L. major*	724	(ABC)2	ART	REG
	PGPA	X17154	*L. tarentolae*	1548	(TMD-ABC)2	OAD	MRP
	PGPB	L29484	*L. tarentolae*	1513	(TMD-ABC)2	OAD	MRP
	PGPC	–	*L. tarentolae*	–	(TMD-ABC)2[e]	OAD	MRP
	PGPD	–	*L. tarentolae*	–	(TMD-ABC)2[e]	OAD	MRP
	PGPE	L29485	*L. tarentolae*	1724	(TMD-ABC)2	OAD	MRP
	PGPE	U55381	*L. tropica*	1677	(TMD-ABC)2	OAD	MRP
	L673.01	AL135898[a]	*L. major*	1571	(TMD-ABC)2	OAD	MRP
	L673.02	AL135898[a]	*L. major*	1824	(TMD-ABC)2	OAD	MRP
	L8329.03	AL446004[a]	*L. major*	659	(ABC)2	RLI	–
Plasmodium							
	PfMDR1	M29154	*P. falciparum*	1419	(TMD-ABC)2	DPL	Pgp
	MAL3P1.7[c]	Z97348	*P. falciparum*	1365	(TMD-ABC)2	DPL	Pgp
	PfMDR2	U04640	*P. falciparum*	1025	TMD-ABC	DPL	HMT
	MAL1P3.03[c]	AL031746	*P. falciparum*	1822	(TMD-ABC)2	OAD	MRP
	GCN20	U37225	*P. falciparum*	816	(ABC)2	ART	REG

[a]Zimmermann, W., Wambutt, R., Ivens, A.C., Murphy, L., Quail, M., Rajandream, M.A. and Barrell, B.G. European *Leishmania major* Friedlin genome sequencing project, Sanger Centre, The Wellcome Trust Genome Campus, http://www.sanger.ac.uk/Projects/L_major/.
[b]Myler, P.J., Sisk, E., Hixson, G., Kiser, P., Rickel, E., Hassebrock, M., Cawthra, J., Marsolini, F., Sunkin, S. and Stuart, K.D. Seattle Biomedical Research Institution, 4 Nickerson Street, Seattle, WA 98109-1651, USA.
[c]The *Plasmodium* Genome Database Collaborative. 2001. (PlasmoDB, 2001).
[d]Transmembrane spans and therefore transmembrane domains were predicted using SOSUI (http://sosui.proteome.bio.tuat.ac.jp/) and TMPRED (http://www.ch.embnet.org/software/) algorithms.
[e]Deduced from hybridization experiments.
[f]http://www.pasteur.fr/recherche/unites/pmtg/abc/species.html; see Chapter 1.
Abbreviations: DPL, drugs, peptides, lipids; EPD, eye pigment precursors and drugs; ART, antibiotic resistance and translation regulation; OAD, organic anion conjugates, anions, drugs; RLI, Rnase L inhibitor; Pgp, Eukaryote multiple drug resistance and lipid export; White, eye pigment precursors and drugs; REG, translation regulation; MRP, conjugate drug exporters; HMT, mitochondrial and bacterial transporters II.

ABC transporters for which there is experimental evidence for a cellular function with particular emphasis on their contribution to drug resistance.

Role of Parasitic ABC Transporters Including Their Involvement in Drug Resistance

ABC transporters in *Plasmodium*

Malaria is the most widespread protozoan parasitic disease and *Plasmodium falciparum*, the etiological agent of the most severe form of malaria, is often resistant to most commonly used antimalarials (White, 1998). Chloroquine (CQ) has long been the drug of choice in the treatment of malaria. Since 1959, when it was first described, resistance to CQ has steadily increased and is now widespread. Chloroquine acts by inhibiting polymerization of the toxic heme that is released during hemoglobin degradation within the digestive vacuole of the parasite (Slater and Cerami, 1992; Sullivan *et al.*, 1996). Active efflux of the drug has long been thought to be the mechanism of resistance (Krogstad *et al.*, 1987) and the demonstration that CQ resistance could be reversed by verapamil (Martin *et al.*, 1987), a phenotype reminiscent of the multidrug resistance phenotype of mammalian cells, has led to the search for a malaria P-glycoprotein homologue by DNA hybridization and polymerase chain reaction (PCR) strategies. A number of ABC transporter genes were isolated and amplification or overexpression of a gene called *pfmdr1* was observed in CQ- or mefloquine-resistant isolates (Foote *et al.*, 1989; Wilson *et al.*, 1989).

The gene *pfmdr1*

The gene *pfmdr1* codes for a protein Pgh1 that is structurally similar to P-glycoproteins **(Table 16.1)**, and Pgh1 was initially proposed to correspond to an efflux pump. This hypothesis received some support from the preferential association of CQ resistance with specific point mutations in Pgh1 (Foote *et al.*, 1990). This was not supported, however, by a genetic cross indicating that the main CQ resistance gene was on chromosome 7 while *pfmdr1* is on chromosome 5 (Wellems *et al.*, 1990, 1991). Moreover, Pgh1 is located in the digestive vacuole of the parasite and its topology would suggest that it transports molecules into the vacuole (Cowman *et al.*, 1991), the site of action of CQ **(Figure 16.1)**. The gene present on chromosome 7 named *pfcrt* was isolated recently and found to be a transmembrane protein that localizes, similarly to Pgh1, to the parasite digestive vacuole (Fidock *et al.*, 2000). Epidemiological studies have found a strong link between mutations in *pfcrt* and CQ resistance in *P. falciparum* (Djimde *et al.*, 2001; Durand *et al.*, 2001) but not in other malaria species (Nomura *et al.*, 2001).

Considerable (and controversial) work has revolved around the issue of CQ resistance and the role played by Pgh1. An inverse correlation was found between the *pfmdr1* copy number and CQ resistance. Indeed, *in vitro* studies indicated that when cells in which *pfmdr1* was amplified were selected for higher CQ resistance, deamplification of *pfmdr1* resulted. This deamplification of *pfmdr1* is associated with collateral sensitivity to mefloquine and halofantrine (Barnes *et al.*, 1992). Conversely, selection for increased mefloquine resistance *in vitro* will lead to an increased copy number of *pfmdr1* and increased collateral sensitivity to CQ (Cowman *et al.*, 1994). The role of Pgh1 in resistance was established recently by gene transfection studies. A number of mutations, S1034C, N1042D, D1246Y, in Pgh1 were known to correlate with CQ resistance (Foote *et al.*, 1990). Allelic exchange at the endogenous *pfmdr1* locus demonstrated that mutations at position 1034, 1042 and 1246 can lead to quinine resistance in various cell backgrounds and also to CQ resistance, although the latter depends on the strain background and other mutated proteins (Reed *et al.*, 2000), possibly PfCRT. The introduction of mutations by allelic replacement in *pfmdr1* will lead to mefloquine and halofantrine sensitivity (Reed *et al.*, 2000) and this is consistent with the result of a genetic cross associating mutations in the *pfmdr1* gene with increased sensitivity to mefloquine (Duraisingh *et al.*, 2000). Thus wild-type *pfmdr1* is a CQ sensitivity gene and a mefloquine resistance gene while point mutations are associated with less susceptibility to CQ and more susceptibility to mefloquine.

The mechanism(s) by which Pgh1 confers resistance and verapamil reverts CQ resistance

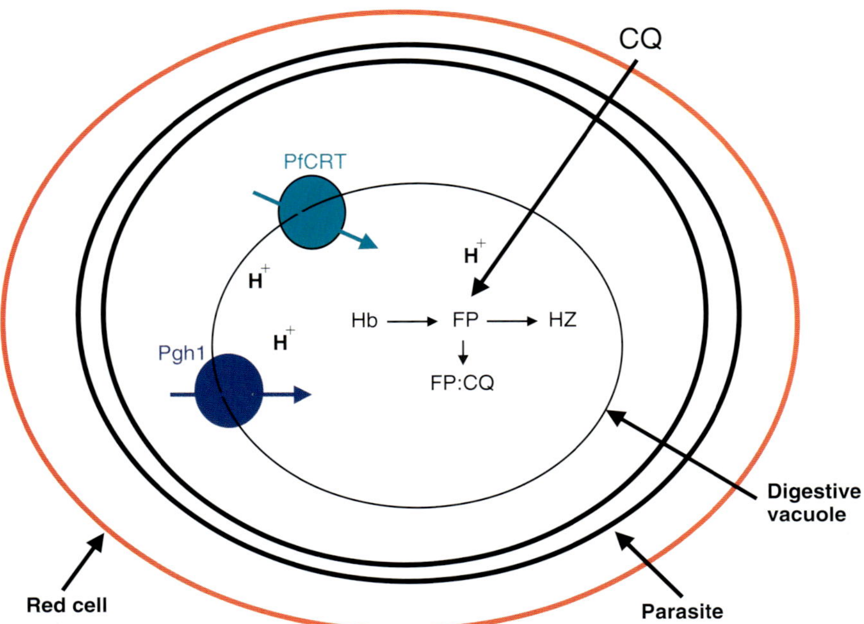

Figure 16.1. ***Possible mechanism of chloroquine resistance in the malaria parasite. Chloroquine (CQ) is a weak base that possibly penetrates by diffusion and is trapped in the acidic digestive vacuole down the pH gradient. Hemoglobin (Hb) breakdown will ultimately lead to globin fragments and to the cellular toxic ferriprotoporphyrin IX (FP). The latter is polymerized to the insoluble polymer hemazoin (HZ). CQ can interact with FP to prevent its detoxification by polymerization. At least two vacuolar membrane proteins are involved in CQ resistance: the ABC protein Pgh1 and the protein PfCRT. Point mutations in these two proteins are correlated with CQ resistance. It is not clear if either of these two proteins transport CQ directly or if these proteins can modulate the vacuolar pH, which in turn will modulate CQ uptake.***

are still unclear. Heterologous transfection of *pfmdr1* in CHO cells indicated that Pgh1 can affect the pH of the lysosomal compartment (van Es *et al.*, 1994), suggesting that mutations in the Pgh1 protein modulate the pH of the digestive vacuole of the parasite and affect the accumulation of antimalarials. Similarly PfCRT is capable of modulating the pH of the digestive vacuole, which may thus confer resistance by altering CQ transport or binding to hemazoin (Fidock *et al.*, 2000). Indeed, decreased accumulation of CQ in resistant parasites was proposed to be due to altered CQ-hemazoin binding parameters (Bray *et al.*, 1998).

The ability of verapamil to reverse CQ resistance was first thought to be due to inhibition of CQ efflux (Krogstad *et al.*, 1987). It is possible, however, that verapamil by either direct or indirect means alters the pH of the vacuole, which alters the ability of CQ to interact with hemazoin (Bray *et al.*, 1998). Interestingly, the food vacuole pH appears to be more acidic in CQ-resistant parasites (Dzekunov *et al.*, 2000; Ursos *et al.*, 2000). Thus resistance to CQ is a complex matter with several proteins involved, including Pgh1 and PfCRT, and changes in pH appear to be key to the modulation of the accumulation of CQ **(Figure 16.1)**. The exact role of Pgh1 in modulating the pH still needs to be defined. One possibility is that its expression will modulate the activity of nearby transporters that will also influence the pH of the vacuole. ABC transporters are well known to modulate the activity of a number of nearby channels or transporters (Higgins, 1995).

The gene *pfmdr2*

The gene *pfmdr2*, which was isolated soon after *pfmdr1*, has a single ATP-binding domain with 10 predicted transmembrane segments and is expressed in a stage-specific manner (Zalis *et al.*, 1993). It is possibly located at the level of the plasma membrane (Rubio and Cowman, 1994). It is related to the fission yeast HMT-1, an ABC transporter that mediates tolerance to cadmium by sequestering the metal conjugated to phytochelatins into the vacuole (Ortiz *et al.*, 1995). Although *pfmdr2* transcripts were found overexpressed in some CQ-resistant parasites (Ekong *et al.*, 1993), it is generally agreed that *pfmdr2* is not implicated in CQ resistance (Rubio and Cowman, 1994; Zalis *et al.*, 1993).

PfGCN20

Considerable work has been done on a third malaria ABC transporter named PfGCN20. This malaria protein was found to be similar to the yeast Gcn20p, which is part of the yeast translation regulatory pathway. This protein is localized to the cytosol of the parasite and in various membranous and non-membranous compartments in the infected erythrocyte (Bozdech *et al.*, 1998). PfGCN20 can complement a yeast GCN20 mutant, suggesting that it may be involved in plasmodial translation regulation. Its localization also suggests that it may act as a molecular chaperone contributing to protein translocation across multiple membranes in infected erythrocytes (Bozdech and Schurr, 1999). Another example of a parasite-encoded protein localized at the parasite–host interface is the *Cryptosporidium parvum* CpABC protein (Perkins *et al.*, 1999). This protein is located at the feeder organelle, the major host–parasite boundary. The CpABC has significant sequence and structural similarities with the MRP subfamily of ABC proteins. Its homology to MRP may suggest that it could be capable of transporting large organic anions and may function as a transporter of endogenous or xenobiotic conjugates. *C. parvum* is intrinsically resistant to several antimicrobial agents and it was proposed that this ABC transporter could contribute to this intrinsic resistance (Perkins *et al.*, 1999). Interestingly, cyclosporin analogues, which bind the mammalian ABC transporters, were shown to be effective against experimental *C. parvum* infection (Perkins *et al.*, 1998).

Other ABC transporters in *Plasmodium*

The analysis of the ongoing *Plasmodium* genome project has revealed two additional full-length ABC transporter genes **(Table 16.1)** in addition to *pfmdr1*, *pfmdr2* and *pfGCN20*. However, the data related to the function of most of these ABC proteins, either in drug resistance or in other functions, are unavailable. Sequence comparison is suggesting that one of these additional ABC transporters is part of the P-glycoprotein gene family while the other one could be part of the organic conjugate pumps of the MRP type **(Table 16.1)**.

ABC transporters in *Leishmania*

Leishmania are intracellular protozoan parasites that cause a wide spectrum of diseases ranging from self-healing cutaneous lesions to visceral infections that can be fatal. It is estimated that there are over 2 million new cases of leishmaniasis each year in 88 countries (Herwaldt, 1999). The first therapeutic choices are in the pentavalent antimony-containing compounds (SbV) sodium stibogluconate (Pentostam) or *N*-methylglucamine (Glucantime) (Berman, 1997; Herwaldt, 1999). The mechanism of action of antimonials is unknown. Cases refractory to treatment were described more than 40 years ago but more recently the incidence of antimony-resistant parasites has increased significantly (Faraut-Gambarelli *et al.*, 1997; Lira *et al.*, 1999; Sundar *et al.*, 2000). The underlying mechanisms that contribute to drug resistance in field isolates are poorly understood but *in vitro* work incriminates ABC proteins.

In vitro metal-resistant *Leishmania* and the ABC transporter PGPA

Analysis of *Leishmania* antimony-resistant mutants indicated that resistance to metals is multifactorial and consistent with the step-by-step mode of selection for mutants. The resistance model is illustrated in **Figure 16.2**. We found that trypanothione is increased in metal-resistant *Leishmania* (Haimeur *et al.*, 2000; Légaré *et al.*, 1997; Mukhopadhyay *et al.*, 1996). Trypanothione (TSH) is the major reduced thiol in *Leishmania* and is composed of a bisglutathione–spermidine conjugate (Fairlamb and Cerami, 1992). The basis for increased TSH in AsIII- and SbIII-resistant cell lines is well understood. The gene *GSH1*, coding for γ-glutamylcysteine synthase (γ-GCS), the rate-limiting step in glutathione (GSH) biosynthesis, is amplified (Grondin *et al.*, 1997; Haimeur *et al.*, 2000). In addition, the gene coding for ornithine decarboxylase (ODC), the rate-limiting step in spermidine biosynthesis, is overexpressed in AsIII-resistant mutants (Haimeur *et al.*, 1999). A dual increase in GSH and spermidine levels, the two building blocks of TSH, leads to an increase in TSH levels in drug-resistant mutants. We found that TSH is essential for resistance but elevated levels of TSH alone are not sufficient for resistance. Indeed, transfection of either *GSH1* or *ODC* leads to an increase in TSH levels in wild-type cells that is even higher than TSH

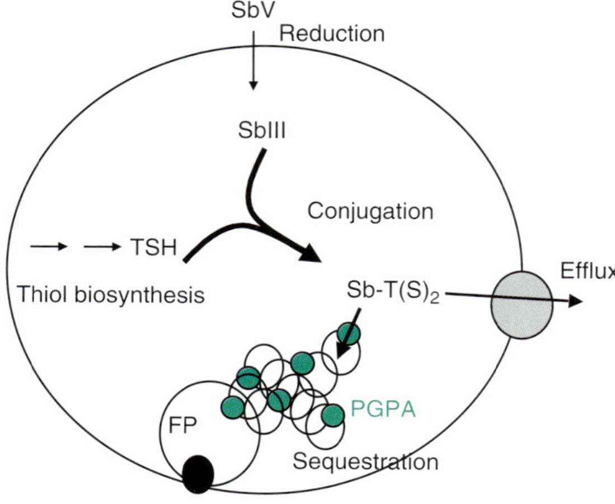

Figure 16.2. Model for metal resistance in **Leishmania**. *Pentavalent metals are probably reduced to the trivalent form, which is thought to be the active form of the metals. The site of reduction is uncertain and could be either in the macrophage or in the parasite. Resistance could arise if the reductase activity were lost and this idea has received support from the analysis of Pentostam-resistant* **L. donovani** *amastigote cells that lost their reductase activity (Shaked-Mishan* et al., *2001). Elevated levels of the bisglutathione–spermidine conjugate trypanothione (TSH) are essential for resistance. This is achieved by amplification of* **GSH1** *(Grondin* et al., *1997) coding for γ-glutamylcysteine synthase and by overexpression of the* **ODC** *gene (Haimeur* et al., *1999, 2000) coding for the enzyme ornithine decarboxylase, which are responsible for the rate-limiting steps in glutathione and spermidine biosynthesis, respectively. A reduction in TSH levels, using specific inhibitors of glutathione and spermidine biosynthesis, will reverse resistance (Haimeur* et al., *1999). Although arsenite–TSH conjugates can form spontaneously in the test tube (Mukhopadhyay* et al., *1996), a putative TSH–conjugase might be necessary inside the cell to increase the rate of generation of the substrate for the various X-thiol transporters. The metal–TSH conjugate can then be sequestered into the intracellular vesicular and tubular membrane organelle by PGPA (Légaré* et al., *2001). These conjugates may then move outside the cell by exocytosis, which occurs exclusively through the flagellar pocket (FP). Alternatively, the metal–TSH conjugate might be extruded directly outside the cell by a plasma membrane thiol-X-efflux pump.*

levels encountered in resistant cells. However, no increase in resistance is observed in wild-type transfectants (Grondin *et al.*, 1997; Haimeur *et al.*, 1999). The γ-GCS- and ODC-specific inhibitors buthionine sulfoximine (BSO) and difluoromethyl-ornithine (DFMO) can reduce the level of TSH in the resistant cells and reverse the resistance phenotype in these mutants (Haimeur *et al.*, 1999, 2000). A strong correlative link therefore exists between TSH levels and resistance but other gene products are implicated in the resistance phenotype.

The gene coding for the ABC transporter PGPA is frequently amplified in metal-resistant *Leishmania* (Ouellette *et al.*, 1998). When discovered, PGPA was found to be the most divergent of eukaryotic ABC transporters (Ouellette *et al.*, 1990). When the MRP sequence became available, PGPA was found to be its closest homologue (Cole *et al.*, 1992). PGPA is now included in the MRP subfamily of ABC transporters **(Table 16.1)**. The results of *PGPA* gene transfection indicated clearly that this gene can contribute to AsIII and SbIII resistance (Callahan and Beverley, 1991; Légaré *et al.*, 1997; Papadopoulou *et al.*, 1994). The level of resistance conferred by PGPA depended on the *Leishmania* species into which the gene was transfected. In *Leishmania tarentolae,* only low level resistance was observed and it was not possible to reach resistance levels observed in drug-resistant mutants (Légaré *et al.*, 1997; Papadopoulou *et al.*, 1994). This led to the suggestion that PGPA requires other factors for conferring high levels of resistance and that the availability of these factors may differ in various *Leishmania* species. The GS-X-mediated resistance pathway of mammalian cells requires sustained elevated GSH levels, increased activity of the GS-X transporter, and increased conjugase activity (Ishikawa, 1992). By analogy to the GS-X pathway, we proposed that PGPA recognizes metals conjugated to TSH. In order to test this hypothesis we have performed co-transfection experiments with *PGPA* and *GSH1* or *ODC*. When these genes were transfected into wild-type cells, we found only the low resistance levels mediated by PGPA. However, when the combination of genes was used to transfect revertant cells (mutants grown in the absence of the drug for prolonged periods) we observed a strong synergy leading to high levels of resistance (Grondin *et al.*, 1997; Haimeur *et al.*, 1999) suggesting indeed that PGPA recognizes metals conjugated to TSH **(Figure 16.2)**. Since this synergy only occurs in revertant cells, it is clear that at least one other mutation is present in the mutant and by analogy to the GS-X system, we are proposing that the missing mutation is a trypanothione-*S*-transferase.

ABC transporters often mediate resistance by increased extrusion of the drug outside the cell and PGPA was initially proposed to correspond to an efflux pump. Transport experiments indeed indicated that there was an active efflux of the metal outside resistant cells. However, this efflux system seemed unrelated to PGPA (Dey *et al.*, 1994). Everted vesicles of fractions enriched for plasma membranes suggested that this efflux system recognizes metal–thiol conjugates (Dey *et al.*, 1996). The activity of this transporter is not increased in membranes derived from mutants or in cells overexpressing PGPA, suggesting that it corresponds to another gene product and that this transporter itself is not rate limiting. PGPA may therefore correspond to an intracellular ABC transporter and this was verified by making a PGPA–green fluorescent protein (GFP) fusion. The PGPA–GFP fusion was totally active and conferred metal resistance in a TSH-dependent manner. The active fusion was indeed shown to be located in an intracellular membrane close to the flagellar pocket (Légaré *et al.*, 2001). Using electron microscopy PGPA was located at the level of the recently described vesicular and tubular membranes (Weise *et al.*, 2000) close to the flagellar pocket (Légaré *et al.*, 2001). Transport experiments using these PGPA-enriched vesicles proved that PGPA transports metal–thiol conjugates in an ATP-dependent fashion (Légaré *et al.*, 2001).

PGPA therefore appears to confer resistance by sequestering thiol–metal conjugates in vesicles close to the flagellar pocket. Several other ABC transporters appear also to confer metal resistance by such a sequestration (reviewed in Ishikawa *et al.*, 1997). The ABC transporter HMT1 confers cadmium tolerance by sequestering phytochelatine (a glutathione-like molecule) cadmium complexes in the fission yeast vacuole (Ortiz *et al.*, 1995). The yeast ABC transporter YCF1 confers cadmium and arsenite resistance by mediating the vacuolar accumulation of metal–glutathione complexes (Ghosh *et al.*, 1999; Li *et al.*, 1996; Tommasini *et al.*, 1996).

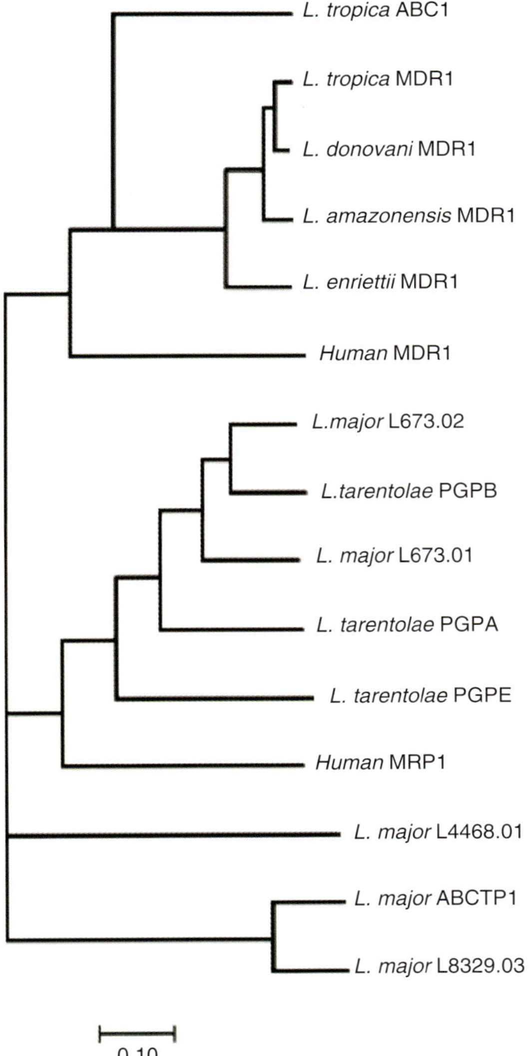

Figure 16.3. Phylogenetic tree of ABC proteins in **Leishmania.** *Only the proteins that are completely sequenced were considered in this analysis. The accession number of the proteins can be found in Table 16.1. The deduced amino acid sequences of the putative ABC proteins were aligned using ClustalW (Thompson* **et al.**, *1994) and subjected to phylogenetic analysis by the neighbor-joining algorithm; Kimura 2-parameters were used to construct the tree. Bootstrap analysis was calculated based on 100 replicates. The scale bar represents 10% changes in amino acid sequences when adding the length of all horizontal lines connecting the two species.*

An MRP-like gene family in *Leishmania*

Since the discovery of MRP1, several other mammalian MRP isoforms have been found, with now at least six members (Borst *et al.*, 1999; see Chapters 19–21). PGPA, which is part of the MRP subfamily of ABC proteins **(Figure 16.3)**, is also part of a large gene family in *Leishmania* with at least four other members termed *PGPB*, *PGPC*, *PGPD* and *PGPE* (Légaré *et al.*, 1994). The nucleotide sequences of *PGPB* and *PGPE* are known and the gene products are highly similar to PGPA (Légaré *et al.*, 1994). PGPA, B and C are linked on chromosome 23 while PGPD and E are linked on a large chromosome (Légaré *et al.*, 1994). The genome sequencing effort has provided the sequence of PGPC (**Table 16.1**, sequence L673.01). Transfection experiments failed to show a role in resistance for any of the four (PGPB–E) novel genes (Légaré *et al.*, 1994) although co-transfection with *GSH1* has never been done and a limited number of drugs were tested. In a methotrexate (MTX)-resistant *Leishmania tropica* cell line, a *PGPE* homologue was shown to be overexpressed (Gamarro *et al.*, 1994). With the recent demonstration that some members of the MRP family have the ability to produce MTX resistance (Hooijberg *et al.*, 1999), the role of PGPE in MTX resistance merits reinvestigation, although transfection of *PGPE* into *L. tarentolae* is not associated with resistance to MTX (Légaré *et al.*, 1994). Owing to their sequence similarities to PGPA and MRP, it is likely that PGPB, C, D and E are organic anion transporters and one of these may correspond to the non-PGPA thiol-X pump located in the plasma membrane and responsible for metal efflux (Dey *et al.*, 1994, 1996) **(Figure 16.2)**.

P-glycoprotein

Leishmania contains in its genome at least one P-glycoprotein homologue. The *Leishmania* gene product is highly homologous to the mammalian MDR1 protein (Hendrickson *et al.*, 1993) **(Figure 16.3)** and it was characterized in several *Leishmania* species **(Table 16.1)**. The *Leishmania MDR1* gene was amplified in *Leishmania* mutants selected for vinblastine or daunomycin resistance and transfection experiments indeed indicated that this *MDR1* gene can cause multidrug resistance (Chiquero *et al.*, 1998; Chow *et al.*, 1993; Gueiros-Filho *et al.*, 1995; Henderson *et al.*, 1992; Katakura *et al.*, 1999). The interactions between flavenoids and the ABC domain of the *Leishmania* MDR1 were characterized and some derivatives with high affinity for the nucleotide-binding domain reversed the multidrug resistance phenotype of resistant cells (Perez-Victoria *et al.*, 1999, 2001).

The high degree of homology between *Leishmania* and human MDR1 suggests that the former could confer resistance by active extrusion of the drug. The efflux of rhodamine 123 in *Leishmania amazonensis*-resistant cells (Gueiros-Filho *et al.*, 1995), the absence of accumulation of puromycin in vinblastine-resistant *Leishmania donovani* (Henderson *et al.*, 1992) and the reduction of daunomycin accumulation in resistant *L. tropica* (Perez-Victoria *et al.*, 1999) were all consistent with this hypothesis. This putative transport defect due to an efflux pump does not fit, however, with subcellular localization studies done in the laboratory of D. Wirth at Harvard. Their studies suggest that the majority of *Leishmania* MDR1 protein is not located in the plasma membrane but in an organelle close to the mitochondria of *Leishmania enriettii* (Chow and Volkman, 1998). Further work is required to understand how MDR1 confers drug resistance in *Leishmania* and to determine its exact cellular location. Recently, it was suggested that *MDR1* could confer resistance to miltefosine (F. Gamarro, personal communication), a promising alkyl-lysophospholipid that can be taken orally and is highly active against *Leishmania* (Jha *et al.*, 1999). Thus *MDR1* has the potential for conferring resistance against useful anti-leishmanial compounds.

Other ABC transporters

The sequencing of the *Leishmania major* genome is well underway and an international consortium of laboratories and institutes (http://www.ebi.ac.uk/parasites/LGN) is now sequencing its 36 chromosomes. A recent survey of the available sequences, as part of sequenced chromosomes, cosmids or genome survey sequences, revealed that *Leishmania* is likely to contain several ABC proteins and to date the sequences of nine full-length ABC genes are known **(Table 16.1)**.

A gene coding for a protein (ABCTP1) with two ABC domains and no apparent transmembrane domains is present on chromosome 3 of *L. major*. A gene coding for a protein with a similar organization belonging to another family is present on another chromosome **(Table 16.1)**. ABCTP1 shares extensive similarities with several other putative ABC transporters found in diverse organisms. The yeast YEF3 and GCN20 ABC proteins also contain duplicated ABC domains without transmembrane domains and are involved in translation (Bauer *et al.*, 1999; Decottignies and Goffeau, 1997; Taglicht and Michaelis, 1998). ABCTP1 may serve a similar function. As part of an ongoing project to determine the function of ABC transporters in *Leishmania,* we are attempting to disrupt several ABC genes by homologous recombination. One of the two alleles of the *ABCTP1* gene was inactivated and no effect on growth properties of the mutants was observed (unpublished observation). Work is in progress to generate a null mutant.

Leishmania has an ABCA1 gene homologue (**Table 16.1;** F. Gamarro, Grenada, personal communication). The ABCA subfamily is absent in the yeast genome and was thought to be restricted to multicellular organisms (Broccardo *et al.*, 1999). The presence of these transporters in the unicellular parasite *Leishmania* is interesting. ABCA1 appears to be involved in the control of membrane lipid composition and in a recessive disorder of lipid metabolism in humans called Tangier disease (see Chapter 23). The same protein has been implicated in the engulfment of apoptotic cells by macrophages (Luciani and Chimini, 1996). The presence of an ABC1-like protein in a parasite engulfed by and living within macrophages is noteworthy and may suggest a role for this ABC protein in host–pathogen interactions possibly in the scavenging of host lipids.

It was already known that *PGPA*, *PGPB* and *PGPC* were linked on the same chromosome (Légaré *et al.*, 1994), and the sequencing effort has indicated that these three genes are part of chromosome 23. A fourth ABC transporter was also found on chromosome 23. It contains one ABC domain and sequence similarities suggest a possible role as an ATP-dependent permease precursor. BLAST analysis of random sequences at Washington University in St Louis (N.S. Akopyants and S.M. Beverley 'A survey of the *Leishmania major* Friedlin strain V1 genome by shotgun sequencing' and the Washington University Genome Sequencing Center) and at the Sanger Centre (*Leishmania major* Friedlin genome sequencing project, Sanger Centre, The Wellcome Trust Genome Campus) clearly indicated the presence of several novel ABC transporters. Once translated, at least eight sequences have clearly recognizable and significant portions of ABC domains that are different from the ABC transporters of **Table 16.1**. As these sequences are partial it is difficult at this point to determine to which subfamily of ABC transporters these proteins belong.

The number of ABC transporters in *Leishmania* is starting to be large enough to carry out phylogenetic analysis. The *Leishmania MDR1* gene was sequenced in four species and as expected these genes cluster together with the human MDR1 gene **(Figure 16.3)**. The PGPA-E proteins cluster together with the human MRP1 protein. From our known genomic organization of the PGPA-B-C locus (Légaré *et al.*, 1994), from the available genomic sequences and from this phylogenetic analysis **(Figure 16.3)** we are confident that the *L. major* L673.01 and L673.02 correspond to the *L. tarentolae* PGPC and PGPB homologues. The two proteins with duplicated ABC domains without transmembrane domains **(Table 16.1)** cluster together **(Figure 16.3)**, although additional sequences may eventually lead to a better discrimination. Similarly, ABCA1 presently stands alone **(Figure 16.3)** but when more *Leishmania* sequences are available for comparison we should obtain a more precise phylogeny.

ABC TRANSPORTERS IN *TRYPANOSOMA* SP.

The African trypanosomes, responsible for sleeping sickness, are coming back with a vengeance and the last WHO statistics indicate that the parasite infects millions of individuals and is responsible for several thousand deaths a year. A number of old drugs are available against *Trypanosoma brucei* infection but in late-stage infection, when the parasite has crossed the blood–brain barrier, the trivalent arsenical melarsoprol is the drug of choice (Pepin and Milord, 1994). The mode of action of arsenicals is not understood. *Trypanosoma cruzi*, the etiologic agent of Chagas' disease infects 16–18 million people in South America. The current drugs nifurtimox and benznidazole are active in the acute phase of the disease but much less in the chronic phase (de Castro, 1993). The genomes of these two trypanosome species are currently being sequenced.

Since an ABC transporter was found implicated in antimony resistance in *Leishmania*, ABC transporters were searched for in *T. brucei*. Several ABC transporters have been described in *T. brucei* (Maser and Kaminsky, 1998); one is highly similar to the *Leishmania* PGPA protein while another is related to the *Leishmania* MDR1. The expression of these genes is similar in resistant and sensitive isolates (Maser and Kaminsky, 1998). It would nonetheless be of interest to test whether the homologous *PGPA* gene of *T. brucei* was capable of conferring resistance to arsenicals. This was recently tested by gene transfection and the *T. brucei PGPA* homologue *TbMRPA* was indeed found to increase the IC_{50} of melarsoprol by 10-fold (Shahi *et al.*, 2002). One other important resistance gene for arsenical resistance was recently isolated. Wild-type trypanosomes have two adenosine transporters, P1 and P2, and arsenical resistant parasites lack P2 (Carter and Fairlamb, 1993). The P2 adenosine transporter gene of *T. brucei* was cloned and drug-resistant trypanosomes harbored a defective transporter (Maser *et al.*, 1999).

A *T. cruzi* ABC transporter TcPGP2, most probably the *Leishmania* PGPA homologue, has also been characterized (Dallagiovanna *et al.*, 1996), although its role in resistance is unknown. A second gene, *TcPGP1*, homologous to *PGPA* but interrupted by the insertion of a retrotransposon, has also been observed in *T. cruzi* (Torres *et al.*, 1999). A probe derived from *TcPGP2* was used as a polymorphic marker in an attempt to discriminate drug-susceptible and drug-resistant strains of *T. cruzi*. An imperfect correlation was observed between drug susceptibility and the TcPGP polymorphism (Murta *et al.*, 1998).

ABC TRANSPORTERS IN ANAEROBIC PROTOZOAN PARASITES

A group of three anaerobic unrelated parasites, *Giardia duodenelis*, *Trichomonas vaginalis* and *Entamoeba histolytica*, are the cause of considerable human suffering. The drug of choice against all three parasites is metronidazole and resistance to this drug has been described, although the resistance mechanisms do not appear to involve ABC transporters (Upcroft and Upcroft, 2001).

E. histolytica is a widely distributed parasite causing dysentery and liver abscesses. Emetine, a protein synthesis inhibitor, was for many years, before the advent of metronidazole, an important drug in the treatment of human amoebiasis. Emetine resistance was induced under laboratory conditions in *E. histolytica* and cells were cross resistant to colchicine, accumulating less radioactive drugs. Verapamil could reverse the defective accumulation (Orozco *et al.*, 1999). Overall, these results were suggestive of the involvement of ABC transporters in the resistance phenotype. A large gene family of P-glycoproteins with at least four genes and two pseudogenes was discovered in *E. histolytica* (Descoteaux *et al.*, 1995), hence constituting the largest currently known P-glycoprotein gene family in a protozoan parasite. None of these genes were amplified but two genes, *Ehpgp1* and *Ehpgp6*, were overexpressed at all drug concentrations while *Ehpgp5* was overexpressed at the highest drug concentration (Descoteaux *et al.*, 1995).

It was proposed that *Ehpgp5* expression is regulated by transcriptional factors induced by the presence of emetine (Perez *et al.*, 1998). The role of Ehpgp1 in emetine resistance was confirmed by gene transfection in *E. histolytica* (Ghosh *et al.*, 1996). It is thus possible that several P-glycoproteins cooperate together to confer high-level resistance to multiple drugs in *E. histolytica* (Orozco et al., 1999). An ABC transporter has been described in the sexually transmitted protozoan parasite *T. vaginalis*. This ABC transporter has six putative transmembrane segments and a carboxy-terminal ABC domain (Johnson *et al.*, 1994). The level of RNA and the copy number of this gene varied greatly between metronidazole-resistant and -sensitive isolates but overall no strict correlation was found between levels of Tvpgp expression and levels of resistance (Johnson *et al.*, 1994).

ABC TRANSPORTERS OF PARASITIC WORMS

Bona fide drug resistance in common worm infections is not yet common in humans but since it is in veterinary medicine (Geerts and Gryseels, 2000), the potential for resistance is important. ABC transporters have been found in a number of human parasitic worms such as *Schistosoma* (Bosch *et al.*, 1994) and *Onchocerca volvulus* (Huang and Prichard, 1999), although the role of any of these proteins in drug resistance needs to be established. The situation seems to differ in the sheep nematode parasite *Haemonchus contortus*, in which resistance to ivermectin and related drugs is an increasing problem. Ivermectin opens the chloride channels of worms, which leads to starvation or paralysis. The expression of a P-glycoprotein from *H. contortus* was higher in ivermectin-selected than in unselected strains (Xu *et al.*, 1998) and the multidrug resistance reversing agent verapamil increased the efficacy of ivermectin in resistant strains (Molento and Prichard, 1999). Ivermectin is a likely substrate for a P-glycoprotein since disruption of the P-glycoprotein gene in mice results in hypersensitivity to ivermectin (Schinkel *et al.*, 1994).

In the nematode *Caenorhabditis elegans*, simultaneous mutations of three genes encoding glutamate-gated chloride channel alpha-type subunits confer high-level resistance to ivermectin (Dent *et al.*, 2000), suggesting that both target mutation and transport alteration can lead to ivermectin resistance in worms. At least 56 ABC transporters have been found in the fully sequenced non-pathogenic *C. elegans* nematode. *C. elegans* contains at least four P-glycoproteins (Lincke *et al.*, 1992), and pgp-3 deletion mutants were found to be more sensitive to both colchicine and chloroquine (Broeks *et al.*, 1995). At least four MRP homologues are also present in *C. elegans* and disruption of mrp-1 causes cells to be more sensitive to heavy metals (Broeks *et al.*, 1996).

Clinical Implications and Outlook

Drug resistance is an important problem in parasitic diseases, which is exacerbated by the limited number of drugs available. Studies on drug resistance can help to find strategies to increase the efficacy or the life span of the few drugs available. The isolation of genes involved in drug resistance can allow the development of tests to detect resistance rapidly. Specific tests have been developed for the genotypic diagnosis of antifolate (Basco *et al.*, 2000; Nzila *et al.*, 2000) and chloroquine resistance (Djimde *et al.*, 2001) in malaria. Tests to detect resistance determinants in other parasites are likely to emerge once resistance mechanisms are better understood. Our ability to detect resistance can be useful to determine the prevalence of resistance in specific geographical regions and can also prevent the use of ineffective or toxic drugs for patients infected with resistant parasites. It may also limit the use of last-resort drugs to only when absolutely required when dealing with resistant parasites.

The understanding of resistance mechanisms including those involving ABC transporters can also lead to new strategies for chemotherapeutic interventions. The dichotomous action of *pfmdr1* mutations on two categories of drugs (quinine and chloroquine on the one hand, and halofantrine, mefloquine and artemisinin on the other) may suggest that a combination with a member of each group of drugs could lead to a reversal of Pghl-mediated resistance. Strategies and inhibitors to modulate the activity of efflux pumps are being developed and these could be useful in the treatment of parasitic diseases when a transport-related mechanism is the main resistance mechanism. In the PGPA-mediated antimony resistance in *Leishmania*, thiols are important and we found that we can revert PGPA-mediated resistance *in vitro* when reducing thiol biosynthesis using specific inhibitors (Haimeur *et al.*, 1999, 2000; Légaré *et al.*, 2001). If warranted, similar strategies may eventually be attempted in patients. Inhibitors of ABC transporters may also turn out to be useful drugs. Inactivation of Pgh1 may alter the pH of the food vacuole of the parasite hence leading to altered function and death. A *Leishmania* mutant deleted for PGPA diminished survival inside murine macrophages (Papadopoulou *et al.*, 1996). Several ABC transporters are located at strategic positions within the host–pathogen interface (e.g. PfGCN20, CpABC) and inactivation of these transporters may indeed decrease parasite survival.

Inventories of ABC transporters are currently being made for the organisms for which the genome is completed. Phylogenetic analysis with ABC protein orthologues from different organisms will help in functional assignment. Considerable work will be required to assess the function of the vast majority of parasitic ABC proteins that are currently being revealed by genome efforts. Episomal transfection, gene disruption by homologous recombination, protein localization using GFP fusions, microarray and proteomic analysis are techniques that are now available for all major parasites. These techniques should be useful to determine the function of parasite ABC proteins and to study their putative role in host–pathogen interactions and in drug resistance.

Acknowledgments

The authors wish to thank Dr Eric Leblanc for assistance with the phylogenetic analyses. This work was supported by the Canadian Institutes for Health Research (CIHR). MO is a CIHR Investigator and a Burroughs Wellcome Fund Scholar in Molecular Parasitology. We wish to thank the scientists and funding agencies comprising the international Malaria Genome Project for making sequence data from the genome of *P. falciparum* (3D7) public prior to publication of the completed sequence. The Sanger Centre (UK) provided sequences for chromosomes 1, 3–9, and 13, with financial support from the Wellcome Trust. A consortium composed of The Institute for Genome Research, along with the Naval Medical Research Center (USA), sequenced chromosomes 2, 10, 11 and 14, with support from NIAID/NIH, the Burroughs Wellcome Fund, and the Department of Defense. The Stanford Genome Technology Center (USA) sequenced chromosome 12, with support from the Burroughs Wellcome Fund.

The *Plasmodium* Genome Database is a collaborative effort of investigators at the University of Pennsylvania (USA) and Monash University (Melbourne, Australia), supported by the Burroughs Wellcome Fund.

References

Barnes, D.A., Foote, S.J., Galatis, D., Kemp, D.J. and Cowman, A.F. (1992) Selection for high-level chloroquine resistance results in deamplification of the pfmdr1 gene and increased sensitivity to mefloquine in *Plasmodium falciparum*. EMBO J. 11, 3067–3075.

Basco, L.K., Tahar, R., Keundjian, A. and Ringwald, P. (2000) Sequence variations in the genes encoding dihydropteroate synthase and dihydrofolate reductase and clinical response to sulfadoxine-pyrimethamine in patients with acute uncomplicated falciparum malaria. J. Infect. Dis. 182, 624–628.

Bauer, B.E., Wolfger, H. and Kuchler, K. (1999) Inventory and function of yeast ABC proteins: about sex, stress, pleiotropic drug and heavy metal resistance. Biochim. Biophys. Acta 1461, 217–236.

Berman, J.D. (1997) Human leishmaniasis: clinical, diagnostic, and chemotherapeutic developments in the last 10 years. Clin. Infect. Dis. 24, 684–703.

Borst, P., Evers, R., Kool, M. and Wijnholds, J. (1999) The multidrug resistance protein family. Biochim. Biophys. Acta 1461, 347–357.

Bosch, I.B., Wang, Z.X., Tao, L.F. and Shoemaker, C.B. (1994) Two *Schistosoma mansoni* cDNAs encoding ATP-binding cassette (ABC) family proteins. Mol. Biochem. Parasitol. 65, 351–356.

Bozdech, Z. and Schurr, E. (1999) Protein transport in the host cell cytoplasm and ATP-binding cassette proteins in *Plasmodium falciparum*-infected erythrocytes. Novartis Found. Symp. 226, 231–241; discussion 241–245.

Bozdech, Z., VanWye, J., Haldar, K. and Schurr, E. (1998) The human malaria parasite *Plasmodium falciparum* exports the ATP-binding cassette protein PFGCN20 to membrane structures in the host red blood cell. Mol. Biochem. Parasitol. 97, 81–95.

Bray, P.G., Mungthin, M., Ridley, R.G. and Ward, S.A. (1998) Access to hematin: the basis of chloroquine resistance. Mol. Pharmacol. 54, 170–179.

Broccardo, C., Luciani, M. and Chimini, G. (1999) The ABCA subclass of mammalian transporters. Biochim. Biophys. Acta 1461, 395–404.

Broeks, A., Janssen, H.W., Calafat, J. and Plasterk, R.H. (1995) A P-glycoprotein protects *Caenorhabditis elegans* against natural toxins. EMBO J. 14, 1858–1866.

Broeks, A., Gerrard, B., Allikmets, R., Dean, M. and Plasterk, R.H. (1996) Homologues of the human multidrug resistance genes MRP and MDR contribute to heavy metal resistance in the soil nematode *Caenorhabditis elegans*. EMBO J. 15, 6132–6143.

Callahan, H.L. and Beverley, S.M. (1991) Heavy metal resistance: a new role for P-glycoproteins in *Leishmania*. J. Biol. Chem. 266, 18427–18430.

Carter, N.S. and Fairlamb, A.H. (1993) Arsenical-resistant trypanosomes lack an unusual adenosine transporter [published erratum appears in Nature 1993 Jan 28;361(6410):374]. Nature 361, 173–176.

Chiquero, M.J., Perez-Victoria, J.M., O'Vallee, F., Gonzalez-Ros, J.M., del Moral, R.G., Ferragut, J.A., Castanys, S. and Gamarro, F. (1998) Altered drug membrane permeability in a multidrug-resistant *Leishmania tropica* line. Biochem. Pharmacol. 55, 131–139.

Chow, L.M.C. and Volkman, S.K. (1998) *Plasmodium* and *Leishmania*: the role of mdr genes in mediating drug resistance. Exp. Parasitol. 90, 135–141.

Chow, L.M., Wong, A.K., Ullman, B. and Wirth, D.F. (1993) Cloning and functional analysis of an extrachromosomally amplified multidrug resistance-like gene in *Leishmania enriettii*. Mol. Biochem. Parasitol. 60, 195–208.

Cole, S.P., Bhardwaj, G., Gerlach, J.H., Mackie, J.E., Grant, C.E., Almquist, K.C., Stewart, A.J., Kurz, E.U., Duncan, A.M. and Deeley, R.G. (1992) Overexpression of a transporter gene in a multidrug-resistant human lung cancer cell line [see comments]. Science 258, 1650–1654.

Cowman, A.F., Karcz, S., Galatis, D. and Culvenor, J.G. (1991) A P-glycoprotein homologue of *Plasmodium falciparum* is localized on the digestive vacuole. J. Cell Biol. 113, 1033–1042.

Cowman, A.F., Galatis, D. and Thompson, J.K. (1994) Selection for mefloquine resistance in *Plasmodium falciparum* is linked to amplification of the pfmdr1 gene and cross-resistance to halofantrine and quinine. Proc. Natl Acad. Sci. USA, 91, 1143–1147.

Dallagiovanna, B., Gamarro, F. and Castanys, S. (1996) Molecular characterization of a P-glycoprotein-related tcpgp2 gene in *Trypanosoma cruzi*. Mol. Biochem. Parasitol. 75, 145–157.

Dassa, E. and Bouige, P. (2001) The ABC of ABCS: a phylogenetic and functional classification of ABC systems in living organisms. Res. Microbiol. 152, 211–229.

de Castro, S.L. (1993) The challenge of Chagas' disease chemotherapy: an update of drugs assayed against *Trypanosoma cruzi*. Acta Trop. 53, 83–98.

Decottignies, A. and Goffeau, A. (1997) Complete inventory of the yeast ABC proteins. Nat. Genet. 15, 137–145.

Dent, J.A., Smith, M.M., Vassilatis, D.K. and Avery, L. (2000) The genetics of ivermectin resistance in *Caenorhabditis elegans*. Proc. Natl Acad. Sci. USA, 97, 2674–2679.

Descoteaux, S., Ayala, P., Samuelson, J. and Orozco, E. (1995) Increase in mRNA of multiple Eh pgp genes encoding P-glycoprotein homologues in emetine-resistant *Entamoeba histolytica* parasites. Gene 164, 179–184.

Dey, S., Papadopoulou, B., Haimeur, A., Roy, G., Grondin, K., Dou, D., Rosen, B.P. and Ouellette, M. (1994) High level arsenite resistance in *Leishmania tarentolae* is mediated by an active extrusion system. Mol. Biochem. Parasitol. 67, 49–57.

Dey, S., Ouellette, M., Lightbody, J., Papadopoulou, B. and Rosen, B.P. (1996) An ATP-dependent As(III)-glutathione transport system in membrane vesicles of *Leishmania tarentolae*. Proc. Natl Acad. Sci. USA, 93, 2192–2197.

Djimde, A., Doumbo, O.K., Cortese, J.F., Kayentao, K., Doumbo, S., Diourte, Y., *et al.* (2001) A molecular marker for chloroquine-resistant falciparum malaria. N. Engl. J. Med. 344, 257–263.

Duraisingh, M.T., Roper, C., Walliker, D. and Warhurst, D.C. (2000) Increased sensitivity to the antimalarials mefloquine and artemisinin is conferred by mutations in the pfmdr1 gene of *Plasmodium falciparum*. Mol. Microbiol. 36, 955–961.

Durand, R., Jafari, S., Vauzelle, J., Delabre, J., Jesic, Z. and Le Bras, J. (2001) Analysis of pfcrt point mutations and chloroquine susceptibility in isolates of *Plasmodium falciparum*. Mol. Biochem. Parasitol. 114, 95–102.

Dzekunov, S.M., Ursos, L.M. and Roepe, P.D. (2000) Digestive vacuolar pH of intact intraerythrocytic *P. falciparum* either sensitive or resistant to chloroquine. Mol. Biochem. Parasitol. 110, 107–124.

Ekong, R.M., Robson, K.J., Baker, D.A. and Warhurst, D.C. (1993) Transcripts of the multidrug resistance genes in chloroquine-sensitive and chloroquine-resistant *Plasmodium falciparum*. Parasitology 106, 107–115.

Fairlamb, A.H. and Cerami, A. (1992) Metabolism and functions of trypanothione in the Kinetoplastida. Annu. Rev. Microbiol. 46, 695–729.

Faraut-Gambarelli, F., Piarroux, R., Deniau, M., Giusiano, B., Marty, P., Michel, G., Faugere, B. and Dumon, H. (1997) In vitro and in vivo resistance of *Leishmania infantum* to meglumine antimoniate: a study of 37 strains collected from patients with visceral leishmaniasis. Antimicrob. Agents Chemother. 41, 827–830.

Fidock, D., Nomura, T., Talley, A., Cooper, R., Dzekunov, S., Ferdig, M., *et al.* (2000) Mutations in the *P. falciparum* digestive vacuole transmembrane protein PfCRT and evidence for their role in chloroquine resistance. Mol. Cell 6, 861–871.

Foote, S.J., Thompson, J.K., Cowman, A.F. and Kemp, D.J. (1989) Amplification of the multidrug resistance gene in some chloroquine-resistant isolates of *P. falciparum*. Cell 57, 921–930.

Foote, S.J., Kyle, D.E., Martin, R.K., Oduola, A.M., Forsyth, K., Kemp, D.J. and Cowman, A.F. (1990) Several alleles of the multidrug-resistance gene are closely linked to chloroquine resistance in *Plasmodium falciparum* [see comments]. Nature 345, 255–258.

Gamarro, F., Chiquero, M.J., Amador, M.V., Légaré, D., Ouellette, M. and Castanys, S. (1994) P-glycoprotein overexpression in methotrexate-resistant *Leishmania tropica*. Biochem. Pharmacol. 47, 1939–1947.

Geerts, S. and Gryseels, B. (2000) Drug resistance in human helminths: current situation and lessons from livestock. Clin. Microbiol. Rev. 13, 207–222.

Ghosh, S.K., Lohia, A., Kumar, A. and Samuelson, J. (1996) Overexpression of P-glycoprotein gene 1 by transfected *Entamoeba histolytica* confers emetine-resistance. Mol. Biochem. Parasitol. 82, 257–260.

Ghosh, M., Shen, J. and Rosen, B.P. (1999) Pathways of As(III) detoxification in *Saccharomyces cerevisiae*. Proc. Natl Acad. Sci. USA 96, 5001–5006.

Grondin, K., Haimeur, A., Mukhopadhyay, R., Rosen, B.P. and Ouellette, M. (1997)

Co-amplification of the gamma-glutamylcysteine synthetase gene gsh1 and of the ABC transporter gene pgpA in arsenite-resistant *Leishmania tarentolae*. EMBO J. 16, 3057–3065.

Gueiros-Filho, F.J., Viola, J.P., Gomes, F.C., Farina, M., Lins, U., Bertho, A.L., Wirth, D.F. and Lopes, U.G. (1995) *Leishmania amazonensis*: multidrug resistance in vinblastine-resistant promastigotes is associated with rhodamine 123 efflux, DNA amplification, and RNA overexpression of a *Leishmania* mdr1 gene. Exp. Parasitol. 81, 480–490.

Haimeur, A., Guimond, C., Pilote, S., Mukhopadhyay, R., Rosen, B.P., Poulin, R. and Ouellette, M. (1999) Elevated levels of polyamines and trypanothione resulting from overexpression of the ornithine decarboxylase gene in arsenite-resistant *Leishmania*. Mol. Microbiol. 34, 726–735.

Haimeur, A., Brochu, C., Genest, P., Papadopoulou, B. and Ouellette, M. (2000) Amplification of the ABC transporter gene PGPA and increased trypanothione levels in potassium antimonyl tartrate (SbIII) resistant *Leishmania tarentolae*. Mol. Biochem. Parasitol. 108, 131–135.

Henderson, D.M., Sifri, C.D., Rodgers, M., Wirth, D.F., Hendrickson, N. and Ullman, B. (1992) Multidrug resistance in *Leishmania donovani* is conferred by amplification of a gene homologous to the mammalian mdr1 gene. Mol. Cell. Biol. 12, 2855–2865.

Hendrickson, N., Sifri, C.D., Henderson, D.M., Allen, T., Wirth, D.F. and Ullman, B. (1993) Molecular characterization of the ldmdr1 multidrug resistance gene from *Leishmania donovani*. Mol. Biochem. Parasitol. 60, 53–64.

Herwaldt, B.L. (1999) Leishmaniasis. Lancet 354, 1191–1199.

Higgins, C.F. (1995) The ABC of channel regulation. Cell 82, 693–696.

Hooijberg, J.H., Broxterman, H.J., Kool, M., Assaraf, Y.G., Peters, G.J., Noordhuis, P., Scheper, R.J., Borst, P., Pinedo, H.M. and Jansen, G. (1999) Antifolate resistance mediated by the multidrug resistance proteins MRP1 and MRP2. Cancer Res. 59, 2532–2535.

Huang, Y.J. and Prichard, R.K. (1999) Identification and stage-specific expression of two putative P-glycoprotein coding genes in *Onchocerca volvulus*. Mol. Biochem. Parasitol. 102, 273–281.

Ishikawa, T. (1992) The ATP-dependent glutathione S-conjugate export pump [see comments]. Trends Biochem. Sci. 17, 463–468.

Ishikawa, T., Li, Z.S., Lu, Y.P. and Rea, P.A. (1997) The GS-X pump in plant, yeast, and animal cells: structure, function, and gene expression. Biosci. Rep. 17, 189–207.

Jha, T.K., Sundar, S., Thakur, C.P., Bachmann, P., Karbwang, J., Fischer, C., Voss, A. and Berman, J. (1999) Miltefosine, an oral agent, for the treatment of Indian visceral leishmaniasis. N. Engl. J. Med. 341, 1795–1800.

Johnson, P.J., Schuck, B.L. and Delgadillo, M.G. (1994) Analysis of a single-domain P-glycoprotein-like gene in the early-diverging protist *Trichomonas vaginalis*. Mol. Biochem. Parasitol. 66, 127–137.

Katakura, K., Iwanami, M., Ohtomo, H., Fujise, H. and Hashiguchi, Y. (1999) Structural and functional analysis of the LaMDR1 multidrug resistance gene in *Leishmania amazonensis*. Biochem. Biophys. Res. Commun. 255, 289–294.

Krogstad, D.J., Gluzman, I.Y., Kyle, D.E., Oduola, A.M., Martin, S.K., Milhous, W.K. and Schlesinger, P.H. (1987) Efflux of chloroquine from *Plasmodium falciparum*: mechanism of chloroquine resistance. Science 238, 1283–1285.

Légaré, D., Hettema, E. and Ouellette, M. (1994) The P-glycoprotein-related gene family in *Leishmania*. Mol. Biochem. Parasitol. 68, 81–91.

Légaré, D., Papadopoulou, B., Roy, G., Mukhopadhyay, R., Haimeur, A., Dey, S., Grondin, K., Brochu, C., Rosen, B.P. and Ouellette, M. (1997) Efflux systems and increased trypanothione levels in arsenite-resistant *Leishmania*. Exp. Parasitol. 87, 275–282.

Légaré, D., Richard, D., Mukhopadhyay, R., Stierhof, Y.D., Rosen, B.P., Haimeur, A., Papadopoulou, B. and Ouellette, M. (2001) The *Leishmania* ABC protein PGPA is an intracellular metal-thiol transporter ATPase. J. Biol. Chem. 276, 26301–26307.

Li, Z.S., Szczypka, M., Lu, Y.P., Thiele, D.J. and Rea, P.A. (1996) The yeast cadmium factor protein (YCF1) is a vacuolar glutathione S-conjugate pump. J. Biol. Chem. 271, 6509–6517.

Lincke, C.R., The, I., van Groenigen, M. and Borst, P. (1992) The P-glycoprotein gene family of *Caenorhabditis elegans*. Cloning and characterization of genomic and complementary DNA sequences. J. Mol. Biol. 228, 701–711.

Lira, R., Sundar, S., Makharia, A., Kenney, R., Gam, A., Saraiva, E. and Sacks, D. (1999)

Evidence that the high incidence of treatment failures in Indian kala-azar is due to the emergence of antimony-resistant strains of *Leishmania donovani*. J. Infect. Dis. 180, 564–567.

Luciani, M.F. and Chimini, G. (1996) The ATP binding cassette transporter ABC1, is required for the engulfment of corpses generated by apoptotic cell death. EMBO J. 15, 226–235.

Martin, S.K., Oduola, A.M. and Milhous, W.K. (1987) Reversal of chloroquine resistance in *Plasmodium falciparum* by verapamil. Science 235, 899–901.

Maser, P. and Kaminsky, R. (1998) Identification of three ABC transporter genes in *Trypanosoma brucei* spp. Parasitol. Res. 84, 106–111.

Maser, P., Sutterlin, C., Kralli, A. and Kaminsky, R. (1999) A nucleoside transporter from *Trypanosoma brucei* involved in drug resistance. Science 285, 242–244.

Molento, M.B. and Prichard, R.K. (1999) Effects of the multidrug-resistance-reversing agents verapamil and CL 347,099 on the efficacy of ivermectin or moxidectin against unselected and drug-selected strains of *Haemonchus contortus* in jirds (*Meriones unguiculatus*). Parasitol. Res. 85, 1007–1011.

Mukhopadhyay, R., Dey, S., Xu, N., Gage, D., Lightbody, J., Ouellette, M. and Rosen, B.P. (1996) Trypanothione overproduction and resistance to antimonials and arsenicals in *Leishmania*. Proc. Natl Acad. Sci. USA, 93, 10383–10387.

Murta, S.M., Gazzinelli, R.T., Brener, Z. and Romanha, A.J. (1998) Molecular characterization of susceptible and naturally resistant strains of *Trypanosoma cruzi* to benznidazole and nifurtimox. Mol. Biochem. Parasitol. 93, 203–214.

Nomura, T., Carlton, J.M., Baird, J.K., del Portillo, H.A., Fryauff, D.J., Rathore, D., *et al.* (2001) Evidence for different mechanisms of chloroquine resistance in 2 *Plasmodium* species that cause human malaria. J. Infect. Dis. 183, 1653–1661.

Nzila, A.M., Mberu, E.K., Sulo, J., Dayo, H., Winstanley, P.A., Sibley, C.H. and Watkins, W.M. (2000) Towards an understanding of the mechanism of pyrimethamine-sulfadoxine resistance in *Plasmodium falciparum*: genotyping of dihydrofolate reductase and dihydropteroate synthase of Kenyan parasites. Antimicrob. Agents Chemother. 44, 991–996.

Orozco, E., Gomez, C. and Perez, D.G. (1999) Physiology and molecular genetics of multidrug resistance in *Entamoeba histolytica*. Drug Resistance Updates 2, 188–197.

Ortiz, D.F., Ruscitti, T., McCue, K.F. and Ow, D.W. (1995) Transport of metal-binding peptides by HMT1, a fission yeast ABC-type vacuolar membrane protein. J. Biol. Chem. 270, 4721–4728.

Ouellette, M. (2001) Biochemical and molecular mechanisms of drug resistance in parasites. Trop. Med. Int. Health, 6, 874–882.

Ouellette, M., Fase-Fowler, F. and Borst, P. (1990) The amplified H circle of methotrexate-resistant *Leishmania tarentolae* contains a novel P-glycoprotein gene. EMBO J. 9, 1027–1033.

Ouellette, M., Haimeur, A., Grondin, K., Légaré, D. and Papadopoulou, B. (1998) Amplification of ABC transporter gene pgpA and of other heavy metal resistance genes in *Leishmania tarentolae* and their study by gene transfection and gene disruption. Methods Enzymol. 292, 182–193.

Papadopoulou, B., Roy, G., Dey, S., Rosen, B.P. and Ouellette, M. (1994) Contribution of the *Leishmania* P-glycoprotein-related gene ltpgpA to oxyanion resistance. J. Biol. Chem. 269, 11980–11986.

Papadopoulou, B., Roy, G., Dey, S., Rosen, B.P., Olivier, M. and Ouellette, M. (1996) Gene disruption of the P-glycoprotein related gene pgpa of *Leishmania tarentolae*. Biochem. Biophys. Res. Commun. 224, 772–778.

Pepin, J. and Milord, F. (1994) The treatment of human African trypanosomiasis. Adv. Parasitol. 33, 1–47.

Perez, D.G., Gomez, C., Lopez-Bayghen, E., Tannich, E. and Orozco, E. (1998) Transcriptional analysis of the EhPgp5 promoter of *Entamoeba histolytica* multidrug-resistant mutant. J. Biol. Chem. 273, 7285–7292.

Perez-Victoria, J.M., Chiquero, M.J., Conseil, G., Dayan, G., Di Pietro, A., Barron, D., Castanys, S. and Gamarro, F. (1999) Correlation between the affinity of flavonoids binding to the cytosolic site of *Leishmania tropica* multidrug transporter and their efficiency to revert parasite resistance to daunomycin. Biochemistry 38, 1736–1743.

Perez-Victoria, J.M., Perez-Victoria, F.J., Conseil, G., Maitrejean, M., Comte, G., Barron, D., Di Pietro, A., Castanys, S. and Gamarro, F. (2001) High-affinity binding of silybin derivatives to the nucleotide-binding domain of a *Leishmania tropica* P-glycoprotein-like transporter and chemosensitization of a multidrug-resistant parasite to daunomycin. Antimicrob. Agents Chemother. 45, 439–446.

Perkins, M.E., Wu, T.W. and Le Blancq, S.M. (1998) Cyclosporin analogs inhibit in vitro growth of *Cryptosporidium parvum*. Antimicrob. Agents Chemother. 42, 843–848.

Perkins, M.E., Riojas, Y.A., Wu, T.W. and Le Blancq, S.M. (1999) CpABC, a *Cryptosporidium parvum* ATP-binding cassette protein at the host-parasite boundary in intracellular stages. Proc. Natl Acad. Sci. USA 96, 5734–5739.

PlasmoDB. (2001) PlasmoDB: an integrative database of the *Plasmodium falciparum* genome. Tools for accessing and analyzing finished and unfinished sequence data. The Plasmodium Genome Database Collaborative. Nucleic Acids Res. 29, 66–69.

Quentin, Y. and Fichant, G. (2000) ABCdb: an ABC transporter database. J. Mol. Microbiol. Biotechnol. 2, 501–504.

Quentin, Y., Fichant, G. and Denizot, F. (1999) Inventory, assembly and analysis of *Bacillus subtilis* ABC transport systems. J. Mol. Biol. 287, 467–484.

Reed, M.B., Saliba, K.J., Caruana, S.R., Kirk, K. and Cowman, A.F. (2000) Pgh1 modulates sensitivity and resistance to multiple antimalarials in *Plasmodium falciparum*. Nature 403, 906–909.

Rubio, J.P. and Cowman, A.F. (1994) *Plasmodium falciparum*: the pfmdr2 protein is not overexpressed in chloroquine-resistant isolates of the malaria parasite. Exp. Parasitol. 79, 137–147.

Schinkel, A.H., Smit, J.J., van Tellingen, O., Beijnen, J.H., Wagenaar, E., van Deemter, L., *et al.* (1994) Disruption of the mouse mdr1a P-glycoprotein gene leads to a deficiency in the blood-brain barrier and to increased sensitivity to drugs. Cell 77, 491–502.

Shahi, S.K., Krauth-Siegel, R.L. and Clayton, C.E. (2002) Overexpression of the putative thiol conjugate transporter Tb MRPA causes melarsoprol resistance in *Trypanosoma brucei*. Mol. Microbiol. 43, 1129–1138.

Shaked-Mishan, P., Ulrich, N., Ephros, M. and Zilberstein, D. (2001) Novel intracellular Sb(V) reducing activity correlates with antimony susceptibility in *Leishmania donovani*. J. Biol. Chem. 276, 3971–3976.

Slater, A.F. and Cerami, A. (1992) Inhibition by chloroquine of a novel haem polymerase enzyme activity in malaria trophozoites. Nature 355, 167–169.

Sullivan, D.J., Jr, Gluzman, I.Y., Russell, D.G. and Goldberg, D.E. (1996) On the molecular mechanism of chloroquine's antimalarial action. Proc. Natl Acad. Sci. USA 93, 11865–11870.

Sundar, S., More, D.K., Singh, M.K., Singh, V.P., Sharma, S., Makharia, A., Kumar, P.C. and Murray, H.W. (2000) Failure of pentavalent antimony in visceral leishmaniasis in India: Report from the Center of the Indian Epidemic. Clin. Infect. Dis. 31, 1104–1107.

Taglicht, D. and Michaelis, S. (1998) *Saccharomyces cerevisiae* ABC proteins and their relevance to human health and disease. Methods Enzymol. 292, 130–162.

Thompson, J.D., Higgins, D.G. and Gibson, T.J. (1994) CLUSTAL W: improving the sensitivity of progressive multiple sequence alignment through sequence weighting, position-specific gap penalties and weight matrix choice. Nucleic Acids Res. 22, 4673–4680.

Tommasini, R., Evers, R., Vogt, E., Mornet, C., Zaman, G.J., Schinkel, A.H., Borst, P. and Martinoia, E. (1996) The human multidrug resistance-associated protein functionally complements the yeast cadmium resistance factor 1. Proc. Natl Acad. Sci. USA 93, 6743–6748.

Torres, C., Barreiro, L., Dallagiovanna, B., Gamarro, F. and Castanys, S. (1999) Characterization of a new ATP-binding cassette transporter in *Trypanosoma cruzi* associated to a L1Tc retrotransposon. Biochim. Biophys. Acta 1489, 428–432.

Upcroft, P. and Upcroft, J.A. (2001) Drug targets and mechanisms of resistance in the anaerobic protozoa. Clin. Microbiol. Rev. 14, 150–164.

Ursos, L.M., Dzekunov, S.M. and Roepe, P.D. (2000) The effects of chloroquine and verapamil on digestive vacuolar pH of *P. falciparum* either sensitive or resistant to chloroquine. Mol. Biochem. Parasitol. 110, 125–134.

van Es, H.H., Karcz, S., Chu, F., Cowman, A.F., Vidal, S., Gros, P. and Schurr, E. (1994) Expression of the plasmodial pfmdr1 gene in mammalian cells is associated with increased susceptibility to chloroquine. Mol. Cell Biol. 14, 2419–2428.

Weise, F., Stierhof, Y.D., Kuhn, C., Wiese, M. and Overath, P. (2000) Distribution of GPI-anchored proteins in the protozoan parasite *Leishmania*, based on an improved ultrastructural description using high-pressure frozen cells. J. Cell Sci. 113, 4587–4603.

Wellems, T.E., Panton, L.J., Gluzman, I.Y., do Rosario, V.E., Gwadz, R.W., Walker-Jonah, A. and Krogstad, D.J. (1990) Chloroquine resistance not linked to mdr-like genes in a *Plasmodium falciparum* cross [see comments]. Nature 345, 253–255.

Wellems, T.E., Walker-Jonah, A. and Panton, L.J. (1991) Genetic mapping of the chloroquine-resistance locus on *Plasmodium falciparum* chromosome 7. Proc. Natl Acad. Sci. USA 88, 3382–3386.

White, N.J. (1998) Drug resistance in malaria. Br. Med. Bull. 54, 703–715.

Wilson, C.M., Serrano, A.E., Wasley, A., Bogenschutz, M.P., Shankar, A.H. and Wirth, D.F. (1989) Amplification of a gene related to mammalian mdr genes in drug-resistant *Plasmodium falciparum*. Science 244, 1184–1186.

Xu, M., Molento, M., Blackhall, W., Ribeiro, P., Beech, R. and Prichard, R. (1998) Ivermectin resistance in nematodes may be caused by alteration of P-glycoprotein homolog. Mol. Biochem. Parasitol. 91, 327–335.

Zalis, M.G., Wilson, C.M., Zhang, Y. and Wirth, D.F. (1993) Characterization of the pfmdr2 gene for *Plasmodium falciparum* [published erratum appears in Mol. Biochem. Parasitol. 1994 Feb; 63(2): 311]. Mol. Biochem. Parasitol. 62, 83–92.

The Plant ABC Transporter Superfamily: the Functions of a Few and Identities of Many

CHAPTER 17

Philip A. Rea, Rocío Sánchez-Fernández, Sixue Chen, Mingsheng Peng, Markus Klein, Markus Geisler and Enrico Martinoia

Introduction

Plants are a particularly rich source of ABC proteins. The genome of the model plant *Arabidopsis thaliana* is capable of encoding a total of 129 ABC proteins, which fall into 12 or more subfamilies (Sánchez-Fernández *et al.*, 2001a). This gene count outstrips those for the human genome and any other animal genome sequenced to date. The human genome (31 500 open reading frames, ORFs) is estimated to encode a mere 48 ABC proteins (http://www.humanabc.org) while those of the nematode worm, *Caenorhabditis elegans* (19 000 ORFs) (http://www.proteome.com/cqui-bin/databases/wwwsearch.cqi), and the fruit fly, *Drosophila melanogaster* (13 600 ORFs) (http://flybase.bio.indiana.edu), encode only 58 and 51, respectively.

This characteristic of plants is equally impressive when viewed in another way – from a transport standpoint. Of the total ABC protein ORFs identified in *Arabidopsis*, at least 103 probably encode membrane proteins – proteins possessing membrane spans contiguous with one or two nucleotide-binding domains (NBDs) (Sánchez-Fernández *et al.*, 2001a). When account is taken of this and the rough equivalence of the total transporter ORF complements of *Arabidopsis* (700 ORFs) with those of *C. elegans* (650 ORFs) and *Drosophila* (600 ORFs), it is evident that *Arabidopsis* allocates nearly twice as many of its transporter genes to ABC transporters as *C. elegans* and *Drosophila*.

In the discussion that follows, we use the *Arabidopsis* ABC protein superfamily as a model for plants in general. *Arabidopsis* is the first plant, indeed the first multicellular organism, whose ORFs encoding ABC proteins have been systematically inventoried in their entirety (Sánchez-Fernández *et al.*, 2001a). The nomenclature adopted is one based on the acronyms of the prototypes, when available, of a particular ABC protein subfamily defined from another source. Searches of the earlier literature and comparisons with the ABC proteins found in organisms other than plants are thereby simplified and expedited. Where possible, a nomenclature similar to that applied to the ABC protein-encoding ORFs of yeast (DeCottignies and Goffeau, 1997) has been favored because yeast null mutants for ABC proteins have been, and will probably continue to be, the system of choice for defining the functional characteristics of plant ABC proteins.

ABC Proteins: From Bacteria to Man
ISBN 0-12-352551-9

SUBFAMILIES OF *ARABIDOPSIS* ABC PROTEINS

The 12 subfamilies (compare with Chapters 1 and 3) to which most of the *Arabidopsis* ABC proteins can be assigned are: those consisting of full-molecule transporters, the MDRs (multidrug resistance homologues), MRPs (multidrug resistance-associated protein homologues), PDRs (pleiotropic drug resistance homologues) and AOHs (ABCA1 homologues); those consisting of half-molecule transporters, the PMPs (peroxisomal membrane protein, PXA1, homologues), WBCs (white–brown complex homologues), ATHs (ABCA4 homologues), ATMs (ABC transporters of the mitochondrion homologues) and TAPs (transporter associated with antigenic processing homologues); those consisting of non-membrane proteins, the RLIs (2′,5′-oligoadenylate-activated RNase L inhibitor homologues), GCNs (yeast GCN 20, general control non-repressible, homologues) and SMCs (structural maintenance of chromosomes homologues) **(Figure 17.1)**. The only *Arabidopsis* ABC proteins that cannot be categorized in this way are the NAPs (non-intrinsic ABC proteins), a heterogeneous group of non-membrane or non-intrinsic membrane proteins that do not bear a close resemblance to each other or to representatives of specific ABC protein subfamilies from other organisms **(Figure 17.1)**.

In most cases, the subfamily assignments made on the basis of protein size (full-, half- or quarter-molecule), orientation (forward or reverse), the presence or absence of idiotypic (transmembrane and/or linker) domains, and overall sequence similarity are robust, as judged by the results of both maximum parsimony- or distance with neighbor-based phylogenetic analyses **(Figure 17.2)**. For eight of the *bona fide* ABC protein subfamilies containing two or more members, the members of each subfamily cluster more tightly with each other than with the members of other subfamilies **(Figure 17.2)**. Hence, all of the MDRs, MRPs, PDRs, ATMs, TAPs, RLIs, SMCs and GCNs cluster within their respective subfamilies. Likewise, the WBCs form a tight cluster, with the exception of WBC25 and WBC29, which nonetheless distribute immediately adjacent to the main WBC cluster, and AOH1, the sole AOH in *Arabidopsis*, groups only with itself,

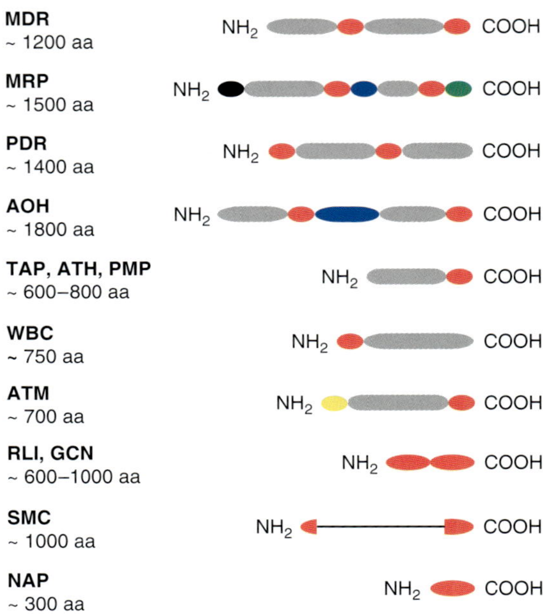

Figure 17.1. Putative domain organizations of representative members of different ABC subfamilies found in **Arabidopsis.** *The subfamilies shown are the MDRs (multidrug resistance homologues), MRPs (multidrug resistance-associated protein homologues), PDRs (pleiotropic drug resistance homologues), AOH (ABCA1 homologue), TAPs (transporter associated with antigenic processing homologues), ATHs (ABC two homologues), PMPs (peroxisomal membrane protein, PXA1, homologues), WBCs (white–brown complex homologues), ATMs (ABC transporters of the mitochondrion homologues), RLIs (2′,5′-oligoadenylate-activated RNase inhibitor homologues), GCNs (yeast GCN 20, general control non-repressible, homologues), SMCs (structural maintenance of chromosomes homologues) and NAPs (non-intrinsic ABC proteins). The different domains are color-coded as follows: red, NBDs 1 and 2; black, hydrophobic N-terminal extension (TMD0); gray, TMDs 1 and 2; blue, linker domain; green, C-terminal extension; yellow, putative targeting peptide.*

further confirming the likely validity of the tenth subfamily designation.

The two exceptions to this pattern are the ATHs and PMPs **(Figure 17.2)**. While 11 members of the ATH subfamily group within the same clade, five (AtATH8, AtATH9, AtATH10, AtATH12, AtATH13) do not, despite their topological equivalence to ABC2. However, with the exception of AtATH12, which groups within the MDR/MRP/TAP/ATM clade, none

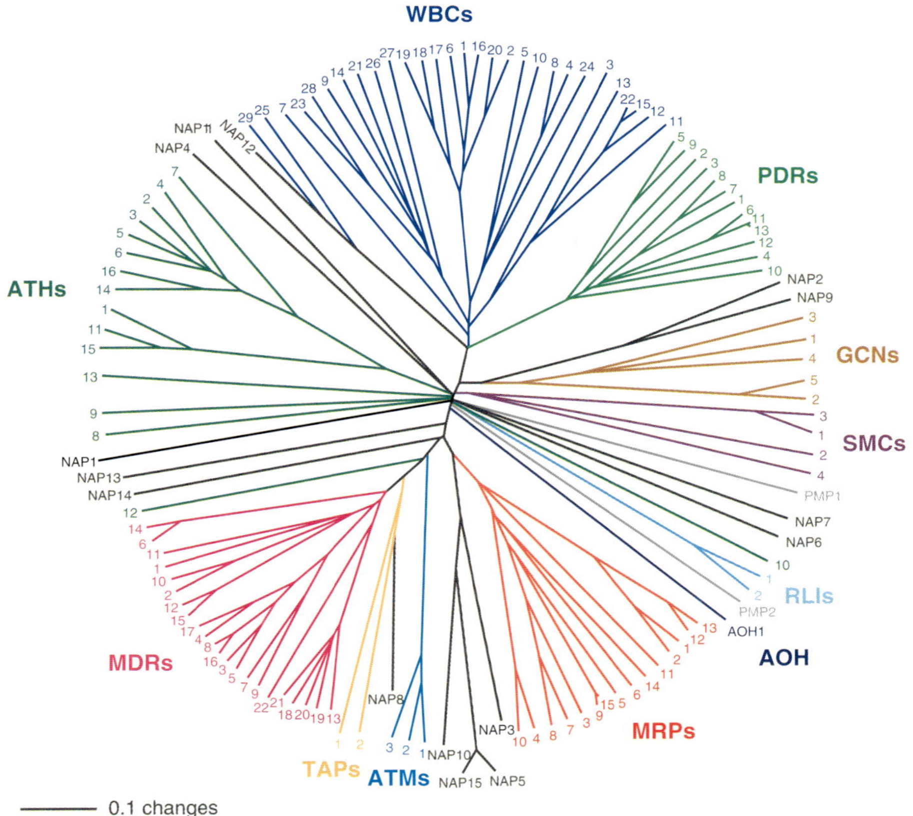

Figure 17.2. Phylogenetic analysis of **Arabidopsis** *ABC protein sequences (after Sánchez-Fernández* **et al.**, *2001a). Protein (ABC domain) sequences were aligned using ClustalX (Thompson* **et al.**, *1994) and subjected to phylogenetic analysis by the distance from neighbor-joining method (Saitou and Nei, 1987) using PAUP4.04a (Swofford, 1999).*

of the outlying AtATHs group within any of the other subfamilies. The lack of coherence within the PMP subfamily – the finding that AtPMP1 and AtPMP2 do not group within any of the other subfamilies or with each other – further confirms that, of the several subfamilies identified, this is the most tentative. The basic difficulty in classifying the *Arabidopsis* PMPs was that while the one that exhibited the weakest alignment with the PMPs from other organisms (AtPMP1) had a PMP-like topology, the one that exhibited the strongest alignment (AtPMP2) has the characteristics of a full-molecule instead of a half-molecule transporter.

Here we will summarize what is and is not known about representatives from these subfamilies not only in *Arabidopsis* but also in those few other plants for which the molecular basis of the ABC transporter-like activities identified are known or have been inferred, at least to a first approximation.

Full-Molecule ABC Transporters

MDRs

The MDRs **(Figure 17.1)**, with 22 members **(Figure 17.2)**, constitute the second largest ABC protein subfamily and the largest full-molecule ABC transporter subfamily in *Arabidopsis*. An *Arabidopsis* MDR (*AtPGP1* alias *AtMDR1*) was the first plant ABC protein gene to be cloned (Dudler and Hertig, 1992) and several other MDR subclass members have since been cloned from *Arabidopsis* and other plant sources (Davies *et al.*, 1996; Noh *et al.*, 2001; Wang *et al.*, 1997).

Despite elegant investigations of transgenic sense plants overexpressing or antisense plants underexpressing *AtPGP1* (alias *AtMDR1* in **Figures 17.1 and 17.2**) (Sidler *et al.*, 1998) and of

T-DNA insertion mutants for *AtMDR1* (alias *AtMDR11* in **Figures 17.1 and 17.2**) and/or *AtPGP1* (Noh *et al.*, 2001), both of which provide strong indications of some of the processes in which plant MDRs might participate, their mode or modes of action and those of their homologues from other plant sources remain obscure. A major obstacle to the ascription of function has been the marked differences in the phenotypes associated with a given *AtMDR* gene depending on how and in which laboratory it has been manipulated. Sidler *et al.* (1998), on the one hand, infer that AtPGP1 is a plasma membrane protein **(Figure 17.3)** involved in low fluence light-dependent hypocotyl elongation. At low photon fluences (10 μ mol m^{-2} s^{-1}), plants ectopically overexpressing the *AtPGP1* sense transcript undergo a 50% increase in hypocotyl length versus wild-type plants, while plants expressing *AtPGP1* antisense transcript undergo an approximately 20% decrease in hypocotyl length. In complete darkness and at high photon fluences

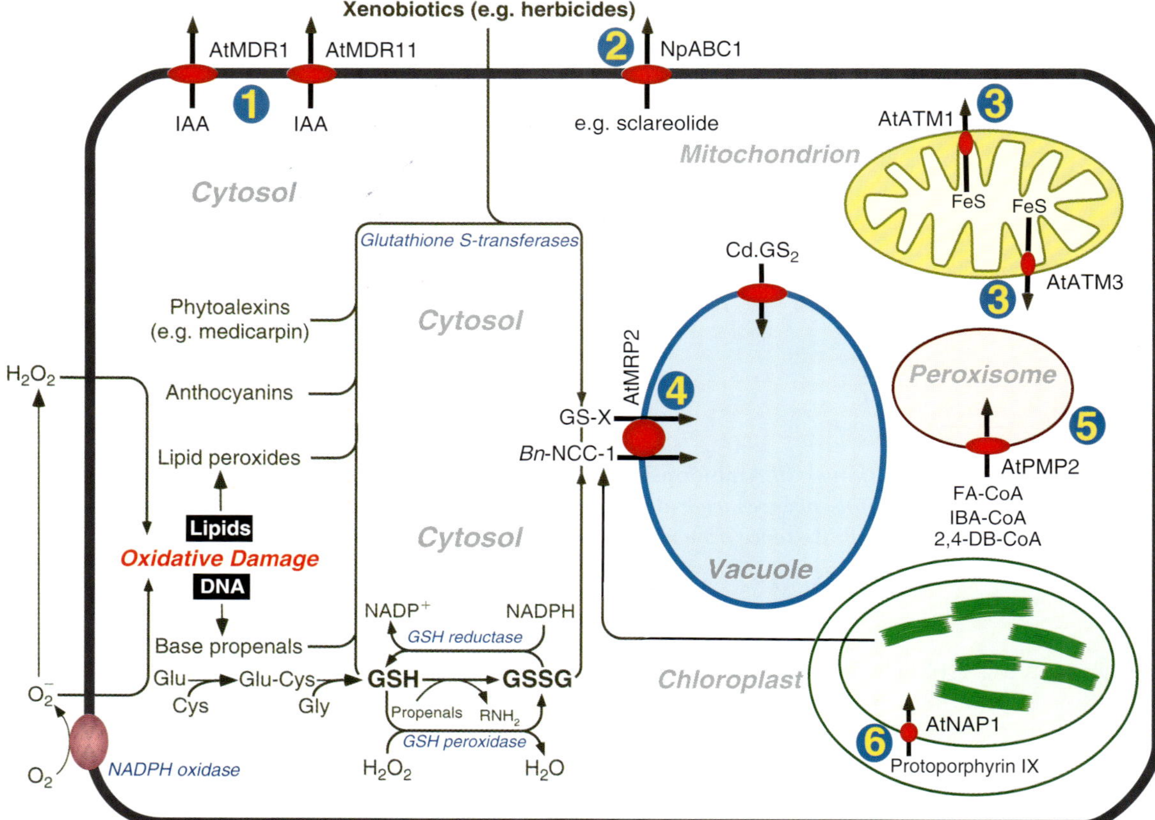

Figure 17.3. Schematic diagram summarizing some of the processes that may converge and depend on plant ABC transporters. The examples shown are: (1) Polar auxin (indole-3-acetic acid, IAA) transport across the plasma membrane by AtMDR1 (alias AtPGP1) and AtMDR11 (alias AtMDR1) (Noh et al.*, 2001). (2) Extrusion of diterpenes, such as the antifungal terpene sclareolide, by NpABC1 (Jasinski* et al.*, 2001) and presumably other plant PDR5 homologues. (3) Export of FeS centers from the mitochondrial matrix into the cytosol by AtATM3 (alias STA1) and also possibly AtATM1 (alias STA2) (Kushnir* et al.*, 2001; R. Sánchez-Fernández, S. Chen and P.A. Rea unpublished). (4) Detoxification of herbicides and other organic xenobiotics, removal of the products of DNA and lipid oxidation, and the vacuolar sequestration of isoflavonoid phytoalexins (e.g. medicarpin), anthocyanins and chlorophyll catabolites (e.g.* Bn-NCC-1*) by some AtMRPs, as exemplified by AtMRP2 (Liu* et al.*, 2001; Lu* et al.*, 1998;). (5) Transport of acyl-CoA derivatives, including fatty acyl-CoA (FA-CoA), indole-3-butyric acyl-CoA (IBA-CoA) and 2,4-dichlorophenoxybutyric acyl-CoA (2,4-DB-CoA) from the cytosol into the peroxisomal matrix by AtPMP2 (alias AtPXA1) (Zolman* et al.*, 2001). (6) Transport of the chlorophyll precursor, protoporphyrin IX, from the chloroplast envelope into the stroma by an ABC transporter in which AtNAP1 is a peripheral nucleotide-binding subunit (Møller* et al.*, 2001). Unlike their yeast counterpart (YCF1), none of the AtMRPs examined to date are competent in the transport of heavy-metal–GS complexes as exemplified by* bis*(glutathionato)cadmium (Cd.GS$_2$).*

(100 μmol $m^{-2}s^{-1}$), by contrast, no differences between the transgenic and wild-type plants are discernible. Noh *et al.* (2001), on the other hand, infer that the contributions of *AtPGP1* alone to morphology are minor in that *atpgp1-1* T-DNA insertion mutants and wild-type plants are indistinguishable, regardless of the developmental stage or illumination conditions. Instead, they attribute a more determinative role to *AtMDR1*. When compared with wild-type plants, *atmdr1-1* but not *atpgp1* single mutants exhibit downward-folded, or epinastic, cotyledons and first true leaves, and diminished apical dominance. Since this phenotype is simulated by the apical application of auxin to wild-type seedlings, Noh *et al.* (2001) propose that *AtMDR1* participates in or is at least required for the transport of auxin from the apical meristematic regions of synthesis to basal tissues **(Figure 17.3)**. Consistent with this hypothesis are the findings that *atmdr1-1* mutants are indeed impaired in the basipetal, but not acropetal, transport of exogenously applied radiolabeled indole-3-acetic acid (IAA) and that 1-naphthylphthalamic (NPA), a chemical inhibitor of polar auxin transport, can phenocopy the *atmdr1-1* mutant phenotype when applied to wild-type plants. On the basis of these findings and the finding that *atmdr1-1 atpgp1-1* double mutants exhibit a more pronounced *atmdr1-1*-type morphology and diminution of polar auxin transport, it has been tentatively concluded that the activity of AtMDR1 affects auxin transport, either directly because it is an auxin transporter or indirectly by an unknown mechanism, and that AtPGP1 may have a similar capability, albeit at a level that is only discernible when the activity of AtMDR1 has been eliminated genetically **(Figure 17.3)** (Noh *et al.*, 2001). Tantalizing, preliminary biochemical results consistent with the idea that AtMDR1 and AtPGP1 might be NPA-binding membrane proteins required for the coordinated transport of auxin from its site of synthesis are twofold: (i) Heterologous expression of AtMDR1 in yeast is associated with a small increase in cellular NPA binding. (ii) Microsomal membranes from *atmdr1-1* and *atpgp1-1* mutants are 35–40% and 20–25% impaired, respectively, in NPA binding by comparison with the same membrane fraction from wild-type plants (Noh *et al.*, 2001).

A recently encountered phenomenon that may in part reconcile the phenotype of *atmdr1-1* and/or *atpgp1-1* mutants with the failure of all attempts to demonstrate direct auxin transport by heterologously expressed AtMDR1 and AtPGP1 is the facility with which the latter interacts with immunophilins. Null mutation of the *Arabidopsis* FKBP- (immunosuppressant FK506 binding protein) like immunophilin *TWISTED DWARF* (*TWD1*) gene yields a pleiotropic phenotype essentially equivalent to that of *atmdr1-1 atpgp1-1* double mutants (M. Geisler, U. Kolukisaoglu, K. Billion, J. Berger, B. Saal, N. Frangne, Z. Konc-Kálmán, C. Koncz, R. Dudler, E. Martinoia and B. Schulz, unpublished). Moreover, and perhaps most importantly, the results of yeast two-hybrid and co-immunoprecipitation experiments indicate a direct physical interaction of TWD1 with AtPGP1 (M. Geisler, U. Kolukisaoglu, K. Billion, J. Berger, B. Saal, N. Frangne, Z. Konc-Kálmán, C. Koncz, R. Dudler, E. Martinoia and B. Schulz, unpublished). Further research is required but there is a distinct possibility that the assembly and/or activation of a transport-active complex by AtPGP1 is contingent on its association with TWD1.

MRPs

The second most highly represented subfamily of *Arabidopsis* full-molecule ABC transporters are the MRPs **(Figure 17.1)**, which number 15 **(Figure 17.2)**. Like the MDRs, the MRPs are forward-orientation full-molecules **(Figure 17.1)**. Unlike the MDRs, however, the MRPs are frequently larger, consisting of at least 1500 amino acid residues, and usually contain three additional subfamily-specific structures. These additional structures are an approximately 200 amino acid residue hydrophobic N-terminal extension, transmembrane domain (TMD0), containing five hydrophilicity minima (putative transmembrane segments); a 'linker' (L) domain contiguous with nucleotide-binding domain 1 (NBD1), rich in charged amino acid residues; and a hydrophilic C-terminal extension **(Figure 17.1)**.

Most studies of MRP-type transport processes in plants have focused on the vacuolar membrane. This is because it was through investigations of the uptake of two model GS-conjugates, *N*-ethylmaleimide-GS (NEM-GS) and *S*-(2,4-dinitrophenyl)-GS (DNP-GS) **(Figure 17.4)**, and the glutathionated chloroacetanilide herbicide, metolachlor (metolachlor-GS) **(Figure 17.4)**, by isolated vacuoles (Martinoia *et al.*, 1993) and vacuolar membrane vesicles from plants (Li *et al.*, 1995) that it was recognized that the transporters responsible bear a close functional resemblance to the GS-conjugate Mg^{2+}-ATPases (GS-X pumps) of mammalian cells and therefore

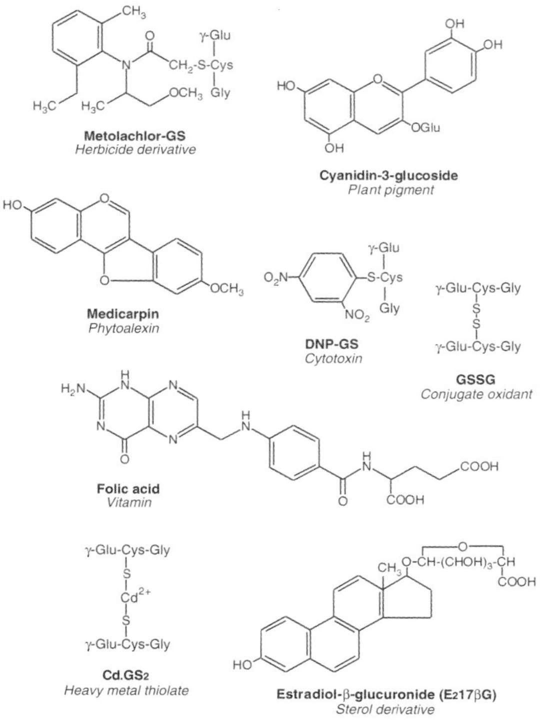

Figure 17.4. ***Examples of some of the compounds known to be transported by plant and/or yeast MRPs. Shown are glutathionated metolachlor (metolachlor-GS), cyanidin-3-glucoside (position of glutathionation unknown), medicarpin (position of glutathionation unknown), the model MRP substrate,*** **S*-(2,4-dinitrophenyl)glutathione (DNP-GS), oxidized glutathione (GSSG), folic acid,*** **bis*(glutathionato)cadmium (Cd.GS$_2$), and the model glucuronide, 17β-estradiol 17-(β-D-glucuronide) (E$_2$17βG).***

probably belong to the MRP subfamily of ABC transporters. These studies, though not precluding the localization of at least some MRP-type transporters on other membranes, have in turn given rise to the notion that plant MRPs are critical for the vacuolar compartmentation of many endogenous compounds that would otherwise be toxic if their accumulation were not in a compartment separate from the cytosol and the sequestration ('excretion storage') of xenobiotics that are subject to glutathione- (GSH-) dependent and/or other modes of conjugation **(Figure 17.3)** (Kreuz *et al.*, 1996; Rea *et al.*, 1998).

It is now established that many GS-conjugates and other amphipathic anions are accumulated in the vacuolar compartment of plant cells. *In situ* cytosolic glutathionation of monochlorobimane to its fluorescent derivative, bimane-GS, and accumulation of the latter in the vacuoles of protoplasts and cell suspension cultures has been demonstrated (Coleman *et al.*, 1997), as has a greater than 50-fold accumulation of the herbicide conjugate, alachlor-GS, in the intact vacuoles of barley leaves (Wolf *et al.*, 1996). Likewise, an MRP-like function is probably responsible for the vacuolar uptake of both glutathionated and non-glutathionated anionic dyes by barley aleurone cells (Rea *et al.*, 1998; Swanson *et al.*, 1998). Moreover, because of an increased appreciation of the range of compounds that might be amenable to glutathionation, primarily as a result of recognition of the operation of glutathione *S*-transferase- (GST-) catalyzed peroxidase reactions, the list of compounds, particularly endogenous compounds, that might undergo MRP-mediated vacuolar compartmentation has been expanded considerably to include products of the phenylpropanoid pathway, namely cinnamic acids, isoflavones and anthocyanins. Many of these compounds, though not susceptible to conjugation with GSH via a canonical GST-catalyzed reaction, electrophilic Michael addition, are amenable to glutathionation by GST-catalyzed radical addition (Dean and Devarenne, 1997).

In the general case of cinnamic acids and isoflavonoids, the implications are clear. An increase in GST transcripts and activities coincides with the production of isoflavonoid antimicrobial agents, phytoalexins, in elicitor-treated or infected cells (Levine *et al.*, 1994). The elicitation of cell cultures with fungal pathogens is considered to result in the partial metabolism of externally applied cinnamic acid to GS-conjugates during the synthesis of isoflavonoids and pterocarpans (Barz and Mackenbrock, 1994). Moreover, medicarpin **(Figure 17.4)**, the major isoflavonoid phytoalexin of alfalfa and several other legumes, is amenable to GST-catalyzed glutathionation *in vitro* and the product of this reaction, medicarpin-GS, is one of the most efficacious (lowest K_m, highest V_{max}) substrates for the vacuolar GS-X pump of the legume mung bean (*Vigna radiata*) (Li *et al.*, 1997a).

In the specific case of anthocyanin plant pigments **(Figure 17.4)**, the implications are more circumstantial and less straightforward, yet warrant further investigation. On the one hand, it has been known for some time that the bronze coloration of maize *Bronze-2* (*bz2*) mutants is due to accumulation of the anthocyanin, cyanidin-3-glucoside (C3G) **(Figure 17.4)**, in the cytosol. In wild-type plants anthocyanins are accumulated in the vacuole as purple or red derivatives, but in *bz2* mutants they are restricted to the cytosol, where they undergo oxidation and crosslinking to generate brown residues. Attributed to a deficiency in the delivery of anthocyanins into the vacuole consequent on mutational inactivation of the type III GST encoded by *Bz2* in *bz2* mutants, the original interpretation of this phenotype was that these mutants cannot deploy a GS-X pump for the delivery of anthocyanins into the vacuole (Marrs *et al.*, 1995). This notion is supported by the finding that chemically synthesized C3G GS-conjugate is subject to *in vitro* GS-X pump-mediated transport at rates four- to sevenfold greater than those measured for the model transport substrate DNP-GS (Lu *et al.*, 1998).

On the other hand, there are two major interpretative and/or technical difficulties associated with the scheme. The first is that all attempts to enzymically glutathionate C3G have failed. Neither Bz2 nor An9 (a sequence-divergent petunia *An9*-encoded type I GST, which is nevertheless able to complement the *Bz2* mutation *in planta* (Alfenito *et al.*, 1998)) glutathionate C3G *in vitro* but are active against the model GST substrate 1-chloro-2,4-dinitrobenzene. The second is that appreciable amounts of glutathionated anthocyanins, or their remnants in the form of sulfur-containing anthocyanin derivatives, either in the vacuole or elsewhere, have yet to be detected *in planta*. The significance of these observations is not known. Do they mean that other activities are operative *in vivo*, activities that are absent *in vitro*, and/or anthocyanin glutathionation, though necessary for transport into the vacuole, is transient *in vivo*, or that the Bz2 and An9 GSTs are primarily responsible for presenting or carrying bound anthocyanin to the pump instead of generating anthocyanin GS-conjugates? A hint that the last explanation, a carrier function for some GSTs, may be applicable is the finding that purified An9 binds flavonoids with moderate affinity but is incapable of catalyzing the formation of GS-flavonoids under the conditions examined (Mueller *et al.*, 2000).

To date, five unique *MRP*s, *AtMRP*s 1–5, have been cloned from *Arabidopsis* and shown to encode functional transporters after heterologous expression in *Saccharomyces cerevisiae*. All are capable of transporting GS-conjugates to different degrees *in vitro* and several are also competent in the transport of other amphipathic anions, including glucuronate conjugates, linearized tetrapyrrole catabolites, and the essential vitamin cofactor folate and its derivatives **(Figure 17.4)** (Gaedeke *et al.*, 2001; Liu *et al.*, 2001; Lu *et al.*, 1997, 1998; Peng *et al.*, 2002; Tommasini *et al.*, 1998). However, none of the AtMRPs that have been characterized so far are active in the transport of glutathione–heavy-metal complexes despite the facility of the first plant-like vacuolar GS-X pump to be identified, *S. cerevisiae* (yeast) cadmium factor 1 (YCF1) for the transport of both organic GS-conjugates (Li *et al.*, 1996; Tommasini *et al.*, 1996) and heavy-metal–GS complexes, as exemplified by *bis* (glutathionato) cadmium ($Cd.GS_2$) **(Figure 17.4)** (Li *et al.*, 1997b) and *tris* (glutathionato)arsenic ($As.GS_3$) (Gosh *et al.*, 1999).

The properties of AtMRPs 2, 4 and 5 are particularly noteworthy. AtMRP2, the sole AtMRP to have been unequivocally localized to the vacuolar membrane *in planta* is a high-capacity multispecific amphipathic anion pump competent in the transport of a broad range of substrates **(Figure 17.3)**. Heterologously expressed AtMRP2, like that in native plant vacuolar membranes (Klein *et al.*, 1998, 2000), is competent in the transport of not only GS-conjugates such as DNP-GS, metolachlor-GS and C3G-GS, but also linearized tetrapyrroles such as *Brassica napus* non-fluorescent chlorophyll catabolite 1 (*Bn*-NCC-1), glucuronides, for example 17β-estradiol 17-(β-D-glucuronide) ($E_2 17\beta G$), GSH and oxidized GSH (GSSG) **(Figure 17.4)** (Liu *et al.*, 2001). Moreover, on the basis of the stimulatory action of DNP-GS on the uptake of $E_2 17\beta G$ and vice versa, together with the results of double-label and preloading experiments demonstrating that the two substrates are subject to simultaneous transport by AtMRP2 consequent on *cis*, not *trans*, interactions, some GS-conjugates and glucuronides are inferred to reciprocally promote each other's transport.

The kinetics of AtMRP2-mediated transport are consistent with the scheme depicted in **Figure 17.5** in which: (i) $E_2 17\beta G$, DNP-GS, GSH and *Bn*-NCC-1 undergo transport via different AtMRP2-dependent pathways; (ii) $E_2 17\beta G$ and DNP-GS promote each other's transport by binding sites distinct from but tightly coupled to the other's transport pathway; (iii) GSH and its redox-inactive analog *S*-methyl-GS, although able to promote $E_2 17\beta G$ transport do so at a site distinct from that responsible for the promotion of $E_2 17\beta G$ transport by DNP-GS. The mechanistic basis of these properties is not known but their existence extends our appreciation of the processes that may converge on transporters of this type. In the case of GS-conjugate and tetrapyrrole transport, the processes that may

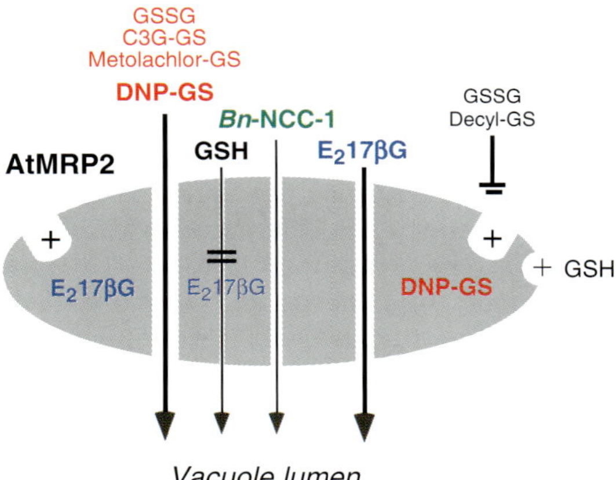

Figure 17.5. Schematic diagram depicting the interactions of $E_2 17\beta G$ and different GS derivatives with AtMRP2 (after Liu et al.*, 2001). DNP-GS, GSH,* Bn*-NCC-1 and $E_2 17\beta G$ are considered to be transported via semi-autonomous pathways. Other GS-conjugates, such as metolachlor-GS and C3G-GS compete with DNP-GS for transport. $E_2 17\beta G$ promotes DNP-GS transport and DNP-GS promotes $E_2 17\beta G$ transport by interacting with sites distinct from but coupled to the DNP-GS and $E_2 17\beta G$ transport pathways, respectively. GSH and* S*-methyl-GS promote $E_2 17\beta G$ transport by interacting with different sites. Decyl-GS and GSSG compete with DNP-GS for binding to the site responsible for promoting $E_2 17\beta G$ transport. $E_2 17\beta G$ blocks the transport of GSH.*

depend on AtMRP2 and its equivalents in plant species other than *Arabidopsis* may include not only xenobiotic detoxification, antimicrobial compound storage, protection from oxidative stress and cell pigmentation but also chlorophyll catabolism **(Figure 17.3)** (Rea, 1999; Rea *et al.*, 1998). In the case of glucuronides, the processes that may depend on AtMRP2 and AtMRP2-like transporters may be broadened to encompass the vacuolar storage of flavonoid antifeedants, UV screening agents and animal attractants. Although the biosynthesis and transport of glucuronate conjugates has been most extensively studied in animals, there is increasing evidence that at least some plants deploy similar mechanisms for the vacuolar sequestration of flavonoids (Martinoia *et al.*, 2000).

The most intriguing property of AtMRP4, one that it shares with AtMRP1, is its facility for the high-capacity transport of folate **(Figure 17.4)** (Peng *et al.*, 2002). In view of the importance of folates in the generation, interconversion and donation of one-carbon units in the metabolism of plants and all other organisms (Hanson and Gregory, 2002), the ability of at least some plant MRPs, like their mammalian counterparts (Zeng *et al.*, 2001), for folate transport may yield new insights into how to manipulate the folate levels of plants, the single most important source of folate for humans.

AtMRP5 is of special interest for other reasons – it is implicated in signal transduction or a process upon which signal transduction depends. As an extension of previous investigations of the effects of sulfonylureas, primarily glibenclamide, on stomatal guard cell function (Leonhardt *et al.*, 1997, 1999), it has been demonstrated that *AtMRP5* (*atmrp5-1*) T-DNA insertion mutants, unlike their wild-type counterparts, are insensitive to these agents (Gaedeke *et al.*, 2001). The stomata of wild-type plants, which together with the vascular system exhibit particularly high levels of expression of *AtMRP5* (Gaedeke *et al.*, 2001), are sensitive to glibenclamide-elicited opening in the dark; a phenomenon attributed to inhibition of the Ca^{2+}-elicited slow anion efflux that ordinarily accompanies closure (Leonhardt *et al.*, 1997, 1999). Stomata of *atmrp5-1* mutants, in striking contrast, are completely insensitive to this agent (Gaedeke *et al.*, 2001). Whether AtMRP5 directly modulates K^+ and/or anion channels, has channel activity itself, is directly involved in the intracellular storage of factors such as auxin-conjugates that might participate in the effects exerted by sulfonylureas, or is some other component of a signal transduction pathway involved in stomatal opening remains to be determined.

Of the 15 putative *MRP* genes in the *Arabidopsis* genome, two, *AtMRP11* and *AtMRP15*, clearly group with the other AtMRPs **(Figure 17.2)** but lack the TMD0 domain **(Figure 17.1)** characteristic of many members of this subfamily. There are two interpretations of this pattern. The first, suggested by the annotators of the *Arabidopsis* genome, is that these are pseudogenes. The second, given that this pattern is not unique to the AtMRPs but is also found among human and yeast MRP transporters, is that it may represent structural divergence within this subfamily. Human MRP4 and MRP5 also lack TMD0s but are nevertheless *bona fide* MRP-type transporters (Klein *et al.*, 1999). Similarly, of the six MRP subfamily members of yeast (YCF1, YOR1 and ORFs YHL035c, YLL048w, YKR103w, YLL015w), one, YOR1, despite its lack of a full-length TMD0, is functional and confers resistance to oligomycin (Katzmann *et al.*, 1995).

PDRs

The third subfamily of full-molecule ABC transporters in *Arabidopsis* is a group of proteins, the PDRs **(Figure 17.1)**, encoded by upwards of 13 ORFs **(Figure 17.2)**. Yeast PDR5, a reverse-orientation 1511-amino acid protein (Balzi *et al.*, 1994), is the prototype of this family. Considered to be a functional equivalent of the mammalian MDRs, albeit one having a reverse orientation and low overall sequence identity, yeast PDR5 is a plasma membrane protein that contributes to the cellular export of all of the anticancer drugs tested (Kolaczkowski *et al.*, 1996). Genes encoding *bona fide* PDR homologues have not been identified in the human genome, despite their being the most common ABC protein subfamily in yeast (9 out of 29 ORFs) and the fifth most common in *Arabidopsis*.

There are two reports concerned with *PDR5*-like genes from plants: *TUR2* in the aquaphyte *Spirodela polyrrhiza* (Smart and Fleming, 1996) and *NpABC1* in tobacco (*Nicotiana plumbaginifolia*) (Jasinski *et al.*, 2001). Initially cloned on the basis of its transcriptional activation by the stress hormone abscisic acid, which in *Spirodela* fronds induces turion (perennating or winter bud) formation, transcription of *TUR2* is enhanced by a broad range of stress factors, including cold and salinity, and diminished by kinetins. By analogy with its yeast homologue, TUR2 is suspected to mediate the excretion or mobilization of toxic metabolites whose production or release is elicited by stress factors.

The tobacco *PDR5* homologue *NpABC1* appears to encode a 160 kDa plasma membrane polypeptide whose levels are enhanced by the treatment of *N. plumbaginifolia* cell cultures with sclareolide, a close analogue of the antifungal diterpene scareol, produced by this plant species (Jasinski *et al.*, 2001). In agreement with the notion that NpABC1 might participate in terpene export **(Figure 17.3)**, the induction of *NpABC1* is associated with an increase in the capacity of cell cultures to exclude or extrude sclareolide derivatives in an energy-dependent manner (Jasinski *et al.*, 2001).

AOH

The *Arabidopsis* genome harbors one ORF (*AtAOH1*) whose translation product **(Figure 17.1)** bears a close resemblance to human ABCA1, a forward-orientation, full-molecule transporter, which localizes to the plasma membrane and Golgi complex (Oram, 2000). As is the case for human and mouse ABCA1, AtAOH1, an 1816-residue protein, is the largest ABC transporter polypeptide encoded by the *Arabidopsis* genome because it possesses an unusually large putative regulatory domain interrupted by a hydrophobic segment in the central linker region of the molecule **(Figure 17.1)**. AtAOH1 is the only full-length ABC transporter in *Arabidopsis* for which there is not a yeast homologue.

Mutation of ABCA1 in humans (Chapter 23) is responsible for Tangier disease, a rare autosomal recessive genetic disorder associated with severe high density lipoprotein deficiency, sterol deposition in tissue macrophages and systemic atherosclerosis (Lawn *et al.*, 1999). Its precise mode of action remains to be determined but ABCA1 appears to be involved in energy-dependent apolipoprotein-mediated lipid efflux (Oram, 2000). Whether AtAOH1 plays an analogous role, for instance in lipid accumulation during seed maturation or lipid mobilization during seed germination, remains to be determined.

HALF-MOLECULE TRANSPORTERS

PMPs

Yeast and animal peroxisomal membranes contain forward-orientation, half-molecule ABC transporters that catalyze the transport of long-chain acyl-CoA substrates (Liu *et al.*, 1999). Suspected to be active as homo- or heterodimers, the PMP subfamily of yeast contains two members (PXA1 and PXA2) and that of humans at least four members. Mutation of two of the human PMPs, ALDP and PNP70, is believed to be the cause of adrenoleukodystrophy and Zellweger syndrome, diseases associated with defective peroxisomal β-oxidation of long-chain fatty acids (Mosser *et al.*, 1994).

As explained in the Introduction, the authenticity of the PMP subfamily in *Arabidopsis* was not clear when the inventory was first compiled (Sánchez-Fernández *et al.*, 2001a), because of the two ORFs identified (*AtPMP1* and *AtPMP2*), one (*AtPMP2*) encoded a full-molecule rather than half-molecule transporter. Moreover, while there was a possibility that AtPMP2 was an annotation error whereby two immediately adjacent ORFs, each encoding a canonical PMP, might have been identified as one, this seemed remote in that all of the ORF prediction programs applied to these sequence data predicted only one (Sánchez-Fernández *et al.*, 2001a). Seminal, therefore, is the isolation and phenotypic characterization of an *Arabidopsis pxa1* mutant (Zolman *et al.*, 2001).

Originally isolated on the basis of the resistance of their roots to growth inhibition and the promotion of lateral root initiation by indole-3-butyric acid (IBA) and its analogue 2,4-dichlorophenoxybutyric acid (2,4-DB), but not IAA, *pxa1* mutants have a G to A mutation at position 5,559 of the coding sequence for AtPMP2 (Zolman *et al.*, 2001). When account is taken of the finding that the two most abundant endogenous auxins are IAA and IBA, the latter of which only differs from the former in having an additional two carbon atoms on its side-chain, and the discovery that *Arabidopsis pxa1* seedlings cannot catabolize fatty acids and consequently arrest developmentally unless sucrose is added to the growth medium, the probable explanation for the mutant phenotype emerges **(Figures 17.3 and 17.6)**. Namely, *Arabidopsis pxa1* mutants lack a functional AtPMP2 and are impaired in the peroxisomal import of fatty acyl-, IBA- and 2,4-DB-CoA derivatives. As a result they are unable to β-oxidize stored fatty acids, for the provision of a substrate (succinate) for respiratory energy production, or IBA or 2,4-DB, to yield the active auxins IAA and 2,4-dichlorophenoxyacetic acid (2,4-D), respectively (Zolman *et al.*, 2001). The discovery that plant peroxisomes appear to convert IBA to IAA and 2,4-DB to 2,4-D using a pathway resembling fatty acid β-oxidation through the deployment of an unusual, internally duplicated, PMP is exciting **(Figure 17.6)**.

Nothing is yet known of *AtPMP1* although, as confirmed by the identification of ESTs corresponding to this gene, it, like *AtPMP2*, is expressed *in vivo*. It will be instructive to determine if both AtPMP2 and AtPMP1 participate in peroxisomal β-oxidation or if the role of AtPMP1 is distinct from that of AtPMP2.

WBCs

One of the biggest surprises to come from the *Arabidopsis* ABC protein superfamily inventory

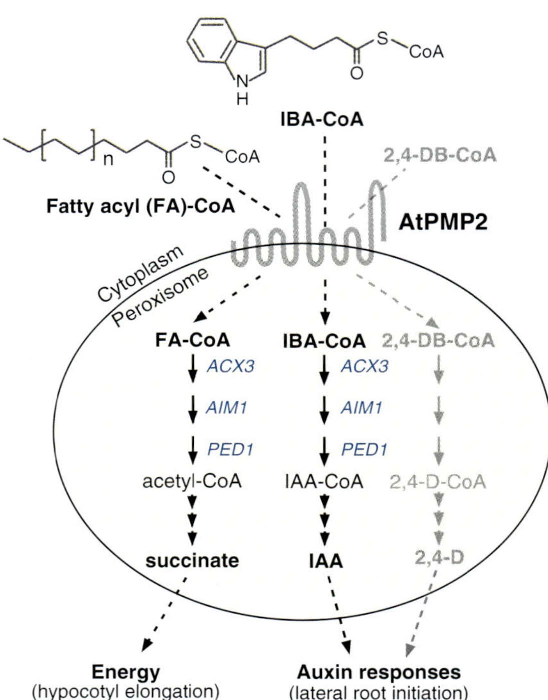

Figure 17.6. Model of the function of the full-molecule PMP, AtPMP2 (alias AtPXA1) proposed to explain the phenotype of **Arabidopsis pxa1** *mutants (after Zolman* **et al., 2001**; *copyright American Society of Plant Biologists, reproduced with permission). AtPMP2 is considered to localize to the peroxisomal membrane and catalyze the transport of FA-CoA, IBA-CoA and 2,4-DB-CoA into the peroxisomal matrix, where they are catabolized to succinate, IAA and 2,4-D, respectively, through the sequential action of acyl-CoA oxidase (ACX3), a multifunctional protein (AIM1) and thiolase (PED1). Succinate serves as an energy source for growth, for instance for seedling hypocotyl elongation, while IAA and 2,4-D, but not their immediate precursors IBA and 2,4-DB, are active as auxins that inhibit root elongation and promote lateral root initiation.*

was the identity of the largest subfamily, the WBCs **(Figure 17.1)**. The large size of the WBC subfamily in *Arabidopsis*, a collection of 29 forward-orientation, half-molecule transporters **(Figure 17.2)**, contrasts sharply with the situation in other organisms whose genomes have been or are being sequenced. For instance, the yeast genome contains only one WBC homologue (ADP1) of unknown function (Decottignies and Goffeau, 1997), and to date only three have been identified in the human genome: ABCA5 and ABCA8, which participate in cholesterol and phospholipid transport

(Berge *et al.*, 2000), and BCRP1, which contributes to multidrug resistance (Doyle *et al.*, 1998). Nothing is known of the function of the *Arabidopsis* WBCs, whether they serve roles analogous to those of their mammalian homologues and/or those of the prototypical *Drosophila* White, Brown and Scarlet subunits of the heterodimeric permease complexes responsible for the transport of pigment precursors such as guanine and tryptophan into eye pigment cells (Dreesen *et al.*, 1988).

Figure 17.7. Chlorotic and stunted phenotype of **AtATM3 (STA1)** *mutants (after Kushnir* **et al.**, *2001). A, Three-week-old* **Arabidopsis** *plants after growth on synthetic medium. Wild-type plants are shown on the left,* **sta1** *mutants on the right. B, Early flowering stage wild-type (left) and* **sta1** *mutant plants (right).*

ATHs

The ATH subfamily of 16 forward-orientation, half-molecule transporters and AtAOH1 (**Figures 17.1 and 17.2**) are the only *Arabidopsis* ABC proteins not represented in yeast. Ascription of a tentative function to the AtATHs is prohibited by the seemingly disparate functions served by their mammalian homologues. The human prototype ABCA2 (alias ABCR) is a retinal transporter, mutation of which is associated with serious eye conditions such as Stargardt disease and fundus flavimaculatus (Allikmets *et al.*, 1997; Sun *et al.*, 1999; Chapter 28), whereas its nearest equivalent ABCA3 is implicated in drug resistance (Klein *et al.*, 1999).

ATMs

The ATM subfamily **(Figure 17.1)** of *Arabidopsis* forward-orientation, half-molecule ABC transporters is small **(Figure 17.2)**. It contains only three ORFs: two that are immediately adjacent to each other on chromosome IV (*AtATM1* and *AtATM2*) and another on chromosome V (*AtATM3*) **(see Figure 17.9)**. All three encode putative proteins possessing canonical N-terminal mitochondrial targeting sequence cleavage sites, specifically ARV/FFF, ARV/MFF and GRL/FST, respectively.

In those that have been investigated in sufficient detail, a basic equivalence of function among the ATMs from different eukaryotes has been demonstrated. The prototypical ATM, yeast ATM1, localizes to the inner mitochondrial membrane and is inferred to participate in the assembly of cytosolic FeS proteins, possibly by mediating the transport of FeS centers from the mitochondrial matrix, where they are synthesized, into the cytosol (Kispal *et al.*, 1999). The human mitochondrial ATM homologues, ABCA7 and MTABC3, mutation of which is associated with iron storage diseases such as X-linked sideroblastic anemia and ataxia, complement yeast *atm1*$^-$ mutants (Csere *et al.*, 1998; Mitsuhashi *et al.*, 2000). Moreover, *Arabidopsis* AtATM3 (alias STA1), whose deficiency causes pronounced dwarfism and chlorosis **(Figure 17.7)** (Babiychuk *et al.*, 1997; Kushnir *et al.*, 2001), restores the maturation of cytosolic FeS proteins in, and suppresses the iron hyperaccumulation phenotype of, *atm1*$^-$ yeast (Kushnir *et al.*, 2001; R. Sánchez-Fernández, S. Chen and P.A. Rea, unpublished), implying functional equivalence and localization to the inner mitochondrial membrane **(Figure 17.3)**.

By comparison with AtATM3, neither AtATM1 nor AtATM2 has been characterized in much detail. Nevertheless, there are indications of at least some functional divergence. *AtATM1* and *AtATM3* are expressed constitutively but *AtATM2* transcripts are evident only in tissues exposed to heavy metals such as cadmium (R. Sánchez-Fernández, S. Chen and P.A. Rea, unpublished). Moreover, whereas AtATM1 (alias STA2) can partially substitute for AtATM3 both in yeast (R. Sánchez-Fernández , S. Chen and P.A. Rea, unpublished) and in the intact plant (Kushnir *et al.*, 2001), its capacity to suppress the iron hyperaccumulation phenotype of yeast *atm1*$^-$ mutants is marginal by comparison with AtATM3 (R. Sánchez-Fernández, S. Chen and P.A. Rea, unpublished).

TAPs

The *Arabidopsis* genome contains two *TAP*-like genes, *AtTAP1* and *AtTAP2* **(Figures 17.1 and**

17.2), both of which are transcriptionally active, as evidenced by the availability of expressed sequence tags (ESTs) for one of them (*AtTAP2*) and the amenability of both to amplification by reverse transcriptase-polymerase chain reaction (RT-PCR) (S. Chen, R. Sánchez-Fernández, M. Peng and P.A. Rea, unpublished). Indeed, both *AtTAP1* and *AtTAP2* have been cloned, heterologously expressed, and shown to have a broad expression pattern and encode proteins that localize to the membranes bounding small intracellular compartments (S. Chen, R. Sánchez-Fernández, M. Peng and P.A. Rea, unpublished). The precise functions of AtTAP1 and AtTAP2, or their equivalents from other plants, remain to be determined but it is suspected that they have capabilities analogous to those inferred for mammalian TAPs 1 and 2 and/or the yeast TAPs, MDL1 and MDL2.

Mammalian TAP1 and TAP2 participate in peptide secretion and translocation across the membranes of the endoplasmic reticulum. TAP1-TAP2 heterodimers actively transport peptide degradation products from the cytosol into the lumen of the endoplasmic reticulum, where they associate with class I molecules of the histocompatibility complex (Reits *et al.*, 2000). Yeast MDL1 and MDL2, by contrast, localize to the inner mitochondrial membrane, function as homodimers and at least in the case of MDL1 promote the mitochondrial export of 6–20 amino acid residue peptides (Young *et al.*, 2001). The physiological significance of the yeast peptide export pathway is a mystery because neither *MDL1* nor *MDL2* is essential for cell viability (Dean *et al.*, 1994).

Phylogenetic analyses reveal that AtTAP1 and AtTAP2 form a subcluster with yeast MDL1 and MDL2 and barley HvID17 (GenBank accession number AAG49003) on a branch distinct from that containing human TAP1 and TAP2 (S. Chen, R. Sánchez-Fernandez, M. Peng and P.A. Rea, unpublished). This might mean that AtTAP1 and/or AtTAP2 are more akin to yeast than mammalian TAPs, although the seemingly greater proximity of the plant TAPs to the MDLs is marginal in that binary comparisons of AtTAP1 and AtTAP2 against human TAP1 and TAP2 and yeast MDL1 and MDL2 yield identities and similarities of only 21–31% and 33–43%, respectively. It is appealing to speculate that AtTAPs 1 and/or 2 also catalyze transmembrane peptide translocation, for instance for storage protein mobilization during seed germination and/or for the delivery of pathogen-derived elicitory peptides to the pathogen response machinery.

'SOLUBLE' ABC PROTEINS

There are 26 ORFs in *Arabidopsis* encoding ABC proteins lacking contiguous transmembrane domains. Eleven of these are beyond the scope of this chapter because they probably encode *bona fide* soluble proteins. As detailed in Sánchez-Fernández *et al.* (2001a), these fall into three well-delineated subfamilies – the RLIs, GCNs and SMCs **(Figures 17.1 and 17.2)**. The remaining 15 so-called 'soluble' ABC proteins, the NAPs **(Figure 17.1)**, do not group in a coherent manner **(Figure 17.2)** and cannot yet be categorized but will almost certainly include a few that are the peripheral subunits of transporters.

NAPs

The NAPs are reminiscent of the peripheral ATP-binding quarter-molecule subunits of the many prokaryotic ABC transporters that have been characterized (Higgins, 1992): they are small, approximately 290 amino acid residue, proteins containing only a single NBD each **(Figure 17.1)**. Whether this is the case or some are genuine non-membrane-associated proteins cannot, however, be decided from the data available with one exception, AtNAP1.

Recent studies have established that *AtNAP1* (alias *LAF6*), when mutated by dissociation (Ds) element insertion, confers a recessive 'long after far red light' phenotype in which plants exhibit a markedly diminished response to continuous far red light. Unlike wild-type plants, *laf6* mutants sustain hypocotyl elongation and undergo diminished cotyledon expansion after irradiation with high intensity far red light (Møller *et al.*, 2001). Attributed to an impairment in import of the chlorophyll precursor protoporphyrin IX from the chloroplast envelope into the stroma **(Figure 17.3)**, the *laf6* mutation is associated with the hyperaccumulation of protoporphyrins, diminution of chlorophyll content and attenuation of far red light-regulated gene expression. This includes genes for chalcone synthase and ferredoxin $NADP^+$ oxidoreductase, possibly consequent on a loss of protoporphyin IX-mediated coordination between the plastidic and nuclear compartments (Møller *et al.*, 2001). In agreement with the notion that *laf6* mutants are deficient in chloroplast protoporphyrin IX import, AtNAP1 possesses an N-terminal plastid transit peptide which alone is sufficient for chloroplast targeting **(Figure 17.8)**. Full-length AtNAP1 and AtNAP1 transit peptide (TP-GFP) fusions, but not truncated AtNAP1 lacking the 63 amino acid TP sequence, localize exclusively to leucoplasts in onion epidermal cells, and anti-AtNAP1 antibodies demonstrate peripheral, envelope-localized immunostaining in wild-type but not *laf6* chloroplasts **(Figure 17.8)**.

The most straightforward explanation of these findings, supported by the capacity of the protoporphyrinogen IX oxidase (PPO) inhibitor, flumioxazin, to phenocopy the *laf6*

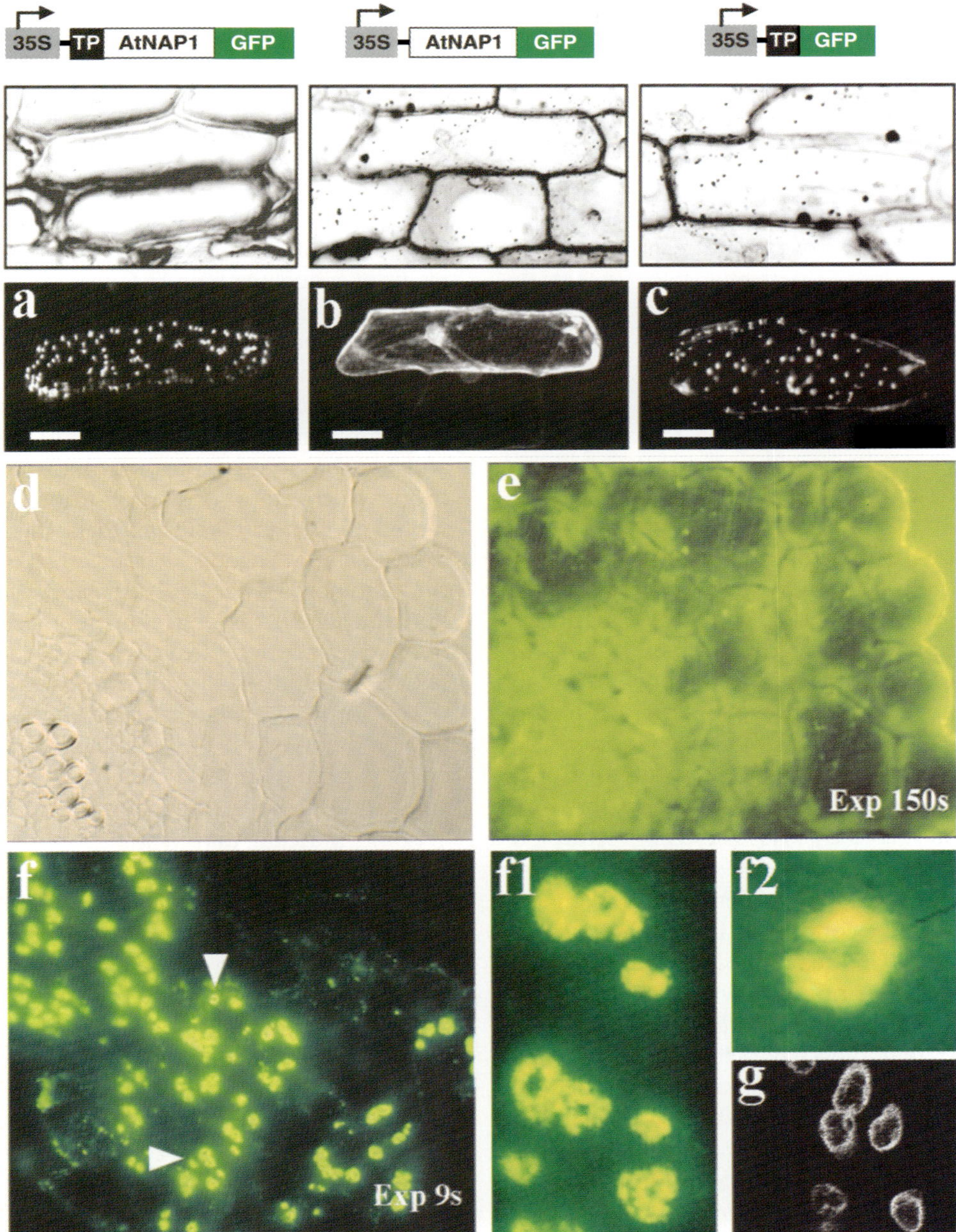

Figure 17.8. Plastid localization of AtNAP1 (alias LAF6) in transiently transformed onion epidermal cells and in **Arabidopsis** *seedlings (after Møller* **et al.**, *2001). a–c, Onion epidermis cells transiently expressing AtNAP1::GFP (a), AtNAP1::GFP lacking transit peptide (TP) (b) or TP::GFP (c). d–f,* **Arabidopsis** *seedlings. Bright-field image of leaf thin section from* **laf6** *mutant (d). Epi-fluorescence image of (d) after incubation with anti-AtNAP1 antibody (e). Epi-fluorescence image of leaf thin section from wild-type after incubation with anti-AtNAP1 antibody (f). f1 and f2, Higher magnification images of the regions indicated by arrows in f. g, Hypocotyl chloroplasts from transgenic* **Arabidopsis** *seedlings overexpressing AtNAP1::GFP. White bars in panels a–c denote 100 μm. Reproduced by permission of Cold Spring Harbor Laboratory Press from Møller* **et al.**, *2001.*

protoporphyrin accumulation phenotype when applied to wild-type plants (Møller *et al.*, 2001), is that AtNAP1 is involved in the transport of protoporphyrin. Specifically, *AtNAP1* is considered to encode a peripheral subunit of a quarter-molecule ABC transporter, whose chloroplast inner envelope localized integral subunit or subunits remain to be discovered, and to catalyze the ATP-energized transport of protoporphyrin into the stroma **(Figure 17.8)**. Another less straightforward explanation, which cannot yet be excluded, is that, rather than being a constituent subunit of a transporter complex directly responsible for protoporphyrin import, AtNAP1 is required for the activation of another transport system whose identity remains to be determined.

WHY SO MANY ABC PROTEIN GENES IN *ARABIDOPSIS*?

The factors responsible for the seemingly disproportionate allocation of coding sequences to ABC proteins – and of the total transporter gene complement to ABC transporters – in *Arabidopsis*, and presumably other plants, can only be speculated. However, there are two major considerations that probably apply. The first is the inherently greater capacity of transporters, such as ABC transporters, that are directly energized by ATP versus H^+ -coupled secondary transporters for establishing very steep concentration gradients across membranes under the conditions that prevail *in vivo* (Kreuz *et al.*, 1996; Rea and Sanders, 1987). The second is what is now known about the chromosome organization of *Arabidopsis*.

The fact that most characterized eukaryotic ABC transporters (albeit a small fraction of those that are known to exist) have been implicated in the transport of secondary metabolites, organic xenobiotics and other complex amphiphiles is pertinent given the extraordinary metabolic versatility of plants (Sánchez-Fernández *et al.*, 2001b). More than 100 000 secondary metabolites have been identified in plants (Harborne, 1993), most of which would be toxic to the cells that produce them, even at pharmacological concentrations, if they were not transported across membranes out of the compartments in which they are synthesized against steep concentration gradients (Martinoia *et al.*, 2000; Rea *et al.*, 1998). Further, while the capacity of green plants for photosynthesis greatly augments their metabolic versatility, this process and its photo-oxidative consequences place even greater demands on the cellular detoxification machinery. As if this is not enough, plants not only manufacture their own secondary products but also have to contend with those of other organisms – allelochemicals and microbial metabolites are examples, as are those associated with humans and their various environmentally destructive activities. In combination, these factors plus the sessile lifestyle of plants – their inability to exercise speedy avoidance behaviors – and their frequent lack of specialized excretory structures, necessitate cellular compartmentation and detoxification mechanisms of exquisite range and sophistication, ones that are reliant on the operation of ABC transporters (Sánchez-Fernández *et al.*, 2001b). By way of extension and validation of this reasoning, similar principles ostensibly apply to another gene family implicated in secondary metabolism and xenobiotic detoxification, the cytochromes P450, which are also disproportionately represented in plants versus animals. The genome of *Arabidopsis* encodes a total of 286 putative cytochromes P450 (The *Arabidopsis* Genome Initiative, 2000), three- to fourfold more than the numbers predicted from the genome sequences of *C. elegans* (73 ORFs) and *Drosophila* (94 ORFs).

The impact of plant chromosome organization on the numbers of genes in superfamilies is evident from **Figures 17.9** and **17.10**. Detailed analyses of the sequence of the *Arabidopsis* genome indicate that whole-genome duplication, followed by gene loss and extensive local tandem gene duplications, were instrumental in the assembly of the five chromosome complement of this organism (The *Arabidopsis* Genome Initiative, 2000). Not only are there estimated to be 24 duplicated segments of 100 kb or more in the *Arabidopsis* genome, which collectively represent 65.6 Mb or 58% of the genome, but alignments between these duplications also show that the number of copies of a gene and its counterparts in a given duplication often differ. For instance, for the one copy of a given gene on one chromosome there may be several on another due to tandem duplications and/ or gene losses after segmental duplication. Therefore the observation that as many as 90 members of the ABC protein gene superfamily probably fall within these duplicated regions is notable. The exact number of ABC protein genes encompassed by the duplications cannot yet be enumerated because the segmental

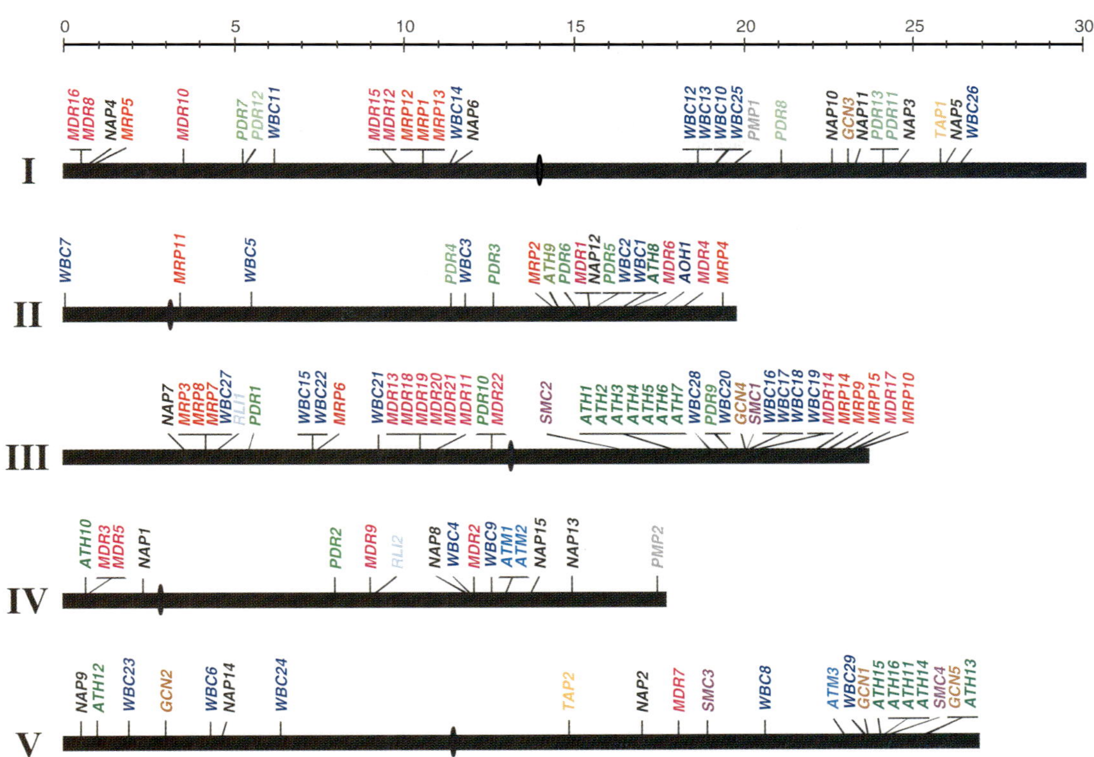

Figure 17.9. ***Chromosome locations of* Arabidopsis *ABC protein-encoding ORFs. Each chromosome is shown as a horizontal black bar intersected by a black ellipse depicting its centromere. Members of the different ABC protein subfamilies are color-coded as in Figure 17.2. ORFs that reside on the same bacterial artificial chromosome (BAC) are grouped together and shown in the order in which they are arranged on the BAC. The scale is in megabases (Mb) as defined by The* Arabidopsis *Information Resource (TAIR).***

duplication map was constructed before the *Arabidopsis* genome was sequenced, and map position (cM) and physical size (Mb) are not strictly collinear between chromosomes and between different segments of the same chromosome. Nevertheless it can be predicted with high confidence that a minimum of 41 and a maximum of about 90 ABC protein genes are encompassed by the duplicated chromosome segments.

By restricting attention to the 41 ABC protein genes that are definitely located within segmental duplications and the overall distribution of ABC protein genes in the genome, two basic patterns are discernible. First, those in which segments of different chromosomes or different segments of the same chromosome seem to have arisen by duplication, so increasing the ABC protein gene count. The direct segmental duplications at the ends of chromosomes II and III illustrate this phenomenon clearly **(Figure 17.10)**. Each of these duplications contains the genes for one AtPDR, two AtWBCs, two AtMDRs and one AtMRP in the same order (in the order AtPDR5-AtWBC2-AtWBC1-AtMDR6-AtMDR4-AtMRP4 on chromosome II; in the order AtPDR9-AtWBC20-AtWBC16-AtMDR14-AtMDR17-AtMRP10 on chromosome III). And, as would be predicted for duplicates, the similarities (identities) between the interchromosomal gene pairs (AtDPR5 with AtPDR9, AtWBC2 with AtWBC20, AtWBC1 with AtWBC16, etc.) exceed 86% (80%) **(Figure 17.10)**, an observation substantiated by their immediate proximity to each other in the unrooted tree for the *Arabidopsis* superfamily **(Figure 17.2)**. Other clear examples are the interchromosomal segmental duplications, some direct some in reverse orientation, encompassing AtMDRs 16 and 8 and 3 and 5, AtMRPs 1 and 2, AtWBCs 12 and 13 and 15 and 22, AtTAPs 1 and 2, AtATHs 1–7 and AtSMC2 and AtATHs 5, 6, 1, 14 and AtSMC4; and the intrachromosomal segmental duplications encompassing AtPDR7 and AtPDR8, and GCN2 and GCN5 **(Figure 17.10)**. The implication is that these segmental duplications, which served to increase the size of the *Arabidopsis* ABC protein gene family, were derived from each of the two

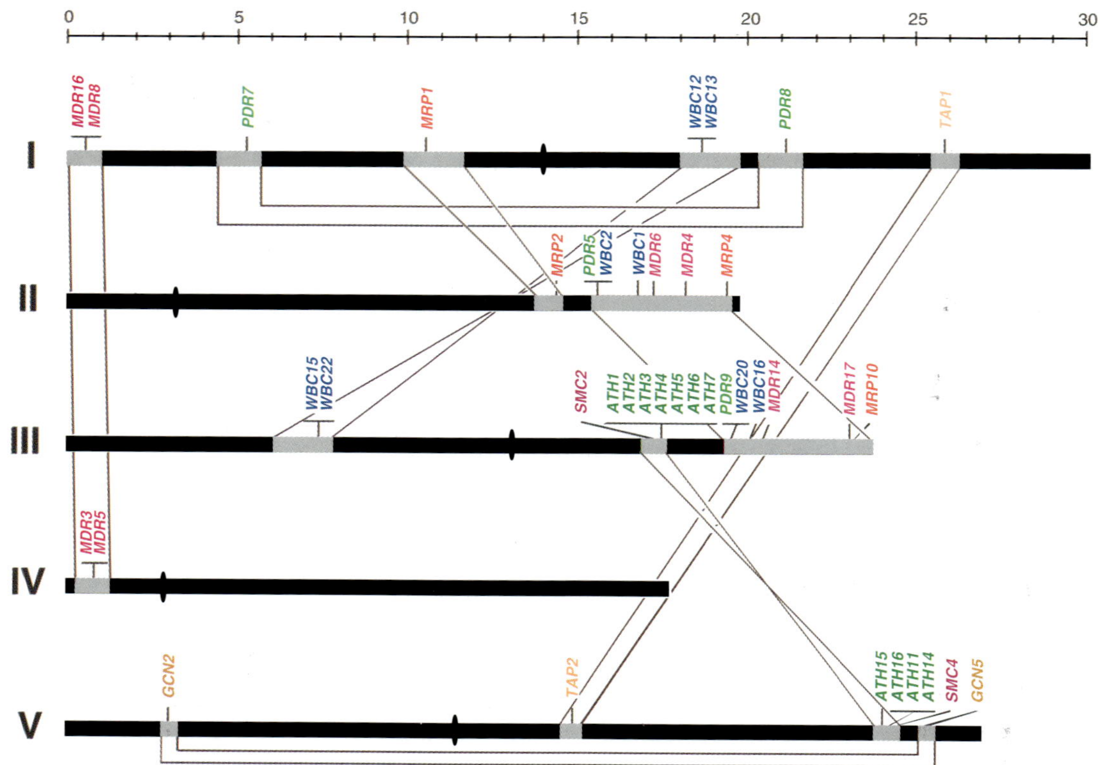

Figure 17.10. ***Examples of chromosome segmental duplications of the* Arabidopsis *genome that encompass ORFs encoding ABC proteins. For clarity, only 41 of a potential 90 or more ABC protein genes encompassed by segmental duplications and only 8 of a total of 24 segmental duplications are shown. The chromosomes are depicted as in Figure 17.9. Parallel lines depict duplications in the same orientation. Diagonal lines depict duplications in the reverse orientation. The scale is in Mb.***

ancestral *Arabidopsis* genomes. The second pattern of duplication, in which multiple ABC protein genes cluster both phylogenetically and chromosomally, presumably results from proximate tandem gene duplication. Examples of this phenomenon are numerous and include the majority of the genes shown in **Figure 17.9**, which are distributed as clusters of genes encoding ABC proteins belonging to the same subclass.

It is not known if the two genomes that fused to give rise to *Arabidopsis* were identical or different. While the high degree of sequence conservation between the ABC proteins encoded by duplicated segments is consistent with autotetraploid origins, there is also the possibility of allotetraploidy in that these genes may have been (and continue to be) subject to pronounced sequence conservation because of the stringent demands placed on the functional repertoire of their translation products. An important corollary therefore follows. When attempting to define the functions of ABC proteins in *Arabidopsis*, and presumably other plants, allowance should be made for functional redundancy while at the same time recognizing that because of the long periods of time, 110 million years or more (Ku *et al.*, 2000), since the ploidy increment there has been ample opportunity for the fixation of functional variants in the gene pool.

CONCLUDING REMARKS

Remarkable features of the *Arabidopsis* ABC protein complement, in addition to its sheer size, are its high content of half-molecule transporters, possession of PDR, AOH and ATH subfamily members and lack of a few subfamilies that have been established to mediate critical transport functions in other organisms.

Because most of the ABC proteins from eukaryotic sources that have been characterized to date are full-molecule transporters, most investigators have assumed that these are the prevalent morphologue. This is certainly not the case for *Arabidopsis,* in which half-molecule and full-molecule ABC transporters are approximately equally (51:52) represented **(Figure 17.2)**.

Arabidopsis, like *S. cerevisiae*, is notable for its high content of PDR homologues and lack of obvious cystic fibrosis transmembrane conductance regulator (CFTR), sulfonylurea receptor (SUR) and heavy metal tolerance factor 1 (HMT1) homologues. It is not known why the PDRs have such a distribution, but this does appear to be a general phenomenon in as much as BLAST searches of GenBank only disclose these proteins in plants and yeast, not in other organisms (Sánchez-Fernández *et al.*, 2001a). By the same token, the significance of the absence of CFTR and SUR homologues, their nearest equivalents being the MRPs, whose ORFs are well represented in the *Arabidopsis* genome, and the presence of AOH and ATH homologues, which are also absent from yeast, cannot yet be decided. The seeming absence of authentic HMT1 homologues, on the other hand, is potentially illuminating. Given that in the fission yeast *Schizosaccharomyces pombe* the half-molecule ABC transporter HMT1 is clearly implicated in the phytochelatin-dependent vacuolar sequestration and detoxification of cadmium (Ortiz *et al.*, 1992, 1995), a process that is also demonstrable in plants (Salt and Rauser, 1995; Vögeli-Lange and Wagner, 1990), a simple yet poignant conclusion follows: namely, that plants employ a subclass of ABC transporter distinct from that employed by *S. pombe* for this purpose. On the one hand, the absence of HMT1 homologues from *Arabidopsis* would explain why all attempts to clone this function have met with failure. On the other hand, it might imply that one or more of the other ABC transporters in the *Arabidopsis* inventory can assume the function of HMT1. The possibility that one or more of the AtMRPs might mediate this reaction has been discussed previously (Rea, 1999). Other possibilities are that completely unexpected ABC transporters serve this function or that one of the AtATMs, which like HMT1 are half-molecule transporters whose homologues in yeast are competent in the transport of metal thiolates (Kispal *et al.*, 1999), serve this function.

Daunting for some but exciting for others is the fact that investigations of the roles played by ABC transporters in plants are still in their infancy. In the last few years exciting new functional data have been gleaned for several plant ABC transporters, especially the MRPs, MDRs, PDRs, ATMs, PMPs and a NAP, but the number investigated is still less than about 10% of the likely total membership of the plant ABC protein superfamily. Regardless, however, investigations of this major class of transporters are not only going to be of deep biological significance but will also have profound biotechnological implications. Why? Because the mechanisms that plants deploy to protect themselves from their own toxic metabolites and those of others are the mechanisms that must be subverted and/or enhanced if plants are to accumulate compounds of high commercial value. A list of the types of compounds that are of interest biotechnologically, all of which are, or are derivatives of analogues of, secondary metabolites, amply illustrates this point. Among these are therapeutic alkaloids (etoposide, taxol, vinblastine, vincristine), neuro-, gluco- and/or lipido-active alkaloids (caffeine, nicotine), flavonoids (antioxidant and steroid mimetic), nutraceuticals (folates and other vitamins), plant pigments (anthocyanins, betanins, chlorophyll derivatives, carotenoids, xanthophylls), sterols, fatty acids, commercial polymer precursors (for instance, phenylpropanoids), herbicides and terpenoids (fragrances and flavor enhancers) to mention a few. If what we have learned from the few plant ABC transporters and ABC transporters from other sources that have been characterized is applicable to many more plant ABC transporters, the vast majority of the compounds of biotechnological interest will prove to be plant ABC transporter substrates.

Acknowledgments

The work from the Rea laboratory was supported by a United States Department of Agriculture National Research Initiative Competitive Grant (99-35304-8094). R.S.-F. was a Department of Energy/National Science Foundation/United States Department of Agriculture Plant Training Grant Research Fellow during the initial phases of this work and sponsored by PlantGenix, Inc., Philadelphia during the final phases. M.P. is a Natural Sciences and Engineering Research Council of Canada Postdoctoral Fellow. The work from the Martinoia laboratory was supported by the Swiss National Foundation and Humboldt Foundation.

References

Alfenito, M.R., Souer, E., Goodman, C.D., Buell, R., Mol, J., Kose, R. and Walbot, V. (1998) Functional complementation of anthocyanin sequestration in the vacuole by widely divergent glutathione *S*-transferases. Plant Cell 10, 1135–1149.

Allikmets, R., Shroyer, N.F., Singh, N., Seddon, J., Lewis, R.A., Bernstein, P.S., *et al.* (1997) Mutation of the Stargardt disease gene (ABCR) in age-related macular degeneration. Science 277, 1805–1807.

Babiychuk, E., Fuangthong, M., Van Montagu, M., Inzé, D. and Kushnir, S. (1997) Efficient gene tagging in *Arabidopsis thaliana* using a gene trap approach. Proc. Natl Acad. Sci. USA 94, 12722–12727.

Balzi, E., Wang, M., Leterme, S., Van Dyck, L. and Goffeau, A. (1994) PDR5, a novel yeast multidrug resistance conferring transporter controlled by the transcription regulator PDR1. J. Biol. Chem. 269, 2206–2214.

Barz, W. and Mackenbrock, U. (1994) Constitutive and elicitation-induced metabolism of isoflavones and pterocarpanes in chickpea (*Cicer arietnum*) cell suspension cultures. Plant Cell Tissue Organ Cult. 38, 199–211.

Berge, K., Tian, H., Graf, G.A., Yu, L., Grishin, N.V., Schultze, J., Kwiterovich, P., Shan, B., Barnes, R. and Hobbs, H.H. (2000) Accumulation of dietary cholesterol in sitosterolemia caused by mutations in adjacent ABC transporters. Science 290, 1771–1775.

Coleman, J.O.D., Blake-Kalff, M.M.A. and Davis, T.G.E. (1997) Detoxification of xenobiotics in plant cells by glutathione conjugation and vacuolar compartmentalization: a fluorescent assay using monochlorobimane. Plant Cell Environ. 20, 449–460.

Csere, P., Lill R. and Kispal, G. (1998) Identification of a human mitochondrial ABC transporter, the functional orthologue of yeast Atm1p. FEBS Lett. 441, 266–270.

Davies, T.G.E., Theodoulou, F.L., Hallahan, D.L. and Forde, B.G. (1996) Cloning and characterization of a novel P-glycoprotein homologue from barley. Gene 199, 195–202.

Dean, J.V. and Devarenne, T.P. (1997) Peroxidase-mediated conjugation of glutathione to unsaturated phenylpropanoids: evidence against glutathione *S*-transferase involvement. Physiol. Plant 99, 271–278.

Dean, M., Allikmets, R., Gerrard, B., Stewart, C., Kistler, A., Shafer, B., Michaelis, S. and Strathern, J. (1994) Mapping and sequencing of two yeast genes belonging to the ATP-binding cassette superfamily. Yeast 10, 377–383.

DeCottignies, A. and Goffeau, A. (1997) Complete inventory of the yeast ABC proteins. Nat. Genet. 15, 137–145.

Doyle, L.A., Yang, W., Abruzzo, L.V., Krogmann, T., Gao, Y., Rishi, A.K. and Ross, D.D. (1998) A multidrug resistance transporter from human MCF-7 breast cancer cells. Proc. Natl Acad. Sci. USA 95, 15665–15670.

Dreesen, T.D., Johnson, D.H. and Henikoff, S. (1988) The brown protein of *Drosophila melanogaster* is similar to the white protein and to components of active transport complexes. Mol. Cell. Biol. 8, 5206–5215.

Dudler, R. and Hertig, C. (1992) Structure of an *mdr*-like gene from *Arabidopsis thaliana*: evolutionary implications. J. Biol. Chem. 267, 5882–5888.

Gaedeke, N., Klein, M., Kolukisaoglu, U., Forestier, C., Muller, A., Ansorge, M., *et al.* (2001) The *Arabidopsis thaliana* ABC transporter AtMRP5 controls root development and stomatal movement. EMBO J. 20, 1875–1887.

Gosh, M., Jian, S. and Rosen, B.P. (1999) Pathways of As(III) detoxification in *Saccharomyces cerevisiae*. Proc. Natl Acad. Sci. USA 96, 5001–5006.

Hanson, A.D. and Gregory, J.F. (2002) Synthesis and turnover of folates in plants. Curr. Opinion Biotechnol. 5, 244–249.

Harborne, J. (1993) Introduction to Ecological Biochemistry, 4th edn. London: Academic Press.

Higgins, C.F. (1992) ABC transporters: from microorganisms to man. Annu. Rev. Cell Biol. 8, 67–113.

Holland, I.B. and Blight, M.A. (1999) ABC-ATPases, adaptable energy generators fueling transmembrane movement of a variety of molecules in organisms from bacteria to humans. J. Mol. Biol. 293, 381–399.

Jasinski, M., Stukkens, Y., Degand, H., Purnelle, B., Marchand-Brynaert, J. and Boutry, M. (2001) A plant plasma membrane ATP-binding cassette-type transporter is involved in antifungal terpenoid secretion. Plant Cell 13, 1095–1107.

Katzmann, D.J., Hallstrom, T., Voet, M., Mysock, W., Golin, G., Volckaert, G. and Moye-Rowley, W. (1995) Expression of an

ATP-binding cassette transporter-encoding gene (*YOR1*) is required for oligomycin resistance in *Saccharomyces cerevisiae*. Mol. Cell. Biol. 15, 6875–6883.

Kispal, G., Csere, P., Prohl, C. and Lill, R. (1999) The mitochondrial proteins Atm1p and Nsf1p are essential for biogenesis of cytosolic Fe/S proteins. EMBO J. 18, 3981–3989.

Klein, I., Sarkadi, B. and Varadi, A. (1999) An inventory of the human ABC proteins. Biochim. Biophys. Acta 1461, 237–262.

Klein, M., Martinoia, E. and Veissenbock, G. (1998) Directly energized uptake of β-estradiol-17-(β-D-glucuronide) in plant vacuoles is strongly stimulated by glutathione conjugates. J. Biol. Chem. 273, 262–270.

Klein, M., Martinoia, E., Hoffmann-Thoma, G. and Veissenbock, G. (2000) A membrane potential-dependent, ABC-like transporter mediates the vacuolar uptake of rye flavone glucuronides. Plant J. 21, 289–304.

Kolaczkowski, M., Van der Rest, M., Cybularz Kolaczkowski, A., Soumillion, J.P., Konings, W.N. and Goffeau, A. (1996) Anticancer drugs, ionophoric peptides and steroids as substrates of the yeast multidrug transporter Pdr5p. J. Biol. Chem. 271, 31543–31548.

Kreuz, K., Tommasini, R. and Martinoia, E. (1996) Old enzymes for a new job. Herbicide detoxification in plants. Plant Physiol. 111, 349–353.

Ku, H.-M., Vision, T., Liu, J. and Tanksley, S.D. (2000) Comparing sequenced segments of the tomato and *Arabidopsis* genome: large-scale duplication followed by selective gene loss creates a network of synteny. Proc. Natl Acad. Sci. USA 97, 9121–9126.

Kushnir, S., Babiychuk, E., Storozhenko, S., Davey, M.W., Papenbrock, J., De Rycke, R., *et al.* (2001) A mutation of the mitochondrial ABC transporter Sta1 leads to dwarfism and chlorosis in the *Arabidopsis* mutant *starik*. Plant Cell 13, 89–100.

Lawn, R.M., Wade, D.P., Garvin, M.R., Wang, X., Schwartz, K., Gordon Porter, J., Seilhamer, J.J., Vaughan, A.M. and Oram, J.F. (1999) The Tangier disease gene product ABC1 controls the cellular apolipoprotein-mediated lipid removal pathway. J. Clin. Invest. 104, R25–R31.

Leonhardt, N., Marin, E., Vavasseur, A. and Forestier, C. (1997) Evidence for the existence of a sulfonylurea receptor-like-protein in plants: modulation of stomatal movements and guard cell potassium channels by sulfonylureas and potassium channel openers. Proc. Natl Acad. Sci. USA 94, 14156–14161.

Leonhardt, N., Vavasseur, A. and Forestier, C. (1999) ATP-binding cassette modulators control abscisic acid-regulated slow anion channels in guard cells. Plant Cell 11, 1141–1152.

Levine, A., Tenhaken, R., Dixon, R. and Lamb, C. (1994) H_2O_2 from the oxidative burst orchestrates the plant hypersensitive disease resistance response. Cell 79, 583–593.

Li, Z.-S., Zhao, Y. and Rea, P.A. (1995) Magnesium adenosine 5′-triphosphate-energized transport of glutathione *S*-conjugates by plant vacuolar membrane vesicles. Plant Physiol. 107, 1257–1268.

Li, Z.-S., Szczypka, M., Lu, Y.-P., Thiele, D.J. and Rea, P.A. (1996) The yeast cadmium factor protein (YCF1) is a vacuolar glutathione *S*-conjugate pump. J. Biol. Chem. 271, 6509–6517.

Li, Z.-S., Alfenito, M., Rea, P.A., Walbot, V. and Dixon, R.A. (1997a) Vacuolar uptake of the phytoalexin medicarpin by the glutathione conjugate pump. Phytochemistry 45, 689–693.

Li, Z.-S., Lu, Y.-P., Zhen, R.-G., Szczypka, M., Thiele, D.J. and Rea, P.A. (1997b) A new pathway for vacuolar cadmium sequestration in *Saccharomyces cerevisiae*: YCF1-catalyzed transport of *bis*(glutathionato)cadmium. Proc. Natl Acad. Sci. USA 94, 42–47.

Liu, L.X., Janiver, K., Berteaux-Lecellier, V., Cartier, N., Benarous, R. and Aubourg, P. (1999) Homo- and heterodimerization of peroxisomal ATP-binding cassette half-transporters. J. Biol. Chem. 274, 32738–32743.

Liu, G.S., Sánchez-Fernández, R., Li, Z.-S. and Rea, P.A. (2001) Enhanced multispecificity of *Arabidopsis* vacuolar multidrug resistance-associated protein-type ATP-binding cassette transporter, AtMRP2. J. Biol. Chem. 276, 8648–8656.

Lu, Y.-P., Li, Z.-S. and Rea, P.A. (1997) *AtMRP1* gene of *Arabidopsis* encodes a glutathione *S*-conjugate pump: isolation and functional definition of a plant ABC-binding cassette transporter gene. Proc. Natl Acad. Sci. USA 84, 8243–8248.

Lu, Y.-P., Li, Z.-S., Drozdowicz, Y.M., Hörtensteiner, S., Martinoia, E. and Rea, P.A. (1998) AtMRP2, an *Arabidopsis* ATP-binding cassette transporter able to transport glutathione *S*-conjugates and chlorophyll catabolites: functional comparison with AtMRP1. Plant Cell 10, 267–282.

Marrs, K.A., Alfenito, M.R., Lloyd, A.M. and Walbot, V. (1995) A glutathione *S*-transferase involved in vacuolar transfer encoded by the maize gene *Bronze-2*. Nature 375, 397–400.

Martinoia, E., Grill, E., Tommasini, R., Kreuz, K. and Amrhein, N. (1993) ATP-dependent glutathione *S*-conjugate 'export' pumps in the vacuolar membrane of plants. Nature 364, 247–249.

Martinoia, E., Klein, M., Geisler, M., Sánchez-Fernández, R. and Rea, P.A. (2000) Vacuolar transport of secondary metabolites and xenobiotics. In: Vacuolar Compartments. Annual Plant Reviews (ed. D.G. Robinson and J.C. Rogers), Vol. 5, pp. 221–253. Sheffield, UK: Sheffield Academic Press.

Mitsuhashi, N., Miki, T., Senbongi, H., Yokoi, N., Yano, H., Miyazaki, M., *et al.* (2000) MTABC3, a novel mitochondrial ATP-binding cassette protein involved in iron homeostasis. J. Biol. Chem. 275, 17536–17540.

Møller, S.G., Kunkel, T. and Chua, N.-H. (2001) A plastidic ABC protein involved in intercompartmental communication of light signaling. Genes Dev. 15, 90–103.

Mosser, J., Lutz, Y., Stoeckel, M.E., Sarde, C-O., Kretz, C., Douar, A-M., Lopez, J., Auborg, P. and Mandel, J.-E. (1994) Putative X-linked adrenoleukodystrophy gene shares unexpected homology with ABC transporters. Hum. Mol. Genet. 3, 265–270.

Mueller, L.A., Goodman, C.D., Silady, R.A. and Walbot, V. (2000) AN9, a petunia glutathione *S*-transferase required for anthocyanin sequestration, is a flavonoid-binding protein. Plant Physiol. 123, 1561–1570.

Noh, B., Murphy, A.S. and Spalding, E.P. (2001) *Multidrug resistance*-like genes of *Arabidopsis* required for auxin transport and auxin-mediated development. Plant Cell 13, 2441–2454.

Oram, J.F. (2000) Tangier disease and ABCA1. Biochim. Biophys. Acta 1529, 321–330.

Ortiz, D.F., Kreppel, L., Speiser, D.M., Scheel, G., McDonald, G. and Ow, D. (1992) Heavy metal tolerance in the fission yeast requires an ATP-binding cassette-type membrane transporter. EMBO J. 11, 3492–3499.

Ortiz, D.F., Ruscitti, T., McCue, K.F. and Ow, D.W. (1995) Transport of metal-binding peptides by HMT1, a fission yeast ABC-type vacuolar membrane protein. J. Biol. Chem. 270, 4721–4728.

Peng, M., Sánchez-Fernández, R., Chen, S., Liang, S. and Rea, P.A. (2002) AtMRP4 is a high-capacity folate transporter. J. Biol. Chem., in preparation.

Rea, P.A. (1999) MRP subfamily ABC transporters from plants and yeast. J. Exp. Bot. 50, 895–913.

Rea, P.A. and Sanders, D. (1987) Tonoplast energization: two H^+ pumps, one membrane. Physiol. Plant 71, 131–141.

Rea, P.A., Li, Z.-S., Lu, Y.-P., Drozdowicz, Y.M. and Martinoia, E. (1998) From vacuolar GS-X pumps to multispecific ABC transporters. Annu. Rev. Plant Physiol. Plant Mol. Biol. 49, 727–760.

Reits, E.A.J., Vos, J.C., Gromme, M. and Neefjes, J. (2000) The major substrates for TAP *in vivo* and from newly synthesized proteins. Nature 404, 774–778.

Saitou, N. and Nei, M. (1987) The neighbor-joining method: a new method for reconstructing phylogenetic trees. Mol. Biol. Evol. 4, 406–425.

Salt, D.E. and Rauser, W.E. (1995) MgATP-dependent transport of phytochelatins across the tonoplast of oat roots. Plant Physiol. 107, 1293–1301.

Sánchez-Fernández, R., Davies, T.G.E., Coleman, J.O.D. and Rea, P.A. (2001a) The *Arabidopsis thaliana* ABC protein superfamily, a complete inventory. J. Biol. Chem. 276, 30231–30244.

Sánchez-Fernández, R., Davies, T.G.E., Coleman, J.O.D. and Rea, P.A. (2001b) Do plants have more genes than humans? Yes, when it comes to ABC proteins. Trends Plant Sci. 6, 347–348.

Sidler, M., Hassa, P., Hasan, S., Ringli, C. and Dudler, R. (1998) Involvement of an ABC transporter in a developmental pathway regulating hypocotyl cell elongation in the light. Plant Cell 10, 1623–1636.

Smart, C.C. and Fleming, A.J. (1996) Hormonal and environmental regulation of a plant PDR5-like ABC transporter. J. Biol. Chem. 271, 19351–19357.

Sun, H., Molday, R.S. and Nathans, J. (1999) Retinal stimulates ATP hydrolysis by purified and reconstituted ABCR, the photoreceptor-specific ATP-binding cassette transporter responsible for Stargardt disease. J. Biol. Chem. 274, 8269–8281.

Swanson, S.J., Bethke, P.C. and Jones, R.L. (1998) Barley aleurone cells contain two types of vacuoles: characterization of lytic organelles by use of fluorescent probes. Plant Cell 10, 685–698.

Swofford, D.L. (1999) PAUP*. Phylogenetic Analysis Using Parsimony (*and Other Methods). Version 4.02b. Sunderland, MA: Sinauer Associates.

The *Arabidopsis* Genome Initiative (2000) Analysis of the genome sequence of the flowering plant *Arabidopsis thaliana.* Nature 408, 796–815.

Thompson, J.D., Higgins, D.G. and Gibson, T.J. (1994) CLUSTAL W: improving the sensitivity of progressive multiple sequence alignment through sequence weighting, position-specific gap penalties and weight matrix choice. Nucleic Acids Res. 22, 4673–4680.

Tommasini, R., Evers, R., Vogt, E., Mornet, C., Zaman, G., Schinkel, A.H., Borst, P. and Martinoia, E. (1996) The human multidrug resistance-associated protein (MRP) functionally complements the yeast cadmium resistance factor (YCF1). Proc. Natl Acad. Sci. USA 93, 6743–6748.

Tommasini, R., Vogt, E., Fromenteau, M., Hörtensteiner, S., Matile, P., Amrhein, N. and Martinoia, E. (1998) An ABC-transporter of *Arabidopsis thaliana* has both glutathione *S*-conjugate and chlorophyll catabolite transport activity. Plant J. 13, 773–780.

Vögeli-Lange, R. and Wagner, G.J. (1990) Subcellular localization of cadmium and cadmium-binding peptides in tobacco leaves. Implications of a transport function for cadmium-binding peptides. Plant Physiol. 92, 1068–1093.

Wang, W., Takezawa, D. and Pooviah, B.W. (1997) A potato cDNA encodes a homologue of mammalian multidrug resistance P-glycoprotein. Plant Mol. Biol. 31, 683–687.

Wolf, A.E., Dietz, K.-J. and Schröder, P. (1996) Degradation of glutathione *S*-conjugates by a carboxypeptidase in the plant vacuole. FEBS Lett. 384, 31–34.

Young, L., Leonhard, K., Tatsuta, T., Trowsdale, J. and Langer, T. (2001) Role of the ABC transporter Mdl1 in peptide export from mitochondria. Science 219, 2135–2138.

Zeng, H., Chen, Z.-S., Belinsky, M.G., Rea, P.A. and Kruh, G.D. (2001) Transport of methotrexate (MTX) and folates by multidrug resistance protein (MRP) 3 and MRP1: effect of polyglutamylation on MTX transport. Cancer Res. 61, 7225–7232.

Zolman, B.K., Silva, I.D. and Bartel, B. (2001) The *Arabidopsis pxa1* mutant is defective in an ATP-binding cassette transporter-like protein required for peroxisomal fatty acid β-oxidation. Plant Physiol. 127, 1266–1278.

Part V

Mammalian Multidrug Resistance ABC Transporters

CHAPTER 18

Solving the Problem of Multidrug Resistance: ABC Transporters in Clinical Oncology

Susan E. Bates

Introduction

Acquired drug resistance was first observed in a laboratory model in 1950, in mouse leukemic cells passaged in mice treated with 4-amino-N^{10}-methyl-pteroylglutamic acid (Burchenal *et al.*, 1950). In 1972, Dano described drug resistance due to the active outward transport of chemotherapeutic agents (Dano, 1973). Daunorubicin-selected resistant tumor cells were found to have energy-dependent transport of daunorubicin that could be inhibited by vinblastine, vincristine, and other anthracyclines. Further, selection of cells for resistance to vinblastine resulted in the same phenotype. Later, Biedler, Beck and Ling more fully characterized the multidrug resistance phenotype (Beck *et al.*, 1979; Biedler and Peterson, 1981; Riordan and Ling, 1979). Tumor cell lines that were selected in the laboratory for resistance to doxorubicin or vincristine became cross-resistant to structurally unrelated anticancer agents, displayed active outward drug efflux, and were characterized by increased expression of a 170 kDa cell membrane glycoprotein that became known as P170 or P-glycoprotein. As critical as this discovery of the first human ATP-binding cassette (ABC) transporter was, it was the observation that drug resistance could be reversed *in vitro* by several different compounds, including verapamil, that brought Pgp into prominence as a potential target for improving cancer therapy (Tsuruo *et al.*, 1981). The first section of this chapter will briefly review the mammalian ABC transporters linked to multidrug resistance (discussed in more detail in Chapters 5, 6 and 19–21). Subsequently, the progress that has been made in developing ABC transporters as clinical targets in anticancer therapy will be reviewed. To date, 48 human ABC genes have been identified and classified into seven distinct subfamilies (Dean *et al.*, 2001). The Human Gene Nomenclature Committee has designated these subfamilies as ABCA through ABCG (Klein *et al.*, 1999). However, the traditional more familiar names will be used for the majority of the transporters described below.

ABC Transporters with Potential Roles in Multidrug Resistance

P-glycoprotein, MDR1 (ABCB1)

In the decade that followed the cloning of the genes encoding rodent P-glycoprotein (Gerlach *et al.*, 1986; Gros *et al.*, 1986; Scotto *et al.*, 1986) and human P-glycoprotein, *MDR1* (Ueda *et al.*, 1987) studies were aimed at exploring the structure and function of P-glycoprotein, and understanding its importance in human malignancy. P-glycoprotein (Pgp) is considered a 'full' transporter, comprising 12 transmembrane (TM) segments divided between two domains, each linked to an ATP-binding domain. Both ATP-binding domains contain Walker A and Walker B sequences as well as the active transport family signature motif 'C' characteristic of all ABC transporters. Current studies suggest that the high-affinity binding of substrate to Pgp results in ATP hydrolysis, which in turn causes a conformational change in Pgp that shifts the substrate to a lower-affinity binding site on the protein, thereby releasing the substrate into either the outer leaflet of the membrane or the extracellular space (Ramachandra *et al.*, 1998). Hydrolysis at the second ATP-binding domain is required to reset the protein conformation to allow binding of a new substrate molecule (Sauna and Ambudkar, 2001; Senior and Bhagat, 1998; van Veen *et al.*, 2000). Thus, Pgp has been viewed as a 'two-cylinder engine' (see also Chapters 4–6).

In vitro studies have shown that overexpression of Pgp in cancer cells confers high levels of resistance to anthracyclines, *Vinca* alkaloids, taxanes, etoposide, and probably hundreds, if not thousands, of other compounds (Gottesman and Pastan, 1993; Scala *et al.*, 1997). Numerous studies suggest that the principal physiological role for Pgp is to protect the organism from toxic substances. This evidence includes the identification of Pgp expression at sites that are involved in drug excretion or at 'sanctuary sites', including the epithelium of the gastrointestinal tract, the renal proximal tubule, the canalicular surface of the hepatocyte, and the endothelial cell surface comprising the blood–brain barrier (Cordon-Cardo *et al.*, 1989, 1990; Thiebaut *et al.*, 1987). Further evidence is derived from *in vivo* knockout mouse models in which the murine orthologue for Pgp has been deleted or disrupted. These mice are healthy, reproduce normally, but display altered sensitivity to, and excretion of, compounds that are Pgp substrates (Borst and Schinkel, 1996; Schinkel *et al.*, 1994, 1997). In human cancer, Pgp expression appears to be due either to continuation of the phenotype found in the normal tissue of origin or to upregulation following exposure to anticancer agents. Numerous studies have attempted to define the extent of Pgp expression in various tumor types and correlate that information with clinical endpoints such as response to chemotherapy and survival. In addition, evidence establishing the importance of Pgp in cancer has been sought in clinical trials with Pgp inhibitors. As discussed below, these studies have advanced our understanding of how to approach Pgp and other ABC transporters as therapeutic targets, but have not yet generated convincing evidence for the use of inhibitors in clinical oncology.

MULTIDRUG RESISTANCE PROTEIN 1, MRP1 (ABCC1)

In 1992, MRP1 was identified as a second human ABC drug transporter (Cole *et al.*, 1992). Cloned from a multidrug resistant human lung carcinoma cell line, MRP1 has an additional five transmembrane segments (TMD0 or MSD1) located at the NH_2-terminus of the protein connected to a Pgp-like core by a linker region (L0 or CL3) (for further details, see Chapter 19). Mutational analyses have suggested that this linker region may be partly responsible for the organic anion affinity of MRP1 but other regions of the protein clearly participate as well (Bakos *et al.*, 1998; Leslie *et al.*, 2001) (Chapter 19). Disruption of *Mrp1* in murine embryonic stem cells results in a three- to fourfold increase in sensitivity to etoposide and teniposide, and a twofold increase in sensitivity to vincristine, doxorubicin and daunorubicin (Lorico *et al.*, 1996). Overexpression of MRP1 confers resistance to etoposide, doxorubicin and vincristine; and MRP1 has also been shown to transport glutathione conjugates, glucuronides and sulfates (Cole *et al.*, 1994; Jedlitschky *et al.*, 1994, 1996; Leslie *et al.*, 2001). Further, MRP1 is able to co-transport certain natural product substrates, such as vincristine with glutathione, without covalent conjugation of the drug (Borst *et al.*, 2000b; Hipfner *et al.*, 1999; Leslie *et al.*, 2001; Loe *et al.*, 1998). Additional evidence has been presented suggesting that MRP1 is able to transport irinotecan and its active metabolite, 7-ethyl-10-hydroxy camptothecin (SN-38), compounds that are glucuronidated in normal metabolism (Chen *et al.*, 1999). Together, these studies indicate that MRP1 is able to transport both unmodified and modified xenobiotics. Recently, it was also discovered that MRP1 can confer resistance to methotrexate, an antifolate antineoplastic agent not usually associated with the multidrug resistance phenotype (Hooijberg *et al.*, 1999). Like Pgp, MRP1 is thought to provide protection to normal tissues, and to be involved in drug disposition (Wijnholds *et al.*, 2000b). Unlike Pgp, low-level expression of MRP1 is ubiquitous throughout the body with higher levels expressed in the lung and kidney (see Chapter 19).

OTHER MRPs

Multiple MRP (ABCC) family members have been identified (Borst *et al.*, 2000b; Dean *et al.*, 2001) (see Chapters 20 and 21). MRP1, 2, 3 and 6 have the highest homology with one another, with 17 predicted TM segments: a Pgp-like core encoding two ATP-binding domains and two membrane-spanning domains, with an additional NH_2-proximal five TM segment region (TMD0) (described in the previous section) (Borst *et al.*, 2000b; Leslie *et al.*, 2001). In contrast, MRP4 (ABCC4), MRP5 (ABCC5), ABCC11 and ABCC12 lack the TMD0 characteristic of MRP1, MRP2, MRP3 and MRP6. MRP4 and MRP5 have been shown to transport nucleosides (Chen *et al.*, 2001; Dean *et al.*, 2001; Jedlitschky *et al.*, 2000; Wijnholds *et al.*, 2000a), while the functions of ABCC11 and ABCC12 are not yet known. MRP2 (ABCC2), also known as cMOAT (*c*analicular *m*ultispecific *o*rganic *a*nion *t*ransporter), has been identified as the bilirubin glucuronide transporter (Buchler *et al.*, 1996; Paulusma *et al.*, 1996) (see Chapter 20). The Dubin–Johnson syndrome in humans, as in the TR^- and EHBR rat models, is characterized by mutations in *MRP2(ABCC2)* which result in the absence of the protein in the canilicular membranes of the liver (Buchler *et al.*, 1996; Paulusma *et al.*, 1996; Toh *et al.*, 1999). Patients accumulate an excess of bilirubin glucuronide and unconjugated bilirubin, resulting in hyperbilirubinemia and hepatic inflammation. Mutations in *MRP6* (*ABCC6*) have been linked to the connective tissue disorder pseudoxanthoma elasticum but have no known role in drug resistance (see Chapters 21 and 28).

The question of whether *MRP2* can confer multidrug resistance has been addressed by *in vitro* transfection studies, with both sense and antisense *MRP2* cDNA constructs. Both types of studies support the conclusion that MRP2 is able to transport cisplatin as well as the MRP1 substrates etoposide, doxorubicin, vincristine and methotrexate (Cole *et al.*, 1994; Cui *et al.*, 1999; Koike *et al.*, 1997; Masuda *et al.*, 1997). However, the prevalence of increased expression of MRP2 as a mechanism of resistance to cisplatin and other anticancer drugs is not yet known (Kool *et al.*, 1997; Taniguchi *et al.*, 1996) (see Chapter 20).

Like MRP1, MRP3 (ABCC3) has been shown to transport etoposide, doxorubicin, vincristine and methotrexate (Hooijberg *et al.*, 1999; Kool *et al.*, 1999; Zeng *et al.*, 1999). MRP3 is expressed at relatively high levels in human liver, localized to the basolateral surface of the hepatocyte (Konig *et al.*, 1999), where, like MRP1, it may be involved in the transport of organic anions back into the bloodstream. Studies with MRP4 and MRP5 have demonstrated transport of cyclic nucleotides, and resistance to 6-mercaptopurine and 6-thioguanine, two anticancer purine analogues (Chen *et al.*, 2001; Jedlitschky *et al.*, 2000; Wijnholds *et al.*, 2000a). Taken together, the findings suggest that the MRP subfamily of ABC transporters has a role, with some possible built-in redundancy, in drug disposition. That function may be subverted by a cancer cell in becoming drug resistant. However, to date, conclusive links to clinical drug resistance have not been established for MRP family members other than MRP1 (see also Chapter 21).

SPGP/BSEP (ABCB11)

Structurally homologous to MDR1/Pgp, the 'sister of P-glycoprotein' was originally cloned from the hamster in a search for genes with homology to *MDR1* (Childs *et al.*, 1995). Subsequently recognized as the bile salt exporter protein (BSEP), SPGP/BSEP (ABCB11) plays an important role in biliary homeostasis (Gerloff *et al.*, 1998). While evidence for a role for SPGP-BSEP in drug resistance is limited, it is interesting to note that paclitaxel is also a substrate for transport by this protein. Overexpression of SPGP/BSEP in human ovarian SKOV3 cells conferred a fourfold resistance to paclitaxel (Childs *et al.*, 1998). Sensitization by PSC833, cyclosporin A and verapamil (typical Pgp/MDR1 inhibitors) was observed.

ABC2 (ABCA2)

Active outward efflux has also been observed in SKEM cells, a human ovarian carcinoma cell line selected for estramustine resistance (Laing *et al.*, 1998). Estramustine is not known to be a substrate for Pgp, and the resistant SKEM cells have a phenotype distinct from that associated with overexpression of Pgp. Amplification of *ABCA2* was detected in these cells, and antisense-mediated downregulation of ABCA2 sensitized the resistant cells to estramustine (Laing *et al.*, 1998). ABC2/ABCA2 belongs to the ABCA subfamily, which also includes

ABCA1, the transporter linked to cholesterol transport, and ABCR (ABCA4), the transporter linked to retinal integrity (Broccardo *et al.*, 1999) (see Chapters 23 and 28).

MXR/BCRP/ABCP (ABCG2)

A member of the ABCG subfamily, MXR/BCRP/ABCP (ABCG2), is a 'half transporter' able to confer high levels of drug resistance to mitoxantrone, topotecan, CPT-11 and its active metabolite SN38, as well as anthracyclines (Allikmets *et al.*, 1998; Brangi *et al.*, 1999; Doyle *et al.*, 1998; Litman *et al.*, 2000; Miyake *et al.*, 1999). Thus, its substrate specificity appears somewhat more limited than Pgp and MRP1. In addition, flavopiridol, a new cell cycle inhibitor in clinical trials, has been found to be a substrate for ABCG2 (Robey *et al.*, 2001). A single ATP-binding domain followed by six TM segments comprising a single membrane-spanning domain make up the half-size transporter designated ABCG2, which is thought to require dimerization to form a functional unit. Two other members of this subfamily are involved in sterol transport (ABCG5 and ABCG8) (see Chapter 22), but a normal function for ABCG2 is not yet known (Dean *et al.*, 2001). High levels of ABCG2 are found in the syncytiotrophoblast cells of the placenta, where the function could be either transport of toxins out of, or transport of nutrients into, the fetal circulation (Maliepaard *et al.*, 2001). In Pgp-deficient mice, increased bioavailability and fetal penetration of topotecan was observed following coadministration of topotecan and GF120918, a Pgp inhibitor found to also inhibit ABCG2 (Jonker *et al.*, 2000). A murine transporter, *Abcg3*, with high homology to human ABCG2 has been described (Mickley *et al.*, 2001). Its tissue distribution pattern is different from ABCG2, suggesting the two transporters are not coexpressed. Overexpression and amplification of *ABCG2* occurs during *in vitro* selection of cells with mitoxantrone or topotecan (Knutsen *et al.*, 2000; Maliepaard *et al.*, 1999). Recent studies have also shown that the substrate specificity of ABCG2 can be significantly altered by a difference in a single amino acid (Honjo *et al.*, 2001).

OTHER ABC TRANSPORTERS

For many of the ABC transporters listed above, no conclusive direct evidence has been obtained to suggest a role in clinical drug resistance. For some transporters, important endogenous substrates are known to exist, and drug transport is probably a secondary function. One question is whether the function of a transporter can be subverted to serve as a mediator of multidrug resistance in tumor cells. In one scenario, an ABC transporter not normally expressed at high levels may be upregulated, induced, or redistributed to the cell surface, and in doing so, becomes capable of conferring drug resistance. In another scenario, mutation of a transporter protein could result in a gain of function. For example, ABCG2 confers resistance primarily to mitoxantrone and camptothecin analogues; however, mutation of amino acid 482 adds rhodamine and anthracyclines to the list of substrates it can transport (Honjo *et al.*, 2001). Similarly, only minor sequence changes are required to improve the efficiency of drug transport by MDR3/Mdr2 (ABCB4), a phosphatidylcholine flippase or translocator closely related to Pgp (MDR1) that normally transports phospholipids into the bile (Borst *et al.*, 2000a; Smit *et al.*, 1993; Zhou *et al.*, 1999) (see Chapter 22). Mutations such as these have not been demonstrated in clinical cancer to date. With at least 48 ABC transporters encoded in the human genome, this list of transporters with a potential role in drug resistance may yet be incomplete. However, the list of substrates encompassed by the already described transporters is quite extensive, and includes some of the newest agents in the anticancer drug armamentarium. It could be argued with considerable conviction that no anticancer agent could be identified for which a drug transporter could not be found.

MVP/LRP

Not an ABC transporter, but included in many clinical studies of multidrug resistance, MVP (major vault protein) (also known as LRP, *l*ung *r*esistance *p*rotein) is a component of the multimeric vault proteins which are found in the cytoplasm and in the nuclear membrane (Scheffer *et al.*, 2000b). Thought to mediate redistribution of drugs away from the nucleus, the expression of vaults may be coordinately regulated with Pgp or MRP1 although direct evidence that this is the case is lacking. MVP/LRP expression has been detected in lung cancer, acute leukemia and ovarian cancer. In several studies, expression of MVP/LRP has been a better correlate of poor prognosis than Pgp (den Boer *et al.*, 1998; Izquierdo *et al.*, 1995; List *et al.*, 1996).

P-GLYCOPROTEIN (ABCB1) AS THE FIRST MDR TRANSPORTER TARGETED IN CLINICAL TRIALS

Table 18.1 lists many of the compounds found to be inhibitors of Pgp-mediated drug efflux and drug resistance. Characterized as both competitive and non-competitive inhibitors, these agents are able to increase chemosensitivity in *in vitro* models by several orders of magnitude. Early characterization of Pgp inhibitors *in vitro* led to trials with what are now referred to as first-generation inhibitors. These compounds were already used in clinical medicine and found in the laboratory to be inhibitors of Pgp and were used in combination with an anticancer agent known to be a Pgp substrate. Several reviews that catalogue these trials are available (Bradshaw and Arceci, 1998; Ferry *et al.*, 1996; Fisher and Sikic, 1995; Fisher *et al.*, 1996). These trials demonstrated the safety of combining a Pgp inhibitor with a chemotherapeutic agent, but fell far short of the goal of defining a role for Pgp inhibition in clinical oncology. This, in turn, meant that a role for Pgp in conferring clinical drug resistance was also not confirmed.

The failure of the first-generation Pgp inhibitor trials to support a role for inhibition of this ABC transporter in clinical oncology could be ascribed to several factors. First, as Pgp inhibitors, the first-generation agents were not very potent, requiring micromolar concentrations for effective inhibition. Concentrations comparable to those that were effective in laboratory models could seldom be obtained without toxicity in patients. Second, the trials were designed to identify a 'home run'; thus, the inhibitors were administered with the anticancer agents without first requiring either that tumors be clearly refractory to treatment, or that randomization be incorporated into the trial design. Third, the trials never sought physical evidence that Pgp inhibition was occurring *in vivo*. Finally, assays were usually not included to confirm the presence of Pgp expression or function in the tumors.

Second-generation Pgp inhibitors were typically analogues of first-generation agents, developed specifically for the purpose of Pgp inhibition. These included R-verapamil (stereoisomer of verapamil) and PSC 833 (derivative of cyclosporin D). These agents were more potent than many of the first-generation agents but still did not achieve the success sought in terms of efficacy. Nor did they confirm a role for Pgp inhibition in clinical oncology. Trials with these second-generation agents again confirmed the safety of adding a Pgp inhibitor to therapy with conventional agents, with the caveat that pharmacokinetic interactions necessitated a lower dose of the anticancer agent in combinations, including PSC 833. Perhaps the most important outcome of the completed Pgp reversal trials was the recognition that a distinction needed to be made between the efficacy of the inhibitor in blocking Pgp and the efficacy of the inhibitor in improving cancer treatment.

Trials with third-generation agents are now in progress, more than 25 years since the identification of the molecular target, Pgp, and more than 20 years since the identification of the first Pgp inhibitor, verapamil. Several of these compounds are reported to have little or no pharmacokinetic interactions, overcoming a

TABLE 18.1. P-GLYCOPROTEIN INHIBITORS USED IN CLINICAL DEVELOPMENT[a]

First-generation agents
- Verapamil
- Quinidine
- Quinine
- Amiodarone
- Nifedipine

Second-generation agents
- R-verapamil
- PSC 833
- Dexniguldipine

Third-generation agents
- GF120918
- VX710
- R101933
- XR9576
- LY335979
- OC144-093

[a]Agents shown represent only a partial list.

major problem linked to the use of PSC 833. These compounds include XR 9576, R101933, LY335979, OC144-093, and GF120918 (Dantzig *et al.*, 1999; Mistry *et al.*, 2001; Newman *et al.*, 2000; Sparreboom *et al.*, 1999; Starling *et al.*, 1997; van Zuylen *et al.*, 2000a). The compounds, evaluated in studies enlightened by lessons from the first- and second-generation inhibitor trials, offer the potential to finally discover the importance of Pgp in clinical oncology.

DETERMINING THE EXPRESSION OF DRUG TRANSPORTERS IN CANCER

It seems logical that the efficacy of a Pgp inhibitor in the clinic will be linked to the importance of Pgp in drug resistance. So clear is this logic that investigators in this field have largely relied upon the clinical trial process to provide the answer to the question of whether Pgp is significant in clinical oncology. This may have been the central flaw in the past decade of clinical research. In breast cancer, markers such as the estrogen receptor, erbB2, aneuploidy, and S-phase are measured in thousands of patients, with steadily improving uniformity of technique, and correlated with clinical outcome. In contrast, in the field of multidrug resistance, we have relied upon 'drug resistance reversal trials' to answer the question of whether Pgp is important in cancer treatment. If a concerted effort to identify the diseases in which Pgp expression confers a resistant phenotype had been made, we might have set the stage for well-conceived clinical trials. Instead, selection of the tumor types and trial designs for clinical studies has relied as much upon guesswork as upon facts.

Early studies of Pgp demonstrated frequent and high levels of expression in colon, kidney, adrenocortical and hepatocellular cancers (Fojo *et al.*, 1987; Goldstein *et al.*, 1989). Initially, there was hope that Pgp could explain the profound intrinsic drug resistance found in these cancers. However, the failure of these cancers to respond to therapies with drugs not transported by Pgp suggested that Pgp alone could not account for the intrinsically drug-resistant phenotype, and attention has turned to cancers that respond to chemotherapy initially, but ultimately acquire resistance. Numerous clinical studies evaluating or measuring Pgp expression and/or function have appeared, and Pgp expression has been correlated with clinical outcome. However, the studies have been largely retrospective, single institution, small studies with insufficient power to provide a definitive statistical outcome.

One problem with designing a study powered to provide this information is that methods for Pgp detection remain imperfect. We and others have previously delineated these issues (Beck *et al.*, 1996; Herzog *et al.*, 1992), and they can be summarized as follows: (1) mRNA and protein methods that use whole tumor specimens risk contamination with normal tissues, which may increase or decrease the Pgp expression level detected; (2) Northern blot analysis for mRNA and immunoblot analysis for protein expression are not sensitive enough for the low levels frequently detected in clinical samples; (3) polymerase chain reaction (PCR) assays for *MDR1* mRNA detection are commonly performed with methods that fail to take into account the fact that quantitation is most accurate in the exponential phase of amplification; (4) immunohistochemical assays are best for direct examination of individual cancer cells, eliminating problems with normal tissue contamination, but are difficult to quantitate; (5) antibodies used in immunohistochemistry studies are not as specific as needed; (6) Pgp is difficult to detect in formalin-fixed tissue; thus, investigators disagree as to whether monoclonal antibody C219, one of the most commonly used antibodies, can detect Pgp in archival samples.

In an effort to address the discrepancies among reports concerning detection of Pgp expression in clinical samples, Beck and co-workers assembled a workshop at St Jude's Children's Hospital (Memphis, USA) to compare Pgp detection methods in use by investigators from around the world (Beck *et al.*, 1996). While specific recommendations were made, there is still disagreement on several levels. For example, should cancer cells be scored as positive for Pgp if membrane staining cannot be identified? Studies requiring membrane staining often report a far lower frequency of Pgp detection in breast cancer. There are also persistent issues of sensitivity. Studies utilizing the PCR method for *MDR1* mRNA detection have a higher frequency of MDR1/Pgp detection than other mRNA detection methods. This can be ascribed to the ability of the amplification process to detect mRNAs of low abundance.

Another issue discussed at the St Jude's Workshop, and still not resolved, is the development of a uniform standard for measurements. Since different PCR assays may run at different efficiencies, it is difficult to know whether the levels measured by one investigator are comparable to those measured by another, unless uniform controls are run. For example, in breast cancer studies, one investigator reported levels of *MDR1* mRNA in tumors as comparable to levels in normal tissues (Lizard-Nacol *et al.*, 1999). Since *MDR1* mRNA levels in normal breast tissue are very low, the investigators concluded that levels of expression in breast cancer were comparably low. Use of one or more standard positive controls would aid in answering this question across studies.

Detection of MRP1 (ABCC1) and other drug transporters has been less intensively investigated (see Chapters 19–21). MRP1 has been detected by the same methods used for Pgp: immunohistochemistry for protein and reverse transcriptase PCR (RT-PCR) or RNase protection for mRNA expression. Nooter *et al.* (1995) examined 370 human cancer samples by RNase protection. High levels of *MRP1* expression were found in chronic lymphocytic leukemia and prolymphocytic leukemia. Occasionally, high levels of expression were found in esophageal carcinoma, in non-small cell lung cancer, and in acute myelogenous leukemia (AML). Predominantly low but ubiquitous expression of *MRP1* was found in the remaining tumor types. An additional 108 samples evaluated by immunohistochemistry with the monoclonal antibody MRPr1 confirmed these findings. The antibodies most commonly used in immunohistochemical analyses, MRPr1, MRPm6 and QCRL-1, recognize sequences specific for human MRP1 and to date, the cross-reactivity problems that have plagued Pgp detection have not arisen (Hipfner *et al.*, 1998).

For other ABC transporters, there is minimal experience to judge the sensitivity and specificity of detection methods. A panel of specific monoclonal antibodies has been generated for detection of other members of the MRP (ABCC) subfamily but their epitope sequences have not yet been precisely defined (Scheffer *et al.*, 2000a). *ABCG2* mRNA expression has been assayed by RT-PCR in single studies in breast cancer and in leukemia (Kanzaki *et al.*, 2001; Ross *et al.*, 2000). Polyclonal and monoclonal antibodies have been developed to detect ABCG2 (MXR/BCRP), but reports have not yet appeared describing expression in tumor tissue.

Expression of ABC Transporters in Selected Malignancies

Leukemia

The most uniform detection of Pgp/MDR1 expression has been that reported in acute leukemia. Leukemic cells from about one-third of patients with acute myelogenous leukemia (AML) express Pgp at the time of diagnosis, and expression is observed in cells from about 50% of patients at the time of relapse **(Table 18.2)**. Certain subtypes of AML are also noted to have higher frequencies of detection, including secondary leukemias. While not invariable, most trials report that Pgp expression is correlated with a reduced complete remission rate, and a greater incidence of refractory disease (Filipits *et al.*, 1998; Legrand *et al.*, 1999; Leith *et al.*, 1999; Michieli *et al.*, 1999; van der Kolk *et al.*, 2000). Complete response rates in the range of 50–70% are reported in Pgp-negative leukemia, compared to 30–50% in Pgp-positive leukemia. Because of the high correlation between CD34 expression and Pgp expression (Campos *et al.*, 1992), some investigators have argued that Pgp, rather than conferring the resistant phenotype through drug efflux, may instead be a phenotypic marker of a poor prognosis subset of leukemia patients. However, *ex vivo* studies using leukemic cells from patients have shown that Pgp expression does correlate with reduced accumulation of daunorubicin (Broxterman *et al.*, 1999; Michieli *et al.*, 1999). In addition, leukemic cells obtained from patients receiving daunorubicin after administration of a Pgp inhibitor have shown increased daunorubicin accumulation (Tidefelt *et al.*, 2000). In a recently reported trial, Broxterman *et al.* (2000) found that the prognostic value of Pgp could be mitigated by substituting idarubicin, an anthracycline not subject to Pgp-mediated efflux, for daunorubicin. One final observation supporting a role for Pgp in drug resistance in AML is derived from trials in which a Pgp inhibitor was used (cyclosporin A or PSC 833) in combination with chemotherapy. Leukemic cells obtained from patients in relapse following treatment with either cyclosporin A or PSC 833 have decreased expression of

TABLE 18.2. EXPRESSION OF P-GLYCOPROTEIN IN ACUTE MYELOGENOUS LEUKEMIA

Author/year	Method[a]	Population	n	Category[b]	Positive n	Positive (%)	Clinical correlate
Expression studies							
Visani *et al.*, 2001	MRK16	Poor risk	16		13	81	
Dorr *et al.*, 2001	C494	Poor risk	38		19	50	
Han *et al.*, 2000	JSB-1	*De novo*	109		29	27	
		Relapse/ Refractory	36		23	64	Relapse vs. *de novo*: $p < 0.05$
Chauncey *et al.*, 2000	MRK-16	*De novo*	20		13	65	
		Secondary	10		8	80	
Kornblau *et al.*, 1997	4E3	Poor risk	7		5	71	
List *et al.*, 1996		*De novo*	21		8	38	
		Secondary	27		17	63	
		Relapse	29		11	38	
		CML-bl	10		5	50	
Samdani *et al.*, 1996	JSB-1 or MRK16	Favorable	54		16	29	OS: $p = 0.001$
		Unfavorable	35		22	62	
Studies with clinical correlations							CR[c] (%)
Leith *et al.*, 1999	MRK-16	*De novo*/ secondary	351	High	64	18	48
				Intermediate	35	10	46
				Low	24	7	63
				Negative	228	65	54 ($p = 0.012$)[d]
	Rh123 efflux ± CsA		318	High	70	22	37
				Intermediate	39	12	62
				Low	12	4	58
				Negative	197	62	55 ($p = 0.028$)[d]
van der Kolk *et al.*, 2000	Rh123 accum.	*De novo*	104	High	32	32	25
				Intermediate	33	34	61
				Low	33	34	58 ($p = 0.008$)
Legrand *et al.*, 1999	UIC2	*De novo*/ secondary	75	Positive	31	52	51
				Negative	29	48	79 ($p = 0.02$)
	CalceinAM accum. ±CsA		40	Positive	14	35	42
				Negative	26	65	76 ($p = 0.03$)
Michieli *et al.*, 1999	MRK-16	*De novo*	96	Positive	47	49	45
				Negative	49	51	80 ($p = 0.006$)
Filipits *et al.*, 1998	C219	*De novo*	82	High	10	12	44
				Intermediate	24	30	
				Low	48	58	75 ($p = 0.05$)

Studies appearing after the 1994 Consensus Conference on MDR Detection Methods (Beck *et al.*, 1996).
[a]All immunohistochemical assays were performed on fresh cytospins of leukemic cells. Rh123 indicates functional assay with the Pgp substrate rhodamine 123. CsA indicates differences in the functional assay with or without the addition of cyclosporin A. Accum. indicates accumulation of either rhodamine 123 or calcein AM in the functional assay. OS, overall survival.
[b]Each set is listed high to low levels of transporter expression or function.
[c]CR, complete remission.
[d]Expression also correlates with resistant disease, $p < 0.005$.

Pgp or *MDR1* mRNA (Kornblau *et al.*, 1997; List *et al.*, 1993, 1996). While it cannot be absolutely concluded that circumvention of Pgp explained the clinical outcome, the absence of a correlation between clinical response and Pgp expression in this trial stands in contrast to the results obtained by numerous investigators from different institutions **(Table 18.2)**. Taken together, the clinical data support an important role for Pgp in drug resistance in AML.

MRP1 and LRP expression have also been evaluated in leukemia patients. MRP1 has been detected at high levels in chronic lymphocytic leukemia and in prolymphocytic leukemia (Nooter *et al.*, 1996b). Levels in AML are less frequently elevated (10–34%) (Legrand *et al.*, 1999; Leith *et al.*, 1999). These studies are divided as to whether MRP1 confers a poor prognosis in a subset of AML patients. The non-ABC protein LRP/MVP (see above) has been detected in AML and in several series has been found to be of greater prognostic value than Pgp (Dorr *et al.*, 2001; Filipits *et al.*, 1998; List *et al.*, 1996; Xu *et al.*, 1999). In these studies, the well-known prognostic value of Pgp expression in AML is not detectable, thus creating a discrepancy that is difficult to reconcile with earlier data. Two of these studies included patients who had received Pgp inhibitors, which conceivably confounded the analysis (Dorr *et al.*, 2001; List *et al.*, 1996). The largest trial to date, reported by Leith *et al.* (1999), found no correlation between LRP/MVP expression and prognosis in a population of previously untreated patients. Finally, low levels of BCRP/MXR (ABCG2) have been observed in AML samples, with one-third having levels as high as 2.6 times those found in the drug sensitive MCF-7 breast cancer cell line (Ross *et al.*, 2000).

Breast cancer

Detection of Pgp in clinical samples from patients with solid tumors has been much more difficult than in hematologic malignancies. These difficulties relate to the lack of specificity of the antibodies, to the heterogeneity of clinical samples, and to the lack of standard laboratory methods. Studies published after the 1994 St Jude's Workshop (see above) have frequently incorporated the recommendations, particularly relating to the need to use more than one detection methodology (Beck *et al.*, 1996). This includes using multiple antibodies or RT-PCR as a second method for Pgp or *MDR1* mRNA detection, respectively. Despite this effort, the results remain variable as observed by Trock *et al.* (1997) in a meta-analysis of 31 studies. In the meta-analysis study, 41% of breast tumors expressed *MDR1*/Pgp, the frequency of detectable expression increased after therapy, and expression was associated with a greater likelihood of treatment failure. However, there was considerable heterogeneity among the studies, with the reported incidence ranging from 0% to 80%. This heterogeneity persists in studies reported since 1996. As shown in **Table 18.3**, the detection rate using immunohistochemistry still ranges from 0% to 71%, and frustratingly, even when the same antibody is being used (Faneyte *et al.*, 2001; Yang *et al.*, 1999). Most studies report some expression of Pgp in breast cancers, and many report membrane staining (Bodey *et al.*, 1997; Chevillard *et al.*, 1996; Hegewisch-Becker *et al.*, 1998; Schneider *et al.*, 2001), considered by most investigators to be the truest indicator of functional Pgp expression. Results with RT-PCR methods have been much less revealing, with studies suggesting no increase in expression relative to normal tissue (Arnal *et al.*, 2000; Dexter *et al.*, 1998; Faneyte *et al.*, 2001; Lizard-Nacol *et al.*, 1999). The discrepancy of these results with those obtained by immunohistochemical methods may be due to the greater sensitivity of PCR as described earlier.

Several studies have also attempted to relate Pgp expression in breast cancer with clinical drug resistance. Pgp expression has been observed to increase in locally advanced breast cancer following therapy, with the incidence increasing from 26% to 57% in one study (Chung *et al.*, 1997) and from 14% to 43% in another (Chevillard *et al.*, 1996). Among 359 samples, including primary cancer, locally advanced, and recurrent disease, the incidence of Pgp expression was 11% in samples obtained from untreated patients, and 30% in samples from patients who had previously received treatment. Although the 1997 meta-analysis concluded that patients with tumors expressing Pgp were more likely to experience treatment failure, several small recent studies have not been able to confirm a significant impact of Pgp expression on response rate or overall survival (Honkoop *et al.*, 1998; Linn *et al.*, 1997; Wang *et al.*, 1997).

Whether MRP1 is found in breast cancer at levels capable of conferring drug resistance is not resolved. As mentioned previously, *MRP1* mRNA is expressed ubiquitously in normal

TABLE 18.3. EXPRESSION OF P-GLYCOPROTEIN IN BREAST CANCER

Reference	Method	*n*	Fixation method	Ab	% positive	Note	Laboratory or clinical correlate
Schneider, *et al.*, 2001	IHC	52	P	C494	39%	PreRx	IHC with PCR
	PCR	46			47%	PostRx	
Faneyte *et al.*, 2001	IHC	140	P	JSB-1	71%	Cytoplasm	
					0%	Membrane	
	PCR	30		TaqMan	Low levels[b]		No change postRx
Yang *et al.*, 1999	IHC	106	F	JSB-1	0%		
				MRK-16			
	PCR	33			39%		No corr. with RR or OS
Hegewisch-Becker *et al.*, 1998	IHC	61	P	MRK-16	75%	Weak	
					35%	Strong	
			P	UIC2	72%	Weak	
					43%	Strong	
	PCR				No positive samples after T-cells removed		
Dexter, *et al.*, 1998	IHC	31	F	JSB-1	6%		
				UIC2	0%		
	PCR				Low levels[b]		
Wang *et al.*, 1997	IHC	48	C	JSB-1	10%		No corr. with CR
				4E3	19%		
				C494	13%		
	PCR	52			84%		
Filipits *et al.*, 1996	IHC	63	F	C219	57%		IHC with PCR
	PCR	134			60%		
Chevillard *et al.*, 1996	IHC	63	C	JSB-1[a]	14%	PreRx	
					43%	PostRx	
	PCR				27%	PreRx	
					51%	PostRx	
Mechetner *et al.*, 1998	IHC	359	P	JSB-1	11%	PreRx	IHC with *in vitro* resistance
					30%	PriorRx	
Honkoop *et al.*, 1998	IHC	30	P	JSB-1	67%		No corr. with DFS, OS
Linn *et al.*, 1997	IHC	40	P	JSB-1	64%	PreRx	No corr. with CR
					57%	Post Rx	
Bodey *et al.*, 1997	IHC	15	P	JSB-1	33%	Strong	
				C494			
				C219			
Chung *et al.*, 1997	IHC	23	P	JSB-1	26%	PreRx	
					57%	PostRx	
Del Vecchio *et al.*, 1997	IHC	30	F	MRK-16	33%		Corr. with sestamibi
Tolcher *et al.*, 1996	IHC	34	C	JSB-1	19%		Post-Paclitaxel: 76%
				C219	6%		59%
				C494	13%		53%
Lacave *et al.*, 1998	PCR	74			3%	High[c]	
					19%	Moderate	
					61%	Low	
					17%	Negative	

(continued)

TABLE 18.3. *(continued)*

Reference	Method	*n*	Fixation method	Ab	% positive	Note	Laboratory or clinical correlate
Lizard-Nacol *et al.*, 1999	PCR	75			92%	PreRx	Equivalent to nl breast tissue
					96%	PostRx	
Arnal *et al.*, 2000	PCR	40			92%	PreRx	No corr. with RR or OS
					100%	PostRx	

Studies were included if they clearly defined the methodology used for MDR1/Pgp detection, and if they delineated a cut-off for positivity. M, membrane staining required for positivity.
Fixation method: C, cytospin – acetone or paraformaldehyde; F, frozen section; P, paraffin-embedded, formalin fixed.
IHC, immunohistochemistry; PCR, polymerase chain reaction.
RR, response rate; OS, overall survival; DFS, disease-free survival; CR, complete response; Rx, therapy.
[a]JSB-1, C219, C494 gave concordant results.
[b]Low levels equivalent to those in normal breast tissue.
[c]Levels defined in relationship to P-glycoprotein expression levels in KB8-5 cells.

human tissues and, consequently, finding expression in tumor tissue is not surprising. Detection of *MRP1* mRNA in 100% of samples by RT-PCR at levels comparable with normal tissue levels reinforces this point (Dexter *et al.*, 1998; Filipits *et al.*, 1996). One study reported a correlation between relapse-free survival in breast cancer patients and MRP1 expression as detected by immunohistochemistry (Nooter *et al.*, 1997). In this series, which comprised breast cancer samples from 259 patients, MRP1 expression was detected in 34%.

OVARIAN CANCER

The problem of variability continues when expression studies in solid tumors other than breast cancer are reviewed. Thus, for ovarian cancer, the reported incidence of Pgp positivity ranges from 17% to 71%. The methodologies described in these studies are more variable than those in the breast cancer studies. The study reporting 71% positivity is the outlier, and was the only one to use immunoblotting as a detection method with a polyclonal antibody not used in the immunohistochemical studies (Joncourt *et al.*, 1998). Two other groups used antibodies not widely accepted, but potentially deserving of further testing since good detection methods for Pgp in archival material have not been established (Schneider *et al.*, 1998; Yokoyama *et al.*, 1999b). If the true incidence of Pgp positivity in ovarian cancer at diagnosis is less than 20%, it can readily be appreciated that a drug resistance reversal trial would need to either select the subset of patients which would be most likely to benefit from a Pgp inhibitor, or expand the size of the trial sufficiently to detect a difference in fewer than one-fifth of patients.

LUNG CANCER

The majority of breast and ovarian cancer trials have used immunohistochemical methods to evaluate Pgp expression. In contrast, in lung cancer, Pgp/*MDR1* mRNA quantitation methods are more prevalent. Most studies measuring *MDR1* mRNA do so by RT-PCR, and report approximately a 25% incidence of expression **(Table 18.4)**, with a range of 15% to 50%. MRP1 expression is reported at a much higher frequency, 70% to 80% in small cell lung carcinoma (SCLC) and 100% in non-small cell lung carcinoma (NSCLC), perhaps not surprising in view of its relatively high level of expression in normal lung tissue. However, few studies have compared both histologies. In studies of lung cancer cell lines, increased levels of both *MRP1* and *MRP3*, but not *MRP2*, correlated with reduced sensitivity to doxorubicin, vincristine, etoposide and cisplatin (Young *et al.*, 2001), suggesting that these transporters may play a role in the intrinsic resistance of lung cancer.

SARCOMA

Investigators have also considered Pgp expression to be important in sarcomas. However, examination of the literature reveals that different detection methodologies with varying results have been reported. An early study in soft tissue sarcomas noted a marked impact of Pgp expression on relapse-free survival and overall survival (Chan *et al.*, 1990). These investigators used a unique immunohistochemical

TABLE 18.4. EXPRESSION OF P-GLYCOPROTEIN, MRP, AND LRP IN SELECTED SOLID TUMORS

Reference	n	Hist[a]	Method	Pgp (%)	MRP (%)	Laboratory or clinical correlation
Ovarian cancer						
Izquierdo *et al.*, 1995	57		IHC	16	68	RR, OS: $p > 0.3$
van der Zee *et al.*, 1995	89		IHC	15		48% positive postRx: $p < 0.001$
Arts *et al.*, 1999	115		IHC	17	44	Pgp and MRP: $p < 0.001$
Schneider *et al.*, 1998	95		IHC	39		OS: $p = 0.05$
Joncourt *et al.*, 1998	39		Blot	71		
Yokoyama *et al.*, 1999b	58		IHC	27.6	22.4	MRP with RR: $p < 0.01$
Baekelandt *et al.*, 2000	73		IHC	47		OS, RFS: $p < 0.05$
Goff *et al.*, 2001	50		IHC		66	
Fracasso *et al.*, 2001	33		IHC	32		
Lung cancer						
Nooter *et al.*, 1996a	35	NSCLC	IHC		74	Membrane staining in 34%
	22		RNAse		35	Ubiquitous mRNA equiv.nl lung; 32% high
Narasaki *et al.*, 1996	6	SCLC	RT-PCR		100	SCLC levels comparable to nl lung
	11	NSCLC	RT-PCR		100	NSCLC levels below nl lung
Savaraj *et al.*, 1997	31	SCLC	RNAblot	26		RR: $p < 0.01$; OS: 10 mo vs. 2 mo, $p < 0.0007$
Oka *et al.*, 1997	84	NSCLC	RNAblot	15		
Galimberti *et al.*, 1998	51	NSCLC	RT-PCR	27	100	
Young *et al.*, 1999	8	SCLC	RT-PCR	25	88	
	7	NSCLC	RT-PCR	43	100	
Yokoyama *et al.*, 1999a	159	NSCLC	IHC	60		OS: 74% vs. 48%, $p < 0.05$
Wright *et al.*, 1998	109	NSCLC	IHC		87	73% intermediate/high levels
	18	SCLC	IHC		56	45% intermediate/high levels
Oshika *et al.*, 1998	107	NSCLC	IHC		44	Cancer cell cytoplasm/nl bronchial epithelium
Haque *et al.*, 1999	56	NSCLC	IHC	61		OS: $p > 0.05$
Sarcoma						
Kuttesch *et al.*, 1996	76	RMS	IHC[b]	41		CR or OS: $p > 0.05$
Lopes *et al.*, 1997	36	SS	IHC	30		PD: $p > 0.05$
Chan *et al.*, 1990	30	STS	IHC	30		RFS: $p < 0.000000012$ OS: $p < 0.0000267$
Levine *et al.*, 1997	65	STS	IHC	48		DFS: 32% vs. 18% $p = 0.039$; OS: 54% vs.14%, $p = 0.07$
Coley *et al.*, 2000	44	STS	IHC	62		OS, DFS: $p > 0.05$
Baldini *et al.*, 1995	92	Osteo	IHC	30		RFS: 80% vs. 42% $p = 0.002$
Lee *et al.*, 1996	77	Osteo	RT-PCR	70		70% > KB8 mRNA; no increase post-Rx
Chan *et al.*, 1997	62	Osteo	IHC	44		RFS: 87% vs. 0%; OS: 94% vs. 35% $p < 0.00001$
Wunder *et al.*, 2000	123	Osteo	RT-PCR	65		High in 36%; DFS: $p > 0.05$
Perri *et al.*, 2001	53	ES	IHC	64		Pgp 3 + 32%; DFS, OS: $p > 0.05$

Abbreviations: Rx, therapy; RFS, relapse-free survival; OS, overall survival; DFS, disease-free survival; RR, response rate; CR, complete response; PD, progressive disease; NSCLC, non-small cell lung cancer; SCLC, small cell lung cancer; STS, soft tissue sarcoma; Osteo, osteosarcoma; RMS, rhabdomyosarcoma; SS, synovial sarcoma; ES, Ewing's sarcoma; IHC, immunohistochemistry; RT-PCR, reverse transcriptase–polymerase chain reaction.

[a]Pgp positive excluding diffuse weak staining, categorized as Pgp negative.

[b]Detection also by RT-PCR, 51% no correlation with survival.

methodology incorporating detection of Pgp in paraffin-embedded tissue with monoclonal antibody C219. Similar results were reported in osteosarcoma using this same methodology (Chan *et al.*, 1997). However, as shown in **Table 18.4**, these findings have been both substantiated and disputed in studies using other methodologies. Thus the importance of Pgp in this tumor remains uncertain (Baldini *et al.*, 1995; Coley *et al.*, 2000; Kuttesch *et al.*, 1996; Perri *et al.*, 2001; Wunder *et al.*, 2000).

Conclusions

In light of the recent progress in the identification and characterization of new ABC transporters, it seems fitting to re-examine recent literature relating to the role of Pgp in clinical drug resistance in five tumor types in which Pgp has been thought to be important. Few studies emphasize serial clinical samples, which have the potential to show the acquisition of multidrug resistance coincident with increased expression of Pgp. Sadly, recent studies reporting the incidence of expression of Pgp in clinical samples appear to be as discordant as older studies. What can we conclude from the available data? It seems that, despite efforts to bring uniformity to the methods used to measure Pgp levels in clinical samples, it is still difficult to discern which studies are valid and which are not. The most valid data appear to be those obtained in AML, where immunostaining, *MDR1* mRNA measurments, and functional studies all confirm Pgp expression in a subset of patients presenting with this disease. Pgp expression in AML is associated with a decreased complete response rate and overall survival. In solid tumors, carefully performed studies repeatedly find some fraction of samples positive, although correlations with response and survival are more variable. Taken together, the studies reviewed here suggest an incidence of Pgp expression of 30% in *de novo* leukemia, 50% in relapsed/refractory/secondary leukemias, 40% in breast cancer, 20% in ovarian cancer, 25% in lung cancer, and 30% in sarcoma. This rate of positivity can be regarded as sufficient to indicate an important role for Pgp in clinical oncology. However, it suggests that subsets of patients need to be selected for multidrug resistance reversal trials, since some tumors do not develop Pgp as a mechanism of resistance.

P-glycoprotein Inhibitors and Pharmacokinetic Interactions

Aside from the difficulties of selecting the most appropriate tumor type in which to test inhibition of Pgp as a means to improve the success of chemotherapy, one of the most vexing problems in the development of Pgp inhibitors has been the impact of the inhibitor on the clearance of certain anticancer agents. For example, PSC 833 was introduced amid very high hopes that reversal of multidrug resistance in clinical oncology would become a reality. It was more potent than any of the first-generation agents including cyclosporin A, and it appeared to be non-toxic in preclinical development. However, the administration of PSC 833 to patients required reduction of anticancer drug doses by 25–70%, in order to prevent toxicity. This dose reduction was determined empirically as the dose of the inhibitor was increased in phase I trials. The greatest impact appeared to be on dosing with paclitaxel and vinblastine.

Dose reductions were due to a delay in clearance of the anticancer drug, and were initially thought to be innocuous, since it was assumed that the delayed clearance would result in a comparable area under the concentration versus time curve (AUC). If all that mattered in cancer chemotherapy was the duration of drug exposure above a certain threshold, then treating patients with doses that resulted in equivalent toxicity would mean equivalent efficacy. This assumption proved to be entirely wrong, and provided an important pharmacology lesson to a number of clinical scientists working in multidrug resistance. Some studies do report equivalent AUCs. However, decreased clearance means a longer half-life. If the AUC is calculated to infinity, the long terminal half-life may account for a significant portion of the AUC calculation, missing the fact that the maximal concentration – and potentially the effective concentration – is in fact reduced. Indeed, two studies reported actual reductions in the AUC, one with paclitaxel and the other with doxorubicin, associated with 30% and 65% dose reductions, respectively (Advani *et al.*, 2001; Fracasso *et al.*, 2000). The assumption also did not take into account changes in clearance of metabolites. Thus, the AUCs for

6-hydroxy-paclitaxel and doxorubicinol were increased by 222% and 259%, respectively, in a phase I trial in which patients received both doxorubicin and paclitaxel with PSC 833 (Advani *et al.*, 2001). Similarly, the AUC for doxorubicinol increased following administration of the Pgp inhibitor GF120918, while the AUC for doxorubicin was not significantly affected (Sparreboom *et al.*, 1999).

At least three potential mechanisms are thought to underlie the pharmacokinetic interactions observed: (1) liver and renal Pgp inhibition; (2) inhibition of drug-metabolizing cytochrome P450s; and (3) impaired bile flow. The relative contribution of each of these mechanisms to the observed pharmacokinetic interactions is not known. However, an estimate of the magnitude of the Pgp interaction can be obtained by referring to studies of knockout mice in which the murine orthologue of Pgp has been deleted. These studies have shown that the absence of Pgp in the mouse results in a delay in clearance of a number of compounds, which is associated with an increase in serum drug levels. Thus, vinblastine levels were increased 1.7-, 2.4-, 2.3-, and 2.1-fold in plasma, liver, kidney and lung, respectively, in Pgp-deficient mice (van Asperen *et al.*, 1996). Doxorubicin levels were only affected in the liver, where they were 4.5-fold higher than in the wild-type mice (van Asperen *et al.*, 1999). Given the relatively high levels of Pgp that are normally found in the kidney and liver, these results suggest that alternate mechanisms for doxorubicin transport and/or metabolism exist in these tissues. Greater increases in doxorubicin levels were observed in the central nervous system (CNS) due to the absence of Pgp in the endothelial cells in the brain; however, redundancy must exist in the human blood–brain barrier, since no toxicity attributable to increased CNS penetration of anticancer agents has been observed in the clinical trials with Pgp inhibitors.

The cytochrome P450 (CYP) mixed-function oxidases are a multigene family encoding enzymes that play a critical role in the metabolism of many drugs and xenobiotics. PSC 833 and cyclosporin A can inhibit the metabolism of numerous compounds that are substrates for the CYP3A4 isoenzyme, thus contributing to pharmacokinetic interactions (Relling, 1996). Numerous anticancer agents, including paclitaxel, are substrates for this isoform of cytochrome P450 (Kivisto *et al.*, 1995; Wacher *et al.*, 1995).

The final mechanism underlying the altered pharmacokinetics resulting from the addition of PSC 833 may be due to inhibition of ATP-dependent bile salt transport mediated by BSEP (ABCB11) (Bohme *et al.*, 1993) described earlier. Studies of mice lacking *Abcb11* revealed impaired bile salt transport and persistent intrahepatic cholestasis (Wang *et al.*, 2001). Further, mutations in *ABCB11/BSEP* have been found in patients afflicted with persistent familial intrahepatic cholestasis, type 2 (PFIC2) (Strautnieks *et al.*, 1998). Since bile salts are the major driving force for bile flow, inhibition of bile salt export results in cholestasis. Hyperbilirubinemia may be explained by reduced levels of MRP2 (ABCC2), which has been observed in several forms of cholestasis (Kullak-Ublick *et al.*, 2000) (see Chapter 20). Drug excretion is also impaired, requiring dose reduction of drugs excreted primarily in the bile when administered to patients with cholestasis, including doxorubicin, vincristine and paclitaxel (Panday *et al.*, 1997; Rollins and Klaassen, 1979). Notably, inhibition of BSEP by PSC 833 or cyclosporin A results in reduced bile salt transport, and a reduction in bile flow (Bohme *et al.*, 1993, 1994; Stieger *et al.*, 2000). A reduction in oxidized glutathione (GSSG) excretion into the bile was also observed, suggesting some impairment of MRP2 function at the canalicular membrane (Song *et al.*, 1998), and potentially explaining the hyperbilirubinemia observed following administration of PSC 833 or cyclosporin A. Thus, by inhibiting BSEP and reducing bile flow, PSC 833 and cyclosporin A may slow excretion of drugs from the liver.

REVERSAL OF DRUG RESISTANCE MEDIATED BY ABC TRANSPORTERS

CLINICAL TRIALS

The majority of completed trials incorporating a Pgp inhibitor have used PSC 833, as shown in **Table 18.5**. Both the design and the breadth of these trials reflect the enthusiasm and optimism which underlay their initiation. The initial phase I trials were conducted in combination with a single cytotoxic drug such as vinblastine, doxorubicin, etoposide and paclitaxel. Subsequent trials incorporated PSC 833 with

Table 18.5. Clinical results from phase I and II trials with second- and third-generation P-glycoprotein antagonists

Trial design	Drug	Dose[a]	*n*	Responses	Reference
PSC 833 trials					
Phase I	Etoposide	None	34	2 PR	Boote *et al.*, 1996
Phase I	Doxorubicin[b]	30%	38	1 PR	Giaccone *et al.*, 1997
Phase I	Doxorubicin[b]	20%	31	1 PR	Minami *et al.*, 2001
Phase I	Paclitaxel	30%	16	1 PR	Fracasso *et al.*, 2000
	3 h infusion				
Phase I	Paclitaxel	50%	41	1 CR	Chico *et al.*, 2001
	96 h infusion			3 PR	
Phase I	Vinblastine	50–66%	79	3 CR	Bates *et al.*, 2001
	120 h infusion			1 PR	
Phase I	Vincristine	25%	22	8 PR	Sonneveld, *et al.*, 1996
	Doxorubicin	25%			
	Dexamethasone				
Phase I	Mitoxantrone	66%	10	6 CR	Kornblau *et al.*, 1997
	Etoposide	66%			
PSC 833 trials in AML					
Phase I	Mitoxantrone	44%	37	12 CR (32%)	Advani *et al.*, 1999
Poor risk AML[c]	Etoposide	58%			
	AraC				
Phase I	Daunorubicin	33%	66	29 CR (44%)	Lee *et al.*, 1999
Elderly AML	Etoposide	40%			
	AraC				
Phase I	Mitoxantrone	40%	30	15 CR (50%)	Chauncey *et al.*, 2000
Secondary AML	Etoposide	40%			
Phase I/II	Daunorubicin (72 h)	0	43	21 CR (49%)	Dorr *et al.*, 2001
Poor risk AML[c]	AraC				
Phase I	Mitoxantrone	25%	23	6 CR (26%)	Visani *et al.*, 2001
Poor risk AML[c]	Etoposide	62.5%			
	AraC				
Phase I	Paclitaxel	50–60%	58	1 CR	Patnaik *et al.*, 2000
	Carboplatin			8 PR	
Phase I	Doxorubicin	65%	33	5 PR	Advani *et al.*, 2001
	Paclitaxel	53%			
Phase II	Paclitaxel	60%	58	5 PR (8.6%)	Fracasso *et al.*, 2001
Ovarian carcinoma	3 h infusion				
Phase I/II	Doxorubicin[b]	30%	33	1 CR	Baekelandt *et al.*, 2001
Ovarian carcinoma	Cisplatin			4 PR	
Phase I trials with third-generation agents					
VX710	Paclitaxel	55–65%	25	—	Rowinsky *et al.*, 1998
	3 h infusion				
VX710	Doxorubicin[b]	10%	25	1 PR	Peck *et al.*, 2001
GF120918	Doxorubicin	0	46	N/A	Sparreboom *et al.*, 1999
R101933	Docetaxel	0	15	N/A	van Zuylen *et al.*, 2000a
XR9576	Vinorelbine	0	12	1 PR	Fojo *et al.*[d]

Abbreviations: CR, complete response; PR, partial response; AML, acute myelogenous leukemia; N/A, not available.
[a]Dose reduction required at the MTD, compared to MTD in the absence of antagonist.
[b]Dose reduction relative to MTD for doxorubicin, 50 mg m^{-2}; administered on a q 3-week schedule.
[c]Poor risk AML: includes variable proportions of patients with relapsed, refractory, or secondary AML.
[d]Unpublished data.

various combination chemotherapy regimens. These included mitoxantrone and etoposide, with or without cytosine arabinoside (AraC), for acute leukemia; VAD (vinblastine and dexamethasone) for myeloma; and paclitaxel and cisplatin for ovarian cancer. Results from ongoing or recently completed randomized trials are not yet available. The dose reductions required in each trial are shown; for doxorubicin, the results are calculated relative to a dose of $50\,mg\,m^{-2}$ on a 3-weekly schedule. However, single agent doxorubicin has been administered at doses as high as $80\,mg\,m^{-2}$ every 3 weeks (Edmonson *et al.*, 1993).

High response rates are found only in the AML trials. These studies were typically undertaken in poor risk populations, including patients with relapsed or refractory leukemia, or elderly patients with secondary leukemia. Nearly 50% of patients on the trial combining PSC 833 with daunorubicin and AraC experienced a complete response (Dorr *et al.*, 2001). These same investigators had noted a complete response rate of 69% with the same regimen combined with cyclosporin A (List *et al.*, 1993). No dose reductions were made in this trial, and pharmacokinetic studies in the PSC 833 trial revealed that one-half of patients had no detectable pharmacokinetic interaction (Dorr *et al.*, 2001). The authors concluded that systematic dose reductions would potentially have led to undertreatment of half of enrolled patients. Perhaps tellingly, response rates were lower on two trials with mitoxantrone, etoposide and AraC (26% and 32%), where dose reductions were required to prevent severe toxicity (Advani *et al.*, 1999; Visani *et al.*, 2001), although heterogeneity in AML subtypes may have contributed to these differences as well.

Findings with paclitaxel are also illustrative. Dose reduction of paclitaxel was required whether administered as a 3-hour or a 96-hour infusion (Chico *et al.*, 2001; Fracasso *et al.*, 2001). In our study combining a 96-hour infusion of paclitaxel with PSC 833, both clinical evidence and pharmacokinetic studies suggested that one-third of patients were undertreated, one-third were overtreated, and only one-third of patients had appropriate doses of paclitaxel when administered at the maximum tolerated dose determined in combination with PSC 833 (Chico *et al.*, 2001). In the phase I trial with a 3-hour infusion of paclitaxel, reduced doses were given to allow equal toxicity following addition of PSC 833, and it was assumed that the AUCs would be comparable to AUCs in the absence of PSC 833. However, the AUCs were reduced by an average of 41% (range 24–59%) in patients receiving 30–50% dose reductions of paclitaxel (Fracasso *et al.*, 2000). In light of these observations, the 8.6% response rate observed with paclitaxel plus PSC 833 in refractory ovarian cancer may be significant, given that the $70\,mg\,m^{-2}$ dose administered every 3 weeks represented a 60% dose reduction from the standard dose (Fracasso *et al.*, 2001).

From one perspective, it could be argued that the addition of a Pgp inhibitor to a combination chemotherapy regimen could not be expected to have a large impact, particularly on response rates. Indeed, both the effect of Pgp expression on clinical outcome and the impact of PSC 833 have been measured in the penumbra of chemotherapy combinations that frequently include potent additional agents. For example, Pgp inhibitors can only be expected to have an impact on the 30–50% of leukemias expressing Pgp, and can only be expected to enhance the contribution of the anthracycline to the clinical response. The same can be said for the ovarian cancer trials. The Pgp expression studies reported thus far suggest that fewer than 20% of ovarian cancers express this transporter. Thus, when a Pgp inhibitor is added to a regimen combining paclitaxel with cisplatin, the incremental benefit provided by the inhibitor to the combination would only be a fraction of a fraction, and thus could only be detected in a randomized trial encompassing large numbers of patients. Furthermore, benefit may only be seen in survival analyses, if the main role of the inhibitor is to prevent the emergence of a resistant clone.

Hindsight is, of course, 20/20. Given the likelihood that any benefit of Pgp inhibition was lost in the dose reductions required in the PSC 833 trials, a defensible conclusion can be reached that Pgp inhibition has not yet been adequately tested. These trials, as in the first-generation inhibitor studies, provided valuable insight, including convincing evidence that Pgp could be inhibited in patients. As described below, surrogate assays emerged during the course of these trials, confirming increased drug retention in Pgp-bearing normal tissues, and in some tumors.

SURROGATE ASSAYS

In assessing the outcome of drug resistance reversal trials, it became apparent that assays were needed to confirm the inhibition of Pgp function, independent of the effect on tumor responsiveness. Ideally, confirmation of increased uptake of the administered anticancer drug in tumor tissue was needed. Indeed, leukemic cells obtained from patients receiving PSC 833 and daunorubicin displayed a 50% increase in daunorubicin accumulation, after correction for the increased plasma drug concentration (Tidefelt *et al.*, 2000). Given the difficulty of performing such an analysis in any malignancy other than leukemia, surrogate assays have been developed to provide a second approach to confirm the efficacy of Pgp inhibition.

The three major surrogate assays developed include a serum bioassay, the CD56+ rhodamine 123 efflux assay, and ^{99m}Tc-sestamibi imaging. The serum bioassay utilizes serum obtained from patients receiving Pgp inhibitors to either reverse multidrug resistance or increase drug accumulation in a Pgp-overexpressing cell line. This assay addresses the question of whether protein binding might reduce the availability of the Pgp inhibitor, a concern that was raised with the first-generation reversing agents. Serum obtained from patients receiving Pgp inhibitors such as quinine and PSC 833 have been assayed *in vitro*, confirming that concentrations achieved in patients are sufficient to reverse multidrug resistance in laboratory models (Minami *et al.*, 2001; Solary *et al.*, 1991). The second surrogate assay, the CD56+ the rhodamine efflux assay, is based upon the high levels of Pgp expression in circulating CD56+ natural killer cells. The fluorescent dye rhodamine 123 is a substrate for Pgp, and differences in cellular rhodamine 123 fluorescence following incubation in the presence or absence of a specific Pgp inhibitor have been shown to be a reliable functional assay for Pgp in multiple studies **(Figure 18.1)**. Efflux of rhodamine 123 is readily demonstrated in the CD56+ subset of circulating mononuclear cells. CD56+ cells obtained from patients following treatment with an effective Pgp inhibitor show inhibition of rhodamine 123 efflux (Robey *et al.*, 1999; Witherspoon *et al.*, 1996). A disadvantage of both the surrogate serum bioassay and the CD56+ rhodamine 123 efflux assay is that neither measure tumor uptake of the chemotherapeutic agents.

In contrast, the third surrogate assay, ^{99m}Tc-sestamibi imaging, requires that sestamibi be delivered to and penetrate the tumor tissue. ^{99m}Tc-sestamibi is a radionuclide imaging agent marketed as Cardiolyte for the detection of cardiac function. Shown to be a Pgp substrate (Piwnica-Worms *et al.*, 1993), Pgp inhibitors increase the accumulation of sestamibi in

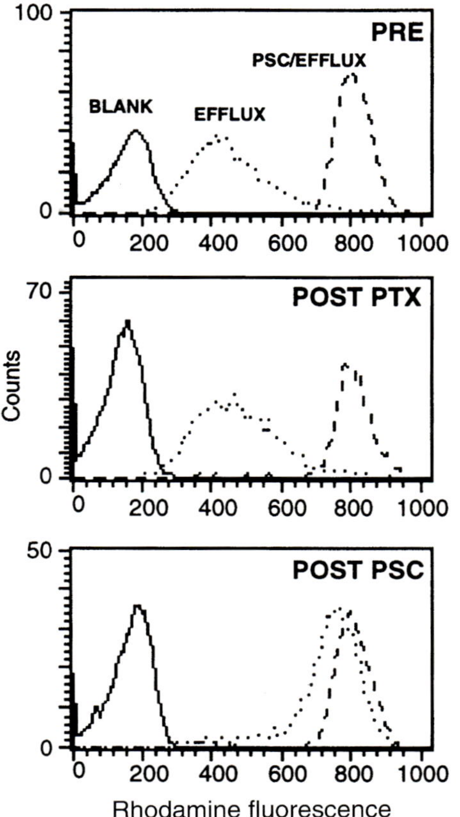

Figure 18.1. Effect of PSC 833 on rhodamine efflux from CD56+ circulating mononuclear cells. Autofluorescence of cells without rhodamine is indicated by the solid line (Blank). Cells incubated in rhodamine for 30 min without or with PSC 833 added in the laboratory and then continuing without (Efflux) or with PSC 833 (PSC/Efflux) in rhodamine-free media for an efflux period are indicated by the dotted and dashed histograms, respectively. PRE: rhodamine fluorescence in CD56+ cells obtained from a patient before the administration of PSC 833. POST PTX: Experiment performed on cells obtained from a patient before the termination of a 96-hour infusion of paclitaxel. POST PSC: Cells obtained 2 hours following oral administration of PSC 833. The results show that rhodamine fluorescence in CD56+ cells obtained from patients following administration of PSC 833 is identical to that in CD56+ cells that have been incubated in exogenous PSC 833, due to the inhibition of P-glycoprotein-mediated rhodamine efflux. In contrast, no inhibition was observed following the administration of paclitaxel alone.

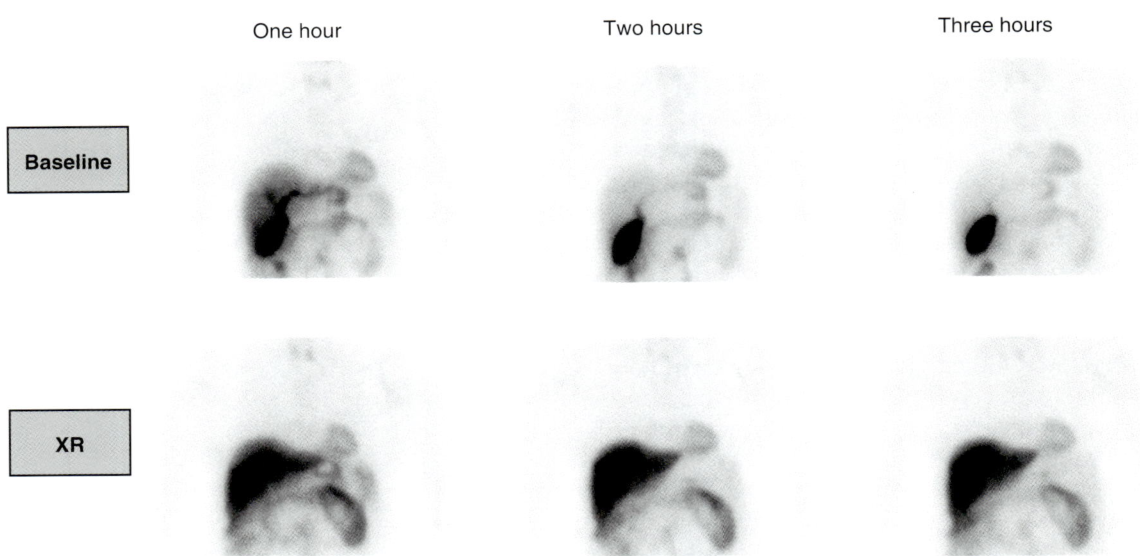

Figure 18.2. ^{99m}Tc-sestamibi images obtained from a patient with renal cell cancer before and after administration of a single intravenous dose of the P-glycoprotein inhibitor XR9576. Images were obtained at 1, 2 and 3 hours. The upper panels show the baseline images. The patient's heart can be seen in the mid to left side of his chest. Sestamibi can be seen in the gall bladder, having been excreted by the liver. The liver is seen only as a faint image. The lower panels show the sestamibi images after administration of XR9576. No significant difference in the cardiac image is observed. The liver is now prominent, due to inhibition of sestamibi excretion by this organ. As a consequence of this inhibition, the gall bladder is no longer visible. Images obtained 48 hours after the single dose show continued sestamibi retention in the liver.

normal organs bearing Pgp and in tumors overexpressing Pgp. Interestingly, no increase in CNS uptake of ^{99m}Tc-sestamibi has been described. Instead of immediate elimination, there is retention of sestamibi in the liver for up to 8 hours when patients receiving PSC 833 are scanned (Chen *et al.*, 1997). Viewed as a surrogate measure of Pgp inhibition, hepatic accumulation of sestamibi is increased in patients receiving both VX710 and PSC 833 (Chen *et al.*, 1997; Luker *et al.*, 1997; Peck *et al.*, 2001). Similar results have been observed with XR9576 **(Figure 18.2)**.

Both the hepatic retention of sestamibi and the inhibition of rhodamine 123 efflux from CD56+ cells are believed to reflect the inhibition of Pgp in a normal tissue. Information regarding tumor uptake is more limited. We studied ^{99m}Tc-sestamibi accumulation following PSC 833 administration (Chen *et al.*, 1997). A small increase in retention in tumor tissue was seen in several patients, after taking into account the small increase observed in heart tissue, which was used to represent the increase in blood levels that probably followed administration of PSC 833. Four of 10 patients showed increases in sestamibi accumulation of 15% or greater over background. To gain perspective on this difference, comparison can be made to the liver scans, which showed increases in sestamibi accumulation greater than 35%, and which are known to have a markedly delayed sestamibi efflux following addition of PSC 833. Similar results were reported from a second group using PSC 833 (Bakker *et al.*, 1999), and from a phase I trial with VX710, a third-generation reversing agent (Peck *et al.*, 2001). In nine patients, tumor enhancement indices ranged from 0.93 to 1.61, compared to liver enhancement, which ranged from 0.8 to 9.85.

In addition to the studies performed with Pgp inhibitors, multiple other studies have correlated either ^{99m}Tc-sestamibi imaging with Pgp expression or ^{99m}Tc-sestamibi imaging with clinical response. In an elegant study by Del Vecchio *et al.* (1997), increasing levels of Pgp (quantitated by ^{125}I-labeled monoclonal antibody MRK-16 binding to tumor tissue) were precisely correlated with reduced ^{99m}Tc-sestamibi retention. This result was confirmed by other investigators using traditional immunohistochemical methods. Taken together, at least five trials evaluating primary and locally advanced breast cancer in 146 patients have demonstrated a correlation between Pgp expression and ^{99m}Tc-sestamibi retention (Fujii *et al.*, 1998; Kao *et al.*, 2001a; Kostakoglu *et al.*, 1998 Sun *et al.*, 2000; Vecchio *et al.*, 1997). Whether reflecting Pgp expression or not, it is interesting to note that several groups have reported that tumors that are not visualized following ^{99m}Tc-sestamibi administration frequently do not respond to chemotherapy (Ciarmiello *et al.*, 1998; Kao *et al.*, 1998, 2000, 2001b; Komori *et al.*, 2000; Nishiyama *et al.*, 2000; Yamamoto *et al.*, 1998). Most of these studies are in lung cancer, but similar findings in breast cancer and in lymphoma have also been described. Since tumor uptake of ^{99m}Tc-sestamibi may involve mechanisms unrelated to drug transport, one must be cautious before concluding that these studies represent expression of Pgp. In addition, this agent is also known to be a substrate for MRP1 (Hendrikse *et al.*, 1998) (see also Chapter 19).

New inhibitors and future trials

With the recognition of the problems of potency and pharmacokinetic interactions, the development of third-generation Pgp inhibitors has been more cautious and strategic. These third-generation inhibitors are more potent than their predecessors. Many (e.g. XR9576, R101933, LY335979, and OC144-093) are also reported to lack significant pharmacokinetic interaction and are free of toxicity (Dantzig *et al.*, 1999; Mistry *et al.*, 2001; Newman *et al.*, 2000; van Zuylen *et al.*, 2000a). A phase I trial combining paclitaxel with VX710 is complete; however, a 55–65% reduction in paclitaxel dose was required, compared to the maximum tolerated dose without VX710, suggesting that many of the problems observed with PSC 833 could recur in development of this agent (Rowinsky *et al.*, 1998). As described in the previous section above, ancillary ^{99m}Tc-sestamibi imaging studies performed with VX710 demonstrated inhibition of Pgp, based on increased drug uptake in both liver and tumor tissue (Peck *et al.*, 2001). The results of a trial with XR9576 with normal volunteers has been reported and inhibition of rhodamine 123 efflux from CD56+ cells was observed in all individuals treated (Coley *et al.*, 2000). Thus, this study demonstrated that XR9576 in a single dose of $2\,mg\,kg^{-1}$ can prevent rhodamine 123 efflux from CD56+ cells for over 24 hours; preliminary results in cancer patients suggest that this inhibition may last as long as 72 hours (unpublished data).

Clinical trials aimed at inhibiting ABC transporters other than Pgp lag far behind. For MRP1, compounds commonly used for inhibition in the laboratory include sulfinpyrazone and probenecid (Evers *et al.*, 1996). The high concentrations required for inhibition of MRP1 function make these agents unlikely candidates for clinical trial although they provide possible leads for the development of more potent inhibitors. Several Pgp inhibitors have been tested for their ability to inhibit MRP1 function: VX710, verapamil, cyclosporin A, MS-209 and GF120918 (Germann *et al.*, 1997). Among these, VX710 was found to be the most potent, at concentrations ranging from 0.5 to 5 μM. However, no trials directly testing VX710 in a tumor type thought to have MRP1-mediated resistance have appeared. Without some evidence for the other MRPs in clinical drug resistance, it would be premature to search for specific inhibitors for these transporters.

Few studies have as yet appeared documenting expression of the ABC half transporter MXR/BCRP (ABCG2) in tumors. A range of expression was documented in human leukemia (Ross *et al.*, 2000), although levels overall appeared to be low. However, two MXR/BCRP inhibitors have been described: GF120918 (also a Pgp inhibitor), and fumitremorgin C (FTC), derived from *Aspergillus fumigatus* (de Bruin *et al.*, 1999; Rabindran *et al.*, 1998). Micromolar concentrations of these inhibitors are required to reverse resistance mediated by BCRP/MXR, and so may not yet be potent enough for clinical application. The potential for GF120918 or a similar compound to inhibit both Pgp and MXR/BCRP could be utilized in leukemia, where mitoxantrone combined with etoposide and ara-C has shown some benefit in poor risk AML **(see Table 18.5)**.

As inhibitors of other ABC transporters move into efficacy trials, lessons learned from the earlier studies of Pgp inhibitors can be applied. One of the most important of these is the critical need to use surrogate assays to confirm the inhibition of transporter-mediated drug efflux in normal tissues (thereby confirming that the inhibitor can work *in vivo*). A second lesson is the importance of choosing a tumor type in which a transporter is understood to play a role in drug resistance. A third is the need to document expression of that transporter in the tumors of patients enrolled on the clinical trial. Predicting that a subset of patients would benefit from inhibition of a transporter will be valid only if the subset can be identified. A fourth lesson is the need to avoid agents with drug interactions requiring dose reduction of the anticancer drugs. Trials incorporating these features may require more effort, but they offer the only chance of preventing history from repeating itself.

The issue of trial design is worth a brief mention. The earliest Pgp inhibitor trials were organized as 'home run' trials, in which it was thought that the benefit of adding a Pgp inhibitor would be so clear that there was no real need to consider including control arms in the trial. The result was outcome data that ranged from 0% to 70% responses in a range of malignancies. Later, several studies were conducted as crossover trials (Lehnert *et al.*, 1998; Miller *et al.*, 1998; Taylor *et al.*, 1997; Thurlimann *et al.*, 1995; Wilson *et al.*, 1995), in which the patient was treated with the chemotherapy regimen until resistance was documented, then the Pgp inhibitor was added. This type of design has the benefit of clearly demonstrating that a patient is resistant to the anticancer drug to be used in combination with the Pgp inhibitor. The disadvantage, however, is that prior exposure to an anticancer agent almost certainly triggers several resistance mechanisms. It is no longer believed that Pgp is likely to operate as a single resistance mechanism. Indeed, the reduction in drug accumulation that results from activity of Pgp might enhance the emergence of other resistance mechanisms. Thus, even though Pgp may be expressed, other resistance mechanisms may be equally or even more important in contributing to the resistance phenotype before the addition of an inhibitor. The design best suited to answer the question of whether inhibition of Pgp (or other transporter) is of clinical benefit in oncology is the randomized trial, coupled with expression, pharmacokinetic, and surrogate studies (see also van Zuylen *et al.*, 2000b). While the number of patients required for a randomized study is daunting, experience with single-arm phase II trials aimed at drug resistance reversal leads to the unavoidable conclusion that such large studies are essential. Even if a phase II trial incorporates surrogate assays and inhibition of the transporter protein is confirmed, the variability of patient populations from study to study will inevitably make the trials difficult to interpret. The randomization controls for variability in the patient population enrolled.

Prevention of Drug Resistance

Finally, it may be argued that 'reversal' of drug resistance will never succeed. Indeed, it has been suggested that increased Pgp expression, as a modulator of intracellular drug concentrations, may well facilitate the induction of other mechanisms of drug resistance. Thus, one strategy to overcome resistance is to prevent it from emerging in the first place. One approach might be to use a Pgp inhibitor at the time of initial therapy to increase intracellular drug concentrations. Major differences in chemotherapy responsiveness would not necessarily be expected from the addition of a Pgp inhibitor at the outset; instead, reduced selection of resistant clones would not become apparent until relapse and treatment failure occurred. *In vitro* models have been described that support the feasibility of such a strategy. In single-step drug selections, co-administration of a Pgp inhibitor has been shown to reduce the mutation rate for doxorubicin-selected resistance by sevenfold from 1.8×10^{-6} to 2.5×10^{-7} per cell generation, while at the same time, suppressing the emergence of resistant cells expressing Pgp (Beketic-Oreskovic *et al.*, 1995). In continuous exposure drug selections, co-administration of PSC 833 with paclitaxel to K562 leukemic cells (Jaffrezou *et al.*, 1995), verapamil with paclitaxel to A2780 ovarian cancer cells (Giannakakou *et al.*, 2000), and verapamil with doxorubicin to both MCF-7 breast cancer cells (Chen *et al.*, 1990) and 8226 multiple myeloma cells (Futscher *et al.*, 1996) prevented emergence of increased levels of Pgp. Finally, studies of murine knockout cell lines can also be cited. Cells lacking Pgp had increased sensitivity to paclitaxel (16-fold), anthracyclines (fourfold) and *Vinca* alkaloids (threefold). Cells lacking both Pgp and Mrp1 had a further increase in sensitivity to anthracyclines (six- to sevenfold) and vincristine (28-fold) (Allen *et al.*, 2000). These studies suggest that the basal levels of Pgp and MRP1 found in unselected cells can contribute to intrinsic resistance.

Clinical evidence for treating patients early with a drug resistance modulator can also be cited. In several studies, leukemia cells obtained at relapse following treatment with daunorubicin and AraC in the presence of a Pgp inhibitor were shown to have reduced levels of Pgp expression (Kornblau *et al.*, 1997; List *et al.*, 1993, 1996). Similarly, biopsies of metastatic sarcoma were obtained during single isolated lung perfusion with doxorubicin (Abolhoda *et al.*, 1999). Pgp levels were measured by RT-PCR in samples obtained before and 50 min after the doxorubicin perfusion. In four of five patients, a three- to 15-fold (median, 6.8) increase in *MDR1* mRNA levels was detected. This evidence, along with enduring evidence that drug transporter expression can be detected in tumor cells at diagnosis from patients with leukemia, lymphoma, and cancer of the ovary, breast, lung, colon, pancreas, kidney and adrenal gland, suggests that inhibitors of drug transport ought to be included early in cancer therapy.

CONCLUSION

As one of the first 'molecular targets' to be addressed in cancer therapy, inhibition of the most intensively studied multidrug transporter, Pgp, has not yet proven its value. However, the weight of evidence suggests that Pgp is expressed in human cancer and is likely to function to promote drug resistance. Efforts to identify potent, non-toxic inhibitors without pharmacokinetic interactions seem to have at last yielded fruit. Surrogate assays have been developed to confirm inhibition of Pgp function in patients. This sets the stage at last for definitive clinical trials testing the efficacy and value of Pgp inhibition. One caveat remains: coexisting transporters may confound efforts to inhibit Pgp. However, with all ABC transporters in the human genome now identified, the ability to identify transporters in individual tumors is within reach. If efforts to measure the expression of ABC transporters in tumors can be renewed and intensified, we should be able to interpret the results of clinical trials testing Pgp inhibition, and determine whether inhibitors of other transporters should be brought forward for future clinical trials.

REFERENCES

Abolhoda, A., Wilson, A.E., Ross, H., Danenberg, P.V., Burt, M. and Scotto, K.W. (1999) Rapid activation of MDR1 gene expression in human metastatic sarcoma after *in vivo* exposure to doxorubicin. Clin. Cancer Res. 5, 3352–3356.

Advani, R., Saba, H.I., Tallman, M.S., Rowe, J.M., Wiernik, P.H., Ramek, J., *et al.* (1999) Treatment of refractory and relapsed acute myelogenous leukemia with combination chemotherapy plus the multidrug resistance modulator PSC 833 (Valspodar). Blood 93, 787–795.

Advani, R., Fisher, G.A., Lum, B.L., Hausdorff, J., Halsey, J., Litchman, M. and Sikic, B.I. (2001) A phase I trial of doxorubicin, paclitaxel, and valspodar (PSC 833), a modulator of multidrug resistance. Clin. Cancer Res. 7, 1221–1229.

Allen, J.D., Brinkhuis, R.F., van Deemter, L., Wijnholds, J. and Schinkel, A.H. (2000) Extensive contribution of the multidrug transporters P-glycoprotein and Mrp1 to basal drug resistance. Cancer Res. 60, 5761–5766.

Allikmets, R., Schriml, L.M., Hutchinson, A., Romano-Spica, V. and Dean, M. (1998) A human placenta-specific ATP-binding cassette gene (ABCP) on chromosome 4q22 that is involved in multidrug resistance. Cancer Res. 58, 5337–5339.

Arnal, M., Franco, N., Fargeot, P., Riedinger, J.M., Brunet-Lecomte, P. and Lizard-Nacol, S. (2000) Enhancement of mdr1 gene expression in normal tissue adjacent to advanced breast cancer. Breast Cancer Res. Treat. 61, 13–20.

Arts, H.J., Katsaros, D., de Vries, E.G., Massobrio, M., Genta, F., Danese, S., *et al.* (1999) Drug resistance-associated markers P-glycoprotein, multidrug resistance-associated protein 1, multidrug resistance-associated protein 2, and lung resistance protein as prognostic factors in ovarian carcinoma. Clin. Cancer Res. 5, 2798–2805.

Baekelandt, M.M., Holm, R., Nesland, J.M., Trope, C.G. and Kristensen, G.B. (2000) P-glycoprotein expression is a marker for chemotherapy resistance and prognosis in advanced ovarian cancer. Anticancer Res. 20, 1061–1067.

Baekelandt, M., Lehne, G., Trope, C.G., Szanto, I., Pfeiffer, P., Gustavssson, B. and Kristensen, G.B. (2001) Phase I/II trial of the multidrug-resistance modulator valspodar combined with cisplatin and doxorubicin in refractory ovarian cancer. J. Clin. Oncol. 19, 2983–2993.

Bakker, M., van der Graaf, W.T., Piers, D.A., Franssen, E.J., Groen, H.J., Smit, E.F., Kool, W., Hollema, H., Muller, E.A. and De Vries, E.G. (1999) 99mTc-Sestamibi scanning with SDZ PSC 833 as a functional detection method for resistance modulation in patients with solid tumours. Anticancer Res. 19, 2349–2353.

Bakos, E., Evers, R., Szakacs, G., Tusnady, G.E., Welker, E., Szabo, K., *et al.* (1998) Functional multidrug resistance protein (MRP1) lacking the N-terminal transmembrane domain. J. Biol. Chem. 273, 32167–32175.

Baldini, N., Scotlandi, K., Barbanti-Brodano, G., Manara, M.C., Maurici, D., Bacci, G., *et al.* (1995) Expression of p-glycoprotein in high-grade osteosarcomas in relation to clinical outcome. N. Engl. J. Med. 333, 1380–1385.

Bates, S., Kang, M., Meadows, B., Bakke, S., Choyke, P., Merino, M., *et al.* (2001) A phase I study of infusional vinblastine in combination with the P-glycoprotein antagonist PSC 833 (Valspodar). Cancer 92, 1577–1590.

Beck, W.T., Mueller, T.J. and Tanzer, L.R. (1979) Altered surface membrane glycoproteins in vinca alkaloid-resistant human leukemic lymphoblasts. Cancer Res. 39, 2070–2076.

Beck, W.T., Grogan, T.M., Willman, C.L., Cordon-Cardo, C., Parham, D.M., Kuttesch, J.F., *et al.* (1996) Methods to detect P-glycoprotein-associated multidrug resistance in patients' tumors: consensus recommendations. Cancer Res. 56, 3010–3020.

Beketic-Oreskovic, L., Duran, G.E., Chen, G., Dumontet, C. and Sikic, B.I. (1995) Decreased mutation rate for cellular resistance to doxorubicin and suppression of mdr1 gene activation by the cyclosporin PSC 833. J. Natl Cancer Inst. 87, 1593–1602.

Biedler, J.L. and Peterson, R.H.F. (1981) Molecular Actions and Targets for Cancer Chemotherapeutic Agents (ed. A. Sartorelli, J. Lazo and J. Bertino), pp. 453–482. New York: Academic Press.

Bodey, B., Bodey, B. Jr, Groger, A.M., Luck, J.V., Siegel, S.E., Taylor, C.R. and Kaiser, H.E. (1997) Immunocytochemical detection of the p170 multidrug resistance (MDR) and the p53 tumor suppressor gene proteins in human breast cancer cells: clinical and therapeutical significance. Anticancer Res. 17, 1311–1318.

Bohme, M., Buchler, M., Muller, M. and Keppler, D. (1993) Differential inhibition by cyclosporins of primary-active ATP-dependent transporters in the hepatocyte canalicular membrane. FEBS Lett. 333, 193–196.

Bohme, M., Muller, M., Leier, I., Jedlitschky, G. and Keppler, D. (1994) Cholestasis caused by inhibition of the adenosine triphosphate-dependent bile salt transport in rat liver. Gastroenterology 107, 255–265.

Boote, D.J., Dennis, I.F., Twentyman, P.R., Osborne, R.J., Laburte, C., Hensel, S., Smyth, J.F., Brampton, M.H. and Bleehen, N.M. (1996) Phase I study of etoposide with SDZ PSC 833 as a modulator of multidrug resistance in patients with cancer. J. Clin. Oncol. 14, 610–618.

Borst, P. and Schinkel, A.H. (1996) What have we learnt thus far from mice with disrupted P-genes? Eur. J. Cancer 32A, 985–990.

Borst, P., Zelcer, N. and van Helvoort, A. (2000a) ABC transporters in lipid transport. Biochim. Biophys. Acta 1486, 128–144.

Borst, P., Evers, R., Kool, M. and Wijnholds, J. (2000b) A family of drug transporters: the multidrug resistance-associated proteins. J. Natl Cancer Inst. 92, 1295–1302.

Bradshaw, D.M. and Arceci, R.J. (1998) Clinical relevance of transmembrane drug efflux as a mechanism of multidrug resistance. J. Clin. Oncol. 16, 3674–3690.

Brangi, M., Litman, T., Ciotti, M., Nishiyama, K., Kohlhagen, G., Takimoto, C., Robey, R., Pommier, Y., Fojo, T. and Bates, S.E. (1999) Camptothecin resistance: role of the ATP-binding cassette (ABC), mitoxantrone-resistance half-transporter (MXR), and potential for glucuronidation in MXR-expressing cells. Cancer Res. 59, 5938–5946.

Broccardo, C., Luciani, M. and Chimini, G. (1999) The ABCA subclass of mammalian transporters. Biochim. Biophys. Acta 1461, 395–404.

Broxterman, H.J., Sonneveld, P., Pieters, R., Lankelma, J., Eekman, C.A., Loonen, A.H., *et al.* (1999) Do P-glycoprotein and major vault protein (MVP/LRP) expression correlate with *in vitro* daunorubicin resistance in acute myeloid leukemia? Leukemia 13, 258–265.

Broxterman, H.J., Sonneveld, P., van Putten, W.J., Lankelma, J., Eekman, C.A., Ossenkoppele, G.J., Pinedo, H.M., Lowenberg, B. and Schuurhuis, G.J. (2000) P-glycoprotein in primary acute myeloid leukemia and treatment outcome of idarubicin/cytosine arabinoside-based induction therapy. Leukemia 14, 1018–1024.

Buchler, M., Konig, J., Brom, M., Kartenbeck, J., Spring, H., Horie, T. and Keppler, D. (1996) cDNA cloning of the hepatocyte canalicular isoform of the multidrug resistance protein, cMrp, reveals a novel conjugate export pump deficient in hyperbilirubinemic mutant rats. J. Biol. Chem. 271, 15091–15098.

Burchenal, J.H., Robinson, E., Johnston, S.F. and Kushida, M.N. (1950) The induction of resistance to 4-amino-N^{10}-methyl-pteroylglutamic acid in a strain of transmitted mouse leukemia. Science 111, 116.

Campos, L., Guyotat, D., Archimbaud, E., Calmard-Oriol, P., Tsuruo, T., Troncy, J., Treille, D. and Fiere, D. (1992) Clinical significance of multidrug resistance P-glycoprotein expression on acute nonlymphoblastic leukemia cells at diagnosis. Blood 79, 473–476.

Chan, H.S.L., Thorner, P.S., Haddad, G. and Ling, V. (1990) Immunohistochemical detection of P-glycoprotein: prognostic correlation in soft tissue sarcoma of childhood. J. Clin. Oncol. 50, 2818–2822.

Chan, H.S., Grogan, T.M., Haddad, G., DeBoer, G. and Ling, V. (1997) P-glycoprotein expression: critical determinant in the response to osteosarcoma chemotherapy. J. Natl Cancer Inst. 89, 1706–1715.

Chauncey, T.R., Rankin, C., Anderson, J.E., Chen, I., Kopecky, K.J., Godwin, J.E., *et al.* (2000) A phase I study of induction chemotherapy for older patients with newly diagnosed acute myeloid leukemia (AML) using mitoxantrone, etoposide, and the MDR modulator PSC 833: a Southwest Oncology Group study 9617. Leuk. Res. 24, 567–574.

Chen, C.C., Meadows, B., Regis, J., Kalafsky, G., Fojo, T., Carrasquillo, J.A. and Bates, S.E. (1997) Detection of *in vivo* p-glycoprotein inhibition by PSC 833 using Tc-99m sestamibi. Clin. Cancer Res. 4, 545–552.

Chen, Y.-N., Mickley, L.A., Schwartz, A.M., Acton, E.M., Hwang, J. and Fojo, A.T. (1990) Characterization of Adriamycin-resistant human breast cancer cells which display overexpression of a novel resistance-related membrane protein. J. Biol. Chem. 265, 10073–10080.

Chen, Z., Lee, K. and Kruh, G.D. (2001) Transport of cyclic nucleotides and estradiol 17-β-D-glucuronide by multidrug resistance protein 4: resistance to 6-mercaptopurine and 6-thioguanine. J. Biol. Chem. 276, 33747–33754.

Chen, Z.S., Furukawa, T., Sumizawa, T., Ono, K., Ueda, K., Seto, K. and Akiyama, S.I. (1999) ATP-dependent efflux of CPT-11 and SN-38 by the multidrug resistance protein (MRP) and its inhibition by PAK-104P. Mol. Pharmacol. 55, 921–928.

Chevillard, S., Pouillart, P., Beldjord, C., Asselain, B., Beuzeboc, P., Magdelenat, H. and Vielh, P. (1996) Sequential assessment of multidrug resistance phenotype and measurement of S-phase fraction as predictive markers of breast cancer response to neoadjuvant chemotherapy. Cancer 77, 292–300.

Chico, I., Kang, M.H., Bergan, R., Abraham, J., Bakke, S., Meadows, B., *et al.* (2001) Phase I study of infusional paclitaxel in combination with the P-glycoprotein antagonist PSC 833. J. Clin. Oncol. 19, 832–842.

Childs, S., Yeh, R.L., Georges, E. and Ling, V. (1995) Identification of a sister gene to P-glycoprotein. Cancer Res. 55, 2029–2034.

Childs, S., Yeh, R.L., Hui, D. and Ling, V. (1998) Taxol resistance mediated by transfection of the liver-specific sister gene of P-glycoprotein. Cancer Res. 58, 4160–4167.

Chung, H.C., Rha, S.Y., Kim, J.H., Roh, J.K., Min, J.S., Lee, K.S., Kim, B.S. and Lee, K.B. (1997) P-glycoprotein: the intermediate end point of drug response to induction chemotherapy in locally advanced breast cancer. Breast Cancer Res. Treat. 42, 65–72.

Ciarmiello, A., Del Vecchio, S., Silvestro, P., Potena, M.I., Carriero, M.V., Thomas, R., Botti, G., D'Aiuto, G. and Salvatore, M. (1998) Tumor clearance of technetium 99m-sestamibi as a predictor of response to neoadjuvant chemotherapy for locally advanced breast cancer. J. Clin. Oncol. 16, 1677–1683.

Cole, S.P.C., Bhardwaj, G., Gerlach, J.H., Mackie, J.E., Grant, C.E., Almquist, K.C., Stewart, A.J., Kurz, E.U., Duncan, A.M. and Deeley, R.G. (1992) Overexpression of a transporter gene in a multidrug-resistant human lung cancer cell line. Science 258, 1650–1654.

Cole, S.P.C., Sparks, K.E., Fraser, K., Loe, D.W., Grant, C.E. and Wilson, G.M. (1994) Pharmacological characterization of multidrug resistant MRP-transfected human tumor cells. Cancer Res. 54, 5902–5910.

Coley, H.M., Verrill, M.W., Gregson, S.E., Odell, D.E., Fisher, C. and Judson, I.R. (2000) Incidence of P-glycoprotein overexpression and multidrug resistance (MDR) reversal in adult soft tissue sarcoma. Eur. J. Cancer 36, 881–888.

Cordon-Cardo, C., O'Brien, J.P., Casals, D., Rittman-Grauer, L., Biedler, J.L., Melamed, M.R. and Bertino, J.R. (1989) Multidrug-resistance gene (P-glycoprotein) is expressed by endothelial cells at blood–brain barrier sites. Proc. Natl Acad. Sci. USA 86, 695–698.

Cordon-Cardo, C., O'Brien, J.P., Boccia, J., Casals, D., Bertino, J.R. and Melamed, M.R. (1990) Expression of the multidrug resistance gene product (P-glycoprotein) in human normal and tumor tissues. J. Histochem. Cytochem. 38, 1277–1287.

Cui, Y., Konig, J., Buchholz, J.K., Spring, H., Leier, I. and Keppler, D. (1999) Drug resistance and ATP-dependent conjugate transport mediated by the apical multidrug resistance protein, MRP2, permanently expressed in human and canine cells. Mol. Pharmacol. 55, 929–937.

Dano, K. (1973) Active outward transport of daunomycin in resistant Ehrlich ascites tumor cells. Biochim. Biophys. Acta 323, 466–483.

Dantzig, A.H., Shepard, R.L., Law, K.L., Tabas, L., Pratt, S., Gillespie, J.S., Binkley, S.N., Kuhfeld, M.T., Starling, J.J. and Wrighton, S.A. (1999) Selectivity of the multidrug resistance modulator, LY335979, for P-glycoprotein and effect on cytochrome P-450 activities. J. Pharmacol. Exp. Ther. 290, 854–862.

de Bruin, M., Miyake, K., Litman, T., Robey, R. and Bates, S.E. (1999) Reversal of resistance by GF120918 in cell lines expressing the ABC half-transporter, MXR. Cancer Lett. 146, 117–126.

Dean, M., Hamon, Y. and Chimini, G. (2001) The human ATP-binding cassette (ABC) transporter superfamily. J. Lipid Res. 42, 1007–1017.

Del Vecchio, S., Ciarmiello, A., Pace, L., Potena, M.I., Carriero, M.V., Mainolfi, C., Thomas, R., D'Aiuto, G., Tsuruo, T. and Salvatore, M. (1997) Fractional retention of technetium-99m-sestamibi as an index of P-glycoprotein expression in untreated breast cancer patients. J. Nucl. Med. 38, 1348–1351.

den Boer, M.L., Pieters, R., Kazemier, K.M., Rottier, M.M., Zwaan, C.M., Kaspers, G.J., *et al.* (1998) Relationship between major vault protein/lung resistance protein, multidrug resistance-associated protein, P-glycoprotein expression, and drug resistance in childhood leukemia. Blood 91, 2092–2098.

Dexter, D.W., Reddy, R.K., Geles, K.G., Bansal, S., Myint, M.A., Rogakto, A., Leighton, J.C. and Goldstein, L.J. (1998) Quantitative reverse transcriptase-polymerase chain reaction measured expression of MDR1 and MRP in primary breast carcinoma. Clin. Cancer Res. 4, 1533–1542.

Dorr, R., Karanes, C., Spier, C., Grogan, T., Greer, J., Moore, J., *et al.* (2001) Phase I/II study of the P-glycoprotein modulator PSC 833 in patients with acute myeloid leukemia. J. Clin. Oncol. 19, 1589–1599.

Doyle, L.A., Yang, W., Abruzzo, L.V., Krogmann, T., Gao, Y., Rishi, A.K. and Ross, D.D. (1998) A multidrug resistance transporter from human MCF-7 breast cancer cells. Proc. Natl Acad. Sci. USA 95, 15665–15670.

Edmonson, J.H., Ryan, L.M., Blum, R.H., Brooks, J.S., Shiraki, M., Frytak, S. and Parkinson, D.R. (1993) Randomized comparison of doxorubicin alone versus ifosfamide plus doxorubicin or mitomycin, doxorubicin, and cisplatin against advanced soft tissue sarcomas. J. Clin. Oncol. 11, 1269–1275.

Evers, R., Zaman, G.J., van Deemter, L., Jansen, H., Calafat, J., Oomen, L.C., Oude Elferink, R.P., Borst, P. and Schinkel, A.H. (1996) Basolateral localization and export activity of the human multidrug resistance-associated protein in polarized pig kidney cells. J. Clin. Invest. 97, 1211–1218.

Faneyte, I.F., Kristel, P.M. and van de Vijver, M.J. (2001) Determining MDR1/P-glycoprotein expression in breast cancer. Int. J. Cancer 93, 114–122.

Ferry, D.R., Traunecker, H. and Kerr, D.J. (1996) Clinical trials of P-glycoprotein reversal in solid tumours. Eur. J. Cancer 32A, 1070–1081.

Filipits, M., Suchomel, R.W., Dekan, G., Haider, K., Valdimarsson, G., Depisch, D. and Pirker, R. (1996) MRP and MDR1 gene expression in primary breast carcinomas. Clin. Cancer Res. 2, 1231–1237.

Filipits, M., Pohl, G., Stranzl, T., Suchomel, R.W., Scheper, R.J., Jager, U., Geissler, K., Lechner, K. and Pirker, R. (1998) Expression of the lung resistance protein predicts poor outcome in *de novo* acute myeloid leukemia. Blood 91, 1508–1513.

Fisher, G.A. and Sikic, B.I. (1995) Clinical studies with modulators of multidrug resistance. Hematol. Oncol. Clin. North Am. 9, 363–382.

Fisher, G.A., Lum, B.L., Hausdorff, J. and Sikic, B.I. (1996) Pharmacological considerations in the modulation of multidrug resistance. Eur. J. Cancer 32A, 1082–1088.

Fojo, A.T., Ueda, K., Slamon, D.J., Poplack, D.G., Gottesman, M.M. and Pastan, I. (1987) Expression of a multidrug resistance gene in human tumors and tissues. Proc. Natl Acad. Sci. USA 84, 265–269.

Fracasso, P.M., Westerveldt, P., Fears, C.L., Rosen, D.M., Zuhowski, E.G., Cazenave, L.A., Litchman, M., Egorin, M.J., Westervelt P. and Fears, C.A. (2000) Phase I study of paclitaxel in combination with a multidrug resistance modulator, PSC 833 (Valspodar), in refractory malignancies. J. Clin. Oncol. 18, 1124–1134.

Fracasso, P.M., Brady, M.F., Moore, D.H., Walker, J.L., Rose, P.G., Letvak, L., Grogan, T.M. and McGuire, W.P. (2001) Phase II study of paclitaxel and valspodar

(PSC 833) in refractory ovarian carcinoma: a gynecologic oncology group study. J. Clin. Oncol. 19, 2975–2982.

Fujii, H., Nakamura, K., Kubo, A., Enomoto, K., Ikeda, T., Kubota, T., Matsuzaki, S.W. and Kitajima, M. (1998) 99mTc-MIBI scintigraphy as an indicator of the chemosensitivity of anthracyclines in patients with breast cancer. Anticancer Res. 18, 4601–4605.

Futscher, B.W., Foley, N.E., Gleason-Guzman, M.C., Meltzer, P.S., Sullivan, D.M. and Dalton, W.S. (1996) Verapamil suppresses the emergence of P-glycoprotein-mediated multidrug resistance. Int. J. Cancer 66, 520–525.

Galimberti, S., Marchetti, A., Buttitta, F., Carnicelli, V., Pellegrini, S., Bevilacqua, G. and Petrini, M. (1998) Multidrug resistance related genes and p53 expression in human non small cell lung cancer. Anticancer Res. 18, 2973–2976.

Gerlach, J.H., Endicott, J.A., Juranka, P.F., Henderson, G., Sarangi, F., Deuchars, K.L. and Ling, V. (1986) Homology between P-glycoprotein and a bacterial haemolysin transport protein suggests a model for multidrug resistance. Nature 324, 485–489.

Gerloff, T., Stieger, B., Hagenbuch, B., Madon, J., Landmann, L., Roth, J., Hofmann, A.F. and Meier, P.J. (1998) The sister of P-glycoprotein represents the canalicular bile salt export pump of mammalian liver. J. Biol. Chem. 273, 10046–10050.

Germann, U.A., Ford, P.J., Shlyakhter, D., Mason, V.S. and Harding, M.W. (1997) Chemosensitization and drug accumulation effects of VX-710, cyclosporin A, MS-209 and GF120918 in multidrug resistant HL60/ADR cells expressing the multidrug resistance-associated protein MRP. Anticancer Drug 8, 141–155.

Giaccone, G., Linn, S.C., Welink, J., Catimel, G., Stieltjes, H., van der Vijgh, W.J., Eeltink, C., Vermorken, J.B. and Pinedo, H.M. (1997) A dose-finding and pharmacokinetic study of reversal of multidrug resistance with SDZ PSC 833 in combination with doxorubicin in patients with solid tumors. Clin. Cancer Res. 3, 2005–2015.

Giannakakou, P., Gussio, R., Nogales, E., Downing, K.H., Zaharevitz, D., Bollbuck, B., Poy, G., Sackett, D., Nicolaou, K.C. and Fojo, T. (2000) A common pharmacophore for epothilone and taxanes: molecular basis for drug resistance conferred by tubulin mutations in human cancer cells. Proc. Natl Acad. Sci. USA 97, 2904–2909.

Goff, B.A., Paley, P.J., Greer, B.E. and Gown, A.M. (2001) Evaluation of chemoresistance markers in women with epithelial ovarian carcinoma. Gynecol. Oncol. 81, 18–24.

Goldstein, L.J., Galski, H., Fojo, A., Willingham, M., Lai, S.L., Gazder, A., *et al.* (1989) Expression of a multidrug resistance gene in human cancers. J. Natl Cancer Inst. 81, 116–124.

Gottesman, M.M. and Pastan, I. (1993) Biochemistry of multidrug resistance mediated by the multidrug transporter. Annu. Rev. Biochem. 62, 385–427.

Gros, P., Croop, J. and Housman, D. (1986) Mammalian multidrug resistance genes: complete cDNA sequence indicates strong homology to bacterial transport proteins. Cell 47, 371–380.

Han, K., Kahng, J., Kim, M., Lim, J., Kim, Y., Cho, B., *et al.* (2000) Expression of functional markers in acute nonlymphoblastic leukemia. Acta. Haematol. 104, 174–180.

Haque, A.K., Adegboyega, P., Al-Salameh, A., Vrazel, D.P. and Zwischenberger, J. (1999) p53 and P-glycoprotein expression do not correlate with survival in nonsmall cell lung cancer: a long-term study and literature review. Mod. Pathol. 12, 1158–1166.

Hegewisch-Becker, S., Staib, F., Loning, T., Pichlmeier, U., Kroger, N., Reymann, A. and Hossfeld, D.K. (1998) No evidence of significant activity of the multidrug resistance gene product in primary human breast cancer. Ann. Oncol. 9, 85–93.

Hendrikse, N.H., Franssen, E.J.F., van der Graaf, W.T.A., Meijer, C., Piers, D.A., Vaalburg, W. and de Vries, E.G.E. (1998) ^{99m}Tc-sestamibi is a substrate for P-glycoprotein and the multidrug resistance-associated protein. Br. J. Cancer 77, 353–358.

Herzog, C.E., Trepel, J.B., Mickley, L.A., Bates, S.E. and Fojo, A.T. (1992). Various methods of analysis of *mdr-1*/P-glycoprotein in human colon cancer cell lines. J. Natl Cancer Inst. 84, 711–716.

Hipfner, D.R., Gao, M., Scheffer, G., Scheper, R.J., Deeley, R.G. and Cole, S.P.C. (1998) Epitope mapping of monoclonal antibodies specific for the 190-kDa multidrug resistance protein (MRP). Br. J. Cancer 78, 1134–1140.

Hipfner, D.R., Deeley, R.G. and Cole, S.P.C. (1999) Structural, mechanistic and clinical aspects of MRP1. Biochim. Biophys. Acta 1461, 359–376.

Honjo, Y., Hrycyna, C.A., Yan, Q.-W., Medina-Perez, W.Y., Robey, R.W., van de Laar, A., Litman, T., Dean, M. and Bates, S.E. (2001) Acquired mutations in the MXR/BCRP/ABCP gene alter substrate specificity in MXR/BCRP/ABCP-overexpressing cells. Cancer Res. 61, 6635–6639.

Honkoop, A.H., van Diest, P.J., de Jong, J.S., Linn, S.C., Giaccone, G., Hoekman, K., Wagstaff, J. and Pinedo, H.M. (1998) Prognostic role of clinical, pathological and biological characteristics in patients with locally advanced breast cancer. Br. J. Cancer 77, 621–626.

Hooijberg, J.H., Broxterman, H.J., Kool, M., Assaraf, Y.G., Peters, G.J., Noordhuis, P., Scheper, R.J., Borst, P., Pinedo, H.M. and Jansen, G. (1999) Antifolate resistance mediated by the multidrug resistance proteins MRP1 and MRP2. Cancer Res. 59, 2532–2535.

Izquierdo, M.A., van der Zee, A.G., Vermorken, J.B., van der Valk, P., Belien, J.A., Giaccone, G., *et al.* (1995) Drug resistance-associated marker Lrp for prediction of response to chemotherapy and prognoses in advanced ovarian carcinoma. J. Natl Cancer Inst. 87, 1230–1237.

Jaffrezou, J.P., Dumontet, C., Derry, W.B., Duran, G., Chen, G., Tsuchiya, E., Wilson, L., Jordan, M.A. and Sikic, B.I. (1995) Novel mechanism of resistance to paclitaxel (Taxol) in human K562 leukemia cells by combined selection with PSC 833. Oncol. Res. 7, 517–527.

Jedlitschky, G., Leier, I., Buchholz, U., Center, M. and Keppler, D. (1994) ATP-dependent transport of glutathion S-conjugates by multidrug resistance-associated protein. Cancer Res. 54, 4833–4836.

Jedlitschky, G., Leier, I., Buchholz, U., Barnouin, K., Kurz, G. and Keppler, D. (1996) Transport of glutathione, glucuronate, and sulfate conjugates by the MRP gene-encoded conjugate export pump. Cancer Res. 56, 988–994.

Jedlitschky, G., Burchell, B. and Keppler, D. (2000) The multidrug resistance protein 5 functions as an ATP-dependent export pump for cyclic nucleotides. J. Biol. Chem. 275, 30069–30074.

Joncourt, F., Buser, K., Altermatt, H., Bacchi, M., Oberli, A. and Cerny, T. (1998) Multiple drug resistance parameter expression in ovarian cancer. Gynecol. Oncol. 70, 176–182.

Jonker, J.W., Smit, J.W., Brinkhuis, R.F., Maliepaard, M., Beijnen, J.H., Schellens, J.H. and Schinkel, A.H. (2000) Role of breast cancer resistance protein in the bioavailability and fetal penetration of topotecan. J. Natl Cancer Inst. 92, 1651–1656.

Kanzaki, A., Toi, M., Nakayama, K., Bando, H., Mutoh, M., Uchida, T., Fukumoto, M. and Takebayashi, Y. (2001) Expression of multidrug resistance-related transporters in human breast carcinoma. Jpn J. Cancer Res. 92, 452–458.

Kao, C.H., ChangLai, S.P., Chieng, P.U. and Yen, T.C. (1998) Technetium-99m methoxy-isobutylisonitrile chest imaging of small cell lung carcinoma: relation to patient prognosis and chemotherapy response – a preliminary report. Cancer 83, 64–68.

Kao, C.H., Hsieh, J.F., Tsai, S.C., Ho, Y.J. and Lee, J.K. (2000) Quickly predicting chemotherapy response to paclitaxel-based therapy in non-small cell lung cancer by early technetium-99m methoxyisobutylisonitrile chest single-photon-emission computed tomography. Clin. Cancer Res. 6, 820–824.

Kao, C.H., Tsai, S.C., Liu, T.J., Ho, Y.J., Wang, J.J., Ho, S.T. and ChangLai, S.P. (2001a) P-glycoprotein and multidrug resistance-related protein expressions in relation to technetium-99m methoxyisobutylisonitrile scintimammography findings. Cancer Res. 61, 1412–1414.

Kao, C.H., Tsai, S.C., Wang, J.J., Ho, Y.J., Ho, S.T. and Changlai, S.P. (2001b) Evaluation of chemotherapy response using technetium-99M-sestamibi scintigraphy in untreated adult malignant lymphomas and comparison with other prognosis factors: a preliminary report. Int. J. Cancer 95, 228–231.

Kivisto, K.T., Kroemer, H.K. and Eichelbaum, M. (1995) The role of human cytochrome P450 enzymes in the metabolism of anticancer agents: implications for drug interactions. Br. J. Clin. Pharmacol. 40, 523–530.

Klein, I., Sarkadi, B. and Varadi, A. (1999) An inventory of the human ABC proteins. Biochim. Biophys. Acta 1461, 237–262.

Knutsen, T., Rao, V.K., Ried, T., Mickley, L., Schneider, E., Miyake, K., *et al.* (2000) Amplification of 4q21-q22 and the MXR gene in independently derived mitoxantrone-resistant cell lines. Genes Chrom. Cancer 27, 110–116.

Koike, K., Kawabe, T., Tanaka, T., Toh, S., Uchiumi, T., Wada, M., Akiyama, S., Ono, M. and Kuwano, M. (1997) A canalicular multi-specific organic anion transporter (cMOAT) antisense cDNA enhances drug sensitivity in human hepatic cancer cells. Cancer Res. 57, 5475–5479.

Komori, T., Narabayashi, I., Matsui, R., Sueyoshi, K., Aratani, T. and Utsunomiya, K. (2000) Technetium-99m MIBI single photon emission computed tomography as an indicator of prognosis for patients with lung cancer – preliminary report. Ann. Nucl. Med. 14, 415–420.

Konig, J., Rost, D., Cui, Y. and Keppler, D. (1999) Characterization of the human multidrug resistance protein isoform MRP3 localized to the basolateral hepatocyte membrane. Hepatology 29, 1156–1163.

Kool, M., de Haas, M., Scheffer, G.L., Scheper, R.J., van Eijk, M.J., Juijn, J.A., Baas, F. and Borst, P. (1997) Analysis of expression of cMOAT (MRP2), MRP3, MRP4, and MRP5, homologues of the multidrug resistance-associated protein gene in human cancer cell lines. Cancer Res. 57, 3537–3547.

Kool, M., van der Linden, M., de Haas, M., Scheffer, G.L., de Vree, J.M., Smith, A.J., *et al.* (1999) MRP3, an organic anion transporter able to transport anti-cancer drugs. Proc. Natl Acad. Sci. USA 96, 6914–6919.

Kornblau, S.M., Estey, E., Madden, T., Tran, H.T., Zhao, S., Consoli, U., *et al.* (1997) Phase I study of mitoxantrone plus etoposide with multidrug blockade by SDZ PSC-833 in relapsed or refractory acute myelogenous leukemia. J. Clin. Oncol. 15, 1796–1802.

Kostakoglu, L., Ruacan, S., Ergun, E.L., Sayek, I., Elahi, N. and Bekdik, C.F. (1998) Influence of the heterogeneity of P-glycoprotein expression on technetium-99m-MIBI uptake in breast cancer. J. Nucl. Med. 39, 1021–1026.

Kullak-Ublick, G.A., Stieger, B., Hagenbuch, B. and Meier, P.J. (2000) Hepatic transport of bile salts. Semin. Liver Dis. 20, 273–292.

Kuttesch, J.F., Parham, D.M., Luo, X., Meyer, W.H., Bowman, L., Shapiro, D.N., Pappo, A.S., Crist, W.M., Beck, W.T. and Houghton, P.J. (1996) P-glycoprotein expression at diagnosis may not be a primary mechanism of therapeutic failure in childhood rhabdomyosarcoma. J. Clin. Oncol. 14, 886–900.

Lacave, R., Coulet, F., Ricci, S., Touboul, E., Flahault, A., Rateau, J.G., Cesari, D., Lefranc, J.P. and Bernaudin, J.F. (1998) Comparative evaluation by semiquantitative reverse transcriptase polymerase chain reaction of MDR1, MRP and GSTp gene expression in breast carcinomas. Br. J. Cancer 77, 694–702.

Laing, N.M., Belinsky, M.G., Kruh, G.D., Bell, D.W., Boyd, J.T., Barone, L., Testa, J.R. and Tew, K.D. (1998) Amplification of the ATP-binding cassette 2 transporter gene is functionally linked with enhanced efflux of estramustine in ovarian carcinoma cells. Cancer Res. 58, 1332–1337.

Lee, E.J., George, S.L., Caligiuri, M., Szatrowski, T.P., Powell, B.L., Lemke, S., Dodge, R.K., Smith, R., Baer, M. and Schiffer, C.A. (1999) Parallel phase I studies of daunorubicin given with cytarabine and etoposide with or without the multidrug resistance modulator PSC-833 in previously untreated patients 60 years of age or older with acute myeloid leukemia: results of cancer and leukemia group B study 9420. J. Clin. Oncol. 17, 2831–2839.

Lee, P.D., Noble-Topham, S.E., Bell, R.S. and Andrulis, I.L. (1996) Quantitative analysis of multidrug resistance gene expression in human osteosarcomas. Br. J. Cancer 74, 1046–1050.

Legrand, O., Simonin, G., Beauchamp-Nicoud, A., Zittoun, R. and Marie, J.P. (1999) Simultaneous activity of MRP1 and P-glycoprotein is correlated with *in vitro* resistance to daunorubicin and with *in vivo* resistance in adult acute myeloid leukemia. Blood 94, 1046–1056.

Lehnert, M., Mross, K., Schueller, J., Thuerlimann, B., Kroeger, N. and Kupper, H. (1998) Phase II trial of dexverapamil and epirubicin in patients with non-responsive metastatic breast cancer. Br. J. Cancer 77, 1155–1163.

Leith, C.P., Kopecky, K.J., Chen, I.M., Eijdems, L., Slovak, M.L., McConnell, T.S., *et al.* (1999) Frequency and clinical significance of the expression of the multidrug resistance proteins MDR1/P-glycoprotein, MRP1, and LRP in acute myeloid leukemia: a Southwest Oncology Group Study. Blood 94, 1086–1099.

Leslie, E.M., Deeley, R.G. and Cole, S.P.C. (2001) Toxicological relevance of the multidrug resistance protein 1, MRP1 (ABCC1) and related transporters. Toxicology 167, 3–23.

Levine, E.A., Holzmayer, T., Bacus, S., Mechetner, E., Mera, R., Bolliger, C., Roninson, I.B. and Das Gupta, T.K. (1997) Evaluation of newer prognostic markers for adult soft tissue sarcomas. J. Clin. Oncol. 15, 3249–3257.

Linn, S.C., Pinedo, H.M., van Ark-Otte, J., van der Valk, P., Hoekman, K., Honkoop, A.H., Vermorken, J.B. and Giaccone, G. (1997) Expression of drug resistance proteins in breast cancer, in relation to chemotherapy. Int. J. Cancer 71, 787–795.

List, A.F., Spier, C., Greer, J., Wolff, S., Hutter, J., Dorr, R., Salmon, S., Futscher, B., Baier, M. and Dalton, W. (1993) Phase I/II trial of cyclosporine as a chemotherapy-resistance modifier in acute leukemia. J. Clin. Oncol. 11, 1652–1660.

List, A.F., Spier, C.S., Grogan, T.M., Johnson, C., Roe, D.J., Greer, J.P., *et al.* (1996) Overexpression of the major vault transporter protein lung-resistance protein predicts treatment outcome in acute myeloid leukemia. Blood 87, 2464–2469.

Litman, T., Brangi, M., Hudson, E., Fetsch, P., Abati, A., Ross, D.D., Miyake, K., Resau, J.H. and Bates, S.E. (2000) The multidrug resistant phenotype associated with overexpression of the new ABC half-transporter, MXR. J. Cell Sci. 113, 2011–2021.

Lizard-Nacol, S., Genne, P., Coudert, B., Riedinger, J.M., Arnal, M., Sancy, C., Brunet-Lecomte, P. and Fargeot, P. (1999) MDR1 and thymidylate synthase (TS) gene expressions in advanced breast cancer: relationships to drug exposure, p53 mutations, and clinical outcome of the patients. Anticancer Res. 19, 3575–3581.

Loe, D.W., Deeley, R.G. and Cole, S.P.C. (1998) Characterization of vincristine transport by the M(r) 190,000 multidrug resistance protein (MRP): evidence for cotransport with reduced glutathione. Cancer Res. 58, 5130–5136.

Lopes, J.M., Bruland, O.S., Bjerkehagen, B., Silva, M.C., Holm, R., Pettersen, E.O., Solheim, O.P., Sobrinho-Simoes, M. and Nesland, J.M. (1997) Synovial sarcoma: immunohistochemical expression of P-glycoprotein and glutathione S transferase-pi and clinical drug resistance. Pathol. Res. Pract. 193, 21–36.

Lorico, A., Rappa, G., Flavell, R.A. and Sartorelli, A.C. (1996) Double knockout of the *MRP* gene leads to increased drug sensitivity *in vitro*. Cancer Res. 56, 5351–5355.

Luker, G.D., Facasso, P.M., Dobkin, J. and Piwnica-Worms, D. (1997) Modulation of the multidrug resistance P-glycoprotein: detection with technetium-99m-sestamibi *in vivo*. J. Nucl. Med. 38, 369–372.

Maliepaard, M., van Gastelen, M.A., de Jong, L.A., Pluim, D., van Waardenburg, R.C., Ruevekamp-Helmers, M.C., Floot, B.G. and Schellens, J.H. (1999) Overexpression of the BCRP/MXR/ABCP gene in a topotecan-selected ovarian tumor cell line. Cancer Res. 59, 4559–4563.

Maliepaard, M., Scheffer, G.L., Faneyte, I.F., van Gastelen, M.A., Pijnenborg, A.C.L.M., Schinkel, A.H., van de Vijver, M.J., Scheper, R.J. and Schellens, J.H.M. (2001) Subcellular localization and distribution of the breast cancer resistance protein transporter in normal human tissues. Cancer Res. 61, 3458–3464.

Masuda, M., I'izuka, Y., Yamazaki, M., Nishigaki, R., Kato, Y., Ni'inuma, K., Suzuki, H. and Sugiyama, Y. (1997) Methotrexate is excreted into the bile by canalicular multispecific organic anion transporter in rats. Cancer Res. 57, 3506–3510.

Mechetner, E., Kyshtoobayeva, A., Zonis, S., Kim, H., Stroup, R., Garcia, R., Parker, R.J. and Fruehauf, J.P. (1998) Levels of multidrug resistance (MDR1) P-glycoprotein expression by human breast cancer correlate with *in vitro* resistance to taxol and doxorubicin. Clin. Cancer Res. 4, 389–398.

Michieli, M., Damiani, D., Ermacora, A., Masolini, P., Raspadori, D., Visani, G., Scheper, R.J. and Baccarani, M. (1999) P-glycoprotein, lung resistance-related protein and multidrug resistance associated protein in *de novo* acute non-lymphocytic leukaemias: biological and clinical implications. Br. J. Haematol. 104, 328–335.

Mickley, L., Jain, P., Miyake, K., Schriml, L.M., Rao, K., Fojo, T., Bates, S. and Dean, M. (2001) An ATP-binding cassette gene (ABCG3) closely related to the multidrug transporter ABCG2 (MXR/ABCP) has an unusual ATP-binding domain. Mamm. Genome 12, 86–88.

Miller, T.P., Chase, E.M., Dorr, R., Dalton, W.S., Lam, K.S. and Salmon, S.E. (1998) A phase I/II trial of paclitaxel for non-Hodgkin's lymphoma followed by paclitaxel plus quinine in drug-resistant disease. Anticancer Drug. 9, 135–140.

Minami, H., Ohtsu, T., Fujii, H., Igarashi, T., Itoh, K., Uchiyama-Kokubu, N., *et al.* (2001) Phase I study of intravenous PSC-833 and doxorubicin: reversal of multidrug resistance. Jpn J. Cancer Res. 92, 220–230.

Mistry, P., Stewart, A.J., Dangerfield, W., Okiji, S., Liddle, C., Bootle, D., Plumb, J.A.,

Templeton, D. and Charlton, P. (2001) *In vitro* and *in vivo* reversal of P-glycoprotein-mediated multidrug resistance by a novel potent modulator, XR9576. Cancer Res. 61, 749–758.

Miyake, K., Mickley, L., Litman, T., Zhan, Z., Robey, R., Cristensen, B., *et al.* (1999) Molecular cloning of cDNAs which are highly overexpressed in mitoxantrone-resistant cells: demonstration of homology to ABC transport genes. Cancer Res. 59, 8–13.

Narasaki, F., Matsuo, I., Ikuno, N., Fukuda, M., Soda, H. and Oka, M. (1996) Multidrug resistance-associated protein (MRP) gene expression in human lung cancer. Anticancer Res. 16, 2079–2082.

Newman, M.J., Rodarte, J.C., Benbatoul, K.D., Romano, S.J., Zhang, C., Krane, S., *et al.* (2000) Discovery and characterization of OC144-093, a novel inhibitor of P-glycoprotein-mediated multidrug resistance. Cancer Res. 60, 2964–2972.

Nishiyama, Y., Yamamoto, Y., Satoh, K., Ohkawa, M., Kameyama, K., Hayashi, E., Fujita, J. and Tanabe, M. (2000) Comparative study of Tc-99m MIBI and TI-201 SPECT in predicting chemotherapeutic response in non-small-cell lung cancer. Clin. Nucl. Med. 25, 364–369.

Nooter, K., Westerman, A.M., Flens, M.J., Zaman, G.J., Scheper, R.J., van Wingerden, K.E., *et al.* (1995) Expression of the multidrug resistance-associated protein (MRP) gene in human cancers. Clin. Cancer Res. 1, 1301–1310.

Nooter, K., Bosman, F.T., Burger, H., van Wingerden, K.E., Flens, M.J., Scheper, R.J., Oostrum, R.G., Boersma, A.W., van der Gaast, A. and Stoter, G. (1996a) Expression of the multidrug resistance-associated protein (MRP) gene in primary non-small-cell lung cancer. Ann. Oncol. 7, 75–81.

Nooter, K., Burger, H. and Stoter, G. (1996b) Multidrug resistance-associated protein (MRP) in haematological malignancies. Leuk. Lymphoma 20, 381–387.

Nooter, K., Brutel de la Riviere, G., Look, M.P., van Wingerden, K.E., Henzen-Logmans, S.C., Scheper, R.J., Flens, M.J., Klijn, J.G., Stoter, G. and Foekens, J.A. (1997) The prognostic significance of expression of the multidrug resistance-associated protein (MRP) in primary breast cancer. Br. J. Cancer 76, 486–493.

Oka, M., Fukuda, M., Sakamoto, A., Takatani, H., Fukuda, M., Soda, H. and Kohno, S. (1997) The clinical role of MDR1 gene expression in human lung cancer. Anticancer Res. 17, 721–724.

Oshika, Y., Nakamura, M., Tokunaga, T., Fukushima, Y., Abe, Y., Ozeki, Y., Yamazaki, H., Tamaoki, N. and Ueyama, Y. (1998) Multidrug resistance-associated protein and mutant p53 protein expression in non-small cell lung cancer. Mod. Pathol. 11, 1059–1063.

Panday, V.R., Huizing, M.T., Willemse, P.H., De Graeff, A., ten Bokkel Huinink, W.W., Vermorken, J.B. and Beijnen, J.H. (1997) Hepatic metabolism of paclitaxel and its impact in patients with altered hepatic function. Semin. Oncol. 24 (Suppl. II), 34–38.

Patnaik, A., Warner, E., Michael, M., Egorin, M.J., Moore, M.J., Siu, L.L., *et al.* (2000) Phase I dose-finding and pharmacokinetic study of paclitaxel and carboplatin with oral valspodar in patients with advanced solid tumors. J. Clin. Oncol. 18, 3677–3689.

Paulusma, C.C., Bosma, P.J., Zaman, G.J.R., Bakker, C.T.M., Otter, M., Scheffer, G.L., Scheper, R.J., Borst, P. and Oude Elferink, R.P.J. (1996) Congenital jaundice in rats with a mutation in a multidrug resistance-associated protein gene. Science 271, 1126–1128.

Peck, R.A., Hewett, J., Harding, M.W., Wang, Y.M., Chaturvedi, P.R., Bhatnagar, A., Ziessman, H., Atkins, F. and Hawkins, M.J. (2001) Phase I and pharmacokinetic study of the novel MDR1 and MRP1 inhibitor biricodar administered alone and in combination with doxorubicin. J. Clin. Oncol. 19, 3130–3141.

Perri, T., Fogel, M., Mor, S., Horev, G., Meller, I., Loven, D., *et al.* (2001) Effect of P-glycoprotein expression on outcome in the Ewing family of tumors. Pediatr. Hematol. Oncol. 18, 325–334.

Piwnica-Worms, D., Chiu, M.L., Budding, M., Kronaauge, J.F., Kramer, R.A. and Croop, J.M. (1993) Functional imaging of multidrug-resistant P-glycoprotein with an organotechnetium complex. Cancer Res. 53, 977–984.

Rabindran, S.K., He, H., Singh, M., Brown, E., Collins, K.I., Annable, T. and Greenberger, L.M. (1998) Reversal of a novel multidrug resistance mechanism in human colon carcinoma cells by fumitremorgin C. Cancer Res. 58, 5850–5858.

Ramachandra, M., Ambudkar, S.V., Chen, D., Hrycyna, C.A., Dey, S., Gottesman, M.M. and Pastan, I. (1998) Human P-glycoprotein exhibits reduced affinity for substrates during a catalytic transition state. Biochemistry 37, 5010–5019.

Relling, M.V. (1996) Are the major effects of P-glycoprotein modulators due to altered pharmacokinetics of anticancer drugs? Therap. Drug Monit. 18, 350–356.

Riordan, J.R. and Ling, V. (1979) Purification of P-glycoprotein from plasma membrane vesicles of Chinese hamster ovary cell mutants with reduced colchicine permeability. J. Biol. Chem. 254, 12701–12705.

Robey, R., Bakke, S., Stein, W., Meadows, B., Litman, T., Patil, S., Smith, T., Fojo, T. and Bates, S. (1999) Efflux of rhodamine from CD56+ cells as a surrogate marker for reversal of P-glycoprotein-mediated drug efflux by PSC 833. Blood 93, 306–314.

Robey, R.W., Medina-Perez, W.Y., Nishiyama, K., Lahusen, T., Miyake, K., Litman, T., Senderowicz, A.M., Ross, D.D. and Bates, S.E. (2001) Overexpression of the ATP-binding cassette half-transporter, ABCG2 (MXR/BCRP/ABCP1), in flavopiridol-resistant human breast cancer cells. Clin. Cancer Res. 7, 145–152.

Rollins, D.E. and Klaassen, C.D. (1979) Biliary excretion of drugs in man. Clin. Pharmacokinet. 4, 368–379.

Ross, D.D., Karp, J.E., Chen, T.T. and Doyle, L.A. (2000) Expression of breast cancer resistance protein in blast cells from patients with acute leukemia. Blood 96, 365–368.

Rowinsky, E.K., Smith, L., Wang, Y.M., Chaturvedi, P., Villalona, M., Campbell, E., *et al.* (1998) Phase I and pharmacokinetic study of paclitaxel in combination with biricodar, a novel agent that reverses multidrug resistance conferred by overexpression of both MDR1 and MRP. J. Clin. Oncol. 16, 2964–2976.

Samdani, A., Vijapurkar, U., Grimm, M.A., Spier, C.S., Grogan, T.M., Glinsmann-Gibson, B.J. and List, A.F. (1996) Cytogenetics and P-glycoprotein (P-GLYCOPROTEIN) are independent predictors of treatment outcome in acute myeloid leukemia (AML). Leuk. Res. 20, 175–180.

Sauna, Z.E. and Ambudkar, S.V. (2001) Characterization of the catalytic cycle of ATP hydrolysis by human P-glycoprotein. The two ATP hydrolysis events in a single catalytic cycle are kinetically similar but affect different functional outcomes. J. Biol. Chem. 276, 11653–11661.

Savaraj, N., Wu, C.J., Xu, R., Lampidis, T., Lai, S., Donnelly, E., Solomon, J. and Feun, L.G. (1997) Multidrug-resistant gene expression in small-cell lung cancer. Am. J. Clin. Oncol. 20, 398–403.

Scala, S., Akhmed, N., Roa, U.S., Paull, K., Dickstein, B., Lee, J.-S., Elgemeie, G.H., Stein, W.D. and Bates, S.E. (1997) P-glycoprotein substrates and antagonists cluster into two distinct groups. Mol. Pharmacol. 51, 1024–1033.

Scheffer, G.L., Kool, M., Heijn, M., de Haas, M., Pijnenborg, A.C., Wijnholds, J., *et al.* (2000a) Specific detection of multidrug resistance proteins MRP1, MRP2, MRP3, MRP5, and MDR3 P-glycoprotein with a panel of monoclonal antibodies. Cancer Res. 60, 5269–5277.

Scheffer, G.L., Schroeijers, A.B., Izquierdo, M.A., Wiemer, E.A. and Scheper, R.J. (2000b) Lung resistance-related protein/major vault protein and vaults in multidrug-resistant cancer. Curr. Opin. Oncol. 12, 550–556.

Schinkel, A.H., Smit, J.J., van Tellingen, O., Beijnen, J.H., Wagenaar, E., van Deemter, L., *et al.* (1994) Disruption of mouse *mdr-1a* p-glycoprotein gene leads to a deficiency in the blood–brain barrier and to increased sensitivity to drugs. Cell 77, 491–502.

Schinkel, A.H., Mayer, U., Wagenaar, E., Mol, C.A., van Deemter, L., Smit, J.J., *et al.* (1997) Normal viability and altered pharmacokinetics in mice lacking mdr1-type (drug-transporting) P-glycoproteins. Proc. Natl Acad. Sci. USA 94, 4028–4033.

Schneider, J., Jimenez, E., Marenbach, K., Marx, D. and Meden, H. (1998) Co-expression of the MDR1 gene and HSP27 in human ovarian cancer. Anticancer Res. 18, 2967–2971.

Schneider, J., Gonzalez-Roces, S., Pollan, M., Lucas, R., Tejerina, A., Martin, M. and Alba, A. (2001) Expression of LRP and MDR1 in locally advanced breast cancer predicts axillary node invasion at the time of rescue mastectomy after induction chemotherapy. Breast Cancer Res. 3, 183–191.

Scotto, K.W., Biedler, J.L. and Melera, P.W. (1986) Amplification and expression of genes associated with multidrug resistance in mammalian cells. Science 232, 751–755.

Senior, A.E. and Bhagat, S. (1998) P-glycoprotein shows strong catalytic cooperativity between the two nucleotide sites. Biochemistry 37, 831–836.

Smit, J.J.M., Schinkel, A.H., Oude Elferink, R.P.J., Groen, A.K., Wagenaar, E., van Deemter, L., *et al.* (1993) Homozygous disruption of the murine *mdr2* P-glycoprotein gene leads to a complete absence of phospholipid from bile and to liver disease. Cell 75, 451–462.

Solary, E., Velay, I., Chauffert, B., Bidan, J.M., Caillot, D., Dumas, M. and Guy, H. (1991) Sufficient levels of quinine in the serum circumvent the multidrug resistance of the human leukemic cell line K562/ADM. Cancer 68, 1714–1719.

Song, S., Suzuki, H., Kawai, R., Tanaka, C., Akasaka, I. and Sugiyama, Y. (1998) Dose-dependent effects of PSC 833 on its tissue distribution and on the biliary excretion of endogenous substrates in rats. Drug Metab. Dispos. 26, 1128–1133.

Sonneveld, P., Marie, J.-P., Huisman, C., Vekhoff, A., Schoester, M., Faussat, A.M., *et al.* (1996) Reversal of multidrug resistance by SDZ PSC 833, combined with VAD (vincristine, doxorubicin, dexamethasone) in refractory multiple myeloma. A phase I study. Leukemia 10, 1741–1750.

Sparreboom, A., Planting, A.S., Jewell, R.C., van der Burg, M.E., van der Gaast, A., de Bruijn, P., *et al.* (1999) Clinical pharmacokinetics of doxorubicin in combination with GF120918, a potent inhibitor of MDR1 P-glycoprotein. Anticancer Drug. 10, 719–728.

Starling, J.J., Shepard, R.L., Cao, J., Law, K.L., Norman, B.H., Kroin, J.S., *et al.* (1997) Pharmacological characterization of LY335979: a potent cyclopropyldibenzosuberane modulator of P-glycoprotein. Adv. Enzyme Regul. 37, 335–347.

Stieger, B., Fattinger, K., Madon, J., Kullak-Ublick, G.A. and Meier, P.J. (2000) Drug- and estrogen-induced cholestasis through inhibition of the hepatocellular bile salt export pump (Bsep) of rat liver. Gastroenterology 118, 422–430.

Strautnieks, S.S., Bull, L.N., Knisely, A.S., Kocoshis, S.A., Dahl, N., Arnell, H., *et al.* (1998) A gene encoding a liver-specific ABC transporter is mutated in progressive familial intrahepatic cholestasis. Nat. Genet. 20, 233–238.

Sun, S.S., Hsieh, J.F., Tsai, S.C., Ho, Y.J., Lee, J.K. and Kao, C.H. (2000) Expression of mediated P-glycoprotein multidrug resistance related to Tc-99m MIBI scintimammography results. Cancer Lett. 153, 95–100.

Taniguchi, K., Wada, M., Kohno, K., Nakamura, T., Kawabe, T., Kawakami, M., Kagotani, K., Okumura, K., Akiyama, S. and Kuwano, M. (1996) A human canalicular multispecific organic anion transporter (cMOAT) gene is overexpressed in cisplatin-resistant human cancer cell lines with decreased drug accumulation. Cancer Res. 56, 4124–4129.

Taylor, C.W., Dalton, W.S., Mosley, K., Dorr, R.T. and Salmon, S.E. (1997) Combination chemotherapy with cyclophosphamide, vincristine, adriamycin, and dexamethasone (CVAD) plus oral quinine and verapamil in patients with advanced breast cancer. Breast Cancer Res. Treat. 42, 7–14.

Thiebaut, F., Tsuruo, T., Hamada, H., Gottesman, M.M., Pastan, I. and Willingham, M.C. (1987) Cellular localization of the multidrug resistance gene product P-glycoprotein in normal human tissues. Proc. Natl Acad. Sci. USA 84, 7735–7738.

Thurlimann, B., Kroger, N., Greiner, J., Mross, K., Schuller, J., Schernhammer, E., *et al.* (1995) Dexverapamil to overcome epirubicin resistance in advanced breast cancer. J. Cancer Res. Clin. Oncol. 121 (Suppl. 3), R3–R6.

Tidefelt, U., Liliemark, J., Gruber, A., Liliemark, E., Sundman-Engberg, B., Juliusson, G., *et al.* (2000) P-Glycoprotein inhibitor valspodar (PSC 833) increases the intracellular concentrations of daunorubicin *in vivo* in patients with P-glycoprotein-positive acute myeloid leukemia. J. Clin. Oncol. 18, 1837–1844.

Toh, S., Wada, M., Uchiumi, T., Inokuchi, A., Makino, Y., Horie, Y., Adachi, Y., Sakisaka, S. and Kuwano, M. (1999) Genomic structure of the canalicular multispecific organic anion-transporter gene (MRP2/cMOAT) and mutations in the ATP-binding-cassette region in Dubin–Johnson syndrome. Am. J. Hum. Genet. 64, 739–746.

Tolcher, A.W., Cowan, K.H., Solomon, D., Ognibene, F., Goldspiel, B., Chang, R., *et al.* (1996) Phase I crossover study of paclitaxel with r-verapamil in patients with metastatic breast cancer. J. Clin. Oncol. 14, 1173–1184.

Trock, B.J., Leonessa, F. and Clarke, R. (1997) Multidrug resistance in breast cancer: a meta-analysis of MDR1/gp170 expression and its possible functional significance [see comments]. J. Natl. Cancer Inst. 89, 917–931.

Tsuruo, T., Iida, H., Tsukagoshi, S. and Sakurai, Y. (1981) Overcoming of vincristine resistance in P388 leukemia *in vivo* and *in vitro* through enhanced cytotoxicity of vincristine and vinblastine by verapamil. Cancer Res. 41, 1967–1972.

Ueda, K., Clark, D.P., Chen, C.-J., Roninson, I.B., Gottesman, M.M. and Pastan, I. (1987) The human multidrug resistance (*mdr*1) gene. J. Biol. Chem. 262, 505–508.

van Asperen, J., Schinkel, A.H., Beijnen, J.H., Nooijen, W.J., Borst, P. and van Tellingen, O. (1996) Altered pharmacokinetics of vinblastine in *Mdr1a* P-glycoprotein-deficient mice. J. Natl Cancer Inst. 88, 994–999.

van Asperen, J., van Tellingen, O., Tijssen, F., Schinkel, A.H. and Beijnen, J.H. (1999) Increased accumulation of doxorubicin and doxorubicinol in cardiac tissue of mice lacking mdr1a P-glycoprotein. Br. J. Cancer 79, 108–113.

van der Kolk, D.M., de Vries, E.G., van Putten, W.J., Verdonck, L.F., Ossenkoppele, G.J., Verhoef, G.E. and Vellenga, E. (2000) P-glycoprotein and multidrug resistance protein activities in relation to treatment outcome in acute myeloid leukemia. Clin. Cancer Res. 6, 3205–3214.

van der Zee, A.G., Hollema, H., Suurmeijer, A.J., Krans, M., Sluiter, W.J., Willemse, P.H., Aalders, J.G. and de Vries, E.G. (1995) Value of P-glycoprotein, glutathione S-transferase pi, c-erbB-2, and p53 as prognostic factors in ovarian carcinomas. J. Clin. Oncol. 13, 70–78.

van Veen, H.W., Margolles, A., Muller, M., Higgins, C.F. and Konings, W.N. (2000) The homodimeric ATP-binding cassette transporter LmrA mediates multidrug transport by an alternating two-site (two-cylinder engine) mechanism. EMBO J. 19, 2503–2514.

van Zuylen, L., Sparreboom, A., van der Gaast, A., van der Burg, M.E., van Beurden, V., Bol, C.J., Woestenborghs, R., Palmer, P.A. and Verweij, J. (2000a) The orally administered P-glycoprotein inhibitor R101933 does not alter the plasma pharmacokinetics of docetaxel. Clin. Cancer Res. 6, 1365–1371.

van Zuylen, L., Nooter, K., Sparreboom, A. and Verweij, J. (2000b) Development of multidrug-resistance convertors: sense or nonsense? Invest. New Drugs 18, 205–220.

Vecchio, S.D., Ciarmiello, A., Potena, M.I., Carriero, M.V., Mainolfi, C., Botti, G., *et al.* (1997) *In vivo* detection of multidrug-resistant (MDR1) phenotype by technetium-99m sestamibi scan in untreated breast cancer patients. Eur. J. Nucl. Med. 24, 150–159.

Visani, G., Milligan, D., Leoni, F., Chang, J., Kelsey, S., Marcus, R., *et al.* (2001) Combined action of PSC 833 (Valspodar), a novel MDR reversing agent, with mitoxantrone, etoposide and cytarabine in poor-prognosis acute myeloid leukemia. Leukemia 15, 764–771.

Wacher, V.J., Wu, C.Y. and Benet, L.Z. (1995) Overlapping substrate specificities and tissue distribution of cytochrome P450 3A and P-glycoprotein: implications for drug delivery and activity in cancer chemotherapy. Mol. Carcinog. 13, 129–134.

Wang, C.S., LaRue, H., Fortin, A., Gariepy, G. and Tetu, B. (1997) mdr1 mRNA expression by RT-PCR in patients with primary breast cancer submitted to neoadjuvant therapy. Breast Cancer Res. Treat. 45, 63–74.

Wang, R., Salem, M., Yousef, I.M., Tuchweber, B., Lam, P., Childs, S.J., Helgason, C.D., Ackerley, C., Phillips, M.J. and Ling, V. (2001) Targeted inactivation of sister of P-glycoprotein gene (spgp) in mice results in nonprogressive but persistent intrahepatic cholestasis. Proc. Natl Acad. Sci. USA 98, 2011–2016.

Wijnholds, J., Mol, C.A.A.M., van Deemter, L., de Haas, M., Scheffer, G.L., Baas, F., *et al.* (2000a) Multidrug resistance protein 5 is a multispecific organic anion transporter able to transport nucleotide analogs. Proc. Natl Acad. Sci. USA 97, 7476–7481.

Wijnholds, J., de Lange, E.C.M., Scheffer, G.L., van den Berg, D.-J., Mol, C.A.A.M., van der Valk, M., Schinkel, A.H., Scheper, R.J., Breimer, D.D. and Borst, P. (2000b) Multidrug resistance protein 1 protects the choroid plexus epithelium and contributes to the blood–cerebrospinal fluid barrier. J. Clin. Invest. 105, 279–285.

Wilson, W.H., Bates, S.E., Fojo, A.T., Bryant, G., Zhan, Z., Regis, J., *et al.* (1995) Controlled trial of dexverapamil, a modulator of multidrug resistance, in lymphomas refractory to EPOCH chemotherapy. J. Clin. Oncol. 13, 1995–2004.

Witherspoon, S.M., Emerson, D.L., Kerr, B.M., Lloyd, T.L., Dalton, W.S. and Wissel, P.S. (1996) Flow cytometric assay of modulation of P-glycoprotein function in whole blood by the multidrug resistance inhibitor GG918. Clin. Cancer Res. 2, 7–12.

Wright, S.R., Boag, A.H., Valdimarsson, G., Hipfner, D.R., Campling, B.G., Cole, S.P.C. and Deeley, R.G. (1998) Immunohistochemical detection of multidrug resistance protein in human lung cancer and normal lung. Clin. Cancer Res. 4, 2279–2289.

Wunder, J.S., Bull, S.B., Aneliunas, V., Lee, P.D., Davis, A.M., Beauchamp, C.P., *et al.* (2000) MDR1 gene expression and outcome in osteosarcoma: a prospective, multicenter study. J. Clin. Oncol. 18, 2685–2694.

Xu, D., Arestrom, I., Virtala, R., Pisa, P., Peterson, C. and Gruber, A. (1999) High levels of lung resistance related protein mRNA in leukaemic cells from patients with acute myelogenous leukemia are associated with inferior response to chemotherapy and prior treatment with mitoxantrone. Br. J. Haematol. 106, 627–633.

Yamamoto, Y., Nishiyama, Y., Satoh, K., Takashima, H., Ohkawa, M., Fujita, J., Kishi, T., Matsuno, S. and Tanabe, M. (1998) Comparative study of technetium-99m-sestamibi and thallium-201 SPECT in predicting chemotherapeutic response in small cell lung cancer. J. Nucl. Med. 39, 1626–1629.

Yang, X., Uziely, B., Groshen, S., Lukas, J., Israel, V., Russell, C., Dunnington, G., Formenti, S., Muggia, F. and Press, M.F. (1999) MDR1 gene expression in primary and advanced breast cancer. Lab. Invest. 79, 271–280.

Yokoyama, H., Ishida, T., Sugio, K., Inoue, T. and Sugimachi, K. (1999a) Immunohistochemical evidence that P-glycoprotein in non-small cell lung cancers is associated with shorter survival. Surg. Today 29, 1141–1147.

Yokoyama, Y., Sato, S., Fukushi, Y., Sakamoto, T., Futagami, M. and Saito, Y. (1999b) Significance of multi-drug-resistant proteins in predicting chemotherapy response and prognosis in epithelial ovarian cancer. J. Obstet. Gynaecol. Res. 25, 387–394.

Young, L.C., Campling, B.G., Voskoglou-Nomikos, T., Cole, S.P.C., Deeley, R.G. and Gerlach, J.H. (1999) Expression of multidrug resistance protein-related genes in lung cancer: correlation with drug response. Clin. Cancer Res. 5, 673–680.

Young, L.C., Campling, B.G., Cole, S.P.C., Deeley, R.G. and Gerlach, J.H. (2001) Multidrug resistance proteins MRP3, MRP1, and MRP2 in lung cancer: correlation of protein levels with drug response and messenger RNA levels. Clin. Cancer Res. 7, 1798–1804.

Zeng, H., Bain, L.J., Belinsky, M.G. and Kruh, G.D. (1999) Expression of multidrug resistance protein-3 (multispecific organic anion transporter-D) in human embryonic kidney 293 cells confers resistance to anticancer agents. Cancer Res. 59, 5964–5967.

Zhou, Y., Gottesman, M.M. and Pastan, I. (1999) Studies of human MDR1-MDR2 chimeras demonstrate the functional exchangeability of a major transmembrane segment of the multidrug transporter and phosphatidylcholine flippase. Mol. Cell. Biol. 19, 1450–1459.

CHAPTER 19

Multidrug Resistance Protein 1 (ABCC1)

Roger G. Deeley and Susan P.C. Cole

Introduction

Historical background of the discovery and initial characterization of MRP1

Multidrug resistance protein 1 (MRP1/ABCC1) was discovered in 1992 following a search for proteins that could confer a form of multidrug resistance (MDR) previously associated exclusively with overexpression of the drug transporter P-glycoprotein (Pgp) (MDR1) (Ambudkar *et al.*, 1999; Cole *et al.*, 1992; Juliano and Ling, 1976). This form of resistance, sometimes referred to as 'classical' MDR, is frequently observed when tumor cell lines are subjected to selection *in vitro* by exposure to increasing concentrations of a single natural product type cytotoxic agent. Cells that survive this type of selection are often resistant to a wide variety of structurally and functionally unrelated, natural product drugs, in addition to the original selecting agent. In general, they also display an enhanced ability to efflux the drugs to which they are resistant and resistance can be reversed by verapamil and other membrane active agents that restore drug accumulation by inhibiting Pgp-mediated drug efflux (Ambudkar *et al.*, 1999; Deeley and Cole, 1997).

The strong association between classical MDR and overexpression of Pgp in many *in vitro* selected cell lines, coupled with the observation that highly conserved isoforms (sometimes referred to as type II Pgps) do not confer drug resistance, led to the conclusion that Pgp was likely to be the only ATP-binding cassette (ABC) transporter capable of conferring a classical MDR phenotype. However, in the late 1980s, approximately a decade after the original biochemical identification of Pgp, several drug-selected cell lines were described that had resistance characteristics similar to classical MDR but without increased expression of Pgp. One of these was derived from the human small cell lung cancer cell line H69, by intermittent exposure to increasing concentrations of the anthracycline antibiotic doxorubicin (Mirski *et al.*, 1987). The multidrug-resistant derivative cell line, designated H69AR, was resistant to a range of natural product drugs including anthracyclines, *Vinca* alkaloids and epipodophyllotoxins (Cole, 1990; Mirski *et al.*, 1987). Differential cDNA screening of the H69AR cell line for mRNAs that were overexpressed relative to the drug-sensitive parental H69 cells yielded cDNA clones that encoded a novel ABC transporter (Cole *et al.*, 1992). The protein, initially termed multidrug resistance-associated protein or MRP and subsequently MRP1, was the founding member of a relatively large subfamily of transporters that is now known to contain eight additional proteins. Phylogenetic analyses of human ABC transporters **(Figure 19.1)**, suggest that the nine MRPs, together with the cystic fibrosis transmembrane conductance regulator (CFTR) and the two sulfonylurea receptors (SURs), all evolved from a common

ABC Proteins: From Bacteria to Man
ISBN 0-12-352551-9

ancestor different from that of the Pgps. Collectively, these proteins comprise the ABCC branch of the human ABC transporter superfamily (Borst *et al.*, 1999) (see next section).

Despite the fact that MRP1 is evolutionarily very distant from the Pgps, gene transfer experiments proved that it was capable of conferring resistance to a similar, although not identical, range of drugs (Cole *et al.*, 1994; Grant *et al.*, 1994). MRP1 overexpressing cells also display increased resistance to antimonial and arsenical oxyanions, a property which is poorly or not shared with cells that express elevated levels of Pgp. Moreover, MRP1-mediated resistance is not reversed by many agents that reverse resistance caused by increased expression of Pgp (Cole *et al.*, 1989, 1994). Elevated levels of MRP1 have now been found in many *in vitro* selected multidrug-resistant cell lines and cells derived from tumors which are intrinsically resistant to many chemotherapeutic agents (Cole and Deeley, 1996; Lautier *et al.*, 1996). The protein has also been detected clinically in a wide range of tumor types (Hipfner *et al.*, 1999a) (see below). Why similar selection protocols sometimes result in cell lines that overexpress Pgp and in other cases MRP1 has not been established (Reeve *et al.*, 1990). Some evidence suggests that in certain cell types, increased expression of MRP1 may occur preferentially following exposure to low concentrations of drug, while selection for higher levels of resistance may favor increased expression of Pgp (Slapak *et al.*, 1990). However, since this does not always occur, there are clearly exceptions to the 'rule'.

Although transfected cells that express elevated levels of MRP1 display reduced accumulation and enhanced efflux of drugs to which they are resistant, attempts to obtain definitive proof that the protein could bind and actively transport chemotherapeutic agents were initially unsuccessful. The first *in vitro* evidence of ATP-dependent transport by MRP1 came from biochemical studies of proteins involved in the active efflux of organic anions. ATP-dependent transport of conjugated organic anions can be demonstrated using inside-out membrane vesicles from many cell types. This activity was attributed to a molecularly uncharacterized protein termed the multispecific organic anion transporter (MOAT). The levels of MOAT activity are particularly high in vesicles prepared from hepatocanalicular membranes and plasma membranes from mast cells. One of the highest-affinity substrates for MOAT was known to be the glutathione-conjugated leukotriene, LTC_4 **(Figure 19.4)**. *In vivo,* this conjugated arachidonic acid derivative is actively effluxed by mast cells following an IgE-mediated inflammatory response and, when metabolized to LTD_4 by the ectoenzyme γ-glutamyl transpeptidase, is a potent activator of receptor-mediated signaling pathways involved in bronchoconstriction and vasoconstriction. Using radiolabeled LTC_4, which is intrinsically photoactivatable, to label the transporter in membranes from a murine mast cell-derived tumor cell line revealed that it was of a size similar to that of MRP1 (Leier *et al.*, 1994). Subsequent LTC_4 binding and transport studies using membrane vesicles from multidrug-resistant human tumor cells known to overexpress MRP1 and from MRP1 transfected cells confirmed that MRP1 could also be photolabeled with LTC_4 and was capable of transporting the conjugated leukotriene (Jedlitschky *et al.*, 1994; Leier *et al.*, 1996; Loe *et al.*, 1996b; Muller *et al.*, 1994). The fact that antibodies specific for MRP1 inhibited transport confirmed that MRP1 and MOAT were the same proteins (Hipfner *et al.*, 1999b; Loe *et al.*, 1996b).

Because of apparently similar substrate specificities, it was originally proposed that the MOAT in hepatocanalicular membranes (designated cMOAT) and in cells from non-hepatic tissues may be attributable to the same, or closely related, proteins. The substrate specificity of cMOAT had been defined by *in vivo* studies of mutant rats in which the transporter was inactive. The mutant rats are animal models of the human condition known as Dubin–Johnson syndrome, a form of heritable conjugated hyperbilirubinemia. The defect in Dubin–Johnson syndrome is now known to be caused by a lack of functional MRP2, the first human homologue of MRP1 to be identified (Kartenbeck *et al.*, 1996; Paulusma *et al.*, 1997). The information derived from studies of bile transport in the mutant rat strains guided screening of potential MRP1 substrates, which resulted in the demonstration that MRP1, like MRP2, could transport not only glutathione conjugates, including oxidized glutathione (GSSG), but also a range of glucuronidated and sulfated compounds, as well as some unconjugated molecules (Hipfner *et al.*, 1999a; Leslie *et al.*, 2001a) (see later section on substrate specificity). These include anionic conjugates of steroid hormones, such as 17β-estradiol-17βD-glucuronide and estrone-3-sulfate, that have no structural similarity to LTC_4 (Loe *et al.*, 1996a;

Qian *et al.*, 2001b). The physiological importance of MRP1-mediated transport of the vast majority of potential substrates identified by *in vitro* studies remains to be established. However, it is clear that despite a very high degree of amino acid identity, considerable species variation exists among mammalian MRP1 orthologues in their ability to transport some of them, e.g. 17β-estradiol-17βD-glucuronide (Stride *et al.*, 1997, 1999). While this may cast doubt on the physiological relevance of MRP1's ability to transport some of these compounds, the functional differences between closely related orthologues have proven useful for investigating structure/substrate specificity relationships of the protein (see later section on substrate specificity).

Although the ability of MRP1 to transport organic anionic conjugates was readily demonstrable, initial *in vitro* attempts to demonstrate that the protein could directly transport unmodified chemotherapeutic agents failed. Thus it was proposed that MRP1, in contrast to Pgp, conferred resistance by transporting drug conjugates rather than the unmodified xenobiotic. This suggestion provided a possible explanation for the inability to detect transport of unmodified drugs by MRP1-enriched membrane vesicles and was supported by the demonstration that the protein was capable of transporting several synthetic glutathione and glucuronide drug conjugates (Jedlitschky *et al.*, 1996; Priebe *et al.*, 1998). However, MRP1 can confer a drug resistance phenotype in cells that lack the necessary complement of drug-conjugating enzymes, and many drugs to which the protein confers resistance are not extensively conjugated *in vivo* prior to elimination. The apparent conflict between data obtained with intact cells and from *in vitro* transport studies has been at least partially resolved by the demonstration that MRP1-enriched membrane vesicles are capable of direct transport of at least some unmodified natural product xenobiotics providing that they are supplemented with physiological concentrations of glutathione (Loe *et al.*, 1996b, 1997, 1998; Renes *et al.*, 1999).

PHYLOGENETIC RELATIONSHIPS BETWEEN MRP1 AND OTHER HUMAN PROTEINS IN THE 'C' BRANCH OF THE SUPERFAMILY

When the predicted amino acid sequence of MRP1 was first compared with the sequences of known ABC proteins, it was found to be most similar (30% amino acid identity) to a protein from *Leishmania tarentolae*. The protein, LtPgpA, confers resistance to arsenite and animonials, as does MRP1, but not to natural product drugs. The most closely related human ABC protein identified at the time was the ATP-gated chloride channel, CFTR (19% identity). All three proteins were noted to share certain structural features in their nucleotide-binding domains (NBDs) that distinguished them from more distantly related members of the superfamily (Cole *et al.*, 1992). Thus, although both NH_2- and COOH-proximal NBDs in all three proteins contain variations of the three conserved motifs typical of ABC proteins, within each protein, the two NBDs are relatively divergent. Furthermore, inter-protein similarity between corresponding NBDs is higher than between the two NBDs within a single protein. This feature provides strong evidence for their evolution from a common ancestral protein containing both domains (Grant *et al.*, 1997). The major distinguishing characteristic present in all MRP-related proteins identified to date, which is also present in CFTR, is the spacing between the characteristic Walker A and ABC signature motifs in their NH_2-proximal NBDs. Relative to a protein such as Pgp, the MRPs (with the exception of MRP7/ABCC10) and CFTR appear to be missing 13 amino acids at exactly the same location between the two motifs. In the case of ABCC10, the gap is 10 rather than 13 amino acids. It has now been established for several members of the ABCC branch that the structural divergence between NH_2- and COOH-proximal NBDs is associated with differences in their ability to bind and hydrolyze ATP and the role that each NBD may play in the functional cycle of the proteins.

Alignment of the amino acid sequences of MRP1 and LtPgPA also revealed that the two proteins shared a major structural feature not present in CFTR, namely a relatively hydrophobic NH_2-terminal region (TMD1, also designated TMD0 in other chapters) of approximately 200 amino acids. Similar regions are found in MRPs 2, 3, 6 and 7, and in the SURs, but not in MRPs 4 and 5 or in the two most recently identified MRP-related proteins designated ABCC11 and ABCC12. MRPs 2, 3 and 6 are relatively closely related to MRP1 having 49%, 58% and 45% overall amino acid identity, while MRP7(ABCC10), despite the presence of an NH_2-terminal extension, is relatively distantly related (28% identity). Phylogenetically, MRP4, MRP5, ABCC11 and ABCC12, all of which lack the additional NH_2-terminal region, are almost as closely related to CFTR as they are to MRP1 **(Figure 19.1)**.

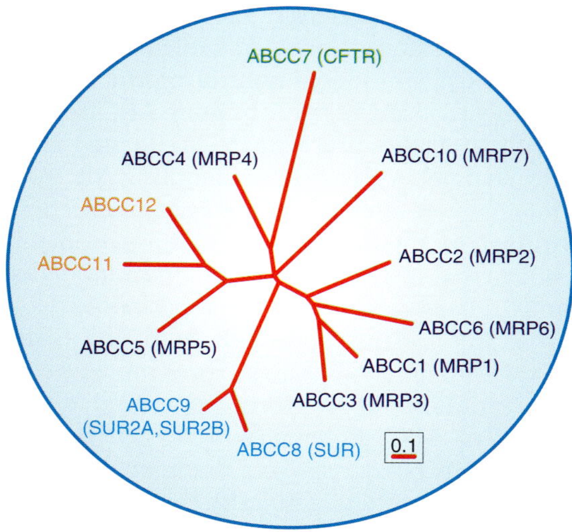

Figure 19.1. ***Evolutionary relationships among human members of the ABCC branch of the ABC superfamily. Multiple amino acid sequence alignments were performed and analyzed using the ClustalX (version 1.8) Windows interface for ClustalW. Alignments were examined using GeneDoc 2.6. These alignments were used to generate unrooted phylogenetic trees using the neighbor-joining method of Saitou and Nei (1987) and were corrected for multiple substitutions at a single site. The reliability of the trees was estimated using a bootstrap procedure with 1000 trial runs. The horizontal bar in the lower right quadrant gives a scale for the fractional divergence between amino acid sequences following correction for multiple substitutions at a single site. See text for additional details.***

Primary structure conservation of the NH_2-terminal 'extension' present in some of the MRPs is relatively low, but the predicted hydropathy profiles are similar and, depending on the algorithm used, suggestive of the presence of four to six transmembrane segments or helices. Alignment of the amino acid sequences of the proteins, together with conserved features of the intron–exon organization of their respective genes where known, suggests that the additional membrane-spanning domains were acquired by gene fusion events (Grant *et al.*, 1997; Konig *et al.*, 2000). It is not clear whether the low level of amino acid identity between the additional TMDs reflects relatively relaxed functional constraints on these regions, or is indicative of several independent gene fusion events. In the case of MRP1 and MRP2, experimental evidence indicates that this region contains five TM helices and that the proteins have extracellular NH_2-termini (Hipfner *et al.*, 1997; Kast and Gros, 1997; Konig *et al.*, 1999) **(see Figure 19.2)**. The functional roles of the additional domains present in MRPs 1, 2, 3, 6 and 7 remain to be fully defined (see below).

PROPERTIES OF MRP1

POST-TRANSLATIONAL MODIFICATION AND MEMBRANE LOCALIZATION

Based on the amino acid sequence of MRP1, the protein has a predicted M_r of 170 000. Biochemical studies have shown that the protein is both N-glycosylated and phosphorylated primarily on serine residues (Almquist *et al.*, 1995; Bakos *et al.*, 1996; Ma *et al.*, 1995). The extent of glycosylation varies among cell lines and the mature protein has an apparent molecular mass of 180–190 kDa. To date, no major functional consequences of these modifications have been identified. However, determination of the sites used for N-glycosylation provided important experimental evidence with respect to the topology of the protein. Mutation of selected asparagine residues has identified which of the 14 potential N-glycosylation sites in the MRP1 sequence are actually used. Two are located close to the NH_2-terminus at Asn 19 and 23, providing strong experimental evidence that the NH_2-terminus is extracellular, as predicted by one topology algorithm (Hipfner *et al.*, 1997) **(see Figure 19.2)**. The extracellular location of the NH_2-terminus has also been confirmed by epitope insertion studies (Kast and Gros, 1997). The third N-glycosylation site is at Asn^{1006}, confirming the location of the first extracellular loop of the COOH-proximal TMD of the protein **(Figure 19.2)** (Hipfner *et al.*, 1997). Processing of the 170 kDa precursor in the endoplasmic reticulum to the mature 190 kDa protein present in post-Golgi membrane vesicles and in the plasma membrane takes approximately 90 min and the half-life of the mature protein is approximately 20 h. Interestingly, 80–90% of immature MRP1 in lung cancer cells that express large amounts of the protein is rapidly degraded, presumably in the endoplasmic reticulum (Almquist *et al.*, 1995). MRP1 shares this behavior with CFTR but it is not characteristic of other ABC transporters, such as Pgp. Whether the degradation is attributable to

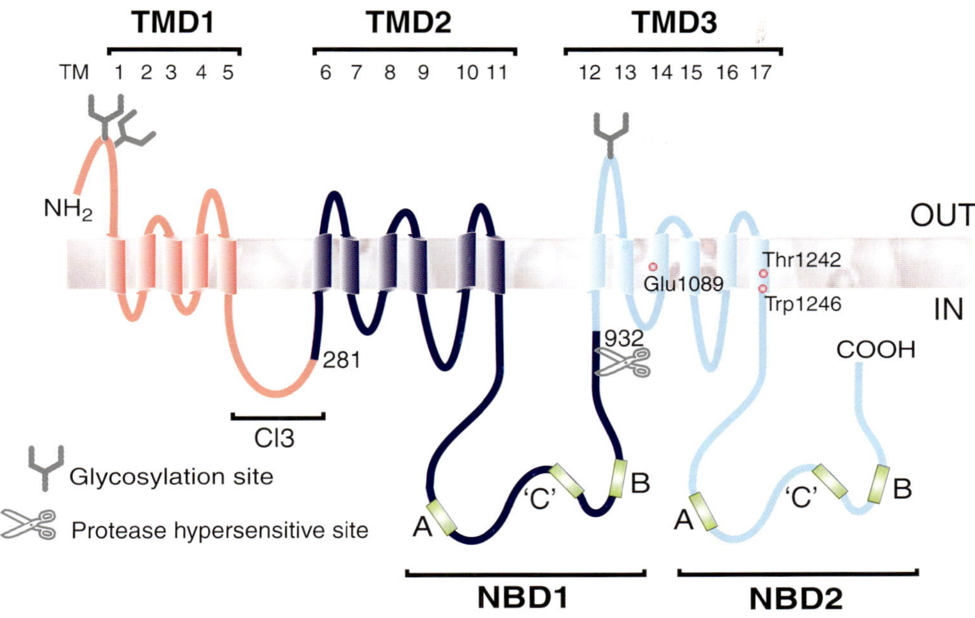

Figure 19.2. A predicted membrane topology of MRP1. The topology shown is based on the predictions of several computer algorithms and is supported by experimental evidence obtained from mapping the location of functional N-glycosylation sites and epitope insertion studies (described in more detail in the section on topology and higher-order structure of MRP1). The figure indicates the approximate location of MRP1's 17 transmembrane helices organized into three transmembrane domains (TMDs) and the protein's two cytoplasmic nucleotide-binding domains (NBDs). Within each NBD, the approximate locations of conserved Walker A (A) and B (B) motifs are shown, as well as the so-called ABC signature sequence (C). Also shown are cytoplasmic loop 3 (Cl3), residues known to be important for determining substrate specificity (Glu1089, Thr1242 and Trp1246) and the location of a protease hypersensitive site in the cytoplasmic region linking NBD1 with MSD3. Amino acid locations in Cl3 and the NBD1/TMD3 linker that have been used to divide the protein into fragments capable of reassociating to form a functional transporter are indicated as 281 and 932, respectively.

some common structural characteristics of MRP1 and CFTR has not been established.

In most cells, mature MRP1 is present primarily in the plasma membrane although the protein can also be detected in intracellular vesicular membranes in some cell types and under some culture conditions. In polarized cells, MRP1 is restricted to basolateral segments of the plasma membrane **(Figure 19.3A)**, with the possible exception of the placental synciytiotrophoblast, where MRP1 has been reported to be apically located (St Pierre *et al.*, 2000). The partitioning of the mature protein between intracellular membrane vesicles and the plasma membrane appears to vary with cell type and growth state. It is clear that in human multidrug-resistant H69AR and GLC4-Adr small cell lung cancer cell lines a substantial portion of the protein is present in intracellular membrane vesicles and is capable of sequestering drugs within the intravesicular space (Cole, 1992; van Luyn *et al.*, 1998) **(Figure 19.3B)**. The contribution that vesicular MRP1 makes to the drug resistance profile of these cells has not been established, but it may provide an explanation for the relatively modest or non-detectable increase in drug efflux observed in some cells that express high levels of the protein. Interestingly, the *L. tarentolae* MRP1 homologue LtPpgA has recently been shown to be present on intracellular membranes of the parasite rather than the plasma membrane and appears likely to contribute to antimonial and arsenical resistance by sequestering metal-thiol conjugates in vesicles that may be part of an exocytic pathway (Legare *et al.*, 2001). Similarly, the *Saccharomyces cerevisiae* MRP1 orthologue, YCF1, is a vacuolar membrane protein which confers cadmium resistance and, like MRP1, can transport a wide range of GSH-conjugates (Li *et al.*, 1996). When expressed in insect cells, YCF1 is found in the plasma membrane and transports a range of organic anions similar to MRP1 (Ren *et al.*, 2000). Conversely, when MRP1 is expressed in

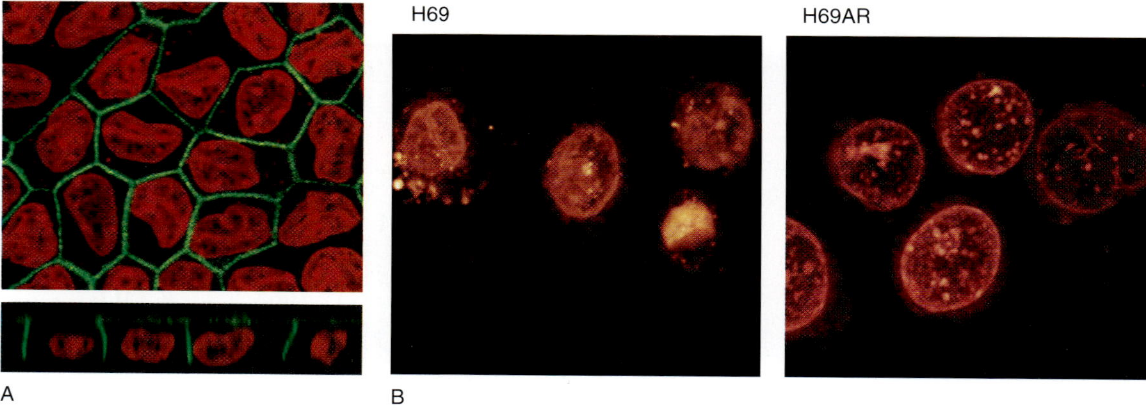

Figure 19.3. ***A, Basolateral plasma membrane location of MRP1 in a polarized monolayer of Madine Darby canine kidney (MDCK) I cells detected by indirect immunofluorescence. The cells were stably transfected with an MRP1 expression vector, pCDNA 3.1-MRP1 DNA, and MRP1 was visualized by confocal laser scanning fluorescence microscopy. Cells were fixed with paraformaldehyde and permeabilized with digitonin prior to incubation with the MRP1 specific MAb QCRL-3 and a fluorescein-labeled second antibody (green). Nuclei were then stained with propidium iodide (red). In the X/Z cross-section shown in the lower part of the figure the apical face of the cells is uppermost. B, Subcellular distribution of daunorubicin in the human multidrug-resistant small cell lung cancer cell line, H69AR, and drug-sensitive parental H69 cells detected by fluorescence microscopy. H69 and H69AR cells were treated for 20 min with daunorubicin (2 μM), which is naturally fluorescent. Cells were then examined by confocal laser scanning fluorescence microscopy. Fluorescence intensity was color graded from magenta (low) to yellow (high). Strong nuclear fluorescence is evident in the drug-sensitive H69 cells (left) while the H69AR cells (right) display low nuclear fluorescence with drug accumulation in vesicles and at the periphery of the nucleus.***

yeast, it can complement a defect in YCF1 and confers cadmium resistance despite the fact that it has not been possible to demonstrate that MRP1 increases cadmium resistance in mammalian cell transfectants (Cole *et al.*, 1994). However, unlike YCF1, MRP1 expressed in yeast is found in both vacuolar and other internal membranes (Tommasini *et al.*, 1996). Thus despite the functional complementation, the signals that direct the trafficking of YCF1 to the yeast vacuole have not been completely conserved in MRP1. Presently, the structural features of MRP1 that target it to the plasma membrane in mammalian cells, and specifically the basolateral membrane in most polarized cells, are not known.

TOPOLOGY AND HIGHER-ORDER STRUCTURE OF MRP1

A number of different topologies have been proposed for MRP1 with most models predicting three transmembrane domains (TMDs), two NBDs and 17 TM helices (Bakos *et al.*, 1996; Cole and Deeley, 1998; Hipfner *et al.*, 1997) **(Figure 19.2)**. Different conventions have been followed in the literature with respect to numbering of the protein's three TMDs. The convention followed here is to number the TMDs 1, 2 and 3 proceeding from the NH_2-terminus to the COOH-terminus of the protein. Several topology algorithms support models in which TMD1 and TMD2 contain five and six TM helices, respectively, although the precise positioning of the helices is somewhat variable. These models are consistent with all data derived from analyses of the utilization of N-glycosylation sites, as well as epitope localization studies using antibodies against both naturally occurring and inserted epitopes. The topology of TMD3 remains less certain. The commonly used PredictProtein algorithm suggests that TMD3 in MRP1 and some related proteins contains only four TM helices (Hipfner *et al.*, 1999a). However, there is presently no experimental evidence to support this model and the mapping of inserted epitopes and the use of inserted glycosylation sequences favors a model with six TM helices (Kast and Gros, 1998). It is presently not known how the three

TMDs are oriented with respect to each other, nor which TM helices are involved in inter-domain contacts.

Physical studies of the structure of MRP1 are at an early stage. The structure of purified native MRP1 isolated from membranes of H69AR cells has been examined by electron microscopy (Rosenberg *et al.*, 2001). Single particle analysis of negatively stained preparations yields a predominant 'picture' of an MRP1 monomer as an approximately pentagonal ring of protein surrounding a pore with a diameter of ≈35 Å. In addition to the 'doughnut' structure of MRP1 similar to that observed previously for Pgp (Rosenberg *et al.*, 1997), two small electron-dense projections on the outside of the protein ring are apparent, one or the other of which may represent the location of TMD1. The structure obtained from 2-D crystals is consistent in some respects with results of single particle analysis, but there are significant differences between them. The unit cell of the crystal consists of a dimer of MRP1 molecules, each of which appears elliptical rather than pentagonal, with a long axis similar to the diameter of the single particles. In addition, no pore-like structure is visible in either of the MRP1 molecules in the unit cell. These discrepancies might be explained by the possibility that the preferred orientation of the protein bound to the mica sample support grid used for single particle microscopy favors predominant staining of one face of the protein while the other face is stained in the 2-D crystals. By analogy with the model proposed for Pgp, the open pore-like structure may then represent a view of the extracellular face of the protein while the pore on the cytoplasmic side is occluded by the NBDs and/or cytoplasmic loops. It remains to be determined whether the MRP1 dimers exist in native plasma membranes (Soszynski *et al.*, 1998).

Identifying regions of MRP1 required for transport activity

In contrast to many bacterial ABC transporters (importers) in which transmembrane and nucleotide-binding domains are each encoded by separate polypeptides, the four-domain core structure of eukaryotic transporters is typically contained in just one or two polypeptides. Thus the proteins have evolved to include cytoplasmic regions which physically connect the individual domains. Whether these regions have acquired some functional purpose other than to link domains has, for the most part, not been determined. Defining the essential core regions of transporters such as MRP1 is complicated by the difficulty of creating mutations between domains, particularly deletions of large numbers of amino acids, without interfering with the folding and overall topology of the protein. Fortunately, the individual domains of eukaryotic transporters appear to have retained the ability to associate without being physically linked. Thus it is possible to coexpress variously modified domain modules of proteins such as MRP1 and to monitor their ability to associate into a functional transporter. For a number of practical reasons, the preferred methodology has been to use insect cells (e.g. *Spodoptera fugiperda* Sf21), cells which have been co-infected with baculovirus vectors so that they express two or three regions of the protein as separate polypeptides (Gao *et al.*, 1996).

The approach described above has been used to investigate the functional importance of regions linking NBD1 and TMD3 and TMD1 and TMD2, as well as TMD1 itself. These studies have established that much of the linker between NBD1 and TMD3 of MRP1 can be removed without altering the kinetics of LTC_4 transport and thus this region is clearly not required for binding, or any obligatory step in the transport of this substrate (Gao *et al.*, 1998). Similarly, although initial studies suggested that TMD1 was essential for activity, it is now apparent that this additional TMD is also not required for LTC_4 transport (Bakos *et al.*, 1998). To date the transport activity of MRP1 lacking either TMD1 or the NBD1/TMD3 linker has been characterized with very few substrates. Consequently, it is too early to say whether these regions are important for recognition and transport of only certain compounds, or whether they perhaps fulfill a completely different function, such as mediating the interaction between MRP1 and other proteins not required for basal transport activity.

Although removal of TMD1 and six to seven amino acids of the TMD1/TMD2 linker by truncation of MRP1 to amino acid 204 has relatively little effect on LTC_4 transport, truncation to amino acids 228 or 281 completely eliminates activity (Bakos *et al.*, 1998; Gao *et al.*, 1998). Deletion of the region between amino acids 204 and 281, or substitution with the comparable region of MRP2, also inactivates the protein. However, activity can be restored by coexpression of fragments corresponding to amino acids 1–281 or 204–281 but not amino acids 1–228 (Bakos *et al.*, 2000a; Gao *et al.*, 1998). Overall, these studies indicate that the physical integrity of all or part of the NH_2-proximal segment of the TMD1/TMD2 linker of MRP1, between amino acids 204 and 281, is essential for activity. Although topology predictions and epitope mapping indicate that amino acids 204–281 should be cytoplasmic, this region of the linker appears to form a relatively stable association with the plasma membrane independent of the presence of the remainder of the protein (Bakos *et al.*, 2000a; Hipfner *et al.*, 1998). One element within the linker that appears to be necessary for membrane association and correct trafficking of the protein is a predicted amphipathic helix in a relatively conserved segment between amino acids 223 and 232 (Bakos *et al.*, 2000a).

SUBSTRATE SPECIFICITY

The substrate specificity of MRP1 has been defined primarily by three experimental approaches. The first, and most direct, has been to monitor ATP-dependent, osmotically sensitive uptake of radiolabeled substrates by inside-out membrane vesicles prepared from cells that express relatively high levels of MRP1, preferably as a result of gene transfer. The use of transfected cells provides considerable assurance that any increase in transport relative to vesicles from control, reference cells is attributable to MRP1. Additional assurances that transport is indeed MRP1 mediated can be obtained by using MAbs that specifically inhibit MRP1 transport activity, or small molecules that are known to inhibit MRP1-mediated transport. However, the availability of chemical inhibitors that are truly MRP1 specific is limited. A second approach is to use intact cells to monitor the rate at which a radiolabeled or fluorescent substrate is effluxed. This approach, particularly in the case of MRP1, has provided a useful complement to vesicle transport studies. However, definitive proof that efflux is attributable to a specific transporter is more difficult to obtain and the possibility that a metabolite rather than the parental compound may be the actual substrate has to be considered. On the other hand, an advantage of using intact cells is that any co-factors or activators required for optimal transport are likely to be present. The third approach, and the least direct, is most frequently used with compounds that are cytotoxic. In these assays, the viability (or survival) of MRP1-expressing cells is compared with that of a parental or non-transfected cell line in the presence of increasing concentrations of cytotoxic agent. In practice, the ratio of the drug concentrations which kill 50% of the MRP1-expressing cells compared with the reference cells is typically used to provide a relative level of resistance, and by extrapolation, an indication of how effectively the protein may transport the test compound. The manner in which these three approaches have been integrated to define the substrate specificity of MRP1 is summarized below.

MRP1 displays an amazingly broad substrate specificity while retaining the ability to discriminate between related compounds with very similar structures. As a consequence, it is important to recognize the limitations of any generalization about the structural characteristics that define a potential substrate. MRP1 substrates include structurally diverse amphiphilic organic anions, many of which are conjugated with glutathione, glucuronate or sulfate. Examples of conjugated and non-conjugated compounds shown by vesicle transport studies to be directly transported by MRP1 are given in **Table 19.1**. These include examples of substrates that display a dependence on GSH for their transport. The list is not exhaustive and many more potential substrates have been identified by virtue of their ability to inhibit the transport of a well-characterized substrate, such as LTC_4, which remains the highest-affinity MRP1 substrate identified to date. Clinically relevant anionic substrates that are not conjugated and which can be directly transported by MRP1 include the anti-metabolite methotrexate and anionic fluorescent dyes such as Fluo-3 or calcein, which can also be used to monitor MRP1 (and Pgp) expression in cultured cells and cells derived from hematological tumors (Hollo *et al.*, 1994; Hooijberg *et al.*, 1999; Lautier *et al.*, 1996; Lohoff *et al.*, 1998). In addition, MRP1 has recently been shown to confer resistance to some cytotoxic hydrophobic peptides such as N-acetyl-Leu-Leu-norleucinal (ALLN) and various derivatized peptides with a Thr-His-Thr-Nle-Glu-Gly backbone. The resistance is associated with a decreased accumulation of the peptides and in the case of ALLN, but not the other peptides, can be reversed by depletion of GSH (de Jong *et al.*, 2001).

The structural diversity of the GSH conjugates that are known MRP1 substrates highlights the

TABLE 19.1. COMPOUNDS SHOWN TO BE MRP1 SUBSTRATES BY DIRECT TRANSPORT STUDIES USING INSIDE-OUT MEMBRANE VESICLES

Substrate	Reference
(A) GSH-dependent/stimulated	
Vincristine	Loe *et al.* (1996b, 1998); Mao *et al.* (2000)
Aflatoxin B_1	Loe *et al.* (1997)
Estrone-3-sulfate	Qian *et al.* (2001b)
NNAL-*O*-glucuronide	Leslie *et al.* (2001b)
Daunorubicin	Renes *et al.* (1999)
(B) GSH-conjugated	
Leukotriene C_4	Leier *et al.* (1994)
	Loe *et al.* (1996b)
	Jedlitschky *et al.* (1996)
PGA_2-GS	Evers *et al.* (1997)
Ethacrynic acid-GS	Zaman *et al.* (1996)
(*S*)-2,4-Dinitrophenol-GS	Muller *et al.* (1994)
Aflatoxin B_1-epoxide-GS	Loe *et al.* (1997)
Metolachlor-GS	Leslie *et al.* (2001a)
4-Hydroxynonenol-GS	Renes *et al.* (2000)
Chlorambucil-GS	Barnouin *et al.* (1998)
	Paumi *et al.* (2001)
Melphalan-GS	Barnouin *et al.* (1998)
	Paumi *et al.* (2001)
N-ethylmaleimide-GS	Bakos *et al.* (1998)
Glutathione disulfide	Leier *et al.* (1996)
	Heijn *et al.* (1997)
(C) Glucuronide-conjugated	
17β-estradiol-17-(βD-glucuronide)	Loe *et al.* (1996a)
	Jedlitschky *et al.* (1996)
Etoposide glucuronide	Jedlitschky *et al.* (1996)
	Sakamoto *et al.* (1999)
Bilirubin monoglucuronide	Jedlitschky *et al.* (1997)
Bilirubin diglucuronide	Jedlitschky *et al.* (1997)
Hyodeoxycholate-6-α-glucuronide	Jedlitschky *et al.* (1996)
(D) Others	
Methotrexate	Hooijberg *et al.* (1999)
	Bakos *et al.* (2000b)
	Zeng *et al.* (2001)
3-α-sulfatolithocholyl-taurine	Jedlitschky *et al.* (1996)
Leukotriene D_4	Jedlitschky *et al.* (1996)
Leukotriene E_4	Jedlitschky *et al.* (1996)
N-acetyl-leukotriene E_4	Jedlitschky *et al.* (1996)
Folate	Keppler *et al.* (1998)
	Zeng *et al.* (2001)
Reduced glutathione	Qian *et al.* (2001b)
	Leslie *et al.* (2001b)

importance of the glutathione moiety in determining their ability to be transported by the protein. However, their affinity for MRP1 is clearly also markedly influenced by the structure of the parental compound and ranges over almost three orders of magnitude (e.g. K_m $LTC_4 \approx 0.1\,\mu M$, K_m GSSG $\approx 100\,\mu M$) (Leier *et al.*, 1996; Loe *et al.*, 1996b). It is presently impossible to predict structure/affinity relationships among the GSH-conjugated substrates. For example, LTC_4 and the GSH conjugates of the complex heterocyclic compound aflatoxin B_1, AFB_1, which are structurally unrelated, have similar, high affinities (K_m AFB_1-SG $\sim 0.2\,\mu M$) (Loe *et al.*, 1997) **(Figure 19.4)**. LTC_4, which is presently the only established physiological substrate of MRP1, is exceptional in that it is metabolized to physiologically active products, LTD_4 and LTE_4, which have lost γ-glutamate or γ-glutamate and glycine, respectively, from the glutathionyl moiety. LTD_4 and LTE_4 can be transported by MRP1 but with much lower

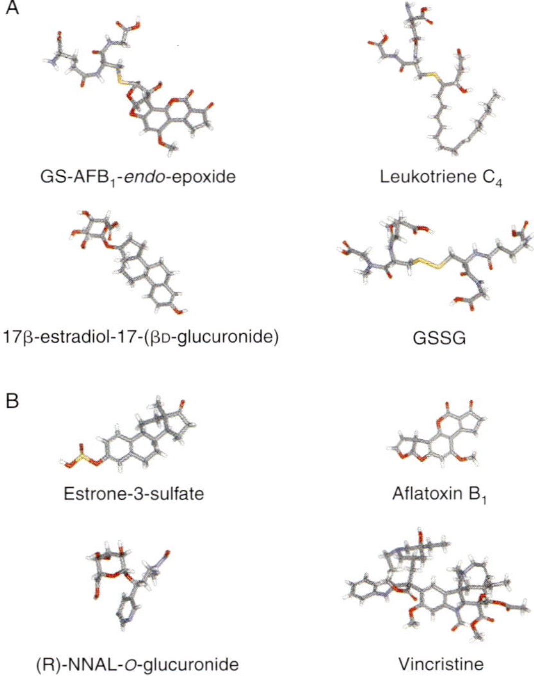

Figure 19.4. Chemical structures of several well-characterized MRP1 substrates. The compounds shown have all been demonstrated to be MRP1 substrates by direct transport studies. Using inside-out membrane vesicles from cells that overexpress protein following drug selection or transfection with an appropriate MRP1 expression vector. A, Substrates transported by MRP1 in the absence of GSH. B, Substrates whose transport is stimulated by or dependent upon GSH.

efficiencies (Jedlitschky *et al.*, 1996). Thus the integrity of the GSH moiety is not absolutely essential for transport of these compounds. However, the unconjugated leukotriene LTB_4 displays very little affinity for the protein, suggesting that the remaining cysteine residue (possibly the additional carboxyl group) contributes significantly to the ability of the protein to bind and transport LTE_4.

Analyses of the ability of various steroid conjugates to inhibit the transport activity of MRP1 illustrate the surprising specificity that the protein can display. The estrogen conjugate, 17β-estradiol-17-(βD-glucuronide) ($E_2$17βG) is a major physiological conjugate of estradiol and is transported by MRP1 with a K_m of 2–3 μM (Jedlitschky *et al.*, 1996; Loe *et al.*, 1996a) **(Figure 19.4)**. It is also transported by MRP2, MRP3 and MRP4, although with approximately 10-fold higher K_m values (Cui *et al.*, 1999; Hirohashi *et al.*, 2000). Studies of the ability of other conjugated steroids to competitively inhibit the MRP1-mediated transport of $E_2$17βG and other substrates have revealed stringent structural requirements with respect to the position of the glucuronide moiety (Loe *et al.*, 1996a, 1997). Although estrogens glucuronidated at the 17β position of the D-ring of the steroid nucleus are effective competitors for $E_2$17βG transport, a shift of the glucuronide to the 16α position of the D-ring decreases inhibitory potency more than 20-fold and naturally occurring estrogen conjugates glucuronidated on the A-ring, such as 17β-estradiol-3-(βD-glucuronide) are essentially inactive. However, a conjugated bile salt sulfated at the 3-position of the A-ring, 3α-sulfatolithocholyltaurine, is a potent inhibitor. Consequently, both the position and nature of the anionic substituent can be important in determining binding affinity. In contrast to the ability of MRP1 to discriminate between structural isomers of estradiol glucuronide, the protein displays similar affinities for both stereoisomers of AFB_1-GS and prostaglandin A_2-GS and transports them with comparable efficiency (Evers *et al.*, 1997; Loe *et al.*, 1997). Thus MRP1 does not exhibit strict stereospecificity with respect to the glutathione moiety.

As mentioned previously, enhanced efflux of at least some of the natural product drugs to which MRP1 confers resistance can be readily demonstrated in MRP1 transfected cells, but initial attempts to demonstrate transport *in vitro* were unsuccessful (Jedlitschky *et al.*, 1996; Loe *et al.*, 1996b; Muller *et al.*, 1994). *Vinca* alkaloids, such as vincristine, and anthracyclines, such as daunorubicin, also proved to be extremely poor inhibitors of vesicular transport of both LTC_4 and $E_2$17βG (Loe *et al.*, 1996b; Muller *et al.*, 1994). At physiological pH, many of these compounds are neutral or cationic in an unmodified form, rather than anionic as are most MRP1 substrates. Two lines of evidence provided an explanation for the initial failure of *in vitro* transport studies. In certain drug-selected MRP1 overexpressing cells, depletion of intracellular GSH was shown to decrease efflux of daunorubicin and partially reversed resistance both to daunorubicin and vincristine (Versantvoort *et al.*, 1995a, 1995b). Although no increase in GSH efflux was detected in response to exposure to daunorubicin, other compounds such as the calcium channel blocker verapamil and the bioflavonoid genistein did stimulate GSH efflux from MRP1 overexpressing cells. Direct evidence of the involvement of GSH in MRP1-mediated transport of unmodified drugs came from vesicle transport studies which demonstrated that the inhibitory potency of agents such as vincristine and vinblastine was dramatically increased in the presence of physiological concentrations of GSH. It was also possible to detect ATP-dependent vesicle uptake of vincristine providing that GSH was present (Loe *et al.*, 1996b, 1998). Subsequent studies demonstrated a similar GSH dependence for the transport of daunorubicin and the unmodified form of AFB_1 (Loe *et al.*, 1997; Renes *et al.*, 1999). The possibility that GSH stimulates transport by altering the redox state of MRP1 or by glutathionylating the protein was excluded because non-reducing short-chain alkyl derivatives of GSH (e.g. *S*-methyl- and *S*-ethyl-GSH) also stimulated transport of vincristine (Loe *et al.*, 1998). More recently, the naturally occurring GSH analogue ophthalmic acid, which contains α-aminobutyrate instead of cysteine, has also been shown to stimulate transport of some conjugated and unconjugated MRP1 substrates (Leslie *et al.*, 2001b; Sethna *et al.*, 1984). Thus there is no requirement for either an available sulfhydryl group or the sulfur atom itself.

Although GSH was first shown to stimulate transport of certain unconjugated compounds by MRP1, it is now apparent that it can also enhance the efficiency with which some conjugated anions are transported. The A-ring conjugated estrogen, estrone-3-sulfate, is a very poor substrate for MRP1 but in the presence of GSH or *S*-methyl GSH, its affinity for the protein and its maximum rate of transport are increased approximately fivefold (Qian *et al.*, 2001b). Despite the ability of GSH to stimulate transport of estrone-3-sulfate, it appears to have no effect on the affinity of MRP1 for estrogens glucuronidated at the same position (unpublished). However, GSH does stimulate the transport of other glucuronide conjugates. For example, transport of etoposide glucuronide is stimulated approximately threefold by GSH (Sakamoto *et al.*, 1999). In addition, transport of the glucuronide conjugate of the potent tobacco-derived carcinogen, 4-(methylnitrosamino)-1-(3-pyridyl)-butanol (NNAL), NNAL-*O*-glucuronide, appears completely dependent on the presence of GSH or certain of its analogues (Leslie *et al.*, 2001b). Consequently, it is currently not possible to predict the structural requirements of conjugated compounds which will display GSH-dependent transport. It is of additional interest to note that NNAL-*O*-glucuronide is also transported by MRP2 (ABCC2) but remarkably, rather than stimulate uptake, GSH is inhibitory (Leslie *et al.*, 2001b). At present, there is no obvious explanation for this difference in GSH dependence/inhibition between MRP1 and MRP2 since GSH itself is transported by both of these proteins with a similar low affinity, and transport of this tripeptide can be stimulated by the same xenobiotics such as verapamil (Loe *et al.*, 2000a; Paulusma *et al.*, 1999). The amino acids in MRP1 and MRP2 that are responsible for their differences in substrate and inhibitor/stimulator specificity and affinity remain to be identified. Thus although the ability to transport NNAL-*O*-glucuronide is a common characteristic of both MRP1 and MRP2, it cannot simply be attributed to conservation of primary structure because the mechanism by which each protein transports this compound clearly differs.

The mechanism by which GSH and certain analogues enhance MRP1-mediated transport of conjugated and unconjugated compounds appears complex. In the case of vincristine, it has been shown that the drug reciprocally stimulates GSH transport (Loe *et al.*, 1998). Furthermore, the presence of GSH increases the affinity of vincristine for the protein approximately 50-fold. Similarly, GSH alone has a K_m greater than 1 mM but in the presence of vincristine this decreases to approximately 100 μM. However, it has not been possible to demonstrate stimulation of GSH transport by estrone-3-sulfate and NNAL-*O*-glucuronide (Leslie *et al.*, 2001b; Qian *et al.*, 2001b). On the other hand, verapamil, as well as a number of its dithiane analogues, potently stimulates MRP1-mediated GSH transport *in vitro* while there is no net apparent vesicular uptake of verapamil itself (Loe *et al.*, 2000a, 2000b). Overall, the data are consistent with the existence of a site or sites on MRP1 capable of binding free GSH and certain non-sulfhydryl-reducing analogues. They also indicate that there is positive cooperativity between the interaction of GSH (or related molecule) and the second substrate with the protein. What remains unclear at present is why in some cases both the second substrate and GSH are transported, while in others, it appears that transport of only one or the other substrate occurs.

SUBSTRATE RECOGNITION

Despite the structural diversity of MRP1 substrates, in most but not all cases, they compete reciprocally with each other for transport, even if there is no structural similarity between parental compounds and they are conjugated to different anionic moieties, e.g. LTC_4 and $E_217\beta G$ compete reciprocally for transport with K_i's of 0.53 μM and 22 μM, respectively (Loe *et al.*, 1996a, 1996b). This suggests that each substrate establishes mutually exclusive interactions with the protein. To what extent these interactions are shared at the atomic level has not yet been resolved, although individual amino acids critical for the binding of some substrates are beginning to be identified. Several approaches have been taken to identify the regions and specific amino acids in MRP1 that participate in substrate recognition and transport. These have involved the use of compounds capable of photolabeling amino acids presumed to be in or close to the site of substrate binding, as well as investigation of the functional consequences of mutating amino acids that are conserved among MRP-related proteins. In addition, it has been possible to exploit differences in substrate specificity between the human protein and its highly conserved murine orthologue to identify non-conserved amino acids that contribute to recognition of particular compounds. Although these studies have been informative with respect to identifying critical regions and individual amino acids involved in substrate recognition and binding, they have also revealed that conservation of substrate specificity is not necessarily simply attributable to a conservation of primary structure. This makes it extremely difficult to extrapolate structure/function relationships from one protein to another, even between highly conserved orthologues.

MRP1 has been photolabeled in crude membranes with iodinated azido derivatives of a quinoline-based drug, *N*-(hydrocinchonidin-8′-yl)-4-azido-2-hydroxybenzamide (IACI), and rhodamine123 (IAARh123) (Daoud *et al.*, 2000). These azido compounds, both of which also label Pgp, are not anionic conjugates and the affinity with which they bind to the protein is not known. However, photolabeling of MRP1 by IACI and IAARh123 can be inhibited by LTC_4, suggesting that they do interact with similar regions of the protein. By using MRP1 that had been modified by insertion of a known epitope at various locations, it was possible to identify two tryptic fragments, containing either TM helices 10 and 11 in TMD2, or TM helices 16 and 17 in TMD3 **(Figure 19.2)**, that were preferentially photolabeled by both compounds. More recently, using a baculovirus coexpression system and limited proteolysis of the native protein, two sites have been detected, one in TMD2 and the other in TMD3 that are also labeled following photoactivation of bound LTC_4 (Qian *et al.*, 2001a). However, it has not been established whether the sites of labeling are the same as those determined for IACI and IAARh123. Nevertheless, labeling with all three compounds favors the existence of either two substrate-binding sites on the protein, or a single site that can exist in two different conformations.

Because of the additional five TM helices in TMD1, TM helices 10, 11, 16 and 17 of MRP1 correspond topographically to TM helices 5, 6, 11 and 12 of Pgp, respectively. Considerable evidence exists that TM helices 6 and 12 of Pgp play a major role in substrate binding, and crosslinking studies using a cysteineless mutant of Pgp also indicate that these two TM helices are in close proximity (Ueda *et al.*, 1997). TM helix 17 of MRP1 is predicted to be much more amphipathic than TM12 of Pgp and is exceptionally well conserved among members of the ABCC subfamily (Ito *et al.*, 2001b). Substitution of a highly conserved tryptophan residue (Trp1246), predicted to be on the hydrophilic face of TM helix 17 close to the membrane cytosol interface, has been shown to eliminate the ability of MRP1 to confer resistance to cationic and electroneutral drugs, and to alter sodium arsenite resistance, regardless of whether or not it is substituted with conserved or non-conserved amino acids (Ito *et al.*, 2001b) **(Figure 19.2)**. Mutation of this residue also essentially eliminates $E_217\beta G$, as well as NNAL-*O*-glucuronide and GSH-stimulated estrone-3-sulfate transport (unpublished). On the other hand, it has only a minor effect on LTC_4 transport and no effect on GSH transport. The exact role of this highly conserved tryptophan residue is not yet clear. While it may function to maintain or form part of a MRP1 substrate-binding pocket, possibly by ensuring the correct positioning of the TM segments, it may also serve a membrane anchoring role since it is predicted to be at the membrane/cytosol interface. It is also possible that it may interact directly with at least some MRP1 substrates because the tryptophan indole ring could contribute to cation-π bond formation with substrates containing positive charges or polarities, π–π stacking interactions with substrates containing aromatic rings, as well as hydrogen bonding interactions with polar substituents. Interestingly, mutations of the comparable tryptophan residue in MRP2 (Trp1254) have a different effect on substrate specificity (Ito *et al.*, 2001a). Thus only non-conservative substitutions of Trp1254 eliminates $E_217\beta G$ transport while more conservative substitutions have no effect. On the other hand, and in contrast to MRP1, only the most conservatively substituted MRP2 Trp1254 mutant transported LTC_4. Thus conserved residues in the two proteins do not necessarily make similar contributions to their shared substrate specificity.

The results obtained from photolabeling studies described earlier with radiolabeled IACI and IAARh123, as well as LTC_4, differ somewhat from those obtained using an azido derivative of the marine sponge polyhydroxylated sterol acetate, Agosterol-A (AG-A) (Ren *et al.*, 2001). In contrast to the other three photolabeled compounds described above, binding of azido-AG-A is GSH dependent. Nevertheless, in the presence of GSH it competitively inhibits LTC_4 transport with a K_i of 30 μM. Using a baculovirus-mediated coexpression system, it was found that azido-AG-A labeling of MRP1 was confined to a fragment containing TMD3 and the cytoplasmic COOH-proximal region of the protein, consistent with the possible involvement of TM helices 16 and/or 17. On the basis of these results, it has been suggested that the site labeled by AG-A has a high affinity for drug and that GSH is binding to a site presumed to be in TMD2. However, this has not been proven by direct binding studies and the possibility remains that azido-AG-A may bind but fail to label a site in TMD2, either because of the presence of GSH or because the geometry of binding of the drug itself is such that the azido group is unable to crosslink to a reactive side-chain in this region of the protein.

The murine and human orthologues of MRP1 are 88% identical, but they display major differences in substrate specificity (Stride *et al.*, 1996, 1997). Both proteins transport LTC_4 with similar kinetic characteristics and confer comparable levels of resistance to drugs such as vincristine and VP-16. However, murine mrp1 does not confer resistance to commonly used anthracyclines and it transports $E_217\beta G$ very poorly relative to MRP1 (Stride *et al.*, 1997). By constructing hybrid mrp1/MRP1 proteins, regions of the human protein that can at least partially 'restore' the ability of the mouse mrp1 to confer resistance to anthracyclines and to transport $E_217\beta G$ were identified and were localized in different parts of TMD3 (Stride *et al.*, 1999). Subsequent systematic conversion of variant amino acids in the comparable regions of mrp1 to those present at the corresponding location in the human protein has identified two non-conserved residues, one of which is critical for the ability of MRP1 to confer resistance to anthracyclines and another which is important for the efficient transport of $E_217\beta G$. Thus, the conversion of Gln1046 in TM helix 14 of mrp1 to Glu as present in MRP1 **(Figure 19.2)** results in a protein capable of conferring resistance to anthracyclines with approximately 40% of the efficiency of wild-type MRP1 (Zhang *et al.*, 2001b). This mutation does not affect the transport of LTC_4 or $E_217\beta G$, nor the ability of the mutant protein to confer resistance to drugs such as vincristine and VP-16. The reciprocal mutation in MRP1 does not affect transport of LTC_4 and $E_217\beta G$ but completely abolishes resistance to anthracyclines. Additional studies indicate that it is the presence of a negatively charged amino acid side-chain at this location that is essential for transport of the cationic anthracyclines. However, the mutation in MRP1, unlike mrp1, also decreases resistance to the weakly cationic vincristine and electroneutral VP-16.

The non-conserved amino acid residue identified as being important for $E_217\beta G$ transport is located in TM helix 17, very close to the conserved Trp1246 described earlier, which is also critical for the transport of this and other substrates (Ito *et al.*, 2001b; Zhang *et al.*, 2001a). Conversion of mrp1 Ala1239 to Thr, as in the corresponding position (1242) in MRP1, increases $E_217\beta G$ transport threefold without affecting athracycline resistance **(Figure 19.2)**. However, the mrp1 Ala1239 to Thr mutation also decreases resistance to vincristine and VP-16 despite the fact that human and murine proteins confer similar levels of resistance to these drugs. The reciprocal Thr1242 to Ala mutation in MRP1 decreases $E_217\beta G$ transport, as expected, but again decreases drug resistance. Surprisingly, the effects on vincristine and VP-16 resistance resulting from mutation of MRP1 Thr1242 to Ala and mrp1 Ala1239 to Thr in TM helix 17 could be suppressed by introducing a second reciprocal mutation in TM helix 14 at the non-conserved position shown to be involved in anthracycline resistance. Thus the apparently conserved ability of mrp1 and MRP1 to transport drugs such as vincristine and VP-16 is not simply attributable to a conservation of primary structure, but also involves compensatory changes in non-conserved amino acids in TM helices 14 and 17. Overall, the results of mutation of both conserved and non-conserved amino acids are consistent with a model in which substrates establish overlapping but non-identical interactions with the protein in one or more shared binding sites.

ATPase activity of MRP1 and models of the mechanism of transport

The ATPase activity of MRP1 has proven difficult to study in crude membrane preparations, because both the basal and substrate stimulated levels of activity are relatively low when compared with a transporter such as Pgp. Detection of ATPase activity attributed to MRP1 has been reported using membranes from MRP1 overexpressing lung cancer cells and insect cells infected with an MRP1 expression vector (Bakos *et al.*, 2000b; Hooijberg *et al.*, 1997, 2000). In the latter case, vanadate inhibitable ATPase activity attributed to MRP1 was estimated to be in the range of 4–5 nmol mg^{-1} membrane protein min^{-1} and to be stimulated two- to threefold by the substrate GS-*N*-ethylmaleimide. However, more detailed characterization of the ATPase activity of MRP1 has required purification of the protein.

MRP1 has been purified by immunoaffinity chromatography from the human small cell lung cancer cell line, H69AR, from which the protein was initially cloned, as well as from transfected baby hamster kidney cells (Chang *et al.*, 1997; Mao *et al.*, 1999). In the latter case, the protein was tagged with a histidine tail to facilitate purification by nickel affinity chromatography. The ATPase activity of native purified MRP1 reconstituted into proteoliposomes is 5–10 nmols mg^{-1} protein min^{-1}, compared with approximately 1 μmol mg^{-1} min^{-1} for Pgp (Mao *et al.*, 1999). Initial reports of the activity of histidine-tagged, detergent-solubilized MRP1 suggested that the protein had an ATPase activity comparable to that of Pgp with respect to both V_{max} and K_m (Chang *et al.*, 1997). However, when reconstituted into proteoliposomes, the ATPase activity of the protein was subsequently reported to be similar to that of native MRP1 (Manciu *et al.*, 2000). Purified reconstituted native MRP1 has been determined to have a K_m for ATP in the range of 100–300 μM (Mao *et al.*, 1999). This is similar to values derived from transport studies using native membrane vesicles and is considerably lower than that of Pgp, which is estimated to have a K_m of approximately 1 mM. Thus the affinity of MRP1 for ATP is several-fold higher than that of Pgp while the maximal basal ATPase activity is approximately 100-fold lower.

Like Pgp, the basal ATPase activity of purified MRP1 can be stimulated by certain substrates and inhibitors (Chang *et al.*, 1997; Leslie *et al.*, 2001c; Mao *et al.*, 1999). However, in general, the stimulation by substrates such as LTC_4 is only twofold or less and the stoichiometry of coupling between ATP hydrolysis and transport of different substrates has not been firmly established. Nevertheless, the purified reconstituted protein possesses many of the transport characteristics determined by studies of MRP1-enriched plasma membrane vesicles, including the reciprocal co-stimulation of transport of GSH and drugs such as vincristine (Loe *et al.*, 1998; Mao *et al.*, 2000). These studies confirm that the basal transport characteristics of MRP1 observed with plasma membrane vesicles are intrinsic properties of the protein and are not markedly influenced by interaction with other components of the membrane.

Another approach to investigating the ATPase properties of MRP1 and the role that each NBD plays in the transport process has been to study the interaction of the protein with radiolabeled photoactivatable ATP analogues such as 8-azido-[^{32}P] ATP. By comparing results obtained using azido-ATP labeled with γ or α [^{32}P] phosphate, it possible to ascertain whether the nucleotide is bound in the form of ATP or ADP. The binding of ADP can also be greatly stabilized by the presence of vanadate, which effectively locks the NBD in a transition state equivalent to that formed in the presence of bound ADP, Mg^{2+} and inorganic phosphate. When analyses of this type are carried out with Pgp, it is technically difficult to demonstrate binding of ATP by either NBD probably because of their relatively low affinity for the nucleotide and high ATPase activity. However, in the presence of vanadate both NBDs display strong equivalent trapping of ADP (Hrycyna *et al.*, 1998; Szabo *et al.*, 1998). Furthermore, trapping at either NBD requires that both be functional, indicating an obligatory interdependency between them. These and other lines of evidence support a model of Pgp transport in which each NBD is functionally equivalent and hydrolyzes ATP alternately, with each hydrolysis event being coupled to the transport of a molecule of substrate (Senior *et al.*, 1995).

In contrast to the results obtained with Pgp, the two NBDs of MRP1 behave quite differently with respect to their ability to bind ATP and to trap ADP in the presence of vanadate. High-affinity binding of ATP to NBD1 is readily demonstrable but very little binding can be detected by NBD2 (Gao *et al.*, 2000; Hou *et al.*, 2000; Nagata *et al.*, 2000). Under vanadate trapping conditions, the reverse is observed, with ADP being trapped almost exclusively by NBD2. Furthermore, although the trapping of ADP by NBD2 requires that NBD1 be capable of binding ATP, the binding of ATP by NBD1 is actually enhanced when NBD2 is inactivated by mutation of essential residues in the conserved Walker A motif. These observations suggest that NBD1 hydrolyzes ATP very poorly relative to NBD2. That the two NBDs have different functional roles is also supported by mutational studies showing that inactivation of each of the NBDs by similar mutations of their Walker A motifs has a different effect on transport activity. Mutation of NBD1 decreases transport activity by 60–70% while mutation of NBD2 essentially inactivates the protein (Gao *et al.*, 2000).

A third line of evidence also indicates that the two NBDs of MRP1 are not functionally equivalent. The ability of Pgp to bind substrate is markedly diminished when ADP is trapped at either NBD (Ramachandra *et al.*, 1998). This is attributed to a conformational change that occurs during the transition state which decreases the affinity of the protein for substrate, enabling the release or transfer of substrate to a low-affinity binding site. Under similar vanadate trapping conditions, MRP1 also displays a marked reduction in the affinity for LTC_4 (Qian *et al.*, 2001a). However, this reduction in affinity is exclusively dependent on the ability of NBD2 to hydrolyze ATP and it is unaffected by inactivation of NBD1. Thus it is clear that hydrolysis of ATP by NBD2 is necessary for MRP1-mediated transport, but it remains to be established whether hydrolysis by NBD1 is an essential component of the transport cycle, or whether nucleotide binding by this domain (and possibly hydrolysis) serves to regulate the ATPase activity of NBD2.

MODEL OF MRP1 TRANSPORT

Most of the available data, based on the ability of structurally different, GSH dependent and independent MRP1 substrates to compete reciprocally for transport, coupled with the effect that mutation of single amino acids has on substrate specificity, suggests that initial interaction of substrates with MRP1 involves a common site, or pocket, on the protein with which each substrate establishes its own set of distinct but partially overlapping atomic contacts. The reciprocal positive cooperativity observed between the binding of GSH and some compounds, such as vincristine, suggests that a conformational change occurs following binding of either GSH or the second substrate. It is presumed that the binding of conjugated substrates that is independent of the presence of GSH induces a similar conformational change. Consistent with this suggestion, LTC_4 has been shown to enhance the binding of ATP by NBD1 of MRP1 (Gao *et al.*, 2000). The fact that high concentrations of GSH compete with the binding of substrates such as LTC_4 also suggests that the site of initial low-affinity GSH binding overlaps with at least part of the site that is occluded by the high-affinity binding of some conjugated substrates (Qian *et al.*, 2001a). By analogy with other drug transporters such as Pgp and the bacterial transporter LmrA (van Veen *et al.*, 2000), stimulation of ATP hydrolysis, possibly only by NBD2 in the case of MRP1, following substrate binding may induce a second conformational change that favors substrate release either directly to the outside of the cell or, more likely, by transfer to a second low-affinity site, from which the substrate is ultimately released. This is supported by the effects of mimicking a transition state by the vanadate-induced trapping of ADP, which decreases the affinity with which LTC_4 binds to the site photolabeled in TMD2. The existence of two cooperatively linked binding sites on MRP1, as proposed for Pgp and LmrA, is favored by data showing that compounds which compete with each other for transport at high concentrations, such as estrone sulfate and LTC_4, can reciprocally stimulate transport at low concentrations (Dey *et al.*, 1997; Shapiro and Ling, 1997; Sharom *et al.*, 1999; van Veen *et al.*, 2000). In addition, although LTC_4 can compete for transport of NNAL-*O*-glucuronide in the presence of GSH, the glucuronide actually increases the V_{max} for LTC_4 transport (Leslie *et al.*, 2001b). A hypothetical model of MRP1-mediated transport that is consistent with current data is illustrated in **Figure 19.5**. In this model, interaction of substrate(s) with a shared high-affinity binding pocket on the protein induces a conformational change stimulating ATP binding by NBD1 and hydrolysis of ATP by NBD2. During the transition state formed at NBD2 a second conformational change occurs, which favors transfer of substrate from the initial binding site to a low-affinity site, from which the substrate is released. Return of MRP1 to the original high-affinity binding state could be coupled to release of ADP from NBD2 and/or release of ATP (or possibly ADP) from NBD1, as well as the release of substrate(s).

BIOLOGICAL RELEVANCE AND PHYSIOLOGICAL ROLE OF MRP1

MRP1 mRNA or protein is readily detectable in many normal tissues with the exception of healthy adult liver, where, if the protein is expressed at all, levels are extremely low. The levels of MRP1 are highest in the lung, testis, kidney, skeletal muscle and heart (Cole *et al.*, 1992; Flens *et al.*, 1996). However, expression within a tissue is often cell type specific, such as bronchial epithelium and hyperplastic type II pneumocytes in the lung, and proliferating Paneth cells in the colon (Brechot *et al.*, 1998; Peng *et al.*, 1999; Wright *et al.*, 1998). There is also some *in vivo* and *in vitro* evidence indicating that levels of MRP1 are increased in other proliferating cells, including hepatocyte-derived cells, and in the liver itself under conditions of induced cholestasis (Roelofsen *et al.*, 1997; Vos *et al.*, 1998).

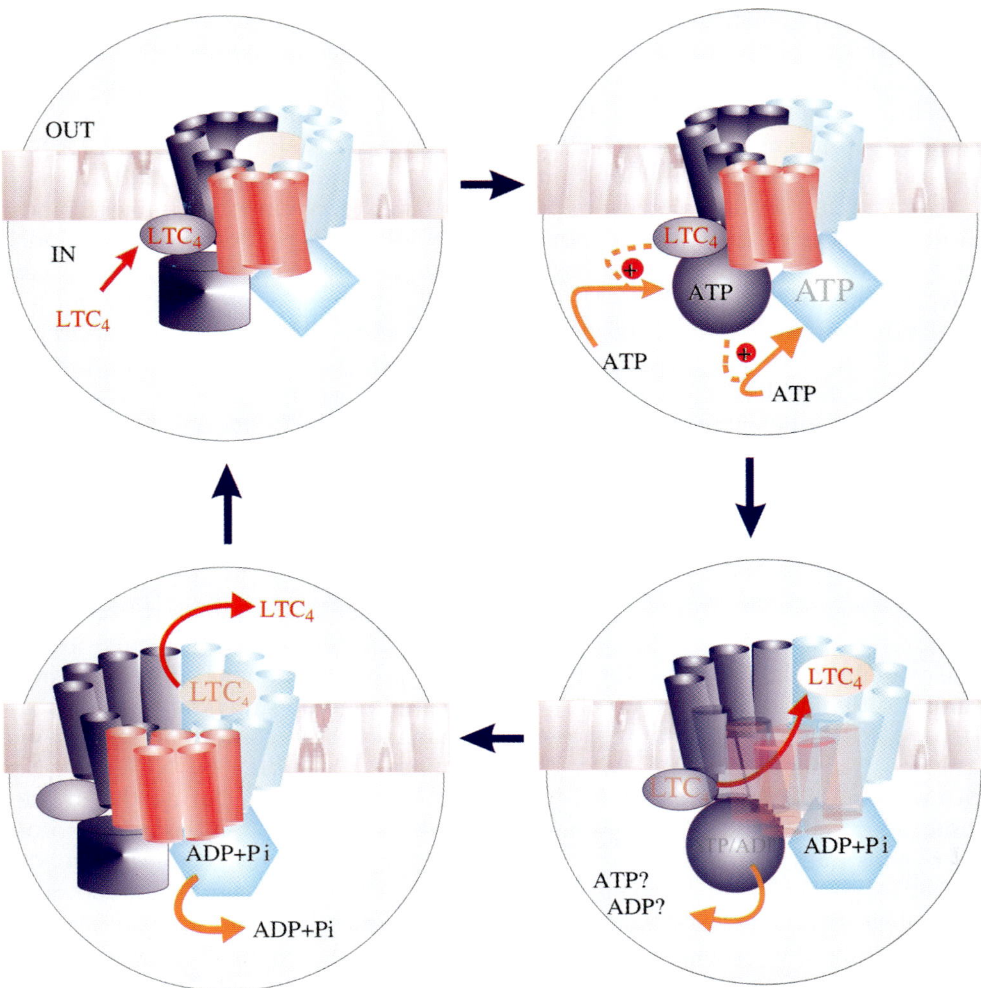

Figure 19.5. ***Hypothetical model of MRP1-mediated transport. The figure illustrates the possible steps involved in MRP1-mediated transport of a GSH-conjugated substrate such as LTC_4. By analogy with models proposed for other transporters, two binding sites are shown, a high-affinity site proximal to the cytoplasmic side of the membrane and a lower-affinity site proximal to the extracellular side. Each of the sites shown may be comprised of a region to which GSH can bind with low affinity and a region with which the hydrophobic component of the substrate interacts. An explanation of the mechanistic steps involved is provided in the section on the model of MRP1 transport.***

The *MRP1* gene is located on chromosome 16 at band p13.1 (Cole *et al.*, 1992; Slovak *et al.*, 1993). It comprises 31 exons and spans over 200 kb **(Figure 19.6)**. A number of splicing variants of MRP1 mRNA with continuous open reading frames have been identified. However, they are generated by 'skipping' one or two exons, primarily in one or the other NBD, and it appears unlikely that the proteins they encode retain any function (Grant *et al.*, 1997). In many but not all drug-selected cell lines that overexpress MRP1 protein, the gene is amplified. In some cases, the *MRP6* gene, which is located only a few kilobases away, is co-amplified but there is presently no evidence that MRP6 contributes to drug resistance (Kool *et al.*, 1999). However, mutations in the gene appear to be the cause of a rare autosomal recessive disorder, pseudoxanthoma pigmentosum (Ringpgeil *et al.*, 2000). Relatively little is known of the mechanisms that regulate expression of *MRP1* and recent comparisons with the regulatory regions of the orthologous gene in mice and rat have revealed little sequence conservation in promoter regions, other than binding sites for the transacting factor SP1 or a related protein (Kurz *et al.*, 2001; Zhu and Center, 1994, 1996; M. Muredda, K.I. Nunoya, S.P.C. Cole and R. Deeley, unpublished). There is evidence suggesting that the expression of *MRP1* may be

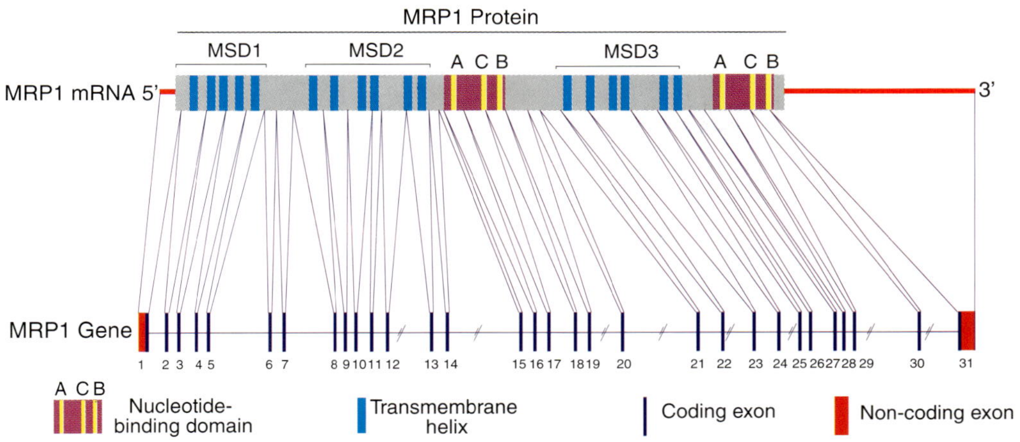

Figure 19.6. ***Alignment of the exons of the MRP1 gene with MRP1 mRNA and protein. The MRP1 gene consists of 31 exons and spans at least 200 kb. The intron/exon organization of MRP1 is depicted in the lower part of the figure and the regions of MRP1 mRNA and protein encoded by each exon are illustrated above. As is apparent from the figure, the first five exons of the gene encode amino acids 1–205, which comprise MSD1. As described in the text, this region can be deleted without loss of the ability to transport at least some MRP1 substrates such as LTC_4.***

influenced by the p53 status of the cell and oxidative stress (Yamane *et al.*, 1998) but it remains to be firmly established whether the response involves transcriptional or post-transcriptional mechanisms.

Several polymorphisms in *MRP1* and *MRP2* have recently been described, but whether they result in differences in the substrate specificity of the encoded proteins is not yet known (Conrad *et al.*, 2001; Ito *et al.*, 2001c). However, there is now convincing evidence that polymorphisms in the *MDR1* (*ABCB1*) gene encoding Pgp can play a clinically important role in the bioavailibility and disposition of a variety of important drugs (Hoffmeyer *et al.*, 2000). Thus some polymorphisms in MRP1 and its related proteins may also be expected to contribute to the pharmacokinetic properties of certain anticancer agents, which in turn could affect their efficacy and/or the incidence of side effects associated with their use. Such pharmacogenetic variation might also affect an individual's susceptibility to damage incurred by exposure to endo- and xenotoxins, carcinogens and their metabolites, whose transport is mediated by these proteins.

Studies with knockout mice have shown that mrp1 is not necessary for normal development (Lorico *et al.*, 1996, 1997; Rappa *et al.*, 1999; Wijnholds *et al.*, 1997, 1998). However, as suggested by *in vitro* studies, the mrp1$^{-/-}$ mice have an impaired response to an IgE-mediated inflammatory stimulus consistent with an inability to efflux LTC_4 from mast cells and eosinophils. Interestingly, the mrp1$^{-/-}$ mice are also more resistant to bacterial lung infection, as exemplified by the increased survival of the knockout mice following infection with *Streptococcus pneumoniae* (Schultz *et al.*, 2001). The relative resistance to infection is attributable to a perturbation in leukotriene metabolism in the lung caused by the inability of alveolar macrophages to efflux LTC_4. The accumulated intracellular LTC_4 is thought to cause product inhibition of LTC_4 synthase with a resultant shift in favor of production of LTB_4 from the common precursor, LTA_4. LTB_4 is a potent stimulator of the microbicidal activity of phagocytic cells in the lung.

The presence of MRP1 in cells at a number of blood–organ interfaces that create pharmacological sanctuary sites in the body, such as the central nervous system, the testis and the placenta, as well as in mucosal epithelium, suggests that the protein may have an important role in protecting these tissues from drug- or toxin-induced injury (St Pierre *et al.*, 2000; Wijnholds *et al.*, 1998, 2000). This suggestion is supported by a considerable body of evidence derived from studies of the mrp1$^{-/-}$ mice and by *in vitro* characterization of the protein's substrate specificity (reviewed in Leslie *et al.*, 2001a).

The mrp1$^{-/-}$ mice have provided valuable models for studying the *in vivo* importance of MRP1/mrp1 in drug disposition. The mice have an increased sensitivity to etoposide phosphate accompanied by increased bone marrow toxicity (Lorico *et al.*, 1997; Wijnholds *et al.*, 1997). Furthermore, these animals sustain increased etoposide-induced damage to the mucosa of the oropharyngeal cavity and exhibit polyuria. In addition, the mice display marked aberrations in spermatogenesis with an almost complete lack of meiotic divisions (Wijnholds *et al.*, 1998). Consistent with a protective role for MRP1/mrp1 in the testis, the protein is expressed at high levels in epididymal epithelium, as well as in Leydig and Sertoli cells, which also contain high levels of GSH *S*-transferases (GST) **(Figure 19.7)**. High levels of MRP1 expression can also be detected in various forms of testicular cancer **(Figure 19.7)**. In addition, mrp1 has been shown to be important in preventing the accumulation of drugs such as vincristine and etoposide in the CNS, as well as in decreasing their absorption across the gastrointestinal mucosa (Wijnholds *et al.*, 2000). Thus while MRP1 may serve to protect certain normal tissues from oncolytic drugs such as etoposide, there is legitimate concern that the use of MRP1 inhibitors clinically may

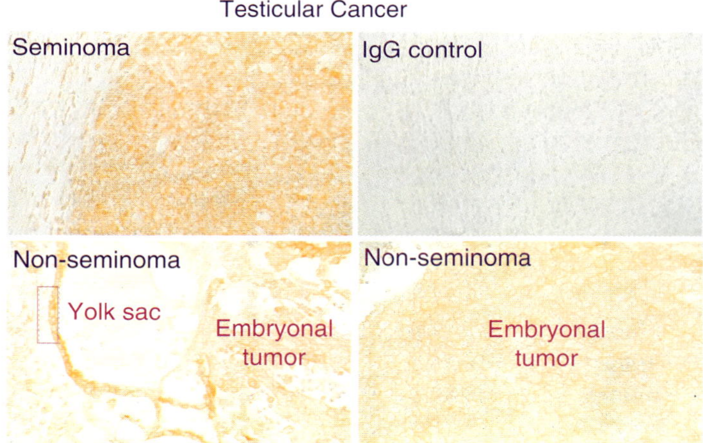

Figure 19.7. Immunohistochemical detection of MRP1 expression in normal human testis and testicular carcinomas. Fixed and paraffin-embedded sections were subjected to antigen recovery as described (Wright et al., *1998) and MRP1 was detected using the highly specific antibody QCRL1 (Hipfner* et al., *1994, 1996). In normal testis (upper panel), MRP1 is readily detectable in epididymal epithelium, spermatogenic cells and Leydig cells. MRP1 is also expressed in Sertoli cells, as evidenced by the staining profile obtained in sections from the testis of an infertile patient lacking spermatogenic cells. MRP1 can also be detected in a high proportion of germ cell carcinomas (lower panel), both seminomas and non-seminomas, regardless of the germ cell type of origin.*

lead to an increased incidence of toxic side effects from at least some chemotherapeutic agents.

Active efflux of unmodified xenobiotics has obvious implications for detoxification, while conjugated metabolites of the parent drugs or toxins, which are more likely substrates for MRP1, are typically less reactive and are usually considered to have been 'detoxified' prior to their elimination. However, in some cases the conjugates remain directly toxic by virtue of their ability to inhibit enzymes important for cell viability. Furthermore, if the metabolites are allowed to accumulate, they may be converted back to the parent compound by hydrolytic enzymes, or cause product inhibition of the conjugating transferases involved in their formation. Thus, MRP1 may act in concert with phase III conjugating enzymes to maintain low steady state levels of their potentially toxic products. For example, drug-selected breast cancer cells that overexpress MRP1 but which have low GST activity were protected from the carcinogen 4-nitroquinoline-1-oxide (Morrow *et al.*, 1998) and the toxin 1-chloro-2,4-dinitrobenzene (Diah *et al.*, 1999) by transfection with a cDNA encoding the P1-1 isoform of GST. Conversely, transfection with a cDNA encoding MRP1 into cells that lack the protein prevented toxicity resulting from product inhibition of the A1-1 isoform of GST, GSTA1-1, caused by accumulation of the GS conjugate of the alkylating agent, chlorambucil (Paumi *et al.*, 2001).

The potent carcinogen, aflatoxin B_1 (AFB_1), a mycotoxin produced by certain *Aspergillus* species, was among the first environmental toxins shown to be a substrate for MRP1 (Loe *et al.*, 1997; Massey *et al.*, 1995). Unmodified AFB_1 is transported by MRP1 in the presence of physiological concentrations of GSH and the protein also transports the GSH conjugate of AFB_1 epoxide with very high affinity (K_m 189 nM). The major tobacco-derived pulmonary carcinogens nitrosamine 4-(methylnitrosamino)-1-(3-pyridyl)-1-butanone (NNK) and its reductive metabolite, nitrosamine 4-(methylnitrosamino)-1-(3-pyridyl)-1-butanol (NNAL), are primarily detoxified by glucuronidation. The product is [4-(methylnitrosamino)-1-(3-pyridyl) but-1-yl]-β-*O*-D-glucosiduronic acid (NNAL-*O*-glucuronide), which, as mentioned previously, has been identified as a GSH-dependent substrate of MRP1 of moderate affinity (K_m 39 μM) (Leslie *et al.*, 2001b). Low levels of two additional glucuronide conjugates are also formed during NNK metabolism, but whether these are substrates of MRP1 or related proteins and if so, whether their transport is inhibited or stimulated by GSH, remains to be determined (Murphy *et al.*, 1995).

The conjugates of some herbicides and pesticides are also likely to be MRP1 substrates. GSH conjugates of certain agrochemicals have been shown to be substrates for an MRP1 homologue expressed in vacuolar membranes of the vascular plant *Arabidopsis thaliana*, suggesting that these compounds might also interact with the human protein (Martinoia *et al.*, 1993; Rea *et al.*, 1998). Consistent with this possibility, MRP1 transports the GSH conjugate of the chloracetanilide herbicide, metolachlor (Leslie *et al.*, 2001a). However, it is important to remember that plant MRP1 homologues are located in intracellular vacuolar membranes and while they may protect the plant by effectively sequestering their toxic substrates, the toxins may nevertheless pass directly up the food chain.

MRP1 may also be involved in protection against exposure to certain forms of heavy metals. Transfection studies demonstrate that MRP1 decreases the cytotoxicity of some arsenic and antimonial centered oxyanions (Cole *et al.*, 1994; Stride *et al.*, 1997) and, conversely, heavy metal-selected tumor cell lines have been shown to overexpress MRP1 (Vernhet *et al.*, 1999). Antimony (Sb) and arsenic (As) toxicity has been attributed to their ability to modify sulfhydryl-containing proteins and enzyme systems including mitochondrial enzymes, which as a result impairs tissue respiration. In several instances, protection against these toxic metals has been shown to be associated with GSH efflux from the cell and GSH depletion has been shown to increase sensitivity (Chen *et al.*, 1997; Vernhet *et al.*, 1999; Zaman *et al.*, 1995). Consequently, MRP1 is assumed to transport metal–GSH complexes or possibly to co-transport metal oxyanions and GSH. However, direct evidence of such transport *in vitro* remains elusive.

In addition to its involvement in protecting cells and tissues from exposure to endo- and xenotoxins, MRP1 may also play a role in protection against oxidative stress. Under conditions of oxidative stress, the concentration of GSSG increases, resulting in glutathionylation of reactive sulfydryl groups, some of which are on regulatory proteins controlling the activation of a variety of cellular stress response mechanisms. Since MRP1 transports GSSG with a K_m (100 μM) comparable to the concentration of oxidized glutathione in many cell types, the protein may contribute to 'buffering' the cell against fluctuations in GSSG levels that could trigger a stress response. In addition, many compounds capable of generating reactive oxygen species are conjugated with GSH as part of their detoxification mechanism and are likely substrates for MRP1.

For reasons cited above, the ability of MRP1 to efflux the conjugates from the cell and thus prevent their accumulation may be an important step in the defense against oxidative stress. This may be of particular relevance for lipid peroxidation products, which are formed endogenously during periods of oxidative stress. These compounds, such as 4-hydroxynonenal, are aldehydes produced by the oxidation of polyunsaturated fatty acids and they are extremely toxic. They are known to be conjugated to GSH and *in vitro* evidence suggests that their conjugates are MRP1 substrates (Renes *et al.*, 2000). The ability of MRP1 to transport some compounds capable of generating reactive oxygen species, such as anthracyclines, in a GSH-dependent manner also indicates that the protein may play a protective role even in the absence of conjugate formation. Finally, although GSH alone is a poor substrate for MRP1 *in vitro*, its transport can be stimulated by a number of compounds including certain bioflavonoids that are common dietary constituents, as well as possibly unidentified endogenous cellular constituents (Leslie *et al.*, 2001c; Loe *et al.*, 2000a, 2000b). Thus the MRP1-mediated efflux of free GSH may contribute to the high levels of GSH found in extracellular fluids thought to offer protection for some tissues against exogenous toxic electrophiles capable of spontaneously forming GS conjugates.

CLINICAL RELEVANCE OF MRP1

MRP1 has been detected in a wide variety of solid and hematological tumors, including pediatric tumors such as neuroblastoma and retinoblastoma. However, evaluation of the clinical significance of the presence of MRP1 or its cognate mRNA in a tumor sample is complicated by the fact that the *MRP1* gene is expressed in many of the tissues from which these tumors originate. Consequently, it is important that bulk analyses of tumor tissue by Northern and Western blotting or RT-PCR be carried out with carefully paired tumor and normal tissue samples and, ideally, should be accompanied by immunohistological verification of the level and pattern of expression of the protein in malignant and adjacent normal tissue. These caveats notwithstanding, there is a significant body of evidence that MRP1 expression is elevated in a wide range of solid tumors including: lung, gastrointestinal and urothelial carcinomas, neuroblastoma, mesothelioma, glioma, retinoblastoma, melanoma, and cancers of the breast, endometrium, ovary, testis, prostate and thyroid. In some cases, there is also evidence that MRP1 expression may be a negative prognostic indicator of disease outcome.

A number of studies have reported high levels of MRP1 expression in lung cancer, particularly non-small cell lung cancer (NSCLC), which accounts for approximately 80% of all lung cancer cases. NSCLC, unlike small cell lung cancer (SCLC), is inherently multidrug resistant. Moderate to high levels of expression of MRP1 are frequently found in both adenocarcinoma and squamous cell carcinoma, the two major forms of NSCLC (Wright *et al.*, 1998). In the latter form, MRP1 has been reported to be a negative indicator of patient survival (Ota *et al.*, 1995). MRP1 has also been found in carcinoma *in situ* and in hyperplastic alveolar type II cells, which may be the progenitor cells of some adenocarcinomas. Consequently, increased expression of MRP1 may be a very early event in the progression of NSCLC. Unlike NSCLC, SCLC generally responds well to initial chemotherapy but within a 2-year period approximately 80% of patients relapse with a multidrug resistant form of the disease. MRP1 expression in untreated small cell lung cancer (SCLC) is less prevalent than in non-small cell lung cancer (NSCLC) and when present appears to be restricted to foci involving a small number of cells (Wright *et al.*, 1998). This is consistent with the possibility that MRP1 positive cells may survive initial chemotherapy and subsequently contribute to the multidrug-resistant character of the disease at relapse. Establishing whether MRP1 expression and the development of multidrug resistance in SCLC are indeed correlated requires longitudinal studies of changes in tumor biology in individual patients and such studies are rarely performed for ethical reasons. However, one small study in which tumor samples were taken from patients at the time of both diagnosis and relapse did find that MRP1 levels, as well as the levels of some other drug resistance markers, were increased following treatment (Kreisholt *et al.*, 1998).

Among other common solid tumors, MRP1 expression has been reported to be frequently increased in breast and prostate cancer and to be positively correlated in both cases with disease stage (Filipits *et al.*, 1996; Kim *et al.*, 2001; Ostlund Farrants *et al.*, 1987). In breast cancer, elevated levels of MRP1 have also been found to be strongly associated with shorter times to relapse following post-surgical adjuvant chemotherapy of some relatively early stage tumors (Nooter *et al.*, 1997a, 1997b) and in prostate cancer, to be positively correlated with p53 status (Sullivan *et al.*, 1998).

MRP1 is also expressed in the two most common solid tumors of childhood, neuroblastoma and retinoblastoma. In the former, two studies have found MRP1 to be a strong independent negative prognostic indicator of survival and to be positively associated with amplification of the *N-Myc* oncogene, also a strong negative prognostic indicator (Bader *et al.*, 1999; Norris *et al.*, 1996). A third study found that the presence of MRP1 in the primary tumor was a negative indicator of outcome, but in contrast to the other two, found no correlation between the levels of MRP1 and *N-myc* amplification, nor between disease outcome and the level of MRP1 expression (Goto *et al.*, 2000). In retinoblastoma, MRP1 expression in untreated tumors appears to be relatively infrequent, but the protein has been detected in tumors that fail to respond to chemotherapy alone or chemotherapy in combination with a Pgp reversing agent such as cyclosporin (Chan *et al.*, 1997).

Although MRP1 has been detected in various forms of leukemia, the protein is also expressed in all lineages of normal hematopoietic cells, and the clinical significance of MRP1 expression remains controversial (Abbaszadegan *et al.*, 1994). One of the difficulties encountered in correlating the expression of MRP1 with clinical outcome is the possibility that the protein may be expressed together with other drug efflux pumps, such as Pgp and the recently identified BCRP (ABCG2) (Doyle *et al.*, 1998; Miyake *et al.*, 1999). Consequently, functional assays capable of detecting the total activity of drug efflux in leukemic blasts may be more informative than analyses that attempt to correlate the levels of individual transporters with disease outcome. These assays are based on the efflux of fluorescent dyes, such as calcein and Fluo-3, in the presence and absence, when available, of specific inhibitors of individual drug pumps. Using this type of analysis, two studies have found that the combined activity of MRP1 and Pgp in acute myeloid leukemia is a far better predictor of response and outcome than either protein alone (Legrand *et al.*, 1996; van der Kolk *et al.*, 2000).

Overall, currently available data strongly suggest that MRP1 plays a role in the clinical multidrug resistance of some tumor types. It is anticipated that in the near future highly specific reversing agents will be available that are capable of selectively targeting MRP1 to the exclusion of other drug efflux pumps. Clinical trials of such agents should establish the contribution of MRP1 to the complex mechanisms likely to be responsible for the multidrug resistance characteristics of some common malignancies, such as NSCLC and refractory cancers of the prostate and breast.

REFERENCES

Abbaszadegan, M.R., Futscher, B.W., Klimecki, W.T., List, A. and Dalton, W.S. (1994) Analysis of multidrug resistance-associated protein (MRP) messenger RNA in normal and malignant hematopoietic cells. Cancer Res. 54, 4676–4679.

Almquist, K.C., Loe, D.W., Hipfner, D.R., Mackie, J.E., Cole, S.P.C. and Deeley, R.G. (1995) Characterization of the 190 kDa multidrug resistance protein (MRP) in drug-selected and transfected human tumor cells. Cancer Res. 55, 102–110.

Ambudkar, S.V., Dey, S., Hyrcyna, C.A., Ramachandra, M., Pastan, I. and Gottesman, M.M. (1999) Biochemical, cellular, and pharmacological aspects of the multidrug transporter. Annu. Rev. Pharmacol. Toxicol. 39, 361–398.

Bader, P., Schilling, F., Schlaud, M., Girgert, R., Handgretinger, R., Klingebiel, T., Treuner, J., Liu, C., Niethammer, D. and Beck, J.F. (1999) Expression analysis of multidrug resistance associated genes in neuroblastomas. Oncology Reports 6, 1143–1146.

Bakos, E., Hegedus, T., Hollo, Z., Welker, E., Tusnady, G.E., Zaman, G.J.R., Flens, M.J., Varadi, A. and Sarkadi, B. (1996) Membrane topology and glycosylation of the human multidrug resistance-associated protein. J. Biol. Chem. 271, 12322–12326.

Bakos, E., Evers, R., Szakacs, G., Tusnady, G.E., Welker, E., Szabo, K., *et al.* (1998) Functional multidrug resistance protein (MRP1) lacking the N-terminal transmembrane domain. J. Biol. Chem. 273, 32167–32175.

Bakos, E., Evers, R., Calenda, G., Tusnady, G.E., Szakacs, G., Varadi, A. and Sarkadi, B. (2000a) Characterization of the amino-terminal regions in the human multidrug resistance protein (MRP1). J. Cell Sci. 113, 4451–4461.

Bakos, E., Evers, R., Sinkule, J.A., Varadi, A., Borst, P. and Sarkadi, B. (2000b) Interactions of the human multidrug resistance proteins MRP1 and MRP2 with organic anions. Mol. Pharmacol. 57, 760–768.

Barnouin, K., Leier, I., Jedlitschky, G., Pourtier-Manzanedo, A., Konig, J., Lehmann, W.-D. and Keppler, D. (1998) Multidrug resistance protein-mediated transport of chlorambucil and melphalan conjugated to glutathione. Br. J. Cancer 77, 201–209.

Borst, P., Evers, R., Kool, M. and Wijnholds, J. (1999) The multidrug resistance protein family. Biochim. Biophys. Acta 1461, 347–357.

Brechot, J.-M., Hurgain, I., Fajac, A., Daty, N. and Bernaudin, J.-F. (1998) Different pattern of MRP localization in ciliated and basal cells from human bronchial epithelium. J. Histochem. Cytochem. 46, 513–517.

Chan, H.S.L., Lu, Y., Grogan, T.M., Haddad, G., Hipfner, D.R., Cole, S.P.C., Deeley, R.G., Ling, V. and Gallie, B.L. (1997) Multidrug resistance protein (MRP) expression in retinoblastoma correlates with rare failure of chemotherapy despite cyclosporine for reversal of P-glycoprotein. Cancer Res. 57, 2325–2330.

Chang, X.-B., Hou,Y.-X. and Riordan, J.R. (1997) ATPase activity of purified multidrug resistance-associated protein. J. Biol. Chem. 272, 30962–30968.

Chen, Z., Mutoh, M., Sumizawa, T., Furukawa, T., Haraguchi, M., Tani, A. and Akiyama, S. (1997) Reversal of heavy metal resistance in multidrug-resistant human KB carcinoma cells. Biochem. Biophys. Res. Commun. 236, 586–590.

Cole, S.P.C. (1990) Patterns of cross-resistance in a multidrug-resistant small-cell lung carcinoma cell line. Cancer Chemother. Pharmacol. 26, 250–256.

Cole, S.P.C. (1992) Drug resistance and lung cancer. In: Cancer: Concept to Clinic (ed. J. Wood), pp. 15–21. Fairlawn, NJ: Medical Publishing Enterprises.

Cole, S.P.C. and Deeley, R.G. (1996) Multidrug resistance associated with overexpression of MRP. In: Drug Resistance (ed. W.N. Hait), pp. 39–62. Norwell, MA: Kluwer Academic Press.

Cole, S.P.C. and Deeley, R.G. (1998) Multidrug resistance mediated by the ATP-binding cassette transporter protein, MRP. BioEssays 20, 931–940.

Cole, S.P.C., Downes, H.F. and Slovak, M.L. (1989) Effect of calcium antagonists on the chemosensitivity of two multidrug resistant human tumour cell lines which do not overexpress P-glycoprotein. Br. J. Cancer 59, 42–46.

Cole, S.P.C., Bhardwaj, G., Gerlach, J.H., Mackie, J.E., Grant, C.E., Almquist, K.C., Stewart, A.J., Kurz, E.U., Duncan, A.M.V. and Deeley, R.G. (1992) Overexpression of a transporter gene in a multidrug-resistant human lung cancer cell line. Science 258, 1650–1654.

Cole, S.P.C., Sparks, K.E., Fraser, K., Loe, D.W., Grant, C.E., Wilson, G.M. and Deeley, R.G. (1994) Pharmacological characterization of multidrug resistant MRP-transfected human tumor cells. Cancer Res. 54, 5902–5910.

Conrad, S., Kaufmann, H-M., Ito, K.-I., Deeley, R., Cole, S.P.C. and Schrenk, D. (2001) Identification of human multidrug resistance protein 1 (*MRP1*) mutations and characterization of a G671V substitution. J. Hum. Genet. 46, 656–663.

Cui, Y., Konig, J., Buchholz, U., Spring, H., Leier, I. and Keppler, D. (1999) Drug resistance and ATP-dependent conjugate transport mediated by the apical multidrug resistance protein, MRP2, permanently expressed in human and canine cells. Mol. Pharmacol. 55, 929–937.

Daoud, R., Desneves, J., Deady, L.W., Tilley, L., Scheper, R.J., Gros, P. and Georges, E. (2000) The multidrug resistance protein is photoaffinity labeled by a quinoline-based drug at multiple sites. Biochemistry 39, 6094–6102.

de Jong, M.C., Slootstra, J.W., Scheffer, G.L., Schroeijers, A.B., Puijk, W.C., Dinkelberg, R., Kool, M., Broxterman, H.J., Meloen, R.H. and Scheper, R.J. (2001) Peptide transport by the multidrug resistance protein MRP1. Cancer Res. 61, 2552–2557.

Deeley, R.G. and Cole, S.P.C. (1997) Multidrug resistance in mammalian cells mediated by members of the ATP-binding cassette superfamily: the P-glycoproteins and MRP. In: Molecular Genetics of Drug Resistance, pp. 247–298. Harwood Academic Press.

den Boer, M.L., Zwaan, C.M., Pieters, R., Kazemier, K.M., Rottier, M.M.A., Flens, M.J., Scheper, R.J. and Veerman, A.J.P. (1997) Optimal immunocytochemical and flow cytometric detection of P-gp, MRP and LRP in childhood acute lymphoblastic leukemia. Leukemia 11, 1078–1085.

Dey, S., Ramachandra, M., Pastan, I., Gottesman, M.M. and Ambudkar, S.V. (1997) Evidence for two nonidentical drug-interaction sites in the human P-glycoprotein. Proc. Natl Acad. Sci. USA 94, 10594–10599.

Diah, S.K., Smitherman, P.K., Townsend, A.J. and Morrow, C.S. (1999) Detoxification of 1-chloro-2,4-dinitrobenzene in MCF7 breast cancer cells expressing glutathione *S*-transferase P1-1 and/or multidrug resistance protein 1. Tox. Appl. Pharmacol. 157, 85–93.

Doyle, L.A., Yang, W., Abruzzo, L.V., Krogmann, T., Gao, Y., Rishi, A.K. and Ross, D.D. (1998) A multidrug resistance transporter from human MCF-7 breast cancer cells. Proc. Natl Acad. Sci. USA 95, 15665–15670.

Evers, R., Cnubben, N.H.P., Wijnholds, J., van Deemter, L., van Bladeren, P.J. and Borst, P. (1997) Transport of glutathione prostaglandin A conjugates by the multidrug resistance protein 1. FEBS Lett. 419, 112–116.

Filipits, M., Suchomel, R.W., Dekan, G., Haider, K., Valdimarsson, G., Depisch, D. and Pirker, R. (1996) *MRP* and *MDR1* gene expression in primary breast carcinoma. Clin. Cancer Res. 2, 1231–1237.

Flens, M.J., Zaman, G.J.R., van der Valk, P., Izquierdo, M.A., Schroeijers, A.B., Scheffer, G.L., van der Groep, P., de Haas, M., Meijer, C.J.L.M. and Scheper, R.J. (1996) Tissue distribution of the multidrug resistance-associated protein. Am. J. Pathol. 148, 1237–1247.

Gao, M., Loe, D.W., Grant, C.E., Cole, S.P.C. and Deeley, R.G. (1996) Reconstitution of ATP-dependent LTC_4 transport by co-expression of both half-molecules of human MRP in insect cells. J. Biol. Chem. 271, 27782–27787.

Gao, M., Yamazaki, M., Loe, D.W., Westlake, C.J., Grant, C.E., Cole, S.P.C. and Deeley, R.G. (1998) Multidrug Resistance Protein: Identification of regions required for active transport of leukotriene C_4. J. Biol. Chem. 273, 10733–10740.

Gao, M., Cui, H.-R., Loe, D.W., Grant, C.E., Almquist, K.C., Cole, S.P.C. and Deeley, R.G. (2000) Comparison of the functional characteristics of the nucleotide binding domains of multidrug resistance protein 1. J. Biol. Chem. 275, 13098–13108.

Goto, H., Keshelava, N., Matthay, K.K., Lukens, J.N., Gerbing, R.B., Stram, D.O., Seeger, R.C. and Reynolds, C.P. (2000) Multidrug resistance-associated protein 1 (MRP1) expression in neuroblastoma cell lines and primary tumors. Med. Pediatr. Oncol. 35, 619–622.

Grant, C.E., Valdimarsson, G., Hipfner, D.R., Almquist, K.C., Cole, S.P.C. and Deeley, R.G. (1994) Overexpression of multidrug resistance-associated protein (MRP) increases resistance to natural product drugs. Cancer Res. 54, 357–361.

Grant, C.E., Kurz, E.U., Cole, S.P.C. and Deeley, R.G. (1997) Analysis of the intron-exon organization of the multidrug resistance protein (MRP) gene and alternative splicing of its mRNA. Genomics 45, 368–378.

Heijn, M., Hoojiberg, J.H., Scheffer, G.L., Szabo, G., Westerhoff, H.V. and Lankelma, J. (1997) Anthracyclines modulate multidrug resistance protein (MRP) mediated organic anion transport. Biochim. Biophys. Acta 1326, 12–22.

Hipfner, D.R., Gauldie, S.D., Deeley, R.G. and Cole, S.P.C. (1994) Detection of the M_r 190,000 multidrug resistance protein, MRP, with monoclonal antibodies. Cancer Res. 54, 5788–5792.

Hipfner, D.R., Almquist, K.C., Stride, B.D., Deeley, R.G. and Cole, S.P.C. (1996) Location of a protease- hypersensitive region in the Multidrug Resistance Protein (MRP) by mapping of the epitope of MRP-specific monoclonal antibody QCRL-1. Cancer Res. 56: 3307–3314.

Hipfner, D.R., Almquist, K.C., Leslie, E.M., Gerlach, J.H., Grant, C.E., Deeley, R.G. and Cole, S.P.C. (1997) Membrane topology of the multidrug resistance protein, MRP: a study of glycosylation-site mutants reveals an extracytosolic NH_2-terminus. J. Biol. Chem. 272, 23623–23630.

Hipfner, D.R., Gao, M., Scheffer, G., Scheper, R., Deeley, R.G. and Cole, S.P.C. (1998) Epitope mapping of monoclonal antibodies specific for the 190-kDa multidrug resistance protein (MRP). Br. J. Cancer 78, 1134–1140.

Hipfner, D.R., Deeley, R.G. and Cole, S.P.C. (1999a) Structural, mechanistic and clinical aspects of MRP1. Biochim. Biophys. Acta 1461, 359–376.

Hipfner, D.R., Mao, Q., Qiu, W., Leslie, E.M., Deeley, R.G. and Cole, S.P.C. (1999b)

Monoclonal antibodies that inhibit the transport function of the 190 kDa multidrug resistance protein, MRP: localization of their epitopes to the nucleotide binding domains of the protein. J. Biol. Chem. 274, 15420–15426.

Hirohashi, T., Suzuki, H., Takikawa, H. and Sugiyama, Y. (2000) ATP-dependent transport of bile salts by rat multidrug resistance-asssociated protein 3 (Mrp3). J. Biol. Chem. 275, 2905–2910.

Hoffmeyer, S., Burk, O., von Richter, O., Arnold, H.P., Brockmoller, J., Johne, A., *et al.* (2000) Functional polymorphisms of the human multidrug resistance gene: multiple sequence variations and correlation of one allele with P-glycoprotein expression and activity *in vivo*. Proc. Natl Acad. Sci. USA 97, 3473–3478.

Hollo, Z., Homolya, L., Davis, C.W. and Sarkadi, B. (1994) Calcein accumulation as a fluorometric functional assay of the multidrug transporter. Biochim. Biophys. Acta 1191, 384–388.

Hooijberg, J.H., Broxterman, H.J., Heijn, M., Fles, D.L.A., Lankelma, J. and Pinedo, H.M. (1997) Modulation by (iso)flavonoids of the ATPase activity of the multidrug resistance protein. FEBS Lett. 413, 344–348.

Hooijberg, J.H., Broxterman, H.J., Kool, M., Assaraf, Y.G., Peters, G.J., Scheper, R.J., Borst, P., Pinedo, H.M. and Jansen, G. (1999) Antifolate resistance mediated by the multidrug resistance proteins MRP1 and MRP2. Cancer Res. 59, 2532–2535.

Hooijberg, J.H., Pinedo, H.M., Vrasdonk, C., Priebe, W., Lankelma, J. and Broxterman, H.J. (2000) The effect of glutathione on the ATPase activity of MRP1 in its natural membranes. FEBS Lett. 469, 47–51.

Hou, Y.-X., Cui, L., Riordan, J.R. and Chang, X.-B. (2000) Allosteric interactions between the two non-equivalent nucleotide binding domains of multidrug resistance protein MRP1. J. Biol. Chem. 275, 20280–20287.

Hrycyna, C.A., Ramachandra, M., Ambudkar, S.V., Ko, Y.H., Peebles, C.L., Pastan, I. and Gottesman, M.M. (1998) Mechanism of action of human P-glycoprotein ATPase activity. Photochemical cleavage during a catalytic transition state using orthovanadate reveals cross-talk between the two ATP sites. J. Biol. Chem. 273, 16631–16634.

Ito, K., Oleschuk, C.J., Westlake, C.J., Vasa, M., Deeley, R.G. and Cole, S.P.C. (2001a) Mutation of Trp1254 in the multispecific organic anion transporter, MRP2 (ABCC2), alters substrate specificity and results in loss of methotrexate transport activity. J. Biol. Chem. 276, 38108–38114.

Ito, K., Olsen, S.L., Qiu, W., Deeley, R.G. and Cole, S.P.C. (2001b) Mutation of a single conserved tryptophan in multidrug resistance protein 1 (MRP1/ABCC1) results in loss of drug resistance and selective loss of organic anion transport. J. Biol. Chem. 276, 15616–15624.

Ito, S., Ieiri, I., Tanabe, M., Suzuki, A., Higuchi, S. and Otsubo, K. (2001c) Polymorphism of the ABC transporter genes, *MDR1*, *MRP1* and *MRP2/cMOAT*, in healthy Japanese subjects. Pharmacogenetics 11, 175–184.

Jedlitschky, G., Leier, I., Buchholz, U., Center, M. and Keppler, D. (1994) ATP-dependent transport of glutathione *S*-conjugates by the multidrug resistance-associated protein. Cancer Res. 54, 4833–4836.

Jedlitschky, G., Leier, I., Buchholz, U., Barnouin, K., Kurz, G. and Keppler, D. (1996) Transport of glutathione, glucuronate, and sulfate conjugates by the *MRP* gene-encoded conjugate export pump. Cancer Res. 56, 988–994.

Jedlitschky, G., Leier, I., Buchholz, U., Hummel-Eisenbeiss, J., Burchell, B. and Keppler, D. (1997) ATP-dependent transport of bilirubin glucuronides by the multidrug resistance protein MRP1 and its hepatocytes canalicular isoform MRP2. Biochem. J. 327, 305–310.

Juliano, R.L. and Ling, V. (1976) A surface glycoprotein modulating drug permeability in Chinese hamster ovary cell mutants. Biochim. Biophys. Acta 455, 152–162.

Kartenbeck, J., Leuschner, U., Mayer, R. and Keppler, D. (1996) Absence of the canalicular isoform of the *MRP* gene-encoded conjugate export pump from the hepatocytes in Dubin-Johnson syndrome. Hepatology 23, 1061–1066.

Kast, C. and Gros, P. (1997) Topology mapping of the amino-terminal half of multidrug resistance-associated protein by epitope insertion and immunofluorescence. J. Biol. Chem. 272, 26479–26487.

Kast, C. and Gros, P. (1998) Epitope insertion favors a six transmembrane domain model for the carboxy-terminal portion of the multidrug resistance-associated protein. Biochemistry 37, 2305–2313.

Keppler, D., Jedlitschky, G. and Leier, I. (1998) Transport function and substrate specificity of multidrug resistance protein. Methods Enzymol. 292, 607–616.

Kim, R., Nishimoto, N., Osaki, A., Kojima, J., Fujii, S., Inai, K. and Togo, T. (2001) An analysis of relapsed breast cancer in patients previously treated with breast conserving surgery. Breast Cancer 8, 146–152.

Konig, J., Nies, A.T., Cui, Y., Leier, I. and Keppler, D. (1999) Conjugate export pumps of the multidrug resistance protein (MRP) family: localization, substrate specificity and MRP2-mediated drug resistance. Biochim. Biophys. Acta 1461, 377–394.

Konig, J., Cui, Y., Nies, A.T. and Keppler, D. (2000) Localization and genomic organization of a new hepatocellular organic anion transporting polypeptide. J. Biol. Chem. 275, 23161–23168.

Kool, M., van der Linden, M., de Haas, M., Baas, F. and Borst, P. (1999) Expression of human *MRP6*, a homologue of the multidrug resistance protein gene *MRP1*, in tissues and cancer cells. Cancer Res. 59, 175–182.

Kreisholt, J., Sorensen, M., Jensen, P.B., Nielsen, B.E., Andersen, C.B. and Sehested, M. (1998) Immunohistochemical detection of DNA topoisomerase IIα, P-glycoprotein and multidrug resistance protein (MRP) in small cell and non-small cell lung cancer. Br. J. Cancer 77, 1469–1473.

Kurz, E.U., Cole, S.P.C. and Deeley, R.G. (2001) Identification of DNA-protein interactions in the 5′ flanking and 5′ untranslated regions of the human multidrug resistance protein (MRP) gene: Evaluation of a putative antioxidant response element. Biochem. Biophys. Res. Commun. 285, 981–990.

Lautier, D., Canitrot, Y., Deeley, R.G. and Cole, S.P.C. (1996) Multidrug resistance mediated by the Multidrug Resistance Protein (*MRP*) gene. Biochem. Pharmacol. 52, 967–977.

Legare, D., Richard, D., Mukhopadhyay, R., Stierhof, Y.D., Rosen, B.P., Haimeur, A., Papadopoulou, B. and Ouellette, M. (2001) The *Leishmania* ATP-binding cassette protein PGPA is an intracellular metal-thiol transporter ATPase. J. Biol. Chem. 276, 26301–26307.

Legrand, O., Perrot, J.-Y., Tang, R., Simonin, G., Gurbuxani, S., Zittoun, R. and Marie, J.-P. (1996) Expression of the multidrug resistance-associated protein (MRP) mRNA and protein in normal peripheral blood and bone marrow haemopoietic cells. Br. J. Haematol. 94, 23–33.

Leier, I., Jedlitschky, G., Buchholz, U. and Keppler, D. (1994) Characterization of the ATP-dependent leukotriene C_4 export carrier in mastocytoma cells. Eur. J. Biochem. 220, 599–606.

Leier, I., Jedlitschky, G., Buchholz, U., Center, M., Cole, S.P.C., Deeley, R.G. and Keppler, D. (1996) ATP-dependent glutathione disulfide transport mediated by the *MRP* gene-encoded conjugate export pump. Biochem. J. 314, 433–437.

Leslie, E.M., Deeley, R.G. and Cole, S.P.C. (2001a) Toxicological relevance of the multidrug resistance protein 1, MRP1 (ABCC1), and related transporters. Toxicology 167, 3–23.

Leslie, E.M., Ito, K., Upadhyaya, P., Hecht, S.S., Deeley, R.G. and Cole, S.P.C. (2001b) Transport of the β-*O*-glucuronide conjugate of the tobacco-specific carcinogen 4-(methylnitrosamino)-1-(3-pyridyl)-1-butanol (NNAL) by the multidrug resistance protein 1 (MRP1): Requirement for glutathione or a non-sulfur-containing analog. J. Biol. Chem. 276, 27846–27854.

Leslie, E.M., Mao, Q., Oleschuk, C.J., Deeley, R.G. and Cole, S.P.C. (2001c) Modulation of multidrug resistance protein 1 (MRP1/ABCC1) transport and ATPase activities by interaction with dietary flavonoids. Mol. Pharmacol. 59, 1171–1180.

Li, Z., Szczypka, M.S., Lu, Y., Thiele, D.J. and Rea, P.A. (1996) The yeast cadmium factor (YCF1) is a vacuolar glutathione *S*-conjugate pump. J. Biol. Chem. 271, 6509–6517.

Loe, D.W., Almquist, K.C., Cole, S.P.C. and Deeley, R.G. (1996a) ATP-dependent 17β-estradiol 17-(β-D-glucuronide) transport by multidrug resistance protein: inhibition by cholestatic steroids. J. Biol. Chem. 271, 9683–9689.

Loe, D.W., Almquist, K.C., Deeley, R.G. and Cole, S.P.C. (1996b) Multidrug Resistance Protein (MRP)-mediated transport of leukotriene C_4 and chemotherapeutic agents in membrane vesicles: Demonstration of glutathione-dependent vincristine transport. J. Biol. Chem. 271, 9675–9682.

Loe, D.W., Stewart, R.K., Massey, T.E., Deeley, R.G. and Cole, S.P.C. (1997) ATP-dependent transport of aflatoxin B_1 and its glutathione conjugates by the product of the *MRP* gene. Mol. Pharmacol. 51, 1034–1041.

Loe, D.W., Deeley, R.G. and Cole, S.P.C. (1998) Characterization of vincristine transport by the 190 kDa multidrug resistance protein, MRP: Evidence for co-transport with reduced glutathione. Cancer Res. 58, 5130–5136.

Loe, D.W., Deeley, R.G. and Cole, S.P.C. (2000a) Verapamil stimulates glutathione transport by the multidrug resistance protein 1 (MRP1). J. Pharmacol. Exp. Therap. 293, 530–538.

Loe, D.W., Oleschuk, C.J., Deeley, R.G. and Cole, S.P.C. (2000b) Structure-activity studies of verapamil analogs that modulate transport of leukotriene C_4 and reduced glutathione by multidrug resistance protein MRP1. Biochem. Biophys. Res. Commun. 275, 795–803.

Lohoff, M., Prechtl, S., Sommer, F., Roellinghoff, M., Schmitt, E., Gradehandt, G., Rohwer, P., Stride, B.D., Cole, S.P.C. and Deeley, R.G. (1998) A multidrug-resistance protein (MRP)-like transmembrane pump is highly expressed by resting murine T helper (Th) 2, but not Th1 cells, and is induced to equal expression levels in Th1 and Th2 cells after antigenic stimulation *in vivo*. J. Clin. Invest. 101, 703–710.

Lorico, A., Rappa, G., Flavell, R.A. and Sartorelli, A.C. (1996) Double knockout of the *MRP* gene leads to increased drug sensitivity *in vitro*. Cancer Res. 56, 5351–5355.

Lorico, A., Rappa, G., Finch, R.A., Yang, D., Flavell, R.A. and Sartorelli, A.C. (1997) Disruption of the murine *MRP* (multidrug resistance protein) gene leads to increased sensitivity to etoposide (VP-16) and increased levels of glutathione. Cancer Res. 57, 5238–5242.

Ma, L., Krishnamachary, N. and Center, M.S. (1995) Phosphorylation of the multidrug resistance associated protein gene encoded protein P190. Biochemistry 34, 3338–3343.

Manciu, L., Chang, X.-B., Riordan, J.R. and Ruysschaert, J.M. (2000) Multidrug resistance protein MRP1 reconstituted into lipid vesicles: secondary structure and nucleotide-induced tertiary structure changes. Biochemistry 39, 13026–13033.

Mao, Q., Leslie, E.M., Deeley, R.G. and Cole, S.P.C. (1999) ATPase activity of purified and reconstituted multidrug resistance protein MRP1 from drug-selected H69AR cells. Biochim. Biophys. Acta 1461, 69–82.

Mao, Q., Deeley, R.G. and Cole, S.P.C. (2000) Functional reconstitution of substrate transport by purified multidrug resistance protein MRP1 (ABCC1) in phopholipid vesicles. J. Biol. Chem. 275, 34165–34172.

Martinoia, E., Grill, E., Tommasini, R., Kreuz, K. and Amrhein, N. (1993) ATP-dependent glutathione *S*-conjugate 'export' pump in the vacuolar membrane of plants. Nature 364, 247–249.

Massey, T.E., Stewart, R.K., Daniels, J.M. and Liu, L. (1995) Biochemical and molecular aspects of mammalian susceptibility to aflatoxin B_1 carcinogenicity. Proc. Soc. Exp. Biol. Med. 208, 1–15.

Mirski, S.E.L., Gerlach, J.H. and Cole, S.P.C. (1987) Multidrug resistance in a human small cell lung cancer cell line selected in adriamycin. Cancer Res. 47, 2594–2598.

Miyake, K., Mickley, L., Litman, T., Zhan, Z., Robey, R., Cristensen, B., *et al.* (1999) Molecular cloning of cDNAs which are highly overexpressed in mitoxantrone-resistant cells: demonstration of homology to ABC transport genes. Cancer Res. 59, 8–13.

Morrow, C.S., Diah, S., Smitherman, P.K., Schneider, E. and Townsend, A.J. (1998) Multidrug resistance protein and glutathione *S*-transferase P1-1 act in synergy to confer protection from 4-nitroquinoline 1-oxide toxicity. Carcinogenesis 19, 109–115.

Muller, M., Meijer, C., Zaman, G.J.R., Borst, P., Scheper, R.J., Mulder, N.H., de Vries, E.G.E. and Jansen, P.L.M. (1994) Overexpression of the gene encoding the multidrug resistance-associated protein results in increased ATP-dependent glutathione *S*-conjugate transport. Proc. Natl Acad. Sci. USA 91, 13033–13037.

Murphy, S.E., Spina, D.A., Nunes, M.G. and Pullo, D.A. (1995) Glucuronidation of 4-((hydroxymethyl)nitrosamino)-1-(3-pyridyl)-1-butanone, a metabolically activated form of 4-(methylnitrosamino)-1-(3-pyridyl)-1-butanone, by phenobarbital-treated rats. Chem. Res. Toxicol. 8, 772–779.

Nagata, K., Nishitani, M., Matsuo, M., Kioka, N., Amachi, T. and Ueda, K. (2000) Nonequivalent nucleotide trapping in the two nucleotide binding folds of the human multidrug resistance protein MRP1. J. Biol. Chem. 275, 17626–17630.

Nooter, K., Brutel de la Riviere, G., Klijn, J.G.M., Stoter, G. and Foekens, J.A. (1997a) Multidrug resistance protein in recurrent breast cancer. Lancet 349, 1885–1886.

Nooter, K., Brutel de la Riviere, G., Look, M.P., van Wingerden, K.E., Henzen-Logmans, S.C., Scheper, R.J., Flens, M.J., Klijn, J.G.M., Stoter, G. and Foekens, J.A. (1997b) The prognostic significance of expression of the multidrug resistance-associated protein (MRP) in primary breast cancer. Br. J. Cancer 76, 486–493.

Norris, M.D., Bordow, S.B., Marshall, G.M., Haber, P.S., Cohn, S.L. and Haber, M. (1996) Expression of the gene for multidrug-resistance-associated protein and outcome in patients with neuroblastoma. N. Engl. J. Med. 334, 231–238.

Ostlund Farrants, A.-K., Meyer, D.J., Coles, B., Southan, C., Aitken, A., Johnson, P.J. and Ketterer, B. (1987) The separation of glutathione transferase subunits by using reverse-phase high-pressure liquid chromatography. Biochem. J. 245, 423–428.

Ota, E., Abe, Y., Oshika, Y., Ozeki, Y., Iwasaki, M., Inoue, H., *et al.* (1995) Expression of the multidrug resistance-associated protein (*MRP*) gene in non-small-cell lung cancer. Br. J. Cancer 72, 550–554.

Paulusma, C.C., Kool, M., Bosma, P.J., Scheffer, G.L., ter Borg, F., Scheper, R.J., Tytgat, G.N.J., Borst, P., Baas, F. and Oude Elferink, R.P.J. (1997) A mutation in the human canalicular multispecific organic anion transporter gene causes the Dubin-Johnson syndrome. Hepatology 25, 1539–1542.

Paulusma, C.C., van Geer, M.A., Evers, R., Heijn, M., Ottenhoff, R., Borst, P. and Oude Elferink, R.P.J. (1999) Canalicular multispecific organic anion transporter/multidrug resistance protein 2 mediates low-affinity transport of reduced glutathione. Biochem. J. 338, 393–401.

Paumi, C.M., Ledford, B.G., Smitherman, P.K., Townsend, A.J. and Morrow, C.S. (2001) Role of the multidrug resistance protein 1 (MRP1) and glutathione *S*-transferase A1-1 in alkylating agent resistance: kinetics of gluathione conjugate formation and efflux govern differential cellular sensitivity to chlorambucil versus melphalan toxicity. J. Biol. Chem. 276, 7952–7956.

Peng, K.C., Cluzeaud, F., Bens, M., Van Huyen, J.P., Wioland, M.A., Lacave, R. and Vandewalle, A. (1999) Tissue and cell distribution of the multidrug resistance-associated protein (MRP) in mouse intestine and kidney. J. Histochem. Cytochem. 47, 757–768.

Priebe, W., Krawczyk, M., Kuo, M.T., Yamane, Y., Savaraj, N. and Ishikawa, T. (1998) Doxorubicin- and daunorubicin-glutathione conjugates, but not unconjugated drugs, competitively inhibit leukotriene C4 transport mediated by MRP/GS-X pumpl. Biochem. Biophys. Res. Commun. 247, 859–863.

Qian, Y.M., Qiu, W., Gao, M., Westlake, C.J., Cole, S.P.C. and Deeley, R.G. (2001a) Characterization of binding of leukotriene C_4 by human multidrug resistance protein 1: evidence of differential interactions with NH_2- and COOH-proximal halves of the protein. J. Biol. Chem. 276, 38636–38644.

Qian, Y.M., Song, W.C., Cui, H.-R., Cole, S.P.C. and Deeley, R.G. (2001b) Glutathione stimulate sulfated estrogen transport by multidrug resistance protein 1. J. Biol. Chem. 276, 6404–6411.

Ramachandra, M., Ambudkar, S.V., Chen, D., Hrycyna, C.A., Dey, S., Gottesman, M.M. and Pastan, I. (1998) Human P-glycoprotein exhibits reduced affinity for substrates during a catalytic transition state. Biochemistry 37, 5010–5019.

Rappa, G., Finch, R.A., Sartorelli, A.C. and Lorico, A. (1999) New insights into the biology and pharmacology of the multidrug resistance protein (MRP) from gene knockout models. Biochem. Pharmacol. 58, 557–562.

Rea, P.A., Li, Z., Lu, Y., Drozdowicz, Y.M. and Martinoia, E. (1998) From vacuolar GS-X pumps to multispecific ABC transporters. Annu. Rev. Plant Physiol. Plant Mol. Biol. 49, 727–760.

Reeve, J.G., Rabbitts, P.H. and Twentman, P.R. (1990) Non-P-glycoprotein-mediated multidrug resistance with reduced EGF receptor expression in a human large cell lung cancer cell line. Br. J. Cancer 61, 851–855.

Ren, X.Q., Furukawa, T., Chen, Z., Okumura, H., Aoki, S., Sumizawa, T., Tani, A., Komatsu, M., Mei, X.D. and Akiyama, S. (2000) Functional comparison between YCF1 and MRP1 expressed in Sf21 insect cells. Biochem. Biophys. Res. Commun. 270, 608–615.

Ren, X.Q., Furukawa, T., Aoki, S., Nakajima, T., Sumizawa, T., Haraguchi, M., Chen, Z., Kobayashi, M. and Akiyama, S. (2001) Glutathione-dependent binding of a photoaffinity analog of Agosterol A to the C-terminal half of human multidrug resistance protein. J. Biol. Chem. 276, 23197–23206.

Renes, J., de Vries, E.G.E., Nienhuis, E.F., Jansen, P.L.M. and Muller, M. (1999) ATP- and glutathione-dependent transport of chemotherapeutic drugs by the multidrug resistance protein MRP1. Br. J. Pharmacol. 126, 681–688.

Renes, J., de Vries, E.G.E., Hooiveld, G.J.E.J., Krikken, I., Jansen, P.L.M. and Muller, M. (2000) Multidrug resistance protein MRP1 protects against the toxicity of the major lipid peroxidation product 4-hydroxynonenal. Biochem. J. 350, 555–561.

Ringpgeil, F., Lebwohl, M.G., Christiano, A.M. and Uitto, J. (2000) Pseudoxanthoma elasticum: Mutations in the *MRP6* gene encoding a transmembrane ATP-binding cassette (ABC) transporter. Proc. Natl Acad. Sci. USA 97, 6001–6006.

Roelofsen, H., Vos, T.A., Schippers, I.J., Kuipers, F., Koning, H., Moshage, H., Jansen, P.L.M. and Muller, M. (1997) Increased levels of the multidrug resistance protein in lateral membranes of proliferating hepatocyte-derived cells. Gastroenterology 112, 511–521.

Rosenberg, M.F., Callaghan, R., Ford, R.C. and Higgins, C.F. (1997) Structure of the multidrug resistance P-glycoprotein to 2.5 nm resolution determined by electron microscopy and image analysis. J. Biol. Chem. 272, 10685–10694.

Rosenberg, M.F., Mao, Q., Holzenburg, A., Ford, R.C., Deeley, R.G. and Cole, S.P.C. (2001) The structure of the multidrug resistance protein 1 (MRP1/ABCC1): crystallization and single-particle analysis. J. Biol. Chem. 276, 16076–16082.

Saitou, N. and Nei, M. (1987) The neighbour-joining method: a new method for reconstruction of phylogenetic trees. Mol. Biol. Evol. 4, 406–425.

Sakamoto, H., Hara, H., Hirano, K. and Adachi, T. (1999) Enhancement of glucuronosyl etoposide transport by glutathione in multidrug resistance-associated protein-overexpressing cells. Cancer Lett. 135, 113–119.

Schultz, M.J., Wijnholds, J., Peppelenbosch, M.P., Vervoordeldonk, M.J.B.M., Speelman, P., van Deventer, S.J.H., Borst, P. and van der Poll, T. (2001) Mice lacking the multidrug resistance protein 1 are resistant to *Streptococcus pneumoniae*-induced pneumonia. J. Immunol. 166, 4059–4064.

Senior, A.E., Al-Shawi, M.K. and Urbatsch, I.L. (1995) The catalytic cycle of P-glycoprotein. FEBS Lett. 377, 285–289.

Sethna, S.S., Gander, J.E. and Rathbun, W.B. (1984) Glutathione synthetase of bovine lens: anomalies of the enzyme-catalyzed formation of ophthalmic acid. Curr. Eye Res. 3, 923–928.

Shapiro, A.B. and Ling, V. (1997) Positively cooperative sites for drug transport by P-glycoprotein with distinct drug specificities. Eur. J. Biochem. 250, 130–137.

Sharom, F.J., Liu, R., Romsicki, Y. and Lu, P. (1999) Insights into the structure and substrate interactions of the P-glycoprotein multidrug transporter from spectroscopic studies. Biochim. Biophys. Acta 1461, 327–345.

Slapak, C.A., Daniel, J.C. and Levy, S.B. (1990) Sequential emergence of distinct resistance phenotypes in murine erythroleukemia cells under adriamycin selection: decreased anthracycline uptake precedes increased P-glycoprotein expression. Cancer Res. 50, 7895–7901.

Slovak, M.L., Ho, J.P., Bhardwaj, G., Kurz, E.U., Deeley, R.G. and Cole, S.P.C. (1993) Localization of a novel multidrug resistance-associated gene in the HT1080/DR4 and H69AR human tumor cell lines. Cancer Res. 53, 3221–3225.

Soszynski, M., Kaluzna, A., Rychlik, B., Sokal, A. and Bartosz, G. (1998) Radiation inactivation suggests that human multidrug resistance-associated protein 1 occurs as a dimer in the human erythrocyte membrane. Arch. Biochem. Biophys. 354, 311–316.

St Pierre, M.V., Serrano, M.E., Macias, R.I.R., Dubs, U., Hoechli, M., Lauper, U., Meier, P.J. and Marin, J.J.G. (2000) Expression of members of the multidrug resistance protein family in human term placenta. Am. J. Physiol. 279, R1495–R1503.

Stride, B.D., Valdimarsson, G., Gerlach, J.H., Wilson, G.M., Cole, S.P.C. and Deeley, R.G. (1996) Structure and expression of the mRNA encoding the murine multidrug resistance protein (MRP), an ATP-binding cassette transporter. Mol. Pharmacol. 49, 962–971.

Stride, B.D., Grant, C.E., Loe, D.W., Hipfner, D.R., Cole, S.P.C. and Deeley, R.G. (1997) Pharmacological characterization of the murine and human orthologs of multidrug resistance protein in transfected human embryonic kidney cells. Mol. Pharmacol. 52, 344–353.

Stride, B.D., Cole, S.P.C. and Deeley, R.G. (1999) Localization of a substrate specificity domain in the multidrug resistance protein. J. Biol. Chem. 274, 22877–22883.

Sullivan, G.F., Amenta, P.S., Villanueva, J.D., Alvarez, C.J., Yang, J. and Hait, W.N. (1998) The expression of drug resistance gene products during the progression of human prostate cancer. Clin. Cancer Res. 4, 1393–1403.

Szabo, K., Welker, E., Bakos, E., Muller, M., Roninson, I., Varadi, A. and Sarkadi, B. (1998) Drug-stimulated nucleotide trapping in the human multidrug transporter MDR1.

Cooperation of the nucleotide binding domains. J. Biol. Chem. 273, 10132–10138.

Tommasini, R., Evers, R., Vogt, E., Mornet, C., Zaman, G.J.R., Schinkel, A.H., Borst, P. and Martinoia, E. (1996) The human multidrug resistance-associated protein functionally complements the yeast cadmium resistance factor 1. Proc. Natl Acad. Sci. USA 93, 6743–6748.

Ueda, K., Taguchi, Y. and Morishima, M. (1997) How does P-glycoprotein recognize its substrates? Semin. Cancer Biol. 8, 151–159.

van der Kolk, D.M., de Vries, E.G.E., van Putten, W.J., Verdonck, L.F., Ossenkoppele, G.J., Verhoef, G.E. and Vellenga, E. (2000) P-glycoprotein and multidrug resistance protein activities in relation to treatment outcome in acute myeloid leukemia. Clin. Cancer Res. 6, 3205–3214.

van Luyn, M.J.A., Muller, M., Renes, J., Meijer, C., Scheper, R.J., Nienhuis, E.F., Mulder, N.H., Jansen, P.L.M. and de Vries, E.G.E. (1998) Transport of glutathione conjugates into secretory vesicles is mediated by the multidrug-resistance protein 1. Int. J. Cancer 76, 55–62.

van Veen, H.W., Margolles, A., Muller, M., Higgins, C.F. and Konings, W.N. (2000) The homodimeric ATP-binding cassette transporter LmrA mediates multidrug transport by an alternating two-site (two-cylinder engine) mechanism. EMBO J. 19, 2503–2514.

Vernhet, L., Courtois, A., Allain, N., Payen, L., Anger, J.-P., Guillouzo, A. and Fardel, O. (1999) Overexpression of the multidrug resistance-associated protein (MRP1) in human heavy metal-selected tumor cells. FEBS Lett. 443, 321–325.

Versantvoort, C.H.M., Bagrij, T., Wright, K.A. and Twentyman, P.R. (1995a) On the relationship between the probenecid-sensitive transport of daunorubicin or calcein and the glutathione status of cells overexpressing the multidrug resistance-associated protein (MRP). Int. J. Cancer 63, 855–862.

Versantvoort, C.H.M., Broxterman, H.J., Bagrij, T., Scheper, R.J. and Twentyman, P.R. (1995b) Regulation by glutathione of drug transport in multidrug-resistant human lung tumour cell lines overexpressing multidrug resistance-associated protein. Br. J. Cancer 72, 82–89.

Vos, T.A., Hooiveld, G.J.E.J., Koning, H., Childs, S., Meijer, D.K.F., Moshage, H., Jansen, P.L.M. and Muller, M. (1998) Up-regulation of the multidrug resistance genes, *Mrp1* and *Mdr1b*, and down-regulation of the organic anion transporter, Mrp2, and the bile salt transporter, Spgp, in endotoxemic rat liver. Hepatology 28, 1637–1644.

Wijnholds, J., Evers, R., Van Leusden, M.R., Mol, C.A.A.M., Zaman, G.J.R., Mayer, U., Beijnen, J.H., van der Valk, M., Krimpenfort, P. and Borst, P. (1997) Increased sensitivity to anticancer drugs and decreased inflammatory response in mice lacking the multidrug resistance-associated protein. Nat. Med. 3, 1275–1279.

Wijnholds, J., Scheffer, G.L., van der Valk, M., van der Valk, P., Beijnen, J., Scheper, R.J. and Borst, P. (1998) Multidrug resistance protein 1 protects the oropharyngeal mucosal layer and the testicular tubules against drug-induced damage. J. Exp. Med. 188, 797–808.

Wijnholds, J., de Lange, E.C.M., Scheffer, G.L., van den Berg, D.-J., Mol, C.A.A.M., van der Valk, M., Schinkel, A.H., Scheper, R.J., Breimer, D.D. and Borst, P. (2000) Multidrug resistance protein 1 protects the choroid plexus epithelium and contributes to the blood-cerebrospinal fluid barrier. J. Clin. Invest. 105, 279–285.

Wright, S.R., Boag, A.H., Campling, B.G., Valdimarsson, G., Hipfner, D.R., Cole, S.P.C. and Deeley, R.G. (1998) Immunohistochemical detection of multidrug resistance protein (MRP) in human lung cancer and normal lung. Clin. Cancer Res. 4, 2279–2289.

Yamane, Y., Furuichi, M., Song, R., Van, N.T., Mulcahy, R.T., Ishikawa, T. and Kuo, M.T. (1998) Expression of multidrug resistance protein/GS-X pump and gamma-glutamylcysteine synthetase genes is regulated by oxidative stress. J. Biol. Chem. 273, 31075–31085.

Zaman, G.J.R., Lankelma, J., van Tellingen, O., Beijnen, J., Dekker, H., Paulusma, C., Oude Elferink, R.P.J., Baas, F. and Borst, P. (1996) Role of glutathione in the export of compounds from cells by the multidrug-resistance-associated protein. Proc. Natl. Acad. Sci. USA 92, 7690–7694.

Zeng, H., Chen, Z.-S., Belinsky, M.G., Rea, P.A. and Kruh, G.D. (2001) Transport of methotrexate (MTX) and folates by multidrug resistance protein (MRP) 3 and MRP1: effect of polyglutamylation on MTX transport. Cancer Res. 61, 7225–7232.

Zhang, D., Cole, S.P.C. and Deeley, R.G. (2001a) Identification of a non-conserved amino acid residue in Multidrug Resistance

Protein (MRP) 1 important for determining substrate specificity: evidence for functional interaction between transmembrane helices 14 and 17. J. Biol. Chem. 276, 34966–34974.

Zhang, D., Cole, S.P.C. and Deeley, R.G. (2001b) Identification of an amino acid residue in multidrug resistance protein (MRP) 1 critical for conferring resistance to anthracyclines. J. Biol. Chem. 276, 13231–13239.

Zhu, Q. and Center, M.S. (1994) Cloning and sequence analysis of the promoter region of the *MRP* gene of HL60 cells isolated for resistance to Adriamycin. Cancer Res. 54, 4488–4492.

Zhu, Q. and Center, M.S. (1996) Evidence that SP1 modulates transcriptional activity of the multidrug resistance-associated protein gene. DNA Cell Biol. 15, 105–111.

MRP2, THE APICAL EXPORT PUMP FOR ANIONIC CONJUGATES

JÖRG KÖNIG, ANNE T. NIES, YUNHAI CUI AND DIETRICH KEPPLER

CHAPTER 20

INTRODUCTION

The ATP-dependent unidirectional transport of anionic conjugates, such as bilirubin glucuronosides and leukotriene C_4 (LTC_4), across the apical membrane domain of polarized cells plays an important role in the elimination and detoxification of endogenous and xenobiotic substances. This process has been functionally characterized by measurements of ATP-dependent transport of labeled conjugates into inside-out membrane vesicles prepared from the apical membranes of hepatocytes (Ishikawa *et al.*, 1990; Kitamura *et al.*, 1990; Kobayashi *et al.*, 1988). This transport function was originally described as a glutathione *S*-conjugate transport system (Ishikawa *et al.*, 1990) and as a canalicular multispecific organic anion transporter (abbreviated cMOAT) (Oude Elferink *et al.*, 1993). Subsequent cloning, expression and functional analysis of the recombinant protein has established that the apical conjugate export pump is encoded by the *MRP2* (*ABCC2*) gene (Büchler *et al.*, 1996; Cui *et al.*, 1999; Evers *et al.*, 1998; Paulusma *et al.*, 1996; Taniguchi *et al.*, 1996). Antibodies raised against various epitopes of MRP2 from several species, including human, monkey, dog, rabbit, rat and mouse, served to localize the MRP2 glycoprotein to the apical membrane of polarized cells, including hepatocytes (Büchler *et al.*, 1996; Keppler and Kartenbeck, 1996), kidney proximal tubules (Schaub *et al.*, 1997, 1999), intestinal epithelia (Fromm *et al.*, 2000; Mottino *et al.*, 2000; van Aubel *et al.*, 2000), gallbladder (Rost *et al.*, 2001) and lung. The apical localization of MRP2 and its broad substrate specificity for various conjugates qualify this ATP-binding cassette (ABC) transporter as an important terminal component in detoxification, subsequent to the phase I and phase II reactions of xenobiotic metabolism. The latter is comprised predominantly of cytochrome P450-catalyzed oxidations and conjugation reactions catalyzed by various transferases. Hepatocytes and kidney proximal tubule epithelia are the major sites for detoxification and excretion of xenobiotics, and in both cell types MRP2 contributes to the vectorial transport of these substances. In addition, in the liver, MRP2 contributes to the bile-salt-independent bile flow, as evidenced by the strong reduction in bile flow in mutant rats lacking the Mrp2 protein (Jansen *et al.*, 1985; Keppler and König, 1997). In the intestine, the apical localization of MRP2 may counteract the entry of toxic or carcinogenic MRP2 substrates from the intestinal lumen into the epithelia and into the blood circulation (Dietrich *et al.*, 2001). Thus, in the intestinal tract, MRP2 may have a similar protective role as proposed previously for MDR1 P-glycoprotein (Benet *et al.*, 1999). In addition to this protective role, MRP2 has been shown directly to confer resistance to several chemotherapeutic agents including cisplatin (Cui *et al.*, 1999).

A hereditary defect of the hepatobiliary elimination of anionic conjugates has long been

ABC Proteins: From Bacteria to Man
ISBN 0-12-352551-9

known in human Dubin–Johnson syndrome (Dubin and Johnson, 1954; Sprinz and Nelson, 1954), and in several animal species (for review see Roy Chowdhury *et al.*, 1994), including two different mutant rat strains (Jansen *et al.*, 1985; Takikawa *et al.*, 1991). These mutant rat strains served as models which facilitated the localization of the defect in ATP-dependent conjugate transport to the hepatocyte canalicular membrane (Ishikawa *et al.*, 1990; Kitamura *et al.*, 1990) and enabled the cloning of the rat *Mrp2* cDNA (Büchler *et al.*, 1996; Ito *et al.*, 1997; Mayer *et al.*, 1995; Paulusma *et al.*, 1996). The absence of functional MRP2 protein from human liver has been recognized as the cause of Dubin–Johnson syndrome (Kartenbeck *et al.*, 1996; Keppler and Kartenbeck, 1996) and many naturally occurring mutations in the *MRP2* (*ABCC2*) gene have been identified (Mor-Cohen *et al.*, 2001; Paulusma *et al.*, 1997; Tsujii *et al.*, 1999; Wada *et al.*, 1998). Mutations and polymorphisms in the human *MRP2* gene that affect MRP2 function may be relevant for adverse drug reactions because of an impaired hepatobiliary and renal clearance of anionic drug conjugates. Moreover, such polymorphisms may also affect the oral bioavailability of drugs that are substrates for intestinal MRP2 or become substrates following conjugation inside intestinal epithelia.

MOLECULAR CHARACTERIZATION OF THE APICAL CONJUGATE EXPORT PUMP MRP2

The first cDNA fragment of *Mrp2* (Abcc2; formerly described as cMrp and cMoat) was identified in 1995 in a comparative analysis of normal and transport-deficient GY/TR$^-$ mutant rat liver (Mayer *et al.*, 1995). Using degenerate oligonucleotides complementary to human *MRP1* mRNA, an *MRP1*-related 347 bp cDNA fragment was amplified from normal rat liver but not from RNA from transport-deficient liver (Mayer *et al.*, 1995). Subsequently, the full-length cDNA encoding an MRP1-related protein now known as Mrp2 was cloned and further analyzed (Büchler *et al.*, 1996; Paulusma *et al.*, 1996). At present, the full-length *MRP2* cDNA sequences and deduced amino acid sequences from five mammalian species are known, including the orthologues from human, rat, rabbit, mouse and dog (Büchler *et al.*, 1996; Conrad *et al.*, 2001; Fritz *et al.*, 2000; Paulusma *et al.*, 1996; Taniguchi *et al.*, 1996; van Aubel *et al.*, 1998). These five mammalian MRP2 proteins are highly homologous, with amino acid identities ranging from 77% for the identity between the MRP2 proteins from rat and dog, to 87% identity for the proteins from rat and mouse. Furthermore, MRP2-related sequences from other organisms including *Caenorhabditis elegans* (Broeks *et al.*, 1996) and the plant *Arabidopsis thaliana* (Rea et al., 1998) (Chapter 17) have been described and, in part, functionally characterized. Within the human MRP (ABCC) subfamily, MRP2 shows the highest degree of similarity to MRP1 with 48% identity (Cole *et al.*, 1992), followed by MRP3 with 47% identity (Kiuchi *et al.*, 1998), and MRP6 with 38% identity (Kool *et al.*, 1999). The lowest degree of amino acid identity was found between MRP2 and MRP8 (GenBank accession XM_040766) and CFTR (Riordan *et al.*, 1989) with 29% and 26% identity, respectively.

The identity of human MRP2 with respect to MDR1 (Ambudkar *et al.*, 1999), a member of the P-glycoprotein (ABCB) subfamily, is only 18%, underlining a major difference between the ABCB and the ABCC transporter subfamilies. Differences between the proteins belonging to the two subfamilies are also apparent based on studies of the membrane topology of these transporters. In contrast to the typical organization described for members of the ABCB subfamily with two transmembrane domains and two ATP-binding domains, MRP2, as well as MRP1, MRP3, MRP6 and MRP7, contains an additional NH_2-proximal membrane-spanning domain **(Figure 20.1)** (Borst *et al.*, 1999; Büchler *et al.*, 1996; Hipfner *et al.*, 1997; König *et al.*, 1999a). This additional domain is represented by an extension of approximately 200 amino acids when compared with the length of the ABCB subfamily members. Another striking feature of MRP2 was found in studies on the localization of the NH_2-terminus. Owing to an odd number of predicted transmembrane helices, the NH_2-terminus was predicted to be extracytosolic on the basis of computational analysis by the TMAP program (Büchler *et al.*, 1996) (see section on mutations in the *MRP2* gene). This was recently directly established by immunofluorescence microscopy studies using an antibody directed against the NH_2-terminus of MRP2 (Cui *et al.*, 1999). The extracellular

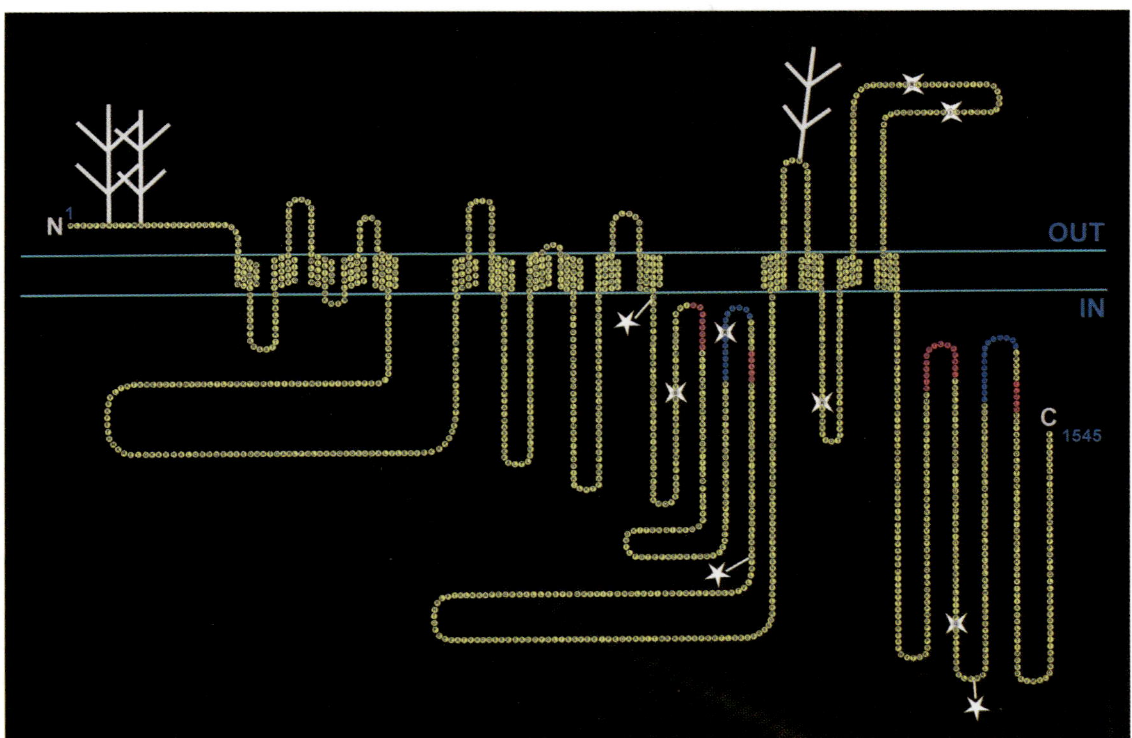

Figure 20.1. ***A predicted membrane topology model for human MRP2. Amino acids in the nucleotide-(ATP)-binding domains are indicated, with the Walker A and B motifs in red and the ABC transporter family signature in blue. Mutated amino acids in patients with Dubin–Johnson syndrome are indicated as white stars on the polypeptide chain, whereas splice site mutations are indicated as pentagonal stars near the polypeptide chain. (**See Chapter 3** for detailed discussion on topologies).***

localization of the NH_2-terminus of MRP1 was also demonstrated by glycosylation site mutational studies and epitope insertion experiments (Hipfner *et al.*, 1997; Kast and Gros, 1998) and, based on sequence similarities, it is expected that MRP3 and MRP6 will be the same.

The human *MRP2* gene has been localized to chromosome 10q23–q24 (Taniguchi *et al.*, 1996). It spans approximately 65 kbp and contains 32 exons with a high proportion of class 0 introns (Tsujii *et al.*, 1999). The size of coding exons ranges from 56 bp (exon 6) to 255 bp (exon 10), and each nucleotide-binding domain is encoded by three exons (Toh *et al.*, 1999; Tsujii *et al.*, 1999). Comparison of the genomic organization of the human *MRP2* and *MRP1* genes shows that they display remarkable similarities as indicated by size and number of exons (Grant *et al.*, 1997). Furthermore, human *MRP1*, *MRP2* and *MRP3* (GenBank accession AC004590) have 21 identical splice sites based on an amino acid alignment of the three cognate proteins (Tsujii *et al.*, 1999). Despite the fact that these three human MRP family members share a relatively moderate degree of amino acid identity, their similar genomic organization suggests a close evolutionary relationship, possibly originating from gene duplication (Keppler *et al.*, 2000).

Tissue Distribution and Localization of MRP2 in Polarized Cells

Antibodies of high affinity and specificity have been useful for the determination of the localization and tissue distribution of MRP2. Initially, MRP2 was localized to the apical membrane of rat (Büchler *et al.*, 1996) and human hepatocytes (Keppler and Kartenbeck, 1996; Paulusma *et al.*, 1997), and was, therefore, termed canalicular MRP (cMRP) (Büchler *et al.*, 1996), or canalicular multispecific organic anion transporter (cMOAT) (Paulusma *et al.*, 1996, 1997; Taniguchi *et al.*, 1996). Soon after the cloning

TABLE 20.1. EXPRESSION AND LOCALIZATION OF MRP2 IN NORMAL TISSUES

Tissue	Species	mRNA	Protein		References
			Immunoblot	Immunolocalization	
Liver	Human			+	Keppler and Kartenbeck, 1996
	Monkey	+	+	+	Kauffmann *et al.*, 1998
	Dog	+	+		Conrad *et al.*, 2001
	Rabbit		+		van Aubel *et al.*, 1998
	Rat	+	+	+	Büchler *et al.*, 1996
	Mouse	+	+		Wielandt *et al.*, 1999
Kidney	Human	+	+	+	Schaub *et al.*, 1999
	Dog	+	+		Conrad *et al.*, 2001
	Rabbit		+		van Aubel *et al.*, 1998
	Rat	+	+	+	Schaub *et al.*, 1997
	Mouse	+			Fritz *et al.*, 2000
Small intestine	Human	+		+	Fromm *et al.*, 2000
	Dog	+	+		Conrad *et al.*, 2001
	Rabbit		+	+	van Aubel *et al.*, 2000
	Rat	+	+	+	Mottino *et al.*, 2000
Colon	Human			+ [a]	
Gallbladder	Human	+		+	Rost *et al.*, 2001
Lung	Human			+ [a]	
Placenta	Human	+	+	+	St-Pierre *et al.*, 2000

[a]See Figure 20.2.

of MRP2 from rat and human liver, *MRP2* mRNA was also detected in kidney, duodenum and peripheral nerve (Kool *et al.*, 1997; Schaub *et al.*, 1997; van Aubel *et al.*, 1998). Kidney proximal tubule epithelia are a particularly good example for the extrahepatic apical localization of MRP2 (Schaub *et al.*, 1997, 1999). **Table 20.1** summarizes the tissues from different mammalian species in which MRP2 mRNA and/or protein has been detected so far. Localization of MRP2 in the apical membrane of hepatocytes, kidney proximal tubules, and epithelial cells of gallbladder, small intestine, colon and lung was confirmed by immunofluorescence microscopy or immunohistochemistry **(Table 20.1 and Figure 20.2)**. In the placenta, MRP2 is localized to the apical syncytiotrophoblast membrane (St-Pierre *et al.*, 2000). In addition to the apically localized MRP2, polarized cells express other MRP homologues in the basolateral membrane (Keppler *et al.*, 2001). For example, MRP6 is highly expressed in rat liver (Hirohashi *et al.*, 1998; Madon *et al.*, 2000), and in human liver and kidney (Kool *et al.*, 1999), and is localized to the basolateral membrane of human hepatocytes and kidney proximal tubule epithelia **(Figure 20.2)**.

The distribution of MRP2 within an organ may change during different pathophysiological conditions. For example, Mrp2 is homogeneously distributed throughout a lobule in normal rat liver; however, cholestasis causes Mrp2 to concentrate near the central (perivenous) area of the liver lobule (Paulusma *et al.*, 2000). On the subcellular level, selective retrieval of Mrp2 from the canalicular membrane to pericanalicular vesicles of rat hepatocytes has been observed as an early event of cholestasis by immunofluorescence microscopy (Dombrowski *et al.*, 2000; Kubitz *et al.*, 1999; Rost *et al.*, 1999; Trauner *et al.*, 1997) and immunogold electron microscopy (Beuers *et al.*, 2001; Dombrowski *et al.*, 2000).

Several cell lines have been used for studies of MRP2 function in intact cells. Rat and human hepatoma cells express endogenous MRP2 in the apical membrane surrounding apical vacuoles or bile canaliculus-like structures between adjacent cells (Cantz *et al.*, 2000; Nies *et al.*, 1998). Secretion of fluorescent MRP2 substrates into these apical vacuoles can be observed by fluorescence microscopy **(Figure 20.3)**. Caco-2 cells, derived from a human colon carcinoma, also express endogenous MRP2 in the apical

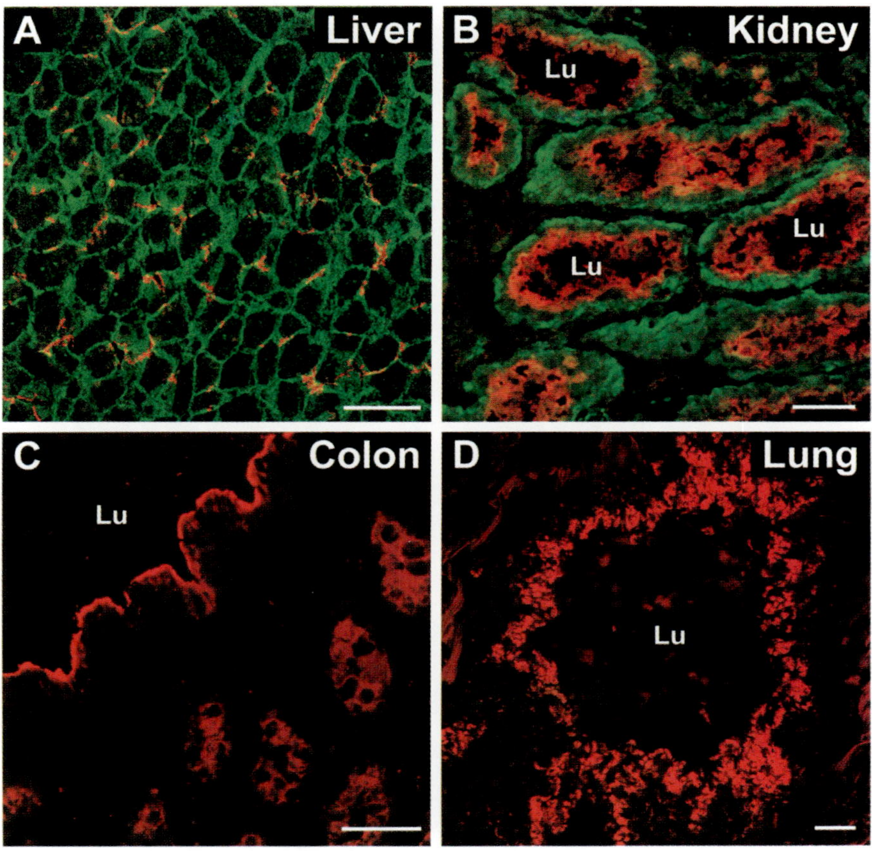

Figure 20.2. Localization of MRP2 (red in A–D) and MRP6 (green in A and B) by immunofluorescence in different human tissues. Double-label (A, B) and single-label (C, D) immunofluorescence microscopy of frozen tissue sections (5 μm thickness) was performed as described (König et al.*, 1999b). Pictures were taken by confocal laser scanning microscopy (A, B, D) or by conventional fluorescence microscopy (C). MRP2 is localized to the apical membrane of human hepatocytes (A), proximal tubule epithelia in the kidney (B), epithelia of the colon (C), and of bronchial epithelia (D) as detected either with the monoclonal antibody M_2III-6 (Paulusma* et al.*, 1997) in A, B, D or with the polyclonal antiserum EAG5 in C (Cui* et al.*, 1999; Schaub* et al.*, 1999). The isoform MRP6 is localized to the basolateral membrane of hepatocytes (A) and proximal tubule epithelia (B) as detected with the antiserum AQL (König* et al.*, 1999b) Lu, lumen. Bars in A–D, 50 μm.*

membrane (Bock *et al.*, 2000; Walgren *et al.*, 2001). Transfection of rat or human *MRP2* cDNA into Madine–Darby canine kidney (MDCKII) cells also results in apical localization of the respective recombinant MRP2 protein (Cui *et al.*, 1999; Evers *et al.*, 1998). Because Caco-2 cells and MRP2-expressing MDCKII cells grow in a polarized fashion on special membrane filters, these cells are often used as a model system for studies on the uptake, transcellular transport, and MRP2-mediated export of substances by intact cells (Cui *et al.*, 2001).

The clinical relevance of MRP2 as an ATP-dependent export pump for chemotherapeutic agents has not yet been extensively investigated but has been supported by the localization of MRP2 in the plasma membrane of renal clear-cell (Schaub *et al.*, 1999), ovarian (Arts *et al.*, 1999), colorectal (Hinoshita *et al.*, 2000), and hepatocellular (Nies *et al.*, 2001) carcinoma cells. MRP2 expression was also detected by reverse transcriptase (RT)-PCR and immunoblotting in cell lines from lung, gastric, uterine and colorectal cancers (Kool *et al.*, 1997; Minemura *et al.*, 1999; Narasaki *et al.*, 1997; Young *et al.*, 1999, 2001). Thus, MRP2 is expressed in several malignant tumor types and may contribute to their resistance to a wide variety of antitumor drugs, as demonstrated *in vitro* in MRP2-transfected cells (Cui *et al.*, 1999).

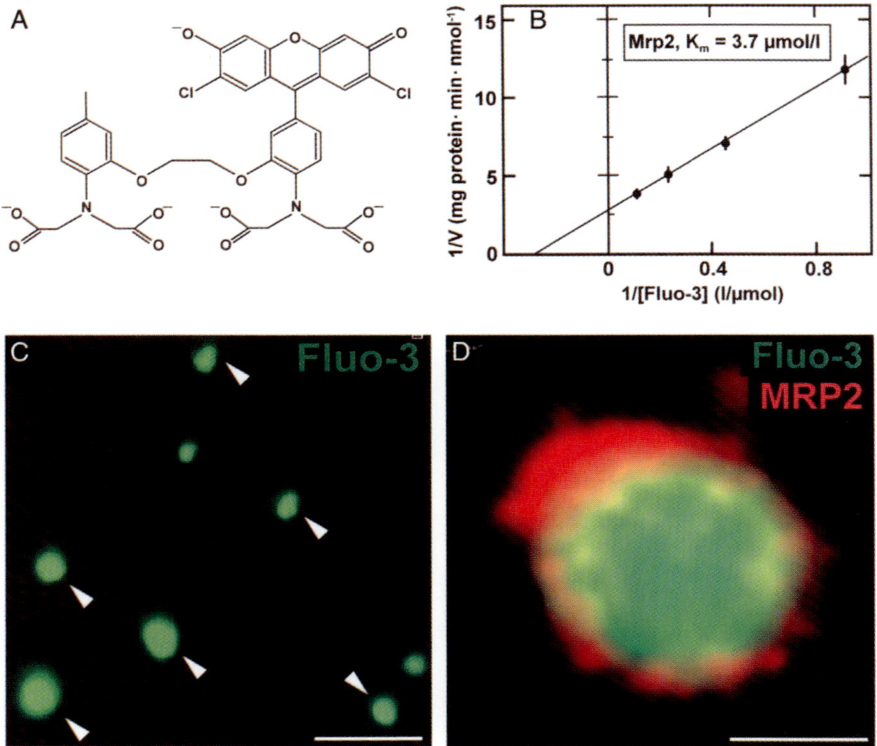

Figure 20.3. Fluo-3 as a fluorescent substrate for MRP2. A, Structure of the fluorescent organic anion Fluo-3 (Kao et al.*, 1989; Minta* et al.*, 1989). B, ATP-dependent transport of Fluo-3 into Mrp2-containing canalicular membrane vesicles isolated from rat hepatocytes. Vesicle-associated fluorescence was determined fluorometrically as described; a* Km *value of 3.7 µM was obtained for Fluo-3 as the substrate for rat Mrp2 (Nies* et al.*, 1998). C, Vectorial transport of Fluo-3 into apical vacuoles of human HepG2 cells. Polarized HepG2 cells, derived from hepatocellular carcinoma, were incubated with the non-fluorescent acetoxymethyl ester of Fluo-3, which was taken up by the cells and hydrolyzed to the fluorescent Fluo-3. Fluo-3-filled apical vacuoles (arrowheads pointing to green fluorescence) were observed by fluorescence microscopy as described (Cantz* et al.*, 2000). D, Single Fluo-3-filled vacuole (green) is shown after immunostaining of MRP2 (red). Merging both fluorescences demonstrates secretion of Fluo-3 into the apical vacuole of polarized HepG2 cells. Bar in C, 25 µm; in D, 5 µm.*

FUNCTIONAL ANALYSIS AND SUBSTRATE SPECIFICITY OF MRP2

Elucidation of the physiological function of MRP1, the first identified member of the MRP family (see Chapter 19), was closely linked to the characterization of the membrane proteins mediating the ATP-dependent transport of the endogenous glutathione *S*-conjugate, LTC_4 (Keppler, 1992; Leier *et al.*, 1994b). The search for the molecular identity of the ATP-dependent LTC_4 transporter localized to the hepatocyte canalicular membrane (Ishikawa *et al.*, 1990) led subsequently to the identification of MRP2 (see above) (Büchler *et al.*, 1996; Paulusma *et al.*, 1996; Taniguchi *et al.*, 1996). Previously, ATP-dependent transport measurements using inside-out hepatocyte canalicular membrane vesicles from normal and Mrp2-deficient GY/TR$^-$ rats (Ishikawa *et al.*, 1990) or EHBR rats (Fernandez-Checa *et al.*, 1992; Takenaka *et al.*, 1995) have been invaluable in elucidating the substrate specificity of Mrp2 (Keppler and Kartenbeck, 1996; König *et al.*, 1999a). Since cell lines stably expressing recombinant rat or human MRP2 became available (Chen *et al.*, 1999; Cui *et al.*, 1999; Evers *et al.*, 1998; Ito *et al.*, 1998), the substrate specificity of MRP2 has been studied under more defined conditions using inside-out membrane vesicles prepared from these transfected cells. Human MRP2 has also been purified to homogeneity, and shown to exhibit substrate-stimulated ATPase and transport activity when reconstituted in proteoliposomes (Hagmann *et al.*, 1999, 2002).

Various labeled substrates for MRP2 as well as for MRP1 are useful in assessing the transport function of both proteins. $[^3H]LTC_4$ has become the preferred substrate for transport measurements because of its high affinity for both MRP1 and MRP2 (Cui *et al.*, 1999; Leier *et al.*, 1994a), and its commercial availability. ATP-dependent transport of $[^3H]LTC_4$ into membrane vesicles is measured by incubating the labeled substrate with the inside-out membrane vesicles for the desired time period and subsequently separating membrane vesicles from extravesicular labeled substrate by rapid filtration through nitrocellulose filter membranes (Keppler *et al.*, 1998). With substrates more hydrophobic than $[^3H]LTC_4$, which bind strongly to the filters and to the membrane vesicles, small Sephadex G-50 columns (Böhme *et al.*, 1993) or glass filters (Loe *et al.*, 1996) have been used for the separation of vesicles and labeled substrate.

The apical export pump MRP2 shares a very similar substrate spectrum with MRP1. High-affinity substrates for MRP2 include amphiphilic anions, particularly those conjugated with glutathione and glucuronate, such as LTC_4, bilirubin glucuronosides, and 17β-glucuronosyl estradiol **(Table 20.2)**. The comparison of both recombinant proteins shows that MRP1 has a 10-fold higher affinity for LTC_4 and a 5-fold higher affinity for 17β-glucuronosyl estradiol than MRP2 (Cui *et al.*, 1999), whereas mono- and bisglucuronosyl bilirubin are preferred substrates for MRP2 (Kamisako *et al.*, 1999). MRP2 is also able to transport non-conjugated compounds such as the penta-anionic fluorescent dye Fluo-3, the model compound for hepatic transport studies sulfobromophthalein, the anionic anticancer drug methotrexate, the HMG-CoA reductase inhibitor pravastatin, and the angiotensin-converting enzyme inhibitor temocaprilat **(Table 20.2)**.

Fluorescent substrates, such as the Ca^{2+}-indicator Fluo-3, are useful for studies of MRP2 function in intact cells (Cantz *et al.*, 2000; Nies *et al.*, 1998). The fluorescent glutathione *S*-conjugates glutathione bimane (Oude Elferink *et al.*, 1993; Roelofsen *et al.*, 1995) and glutathione methylfluorescein (Roelofsen *et al.*, 1998) are also likely substrates for Mrp2 because they are not transported from hepatocytes into bile of Mrp2-deficient mutant rats. The ATP-dependent transport of Fluo-3 by Mrp2 was demonstrated using inside-out membrane vesicles from normal and Mrp2-deficient rat hepatocytes (Nies *et al.*, 1998) **(Figure 20.3)** and shown to represent the predominant export pump mediating extrusion of Fluo-3 across the apical membrane of polarized cells. Polarized rat (Ihrke *et al.*, 1993) and human (Sormunen *et al.*, 1993) hepatoma cells form apical vacuoles between adjacent cells and express MRP2 in the apical membrane (Cantz *et al.*, 2000; Nies *et al.*, 1998). Secretion of Fluo-3 and other fluorescent anions into these apical vacuoles is readily observed by fluorescence microscopy **(Figure 20.3)**. MRP2-mediated Fluo-3 secretion into apical vacuoles is inhibited by cyclosporin A but not by the selective MDR1 P-glycoprotein inhibitor LY335979. Recently, probenecid-sensitive efflux of carboxyfluorescein has been used for estimation of MRP2 transport activity in intact cells (Mor-Cohen *et al.*, 2001).

DETOXIFICATION AND DRUG RESISTANCE CONFERRED BY MRP2

The strategic localization of MRP2 to the apical membrane of hepatocytes, renal proximal tubule epithelia, and epithelia of the small intestine, colon and bronchia **(Table 20.1, Figure 20.2)** suggests that a physiological function of MRP2 is to excrete endogenous metabolites and xenobiotics, and to prevent toxic compounds from entering the body. It has been reported that the Mrp2-deficient GY/TR^- rats have a much lower excretion rate for the food-derived carcinogen 2-amino-1-methyl-6-phenylimidazo[4,5-b] pyridine (PhIP) and its glucuronate conjugate compared to wild-type Wistar rats (Dietrich *et al.*, 2001). MRP2 has also been shown to act synergistically with the glutathione *S*-transferase GSTπ1-1 in the detoxification of the cytotoxic and genotoxic agent 4-nitroquinoline 1-oxide (4-NQO) (Morrow *et al.*, 2001).

Because of the similar substrate spectrum of MRP2 and MRP1, which has been shown to confer resistance to different anticancer drugs when overexpressed in mammalian cells **(see Chapter 19)** (Cole *et al.*, 1994; Grant *et al.*, 1994), it has been proposed that MRP2 may also confer drug resistance by pumping drug conjugates or drug–glutathione complexes out of the cell. Northern blot analyses and RNase protection

TABLE 20.2. SUBSTRATE SPECIFICITY OF HUMAN AND RAT MRP2

Substrate	K_m value (μM)	References
MRP2 (human recombinant)		
LTC_4	1	Cui *et al.*, 1999
S-Glutathionyl 2,4-dinitrobenzene	6.5	Evers *et al.*, 1998
S-Glutathionyl ethacrynic acid		Evers *et al.*, 1998
Bilirubin		
Monoglucuronosyl	0.7	Kamisako *et al.*, 1999
Bisglucuronosyl	0.9	Kamisako *et al.*, 1999
17β-Glucuronosyl estradiol	7.2	Cui *et al.*, 1999
Sulfobromophthalein	11	Cui *et al.*, 2001
p-Aminohippurate	880	Leier *et al.*, 2000
Ochratoxin A		Leier *et al.*, 2000
Mrp2 (rat recombinant)		
LTC_4	1.1	Cui *et al.*, 1999
S-Glutathionyl 2,4-dinitrobenzene	0.2	Ito *et al.*, 1998
Bilirubin		
Monoglucuronosyl	0.8	Kamisako *et al.*, 1999
Bisglucuronosyl	0.5	Kamisako *et al.*, 1999
17β-Glucuronosyl estradiol	6.9	Cui *et al.*, 1999
Sulfatolithocholyl taurine	3.9	Akita *et al.*, 2001
Mrp2 (rat; normal/mutant BCM)[a]		
LTC_4	0.3	Ishikawa *et al.*, 1990
LTD_4	1.5	Ishikawa *et al.*, 1990
LTE_4		Ishikawa *et al.*, 1990
N-Acetyl LTE_4	5.2	Ishikawa *et al.*, 1990
S-Glutathionyl 2,4-dinitrobenzene		Ishikawa *et al.*, 1990
S-Glutathionyl sulfobromophthalein		Ishikawa *et al.*, 1990
Glutathione disulfide		Fernandez-Checa *et al.*, 1992
Bilirubin		
Monoglucuronosyl		Jedlitschky *et al.*, 1997
Bisglucuronosyl		Nishida *et al.*, 1992
		Jedlitschky *et al.*, 1997
Glucuronosyl nafenopin		Jedlitschky *et al.*, 1994
Glucuronosyl E3040[a]	5.7	Niinuma *et al.*, 1997
Glucuronosyl grepafloxacin	7.2	Sasabe *et al.*, 1998
Glucuronosyl SN38 carboxylate[a]		Chu *et al.*, 1997
Glucuronosyl SN38 lactone[a]		Chu *et al.*, 1997
SN38 carboxylate[a]		Chu *et al.*, 1997
Sulfobromophthalein	31	Nishida *et al.*, 1992
Fluo-3[a]	3.7	Nies *et al.*, 1998
Methotrexate	295	Masuda *et al.*, 1997
Temocaprilat	93	Ishizuka *et al.*, 1997
Pravastatin	220	Yamazaki *et al.*, 1997
Sulfatolithocholyl taurine	1.5	Akita *et al.*, 2001
Sulfatochenodeoxycholyl taurine	8.8	Akita *et al.*, 2001

Compounds listed have been identified as substrates by measurement of their ATP-dependent transport into inside-out membrane vesicles from cells expressing the recombinant MRP2/Mrp2 in comparison with membrane vesicles from control vector-expressing cells. In addition, measurement of ATP-dependent transport into hepatocyte canalicular membranes vesicles from Mrp2-deficient mutant rats (GY/TR$^-$ and EHBR) compared with those from normal rats are presented (summarized by König *et al.*, 1999a).

[a]Abbreviations: BCM, bile (hepatocyte) canalicular membranes; E3040, 6-hydroxy-5,7-dimethyl-2-methylamino-4-(3-pyridylmethyl)benzothiazole; Fluo-3, 1-[2-amino-5-(2,7-dichloro-6-hydroxy-3-oxo-3H-xanthen-9-yl)]-2-(2′-amino-5′-methyl-phenoxy)-ethane-*N*,*N*,*N*′,*N*′,-tetraacetic acid penta ammonium salt; SN38, de-esterified metabolite of CPT11 (7-ethyl-10-hydroxycamptothecin).

TABLE 20.3. MRP2-MEDIATED RESISTANCE TO ANTICANCER DRUGS IN TRANSFECTED CELLS

I. Stably transfected MDCKII cells

Drug	MDCK-MRP2 (human)		MDCK-Mrp2 (rat)		MDCK-Con	
	IC_{50}	RR	IC_{50}	RR	IC_{50}	RR
Etoposide (μM)	612	3.8[a]	809	5.0[a]	163	1.0
Vincristine (μM)	49	6.0[a]	19	2.3[a]	8.2	1.0

II. Stably transfected HEK293 cells

Drug	HEK-MRP2 (human)	HEK-Con	RR
	IC_{50}	IC_{50}	
Cisplatin (μM)	24	2.4	10.0[a]
Etoposide (μM)	1.2	0.3	4.0[a]
Doxorubicin (nM)	346	44	7.8[a]
Epirubicin (nM)	19	3.8	5.0[a]

[a]$p < 0.01$. Sensitivity to antitumor drugs was determined by a tetrazolium salt-based cell viability assay. The relative resistance factor (RR) was calculated by dividing the IC_{50} value of cells transfected with human or rat *MRP2* cDNA by the IC_{50} value of cells transfected with control vector. Modified from Cui *et al.* (1999).

assays indicate that a correlation between MRP2 expression and multidrug resistance may exist (Kool *et al.*, 1997; Taniguchi *et al.*, 1996). MRP2-related drug resistance was also suggested by its cloning from the cisplatin-resistant human cancer cell lines KB-CDP4 and P-CDP5 (Taniguchi *et al.*, 1996). In both cisplatin-resistant cell lines, *MRP2* mRNA was overexpressed relative to non-resistant parental cell lines. A correlation between cisplatin resistance and *MRP2* expression was also demonstrated for several additional cell lines by RNase protection assays and Northern blot analyses (Kool *et al.*, 1997; Minemura *et al.*, 1999). Finally, drug resistance conferred by MRP2 has also been demonstrated by the use of antisense cDNA, studied in the human hepatoma cell line HepG2 (Koike *et al.*, 1997). The amount of *MRP2* mRNA was reduced, leading to elevated intracellular glutathione levels and enhanced sensitivity to anticancer drugs including cisplatin, vincristine, doxorubicin and the camptothecin derivatives CPT11 and SN38.

Direct evidence for MRP2-mediated multidrug resistance was obtained by transfection studies with *MRP2* cDNA (Cui *et al.*, 1999; Hooijberg *et al.*, 1999). In *MRP2*-transfected MDCKII cells and HEK293 cells, MRP2 was localized to the plasma membrane, which is a prerequisite for the measurement of MRP2-mediated resistance (Cui *et al.*, 1999). Expression of both recombinant rat and human MRP2 in MDCKII and HEK293 cells leads to significant resistance to cisplatin, etoposide, vincristine, doxorubicin and epirubicin **(Table 20.3)**. The ability of MRP2 to confer resistance to cisplatin has also been shown in MRP2-transfected LLC-PK1 cells (Chen *et al.*, 1999). Moreover, MRP2 confers resistance to the antifolate methotrexate in transfected human ovarian carcinoma 2008 cells (Hooijberg *et al.*, 1999). The possible clinical relevance of MRP2-mediated drug resistance was further suggested by the detection of MRP2 in various carcinoma samples.

MUTATIONS IN THE *MRP2* GENE

Naturally occurring mutations in the *MRP2* gene have been discovered in humans (Mor-Cohen *et al.*, 2001; Paulusma *et al.*, 1997; Toh *et al.*, 1999; Tsujii *et al.*, 1999; Wada *et al.*, 1998) and rat (Ito *et al.*, 1997; Paulusma *et al.*, 1996). Some of these mutations were shown to be associated with the absence of the MRP2 protein from the hepatocyte canalicular membrane (Büchler *et al.*, 1996; Kartenbeck *et al.*, 1996). In humans, the Dubin–Johnson syndrome, originally described in 1954, is an autosomal recessively inherited disorder characterized by conjugated hyperbilirubinemia (Dubin and Johnson, 1954; Sprinz and Nelson, 1954). The liver of patients with Dubin–Johnson syndrome appears dark blue or black because of deposition of a dark pigment in the pericanalicular area (Roy Chowdhury *et al.*, 1994). The deficient transport of anionic conjugates, including monoglucuronosyl bilirubin and bisglucuronosyl bilirubin, from hepatocytes into bile is caused by the absence of the MRP2 protein from the canalicular membrane (Kartenbeck *et al.*, 1996; Keppler and Kartenbeck, 1996; Paulusma *et al.*, 1997; Tsujii *et al.*, 1999). Established mutations identified in patients with Dubin–Johnson syndrome include splice site mutations leading to exon deletions with subsequent premature termination codons (Kajihara *et al.*, 1998; Toh *et al.*, 1999; Wada *et al.*, 1998), missense mutations (Mor-Cohen *et al.*, 2001; Toh *et al.*, 1999), a nonsense mutation leading to a premature termination codon (Paulusma *et al.*, 1997; Tsujii *et al.*, 1999), and a deletion mutation leading to the loss of two amino acids in the second nucleotide-binding domain, (Tsujii *et al.*, 1999) **(Figure 20.4 and Table 20.4)**. Interestingly, all mutations identified so far are located in the COOH-proximal half of the MRP2 protein and only two of them are located in a predicted extracellular loop **(Figures 20.1 and 20.4).**

The MRP2 membrane topology has been predicted using several different algorithms including TMpred (http://www.ch.embnet.org/software/TMPRED_form.html), TopPred2 (http://bioweb.pasteur.fr/seqanal/interfaces/toppred.html), and SOSUI (http://sosui.proteome.bio.tuat.ac.jp/sosuiframe0.html). These algorithms predict four, or at most five, transmembrane helices for the region between the first and the second ATP-binding domain of human MRP2. None of the programs predict six transmembrane helices for this domain of MRP2. Because an even number of transmembrane helices is required between the two cytosolic ATP-binding domains, we have used the four-transmembrane-helix topology for the prediction of the location of mutations and polymorphisms in the COOH-proximal portion of the protein **(Figures 20.1 and 20.4)**.

In addition to the known mutations in Dubin–Johnson syndrome, seven base pair changes, of which five are in the coding region of the *MRP2*

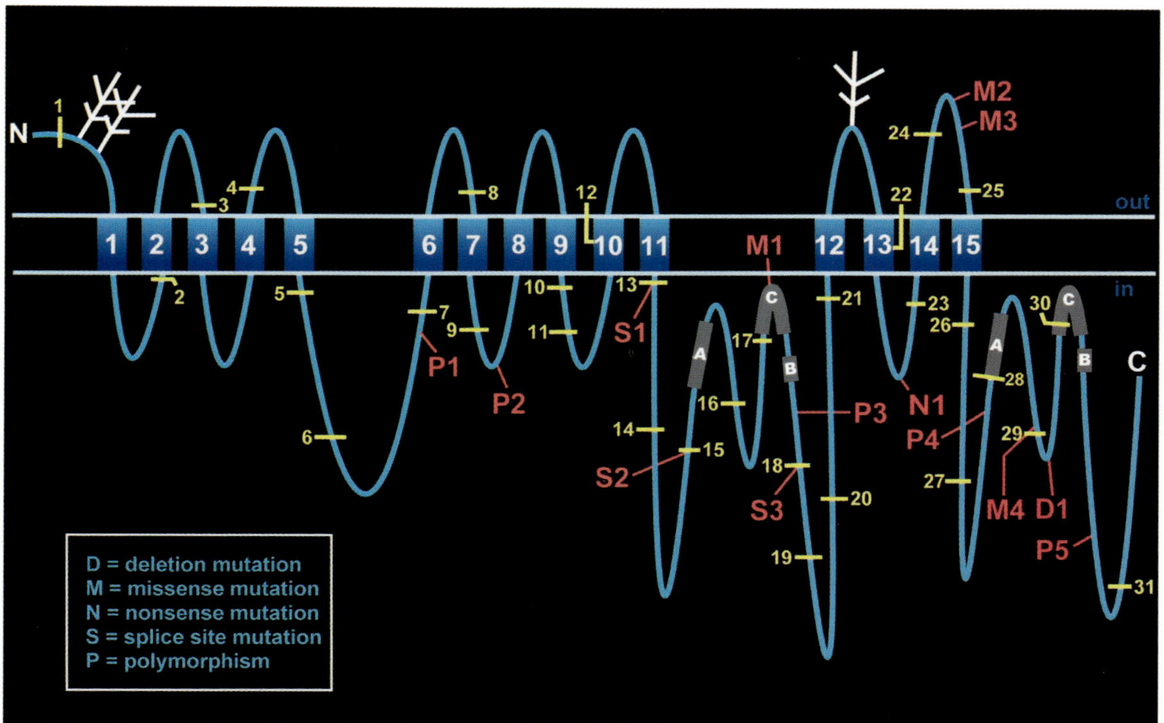

Figure 20.4. Schematic membrane topology of human MRP2 with the locations of intron–exon boundaries of the MRP2 *gene indicated in yellow. Yellow numbers indicate the number of the exon encoding the* NH_2*-proximal amino acid sequence. Thus, the COOH-terminal amino acid sequence is encoded by exon 32. Mutations causing Dubin–Johnson syndrome or polymorphisms are indicated in red as detailed in Table 20.4.*

gene and which are not associated with the Dubin–Johnson syndrome phenotype, have been recently characterized as polymorphisms (Ito *et al.*, 2001c; Mor-Cohen *et al.*, 2001) **(Figure 20.4 and Table 20.4)**. In the non-coding region of the *MRP2* gene, one C → T transition in the MRP2 promoter (−24 C → T; frequency of 12.5%) and one G → A transition in intron 29 (frequency of 3.5–5.4%, depending on the ethnic group), have been described (Mor-Cohen *et al.*, 2001). Polymorphisms in the coding region include one G → A transition in exon 7 (842G → A, predicting Ser^{281} → Asn; frequency of 0.6% in Iranian Jews and 5.6% in Moroccan Jews) (Mor-Cohen *et al.*, 2001), one G → A transition in exon 10 (1249G → A, leading to Val^{417} → Ile; frequency of 12.5%) (Ito *et al.*, 2001c), one C → T transition in exon 18 (2366C → T, predicting Ser^{789} → Phe; frequency of 1%) (Ito *et al.*, 2001c), one C → T transition in exon 28 (3972C → T, with no change in the amino acid sequence; frequency of 22%), and one G → A transition in exon 31 (4348G → A, resulting in Ala^{1450} → Thr) with a frequency of 1% (Ito *et al.*, 2001c). So far, no mutations in the *MRP2* gene have been found that lead to the expression of a truncated but apically localized MRP2 protein. We have identified a Dubin–Johnson syndrome mutation in exon 30 of the *MRP2* gene **(Figure 20.4)**, leading to the loss of arginine 1392 and methionine 1393 in the second ATP-binding domain (Tsujii *et al.*, 1999). Transfection and expression of the corresponding mutated *MRP2* cDNA showed that the mutant MRP2 protein is expressed in polarized human HepG2 cells; however, it is retained in the endoplasmic reticulum and is not sorted to the apical membrane **(Figure 20.5)**. The mutant MRP2 protein does not mature correctly and is thus recognized by the cellular quality control machinery and degraded by proteasomes (Keitel *et al.*, 2000). Recently, the Dubin–Johnson missense mutation 3449G → A resulting in the amino acid substitution Arg^{1150} → His was analyzed on the molecular level. This mutant protein was located in the membrane of unpolarized human embryonic kidney HEK293 cells but was transport deficient (Mor-Cohen *et al.*, 2001).

Mutations have also been identified in two well-characterized hyperbilirubinemic rat strains, which, as mentioned previously, are

TABLE 20.4. POLYMORPHISMS IN MRP2 AND MUTATIONS CAUSING DUBIN–JOHNSON SYNDROME

Designation[a]	Nucleotide change	Predicted consequence or amino acid change	References
Polymorphisms			
P1	842G → A	S281N	Mor-Cohen *et al.*, 2001
P2	1249G → A	V417I	Ito *et al.*, 2001c
P3	2366C → T	S789F	Ito *et al.*, 2001c
P4	3972C → T	I1324I	Ito *et al.*, 2001c
P5	4348G → A	A1450T	Ito *et al.*, 2001c
Splice site mutations			
S1	1815 + 2T → A	Splice donor, loss of exon 13	Toh *et al.*, 1999
S2	1967 + 2T → C	Splice donor, loss of exon 15	Kajihara *et al.*, 1998
S3	2439 + 2T → C	Splice donor, loss of exon 18	Toh *et al.*, 1999
Missense mutations			
M1	2302C → T	R768W	Toh *et al.*, 1999
M2	3449G → A	R1150H	Mor-Cohen *et al.*, 2001
M3	3517A → T	I1173F	Mor-Cohen *et al.*, 2001
M4	4145A → G	Q1382R	Toh *et al.*, 1999
Nonsense mutations			
N1	3196C → T	R1066X	Paulusma *et al.*, 1997
Deletion mutations			
D1	4175–4180DelGGATGA	R1392 + M1393del	Tsujii *et al.*, 1999

[a]For location, see **Figures 20.1 and 20.4.**

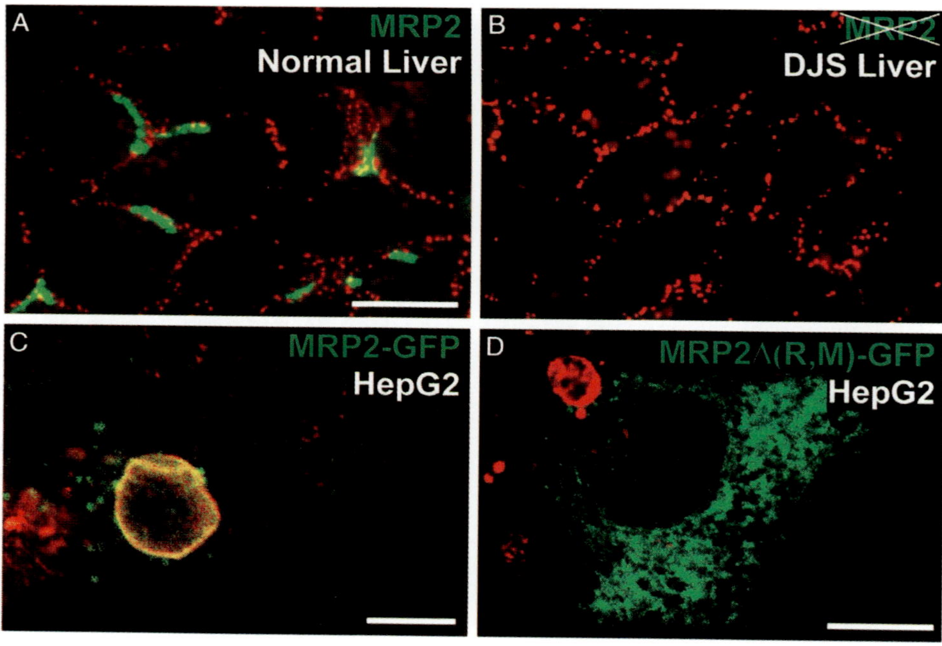

Figure 20.5. Mutations in the **MRP2** *gene lead to the loss of MRP2 protein in Dubin–Johnson syndrome because of impaired maturation of the mutant protein. A, Immunofluorescence microscopy of a normal human liver stained for MRP2 (green) and for desmoplakin (red). MRP2 is localized to the apical (canalicular) membrane of hepatocytes in A. B, Immunofluorescence microscopy of a human liver biopsy from a patient with Dubin–Johnson syndrome (DJS) (Tsujii* et al., *1999) shows the complete absence of MRP2 from the canalicular domain, whereas staining of desmoplakin (red) is not affected. C, Polarized HepG2 cells express a fusion protein of MRP2 and the green fluorescent protein (MRP2-GFP; green) in the apical membrane. This domain also stains positive for the apical marker protein dipeptidylpeptidase IV (red) yielding a yellow color where both proteins co-localize. D, cDNA carrying the 6-nucleotide deletion within the second nucleotide-binding domain, leading to loss of Arg1392 and Met1393 (Tsujii* et al., *1999), is expressed in HepG2 cells; however, the mutant MRP2 protein is retained in the endoplasmic reticulum (MRP2Δ(R,M)-GFP, green). MRP2Δ(R,M)-GFP is not detected in the apical membrane as is evident from the lack of co-localization with the apical marker protein, dipeptidylpeptidase IV (red) (Keitel* et al., *2000). Bars, 10 μm.*

long-established animal models of the human Dubin–Johnson syndrome, the Eisai hyperbilirubinemic rat (EHBR) (Ito *et al.*, 1997) and the GY/TR$^-$ mutant rat (Paulusma *et al.*, 1996). In both strains, premature termination codons were detected which lead to the absence of the Mrp2 protein from the hepatocyte canalicular membrane. Interestingly, as with corresponding human mutations, no truncated Mrp2 protein was detected and the *Mrp2* mRNA was below detectable levels by Northern blotting analysis (Büchler *et al.*, 1996; Ito *et al.*, 1997; Paulusma *et al.*, 1996). The absence of *Mrp2* mRNA may be explained by a mechanism termed 'nonsense mediated decay' (Thermann *et al.*, 1998), and it is likely that the absence of the MRP2 protein is a consequence of the rapid degradation of the mRNA. Other mutations in the *MRP2* gene may influence apical sorting of the protein or may lead to altered protein stability. These alternatives should be considered for each of the newly identified mutations and polymorphisms.

Recently, experimentally induced mutations have been described both in human and rat MRP2. The non-conservative substitution of tryptophan 1254 in human MRP2 alters the substrate specificity of the protein and results in the loss of methotrexate transport (Ito *et al.*, 2001a). In rat Mrp2, it was demonstrated that the charged amino acids within predicted transmembrane helices 6, 11, 13, 16 and 17 (based on a secondary structure prediction with a total of 17 transmembrane helices for rat Mrp2) play an important role in substrate recognition and determination of substrate specificity (Ito *et al.*, 2001b; Ryu *et al.*, 2000).

REGULATION OF MRP2 GENE EXPRESSION

The regulation of *MRP2* gene expression has been studied under a variety of conditions associated with changes in mRNA and protein levels. Induction of human *MRP2* mRNA expression was found after treatment of primary human hepatocytes with arsenite (Vernhet *et al.*, 2001) and in liver-derived HepG2 cells exposed to the chemical carcinogen, 2-acetylaminofluorene, to phenobarbital, or to cisplatin (Schrenk *et al.*, 2001). In addition to the increase in *MRP2* mRNA, an increase in the MRP2 protein was observed in all of these studies. Furthermore, in human duodenal biopsies obtained after oral rifampin treatment, an induction of *MRP2* mRNA and protein was observed (Fromm *et al.*, 2000). One possible regulatory pathway responsible for the upregulation of human *MRP2* is mediated by the orphan nuclear receptor SXR (Dussault *et al.*, 2001). Several ligands, such as the HIV protease inhibitor ritonavir, bind SXR and activate its target genes including the human *MRP2* gene. The ability of SXR to activate *MRP2* implies that SXR may regulate the biliary excretion of xenobiotic compounds. Downregulation of human *MRP2* mRNA was found in livers from patients with primary sclerosing cholangitis as demonstrated by quantitative RT-PCR (Oswald *et al.*, 2001).

The promoter region of the human *MRP2* gene has been cloned and partially characterized. Sequence analysis of the 5′-flanking region of the human *MRP2* gene identified a number of consensus binding sites for both liver-specific and ubiquitous transcription factors (Stöckel *et al.*, 2000; Tanaka *et al.*, 1999). Using promoter deletion constructs in reporter gene analysis, the region between nucleotides −517 and −197 in front of the start codon was identified to be critical for basal *MRP2* expression (Stöckel *et al.*, 2000). This region also contains the transcriptional start site at bp −247. Interestingly, reporter gene analysis in HepG2 cells suggests that human *MRP2* and *MRP3* may be inversely regulated. MRP3 is normally localized to the basolateral membrane (König *et al.*, 1999b). Under normal conditions *MRP3* expression was only 4% of that measured for *MRP2* (Stöckel *et al.*, 2000). However, disruption of microtubules with nocodazole decreased the amount of *MRP2* mRNA and protein but increased the expression of *MRP3* mRNA and protein. This inverse regulation of these two MRP family members is consistent with their overlapping substrate spectrum and their different localization and direction of substrate transport under normal and pathophysiological conditions. However, the molecular mechanisms of how such inverse regulation occurs remain to be elucidated.

Detailed analyses of rat *Mrp2* gene expression have been performed in cholestasis models including endotoxin treatment, ethinylestradiol treatment, and common bile duct ligation (Kubitz *et al.*, 1999; Paulusma *et al.*, 2000; Trauner *et al.*, 1997; Vos *et al.*, 1998). Under these conditions, *Mrp2* mRNA and protein were downregulated. In contrast, treatment of primary rat hepatocytes with dexamethasone (Courtois *et al.*, 1999), as well as 2-acetylaminofluorene, phenobarbital and cisplatin (Schrenk *et al.*, 2001), increased both rat *Mrp2* mRNA and protein expression. In addition to the regulatory effects on gene expression, short-term regulation of Mrp2 protein expression has been described. Thus, during hypo-osmotic exposure, rat Mrp2 is localized largely in the canalicular membrane whereas during hyperosmotic exposure, Mrp2 becomes detectable in intracellular vesicles (Kubitz *et al.*, 1997). The osmotic dependence of the subcellular distribution of Mrp2 is fully reversible, suggesting the possibility of short-term regulation of protein function by changing the localization of the protein by endocytic retrieval and exocytic insertion. Recently, some regulatory properties of mouse Mrp2 expression have been investigated, showing that several bile acids, such as ursodeoxycholic acid and cholic acid, can induce *Mrp2* mRNA and protein in mouse liver. This increased expression of Mrp2 may serve to prevent hepatocellular accumulation of potentially toxic bile acids (Fickert *et al.*, 2001).

CONCLUSIONS AND PERSPECTIVES

MRP2 (ABCC2) is located in the apical membrane of many polarized cells, including hepatocytes, kidney proximal tubules, intestinal epithelia and bronchial epithelia. MRP2 shares only 48% amino acid identity with MRP1, but has a similar substrate specificity. Prototypic substrates include the glutathione *S*-conjugate LTC_4, monoglucuronosyl and bisglucuronosyl bilirubin, and 17β-glucuronosyl estradiol. Hereditary mutations leading to the absence of functional MRP2 protein from the apical membrane cause Dubin–Johnson syndrome in humans, which is associated with conjugated hyperbilirubinemia because of the impaired ATP-dependent transport of bilirubin glucuronosides across the hepatocyte canalicular (apical) membrane. Several missense mutations in the *MRP2* gene lead to impaired maturation of the MRP2 glycoprotein, deficient trafficking to the apical membrane, and to degradation of the mutant protein in proteasomes. MRP2-mediated transport of anionic conjugates into bile, urine, and into the intestinal lumen represents a important final step in the detoxification of drugs, toxins and endogenous substances. Polymorphisms in the human *MRP2* gene may affect the pathways of drug elimination and detoxification. Enhanced expression of MRP2 in cancer cells and its localization to the plasma membrane confers resistance to multiple chemotherapeutic agents. In normal tissues and epithelia, however, MRP2 has its role in detoxification and chemoprotection.

ACKNOWLEDGMENTS

The studies in the authors' laboratory were supported by Deutsches Krebsforschungszentrum and by the Deutsche Forschungsgemeinschaft, Bonn, Germany. We acknowledge the contributions to the work described here from past and present members of our laboratory, particularly from Inka Leier, Hiroyuki Tsujii, Gabriele Jedlitschky, Wolfgang Hagmann, Daniel Rost, Thomas Schaub, Tobias Cantz, Markus Donner and Verena Keitel, as well as the collaboration with Jürgen Kartenbeck and Herbert Spring from the Cell Biology Division of Deutsches Krebsforschungszentrum.

REFERENCES

Akita, H., Suzuki, H., Ito, K., Kinoshita, S., Sato, N., Takikawa, H. and Sugiyama, Y. (2001) Characterization of bile acid transport mediated by multidrug resistance associated protein 2 and bile salt export pump. Biochim. Biophys. Acta 1511, 7–16.

Ambudkar, S.V., Dey, S., Hrycyna, C.A., Ramachandra, M., Pastan, I. and Gottesman, M.M. (1999) Biochemical, cellular, and pharmacological aspects of the multidrug transporter. Annu. Rev. Pharmacol. Toxicol. 39, 361–398.

Arts, H.J., Katsaros, D., de Vries, E.G., Massobrio, M., Genta, F., Danese, S., *et al.* (1999) Drug resistance-associated markers P-glycoprotein, multidrug resistance-associated protein 1, multidrug resistance-associated protein 2, and lung resistance protein as prognostic factors in ovarian carcinoma. Clin. Cancer Res. 5, 2798–2805.

Benet, L.Z., Izumi, T., Zhang, Y., Silverman, J.A. and Wacher, V.J. (1999) Intestinal MDR transport proteins and P-450 enzymes as barriers to oral drug delivery. J. Control. Release 62, 25–31.

Beuers, U., Bilzer, M., Chittattu, A., Kullak-Ublick, G.A., Keppler, D., Paumgartner, G. and Dombrowski, F. (2001) Tauroursodeoxycholic acid inserts the apical conjugate export pump, Mrp2, into canalicular membranes and stimulates organic anion secretion by protein kinase C-dependent mechanisms in cholestatic rat liver. Hepatology 33, 1206–1216.

Bock, K.W., Eckle, T., Ouzzine, M. and Fournel-Gigleux, S. (2000) Coordinate induction by antioxidants of UDP-glucuronosyltransferase UGT1A6 and the apical conjugate export pump MRP2 (multidrug resistance protein 2) in Caco-2 cells. Biochem. Pharmacol. 59, 467–470.

Böhme, M., Büchler, M., Müller, M. and Keppler, D. (1993) Differential inhibition by cyclosporins of primary-active ATP-dependent transporters in the hepatocyte canalicular membrane. FEBS Lett. 333, 193–196.

Borst, P., Evers, R., Kool, M. and Wijnholds, J. (1999) The multidrug resistance protein family. Biochim. Biophys. Acta 1461, 347–357.

Broeks, A., Gerrard, B., Allikmets, R., Dean, M. and Plasterk, R.H. (1996) Homologues of the human multidrug resistance genes MRP and MDR contribute to heavy metal resistance in the soil nematode *Caenorhabditis elegans*. EMBO J. 15, 6132–6143.

Büchler, M., König, J., Brom, M., Kartenbeck, J., Spring, H., Horie, T. and Keppler, D. (1996) cDNA cloning of the hepatocyte canalicular isoform of the multidrug resistance protein, cMRP, reveals a novel conjugate export pump deficient in hyperbilirubinemic mutant rats. J. Biol. Chem. 271, 15091–15098.

Cantz, T., Nies, A.T., Brom, M., Hofmann, A.F. and Keppler, D. (2000) MRP2, a human conjugate export pump, is present and transports fluo 3 into apical vacuoles of Hep G2 cells. Am. J. Physiol. 278, G522–G531.

Chen, Z.S., Kawabe, T., Ono, M., Aoki, S., Sumizawa, T., Furukawa, T., Uchiumi, T., Wada, M., Kuwano, M. and Akiyama, S.I. (1999) Effect of multidrug resistance-reversing agents on transporting activity of human canalicular multispecific organic anion transporter. Mol. Pharmacol. 56, 1219–1228.

Chu, X.Y., Kato, Y., Niinuma, K., Sudo, K.I., Hakusui, H. and Sugiyama, Y. (1997) Multispecific organic anion transporter is responsible for the biliary excretion of the camptothecin derivative irinotecan and its metabolites in rats. J. Pharmacol. Exp. Ther. 281, 304–314.

Cole, S.P.C., Bhardwaj, G., Gerlach, J.H., Mackie, J.E., Grant, C.E., Almquist, K.C., Stewart, A.J., Kurz, E.U., Duncan, A.M. and Deeley, R.G. (1992) Overexpression of a transporter gene in a multidrug-resistant human lung cancer cell line. Science 258, 1650–1654.

Cole, S.P.C., Sparks, K.E., Fraser, K., Loe, D.W., Grant, C.E., Wilson, G.M. and Deeley, R.G. (1994) Pharmacological characterization of multidrug resistant MRP-transfected human tumor cells. Cancer Res. 54, 5902–5910.

Conrad, S., Viertelhaus, A., Orzechowski, A., Hoogstraate, J., Gjellan, K., Schrenk, D. and Kauffmann, H.M. (2001) Sequencing and tissue distribution of the canine MRP2 gene compared with MRP1 and MDR1. Toxicology 156, 81–91.

Courtois, A., Payen, L., Guillouzo, A. and Fardel, O. (1999) Up-regulation of multidrug resistance-associated protein 2 (MRP2) expression in rat hepatocytes by dexamethasone. FEBS Lett. 459, 381–385.

Cui, Y., König, J., Buchholz, U., Spring, H., Leier, I. and Keppler, D. (1999) Drug resistance and ATP-dependent conjugate transport mediated by the apical multidrug resistance protein, MRP2, permanently expressed in human and canine cells. Mol. Pharmacol. 55, 929–937.

Cui, Y., König, J. and Keppler, D. (2001) Vectorial transport by double-transfected cells expressing the human uptake transporter SLC21A8 and the apical export pump ABCC2. Mol. Pharmacol. 60, 934–943.

Dietrich, C.G., de Waart, D.R., Ottenhoff, R., Bootsma, A.H., van Gennip, A.H. and Elferink, R.P. (2001) Mrp2-deficiency in the rat impairs biliary and intestinal excretion and influences metabolism and disposition of the food-derived carcinogen 2-amino-1-methyl-6-phenylimidazo. Carcinogenesis 22, 805–811.

Dombrowski, F., Kubitz, R., Chittattu, A., Wettstein, M., Saha, N. and Häussinger, D. (2000) Electron-microscopic demonstration of multidrug resistance protein 2 (Mrp2) retrieval from the canalicular membrane in response to hyperosmolarity and lipopolysaccharide. Biochem. J. 348, 183–188.

Dubin, I.N. and Johnson, F.B. (1954) Chronic idiopathic jaundice with unidentified pigment in liver cells; a new clinicopathologic entity with report of 12 cases. Medicine 33, 155–179.

Dussault, I., Lin, M., Hollister, K., Wang, E.H., Synold, T.W. and Forman, B.M. (2001) Peptide mimetic HIV protease inhibitors are ligands for the orphan receptor SXR. J. Biol. Chem. 276, 33309–33312.

Evers, R., Kool, M., van Deemter, L., Janssen, H., Calafat, J., Oomen, L.C., *et al.* (1998) Drug export activity of the human canalicular multispecific organic anion transporter in polarized kidney MDCK cells expressing cMOAT (MRP2) cDNA. J. Clin. Invest. 101, 1310–1319.

Fernandez-Checa, J.C., Takikawa, H., Horie, T., Ookhtens, M. and Kaplowitz, N. (1992) Canalicular transport of reduced glutathione in normal and mutant Eisai hyperbilirubinemic rats. J. Biol. Chem. 267, 1667–1673.

Fickert, P., Zollner, G., Fuchsbichler, A., Stumptner, C., Pojer, C., Zenz, R., *et al.* (2001) Effects of ursodeoxycholic and cholic acid feeding on hepatocellular transporter expression in mouse liver. Gastroenterology 121, 170–183.

Fritz, F., Chen, J., Hayes, P. and Sirotnak, F.M. (2000) Molecular cloning of the murine cMOAT ATPase. Biochim. Biophys. Acta 1492, 531–536.

Fromm, M.F., Kauffmann, H.M., Fritz, P., Burk, O., Kroemer, H.K., Warzok, R.W., Eichelbaum, M., Siegmund, W. and Schrenk, D. (2000) The effect of rifampin treatment on intestinal expression of human MRP transporters. Am. J. Pathol. 157, 1575–1580.

Grant, C.E., Valdimarsson, G., Hipfner, D.R., Almquist, K.C., Cole, S.P. and Deeley, R.G. (1994) Overexpression of multidrug resistance-associated protein (MRP) increases resistance to natural product drugs. Cancer Res. 54, 357–361.

Grant, C.E., Kurz, E.U., Cole, S.P. and Deeley, R.G. (1997) Analysis of the intron–exon organization of the human multidrug-resistance protein gene (MRP) and alternative splicing of its mRNA. Genomics 45, 368–378.

Hagmann, W., Nies, A.T., König, J., Frey, M., Zentgraf, H. and Keppler, D. (1999) Purification of the human apical conjugate export pump MRP2. Reconstitution and functional characterization as substrate-stimulated ATPase. Eur. J. Biochem. 265, 281–289.

Hagmann, W., Schubert, J., König, J. and Keppler, D. (2002) Reconstitution of transport-active multidrug resistance protein 2 (MRP2; ABCC2) in proteoliposomes. Biol. Chem. 383, 1001–1009.

Harris, M.J., Kuwano, M., Webb, M. and Board, P.G. (2001) Identification of the apical membrane-targeting signal of the multidrug resistance-associated protein 2 (MRP2/MOAT). J. Biol. Chem. 276, 20876–20881.

Hinoshita, E., Uchiumi, T., Taguchi, K., Kinukawa, N., Tsuneyoshi, M., Maehara, Y., Sugimachi, K. and Kuwano, M. (2000) Increased expression of an ATP-binding cassette superfamily transporter, multidrug resistance protein 2, in human colorectal carcinomas. Clin. Cancer Res. 6, 2401–2407.

Hipfner, D.R., Almquist, K.C., Leslie, E.M., Gerlach, J.H., Grant, C.E., Deeley, R.G. and Cole, S.P. (1997) Membrane topology of the multidrug resistance protein (MRP). A study of glycosylation-site mutants reveals an extracytosolic NH_2 terminus. J. Biol. Chem. 272, 23623–23630.

Hirohashi, T., Suzuki, H., Ito, K., Ogawa, K., Kume, K., Shimizu, T. and Sugiyama, Y. (1998) Hepatic expression of multidrug resistance-associated protein-like proteins maintained in Eisai hyperbilirubinemic rats. Mol. Pharmacol. 53, 1068–1075.

Hooijberg, J.H., Broxterman, H.J., Kool, M., Assaraf, Y.G., Peters, G.J., Noordhuis, P., Scheper, R.J., Borst, P., Pinedo, H.M. and Jansen, G. (1999) Antifolate resistance mediated by the multidrug resistance proteins MRP1 and MRP2. Cancer Res. 59, 2532–2535.

Ihrke, G., Neufeld, E.B., Meads, T., Shanks, M.R., Cassio, D., Laurent, M., Schroer, T.A., Pagano, R.E. and Hubbard, A.L. (1993) WIF-B cells: an *in vitro* model for studies of hepatocyte polarity. J. Cell Biol. 123, 1761–1775.

Ishikawa, T., Müller, M., Klünemann, C., Schaub, T. and Keppler, D. (1990) ATP-dependent primary active transport of cysteinyl leukotrienes across liver canalicular membrane. Role of the ATP-dependent transport system for glutathione *S*-conjugates. J. Biol. Chem. 265, 19279–19286.

Ishizuka, H., Konno, K., Naganuma, H., Sasahara, K., Kawahara, Y., Niinuma, K., Suzuki, H. and Sugiyama, Y. (1997) Temocaprilat, a novel angiotensin-converting enzyme inhibitor, is excreted in bile via an ATP-dependent active transporter (cMOAT) that is deficient in Eisai hyperbilirubinemic mutant rats (EHBR). J. Pharmacol. Exp. Ther. 280, 1304–1311.

Ito, K., Suzuki, H., Hirohashi, T., Kume, K., Shimizu, T. and Sugiyama, Y. (1997) Molecular cloning of canalicular multispecific organic anion transporter defective in EHBR. Am. J. Physiol. 272, G16–G22.

Ito, K., Suzuki, H., Hirohashi, T., Kume, K., Shimizu, T. and Sugiyama, Y. (1998) Functional analysis of a canalicular multispecific organic anion transporter cloned from rat liver. J. Biol. Chem. 273, 1684–1688.

Ito, K., Oleschuk, C.J., Westlake, C., Vasa, M.Z., Deeley, R.G. and Cole, S.P.C. (2001a) Mutation of Trp^{1254} in the multispecific organic anion transporter, MRP2 (ABCC2), alters substrate specificity and results in loss of methotrexate transport activity. J. Biol. Chem. 276, 38108–38114.

Ito, K., Suzuki, H. and Sugiyama, Y. (2001b) Charged amino acids in the transmembrane domains are involved in the determination of the substrate specificity of rat Mrp2. Mol. Pharmacol. 59, 1077–1085.

Ito, S., Ieiri, I., Tanabe, M., Suzuki, A., Higuchi, S. and Otsubo, K. (2001c) Polymorphism of the ABC transporter genes, MDR1, MRP1 and MRP2/cMOAT, in healthy Japanese subjects. Pharmacogenetics 11, 175–184.

Jansen, P.L., Peters, W.H. and Lamers, W.H. (1985) Hereditary chronic conjugated hyperbilirubinemia in mutant rats caused by defective hepatic anion transport. Hepatology 5, 573–579.

Jedlitschky, G., Leier, I., Böhme, M., Buchholz, U., Bar-Tana, J. and Keppler, D. (1994) Hepatobiliary elimination of the peroxisome proliferator nafenopin by conjugation and subsequent ATP-dependent transport across the canalicular membrane. Biochem. Pharmacol. 48, 1113–1120.

Jedlitschky, G., Leier, I., Buchholz, U., Hummel-Eisenbeiss, J., Burchell, B. and Keppler, D. (1997) ATP-dependent transport of bilirubin glucuronides by the multidrug resistance protein MRP1 and its hepatocyte canalicular isoform MRP2. Biochem. J. 327, 305–310.

Kajihara, S., Hisatomi, A., Mizuta, T., Hara, T., Ozaki, I., Wada, I. and Yamamoto, K. (1998) A splice mutation in the human canalicular multispecific organic anion transporter gene causes Dubin–Johnson syndrome. Biochem. Biophys. Res. Commun. 253, 454–457.

Kamisako, T., Leier, I., Cui, Y., König, J., Buchholz, U., Hummel-Eisenbeiss, J. and Keppler, D. (1999) Transport of monoglucuronosyl and bisglucuronosyl bilirubin by recombinant human and rat multidrug resistance protein 2. Hepatology 30, 485–490.

Kao, J.P., Harootunian, A.T. and Tsien, R.Y. (1989) Photochemically generated cytosolic calcium pulses and their detection by fluo-3. J. Biol. Chem. 264, 8179–8184.

Kartenbeck, J., Leuschner, U., Mayer, R. and Keppler, D. (1996) Absence of the canalicular isoform of the MRP gene-encoded conjugate export pump from the hepatocytes in Dubin–Johnson syndrome. Hepatology 23, 1061–1066.

Kast, C. and Gros, P. (1998) Epitope insertion favors a six transmembrane domain model for the carboxy-terminal portion of the multidrug resistance-associated protein. Biochemistry 37, 2305–2313.

Kauffmann, H.M., Keppler, D., Gant, T.W. and Schrenk, D. (1998) Induction of hepatic *mrp2 (cmrp/cmoat)* gene expression in nonhuman primates treated with rifampicin or tamoxifen. Arch. Toxicol. 72, 763–768.

Keitel, V., Kartenbeck, J., Nies, A.T., Spring, H., Brom, M. and Keppler, D. (2000) Impaired protein maturation of the conjugate export pump MRP2 as a consequence of a deletion mutation in Dubin–Johnson syndrome. Hepatology 32, 1317–1328.

Keppler, D. (1992) Leukotrienes: biosynthesis, transport, inactivation, and analysis. Rev. Physiol. Biochem. Pharmacol. 121, 1–30.

Keppler, D. and Kartenbeck, J. (1996) The canalicular conjugate export pump encoded by the *cmrp/cmoat* gene. Prog. Liver Dis. 14, 55–67.

Keppler, D. and König, J. (1997) Expression and localization of the conjugate export pump encoded by the *MRP2 (cMRP/cMOAT)* gene in liver. FASEB J. 11, 509–516.

Keppler, D., Jedlitschky, G. and Leier, I. (1998) Transport function and substrate specificity of multidrug resistance protein. Methods Enzymol. 292, 607–616.

Keppler, D., Kamisako, T., Leier, I., Cui, Y., Nies, A.T., Tsujii, H. and Konig, J. (2000) Localization, substrate specificity, and drug resistance conferred by conjugate export pumps of the MRP family. Adv. Enzyme Regul. 40, 339–349.

Keppler, D., König, J. and Nies, A.T. (2001) Conjugate export pumps of the multidrug resistance protein (MRP) family in liver. In: The Liver. Biology and Pathobiology (ed. I.M. Arias, J.L. Boyer, F.V. Chisari, N. Fausto, D. Schachter and D.A. Shafritz), pp. 373–382. New York: Lippincott/Williams & Wilkins.

Kitamura, T., Jansen, P., Hardenbrook, C., Kamimoto, Y., Gatmaitan, Z. and Arias, I.M. (1990) Defective ATP-dependent bile canalicular transport of organic anions in mutant (TR^-) rats with conjugated hyperbilirubinemia. Proc. Natl Acad. Sci. USA 87, 3557–3561.

Kiuchi, Y., Suzuki, H., Hirohashi, T., Tyson, C.A. and Sugiyama, Y. (1998) cDNA cloning and inducible expression of human multidrug resistance associated protein 3 (MRP3). FEBS Lett. 433, 149–152.

Kobayashi, K., Sogame, Y., Hayashi, K., Nicotera, P. and Orrenius, S. (1988) ATP stimulates the uptake of *S*-dinitrophenylglutathione by rat liver plasma membrane vesicles. FEBS Lett. 240, 55–58.

Koike, K., Kawabe, T., Tanaka, T., Toh, S., Uchiumi, T., Wada, M., Akiyama, S., Ono, M. and Kuwano, M. (1997) A canalicular multispecific organic anion transporter (cMOAT) antisense cDNA enhances drug sensitivity in human hepatic cancer cells. Cancer Res. 57, 5475–5479.

König, J., Nies, A.T., Cui, Y., Leier, I. and Keppler, D. (1999a) Conjugate export pumps of the multidrug resistance protein (MRP) family: localization, substrate specificity, and MRP2-mediated drug resistance. Biochim. Biophys. Acta 1461, 377–394.

König, J., Rost, D., Cui, Y. and Keppler, D. (1999b) Characterization of the human multidrug resistance protein isoform MRP3 localized to the basolateral hepatocyte membrane. Hepatology 29, 1156–1163.

Kool, M., de Haas, M., Scheffer, G.L., Scheper, R.J., van Eijk, M.J., Juijn, J.A., Baas, F. and Borst, P. (1997) Analysis of expression of cMOAT (MRP2), MRP3, MRP4, and MRP5, homologues of the multidrug resistance-associated protein gene (MRP1), in human cancer cell lines. Cancer Res. 57, 3537–3547.

Kool, M., van der Linden, M., de Haas, M., Baas, F. and Borst, P. (1999) Expression of human MRP6, a homologue of the multidrug resistance protein gene MRP1, in tissues and cancer cells. Cancer Res. 59, 175–182.

Kubitz, R., D'Urso, D., Keppler, D. and Häussinger, D. (1997) Osmodependent dynamic localization of the multidrug resistance protein 2 in the rat hepatocyte canalicular membrane. Gastroenterology 113, 1438–1442.

Kubitz, R., Wettstein, M., Warskulat, U. and Häussinger, D. (1999) Regulation of the multidrug resistance protein 2 in the rat liver by lipopolysaccharide and dexamethasone. Gastroenterology 116, 401–410.

Leier, I., Jedlitschky, G., Buchholz, U., Cole, S.P., Deeley, R.G. and Keppler, D. (1994a) The MRP gene encodes an ATP-dependent export pump for leukotriene C_4 and structurally related conjugates. J. Biol. Chem. 269, 27807–27810.

Leier, I., Jedlitschky, G., Buchholz, U. and Keppler, D. (1994b) Characterization of the ATP-dependent leukotriene C_4 export carrier in mastocytoma cells. Eur. J. Biochem. 220, 599–606.

Leier, I., Hummel-Eisenbeiss, J., Cui, Y. and Keppler, D. (2000) ATP-dependent para-aminohippurate transport by apical multidrug resistance protein MRP2. Kidney Int. 57, 1636–1642.

Loe, D.W., Almquist, K.C., Deeley, R.G. and Cole, S.P. (1996) Multidrug resistance protein (MRP)-mediated transport of leukotriene C_4 and chemotherapeutic agents in membrane vesicles. Demonstration of glutathione-dependent vincristine transport. J. Biol. Chem. 271, 9675–9682.

Madon, J., Hagenbuch, B., Landmann, L., Meier, P.J. and Stieger, B. (2000) Transport function and hepatocellular localization of mrp6 in rat liver. Mol. Pharmacol. 57, 634–641.

Masuda, M., I'izuka, Y., Yamazaki, M., Nishigaki, R., Kato, Y., Ni'inuma, K., Suzuki, H. and Sugiyama, Y. (1997) Methotrexate is excreted into the bile by canalicular multispecific organic anion transporter in rats. Cancer Res. 57, 3506–3510.

Mayer, R., Kartenbeck, J., Büchler, M., Jedlitschky, G., Leier, I. and Keppler, D. (1995) Expression of the MRP gene-encoded conjugate export pump in liver and its selective absence from the canalicular membrane in transport-deficient mutant hepatocytes. J. Cell Biol. 131, 137–150.

Minemura, M., Tanimura, H. and Tabor, E. (1999) Overexpression of multidrug resistance genes MDR1 and cMOAT in human hepatocellular carcinoma and hepatoblastoma cell lines. Int. J. Oncol. 15, 559–563.

Minta, A., Kao, J.P.Y. and Tsien, R.Y. (1989) Fluorescent indicators for cytosolic calcium based on rhodamine and fluorescein chromophores. J. Biol. Chem. 264, 8171–8178.

Mor-Cohen, R., Zivelin, A., Rosenberg, N., Shani, M., Muallem, S. and Seligsohn, U. (2001) Identification and functional analysis of two novel mutations in the multidrug-resistance protein 2 gene in Israeli patients with Dubin–Johnson syndrome. J. Biol. Chem. 276, 36923–36930.

Morrow, C.S., Smitherman, P.K. and Townsend, A.J. (2001) Role of multidrug-resistance protein 2 in glutathione S-transferase P1-1-mediated resistance to 4-nitroquinoline 1-oxide toxicities in HepG2 cells. Mol. Carcinog. 29, 170–178.

Mottino, A.D., Hoffman, T., Jennes, L. and Vore, M. (2000) Expression and localization of multidrug resistant protein mrp2 in rat small intestine. J. Pharmacol. Exp. Ther. 293, 717–723.

Narasaki, F., Oka, M., Nakano, R., Ikeda, K., Fukuda, M., Nakamura, T., Soda, H., Nakagawa, M., Kuwano, M. and Kohno, S. (1997) Human canalicular multispecific organic anion transporter (cMOAT) is expressed in human lung, gastric, and colorectal cancer cells. Biochem. Biophys. Res. Commun. 240, 606–611.

Nies, A.T., Cantz, T., Brom, M., Leier, I. and Keppler, D. (1998) Expression of the apical conjugate export pump, Mrp2, in the polarized hepatoma cell line, WIF-B. Hepatology 28, 1332–1340.

Nies, A.T., König, J., Pfannschmidt, M., Klar, E., Hofmann, W.J. and Keppler, D. (2001) Expression of the multidrug resistance proteins MRP2 and MRP3 in human hepatocellular carcinoma. Int. J. Cancer 94, 492–499.

Niinuma, K., Takenaka, O., Horie, T., Kobayashi, K., Kato, Y., Suzuki, H. and Sugiyama, Y. (1997) Kinetic analysis of the primary active transport of conjugated metabolites across the bile canalicular membrane: comparative study of S-(2, 4-dinitrophenyl)-glutathione and 6-hydroxy-5,7-dimethyl-2-methylamino-4-(3-pyridylmethyl)benzothiazole glucuronide. J. Pharmacol. Exp. Ther. 282, 866–872.

Nishida, T., Hardenbrook, C., Gatmaitan, Z. and Arias, I.M. (1992) ATP-dependent organic anion transport system in normal and TR^- rat liver canalicular membranes. Am. J. Physiol. 262, G629–G635.

Oswald, M., Kullak-Ublick, G.A., Paumgartner, G. and Beuers, U. (2001)

Expression of hepatic transporters OATP-C and MRP2 in primary sclerosing cholangitis. Liver 21, 247–253.

Oude Elferink, R.P.J., Bakker, C.T.M., Roelofsen, H., Ottenhoff, R., Heijn, M. and Jansen, P.L.M. (1993) Accumulation of organic anion in intracellular vesicles of cultured rat hepatocytes is mediated by the canalicular multispecific organic anion transporter. Hepatology 17, 434–444.

Paulusma, C.C., Bosma, P.J., Zaman, G.J., Bakker, C.T., Otter, M., Scheffer, G.L., Scheper, R.J., Borst, P. and Oude Elferink, R.P. (1996) Congenital jaundice in rats with a mutation in a multidrug resistance-associated protein gene. Science 271, 1126–1128.

Paulusma, C.C., Kool, M., Bosma, P.J., Scheffer, G.L., ter Borg, F., Scheper, R.J., Tytgat, G.N., Borst, P., Baas, F. and Oude Elferink, R.P. (1997) A mutation in the human canalicular multispecific organic anion transporter gene causes the Dubin–Johnson syndrome. Hepatology 25, 1539–1542.

Paulusma, C.C., Kothe, M.J., Bakker, C.T., Bosma, P.J., van Bokhoven, I., van Marle, J., Bolder, U., Tytgat, G.N. and Oude Elferink, R.P. (2000) Zonal down-regulation and redistribution of the multidrug resistance protein 2 during bile duct ligation in rat liver. Hepatology 31, 684–693.

Rea, P.A., Li, Z., Lu, Y. and Drozdowicz, Y.M. (1998) From vacuolar GS-X pumps to multispecific ABC transporters. Annu. Rev. Plant Physiol. Plant Mol. Biol. 49, 727–760.

Riordan, J.R., Rommens, J.M., Kerem, B., Alon, N., Rozmahel, R., Grzelczak, Z., *et al.* (1989) Identification of the cystic fibrosis gene: cloning and characterization of complementary DNA. Science 245, 1066–1073.

Roelofsen, H., Bakker, C.T., Schoemaker, B., Heijn, M., Jansen, P.L. and Oude Elferink, R.P.J. (1995) Redistribution of canalicular organic anion transport activity in isolated and cultured rat hepatocytes. Hepatology 21, 1649–1657.

Roelofsen, H., Soroka, C.J., Keppler, D. and Boyer, J.L. (1998) Cyclic AMP stimulates sorting of the canalicular organic anion transporter (Mrp2/cMoat) to the apical domain in hepatocyte couplets. J. Cell Sci. 111, 1137–1145.

Rost, D., Kartenbeck, J. and Keppler, D. (1999) Changes in the localization of the rat canalicular conjugate export pump Mrp2 in phalloidin-induced cholestasis. Hepatology 29, 814–821.

Rost, D., König, J., Weiss, G., Klar, E., Stremmel, W. and Keppler, D. (2001) Expression and localization of the multidrug resistance proteins MRP2 and MRP3 in human gallbladder epithelia. Gastroenterology 121, 1203–1208.

Roy Chowdhury, J., Roy Chowdhury, N., Wolkoff, A.W. and Arias, I.M. (1994) Heme and bile pigment metabolism. In: The Liver: Biology and Pathology (ed. I.M. Arias, J.L. Boyer, N. Fausto, W.B. Jakoby, D.A. Schachter and D.A. Shafritz), pp. 471–504. New York: Raven.

Ryu, S., Kawabe, T., Nada, S. and Yamaguchi, A. (2000) Identification of basic residues involved in drug export function of human multidrug resistance-associated protein 2. J. Biol. Chem. 275, 39617–39624.

Sasabe, H., Tsuji, A. and Sugiyama, Y. (1998) Carrier-mediated mechanism for the biliary excretion of the quinolone antibiotic grepafloxacin and its glucuronide in rats. J. Pharmacol. Exp. Ther. 284, 1033–1039.

Schaub, T.P., Kartenbeck, J., König, J., Vogel, O., Witzgall, R., Kriz, W. and Keppler, D. (1997) Expression of the conjugate export pump encoded by the mrp2 gene in the apical membrane of kidney proximal tubules. J. Am. Soc. Nephrol. 8, 1213–1221.

Schaub, T.P., Kartenbeck, J., König, J., Spring, H., Dörsam, J., Staehler, G., Störkel, S., Thon, W.F. and Keppler, D. (1999) Expression of the MRP2 gene-encoded conjugate export pump in human kidney proximal tubules and in renal cell carcinoma. J. Am. Soc. Nephrol. 10, 1159–1169.

Schrenk, D., Baus, P.R., Ermel, N., Klein, C., Vorderstemann, B. and Kauffmann, H.M. (2001) Up-regulation of transporters of the MRP family by drugs and toxins. Toxicol. Lett. 120, 51–57.

Sormunen, R., Eskelinen, S. and Lehto, V. (1993) Bile canaliculus formation in cultured HepG2 cells. Lab. Invest. 68, 652–662.

Sprinz, H. and Nelson, R.S. (1954) Persistent nonhemolytic hyperbilirubinemia associated with lipochrome-like pigment in liver cells; report of four cases. Ann. Intern. Med. 41, 952–962.

St-Pierre, M.V., Serrano, M.A., Macias, R.I., Dubs, U., Hoechli, M., Lauper, U., Meier, P.J. and Marin, J.J. (2000) Expression of members of the multidrug resistance protein family in human term placenta. Am. J. Physiol. Regul. Integr. Comp. Physiol. 279, R1495–R1503.

Stöckel, B., König, J., Nies, A.T., Cui, Y., Brom, M. and Keppler, D. (2000) Characterization of the 5′-flanking region of the human multidrug resistance protein 2 (MRP2) gene and its regulation in comparison with the multidrug resistance protein 3 (MRP3) gene. Eur. J. Biochem. 267, 1347–1358.

Takenaka, O., Horie, T., Kobayashi, K., Suzuki, H. and Sugiyama, Y. (1995) Kinetic analysis of hepatobiliary transport for conjugated metabolites in the perfused liver of mutant rats (EHBR) with hereditary conjugated hyperbilirubinemia. Pharm. Res. 12, 1746–1755.

Takikawa, H., Sano, N., Narita, T., Uchida, Y., Yamanaka, M., Horie, T., Mikami, T. and Tagaya, O. (1991) Biliary excretion of bile acid conjugates in a hyperbilirubinemic mutant Sprague-Dawley rat. Hepatology 14, 352–360.

Tanaka, T., Uchiumi, T., Hinoshita, E., Inokuchi, A., Toh, S., Wada, M., Takano, H., Kohno, K. and Kuwano, M. (1999) The human multidrug resistance protein 2 gene: functional characterization of the 5′-flanking region and expression in hepatic cells. Hepatology 30, 1507–1512.

Taniguchi, K., Wada, M., Kohno, K., Nakamura, T., Kawabe, T., Kawakami, M., Kagotani, K., Okumura, K., Akiyama, S. and Kuwano, M. (1996) A human canalicular multispecific organic anion transporter (cMOAT) gene is overexpressed in cisplatin-resistant human cancer cell lines with decreased drug accumulation. Cancer Res. 56, 4124–4129.

Thermann, R., Neu-Yilik, G., Deters, A., Frede, U., Wehr, K., Hagemeier, C., Hentze, M.W. and Kulozik, A.E. (1998) Binary specification of nonsense codons by splicing and cytoplasmic translation. EMBO J. 17, 3484–3494.

Toh, S., Wada, M., Uchiumi, T., Inokuchi, A., Makino, Y., Horie, Y., Adachi, Y., Sakisaka, S. and Kuwano, M. (1999) Genomic structure of the canalicular multispecific organic anion-transporter gene (MRP2/cMOAT) and mutations in the ATP-binding-cassette region in Dubin–Johnson syndrome. Am. J. Hum. Genet. 64, 739–746.

Trauner, M., Arrese, M., Soroka, C.J., Ananthanarayanan, M., Koeppel, T.A., Schlosser, S.F., Suchy, F.J., Keppler, D. and Boyer, J.L. (1997) The rat canalicular conjugate export pump (Mrp2) is down-regulated in intrahepatic and obstructive cholestasis. Gastroenterology 113, 255–264.

Tsujii, H., König, J., Rost, D., Stöckel, B., Leuschner, U. and Keppler, D. (1999) Exon–intron organization of the human multidrug resistance protein 2 (MRP2) gene mutated in Dubin–Johnson syndrome. Gastroenterology 117, 653–660.

van Aubel, R.A., van Kuijck, M.A., Koenderink, J.B., Deen, P.M., van Os, C.H. and Russel, F.G. (1998) Adenosine triphosphate-dependent transport of anionic conjugates by the rabbit multidrug resistance-associated protein Mrp2 expressed in insect cells. Mol. Pharmacol. 53, 1062–1067.

van Aubel, R.A., Hartog, A., Bindels, R.J., van Os, C.H. and Russel, F.G. (2000) Expression and immunolocalization of multidrug resistance protein 2 in rabbit small intestine. Eur. J. Pharmacol. 400, 195–198.

Vernhet, L., Seite, M.P., Allain, N., Guillouzo, A. and Fardel, O. (2001) Arsenic induces expression of the multidrug resistance-associated protein 2 (MRP2) gene in primary rat and human hepatocytes. J. Pharmacol. Exp. Ther. 298, 234–239.

Vos, T.A., Hooiveld, G.J., Koning, H., Childs, S., Meijer, D.K., Moshage, H., Jansen, P.L. and Muller, M. (1998) Up-regulation of the multidrug resistance genes, Mrp1 and Mdr1b, and down-regulation of the organic anion transporter, Mrp2, and the bile salt transporter, Spgp, in endotoxemic rat liver. Hepatology 28, 1637–1644.

Wada, M., Toh, S., Taniguchi, K., Nakamura, T., Uchiumi, T., Kohno, K., *et al.* (1998) Mutations in the canalicular multispecific organic anion transporter (cMOAT) gene, a novel ABC transporter, in patients with hyperbilirubinemia II/Dubin–Johnson syndrome. Hum. Mol. Genet. 7, 203–207.

Walgren, R.A., Karnaky, K.J., Jr, Lindenmayer, G.E. and Walle, T. (2001) Efflux of dietary flavonoid quercetin 4′-beta-glucoside across human intestinal Caco-2 cell monolayers by apical multidrug resistance-associated protein-2. J. Pharmacol. Exp. Ther. 294, 830–836.

Wielandt, A.M., Vollrath, V., Manzano, M., Miranda, S., Accatino, L. and Chianale, J. (1999) Induction of the multispecific organic anion transporter (cMoat/mrp2) gene and biliary glutathione secretion by the herbicide 2,4,5-trichlorophenoxyacetic acid in the mouse liver. Biochem. J. 341, 105–111.

Yamazaki, M., Akiyama, S., Ni'inuma, K., Nishigaki, R. and Sugiyama, Y. (1997) Biliary excretion of pravastatin in rats: contribution of the excretion pathway mediated by canalicular multispecific organic anion transporter. Drug Metab. Dispos. 25, 1123–1129.

Young, L.C., Campling, B.G., Voskoglou-Nomikos, T., Cole, S.P.C., Deeley, R.G. and Gerlach, J.H. (1999) Expression of multidrug resistance protein-related genes in lung cancer: correlation with drug response. Clin. Cancer Res. 5, 673–680.

Young, L.C., Campling, B.G., Cole, S.P.C., Deeley, R.G. and Gerlach, J.H. (2001) Multidrug resistance proteins MRP3, MRP1, and MRP2 in lung cancer: correlation of protein levels with drug response and messenger RNA levels. Clin. Cancer Res. 7, 1798–1804.

CHAPTER 21

THE MULTIDRUG RESISTANCE PROTEINS 3–7

PIET BORST, GLEN REID, TOHRU SAEKI, PETER WIELINGA AND NOAM ZELCER

INTRODUCTION

Following the discovery of the multidrug resistance protein 1 (MRP1) in 1992 (Cole *et al.*, 1992) (Chapter 19) and the subsequent demonstration that the well-known liver canalicular multispecific organic anion transporter, cMOAT (now known as MRP2), was a transporter closely related to MRP1 (Chapter 20), several other related sequences were uncovered by Allikmets *et al.* (1996) and Kool *et al.* (1997) in a database search. The existence of a family of MRP-related transporters has been confirmed in subsequent work and these transporters are now assembled in the C group of ABC transporters (see http://www.nutrigene.4t.com/humanabc.htm) together with CFTR (Chapter 29) and the sulfonylurea receptors (Chapter 27). The present count of MRPs stands at nine and it is unlikely that there are more to come. MRP1 and MRP2 are discussed in separate chapters of this book and very little is known yet about MRP8 and MRP9 (Tammur *et al.*, 2001). Here we focus on MRP3–7. Other recent reviews of the MRP family can be found in (Borst *et al.*, 1999, 2000; Borst and Oude Elferink, 2002; Ishikawa *et al.*, 1994; Keppler, 1999; Renes *et al.*, 1999).

MRPs come in two types of structures, as illustrated in **Figure 21.1**: the MRP1 type, shared by MRP2, 3, 6 and 7; and the MRP4 type, shared by MRP5 (and probably MRP8 and MRP9). The MRP1 type has an additional NH_2-terminal domain, which is thought to have five transmembrane segments and is not present in the MRP4 type. In MRP1, this domain is dispensable for transport function, explaining why proteins that differ substantially in size and putative structure, like MRP1 and MRP4, can still have similar functions. The sequence differences between the known MRPs are summarized in **Table 21.1**.

All MRPs studied thus far are organic anion pumps, but they differ widely in their preferred substrate or tissue location, as illustrated by **Tables 21.2 and 21.3**. MRP1–5 are inhibited by sulfinpyrazone, a classical inhibitor of organic anion transport. Note, however, that sulfinpyrazone is also a substrate of MRP2 (Evers *et al.*, 2000) and that it can actually stimulate transport of *S*-(2,4-dinitrophenyl) glutathione (GS-DNP) (Evers *et al.*, 1998) or glutathione (GSH) (Evers *et al.*, 2000) at lower concentrations.

MRP3

Human MRP3, also known as MOAT-D or cMOAT-2, and formally designated ABCC3, was first spotted by Allikmets and co-workers (1996) in their initial inventory of the human ABC superfamily. Kool *et al.* (1997) recruited this putative transporter to the MRP family and showed that *MRP3* RNA has a rather restricted tissue distribution, with high concentrations in liver, intestine and adrenal gland. The *MRP3* gene is located at human chromosome 17q21.3 and the corresponding protein is 1527 amino acids long (Belinsky *et al.*, 1998; Kiuchi *et al.*,

ABC Proteins: From Bacteria to Man
ISBN 0-12-352551-9

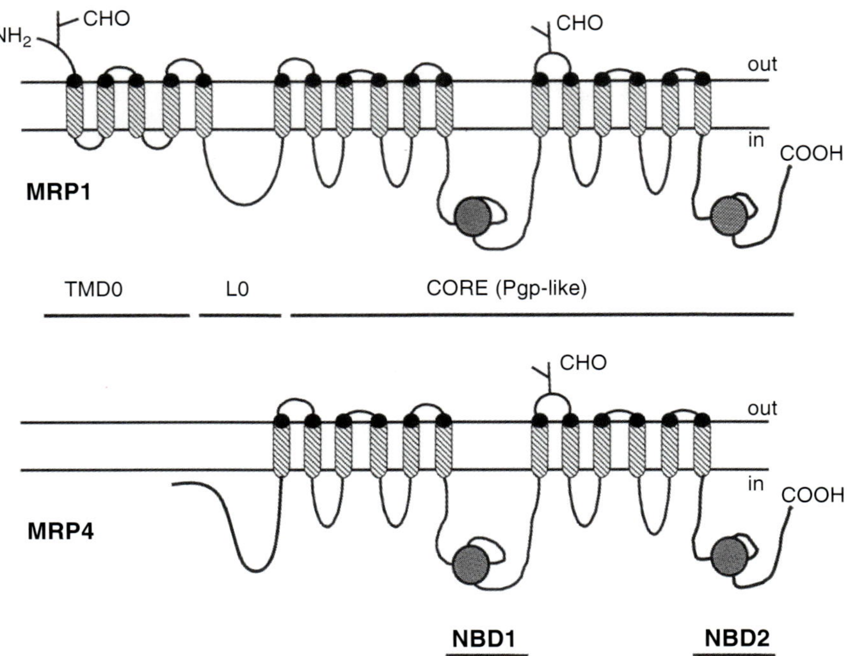

Figure 21.1. ***Predicted membrane (secondary) structure of the two types of MRP-related proteins represented by their prototypic members, MRP1 (MRP1, 2, 3, 6 and 7) and MRP4 (MRP4 and MRP 5). Shown are the TMD0 (extra transmembrane domain), the linker (L0), and the P-glycoprotein-like MDR core; NBD, Nucleotide-binding domain; CHO, glycosylation site. (Adapted from Borst*** **et al.*, 2000.)***

TABLE 21.1. PERCENTAGE AMINO ACID IDENTITY BETWEEN FULLY SEQUENCED HUMAN MULTIDRUG RESISTANCE PROTEINS (MRPS)[a]

	MRP1 1531 aa	MRP2 1545 aa	MRP3 1527 aa	MRP4 1325 aa	MRP5 1437 aa	MRP6 1503 aa	MRP7 1492 aa
MRP1 (ABCC1)	–						
MRP2 (ABCC2)	49	–					
MRP3 (ABCC3)	58	48	–				
MRP4 (ABCC4)	39	37	36	–			
MRP5 (ABCC5)	34	35	33	36	–		
MRP6 (ABCC6)	45	38	43	34	31	–	
MRP7 (ABCC10)	34	34	36	36	36	34	–

[a]Homology between human MRPs, expressed as % amino acid (aa) identity; Borst (1999). For the multiple sequence alignment the GAP program from the University of Wisconsin Genetics Group (GCG) package (version 9.1) was used. The following accession numbers were used: MRP1, L05628; MRP2, U49248; MRP3, AF009670; MRP4, AF071202; MRP5, AF104942; MRP6, AF076622; MRP7, BAA92227.

1998; König *et al.*, 1999; Kool *et al.*, 1999b). Within the MRP family, MRP3 is the MRP most closely related to MRP1 **(Table 21.1)**.

Studies on rat Mrp3 have contributed substantially to our present understanding of the substrate specificity of human MRP3. An MRP-like protein, called MLP-2, was first identified by the group of Suzuki and Sugiyama in the liver of *Mrp2* (−/−) rats, in which it is highly upregulated (Hirohashi *et al.*, 1998). This protein proved to be the rat homologue of human MRP3, and its substrate specificity and inducbility have been studied in some detail (Hirohashi *et al.*, 1999, 2000; Ogawa *et al.*, 2000). A gene knockout (KO) of the mouse *Mrp3* gene was recently produced (our unpublished results). The mice are healthy and fertile, but remain to be characterized.

TABLE 21.2. SUBSTRATE SPECIFICITY OF MRPS

MRP	GS-X pump	Preferred substrates	Transport of MDR drugs	Transport of MTX	Transport of GSH	Sulfinpyrazone inhibition
MRP1	+	GS-X Gluc-X	+	+	+	+
MRP2	+	GS-X Gluc-X	+	+	+	+
MRP3	+	Gluc-X Sulf-X	+	+	−	+
MRP4	?	cGMP, cAMP, NMP analogues, Gluc-X	−	+	?	+
MRP5	+	cGMP, NMP analogues	−?	−	+	+
MRP6	−	Peptides?	−	−	?	−
MRP7	?	?	?	?	?	?

Abbreviations: GS-X, Gluc-X, Sulf-X are conjugates of an organic compound (X) with glutathione (GS), glucuronide (Gluc) or sulfate (Sulf), respectively; MDR drugs are drugs belonging to the multidrug resistance (MDR) spectrum; MTX, methotrexate; NMP, nucleoside monophosphate. See text for details and references.

TABLE 21.3. TISSUE DISTRIBUTION AND LOCATION IN THE PLASMA MEMBRANE OF POLARIZED EPITHELIA OF HUMAN MULTIDRUG RESISTANCE PROTEINS (MRPS)

	Tissue distribution	Plasma membrane location
MRP1	Ubiquitous (low in liver)	Basolateral
MRP2	Liver, kidney, gut	Apical
MRP3	Liver, adrenals, pancreas, kidney, gut, gallbladder	Basolateral
MRP4	Prostate, lung, muscle, pancreas, testis, ovary, bladder, gallbladder	Apical?
MRP5	Ubiquitous	Basolateral
MRP6	Liver, kidney	Basolateral?
MRP7	Ubiquitous (low)	?

Adapted from Borst *et al.* (2000).

SUBSTRATE SPECIFICITY OF MRP3 AND rMRP3

The substrate specificity of MRP3 is summarized in **Tables 21.4 and 21.5**. Drug resistance of cells transfected with *MRP3* cDNA constructs is limited, as shown by the results presented in **Table 21.4**. Substantial resistance of the transfected cells is found only against etoposide and teniposide, and against methotrexate (MTX) in short-time (4 h) exposures with high MTX concentrations. Low vincristine resistance was only found in *MRP3*-transfected HEK293 cells (Zeng *et al.*, 2000), but not in transfected 2008 human ovarian cells (Kool *et al.*, 1999b), pig kidney cells (Haga *et al.*, 2001), or mouse fibroblast cells isolated from triple KO (TKO) mice lacking P-glycoprotein and Mrp1 (Allen *et al.*, 2000). As the transfected TKO cells have the highest etoposide resistance of any *MRP3*-transfected cell analyzed **(Table 21.4)**, we think that more work is required to establish unambiguously transport of *Vinca* alkaloids by MRP3. A problem in all these experiments is that expression of MRP3 is relatively low in most transfected cells (Kool *et al.*, 1999b), complicating the analysis of the resistance spectrum associated with the presence of MRP3. Polarized MDCKII cell transfectants were used to show that MRP3 can transport a classical substrate of organic anion pumps, GS-DNP (Kool *et al.*, 1999b).

We have shown that etoposide resistance in TKO cells is associated with diminished drug accumulation and increased drug extrusion (Zelcer *et al.*, 2001), but the mechanism of etoposide transport is not yet clear. Unlike MRP1 and MRP2, MRP3 does not detectably transport reduced glutathione (GSH) (Kool *et al.*, 1999b) and does not co-transport etoposide with GSH (Zelcer *et al.*, 2001). Resistance

TABLE 21.4. DRUG RESISTANCE INDUCED BY MRP3 EXPRESSION IN TRANSFECTED CELLS (EXPRESSED AS RESISTANCE RELATIVE TO THE UNTRANSFECTED PARENTAL CELL)

Drug	Resistance of MRP3 cells relative to wild-type		
	Human ovarian 2008 carcinoma[a]	HEK293 cells[b]	TKO mouse cells[c]
Etoposide	3	4	8
Teniposide	3	2	5
Podophyllotoxin	1		1
Methotrexate (high)	50		
Methotrexate (low)	1	2	
Vincristine	1	2	1
Daunorubicin	1	1	1
Paclitaxel	1	1	
Mitoxantrone	1	1	
SN-38	1		1
Cisplatin	1	1	1
Arsenite	1		

[a]Kool *et al.* (1999b).
[b]Zeng *et al.* (2000).
[c]Zelcer *et al.* (2001).

TABLE 21.5. SUBSTRATE SPECIFICITY OF RAT MRP3 AND HUMAN MRP3 IN VESICULAR TRANSPORT

Substrates	Hirohashi *et al.*, 1999, 2000 Rat Mrp3 — LLC-PK1[a] or HeLa		Zeng *et al.*, 2000 Human MRP3 — HEK293[a]		Zelcer *et al.*, 2001 Human MRP3 — Sf9[a]	
	K_m	V_{max}	K_m	V_{max}	K_m	V_{max}
Glucuronides						
Estradiol-17β-glucuronide	67	415	26	76	18	474
Etoposide glucuronide					11	~138
E 3040 glucuronide	+					
Glutathione conjugates						
LTC_4	±		5	20	+	
GS-DNP	±		6	4	+	
Bile salts (and conjugates)						
Taurocholate	16	50	–		+	
Glycocholate	+		248	183	+	
TLC-sulfate	3	162				
Other compounds						
Methotrexate	+		776	288	+	
DHEA-sulfate					–	
GSH					–	

[a]Transport was measured using membrane vesicles from transfected cells. The LLC-PK1 cells are pig kidney cells; the HeLa cells are human cervical tumor cells; the HEK293 cells are immortalized human embryonic kidney cells; and the Sf9 cells are *Spodoptera frugiperda* insect cells infected with a recombinant baculovirus *MRP3* gene construct. Values for K_m and V_{max} are μM and pmol mg^{-1} (protein) min^{-1}, respectively. For some substrates, no K_m or V_{max} was determined and are presented as follows: (+) = transport, (±) = marginal transport, (−) = no transport. Abbreviations: DHEA, dihydroepiandrosterone; GSH, glutathione; GS-DNP, *S*-(2,4-dinitrophenyl)glutathione; LTC_4, leukotriene C_4.

is not due to intracellular conversion of etoposide to glucuronosyl-etoposide (which is a good substrate for MRP3, see below) and the simplest interpretation of the available results is that MRP3 transports unmodified etoposide by itself.

Vesicular transport studies, summarized in **Table 21.5**, confirmed that the substrate specificity of MRP3 differs from that of MRP1 and MRP2 in that glutathione conjugates are relatively poor substrates for MRP3 (Hirohashi *et al.*, 1999). Hirohashi *et al.* (2000) found that rat Mrp3 also transports several bile salts at a high rate and with high affinity **(Table 21.5)**. The best substrates were taurolithocholate-3-sulfate and taurochenodeoxycholate-3-sulfate (also substrates of MRP2) (Chapter 20), and taurocholate and glycocholate (also substrates of BSEP, the bile salt export pump). Competition experiments suggest that rat Mrp3 may also interact with other organic sulfate compounds, such as estrone sulfate (Hirohashi *et al.*, 1999), but MRP3 does not transport a prominent human steroid derivative, dehydroepiandrosterone sulfate (DHEA-sulfate).

The results of Hirohashi *et al.* (1999) with rat Mrp3 have been reproduced to some extent with human MRP3 **(Table 21.5)**. Differences are that the human MRP3 appears to transport bile salts more sluggishly than rat Mrp3, and glutathione conjugates more briskly. Although Zeng *et al.* (2000) found a low V_{max} for transport of glutathione conjugates by vesicles derived from MRP3-overexpressing HEK293 cells, much higher rates were observed in the baculovirus system, in which the transport rate for DNP-GS is similar to that of estradiol-17β-glucuronide ($E_2$17βG), in agreement with the substantial transport of DNP-GS observed in MDCKII cells transfected with *MRP3* (Kool *et al.*, 1999b). Like MRP1 and MRP2, MRP3 is inhibited by common organic anion transport inhibitors, such as sulfinpyrazone (1 mM), benzbromarone (250 μM), and indomethacin (250 μM), and less efficiently by probenecid (1 mM) (Zelcer *et al.*, 2001).

In summary, MRP3 is a typical organic anion pump, able to transport acidic drugs, such as MTX, and conjugates of organic compounds with GSH, glucuronate or sulfate. It is inhibited by sulfinpyrazone and other inhibitors of organic anion transport. A major difference compared with MRP1 and MRP2 is the inability of MRP3 to transport free GSH. This may limit its ability to transport unconjugated drugs and may explain, at least in part, the very restricted drug resistance spectrum associated with MRP3 **(Table 21.4)**. Another remarkable property of rat Mrp3 is its ability to transport a range of bile salts at a high rate. Whether human MRP3 has this same property remains to be verified.

Tissue distribution and regulation of MRP3 expression

Kool *et al.* (1997) found substantial amounts of *MRP3* RNA in adrenal gland, colon, small intestine and liver, and low amounts in kidney, bladder, pancreas, stomach, lung, spleen and tonsil. Similar results were obtained in a more limited survey by Belinsky *et al.* (1998), König *et al.* (1999) and Uchiumi *et al.* (1998), and for rat tissues by Kiuchi *et al.* (1998). The only other tissues found to be weakly positive were placenta and prostate. Noteworthy is the absence of detectable *MRP3* expression in muscle, heart, brain, mammary gland, thyroid, salivary gland, testis and ovary (see overview in Scheffer *et al.*, 2002). The presence of MRP3 has been verified at the protein level in gut, pancreas, gallbladder, liver, spleen and adrenal gland (Scheffer *et al.*, 2002). In the adrenals, MRP3 is only present in the cortex, and staining was restricted to the two innermost zones, the zona fasciculata and the zona reticularis. In the kidney, MRP3 is only seen in the distal convoluted tubules and the ascending loops of Henle (Scheffer *et al.*, 2002). In denaturing acrylamide gels, MRP3 from most tissues and cell lines migrates as two separate bands with apparent masses of 170 kDa and 190 kDa (Kool *et al.*, 1999b; Scheffer *et al.*, 2002). The two bands reduce to a single 150 kDa band in cells incubated with tunicamycin, an inhibitor of N-linked glycosylation. Why glycosylation results in two distinct MRP3 protein bands rather than a smear is not known.

In polarized epithelia, MRP3 is located in the basolateral membrane (König *et al.*, 1999; Kool *et al.*, 1999b). Some confusion was created when Ortiz *et al.* (1999) reported that a polyclonal antibody raised against MRP3 stained the canalicular (apical) membrane of the hepatocyte. In view of the unambiguous basolateral localization of MRP3 with several independent monoclonal antibodies in hepatocytes and in other epithelia (König *et al.*, 1999; Kool *et al.*, 1999b; Scheffer *et al.*, 2002), the result of Ortiz *et al.* (1999) must be an artifact.

The inducibility of *MRP3* in the liver has generated intense interest. Hirohashi *et al.* (1998) discovered that *Mrp3* RNA is very low in normal rat liver and upregulated in *Mrp2* (−/−) rats or after bile duct ligation. Less dramatic induction was obtained by feeding rats phenobarbital or α-naphthylisothiocyanate, a compound that induces cholestasis in rats (Ogawa *et al.*, 2000). Initially, these results seemed in contradiction with results reported for human liver RNA. Kool *et al.* (1997) found high levels of *MRP3* RNA in liver samples and this was also observed in some other laboratories (Belinsky *et al.*, 1998; König *et al.*, 1999; Uchiumi *et al.*, 1998). Immunohistochemistry of normal human liver showed, however, only prominent staining of the intra-hepatic bile duct epithelial cells (cholangiocytes) and weak staining of a small subset of hepatocytes surrounding the portal tracts. This resembles the staining seen in normal rat liver (Soroka *et al.*, 2001). Presumably, some of the initial human liver RNA samples with high *MRP3* RNA came from damaged livers.

Absence of MRP2 also leads to strong induction of *MRP3* in humans (König *et al.*, 1999; Scheffer *et al.*, 2002), and increased MRP3 in hepatocytes was also seen in patients with hepatitis, biliary atresia, and especially patients lacking the *MDR3* P-glycoprotein (Scheffer *et al.*, 2002). However, not all cholestatic patients upregulate *MRP3* in their hepatocytes, as a patient with obstructive cholestasis and patients lacking BSEP (ABCB2) had no increase in MRP3 in their hepatocytes, even though they were highly cholestatic and had high concentrations of MRP3 in their proliferating bile ducts (Scheffer *et al.*, 2002). Induction of *MRP3* in the liver is therefore not a simple consequence of cholestasis, but requires a more specific signal generated in some, but not all, cholestatic conditions. The nature of this signal is not known.

PHYSIOLOGICAL FUNCTION OF MRP3 AND ITS POSSIBLE ROLE IN DRUG RESISTANCE OF TUMOR CELLS

The physiological function of MRP3 is not known, but on the basis of its substrate distribution and location in the body, several functions have been proposed. At the top of the list is a function in the cholehepatic and enterohepatic circulation of bile salts (Hirohashi *et al.*, 2000; König *et al.*, 1999; Kool *et al.*, 1999b; Ogawa *et al.*, 2000). Bile salts are secreted in the liver. On their way to the gut they may enter epithelial cells lining the bile ducts, either actively through the apical bile salt transporter (ASBT) or passively. MRP3 may help the bile salts to leave the epithelial cells at the basolateral side. The presence of MRP3 in the ductules of the pancreas and induction of MRP3 in the hepatocyte may serve an analogous function. In the gut, bile salts are taken up, passively or through ASBT, and MRP3 may be the basolateral transporter allowing exit of the bile salts from the enterocyte.

A second possibility is that MRP3 has a defense function and contributes to the elimination of toxic organic anions, notably glucuronosyl derivatives. Humans produce at least 15 UDP-glucuronosyltransferases and these enzymes can glucuronidate a wide range of endogenous and exogenous toxic compounds, not only in the liver and gastrointestinal tract, but also in many other tissues in the body (reviewed by Tukey and Strassburg, 2000). MRP3 and MRP1 may allow cells to export the glucuronosyl derivatives produced intracellularly. The relatively high affinity of MRP3 for $E_217\beta G$ **(Table 21.5)**, and the high concentration of MRP3 in the adrenal cortex, suggest a role in steroid metabolism, but the physiological substrates transported are not yet known. The *Mrp3* (−/−) mouse, recently generated, should provide a test model for these speculations.

Whether MRP3 can contribute to clinical drug resistance is also still unclear. Kool *et al.* (1997) found no association between drug resistance and MRP3 in a diverse panel of cell lines. Young *et al.* (1999, 2001) studied a series of 30 lung cancer cell lines and observed that MRP3 was increased in many of the non-small cell lung cancers, but not in the small cell lung cancers. They found a significant correlation between MRP3 levels and doxorubicin resistance, and a weaker association with resistance to vincristine, etoposide and cisplatin. This does not fit the resistance spectrum of the *MRP3*-transfected cells described in **Table 21.4**. A further complication is the positive correlation between overexpression of *MRP3* and *MRP1* in these cell lines (Young *et al.*, 2001). It is therefore difficult to assess whether the association between MRP3 levels and resistance is not due to the association between *MRP3* and *MRP1* overexpression.

MRP4

MRP4 first appeared in the literature as one of the 21 new ABC genes found by Allikmets *et al.* (1996) by screening the EST database. Kool *et al.* (1997) then showed that a cDNA

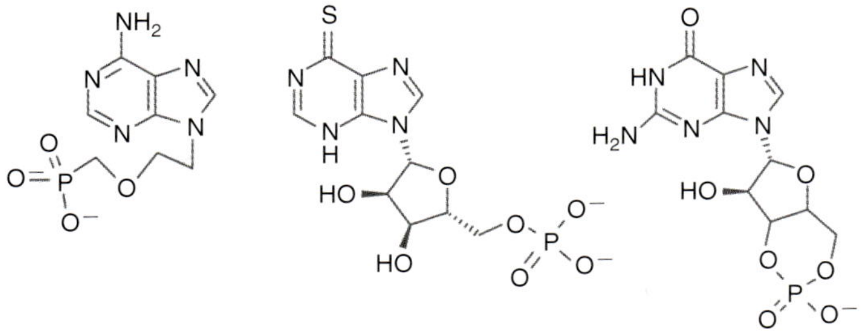

Figure 21.2. ***Chemical structures of the MRP4 and MRP5 substrates PMEA, thio-IMP and cGMP.***

corresponding to the 3′-half of *MRP4* is expressed at low levels in several organs, and that the *MRP4* gene is located on chromosome 13, a location that has since been refined to 13q32 (Schuetz *et al.*, 1999). The *MRP4* cDNA, first reported by Lee *et al.* (1998), has a reading frame encoding a protein of 1325 amino acids and a predicted secondary structure most closely resembling that of MRP5 **(Figure 21.1)**.

The first substrates of MRP4 were deduced with the human T-lymphoid cell line CEM-r1. This cell line, generated by continual selection on the antiviral agent 9-(2-phosphonylmethoxyethyl) adenine (PMEA) **(Figure 21.2)**, an analogue of AMP, is highly resistant to PMEA and related compounds, as well as other nucleoside analogues, but shows no cross-resistance to typical MRP1 substrates such as vinblastine (Robbins *et al.*, 1995). This cell line was shown to rapidly efflux PMEA and other nucleoside monophosphates, such as AZTMP. The finding that the CEM-r1 cells have an amplification of the *MRP4* gene led to the conclusion that MRP4 can transport nucleoside monophosphate analogues (Schuetz *et al.*, 1999).

SUBSTRATE SPECIFICITY OF MRP4

Initial studies with MRP4 using the PMEA-selected CEM-r1 line suggested a relatively broad spectrum of transportable substrates. However, as was apparent at the time, the other genetic changes present in this cell line have a considerable influence on the drug resistance phenotype. This is especially true for the downregulation of adenylate kinase activity in these cells, which increases the pool of transportable nucleoside monophosphates (Robbins *et al.*, 1995; Schuetz *et al.*, 1999).

Lee *et al.* (2000) transfected NIH3T3 cells with *MRP4* cDNA and, somewhat surprisingly, found these cells exhibited resistance only against PMEA and short-term MTX exposure. In contrast to the cross-resistance of CEM-r1 cells to a variety of nucleoside analogues, the NIH3T3/MRP4 cells showed no resistance against AZT, 3TC, ddC or d4T. Further work by Schuetz and co-workers (presented at the FEBS 2001 ABC Meeting in Gosau), however, broadened the substrate specificity to include thiopurine derivatives. Resistance to 6-mercaptopurine (6-MP) and thioguanine (TG) was recently found in the *MRP4*-transfected NIH3T3 cells studied by the group of Kruh (Chen *et al.*, 2001).

We have transfected a variety of cell lines with the MRP4 *cDNA* (kindly provided by J. Schuetz), and found the highest *MRP4* expression in HEK293 cells. Initial experiments with these HEK293/MRP4 cells showed that they efflux PMEA when loaded with bis-POM-PMEA (a membrane permeable form of PMEA), and show resistance under continuous exposure to PMEA, to 6-MP and to the antiviral nucleoside analogue abacavir (our unpublished results). In common with the exogenous expression of other MRPs, it appears relatively difficult to obtain substantial levels of the protein after transduction. A further complication is the endogenous expression of MRP4 and MRP5 in many of the cell lines used for transfection studies. For example, HEK293 cells contain levels of *MRP4* mRNA comparable to those found in the most MRP4-rich tissues (our unpublished results).

A recently published study suggests that expression of *MRP4* in insect cells could be the best way to overcome problems of endogenous transporter background. Chen *et al.* (2001) used *MRP4*-containing baculovirus to infect insect cells from which they made inside-out membrane vesicles. Taking into account the structural and substrate similarities between MRP4 and MRP5, and a previous study on MRP5 (Jedlitschky *et al.*, 2000) (see later section on substrate specificity of human MRP5 for details), they demonstrated that cyclic nucleotides are substrates for this transporter. Low rates of transport were observed for 3′,5′-cyclic GMP (cGMP) **(Figure 21.2)** (K_m and V_{max} values of 10 μM and 2 pmol mg^{-1} min^{-1}) and 3′,5′-cyclic AMP (cAMP) (K_m and V_{max} values of 45 μM and 4 pmol mg^{-1} min^{-1}). Remarkably, estradiol $E_2$17βG was a relatively good substrate (K_m and V_{max} values of 30 μM and 102 pmol mg^{-1} min^{-1}).

TISSUE DISTRIBUTION OF *MRP4* EXPRESSION

Initial reports concerning the expression of *MRP4* described a gene with a restricted tissue distribution, as determined by polymerase chain reaction (PCR) mapping (Allikmets *et al.*, 1996) and RNase protection assays (Kool *et al.*, 1997). More recent data suggest that the gene is more widely expressed. Lee *et al.* (1998) detected MRP4 protein in most of the tissues examined, with levels ranging from very high in the prostate to barely detectable in the liver. Using a semi-quantitative reverse transcriptase PCR (RT-PCR) method to standardize *MRP4* transcript levels to β-actin, we find high levels of expression in the kidney, with lower but substantial expression in the gallbladder, testis and prostate. We also find *MRP4* mRNA in all cell lines tested. More recently, we generated a new monoclonal antibody against human MRP4, which we have used to confirm the high expression of *MRP4* in the kidney, as well as its presence in cell lines. Lee *et al.* (2000) found MRP4 in the basolateral membrane of the acinar cells in the prostate. In contrast, Van Aubel reported at the FEBS 2001 ABC Meeting in Gosau that MRP4 is in the apical membrane, not the basolateral membrane, of rat and human kidney cells. Whether MRP4 is indeed localized to different membranes in different epithelial tissues needs verification with antibodies that enable more conclusive immunohistochemistry.

PHYSIOLOGICAL FUNCTION OF MRP4 AND ROLE IN DRUG RESISTANCE

There are still very few clues as to the normal physiological function of MRP4, and the role, if any, played by MRP4 in anticancer drug resistance. The recent discovery by Chen *et al.* (2001) that MRP4, like MRP5 (Jedlitschky *et al.*, 2000), can serve as an efflux pump for cGMP and cAMP indicates that MRP4 is able to remove physiologically relevant (cyclic) nucleoside monophosphates from the cell. Any role MRP4 may have in drug resistance is also under investigation. As nucleobase and nucleoside analogues are used extensively in anticancer and antiviral therapies, there is potential for MRP4 to mediate resistance to these compounds. As pointed out by Chen *et al.* (2001), 6-MP and MTX are both used in the treatment of childhood leukemias and MRP4 is the only drug transporter known thus far that can transport both drugs. In a screen of drug-selected human cancer cell lines by RNase protection assays, Kool *et al.* (1997) found MRP4 to be expressed at low levels in all cell lines, but this did not correlate with resistance. However, the cell lines tested were not selected by nucleobase or nucleoside analogues nor tested for resistance against these compounds.

MRP5

Human *MRP5*, also known as *MOAT-C*, and formally called *ABCC5* (GenBank: AF146074), was cloned by several groups (Belinsky *et al.*, 1998; Jedlitschky *et al.*, 2000; McAleer *et al.*, 1999; Wijnholds *et al.*, 2000). The mouse homologue of human *MRP5*, called *mrp5* or *Mrp5* (GenBank AB019003), turned out to be the same as the previously identified *sMRP* (Suzuki *et al.*, 1997; Tusnady and Varadi, 1998), a cloning artifact missing the part encoding the first transmembrane domain (Suzuki *et al.*, 2000). Like MRP4, MRP5 is an organic anion pump with a high affinity for nucleotide analogues and cyclic nucleotides (Jedlitschky *et al.*, 2000; Wijnholds *et al.*, 2000). No known human disease is associated with MRP5 defects and the Mrp5 KO mouse has no phenotype thus far (Wijnholds *et al.*, 2000) (our unpublished results).

SUBSTRATE SPECIFICITY OF HUMAN MRP5

Cells transfected with *MRP5* cDNA constructs were used by McAleer *et al.* (1999) and Wijnholds *et al.* (2000) to study the substrate specificity of MRP5. McAleer *et al.* (1999) found reduced accumulation in MRP5 cells for the acidic organic dyes, 5-chloromethylfluorescein diacetate (CMFDA), 5-fluorescein diacetate (FDA), and 2′,7′-bis-(2-carboxyethyl)-5 (and-6)-carboxyfluorescein acetoxymethyl ester (BCECF-AM), but not for structural analogues, the anionic calcein and the cationic tetramethylrosamine. Further evidence that MRP5 is a typical organic pump, like other MRPs, came from Wijnholds *et al.* (2000), who showed that MRP5 transports DNP-GS and GSH, and that MRP5 is inhibited by nonspecific organic anion transport inhibitors, such as sulfinpyrazone and benzbromarone. MRP5 does not seem to transport MTX, in contrast to MRP1-4. There is also no evidence that MRP5 can mediate resistance to any of the anticancer drugs belonging to the MDR spectrum and transported by MRP1 with one exception. Wijnholds *et al.* (2000) found a low level of etoposide resistance but this was not found by McAleer *et al.* (1999). Conversely, resistance against cadmium chloride and potassium antimonyl tartrate was found by McAleer *et al.* (1999) but this could not be reproduced by Wijnholds *et al.* (2000).

Like MRP4, MRP5 can also cause resistance against the nucleoside monophosphate analogue PMEA **(Figure 21.2)**. When cells are loaded with the membrane-permeable PMEA precursor bis-POM-PMEA, MRP5 mediates excretion of PMEA, but not of the di-(PMEAp) and triphosphate (PMEApp) of PMEA (Wijnholds *et al.*, 2000), which are formed intracellularly (Balzarini *et al.*, 1991). The ability of MRP5 to transport nucleotide analogues may also explain the resistance of MRP5-transfected cells to the thiopurines 6-MP and TG. As shown in **Figure 21.3**, these thiopurines are converted into the corresponding nucleoside monophosphates (e.g. thio-IMP) **(Figure 21.2)** and these are excreted via MRP5 (Wijnholds *et al.*, 2000) (our unpublished results). Whether MRP5 can mediate excretion of methylated thiopurine derivatives, as reported for MRP4 by J. Schuetz at the 2001 FEBS 2001 ABC Meeting in Gosau, remains to be tested.

Using vesicular transport by plasma membrane vesicles made from hamster V79 cells overexpressing MRP5, Jedlitschky *et al.* (2000) identified cGMP as a MRP5 substrate with a micromolar affinity for the pump. cAMP was also transported, but with a lower affinity. Interestingly, they also found that the phosphodiesterase (PDE) inhibitors sildenafil (better known as Viagra), trequinsin and zaprinast, which prevent intracellular breakdown of cGMP, inhibited the MRP5-mediated cGMP transport as well.

MRP5 does not only transport purine-based compounds. In unpublished experiments (J. Wijnholds, P.W. and P.B.), we have also found transport of a pro-drug of 3′-deoxy 2′,3′-didehydrothymidine 5′monophosphate (d4TMP), alaninyl-d4TMP, an antiviral agent (Balzarini *et al.*, 1996). MRP5 can therefore also transport nucleotide analogues with a normal pyrimidine (thymine) ring.

Expression of MRP5 in tissues and cell lines

Analysis of tissue RNA suggested that *MRP5* is ubiquitously expressed **(Table 21.3)**. The highest levels are found in skeletal muscle and brain (Belinsky *et al.*, 1998; Kool *et al.*, 1997; McAleer *et al.*, 1999; Zhang *et al.*, 2000). All attempts to generate antibodies that allow the localization of MRP5 in tissues have failed thus

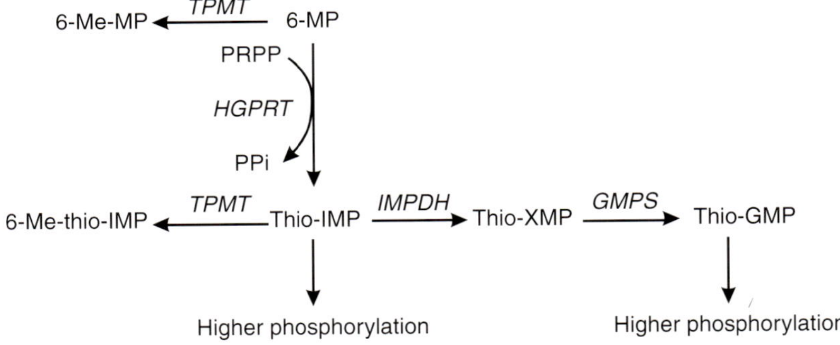

Figure 21.3. *A simplified schematic diagram of 6-mercaptopurine metabolism. With the metabolites: 6-MP, 6-mercaptopurine; thio-IMP, 6-thio-inosine monophosphate; thio-XMP, 6-thio-xanthidine monophosphate; thio-GMP, thio-guanosine monophosphate; 6-Me-MP, 6-methyl-mercaptopurine; 6-Me-thioIMP, 6-methyl-thio-inosine monophosphate; PRPP, phosphoribosyl pyrophosphate; PPi, pyrophosphate; and the enzymes:* **HGPRT**, *hypoxanthine-guanosine phosphoribosyltransferase,* **TPMT**, *thiopurine methyltransferase;* **IMPDH**, *inosine monophosphate dehydrogenase;* **GMPS**, *guanosine monophosphate synthetase. (Adapted from Zimm* et al.*, 1985.)*

far. This is presumably because the expression levels are too low, as these antibodies readily detect MRP5 in transfected cells (Wijnholds *et al.*, 2000). On Western blots, MRP5 can be detected in human and murine erythrocytes (Jedlitschky *et al.*, 2000) (our unpublished results) and MRP5 might be the cGMP pump detected in erythrocytes (Schultz *et al.*, 1998), although this needs to be shown using red cells from Mrp5 (−/−) mice. Levels of MRP5 protein were also high in brain extracts, but not in skeletal muscle, in contrast to the MRP5 RNA levels in this tissue.

The negative results obtained on intact tissues using antibodies against human or murine MRP5 make it impossible to decide whether MRP5 is present in many cell types or only in some cell types present in all tissues, for example, endothelial cells. In a recent survey using Western blots, we found MRP5 in all human tumor cell lines analyzed, including colon, breast, ovarian, lung carcinoma lines, leukemia and embryonic kidney cell lines. This suggests, but does not prove, that MRP5 is present in many different normal cell types.

THE POSSIBLE INVOLVEMENT OF MRP5 IN DRUG RESISTANCE OR DISEASE

Base, nucleoside and nucleotide analogues are used in antiviral and in anticancer therapy. Potentially, elevated levels of MRP5 could therefore contribute to clinical resistance to these agents. In MRP5-transfected cells, unambiguous resistance has only been found thus far against 6-MP, TG, PMEA, 5-hydroxypyrimidine-2-carboxaldehyde thiosemicarbazone (an experimental anticancer drug), and an aryloxyphosphoramidate derivative of 2′,3′-dideoxyadenosine (Cf 1093). Whether clinical resistance against these compounds is ever associated with elevated MRP5 levels remains to be studied. No resistance was observed thus far in MRP5 cells for other base or nucleoside analogues used in cancer or antiviral chemotherapy, such as ara-C, 5-fluorouracil, cidofovir and fludarabine (our unpublished results).

No human disease has been associated with alterations in MRP5, and the Mrp5 KO mouse, generated by Wijnholds *et al.* (2000), has no obvious phenotype. It is possible, however, that the overlapping substrate specificities of MRP5 and MRP4 (and possibly MRP8 and MRP9) may hide the physiological function of Mrp5 (e.g. in cyclic nucleotide transport), and that the generation of mice lacking all these transporters may lead to an understanding of the physiological function of each of them.

MRP6

MRP6 sprung into prominence when defects in the *MRP6* gene were identified as the cause of pseudoxanthoma elasticum (Bergen *et al.*, 2000; Le Saux *et al.*, 2000; Ringpfeil *et al.*, 2000), a connective tissue disease affecting multiple organs. MRP6 is a protein of 1503 amino acids (Belinsky and Kruh, 1999; Kool *et al.*, 1999a), 45% identical to *MRP1*, and its gene is located next to *MRP1* on chromosome 16 in a tail-to-tail configuration (Kool *et al.*, 1999a). Human *MRP6* is mainly expressed in liver and kidney (Belinsky and Kruh, 1999; Kool *et al.*, 1997, 1999a), like *Mrp6* (*MLP-1*), its rat homologue (Hirohashi *et al.*, 1998, 1999; Madon *et al.*, 2000), but low RNA levels have also been detected in other tissues. In initial immunofluorescence studies, Madon *et al.* (2000) localized rat Mrp6 in the basolateral and apical membranes of hepatocytes. More recent work reported at the FEBS 2001 ABC Meeting at Gosau strongly indicates, however, that MRP6 is in the basolateral membrane of polarized cells. In contrast to some other MRPs, expression of *Mrp6* appears stable, whatever damage is inflicted on the liver (Madon *et al.*, 2000).

The substrate specificity of MRP6 is still a mystery. Madon *et al.* (2000) tested a series of typical MRP substrates in vesicular transport studies and found only transport of BQ-123, an anionic cyclopentapeptide and endothelin A receptor antagonist. Endothelin-1 itself was transported by Mrp2, but not by Mrp6. These results suggest that MRP6 could be a highly selective organic anion pump. It should be noted, however, that Madon *et al.* (2000) only tested radioactive substrates at relatively low concentrations. No competition experiments were done with high competitor concentrations, substrates such as MTX were not tested, and standard inhibitors of MRPs were not tested either.

MRP6 AND DRUG RESISTANCE

Amplification of the 3′-part of the *MRP6* gene was found in leukemia cells selected for anthracycline (epirubicin) resistance (Kuss *et al.*, 1998; Longhurst *et al.*, 1996; O'Neill *et al.*, 1998). The anthracycline resistance was initially thought to be due to a new resistance determinant, called the anthracycline resistance gene *ARA*. Subsequent work has shown, however, that the epirubicin resistance of cell lines with *ARA* gene amplification can be explained by co-amplification of the *MRP1* gene with the 3′ half of the adjacent *MRP6* gene (Belinsky and Kruh, 1999; Kool *et al.*, 1999a).

There is no indication that the *MRP6* gene is ever associated with anticancer drug resistance. *MRP6* is expressed at low or undetectable levels in all cancer cell lines tested (Kool *et al.*, 1997, 1999a; Madon *et al.*, 2000) and no correlation between expression and drug resistance was observed (Kool *et al.*, 1997).

MRP6 AND PXE

Pseudoxanthoma elasticum is a heritable disorder characterized by calcification of elastic fibers in skin, arteries and retina, resulting in loss of elasticity of the skin, arterial insufficiency and retinal hemorrhage. Why loss of a highly specialized pump located in the basolateral membrane of liver and kidney cells would lead to such a generalized connective tissue disease is unclear. Speculations include indirect effects on Ca^{2+} metabolism or elastic fiber assembly, through excretion of cytokine-like organic anionic peptides (Bergen *et al.*, 2000; Le Saux *et al.*, 2000; Ringpfeil *et al.*, 2000) (see Chapter 28).

MRP7

Hopper *et al.* (2001) identified an MRP homologue designated MRP7 by a database search. *MRP7* cDNA encodes a protein of 1492 amino acids and, translated in a reticulocyte lysate system, the cDNA produces a protein of approximately 158 kDa. MRP7 has a predicted secondary structure and topology similar to that of MRP1, with an extra NH_2-terminal transmembrane domain. Although clearly an MRP homologue, MRP7 has the lowest overall homology with other members of the MRP family (33–36% sequence identity). *MRP7* seems to be ubiquitously expressed as assessed by RT-PCR, but at a low level, since the mRNA transcript could not be detected by RNA blot analysis. The substrate specificity and mechanism of transport of MRP7 have not yet been studied. Whether or not MRP7 is an organic anion transporter also remains to be tested.

ACKNOWLEDGMENTS

We thank Drs Marcel Kool, Alfred Schinkel and Jan Wijnholds for their helpful comments on the manuscript. T.S. was supported by a postdoctoral fellowship from the Japanese Society for the Promotion of Science and our research was supported by grants (NKI 2001-2474 and 1998-1794) of the Dutch Cancer Society to P.B.

REFERENCES

Allen, J.D., Brinkhuis, R.F., Van Deemter, L., Wijnholds, J. and Schinkel, A.H. (2000) Extensive contribution of the multidrug transporters P-glycoprotein and Mrp1 to bassal drug resistance. Cancer Res. 60, 5761–5766.

Allikmets, R., Gerrard, B., Hutchinson, A. and Dean, M. (1996) Characterization of the human ABC superfamily: isolation and mapping of 21 new genes using the expressed sequence tags database. Hum. Mol. Genet. 5, 1649–1655.

Balzarini, J., Hao, Z., Herdewijn, P., Johns, D.G. and De Clercq, E. (1991) Intracellular metabolism and mechanism of anti-retrovirus action of 9-(2-phosphonylmethoxyethyl) adenine, a potent anti-human immunodeficiency virus compound. Proc. Natl Acad. Sci. USA 88, 1499–1503.

Balzarini, J., Karlsson, A., Aquaro, S., Perno, C.F., Cahard, D., Naesens, L., De Clercq, E. and McGuigan, C. (1996) Mechanism of anti-HIV action of masked alaninyl D4T-MP derivatives. Proc. Natl Acad. Sci. USA 93, 7295–7299.

Belinsky, M.G. and Kruh, G.D. (1999) MOAT-E (ARA) is a full-length MRP/cMOAT subfamily transporter expressed in kidney and liver. Br. J. Cancer 80, 1342–1349.

Belinsky, M.G., Bain, L.J., Balsara, B.B., Testa, J.R. and Kruh, G.D. (1998) Characterization of MOAT-C and MOAT-D, new members of the MRP/cMOAT subfamily of transporter proteins. J. Natl. Cancer Inst. 90, 1735–1741.

Bergen, A.A., Plomp, A.S., Schuurman, E.J., Terry, S., Breuning, M., Dauwerse, H., *et al.* (2000) Mutations in ABCC6 cause *Pseudoxanthoma elasticum*. Nat. Genet. 25, 228–231.

Borst, P. and Oude Elferink, R. (2002) Mammalian ABC transporters in health and disease. Ann. Rev. Biochem. 71, 537–592.

Borst, P., Evers, R., Kool, M. and Wijnholds, J. (1999) The multidrug resistance protein family. Biochim. Biophys. Acta 1461, 347–357.

Borst, P., Evers, R., Kool, M. and Wijnholds, J. (2000) A family of drug transporters, the MRP's. J. Natl Cancer Inst. 92, 1295–1302.

Chen, Z.-S., Lee, K. and Kruh, G.D. (2001) Transport of cyclic nucleotides and estradiol 17-β-D-glucuronide by multidrug resistance protein 4: resistance to 6-mercaptopurine and 6-thioguanine. J. Biol. Chem. 276, 33747–33754.

Cole, S.P.C., Bhardwaj, G., Gerlach, J.H., Mackie, J.E., Grant, C.E., Almquist, K.C., Stewart, A.J., Kurz, E.U., Duncan, A.M.V. and Deeley, R.G. (1992) Overexpression of a transporter gene in a multidrug-resistant human lung cancer cell line. Science 258, 1650–1654.

Evers, R., Kool, M., Van Deemter, L., Jansen, H., Calafat, J., Oomen, L.C.J.M., *et al.* (1998) Drug export activity of the human canalicular multispecific organic anion transporter in polarized kidney MDCK cells expressing *cMOAT (MRP2)* cDNA. J. Clin. Invest. 101, 1310–1319.

Evers, R., De Haas, M., Sparidans, R., Beijnen, J., Wielinga, P.R., Lankelma, J. and Borst, P. (2000) Vinblastine and sulfinpyrazone export by the multidrug resistance protein MRP2 is associated with glutathione export. Br. J. Cancer 83, 375–383.

Haga, S., Hinoshita, E., Ikezaki, K., Fukui, M., Scheffer, G.L., Uchiumi, T. and Kuwano, M. (2001) Involvement of the multidrug resistance protein 3 in drug sensitivity and its expression in human glioma. Jpn J. Cancer Res. 92, 211–219.

Hirohashi, T., Suzuki, H., Ito, K., Ogawa, K., Kume, K., Shimizu, T. and Sugiyama, Y. (1998) Hepatic expression of multidrug resistance-associated protein-like proteins maintained in eisai hyperbilirubinemic rats. Mol. Pharmacol. 53, 1068–1075.

Hirohashi, T., Suzuki, H. and Sugiyama, Y. (1999) Characterization of the transport properties of cloned rat multidrug resistance-associated protein 3 (MRP3). J. Biol. Chem. 274, 15181–15185.

Hirohashi, T., Suzuki, H., Takikawa, H. and Sugiyama, Y. (2000) ATP-dependent transport of bile salts by rat multidrug resistance-associated protein 3 (Mrp3). J. Biol. Chem. 275, 2905–2910.

Hopper, E., Belinsky, M.G., Zeng, H., Tosolini, A., Testa, J.R. and Kruh, G.D. (2001) Analysis of the structure and expression pattern of MRP7 (ABCC10), a new member of the MRP subfamily. Cancer Lett. 162, 181–191.

Ishikawa, T., Wright, C.D. and Ishizuka, H. (1994) GS-X pump is functionally overexpressed in *cis*-diamminedichloroplatinum(II)-resistant human leukemia HL-60 cells and downregulated by cell differentiation. J. Biol. Chem. 269, 29085–29093.

Jedlitschky, G., Burchell, B. and Keppler, D. (2000) The multidrug resistance protein 5 (MRP5) functions as an ATP-dependent export pump for cyclic nucleotides. J. Biol. Chem. 275, 30069–30074.

Keppler, D. (1999) Export pumps for glutathione S-conjugates. Free Radical Biol. Med. 27, 985–991.

Kiuchi, Y., Suzuki, H., Hirohashi, T., Tyson, C.A. and Sugiyama, Y. (1998) cDNA cloning and inducible expression of human multidrug resistance associated protein 3 (MRP3). FEBS Lett. 433, 149–152.

König, J., Rost, D., Cui, Y. and Keppler, D. (1999) Characterization of the human multidrug resistance protein isoform MRP3 localized to the basolateral hepatocyte membrane. Hepatology 29, 1156–1163.

Kool, M., De Haas, M., Scheffer, G.L., Scheper, R.J., Van Eijk, M.J.T., Juijn, J.A., Baas, F. and Borst, P. (1997) Analysis of expression of *cMOAT (MRP2), MRP3, MRP4,* and *MRP5,* homologs of the multidrug resistance-associated protein gene *(MRP1),* in human cancer cell lines. Cancer Res. 57, 3537–3547.

Kool, M., Van der Linden, M., De Haas, M., Baas, F. and Borst, P. (1999a) Expression of human *MRP6*, a homologue of the multidrug resistance protein gene *MRP1*, in tissues and cancer cells. Cancer Res. 59, 175–182.

Kool, M., Van der Linden, M., De Haas, M., Scheffer, G.L., De Vree, J.M.L., Smith, A.J., *et al.* (1999b) MRP3, an organic anion transporter able to transport anti-cancer drugs. Proc. Natl Acad. Sci. USA 96, 6914–6919.

Kuss, B.J., O'Neill, G.M., Eyre, H., Doggett, N.A., Callen, D.F. and Davey, R.A. (1998) *ARA*, a novel ABC transporter, is located at 16p13.1, is deleted in inv(16) leukemias, and is shown to be expressed in primitive hematopoietic precursors. Genomics 51, 455–458.

Le Saux, O., Urban, Z., Tschuch, C., Csiszar, K., Bacchelli, B., Quaglino, D., *et al.* (2000) Mutations in a gene encoding an ABC transporter cause *Pseudoxanthoma elasticum*. Nat. Genet. 25, 223–227.

Lee, K., Belinsky, M.G., Bell, D.W., Testa, J.R. and Kruh, G.D. (1998) Isolation of MOAT-B, a widely expressed multidrug resistance-associated protein canalicular multispecific organic anion transporter-related transporter. Cancer Res. 58, 2741–2747.

Lee, K., Klein-Szanto, A.J.P. and Kruh, G.D. (2000) Analysis of the MRP4 drug resistance profile in transfected NIH3T3 cells. J. Natl Cancer Inst. 92, 1934–1940.

Longhurst, T.J., O'Neill, G.M., Harvie, R.M. and Davey, R.A. (1996) The anthracycline resistance-associated (*ara*) gene, a novel gene associated with multidrug resistance in a human leukaemia cell line. Br. J. Cancer 74, 1331–1335.

Madon, J., Hagenbuch, B., Landmann, L., Meier, P.J. and Stieger, B. (2000) Transport function and hepatocellular localization of mrp6 in rat liver. Mol. Pharmacol. 57, 634–641.

McAleer, M.A., Breen, M.A., White, N.L. and Matthews, N. (1999) pABC11 (also known as MOAT-C and MRP5), a member of the ABC family of proteins, has anion transporter activity but does not confer multidrug resistance when overexpressed in human embryonic kidney 293 cells. J. Biol. Chem. 274, 23541–23548.

Ogawa, K., Suzuki, H., Hirohashi, T., Ishikawa, T., Meier, P.J., Hirose, K., Akizawa, T., Yoshioka, M. and Sugiyama, Y. (2000) Characterization of inducible nature of MRP3 in rat liver. Am. J. Physiol. Gastrointest. Liver Physiol. 278, G438–G446.

O'Neill, G.M., Peters, G.B., Harvie, R.M., MacKenzie, H.B., Henness, S. and Davey, R.A. (1998) Amplification and expression of the ABC transporters ARA and MRP in a series of multidrug-resistant leukaemia cell sublines. Br. J. Cancer 77, 2076–2080.

Ortiz, D.F., Li, S., Iyer, R., Zhang, X., Novikoff, P. and Arias, I.M. (1999) MRP3, a new ATP-binding cassette protein localized to the canalicular domain of the hepatocyte. Am. J. Physiol. 276, G1493–G1500.

Renes, J., De Vries, E.G.E., Nienhuis, E.F., Jansen, P.L.M. and Müller, M. (1999) ATP- and glutathione-dependent transport of chemotherapeutic drugs by the multidrug resistance protein MRP1. J. Pharmacol. 126, 681–688.

Ringpfeil, F., Lebwohl, M.G., Christiano, A.M. and Uitto, J. (2000) *Pseudoxanthoma elasticum:* mutations in the MRP6 gene encoding a transmembrane ATP-binding cassette (ABC) transporter. Proc. Natl Acad. Sci. USA 97, 6001–6006.

Robbins, B.L., Connelly, M.C., Marshall, D.R., Srinivas, R.V. and Fridland, A. (1995) A human T lymphoid cell variant resistant to the acyclic nucleoside phosphonate 9-(2-phosphonylmethoxyethyl)adenine shows a unique combination of a phosphorylation defect and increased efflux of the agent. Mol. Pharmacol. 47, 391–397.

Scheffer, G.L., Kool, M., De Haas, M., De Vree, J.M.L., Pijnenborg, A.C.L.M., Bosman, D.K., Oude Elferink, R.P.J., Van der Valk, P., Borst, P. and Scheper, R.J. (2002) Tissue distribution and induction of human MRP3. Lab. Invest. 82, 193–201.

Schuetz, J.D., Connelly, M.C., Sun, D., Paibir, S.G., Flynn, P.M., Srinivas, R.V., Kumar, A. and Fridland, A. (1999) MRP4: A previously unidentified factor in resistance to nucleoside-based antiviral drugs. Nat. Med. 5, 1048–1051.

Schultz, C., Vaskinn, S., Kildalsen, H. and Sager, G. (1998) Cyclic AMP stimulates the cyclic GMP-egression pump in human erythrocytes: effects of probenecid, verapamil, progresterone, theophylline, IBMX, forskolin, and cyclic AMP on cyclic GMP uptake and association to inside-out vesicles. Biochemistry 37, 1161–1166.

Soroka, C.J., Lee, J.M., Azzaroli, F. and Boyer, J.L. (2001) Cellular localization and up-regulation of multidrug resistance-associated protein 3 in hepatocytes and cholangiocytes during obstructive cholestasis in rat liver. Hepatology 33, 783–791.

Suzuki, T., Nishio, K., Sasaki, H., Kurokawa, H., Saito-Ohara, F., Ikeuchi, T., Tanabe, S., Terada, M. and Saijo, N. (1997) cDNA cloning of a short type of multidrug resistance protein homologue, SMRP, from a human lung cancer cell line. Biochem. Biophys. Res. Comm. 238, 790–794.

Suzuki, T., Sasaki, H., Ku, H.-J., Agui, M., Tatsumi, Y., Tanabe, S., Terada, M., Saijo, N. and Nishio, K. (2000) Detailed structural analysis on both human *MRP5* and mouse *mrp5* transcripts. Gene 242, 167–173.

Tammur, J., Prades, C., Arnould, I., Rzhetsky, A., Hutchinson, A., Adachi, M., *et al.* (2001) Two new genes from the human ATP-binding cassette transporter superfamily, *ABCC11* and *ABCC12*, tandemly duplicated on chromosome 16q12. Gene 273, 89–96.

Tukey, R.H. and Strassburg, C.P. (2000) Human UDP-glucuronosyltransferases: metabolism, expression, and disease. Annu Rev. Pharmacol. Toxicol. 40, 581–616.

Tusnady, G.E. and Varadi, A. (1998) Short MRP may not be short. Biochem. Biophys. Res. Commun. 242, 465–466.

Uchiumi, T., Hinoshita, E., Haga, S., Nakamura, T., Tanaka, T., Toh, S., *et al.* (1998) Isolation of a novel human canalicular multispecific organic anion transporter, cMOAT2/MRP3, and its expression in cisplatin-resistant cancer cells with decreased ATP-dependent drug transport. Biochem. Biophys. Res. Commun. 252, 103–110.

Wijnholds, J., Mol, C.A.A.M., Van Deemter, L., De Haas, M., Scheffer, G.L., Baas, F., *et al.* (2000) Multidrug-resistance protein 5 is a multispecific organic anion transporter able to transport nucleoside analogs. Proc. Natl Acad. Sci. USA 97, 7476–7481.

Young, L.C., Campling, B.G., Voskoglou-Nomikos, T., Cole, S.P.C., Deeley, R.G. and Gerlach, J.H. (1999) Expression of multidrug resistance protein-related genes in lung cancer: correlation with drug response. Clin. Cancer Res. 5, 673–680.

Young, L.C., Campling, B.G., Cole, S.P.C., Deeley, R.G. and Gerlach, J.H. (2001) Multidrug resistance proteins MRP3, MRP1, and MRP2 in lung cancer: Correlation of protein levels with drug response and messenger RNA levels. Clin. Cancer Res. 7, 1798–1804.

Zelcer, N., Saeki, T., Reid, G., Beijnen, J.H. and Borst, P. (2001) Characterization of drug transport by the human multidrug resistance protein 3 (ABCC3). J. Biol. Chem. 276, 46400–46407.

Zeng, H., Liu, G., Rea, P.A. and Kruh, D. (2000) Transport of amphipathic anions by human multidrug resistance protein. Cancer Res. 60, 4779–4784.

Zhang, Y., Han, H., Elmquist, W.F. and Miller, D.W. (2000) Expression of various multidrug resistance-associated protein (MRP) homologues in brain microvessel endothelial cells. Brain Res. 876, 148–153.

Zimm, S., Johnson, G.E., Chabner, B.C. and Poplack, D.G. (1985) Cellular pharmacokinetics of mercaptopurine in human neoplastic cells and cell lines. Cancer Res. 45, 4156–4161.

Part VI

ABC Transporters, Physiological Roles and Human Disease

CHAPTER 22

Lipid Transport by ABC Transporters

Piet Borst, Gerrit van Meer and Ronald Oude Elferink

Introduction

ABC transporters, like fat, are embedded in a lipid bilayer and some of them are good at transporting lipids. This could already be inferred from the fact that classical multidrug resistance (MDR) of cancer cells can be caused by the ABC transporter, the P-glycoprotein (MDR1, ABCB1). The drugs belonging to the MDR spectrum are rather hydrophobic and MDR1 P-glycoprotein must therefore have affinity for lipophilic compounds and lipids. The importance of ABC transporters for lipid transport was firmly established by the discovery in 1993 that the human MDR3 P-glycoprotein (ABCB4; sometimes referred to as PGY3) is a dedicated phosphatidylcholine (PC) transporter, indispensable for normal bile formation.

Since 1993 many additional ABC transporters have been shown to be involved in lipid transport, as illustrated by the overview presented in **Table 22.1**. The list is undoubtedly incomplete. We know only for a fraction of the 48 human ABC transporters what their physiological substrates are. We expect that some of the new ones that have recently turned up in the human genome will also be involved in lipid transport.

It should be clear from **Table 22.1** that we use a rather broad definition of lipid transport. Included are not only proteins that transport indisputable lipids, such as the MDR3 P-glycoprotein (ABCB4) or ABC1 (ABCA1), but also proteins that transport acidic lipid conjugates, such as MRP1 (ABCC1) and MRP2 (ABCC2), or ALDP (ABCD1). This serves to highlight the diverse roles of the ABC transporters in lipid disposition. The transporters listed in **Table 22.1** differ widely in their contribution to lipid transport. Most of them, notably MDR3 P-glycoprotein, BSEP (ABCB11), MRP2 (ABCC2), ABC1 (ABCA1), ABCG5 and ABCG8, ALDP (ABCD1) and ABCR (ABCA4), are indispensable and their absence or disruption results in disease. For other transporters, natural substrates are known, but transporter absence does not seem to lead to significant alterations in lipid disposition. Examples are MDR1 P-glycoprotein and MRP1. These proteins may mainly serve to defend the body against amphipathic xenotoxins. This may also be the case for BCRP1 (ABCG2), which has a clearly defined protective function, but no physiological substrates are known yet for this drug transporter. The substrate specificity and function of the recently identified large transporter ABCA2 (Vulevic *et al.*, 2001) is also still unclear. ABCA2 appears to be present in lysosomes, but it was also found in the endoplasmic reticulum, Golgi and some peroxisomes. The *ABCA2* gene was overexpressed in a cell line selected for estramustine resistance, but the resistance against this drug of cells transfected with *ABCA2* cDNA constructs was only marginal. Estramustine is a synthetic nitrogen mustard

ABC Proteins: From Bacteria to Man
ISBN 0-12-352551-9

TABLE 22.1. HUMAN AND RODENT ABC TRANSPORTERS INVOLVED IN LIPID TRANSPORT[a]

Nr	Name	Symbol	Size[b] (TMS)	Rodent gene	Main locations		Physiological lipid substrates	Other substrates	Disease[c]		Miscellaneous
					Tissue	Subcellular			Human	Rodent	
1	ABC1 TGD HDLDT1 CERP	ABCA1	2261 (12?)	mAbc1	Ubiquitous	Plasma membrane (apical)	P-lipid, cholesterol?		Tangier disease	Hemorrhage, defective apoptosis	
2	ABC2	ABCA2	2436 (12?)	mAbc2 rAbc2	Brain, kidney, lung, heart	Lysosomal membrane	None known	Estramustine	–	–	Steroid transport?
3	ABCR RmP ABC10 STGD1 STGD	ABCA4	2273 (12?)	mAbcr	Retina	Rim of outer segment disks	*N*-retinylidene-phosphatidyl-ethanolamine	–	Stargardt disease	As humans	
4	PGY1 MDR1 Pgp GP170	ABCB1	1279 (12)	mMdr1a (Mdr3) mMdr1b (Mdr1) (and rat homologue)	Many epithelia, blood–brain barrier	Plasma membrane (apical)	Glucosylceramide, platelet-activating factor	Amphipathic drugs	–	Drug hypersensitivity	Major defense function against xenotoxins
5	PGY3 MDR3(2) PFIC-3	ABCB4	1279 (12)	mMdr2 rMdr2	Liver hepatocytes	Plasma membrane (apical)	Long-chain phosphatidylcholine	Some amphipathic drugs	PFIC-3, cholestasis of pregnancy	Liver disease	Also defense function against xenotoxins?
6	BSEP sPGP PFIC-2 PGY4	ABCB11	1321 (12?)	mBsep rBsep	Liver hepatocytes	Plasma membrane (apical)	Bile salts	Paclitaxel	PFIC-2	Liver disease	Drug resistance is low
7	MRP1 MRP	ABCC1	1531 (17)	mMrp1 rMrp1	Ubiquitous	Plasma membrane (basolateral) and endosomes	LTC_4	Anionic drug conjugates, GSSG, GSH	–	Drug hypersensitivity	Also co-transports drugs with GSH

8	MRP2 cMOAT	ABCC2	1545 (17)	rMrp2	Liver, intestine, kidney	Plasma membrane (apical)	Bilirubin-glucuronides, GSSG and GSH; acidic bile salts	Anionic drug conjugates	Dubin–Johnson syndrome	Altered drug handling[d]	Also co-transports drugs with GSH
9	MRP3	ABCC3	1527 (17?)	rMrp3 mMrp3	Liver, bile ducts, gut, adrenal cortex	Plasma membrane (basolateral)	Bile salts	Anionic drug conjugates[e]	–	?	Strongly upregulated in cholestasis
10	ALD ALDP	ABCD1	745 (6?)	mAld	Many	Peroxisomal membrane	Very long-chain saturated fatty acyl-CoA	–	Adrenoleuko-dystrophy	As humans	Probably heterodimer with 11/12/13
11	ALDL1 ALDR	ABCD2	740 (6?)	rAbcd2	Many	Peroxisomal membrane	As 10?	?	–	–	As 10?
12	PMP70 PXMP1	ABCD3	659 (6?)	mPmp70 rPmp70	Many	Peroxisomal membrane	As 10?	?	–	–	As 10?
13	PMP69 P70R PXMPIL	ABCD4	606 (6?)	mP69r	Many	Peroxisomal membrane	As 10?	?	–	–	As 10?
14	BCRP1 MXR1 ABCP	ABCG2	655 (6)	mBcrp1	Placenta, gut, liver, endothelium	Plasma membrane (apical)	None known	Amphipathic drugs	–	Drug hypersensitivity	Major defense function against xenotoxins
15	ABCG5	ABCG5	651 (6?)	–	Liver, intestine	Plasma membrane? (apical?)	Plant sterols	Cholesterol?	Sitosterolemia	?	Probably heterodimer with ABCG8
16	ABCG8	ABCG8	673 (6?)	–	Liver, intestine	Plasma membrane? (apical?)	Plant sterols	Cholesterol?	Sitosterolemia	–	Probably heterodimer with ABCG5

[a]See also http://nutrigene.4t.com/humanabc.htm.
This is the website of Michael Müller, University of Wageningen, The Netherlands.
[b]Size in number of amino acids and topology as the most probable number of transmembrane segments.
[c]A dash means that no homozygous null alleles have been observed (humans, rats) or constructed (KO mice). A question mark means that no phenotype has (yet) been found.
[d]Decreased biliary drug clearance and increased oral drug availability; decreased biliary excretion of bilirubin glucuronides.
[e]Preference for glucuronosyl derivatives of drugs and steroids; does not transport GSH.

derivative of estradiol and it is therefore possible that ABCA2 is a steroid transporter. This remains to be demonstrated, however.

Many transporters listed in **Table 22.1** are discussed in detail in other chapters of this volume. The involvement of the MDR1 P-glycoprotein and MRP1 in drug transport is discussed in Chapters 18 and 19, MRP2 (ABCC2) in Chapter 20, MRP3 (ABCC3) in Chapter 21, ABCA1 in Chapter 23, ABCA4 in Chapter 28, and the peroxisomal ABC transporters in Chapter 24. Here we concentrate on four topics: (1) the MDR1 P-glycoprotein (ABCB1) and its role in transporting physiological lipids; (2) the transport of lipid analogues by ABC transporters; (3) the MDR3 (ABCB4) P-glycoprotein (the phosphatidylcholine transporter, murine Mdr2); and (4) the role of ABC transporters in sterol transport with an emphasis on transporters not discussed in detail in other chapters.

The MDR1 (ABCB1) P-Glycoprotein and Its Role in Transporting Physiological Lipids

MDR1 P-glycoprotein substrates are rather hydrophobic molecules (Seelig *et al.*, 2000) and current models for MDR1 activity propose that the substrate is recognized within the membrane. In the vacuum cleaner model (Raviv *et al.*, 1990), the substrate molecule enters a hydrophobic cavity and is pumped into the extracellular space (Bolhuis *et al.*, 1996). In the flippase model (Higgins and Gottesman, 1992), the substrate enters the MDR1 P-glycoprotein from the cytosolic leaflet of the plasma membrane and is subsequently moved into the exoplasmic leaflet. From there, it can freely diffuse into the extracellular space. Not surprisingly, the question was raised as to whether MDR1 P-glycoprotein would be capable of moving natural membrane lipids from the cytosolic into the exoplasmic leaflet of the plasma membrane, a process termed flop as opposed to lipid flip in the opposite direction (Devaux and Zachowski, 1994). In this case, pumping of the substrate lipid into the aqueous phase would be unlikely, as the change in free energy between the monomer in aqueous solution and the membrane form (70 kJ mol^{-1} for PC) is more than the energy released from hydrolysis of an ATP molecule (30 kJ mol^{-1}) (McLean and Phillips, 1984). This problem would be solved by the flippase model, in which the substrate is only moved from the cytosolic to the exoplasmic leaflet of the plasma membrane (Higgins and Gottesman, 1992). Less hydrophobic compounds might be translocated into the exoplasmic leaflet by flippase action and readily equilibrate with the extracellular water phase.

One natural lipid found to be a substrate for MDR1 P-glycoprotein is platelet-activating factor (PAF) (Ernest and Bello-Reuss, 1999; Raggers *et al.*, 2001). This bioactive lipid is synthesized by inflammatory cells upon cell activation by a number of physiological stimuli. Some activated cells release PAF upon specific induction whereas other cells need no additional stimulation (Prescott *et al.*, 2000). PAF release from these cells may be the consequence of a process that scrambles the asymmetric distribution of the bulk membrane lipids (Bratton, 1993).

Ernest and Bello-Reuss (1999) found, unexpectedly, that PAF release from ionophore-stimulated cells is inhibited by MDR1 inhibitors. This was followed up by Raggers *et al.* (2001), who found that transfection of kidney epithelial cell monolayers with human *MDR1* selectively increased PAF transport across the apical plasma membrane domain. PAF transport was independent of vesicular traffic and was inhibited by the MDR1 P-glycoprotein inhibitors PSC833 and cyclosporin A (CsA). It is very likely that this system mimics the *in vivo* situation of constitutive PAF secretion since kidney cells possess a cholinephosphotransferase that is specific for PAF synthesis (Woodard *et al.*, 1987), and have a high endogenous level of MDR1.

Like PAF, its structural analogue, the antineoplastic agent edelfosine (the di-ether PC analogue 1-*O*-octadecyl-2-*O*-methyl-sn-glycero-3-phosphocholine) may be a substrate for translocation by MDR1. However, a report on such an activity for murine Mdr3 (also known as Mdr1a) P-glycoprotein (Abcb1) was not consistently confirmed in subsequent studies (Ruetz *et al.*, 1997).

The fact that Mdr1 (Abcb1) does not functionally replace Mdr2 (Abcb4) in the *Mdr2* murine knockout (KO) model predicts that natural long-chain PC is not a substrate for human MDR1 (Smit *et al.*, 1993), but a direct test must still be done to prove this. A long-chain membrane lipid that may be a substrate for MDR1 is sphingomyelin (SM), as one study claimed that the MDR1 inhibitor PSC833 caused an increase in the fraction of SM in the inner leaflet of the plasma membrane (Bezombes *et al.*, 1998). Another sphingolipid that appears to be translocated by MDR1 is glucosylceramide. From its site of synthesis on the cytosolic surface of the Golgi, glucosylceramide did not reach the surface of fibroblasts derived from an *Mdr1* null mutant mouse. In control fibroblasts, which express *Mdr1*, transport was inhibited by Mdr1/MDR1 inhibitors. Translocation of glucosylceramide across the Golgi membrane, as measured from the synthesis of higher glycolipids in the Golgi lumen, continued in the null fibroblasts, indicating the presence of an Mdr1-independent translocator (Raggers *et al.*, unpublished results). Still, the results do not exclude the possibility that Mdr1 can also be active in membranes of the Golgi. This has been suggested from the observations that transfection of Madine Darby canine kidney (MDCK) cells with human *MDR1* cDNA dramatically increased synthesis of the (lumenal) glycolipid globotriaosylceramide, and that this increase could be inhibited by MDR1 inhibitors (Lala *et al.*, 2000).

Finally, sterols have been found to act as substrates for MDR1 with variable efficiencies. Whereas dexamethasone and cortisol are relatively good substrates, progesterone binds to the substrate-binding site without being transported, and is a good inhibitor. MDR1 has also been proposed to be involved in the intracellular trafficking of cholesterol but it is unclear whether the effect is related to MDR1 P-glycoprotein drug transport activity (Debry *et al.*, 1997; Field *et al.*, 1995; Luker *et al.*, 1999). Alternatively, it might be linked to translocation of glucosylceramide or another sphingolipid as cholesterol preferentially interacts with sphingolipids. Whether cholesterol, the major mammalian membrane sterol, is translocated awaits a direct transport experiment (Barnes *et al.*, 1996; Wang *et al.*, 2000).

TRANSPORT OF LIPID ANALOGUES BY ABC TRANSPORTERS

Traditional assays for measuring the fraction of a lipid on one side of the membrane relied on the use of phospholipases, labeling reagents and lipid exchange techniques, and were not well suited as sensitive assays for protein-mediated lipid translocation (Op den Kamp, 1979; Sillence *et al.*, 2000). Measurements have been greatly facilitated by the use of analogues of membrane lipids, in which one long lipid chain has been replaced by a short C5-C6 acyl chain (Seigneuret and Devaux, 1984; Sleight and Pagano, 1985). The lower hydrophobicity enhances the off-rate from the membrane, allowing the analogues to exchange via the aqueous phase with half-times of seconds, whereas the natural lipids need hours. As a first step, the analogues can thus be easily inserted into the surface of the membrane of interest. For detection, the analogues are labeled on the short chain with a spin-label, a fluorescent moiety or a radiolabel. Translocation can then be monitored in various ways. One general principle is the 'back-exchange'. After the incubation, exogenous bovine serum albumin (BSA) is used to selectively extract the short-chain analogue from the outer leaflet. Alternatively, the spin-labeled or fluorescent analogues in the outer leaflet can be chemically quenched (Bosch *et al.*, 1997; Margolles *et al.*, 1999; Romsicki and Sharom, 2001). The activity of a translocator can be measured as a change in the amount of analogue that has been translocated. One can also compare the rate of analogue uptake from an exogenous source (or the resulting accumulation), which is a combination of inward and outward translocation. Finally, a change in the equilibrium distribution of a natural lipid across the bilayer may reflect the activity of a translocator (Dekkers *et al.*, 2000). Still, it must be realized that results obtained with analogues cannot be extrapolated to the natural lipids without further experiments.

The ability of ABC transporters to transport lipid analogues was first shown for the PC translocator encoded by the mouse *Mdr2* P-glycoprotein (Abcb4) and human *MDR3* P-glycoprotein (ABCB4) genes. Translocation of C6-NBD-PC (*N*-6[7-nitro-2,1,3-benzoxadiazol-4-yl]-amino-hexanoyl-phosphatidylcholine) from the cytosolic leaflet to the lumen (or lumenal leaflet) of secretory vesicles isolated from yeast transformed with *Mdr2* was higher than in vesicles from control cells (Ruetz and Gros, 1994). Translocation was ATP- and Mg^{2+}-dependent, sensitive to the inhibitor verapamil, and appeared selective for Mdr2 since Mdr3 was inactive. The activity of human MDR3 towards C6-NBD-PC was confirmed in *MDR3* transfected cells, where transport of intracellularly synthesized C6-NBD-PC to the cell surface was measured by the BSA assay described earlier (van Helvoort *et al.*, 1996). However, in contrast to the results of fluorescence quenching experiments (Ruetz and Gros, 1994), the experiments on *MDR1* transfected cells demonstrated that human MDR1 (and mouse Mdr3) possess the same capability as human MDR3 to translocate C6-NBD-PC. Unexpectedly, MDR1 was found to translocate a wide variety of short-chain analogues **(Table 22.2)**. This finding was corroborated by subsequent work on intact cells (Bosch *et al.*, 1997), and in reconstituted proteoliposomes with hamster Abcb1 (Romsicki and Sharom, 2001). The hamster P-glycoprotein displayed even less specificity than the human MDR1, because in the fluorescence quenching assay of Ruetz and Gros (1994), it transported various analogues of phosphatidylserine which were not recognized by human MDR1 in the intact cell system. Hamster Abcb1 even transported phospholipid analogues with long acyl chains and a phosphoethanolamine-NBD headgroup, *N*-NBD-PE (Romsicki and Sharom, 2001). These analogues may resemble the physiological lipid *N*-retinylidene-PE, a presumed substrate of ABCR (ABCA4), the rod outer segment disk ABC transporter (see Chapter 28) (Sun *et al.*, 1999; Weng *et al.*, 1999).

Like MDR1, human MRP1 (ABCC1) was found to transport only C6-NBD sphingolipid analogues, but translocation by MRP1 depended on the presence of the NBD moiety (Raggers *et al.*, 1999). C6-NBD-PS is also transported by human and mouse MRP1/Mrp1 (Dekkers *et al.*, 1998; Kamp and Haest, 1998). Most convincingly, outward transport of C6-NBD-PS across the erythrocyte membrane was completely absent in erythrocytes from *Mrp1*(−/−) mice (Dekkers *et al.*, 1998). The situation for PC is more complex. The PCs PAF and C16:0/C6-NBD-PC were not substrates (Ernest and Bello-Reuss, 1999; Raggers *et al.*, 1999). However, in human erythrocytes, the outward movement of C18:1/C6-NBD-PC and C14:0/C12-NBD-PC required reduced glutathione (GSH) and was strongly inhibited by MRP1/Mrp1 inhibitors (Dekkers *et al.*, 1998; Kamp and Haest, 1998). One explanation for the difference could be the difference in fatty acid at the sn-1 position. As might be expected from the similarities in substrate specificity, ABC transporters from yeasts and bacteria have also been found to be capable of translocating lipid analogues. The available data are summarized in **Table 22.2**.

THE PC TRANSLOCATOR (MDR3, MDR2, ABCB4)

MDR3/MDR2 TRANSLOCATES PHOSPHATIDYLCHOLINE

The *MDR3* (*ABCB4*) gene for the human PC translocator is located on chromosome 7 (q 21.1), only 34 kb downstream of the *MDR1* (*ABCB1*) gene (Lincke *et al.*, 1991). A similar gene cluster on chromosome 5 of the mouse contains the mouse orthologue, *Mdr2* (Kirschner, 1995; Raymond *et al.*, 1990). Disruption of this gene led to the discovery that the Mdr2 P-glycoprotein is essential for secretion of long-chain PC into bile (Smit *et al.*, 1993). *Mdr2*(+/−)-heterozygotes have no defects, but secrete only about half as much PC as wild-type *Mdr2*(+/+) mice (Smit *et al.*, 1993). The absence of PC in the bile of *Mdr2* (−/−) mice leads to a mild liver disease, because bile salt secretion is normal in these mice and the high bile salt concentrations, without accompanying PC, damage the canalicular membrane of the hepatocyte and the small bile ducts. This causes extensive bile duct proliferation and some hepatocyte damage (Mauad *et al.*, 1994; Oude Elferink *et al.*, 1997). All defects in these mice are due to the absence of Mdr2 in the liver, as they can be completely prevented in the KO mice by the liver-specific expression of the human *MDR3* gene under the control of an albumin promoter active only in the liver (Smith *et al.*, 1998). The murine, rat and human PC translocator genes are also expressed at low levels in adrenal glands, skeletal and heart muscle, tonsil and spleen (see Smit *et al.*, 1994), and the protein has been detected in murine erythrocytes (Vermeulen, 1996). No function for the PC translocator has been found, however, in any other tissue than liver.

Our present ideas about PC secretion from hepatocytes into bile are summarized in **Figure 22.1**. PC secretion depends both on bile salts and on the Mdr2 P-glycoprotein. If either is lacking, no PC secretion is detectable. If Mdr2 is present in the hepatocyte canalicular membrane, the rate of PC secretion is hyperbolically dependent on the bile salt concentration. Interestingly, PC secretion is dependent on the Mdr2 levels at all bile salt concentrations **(Figure. 22.2)**. PC secretion is higher in wild-type *Mdr2*(+/+) mice than in *Mdr2*(+/−) heterozygotes. Secretion is

TABLE 22.2. TRANSPORT OF MEMBRANE LIPID ANALOGUES BY MULTIDRUG TRANSPORTERS

	Species[a]	Glycerophospholipids			Sphingolipids		References
		PC	PE	PS	GlcCer	SM	
MDR1 Pgp	H	PAF					1, 2
	H, M				C6-NBD		3–5
	H, M, CH	C16:0/C6-NBD	C16:0/C6-NBD			C6-NBD	3, 4, 6
	CH	C18:1/C6-NBD	N-NBD-diC16:0	C16:0/C6-NBD			6[b]
	H	C8:0/C8:0	C8:0/C8:0		C8:1/C8:0[c], C6	C8:1/C8:0[c]	3
	H	C16:0/C12-NBD	C12-NBD				7
				not C12-NBD			
MDR3 Pgp	M	Cx/C6-NBD					8
	H	C16:0/C6-NBD	*not* C16:0/C6-NBD		*not* C6-NBD	*not* C6-NBD	5
MRP1	H	*not* PAF					1, 2
	H, M	C18:1/C6-NBD		C18:1/C6-NBD			9–11
	H	C14:0/C12-NBD					10
	H					C6-NBD	11
	H	*not* C16:0/C6-NBD	*not* C16:0/C6-NBD		C6-NBD *not* C6	C6-NBD	5
ABC1	M			C16:0/C5-doxyl[d]			12
YOR1p and Pdr5p	S		C14:0/C6-NBD				13
LmrA	L	*not* C14:0/C6-NBD	C14:0/C6-NBD				14

[a]CH, Chinese hamster; H, human; L, *Lactococcus lactis*; M, mouse; S, *Saccharomyces cerevisiae*.
[b]Similar data for C12-NBD-PC and –PS and for N-NBD-diC18:1.
[c]Contain a truncated ceramide, consisting of C8 sphingosine amide-linked to C8:0 fatty acid.
[d]In addition, evidence was provided for the translocation of natural PS.
1, Ernest and Bello-Reuss (1999); 2, Raggers *et al.* (2001); 3, van Helvoort *et al.* (1996); 4, van Helvoort *et al.* (1997); 5, Raggers *et al.* (1999); 6, Romsicki and Sharom (2001); 7, Bosch *et al.* (1997); 8, Ruetz and Gros (1994); 9, Dekkers *et al.* (1998); 10, Kamp and Haest (1998); 11, Dekkers *et al.* (2000); 12, Hamon *et al.* (2000); 13, Decottignies *et al.* (1998); 14, Margolles *et al.* (1999).

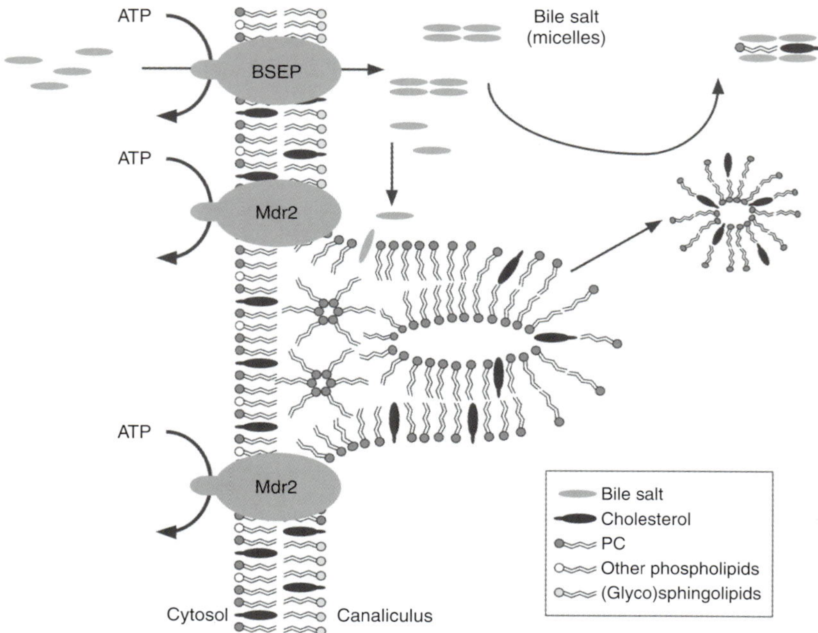

Figure 22.1. The role of Mdr2/MDR3 P-glycoprotein in bile formation (modified from Borst et al.*, 2000). cBAT is the canalicular bile acid transporter, BSEP (ABCB11). Phosphatidylcholine (PC) from the inner leaflet of the canalicular membrane is flipped by Mdr2/MDR3 (ABCB4) to the exoplasmic leaflet where it is accessible to extraction by bile salts.*

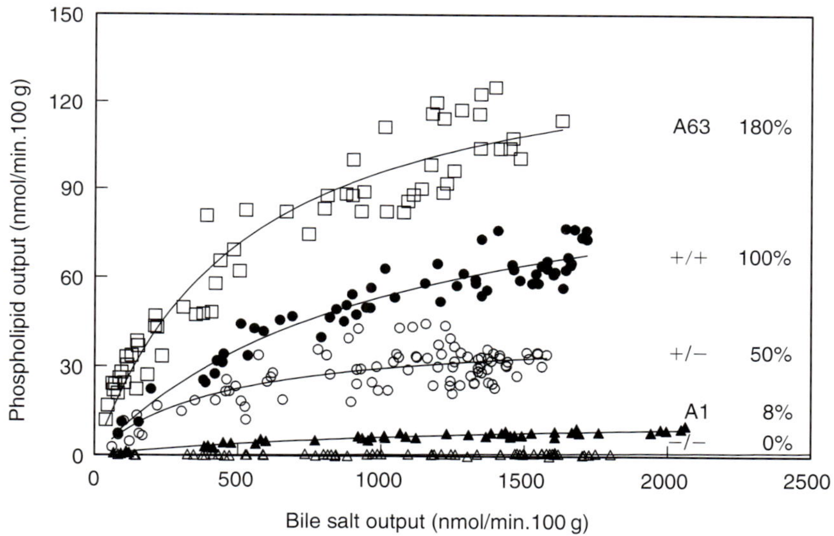

*Figure 22.2. Relation between bile salt and phospholipid transport in mice with different expression levels of the PC translocator gene (*MDR3/Mdr2/ABCB4*). Normal wild-type mice (+/+; filled circles), mice with a homozygous (−/−; open triangles) or heterozygous (+/−; open circles) disruption of the* Mdr2 *gene as well as A63 mice (transgenic for the* MDR3 *gene against wild-type background; open squares) and A1 mice (transgenic for* MDR3 *against* Mdr2(−/−) *background; closed triangles) were infused with increasing amounts of the bile salt tauroursodeoxycholate, while bile was continuously collected. Phospholipid secretion is hyperbolically dependent on bile secretion and the maximal phospholipid secretion capacity is strictly dependent on the expression levels of* Mdr2/MDR3*. The total expression levels are given in the figure, expressed as percentages of that in wild-type mice. Modified from Oude Elferink* et al. *(1998).*

highest in mice transgenic for the human *MDR3* gene, which have a supraphysiological PC translocator concentration in their livers, and very low in the A1 *Mdr2*(−/−) homozygote with low *MDR3* transgene expression (Smith *et al.*, 1998). Crawford and co-workers (1997) have shown by electron microscopy that adherent monolamellar vesicles on the outer canalicular membrane may represent an intermediate structure in the secretion process of biliary lipid. Formation of these vesicles requires the presence of functional Mdr2. Interestingly, the appearance of lipoprotein X (LpX) in the blood of cholestatic mice is also completely dependent on functional Mdr2 (Oude Elferink *et al.*, 1998). LpX consists of 40–100 nm vesicles comprising phospholipid and cholesterol with an aqueous lumen. They appear shortly after bile duct ligation in wild-type mice, but not in *Mdr2*(−/−) mice. How LpX reaches the blood is not known. Oude Elferink *et al.* (1998) favor a model in which biliary vesicles continue to be formed at the canalicular membrane after ligation and the LpX vesicles reach the blood by transcytosis through the hepatocyte. Another possibility is that the increased pressure in the biliary compartment after ligation is released from time to time by opening of the tight junctions between hepatocytes and a paracellular flux of bile into the blood.

THE PC TRANSLOCATOR (MDR3/MDR2, ABCB4) CAN ALSO TRANSPORT DRUGS

Initial experiments on the substrate specificity of the MDR3 PC translocator were done with membrane vesicles from transgenic yeast overexpressing the murine orthologue Mdr2. With this system, Ruetz and Gros (1994) showed that the translocator is highly specific for phospholipid analogues with a choline head group and this was confirmed by van Helvoort *et al.* (1996) in animal cells. In addition, they found that the protein is selective towards the PC fatty acid moieties.

All these results suggested that the PC translocator had evolved for the specific purpose of transporting natural membrane PC into bile. It therefore came as a surprise that MDR3/Mdr2 is inhibited by verapamil, a classical inhibitor of P-glycoprotein (Ruetz and Gros, 1994; van Helvoort *et al.*, 1996). The group of Ueda (Kino *et al.*, 1996) then showed that transfection of yeast cells with an *MDR3* cDNA construct resulted in low-level resistance to the antifungal agent aureobasidin, also a substrate of MDR1. No multidrug resistance had ever been seen in animal cells transfected with *MDR3* or *Mdr2* cDNA constructs, but this is a rather insensitive assay for drug transport. Drug transport through epithelial monolayers provides a more sensitive assay, and Smith *et al.* (2000), using pig kidney cell monolayers expressing *MDR3,* found substantial transport of digoxin, lower rates of transport of paclitaxel, daunorubicin and vinblastine, but no significant transport of other drugs that are transported at high rate by the drug-transporting P-glycoproteins, such as CsA and dexamethasone. Digoxin transport by MDR3 was efficiently inhibited by P-glycoprotein inhibitors, including CsA, PSC833 and verapamil. To exclude the unlikely possibility that MDR3 had activated an endogenous drug transporter, Smith *et al.* (2000) verified that the protein interacts with drugs by studying nucleotide trapping by MDR3. They found that the substrates paclitaxel and vinblastine, and the inhibitors CsA and PSC833, were able to decrease nucleotide trapping by 90% or more in concentrations similar to those used for inhibiting MDR1.

As we have pointed out elsewhere (Borst *et al.*, 2000; Smith *et al.*, 2000), it is puzzling that the MDR3/Mdr2 PC translocator can bind and transport drugs in a membrane environment full of long-chain PC and that transport of PC analogues by the protein is efficiently inhibited by drugs. Whatever the explanation for this apparent paradox, it is important to acknowledge that the PC translocator *can* be inhibited by drugs. There is now ample evidence that patients with a diminished level of MDR3 are at risk of intrahepatic cholestasis (see below). Some drugs might increase that risk.

The possibility also remains that the MDR3 PC translocator might contribute to the MDR phenotype in some types of human cancer. This was first suggested by studies of Nooter and co-workers on drug-resistant B-cell leukemias. They noted that cells with substantial *MDR3* expression had diminished daunorubicin uptake that was reversed by CsA (Herweijer *et al.*, 1990; Nooter *et al.*, 1990). *MDR3* expression also correlated negatively with clinical outcome. This was confirmed in a larger group of patients by Arai *et al.* (1997). Although these results are compatible with a contribution of the MDR3 PC translocator to resistance through its ability to transport drugs, intervention studies with specific blockers will be required to prove the point.

DEFECTS IN THE *MDR2/MDR3 (ABCB4)* GENE LEAD TO LIVER DISEASE IN MICE AND HUMANS

Mdr2(−/−) mice, unable to make any PC translocator, are normal at birth. Only when bile flow starts do the symptoms of a non-suppurative inflammatory cholangitis appear, with portal inflammation and proliferation of the bile ducts (Mauad *et al.*, 1994; Smit *et al.*, 1993). At the age of 4–6 months, the mice start to develop multiple foci in their liver parenchyma, which eventually progress to tumors, often with necrosis and hemorrhage. These tumors may metastasize to the lung (Mauad *et al.*, 1994). The liver disease in the *Mdr2*(−/−) mice is accompanied by an increased bile flow and a strongly decreased cholesterol and GSH secretion into bile (Smit *et al.*, 1993). These abnormalities are the consequence of the inflammatory cholangitis caused by the (normal) secretion of bile salts without accompanying PC. One would expect the severity of the cholangitis to be affected by the nature of the bile salts secreted and this has been verified. Thus, if the diet is supplemented with the more hydrophobic bile salt cholate, liver pathology worsens, while ingestion of a more hydrophilic bile salt, ursodeoxycholate, results in decreased liver damage (Van Nieuwkerk *et al.*, 1996). The severity of liver damage in the *Mdr2*(−/−) mice is therefore dependent on bile salt hydrophobicity. A level of 15% of normal PC secretion, induced by expression of a *MDR3* transgene in the liver, prevented liver defects on a standard diet in male mice and mitigated pathology in female mice (De Vree, 1999). The fact that female mice suffer more from defects in the Mdr2 PC translocator than male mice correlates with females having a more hydrophobic bile salt composition and, hence, increased cytotoxicity of bile salts (Van Nieuwkerk *et al.*, 1997).

The *Mdr2* gene in the liver is relatively impervious to drastic treatments that strongly affect other liver ABC transporters, such as bile duct ligation, partial hepatectomy or endotoxin treatment (Vos *et al.*, 1998, 1999). However, when mice were fed a diet supplemented with cholate, their liver *Mdr2* mRNA and PC secretion capacity increased by approximately 50% (Frijters *et al.*, 1997). A much larger induction of the *Mdr2* gene was observed in mice exposed to compounds that induce peroxisome proliferation (Chianale *et al.*, 1996; Miranda *et al.*, 1997). Thus, a nearly sixfold increase in *Mdr2* mRNA was induced by 2,4,5-trichlorophenoxyacetic acid and this was accompanied by a fivefold increase in *Mdr2* mRNA synthesis. Biliary PC secretion went up only twofold and it is possible that Mdr2 protein levels were less elevated than the mRNA level, or that PC supply to the protein becomes rate limiting at these extreme levels of *Mdr2* induction. Two groups (Carralla *et al.*, 1999; Hooiveld *et al.*, 1999) have independently shown that *Mdr2* expression is induced by treatment of rats with statins (inhibitors of HMG-CoA reductase). Hooiveld *et al.* (1999) proposed that this induction is mediated by the sterol regulatory binding protein, SREBP, because the 5′ flanking region of *Mdr2* contains a potential sterol responsive element. Statin treatment causes a transient induction of SREBP expression, due to the depletion of cholesterol. It is not clear what the physiological function of increased *Mdr2* expression during sterol depletion would be.

Whereas overexpression of the MDR3/Mdr2 PC translocator gene in liver had no obvious deleterious effects, mice expressing an *MDR3* gene under a vimentin promoter develop cataracts (Dunia *et al.*, 1996) and a peripheral neuropathy (Smit *et al.*, 1996). It is not clear whether these pathological consequences are due to a redistribution of PC in membranes or only to the presence of a bulky glycoprotein in membranes where it does not normally belong.

Although it was obvious that there should be a human counterpart of the *Mdr2*(−/−) mouse, it took until 1996 before Deleuze *et al.* (1996) found that expression of *MDR3* was absent in a patient with progressive familial intrahepatic cholestasis (PFIC). This patient belonged to a subgroup of PFIC patients characterized by a high serum γ-glutamyltransferase (GGT), strong bile duct proliferation, and eventual liver cirrhosis requiring liver transplantation. This subgroup is now called PFIC, type 3. In subsequent work, 17 out of 31 patients with high GGT PFIC were found to have a mutation in the *MDR3* (*ABCB4*) gene (De Vree *et al.*, 1998; Jacquemin *et al.*, 1999, 2001). Since the gene was not completely sequenced in all these patients, it is not clear whether the remaining 14 patients have as yet unidentified *MDR3* mutations, or that another gene might also be involved in this form of PFIC.

Defects in the *MDR3* gene do not only give rise to pediatric liver disease. Jacquemin *et al.* (1999) reported that the mother of a patient with PFIC type 3 and several other women from this family suffered from intrahepatic cholestasis of pregnancy. These women turned out to be heterozygotes for the mutation in the *MDR3* gene, which caused PFIC type 3 in the homozygous index patient. Obviously bile formation is compromised during the last trimester of pregnancy and this unmasks heterozygous *MDR3* mutations. The mechanism of reduced transport during pregnancy has not been solved yet, but it could be due either to reduced expression of canalicular transporters (Bossard *et al.*, 1993), or to enhanced biliary disposition of steroid hormones, which may cause *trans*-inhibition of bile salt transport (Chambenoit *et al.*, 2001).

Rosmorduc *et al.* (2001) reported on six patients with gallstones, in whom mutations in the *MDR3* gene were found. This included homozygous and heterozygous missense mutations as well as a heterozygous insertion leading to frameshift and protein truncation. None of these mutations were observed in more than 100 control subjects, suggesting that they may be associated with the disease phenotype. Finally, Strautnicks and co-workers reported at the FEBS 2001 ABC Meeting at Gosau that three members of a Chinese family with idiopathic adulthood ductopenia were compound heterozygotes for mutations in the *MDR3* gene, whereas these mutations were absent in controls.

These reports indicate that the range of disease associated with defects of the human *MDR3/ABCB4* gene may be much larger than initially thought. Apart from the severe phenotype associated with the rare complete or nearly complete absence of MDR3 function, milder phenotypes also exist that are associated with reduction but not complete absence of function.

Sterol Transport by ABC Transporters

Whether cholesterol flip-flop through lipid bilayers requires proteins is a long-standing issue. Very different values have been reported for the rate of spontanous flip-flop of cholesterol with half-times ranging from seconds to hours (Brasaemle *et al.*, 1988; Lange *et al.*, 1981). This seems to depend mostly on the experimental system used. Clearly, if the lower rates are correct, protein-mediated translocation is necessary. Recently, considerable progress was made in the recognition of transporters that play a role in sterol transport since it turns out that ABC transporters are crucial for this type of lipid as well. We will describe the available evidence here rather briefly, because this topic will also be partly dealt with in other chapters.

Cholesterol transport from peripheral cells secondarily depends on ABCA1

The ABC gene responsible for Tangier disease was recently identified. Tangier disease is a rare inherited disorder characterized by the virtual absence of high-density lipoprotein (HDL). Patients with this disorder have mutations in the *ABCA1* gene (Bodzioch *et al.*, 1999; Brooks-Wilson *et al.*, 1999; Drobnik *et al.*, 1999; Rust *et al.*, 1999). Before the function of this protein was discovered, it was already known that the defect in Tangier disease involves the absence of cholesterol efflux from peripheral tissues. Fibroblasts from these patients are incapable of donating cholesterol and phospholipid to lipid-poor apoA-I, a step that is essential for maturation of the lipoprotein. When the association between *ABCA1* defects and Tangier disease was discovered, the initial suggestion was that the ABCA1 protein transports cholesterol, but recent evidence demonstrates that this is not the case. In an elegant study, Wang *et al.* (2001) investigated the molecular mechanism of ABCA1-mediated cholesterol efflux from transfected cells. They demonstrated that apoA-I binds to the cell membrane during this process and that this binding depends on the activity of the ABCA1 pump. Inhibition of the pump with glibenclamide not only prevented cholesterol and phospholipid transfer, but also eliminated binding of apoA-I to the cells. Depletion of cholesterol from the cells by extraction with the cholesterol acceptor cyclodextrin resulted in a marked decrease of ABCA1-mediated efflux, while phospholipid transfer continued to take place. Conversely, cholesterol efflux to mature HDL, which contains phospholipid, or to cyclodextrin, which does not need phospholipid for cholesterol extraction, is ABCA1 independent. This led Wang *et al.* (2001) to the conclusion that ABCA1 functions primarily as an outward phospholipid translocase rather than a cholesterol transporter. The authors propose the following model: ATP binding/hydrolysis by ABCA1 probably induces a conformational change of the transporter, which leads to the binding of apoA-I and phospholipid translocation. The binding site may consist of phospholipids together with ABCA1 itself. Once apoA-I has recruited the phospholipids translocated by ABCA1, the nascent complexes promote cholesterol efflux. Apparently, phospholipid-free apoA-I has little affinity for cholesterol and addition of phospholipid to the complex enhances the affinity for the sterol. ABCA1 is not only a transporter but also a receptor for apoA-I, as transfection of *ABCA1* in cells confers the competence to bind apoA-I (Chambenoit *et al.*, 2001). This binding competence depends on the activity of the pump, because transfection with mutant, ATPase-deficient *ABCA1* did not elicit apoA-I binding.

This model is highly reminiscent of the situation with cholesterol secretion at the canalicular membrane of the hepatocyte. It was shown in the *Mdr2(−/−)* mouse that both phospholipid and cholesterol secretion into bile are impaired. However, in transgenic rescue mice with a very low expression level of *Mdr2* PC translocator (15% of controls), cholesterol secretion is normal (Smith *et al.*, 1998). This shows that the presence of a small amount of phospholipid strongly enhances the affinity of bile salt micelles for cholesterol. In addition, infusion and subsequent biliary secretion of more hydrophobic bile salts in the *Mdr2(−/−)* mouse induced normal cholesterol secretion (Oude Elferink *et al.*, 1995). Thus, the extraction of cholesterol seems to be secondarily driven by the Mdr2-mediated extrusion of phospholipid from the plasma membrane.

It should be stressed, however, that both models for phospholipid-dependent cholesterol secretion leave open the possibility that cholesterol translocation across the plasma membrane could be protein-mediated as well. Although the final extraction from the membrane depends on the loading of acceptors with phospholipid (i.e. nascent HDL in the case of ABCA1 and mixed bile salt micelles in the case of MDR3/Mdr2/ABCB4), this might be preceded by protein-mediated transport of cholesterol from the inner leaflet to the outer leaflet of the membrane. If the protein involved is crucial for many processes involving cholesterol translocation, mutations in the gene for this protein might be lethal and therefore not be found.

Recently, Vaisman *et al.* (2001) reported a study on transgenic mice that overexpress *ABCA1*. As expected, these mice had increased plasma levels of HDL. The authors also reported that the concentration of cholesterol in bile was increased. This could mean that ABCA1 also fulfills a function in lipid secretion in the canalicular membrane. To investigate this aspect in a more rigorous model, Groen *et al.* (2001) measured cholesterol secretion rates into bile in *Abca1(−/−)* mice during bile diversion. They observed no difference between *Abca1(−/−)* mice and controls, suggesting that Abca1 does not contribute to biliary cholesterol efflux. Localization of ABCA1/Abca1 by immunohistochemistry is required to demonstrate whether ABCA1/Abca1 is present in the canalicular membrane of normal liver and whether a canalicular function is possible.

The *ABCA1* gene was initially cloned in 1994 by Chimini and her collaborators as a novel ABC transporter gene of unknown function (Luciani *et al.*, 1994). These investigators have suggested that ABCA1 can affect the distribution of phospholipids in the plasma membrane and thereby promote engulfment of apoptotic cells by macrophages (Hamon *et al.*, 2000; Luciani and Chimini, 1996 (Chapter 23)). Neither Tangier disease patients nor *Abca1(−/−)* mice have defects in developmental pathways requiring apoptosis, however (McNeish *et al.*, 2000). This indicates that the role of ABCA1 in dealing with apoptotic cells is not an essential one.

TRANSPORT OF PHYTOSTEROLS BY ABCG5/ABCG8 AND ITS DEFECT IN PATIENTS WITH SITOSTEROLEMIA

Sitosterolemia is a very rare, recessively inherited disease. The clinical presentation includes tendon xanthomas, accelerated atherosclerosis particularly affecting males at young age, hemolytic episodes, and arthritis and arthralgias (Salen *et al.*, 1992). The hallmark biochemical feature of the disease is the elevated concentration of plant sterols in plasma. Because sitosterol (24-ethyl cholesterol) is the most important accumulated sterol in plasma of these patients (as well as the most abundant plant sterol in the diet), this disease is referred to as sitosterolemia. A host of other sterol variants, such as campesterol, stigmasterol and brassicasterol, are present in plants and other components of the diet such as shellfish. Several studies have indicated that absorption of all these plant sterols in the intestine is strongly increased in sitosterolemic patients. Control subjects absorb approximately 50–70% of dietary cholesterol, but only less than 5% of the ingested plant sterols. In stark contrast, sitosterolemic patients absorb plant sterols to about the same extent (30–60%) as cholesterol (Salen *et al.*, 1992). This leads to a marked accumulation of these sterols in plasma, and in tissues such as liver, lung, heart and red blood cells. The brain content of plant sterols is low, indicating that the blood–brain barrier for these exogenous sterols is intact (Salen *et al.*, 1992). Apparently the blood–brain barrier prevents all sterols from being taken up, as defective cholesterol synthesis in patients with the Smith–Lemli–Opitz syndrome leads to a deprivation of cholesterol in brain tissue while other tissues can be partially rescued by dietary cholesterol (Salen *et al.*, 1996).

In normal subjects, the low amount of absorbed sitosterol is quickly secreted into bile so that only trace amounts of sitosterol can be found in blood (Salen *et al.*, 1992). In sitosterolemic patients, this biliary secretion of sitosterol and other phytosterols is impaired (Gregg *et al.*, 1986; Lutjohann *et al.*, 1995; Miettinen, 1980). Bile and plasma analysis in two patients revealed a more than 30-fold reduced bile/plasma ratio of total plant sterols compared with controls (Gregg *et al.*, 1986). Phytosterol handling in these patients is therefore impaired at two levels: absorption in the intestine and secretion into bile.

The disease locus for sitosterolemia was initially localized to chromosome 2p21 (Patel *et al.*, 1998) and more recently the genes involved in this disease were identified as *ABCG5* and *ABCG8* (Berge *et al.*, 2000; Lee *et al.*, 2001). These genes are separated by less than 140 bp and are located in opposite orientations on the chromosome (Lu *et al.*, 2001). *ABCG5* and *ABCG8* encode two ABC half transporters that are thought to dimerize into functional pumps. The fact that mutations in either *ABCG5* or *ABCG8* cause sitosterolemia (Lu *et al.*, 2001) is consistent with the idea that these two half transporters do indeed form a heterodimer that is essential for function. Both genes are expressed in liver and intestine, in keeping with their elimination function in both organs (Berge *et al.*, 2000). No studies have been reported yet on the transport of sterols by cells transfected with *ABCG5* and/or *ABCG8* gene constructs. Such studies should answer the important question whether ABCG5 and/or ABCG8 also transport cholesterol itself, or only plant sterols. Although most investigators think that ABCG5/ABCG8 does transport cholesterol, this remains a speculation in the absence of genuine transport experiments. The patient data are scarce but seem to indicate that the defect is much more pronounced for plant sterols than for cholesterol. The relative increase in plasma phytosterols is much greater than that of cholesterol, and the increase in intestinal absorption of phytosterols is greater than that of cholesterol (Salen *et al.*, 1992). The decrease in biliary elimination is also probably greater for plant sterols than for cholesterol since the ratio of phytosterol over cholesterol in bile is significantly decreased in sitosterolemic patients compared with controls (Björkhem and Boberg, 1995). The localization of the proteins in normal human tissues has also not been described yet. However, the increased sterol absorption and decreased biliary secretion in patients suggests that ABCG5/ABCG8 is an outward pump for plant sterols present in the apical membrane. In the intestine, this pump reduces absorption and in the liver, it mediates canalicular secretion of these unwanted sterols. Whether or not cholesterol is a substrate remains to be determined.

ACKNOWLEDGMENTS

The experimental work in our laboratories is supported by grants from the Dutch Cancer Society and The Netherlands Organization for Scientific Research and by a grant from the Mizutani Foundation for Glycoscience to G.v.M. We are grateful to Drs Alfred Schinkel and Bert Groen for suggestions to improve the manuscript.

REFERENCES

Arai, Y., Masuda, M., Sugawara, I., Arai, T., Motoji, T., Tsuruo, T., Oshimi, K. and Mizoguchi, H. (1997) Expression of the mdr1 and mdr3 gene products in acute and chronic leukemias. Leuk. Res. 21, 313–319.

Barnes, K.M., Dickstein, B., Cutler, J.G.B., Fojo, T., Bates, S.E. (1996) Steroid transport, accumulation, and antagonism of P-glycoprotein in multidrug-resistant cells. Biochemistry 35, 4820–4827.

Berge, K.E., Tian, H., Graf, G.A., Yu, L., Grishin, N.V., Schultz, J., Kwiterovich, P., Shan, B., Barnes, R. and Hobbs, H.H. (2000) Accumulation of dietary cholesterol in sitosterolemia caused by mutations in adjacent ABC transporters. Science 290, 1771–1775.

Bezombes, C., Maestre, N., Laurent, G., Levade, T., Bettaïeb, A. and Jaffrézou, J.-P. (1998) Restoration of TNF-α-induced ceramide generation and apoptosis in resistant human leukemia KG1a cells by the P-glycoprotein blocker PSC833. FASEB J. 12, 101–109.

Björkhem, I. and Boberg, K.M. (1995) Inborn errors in bile acid biosynthesis and storage of sterols other than cholesterol. In: The Metabolic and Molecular Basis of Inherited Disease (ed. C.R. Scriver). New York: McGraw-Hill.

Bodzioch, M., Orsó, E., Klucken, J., Langmann, T., Böttcher, A., Diederich, W., *et al.* (1999) The gene encoding ATP-binding cassette transporter 1 is mutated in Tangier disease. Nat. Genet. 22, 347–351.

Bolhuis, H., Van Veen, H.W., Molenaar, D., Poolman, B., Driessen, A.J.M. and Konings, W. N. (1996) Multidrug resistance in *Lactococcus lactis*: evidence for ATP-dependent drug extrusion from the inner leaflet of the cytoplasmic membrane. EMBO J. 15, 4239–4245.

Borst, P., Zelcer, N. and van Helvoort, A. (2000) ABC-transporters in lipid transport. Biochim. Biophys. Acta 1486, 128–144.

Bosch, I., Dunussi-Joannopoulos, K., Wu, R.-L., Furlong, S.T. and Croop, J. (1997) Phosphatidylcholine and phosphatidylethanolamine behave as substrates of the human MDR1 P-glycoprotein. Biochemistry 36, 5685–5694.

Bossard, R., Stieger, B., O'Neill, B., Fricker, G. and Meier, P.J. (1993) Ethinylestradiol treatment induces multiple canalicular membrane transport alterations in rat liver. J. Clin. Invest. 91, 2714–2720.

Brasaemle, D.L., Robertson, A.D. and Attie, A.D. (1988) Transbilayer movement of cholesterol in the human erythrocyte membrane. J. Lipid Res. 29, 481–489.

Bratton, D.L. (1993) Release of platelet activation factor from activated neutrophils. Transglutaminase-dependent enhancement of transbilayer movement across the plasma membrane. J. Biol. Chem. 268, 3364–3373.

Brooks-Wilson, A., Marcil, M., Clee, S.M., Zhang, L.-H., Roomp, K., Van Dam, M., *et al.* (1999) Mutations in *ABC1* in Tangier disease and familial high-density in lipoprotein deficiency. Nat. Genet. 22, 336–345.

Carralla, M., Feldman, D., Cogoi, S., Csillaghy, A. and Weinhold, P.A. (1999) Enhancement of mdr2 gene transcription mediates the biliary transfer of phosphatidylcholine supplied by an increased biosynthesis in the pravastarin-treated rat. Hepatology 29, 1825–1832.

Chambenoit, O., Hamon, Y., Marguet, D., Rigneault, H., Rosseneu, M. and Chimini, G. (2001) Specific docking of apolipoprotein A-1 at the cell surface requires a functional ABCA1 transporter. J. Biol. Chem. 276, 9955–9960.

Chianale, J., Vollrath, V., Wielandt, A.M., Amigo, L., Rigotti, A., Nervi, F., Gonzalez, S., Andrade, L., Pizarro, M. and Accatino, L. (1996) Fibrates induce *mdr2* gene expression and biliary phospholipid secretion in the mouse. Biochem. J. 314, 781–786.

Crawford, A.R., Smith, A.J., Hatch, V.C., Oude Elferink, R.P.J., Borst, P. and Crawford, J.M. (1997) Hepatic secretion of phospholipid vesicles critically depends on mdr2 or MDR3 P-glycoprotein expression. J. Clin. Invest. 100, 2562–2567.

De Vree, M. (1999) Defects in hepatogiliary lipid transport. Academic Thesis, Amsterdam.

De Vree, J.M.L., Jacquemin, E., Sturm, E., Cresteil, D., Bosma, P.J., Aten, J., *et al.* (1998) Mutations in the *MDR3* gene cause progressive familial intrahepatic cholestasis. Proc. Natl Acad. Sci. USA 95, 282–287.

Debry, P., Nash, E.A., Neklason, D.W. and Metherall, J.E. (1997) Role of multidrug resistance P-glycoproteins in cholesterol esterification. J. Biol. Chem. 272, 1026–1031.

Decottignies, A., Grant, A.M., Nichols, J.W., de Wet, H., McIntosh, D.B. and Goffeau, A. (1998) ATPase and multidrug transport activities of the overexpressed yeast ABC protein Yorp1. J. Biol. Chem. 273, 12612–12622.

Dekkers, D.W., Comfurius, P., Schroit, A.J., Bevers, E.M. and Zwaal, R.F. (1998) Transbilayer movement of NBD-labeled phospholipids in red blood cell membranes: outward-directed transport by the multidrug resistance protein 1 (MRP1). Biochemistry 37, 14833–14837.

Dekkers, D.W.C., Comfurius, P., Van Gool, R.G.J., Bevers, E.M. and Zwaal, R.F.A. (2000) Multidrug resistance protein 1 regulates lipid asymmetry in erythrocyte membranes. Biochem. J. 350, 531–535.

Deleuze, J.-F., Jacquemin, E., Dubuisson, C., Cresteil, D., Dumont, M., Erlinger, S., Bernard, O. and Hadchouel, M. (1996) Defect of multidrug-resistance 3 gene expression in a subtype of progressive familial intrahepatic cholestasis. Hepatology 23, 904–908.

Devaux, P.F. and Zachowski, A. (1994) Maintenance and consequences of membrane phospholipid asymmetry. Chem. Phys. Lipids 73, 107–120.

Drobnik, W., Liebisch, G., Biederer, C., Trumbach, B., Rogier, G., Muller, P. and Schmitz, G. (1999) Growth and cell cycle abnormalities of fibroblasts from Tangier disease patients. Arterioscl. Thromb. Vasc. Biol. 19, 28–38.

Dunia, I., Smit, J.J.M., Van der Valk, M.A., Bloemendal, H., Borst, P. and Benedetti, E.L. (1996) Human MDR3-P-glycoprotein expression in transgenic mice induces lens membrane alterations leading to cataract. J. Cell Biol. 132, 701–706.

Ernest, S. and Bello-Reuss, E. (1999) Secretion of platelet-activating factor is mediated by MDR1 P-glycoprotein in cultured human mesangial cells. J. Am. Soc. Nephrol. 10, 2306–2313.

Field, F.J., Born, E., Chen, H., Murthy, S. and Mathur, S.N. (1995) Esterification of plasma membrane cholesterol and triacylglycerol-rich lipoprotein secretion in CaCo-2 cells: possible role of P-glycoprotein. J. Lipid Res. 36, 1533–1543.

Frijters, C.M.G.G., Ottenhoff, R., Van Wijland, M.J., Van Nieuwkerk, C.M.J., Groen, A.K. and Oude Elferink, R.P.J. (1997) Regulation of mdr2 P-glycoprotein expression by bile salts. Biochem. J. 321, 389–395.

Gregg, R.E., Connor, W.E., Lin, D.S. and Brewer, H.B.J. (1986) Abnormal metabolism of shellfish sterols in a patient with sitosterolemia and xanthomatosis. J. Clin. Invest. 1864–1872.

Groen, A.K., Brandsma, R.H., Ottenhoff, R., Chimini, G. and Kuipers, F. (2001) Hepatobiliary cholesterol transport is not impaired in ABCA1 null mice lacking high density lipoproteins. J. Clin. Invest. 108, 843–850.

Hamon, Y., Broccardo, C., Chambenoit, O., Luciani, M.F., Toti, F., Chaslin, S., *et al.* (2000) ABC1 promotes engulfment of apoptotic cells and transbilayer redistribution of phosphatidylserine. Nat. Cell Biol. 2, 399–406.

Herweijer, H., Sonneveld, P., Baas, F. and Nooter, K. (1990) Expression of mdr1 and mdr3 multidrug-resistance genes in human acute and chronic leukemias and association with stimulation of drug accumulation by cyclosporine. J. Natl Cancer Inst. 82, 1133–1140.

Higgins, C.F. and Gottesman, M.M. (1992) Is the multidrug transporter a flippase? Trends Biochem. Sci. 17, 18–21.

Hooiveld, G.J.E.J., Vos, T.A., Scheffer, G.L., Van Goor, H., Koning, H., Bloks, V., *et al.* (1999) 3-Hydroxy-3-methylglutaryl-coenzyme A reductase inhibitors (statins) induce hepatic expression of the phospholipid translocase mdr2 in rats. Gastroenterology 117, 678–687.

Jacquemin, E., Cresteil, D., Manouvrier, S., Boute, O. and Hadchouel, M. (1999) Heterozygous non-sense mutation of the *MDR3* gene in familial intrahepatic cholestatis of pregnancy. Lancet 353, 210–211.

Jacquemin, E., De Vree, J.M.L., Cresteil, D., Sokal, E.M., Sturm, E., Dumont, M., *et al.* (2001) The wide spectrum of multidrug resistance 3 deficiency: from neonatal cholestasis to cirrhosis of adulthood. Gastroenterology 120, 1448–1458.

Kamp, D. and Haest, C.W.M. (1998) Evidence for a role of the multidrug resistance protein (MRP) in the outward translocation of NBD-phospholipids in the erythrocyte membrane. Biochim. Biophys. Acta 1372, 91–101.

Kino, K., Taguchi, Y., Yamada, K., Komano, T. and Ueda, K. (1996) Aureobasidin A, an antifungal cyclic depsipeptide antibiotic, is a substrate for both human MDR1 and MDR2/P-glycoproteins. FEBS Lett. 399, 29–32.

Kirschner, L.S. (1995) Amplification of the murine *mdr2* gene and a reconsideration of the structure of the murine *mdr* gene locus. DNA Cell Biol. 14, 47–59.

Lala, P., Ito, S. and Lingwood, C.A. (2000) Retroviral transfection of Madin-Darby canine kidney cells with human *MDR1* results in a major increase in globotriaosylceramide and 10^5- to 10^6-fold increased cell sensitivity to verocytotoxin. J. Biol. Chem. 275, 6246–6251.

Lange, Y., Dolde, J. and Steck, T.L. (1981) The rate of transmembrane movement of cholesterol in the human erythrocyte. J. Biol. Chem. 256, 5321–5323.

Lee, M.H., Lu, K.L., Hazard, S., Yu, H., Shulenin, S., Hidaka, H., *et al.* (2001) Identification of a gene, *ABCG5*, important in the regulation of dietary cholesterol absorption. Nat. Genet. 27, 79–83.

Lincke, C.R., Smit, J.J., Van der Velde-Koerts, T. and Borst, P. (1991) Structure of the human MDR3 gene and physical mapping of the human MDR locus. J. Biol. Chem. 266, 5303–5310.

Lu, K., Lee, M.H., Hazard, S., Brooks-Wilson, A., Hidaka, H., Kojima, H., Ose, L., Stalenhoef, A.F., Mietinnen, T. and Bjorkhem, I. (2001) Two genes that map to the sts1 locus cause sitosterolemia: genomic structure and spectrum of mutations involving sterolin-1 and sterolin-2, encoded by abcg5 and abcg8, respectively. Am. J. Hum. Genet. 69, 278–290.

Luciani, M.-F. and Chimini, G. (1996) The ATP binding cassette transporter ABC1, is required for the engulfment of corpses generated by apoptotic cell death. EMBO J. 15, 226–235.

Luciani, M.F., Denizot, F., Savary, S., Mattei, M.G. and Chimini, G. (1994) Cloning of two novel ABC transporters mapping on human chromosome 9. Genomics 21, 150–159.

Luker, G.D., Nilsson, K.R., Covey, D.F. and Piwnica-Worms, D. (1999) Multidrug resistance *(MDR1)* P-glycoprotein enhances esterification of plasma membrane cholesterol. J. Biol. Chem. 274, 6979–6991.

Lutjohann, D., Bjorkhem, I., Beil, U.F. and Von Bergmann, K. (1995) Sterol absorption and sterol balance in phytosterolemia evaluated by deuterium-labeled sterols: effect of sitostanol treatment. J. Lipid Res. 36, 1763–1773.

Margolles, A., Putman, M., Van Veen, H.W. and Konings, W.N. (1999) The purified and functionally reconstituted multidrug transporterLmrA of *Lactococcus lactis* mediates the transbilayer movement of specific fluorescent phospholipids. Biochemistry 38, 16298–16306.

Mauad, T.H., Van Nieuwkerk, C.M.J., Dingemans, K.P., Smit, J.J.M., Schinkel, A.H., Notenboom, R.G.E., *et al.* (1994) Mice with homozygous disruption of the mdr2 P-glycoprotein gene: a novel animal model for studies of nonsuppurative inflammatory cholangitis and hepatocarcinogenesis. Am. J. Pathol. 145, 1237–1245.

McLean, L.R. and Phillips, M.C. (1984) Kinetics of phosphatidylcholine and lysophosphatidylcholine exchange between unilamellar vesicles. Biochemistry 23, 4624–4630.

McNeish, J., Aiello, R.J., Guyot, D., Turi, T., Gabel, C., Aldinger, C., *et al.* (2000) High density lipoprotein deficiency and foam cell accumulation in mice with targeted disruption of ATP-binding cassette transporter-1. Proc. Natl Acad. Sci. USA 97, 4245–4250.

Miettinen, T.A. (1980) Phytosterolaemia, xanthomatosis and premature atherosclerotic arterial disease: a case with high plant sterol absorption, impaired sterol elimination and low cholesterol synthesis. Eur. J. Clin. Invest. 27–35.

Miranda, S., Vollrath, V., Wielandt, A.M., Loyola, G., Bronfman, M. and Chianale, J. (1997) Overexpression of mdr2 gene by peroxisome proliferators in the mouse liver. J. Hepatol. 26, 1331–1339.

Nooter, K., Sonneveld, P., Janssen, A., Oostrum, R., Boersma, T., Herweijer, H., Valerio, D., Hagemeijer, A. and Baas, F. (1990) Expression of the mdr3 gene in prolymphocytic leukemia: association with cyclosporin-A-induced increase in drug accumulation. Int. J. Cancer 45, 626–631.

Op den Kamp, J.A.F. (1979) Lipid asymmetry in membranes. Annu. Rev. Biochem. 48, 47–71.

Oude Elferink, R.P.J., Ottenhoff, R., Van Wijland, M., Smit, J.J.M., Schinkel, A.H. and Groen, A.K. (1995) Regulation of biliary lipid secretion by mdr2 P-glycoprotein in the mouse. J. Clin. Invest. 95, 31–38.

Oude Elferink, R.P.J., Tytgat, G.N.J. and Groen, A.K. (1997) The role of mdr2 P-glycoprotein in hepatobiliary lipid transport. FASEB J. 11, 19–28.

Oude Elferink, R.P.J., Ottenhoff, R., Van Marle, J., Frijters, C.M.G., Smith, A.J. and Groen, A.K. (1998) Class III P-glycoproteins mediate the formation of lipoprotein X in the mouse. J. Clin. Invest. 102, 1749–1757.

Patel, S.B., Salen, G., Hidaka, H., Kwiterovich, P.O., Stalenhoef, A.F., Mietinnen, T.A., Grundy, S.M., Lee, M.H., Rubenstein, S.J. and Polymeropoulos, M.H. (1998) Mapping a gene involved in regulating dietary cholesterol absorption: the sitosterolemia locus is found at chromosome 2p21. J. Clin. Invest. 102, 1041–1044.

Prescott, S.M., Zimmerman, G.A., Stafforini, D.M. and McIntyre, T.M. (2000) Platelet-activating factor and related lipid mediators. Annu. Rev. Biochem. 69, 419–445.

Raggers, R.J., van Helvoort, A., Evers, R. and van Meer, G. (1999) The human multidrug resistance protein MRP1 translocates sphingolipid analogs across the plasma membrane. J. Cell Sci. 112, 415–422.

Raggers, R.J., Vogels, I. and van Meer, G. (2001) MDR1 P-glycoprotein (ABCB1) secretes platelet activating factor (PAF). Biochem. J. 357, 859–865.

Raviv, Y., Pollard, H.B., Bruggemann, E.P., Pastan, I. and Gottesman, M.M. (1990) Photosensitized labeling of a functional multidrug transporter in living drug-resistant tumor cells. J. Biol. Chem. 265, 3975–3980.

Raymond, M., Rose, E., Housman, D.E. and Gros, P. (1990) Physical mapping, amplification, and overexpression of the mouse *mdr* gene family in multidrug-resistant cells. Mol. Cell Biol. 10, 1642–1651.

Romsicki, Y. and Sharom, F.J. (2001) Phospholipid flippase activity of the reconstituted P-glycoprotein multidrug transporter. Biochemistry 40, 6937–6947.

Rosmorduc, O., Hermelin, B. and Poupon, R. (2001) MDR3 gene defect in adults with symptomatic intrahepatic and gallbladder cholesterol cholelithiasis. Gastroenterology 120, 1459–1467.

Ruetz, S. and Gros, P. (1994) Phosphatidylcholine translocase: a physiological role for the *mdr*2 gene. Cell 77, 1071–1083.

Ruetz, S., Brault, M., Dalton, W. and Gros, P. (1997) Functional interactions between synthetic alkyl phospholipids and the ABC transporters P-glycoprotein, Ste-6, MRP, and Pgh-1. Biochemistry 36, 8180–8188.

Rust, S., Rosier, M., Funke, H., Real, J., Amoura, Z., Piette, J.C., *et al.* (1999) Tangier disease is caused by mutations in the gene encoding ATP-binding cassette transporter 1. Nat. Genet. 22, 352–355.

Salen, G., Shefer, S., Nguyen, L., Ness, G.C., Tint, G.S. and Shore, V. (1992) Sitosterolemia. J. Lipid Res. 33, 945–955.

Salen, G., Shefer, S., Batta, A.K., Tint, G.S., Xu, G., Honda, A., Irons, M. and Elias, E.R. (1996) Abnormal cholesterol biosynthesis in the Smith-Lemli-Opitz syndrome. J. Lipid Res. 37, 1169–1180.

Seelig, A., Blatter, X.L. and Wohnsland, F. (2000) Substrate recognition by P-glycoprotein and the multidrug resistance-associated protein MRP1: a comparison. Int. J. Clin. Pharmacol. Therapeut. 38, 111–121.

Seigneuret, M. and Devaux, P.F. (1984) ATP-dependent asymmetric distribution of spin-labeled phospholipids in the erythrocyte membrane: relation to shape changes. Proc. Natl Acad. Sci. USA 81, 3751–3755.

Sillence, D.J., Raggers, R.J. and van Meer, G. (2000) Assays for transmembrane movement of sphingolipids. Methods Enzymol. 312, 562–579.

Sleight, R.G. and Pagano, R.E. (1985) Transbilayer movement of a fluorescent phosphatidylethanolamine analogue across the plasma membranes of cultured mammalian cells. J. Biol. Chem. 260, 1146–1154.

Smit, J.J.M., Schinkel, A.H., Oude Elferink, R.P.J., Groen, A.K., Wagenaar, E., Van Deemter, L., *et al.* (1993) Homozygous disruption of the murine *mdr2* P-glycoprotein gene leads to a complete absence of phospholipid from bile and to liver disease. Cell 75, 451–462.

Smit, J.J.M., Schinkel, A.H., Mol, C.A.A.M., Majoor, D., Mooi, W.J., Jongsma, A.P.M., Lincke, C.R. and Borst, P. (1994) Tissue distribution of the human MDR3 P-glycoprotein. Lab. Invest. 71, 638–649.

Smit, J.J.M., Baas, F., Hoogendijk, J.E., Jansen, G.H., Van der Valk, M.A., Schinkel, A.H., *et al.* (1996) Peripheral neuropathy in mice transgenic for a human MDR3 P-glycoprotein mini-gene. J. Neurosci. 16, 6386–6393.

Smith, A.J., De Vree, J.M.L., Ottenhof, R., Oude Elferink, R.P.J., Schinkel, A.H. and Borst, P. (1998) Hepatocyte-specific expression of the human *MDR3* P-glycoprotein gene restores the biliary phosphatidylcholine excretion absent in *Mdr2*(−/−) mice. Hepatology 28, 530–536.

Smith, A.J., van Helvoort, A., van Meer, G., Borst, P., Szabó, K., Welker, E., Szakács, G., Váradi, A. and Sarkadi, B. (2000) MDR3 P-glycoprotein, a phosphatidylcholine translocase, transports several cytotoxic drugs and directly interacts with drugs as judged by interference with nucleotide trapping. J. Biol. Chem. 275, 23530–23539.

Sun, H., Molday, R.S. and Nathans, J. (1999) Retinal stimulates ATP hydrolysis by purified and reconstituted ABCR, the photoreceptor-specific ATP-binding cassette transporter responsible for Stargardt disease. J. Biol. Chem. 274, 8269–8281.

Vaisman, B.L., Lambert, G., Amar, M., Joyce, C., Ito, T., Shamburek, R.D., Cain, W.J., Fruchart-Najib, J., Neufeld, E.D. and Remaley, A.T. (2001) ABCA1 overexpression leads to hyperalphalipoproteinemia and increased biliary cholesterol excretion in transgenic mice. J. Clin. Invest. 108, 303–309.

van Helvoort, A., Smith, A.J., Sprong, H., Fritzsche, I., Schinkel, A.H., Borst, P. and van Meer, G. (1996) MDR1 P-glycoprotein is a lipid translocase of broad specificity, while MDR3 P-glycoprotein specifically translocates phosphatidylcholine. Cell 87, 507–517.

van Helvoort, A., Giudici, M.L., Thielemans, M. and van Meer, G. (1997) Transport of sphingomyelin to the cell surface is inhibited by brefeldin A and in mitosis, where C_6-NBD-sphingomyelin is translocated across the plasma membrane by a multidrug transporter activity. J. Cell Sci. 110, 75–83.

Van Nieuwkerk, C.M.J., Oude Elferink, R.P.J., Groen, A.K., Ottenhoff, R., Tytgat, G.N.J., Dingemans, K.P., Van den Bergh Weerman, M.A. and Offerhaus, G.J.A.

(1996) Effects of ursodeoxycholate and cholate feeding on liver disease in FVB mice with a disrupted *mdr2* P-glycoprotein gene. Gastroenterology 111, 165–171.

Van Nieuwkerk, C.M.J., Groen, A.K., Van Wijland, M., Van den Bergh Weerman, M.A., Tytgat, G.N.J., Offerhaus, G.J.A. and Oude Elferink, R.P.J. (1997) The role of bile salt composition in liver pathology of *mdr2* (−/−) mice: differences between males and females. J. Hepatol. 26, 138–145.

Vermeulen, W.P. (1996) Effects of ursodeoxycholate and cholate feeding on liver disease in FVB mice with a disrupted *mdr2* P-glycoprotein gene. Academic Thesis, Utrechtl, 1–120.

Vos, T.A., Hooiveld, G.J.E.J., Koning, H.H., Childs, S., Meijer, D.K.F., Moshage, H., Jansen, P.L.M. and Müller, M. (1998) Up-regulation of the multidrug resistance genes, mrp1 and mdr1b, and down-regulation of the organic anion transporter, Mrp2, and the bile salt transporter, spgp, in endotoxemic rat liver. Hepatology 28, 1637–1644.

Vos, T.A., Ros, J.E., Havinga, R., Moshage, H., Kuipers, F., Jansen, P.L.M. and Müller, M. (1999) Regulation of hepatic transport systems involved in bile secretion during liver regeneration in rats. Hepatology 29, 1833–1839.

Vulevic, B.C.Z., Boyd, J.T., Davis, W., Walsh, E.S., Belinsky, M.G. and Tew, K.D. (2001) Cloning and characterization of human adenosine 5′-triphosphate-binding cassette, sub-family A, transporter 2 (*ABCA2*). Cancer Res. 61, 3339–3347.

Wang, E., Casciano, C.N., Clement, R.P. and Johnson, W.W. (2000) Cholesterol interaction with the daunorubicin binding site of P-glycoprotein. Biochem. Biophys. Res. Comm. 276, 909–916.

Wang, N., Silver, D.L., Thiele, C. and Tall, A.R. (2001) Atp-binding cassette transporter a1 (abca1) functions as a cholesterol efflux regulatory protein. J. Biol. Chem. 276, 23742–23747.

Weng, J., Mata, N.L., Azarian, S.M., Tzekow, R.T., Birch, D.G. and Travis, G.H. (1999) Insights into the function of rim protein in photoreceptors and etiology of Stargardt's disease from the phenotype in *abcr* knockout mice. Cell 98, 13–23.

Woodard, D.S., Lee, T.C. and Snyder, F. (1987) The final step in the *de novo* biosynthesis of platelet-activating factor. Properties of a unique CDP-choline: 1-alkyl-2-acetyl-sn-glycerol choline-phosphotransferase in microsomes from the renal inner medulla of rats. J. Biol. Chem. 262, 2520–2527.

Role of ABCA1 in Cell Turnover and Lipid Homeostasis

CHAPTER 23

Giovanna Chimini, Olivier Chambenoit and Christopher Fielding

Structural Traits of ABCA1 and Related Transporters

The gene and its expression

The discovery of ABC1 in 1994 stemmed from an effort to identify novel ATP-binding cassette (ABC) transporters in mouse macrophages, based on the selective amplification of consensus motifs in the nucleotide-binding domain (NBD) (Luciani *et al.*, 1994; Savary *et al.*, 1996, 1997). Among the new genes discovered, it soon became evident that ABC1 bore structural features distinct from those of known transporters and that it was not an isolated example in the mammalian genome (Broccardo *et al.*, 1999; Dean *et al.*, 2001). The group of transporters most closely related to ABC1 has been recently renamed the A subclass (ABC1 becoming ABCA1) and, to date, includes 12 transporters (see Chapter 3). All the ABCA genes encode complete transporters with four domains organized in the following fashion: TMD1/NBD1/TMD2/NBD2. ABCA genes are highly conserved in mammals and are present in *Drosophila melanogaster* and in the nematode *Caenorhabditis elegans*, but are absent from yeast (Dean *et al.*, 2001; Decottignies and Goffeau, 1997). In contrast to mammals, insects and nematodes, most of the ABCA genes expressed in *Arabidopsis thaliana* appear to encode transporters with only two domains: TMD/NBD, so-called hemi- or half transporters (Sanchez-Fernandez *et al.*, 2001).

Of the 12 members of the ABCA subclass, ABCA1, ABCA2, ABCA3, ABCA4 and ABCA7 have been well characterized and identify a closely related group. Two additional genes, ABCA12 and ABCA13, (M. Dean, personal communication) also belong to this cluster but have only been partially characterized so far. The remaining five ABCA genes are clustered on chromosome 17 in humans and, on the basis of sequence alignments, define a subgroup distinct from that defined by ABCA1, 2, 3, 4 and 7. The ABCA1–4 and ABCA7 genes probably originated by duplications before speciation, as suggested by their localization on different chromosomes and their mapping in syntenic regions in the human and mouse genome **(Figure 23.1)**. In spite of this remote evolutionary origin, they retain a very similar exon–intron structure exemplified by that of ABCA1 **(Figure 23.2)** (Allikmets *et al.*, 1998; Azarian *et al.*, 1998; Broccardo *et al.*, 2001; Kaminski *et al.*, 2000, 2001; Remaley *et al.*, 1999; Santamarina-Fojo *et al.*, 2000; Vulevic *et al.*, 2001). In fact, the most divergent characteristic concerns the shrinking or expansion of intervening sequences. This latter feature leads to genomic loci spanning more than 100 kb for ABCA1 and ABCA4 whereas ABCA2, ABCA3

ABC Proteins: From Bacteria to Man
ISBN 0-12-352551-9

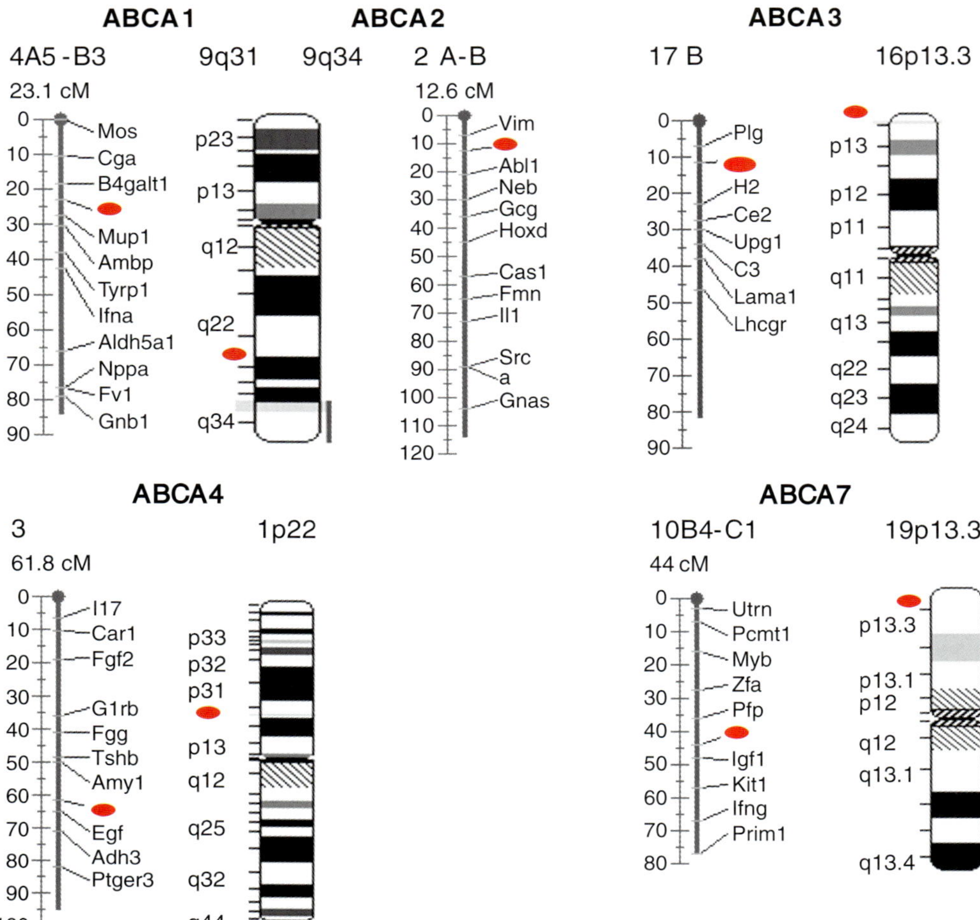

Figure 23.1. Schematic diagram of chromosomal mapping in the mouse or human genome of ABCA1, ABCA2, ABCA3, ABCA4 and ABCA7. Centimorgans (cM) are shown on mouse chromosomes whereas cytogenetic banding is represented on the ideograms of human chromosomes.

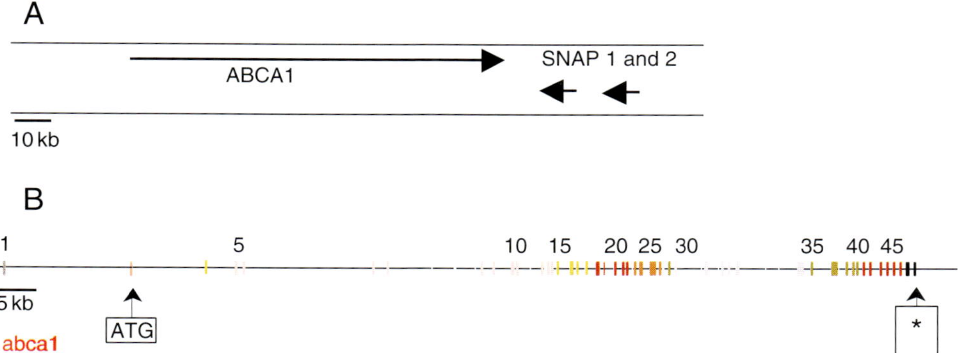

Figure 23.2. A, Schematics of the 200 kb spanning the ABCA1 locus on mouse chromosome 4 as reported by Qiu et al., 2001. *The exon–intron structure of the ABCA1 gene is shown in part B. Color-coded boxes identify individual exons and the encoded protein domains: green, transmembrane segments; rose and light blue, N- and C-terminal extracellular loops; red, nucleotide-binding domains; orange, intervening domain with putative regulatory function. The position of the starting methionine (ATG) and of the stop codon (*) are shown. Figures on top indicate exon numbering. ABCA2, ABCA4 and ABCA7 show a largely superposable gene structure.*

(C. Broccardo, unpublished results) and ABCA7 loci span a genomic region of 20–30 kb.

Most of these five ABCA genes have been cloned from different species (mouse, man, rat and cow) and a close homologue to ABCA1 has been identified in the chicken (Schreyer *et al.*, 1994). In all species, the ABCA1 coding region is spread over 50 exons and the translational start site is in exon 2 (Qiu *et al.*, 2001; Santamarina-Fojo *et al.*, 2000). The relative relationship between exons and protein domains is schematized in **Figure 23.2B** and is closely conserved in ABCA1, ABCA2, ABCA4 and ABCA7. The exon–intron structure of ABCA3 has not been reported so far but is likely to be similar. A recent survey of the genomic region spanning the ABCA1 locus both in mouse and in man failed to identify genes in the regions upstream of ABCA1 (87 kb in the mouse and 34 kb in man) but detected one gene in man and two genes in the mouse, in close proximity to the ABCA1 polyadenylation site, that are encoded in the opposite orientation to ABCA1 itself (Qiu *et al.*, 2001). These were named hSNAP, mSNAP1 and mSNAP2. hSNAP and mSNAP1 are located 9 kb and 8 kb downstream of ABCA1, respectively. In mouse, SNAP2 is located 12 kb downstream of SNAP1.

The transcriptional regulation of ABCA1 appears to be exceptionally complex, and at present, poorly understood. Three clusters of transcriptional start sites for ABCA1 have been identified. The first type (class 1 in **Figure 23.3**), identified in placenta, is 40 bp downstream from a modified TATA box (Pullinger *et al.*, 2000; Schwartz *et al.*, 2000). Six G/C-rich sequences, potential binding sites for Sp1 and/or SREBP, as well as AP1 and NFkB sites were identified in the same region. The second start site (class 2, **Figure 23.3**) is approximately 90 bp downstream of the start sites for class 1 transcripts (Santamarina-Fojo *et al.*, 2000). A weak TATA box is present at 32 bp 5′ of this start site. A LXR/RXR site is present between −70 and −55 bp of this start site (that is, at +19 to +44 relative to start site 1) (Costet *et al.*, 2000). Transcripts with the second start site were reported to predominate in HepG2 and THP-1 cells, and transformed human lines originating from liver and monocytic cells respectively. The third group of transcripts (class 3, **Figure 23.3**) are initiated within intron 1 of the full-length gene. This leads to formation of a novel first exon (exon 1a) with the loss of 28 amino acids from ABCA1 (Cavelier *et al.*, 2001; Singaraja *et al.*, 2001). This group of transcripts is initiated

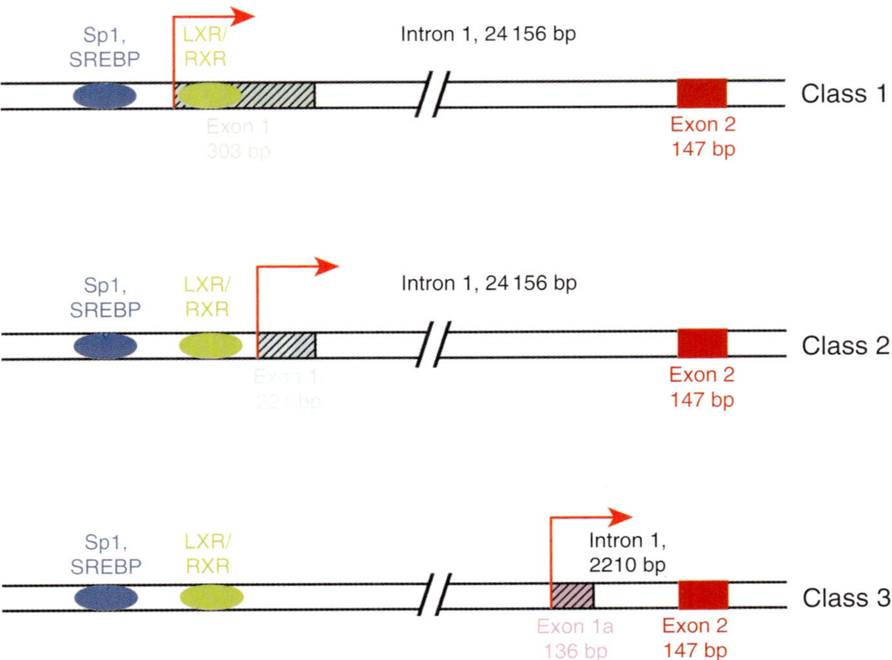

Figure 23.3. Alternative start sites for ABCA1 gene transcription. Transcript classes 1–3 are defined in the text. In different analyses of the same transcript class, slight differences in length have been reported. The data shown correspond to those in the first report of each class. Sp1, SREBP, LXR/RXR, consensus binding sites for these transcription factors.

downstream of classical TATA and CAAT sequences and a variety of potential lipid-dependent binding sites for transcription factors. Class 3 transcripts predominate in liver tissue in mice expressing a human ABCA1 gene construct lacking wild-type exon 1. According to classical concepts, the basic transcription machinery, assembled at the start site, forms an activated complex with DNA-binding proteins generally within 300–400 bp upstream. On this basis, transcript classes 1–3 could each respond to a different set of regulatory proteins. The functional effect of a given inducer (e.g. oxysterol, free cholesterol – FC) on the overall ABCA1 mRNA levels would thus also depend on the tissue-specific proportions of each transcript present under baseline conditions.

Finally, alternative ABCA1 transcripts lacking part of exon 3 and all of exon 4 were recently detected in human fibroblasts, endothelial and smooth muscle cells, and HepG2 cells (Bellincampi *et al.*, 2001). This variant does not affect the promoter sequence. As a result, and predictably, induction of ABCA1 mRNA with FC did not change the proportion of full-length and shorter transcripts.

The best-studied regulatory element controlling ABCA1 expression is the class 2 start site LRR/RXR (Costet *et al.*, 2000). This is controlled by the transcription factor PPAR gamma/delta (Chawla *et al.*, 2001; Oliver *et al.*, 2001; Venkateswaran *et al.*, 2000). Oxysterols and retinoic acid strongly upregulated the expression of luciferase constructs linked to such ABCA1 type 2 promoter constructs. The *in vivo* relevance of the site is indicated by upregulation of ABCA1-mediated lipid efflux by a PPAR-delta agonist in monkeys (Oliver *et al.*, 2001). An FC-sensitive promoter region was also identified 100–200 bp upstream of the class 1 start site (Santamarina-Fojo *et al.*, 2000). This may be functional for the production of class 2 transcripts but probably not for those in class 3. Finally, cAMP-dependent expression has been described in transformed rodent monocyte-derived cell lines (RAW264 and J774 cells) but not in human-derived THP-1, CaCo-2 or HepG2 cells, or normal skin fibroblasts (Bortnick *et al.*, 2000). The target for cAMP-mediated upregulation in responsive cells has not been identified, nor the significance of this activity established.

The expression pattern of ABCA1 has been extensively studied. Early studies by *in situ* RNA hybridization revealed a tight spatio-temporal correlation between the expression of the ABCA1 transcript and the occurrence of cell death during embryo development (Luciani and Chimini, 1996). This was further interpreted as due to the local recruitment of macrophages responsible for clearing the corpses of the cells committed to die. The exclusive expression of ABCA1 by phagocytes in the areas of developmental cell death has recently been formally proven by the undetectability of ABCA1 transcript in these areas in PU1 null embryos (Wood *et al.*, 2000). These embryos, owing to the lack of this transcription factor crucial for the differentiation of hematopoietic cell lineages, are in fact virtually devoid of macrophages (Wood *et al.*, 2000). The expression of ABCA1 in cells of myeloid lineage is unequivocal. It has indeed been assessed in several cell lines and in primary cellular systems in mouse, such as those of resident or elicited peritoneal and bone marrow derived macrophages and in humans in activated monocytes, macrophages and foam cells (Christiansen-Weber *et al.*, 2000; Langmann *et al.*, 1999; Lawn *et al.*, 2001; Luciani *et al.*, 1994). Dendritic cells, whilst sharing a common precursor with monocytes in myeloid lineages, lack ABCA1 transcripts and instead express ABCA7 (C. Broccardo, unpublished). ABCA1 expression by tissue macrophages can account for the detection of low/medium levels of ABCA1 transcripts in many adult tissues. In addition, however, some parenchymal cells, such as liver and adrenal cells, do also express significant levels of ABCA1 in the mouse (Luciani *et al.*, 1994, and unpublished observations). Northern blot analysis of human tissues indicated that kidney, lung and spleen were among the major sites of ABCA1 expression (Langmann *et al.*, 1999). A similar tissue distribution in baboon was observed by *in situ* hybridization studies (Lawn *et al.*, 2001). This study also reported that although normal veins and arteries did not express ABCA1 mRNA, this was upregulated in the setting of atherosclerosis, where widespread expression was found in macrophages within atherosclerotic lesions. In contrast, the expression of ABCA1 by intestinal epithelium cells has been repeatedly suggested but not yet formally demonstrated (Lawn *et al.*, 2001). A massive upregulation of ABCA1 transcription has been demonstrated in the uterus and the developing placenta, where the transcript has been detected in both the labyrinthine and decidual layers (Christiansen-Weber *et al.*, 2000; Hamon *et al.*, 2000; Luciani *et al.*, 1994).

It is important to note that the five ABCA1-like genes show non-overlapping territories of preferred expression. This may indicate that they exert similar functions in diverse cell specific contexts (Broccardo *et al.*, 1999). ABCA1, ABCA2, ABCA3 and ABCA7 are expressed during embryonic development as witnessed by the detection of specific transcripts in whole embryo RNA. However, no detailed morphological assessment of the developmental expression pattern of ABCA 2, 3 or 7 has been reported as yet.

The protein: topology, ATPase activity and subcellular localization

The protein encoded by the ABCA1 gene is 2261 amino acids long in both mouse and man. This corresponds to a product 60 amino acids longer than that of the one originally described (Costet *et al.*, 2000; Luciani *et al.*, 1994; Pullinger *et al.*, 2000; Tanaka *et al.*, 2001). Both the methionines +1 and +61 are able to support the production of a protein but only translation from the first methionine produces an active protein (Fitzgerald *et al.*, 2001; Wang *et al.*, 2000). The shorter product is retained in the endoplasmic reticulum (Y. Hamon, unpublished result), possibly as a result of improper folding.

As is frequently the case for ABC transporters and more generally for other polytopic membrane proteins, determination of a precise topological organization is extremely difficult. In spite of the unambiguous identification of large blocks of hydrophobic residues, their precise distribution into the succession of individual transmembrane segments is rather uncertain on the basis of computer predictions alone.

Experimental topological analyses are underway in many laboratories. Based on both direct evidence and on the analogy with ABCA4 (Bungert *et al.*, 2001; Fitzgerald *et al.*, 2002), it seems reasonable, at present, to favor the topological model shown in **Figure 23.4** (see also Chapter 2). This predicts two large extracellular

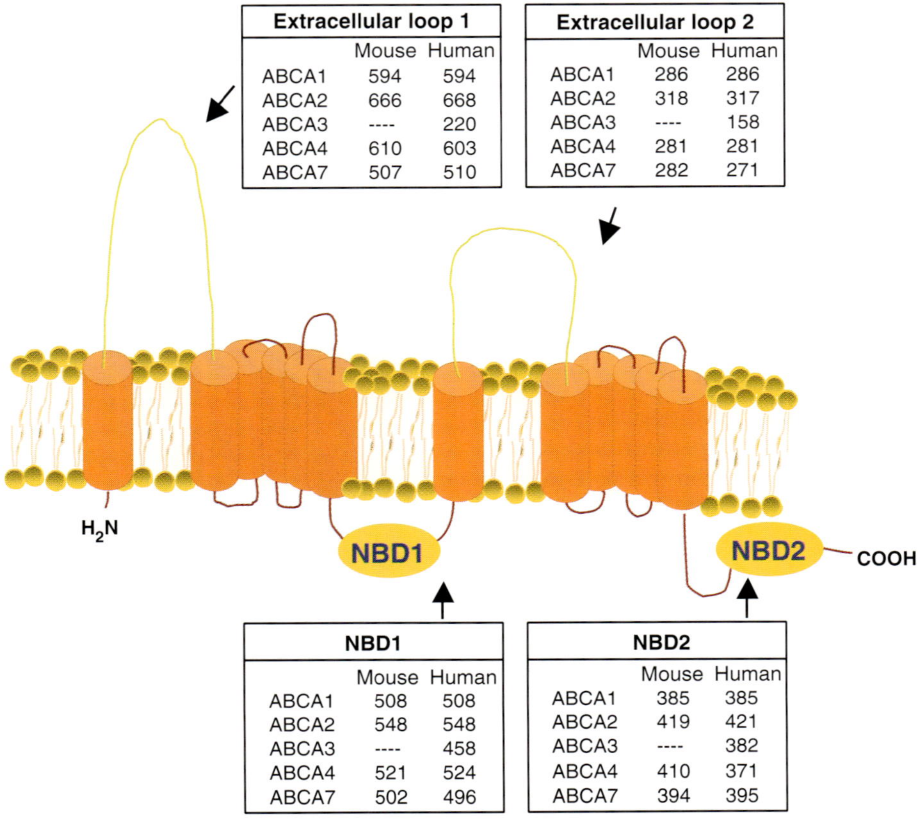

Extracellular loop 1

	Mouse	Human
ABCA1	594	594
ABCA2	666	668
ABCA3	----	220
ABCA4	610	603
ABCA7	507	510

Extracellular loop 2

	Mouse	Human
ABCA1	286	286
ABCA2	318	317
ABCA3	----	158
ABCA4	281	281
ABCA7	282	271

NBD1

	Mouse	Human
ABCA1	508	508
ABCA2	548	548
ABCA3	----	458
ABCA4	521	524
ABCA7	502	496

NBD2

	Mouse	Human
ABCA1	385	385
ABCA2	419	421
ABCA3	----	382
ABCA4	410	371
ABCA7	394	395

Figure 23.4. Model of ABCA1 membrane topology. The type I or type II orientation of the N-terminus is still controversial (see text). The length of the major extracellular loops (green) and of NBDs (yellow) in five distinct ABCA transporters is shown for comparison.

loops between TM1 (amino acids 25–45) and 2 (starting at amino acid 640 in ABCA1) and TMS7 (amino acids 1350–1370) and 8 (starting at amino acid 1668). A similar topology is probably shared by the other ABCA members, whose hydrophobicity plots are largely superimposable on that of ABCA1. The length of the predicted extracellular loops, however, varies greatly among the individual transporters. ABCA2 and ABCA3 are the most divergent and show respectively the longest and shortest extracellular loops **(Figure 23.4)**. Other aspects of the topological model remain ambiguous. Thus, contradictory results have been reported as to whether the first hydrophobic segment (amino acids 24–48) serves as a signal peptide, or as a signal anchor sequence. In the former situation, processing of the peptide would lead to an externally exposed N-terminus, commencing at position 49 (potential cleavage site by the algorithm SignalP V1.1 World Wide Web Prediction Server (Nielsen *et al.*, 1997)) and to an asymmetrical number of transmembrane segments in the two halves of the transporter. Alternatively, if this segment acts as a signal anchor, a type II orientation of the short free N-terminus will result, and the two halves of the transporter will show a symmetrical architecture, each with six predicted transmembrane segments. The first option is favored by Ueda and co-workers and is based on the inability to detect an HA epitope fused to the N-terminus of ABCA1 by Western blotting (Tanaka *et al.*, 2001). In the hands of other investigators (Fitzgerald *et al.*, 2001), the analysis of a similar EGFP/ABCA1 chimera suggested, in contrast, its function as signal anchor. Moreover, Fitzgerald *et al.* not only detected EGFP in the final product, but also reported the inability of the ABCA1 'signal peptide' to support the secretion of rhodopsine fused to its C-terminus. Taking into account also the formal biochemical evidence that in ABCA4 the putative signal peptide is not processed (Illing *et al.*, 1997), we favor the hypothesis of a type II orientation. Constrained folding within the hydrophobic membrane environment of the very short N-terminus and/or the technical inability to completely denature the transporter may account for the lack of detection of the HA epitope reported by Ueda and co-workers (Tanaka *et al.*, 2001). In line with that, we have observed an inability to detect an HA epitope inserted into the short loops separating TMSs, that is in positions where a tight interaction with the membrane bilayer is likely (Rigot *et al.*, in preparation). In the case of ABCA4, disulfide bridging between the large extracellular loops has been reported (Bungert *et al.*, 2001). Similar molecular interactions may exist in the case of ABCA1, on the basis of the observed conservation of cysteine residues in the extracellular loops of both proteins.

Ueda and co-workers (Tanaka *et al.*, 2001) showed that ABCA1 is able to bind and hydrolyze ATP, although with low efficiency. This is in line with the presence of the two conserved Walker motifs in the NBDs and is also consistent with the known ATPase activity of ABCA4 (Ahn and Molday, 2000; Biswas and Biswas, 2000; Sun *et al.*, 1999).

ABCA1 has been shown to reach the plasma membrane in a variety of transfected cell lines (Hamon *et al.*, 2000; Neufeld *et al.*, 2001; Wang *et al.*, 2000) **(Figure 23.5)**. The staining at the membrane is not homogeneous but rather punctate. The localization of ABCA1 in plasma membrane domains enriched in cholesterol and sphingolipids (lipid rafts or caveolae) is suggested by the partial coalescence with the membrane staining of CD14, the GPI-linked lipopolysaccharide (LPS) receptor and with GM1 distribution (O. Chambenoit, unpublished). It has been reported (Drobnik *et al.*, 2002; Mendez *et al.*, 2000), however, that in

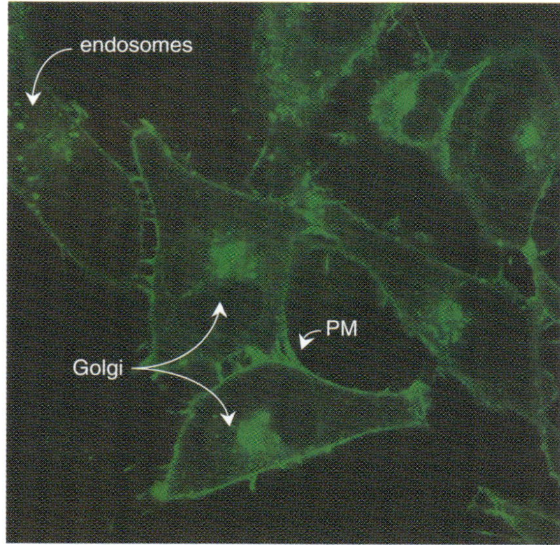

Figure 23.5. Subcellular distribution of a transfected ABCA1/EGFP chimera as determined by confocal microscopy. The Golgi and endo-lysosomal compartments were identified by costaining with known markers (Hamon et al.*, 2000). The discrete staining at the plasma membrane is clearly visible.*

standard purification procedures ABCA1 does not partition with Triton X-100 insoluble membrane domains. It has to be noted, however, that the cell model chosen for this study (immortalized skin fibroblasts) expresses few rafts or caveolae and that also the isolation procedure used was relatively nonspecific. Thus, it seems possible that ABCA1 is preferentially localized at the periphery of these domains, from which it dissociates in the presence of detergents, under standard purification procedures. The transfected ABCA1 protein has also been localized to intracellular vesicles belonging to the endo-lysosomal compartment and in the Golgi stack (Hamon *et al.*, 2000; Neufeld *et al.*, 2001; Wang *et al.*, 2000). An ABCA1-specific staining in the latter compartment is detected also in untransfected macrophages. At present it is not known whether ABCA1 is functional as a transporter in these intracellular compartments. An interesting feature associated with the stable expression of ABCA1 in transfected cells is a significant delay in cell growth (Y. Hamon, unpublished). This may be related to an ability of the transporter to modulate the composition and dynamics of the membrane, which in turn may alter growth parameters. It is interesting to note that the forced and stable expression of CED-7, the *C. elegans* homologue of ABCA1, has a similar impact on cell growth (Y. Hamon unpublished).

THE IMPACT OF ABCA1 ON PHYSIOLOGICAL FUNCTIONS

ABCA1 AND CELL TURNOVER: THE CLEARANCE OF CELLS DYING BY APOPTOSIS

Apoptosis or programmed cell death is a genetically controlled and highly regulated event responsible for cell turnover in healthy adult tissues and of focal elimination of cells during embryonic development (Kerr *et al.*, 1972). The apoptotic process itself consists of the systematic dismantling of the cell factory orchestrated by the activation of caspases. From a morphological standpoint, apoptosis can be easily distinguished from other forms of cell elimination. The structural changes during apoptosis take place in two distinct steps. The first involves the generation of cell fragments still with an intact cell membrane (apoptotic bodies). This is then swiftly followed by their uptake and degradation by phagocytes, most frequently macrophages, recruited locally in large numbers by mechanisms yet unknown. Typically, a cell committed to die, rounds up and detaches from its neighbors, then it undergoes nuclear condensation and shrinkage of the cytoplasm without major morphological alteration of intracellular organelles. Membrane blebs are now formed, which progressively lead to the generation of membrane-bound, compact but otherwise well-preserved cell remnants, the apoptotic bodies.

Many of the morphological aspects of the apoptotic process have now been linked to precise biochemical events and depend on the proteolytic cleavage of one or more of the molecules targeted by the effector caspases (Hengartner, 2000; Leverrier and Ridley, 2001a). A similar orchestration regulates the clearance of corpses by phagocytes (Savill and Fadok, 2000). In physiological situations virtually no free dying cells are detectable in the body. Indeed the persistence of self cells undergoing progressive disintegration is to be avoided at any cost for two main reasons. Their slow removal would be immediately harmful as a consequence of the leakage of noxious intracellular contents. It would also be expected to be dangerous in the longer term in view of the ability of cell fragments or their contents to trigger immune responses against self antigens persistently exposed to antigen-presenting cells. The molecular circuits controlling recognition and ingestion of corpses by phagocytes are far from being elucidated. Most of the available clues come from the model system *C. elegans*. There, genetic dissection has highlighted the 14 genes controlling the process of programmed cell death from the initial commitment to the final degradation inside the phagocyte. Those are designated as *ced* for the *c*ell *d*eath abnormal phenotypes deriving from their mutation (Ellis *et al.*, 1991a, 1991b). As far as the engulfment phase is concerned at least two genetic pathways exist in the worm that are conserved in mammals. The ced-2, ced-5, ced-10 and ced-12 group of genes controls the first clearance pathway, which corresponds, as shown in **Figure 23.6**, to an integrin-triggered signaling cascade in mammalian phagocytes (Albert *et al.*, 2000; Reddien and Horvitz, 2000; Tosello-Trampont *et al.*, 2001; Wu and Horvitz, 1998a; Wu *et al.*, 2001; Zhou *et al.*, 2001b). The second pathway is less defined but

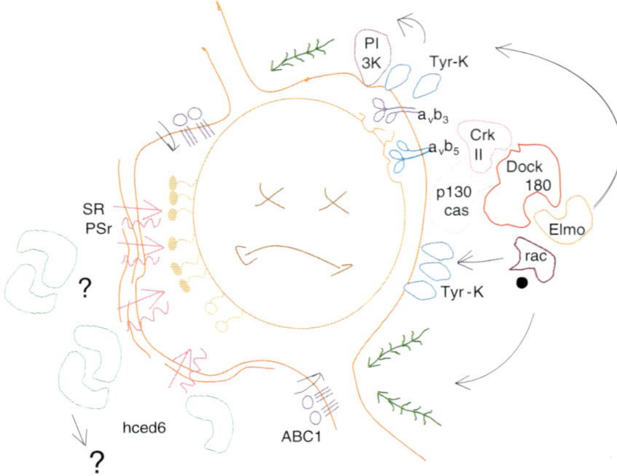

Figure 23.6. ***Two parallel pathways are responsible for the recognition and uptake of corpses by phagocytes. This scheme combines data from both the nematode and mammalian systems. In the left part the pathway involving the ABC transporter (ABC1 or CED-7), a scavenger receptor (SR = CD36 or CED-1) and the adaptor ced-6 is represented. Note that so far a direct molecular interaction between the partners has not been demonstrated. In the right section, the integrin-mediated triggering of the signaling cascade involving p130cas, CrkII, Dock 180, Elmo and rac1 is represented. In*** **C. elegans** ***the triggering receptor has not been identified but the membrane recruitment and molecular interaction of CED-2, CED-5, CED10 and CED-12 was demonstrated. Membrane recruitment of tyrosine kinases of unknown identity has also been reported (Albert*** **et al.,** ***2000; Leverrier and Ridley, 2001b; Reddien and Horvitz, 2000; Zhou*** **et al.,** ***2001a).***

includes the ABC transporter CED-7 (Wu and Horvitz, 1998b), which bears a high sequence similarity to ABCA1 and belongs to the nematode ABCA class of transporters. CED-7 works in concert with a membrane scavenger receptor, CED-1 (Zhou *et al.*, 2001a), and a downstream signaling protein, ced-6, which is also conserved in mouse and man (Liu and Hengartner, 1998, 1999; Su *et al.*, 2000). Apart from their clustering in the same epistatic group, a cascade of molecular interactions has not been determined so far.

In the mammalian system, where investigations have mainly relied on *in vitro* assays, a number of well-known phagocytic receptors are involved in the uptake of apoptotic bodies (Gregory, 2000; Platt *et al.*, 1998; Ren and Savill, 1998; Savill and Fadok, 2000). Among these are the LPS receptor, CD14, members of the family of scavenger receptors (CD36, SRA) and of integrins (αvβ3 and αvβ5). Unfortunately, neither the molecular entities any of these receptors recognize on the surface of the apoptotic prey nor what molecular modifications occur on the surface of the prey to be during apoptosis are known. These are globally indicated as ACAMP (apoptotic cell-associated molecular patterns) to underline the lack of molecular data. Recently two new molecules on the phagocyte surface able to engage prey ingestion have been identified: a specific receptor for phosphatidylserine (PSR) and the tyrosine kinase receptor MER. Both are expected to participate in the recognition of the unusual amounts of phosphatidylserine exposed on the dying cells, at present the only available hallmark of membrane modifications during apoptosis. The PSR acting alone or in concert with CD36 provides the required stereospecificity for phosphatidylserine (PS) recognition, whereas the second could act as a receptor for gas-6 (the product of growth arrested specific gene 6), a soluble protein previously implicated as a mediator of macrophage binding to PS.

A thorough overview of these receptors is beyond the scope of this chapter and is provided elsewhere (Ren and Savill, 1998; Savill, 1998). It is, however, worth underlining the redundancy of surface molecules implicated in prey recognition by the macrophages; however, none of them seem to be used exclusively for the engulfment. This underscores the high physiological impact of the phenomenon, whose major goal is to avoid any escape of apoptotic prey from their fate.

THE ROLE OF ABCA1 DURING THE ENGULFMENT OF APOPTOTIC CORPSES

In mammals, the participation of an ABC transporter in engulfment was established in the mid-1990s by the description of an upregulation of ABCA1 transcripts in the macrophages recruited to areas of developmental cell death (Luciani and Chimini, 1996). The functional meaning of this upregulation was suggested by *in vitro* results where an antibody-mediated block of ABCA1 function led to a reduced phagocytic performance of peritoneal macrophages exclusively when the prey consisted of an apoptotic cell. The subsequent molecular identification of CED-7 in the worm as an ABC transporter (Wu and Horvitz, 1998b) reinforced, by analogy, the hypothesis of an active role for ABCA1 during clearance of apoptotic cells by professional macrophages.

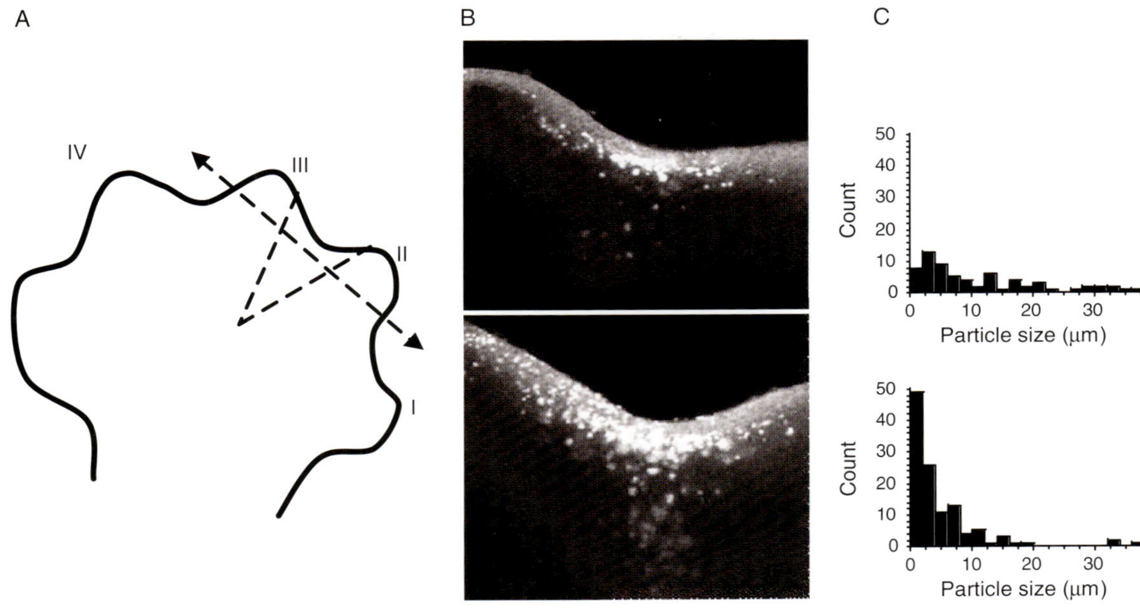

Figure 23.7. Delayed engulfment of apoptotic cells is detected during embryonic development in ABCA1 null animals. A microscopic analysis of limb buds at embryonic day 13.5 is shown. A schematic representation of the virtual section in the limb bud is shown in panel A. Apoptotic cells are detected by neutral red staining (panel B) in wild-type (upper) or ABCA1 null (lower) animals. An increased number of particles of larger size is clearly detected in the ABCA1 −/− mice. In panel C, the distribution and size of particles corresponding to apoptotic corpses in wild-type and null animals is shown. The graph was derived from the microscopic analysis of buds stained for apoptotic corpses by the TUNEL technique (Hamon et al., *2000).*

The development and combined analysis of an *in vivo* loss of function and an *in vitro* gain of function model (Hamon *et al.*, 2000) allowed us to establish unambiguously that ABCA1 is able to promote the engulfment function of macrophages both during embryonic development and in adult life **(Figure 23.7)**. However, this is likely to be a consequence of the ability of ABCA1 to influence the distribution of lipids on both the transversal and lateral dimension of the membrane **(Figure 23.8**; see also **Figure 23.9)**.

Indeed the loss or gain of ABCA1 function has been directly correlated with a reduction or an increase, respectively, in the outward flip of PS from the inner leaflet of the plasma membrane. This is not an unusual activity among ABC transporters (Higgins, 1994; Higgins and Gottesman, 1992). It is, however, important to stress that, in spite of clear evidence for the participation of ABCA1 in the distribution of lipid species across the bilayer, we cannot formally consider PS as the sole or direct substrate of ABCA1. Indeed the interrelationship between the different lipid species in the environment of the membrane is complex and highly dynamic and we cannot 'a priori' exclude the possibility that the observed movement of PS is balanced

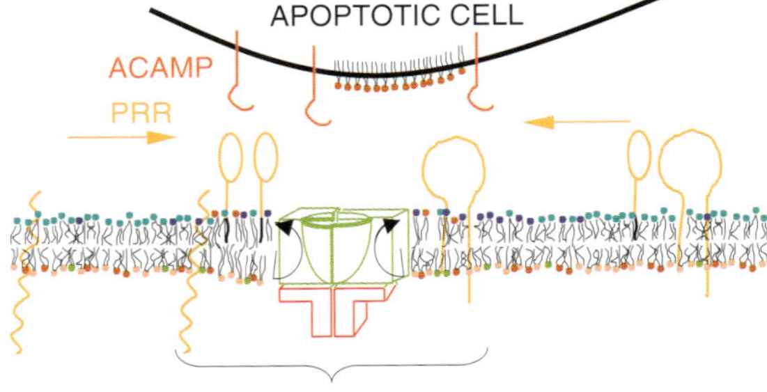

Figure 23.8. Proposed model of ABCA1 function during engulfment. The ABCA1-generated domains increase the efficiency of engulfment by promoting clustering, by lateral diffusion, of receptors engaged on the phagocyte membrane in the recognition of the apoptotic prey, through modulation of the properties of the bilayer. CD36 (scavenger receptor of B class) and the PSR (PS receptor) mobilization along the lateral axis of the membrane are indicated by arrows. This may allow the generation on the phagocyte surface of specific molecular arrays (PRR: pattern recognition receptors), which then efficiently detect patterns on the surface of the apoptotic prey (ACAMP: apoptotic cell-associated molecular patterns) (Franc et al., *1999).*

by a primary flip of an as yet unknown substrate. We can nonetheless conclude that ABCA1 exerts an indirect control on the biophysical properties of the membrane, which in a cascade facilitates the engulfment. Indeed in *C. elegans* it has been reported that the absence of CED-7 (the ABCA1 orthologue) hampers the lateral mobility of protein involved in the recognition of the prey. It has been shown that the redistribution and clustering on the phagocyte membrane of the CED-1 scavenger receptor, an event normally triggered by the contact with a dying cell, is absent in ced-7 mutants (Zhou *et al.*, 2001a).

ABCA1 AND LIPID HOMEOSTASIS

Normal cellular lipid homeostasis

In normal metabolism, there is continuous traffic of cell phospholipid (PL) and free cholesterol (FC) to and from their external milieu – interstitial fluids, large vessel lymph, and plasma (Fielding *et al.*, 1998). The extracellular acceptors of cell-derived lipids are mainly high density lipoproteins (HDL), whose major protein component is apolipoprotein A-I (apo A-I) (Frank and Marcel, 2000). Of these acceptors, a lipid-poor fraction (normally representing about 5% of total HDL), appears to play the major role (Castro and Fielding, 1988). These particles are distinguished by their prebeta-electrophoretic mobility, which contrasts with the alpha-mobility of the major, lipid-rich HDL fraction. *In vivo*, it is not clear if prebeta-HDL originates from lipid-free apo A-I, or from a lipoprotein precursor. However, the apoprotein is widely available commercially and has been used as a convenient surrogate for the, as yet unidentified, precursor of physiological *de novo* HDL formation (Hara and Yokoyama, 1992; Oram and Yokoyama, 1996).

Peripheral cells, even when quiescent, synthesize PL at significant rates. Part of this PL is transferred out of the cell onto lipoprotein acceptors. PL is also internalized from extracellular lipoproteins and degraded by lysosomal phospholipases (Waite, 1996). In contrast, FC is neither synthesized nor catabolized in most peripheral cells at rates that are significant in comparison to those of either FC efflux or the uptake of preformed FC from lipoproteins (Fielding *et al.*, 1998). As a result, most FC leaving the cell has been recycled from lipoproteins, while most PL is newly synthesized.

It was recently shown that FC, internalized from extracellular lipoproteins, recycles within the cell in extra-lysosomal, weakly acidified recycling endosomes which are rich in FC, sphingolipids and caveolin, the major structural protein of cell surface caveolae (Pol *et al.*, 2001). Caveolin also plays a key role in FC efflux and in returning both recycling and newly synthesized FC to the cell surface (Fielding and Fielding, 1995; Smart *et al.*, 1996). In contrast, phosphatidylcholine (PC), the major PL of the plasma membrane and probably other glycerophospholipids, newly synthesized in the endoplasmic reticulum, are transported to the cell surface by PL transfer proteins (Voelker, 1996).

In summary, while FC and PC both transfer from the cell surface to the same lipoprotein acceptor (lipid-poor HDL), their respective precursor pathways in the cell appear to differ. The efflux of cellular FC, and its subsequent metabolism within the plasma compartments and catabolism in the liver, have been termed the *reverse cholesterol transport* pathway (Castro and Fielding, 1988). In this way it is possible to distinguish this flux from the equivalent but opposite 'forward' transport of FC, synthesized in the liver, to peripheral cells (Castro and Fielding, 1988).

Cellular lipid homeostasis in the context of ABCA1 deficiency

Spontaneous ABCA1 deficiency (Tangier disease) is characterized by the complete absence of alpha-migrating HDL from the plasma of affected human subjects and storage of cholesteryl esters (CE) within focal accumulations of macrophages, notably the tonsils (Assmann *et al.*, 2001). The low levels of HDL molecules present in the plasma of Tangier disease patients are almost all lipid-poor particles (Asztalos *et al.*, 2001). These must differ structurally from the prebeta-HDL of normal plasma, which are effectively converted into alpha-HDL in the presence of lecithin:cholesterol acyltransferase (LCAT), which is decreased but not absent in Tangier disease (Assmann *et al.*, 2001). However, the composition and properties of Tangier lipid-poor particles have been little investigated. One intriguing recent study reported that they were ineffective as a substrate for plasma phospholipid transfer protein activity (von Eckardstein *et al.*, 1998).

Using 'knockout' technology, ABCA1-deficient mice have been generated and characterized

(McNeish *et al.*, 2000; Orso *et al.*, 2000). These share the human Tangier phenotype – absence of normal HDL, and presence of macro-phage CE deposits. Additionally, as mentioned previously, ABCA1-deficient mice also fail to clear cell corpses normally during embryonic development (Hamon *et al.*, 2000). The defect, still detectable in isolated adult macrophages, does not hamper the normal development of the animal. It is likely, however, to induce, in adult life, perturbations of the immune responses and propensity to develop autoimmune diseases. It cannot be determined at present if similar developmental abnormalities occur in Tangier fetuses, although this seems likely, given the conserved structure of the ABCA1 gene, and the similar effect of the mutations on HDL metabolism in humans and mice.

Lipid efflux in Tangier disease

In 1994, Walter and colleagues first reported defective efflux of FC from Tangier skin fibroblasts in culture (Walter *et al.*, 1994). The observation has been confirmed by many different laboratories (Francis *et al.*, 1995; Remaley *et al.*, 1997; Rogler *et al.*, 1995). Further analysis showed that the efflux of newly synthesized FC onto HDL was also affected. Thus, efflux of FC from uniformly labeled cells onto lipid-free apoA-I was found to be defective, whilst the efflux of FC from uniformly labeled cells to HDL was normal. These findings were consistent with other data suggesting that FC efflux was heterogeneous (Castro and Fielding, 1988; Hara and Yokoyama, 1992; Oram and Yokoyama, 1996), and also with the concept that the role of ABCA1 in FC efflux was limited to the early steps of HDL formation (Fielding *et al.*, 2000).

PL efflux from ^{3}H-choline labeled cells to lipid-free apo A-I is also reduced in cells lacking ABCA1 (Francis *et al.*, 1995). More limited data along the same lines has been reported using Tangier disease monocyte/macrophages (Hirano *et al.*, 2000). The results of these findings led to the conclusion that a single pathway, defective in Tangier disease, promoted the efflux of both FC and PL to lipid-poor lipoproteins.

Following localization of the Tangier disease defect to within the ABCA1 gene in 1999 (Bodzioch *et al.*, 1999; Brooks-Wilson *et al.*, 1999; Lawn *et al.*, 1999; Rust *et al.*, 1998), it was generally concluded, based on all the findings above, that ABCA1 was indeed a molecular pump required to transport both FC and PL to the cell surface for efflux. By analogy with the structure and function of the better-characterized multidrug resistance 1 (MDR1) transporter, also a member of the ABC family, ABCA1 was hypothesized to contain a central cavity accommodating both PL and FC, and to transport these lipids across the plasma membrane to the cell surface in a reaction driven by ATP (Rosenberg *et al.*, 2001). The proportions of different lipids transported would then depend on their local concentrations in the membrane bilayer, and probably also on a selectivity expressed at the level of the transporter ligand-binding site. Consistent with this notion, direct binding of apo A-I to the ABCA1 protein was reported (Rust *et al.*, 1999), suggesting that the initial formation of HDL lipid was directly linked to ABCA1 transporter activity. However, in the succeeding two years, data has been obtained by a number of different laboratories that points to different conclusions, as discussed in the following section.

Role of ABCA1 in PL efflux: direct transport of PS?

The major PL of the mammalian plasma membrane is PC (Voelker 1996). Smaller amounts of phosphatidylethanolamine (PE) and sphingomyelin (SPH) are present, but only very low levels of other classes, such as PS. Most PC and SPH is in the external leaflet of the bilayer; PE and PS are mainly restricted to the internal leaflet. The distribution of PS in the membrane bilayer is regulated by the balance between an endogenous PS transfer activity, promoting transfer from the inner to the outer leaflet; and aminophospholipid translocase activity, which catalyzes the reverse reaction (Daleke and Lyles, 2000). The factors involved in maintaining the asymmetric distribution of other PL classes are less well understood, but may involve several different PL-specific transfer activities.

A role for ABCA1 in PS movement was shown experimentally, by analysis of the distribution of PS in the bilayer as expression of this transporter was decreased or increased (Marguet *et al.*, 1999). Nevertheless, ABCA1 activity was not linked to the efflux of any PS in the initial formation of HDL from lipid-free apo A-I; the major PL class transferred out of the cell was PC (Fielding *et al.*, 2000; Wang *et al.*, 2000). These data indicate that PL efflux from the cell is unlikely to be the direct consequence

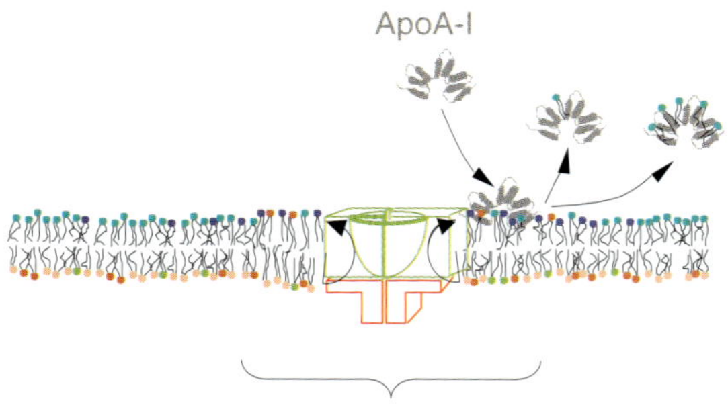

Figure 23.9. ***The ABCA1-generated domains in the bilayer promote ApoA-I docking at the cell membrane. This, whilst certainly occurring in close proximity to ABCA1 molecules, may actually involve PC-rich areas.***

of ABCA1 pumping activity. Indeed, it is possible that the transfer of PS between the leaflets of the bilayer is itself an indirect result of ABCA1 activity, and that the primary transport substrate of ABCA1 is presently unrecognized. Other recent experiments suggest that the cross-linking of apo A-I to ABCA1 reported earlier is also an indirect consequence of such secondary changes induced in membrane lipid domains (Chambenoit *et al.*, 2001). While details at the molecular level remain to be worked out it seems likely, based on the most recent evidence, that ABCA1 induces local rearrangement of PL within the bilayer that favors apo A-I binding to an adjacent but non-identical PC-rich microdomain; and that PC–apo A-I interaction is followed by dissociation of the activated apo A-I/PC complex **(Figure 23.9)**. At this point there is no compelling evidence for the simultaneous transfer, under physiological conditions, of more than a single PC molecule per binding event, but the possibility that more than one molecule of PC associates to bound ApoA-I has not been excluded.

Role of ABCA1 in the efflux of FC

Three recent studies by different laboratories challenge the earlier conclusion that ABCA1 is directly involved in FC efflux (Arakawa *et al.*, 2000; Fielding *et al.*, 2000; Wang *et al.*, 2001). These earlier studies showed that glyburide, an inhibitor of ABCA1 activity (Becq *et al.*, 1997), inhibited the efflux of both PL and FC from normal human smooth muscle cells and fibroblasts. Polyorthovanadate, an inhibitor of caveolar function (Aoki *et al.*, 1999), blocked FC efflux to apo A-I but had little effect on PL efflux. When smooth muscle cells were pretreated with vanadate, apo A-I/PL complexes were formed as a result of ABCA1 activity. These activated complexes bound caveolar FC from endothelial cells which lacked significant ABCA1 protein or activity. This FC efflux was blocked by vanadate, but resistant to glyburide. Lipid-free apo A-I added directly to endothelial cells was unable to form any HDL. These data suggested that HDL formation from apo A-I was a two-step process, in which ABCA1-dependent PL efflux was followed by an ABCA1-independent efflux of FC (Fielding *et al.*, 2000) **(Figure 23.10)**.

Consistent with this model, unactivated human THP-1 macrophage-like cells formed a PL-rich complex with apo A-I, whose synthesis was dependent on ABCA1 activity (Arakawa *et al.*, 2000). After activation with phorbol esters and cholesterol loading with acetylated low density lipoprotein (LDL), which upregulated caveolae, these cells formed FC-rich particles with apo A-I. The FC content of these HDLs was significantly decreased if the cells had been transfected with caveolin antisense DNA, which significantly reduces the expression of caveolae (Arakawa *et al.*, 2000; Fielding *et al.*, 1997). Sham transfection was without effect on HDL FC content. These data are, therefore, consistent with an origin of the major part of FC efflux from cell-surface caveolae.

Finally, in another recent study, HEK293 cells, which normally do not express ABCA1, were transfected with ABCA1 cDNA (Wang *et al.*, 2001) and transfected cells produced HDL containing both PL and FC. Mock-transfected cells produced no HDL in the presence of apo A-I. Treatment of the transfected cells with extracellular cyclodextrin, a sequestrant of FC, was without effect on PL efflux but greatly decreased the content of FC in the particles formed. Cyclodextrin exposure had previously been shown to downregulate the expression of caveolae at the cell surface (Parpal *et al.*, 2001). Apo A-I/PL particles without FC, produced in the presence of cyclodextrin, promoted FC efflux from non-transfected HEK293 cells without ABCA1 expression. These data are also consistent with a two-step model of HDL formation. One study, using immortalized human skin fibroblasts, concluded that none of the FC transferred out of the cell in response to ABCA1 activity was derived from caveolae (Mendez *et al.*, 2000). However, such transformed cells express few caveolae (Engelman *et al.*, 1997; Koleske *et al.*, 1995). If fact, overexpression of caveolae in

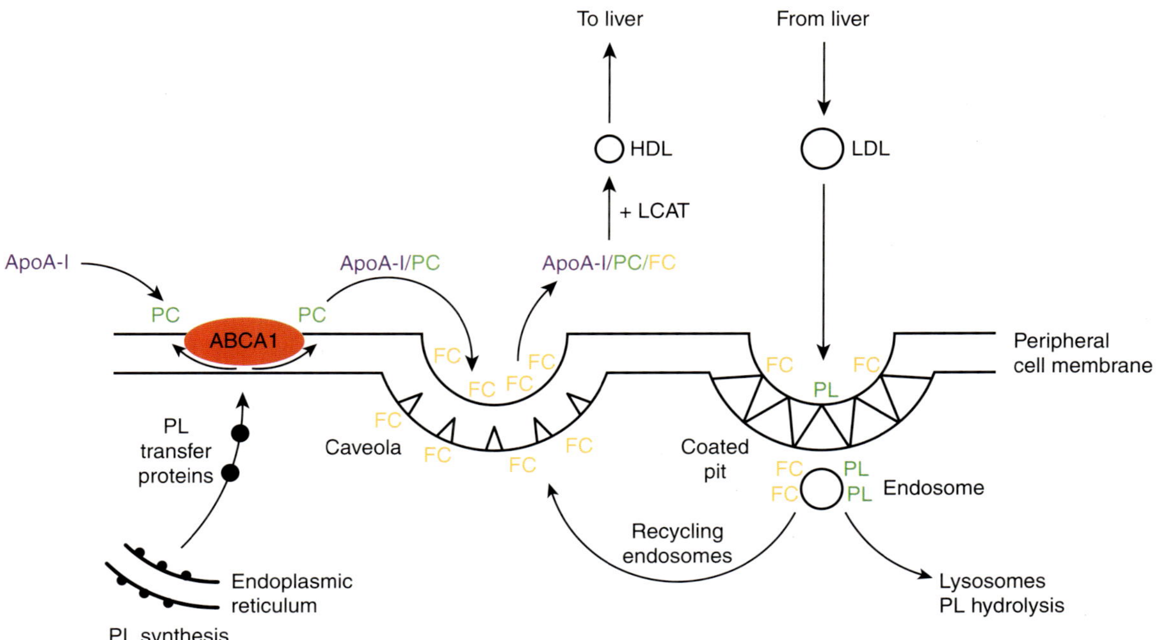

Figure 23.10. ***FC and PL homeostasis at the surface of peripheral cells. The endocytosis of lipids from LDL, the recycling pathway for LDL-derived FC, and the reincorporation of FC with PC–apo A-I, via a two-step model to generate HDL are shown. ApoA-I, apolipoprotein A-I, LCAT, lecithin:cholesterol acyltransferase; FC, free cholesterol; PC, phosphatidylcholine; PL, phospholipid.***

transformed cells effectively inhibits growth (Engelman *et al.*, 1997). Caveolin when present in these cells is primarily in intracellular vesicles. In immortalized cells, in contrast to primary cells, FC efflux is probably mediated by other mechanisms, such as simple diffusion.

In summary, we suggest that the major role of ABCA1 in cellular lipid homeostasis **(Figure 23.10)** is to facilitate, albeit indirectly, the transfer of PL from PC-rich membrane domains to a precursor particle of circulating lipid-poor (pre-beta-) HDL. In cells expressing significant levels of both ABCA1 and caveolae (for example, smooth muscle cell), we conclude that PL efflux is facilitated, either directly or indirectly, by ABCA1, but FC efflux is not. In cells lacking ABCA1 but expressing caveolae (such as endothelial cells), FC efflux depends mainly on apo A-I/PL complexes preformed at other sites.

ACKNOWLEDGMENTS

Research by C.J.F. was supported by the National Institutes of Health through HL 57976 and HL 67294. G.C. and O.C. wish to thank the members of the group for discussion and the Association Nationale pour la Recherche sur le Cancer , the Ligue Nationale Contre le Cancer, la Fondation de France and the association Vaincre la Mucoviscidose for financial support.

The authors acknowledge all the scientists whose work has contributed to the advancement of the field.

REFERENCES

Ahn, J. and Molday, R.S. (2000) Purification and characterization of ABCR from bovine rod outer segments. Methods Enzymol. 315, 864–879.

Albert, M., Kim, J. and Birge, R.B. (2000) αvβ5 integrin recruits the CrkII-Dock180-Rac1 complex for phagocytosis of apoptotic cells. Nature Cell. Biol. 2, 899–906.

Allikmets, R., Wasserman, W.W., Hutchinson, A., Smallwood, P., Nathans, J.K.R., Schneider, T.D. and Dean, M. (1998) Organization of the ABCR gene: analysis of promoter and splice junction sequences. Gene 215, 111–122.

Aoki, T., Nomura, R. and Fujimoto, T. (1999) Tyrosine phosphorylation of caveolin-1 in the endothelium. Exp. Cell Res. 253, 629–636.

Arakawa, R., Abe-Dohmae, S., Asai, M., Ito, J.I. and Yokoyama, S. (2000) Involvement of caveolin-1 in cholesterol enrichment of high density lipoprotein during its assembly by apolipoprotein and THP-1 cells. J. Lipid Res. 41, 1952–1962.

Assmann, G., von Eckardstein, A. and Brewer, H.B. (2001) Familial analphalipoproteinemia: Tangier disease. In: The Metabolic Basis of Inherited Disease, (ed. C.R. Scriver, A.L. Beaudet, W.S. Sly and D. Valle), pp. 2937–2960. New York: McGraw-Hill.

Asztalos, B.F., Brousseau, M.E., McNamara, J.R., Horvath, K.V., Roheim, P.S. and Schaefer, E.J. (2001) Subpopulations of high density lipoproteins in homozygous and heterozygous Tangier disease. Atherosclerosis 156, 217–225.

Azarian, S.M., Megarity, C.F., Weng, J., Horvath, D.H. and Travis, G.T. (1998) The human photoreceptor rim protein gene (ABCR): genomic structure and primer set information for mutation analysis. Hum. Genet. 102, 699–705.

Becq, F., Hamon, Y., Bajetto, A., Gola, M., Verrier, B. and Chimini, G. (1997) ABC1, an ATP binding cassette transporter required for phagocytosis of apoptotic cells, generates a regulated anion flux after expression in *Xenopus laevis* oocytes. J. Biol. Chem. 272, 2695–2699.

Bellincampi, L., Simone, M.L., Motti, C., Cortese, C., Bernardini, S., Bertolini, S. and Calandra, S. (2001) Identification of an alternative transcript of ABCA1 gene in different human cell types. Biochem. Biophys. Res. Commun. 283, 590–597.

Biswas, E.E. and Biswas, S.B. (2000) The C-terminal nucleotide binding domain of the human retinal ABCR protein is an adenosine triphosphatase. Biochemistry 39, 15879–15886.

Bodzioch, M., Orso, E., Klucken, J., Langmann, T., Böttcher, A., Diederich, W., *et al.* (1999) The gene encoding ATP-binding cassette transporter 1 is mutated in Tangier disease. Nat. Genet. 22, 347–351.

Bortnick, A.E., Rothblat, G.H., Stoudt, G., Hoppe, K.L., Royer, L.J., McNeish, J. and Francone, O.L. (2000) The correlation of ATP-binding cassette 1 mRNA levels with cholesterol efflux from various cell lines. J. Biol. Chem. 275, 28634–28640.

Broccardo, C., Luciani, M.F. and Chimini, G. (1999) The ABCA subclass of mammalian transporters. Biochim. Biophys. Acta 1461, 395–404.

Broccardo, C., Osorio, J., Luciani, M.F., Schriml, L.M., Prades, C., Shulenin, S., *et al.* (2001) Comparative analysis of the promoter structure and genomic organization of the human and mouse ABCA7 gene encoding a novel ABCA transporter. Cytogenet. Cell Genet. 92, 264–270.

Brooks-Wilson, A., Marcil, M., Clee, S.M., Zhang, L.H., Roomp, K., van Dam, M., *et al.* (1999) Mutations in ABC1 in Tangier disease and familial high-density lipoprotein deficiency. Nat. Genet. 22, 336–345.

Bungert, S., Molday, L.L. and Molday, R.S. (2001) Membrane topology of the ATP binding cassette transporter ABCR and its relationship to ABC1 and related ABCA transporters. J. Biol. Chem. 276, 23539–23546.

Castro, G.R. and Fielding, C.J. (1988) Early incorporation of cell-derived cholesterol into pre-beta-migrating high-density lipoprotein. Biochemistry 27, 25–29.

Cavelier, L.B., Qiu, Y., Bielicki, J.K., Afzal, V., Cheng, J.F. and Rubin, E.M. (2001) Regulation and activity of the human abca1 gene in transgenic mice. J. Biol. Chem. 276, 18046–18051.

Chambenoit, O., Hamon, Y., Marguet, D., Rigneault, H., Rosseneu, M. and Chimini, G. (2001) Specific docking of apolipoprotein A-I at the cell surface requires a functional ABCA1 transporter. J. Biol. Chem. 276, 9955–9960.

Chawla, A., Boisvert, W.A., Lee, C.H., Laffitte, B.A., Barak, Y., Joseph, S.B., *et al.* (2001) A PPAR gamma-LXR-ABCA1 pathway in macrophages is involved in cholesterol efflux and atherogenesis. Mol. Cell 7, 161–171.

Christiansen-Weber, T.A., Voland, J.R., Wu, Y., Ngo, K., Roland, B.L., Nguyen, S., Peterson, P.A. and Fung-Leung, W.P. (2000) Functional loss of ABCA1 in mice causes severe placental malformation, aberrant lipid distribution, and kidney glomerulonephritis as well as high-density lipoprotein cholesterol deficiency. Am. J. Pathol. 157, 1017–1029.

Costet, P., Luo, Y., Wang, N. and Tall, A.R. (2000) Sterol-dependent transactivation of the human ABC1 promoter by LXR/RXR. J. Biol. Chem. 275, 28240–28245.

Daleke, D.L. and Lyles, J.V. (2000) Identification and purification of aminophospholipid flippases. Biochim. Biophys. Acta 1486, 108–127.

Dean, M., Hamon, Y. and Chimini, G. (2001) The human ATP-binding cassette (ABC) transporter superfamily. J. Lipid Res. 42, 1007–1017.

Decottignies, A. and Goffeau, A. (1997) Complete inventory of the yeast ABC proteins, Nat. Genet. 15, 137–145.

Drobnik, W., Borsukova, H., Bottcher, A., Pfeiffer, A., Liebisch, G., Schultz, G.J., Schindler, H. and Schmitz, G. (2002) Apo A-I/ABCA1-dependent and HDL3-mediated lipid efflux from compositionally distinct cholesterol-based microdomains. Traffic 4, 268–278.

Ellis, R.E., Jacobson, D.M. and Horvitz, H.R. (1991a) Genes required for the engulfment of cell corpses during programmed cell death in *Caenorhabditis elegans*. Genetics 129, 79–94.

Ellis, R.E., Yuan, J. and Horvitz, H.R. (1991b) Mechanisms and functions of cell death. Annu. Rev. Cell Biol. 7, 663–698.

Engelman, J.A., Wykoff, C.C., Yasuhara, S., Song, K.S., Okamoto, T. and Lisanti, M.P. (1997) Recombinant expression of caveolin-1 in oncogenically transformed cells abrogates anchorage-independent growth. J. Biol. Chem. 272, 16374–16381.

Fielding, C.J. and Fielding, P.E. (1995) Molecular physiology of reverse cholesterol transport. J. Lipid Res. 36, 211–228.

Fielding, C.J., Bist, A. and Fielding, P.E. (1997) Caveolin mRNA levels are up-regulated by free cholesterol and down-regulated by oxysterols in fibroblast monolayers. Proc. Natl Acad. Sci. USA 94, 3753–3758.

Fielding, C.J., Bist, A. and Fielding, P.E. (1998) Uptake of lipoprotein free cholesterol and its intracellular transport. In: Intracellular Cholesterol Trafficking (ed. T.Y. Chjang and D.A. Freeman), pp. 273–288. Boston: Kluwer.

Fielding, P.E., Nagao, K., Hakamata, H., Chimini, G. and Fielding, C.J. (2000) A two-step mechanism for free cholesterol and phospholipid efflux from human vascular cells to apolipoprotein A-1. Biochemistry 39, 14113–14120.

Fitzgerald, M.L., Mendez, A.J., Moore, K.J., Andersson, L.P., Panjeton, H.A. and Freeman, M.W. (2001) ABCA1 contains an N-terminal signal-anchor sequence that translocates the protein's first hydrophilic domain to the exoplasmic space. J. Biol. Chem. 276, 15137–15145.

Fitzgerald, M.L., Morris, A.L., Rhee, J.S., Anderson, L.P., Mendez, A.J. and Freeman, M.W. (2002) Naturally occurring mutations in ABCA1's largest extracellular loops can disrupt its direct interactions with apolipoprotein A-I. J. Biol. Chem. Jun 25; [ed pub ahead of print].

Franc, N.C., White, K. and Ezekowitz, R.A. (1999) Phagocytosis and development: back to the future. Curr. Opin. Immunol. 11, 47–52.

Francis, G.A., Knopp, R.H. and Oram, J.F. (1995) Defective removal of cellular cholesterol and phospholipids by apolipoprotein A-I in Tangier disease. J. Clin. Invest. 96, 78–87.

Frank, P.G. and Marcel Y.L. (2000) Apolipoprotein A-I: structure–function relationships. J. Lipid Res. 41, 853–872.

Gregory, C.D. (2000) CD14-dependent clearance of apoptotic cells: relevance to the immune system. Curr. Opin. Immunol. 12, 27–34.

Hamon, Y., Broccardo, C., Chambenoit, O., Luciani, M.F., Toti, F., Chaslin, S., *et al.* (2000) ABC1 promotes engulfment of apoptotic cells and transbilayer redistribution of phosphatidylserine. Nature Cell Biol. 2, 399–406.

Hara, H. and Yokoyama S. (1992) Role of apolipoproteins in cholesterol efflux from macrophages to lipid microemulsion: proposal of a putative model for the pre-beta high-density lipoprotein pathway. Biochemistry 31, 2040–2046.

Hengartner, M.O. (2000) The biochemistry of apoptosis. Nature 407, 770–776.

Higgins, C.F. (1994) Flip-flop: the transmembrane translocation of lipids. Cell 79, 393–395.

Higgins, C.F. and Gottesman, M.M. (1992) Is the multidrug transporter a flippase? TIBS 17, 18–21.

Hirano, K., Matsuura, F., Tsukamoto, K., Zhang, Z., Matsuyama, A., Takaishi, K., *et al.* (2000) Decreased expression of a member of the Rho GTPase family, Cdc42Hs, in cells from Tangier disease – the small G protein may play a role in cholesterol efflux. FEBS Lett. 484, 275–279.

Illing, M., Molday, L.L. and Molday, R.S. (1997) The 220-kDa rim protein of retinal rod outer segments is a member of the ABC transporter superfamily. J. Biol. Chem. 272, 10303–10310.

Kaminski, W.E., Piehler, A. and Schmitz, G. (2000) Genomic organization of the human cholesterol-responsive ABC transporter ABCA7: tandem linkage with the minor histocompatibility antigen HA-1 gene. Biochem. Biophys. Res. Commun. 278, 782–789.

Kaminski, W.E., Piehler, A., Pullmann, K., Porsch-Ozcurumez, M., Duong, C., Bared, G.M., Buchler, C. and Schmitz, G. (2001)

Complete coding sequence, promoter region, and genomic structure of the human ABCA2 gene and evidence for sterol-dependent regulation in macrophages. Biochem. Biophys. Res. Commun. 281, 249–258.

Kerr, J.F.R., Wyllie, A.H. and Currie, A.R. (1972) Apoptosis: a basic biological phenomenon with wide-ranging implication in tissue kinetics. Br. J. Cancer 26, 239–257.

Koleske, A.J., Baltimore, D. and Lisanti, M.P. (1995) Reduction of caveolin and caveolae in oncogenically transformed cells. Proc. Natl Acad. Sci. USA 92, 1381–1385.

Langmann, T., Klucken, J., Reil, M., Liebisch, G., Luciani, M.F., Chimini, G., Kaminski, W.E. and Schmitz, G. (1999) Molecular cloning of the human ATP-binding cassette transporter 1 (hABC1): evidence for sterol-dependent regulation in macrophages. Biochem. Biophys. Res. Commun. 257, 29–33.

Lawn, R.M., Wade, D.P., Garvin, M.R., Wang, X., Schwartz, K., Porter, J.G., Seilhamer, J.J., Vaughan, A.M. and Oram, J.F. (1999) The Tangier disease gene product ABC1 controls the cellular apolipoprotein-mediated lipid removal pathway. J. Clin. Invest. 104, R25–31.

Lawn, R.M., Wade, D.P., Couse, T.L. and Wilcox, J.N. (2001) Localization of human ATP-binding cassette transporter 1 (ABC1) in normal and atherosclerotic tissues. Arterioscler. Thromb. Vasc. Biol. 21, 378–385.

Leverrier, Y. and Ridley, A.J. (2001a) Apoptosis: caspases orchestrate the ROCK 'n' bleb. Nat. Cell Biol. 3, E91–E92.

Leverrier, Y. and Ridley, A.J. (2001b) Requirement for Rho GTPases and PI 3-kinases during apoptotic cell phagocytosis by macrophages. Curr. Biol. 11, 195–199.

Liu, Q. and Hengartner, M. (1998) Candidate adaptor protein CED-6 promotes the engulfment of apoptotic cells in *C. elegans*. Cell 93, 961–972.

Liu, Q.A. and Hengartner, M.O. (1999) Human CED-6 encodes a functional homologue of the *Caenorhabditis elegans* engulfment protein CED-6. Curr. Biol. 9, 1347–1350.

Luciani, M.F. and Chimini, G. (1996) The ATP binding cassette transporter ABC1, is required for the engulfment of corpses generated by apoptoic cell death. EMBO J. 15, 226–235.

Luciani, M.F., Denizot, F., Savary, S., Mattei, M.G. and Chimini, G. (1994) Cloning of two novel ABC transporters mapping on human chromosome 9. Genomics 21, 150–159.

Marguet, D., Luciani, M.F., Moynault, A., Williamson, P. and Chimini, G. (1999) Engulfment of apoptotic cells involves the redistribution of membrane phosphatidylserine on both phagocyte and prey. Nature Cell Biol. 1, 454–456.

McNeish, J., Aiello, R.J., Guyot, D., Turi, T., Gabel, C., Aldinger, C., *et al.* (2000) High density lipoprotein deficiency and foam cell accumulation in mice with targeted disruption of ATP-binding cassette transporter-1. Proc. Natl Acad. Sci. USA 97, 4245–4250.

Mendez, A.J., Lin, G., Wade, D.P., Lawn, R.M. and Oram, J.F. (2000) Membrane lipid domains distinct from cholesterol/sphingomyelin-rich rafts are involved in the ABCA1-mediated lipid secretory pathway. J. Biol. Chem. 276, 3158–3166.

Neufeld, E.B., Remaley, A.T., Demosky, S.J., Stonik, J.A., Cooney, A.M., Comly, M., *et al.* (2001) Cellular localization and trafficking of the human ABCA1 transporter. J. Biol. Chem. 276, 27584–27590.

Nielsen, H., Engelbrecht, J., Brunak, S. and von Heijne, G. (1997) Identification of prokaryotic and eukaryotic signal peptides and prediction of their cleavage sites. Protein Eng. 10, 1–6.

Oliver, W.R., Jr, Shenk, J.L., Snaith, M.R., Russell, C.S., Plunket, K.D., Bodkin, N.L., *et al.* (2001) A selective peroxisome proliferator-activated receptor delta agonist promotes reverse cholesterol transport. Proc. Natl Acad. Sci. USA 98, 5306–5311.

Oram, J.F. and Yokoyama, S. (1996) Apolipoprotein-mediated removal of cellular cholesterol and phospholipids. J. Lipid Res. 37, 2473–2491.

Orso, E., Broccardo, C., Böttcher, A., Liebisch, G., Drobnik, W., Kaminski, W., *et al.* (2000) ABC1 mediates the cellular export of cholesterol and phospholipides: defective golgi to plasma membrane lipid transport in ABC1 null mice and Tangier Disease. Nat. Genet. 24, 192–196.

Parpal, S., Karlsson, M., Thorn, H. and Stralfors, P. (2001) Cholesterol depletion disrupts caveolae and insulin receptor signaling for metabolic control via insulin receptor substrate-1, but not for mitogen-activated protein kinase control. J. Biol. Chem. 276, 9670–9678.

Platt, N., da Silva, R.P. and Gordon, S. (1998) Recognizing death: the phagocytosis of apoptotic cells. Trends Cell Biol. 9, 365–372.

Pol, A., Luetterforst, R., Lindsay, M., Heino, S., Ikonen, E. and Parton, R.G. (2001) A caveolin dominant negative mutant associates with lipid bodies and induces intracellular cholesterol imbalance. J. Cell Biol. 152, 1057–1070.

Pullinger, C.R., Hakamata, H., Duchateau, P.N., Eng, C., Aouizerat, B.E., Cho, M.H., Fielding, C.J. and Kane, J.P. (2000) Analysis of hABC1 gene 5′ end: additional peptide sequence, promoter region, and four polymorphisms. Biochem. Biophys. Res. Commun. 271, 451–455.

Qiu, Y., Cavelier, L., Chiu, S., Yang, X., Rubin, E. and Cheng, J.F. (2001) Human and mouse abca1 comparative sequencing and transgenesis studies revealing novel regulatory sequences. Genomics 73, 66–76.

Reddien, P.W. and Horvitz, H.R. (2000) CED-2/CrkII and CED-10/Rac control phagocytosis and cell migration in *Caenorhabditis elegans*. Nat. Cell Biol. 2, 131–136.

Remaley, A.T., Schumacher, U.K., Stonik, J.A., Farsi, B.D., Nazih, H. and Brewer, H.B., Jr (1997) Decreased reverse cholesterol transport from Tangier disease fibroblasts. Acceptor specificity and effect of brefeldin on lipid efflux. Arterioscler. Thromb. Vasc. Biol. 17, 1813–1821.

Remaley, A.T., Rust, S., Rosier, M., Knapper, C., Naudin, L., Broccardo, C., *et al.* (1999) Human ATP-binding cassette transporter 1 (ABC1): genomic organization and identification of the genetic defect in the original Tangier disease kindred. Proc. Natl Acad. Sci. USA 96, 12685–12690.

Ren, Y. and Savill, J. (1998) Apoptosis: the importance of being eaten. Cell Death Differ. 5, 563–568.

Rogler, G., Trumbach, B., Klima, B., Lackner, K.J. and Schmitz, G. (1995) HDL-mediated efflux of intracellular cholesterol is impaired in fibroblasts from Tangier disease patients. Arterioscler. Thromb. Vasc. Biol. 15, 683–690.

Rosenberg, M.F., Mao, Q., Holzenburg, A., Ford, R.C., Deeley, R.G. and Cole, S.P. (2001) The structure of the multidrug resistance protein 1 (MRP1/ABCC1). Crystallization and single-particle analysis. J. Biol. Chem. 276, 16076–16082.

Rust, S., Walter, M., Funke, H., von Eckardstein, A., Cullen, P., Kroes, H., *et al.* (1998) Assignment of Tangier disease to chromosome 9q31 by a graphical linkage exclusion strategy. Nat. Genet. 20, 96–98.

Rust, S., Rosier, M., Funke, H., Real, J., Amoura, Z., Piette, J.C., *et al.* (1999) Tangier disease is caused by mutations in the gene encoding ATP-binding cassette transporter 1. Nat. Genet. 22, 352–355.

Sanchez-Fernandez, R., Davies, T.G., Coleman, J.O. and Rea, P.A. (2001). The *Arabidopsis thaliana* ABC protein superfamily: a complete inventory. J. Biol. Chem. 276, 30231–30244.

Santamarina-Fojo, S., Peterson, K., Knapper, C., Qiu, Y., Freeman, L., Cheng, J.F., *et al.* (2000) Complete genomic sequence of the human ABCA1 gene: analysis of the human and mouse ATP-binding cassette A promoter. Proc. Natl Acad. Sci. USA 97, 7987–7992.

Savary, S., Denizot, F., Luciani, M.F., Mattei, M.G. and Chimini, G. (1996) Molecular cloning of a mammalian ABC transporter homologous to *Drosophila white* gene. Mammalian Genome 7, 673–676.

Savary, S., Allikmets, R., Denizot, F., Luciani, M.F., Mattei, M.G., Dean, M. and Chimini, G. (1997) Isolation and chromosomal mapping of a novel ATP-binding cassette transporter conserved in mouse and human. Genomics 41, 275–278.

Savill, J. (1998) Phagocytic docking without shocking. Nature 392, 442–443.

Savill, J. and Fadok V. (2000) Corpse clearance defines the meaning of cell death. Nature 407, 784–788.

Schreyer, S.A., Hart, L.K. and Attie, A.D. (1994) Hypercatabolism of lipoprotein-free apolipoprotein A I in HDL-deficient mutant chickens. Arterioscler. Thromb. 14, 2053–2059.

Schwartz, K., Lawn, R.M. and Wade, D.P. (2000) ABC1 gene expression and ApoA-I-mediated cholesterol efflux are regulated by LXR. Biochem. Biophys. Res. Commun. 274, 794–802.

Singaraja, R.R., Bocher, V., James, E.R., Clee, S.M., Zhang, L.H., Leavitt, B.R., *et al.* (2001) Human ABCA1 BAC transgenic mice show increased HDL-C and ApoAI dependant efflux stimulated by an internal promoter containing LXREs in intron 1. J. Biol. Chem. 276, 33969–33979.

Smart, E.J., Ying, Y., Donzell, W.C. and Anderson, R.G. (1996) A role for caveolin in transport of cholesterol from endoplasmic reticulum to plasma membrane. J. Biol. Chem. 271, 29427–29435.

Su, H.P., Brugnera, E., Van Criekinge, W., Smits, E., Hengartner, M., Bogaert, T. and Ravichandran, K.S. (2000) Identification and characterization of a dimerization domain in CED-6, an adapter protein involved in engulfment of apoptotic cells. J. Biol. Chem. 275, 9542–9549.

Sun, H., Molday, R.S. and Nathans, J. (1999) Retinal stimulates ATP hydrolysis by purified and reconstituted ABCR, the photoreceptor specific ABC transporter responsible for Stargardt disease. J. Biol. Chem. 274, 8269–8281.

Tanaka, A.R., Ikeda, Y., Abe-Dohmae, S., Arakawa, R., Sadanami, K., Kidera, A., *et al.* (2001) Human ABCA1 contains a large amino-terminal extracellular domain homologous to an epitope of Sjogren's syndrome. Biochem. Biophys. Res. Commun. 283, 1019–1025.

Tosello-Trampont, A.C., Brugnera, E. and Ravichandran, K.S. (2001) Evidence for a conserved role for CrkII and Rac in engulfment of apoptotic cells. J. Biol. Chem. 276, 13797–13802.

Venkateswaran, A., Laffitte, B.A., Joseph, S.B., Mak, P.A., Wilpitz, D.C., Edwards, P.A. and Tontonoz, P. (2000) Control of cellular cholesterol efflux by the nuclear oxysterol receptor LXR alpha. Proc. Natl Acad. Sci. USA 97, 12097–12102.

Voelker, D.R. (1996) Lipid assembly into cell membranes. In: Biochemistry of Lipids, Lipoproteins and Membranes (ed. D.E. Vance and J. Vance), pp. 391–424. New York: Elsevier.

von Eckardstein, A., Chirazi, A., Schuler-Luttmann, S., Walter, M., Kastelein, J.J.P., Geisel, J., *et al.* (1998) Plasma and fibroblasts of Tangier disease patients are disturbed in transferring phospholipids onto apoA-I. J. Lipid Res. 39, 987–998.

Vulevic, B., Chen, Z., Boyd, J.T., Davis, W., Jr, Walsh, E.S., Belinsky, M.G. and Tew, K.D. (2001) Cloning and characterization of human adenosine 5′-triphosphate-binding cassette, sub-family A, transporter 2 (ABCA2). Cancer Res 61, 3339–3347.

Waite, M. (1996) Phospholipases. In: Biochemistry of Lipids, Lipoproteins and Membranes (ed. D.E. Vance and J. Vance), pp. 211–236. New York: Elsevier.

Walter, M., Gerdes, U., Seedorf, U. and Assmann, G. (1994) The high density lipoprotein- and apolipoprotein A-I-induced mobilization of cellular cholesterol is impaired in fibroblasts from Tangier disease subjects. Biochem. Biophys. Res. Commun. 205, 850–856.

Wang, N., Silver, D.L., Costet, P. and Tall, A.R. (2000) Specific binding of ApoA-I, enhanced cholesterol efflux, and altered plasma membrane morphology in cells expressing ABC1. J. Biol. Chem. 275, 33053–33058.

Wang, N., Silver, D.L., Thiele, C. and Tall, A.R. (2001) ABCA1 functions as a cholesterol efflux regulatory protein. J. Biol. Chem. 276, 23742–23747.

Wood, W., Turmaine, M., Weber, R., Camp, V., Maki, R.A., McKercher, S.R. and Martin, P. (2000) Mesenchymal cells engulf and clear apoptotic footplate cells in macrophageless PU.1 null mouse embryos. Development 127, 5245–5252.

Wu, Y. and Horvitz, H.R. (1998a) *C. elegans* phagocytosis and cell-migration protein CED-5 is similar to human DOCK180. Nature 329, 501–504.

Wu, Y. and Horvitz, R.H. (1998b) The *C. elegans* cell corpse engulfment gene ced-7 encodes a protein similar to ABC transporters. Cell 93, 951–960.

Wu, Y.C., Tsai, M.C., Cheng, L.C., Chou, C.J. and Weng, N.Y. (2001) *C. elegans* CED-12 acts in the conserved crkII/DOCK180/Rac pathway to control cell migration and cell corpse engulfment. Dev. Cell 4, 491–502.

Zhou, Z., Hartwieg, E. and Horvitz, H.R. (2001a) CED-1 is a transmembrane receptor that mediates cell corpse engulfment in *C. elegans*. Cell 104, 43–56.

Zhou, Z., Caron, E., Hartwieg, E., Hall. A. and Horvitz, H.R. (2001b) The *C. elegans* PH domain protein CED-12 regulates cytoskeletal reorganization via a Rho/Rac GTPase signaling pathway. Dev. Cell 4, 477–489.

CHAPTER 24

Peroxisomal ABC Transporters

Shlomo Almashanu and David Valle

Introduction

The set of proteins found in the membranes of mammalian peroxisomes includes four half ABC transporters (ALDP, ALDR, PMP70, P70R) comprising a distinct subset of the superfamily of ABC transporters designated subfamily D. Each has an N-terminal hydrophobic transmembrane domain with multiple transmembrane segments (TMS) and a hydrophilic C-terminal half containing a nucleotide-binding domain (NBD) with Walker A and B motifs. Limited topology studies indicate that the C-terminal hydrophilic halves of ALDP and PMP70 extend into the cytosol (Contreras *et al.*, 1996; Kamijo *et al.*, 1990). To provide context in what follows, we first briefly review peroxisome function, genetic diseases and biogenesis; we then focus on the molecular, cellular and evolutionary biology of the mammalian peroxisomal ABC transporters.

Overview of Peroxisome Biology

Peroxisomes are typically spherical (0.1–1 μm diameter), single-membrane-bound organelles present in numbers ranging from a few hundred to a few thousand in most mammalian cells **(Figure 24.1)** (Gould *et al.*, 2001; Purdue and Lazarow, 2001; Tabak *et al.*, 1999). The pH of the mammalian peroxisome matrix has variously been estimated to be more basic (Dansen *et al.*, 2000) or similar to (Jankowksi *et al.*, 2001) that of cytosol. The dense, proteinaceous peroxisome matrix contains 50 or more enzymes, which participate in a variety of metabolic pathways including β-oxidation of straight and branched, very long ($\geq C_{24}$) and long (C_{14-22}) chain fatty acids (VLCFA and LCFA, respectively), synthesis of cholesterol and ether-lipids (e.g. plasmalogens) and oxidation of polyamines, D-amino acids and, in non-primates, uric acid (Gould *et al.*, 2001; Sacksteder and Gould, 2000; Wanders *et al.*, 2001). Many of the peroxisomal oxidation reactions liberate H_2O_2, which is detoxified by peroxisomal catalase. The peroxisome membrane contains a characteristic set of peroxisomal membrane proteins (PMPs) that, in addition to the peroxisomal ABC transporters, includes other small molecule transporters, enzymes and proteins required for import of peroxisomal matrix proteins and peroxisomal membrane biogenesis, plus many whose function is uncertain (Schäfer *et al.*, 2001).

Peroxisomes and genetic disease

The rapid growth of our understanding of peroxisome biogenesis and function over the last decade has depended in part on careful analysis of cells from patients with inherited defects in these processes. Two categories of peroxisomal genetic disorders are recognized, both with profound phenotypic consequences. The first includes the peroxisomal biogenesis disorders (PBDs), a genetically heterogeneous group of

ABC Proteins: From Bacteria to Man
ISBN 0-12-352551-9

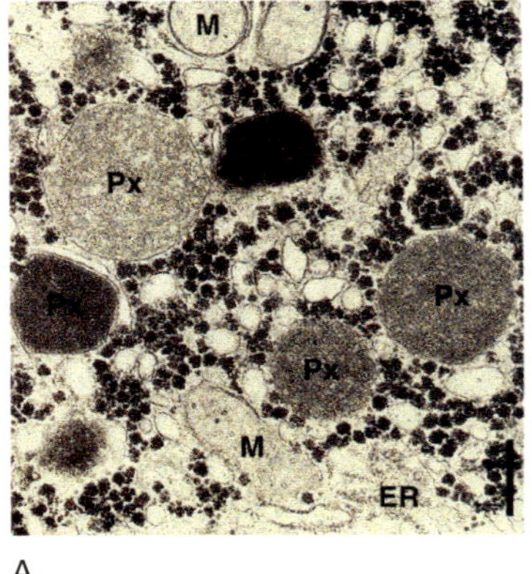

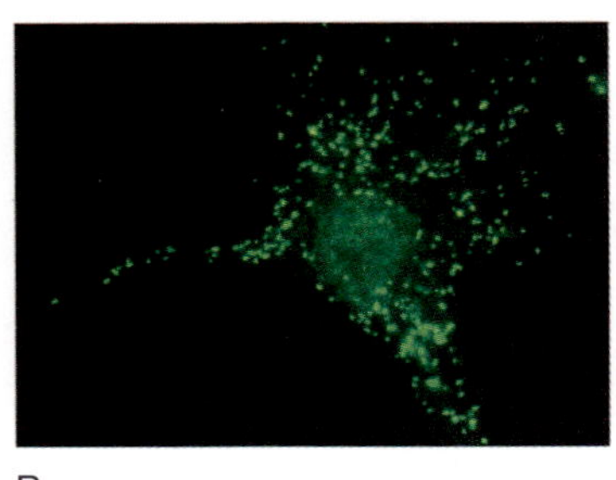

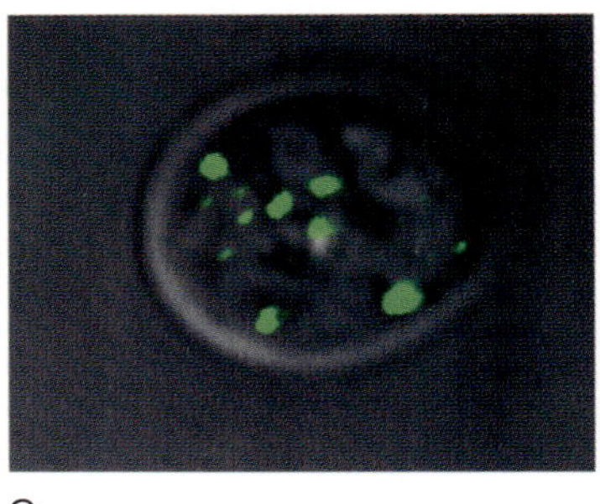

A B C

Figure 24.1. Peroxisome morphology. A, Electron micrograph of peroxisomes in human liver visualized by diaminobenzidine staining for catalase activity (image kindly supplied by M. Espeel and F. Roels, University of Gent, Belgium); scale bar is 0.5 μm. Px, peroxisome; M, mitochondrion; ER, endoplasmic reticulum. B, Immunofluorescence of human fibroblasts stained with anti-PMP70 antibody. Note the punctate appearance of normal peroxisomes. C, Confocal image of **S. cerevisiae**, *grown on oleic acid to induce peroxisomes and expressing C-terminal PTS1 tagged GFP, which localizes to peroxisomes.*

autosomal recessive disorders characterized by deficiency of multiple peroxisomal functions (Gould and Valle, 2000b; Gould *et al.*, 2001). Zellweger syndrome (MIM#214100), a lethal developmental and metabolic disorder, is the phenotypic paradigm for the PBD. Cell fusion and molecular studies have identified at least 12 genes responsible for these disorders, none of which encode peroxisomal ABC transporters (Gould and Valle, 2000b; Gould *et al.*, 2001). The second category includes disorders in which a single peroxisomal function is deficient (Wanders *et al.*, 2001). More than a dozen have been recognized. The exemplar is X-linked adrenoleukodystrophy (X-ALD)(MIM#300100), a progressive neurological disorder caused by mutations in *ABCD1*, the gene that encodes ALDP, a peroxisomal ABC transporter (see below) (Moser *et al.*, 2001). The principal biochemical abnormality of X-ALD, accumulation of VLCFA in plasma and tissues, together with the existence of a VLCFA β-oxidation pathway in the peroxisome, has implicated ALDP in the transport of VLCFA or VLCF acyl-CoAs into the peroxisome. Tissues and cultured cells from X-ALD patients have a 60–80% reduction in β-oxidation of VLCFA.

PEROXISOME BIOGENESIS

Over the last decade at least 23 genes (designated *PEX* genes) have been identified that encode proteins (peroxins) necessary for peroxisome biogenesis (Gould and Valle, 2000b; Gould *et al.*, 2001; Purdue and Lazarow, 2001). The nomenclature convention is that orthologous *PEX* genes in different species are indicated by the same number (Distel *et al.*, 1996). Thus, *Saccharomyces cerevisiae PEX1* is orthologous to human *PEX1*.

Regulation

The number of peroxisomes per cell is dynamic and varies with the metabolic state (Chang *et al.*, 1999; Gould *et al.*, 2001; Purdue and Lazarow, 2001; Subramani, 1998). In rodents, certain hypolipidemic drugs, plasticizers and naturally occurring lipids induce higher peroxisome numbers and the expression of genes encoding many matrix proteins and peroxisome membrane proteins (PMPs) including the ABC transporters (Reddy *et al.*, 1986; Zomer *et al.*, 2000). This coordinated induction is mediated primarily by activation of peroxisome proliferator-activated receptor α (PPARα), a member of the nuclear hormone receptor superfamily (Kersten *et al.*, 2000; Wahli *et al.*, 1999). Activated PPARα heterodimerizes with a second nuclear hormone receptor, retinoid X-responsive receptor α, to form an active transcript factor that recognizes *cis*-acting sequences, peroxisome proliferator responsive elements (PPRE, consensus two direct AGGA/ TCA separated by a single base pair), in the promoters of its target genes (Juge-Aubry *et al.*, 1997; Kersten *et al.*, 2000).

In normal cells, an increase in peroxisome number results mainly from maturation and enlargement of existing peroxisomes with uptake of both membrane and matrix components followed by fission into daughter organelles (Gould and Valle, 2000a; Purdue and Lazarow, 2001). *De novo* synthesis of peroxisomes is also possible (South and Gould, 1999).

Matrix proteins

Peroxisomal matrix proteins are synthesized on free cytosolic ribosomes and targeted post-translationally to the organelle by specific cytosolic receptors that recognize *cis*-acting sequences (peroxisomal targeting signals or PTSs) in the primary peptide sequence (Gould and Valle, 2000b; Gould *et al.*, 2001; Purdue and Lazarow, 2001). Most matrix proteins are targeted by PTS1, a C-terminal -SKL or conservative variant thereof. A few matrix proteins are targeted by PTS2, a degenerate sequence (-R/KX$_5$Q/HL-) located near the N-terminus. A few matrix proteins appear to be targeted by as yet unrecognized PTSs (Purdue and Lazarow, 2001). The PTS1 and 2 receptors have been cloned and characterized. The former, PEX5, is a tetratricopeptide repeat protein (Dodt *et al.*, 1995; Fransen *et al.*, 1995; Wiemer *et al.*, 1995); the latter, PEX7, is a WD40 repeat protein (Braverman *et al.*, 1997; Motley *et al.*, 1997; Purdue *et al.*, 1997). The structure of the PEX5 tetratricopeptide repeat domain complexed with a PTS1 peptide has been mapped in *S. cerevisiae* (Klein *et al.*, 2001) and solved for human PEX5 (Gatto *et al.*, 2000a, 2000b). Both PEX5 and PEX7 bind their cargo proteins in the cytosol and transport them to the peroxisome, where they interact with specific docking proteins in the peroxisomal membrane, release their cargo and recycle to the cytosol (Dodt and Gould, 1996). In mammalian cells, the long isoform of PEX5 interacts with PEX7 and is necessary for its function (Braverman *et al.*, 1998; Otera *et al.*, 1998, 2000).

Membrane proteins

Like peroxisomal matrix proteins, PMPs are synthesized on free cytosolic ribosomes and targeted to the organelle by *cis*-acting targeting sequences (membrane peroxisome targeting signals or mPTS). In contrast to the discrete, well-defined, single targeting sequences found in matrix proteins, the model emerging for several PMPs includes two non-overlapping targeting segments for each peptide, either of which is sufficient for localization and insertion into the peroxisome membrane. In most cases these targeting segments are relatively long and include one or two TM segments. This model derives from recent work on several integral PMPs including: human PMP34 (Jones *et al.*, 2001) and its fungal orthologue PMP47 (Dyer *et al.*, 1996; Wang *et al.*, 2001), a peroxisomal member of the mitochondrial carrier family that probably functions as an ATP/ADP transporter (van Roermund *et al.*, 2001); human PMP22, an abundant PMP of unknown function (Brosius *et al.*, 2002; Pause *et al.*, 2000); and PEX13, the docking protein for matrix protein receptors (Jones *et al.*, 2001). Attempts to define specific and necessary sequence motifs in these targeting segments have not been successful. A five-residue sequence of basic residues implicated in initial studies of PMP47 is not a consistent feature and some targeting persists when these residues are entirely replaced by alanines (Biery and Valle, unpublished; Wang *et al.*, 2001). Work on PMP70 by ourselves (Almashanu and Valle, unpublished) and others (Sacksteder and Gould, 2000) indicates that the peroxisomal membrane ABC transporters also use this two targeting segment mechanism.

A related question is how the newly synthesized, hydrophobic PMPs traverse the aqueous cytosol to acquire their proper location in the peroxisomal membrane. Recent studies with cells from patients with a peroxisome biogenesis disorder (PBD) have implicated PEX19, a farnesylated, mainly cytosolic protein, as a possible receptor for PMPs analogous to the role of PEX5 and PEX7 in the targeting of matrix proteins (Gotte *et al.*, 1998; James *et al.*, 1994; Sacksteder *et al.*, 2000; Snyder *et al.*, 2000). Binding studies show that the multiple PMP targeting segments described above are recognized by PEX19 (Brosius *et al.*, 2002; Jones *et al.*, 2001; Sacksteder *et al.*, 2000). Additionally, two integral PMPs, PEX3 and PEX16, are probably involved in this process (Honsho *et al.*, 1998; Muntau *et al.*, 2000; Snyder *et al.*, 1999; South and Gould, 1999; South *et al.*, 2000). Mutations in any of these three genes not only cause a PBD phenotype but also are associated with the distinct cellular phenotype of absence of peroxisome membranes (Ghaedi *et al.*, 2000; Hettema *et al.*, 2000; Honsho *et al.*, 1998; Matsuzono *et al.*, 1999; Muntau *et al.*, 2000; South and Gould, 1999).

PEROXISOME ABC TRANSPORTERS

The peroxisome ABC transporters are all half ABC transporters and, as a group, are the most thoroughly studied integral PMPs. One, PMP70, is routinely used as the standard marker protein for peroxisomal membranes **(Figure 24.1)**. Despite this wealth of information, much remains to be learned about these molecules. In what follows, we review the current state of knowledge of these interesting proteins.

GENES

ALDP is encoded by *ABCD1*, the gene responsible for X-ALD. *ABCD1* spans ~21 kb, contains 10 exons (Sarde *et al.*, 1994) and maps to Xq28 **(Table 24.1)** (Mosser *et al.*, 1994). *ABCD1* mutations, including frameshifting insertions and deletions and nonsense mutations identified in X-ALD patients, provide compelling evidence that it is the gene responsible for this neurodegenerative disorder. Four autosomal *ABCD1* non-processed pseudogenes with 92–96% nucleotide identity located elsewhere in the genome complicate the molecular diagnosis of X-ALD **(Table 24.1)**. An online X-ALD database (www.x-ald.nl) contains useful information for patients and professionals and is maintained in a collaborative effort between the Peroxisomal Diseases Laboratory at the Kennedy Krieger Institute, Baltimore, MD, USA and the Laboratory of Genetic Metabolic Diseases at the Academic Medical Center, Amsterdam, the Netherlands.

ALDR is encoded by *ABCD2* comprising 10 exons distributed over 33 kb on chromosome 12q12 **(Table 24.1)**. The genomic organization of *ABCD2* closely resembles that of *ABCD1* consistent with the conclusion, based on sequence similarity (64% identity), that these two members of the peroxisome ABC transporter subfamily diverged recently from a common ancestor **(see Figure 24.4)**. Moreover, this high sequence similarity suggests that the function between ALDP and ALDR may be partially redundant. Thus, ALDR may have the potential to modify the phenotype of X-ALD (Holzinger *et al.*, 1997a). Induction of the *ABCD2/Abcd2* gene in

TABLE 24.1. MAMMALIAN PEROXISOMAL ABC TRANSPORTERS

The nomenclature is based on the guidelines for the human and the mouse ABC transporter gene nomenclature (http://www.gene.ucl.ac.uk/nomenclature/genefamily/abc.html). The information on the localization, nucleotide sequence, locus and PubMed IDs was derived from the National Center for Biotechnology Information (http://www.ncbi.nlm.nih.gov).

Symbol	Previous symbol, ABC #	Alias	Map location	Number of exons	Genomic NT	mRNA NM	Locus ID	OMIM	PMID
Hs ABCD1	ALD ABC42	ALDp ALD AMA	Xq28	10	025965	000033	215	300100	8441467
Hs ABCD1P1		Pseudogene 1	2p11		–	–	23785		8829626
Hs ABCD1P2		Pseudogene 2	10p11		–	–	26983		9215666
Hs ABCD1P3		Pseudogene 3	16p11.2		024779	–	26982		9215666
Hs ABCD1P4		Pseudogene 4	22q11		–	–	26957		9215666
Hs ABCD2	ALDL1 ABC39	ALDR ALDRp	12q11	10	–	005164	225	601081	8577752
Hs ABCD3	PXMP1 ABC43	PMP70 PXMP1	1p21–p22	23	029227	002858	5825	170995	1536884
Hs ABCD4	PXMP1L ABC41	P70R PMP69	14q24.3	19	–	005050	5826	603214	9266848
Mm Abcd1		ALD	X 29.5 cM		–	007435	11666		11076861
Mm Abcd2		ALDR	15 E–F		–	011994	26874		10504404
Mm Abcd3		PMP70	3 56.6cM		–	008991	19299		11076861
Mm Abcd4		P70R	12 39.0 cM		–	008992	19300		10708515

OMIM, Online Mendelian Inheritance in Man. PMID, Pubmed entry or Pubmed indexed for MEDLINE.

cells from X-ALD patients or in the *Abcd1* knockout mice by exposure to 4-phenylbutyrate temporarily corrected the deficiency of VLCFA β-oxidation (Kemp *et al.*, 1998). If this response could be maintained, it would be a promising therapeutic strategy for X-ALD. Expression of *ABCD2* is induced by fibrates and other xenobiotics as well as certain endogenous lipids in a response that depends on PPARα (Fourcade *et al.*, 2001). Survey of 2 kb of 5′ flanking sequence of rat *Abcd2* identified several candidate PPREs (see above) but none of these were functional in transient transfection assays with chimeric promoter/reporter constructs. Thus, either the responsible PPRE is located more remotely or the PPARα-dependent regulation of the *Abcd2* involves a different mechanism (Fourcade *et al.*, 2001). An earlier study showed that the human and murine *ABCD2* genes share more than 500 bp of conserved 5′ flanking sequence with potential Sp1- and AP-2-binding sites but no TATAA box. Moreover, in transient transfection assays with chimeric promoter/reporter constructs, 1.3 kb of the 5′-flanking region of human and murine *ABCD2* genes was shown to be necessary for upregulation by 9-*cis*-retinoic acid and forskolin, while no effect of PPARα could be detected (Pujol *et al.*, 2000).

PMP70, encoded by *ABCD3*, is abundantly and widely expressed and, like *ABCD2*, is induced in mammalian liver following the administration of fibrates (Kamijo *et al.*, 1990). A rat *Abcd3* cDNA was initially identified by screening an expression library (Kamijo *et al.*, 1990) and the sequence of the rat cDNA was used, in turn, to clone the human PMP70 cDNA and structural gene (Gärtner *et al.*, 1992, 1998). Human *ABCD3* maps to chromosome 1p21–p22 (Gärtner *et al.*, 1993) and comprises 23 exons distributed over 65 kb of genomic DNA **(Table 24.1)**. The proximal promoter region of human and murine *ABCD3* genes have a high GC content and multiple consensus Sp1-binding sites, features consistent with its broad tissue expression (Gärtner *et al.*, 1998). Several PPRE-like sequences with the correct spacing are present in the 5′ flanking sequence of *ABCD3* (Berger *et al.*, 1999; Fourcade *et al.*, 2001). The organization of *ABCD3* differs from *ABCD1* and *ABCD2* genes with only 2 of 22 introns falling in positions corresponding to *ABCD1* introns. This observation plus the fact that the first exon of *ABCD1* is exceptionally large (⩾1286 bp) lead to the speculation that the modern *ABCD1* gene may have arisen from an ancient retrotransposition of a cDNA derived from a *ABCD3*-like gene followed by intron acquisition in the 3′ half of the gene (Gärtner *et al.*, 1998).

P70R, also known as PMP69, is encoded by *ABCD4*. This gene maps to chromosome 14q24.3, covers 16 kb and has 19 exons (Holzinger *et al.*, 1998; Shani *et al.*, 1997). The position of several *ABCD4* introns corresponds to those in *ABCD3*, consistent with the suggestion, based on sequence similarity, that these two genes are more closely related to each other than to *ABCD1* and *2*. Also, as in *ABCD3*, the 5′ flanking sequence of *ABCD4* has a high GC content, contains several consensus Sp1-binding sites and lacks a TATAA box. No PPREs were identified in the 5′ 1.8 kb. Variant *ABCD4* transcripts result from the use of alternative polyadenylation sites and alternative exon splicing events including one that confers an alternative C-terminus (Holzinger *et al.*, 1997b, 1998; Shani *et al.*, 1997).

EXPRESSION PATTERNS

Northern blot studies of RNA harvested from adult mice have shown different expression patterns for the four peroxisomal ABC transporters (Albet *et al.*, 1997; Berger *et al.*, 1999). *Abcd1* is mainly expressed in heart, lung, intestine and spleen; *Abcd2* in skeletal muscle and brain; *Abcd3* in all tissues studied with greatest abundance in liver and kidney; while *Abcd4* mRNA is 10-fold more abundant in kidney than in other tissues analyzed.

The expression of the four peroxisomal ABC transporter genes is also regulated differentially during mouse brain development. *Abcd1* mRNA is most abundant in embryonic brain and gradually decreases during maturation; *Abcd2* and *Abcd4* mRNA accumulates in the early postnatal period; and *Abcd3* transcripts increase during the second and third postnatal weeks (Albet *et al.*, 1997; Berger *et al.*, 1999). Similarly, *in situ* hybridization studies in rat brain showed different spatial and temporal expression patterns of *Abcd1* and *Abcd3* during postnatal development (Pollard *et al.*, 1995). *Abcd3* expression was low at birth and increased to a peak between the second and third week in hippocampus and cerebellum. By contrast, *Abcd1* expression was maximal at birth in all areas of the brain and decreased thereafter (Pollard *et al.*, 1995). Administration of fenofibrate strongly increased the expression of the *Abcd2* and *Abcd3* in rat intestine and liver, respectively, but not the expression of

Abcd1 (Albet *et al.*, 1997). These observations suggest that transcriptional regulation is an important variable in the expression of the peroxisomal half ABC transporters and must be taken into account when considering possible *in vivo* heterodimerization partners.

Assembly of the four half ABC transporters

All known peroxisomal ABC transporters are half transporters that must dimerize to be functional. The partially overlapping patterns of developmental and tissue expression suggest that both hetero- and homodimerization are possible. Using a yeast two-hybrid system, Liu *et al.* found hetero- and homodimerization of the C-terminal halves of ALDP, ALDR and PMP70 (Liu *et al.*, 1999). P70R was not tested. Two mutations in ALDP (P484R and R591Q) known to cause X-ALD impaired both hetero- and homodimerization. These results were supported by co-immunoprecipitation experiments showing homodimerization of ALDP and heterodimerization of ALDP with either ALDR or PMP70. ALDR also heterodimerized with PMP70. Formation of ALDP homodimers and ALDP/PMP70 heterodimers was also demonstrated by co-immunoprecipitation of *in vitro* synthesized proteins (Smith *et al.*, 1999). Considering the protein interaction and expression studies, it seems likely that both homodimerization and heterodimerization of certain peroxisomal half ABC transporters takes place and that this may vary from tissue to tissue. The functional consequences of the choice of partners are not known.

Targeting to peroxisomal membrane

As discussed above, our understanding of how PMPs achieve their proper and specific location in the peroxisomal membrane is not well understood. Nevertheless, recent work suggests that PEX19, a farnesylated protein located primarily in the cytosol, appears to have a major role in PMP import. PEX19 interacts *in vitro* with numerous PMPs including ALDP, PMP70 and ALDR (Gloeckner *et al.*, 2000; Sacksteder *et al.*, 2000; Snyder *et al.*, 2000). More specifically, a region of ALDP located towards the N-terminus (between amino acids 67–186) has been shown to be important for proper peroxisomal targeting and an overlapping fragment (residues 1–203) interacts in a two-hybrid system with PEX19 (Gloeckner *et al.*, 2000). Similarly, expressing a series of N-terminal and C-terminal deletions of PMP70 tagged with a C-terminal green fluorescent protein (GFP) in Chinese hamster ovary (CHO) cells, we localized a peroxisomal membrane targeting signal to the N-terminal 80 amino acids of PMP70 as well as a second site influencing targeting in the C-terminal 100 amino acids (Almashanu and Valle, unpublished observations). Additionally, we made a chimeric protein with the 183 N-terminal residues of PMP70 followed by the 428 C-terminal residues of its *Escherichia coli* homologue (YDDA) tagged with GFP. When expressed in CHO cells, the chimeric PMP70/YDDA-GFP localized to peroxisomes while YDDA-GFP alone showed a non-peroxisomal pattern. Mutagenesis studies of conserved residues in the N-terminal 80 amino acids of PMP70 failed to detect a discrete targeting sequence. This suggests that recognition of PMP70 by the targeting apparatus depends on more general secondary structure features rather than on specific residues.

Topology

By analogy to other ABC transporters, the N-terminal hydrophobic half of the peroxisomal ABC transporters is expected to have six TM segments. However, the location and number of these hydrophobic segments has been difficult to establish with certainty (see also Chapter 2). Analyses with multiple protein motif prediction algorithms failed to identify six unequivocal TM segments (TMS) in any of the peroxisomal half ABC transporters **(Figure 24.2)**. Moreover, there was some variation in the location of the predicted TMS. In **Figure 24.2**, we present the TMSs predicted by several algorithms for 11 PMP70 homologues, from bacteria to mammals. The location of five TMSs is clear; the position of the sixth is less certain.

The initial description of rat PMP70 by Kamijo *et al.* included a protease sensitivity study indicating that the C-terminal hydrophobic half of PMP70 faces the cytosol (Kamijo *et al.*, 1990). Similarly, immunohistochemical studies with antibodies directed at specific segments of ALDP indicated that the C-terminal half of the protein projects into the cytosol (Contreras *et al.*, 1996). Additionally, protease treatment followed by immunoblot analysis showed that the C-terminal segment of ALDP could be released from the surface of intact rat liver

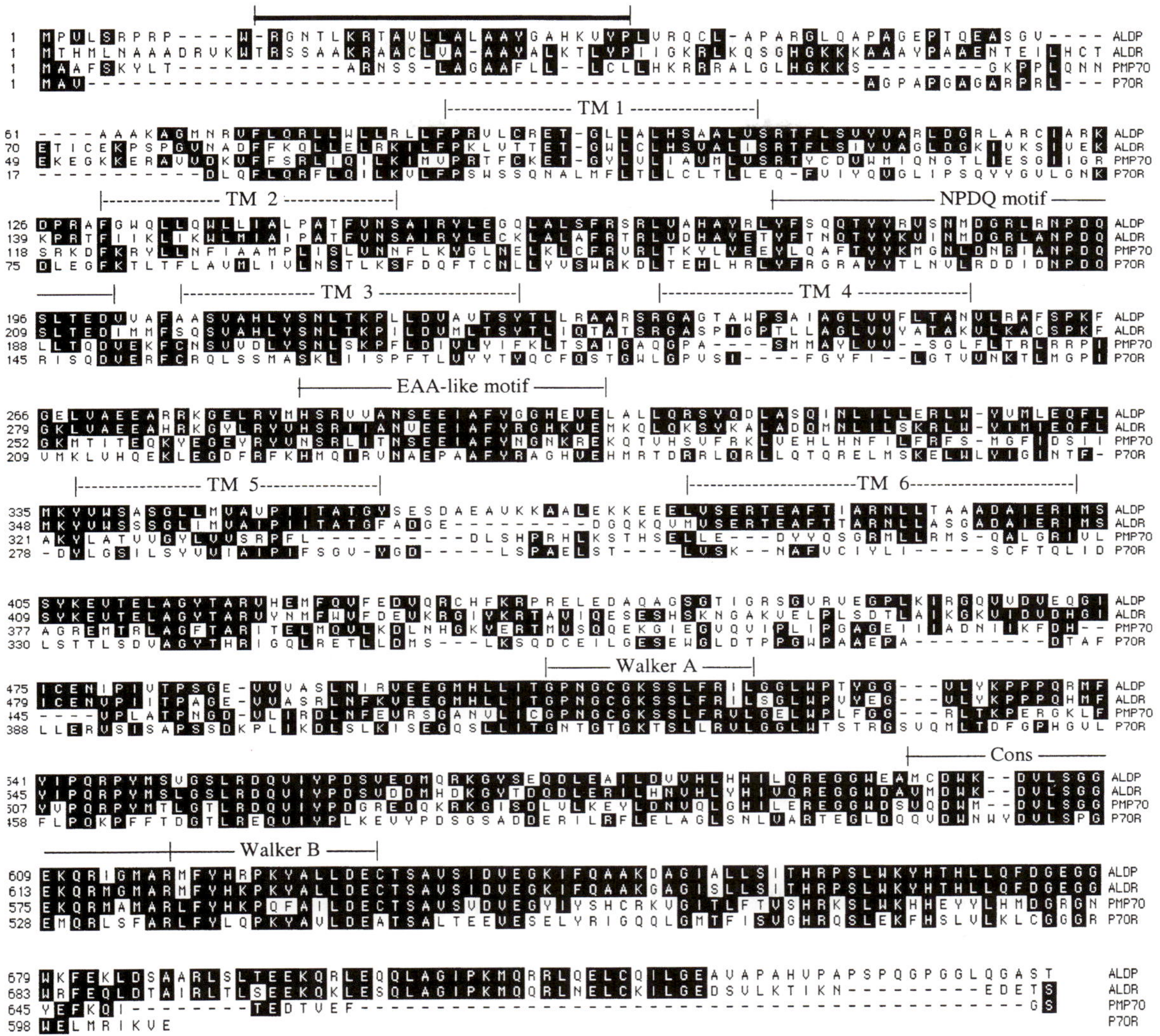

Figure 24.2. Alignment of the human peroxisomal ABC transporters. Amino acid sequences were aligned using the MegAlign program (DNASTAR Inc.). Identical residues in two or more polypeptides are boxed in black. The indicated transmembrane segments were predicted using the transmembrane region detection programs available at http://www.us.expasy.org. The labeled solid overlines designate the EAA-like motif and the NPDQ motif (see text). The heavy solid overline designates an N-terminal helical hydrophobic region of unknown function present in ALDP, ALDR and PMP70.

peroxisomes and retain its ability to bind ATP *in vitro* (Contreras *et al.*, 1996). We are unaware of any topology studies of ALDR or P70R.

NON-MAMMALIAN HOMOLOGUES OF THE PEROXISOMAL MEMBRANE ABC TRANSPORTER

Homologues from several non-mammalian species have been identified (Smith *et al.*, 1999); those from *S. cerevisiae* have been well characterized.

PXA1/PAL1/PAT2 (Hettema *et al.*, 1996; Shani *et al.*, 1995; Swartzman *et al.*, 1996) and *PXA2/PAT1* (Hettema *et al.*, 1996; Shani and Valle, 1996) are the two yeast homologues of the mammalian half ABC transporters. Taking advantage of the induction of PMPs and peroxisomal matrix proteins by growth on oleic acid as a sole carbon source, Shani *et al.* used degenerate PCR primers corresponding to the conserved Walker A and Walker B sequences of *ABCD1* and *ABCD3* to clone *PXA1*. The conceptual protein product Pxa1p has 758 amino acids, a predicted molecular mass of 87 kDa and is slightly more similar to ALDP than to PMP70 (Shani *et al.*, 1995). Swartzman *et al.* used an alternate strategy to clone a cDNA identical with *PXA1* over the 758 C-terminal amino acids but with an additional 112 N-terminus codons with a predicted protein mass of 100 kDa (Swartzman *et al.*, 1996). The shorter Pxa1p is clearly functional in that it rescues growth on oleic acid in mutant yeast lacking *PXA1* (Shani *et al.*, 1995). Functional studies with Pal1p were not reported and the N-terminal extension of Pal1p is not homologous with the peroxisomal ABC transporters of other species. Thus, the significance of this N-terminal extension is uncertain.

PXA2/PAT1, originally identified as YKL741, encodes a half ABC transporter with highest similarity to mammalian PMP70 and ALDP. Shani *et al.* showed that the combined disruption of *PXA1* and *PXA2* gave a growth phenotype identical to that of either single disruption (Shani and Valle, 1996) and that the stability of Pxa1p was reduced in yeast with a *PXA2* deletion. Finally, in co-immunoprecipitation studies with either Pxa1p or Pxa2p, they showed direct physical evidence for the heterodimerization of the two proteins to assemble a full peroxisomal half ABC transporter. Thus, the functional transporter in yeast is a heterodimer of Pxa1p/Pxa2p.

CONSERVED SEQUENCE MOTIFS

Availability of sequence information of peroxisomal ABC transporters from yeast and mammals provides the opportunity to look for conserved sequence motifs of possible functional significance. In addition to the TM segments and the Walker A and B and C segments of the NBD, at least two additional conserved motifs can be identified in the peroxisomal ABC transporters. The first, the EAA-like motif, is a 15 amino acid sequence (- N S E E I A F Y X G X K X E X where, in a comparison of Pxa1p, Pxa2p, PMP70 and ALDP, the designated residues are present in three out of four and the underlined residues are present in all four proteins). The EAA-like motif is present between TMS 4 and 5 and resembles the central core of a 30-residue sequence (the EAA motif) found in a similar location in many prokaryotic ABC transporters (Saurin *et al.*, 1994; Shani *et al.*, 1996b). Mutagenesis studies in Pxa1 showed that conservative missense mutations at E294 and G301 reduce the function of Pxa1p but do not alter stability or targeting (Shani *et al.*, 1996a). Recent mutagenesis and crosslinking studies of the prokaryotic EAA motif suggest that this sequence may interact with certain regions of the ATP-binding cassette (Hunke *et al.*, 2000; Mourez *et al.*, 1997). Photoaffinity labeling and mass spectrometry study of the human P-glycoprotein (Pgp, ABCB1) showed that its EAA-like motif (cytoplasmic loop 2) was one of nine tryptic peptides that bind to photoaffinity-labeled substrate analogues (Ecker *et al.*, 2002). These results suggest that the EAA-like motif participates in substrate binding/translocation or in the interaction of these processes with ATP binding/hydrolysis.

The second conserved motif, which we designate the NPDQ motif **(see Figure 24.3 for sequence consensus)**, is located between TMS 2 and 3. Preliminary studies replacing one, two or four of the residues in the conserved NPDQ core with alanines did not affect targeting (Almashanu and Valle, unpublished). We could not identify this motif in other categories of ABC transporters including other mammalian half ABC transporters like the TAP proteins. Therefore, we considered that it might be specific for the peroxisomal ABC transporters and used it to search for the subset of ABC transporters with homology to the known peroxisome members of the superfamily. Interestingly, our results suggest that the NPDQ motif is a specific signature for peroxisomal transporters among eukaryotes but that it is also present in a small subset of prokaryotic transporters (see section on evolution, below).

FUNCTION OF THE PEROXISOME ABC TRANSPORTERS

The spectrum of physiological ligands and the direction of transport is not known with certainty for any of the mammalian peroxisomal half ABC transporters. A variety of studies suggest, however, that one or more of the peroxisomal ABC transporters transport straight or branched LCFA or VLCFA or their acyl-CoA derivatives into the peroxisome. Two important variables in the transport of fatty acids across membranes are the chain length and the site of activation of the fatty acid to its acyl-CoA derivative. The latter is accomplished by acyl-CoA synthetases that differ in chain length specificity and subcellular location. In general, the longer the chain length, the more difficult it becomes for fatty acids to move across the peroxisomal membrane; medium-chain fatty acids do not require transporters while LCFAs do. Activation of fatty acids makes them more polar and impairs movement across the peroxisomal membrane (Hettema and Tabak, 2000).

The most detailed studies of the functions of the peroxisomal ABC transporters have utilized *S. cerevisiae* as a model system. In this regard, yeast offer several advantages over mammalian cells for the study of peroxisome biogenesis and function (Kunau *et al.*, 1993). In contrast to mammalian peroxisomes, the peroxisomes of yeast are more easily isolated and their components more easily induced. Moreover, fatty acid β-oxidation in yeast is limited to peroxisomes, while in mammalian cells, β-oxidation of fatty acids occurs in both peroxisomes and mitochondria. In *S. cerevisiae,* Pxa1p and Pxa2p heterodimerize to form the functional peroxisome ABC transporter that is essential for growth on LCFAs, especially oleic acid, $C_{18:1}$, as a sole carbon source (Hettema *et al.*, 1996; Shani and Valle, 1996). β-Oxidation of LCFA in intact cells deleted for either the *PXA1* or *PXA2* gene is reduced to approximately 20% of the wild-type level. In detergent lysates of these same mutant yeasts, β-oxidation of LCFA is unaffected, indicating that the peroxisome membrane is a barrier in the intact mutant yeast (Hettema and Tabak, 2000; Hettema *et al.*, 1996). Using protoplasts in which the plasma membrane has been selectively permeabilized by digitonin, it was shown that $C_{18:1}$-CoA, but not $C_{8:0}$-CoA, enters peroxisomes in a Pxa1p/Pxa2p and ATP-dependent process (Hettema and Tabak, 2000; Verleur *et al.*, 1997). This result is consistent with the observation that the acyl-CoA synthetase with activity towards long-chain fats is extraperoxisomal in yeast (Hettema *et al.*, 1996). Thus, the available evidence indicates that the yeast PXA transporter is necessary for the transport of LCF acyl-CoAs into the peroxisome.

Our understanding of peroxisomal ABC transporter function in mammalian cells derives largely from observations made in cells or tissues with mutations in one or more of the transporters. Mutations in *ABCD1* cause X-ALD, a neurodegenerative disorder with a highly variable clinical phenotype (Moser *et al.*, 2001). Biochemically, X-ALD patients accumulate VLCFA in plasma and tissues and exhibit deficient VLCFA β-oxidation with decreased activity of the peroxisomal VLCFA acyl-CoA synthetase (VLCS) that activates VLCFAs to their CoA thioesters (Moser *et al.*, 2001). The latter is thought to be localized on the matrix side of the peroxisome membrane (Lazo *et al.*, 1990; Steinberg *et al.*, 1999b). Functional interaction between ALDP and peroxisomal VLCS is also implied by the observation of a synergistic effect on VLCFA β-oxidation when both VLCS and ALDP were overexpressed in humans and mouse fibroblasts (Steinberg *et al.*, 1999a; Yamada *et al.*, 1999). Several hypotheses regarding the functional relationship between VLCS and ALDP and their role in the pathogenesis of X-ALD have been proposed but the mechanism remains obscure. A relationship between fatty acid β-oxidation and PMP70 has also been suggested. Imanaka *et al.* (1999) demonstrated a two- to threefold increase in β-oxidation of palmitic acid ($C_{16:0}$) in CHO cell line overexpressing PMP70, whereas the oxidation of lignoceric acid ($C_{24:0}$) decreased about 30–40%. In summary, current data suggest that ALDP and PMP70 are involved in the transport of LCFA and VLCFA or their CoA derivatives across the peroxisomal membrane.

Creation of mouse models, each lacking one of the four peroxisomal half ABC transporters, also has the promise of providing functional insight. Targeted knockouts for the genes encoding ALDP, ALDR and PMP70 have been produced (Forss-Petter *et al.*, 1997; Lu *et al.*, 1997; Yamada *et al.*, 2000). The X-ALD mouse model has some of the human biochemical features including high levels of VLCFA in brain and adrenal gland. However, the mice appear to lack the neurological phenotype of humans as they do not develop symptoms or evidence of cerebral or spinal cord demyelination by up to 2 years of age. The phenotypes of both the ALDR and the PMP70 knockout mice do not correspond to a recognizable human disease. Mice lacking PMP70 have impaired metabolism of very long branched-chain fatty acids including pristanic acid, phytanic acid and bile acid precursors (Jimenez-Sanchez *et al.*, 2000).

Biochemical studies of purified peroxisomal ABC transporters in reconstituted lipid vesicles should help to clarify our understanding of their function but have not yet been described.

EVOLUTION OF THE PEROXISOMAL HALF ABC TRANSPORTERS

The superfamily of ABC proteins is large and diverse. The availability of whole genome sequences from an increasing number of organisms has led to the identification of many new members and to a better understanding of the set of ABC transporters characteristic of each species. Taking advantage of this sequence information, we searched for a specific signature sequence for the ABCD subfamily that would identify all known peroxisomal ABC

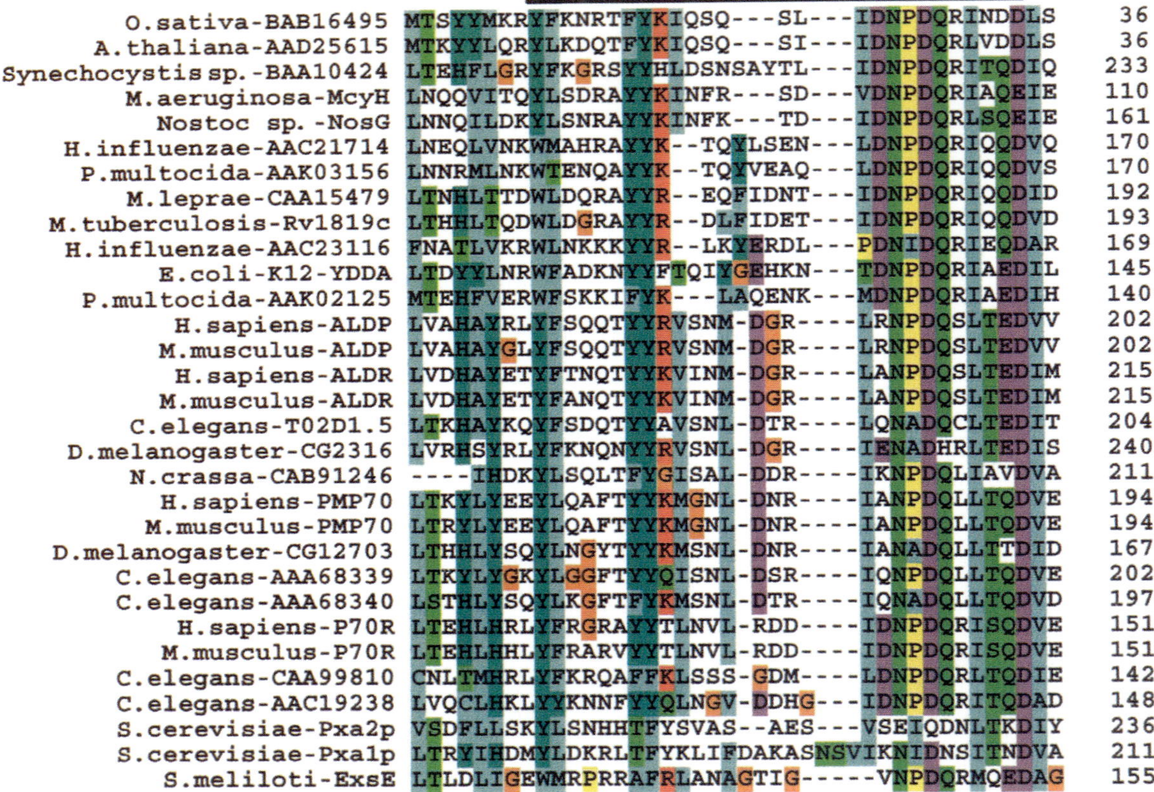

Figure 24.3. *Sequence alignment of the NPDQ motif from 30 homologues of human ALDP. We used ClustalX 1.81 to align the protein sequences listed in Table 24.2. Color code for residues: G, brown; P, yellow; conserved K, R, red; conserved D, E, purple; conserved neutrals, green, blue, teal. Residue number is indicated on the right.*

transporters in different species. We used a short segment of a highly conserved motif located at the second predicted loop of the mammalian peroxisomal half ABC transporters (the NPDQ motif; **Figure 24.3**) and used it to perform BLAST searches of different NCBI databases. This specifically identified all the known peroxisomal half ABC transporters as well as their apparent orthologues whose subcellular localization is yet to be determined in other eukaryotes. Additionally this search identified a subset of prokaryotic half ABC transporters. Interestingly, these prokaryotic transporters are all half ABC transporters, each is encoded by a single gene without subdivision into an operon. We used more than 30 of the hits **(Table 24.2)** to run the ClustalX algorithm, producing a multiple sequence alignment, and we analyzed this comparison to generate a phylogenetic relationship by maximum parsimony **(Figure 24.4)**. This analysis indicates that among the four mammalian peroxisomal half ABC proteins P70R represents the ancestral gene, more closely related to the bacterial and the plant homologues. Pxa1p and Pxa2p, the two *S. cerevisiae* homologues, are more divergent from the ancestral gene and closer to the PMP70/ALDP branch. Interestingly, two of the *Caenorhabiditis elegans* homologues are closely related to P70R, two to PMP70 and only one to ALDP/ALDR. This supports the hypothesis that genes encoding these two transporters diverged relatively recently, a hypothesis that is also supported by the genomic organization of these two genes (see section on genes, above). In addition, the genomic organization of *ABCD3* gene, when compared to that of *ABCD1* and *ABCD2*, suggests that the modern *ABCD1* gene may have arisen from an ancient retrotransposition event followed by intron acquisition (Gärtner *et al.*, 1998). The prokaryotic homologues identified in this search number one per species but *Haemophilus influenzae* Rd and possibly *E. coli* have two.

A separate and additional aspect of the evolutionary history of the human *ABCD1* gene is the presence of four autosomal *ABCD1* pseudogenes

TABLE 24.2. HOMOLOGUES FOR THE HUMAN PEROXISOMAL ABC TRANSPORTERS

The human ALDP NPDQ motif (Figure 24.3) was used as the query for the PSI- and PHI-blast algorithms, National Center for Biotechnology Information (http://www.ncbi.nlm.nih.gov/BLAST/) and for the CMR-blast at TIGR (The Institute for Genomic Research) website (http://tigrblast.tigr.org/cmr-blast).

Species	Gene/Protein name	Number of amino acids	Protein accession #	PID
Homo sapiens[a]	ALDp	745	NP_000024	7262393
Homo sapiens[a]	ALDRp	740	NP_005155	9945308
Homo sapiens[a]	PMP70p	659	NP_002849	4506341
Homo sapiens[a]	P70Rp	606	NP_005041	4826958
Mus musculus[a]	ALDp	736	NP_031461	6671497
Mus musculus[a]	ALDRp	741	NP_036124	6752942
Mus musculus[a]	PMP70p	659	NP_033017	6680612
Mus musculus	P70Rp	606	NP_033018	6680614
Drosophila melanogaster	CG2316	730	AAF59365	7304333
Drosophila melanogaster	CG12703	618	AAF49018	7293647
Caenorhabditis elegans	C54G10.3	660	CAA99810	3875241
Caenorhabditis elegans	T10H9.5	598	AAC19238	3193212
Caenorhabditis elegans	C44B7.8	665	AAA68339	861307
Caenorhabditis elegans	C44B7.9	661	AAA68340	861308
Caenorhabditis elegans	T02D1.5	734	T24357	7506941
Neurospora crassa	B17C10.260	700	CAB91246	7800888
Saccharomyces cerevisiae[a]	Pxalp	758	AAC49009	619668
Saccharomyces cerevisiae[a]	Pxa2p	853	NP_012733	6322660
Oryza sativa		514	BAB16495	10800075
Arabidopsis thaliana		514	AAD25615	4585979
Arabidopsis thaliana		1383	CAB38898	4490736
Microcystis aeruginosa	McyH	538	AAF00956	6007549
Nostoc sp.	NosG	663	AAF17285	6563404
Synechocystis sp.		661	BAA10424	1001688
Haemophilus influenzae Rd		592	AAC21714	1572982
Haemophilus influenzae Rd		589	AAC23116	1574308
Mycobacterium leprae		638	CAA15479	2578388
Mycobacterium tuberculosis		639	CAB01460	1483552
Escherichia coli	YDDA	561	AAC74569	1787772
Sinorhizobium meliloti	ExsE	606	CAA12533	2808503
Pasteurella multocida	PM1072	585	AAK03156	12721409
Pasteurella multocida	PM0041	564	AAK02125	12720245

[a]Experimental evidence for peroxisomal localization.

identified using the 3′ exons as a hybridization probe (Sarde *et al.*, 1994). These were mapped to pericentric regions of chromosomes 2, 10, 16 and 22. Sarde *et al.* proposed interchromosomal duplications of a genomic segment containing the *ABCD1* locus as the mechanism leading to these pericentromeric pseudogenes. The breakpoint sequences and phylogenetic analysis of the duplicated segments predicted a two-step transposition model in which a duplication from Xq28 to pericentromeric 2p11 occurred once followed by a rapid distribution of a large duplication cassette among the other pericentromeric regions, 10p11, 16p11, and 22p11 (Eichler *et al.*, 1997). Fluorescence *in situ* hybridization (FISH) analysis using cloned genomic fragments of the autosomal pseudogenes identified two additional paralogs on chromosomes 1 and 20, which may represent more divergent sequences arising from an earlier duplication (Smith *et al.*,

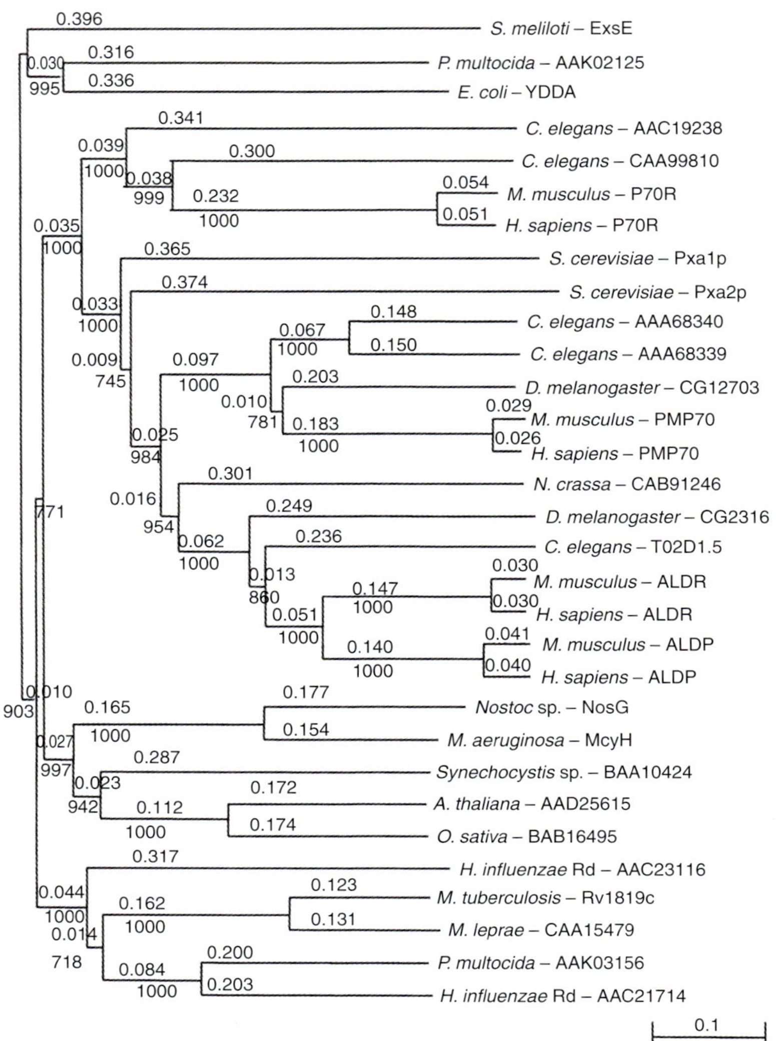

Figure 24.4. ***Unrooted phylogenetic tree of ALDP homologues. We used the ClustalX 1.81 alignment of the protein sequences listed in Table 24.2 as the input for calculating the tree, and the neighbor-joining method to calculate the distances (percent divergence) indicated above each line and the bootstrap values (1000 replicates) below each line. The scale at the bottom relates the length of the branch to the number of substitutions.***

1999). Southern blot analysis of murine and primate genomic DNA indicated that there might have been two duplication events of the X-ALD locus in higher primates, supporting the two-step transposition model. An initial expansion appears to have occurred on the evolutionary line leading to orang utans with a subsequent or independent expansion in the great apes (Braun *et al.*, 1996; Smith *et al.*, 1999).

ACKNOWLEDGMENTS

We thank Amir Rattner for help with the phylogenetic analysis, and Sandy Muscelli for help with manuscript preparation. A portion of this work was supported by a grant from NICHD (2PO1HD 10981) (D.V.). D.V. is an Investigator and S.A. is a Research Specialist in the Howard Hughes Medical Institute.

REFERENCES

Albet, S., Causeret, C., Bentejac, M., Mandel, J.-L., Aubourg, P. and Maurice, B. (1997) Fenofibrate differently alters expression of genes encoding ATP-binding transporter proteins of the peroxisomal membrane. FEBS Lett. 405, 394–397.

Berger, J., Albet, S., Bentejac, M., Netik, A., Holzinger, A., Roscher, A.A., Bugaut, M. and Forss-Petter, S. (1999) The four murine peroxisomal ABC-transporter genes differ in constitutive, inducible and developmental expression. Eur. J. Biochem. 265, 719–727.

Braun, A., Kammerer, S., Ambach, H. and Roscher, A.A. (1996) Characterization of a partial pseudogene homologous to the adrenoleukodystrophy gene and application of mutation detection. Hum. Mutat. 7, 105–108.

Braverman, N., Steel, G., Obie, C., Moser, A., Moser, H., Gould, S.J. and Valle, D. (1997) Human *PEX7* encodes the peroxisomal PTS2 receptor and is responsible for rhizomelic chondrodysplasia punctata. Nat. Genet. 15, 369–376.

Braverman, N., Dodt, G., Gould, S.J. and Valle, D. (1998) An isoform of Pex5p, the human PTS1 receptor, is required for the import of PTS2 proteins into peroxisomes. Hum. Mol. Genet. 7, 1195–1205.

Brosius, U., Dehmel, T. and Gärtner, J. (2002) Two different targeting signals direct human peroxisomal membrane protein 22 to peroxisomes. J. Biol. Chem. 277, 774–784.

Chang, C.C., South, S., Warren, D., Jones, J., Moser, A.B., Moser, H.W. and Gould, S.J. (1999) Metabolic control of peroxisome abundance. J. Cell Sci. 112, 1579–1590.

Contreras, M., Sengupta, T.K., Sheikh, F., Aubourg, P. and Singh, I. (1996) Topology of ATP-binding domain of adrenoleukodystrophy gene product in peroxisomes. Arch. Biochem. Biophys. 334, 369–379.

Dansen, T.B., Wirtz, K.W., Wanders, R.J. and Pap, E.H. (2000) Peroxisomes in human fibroblasts have a basic pH. Nat. Cell Biol. 2, 51–53.

Distel, B., Erdmann, R., Gould, S.J., Blobel, G., Crane, D.I., Cregg, J.M., *et al.* (1996) A unified nomenclature for peroxisome biogenesis factors. J. Cell Biol. 135, 1–3.

Dodt, G. and Gould, S.J. (1996) Multiple PEX genes are required for proper subcellular distribution and stability of Pex5p, the PTS1 receptor: evidence that PTS1 protein import is mediated by a cycling receptor. J. Cell Biol. 135, 1763–1774.

Dodt, G., Braverman, N., Wong, C., Moser, A., Moser, H.W., Watkins, P., Valle, D. and Gould, S.J. (1995) Mutations in the PTS1 receptor gene, PXR1, define complementation group 2 of the peroxisome biogenesis disorders. Nat. Genet. 9, 115–124.

Dyer, J.M., McNew, J.A. and Goodman, J.M. (1996) The sorting sequence of the peroxisomal integral membrane protein PMP47 is contained within a short hydrophilic loop. J. Cell Biol. 133, 269–280.

Ecker, G.F., Csaszar, E., Kopp, S., Plagens, B., Holzer, W.E., Ernst, W. and Chiba, P. (2002) Identification of ligand-binding regions of P-glycoprotein by activated-pharmacophore photoaffinity labeling and matrix-assisted laser desorption/ionization time-of-flight mass spectrometry. Mol. Pharmacol. 61, 637–648.

Eichler, E.E., Budarf, M.L., Rocchi, M., Deaven, L.L., Doggett, N.A., Baldini, A., Nelson, D.L. and Mohrenweiser, H.W. (1997) Interchromosomal duplications of the adrenoleukodystrophy locus: a phenomenon of pericentromeric plasticity. Hum. Mol. Genet. 6, 991–1002.

Forss-Petter, S., Werner, H., Berger, J., Lassmann, H., Molzer, B., Schwab, M.H., Bernheimer, H., Zimmermann, F. and Nave, K.A. (1997) Targeted inactivation of the X-linked adrenoleukodystrophy gene in mice. J. Neurosci. Res. 50, 829–843.

Fourcade, S., Savary, S., Albet, S., Gauthe, D., Gondcaille, C., Pineau, T., *et al.* (2001) Fibrate induction of the adrenoleukodystrophy-related gene (ABCD2) promoter analysis and role of the peroxisome proliferator-activated receptor PPARalpha. Eur. J. Biochem. 268, 3490–3500.

Fransen, M., Brees, C., Baumgart, E., Vanhooren, J.C.T., Baes, M., Mannaerts, G.P. and Van Veldhoven, P.P. (1995) Identification and characterization of the putative human peroxisomal C-terminal targeting signal import receptor. J. Biol. Chem. 270, 7731–7736.

Gärtner, J., Moser, H. and Valle, D. (1992) Mutations in the 70 kD peroxisomal membrane protein gene in Zellweger syndrome. Nat. Genet. 1, 16–23.

Gärtner, J., Kearns, W., Rosenberg, C., Pearson, P., Copeland, N.G., Gilbert, D.J., Jenkins, N. and Valle, D. (1993) Localization of the 70 kD peroxisomal membrane protein to human 1p21-22 and mouse 3. Genomics 15, 412–414.

Gärtner, J., Jimenez-Sanchez, G., Roerig, P. and Valle, D. (1998) Genomic organization of the 70-kDa peroxisomal membrane protein gene (PXMP1). Genomics 48, 203–208.

Gatto, G.J., Jr, Geisbrecht, B.V., Gould, S.J. and Berg, J.M. (2000a) A proposed model for the PEX5-peroxisomal targeting signal-1 recognition complex. Proteins 38, 241–246.

Gatto, G.J., Jr, Geisbrecht, B.V., Gould, S.J. and Berg, J.M. (2000b) Peroxisomal targeting signal-1 recognition by the TPR domains of human PEX5. Nat. Struct. Biol. 7, 1091–1095.

Ghaedi, K., Honsho, M., Shimozawa, N., Suzuki, Y., Kondo, N. and Fujiki, Y. (2000) PEX3 is the causal gene responsible for peroxisome membrane assembly-defective Zellweger syndrome of complementation group G. Am. J. Hum. Genet. 67, 976–981.

Gloeckner, C.J., Mayerhofer, P.U., Landgraf, P., Muntau, A.C., Holzinger, A., Gerber, J.-K., Kammerer, S., Adamski, J. and Roscher, A.A. (2000) Human adrenoleukodystrophy protein and related peroxisomal ABC transporters interact with the peroxisomal assembly protein PEX19p. Biochem. Biophys. Res. Commun. 271, 144–150.

Gotte, K., Girzalsky, W., Linkert, M., Baumgart, E., Kammerer, S., Kunau, W.H. and Erdmann, R. (1998) Pex19p, a farnesylated protein essential for peroxisome biogenesis. Mol. Cell Biol. 18, 616–628.

Gould, S.J. and Valle, D. (2000a) The genetics and cell biology of the peroxisome biogenesis disorders. Trends Genet. 16, 340–344.

Gould, S.J. and Valle, D. (2000b) Peroxisome biogenesis disorders: genetics and cell biology. TIGS 16, 340–344.

Gould, S.J., Raymond, G.V. and Valle, D. (2001) The peroxisome biogenesis disorders. In: The Metabolic and Molecular Basis of Inherited Disease (ed. C.R. Scriver, A.L. Beaudet, W.S. Sly and D. Valle), pp. 3181–3217. New York: McGraw-Hill.

Hettema, E.H. and Tabak, H.F. (2000) Transport of fatty acids and metabolites across the peroxisomal membrane. Biochim. Biophys. Acta 1486, 18–27.

Hettema, E.H., van Roermund, C.W.T., Distel, B., van den Berg, M., Vilela, C., Rodrigues-Pousada, C., Wanders, R.J.A. and Tabak, H.F. (1996) The ABC transporter proteins Pat1 and Pat2 are required for import of long-chain fatty acids into peroxisomes of *Saccharomyces cerevisiae*. EMBO J. 15, 3813–3822.

Hettema, E.H., Girzalsky, W., van Den Berg, M., Erdmann, R. and Distel, B. (2000) *Saccharomyces cerevisiae* pex3p and pex19p are required for proper localization and stability of peroxisomal membrane proteins. EMBO J. 19, 223–233.

Holzinger, A., Kammerer, S., Berger, J. and Roscher, A.A. (1997a) cDNA cloning and mRNA expression of the human adrenoleukodystrophy related protein (ALDP), a peroxisomal ABC transporter. Biochem. Biophys. Res. Commun. 239, 261–264.

Holzinger, A., Kammerer, S. and Roscher, A.A. (1997b) Primary structure of human PMP69, a putative peroxisomal ABC-transporter. Biochem. Biophys. Res. Commun. 237, 152–157.

Holzinger, A., Roscher, A.A., Landgraf, P., Lichtner, P. and Kammerer, S. (1998) Genomic organization and chromosomal localization of the human peroxisomal membrane protein-1-like protein (PXMP1-L) gene encoding a peroxisomal ABC transporter. FEBS Lett. 426, 238–242.

Honsho, M., Tamura, S., Shimozawa, N., Suzuki, Y., Kondo, N. and Fujiki, Y. (1998) Mutation in PEX16 is causal in the peroxisome-deficient Zellweger syndrome of complementation group D. Am. J. Hum. Genet. 63, 1622–1630.

Hunke, S., Mourez, M., Jéhanno, M., Dassa, E. and Schneider, E. (2000) ATP modulates subunit-subunit interactions in an ATP-binding cassette transporter ($MalFGK_2$) determined by site-directed chemical cross-linking. J. Biol. Chem. 275, 15526–15534.

Imanaka, T., Aihara, K., Takano, T., Yamashita, A., Sato, R., Suzuki, Y., Yokota, S. and Osumi, T. (1999) Characterization of the 70 kDa peroxisomal membrane protein, an ATP binding cassette transporter. J. Biol. Chem. 274, 11968–11976.

James, G.L., Goldstein, J.L., Pathak, R.K.W., Anderson, R.G. and Brown, M.S. (1994) PxF, a prenylated protein of peroxisomes. J. Biol. Chem. 269, 14182–14190.

Jankowksi, A., Kim, J.H., Collins, R.F., Daneman, R., Walton, P. and Grinstein, S. (2001) *In situ* measurements of the pH of mammalian peroxisomes using the fluorescent protein pHluorin. J. Biol. Chem. 276, 48748–48753.

Jimenez-Sanchez, G., Hebron, K., Silva-Zolezzi, I., Mihalik, S.J., Watkins, P.A., Espeel, M., Moser, A., Thomas, G., Röels, F. and Valle, D. (2000) Fasting fuel homeostasis triggered by defective phytanic and pristanic acids metabolism in the 70 kDa peroxisomal protein deficient mice. Am. J. Hum. Genet. 67, A65.

Jones, J.M., Morrell, J.C. and Gould, S.J. (2001) Multiple distinct targeting signals in integral peroxisomal membrane proteins. J. Cell Biol. 153, 1141–1149.

Juge-Aubry, C., Pernin, A., Favez, T., Burger, A.G., Wahli, W., Meier, C.A. and Desvergnee, B. (1997) DNA binding properties of peroxisome proliferator-activated receptor subtypes on various natural peroxisome proliferator response elements. J. Biol. Chem. 272, 25252–25259.

Kamijo, K., Taketani, S., Yokata, S., Osumi, T. and Hashimoto, T. (1990) The 70 kDa peroxisomal membrane protein is a member of the Mdr (P-glycoprotein)-related ATP-binding protein superfamily. J. Biol. Chem. 265, 4534–4540.

Kemp, S., Wei, H.M., Lu, J.F., Braiterman, L.T., McGuinness, M.C., Moser, A.B., Watkins, P.A. and Smith, K.D. (1998) Gene redundancy and pharmacologic gene therapy: Implications for X-linked adrenoleukodystrophy. Nat. Med. 4, 1261–1268.

Kersten, S., Desvergne, B. and Wahli, W. (2000) Roles of PPARs in health and disease. Nature 405, 421–424.

Klein, A.T.J., Barnett, P., Bottger, G., Konings, D., Tabak, H.F. and Distel, B. (2001) Recognition of peroxisomal targeting signal type 1 by the import receptor Pex5p. J. Biol. Chem. 276, 15034–15041.

Kunau, W.H., Beyer, A., Franken, T., Gotte, K., Marzioch, M., Saidowsky, J., Skaletz-Rorowski, A. and Wiebel, F.F. (1993) Two complementary approaches to study peroxisome biogenesis in *Saccharomyces cerevisiae*: forward and reversed genetics. Biochimie 75, 209–224.

Lazo, O., Contreras, M. and Singh, I. (1990) Topographical localization of peroxisomal acyl-CoA ligases: differential localization of palmitoyl-CoA and lignoceroyl-CoA ligases. Biochemistry 29, 3981–3986.

Liu, L.-S., Janvier, K., Berteaux-Lecellier, V., Cartier, N., Benarous, R. and Aubourg, P. (1999) Homo- and heterodimerization of peroxisomal ATP-binding cassette half-transporters. J. Biol. Chem. 274, 32738–32743.

Lu, J.F., Lawler, A.M., Watkins, P.A., Powers, J.M., Moser, A.B., Moser, H.W. and Smith, K.D. (1997) A mouse model for X-linked adrenoleukodystrophy. Proc. Natl Acad. Sci. USA 94, 9366–9371.

Matsuzono, Y., Kinoshita, N., Tamura, S., Shimozawa, N., Hamasaki, M., Ghaedi, K., Wanders, R.J.A., Suzuki, Y., Kondo, N. and Fujiki, Y. (1999) Human *PEX19:* cDNA cloning by functional complementation, mutation analysis in a patient with Zellweger syndrome and potential role in peroxisomal membrane assembly. Proc. Natl Acad. Sci. USA 96, 2116–2121.

Moser, H.W., Smith, K.D., Watkins, P.A., Powers, J.M. and Moser, A.B. (2001) X-linked adrenoleukodystrophy. In: The Metabolic and Molecular Basis of Inherited Disease (ed. C.R. Scriver, A.L. Beaudet, W.S. Sly and D. Valle), pp. 3257–3302. New York: McGraw-Hill.

Mosser, J., Lutz, Y., Stoeckel, M.E., Sarde, C.O., Kretz, C., Douar, A.M., Lopez, J., Aubourg, P. and Mandel, J.L. (1994) The gene responsible for adrenoleukodystrophy encodes a peroxisomal membrane protein. Hum. Mol. Genet. 3, 265–271.

Motley, A.M., Hettema, E.H., Hogenhout, E.M., Brites, P., ten Asbroek, A.L.M.A., Wijburg, F.A., *et al.* (1997) Rhizomelic chondrodysplasia punctata is a peroxisomal protein targeting disease caused by a non-functional PTS2 receptor. Nat. Genet. 15, 377–380.

Mourez, M., Hofnung, M. and Dassa, E. (1997) Subunit interactions in ABC transporters: a conserved sequence in hydrophobic membrane proteins of periplasmic permeases defines an important site of interaction with the ATPase subunits. EMBO J. 16, 3066–3077.

Muntau, A.C., Mayerhofer, P.U., Paton, B.C., Kammerer, S. and Roscher, A.A. (2000) Defective peroxisome membrane synthesis due to mutations in human *PEX3* causes Zellweger Syndrome, complementation group G. Am. J. Hum. Genet. 67, 967–975.

Otera, H., Okumoto, K., Tateishi, K., Ikoma, Y., Matsuda, E., Nishimura, M., *et al.* (1998) Peroxisome targeting signal type 1 (PTS1) receptor is involved in import of both PTS1 and PTS2: Studies with *PEX5*-defective CHO cell mutants. Mol. Cell Biol. 18, 388–399.

Otera, H., Harano, T., Honsho, M., Ghaedi, K., Mukai, S., Tanaka, A., Kawai, A., Shimizu, N. and Fujiki, Y. (2000) The mammalian peroxin Pex5pL, the longer isoform of the mobile peroxisome targeting signal (PTS) type 1 transporter, translocates the Pex7p PTS2 protein complex into peroxisomes via its initial docking site, Pex14p. J. Biol. Chem. 275, 21703–21714.

Pause, B., Saffrich, R., Hunziker, A., Ansorge, W. and Just, W.W. (2000) Targeting of the 22 kDa integral peroxisomal membrane protein. FEBS Lett. 471, 23–28.

Pollard, H., Moreau, J. and Aubourg, P. (1995) Localization of mRNAs for adrenoleukodystrophy and the 70 kDa peroxisomal (PMP70) proteins in the rat brain during post-natal development. J. Neurosci. Res. 42, 433–437.

Pujol, A., Troffer-Charlier, N., Metzger, E., Chimini, G. and Mandel, J.L. (2000) Characterization of the adrenoleukodystrophy-related (ALDR, ABCD2) gene promoters: Inductibility by retinoic acid and forskolin. Genomics 70, 131–139.

Purdue, P.E. and Lazarow, P.B. (2001) Peroxisome biogenesis. Annu. Rev. Cell Dev. Biol. 17, 701–752.

Purdue, P.E., Zhang, J.W., Skoneczny, M. and Lazarow, P.B. (1997) Rhizomelic chondrodysplasia punctata is caused by deficiency of human PEX7, a homologue of the yeast PTS2 receptor. Nat. Genet. 15, 381–384.

Reddy, J.K., Goel, S.K., Nemali, M.R., Carrino, J.J., Laffer, T.G., Reddy, M.K., *et al.* (1986) Transcriptional regulation of peroxisomal fatty acyl-CoA oxidase and enoyl-CoA hydratase/3-hydroxyacyl-CoA dehydrogenase in rat liver by peroxisome proliferators. Proc. Natl Acad. Sci. USA 83, 1747–1751.

Sacksteder, K.A. and Gould, S.J. (2000) The genetics of peroxisome biogenesis. Annu. Rev. Genet. 34, 623–652.

Sacksteder, K.A., Jones, J.M., South, S.T., Li, X., Liu, Y. and Gould, S.J. (2000) *PEX19* binds multiple peroxisomal membrane proteins, is predominantly cytoplasmic, and is required for peroxisome membrane synthesis. J. Cell Biol. 148, 931–944.

Sarde, C.O., Mosser, J., Kioschis, P., Kretz, C., Vicaire, S., Aubourg, P., Poustka, A. and Mandel, J.L. (1994) Genomic organization of the adrenoleukodystrophy gene. Genomics 22, 13–20.

Saurin, W., Koster, W. and Dassa, E. (1994) Bacterial binding protein-dependent permeases: characterization of distinctive signatures for functionally related integral cytoplasmic membrane proteins. Mol. Microbiol. 12, 993–1004.

Schäfer, H., Nau, K., Sickmann, A., Erdmann, R. and Meyer, H.E. (2001) Identification of peroxisomal membrane proteins of *Saccharomyces cerevisiae* by mass spectrometry. Electrophoresis 22, 2955–2968.

Shani, N. and Valle, D. (1996) A *Saccharomyces cerevisiae* homolog of the human adrenoleukodystrophy transporter is a heterodimer of two half ATP-binding cassette transporter. Proc. Natl Acad. Sci. USA 93, 11901–11906.

Shani, N., Watkins, P.A. and Valle, D. (1995) PXA1, a putative *S. cerevisiae* homolog of the human adrenoleukodystrophy gene. Proc. Natl Acad. Sci. USA 92, 6012–6016.

Shani, N., Sapag, A. and Valle, D. (1996a) Characterization and analysis of conserved motifs in a peroxisomal ATP-binding cassette transporter. J. Biol. Chem. 271, 8725–8730.

Shani, N., Steel, G., Dean, M. and Valle, D. (1996b) Four half ABC transporters may heterodimerize in the peroxisome membrane. Am. J. Hum. Genet. 59, A42.

Shani, N., Jimenez-Sanchez, G., Steel, G., Dean, M. and Valle, D. (1997) Identification of a fourth half ABC transporter in the human peroxisomal membrane. Hum. Mol. Genet. 6, 1925–1931.

Smith, K.D., Kemp, S., Braiterman, L.T., Lu, J.-F., Wei, H.-M., Geraghty, M., Stetten, G., Bergin, J.S., Pevsner, J. and Watkins, P.A. (1999) X-linked adrenoleukodystrophy: Genes, mutations and phenotypes. Neurochem. Res. 24, 521–535.

Snyder, W.B., Faber, K.N., Wenzel, T.J., Koller, A., Luers, G.H., Rangell, L., Keller, G.A. and Subramani, S. (1999) Pex19p interacts with pex3p and pex10p and is essential for peroxisome biogenesis in *Pichia pastoris*. Mol. Biol. Cell 10, 1745–1761.

Snyder, W.B., Koller, A., Choy, A.J. and Subramani, S. (2000) The peroxin Pex19p interacts with multiple, integral membrane proteins at the peroxisomal membrane. J. Cell Biol. 149, 1171–1178.

South, S.T. and Gould, S.J. (1999) Peroxisome synthesis in the absence of preexisting peroxisomes. J. Cell Biol. 144, 255–266.

South, S.T., Sacksteder, K.A., Li, X., Liu, Y., Santos, M. and Gould, S.J. (2000) Inhibitors of COPI and COPII do not block *PEX3*-mediated peroxisome synthesis. J. Cell Biol. 149, 1345–1360.

Steinberg, S.J., Kemp, S., Braiterman, L.T. and Watkins, P.A. (1999a) Role of very long chain acyl-coenzyme A synthetase in X-linked adrenoleukodystrophy. Ann. Neurol. 46, 409–412.

Steinberg, S.J., Wang, S.J., McGuinness, M.C. and Watkins, P.A. (1999b) Human liver-specific very long chain acyl-coenzyme A synthetase: cDNA cloning and characterization of a second enzymatically active protein. Mol. Genet. Metab. 68, 32–42.

Subramani, S. (1998) Components involved in peroxisome import, biogenesis, proliferation, turnover and movement. Physiol. Rev. 78, 171–188.

Swartzman, E.E., Viswanathan, M.N. and Thorner, J. (1996) The *PAL1* gene product is a peroxisomal ATP-binding cassette transporter in the yeast *Saccharomyces cerevisiae*. J. Cell Biol. 132, 549–563.

Tabak, H.F., Braakman, I. and Distel, B. (1999) Peroxisomes: simple function but complex in maintenance. Trends Cell Biol. 9, 447–453.

van Roermund, C.W., Drissen, R., van den Berg, M., Ijlst, L., Hettema, E.H., Tabak, H.F., Waterham, H.R. and Wanders, R.J. (2001) Identification of a peroxisomal ATP carrier required for medium-chain fatty acid β-oxidation and normal peroxisome proliferation in *Saccharomyces cerevisiae*. Mol. Cell Biol. 21, 4321–4329.

Verleur, N., Hettema, E.H., Van Roermund, C.W.T., Tabak, H.F. and Wanders, R.J.A. (1997) Transport of activated fatty acids by the peroxisomal ATP-binding cassette transporter Pxa2 in a semi-intact yeast system. Eur. J. Biochem. 249, 657–661.

Wahli, W., Devchand, P.R., Ijpenberg, A. and Desvergne, B. (1999) Fatty acids, eicosanoids, and hypolipidemic agents regulate gene expression through direct binding to peroxisome proliferator-activated receptors. Adv. Exp. Med. Biol. 447, 199–209.

Wanders, R.J., Barth, P.G. and Heymans, H.S.A. (2001) Single peroxisomal enzyme deficiencies. In: The Metabolic and Molecular Basis of Inherited Disease (ed. C.R. Scriver, A.L. Beaudet, W.S. Sly and D. Valle), pp. 3219–3256. New York: McGraw-Hill.

Wang, X., Unruh, M.J. and Goodman, J.M. (2001) Discrete targeting signals direct Pmp47 to oleate-induced peroxisomes in *Saccharomyces cerevisiae*. J. Biol. Chem. 276, 10897–10905.

Wiemer, E.A.C., Nuttley, W.M., Bertolaet, B.L., Li, X., Franke, U., Wheelock, M.J., Anné, U.K., Johnson, K.R. and Subramani, S. (1995) Human peroxisomal targeting signal-1 receptor restores peroxisomal protein import in cells from patients with fatal peroxisomal disorders. J. Cell Biol. 130, 51–65.

Yamada, T., Taniwaki, T., Shinnoh, N., Uchiyama, A., Shimozawa, N., Ohyagi, Y., Asahara, H. and Kira, J. (1999) Adrenoleukodystrophy protein enhances association of very long chain acyl-coenzyme A synthetase with the peroxisome. Neurology 52, 614–616.

Yamada, T., Shinnoh, N., Kondo, A., Uchiyama, A., Shimozawa, N., Kira, J. and Kobayashi, T. (2000) Very long chain fatty acid metabolism in adrenoleukodystrophy protein-deficient mice. Cell Biochem. Biophys. 32, 239–246.

Zomer, A.W.M., van der Burg, B., Jansen, G.A., Wanders, R.J., Poll-The, B.T. and van der Saag, P.T. (2000) Pristanic acid acid and phytanic acid: Naturally occurring ligands for the nuclear receptor peroxisome proliferator-activated receptor α. J. Lipid Res. 41, 1801–1807.

CHAPTER 25

ABC Transporters in Mitochondria

Roland Lill and Gyula Kispal

Introduction

Mitochondria are essential organelles of most eukaryotic cells including fungi, invertebrates, vertebrates and plants. They perform various processes such as oxidative phosphorylation, the tricarboxylic acid cycle, fatty acid oxidation, the biosynthesis of various amino acids, the generation of iron-sulfur (Fe/S) clusters and their insertion into apoproteins, as well as partial reactions of heme biosynthesis and the urea cycle. According to the endosymbiont hypothesis, virtually all of these functions have been inherited from the bacterial ancestor of the present-day mitochondrion, an α-proteobacterium. Hence, both the components and mechanisms of the shared processes are highly related in mitochondria and bacteria.

In contrast to the aforementioned functions, reactions including membrane transport of proteins, peptides, sugars, metabolites, vitamins and lipids into and out of the organelle differ quite significantly from those operating in bacteria. For instance, the mitochondrial protein import system involving the TOM and TIM preprotein translocases does not exist in bacteria (Neupert, 1997; Pfanner and Geissler, 2001). Likewise, only one of the bacterial protein export systems has been maintained in mitochondria, namely the Oxa1/YidC complex (Dalbey and Kuhn, 2000). Striking differences between mitochondria and bacteria also exist with respect to trafficking small molecules. To facilitate this task, mitochondria contain more than 30 so-called 'carrier' proteins, which transport a variety of compounds (e.g. nucleotides, di- and tricarboxylates, vitamins and amino acids) across the inner membrane (reviewed by El Moualij *et al.*, 1997; Nelson *et al.*, 1998; Palmieri *et al.*, 2000).

No bacterial counterparts of these carrier proteins are known. Apparently, mitochondrial carrier proteins have replaced most of the versatile membrane transport functions performed by ATP-binding cassette (ABC) transporters of the bacterial ancestors of mitochondria. In present-day bacteria such as *Escherichia coli*, more than 50 members of this large protein family are found, and they are crucial for transport into and out of the bacterial cytosol (Linton and Higgins, 1998). In comparison, only a small number of ABC transporters exist in mitochondria. Strikingly, both structural and functional evidence suggests that these mitochondrial transporters do not closely resemble any of the bacterial counterparts, but rather represent proteins with a role specifically adapted for eukaryotic cells. Today, we can distinguish different types of mitochondrial ABC transporters. Two types belong to subclass B of the ABC transporter superfamily (MDR-like proteins) (Bauer *et al.*, 1999; Taglicht and Michaelis, 1998) and are distinguished according to their degree of homology to the three ABC transporters present in the yeast *Saccharomyces cerevisiae*, namely the Atm1p-like proteins and the Mdl1p/Mdl2p-like

ABC Proteins: From Bacteria to Man
ISBN 0-12-352551-9

proteins. An additional type of ABC transporter, termed CcmAB, may exist in plant mitochondria, but to date only its membrane-spanning domain (CcmB) has been identified.

This review will summarize our current knowledge of mitochondrial ABC transporters. We shall first address the properties and functions of the mitochondrial ABC transporters in *S. cerevisiae*. Then, we shall introduce the ABC transporters of mammalian cells and discuss their (putative) functions in comparison to those defined for the yeast proteins. Finally, we shall briefly review recent insights into plant mitochondrial ABC transporters and their (putative) functions.

MITOCHONDRIAL ABC TRANSPORTERS IN *S. CEREVISIAE*

IDENTIFICATION OF THE FIRST MITOCHONDRIAL ABC TRANSPORTER, YEAST ATM1P

Based on the bacterial origin of mitochondria, Leighton and Schatz (1995) predicted the existence of ABC transporters in these organelles. By using a polymerase chain reaction

TABLE 25.1. MITOCHONDRIAL ABC TRANSPORTERS

Name of ABC transporter	Chromosomal localization	Amino acid residues	Molecular mass (kDa)	Homologous to yeast protein	(Putative) Function
Yeast (*Saccharomyces cerevisiae*)					
Atm1p	XIII	690	78	–	Maturation of cytosolic Fe/S proteins, Iron homeostasis
Mdl1p	XII	695	76	–	Peptide export
Mdl2p	XVI	820	91	–	?
Man (*Homo sapiens*)					
hABC7	Xq13.1–q13.3	752	83	Atm1p (47%)	Maturation of cytosolic Fe/S proteins, Iron homeostasis
MTABC3	2q36	842	94	Atm1p (38%)	Iron homeostasis
M-ABC1	7q35–q36	718	78	Mdl2p (34%) Mdl1p (32%)	?
M-ABC2	1q42	738	79	Mdl1p (42%) Mdl2p (38%)	?
Mouse (*Mus musculus*)					
ABC-me	–	715	77	Mdl1p (39%) Mdl2p (37%)	Heme transport ?
Plants					
Sta1 (*A. thaliana*)	V	728	80	Atm1p (45%)	Maturation of cytosolic Fe/S proteins
Sta2 (*A. thaliana*)	IV	680	76	Atm1p (44%)	?
Sta3 (*A. thaliana*)	IV	678	75	Atm1p (45%)	?
CcmB (*Triticum aestivum*) (Membrane domain)	–	206	24	*E. coli* CcmB (27%)	*c*-type cytochrome biogenesis?

In all cases, a (putative) N-terminal mitochondrial presequence might be cleaved from the proteins, thus resulting in slightly shorter mature forms. The highest sequence homology between the listed mammalian or plant proteins and the *Saccharomyces cerevisiae* proteins is given as the fraction of identical amino acid residues in both proteins. For references see text.

(PCR) approach, they identified genes for several of the *S. cerevisiae* ABC transporters. The first mitochondrial representative, termed Atm1p, was identified by virtue of an N-terminal sequence resembling a mitochondrial targeting signal (presequence). In a parallel genetic screen originally intended to isolate new components of the biogenesis of *c*-type cytochromes (Kranz *et al.*, 1998), a temperature-sensitive mutant of the yeast *ATM1* gene (Kispal *et al.*, 1997) was found. This encodes a protein comprising 690 amino acid residues with six putative transmembrane segments and a C-terminal ATP-binding domain **(Table 25.1)** exhibiting the characteristic features of ABC transporter proteins. Atm1p therefore belongs to the group of 'half transporters'. It should be mentioned that no attempts have been made so far to determine precisely the structural mode of membrane integration of Atm1p (or of the other mitochondrial ABC transporters). Different algorithms used to predict transmembrane helices have identified five to six hydrophobic sequences that fulfill the criteria for membrane integration. Thus, by analogy with classical ABC transporters (Higgins, 1992), the Atm1p polypeptide chain may be expected to span the membrane six times and the functional protein may be a homodimer consisting of two molecules of Atm1p **(Figure 25.1)**.

The function of the N-terminus of Atm1p as a mitochondrial presequence was verified by its ability to target attached proteins to mitochondria (Leighton and Schatz, 1995). The precise localization of the Atm1p presequence cleavage site is not known but, based on the consensus sequence recognized by matrix processing peptidase (MPP), it is predicted to be after amino acid residues 25 or 41. Subcellular localization of Atm1p was demonstrated by immunostaining of cell fractions and by immunofluorescence. Atm1p is localized in the mitochondrial inner membrane with the nucleotide-binding domain facing the matrix space **(Figure 25.1)**. We presume, as will be developed in later sections, that Atm1p is predicted to function as an exporter of compounds from the matrix to the intermembrane space.

DELETION OF THE YEAST *ATM1* GENE

Cells deficient in the *ATM1* gene (strain Δatm1) display a strong growth defect on rich media containing glucose (Kispal *et al.*, 1997; Leighton and Schatz, 1995) and do not grow on non-fermentable carbon sources such as glycerol. The rate of growth of Δatm1 cells in the presence of glucose is much slower than that of cells harboring mitochondria defective in respiration. Thus, Atm1p plays a role that goes beyond the formation of respiratory competent mitochondria. Another phenotype resulting from the deletion of *ATM1* is a large reduction in the level of holocytochromes (Kispal *et al.*, 1997; Leighton and Schatz, 1995). Immunostaining analysis showed that this is not due to the defective biosynthesis of the apoforms of the *c*-type cytochromes in Δatm1 cells (Kispal *et al.*, 1997).

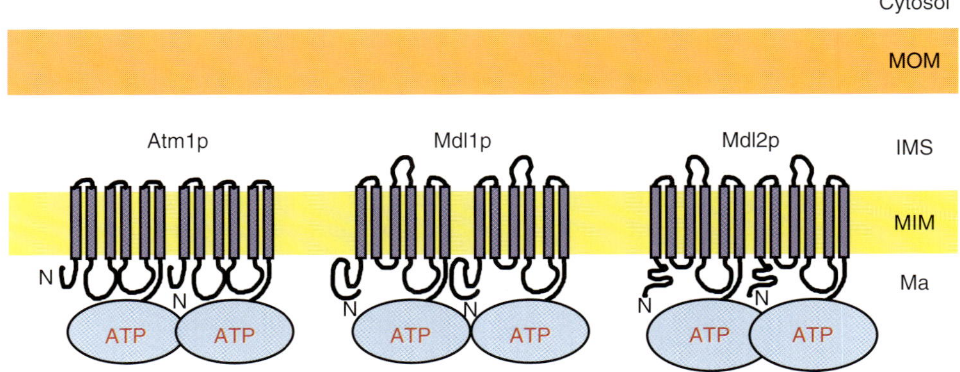

Figure 25.1. Model for the membrane orientation of the yeast mitochondrial ABC transporters. All three known yeast ABC transporters, Atm1p, Mdl1p and Mdl2p, share a similar membrane orientation with the N-terminus (N) facing the matrix space, an N-terminal ATP-binding domain and a C-terminal membrane-spanning domain with six putative transmembrane helices. The drawing represents the predicted size of the loops between the membrane segments and indicates the formation of possible homodimers. MOM, mitochondrial outer membrane; IMS, intermembrane space; MIM, mitochondrial inner membrane; Ma, matrix.

Moreover, heme biosynthesis occurs at wild-type rates in these cells (H. Lange, unpublished). Consequently, cells defective in Atm1p seem to face a condition leading to the degradation of protein-bound heme, which can most probably be explained by the oxidative stress prevailing in Δatm1 cells (Kispal *et al.*, 1997). Reduced and oxidized glutathione, which are the most important compounds required to balance the cellular redox level in yeast, are substantially increased in Δatm1 cells (Kispal *et al.*, 1997). The state of oxidative stress itself may be a consequence of the dramatic increase in the concentration of 'free' iron (i.e. non-heme and non Fe/S iron), which appears to be an early phenotype resulting from the loss of Atm1p function (Kispal *et al.*, 1999). Together with the mitochondrial matrix protein Yfh1p (frataxin), Atm1p was the first protein for which a function in mitochondrial iron homeostasis could be demonstrated (Babcock *et al.*, 1997; Foury and Cazzalini, 1997; Kispal *et al.*, 1997).

In some genetic backgrounds, Δatm1 cells lose mitochondrial DNA to yield so-called ρ^0 cells (Leighton and Schatz, 1995; Senbongi *et al.*, 1999). This phenomenon is not an obligatory consequence of the inactivation of *ATM1*; for example, deletion of the gene in strain W303 does not result in ρ^0 cells (Kispal *et al.*, 1997). Thus, loss of mitochondrial DNA may well be an indirect consequence of the oxidative damage resulting from iron overload and impairment of components of the machinery involved in mitochondrial DNA maintenance (Kaufman *et al.*, 2000).

ROLE OF ATM1P IN THE MATURATION OF CYTOSOLIC FE/S PROTEINS

Of all these pleiotropic phenotypes associated with Δatm1 cells none provide any clues towards the understanding of the function of the ABC transporter. Initial insight into the process in which Atm1p is involved came from the observation that Δatm1 cells fail to grow without added leucine (Kispal *et al.*, 1999). In yeast, leucine is synthesized from the common leucine/valine precursor α-ketoisovalerate by three specific steps catalyzed by the enzymes α-isopropyl malate synthase (Leu4p and Leu9p), isopropyl malate isomerase (Leu1p) and β-isopropyl malate dehydrogenase (Leu2p) (see reaction schemes in Hinnebusch, 1992; Jones and Fink, 1982; Prohl *et al.*, 2001). These enzymes are compartmentalized, distributed between the mitochondrial matrix (Leu4p and Leu9p) and the cytosol (Leu4p, Leu1p and Leu2p) (Beltzer *et al.*, 1988; Casalone *et al.*, 2000; Kohlhaw, 1988a, 1988b). Measurements of individual enzymatic activities showed a quantitative deficiency of isopropyl malate isomerase (Leu1p) in Δatm1 cells while the other enzymes were active at wild-type levels (Kispal *et al.*, 1999).

What is the reason for these observations? Leu1p is a cytosolic protein that requires an Fe/S cluster, generated in this mitochondrial matrix, for activity. Leu1p closely resembles aconitase of the mitochondrial matrix (Kohlhaw, 1988b). However, in contrast to Leu1p, mitochondrial aconitase, another Fe/S protein, exhibits almost wild-type activity in Δatm1 cells, rendering a general defect in cellular Fe/S proteins unlikely (Kispal *et al.*, 1997). Rather, the specific defect in Leu1p indicated that Atm1p may perform a function in the maturation of extra-mitochondrial Fe/S proteins (Kispal *et al.*, 1999).

To investigate the immediate effects of Atm1p deficiency, as opposed to long-term consequences (see above), a yeast mutant in which expression of the *ATM1* gene was under the control of a galactose-regulatable promoter (Gal-ATM1 cells) was created (Kispal *et al.*, 1999). These cells can readily be depleted of Atm1p when grown in the absence of galactose. Nevertheless, in the presence of galactose they do not exhibit a dramatic growth defect nor do they display any of the pleiotropic phenotypes reported above (e.g. cytochrome deficiency, oxidative stress). Upon depletion of Atm1p, the activity of Leu1p decreased at least 10-fold, indicating that incorporation of the Fe/S cluster into the cytosolic Leu1p apoprotein is an early consequence of Atm1p deficiency. A direct function of Atm1p in the assembly of the Fe/S cluster holoprotein, Leu1p, could be shown by briefly radiolabeling wild-type cells with ferrous iron (^{55}Fe), followed by immunoprecipitation of Leu1p from cell extracts using specific antibodies (Kispal *et al.*, 1999). The radioactive iron associated with Leu1p served as a direct measure of the formation of the Fe/S cluster in Leu1p. Cells lacking Atm1p did not incorporate any significant ^{55}Fe radioactivity into Leu1p. These results provided convincing evidence for the involvement of Atm1p in the maturation of a cytosolic Fe/S protein.

Recently, these results have been supported and extended by the analysis of another cytosolic Fe/S protein, namely the essential protein Rli1p, which harbors an Fe/S cluster domain at its N-terminus (G. Kispal, unpublished). Assembly of the Fe/S cluster in Rli1p also involves the function of Atm1p, suggesting a general role for this ABC transporter in the biogenesis of extra-mitochondrial Fe/S proteins. Atm1p function in cytosolic Fe/S protein maturation is highly specific because no defects were observed in Fe/S proteins localized inside mitochondria upon depletion of Atm1p (Kispal *et al.*, 1999). For a better understanding of the distinct function of Atm1p in the maturation of cytosolic Fe/S proteins, it is necessary to provide a brief outline of the biogenesis of Fe/S proteins in a eukaryotic cell. For a more comprehensive discussion of this recently discovered process, the reader is referred to several detailed reviews (Craig *et al.*, 1999; Lill *et al.*, 1999; Lill and Kispal, 2000; Mühlenhoff and Lill, 2000).

Biogenesis of eukaryotic Fe/S proteins

Assembly of mitochondrial Fe/S proteins

Many studies over the past four years have led to the identification of some ten proteins of the mitochondrial matrix, which play a role in the formation of the Fe/S clusters and their incorporation into mitochondrial apoproteins (for examples see Garland *et al.*, 1999; Jensen and Culotta, 2000; Kaut *et al.*, 2000; Kim *et al.*, 2001; Kispal *et al.*, 1999; Lange *et al.*, 2000; Li *et al.*, 2001; Pelzer *et al.*, 2000; Schilke *et al.*, 1999; Strain *et al.*, 1998; Voisine *et al.*, 2001). These proteins are highly homologous to bacterial proteins encoded by the *isc* (iron sulfur cluster) operons (Zheng *et al.*, 1998), and were therefore defined as compounds of the 'ISC assembly machinery' (Lill and Kispal, 2000). Even though virtually all of these proteins have been shown to participate in the assembly of Fe/S clusters, comparatively little is known about the precise roles of individual proteins or the overall molecular mechanism of the pathway. Nevertheless, a number of functional studies have been performed on the bacterial Isc proteins (for examples see Agar *et al.*, 2000b, 2000c; Hoff *et al.*, 2000; Krebs *et al.*, 2001; Ollagnier-de-Choudens *et al.*, 2001; Silberg *et al.*, 2000; Yuvaniyama *et al.*, 2000; Zheng *et al.*, 1993, 1994). Thus, the following model combines knowledge gained from studies on both mitochondrial and bacterial Isc proteins, assuming that the process is highly similar in both environments. However, it should be emphasized that we are just beginning to understand Fe/S cluster biogenesis, and any putative mechanistic pathways are based on rather limited experimental evidence.

According to a present working model, shown in **Figure 25.2**, iron, after its membrane potential-dependent import into mitochondria (Lange *et al.*, 1999), binds to the two proteins, Isu1p and Isu2p. The cysteine desulfurase Nfs1p generates elemental sulfur (S^0) from cysteine, which is then used to form an 'intermediate'

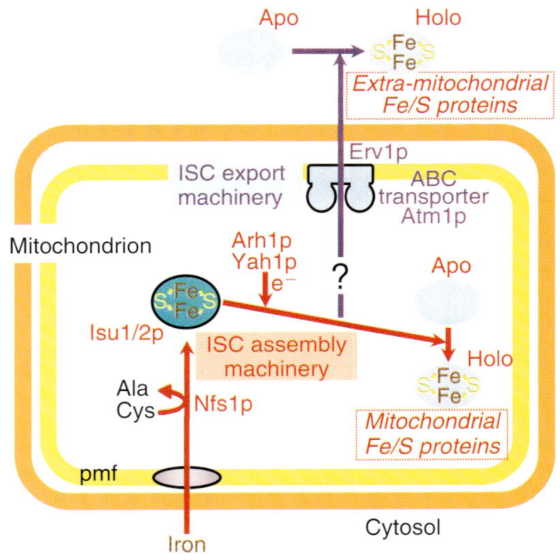

Figure 25.2. Working model for the function of Atm1p in cytosolic Fe/S protein assembly in eukaryotic cells. The assembly of Fe/S clusters, for both mitochondrial and cytosolic Fe/S proteins, is achieved by the ISC assembly machinery. First, ferrous iron enters the mitochondrial matrix in a membrane potential (pmf)-dependent step. Iron binds to the Isu proteins which provide a scaffold for the assembly of the Fe/S clusters. The cysteine desulfurase, Nfs1p, generates elemental sulfur (S^0) from cysteine needed for Fe/S cluster formation on the Isu proteins. The nascent Fe/S clusters are released from the Isu proteins upon reduction by the electron transfer chain shuttling electrons from NAD(P)H to the ferredoxin reductase Arh1p and the ferredoxin Yah1p. The Fe/S clusters are then incorporated into the apoforms of mitochondrial Fe/S proteins or exported to the cytosol, a step most likely involving Atm1p. The exact nature of the substrate of Atm1p is not known yet, but a likely compound is a chelated Fe/S cluster. The export process may be assisted by Erv1p, a sulfhydryl oxidase in the intermembrane space. It should be noted that many of the proposed steps of this model need further experimental verification.

[2Fe-2S] cluster on Isu1p/Isu2p. (Yuvaniyama *et al.*, 2000). This cluster may further be modified to generate a [4Fe-4S] cluster (Agar *et al.*, 2000a). The next steps of Fe/S cluster release and incorporation into apoproteins have not been defined experimentally, leaving us to speculate about the possible mechanism. *In vitro*, the intermediate Fe/S cluster can be released from the Isu proteins upon the addition of reducing agents. Therefore, the ferredoxin reductase Arh1p and the ferredoxin Yah1p may form an electron transfer chain that provides the reducing electrons for the release of the Fe/S cluster from the Isu proteins (Lange *et al.*, 2000; Li *et al.*, 2001).

The fate of the released Fe/S cluster is unknown. It may be transferred to and incorporated into the apoproteins spontaneously, or the process may need the help of accessory proteins. It is tempting to speculate that the insertion of the Fe/S cluster into apoproteins is a protein-assisted reaction. Stabilization of the apoproteins before incorporation of the Fe/S cluster could be an obvious task of the two mitochondrial heat shock proteins of the Hsp70/DnaK and Hsp40/DnaJ classes, Ssq1p and Jac1p, respectively (Kim *et al.*, 2001; Lutz *et al.*, 2001; Schilke *et al.*, 1999; Strain *et al.*, 1998; Voisine *et al.*, 2001). However, evidence for an interaction between the chaperones and the apoproteins has not, so far, been reported. On the contrary, the bacterial homologues of the two heat shock proteins have been shown to bind to the Isu proteins, leading to a stimulation of the ATPase activity of the Hsp70 chaperone (Hoff *et al.*, 2000; Silberg *et al.*, 2000). The mechanistic significance of this interaction remains to be discovered.

The Isa proteins have recently been shown to be crucial for Fe/S cluster assembly (Jensen and Culotta, 2000; Kaut *et al.*, 2000; Pelzer *et al.*, 2000) and, according to *in vitro* data, they may provide the necessary scaffold for the assembly of these Fe/S clusters (Krebs *et al.*, 2001; Ollagnier-de-Choudens *et al.*, 2001). Thus, the Isa proteins may represent an alternative to the Isu proteins in the assembly of the Fe/S clusters. Finally, a requirement for frataxin (yeast Yfh1p) for the normal activity of mitochondrial Fe/S proteins has been documented, even though the effects of deleting the frataxin gene were not dramatic (Foury, 1999; Rötig *et al.*, 1997). According to a recent study, frataxin might play a role in the storage of iron in mitochondria (Adamec *et al.*, 2000). Thus, the requirement for frataxin in Fe/S protein maturation might well be an indirect consequence of the impaired delivery of iron to the Isu and Isa proteins.

Maturation of extra-mitochondrial Fe/S proteins

In addition to the assembly of mitochondrial Fe/S proteins, the ISC assembly machinery also plays a crucial role in the maturation of extra-mitochondrial Fe/S proteins. The currently available data suggest that the Fe/S clusters of cytosolic Fe/S proteins are assembled in the mitochondrial matrix and, therefore, need to be exported, in some form, from mitochondria (summarized by Lill and Kispal, 2000). This contention is based on the fact that depletion of the mitochondrial Isc components abolishes cytosolic Fe/S protein maturation. Nevertheless, the molecular moiety leaving the organelle is not known at present. Similarly, we are only just beginning to understand the molecular mechanisms underlying the export process.

Since Atm1p is specifically required for the assembly of cytosolic, but not mitochondrial Fe/S proteins, it is thought to play a central role in the release of a moiety synthesized by the ISC assembly machinery from the organelles and may be required for the assembly of cytosolic Fe/S proteins. Only a few components of the so-called 'ISC export machinery', other than Atm1p, have been identified so far, namely Erv1p and the two homologous proteins Bat1p and Bat2p. Since these proteins appear to be functionally related to Atm1p, the findings that support their involvement in Fe/S protein maturation in the cytosol are briefly summarized as follows.

Erv1p is a component of the intermembrane space and is essential for yeast viability (Lange *et al.*, 2001; Lisowsky, 1992). Inactivation of Erv1p leads to a dramatic reduction in the assembly of cytosolic Fe/S proteins. Similar to what is observed when Atm1p is depleted, mitochondrial Fe/S protein assembly is not affected in Erv1p-defective cells. Erv1p was found to possess sulfhydryl oxidase activity associated with the C-terminal domain of the protein (Lee *et al.*, 2000). Currently, the role of this domain in Fe/S protein assembly in the cytosol is unclear. Nevertheless, the localization of Erv1p in the intermembrane space suggests that it plays a role in the export pathway subsequent to that in which Atm1p is implicated. Whether Erv1p transiently binds directly to the transported molecule, or introduces disulfide bonds into a component of the pathway, remains to be determined. Interestingly, the mammalian homologue of Erv1p, termed ALR ('augmenter of liver regeneration'), can functionally replace the yeast protein and thus the two proteins appear to be orthologues. All of the components of the ISC assembly machinery and Atm1p (see below) are conserved in mammals, suggesting that Fe/S cluster assembly follows similar pathways in virtually all eukaryotes.

The *BAT1* gene was isolated as a high-copy suppressor of a temperature-sensitive mutant of ATM1 (Kispal *et al.*, 1996). *BAT1* and the highly homologous gene *BAT2* encode the mitochondrial and cytosolic forms of *b*ranched-chain *a*mino acid *t*ransaminases, respectively (Eden *et al.*, 1996; Kispal *et al.*, 1996). The Bat proteins catalyze the reversible inter-conversion of branched-chain α-keto acids and amino acids (i.e. leucine, isoleucine and valine). Additionally, they perform a second function unrelated to amino acid synthesis. This is evident from the growth defect of Δbat1 Δbat2 cells, lacking both *BAT* genes, on rich media containing glucose, which occurs even after additional branched-chain amino acids are added to the medium (Kispal *et al.*, 1996). This observation may be explained by the participation of the Bat proteins in the maturation of cytosolic Fe/S proteins (Prohl *et al.*, 2000; C. Prohl, unpublished). The double mutant cells show a threefold reduction in the *de novo* synthesis of both Leu1p and Rli1p Fe/S proteins in the cytosol. Thus, the Bat proteins are not essential for maturation of cytosolic Fe/S proteins, but apparently perform an accessory function, increasing the efficiency of the formation of holoprotein in an, as yet, unknown way. Expression of either *BAT* gene is sufficient for the normal formation of cytosolic Fe/S proteins, indicating that the specific Bat function can be performed either in mitochondria, or in the cytosol. Similar to Atm1p and Erv1p, the Bat proteins are not required for the biogenesis of Fe/S proteins within the mitochondria, suggesting that they participate in the Atm1p-mediated export pathway. One possible function may be the catalytic formation of a compound required for chelation of the Fe/S cluster (or a related compound) during export from the mitochondria.

In summary, there is ample evidence for the involvement of Atm1p in the maturation of cytosolic Fe/S proteins, yet the molecular details underlying its precise function have not been unraveled so far. Future progress in understanding the roles of Atm1p, Erv1p and the Bat proteins will require the identification of the substrate for Atm1p and of any additional components of the ISC export machinery.

MDL1P AND MDL2P, TWO HOMOLOGOUS YEAST MITOCHONDRIAL ABC TRANSPORTERS WITH DIFFERENT FUNCTIONS

Recently, two additional ABC transporters, termed Mdl1p and Mdl2p, have been identified in yeast mitochondria, and were found to be homologues of the human ABCB8 and ABCB10 genes (see below) (Young *et al.*, 2001). Like Atm1p, these proteins are half transporters with an N-terminal membrane-spanning domain **(Figure 25.1)** (Dean *et al.*, 1994). The two proteins show rather high sequence homology (46% identical amino acid residues) and have a molecular mass of 76 kDa (Mdl1p) and 91 kDa (Mdl2p), including a putative N-terminal extension serving as a mitochondrial presequence **(Table 25.1)**. In fact, only the N-terminus of Mdl1p resembles a canonical mitochondrial targeting signal, whereas the N-terminal segment of Mdl2p does not conform to the properties of a presequence. According to biochemical fractionation experiments using specific antibodies, both proteins are localized in the mitochondrial inner membrane with the ABC domains facing the matrix space (Young *et al.*, 2001) **(Figure 25.1)**. Thus, all three yeast mitochondrial ABC transporters appear to exhibit the same membrane orientation and thus are presumed to export substrates from the matrix towards the cytosol.

Deletion of the *MDL1* and *MDL2* genes does not cause major growth defects in *S. cerevisiae* (Dean *et al.*, 1994). However, whilst Δmdl1 cells exhibit normal growth, growth of Δmdl2 cells is retarded on glycerol-containing media. In part, this may be explained by the finding that Δmdl2 cells tend to gradually lose mitochondrial DNA (J. Gerber, unpublished). Double deletion of both *MDL* genes slightly exacerbates the growth defect observed for Δmdl2 cells, suggesting that the proteins may perform non-overlapping functions. This is supported by recent insights into the function of Mdl1p.

Both Mdl1p and Mdl2p are close homologues of the yeast **a**-factor pheromone receptor Ste6p and of another ABC protein, the mammalian TAP transporter (ABCB2/ABCB3). This protein mediates the transfer of antigenic peptides after their generation by the cytosolic proteasome to the class I major histocompatibility complex (MHC class I) in the endoplasmic reticulum (ER) (see Chapter 26). Hence, it was postulated that the Mdl proteins may facilitate the export of peptides from the matrix to the intermembrane space. A direct test of this idea showed that mitochondria derived from Δmdl1 mutant cells displayed a 40% reduction in peptide release (Young *et al.*, 2001). In the assay system used, the peptides were generated by the inner membrane protease Yta10p/Yta12p (also termed Afg3p/Rca1p) from mitochondria-encoded membrane proteins (Arlt *et al.*, 1996; Rep *et al.*, 1996) **(Figure 25.3)**. This member of the family of ATP-dependent AAA proteases exposes its proteolytic domain in the matrix space and forms a large hetero-oligomeric complex (for a recent review on AAA proteases, see Langer, 2000). The rather small decrease in peptide export observed after deletion of *MDL1* is explained by the fact that another set of peptides generated by the inner membrane protease Yme1p can still leave the organelle in the absence of Mdl1p. This second mitochondrial AAA protease forms a homo-oligomer with its proteolytic domain in the intermembrane space (Langer, 2000) **(Figure 25.3)**. When a double mutant Δmdl1 Δyme1 was analyzed, a 75% reduction in peptide release from the organelle was observed. The length of the released peptides varied between 6 and 20 residues, strikingly similar in size to peptides transported by the TAP transporter in the ER (Elliott, 1997; Ritz and Seliger, 2001) (see Chapter 26). The function of Mdl1p in peptide export depended on a conserved motif in the Walker A and B sites of the nucleotide-binding domain and a loop characteristic for ATPases. Peptide export through Mdl1p therefore seems to require the hydrolysis of ATP (Young *et al.*, 2001). For final exit from the organellar intermembrane space, as illustrated in **Figure 25.3**, the peptides possibly pass the outer membrane with the help of mitochondrial porin or the TOM complex, both of which contain large pores **(Figure 25.3)** (Künkele *et al.*, 1998).

On the other hand, deletion of *MDL2* does not result in any alteration of peptide export from the mitochondria, suggesting that only Mdl1p mediates the release of peptides from the mitochondrial matrix. These findings are nicely corroborated by the observation that Mdl1p and Mdl2p appear to be associated with different high molecular mass complexes of

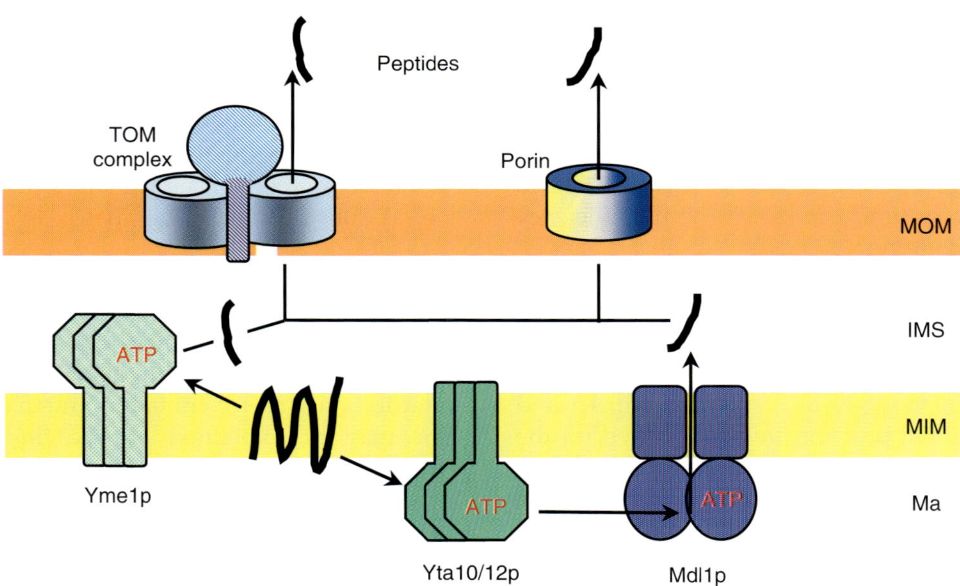

Figure 25.3. Model for the function of Mdl1p in the export of peptides from the mitochondrial matrix. Peptides generated by the inner membrane protease, Yta10p/Yta12p, are exported by the ABC transporter, Mdl1p, in an ATP-dependent fashion. Another pool of proteolytic fragments is formed by the inner membrane protease Yme1p in the intermembrane space. Most likely, the peptides leave the mitochondria via porin or the TOM complex, both of which contain large pores. Currently, it is unknown how peptides generated by the matrix protease Pim1p (not shown) are exported from the organelles. MOM, mitochondrial outer membrane; IMS, intermembrane space; MIM, mitochondrial inner membrane; Ma, matrix.

200 kDa and 300 kDa, respectively (Young *et al.*, 2001). This observation may argue for homodimer, rather than heterodimer, association of the Mdl proteins.

Since deletion of *MDL1* is not associated with any detectable phenotype, the question arises as to what the physiological significance of peptide transport by Mdl1p might be. Moreover, mitochondria can further break down the longer peptides to amino acids or di- and tripeptides, which can be transported independently of Mdl1p. Thus, the purpose of Mdl1p-mediated peptide transport remains unclear, even though the latter observation supports the finding that Mdl1p is dispensable in yeast. In vertebrates, proteins homologous to Mdl1p (see below) might play an important role in the transport of antigenic peptides derived from mitochondrial proteins for presentation on the eukaryotic cell surface. Functional complementation studies with the mammalian homologues expressed in the Δmdl1 background have not yet been conducted to test this attractive hypothesis.

Mitochondrial ABC Transporters in Mammals

The sequencing of the human genome has provided us with a complete inventory of ABC transporters in man (Klein *et al.*, 1999; http://www.humanabc.org). Amongst the 48 proteins with ABC domains, a few qualify as potential mitochondrial components based on the presence of a (putative) presequence at their N-termini. Two of these proteins, termed ABC7 (ABCB7, according to the nomenclature of http://www.humanabc.org) and MTABC3 (ABCB6) are homologous to the yeast Atm1p, whereas another two proteins, namely M-ABC1 (ABCB8) and M-ABC2 (ABCB10), closely resemble yeast Mdl1p and Mdl2p. In mice, another homologue of the latter subclass has been identified and analyzed recently, the protein ABC-me. The properties and functions of these proteins are discussed in the following sections.

ABC7 and MTABC3, Functional Orthologues of Yeast Atm1p

The human ABC transporter ABC7 represents the closest homologue of yeast Atm1p. The cDNA corresponding to the gene has been identified independently by several groups (Allikmets *et al.*, 1999; Csere *et al.*, 1998; Mao *et al.*, 1998; Shimada *et al.*, 1998). Sequencing of the entire *ABC7* gene revealed 16 introns and the promoter structure (Bekri *et al.*, 2000). At the protein level ABC7 shares 47% amino acid sequence identity with yeast Atm1p **(Table 25.1)**. Expression of the human gene in yeast has demonstrated that the human protein is the functional orthologue of Atm1p (Allikmets *et al.*, 1999; Csere *et al.*, 1998). This gene is able to revert growth of Δatm1 mutant cells to almost wild-type rates and to restore normal cytochrome levels. Furthermore, mitochondria harboring ABC7 instead of Atm1p do not accumulate iron. All these observations strongly indicate that upon expression in yeast, ABC7 can replace the primary function of Atm1p in Fe/S cluster formation (Bekri *et al.*, 2000). These findings further suggest that ABC7 performs a similar or identical function in the mammalian cell as that carried out by Atm1p in yeast.

Mutations in the human *ABC7* gene cause X-linked sideroblastic anemia and cerebellar ataxia (XLSA/A) (Allikmets *et al.*, 1999; Bekri *et al.*, 2000). As a result of such mutations, mitochondria accumulate high concentrations of iron and form so-called ring sideroblasts (i.e. iron-loaded ring-shaped tubules which are concentrated around the nucleus). Thus, there exists a striking similarity in phenotypes between yeast and man upon impairment of Atm1p and ABC7 function, respectively. Biochemical studies indicate that yeast serves as an excellent model system to study the effects of the mutations in *ABC7*. When expressed in yeast, mutant ABC7 proteins, or Atm1p bearing the corresponding mutations, are functionally impaired (Allikmets *et al.*, 1999; Bekri *et al.*, 2000). For instance, when the ABC7-(E433K) mutants (mutation localized towards the matrix following TM6), or the corresponding *ATM1*-(D398K) mutants, were expressed in Δatm1 yeast cells, maturation of cytosolic Fe/S proteins was twofold lower as compared to wild-type cells (Bekri *et al.*, 2000). The surprisingly weak consequences of these charge exchange mutations underlines the importance of ABC7 function for a healthy cell. In fact, only slight changes to ABC7 can dramatically affect cellular iron homeostasis and elicit severe phenotypical consequences. These observations are consistent with the fact that, in yeast, Fe/S cluster formation is an indispensable process (Lill *et al.*, 1999). Deletion of many genes encoding components of the ISC assembly machinery is lethal, indicating a central role for Fe/S proteins for life.

The human protein, termed MTABC3 (Taglicht and Michaelis, 1998), represents a second functional orthologue of yeast Atm1p (Mitsuhashi *et al.*, 2000). The *MTABC3* gene is encoded by human chromosome 2 **(Table 25.1)** and has been mapped to within the vicinity of the locus for lethal neonatal metabolic syndrome, a disorder of mitochondrial function associated with iron metabolism. Hence, *MTABC3* is a likely candidate gene for this disorder. The homology of MTABC3 and Atm1p is less than that of ABC7 compared with Atm1p (38% as compared to 47% identical amino acid residues). Nevertheless, expression of MTABC3 in Δatm1 yeast cells restores growth to wild-type levels, reverts the increase in mitochondrial iron, and prevents the loss of mitochondrial DNA. Even though the role of MTABC3 in the biogenesis of cytosolic Fe/S proteins has yet to be analyzed, such a function seems likely.

The relationship of the two human orthologues of yeast Atm1p is unclear. Based on their common role in iron homeostasis it is conceivable that ABC7 and MTABC3 form a heterodimer in the human cell. An alternative hypothesis predicts that both genes may be differentially expressed in human tissues. The presence of two copies of Atm1p-like proteins may offer the possibility to fine-tune the function of the ABC transporter, as found for numerous other mammalian proteins.

M-ABC1 AND M-ABC2, MAMMALIAN HOMOLOGUES OF YEAST MDL1P AND MDL2P

The human genome harbors four candidates with homology to the yeast *MDL* genes. Only two of the encoded proteins, termed M-ABC1 and M-ABC2 (ABCB8 and ABCB10 according to nomenclature of http://www.humanabc.org), have been experimentally localized to mitochondria (Hogue *et al.*, 1999; Zhang *et al.*, 2000a). Another gene product, ABCB9, has been found in the lysosomal compartment (Zhang *et al.*, 2000b). Nevertheless, the protein is not a close homologue of the vacuolar ABC transporters, Ycf1p of *S. cerevisiae* (Li *et al.*, 1996) or Hmt1p of *Schizosaccharomyces pombe* (Ortiz *et al.*, 1995). The fourth mammalian Mdl homologue, ABCB5, has not yet been studied. The sequence identity between the human M-ABC1/M-ABC2 and the yeast Mdl proteins varies between 32% and 42% **(Table 25.1)**. Based on sequence comparisons, M-ABC1 may be the counterpart of Mdl2p, while M-ABC2 is more closely related to Mdl1p. However, the differences in homology may be too small to infer a close functional relationship.

Currently, it is not known whether M-ABC1 and M-ABC2 form homo- or heterodimers in the mitochondrial inner membrane. Similarly, the membrane orientation of these proteins is not yet clear, even though it is likely that it is the same as for Mdl1p and Mdl2p, with the nucleotide-binding domain facing the matrix space (however, see Zhang *et al.*, 2000a). No experimental evidence has been obtained for any function of M-ABC1 and M-ABC2 in the transport of peptides out of the mitochondrial matrix, even though such a role, similar to that of Mdl1p, seems probable (see above).

ABC-ME, A MURINE MITOCHONDRIAL ABC TRANSPORTER WITH A FUNCTION IN HEME METABOLISM

Only one mitochondrial ABC transporter has been identified to date in mice, the protein ABC-me (*m*itochondrial *e*rythroid), and its cellular role has been investigated in some detail (Shirihai *et al.*, 2000). The *ABC-me* gene has been isolated as one factor that is induced upon expression of the erythropoietic transcription factor GATA-1. *ABC-me* is highly expressed in erythroid tissues of embryos and adults. In murine erythroleukemia (MEL) cells, overexpression of *ABC-me* strongly increased the heme concentration. Conversely, *ABC-me* mRNA levels are decreased by physiological concentrations of heme. Together, these findings are consistent with a role for ABC-me in the trafficking of intermediates of heme biosynthesis. The heme biosynthetic steps are partitioned between the mitochondrial matrix and the cytosol, with the first reaction and the last three steps taking place in the matrix. The ABC domains of ABC-me face the mitochondrial matrix and this has been taken to indicate that the protein should function as an exporter (Shirihai *et al.*, 2000). ABC-me could be involved in translocating either δ-aminolevulinate or heme from the mitochondrial matrix to the cytosol. The rather specific expression of ABC-me in erythroid cells may be necessary to satisfy the extraordinarily high needs for transporting heme biosynthetic metabolites across the mitochondrial inner membrane. At present, it is unclear which, if any, human ABC protein may represent the counterpart of murine ABC-me, since there is no known human Mdl-like protein with specific expression in erythroid tissues. Further, all known human candidates exhibit a similar degree of sequence similarity to ABC-me.

Mitochondrial ABC Transporters in Plants

The recent sequencing of the genome of the weed *Arabidopsis thaliana* has allowed access to the inventory of plant ABC transporters (Sanchez-Fernandez *et al.*, 2001). The plant genome contains more than 100 distinct members of this protein family. From the homology with yeast ABC transporters, several counterparts to Atm1p and Mdl1p/Mdl2p can be identified in *Arabidopsis*. While two of the three homologues of Atm1p have been characterized as mitochondrial proteins (Kushnir *et al.*, 2001), the subcellular localization of the two Mdl1p-like proteins is not clear. The latter show high sequence similarity to both yeast mitochondrial Mdl1p/Mdl2p and to the mammalian TAP1/TAP2 (ABCB2/ABCB3) transporters of the ER (see Chapter 26). Thus, in the absence of experimental evidence, it is uncertain whether these members of the ABC transporter family resemble mitochondrial or microsomal constituents. In addition to these plant ABC proteins, another potential ABC transporter, termed CcmAB, may exist. CcmB, the membrane-spanning part of this ABC protein, is encoded by the mitochondrial genome of various plants. Bacterial homologues of the plant CcmB protein play a role in *c*-type cytochrome biogenesis. The Atm1p-like and the CcmAB-like ABC proteins of plant mitochondria will be discussed in more detail in the following sections.

The Sta (Atm) proteins, homologues of yeast Atm1p

A. thaliana contains three genes, termed *STA1*, *STA2* and *STA3* (also known as *ATM3*, *ATM2* and *ATM1*, respectively), the products of which share about 45% amino acid identity with the yeast *ATM1* gene product (Kushnir *et al.*, 2001; Sanchez-Fernandez *et al.*, 2001) **(Table 25.1)**. The sequence identity between the three plant proteins varies from 71% (Sta1/Sta2) to 83% (Sta2/Sta3). Although all three Sta proteins contain a putative mitochondrial presequence at their N-termini, mitochondrial localization has only been experimentally determined for Sta1 and Sta2 (Kushnir *et al.*, 2001).

Inactivation of the *STA1* gene results in chlorosis and dwarfism of mutant plants (Kushnir *et al.*, 2001). The most severe phenotype was seen when plants were grown on synthetic media. Nevertheless, mutant plants are photoautotrophic and fertile. Plant leaves in the mutants exhibit a number of abnormalities such as enlarged cells with more air space in between them. The *STA1* mutant can be partially complemented by ectopic expression of *STA2*, resulting in plants which grow to almost wild-type size and show no signs of chlorosis. Even though the two proteins apparently have overlapping functions, their roles are not entirely redundant. A plausible reason for these observations may be differential and tissue-specific expression of the *STA* genes.

A function for Sta1 in Fe/S protein maturation in the cytosol could be inferred from complementation studies in yeast (Kushnir *et al.*, 2001). In these studies, expression of *STA1* in Δatm1 mutant yeast cells fully complemented the defects of Atm1p deficiency. The Sta1 protein supported cytosolic Fe/S protein maturation with an efficiency comparable to yeast Atm1p. However, despite the apparent functional similarities, biochemical analyses of Sta1-deficient plants revealed striking differences compared to yeast Δatm1 cells. In contrast to the 25- to 30-fold increase in iron concentration in yeast mitochondria derived from Δatm1 cells, plant mitochondria displayed only a small increase in free iron concentration. Furthermore, plant cells showed no obvious signs of oxidative stress. The differences between yeast and plants may be explained by the multiplicity of the *ATM1*-like gene in plants. The functional redundancy of the Sta proteins may lead to comparatively weak phenotypic consequences of the inactivation of a single *STA* gene. These results support the view that the phenotypes observed in yeast are indirect (secondary) consequences of the defect in Atm1p.

The function of Sta1 in cytosolic Fe/S protein maturation, together with the presence of numerous plant genes encoding components of the ISC assembly machinery (Kushnir *et al.*, 2001), indicates that the process of Fe/S protein biogenesis in plants resembles that of the model organism yeast. Thus, the function of the three Sta proteins in this biosynthetic process may be to transport a component required for Fe/S protein assembly outside the mitochondria.

CcmB, a novel component of an ABC transporter of land plants?

The sequencing of various plant mitochondrial genomes has led to the suggestion that an ABC transporter, with homology to bacterial proteins implicated in *c*-type cytochrome biogenesis, might be present in these organelles. This expectation is supported by the recent identification of the independently encoded membrane-spanning protein component, CcmB, of a putative ABC transporter (Faivre-Nitschke *et al.*, 2001). In wheat, CcmB consists of 206 amino acid residues **(Table 25.1)**. The plant CcmB protein has a large number of hydrophobic amino acid residues and shares significant sequence identity with CcmB proteins of various bacteria (24–29%). Both plant and bacterial CcmB proteins have hydrophobicity profiles characteristic of membrane proteins with six predicted transmembrane helices. Typical of many genes encoded by the plant mitochondrial genome, the CcmB transcript is highly edited (42 C to U editing positions), affecting 32 out of the 206 amino acid residues.

CcmB has been identified as a constituent of the mitochondrial inner membrane by employing an antibody raised against CcmB (Faivre-Nitschke *et al.*, 2001). This detects a 28 kDa protein, compared to the calculated molecular mass of 24 kDa **(Table 25.1)**, that is enriched in the mitochondrial membrane fraction. Association of CcmB with its putative ABC domain, CcmA, has not been established so far. This is mainly due to the fact that, in land plants, the expected CcmA protein is not encoded by the mitochondrial genome, but rather is thought to derive from a nuclear gene.

The function of CcmB in land plants has not been addressed experimentally so far. The homology to bacterial CcmB, however, has led to the suggestion that the plant protein, like the bacterial counterparts, performs a function in *c*-type cytochrome biogenesis (Faivre-Nitschke *et al.*, 2001). For a better understanding of the potential function of CcmAB, a brief sketch of *c*-type cytochrome biogenesis follows. For further details, the reader is referred to recent review articles by Kranz *et al.* (1998), Page *et al.* (1998), and Thony-Meyer (2000). *c*-Type cytochromes carry a heme moiety covalently attached to two conserved cysteine residues via a thioether bond. The best-known examples are the cytochromes *c* and c_1, which are located in the mitochondrial intermembrane space, or the bacterial periplasm, where they participate in electron transfer during oxidative phosphorylation.

During evolution, three systems have evolved for the biogenesis of these heme proteins (Kranz *et al.*, 1998). System I is the most complex and is found in α- and γ-proteobacteria and in mitochondria of land plants. System II is used by Gram-positive bacteria, cyanobacteria and chloroplasts, and system III is present in fungal, vertebrate and invertebrate mitochondria. The components of the individual systems differ in both structure and number. The simplest pathway (system III) uses just one protein for biogenesis, the cytochrome heme lyases that attach heme to the apocytochromes. Biogenesis in system I, on the other hand, involves some ten proteins which share no obvious homology with cytochrome heme lyases. Studies in *E. coli* and *Rhodobacter capsulatus* have provided us with a rudimentary view of the individual steps of biogenesis in system I (reviewed by Thony-Meyer, 2000). In brief, apocytochrome *c* is translocated into the periplasm by the canonical Sec translocase. In an unknown way, heme is transferred from the cytosol to a periplasmic heme chaperone, CcmE, where it becomes covalently bound in a transient fashion (Schulz *et al.*, 1998). Earlier genetic studies of the bacterial CcmA, CcmB and CcmC (another integral membrane protein) showed their involvement in cytochrome biogenesis, and it was, therefore, suggested that a complex comprising these three proteins may form an ABC transporter necessary to export heme from its site of synthesis in the cytosol to the periplasm (Goldman and Kranz, 2001; Goldman *et al.*, 1998). However, this view was rendered rather unlikely by the findings in *E. coli* that heme can be transferred to the periplasmic heme chaperone CcmE in the absence of CcmA and CcmB, but not without CcmC (Schulz *et al.*, 1999). Furthermore, CcmC and a periplasmic protein, CcmE, were found to interact tightly with each other and with heme (Ren and Thony-Meyer, 2001). Based on these most recent results, it may be expected that CcmC facilitates transport of heme from the bacterial cytosol to CcmE in the periplasm. The function of CcmA and CcmB is not yet clear, but it has recently been proposed that they export a compound required to maintain the reduced states of apocytochrome cysteines, the vinyl groups of protoheme or heme iron (Faivre-Nitschke *et al.*, 2001). Biosynthesis of the holoform of cytochrome *c* is then completed by the covalent attachment of heme to apocytochrome *c*, a reaction most probably catalyzed by CcmF. Prerequisites for this reaction are a number of redox steps that lead to the reduction of both heme and the disulfide bridges of the apoprotein. In plant mitochondria, further studies are needed to verify the existence of the ABC transporter CcmAB and to examine its potential function in *c*-type cytochrome biogenesis.

Future Directions

This review on mitochondrial ABC transporters clearly shows that we are only beginning to understand the biological roles of these interesting proteins. While it is possible that all of the yeast and human mitochondrial ABC transporters have been identified (Bauer *et al.*, 1999; Decottignies and Goffeau, 1997; Klein *et al.*, 1999; Taglicht and Michaelis, 1998), the biochemical characterization of these components lags behind. Future studies on mitochondrial ABC transporters will include the purification and functional reconstitution of these proteins, the identification of substrates and the elucidation of the molecular mechanisms underlying transport. Further challenges include the discovery of new diseases associated with mutations in these proteins, and understanding the structural/functional relationships between these important proteins. Clearly, the most interesting years of research on mitochondrial ABC transporters lie ahead of us.

Acknowledgments

Our work was supported generously by grants from the Sonderforschungsbereich 286, Deutsche Forschungsgemeinschaft, Deutsches Humangenomprojekt, Volkswagen-Stiftung, Fonds der Chemischen Industrie and the Hungarian Funds OKTA.

References

Adamec, J., Rusnak, F., Owen, W.G., Naylor, S., Benson, L.M., Gacy, A.M. and Isaya, G. (2000) Iron-dependent self-assembly of recombinant yeast frataxin: implications for Friedreich ataxia. Am. J. Hum. Genet. 67, 549–562.

Agar, J.N., Krebs, C., Frazzon, J., Huynh, B.H., Dean, D.R. and Johnson, M.K. (2000a) IscU as a scaffold for iron-sulfur cluster biosynthesis: sequential assembly of [2Fe-2S] and [4Fe-4S] clusters in IscU. Biochemistry 39, 7856–7862.

Agar, J.N., Yuvaniyama, P., Jack, R.F., Cash, V.L., Smith, A.D., Dean, D.R. and Johnson, M.K. (2000b) Modular organization and identification of a mononuclear iron-binding site within the NifU protein. J. Biol. Inorg. Chem. 5, 167–177.

Agar, J.N., Zheng, L., Cash, V.L., Dean, D.R. and Johnson, M.K. (2000c) Role of the IscU protein in iron-sulfur cluster biosynthesis: IscS-mediated assembly of a $[Fe_2S_2]$ cluster in IscU. J. Am. Chem. Soc. 122, 2136–2137.

Allikmets, R., Raskind, W.H., Hutchinson, A., Schueck, N.D., Dean, M. and Koeller, D.M. (1999) Mutation of a putative mitochondrial iron transporter gene (ABC7) in X-linked sideroblastic anemia and ataxia (XLSA/A). Hum. Mol. Gen. 8, 743–749.

Arlt, H., Tauer, R., Feldmann, H., Neupert, W. and Langer, T. (1996) The YTA10-12 complex, an AAA protease with chaperone-like activity in the inner membrane of mitochondria. Cell 85, 875–885.

Babcock, M., De Silva, D., Oaks, R., Davis-Kaplan, S., Jiralerspong, S., Montermini, L., Pandolfo, M. and Kaplan, J. (1997) Regulation of mitochondrial iron accumulation by Yfh1p, a putative homolog of frataxin. Science 276, 1709–1712.

Bauer, B.E., Wolfger, H. and Kuchler, K. (1999) Inventory and function of yeast ABC proteins: about sex, stress, pleiotropic drug and heavy metal resistance. Biochim. Biophys. Acta 1461, 217–236.

Bekri, S., Kispal, G., Lange, H., Fitzsimons, E., Tolmie, J., Lill, R. and Bishop, D.F. (2000) Human ABC7 transporter: Gene structure and mutation causing X-linked sideroblastic anemia with ataxia (XLSA/A) with disruption of cytosolic iron-sulfur protein maturation. Blood 96, 3256–3264.

Beltzer, J.P., Morris, S.R. and Kohlhaw, G.B. (1988) Yeast *LEU4* encodes mitochondrial and nonmitochondrial forms of alpha-isopropylmalate synthase. J. Biol. Chem. 263, 368–374.

Casalone, E., Barberio, C., Cavalieri, D. and Polsinelli, M. (2000) Identification by functional analysis of the gene encoding alpha-isopropylmalate synthase II (*LEU9*) in *Saccharomyces cerevisiae*. Yeast 16, 539–545.

Craig, E.A., Voisine, C. and Schilke, B. (1999) Mitochondrial iron metabolism in the yeast *Saccharomyces cerevisiae*. Biol. Chem. 380, 1167–1173.

Csere, P., Lill, R. and Kispal, G. (1998) Identification of a human mitochondrial ABC transporter, the functional orthologue of yeast Atm1p. FEBS Lett. 441, 266–270.

Dalbey, R.E. and Kuhn, A. (2000) Evolutionarily related insertion pathways of bacterial, mitochondrial, and thylakoid. Annu. Rev. Cell Dev. Biol. 16, 51–87.

Dean, M., Allikmets, R., Gerrard, B., Stewart, C., Kistler, A., Shafer, B., Michaelis, S. and Strathern, J. (1994) Mapping and sequencing of two yeast genes belonging to the ATP-binding cassette superfamily. Yeast 10, 377–383.

Decottignies, A. and Goffeau, A. (1997) Complete inventory of the yeast ABC proteins. Nat. Genet. 15, 137–145.

Eden, A., Simchen, G. and Benvenisty, N. (1996) Two yeast homologs of ECA39, a target for c-Myc regulation, code for cytosolic and mitochondrial branched-chain amino acid aminotransferases. J. Biol. Chem. 271, 20242–20245.

El Moualij, B., Duyckaerts, C., Lamotte-Brasseur, J. and Sluse, F.E. (1997) Phylogenetic classification of the mitochondrial carrier family of *Saccharomyces cerevisiae*. Yeast 13, 573–581.

Elliott, T. (1997) Transporter associated with antigen processing. Adv. Immunol. 65, 47–109.

Faivre-Nitschke, S.E., Nazoa, P., Gualberto, J.M., Grienenberger, J.M. and Bonnard, G. (2001) Wheat mitochondria ccmB encodes the membrane domain of a putative ABC transporter involved in cytochrome *c* biogenesis. Biochim. Biophys. Acta 1519, 199–208.

Foury, F. (1999) Low iron concentration and aconitase deficiency in a yeast frataxin homologue deficient strain. FEBS Lett. 456, 281–284.

Foury, F. and Cazzalini, O. (1997) Deletion of the yeast homologue of the human gene associated with Friedreich's ataxia elicits iron accumulation in mitochondria. FEBS Lett. 411, 373–377.

Garland, S.A., Hoff, K., Vickery, L.E. and Culotta, V.C. (1999) *Saccharomyces cerevisiae ISU1* and *ISU2*: members of a well-conserved gene family for iron-sulfur cluster assembly. J. Mol. Biol. 294, 897–907.

Goldman, B.S. and Kranz, R.G. (2001) ABC transporters associated with cytochrome *c* biogenesis. Res. Microbiol. 152, 323–329.

Goldman, B.S., Beck, D.L., Monika, E.M. and Kranz, R.G. (1998) Transmembrane heme delivery systems. Proc. Natl Acad. Sci. USA 95, 5003–5008.

Higgins, C.F. (1992) ABC transporters: from microorganisms to man. Annu. Rev. Cell Biol. 8, 67–113.

Hinnebusch, A.G. (1992) General and pathway-specific regulatory mechanisms controlling the synthesis of amino acid biosynthetic enzymes in *Saccharomyces cerevisiae.* In: The Molecular and Cellular Biology of the Yeast *Saccharomyces*: Vol. II: Gene Expression (ed. E.W. Jones, J.R. Pringle and J.R. Broach), pp. 319–414. Cold Spring Harbor: CSH Laboratory Press.

Hoff, K.G., Silberg, J.J. and Vickery, L.E. (2000) Interaction of the iron-sulfur cluster assembly protein IscU with the Hsc66/Hsc20 molecular chaperone system of *Escherichia coli*. Proc. Natl Acad. Sci. USA 97, 7790–7795.

Hogue, D.L., Liu, L. and Ling, V. (1999) Identification and characterization of a mammalian mitochondrial ATP-binding cassette membrane protein. J. Mol. Biol. 285, 379–389.

Jensen, L.T. and Culotta, V.C. (2000) Role of *Saccharomyces cerevisiae ISA1* and *ISA2* in iron homeostasis. Mol. Cell. Biol. 20, 3918–3927.

Jones, E.W. and Fink, G.R. (1982) Regulation of amino acid and nucleotide biosynthesis in yeast. In: The Molecular Biology of the Yeast *Saccharomyces.* Metabolism and Gene Expression (ed. J.N. Strathern, E.W. Jones and J.R. Broach), pp. 181–299. Cold Spring Harbor: CSH Laboratory Press.

Kaufman, B.A., Newman, S.M., Hallberg, R.L., Slaughter, C.A., Perlman, P.S. and Butow, R.A. (2000) *In organello* formaldehyde crosslinking of proteins to mtDNA: identification of bifunctional proteins. Proc. Natl Acad. Sci. USA 97, 7772–7777.

Kaut, A., Lange, H., Diekert, K., Kispal, G. and Lill, R. (2000) Isa1p is a component of the mitochondrial machinery for maturation of cellular iron-sulfur proteins and requires conserved cysteine residues for function. J. Biol. Chem. 275, 15955–15961.

Kim, R., Saxena, S., Gordon, D.M., Pain, D. and Dancis, A. (2001) J-domain protein, Jac1p, of yeast mitochondria required for iron homeostasis and activity of Fe-S cluster proteins. J. Biol. Chem. 276, 17524–17532.

Kispal, G., Steiner, H., Court, D.A., Rolinski, B. and Lill, R. (1996) Mitochondrial and cytosolic branched-chain amino acid transaminases from yeast, homologs of the *myc* oncogene-regulated Eca39 protein. J. Biol. Chem. 271, 24458–24464.

Kispal, G., Csere, P., Guiard, B. and Lill, R. (1997) The ABC transporter Atm1p is required for mitochondrial iron homeostasis. FEBS Lett. 418, 346–350.

Kispal, G., Csere, P., Prohl, C. and Lill, R. (1999) The mitochondrial proteins Atm1p and Nfs1p are required for biogenesis of cytosolic Fe/S proteins. EMBO J. 18, 3981–3989.

Klein, I., Sarkadi, B. and Varadi, A. (1999) An inventory of the human ABC proteins. Biochem. Biophys. Res. Commun. 1461, 237–262.

Kohlhaw, G.B. (1988a) Beta-isopropylmalate dehydrogenase from yeast. Methods Enzymol. 166, 429–435.

Kohlhaw, G.B. (1988b) Isopropylmalate dehydratase from yeast. Methods Enzymol. 166, 423–429.

Kranz, R., Lill, R., Goldman, B., Bonnard, G. and Merchant, S. (1998) Molecular mechanisms of cytochrome *c* biogenesis: three distinct systems. Mol. Microbiol. 29, 383–396.

Krebs, C., Agar, J.N., Smith, A.D., Frazzon, J., Dean, D.R., Huynh, B.H. and Johnson, M.K. (2001) IscA, an alternate scaffold for Fe-S cluster biosynthesis. Biochemistry 40, 14069–14080.

Künkele, K.P., Juin, P., Pompa, C., Nargang, F.E., Henry, J.P., Neupert, W., Lill, R. and Thieffry, M. (1998) The isolated complex of the translocase of the outer membrane of mitochondria. Characterization of the cation-selective and voltage-gated preprotein-conducting pore. J. Biol. Chem. 273, 31032–31039.

Kushnir, S., Babiychuck, E., Storozhenko, S., Davey, M.W., Papenbrock, J., De Rycke, R., *et al.* (2001) A mutation of the mitochondrial ABC transporter Sta1 leads to dwarfism and chlorosis in the *Arabidopsis* mutant *starik*. Plant Cell 13, 89–100.

Lange, H., Kispal, G. and Lill, R. (1999) Mechanism of iron transport to the site of heme synthesis inside yeast mitochondria. J. Biol. Chem. 274, 18989–18996.

Lange, H., Kispal, G., Kaut, A. and Lill, R. (2000) A mitochondrial ferredoxin is essential for biogenesis of intra- and extra-mitochondrial Fe/S proteins. Proc. Natl Acad. Sci. USA. 97, 1050–1055.

Lange, H., Lisowsky, T., Gerber, J., Mühlenhoff, U., Kispal, G. and Lill, R. (2001) An essential function of the mitochondrial sulfhydryl oxidase Erv1p/ALR in the maturation of cytosolic Fe/S proteins. EMBO Rep. 2, 715–720.

Langer, T. (2000) AAA proteases: cellular machines for degrading membrane proteins. Trends Biochem. Sci. 25, 247–251.

Lee, J.E., Hofhaus, G. and Lisowsky, T. (2000) Erv1p from *Saccharomyces cerevisiae* is a FAD-linked sulfhydryl oxidase. FEBS Lett. 477, 62–66.

Leighton, J. and Schatz, G. (1995) An ABC transporter in the mitochondrial inner-membrane is required for normal growth of yeast. EMBO J. 14, 188–195.

Li, J., Saxena, S., Pain, D. and Dancis, A. (2001) Adrenodoxin reductase homolog (Arh1p) of yeast mitochondria required for iron homeostasis. J. Biol. Chem. 276, 1503–1509.

Li, Z.S., Szczypka, M., Lu, Y.P., Thiele, D.J. and Rea, P.A. (1996) The yeast cadmium factor protein (YCF1) is a vacuolar glutathione S-conjugate pump. J. Biol. Chem. 271, 6509–6517.

Lill, R. and Kispal, G. (2000) Maturation of cellular Fe/S proteins: The essential function of mitochondria. Trends Biochem. Sci. 25, 352–356.

Lill, R., Diekert, K., Kaut, A., Lange, H., Pelzer, W., Prohl, C. and Kispal, G. (1999) The essential role of mitochondria in the biogenesis of cellular iron-sulfur proteins. Biol. Chem. 380, 1157–1166.

Linton, K.J. and Higgins, C.F. (1998) The *Escherichia coli* ATP-binding cassette (ABC) proteins. Mol. Microbiol. 28, 5–13.

Lisowsky, T. (1992) Dual function of a new nuclear gene for oxidative phosphorylation and vegetative growth in yeast. Mol. Gen. Genet. 232, 58–64.

Lutz, T., Westermann, B., Neupert, W. and Herrmann, J.M. (2001) The mitochondrial proteins Ssq1 and Jac1 are required for the assembly of iron sulfur clusters in mitochondria. J. Mol. Biol. 307, 815–825.

Mao, M., Fu, G., Wu, J.S., Zhang, Q.H., Zhou, J., Kan, L.X., *et al.* (1998) Identification of genes expressed in human CD34(+) hematopoietic stem/progenitor cells by expressed sequence tags and efficient full-length cDNA cloning. Proc. Natl Acad. Sci. USA 95, 8175–8180.

Mitsuhashi, N., Miki, T., Senbongi, H., Yokoi, N., Yano, H., Miyazaki, M., *et al.* (2000) MTABC3, a novel mitochondrial ATP-binding cassette protein involved in iron homeostasis. J. Biol. Chem. 275, 17536–17540.

Mühlenhoff, U. and Lill, R. (2000) Biogenesis of iron-sulfur proteins in eukaryotes: a novel task of mitochondria that is inherited from bacteria. Biochim. Biophys. Acta 1459, 370–382.

Nelson, D.R., Felix, C.M. and Swanson, J.M. (1998) Highly conserved charge-pair networks in the mitochondrial carrier family. J. Mol. Biol. 277, 285–308.

Neupert, W. (1997) Protein import into mitochondria. Ann. Rev. Biochem. 66, 861–915.

Ollagnier-de-Choudens, S., Mattioli, T., Takahashi, Y. and Fontecave, M. (2001) Iron-sulfur cluster assembly: characterization of IscA and evidence for a specific and functional complex with ferredoxin. J. Biol. Chem. 276, 22604–22607.

Ortiz, D.F., Ruscitti, T., McCue, K.F. and Ow, D.W. (1995) Transport of metal-binding peptides by *HMT1*, a fission yeast ABC-type vacuolar membrane protein. J. Biol. Chem. 270, 4721–4728.

Page, M.D., Sambongi, Y. and Ferguson, S.J. (1998) Contrasting routes of *c*-type cytochrome assembly in mitochondria, chloroplasts and bacteria. Trends Biochem. Sci. 23, 103–108.

Palmieri, L., Runswick, M.J., Fiermonte, G., Walker, J.E. and Palmieri, F. (2000) Yeast mitochondrial carriers: bacterial expression, biochemical identification and metabolic significance. J. Bioenerg. Biomemb. 32, 67–77.

Pelzer, W., Mühlenhoff, U., Diekert, K., Siegmund, K., Kispal, G. and Lill, R. (2000) Mitochondrial Isa2p plays a crucial role in the maturation of cellular iron-sulfur proteins. FEBS Lett. 476, 134–139.

Pfanner, N. and Geissler, A. (2001) Versatility of the mitochondrial protein import machinery. Nature Rev. Mol. Cell Biol. 2, 339–349.

Prohl, C., Kispal, G. and Lill, R. (2000) Branched-chain amino acid transaminases of yeast *Saccharomyces cerevisiae*. Methods Enzymol. 324, 365–375.

Prohl, C., Pelzer, W., Diekert, K., Kmita, H., Bedekovics, T., Kispal, G. and Lill, R. (2001) The yeast mitochondrial carrier Leu5p and its human homologue Graves' disease protein are required for accumulation of coenzyme A in the matrix. Mol. Cell Biol. 21, 1089–1097.

Ren, Q. and Thony-Meyer, L. (2001) Physical interaction of CcmC with heme and the heme chaperone CcmE during cytochrome *c* maturation. J. Biol. Chem. 276, 32591–32596.

Rep, M., van Dijl, J.M., Suda, K., Schatz, G., Grivell, L.A. and Suzuki, C.K. (1996) Promotion of mitochondrial membrane complex assembly by a proteolytically inactive yeast Lon. Science 274, 103–106.

Ritz, U. and Seliger, B. (2001) The transporter associated with antigen processing (TAP): structural integrity, expression, function, and its clinical relevance. Mol. Med. 7, 149–158.

Rötig, A., de Lonlay, P., Chretien, D., Foury, F., Koenig, M., Sidi, D., Munnich, A. and Rustin, P. (1997) Aconitase and mitochondrial iron-sulphur protein deficiency in Friedreich ataxia. Nat. Genet. 17, 215–217.

Sanchez-Fernandez, R., Davies, T.G., Coleman, J.O. and Rea, P.A. (2001) The *Arabidopsis thaliana* ABC protein superfamily, a complete inventory. J. Biol. Chem. 276, 30231–30244.

Schilke, B., Voisine, C., Beinert, H. and Craig, E. (1999) Evidence for a conserved system for iron metabolism in the mitochondria of *Saccharomyces cerevisiae*. Proc. Natl Acad. Sci. USA 96, 10206–10211.

Schulz, H., Hennecke, H. and Thony-Meyer, L. (1998) Prototype of a heme chaperone essential for cytochrome *c* maturation. Science 281, 1197–1200.

Schulz, H., Fabianek, R.A., Pellicioli, E.C., Hennecke, H. and Thony-Meyer, L. (1999) Heme transfer to the heme chaperone CcmE during cytochrome *c* maturation requires the CcmC protein, which may function independently of the ABC-transporter CcmAB. Proc. Natl Acad. Sci. USA 96, 6462–6467.

Senbongi, H., Ling, F. and Shibata, T. (1999) A mutation in a mitochondrial ABC transporter results in mitochondrial dysfunction through oxidative damage of mitochondrial DNA. Mol. Gen. Genet. 262, 426–436.

Shimada, Y., Okuno, S., Kawai, A., Shinomiya, H., Saito, A., Suzuki, M., *et al.* (1998) Cloning and chromosomal mapping of a novel ABC transporter gene (hABC7), a candidate for X-linked sideroblastic anemia with spinocerebellar ataxia. J. Hum. Genet. 43, 115–122.

Shirihai, O.S., Gregory, T., Yu, C., Orkin, S.H. and Weiss, M.J. (2000) ABC-me: a novel mitochondrial transporter induced by GATA-1 during erythroid differentiation. EMBO J. 19, 2492–2502.

Silberg, J.J., Hoff, K.G., Tapley, T.L. and Vickery, L.E. (2000) The Fe/S assembly protein IscU behaves as a substrate for the molecular chaperone Hsc66 from *Escherichia coli*. J. Biol. Chem. 276, 1696–1700.

Strain, J., Lorenz, C.R., Bode, J., Garland, S., Smolen, G.A., Ta, D.T., Vickery, L.E. and Culotta, V.C. (1998) Suppressors of superoxide dismutase (SOD1) deficiency in *Saccharomyces cerevisiae*. Identification of proteins predicted to mediate iron-sulfur cluster assembly. J. Biol. Chem. 273, 31138–31144.

Taglicht, D. and Michaelis, S. (1998) *Saccharomyces cerevisiae* ABC proteins and their relevance to human health and disease. Methods Enzymol. 292, 130–162.

Thony-Meyer, L. (2000) Haem-polypeptide interactions during cytochrome *c* maturation. Biochim. Biophys. Acta 1459, 316–324.

Voisine, C., Cheng, Y.C., Ohlson, M., Schilke, B., Hoff, K., Beinert, H., Marszalek, J. and Craig, E.A. (2001) Jac1, a mitochondrial J-type chaperone, is involved in the biogenesis of Fe/S clusters in *Saccharomyces cerevisiae*. Proc. Natl Acad. Sci. USA 98, 1483–1488.

Young, L., Leonhard, K., Tatsuta, T., Trowsdale, J. and Langer, T. (2001) Role of the ABC transporter Mdl1 in peptide export from mitochondria. Science 291, 2135–2138.

Yuvaniyama, P., Agar, J.N., Cash, V.L., Johnson, M.K. and Dean, D.R. (2000) NifS-directed assembly of a transient [2Fe-2S] cluster within the NifU protein. Proc. Natl Acad. Sci. USA 97, 599–604.

Zhang, F., Hogue, D.L., Liu, L., Fisher, C.L., Hui, D., Childs, S. and Ling, V. (2000a) M-ABC2, a new human mitochondrial ATP-binding cassette membrane protein. FEBS Lett. 478, 89–94.

Zhang, F., Zhang, W., Liu, L., Fisher, C.L., Hui, D., Childs, S., Dorovini-Zis, K. and Ling, V. (2000b) Characterization of ABCB9, an ATP binding cassette protein associated with lysosomes. J. Biol. Chem. 275, 23287–23294.

Zheng, L., White, R.H., Cash, V.L., Jack, R.F. and Dean, D.R. (1993) Cysteine desulfurase acitivity indicates a role for NifS in metallocluster biosynthesis. Proc. Natl Acad. Sci. USA 90, 2754–2758.

Zheng, L., White, R.H., Cash, V.L. and Dean, D.R. (1994) Mechanism for the desulfurization of L-cysteine catalyzed by the *nifS* gene product. Biochemistry 33, 4714–4720.

Zheng, L., Cash, V.L., Flint, D.H. and Dean, D.R. (1998) Assembly of iron-sulfur clusters. Identification of an *iscSUA-hscBA-fdx* gene cluster from *Azotobacter vinelandii*. J. Biol. Chem. 273, 13264–13272.

The Transporter Associated with Antigen Processing (TAP): a Peptide Transport and Loading Complex Essential for Cellular Immune Response

CHAPTER 26

Brigitte Lankat-Buttgereit and Robert Tampé

Overview of the MHC Class I Antigen Processing

During evolution, the adaptive immune system has developed to protect the organism against pathogens. This system consists of three interrelated branches of defense, depending on where the first step of elimination of foreign antigens occurs. The humoral system is responsible for recognition and elimination of intact pathogens such as viruses or bacteria in the extracellular space via antibodies. These are produced by B-lymphocytes and, subsequently, they activate the complement system. The cellular immune system is subdivided into two components. In the first, endogenous proteins are degraded in the cytosol by the proteasome and the resulting peptides are transported into the endoplasmic reticulum (ER), where they bind to major histocompatibility complex (MHC) class I molecules. In the second, exogenous proteins are degraded after internalization in specialized late endosomal or pre-lysosomal compartments which contain MHC class II molecules. The antigenic peptides are directly loaded onto the MHC class II complexes. Both peptide-loaded MHC class I and class II molecules are transported on the cell surface and recognized by cytotoxic and helper T-lymphocytes, respectively. The focus of this chapter is on the MHC class I pathway.

In the non-infected state, MHC class I complexes are presented on the cell surface by binding peptides derived from normal cellular proteins. Cytotoxic T-lymphocytes (CTLs) are not activated by this chronic presentation of self-peptides, because T-cells with the ability to respond to these molecules are eliminated during thymus development. The pathways leading to the generation of peptides, their binding to MHC molecules, and their subsequent expression on the cell surface are called antigen processing and presentation **(Figure 26.1A)**.

During viral infection or malignant transformation, a set of 'non-self' peptides of the target cell is presented to the CTLs, which recognize the MHC class I molecule as a self-component loaded with a peptide derived from non-self proteins and which eliminate these cells (for review, see Ljunggren *et al.*, 1990; Townsend *et al.*, 1989). Interference with this antigen presentation pathway is an effective method for pathogens to dodge the immune response.

MHC class I molecules are composed of a polymorphic heavy chain (α-chain), which is encoded within the MHC locus, and an invariant

ABC Proteins: From Bacteria to Man
ISBN 0-12-352551-9

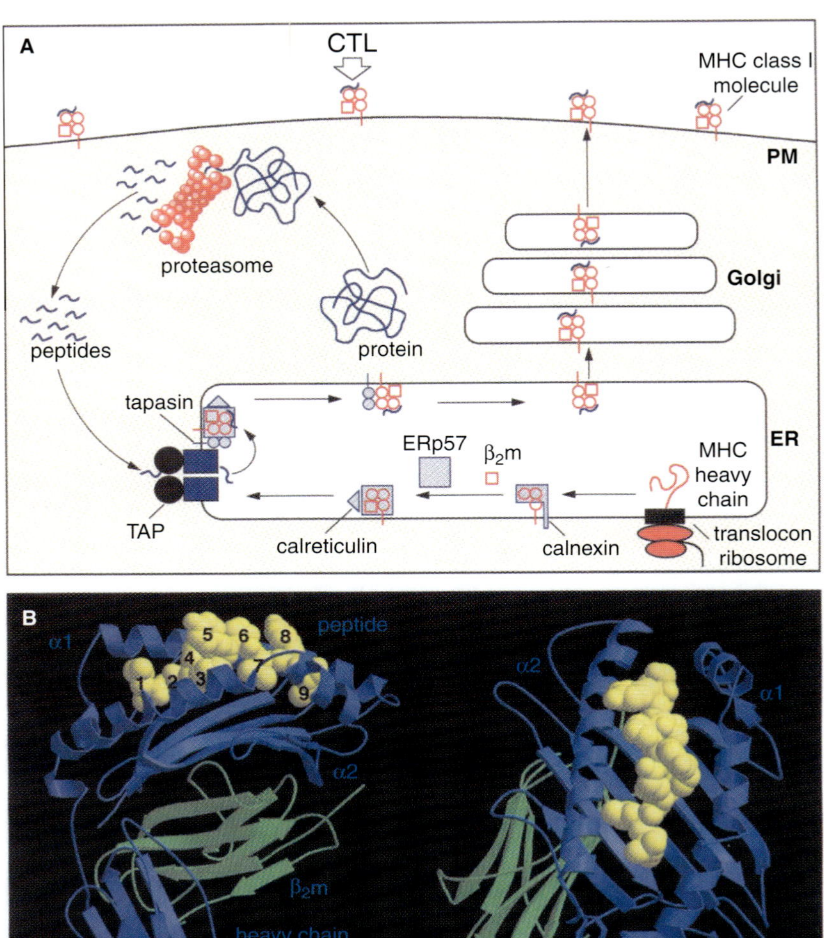

Figure 26.1. A, Antigen processing pathway via MHC class I. In the ER, the MHC class I heavy chain associates with the chaperone calnexin. After assembly with β_2*-microglobulin (*β_2*m) and the thiol reductase ERp57, calnexin is replaced by calreticulin. Subsequently, tapasin mediates the association with TAP, forming a macromolecular transport and loading complex for antigenic peptides. Peptides are generated within the ubiquitin–proteasome pathway in the cytosol. After peptide loading, MHC class I molecules are released from the TAP complex and traffic via the Golgi apparatus and the* trans*-Golgi network to the cell surface. There CTLs recognize the antigenic peptides in complex with MHC class I molecules. B, Structure of the peptide–MHC complex. Side and top view of the MHC class I molecules HLA-A2 with bound antigenic peptide derived from tax protein of human T-cell lymphotropic virus (Madden* et al.*, 1993). The peptide epitope (LLFGYPVYV, yellow, residues are numbered) binds in a groove of the heavy chain of HLA-A2 (blue), which is formed by two* α*-helices on the rim and eight* β*-strands at the bottom of the* α*1 and* α*2 domain. Peptide, heavy chain and* β_2*-microglobulin (green) form a stable complex with a half-life of several days. The groove fits peptides with a length of eight to ten amino acids. Peptides are fixed via their free amino and carboxy termini as well as via side-chain interactions at position two or three and the C-terminus. Side-chains, which point out of the groove, are monitored by the T-cell receptor.*

non-MHC encoded subunit, β_2-microglobulin (for review, see Ljunggren *et al.*, 1990; Townsend *et al.*, 1989). The assembly of the different subunits proceeds in the ER by a folding process that is synchronized in time and space by various chaperones. After co-translational translocation, the α-chain of the MHC class I molecule associates with the chaperone protein, calnexin, and forms a dimer with β_2-microglobulin (for review, see Pamer and Cresswell, 1998). β_2-Microglobulin interacts extensively with domains of the α-chain and, consequently, the correct folding of the α-chain is dependent on the dimerization with β_2-microglobulin. Calnexin is then replaced by another chaperone, calreticulin, and the thiol reductase ERp57 then associates with the complex. For the binding of peptides to the MHC class I heterodimer, a macromolecular loading complex, together with tapasin and an ATP-binding cassette (ABC) heterodimeric protein known as the transporter associated with antigen processing (TAP) (ABCB2/ABCB3), is formed. Tapasin is an ER-resident type I glycoprotein that mediates the efficient interaction of TAP and class I molecules (Sadasivan *et al.*, 1996). The peptide loading onto the MHC class I heterodimer stabilizes the molecule and it is released from the assembly complex for transport to the plasma membrane via the Golgi apparatus and the *trans*-Golgi network **(Figure 26.1B)**.

Normal cellular proteins, as well as viral proteins or proteins that are artificially introduced into the cytosol (Moore *et al.*, 1988; Yewdell *et al.*, 1988), are degraded by an extralysosomal pathway (Morrison *et al.*, 1986). The major sources for antigenic peptides are proteins cleaved by a proteasome-based mechanism, which degrades unfolded or ubiquitinated proteins in the cytoplasmic compartment (Rock and Goldberg, 1999; York *et al.*, 1999). One form of the proteasome is a 20S (700 kDa) cylindrical particle, consisting of 28 subunits arranged in four heptameric rings. The outer rings are composed of seven α subunits with regulatory and structural functions, while the inner rings consist of seven β subunits containing the catalytic sites (Baumeister *et al.*, 1998). The 26S proteasome (1500 kDa) is associated with additional subunits, which have a regulatory function. Approximately one-third of newly synthesized proteins are degraded by proteasomes into peptides with a size distribution of 3–30 amino acid residues (Kisselev *et al.*, 1999; Schubert *et al.*, 2000; Turner and Varshavsky, 2000). The optimal size is 6–11 residues, which overlaps with the size of antigenic peptides (8–11 residues) bound to MHC class I molecules (Kisselev *et al.*, 1999).

The transport of the peptides generated in the cytosol into the ER lumen is executed by TAP (ABCB2/ABCB3). It has been established that any defect in TAP severely impairs antigen presentation. Since peptide binding is necessary for stabilization of the MHC complex, a reduced or abolished transport activity of TAP results in reduced cell surface expression of MHC class I molecules. Thus, TAP function is essential for antigen presentation and, consequently, inhibition of TAP function is an effective strategy for pathogens to avoid immune surveillance, leading to chronic or latent infections. During the last few years, the understanding of the function of TAP has increased significantly. Disturbance of peptide delivery to the MHC class I complex is associated with various human diseases from tumor development to infections. In this chapter, the current knowledge of the mechanisms enabling transport of peptides from the cytosol into the lumen of the ER for antigen presentation is summarized. In addition, TAP serves as an important model system to aid in understanding the topic of multisubstrate specificity, transport mechanism and inhibition of function by natural inhibitors.

The Transporter Associated with Antigen Processing, TAP (ABCB2/ABCB3)

Genomic organization and regulation of *TAP*

Some years ago, it was observed that some tumor cell lines exhibit a low cell surface expression of MHC class I molecules and are deficient in antigen presentation. However, at low temperatures, the expression of the MHC class I α-chain and β_2-microglobulin could be restored to normal levels (Ljunggren *et al.*, 1990; Townsend *et al.*, 1989). The defect was located in the MHC locus and it was concluded that a gene or genes were involved in peptide loading of the class I molecules. In the following years, four groups independently discovered candidate genes for proteins that were implicated in the transport of peptides from the cytosol into the lumen of the ER (Deverson *et al.*, 1990; Monaco *et al.*, 1990; Spies *et al.*, 1990; Trowsdale *et al.*, 1990). Since then, these human, mouse and rat genes have been renamed as the transporter associated with antigen processing, *TAP* (*TAP1/ABCB2* for *RING4*, *PSF1*, *mtp1* and *HAM1*; *TAP2/ABCB3* for *RING11*, *PSF2*, *mtp2* and *HAM2*). Transfection of defective cell lines with *TAP1* (*ABCB2*) and/or *TAP2* (*ABCB3*) cDNAs restores MHC class I surface expression and antigen presentation. These findings indicate that MHC class I molecules are stabilized by binding peptides and that the majority of peptides are transported by TAP from the cytosol into the ER lumen.

The human *TAP* genes are located on chromosome 6 band p21.3 in the MHC II locus. They are 8–12 kb in size and consist of 11 exons each (Hanson and Trowsdale, 1991). The *TAP1* 2.5 kb mRNA encodes a protein of 748 amino acids, while the *TAP2* 2.8 kb mRNA encodes a protein of 686 amino acids. Sequence alignments of the coding region from human to horned shark, the most distant vertebrate class displaying an adaptive immune system, exhibit the expected phylogenetic differences. For example, human TAP1 shares 98.8% amino acid homology with the gorilla TAP1 protein, 69.2% with the hamster protein, and approximately 43% with the horned shark protein. The homology between TAP1 and TAP2 is approximately 35% in all species examined thus far and the two proteins share a similar predicted membrane topology. It is likely that these genes evolved from a common ancestral gene by gene duplication prior to the development of the adaptive immune system in jawed vertebrates.

The human *TAP* genes contain putative GC-rich elements (Sp1-binding sites) in their 5′-flanking sequences, but no TATA box motifs (Beck *et al.*, 1992). It was shown by mutagenesis that the Sp1-binding sites are necessary for the basal promoter activity of *TAP1* (Wright *et al.*, 1995). In addition, several other motifs induce *TAP1* promoter activity such as interferon (IFN)-γ- and p53-responsive elements (Zhu *et al.*, 1999). Interestingly, the *TAP1* gene is coordinately regulated by a bi-directional promoter with the divergently transcribed *LMP2* gene (Israel *et al.*, 1989). *LMP2* encodes the alternative β-type proteasomal subunit, which is important for differential processing of epitopes by constitutive and immunoproteasomes (Gaczynska *et al.*, 1994; Toes *et al.*, 2001). Both genes are stimulated by tumor necrosis factor (TNF)-α. The induced expression of *TAP1* and *LMP2* concordantly with upregulated MHC class I genes suggests a link between generation of peptides and expression levels of the transporter.

Expression of the MHC class I molecules correlates with CTL function and can be increased by cytokines such as IFNs (for review, see Früh and Yang, 1999). *TAP1* and *TAP2* mRNA and protein levels are rapidly upregulated by IFN-γ, whereas MHC class I α-chains and cell surface expression increase more slowly (Ma *et al.*, 1997). A similar enhancement of TAP1 was observed by *in vitro* treatment of tumor samples with TNF-α (Nagy *et al.*, 1998). In contrast to these cytokines, interleukin-10 has a reverse effect on TAP expression and reduces TAP1 and TAP2 levels (Salazar-Onfray *et al.*, 1997).

In addition to the interference of TAP function by cytokines, some other mechanisms are known to regulate TAP activity. In certain breast cancer cell lines, TAP expression was observed to be dependent on the cell cycle, and the overall amounts of *TAP* mRNAs were lower than in normal breast epithelial cells (Alpan *et al.*, 1996). Tumor cells may evade host tumor surveillance by mutations that inhibit TAP function. Because more than 50% of human tumors have no functional p53, the influence of p53 on TAP1 levels was examined (Zhu *et al.*, 1999). Overexpression of p53 increased *TAP1* mRNA and protein levels and, subsequently, MHC class I cell surface expression. The authors suggested that a non-functional p53 cannot induce TAP following genotoxic stress. Thus, p53 may act as a tumor suppressor by inducing TAP and thereby tumor surveillance.

Structural organization of TAP

Like MDR1 (ABCB1) and MDR3 (ABCB4), TAP1 and TAP2 belong to subfamily B of the ABC superfamily. Each TAP protein consists of one ATP-binding domain (nucleotide-binding domain: NBD) and one hydrophobic region of 10 (TAP1) or 9 (TAP2) transmembrane (TM) helices. The homology with other 'half-size' transporters indicated that a functional TAP complex consisted of either a homodimer of TAP1 or TAP2, or a TAP1/TAP2 heterodimer. Heterologous coexpression of TAP1 and/or TAP2 in yeast and insect cells demonstrated that TAP is active as a heterodimer and that no additional factors of the adaptive immune system are needed for TAP function (Meyer *et al.*, 1994; Urlinger *et al.*, 1997). Moreover, immunoprecipitation with antibodies directed against TAP1 co-precipitate

TAP1 and TAP2 (Kelly *et al.*, 1992). The TAP complex is located in the ER as shown by immunoelectron and immunofluorescence microscopy (Kleijmeer *et al.*, 1992; Meyer *et al.*, 1994). Both proteins lack an NH_2-terminal signal sequence for ER targeting. The complex is retarded in the ER by an internal signal sequence. Recent studies with truncated proteins indicate that ER retention of both TAP1 and TAP2 is achieved by multiple signals in the transmembrane regions (Vos *et al.*, 1999). TAP1 has three predicted glycosylation sites, two facing the cytosol and one placed in an ER loop, which is likely to be too short for glycosylation. Consistent with these predictions, it was found that both proteins are predominantly non-glycosylated (Meyer *et al.*, 1994). A minor subpopulation of TAP has been reported to be N-glycosylated, but this may consist of misfolded protein (Russ *et al.*, 1995).

Hydrophobicity analysis predicts that each TAP protein contains 6–10 TM helices depending on the algorithm used **(Figure 26.2A)** (Elliott, 1997; Gileadi and Higgins, 1997; Nijenhuis and Hämmerling, 1996; Tampé *et al.*, 1997). A core domain of six TM-spanning helices, which is found in all other ABC transporters, may possibly serve to align the translocation pore, and the sequence similarity increases from TM1 through to TM6. By sequence alignments, the first 175 and the first 140 NH_2-terminal amino acid residues of TAP1 and TAP2, respectively, which are encoded by exon 1, show no corresponding domains in related ABC transporters. It is assumed that these hydrophobic regions contain an additional four and three TM helices, respectively, as extensions and might be necessary for specialization or assembly of the TAP complex (Tampé *et al.*, 1997). To clarify the topology of the TAP complex, it may be useful to construct cysteine-less mutants of TAP1 and TAP2, which are functionally active. Single cysteines can then be reintroduced in predicted loop regions and the accessibility checked by thiol-specific reagents.

Linked to the hydrophobic domains are the NBDs containing the conserved Walker A and B motifs and the 'C' transport family signature sequence located in the cytosol. According to this model, the complex contains large cytosolic loops, but only a small part passes into the lumen of the ER (Tampé *et al.*, 1997). It was proposed that these TM-spanning domains are arranged in a head–head/tail–tail orientation (Vos *et al.*, 2000), but this alignment contradicts established models of other transporters such

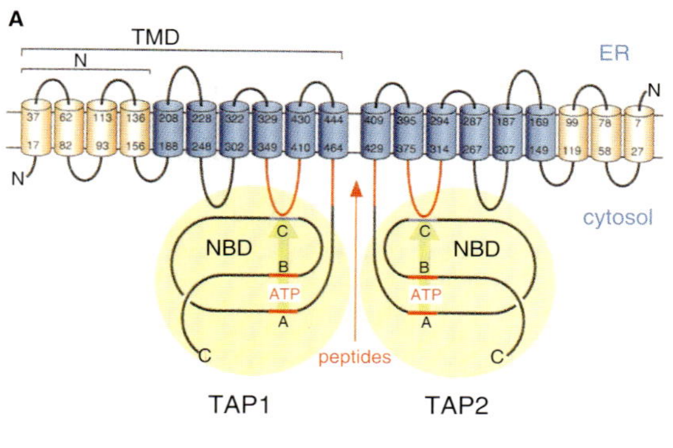

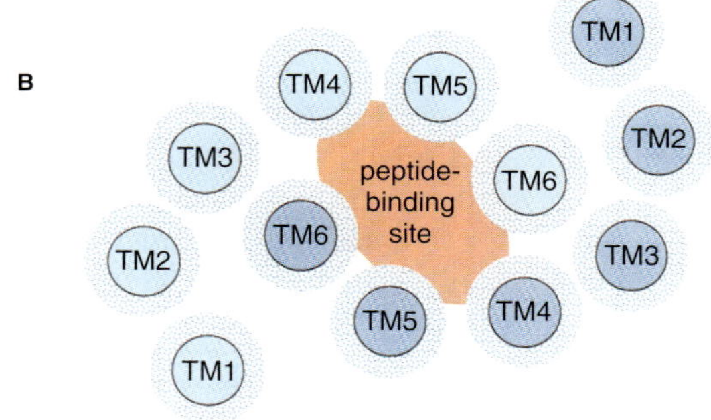

Figure 26.2. ***A, Structural organization of the TAP transporter. Membrane topology was predicted based on sequence alignments with other ABC transporters including MDR1 and hydrophobicity analysis. The translocation pore is framed by 2 × 6 transmembrane helices from TAP1 and TAP2 (blue cylinders). N-terminal regions of TAP1 and TAP2 have no counterparts in other ABC proteins. They putatively contain four and three transmembrane helices, respectively (orange cylinders). The yellow circle encompasses the highly conserved NBD with the Walker A (P loop) and B motifs (red bars) and the C-loop (blue bar). The cytosolic loops following TM6 and within TM5 and TM6 of both subunits delineate the potential peptide-binding site. B, Arrangement of the transmembrane helices. The transmembrane helices (light blue for TAP1 and dark blue for TAP2) are organized according to the model for MDR1 (Loo and Clarke, 2001a).***

as P-glycoprotein (ABCB1) (Loo and Clarke, 1995, 2001a).

The peptide-binding site is shared by both subunits as was shown by peptide photo-crosslinking and binding experiments (Androlewicz and Cresswell, 1994; Androlewicz *et al.*, 1993; van Endert *et al.*, 1994). Digestion of TAP after photo-crosslinking and subsequent immunoprecipitation with antibodies directed against different epitopes of TAP provided more detailed insight into the peptide-binding region of this transporter (Nijenhuis and Hämmerling, 1996). The cytosolic loops between TM4 and TM5 and a COOH-terminal stretch of approximately 15 amino acids following TM6 of TAP1 and TAP2 participate in peptide binding **(Figure 26.2A and B)**. Deletion of some of these potential binding sites of TAP1 resulted in loss of transporter function (Ritz *et al.*, 2001). According to one topological model of TAP (Abele and Tampé, 1999; Tampé *et al.*, 1997), these regions should all be located in the cytosol. A different model was proposed by Vos *et al.* (1999), which is in contrast to the established membrane topology of other ABC transporters (Loo and Clarke, 1995). These authors could find no evidence for membrane integration of the two hydrophobic regions adjacent to the NBDs. This could be a misleading experimental result derived from singularly expressed, non-functional deletion constructs of TAP1 and TAP2.

HOMOLOGUES OF TAP

As mentioned previously, sequence alignments show that both TAP proteins belong to subfamily B of the ABC transporter superfamily. Members of this subfamily may be 'full' transporters like P-glycoprotein/MDR1 (ABCB1) and MDR3 (ABCB4), or half-size transporters like TAP1 (ABCB2), TAP2 (ABCB3) and ABCB9. These transporters translocate a variety of molecules across different biological membranes, e.g. steroids and hydrophobic compounds by P-glycoprotein, phosphatidylcholine by MDR3/ABCB4 (see Chapter 22), possibly iron/glutathione complexes by ABCB6 and ABCB7 (Chapter 25), and monovalent bile salts by BSEP (ABCB11, sPgp). For other members of this subfamily, the substrates are unknown at present. Sequence alignments of the NBDs and a phylogenetic tree of the members of subfamily B reveals that TAP1 and TAP2 are most related to ABCB9, a half-size transporter of unknown function (Zhang *et al.*, 2000) **(Figure 26.3)**. The three genes may have arisen by duplication from an ancestral gene. Because of the close relationship with TAP1 and TAP2, it is likely that ABCB9 may act as a peptide transporter but this remains to be established. At the moment, no partner protein for ABCB9 is known; therefore, the functional complex may be a homo- or a heterodimer. The next closest relatives to the TAP proteins are ABCB8 and ABCB10, two mitochondrial ABC transporters which, due to their relation to the yeast transporter MDL1, putatively transport peptides (see Chapter 25).

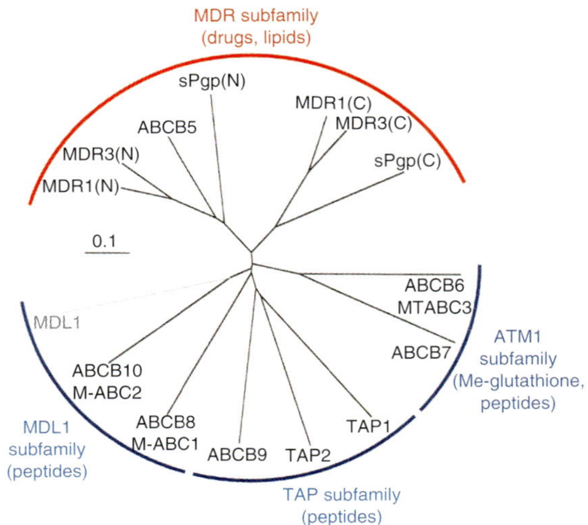

Figure 26.3. Phylogenetic analysis of TAP homologues. The NBDs of all members of the subfamily B of human ABC transporters including the TAP homologue MDL1 in yeast (**S. cerevisiae**) *are aligned by ClustalW. The member of this family most closely related to TAP is ABCB9. Also, the putative peptide transporters ABCB6, ABCB7, ABCB8 and ABCB10, all located in the inner membranes of mitochondria, are close relatives. The yeast mitochondrial peptide transporter MDL1 is included for comparison (gray). The other members of subfamily B transport hydrophobic drugs and lipids (indicated in red). (N) and (C) refer to NBDs of the N- or C-terminal half or full-size transporters.*

SUBSTRATE SELECTION AND SPECIFICITY OF TAP

The first data concerning the character of peptides that are transported by the TAP complex was obtained by trapping peptides in the ER via glycosylation or by binding to MHC class I molecules (Androlewicz *et al.*, 1993; Neefjes *et al.*, 1993). Peptides with a length of 8–16 amino acids were found to have equal affinity for TAP (van Endert *et al.*, 1994), but are most efficiently translocated into the ER when they are 8–12 residues long (Koopmann *et al.*, 1996). Moreover, free NH_2- and COOH-termini are prerequisites for transport (Momburg *et al.*, 1994; Schumacher *et al.*, 1994a; Uebel *et al.*, 1997). By screening combinatorial peptide libraries, the contribution of each amino acid to the

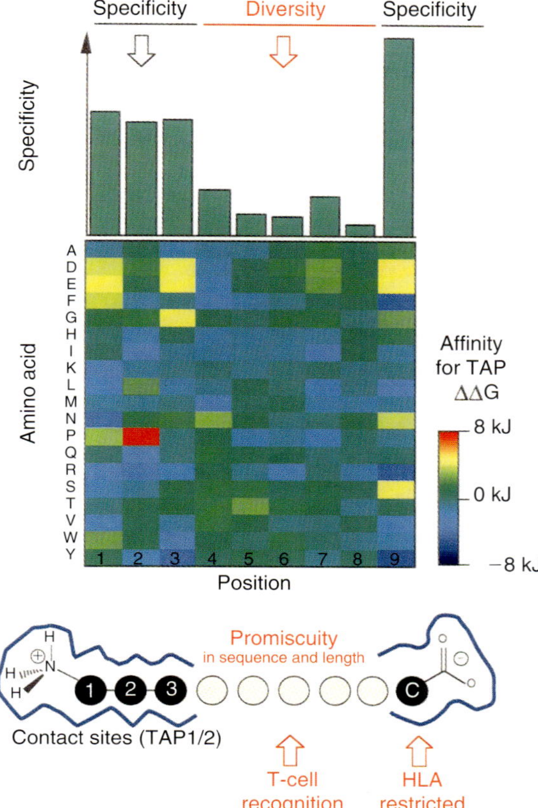

Figure 26.4. Specificity of TAP. By using combinatorial peptide libraries and statistical analysis, human TAP was found to be most specific for the three N-terminal and C-terminal residues (Uebel et al.*, 1997). Favored amino acids (K, N and R in the first, R in the second and W and Y in the third position) are shown in blue (negative ΔΔG values) in the middle panel. At the C-terminus F, L, R or Y are preferred for binding. Disfavored amino acids are shown in red (positive ΔΔG values). In the lower panel a model of the peptide-binding site to TAP is shown with the variable region of the peptide labeled in yellow.*

stabilization of peptide binding to TAP was analyzed (Uebel *et al.*, 1997). A randomized peptide mixture with one defined residue was compared with a totally randomized peptide library, and the influence of each amino acid on the affinity for TAP was determined. The peptide with the best binding characteristics showed a 45-fold higher affinity for TAP than a totally randomized peptide mixture.

The effect of each amino acid was found to be critically dependent on its position in the peptide **(Figure 26.4)**. Thus, the first three NH_2-terminal residues and the COOH-terminal amino acid were most important for substrate specificity. TAP displayed preferences for peptides with Lys, Asn and Arg in the first, Arg in the second, and Trp and Tyr in the third position at the NH_2-terminus. The most profound differences were observed for peptide residues at the COOH-terminus. The highest affinity for TAP binding was found for peptides with hydrophobic or basic amino acids (Phe, Leu, Tyr or Arg) in this position. It is interesting that these residues at the COOH-terminus are also advantageous for binding to MHC class I molecules. Moreover, none of the disfavored amino acids served as a preferred anchor for MHC class I binding. Thus, it is speculated that the recognition and binding principles of TAP and MHC class I molecules coevolved.

The contribution of the peptide backbone to the substrate specificity of TAP was determined by using peptides of different length by exchanging each residue with its D-enantiomer. D-Amino acids in positions 1–3 and the COOH-terminal position resulted in a markedly reduced affinity for TAP. Thus, contact between a peptide and TAP seems to occur via the peptide backbone, the amino acid side-chains and the free NH_2- and COOH-termini, which is fixed by hydrogen bonding (Uebel *et al.*, 1997). The residues in the center of the peptide between positions 1–3 and the COOH-terminal amino acid seem to have only a little or no effect on the substrate specificity of TAP. This binding property can explain how larger peptides can bulge out of the binding pocket and how large amino acid side-chains and even fluorescence labels can be accommodated (Neumann and Tampé, 1999; Uebel *et al.*, 1995). Interestingly, these residues are responsible for the detection of the MHC class I-bound peptide by the T-cell receptor. Therefore, by binding at the termini, TAP transports peptides with maximal diversity in the center of the peptide (positions 5–8), where T-cell recognition occurs. Therefore, a coevolution of the genes involved in antigen presentation seems likely to have taken place in order to optimize the antigen processing and recognition machinery (Uebel and Tampé, 1999).

Polymorphisms in TAP have been found in human, mouse and rat by sequence analysis and restriction length polymorphism analysis (Daniel *et al.*, 1997; Momburg *et al.*, 1994; Powis *et al.*, 1992; Schumacher *et al.*, 1994a, 1994b). Although polymorphisms can contribute to immune diversity, no effect of the amino acid changes on the substrate specificity for human and mouse TAP was observed. However, a rat *TAP* polymorphism has a significant influence on peptide selectivity (Powis *et al.*, 1992). Human TAP and rat TAP from the RT1[a] strain were found to prefer peptides with hydrophobic or basic residues at the COOH-terminus, while mouse TAP and rat TAP from the strain RT1[u] favored peptides with hydrophobic COOH-terminal residues. The TAP complexes from RT1[a] and RT1[u] strains differ by the exchange of 25 amino acids in TAP2: 23 in the transmembrane domain (TMD) and two in the NBD.

In addition to gene polymorphisms, altered substrate specificity can be achieved by alternative splicing. Recently, a variant of the human TAP2 protein, called TAP2iso, was described (Yan *et al.*, 1999). This splice variant lacks exon 11 comprising a part of the coding region and the original 3′-untranslated sequence. Instead, it contains a previously unidentified exon 12. Expression of *TAP2iso* mRNA was found to be coincident with *TAP2* mRNA in several cell lines. Interestingly, heterodimers of TAP1 and TAP2iso more resemble mouse and rat RT1[u] TAP, with preferences for peptides with hydrophobic COOH-termini. Thus, alternative splicing may be a strategy for the organism to acquire broadened epitope diversity. How the variable COOH-terminus of the TAP2 NBD affects TAP substrate specificity remains an open and intriguing question.

TRANSPORT MECHANISM OF TAP

The transport mechanism of TAP is the subject of intensive investigation because of its important role in immune recognition. Translocation into the lumen of the ER is a multistep process, consisting of binding of the peptide to TAP, isomerization of the complex, and transport **(Figure 26.5)** (review by Abele and Tampé, 1999). Peptide interaction with the binding site shared by both TAP subunits is ATP-independent and follows a monophasic 1:1 Langmuir adsorption model (Uebel *et al.*, 1995). In direct binding or competition assays, no evidence for a second interaction site was found. However, it cannot be excluded that a second binding site with a very low affinity, or with a similar affinity, exists. Real-time kinetic analysis of the peptide binding with environmentally sensitive fluorescence labeled peptides revealed that this process could be subdivided into two steps (Neumann and Tampé, 1999). The peptide binding occurs in a fast bimolecular association step and determines the specificity of TAP; subsequently, a slow isomerization of the TAP complex takes place. It is proposed that the conformational change of the molecule triggers ATP hydrolysis and, thereby, peptide transport into the lumen of the ER. The isomerization of the complex also affects its lateral mobility as analyzed by fluorescence recovery after photobleaching (Reits *et al.*, 2000). This increases when TAP is inactive and decreases during peptide translocation, as was shown in studies with TAP1 tagged with green fluorescent protein (GFP) at its cytosolic COOH-terminus. However, owing to the presence of endogenous TAP, the activity of the GFP-tagged complex could not be unequivocally established.

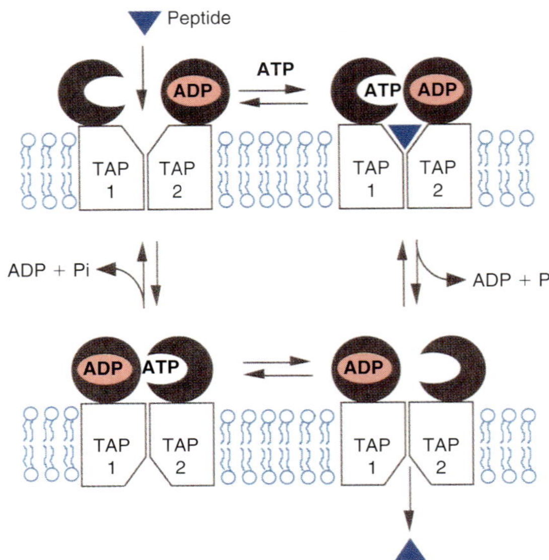

Figure 26.5. Working model of the translocation mechanism by TAP. In the ground state, ATP interacts primarily with the TAP1 subunit, whereas TAP2 most probably contains pre-bound ADP in an occluded state (red) (Alberts et al.*, 2001; Karttunen* et al. *2001; Lapinski* et al.*, 2001; Saveanu* et al.*, 2001). High-affinity peptide binding occurs in a fast reaction followed by a slow isomerization of the TAP complex, promoting allosteric coupling between the two NBDs (Neumann and Tampé, 1999). Peptide binding to TAP triggers ATP hydrolysis and subsequent translocation of the solute (Gorbulev* et al.*, 2001). ATP hydrolysis and peptide binding are tightly coupled. The release of inorganic phosphate and subsequently ADP at TAP1 might catalyze nucleotide exchange at TAP2. ATP hydrolysis at TAP2 finally closes the transport cycle by restoration of the high-affinity peptide-binding pocket. The maximal turnover rate of the transport cycle was determined to be around 5 ATP per second (Gorbulev* et al.*, 2001).*

The transport of peptides from the cytosol to the ER lumen strictly requires hydrolysis and not merely binding of ATP, UTP, CTP or GTP (Androlewicz *et al.*, 1993). Non-hydrolyzable analogues of ATP, nucleotide depletion by apyrase, or competition with ADP, completely abrogate peptide translocation (Meyer *et al.*, 1994; Neefjes *et al.*, 1993; Shepherd *et al.*, 1993). Evidence of the binding of nucleotides to the NBDs was obtained by crosslinking experiments with 8-azido-ATP (Müller *et al.*, 1994; Russ *et al.*, 1995). Nucleoside tri- and diphosphates can compete for binding and have similar affinities for TAP. Thus, for example, ADP inhibits peptide translocation. By developing an enrichment and reconstitution protocol for TAP, it was possible to restore the function in proteoliposomes and to examine the specific ATPase activity (Gorbulev *et al.*, 2001). Nucleotide hydrolysis was found to be strictly dependent on binding of peptides and on crosstalk with the peptide-binding and the translocation sites. The strict correlation between peptide binding and stimulation of ATP hydrolysis may be a strategy to avoid 'wasting' ATP without transport of peptides. A further indication of the tight coupling between ATP hydrolysis and peptide transport is the observation that sterically restricted peptides, which cannot be transported by TAP, do not induce ATP hydrolysis. Maximal ATPase activity of TAP was found to be independent of substrate affinity, because peptides with different K_D values for the transporter exhibited the same V_{max} values (Gorbulev *et al.*, 2001). The two NBDs of the functional TAP complex can both interact with ATP, even if TAP1 or TAP2 are expressed separately (Müller *et al.*, 1994; Wang *et al.*, 1994). However, alone, the NBDs are unable to hydrolyze ATP. Thus, communication between the NBDs and TMDs leading to a conformational change of the NBDs by peptide binding to TAP seems to be a requirement to activate ATPase function. Furthermore, TAP function is dependent on the presence of both NBDs since disruption of one NBD leads to loss of transport (Chen *et al.*, 1996). Thus, it has been speculated that hydrolysis at one NBD is necessary for the beginning of the transport, whereas hydrolysis at the second NBD completes the cycle and may promote the reconversion of the peptide binding site to the initial state (Abele and Tampé, 1999).

But one question remains: are the two NBDs equal in function or do they have distinct functional properties? To address this point, several groups have introduced mutations in the Walker A and/or B motifs in the NBDs of TAP1 and TAP2 and examined the effects on transport function. Lysine mutations in the Walker A sequences affecting nucleotide binding/hydrolysis by TAP1 or TAP2 suggest that each NBD plays a distinct functional role (Karttunen *et al.*, 2001; Lapinski *et al.*, 2001; Saveanu *et al.*, 2001). Even if the data concerning nucleotide and peptide binding from the different groups are in part contradictory, it seems that nucleotide binding to TAP2 maintains a peptide-receptive TAP conformation, and that TAP2-mediated ATP hydrolysis is essential for translocation. The functional consequences of mutations in the Walker A motifs of TAP1 and TAP2 have led to speculations that ATP hydrolysis at TAP1 initiates the transport cycle and is a requirement for binding of ATP to the NBD of TAP2 (Alberts *et al.*, 2001). Hydrolysis by TAP2 might then complete the cycle by restoring the peptide-binding site. The use of chimeric proteins consisting of the TAP1 membrane-spanning domain and the TAP2 NBD, and vice versa, indicate that both membrane-spanning domains of TAP1 and TAP2 are necessary, but that neither NBD encompasses signals unique for peptide binding (Arora *et al.*, 2001). These observations are in agreement with earlier data demonstrating that the TM domains of both TAP1 and TAP2 are needed to form the peptide-binding pocket. The NBD-switched complexes are all transport competent (Arora *et al.*, 2001). Even TAP complexes with two identical NBDs are able to translocate peptides, although with a lower efficiency. The two NBDs of TAP1 and TAP2 appear to possess different functional properties. Further studies will elucidate the exact roles of each NBD within the transport cycle.

TAP AS A CENTRAL PART OF A MACROMOLECULAR PEPTIDE TRANSPORT AND CHAPERONE COMPLEX

TAP not only delivers peptides necessary for antigen presentation into the ER lumen, but is also part of a large macromolecular loading complex which is critical for MHC class I maturation. A number of proteins have been identified which play an important role in this complex (for review, see Cresswell *et al.*, 1999). At least three proteins are involved in the assembly of peptide-loaded TAP and MHC class I heterodimers: the chaperone calreticulin, the thiol reductase ERp57, and the glycoprotein tapasin (TAP-associated glycoprotein). Calreticulin acts as a chaperone to ensure proper folding (Sadasivan *et al.*, 1996), whereas ERp57 probably supports the correct formation of disulfide bridges (Lindquist *et al.*, 1998), and tapasin mediates the efficient interaction of TAP and the MHC class I molecules and stabilizes the loading complex (Ayalon *et al.*, 1998; Ortmann *et al.*, 1997). The stoichiometry of this complex was determined to be four MHC class I heterodimers associated with four tapasins to one TAP heterodimer (Ortmann *et al.*, 1997). In the absence of tapasin, the assembly in the ER is impaired and MHC class I antigen presentation decreases. However, in a tapasin-deficient cell line, the class I cell surface expression and function is restored by a truncated soluble tapasin lacking the transmembrane region, even if the remaining cytosolic tail is not linked to TAP (Lehner *et al.*, 1998). The physical association of TAP and class I molecules leads to the assumption that the peptides are directly loaded from TAP onto the MHC class I complex. However, application of anti-peptide antibodies inhibits peptide binding to class I molecules (Hilton *et al.*, 2001). Therefore, these authors suggested that most TAP-transported peptides diffuse through the ER lumen before being loaded onto MHC class I molecules. The binding of the peptides is necessary for the dissociation of TAP–MHC class I complexes and is dependent on conformational signals from TAP in an ATP-dependent manner (Cresswell *et al.*, 1999; Knittler *et al.*, 1999). Recent observations point to a more pronounced conformational role of TAP1 in the dynamic activity of the loading complex (Alberts *et al.*, 2001). The release of the MHC class I molecules for transport to the cell surface is synchronized with peptide binding and peptide translocation by TAP.

VIRAL STRATEGIES OF IMMUNE EVASION

In non-infected cells, MHC class I molecules are stably expressed on the cell surface presenting peptides derived from intracellular proteins.

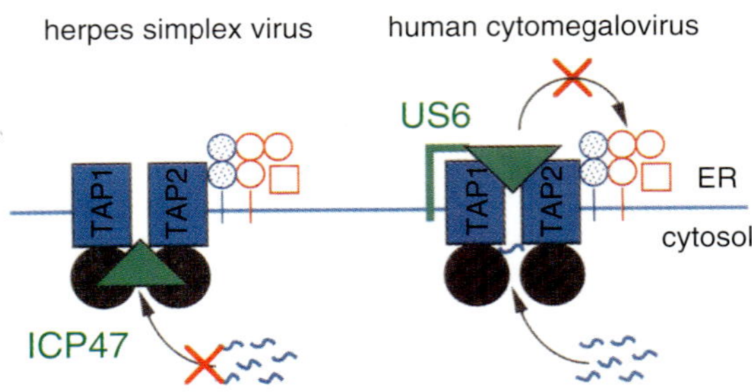

Figure 26.6. ***Inhibition of TAP function by viral proteins. The herpes simplex virus-encoded protein ICP47 blocks the TAP-mediated peptide transport by binding to the cytosolic part of the TAP complex (left side), whereas the human cytomegalovirus protein US6 inhibits TAP function by binding to the ER-luminal side (right side). The association of TAP with tapasin and MHC class I molecules seems to be unaffected by both proteins.***

Presenting peptides from viral proteins enables T-cells to recognize and eliminate infected cells. To propagate in the presence of an active immune system, the virus must develop a strategy to avoid an immune response. One mechanism is to interfere with the antigen presentation pathway at different stages (for review see Ploegh, 1998). Many steps are susceptible to viral disturbances, such as the generation of peptides (Gilbert *et al.*, 1996; Levitskaya *et al.*, 1997), the export of class I molecules to the cell surface (Früh *et al.*, 1999; Hengel *et al.*, 1999), and the transport of peptides by TAP. Here, we will focus on the latter point **(Figure 26.6)**.

Human cytomegalovirus (HCMV) encodes several proteins inhibiting cell surface expression of MHC class I molecules (for review, see Hengel and Koszinowski, 1997). One of these proteins is known to interfere with intracellular peptide transport: gpUS6. gpUS6 is an ER-resident glycoprotein that probably binds to the ER luminal part of the TAP complex (Ahn *et al.*, 1997; Hengel *et al.*, 1997; Lehner *et al.*, 1997). The association of TAP with tapasin, calreticulin and MHC class I molecules seems to be unaffected by gpUS6, as is the binding of peptides to TAP. Recent data point to a binding of gpUS6 to TAP, which results in stabilization of a TAP1 conformation that is unable to bind ATP (Hewitt *et al.*, 2001; Kyritsis *et al.*, 2001). Consequently, the energy for peptide transport is lacking. It is possible to override this inhibition by overexpression of TAP, for example by induction with IFN-γ.

Herpes simplex virus type I (HSV-1) has evolved a completely different strategy to avoid immune recognition. This virus encodes the immediate early protein ICP47 of 88 amino acids. The expression of ICP47 leads to a downregulation of MHC class I antigen presentation in human fibroblasts (York *et al.*, 1994). ICP47 inhibits TAP-mediated peptide transport into the ER lumen by binding with high affinity to the cytosolic side of the TAP complex (Früh *et al.*, 1995; Hill *et al.*, 1995), thereby preventing the binding of peptides and translocation (Ahn *et al.*, 1996; Tomazin *et al.*, 1996). Moreover, it was found that ICP47 has a 100-fold higher affinity for human TAP than for murine TAP and thus it acts in a species-specific manner. The amino acids strictly required for ICP47 function include residues 3–34 (Neumann *et al.*, 1997). This active domain seems to be unstructured in aqueous solution, but in a lipid-like environment it adopts an α-helical structure with two helical regions from amino acid 4–15 and 22–32 (Beinert *et al.*, 1997; Pfänder *et al.*, 1999). The two helices are linked by a flexible loop (helix-loop-helix motif) and the authors propose that one helix binds to TAP, while the other one is attached to the membrane. The elucidation of the ICP47 structure can serve as a template for new specific therapeutic agents, either for use as immune suppressors or for vaccination strategies against HSV.

Viruses from other families can also interfere with immune recognition via TAP impairment. Two of these viruses are the adenovirus and the human papilloma virus (HPV) (Bennett *et al.*, 1999; Vambutas *et al.*, 2000). Adenoviruses express E3/19K, a protein which serves two purposes. On the one hand, it directly binds to MHC class I molecules, trapping them in the ER. On the other hand, it seems to bind to TAP and consequently inhibit TAP–tapasin association and thereby the assembly of the macromolecular loading complex (Bennett *et al.*, 1999). HPV interferes with immune recognition via another mechanism. This virus causes different clinical courses with recurrent papillomatosis and can lead to significant morbidity. Although the precise mechanism is unclear at the moment, it was found that in laryngeal papilloma tissue biopsies and in cell culture of primary explants, TAP levels are markedly downregulated and correlate inversely with the frequency of disease recurrence (Vambutas *et al.*, 2000). The expression of TAP1 is even considered a measure for disease severity. By decreasing the TAP levels and thereby MHC class I levels, HPV may evade immune recognition, which in turn leads to persistent infection of the host cells. For other malignancies induced by HPV, such as HPV-16 and HPV-18 infected carcinomas of the cervix, a similar downregulation of TAP1 and MHC class I molecules has been described (Cromme *et al.*, 1994). One mechanism by which a decrease in TAP can occur is known from studies of the Epstein–Barr virus (EBV) (Zeidler *et al.*, 1997). EBV expresses an interleukin-10-like protein, causing the downregulation of TAP1 and subsequent MHC class I cell surface expression. This strategy may promote persistent EBV infections.

TUMOR ESCAPE STRATEGIES

Although defects in the TAP complex are associated with a subgroup of individuals with a rare genetic disease (bare lymphocyte syndrome type 1) (de la Salle *et al.*, 1994, 1999; Teisserenc *et al.*, 1997), TAP may play a more important role in tumor development. Tumors often avoid immune recognition by ineffective display of the antigen presentation pathway. In some cancers, such as melanomas, it was demonstrated that MHC class I molecules on the cell surface were reduced (Seliger *et al.*, 1997b; Sherman *et al.*, 1998). This effect can be due to low levels of TAP1 and/or TAP2 as was shown for cell lines from human small cell lung cancer (Seliger *et al.*, 1997b) and breast cancer (Alimonti *et al.*, 2000). In various cases, the TAP levels could be restored by treatment of the cells with IFN-γ. Also, TAP1 is downregulated in other human tumors, but the mechanism or mutation responsible is unknown (Chen *et al.*, 1996; Seliger *et al.*, 1997a). A defective or non-existent TAP complex causes a decrease in, or even loss of, MHC class I expression on the cell surface. One way of reducing TAP levels may be via cytokines, such as interleukin-10, which are known to downregulate TAP (Zeidler *et al.*, 1997). A large number of tumors secrete interleukin-10; accordingly, this fact may be of clinical importance and needs further investigation.

In addition to downregulation of TAP expression, mutations in the *TAP* genes can also contribute to tumor escape from immune recognition. In a small cell lung cancer cell line, a mutation was found which introduces an amino acid exchange at position 659 of TAP1 (Chen *et al.*, 1996). This mutation is located between the 'C' motif and the Walker B motif and might interfere with ATP binding or hydrolysis. The resulting TAP complex is non-functional. It is important to stress the fact that mutations in TAP1 occur frequently in a variety of human tumors, such as primary breast cancers, and that the expression of TAP1 can be linked with tumor grading and reduced survival (Kaklamanis *et al.*, 1995; Vitale *et al.*, 1998). In contrast to high-grade breast carcinoma lesions, low-grade lesions exhibited normal TAP levels (Vitale *et al.*, 1998). Based on these findings, the therapeutic potential of exogenous TAP was tested in a TAP-deficient small cell lung carcinoma cell (Alimonti *et al.*, 2000). A clear regression of tumors in mice was shown after infection with vaccinia virus containing the TAP1 gene. Therefore, this method could increase the immune response against tumors and should be kept in mind for implementation in cancer therapies.

Some studies show a correlation between deficiencies in MHC class I antigen presentation and tumor progression (Kaklamanis *et al.*, 1995; Seliger *et al.*, 1997b; Vitale *et al.*, 1998), but the significance for tumor development is still uncertain. To gain a better insight into the problem, matched panels of TAP1-positive and TAP1-negative tumor cell lines were established and inoculated into mice (Johnsen *et al.*, 1999). The TAP-negative cells resulted in large and persistent tumors, but the TAP-positive cells did not produce lasting tumors. As a control, both cell lines were shown to generate tumors in athymic mice, thereby confirming that the tumorigenicity can be attributed to the T-cell immune response. Because of the different types of tumors, the clinical significance of reduced MHC class I cell surface expression varies markedly and needs further studies to explore potential diagnostic and therapeutic applications.

Conclusions

The transport of peptides from the cytosol into the ER lumen is a key step in the MHC class I antigen presentation pathway. This translocation is performed by the TAP complex, a heterodimer of TAP1 and TAP2. It is known that disruption of peptide transport severely interferes with the T-cell-mediated immune response. The TAP transporter may be involved in several (patho) physiological processes; the elucidation of the molecular structure, topology and function of TAP will enhance our knowledge of the underlying mechanisms of various diseases and, thereby, provide the basis for the development of new therapeutic strategies. It is conceivable that new vaccines may be developed by understanding how some viruses escape the immune response. Knowledge about every step in TAP action may even improve the treatment of some cancers, because some studies have shown that loss of TAP function occurs frequently in metastatic tumors. One possible therapeutic strategy may be to restore the immune response against tumor cells which evade immune recognition by disruption of TAP function. However, many questions remain unanswered at the moment. For example, what are the conformational changes upon peptide binding? And how do the molecules involved in the macromolecular peptide transport and loading complex interact? Further investigations will provide new insights about the significance of the loss of TAP function, and about the mechanisms by which viruses and some malignant tumor cells escape immune recognition.

References

Abele, R. and Tampé, R. (1999) Function of the transport complex TAP in cellular immune recognition. Biochim. Biophys. Acta 1461, 405–419.

Ahn, K., Meyer, T.H., Uebel, S., Sempé, P., Djaballah, H., Yang, Y., Peterson, P.A., Früh, K. and Tampé, R. (1996) Molecular mechanism and species-specificity of TAP inhibition by Herpes-Simplex virus protein ICP47. EMBO J. 15, 3247–3255.

Ahn, K., Gruhler, A., Galocha, B., Jones, T.R., Wiertz, E.J.H.J., Ploegh, H.L., Peterson, P.A., Yang, Y. and Früh, K. (1997) The ER-luminal domain of the HCMV glycoprotein US6 inhibits peptide translocation by TAP. Immunity 6, 613–621.

Alberts, P., Daumke, O., Deverson, E.V., Howard, J.C. and Knittler, M.R. (2001) Distinct functional properties of the TAP subunits coordinate the nucleotide-dependent transport cycle. Curr. Biol. 11, 242–251.

Alimonti, J., Zhang, Q.J., Gabathuler, R., Reid, G., Chen, S.S. and Jefferies, W.A. (2000) TAP expression provides a general method for improving the recognition of malignant cells *in vivo*. Nat. Biotechnol. 18, 515–520.

Alpan, R.S., Zhang, M. and Pardee, A.B. (1996) Cell cycle-dependent expression of TAP1, TAP2, and HLA-B27 messenger RNAs in a human breast cancer cell line. Cancer Res. 56, 4358–4361.

Androlewicz, M.J. and Cresswell, P. (1994) Human transporters associated with antigen processing possess a promiscuous peptide binding site. Immunity 1, 7–14.

Androlewicz, M.J., Anderson, K.S. and Cresswell, P. (1993) Evidence that transporter associated with antigen processing translocate a major histocompatibility complex class I-binding peptide into the endoplasmic reticulum in an ATP-dependent manner. Proc. Natl Acad. Sci. USA 90, 9130–9134.

Arora, S., Lapinski, P.E. and Raghavan, M. (2001) Use of chimeric proteins to investigate the role of transporter associated with antigen processing (TAP) structural domains in peptide binding and translocation. Proc. Natl Acad. Sci. USA 98, 7241–7246.

Ayalon, O., Hughes, E.A., Cresswell, P., Lee, J., Odonnell, L., Pardi, R. and Bender, J.R. (1998) Induction of transporter associated with antigen-processing by interferon-gamma confers endothelial-cell cytoprotection against natural killer-mediated lysis. Proc. Natl Acad. Sci. USA 95, 2435–2440.

Baumeister, W., Walz, J., Zühl, F. and Seemüller, E. (1998) The proteasome – paradigm of a self-compartmentalizing protease. Cell 92, 367–380.

Beck, S., Kelly, A., Radley, E., Khurshid, F., Alderton, R.P. and Trowsdale, J. (1992) DNA sequence analysis of 66 kb of the human MHC class II region encoding a cluster of genes for antigen processing. J. Mol. Biol. 228, 433–441.

Beinert, D., Neumann, L., Uebel, S. and Tampé, R. (1997) Structure of the Viral TAP-inhibitior ICP47 induced by membrane association. Biochemistry 36, 4694–4700.

Bennett, E.M., Bennink, J.R., Yewdell, J.W. and Brodsky, F.M. (1999) Cutting edge: adenovirus E19 has two mechanisms for affecting class I MHC expression. J. Immunol. 162, 5049–5052.

Chen, H.L., Gabrilovich, D., Tampé, R., Girgis, K.R., Nadaf, S. and Carbone, D.P. (1996) A functionally defective allele of TAP1 results in loss of MHC class I antigen presentation in a human lung cancer. Nat. Genet. 13, 210–213.

Cresswell, P., Bangia, N., Dick, T. and Diedrich, G. (1999) The nature of the MHC class I peptide loading complex. Immunol. Rev. 172, 21–28.

Cromme, F.V., Airey, J., Heemels, M.T., Ploegh, H.L., Keating, P.J., Stern, P.L., Meijer, J.C.L.M. and Walboomers, C.M.M. (1994) Loss of transporter protein, encoded by the TAP-1 gene, is highly correlated with loss of HLA expression in cervical carcinomas. J. Exp. Med. 179, 335–340.

Daniel, S., Caillatzucman, S., Hammer, J., Bach, J.F. and Van Endert, P.M. (1997) Absence of functional relevance of human transporter associated with antigen processing polymorphism for peptide selection. J. Immunol. 159, 2350–2357.

de la Salle, H., Hanau, D., Fricker, D., Urlacher, A., Kelly, A., Salamero, J., Powis, S.H., Donato, L., Bausinger, H. and Laforet, M. (1994) Homozygous human TAP peptide transporter mutation in hla class-I deficiency. Science 265, 237–241.

de la Salle, H., Zimmer, J., Fricker, D., Angenieux, C., Cazenave, J.P., Okubo, M., *et al.* (1999) HLA class I deficiencies due to mutations in subunit 1 of the peptide transporter TAP1. J. Clin. Invest. 103, R9–R13.

Deverson, E.V., Gow, I.R., Coadwell, W.J., Monaco, J.J., Butcher, G.W. and Howard, J.C. (1990) MHC class II region encoding proteins related to the multidrug resistance family of transmembrane transporters. Nature 348, 738–741.

Elliott, T. (1997) Transporter associated with antigen processing. Adv. Immunol. 65, 47–109.

Früh, K. and Yang, Y. (1999) Antigen presentation by MHC class I and its regulation by interferon gamma. Curr. Opin. Immunol. 11, 76–81.

Früh, K., Ahn, K., Djaballah, H., Sempé, P., van Endert, P.M., Tampé, R., Peterson, P.A. and Yang, Y. (1995) A viral inhibitor of peptide transporters for antigen presentation. Nature 375, 415–418.

Früh, K., Gruhler, A., Krishna, R.M. and Schoenhals, G.J. (1999) A comparison of viral immune escape strategies targeting the MHC class I assembly pathway. Immunol. Rev. 168, 157–166.

Gaczynska, M., Rock, K.L., Spies, T. and Goldberg, A.L. (1994) Peptidase activities of proteasomes are differentially regulated by the major histocompatibility complex-encoded genes for lmp2 and lmp7. Proc. Natl Acad. Sci. USA 91, 9213–9217.

Gilbert, M.J., Riddell, S.R., Plachter, B. and Greenberg, P.D. (1996) Cytomegalovirus selectively blocks antigen-processing and presentation of its immediate-early gene-product. Nature 383, 720–722.

Gileadi, U. and Higgins, C.F. (1997) Membrane topology of the ATP-binding cassette transporter associated with antigen presentation (TAP1) expressed in *Escherichia coli*. J. Biol. Chem. 272, 11103–11108.

Gorbulev, S., Abele, R. and Tampe, R. (2001) Allosteric crosstalk between peptide-binding, transport, and ATP hydrolysis of the ABC transporter TAP. Proc. Natl Acad. Sci. USA 98, 3732–3737.

Hanson, I.M. and Trowsdale, J. (1991) Colinearity of novel genes in the class II regions of the MHC in mouse and human. Immunogenetics 34, 5–11.

Hengel, H. and Koszinowski, U.H. (1997) Interference with antigen-processing by viruses. Curr. Opin. Immunol. 9, 470–476.

Hengel, H., Koopmann, J.O., Flohr, T., Muranyi, W., Goulmy, E., Hämmerling, G.J., Koszinowski, U.H. and Momburg, F. (1997) A viral ER-resident glycoprotein inactivates the MHC-encoded peptide transporter. Immunity 6, 623–632.

Hengel, H., Reusch, U., Gutermann, A., Ziegler, H., Jonjic, S., Lucin, P. and Koszinowski, U.H. (1999) Cytomegaloviral control of MHC class I function in the mouse. Immunol. Rev. 168, 167–176.

Hewitt, E.W., Gupta, S.S. and Lehner, P.J. (2001) The human cytomegalovirus gene product US6 inhibits ATP binding by TAP. EMBO J. 20, 387–396.

Hill, A., Jugovic, P., York, I., Russ, G., Bennink, J., Yewdell, J., Ploegh, H. and Johnson, D. (1995) Herpes simplex virus turns off the TAP to evade host immunity. Nature 375, 411–415.

Hilton, C.J., Dahl, A.M. and Rock, K.L. (2001) Anti-peptide antibody blocks peptide binding to MHC class I molecules in the endoplasmic reticulum. J. Immunol. 166, 3952–3956.

Israel, A., Le Bail, O., Hatat, D., Piette, J., Kieran, M., Logeat, F., Wallach, D., Fellous, M. and Kourilsky, P. (1989) TNF stimulates expression of mouse MHC class I genes by inducing an NF kappa B-like enhancer binding activity which displaces constitutive factors. EMBO J. 8, 3793–3800.

Johnsen, A.K., Templeton, D.J., Sy, M. and Harding, C.V. (1999) Deficiency of transporter for antigen presentation (TAP) in tumor cells allows evasion of immune surveillance and increases tumorigenesis. J. Immunol. 163, 4224–4231.

Kaklamanis, L., Leek, R., Koukourakis, M., Gatter, K.C. and Harris, A.L. (1995) Loss of transporter in antigen-processing-1 transport protein and major histocompatibility complex class-i molecules in metastatic versus primary breast-cancer. Cancer Res. 55, 5191–5194.

Karttunen, J.T., Lehner, P.J., Gupta, S.S., Hewitt, E.W. and Cresswell, P. (2001) Distinct functions and cooperative interaction of the subunits of the transporter associated with antigen processing (TAP). Proc. Natl Acad. Sci. USA 98, 7431–7436.

Kelly, A.P., Powis, S.H., Kerr, L.-A., Mockridge, I., Elliott, T., Bastin, J., Uchanska-Ziegler, B., Ziegler, A., Trowsdale, J. and Townsend, A. (1992) Assembly and function of the two ABC transporter proteins encoded in the human major histocompatibility complex. Nature 355, 641–644.

Kisselev, A.F., Akopian, T.N., Woo, K.M. and Goldberg, A.L. (1999) The sizes of peptides generated from protein by mammalian 26 and 20 S proteasomes – implications for understanding the degradative mechanism and antigen presentation. J. Biol. Chem. 274, 3363–3371.

Kleijmeer, M., Kelly, A., Geuze, H.J., Slot, J.W., Townsend, A. and Trowsdale, J. (1992) Location of MHC-encoded transporters in the endoplasmic reticulum and *cis*-golgi. Nature 357, 342–344.

Knittler, M.R., Alberts, P., Deverson, E.V. and Howard, J.C. (1999) Nucleotide binding by TAP mediates association with peptide and release of assembled MHC class I molecules. Curr. Biol. 9, 999–1008.

Koopmann, J.O., Post, M., Neefjes, J.J., Hämmerling, G.J. and Momburg, F. (1996) Translocation of long peptides by transporters associated with antigen processing (TAP). Eur. J. Immunol. 26, 1720–1728.

Kyritsis, C., Gorbulev, S., Hutschenreiter, S., Pawlitschko, K., Abele, R. and Tampé, R. (2001) Molecular mechanism and structural aspects of transporter associated with antigen processing inhibition by the cytomegalovirus protein US6. J. Biol. Chem. 276, 48031–48039.

Lapinski, P.E., Neubig, R.R. and Raghavan, M. (2001) Walker a lysine mutations of TAP1 and TAP2 interfere with peptide translocation but not peptide binding. J. Biol. Chem. 276, 7526–7533.

Lehner, P.J., Karttunen, J.T., Wilkinson, G.W. and Cresswell, P. (1997) The human cytomegalovirus US6 glycoprotein inhibits transporter associated with antigen processing-dependent peptide translocation. Proc. Natl Acad. Sci. USA 94, 6904–6909.

Lehner, P.J., Surman, M.J. and Cresswell, P. (1998) Soluble tapasin restores MHC class I expression and function in the tapasin-negative cell line 220. Immunity 8, 221–231.

Levitskaya, J., Sharipo, A., Leonchiks, A., Ciechanover, A. and Masucci, M.G. (1997) Inhibition of ubiquitin/proteasome-dependent protein degradation by the Gly-Ala repeat domain of the Epstein-Barr virus nuclear antigen 1. Proc. Natl Acad. Sci. USA 94, 12616–12621.

Lindquist, J.A., Jensen, O.N., Mann, M. and Hammerling, G.J. (1998) ER-60, a chaperone with thiol-dependent reductase activity involved in MHC class I assembly. EMBO J. 17, 2186–2195.

Ljunggren, H.G., Stam, N.J., Ohlen, C., Neefjes, J.J., Hoglund, P., Heemels, M.T., Bastin, J., Schumacher, T.N., Townsend, A., Karre, K., *et al.* (1990) Empty MHC class I molecules come out in the cold. Nature 346, 476–480.

Loo, T.W. and Clarke, D.M. (1995) Membrane topology of a cysteine-less mutant of human P-glycoprotein. J. Biol. Chem. 270, 843–848.

Loo, T.W. and Clarke, D.M. (2001a) Defining the drug-binding site in the human multidrug resistance P-glycoprotein using a methanethiosulfonate analog of verapamil, MTS-verapamil. J. Biol. Chem. 276, 14972–14979.

Loo, T.W. and Clarke, D.M. (2001b) Determining the dimensions of the drug-binding domain of human P-glycoprotein using thiol cross-linking compounds as molecular rulers. J. Biol. Chem. 276, 36877–36880.

Ma, W.L., Lehner, P.J., Cresswell, P., Pober, J.S. and Johnson, D.R. (1997) Interferon-gamma rapidly increases peptide transporter (TAP) subunit expression and peptide-transport capacity in endothelial-cells. J. Biol. Chem. 272, 16585–16590.

Madden, D.R., Garboczi, D.N. and Wiley, D.C. (1993) The antigenic identity of peptide-MHC complexes – a comparison of the conformations of 5 viral peptides presented by HLA-A2. Cell 75, 693–708.

Meyer, T.H., van Endert, P.M., Uebel, S., Ehring, B. and Tampé, R. (1994) Functional expression and purification of the ABC transporter complex-associated with antigen-processing (TAP) in insect cells. FEBS Lett. 351, 443–447.

Momburg, F., Roelse, J., Howard, J.C., Butcher, G.W., Hämmerling, G.J. and Neefjes, J.J. (1994) Selectivity of MHC-encoded peptide transporters from human, mouse and rat. Nature 367, 648–651.

Monaco, J.J., Cho, S. and Attaya, M. (1990) Transport protein genes in the murine MHC: possible implications for antigen processing. Science 250, 1723–1726.

Moore, M.W., Carbone, F.R. and Bevan, M.J. (1988) Introduction of soluble protein into the class I pathway of antigen processing and presentation. Cell 54, 777.

Morrison, L.A., Lukacher, A.E., Braciale, V.L., Fan, D.P. and Braciale, T.J. (1986) Differences in antigen presentation to MHC class I- and class II-restricted influenza virus-specific cytolytic T lymphocyte clones. J. Exp. Med. 163, 903–921.

Müller, K.M., Ebensperger, C. and Tampé, R. (1994) Nucleotide binding to the hydrophilic C-terminal domain of the transporter associated with antigen processing (TAP). J. Biol. Chem. 269, 14032–14037.

Nagy, N., Vanky, F. and Klein, E. (1998) Tumor surveillance: expression of the transporter associated with antigen processing (TAP-1) in ex-vivo human tumor samples and its elevation by *in vitro* treatment with IFN-gamma and TNF-alpha. Immunol. Lett. 64, 153–160.

Neefjes, J.J., Momburg, F. and Hämmerling, G.J. (1993) Selective and ATP-dependent translocation of peptides by the MHC-encoded transporter. Science 261, 769–771.

Neumann, L. and Tampé, R. (1999) Kinetic analysis of peptide binding of the TAP transporter complex: evidence for structural rearrangements induced by substrate binding. J. Mol. Biol. 294, 1203–1213.

Neumann, L., Kraas, W., Uebel, S., Jung, G. and Tampé, R. (1997) The active domain of the herpes-simplex virus protein ICP47 – a potent inhibitor of the transporter associated with antigen-processing (TAP). J. Mol. Biol. 272, 484–492.

Nijenhuis, M. and Hämmerling, G.J. (1996) Multiple regions of the transporter associated with antigen processing (TAP) contribute to its peptide binding site. J. Immunol. 157, 5467–5477.

Ortmann, B., Copeman, J., Lehner, P.J., Sadasivan, B., Herberg, J.A., Grandea, A.G., Riddell, S.R., Tampé, R., Spies, T. and Trowsdale, J. (1997) A critical role for tapasin in the assembly and function of multimeric MHC class I-TAP complexes. Science 277, 1306–1309.

Pamer, E. and Cresswell, P. (1998) Mechanisms of MHC class I-restricted antigen processing. Annu. Rev. Immunol. 16, 323–358.

Pfänder, R., Neumann, L., Zweckstetter, M., Seger, C., Holak, T.A. and Tampé, R. (1999) Structure of the active domain of the herpes simplex virus protein ICP47 in water/sodium dodecyl sulfate solution determined by nuclear magnetic resonance spectroscopy. Biochemistry 38, 13692–13698.

Ploegh, H.L. (1998) Viral strategies of immune evasion. Science 280, 248–253.

Powis, S.J., Deverson, E.V., Coadwell, W.J., Ciruela, A., Huskisson, N.S., Smith, H., Butcher, G.W. and Howard, J.C. (1992) Effect of polymorphism of an MHC-linked transporter on the peptides assembled in a class I molecule. Nature 357, 211–215.

Reits, E.A.J., Vos, J.C., Grommé, M. and Neefjes, J. (2000) The major substrates for TAP *in vivo* are derived from newly synthesized proteins. Nature 404, 774–778.

Ritz, U., Momburg, F., Pircher, H.P., Strand, D., Huber, C. and Seliger, B. (2001) Identification of sequences in the human peptide transporter subunit TAP1 required for transporter associated with antigen processing (TAP) function. Int. Immunol. 13, 31–41.

Rock, K.L. and Goldberg, A.L. (1999) Degradation of cell proteins and the generation of MHC class I-presented peptides. Annu. Rev. Immunol. 17, 739–779.

Russ, G., Esquivel, F., Yewdell, J.W., Cresswell, P., Spies, T. and Bennick, J.R. (1995) Assembly, intracellular localization, and nucleotide binding properties of the human peptide transporters TAP1 and TAP2 expressed by recombinant vaccinia viruses. J. Biol. Chem. 270, 21312–21318.

Sadasivan, B., Lehner, P.J., Ortmann, B., Spies, T. and Cresswell, P. (1996) Roles for calreticulin and a novel glycoprotein, tapasin, in the interaction of MHC class I molecules with TAP. Immunity 5, 103–114.

Salazar-Onfray, F., Charo, J., Petersson, M., Freland, S., Noffz, G., Qin, Z.H., Blankenstein, T., Ljunggren, H.G. and Kiessling, R. (1997) Down-regulation of the expression and function of the transporter associated with antigen-processing in murine tumor-cell lines expressing IL-10. J. Immunol. 159, 3195–3202.

Saveanu, L., Daniel, S. and van Endert, P.M. (2001) Distinct functions of the ATP binding cassettes of transporters associated with antigen processing: a mutational analysis of Walker A and B sequences. J. Biol. Chem. 276, 22107–22113.

Schmitt, L. and Tampé, R. (2000) Affinity, specificity, diversity: a challenge for the ABC transporter TAP in cellular immunity. ChemBioChem. 1, 16–35.

Schubert, U., Anton, L.C., Gibbs, J., Norbury, C.C., Yewdell, J.W. and Bennink, J.R. (2000) Rapid degradation of a large fraction of newly synthesized proteins by proteasomes. Nature 404, 770–774.

Schumacher, T.N., Kantesaria, D.V., Heemels, M.T., Ashton-Rickardt, P.G., Shepherd, J.C., Früh, K., Yang, Y., Peterson, P.A., Tonegawa, S. and Ploegh, H.L. (1994a) Peptide length and sequence specificity of the mouse TAP1/TAP2 translocator. J. Exp. Med. 179, 533–540.

Schumacher, T.N.M., Kantesaria, D.V., Serreze, D.V., Roopenian, D.C. and Ploegh, H.L. (1994b) Transporters from H-2(b), H-2(d), H-2(s), H-2(k), and H-2(g7) (NOD/lt) haplotype translocate similar sets of peptides. Proc. Natl Acad. Sci. USA 91, 13004–13008.

Seliger, B., Hohne, A., Jung, D., Kallfelz, M., Knuth, A., Jaeger, E., Bernhard, H., Momburg, F., Tampé, R. and Huber, C. (1997a) Expression and function of the peptide transporters in escape variants of human renal-cell carcinomas. Exp. Hematol. 25, 608–614.

Seliger, B., Maeurer, M.J. and Ferrone, S. (1997b) TAP off-tumors on. Immunol. Today 18, 292–299.

Shepherd, J.C., Schumacher, T.N., Ashton-Rickardt, P.G., Imaeda, S., Ploegh, H.L., Janeway, C.A.J. and Tonegawa, S. (1993) TAP1-dependent peptide translocation *in vitro* is ATP dependent and peptide selective. Cell 74, 577–584.

Sherman, L.A., Theobald, M., Morgan, D., Hernandez, J., Bacik, I., Yewdell, J., Bennink, J. and Biggs, J. (1998) Strategies for tumor elimination by cytotoxic T lymphocytes. Crit. Rev. Immunol. 18, 47–54.

Spies, T., Bresnahan, M., Bahram, S., Arnold, D., Blanck, G., Mellins, E., Pious, D. and DeMars, R. (1990) A gene in the human major histocompatibility complex class II region controlling the class I antigen presentation pathway. Nature 348, 744–747.

Tampé, R., Urlinger, S., Pawlitschko, K. and Uebel, S. (1997) The transporters associated with antigen processing (TAP). In: Unusual Secretory Pathways: From Bacteria to Man (ed. K. Kuchler, A. Rubartelli, and B. Holland), pp. 115–136. New York: Springer.

Teisserenc, H., Schmitt, W., Blake, N., Dunbar, R., Gadola, S., Gross, W.L., Exley, A. and Cerundolo, V. (1997) A case of primary immunodeficiency due to a defect of the major histocompatibility gene complex class I processing and presentation pathway. Immunol. Lett. 57, 183–187.

Toes, R.E., Nussbaum, A.K., Degermann, S., Schirle, M., Emmerich, N.P., Kraft, M., *et al.* (2001) Discrete cleavage motifs of constitutive and immunoproteasomes revealed by quantitative analysis of cleavage products. J. Exp. Med. 194, 1–12.

Tomazin, R., Hill, A.B., Jugovic, P., York, I., van Endert, P., Ploegh, H.L., Andrews, D.W. and Johnson, D.C. (1996) Stable binding of the Herpes Simplex virus ICP47 protein to the peptide binding-site of TAP. EMBO J. 15, 3256–3266.

Townsend, A., Ohlen, C., Foster, L., Bastin, J., Ljunggren, H.G. and Karre, K. (1989) A mutant cell in which association of class I heavy and light chains is induced by viral peptides. Cold Spring Harb. Symp. Quant. Biol. 54, 299–308.

Trowsdale, J., Hanson, I., Mockridge, I., Beck, S., Townsend, A. and Kelly, A. (1990) Sequences encoded in the class II region of the MHC related to the 'ABC' superfamily of transporters. Nature 348, 741–744.

Turner, G.C. and Varshavsky, A. (2000) Detecting and measuring cotranslational protein degradation *in vivo*. Science 289, 2117–2120.

Uebel, S. and Tampé, R. (1999) Specificity of the proteasome and the TAP transporter. Curr. Opin. Immunol. 11, 203–208.

Uebel, S., Meyer, T.H., Kraas, W., Kienle, S., Jung, G., Wiesmüller, K.H. and Tampé, R. (1995) Requirements for peptide binding to the human transporter associated with antigen processing revealed by peptide scans and complex peptide libraries. J. Biol. Chem. 270, 18512–18516.

Uebel, S., Kraas, W., Kienle, S., Wiesmüller, K.-H., Jung, G. and Tampé, R. (1997) Recognition principle of the TAP-transporter disclosed by combinatorial peptide libraries. Proc. Natl Acad. Sci. USA 94, 8976–8981.

Urlinger, S., Kuchler, K., Meyer, T.H., Uebel, S. and Tampé, R. (1997) Intracellular location, complex-formation, and function of the transporter associated with antigen processing in yeast. Eur. J. Biochem. 245, 266–272.

Vambutas, A., Bonagura, V.R. and Steinberg, B.M. (2000) Altered expression of TAP-1 and major histocompatibility complex class I in laryngeal papillomatosis: correlation of TAP-1 with disease. Clin. Diagn. Lab. Immunol. 7, 79–85.

van Endert, P.M., Tampé, R., Meyer, T.H., Tisch, R., Bach, J.F. and McDevitt, H.O. (1994) A sequential model for peptide binding and transport by the transporters associated with antigen processing. Immunity 1, 491–500.

Vitale, M., Rezzani, R., Rodella, L., Zauli, G., Grigolato, P., Cadei, M., Hicklin, D.J. and Ferrone, S. (1998) Hla class-i antigen and transporter associated with antigen-processing (tap1 and tap2) down-regulation in high-grade primary breast-carcinoma lesions. Cancer Res. 58, 737–742.

Vos, J.C., Spee, P., Momburg, F. and Neefjes, J. (1999) Membrane topology and dimerization of the two subunits of the transporter associated with antigen processing reveal a three-domain structure. J. Immunol. 163, 6679–6685.

Vos, J.C., Reits, E.A., Wojcik-Jacobs, E. and Neefjes, J. (2000) Head-head/tail-tail relative orientation of the pore-forming domains of the heterodimeric ABC transporter TAP. Curr. Biol. 10, 1–7.

Wang, K.N., Früh, K., Peterson, P.A. and Yang, Y. (1994) Nucleotide binding of the C-terminal domains of the major histocompatibility complex-encoded transporter expressed in *Drosophila melanogaster* cells. FEBS Lett. 350, 337–341.

Wright, K.L., White, L.C., Kelly, A., Beck, S., Trowsdale, J. and Ting, J.P.Y. (1995) Coordinate regulation of the human tap1 and lmp2 genes from a shared bidirectional promoter. J. Exp. Med. 181, 1459–1471.

Yan, G., Shi, L.J. and Faustman, D. (1999) Novel splicing of the human MHC-encoded peptide transporter confers unique properties. J. Immunol. 162, 852–859.

Yewdell, J.W., Bennink, J.R. and Hosaka, Y. (1988) Cells process exogenous proteins for recognition by cytotoxic T lymphocytes. Science 239, 637–640.

York, I.A., Roop, C., Andrews, D.W., Riddell, S.R., Graham, F.L. and Johnson, D.C. (1994) A cytosolic herpes simplex virus protein inhibits antigen presentation to CD8+ T lymphocytes. Cell 77, 525–535.

York, I.A., Goldberg, A.L., Mo, X.Y. and Rock, K.L. (1999) Proteolysis and class I major histocompatibility complex antigen presentation. Immunol. Rev. 172, 49–66.

Zeidler, R., Eissner, G., Meissner, P., Uebel, S., Tampé, R., Lazis, S. and Hammerschmidt, W. (1997) Downregulation of TAP1 in B lymphocytes by cellular and Epstein–Barr virus-encoded interleukin-10. Blood 90, 2390–2397.

Zhang, F., Zhang, W., Liu, L., Fisher, C.L., Hui, D., Childs, S., Dorovini-Zis, K. and Ling, V. (2000) Characterization of ABCB9, an ATP binding cassette protein associated with lysosomes. J. Biol. Chem. 275, 23287–23294.

Zhu, K., Wang, J., Zhu, J., Jiang, J., Shou, J. and Chen, X. (1999) p53 induces TAP1 and enhances the transport of MHC class I peptides. Oncogene 18, 7740–7747.

THE SULFONYLUREA RECEPTOR: AN ABCC TRANSPORTER THAT ACTS AS AN ION CHANNEL REGULATOR

CHAPTER 27

MICHINORI MATSUO, KAZUMITSU UEDA, TIMOTHY RYDER AND FRANCES ASHCROFT

INTRODUCTION

ATP-binding cassette (ABC) proteins constitute a large and diverse family of integral membrane proteins, which are found in both prokaryotes and eukaryotes (Dean *et al.*, 2001). Most members of this diverse family are involved in the ATP-dependent transport of solutes across surface or intracellular membranes (Higgins, 1992; Holland and Blight, 1999). Unique functions, however, have been identified for two members of the ABCC subfamily. The cystic fibrosis transmembrane conductance regulator (CFTR) functions as a chloride ion channel, and harnesses the energy of ATP hydrolysis to open and close the channel pore. The sulfonylurea receptor (SUR), the topic of this review, serves as the regulatory subunit of the ATP-sensitive potassium (K_{ATP}) channel, endowing it with the ability to respond to changes in cell metabolism. It is an open question whether or not SUR has an additional classical transport function, and thus in this review we confine ourselves to the role of SUR as a channel regulator.

ATP-sensitive potassium (K_{ATP}) channels are inwardly rectifying potassium channels, which are inhibited by ATP and activated by MgADP. They link the metabolism of the cell to the membrane potential by sensing changes in intracellular adenine nucleotide concentrations. K_{ATP} channels play important functional roles in numerous tissues, including pancreatic β-cells, neurons, cardiac muscle, skeletal muscle and smooth muscle. For example, their activation leads to shortening of the cardiac action potential, relaxation of vascular smooth muscle, and inhibition of insulin secretion and neurotransmitter release.

The K_{ATP} channel is a hetero-octameric complex of four Kir6.x and four SUR subunits. Kir6.x belongs to the family of inwardly rectifying K^+ channels and assembles into tetramers to form the channel **(Figure 27.1)**. Binding of ATP to the intracellular domains of Kir6.x produces channel *inhibition*. Associated with each Kir6.x subunit is a regulatory subunit, the sulfonylurea receptor (SUR). Like other members of the ABC transporter family, SUR has two large intracellular domains, containing consensus sequences for nucleotide binding and hydrolysis, which are known as the nucleotide-binding domains (NBDs). Interaction of Mg-nucleotides with these NBDs mediates *activation* of the K_{ATP} channel. The SUR subunit also binds therapeutic drugs, such as the sulfonylureas, which inhibit K_{ATP} channel activity, and the K_{ATP} channel openers, which stimulate channel activity. More than one isoform exists for both Kir6.x (Kir6.1, Kir6.2) and SUR (SUR1, SUR2A, SUR2B) and variation in the subunit composition of the K_{ATP} channel accounts for the different metabolic and drug sensitivities of K_{ATP} channels in different cells. In most tissues, Kir6.2 serves as

ABC Proteins: From Bacteria to Man
ISBN 0-12-352551-9

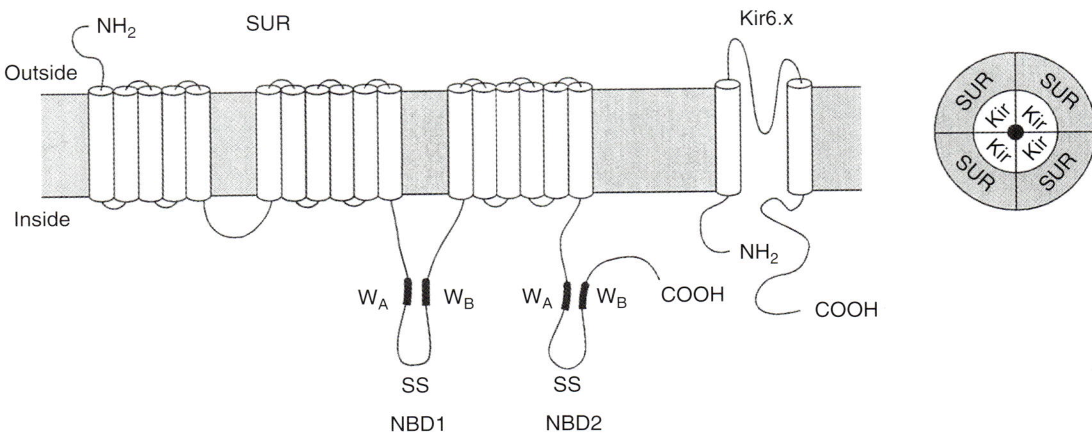

Figure 27.1. K_{ATP} channels are formed from two different types of subunits. The K_{ATP} channel is an octameric complex of four pore-forming Kir6.x subunits and four regulatory sulfonylurea receptor (SUR) subunits. Like other inwardly rectifying K^+ channels, Kir6.2 has only two transmembrane segments, and cytosolic N- and C-termini. The cytosolic domains are involved in channel inhibition by ATP. SUR has 17 transmembrane segments, arranged in groups of 5, 6 and 6, and two cytosolic nucleotide-binding domains (NBDs), each of which contains a Walker A (W_A) motif, a Walker B (W_B) motif and an ABC signature sequence (SS). These motifs are involved in the activation of the K_{ATP} channel by Mg-nucleotides.

the pore-forming subunit, but it associates with different SUR subunits: for example, SUR1 in pancreas and brain, SUR2A in heart and skeletal muscle and SUR2B in a variety of tissues including brain and smooth muscle. In some smooth muscles, the K_{ATP} channel comprises Kir6.1 in association with SUR2B.

In this review, we first focus on the physiological role of the K_{ATP} channel and how this is impaired in disease. We then discuss the molecular composition of the K_{ATP} channel, and detail its regulation by nucleotides, metabolism and other signaling molecules. Finally, we briefly summarize what is known of the pharmacological properties of the channel.

PHYSIOLOGICAL ROLE

K_{ATP} CHANNELS IN THE PANCREAS

The physiological role of the K_{ATP} channel is best understood in the pancreatic β-cell, where metabolically induced changes in K_{ATP} channel activity play a key role in glucose-stimulated insulin secretion. This is illustrated in the cartoon in **Figure 27.2**. At substimulatory glucose concentrations, K_{ATP} channels are open, and their activity serves to maintain the resting membrane potential at a hyperpolarized level. Elevation of the blood glucose concentration increases glucose uptake and metabolism by the β-cell, producing changes in cytosolic nucleotide concentrations that result in closure of the K_{ATP} channels. This leads to depolarization of the β-cell membrane potential, and thus to activation of voltage-gated calcium channels and Ca^{2+} influx. The resulting rise in the intracellular Ca^{2+} concentration triggers insulin release. The physiological importance of the K_{ATP} channel (Kir6.2/SUR1) in regulating insulin secretion is demonstrated by the fact that mutations in the SUR1 subunit (discussed below) have been found in patients with persistent hyperinsulinemic hypoglycemia of infancy (PHHI), a serious disorder characterized by excessive and unregulated insulin secretion. Furthermore, defective metabolic regulation of the K_{ATP} channel results in diabetes mellitus (see **Glossary**, page 566) in both humans and transgenic animals.

K_{ATP} channels have also been described in the glucagon-secreting α-cells (Gopel *et al.*, 2000b) and somatostatin-secreting δ-cells (Gopel *et al.*, 2000a) of the pancreas. In both cell types, glucose metabolism results in K_{ATP} channel closure. In δ-cells, this causes hormonal secretion (as it does in β-cells). In α-cells, however, K_{ATP} channel closure and the resulting membrane depolarization causes inactivation of the voltage-gated channels that participate in action

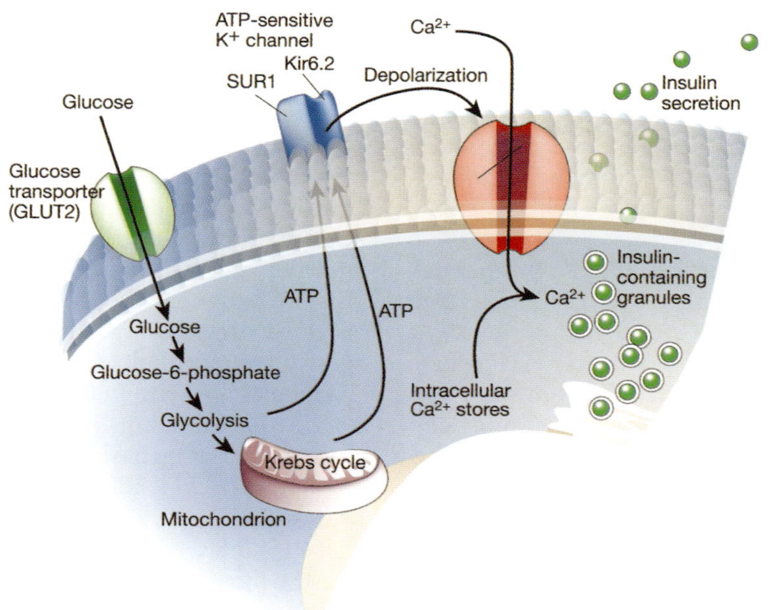

Figure 27.2. Model of a pancreatic β-cell summarizing the roles of major players, including SUR1/Kir6.2, involved in regulating insulin secretion. (Modified and reprinted by permission from **Nature** *(Bell and Polonsky, 2001) copyright 2001 Macmillan Publishers Ltd.) ATP generated by glycolysis and the Krebs cycle can result in inhibition and closure of the SUR1/Kir6.2, ATP-sensitive channel. The consequent reduction in potassium efflux results in membrane depolarization and Ca^{2+} influx. High intracellular Ca^{2+} levels lead to the fusion of secretory granules containing insulin with the membrane and secretion of insulin into the circulation.*

potential generation. Consequently, glucagon secretion is inhibited.

Insights from genetically engineered mice

Transgenic mice have proved to be valuable tools for analyzing the physiological role of the K_{ATP} channel (Seino *et al.*, 2000). Miki and colleagues (1997) showed that mutation of a key glycine in the K^+ channel selectivity sequence to serine (Kir6.2-G132S) has a dominant-negative effect on the K_{ATP} channel. By expressing this transgene under the control of the insulin promoter, they were able to selectively delete K_{ATP} channel function in the pancreatic β-cell. Neonatal transgenic mice developed hypoglycemia with hyperinsulinemia, which progressed to hyperglycemia with hypoinsulinemia in later life. A high frequency of apoptotic β-cells was found prior to the appearance of hyperglycemia, suggesting that K_{ATP} channels may play a role in pancreatic β-cell survival (Miki *et al.*, 2001a).

A Kir6.2 knockout mouse, in which Kir6.2 was disrupted genetically in all tissues, had a different phenotype (Miki *et al.*, 1998). Despite showing defective glucose-induced insulin secretion, these mice had only a mild impairment of glucose tolerance. This appeared to be due to an enhanced insulin sensitivity, which compensated for the reduced insulin secretion. The β-cell number decreased with age, but not as markedly as in the transgenic mice (Miki *et al.*, 2001a). Kir6.2 knockout mice can develop fasting hyperglycemia and glucose intolerance with age, but only if they become obese. This suggests that both the genetic defect and environmental factors are required for the mice to develop diabetes.

Like the Kir6.2 knockout mice, SUR1 knockout mice are normoglycemic, unless stressed (Seghers *et al.*, 2000). However, they are not insulin hypersensitive. They become more hyperglycemic when glucose-loaded, and more hypoglycemic when fasted, than wild-type mice. Isolated islets from SUR1 knockout mice lack first phase insulin secretion and have an attenuated second phase secretion in response to glucose stimulation. As expected, K_{ATP} currents are not found in β-cells of Kir6.2 knockout or SUR1 knockout mice (Miki *et al.*,

2001a; Seghers *et al.*, 2000). Because patients carrying loss-of-function mutations in the SUR1 gene can be severely hypoglycemic, unlike the SUR1 knockout mice, it is hypothesized that K_{ATP}-independent pathways controlling insulin secretion may be regulated differently in man and mouse (Seghers *et al.*, 2000).

The effect of reducing the K_{ATP} channel ATP sensitivity has also been explored. An N-terminal truncated Kir6.2, which forms K_{ATP} channels with about 10-fold lower ATP sensitivity, was expressed under the control of the insulin promoter, so that overexpression was confined to the β-cell (Koster *et al.*, 2000). The mutation is expected to partially prevent the decrease in β-cell K_{ATP} channel activity that occurs when plasma glucose levels rise, and thereby to decrease insulin resistance. As predicted the mice showed a severe diabetic phenotype: they were hyperglycemic, hypoinsulinemic, and most of them died within 5 days of birth. Because histological analysis revealed that islet morphology, insulin localization, and α- and β-cell distribution were normal, the diabetic phenotype may be caused by reduced insulin secretion. An interesting and important finding was that a comparatively small reduction in ATP sensitivity (<10-fold) was sufficient to cause diabetes.

In conclusion, studies of genetically engineered mice demonstrate that K_{ATP} channels play a key role in both glucose-induced and sulfonylurea-induced insulin secretion. They further show that defects in K_{ATP} channel activity can predispose the animal to diabetes, which is exacerbated by obesity.

PHHI

Persistent hyperinsulinemic hypoglycemia of infancy (PHHI), also known as familial hyperinsulinism or nesidioblastosis, is an inherited disorder characterized by abnormally high levels of insulin secretion despite severe hypoglycemia (Glaser, 2000; Sharma *et al.*, 2000). It presents at birth or in early childhood, and in the absence of clinical treatment may be lethal or result in irreversible neurological damage. Mild cases of the disease can be treated with the K_{ATP} channel opener diazoxide (or even by supplementing the diet with glucose), but the more severe forms require subtotal pancreatectomy. The frequency of PHHI in the general population is low but in inbred populations it may be as high as 1 in 2500 live births. In most families, PHHI is inherited in an autosomal recessive manner, but dominant forms of the disease have also been described (Glaser *et al.*, 1998; Huopio *et al.*, 2000). Sporadic cases have also been identified. Histopathologically, PHHI can also be divided into two types (Glaser, 2000). One is a diffuse form of the disease, in which the morphological defect is found within all the islets throughout the pancreas and the size of islets is irregular. The other is a focal type, in which the morphological defect is localized to a particular area within the pancreas.

The disease results from mutations in at least four different genes: the K_{ATP} channel genes Kir6.2 and SUR1, and the enzymes glucokinase (Glaser *et al.*, 1998) and glutamate dehydrogenase (Stanley *et al.*, 1998). Mutations in SUR1 are the most common cause of the disorder and account for about 50% of cases. All mutations are thought to lead to loss of K_{ATP} channel function in the pancreatic β-cell, as a consequence of defects either in the channel itself or in its metabolic regulation. Because the K_{ATP} channel sets the β-cell membrane potential, loss of channel activity produces a persistent membrane depolarization that leads to activation of voltage-gated Ca^{2+} influx and continuous insulin secretion, irrespective of the blood glucose level.

Mutations in Kir6.2 and SUR1 causing PHH1

In this review, we focus on the mutations in K_{ATP} channel subunits responsible for PHHI. To date, three mutations have been identified in the *Kir6.2* gene (Aguilar-Bryan and Bryan, 1999; Nestorowicz *et al.*, 1997; Thomas *et al.*, 1996b) and numerous mutations in the *SUR1* gene (Dunne *et al.*, 1997; Kane *et al.*, 1996; Nestorowicz *et al.*, 1996; Nichols *et al.*, 1996; Otonkoski *et al.*, 1999; Sharma *et al.*, 2000; Shyng *et al.*, 1998; Tanizawa *et al.*, 2000; Thomas *et al.*, 1995, 1996a; Verkarre *et al.*, 1998). The *Kir6.2* mutations comprise two missense mutations, L147P and W91R, and a nonsense mutation that truncates Kir6.2 after 12 amino acids. Kir6.2 carrying these mutations does not reconstitute functional K_{ATP} channels when coexpressed with SUR1 (Aguilar-Bryan and Bryan, 1999; Nestorowicz *et al.*, 1997; Thomas *et al.*, 1996a). The *SUR1* mutations are very heterogeneous. They are found throughout the gene and they include nonsense, missense, frameshift and splice site mutations. Functionally, these mutations can be grouped into two broad classes: those in which the channel is not present in the surface membrane, and those in which the channel is present in the plasma membrane but is always closed, independent of the metabolic state of the cell.

The molecular mechanism of action of several SUR1 mutations has been analyzed in detail, both by examination of β-cells obtained from therapeutic pancreatectomy (Kane *et al.*, 1996) and by heterologous expression of the mutant channels (Cartier *et al.*, 2001; Matsuo *et al.*, 2000b; Nichols *et al.*, 1996; Otonkoski *et al.*, 1999; Shyng *et al.*, 1997). Such studies have provided important insights into the structure–function relationships of the K_{ATP} channel, as well as its physiological role.

A number of mutations are associated with the absence of K_{ATP} channel activity even when exposed to nucleotide-free intracellular solutions, suggesting the channel may not be present in the surface membrane (e.g. Cartier *et al.*, 2001; Otonkoski *et al.*, 1999; Shyng *et al.*, 1997). Direct evidence for defective trafficking to the plasma membrane has been obtained for one of these mutations (deletion of F1388 in SUR1) (Cartier *et al.*, 2001). Interestingly, the ΔF1388 mutant channels have reduced ATP sensitivity and do not respond to stimulation by MgADP or diazoxide, even when expressed at the surface membrane by mutations that correct the trafficking defect.

Other mutations influence nucleotide interactions with the SUR1 subunit. Nichols and colleagues (1996) first reported that the G1479R mutation, which lies within NBD2, just upstream of the signature sequence, strongly reduced channel activation by MgADP. Subsequently, many other mutations have also been found to reduce MgADP activation (e.g. Huopio *et al.*, 2000; Shyng *et al.*, 1997). In these mutant channels, it appears that loss of MgADP activation underlies the inability of the channel to respond to metabolic inhibition. In some cases, the mutant channel retains sensitivity to diazoxide, which accounts for the fact that some PHHI patients respond to this drug. A missense mutation (R1420C) within NBD2 of SUR1 is associated with mild disease (Matsuo *et al.*, 2000b). This mutation slightly reduced the affinity of nucleotide binding to NBD2, and shifted the concentration–response curve for channel activation by MgADP to higher concentrations. It also reduced the surface expression of the channel. These two effects may explain why the mutation causes PHHI. Interestingly, the R1420C mutation impairs the ability of MgADP, acting at NBD2, to stabilize the binding of 8-azido-ATP at NBD1. Because MgADP activation was not abolished, this result suggests that nucleotide stabilization at NBD1 is not required for channel activation by MgADP.

All mutations in which SUR1 does not reach the surface membrane produce a severe form of PHHI. Mutations that are associated with loss or reduction of MgADP activation appear to be somewhat less severe. However, some mutations altered channel function only minimally *in vitro* but were associated with severe clinical disease. Thus, the precise relationship between the individual mutation and the severity of the clinical phenotype is not completely clear.

Diabetes

Because K_{ATP} channels regulate insulin secretion **(see Figure 27.2)**, it seems logical to postulate that gain-of-function mutations in the *SUR1* or *Kir6.2* genes might lead to diabetes mellitus. Enhanced K_{ATP} channel activity in the presence of stimulatory glucose concentrations would impair β-cell depolarization, resulting in reduced Ca^{2+} channel activation, less Ca^{2+} influx and less insulin release. A larger K_{ATP} current could arise from a reduced ATP sensitivity of the channel (as in the case of the transgenic mice), enhanced sensitivity to the stimulatory effects of MgADP, a higher channel density in the plasma membrane, or altered concentrations of channel modulators (e.g. ATP).

Many linkage analyses and mutation screening studies of the *Kir6.2* and *SUR1* genes have been performed, but the results remain confusing. Several groups have reported that genetic variation in *SUR1* (Iwasaki *et al.*, 1996; Lindner *et al.*, 1997; Stirling *et al.*, 1995) or *Kir6.2* (Sakura *et al.*, 1996) does not play a major role in susceptibility to type 2 diabetes in diverse populations (Mexican-American, Japanese, Caucasian or non-Hispanic whites). On the other hand, some *SUR1* (Hart *et al.*, 2000; Inoue *et al.*, 1996; Rissanen *et al.*, 2000) or *Kir6.2* (Hani *et al.*, 1998) gene variants are reported to be associated with type 2 diabetes in different populations, including Caucasians and Finns. Meta-analysis of several studies showed that the polymorphism E23K, which lies in the N-terminus of *Kir6.2*, is associated with type 2 diabetes in Caucasians (Hani *et al.*, 1998). Our unpublished studies indicate that the E23K mutation produces, if anything, only a small reduction in the ATP sensitivity of the K_{ATP} channel. However, it is apparent from studies on genetically engineered mice (Koster *et al.*, 2000) that small changes in K_{ATP} channel ATP sensitivity are sufficient to produce diabetes. This is because the input resistance of the β-cell is very high, so that small changes in K_{ATP} current exert large effects on the membrane potential. In this regard, it is worth remembering that the ATP sensitivity of the K_{ATP} channel can also be reduced by lipid modulators, such as PIP_2 (Baukrowitz *et al.*, 1998; Fan and Makielski, 1997; Shyng and Nichols, 1998) and long-chain acyl-CoA (Bränström *et al.*, 1997, 1998; Gribble *et al.*, 1998a). Because plasma lipids, including long-chain acyl-CoA esters, are elevated in obesity (Prentki and Corkey, 1996; Prentki *et al.*, 1997), this may be one mechanism by which obesity is linked to the development of impaired insulin secretion and diabetes.

Recent studies have suggested that PHHI may be self-limiting since affected β-cells undergo apoptosis (Kassem *et al.*, 2000), and there is accumulating evidence that some forms of PHHI lead to type 2 diabetes in later life. For example, the dominant E1506K mutation in the *SUR1* gene (which lies immediately after the Walker B motif of NBD2), which leads to a reduction in the level, but not complete loss, of K_{ATP} channel function, is associated with the progressive development of a reduced insulin secretory capacity that predisposes to diabetes in adult life (Huopio *et al.*, 2000). This may be because continuous β-cell membrane depolarization, resulting from the loss of functional K_{ATP} channels, produces an increase in the intracellular Ca^{2+} concentration. This, in turn, could activate apoptosis, so reducing β-cell mass (Efanova *et al.*, 1998). Studies of genetically engineered mice support the idea that loss of K_{ATP} channel function gives rise to β-cell apoptosis (Miki *et al.*, 2001a).

Impaired metabolic regulation of K_{ATP} channels resulting from mutations in genes that influence β-cell metabolism can also cause diabetes. Maturity-onset diabetes of the young (MODY) is characterized by early onset and autosomal dominant inheritance. Mutations in six different genes are known to cause MODY. In French families (Froguel *et al.*, 1993), around 60% of all MODY patients carry mutations in the gene encoding glucokinase, the glycolytic enzyme that catalyzes the conversion of glucose to glucose-6-phosphate in liver and β-cells (Randle, 1993). Mutations in the transcription factor, hepatocyte nuclear factor-1α, are the commonest cause of MODY in the UK population (Frayling *et al.*, 1997). Although the molecular mechanism of this MODY variant is not known, there is clear evidence that impaired β-cell metabolism is the basis of the disease, and that this causes reduced metabolic regulation of K_{ATP} channels (Dukes *et al.*, 1998). Impaired mitochondrial metabolism can also give rise to diabetes, as in maternally inherited diabetes with deafness (MIDD), which results from a mutation at position 3243 of the mitochondrial DNA that encodes a leucine transfer RNA (Maassen and Kadowaki, 1996). All these mutations result in reduced K_{ATP} channel inhibition in response to glucose metabolism and consequently impaired insulin secretion.

Uncoupling proteins induce a proton leak in the mitochondria that leads to impaired ATP production. Mice in which the uncoupling protein UCP2 was knocked out had higher islet ATP levels and increased insulin secretion, whereas rodents overexpressing UCP2 had impaired insulin secretion (Chan *et al.*, 2001; Zhang *et al.*, 2001). These results suggest that mutations in uncoupling proteins may also be involved in the development of diabetes and obesity in humans.

K_{ATP} CHANNELS IN THE BRAIN

In situ hybridization and immunocytochemistry suggest that *SUR1* and *Kir6.2* are expressed in several regions of the brain (Karschin *et al.*, 1997). Single-cell polymerase chain reaction (PCR) data are providing finer resolution of neuronal expression. These studies have shown, for example, that gamma-aminobutyric acidergic (GABAergic) neurons in the substantia nigra (SN) selectively express SUR1 and Kir6.2, whereas dopaminergic SN neurons express Kir6.2 together with SUR1 or SUR2B, or both SURs (Liss *et al.*, 1999).

Genetically engineered mice have facilitated our understanding of the role of K_{ATP} channels in the brain. In particular, they point to a role for K_{ATP} channels in seizure prevention. Hypoxia inhibited the activity of substantia nigra reticulata (SNr) neurons in wild-type mice, but enhanced SNr neuronal activity in Kir6.2 knockout mice (Yamada *et al.*, 2001). Furthermore, Kir6.2 knockout mice are susceptible to generalized seizures after brief hypoxia. This suggests that the opening of K_{ATP} channels in SNr neurons protects against seizure propagation during metabolic stress (Yamada *et al.*, 2001). In support of this idea, transgenic mice that overexpress SUR1 in cortex, hippocampus, and striatum are more resistant to kainic acid-induced seizures than wild-type mice (Hernandez-Sanchez *et al.*, 2001). The transgenic animals suffered no loss of hippocampal pyramidal neurons after kainic acid administration, whereas wild-type mice lost 70–80% of their pyramidal neurons. These results indicate that the overexpression of SUR1 in forebrain protects mice from seizures and neuronal damage. It is difficult to assess whether PHHI subjects carrying SUR1 mutations also suffer neurological problems as a result of decreased neuronal K_{ATP} channel function, because any brain damage that is present might be a consequence of low blood glucose levels experienced prior to diagnosis and treatment.

K_{ATP} channels also serve as glucose sensors in glucose-responsive neurons of the ventromedial hypothalamus, a role which is essential for the maintenance of glucose homeostasis (Miki *et al.*, 2001b). Although pancreatic β-cell function is intact in Kir6.2 knockout mice, the animals exhibit a severe defect in glucagon secretion in response to systemic hypoglycemia. In addition, they show a complete loss of glucagon secretion, together with reduced food intake, in response to neuroglucopenia. Thus hypothalamic and β-cell K_{ATP} channels act in concert, as central and peripheral glucose sensors, to regulate glucose homeostasis.

K_{ATP} CHANNELS IN THE CARDIOVASCULAR SYSTEM

In most tissues other than the pancreas, including cardiac and skeletal muscle, K_{ATP} channels are closed under normal conditions. In the heart, K_{ATP} channels open when the intracellular concentration of ATP falls under ischemic stress (Nichols and Lederer, 1990). This serves to shorten the action potential duration and reduce Ca^{2+} influx, so decreasing contractile force and ATP consumption. In this way, K_{ATP} channel activation helps protect the myocardium from ischemic injury. Analysis of Kir6.2 knockout mice confirms that sarcolemmal K_{ATP} channels containing Kir6.2 subunits mediate the depression of cardiac excitability and contractility produced by K_{ATP} channel openers and metabolic inhibition (Suzuki *et al.*, 2001). Vascular smooth muscle K_{ATP} channels are thought to play a role in regulation of vessel tone (Quayle *et al.*, 1997; Yokoshiki *et al.*, 1998). They are regulated by a variety of neurotransmitters, some of which mediate their effects via protein kinase A (which enhances K_{ATP} channel activity) and/ or protein kinase C (which decreases channel activity) (Hayabuchi *et al.*, 2001). Analysis of Kir6.2 knockout mice suggests that Kir6.2 does not contribute to the arterial K_{ATP} channel (Suzuki *et al.*, 2001), and it seems likely that Kir6.1 serves this role (Inagaki *et al.*, 1995b).

In addition to K_{ATP} channels in the plasma membrane, it has been suggested that the mitochondrial membrane contains another type of K_{ATP} channel that may be involved in ischemic preconditioning in the heart (Grover and Garlid, 2000). As the molecular identity of this channel remains unknown, and its relationship to ABC transporters is unclear, it will not be considered further here.

MOLECULAR PROPERTIES

MOLECULAR COMPOSITION

As shown in **Figure 27.1**, the K_{ATP} channel is a hetero-octameric complex of Kir6.x and SUR subunits (Clement *et al.*, 1997; Inagaki *et al.*, 1997; Shyng and Nichols, 1997). Kir6.x belongs to the family of inwardly rectifying K^+ channels and there are two isoforms: Kir6.1 and Kir6.2 (Inagaki *et al.*, 1995a, 1995b; Sakura *et al.*, 1995). Both isoforms have cytosolic N- and C-termini, and two transmembrane segments, linked by a region that is predicted to re-enter the membrane and serve as the selectivity filter for potassium ions. In most tissues, Kir6.2 acts as the pore-forming subunit of the K_{ATP} channel, but in vascular smooth muscle Kir6.1 serves this role. The site at which ATP binds to mediate channel inhibition lies within the cytosolic domains of Kir6.2 (Tanabe *et al.*, 1999; Tucker *et al.*, 1998).

SUR1 is a member of the ABCC subfamily of the ABC proteins (Aguilar-Bryan *et al.*, 1995), which also includes the multidrug resistance-associated protein (MRP), and the cystic fibrosis transmembrane conductance regulator (CFTR) (Dean *et al.*, 2001). Although SUR1 was originally thought to have 13 transmembrane segments (Aguilar-Bryan *et al.*, 1995), sequence alignment of SURs and MRPs suggested that there are 17 transmembrane segments (TMS), arranged in three groups of 5, 6 and 6 TMSs

(Tusnády *et al.*, 1997) **(see Figure 27.1)**. Evidence for the 17 TMS model was obtained subsequently by epitope mapping (Raab-Graham *et al.*, 1999). There are two large intracellular NBDs, one between TMS 11 and 12 and the other at the C-terminus. Like other ABC transporters (Higgins, 1992), each NBD contains several highly conserved motifs. These include a Walker A motif and a Walker B motif, a linker sequence (the ABC signature sequence, LSGGQ), and an invariant glutamine and histidine residue (sometimes equated with the Q-loop and H-loop, respectively).

There are two subtypes of SUR: SUR1 (ABCC8) and SUR2 (ABCC9) (Inagaki *et al.*, 1996). SUR2A shares 68% amino acid identity with SUR1, and Northern blotting has revealed that it is expressed at high levels in heart, skeletal muscle, and ovary. Two splicing variants of SUR2A (SUR2B and SUR2C) have been identified (Chutkow *et al.*, 1999; Yamada *et al.*, 1997). SUR2B is identical to SUR2A except for its C-terminal 42 amino acids, which more closely resemble those of SUR1. SUR2C has a deletion of 35 amino acids near NBD1 of SUR2A (Chutkow *et al.*, 1999).

The Kir6.2 and SUR1 genes map to the same region of human chromosome 11p15.1, and both reading frames are aligned in the same direction (Inagaki *et al.*, 1995a). The Kir6.1 and SUR2 genes are similarly clustered on human chromosome 12p (11.23-12.12) (Chutcow *et al.*, 1996).

STOICHIOMETRY

There is good evidence that Kir6.2 co-assembles as a tetramer, as is the case for other Kir channels (Clement *et al.*, 1997; Shyng and Nichols, 1997) **(see Figure 27.1)**. The stoichiometry of the K_{ATP} channel has been determined using tandem constructs of SUR1 and Kir6.2 subunits (Clement *et al.*, 1997; Inagaki *et al.*, 1997). The SUR1–Kir6.2 construct showed similar properties to native channels, indicating that a 1:1 stoichiometry is sufficient for functional K_{ATP} channels. In contrast, the SUR1–Kir6.2–Kir6.2 construct did not produce functional channels, although it could be rescued by coexpression with SUR1. This suggests that each Kir6.2 subunit requires an SUR partner. The octameric (4 + 4) nature of the K_{ATP} channel complex is supported by biochemical studies showing that the molecular mass of the purified Kir6.2/SUR1 complex is approximately 950 kDa, which corresponds to four SUR1 and four Kir6.2 subunits (Clement *et al.*, 1997).

ASSEMBLY AND TRAFFICKING OF K_{ATP} CHANNELS

Both Kir6.2 and SUR possess an endoplasmic reticulum retention motif (RKR) that prevents their surface expression in the absence of the other type of subunit (Zerangue *et al.*, 1999). This ensures that only fully functional octameric complexes reach the plasma membrane. The RKR motif in Kir6.2 lies within the C-terminal region of the protein and is masked in the presence of SUR1. Truncation of the last 26–36 amino acids of Kir6.2 deletes this retention signal and thus allows plasma membrane expression in the absence of SUR (Tucker *et al.*, 1997). SUR1 also contains an RKR motif in a cytoplasmic loop between TMS11 and NBD1 (Zerangue *et al.*, 1999). In addition to this ER retention signal, the C-terminus of SUR1 has been proposed to contain an anterograde signal, consisting of a dileucine motif and a downstream phenylalanine, which is required for K_{ATP} channels to exit the ER of mammalian cells (Sharma *et al.*, 1999). In contrast, truncation of NBD2 of SUR1 does not prevent functional K_{ATP} channels from reaching the plasma membrane when expressed in *Xenopus* oocytes (Sakura *et al.*, 1999). Whether this reflects a difference in K_{ATP} channel trafficking in oocytes and mammalian cells, or some other process, has not been ascertained.

SUR1 and Kir6.2 must be closely associated, since Kir6.2 can be photoaffinity labeled with a ligand that binds to SUR (Clement *et al.*, 1997), and the subunits can be co-immunoprecipitated (Lorenz *et al.*, 1998). The first transmembrane segment and the N-terminus of Kir6.2 are involved in K_{ATP} assembly (Giblin *et al.*, 1999; Schwappach *et al.*, 2000). A requirement for the C-terminus of Kir6.2 has also been reported (Giblin *et al.*, 1999; Lorenz and Terzic, 1999). The regions of SUR that are involved in assembly and interaction with Kir6.2 have not been fully defined. Using a trafficking-based assay for detection of such interactions, Schwappach and colleagues (2000) concluded that TMS 6–17 of SUR1 are required for interaction with Kir6.2 and that the NBDs are not needed for correct assembly. In a different approach, TMS 12 and 13 were shown not to be required for SUR1 assembly, but TMS12 was found to be involved in interaction with Kir6.2 (Mikhailov *et al.*, 2000).

REGULATION BY NUCLEOTIDES

The nucleotide regulation of the K_{ATP} channel is complex, as channel activity is inhibited by nucleotide binding to Kir6.2 and activated by Mg-nucleotides binding to the two NBDs of SUR. In addition, MgATP may activate lipid and protein kinases, thereby increasing the membrane concentration of phospholipids such as PIP_2 (which modulates channel ATP sensitivity), or altering the phosphorylation state of the channel itself. This makes it more difficult to elucidate the molecular mechanisms underlying nucleotide regulation of the K_{ATP} channel. Two types of experiment have helped to determine the properties of the different binding sites: first, radiolabeled ATP binding to one type of subunit, expressed in the absence of the other; and secondly, electrophysiological studies of K_{ATP} channels containing mutant subunits.

THE KIR6.2 NUCLEOTIDE-BINDING SITE

Truncation of the cytoplasmically located C-terminal 26–36 amino acids of Kir6.2 (Kir6.2ΔC), allows this subunit to reach the surface membrane in the absence of SUR. This construct permitted the first demonstration that ATP binding to Kir6.2 causes channel closure (Tucker *et al.*, 1997). Confirmation that Kir6.2 indeed binds ATP is provided by the fact that this subunit can be photoaffinity labeled with the ATP analogues 8-azido-ATP and ATP-[γ]4-azidoanilido (Tanabe *et al.*, 1999, 2000). Kir6.2ΔC, expressed in the absence of SUR, has proved a useful tool for analyzing the properties of the Kir6.2 ATP-binding site.

Unlike many classical ATP-binding sites, Mg^{2+} is not required for the action of the nucleotide, suggesting that channel inhibition does not require nucleotide hydrolysis. Inhibition is extremely sensitive to the structure of the adenine base moiety. Even ITP, which differs from ATP at only two positions, blocks the channel approximately 50-fold less effectively than ATP (Tucker *et al.*, 1998). In contrast, the terminal (γ)-phosphate group of ATP appears not to be required since ADP blocks almost as effectively as ATP (Tucker *et al.*, 1998). Furthermore, diadenosine polyphosphates, in which extra phosphate groups and an additional adenosine moiety are attached to the γ-phosphate, inhibit channel activity as effectively as ATP. AMP, however, in which the β-phosphate group is removed, is a poor inhibitor of channel activity (Tucker *et al.*, 1998). Hence, it appears that both the adenine base and the β-phosphate group are necessary for high-affinity channel inhibition. Although the site to which ATP binds lies on Kir6.2, coexpression with SUR enhances the potency of ATP inhibition by about 10-fold (Tucker *et al.*, 1997), by a mechanism that is not understood.

There is evidence that each Kir6.2 monomer in the K_{ATP} channel complex has its own ATP-binding site, so that there are four per channel (Markworth *et al.*, 2000). However, binding of ATP to just one subunit is sufficient to induce channel closure. The precise location of the ATP-binding site on Kir6.2 remains unclear. There are no obvious ATP-binding consensus sequences and mutagenesis studies require careful interpretation because a mutation may affect the channel ATP sensitivity in several ways. For example, it may influence ATP binding, impair the ability of the channel to close, or interfere with the transduction mechanism by which ATP binding induces channel closure. If ATP binds to a particular state of the channel (e.g. the closed state), even binding studies may not be able to distinguish between these possibilities. However, several mutations have been found that reduce the channel ATP sensitivity without altering the single-channel kinetics in the absence of ATP. These lie within both the N- and C-terminus and include residues R50, K185, I182, R201A and G334 (Drain *et al.*, 1998; Shyng *et al.*, 2000; Tucker *et al.*, 1997, 1998). Both R50G and K185Q mutations have been directly shown to decrease ATP binding (Tanabe *et al.*, 2000).

THE SUR NUCLEOTIDE-BINDING SITE

It is well established that MgADP stimulates K_{ATP} channel activity. Because MgADP blocks Kir6.2ΔC but stimulates Kir6.2/SUR1 (Tucker *et al.*, 1997), it appears that MgADP stimulation is mediated by the SUR subunit. Mutations within the NBDs of SUR have confirmed that this is the case and demonstrated that MgADP mediates its stimulatory effects by interaction with the NBDs (Gribble *et al.*, 1997a; Nichols *et al.*, 1996; Shyng *et al.*, 1997). Thus, mutation of the conserved Walker A lysine or the Walker B aspartate abolished MgADP activation and unmasked an inhibitory action of the nucleotide on Kir6.2 (Gribble *et al.*, 1997a).

In most ABC transporters, MgATP is the major ligand, and its hydrolysis to MgADP provides the energy required for substrate transport. A difficulty with trying to investigate the effects of MgATP at the NBDs of SUR is that the nucleotide also produces a potent block of the channel via Kir6.2. To circumvent this problem, Gribble *et al.* (1998b) coexpressed SUR1 with an ATP-insensitive pore mutant, Kir6.2-R50G. Addition of MgATP stimulated the activity of these channels (Gribble *et al.*, 1998b), an effect that was abolished by simultaneous mutation of the Walker A lysine in both NBDs of SUR1. Thus, like MgADP, MgATP stimulates K_{ATP} channel activity via interaction with the NBDs of SUR1. There is growing evidence, however, that MgADP is the effective ligand, and that MgATP must be hydrolyzed to MgADP at the NBDs of SUR1 before it is able to enhance K_{ATP} channel activity (Zingman *et al.*, 2001).

Nucleotide binding by the NBDs

The nucleotide-binding properties of the two NBDs of SURs have been examined directly, using the photoaffinity analogue 8-azido-ATP (Ueda *et al.*, 1997). 8-Azido-ATP binding to SUR1 is both high affinity and very stable. This made it possible to investigate the cooperative interaction between the two NBDs (Ueda *et al.*, 1999). Prebound 8-azido-ATP did not dissociate at 0°C and dissociated only slowly at 37°C, provided that Mg^{2+} was present. Addition of MgADP or MgATP markedly stabilized prebound 8-azido-ATP, whereas the slowly hydrolyzable ATP analogue ATP-γS did not. Mutations in the Walker A and B motifs of NBD2 had almost no effect on 8-azido-ATP binding itself, but abolished the ability of MgADP to stabilize prebound 8-azido-ATP. These results suggest that MgADP, either by directly binding to NBD2 or by hydrolysis of bound MgATP, induces a conformational change at NBD2 that results in stabilization of ATP binding at NBD1. A similar cooperativity was found for nucleotide binding to the two NBDs of SUR2A and SUR2B (Matsuo *et al.*, 2000a). The functional significance of this effect is unclear, however, because a mutation in SUR1 (R1420C, located between the Walker A and the LSGG motifs in NBD2) that abolished cooperativity did not impair the ability of either MgATP or MgADP to stimulate channel activity (Matsuo *et al.*, 2000b).

To examine the nucleotide binding of two NBDs in more detail, SURs photoaffinity labeled with 8-azido-[^{32}P]ATP were digested mildly with trypsin and tryptic fragments immunoprecipitated with antibodies against NBD1 or NBD2 (Matsuo *et al.*, 1999, 2000a). The ATP-binding properties of the tryptic fragments indicated that NBD1 of SUR binds 8-azido-ATP in a Mg^{2+}-independent manner, and that NBD2 binds 8-azido-ATP in a Mg^{2+}-dependent manner. Because K_{ATP} channels are activated by ADP only in the presence of Mg^{2+} (Gribble *et al.*, 1997a), NBD2 seems to be primarily responsible for the channel activation by MgADP.

Mutations in the signature sequence of NBD2 of SUR1 (e.g. G1479D, G1479R, G1485D, G1485R, Q1486H and D1506A) abolish K_{ATP} channel activation by MgADP, whereas similar mutations in NBD1 (e.g. G827D, G827R and Q834H) do not (Gribble *et al.*, 1997a; Shyng *et al.*, 1997). Furthermore, mutation of the Walker A lysine in NBD2 of SUR2A (K1348A) abolishes channel activation by MgADP, but the corresponding mutation in NBD1 (K707A) does not (Reimann *et al.*, 2000). These results support the idea that NBD2 of SUR is essential for MgADP activation and that NBD1 may play a secondary role. Although mutation of the Walker A lysine in NBD1 of SUR1 (K719A) abolishes the ability of MgADP to stimulate K_{ATP} channel activity (Gribble *et al.*, 1997a), we have found that the mutation of this residue (K719M) abolishes 8-azido-ATP binding at *both* NBDs of SUR1 (Matsuo *et al.*, unpublished data). Thus, it is not clear yet if ATP binding to NBD1 is required for MgADP activation of the K_{ATP} channel. It remains possible, however, that nucleotide binding to NBD1 facilitates the action of MgATP (or indeed MgADP) at NBD2: for example by enhancing nucleotide hydrolysis or transduction of nucleotide binding/hydrolysis into channel gating.

The nucleotide-binding affinities of the NBD1 of SUR1 are significantly higher than those of SUR2A and SUR2B, particularly in the case of ATP (Matsuo *et al.*, 2000a). Interestingly, the affinity of NBD1 of SUR2B for ATP is higher than that of SUR2A, and the affinities of NBD2 of SUR2B for both ATP and ADP are greater than those of SUR2A. Because SUR2A and SUR2B share the same amino acid sequence except for their C-terminal 42 amino acids, the C-terminal 'tail' may affect the nucleotide-binding properties of both NBDs. The different nucleotide-binding affinities of the various NBDs may explain, in part, the differential nucleotide regulation of K_{ATP} channel subtypes. For example, the fact that higher concentrations of MgADP are needed to activate Kir6.2/SUR2A channels than Kir6.2/SUR1 or Kir6.2/SUR2B channels, when tested in the presence of MgATP (Matsuoka *et al.*, 2000).

ATP hydrolysis by the NBDs

Unlike many other ABC transporters, where both NBDs are believed to hydrolyze ATP, NBD2 of SUR appears to be more efficient at hydrolyzing ATP than NBD1 (but see Chapter 29). Thus, under conditions permitting ATP hydrolysis, NBD2 of all three SUR subtypes was photoaffinity labeled with 8-azido-[α-^{32}P]ATP but not with 8-azido-[γ-^{32}P]ATP, whereas NBD1 was photoaffinity labeled with both ligands (Matsuo *et al.*, 2000a). This suggests that NBD2 has ATPase activity and that NBD1 has little or none. It has also been shown that a construct consisting of NBD2 of SUR2 (lacking the C-terminal tail) fused to the maltose-binding protein has twice the ATPase activity of a similar fusion protein containing NBD1. However, it should be noted that these ATPase activities are very low: $V_{max} = 0.018$ and $0.009\,\mu\text{mol}\,\text{min}^{-1}\text{mg}^{-1}$, for NBD2 and NBD1, respectively (Bienengraeber *et al.*, 2000; Zingman *et al.*, 2001). For comparison, the NBD1 of CFTR fused to the maltose-binding protein and the purified recombinant CFTR from Sf9 cells showed ATPase activity of about 0.03 and $0.05\,\mu\text{mol}\,\text{min}^{-1}\text{mg}^{-1}$ of purified protein, respectively (Bear *et al.*, 1997; Ko and Pederson, 1995; Li *et al.*, 1996). The lower ATPase activity of SUR is consistent with its role as a channel regulator rather than a transporter. Interestingly, the ATPase activity of NBD2 is stimulated by K_{ATP} channel openers, such as rilmakalim, pinacidil, cromakalim, diazoxide and nicorandil (Bienengraeber *et al.*, 2000).

Metabolic Regulation of the K_{ATP} Channel

K_{ATP} channels couple cellular metabolism to electrical excitability, by sensing changes in intracellular ATP and ADP concentrations. The intracellular ATP concentration ($[ATP]_i$) is estimated to lie between 1 and 5 mM, even during metabolic inhibition (Gribble *et al.*, 1997b, 2000; Himmelreich and Dobson, 2000), whereas free MgADP concentrations are thought to be less than 100 μM (Askenasy and Navon, 1997; Ghosh *et al.*, 1991; Himmelreich and Dobson, 2000). Even under metabolic stress, changes in cytosolic ADP concentration only occur within the 100 μM range (Weiss and Venkatesh, 1993). This suggests that metabolically induced changes in MgADP do not modulate K_{ATP} channel activity by directly competing with MgATP for binding at NBD2. Rather, an ATP hydrolysis cycle at NBD2 generates bound MgADP and changes in cell metabolism influence K_{ATP} channel activity by modulating the length of time that NBD2 remains in the MgADP-bound (active) state (Zingman *et al.*, 2001). When metabolic activity is high, cytosolic ADP levels will be low so that MgADP dissociates more rapidly from NBD2, causing channel activity to decrease. In contrast, when metabolic activity declines, the rise in MgADP will slow the off-rate of MgADP, and promote channel opening. In this way, SUR monitors changes in intracellular MgADP concentration.

In addition to their stimulatory effects, mediated via NBD2 of SUR, nucleotides (ATP, ADP) also inhibit the K_{ATP} channel via Kir6.2. The metabolic response of the K_{ATP} channel will reflect all of these processes. It will also be affected by the metabolic rate of the cell, and the extent to which changes in submembrane nucleotide concentrations are buffered, e.g. by creatine kinase and adenylate kinase (Carrasco *et al.*, 2001).

Native K_{ATP} channels exhibit different sensitivities to metabolic inhibition. K_{ATP} channels in pancreatic β-cells and microvascular coronary endothelial cells open when extracellular glucose levels fall (Ashcroft *et al.*, 1984; Langheinrich and Daut, 1997), whereas cardiac channels remain closed in zero glucose solutions and only open in response to ischaemia or metabolic inhibition (Nichols and Lederer, 1991). There is evidence that K_{ATP} channels in vascular smooth muscle contribute to the regulation of coronary flow under both normoxic and hypoxic conditions (Daut *et al.*, 1994). These differences in metabolic sensitivity reflect both differences in cell metabolism and differences in the metabolic sensitivities of the channels themselves. Thus, Kir6.2/SUR2A channels are less activated by MgATP than either Kir6.2/SUR2B or Kir6.2/SUR1 channels

(Reimann *et al.*, 2000). There is evidence that this may relate to differences in nucleotide handling at NBD2 (Song and Ashcroft, unpublished).

COMPARISON WITH OTHER ABC TRANSPORTERS

Most ABC transporters are active transporters that expend the energy of ATP hydrolysis to transport compounds against a chemical gradient (Dean *et al.*, 2001; Higgins, 1992). In these proteins, a catalytic ATP hydrolysis cycle is coupled to substrate transport (Senior and Gadsby, 1997). In contrast, SUR serves as the regulatory subunit of the K_{ATP} channel and it is therefore not perhaps surprising that the nucleotide-binding properties of the NBDs of SUR differ from those of other ABC proteins.

The archetypal ABC transporter MDR1 can be photoaffinity labeled by 8-azido-ATP in the presence of orthovanadate (Senior *et al.*, 1995; Ueda *et al.*, 1997; Urbatsch *et al.*, 1995). This vanadate-induced nucleotide trapping occurs at both NBDs, although not simultaneously, and is abolished by mutations in the Walker A or Walker B motif of either NBD (Azzaria *et al.*, 1989; Loo and Clarke, 1995a; Muller *et al.*, 1996) or by modification of either NBD with *N*-ethylmaleimide (NEM) (Al-Shawi *et al.*, 1994; Liu and Sharom, 1996; Loo and Clarke, 1995b; Takada *et al.*, 1998). This indicates that the ATPase activity of the two NBDs of MDR1 is highly cooperative. Some interaction also occurs between the NBDs of SUR, because MgADP binding to NBD2 stabilizes ATP binding at NBD1. However, the very low ATP hydrolysis rate at NBD1 of SUR in the intact protein (Matsuo *et al.*, 1999, 2000a) suggests that the ATP hydrolysis cycle is not coupled in the same way as found for MDR. Mg^{2+} is required for ATP binding and hydrolysis at both NBDs of MDR1 (Ueda *et al.*, 1997). By analogy, we may assume that although ATP binding to NBD1 of SUR does not require Mg^{2+}, the cation will be necessary if there are conditions under which NBD1 is able to hydrolyze MgATP.

Photoaffinity labeling of MRP1, a xenobiotic exporter that also belongs to the ABCC subfamily, demonstrated that NBD1 binds 8-azido-ATP strongly in the presence of Mg^{2+} and that NBD2 shows vanadate-induced nucleotide trapping (Gao *et al.*, 2000; Hou *et al.*, 2000; Nagata *et al.*, 2000). In contrast, NBD1 of SUR binds 8-azido-ATP strongly even in the absence of Mg^{2+} (Matsuo *et al.*, 1999, 2000a; Ueda *et al.*, 1997). Although NBD2 did not show vanadate-induced nucleotide trapping under the conditions examined in these experiments, other studies are consistent with the presence of nucleotide trapping at NBD2 of SUR2A (Zingman *et al.*, 2001). Beryllium fluoride, which mimics a prehydrolytic ADP-bound state, caused K_{ATP} channel closure, whereas orthovanadate, which mimics a posthydrolytic ADP-bound state, induced channel opening.

ATP hydrolysis has been implicated in the regulation of chloride flux through the CFTR Cl channel (Baukrowitz *et al.*, 1994), although its precise role is somewhat controversial. It was originally proposed that channel gating was entirely dependent on ATP hydrolysis (Hwang *et al.*, 1994; Weinreich *et al.*, 1999). However, the finding that ATP can support channel activity in the absence of Mg suggests this not the case. NBD1 of CFTR is photoaffinity labeled by 8-azido-ATP with high affinity. Binding is Mg^{2+} and temperature dependent, and is very stable, with nucleotide occlusion predicted to take place during the ATP hydrolysis cycle of CFTR (Szabo *et al.*, 1999). ATP binding to NBD1 of SUR1 is also temperature dependent (Matsuo *et al.*, 1999), but does not require Mg^{2+} (Ueda *et al.*, 1997). Thus, although nucleotide occlusion produced by a conformational change at NBD1 of SUR1 might take place, it does not do so as part of an ATP hydrolysis cycle. Szabo *et al.* (1999) predicted that NBD1 of CFTR rapidly enters an occluded nucleotide state, whereas the complete ATP hydrolytic cycle takes much longer. In contrast, both ATP hydrolysis and ADP release are much faster at NBD2, which explains why vanadate is required for stabilizing the transient state of NBD2.

Mg-independent nucleotide binding at NBD1 and slow ATP hydrolysis at NBD2 seem to be unique features of SUR compared with other ABC proteins. However, it is not clear what causes these differences. Hrycyna *et al.* (1999) reported that the context of the NBD, rather than its exact sequence, is an important determinant of ATP binding. This suggests that the three-dimensional structure of the NBDs, which will be determined not only by their amino acid sequence but also by their interaction with membrane domains, serves as a major determinant of the nucleotide-binding properties of SUR.

Effects of other K_{ATP} Channel Regulators

Regulation by phospholipids

Membrane phospholipids such as PIP_2 and PIP_3 interact with K_{ATP} channels to increase their probability of being open and to reduce their ATP sensitivity (Baukrowitz and Fakler, 2000; Baukrowitz *et al.*, 1998; Fan and Makielski, 1997, 1999; Hilgemann and Ball, 1996; Shyng and Nichols, 1998). This has been shown by direct application of PIPs to K_{ATP} channels in excised patches. In addition, overexpression of PI5-kinase, which enhances PIP_2 levels, reduces the ATP sensitivity of the channel in membrane patches (Shyng *et al.*, 2000), whereas break-down of PIP_2 by phospholipase C enhances the K_{ATP} channel ATP sensitivity (Xie *et al.*, 1999).

The effects of PIPs are mediated principally though the Kir6.2 subunit, because Kir6.2ΔC expressed in the absence of SUR also shows a reduced ATP sensitivity to applied PIP_2 (Baukrowitz *et al.*, 1998) and direct binding of PIP_2 to Kir channels has been demonstrated (Huang *et al.*, 1998). However, SUR may also modulate the PIP_2 sensitivity, perhaps by changing the probability of channel opening (Song and Ashcroft, 2001). It appears that PIP_2 may have both a direct effect on the channel ATP sensitivity and an indirect effect that is a consequence of altered channel gating (Fan and Makielski, 1999).

The rundown of both native and cloned K_{ATP} channels that occurs in excised membrane patches is reversed by application of MgATP (Findlay and Dunne, 1986; Ohno-Shosaku *et al.*, 1987; Xie *et al.*, 1999) and PIP_2 has also been implicated in such an effect (Xie *et al.*, 1999). It is postulated that PIP_2 is produced in the plasma membrane by serial phosphorylation of phosphatidylinositol (PI) and that this process is sequentially catalyzed by PI4-kinase and PIP-kinase. MgATP generation of phospholipids therefore has multiple, but related, effects on K_{ATP} channel activity.

Regulation by phosphorylation

Phosphorylation of the R-domain of CFTR by protein kinase A (PKA) is required for channel activity (Gadsby and Nairn, 1999). Interestingly, SUR has a string of negatively charged amino acids in an equivalent position, raising the possibility that they may serve a similar role to the phosphorylated R-domain. The activity of K_{ATP} channels is also regulated by protein kinase phosphorylation, although the effects are not so dramatic as for CFTR. Protein kinase A (PKA) stimulates activity of both native and recombinant K_{ATP} channels (Light, 1996), while protein kinase C (PKC) reduces the activity of smooth muscle K_{ATP} channels. Phosphorylation of Kir6.2 by PKA or PKC has been shown to influence the ATP sensitivity and/or open probability of the K_{ATP} channel (Béguin *et al.*, 1999; Light *et al.*, 2000; Lin *et al.*, 2000). There is also evidence that phosphorylation of SUR may be involved in channel regulation. Human SUR1 has four consensus sites for phosphorylation by PKA (Béguin *et al.*, 1999). PKA phosphorylation of SUR1 (at S1571) decreases the burst duration, interburst interval and open probability of Kir6.2/SUR1 channels (Béguin *et al.*, 1999).

Regulation by G-proteins

There are several reports that K_{ATP} channel activity is upregulated by trimeric GTP-binding (G) proteins (Sánchez *et al.*, 1998; Wada *et al.*, 2000). Thus, G_{α} was found to stimulate both Kir6.2/SUR1 and Kir6.2/SUR2A channels, whereas $G_{\beta\gamma}$ subunits had no effect (Sánchez *et al.*, 1998). In contrast, another study reported that $G_{\beta\gamma}$ subunits cause a reduction of ATP-induced inhibition of Kir6.2/SUR2A channels that is mediated by $G_{\beta\gamma}$ interaction with SUR2A (Wada *et al.*, 2000).

Endogenous ligands

The high affinity with which sulfonylureas bind to SUR1 led to the search for endogenous ligands of the K_{ATP} channel. Two peptides isolated from porcine brain, α- and β-endosulfine, were found to inhibit sulfonylurea binding *in vitro* (Virsolvy-Vergine *et al.*, 1988). Cloning of human α-endosulfine (Heron *et al.*, 1998) and analysis of the recombinant protein revealed that it inhibits Kir6.2/SUR1 channel activity and stimulates insulin secretion in the absence of glucose. Its functional significance has not yet been established.

PHARMACOLOGY

CHANNEL BLOCKERS

Inhibitors of K_{ATP} channel activity fall into two groups: those that interact with Kir6.2 and those that interact with SUR1. All these drugs stimulate insulin secretion. Imidazolines such as phentolamine and cibenzoline block K_{ATP} channels by binding to Kir6.2 (Mukai *et al.*, 1998; Proks and Ashcroft, 1997). In contrast, sulfonylureas (tolbutamide, gliclazide, glimepiride) and benzamido derivatives (meglitinide) close K_{ATP} channels by binding with high affinity to SUR1 (for review, see Ashcroft and Gribble, 1999). Many of these drugs are used therapeutically to stimulate insulin secretion in type 2 diabetes. Sulfonylureas also interact with low affinity with Kir6.2 (Gribble *et al.*, 1997b, 1998c), but this effect is of no clinical relevance, as the concentrations required to inhibit the Kir6.2 subunit are much higher than those found in the plasma of patients treated with the drugs.

Recent studies of cloned channels have revealed that low concentrations of tolbutamide block K_{ATP} channels containing SUR1, but not SUR2A or SUR2B, subunits (Gribble *et al.*, 1998c). In contrast, meglitinide blocks all three types of cloned K_{ATP} channel. This has been interpreted to indicate that sulfonylureas (such as tolbutamide) interact with a binding site that is specific to SUR1, whereas benzamido compounds (such as meglitinide) interact with a site that is common to all SUR subtypes (for review, see Ashcroft and Gribble, 1999). Glibenclamide, which contains both sulfonylurea and benzamido moieties, is postulated to interact with both sites on SUR1 but only a single (benzamido) on SUR2. This may account for the slow off-rate of the drug in electrophysiological experiments (it has to unbind at both sites).

The different affinities of Kir6.2/SUR1 and Kir6.2/SUR2 channels for tolbutamide were exploited to determine the tolbutamide-binding site, using a chimeric approach (Ashfield *et al.*, 1999; Babenko *et al.*, 1999). These studies showed that TMS 14–16 of SUR1 are required for high-affinity tolbutamide inhibition and that serine-1237, which lies in the predicted cytoplasmic loop between TMS 15 and 16, is crucial for high-affinity block. This is consistent with earlier studies suggesting an intracellular site of action for sulfonylureas (Lee *et al.*, 1994; Schwanstecher *et al.*, 1994). Recent studies, using a different approach, have shown that only the TMS 5–6 and TMS 15–16 cytosolic loops are needed for $[^3H]$-glibenclamide binding (Mikhailov *et al.*, 2001). This raises the possibility that the TMS 5–6 loop corresponds to the benzamido-binding site and the TMS 15–16 loop to the tolbutamide-binding site. It also suggests that TMSs 14–16 may be involved in transducing binding at the 'sulfonylurea' site into channel inhibition.

MgADP modulates the inhibitory action of sulfonylureas by enhancing the apparent block of β-cell K_{ATP} channels and reducing that of cardiac channels (Zunkler *et al.*, 1988). Studies of recombinant Kir6.2/SUR1 channels have shown that sulfonylureas prevent the stimulatory effect of MgADP at SUR1, and unmask its inhibitory effect on Kir6.2 (Gribble *et al.*, 1997b). This produces an apparent enhancement of the sulfonylurea block in the presence of MgADP, because the inhibitory effect of MgADP adds to that of the sulfonylurea. It is not known whether the interaction between MgADP and sulfonylureas results because the sulfonylurea displaces MgADP binding to the NBDs, or because it blocks transduction of the stimulatory action of MgADP bound to the NBDs. However, nanomolar concentrations of glibenclamide do not change the affinity of the NBDs of SUR1 for either ATP or ADP (Matsuo *et al.*, unpublished result). Although glibenclamide destabilizes ATP binding at NBD1 of SUR1 in the presence of MgATP or MgADP (Ueda *et al.*, 1999), it is not clear if this is directly connected with high-affinity glibenclamide inhibition of K_{ATP} channels.

The mechanism of the reduced sulfonylurea block of Kir6.2/SUR2 channels in the presence of MgADP is also not certain. However, it seems likely that this is a consequence of the enhanced channel activity produced by MgADP, which results in the channel spending less time in the interburst closed state. Because sulfonylureas appear to act by stabilizing the interburst closed states (Gillis *et al.*, 1989), the blocking potency of the drug will be reduced by MgADP (Alekseev *et al.*, 1997). Thus, one can think of the intraburst state(s) as ligand-insensitive states and the interburst state(s) as ligand-sensitive states. A similar effect would presumably operate in the case of Kir6.2/SUR1 channels, were it not for the ability of sulfonylureas to prevent the stimulatory effect of MgADP in this channel.

POTASSIUM CHANNEL OPENERS

Potassium channel openers (KCOs) are a structurally diverse group of drugs which share the common property that they activate K_{ATP} channels, thereby hyperpolarizing the plasma membrane and reducing electrical excitability (Ashcroft and Gribble, 2000). They include diazoxide, cromakalim, pinacidil and nicorandil. None of these drugs is in widespread clinical use, although diazoxide is used to suppress excessive insulin secretion, minoxidil sulfate is used topically to stimulate hair growth, and nicorandil is currently in clinical trials for the treatment of angina. Different SUR subtypes confer varying sensitivities to K_{ATP} channel openers. Thus, in the presence of cytosolic MgATP, Kir6.2/SUR1 channels are stimulated by diazoxide but not by pinacidil or cromakalim, whereas Kir6.2/SUR2A channels are stimulated by pinacidil and cromakalim but not by diazoxide, and Kir6.2/SUR2B channels are stimulated by pinacidil, cromakalim and diazoxide (Babenko *et al.*, 1998; D'hahan *et al.*, 1999a, 1999b; Gribble *et al.*, 1998c; Inagaki *et al.*, 1995a, 1996; Isomoto *et al.*, 1996).

K_{ATP} channel openers interact with nucleotides in a complex fashion. In particular, ATP is required for binding of the pinacidil analogue [^{3}H]-P1075 (Hambrock *et al.*, 1998, 1999), although channel activity can be stimulated by P1075 in the absence of added nucleotide (Gribble *et al.*, 2000; Reimann *et al.*, 2000; Terzic *et al.*, 1995). The faster off-rate of the drug in the absence of MgATP (Gribble *et al.*, 2000) may mean that it is more difficult to measure P1075 binding in nucleotide-free solutions. Alternatively, ATP may remain bound at NBD1 for some time after patch excision, in electrophysiological experiments. Recent evidence indicates that K_{ATP} channel openers, such as pinacidil, stimulate ATP hydrolysis at NBD2 (Bienengraeber *et al.*, 2000) and that they promote channel opening by stabilizing the channel in the Mg-nucleotide bound state (Zingman *et al.*, 2001). Conversely, agents that enhance the removal of MgADP, such as creatine kinase, promote channel closure (Zingman *et al.*, 2001). The fact that KCOs stimulate the ATPase activity of SURs suggests the possibility that these ABC proteins may transport KCOs across the cell membrane, because MDR1 and MRP1 substrates also induce ATPase activity in those transporters. However, no transport function has yet been demonstrated for SUR.

The binding site for most SUR2-specific K_{ATP} channel openers (e.g. cromakalim, pinacidil, P1076) was first shown to lie within the third set of transmembrane segments (Babenko *et al.*, 2000; D'hahan *et al.*, 1999a; Uhde *et al.*, 1999), and was subsequently narrowed down to TMS 17. Two residues in this TMS were found to be of key importance: L1249 and T1253 of SUR2A (Moreau *et al.*, 2000). Interestingly, TMSs 12–17 are also involved in substrate recognition by MRPs and MDRs (Hafkemeyer *et al.*, 1998; Ito *et al.*, 2001a, 2001b; Loo and Clarke, 2001). The location of the binding site for diazoxide is still uncertain. This drug binds to both SUR1 and SUR2 receptors, as it can activate Kir6.2/SUR1 and Kir6.2/SUR2B channels when MgATP is present (D'hanan *et al.*, 1999b). Moreover, although it does not activate Kir6.2/SUR2A channels under these conditions, it is able to increase channel activity in the presence of MgADP, implying that the drug also binds to SUR2A. This indicates that the diazoxide-binding site is common to all three SURs. It further suggests that MgADP binding, probably at NBD2, is required for action of the drug. The differential effects of MgATP on diazoxide activation are consistent with the idea that this drug stimulates ATPase activity at NBD2 of SUR1 and SUR2B but not SUR2A.

CONCLUSION AND PERSPECTIVES

The sulfonylurea receptor differs from other ABC proteins in that its primary function is that of an ion channel regulator rather than an ATP-dependent transporter. Although SUR retains an ATPase cycle at NBD2, this is not apparently used to power substrate transport against a concentration gradient. Instead, K^+ ions move through the Kir6.2 pore along their electrochemical gradient, (as in other K^+ channels), and the ATPase cycle at NBD2 has been adapted to serve as a sensor of cellular metabolism. This may explain why the rate of ATP hydrolysis is relatively slow, compared to other ABC proteins.

The K_{ATP} channel is a valuable tool for studying the function of the NBDs of an ABC protein, because nucleotide interaction with the NBDs is coupled to gating of Kir6.2. Thus, potassium fluxes through the Kir6.2 pore can be used to monitor NBD function with high time resolution. Furthermore, binding of MgADP to the NBDs of SUR is sufficient to stimulate channel activity, which enables the ATP hydrolysis step to be bypassed, and the effects of ligand binding and transduction examined in isolation.

Many issues remain to be resolved. As in the case of some other ABC proteins, the NBDs of SUR are not functionally equivalent. However, the precise role of NBD1 is not yet clear. Likewise, the role of the ABC signature sequence in SUR, as in other ABC proteins, remains controversial and it is not certain whether it is involved in ATP binding and/or hydrolysis or in signal transduction. Nucleotide-binding studies, coupled with electrophysiological recordings, may help to resolve these questions. The way in which conformational changes at the NBDs are transduced into changes in the transmembrane domains is not yet known. Indeed, it is not clear whether regulation of Kir6.2 gating by SUR is mediated via the TMDs of SUR or through the cytosolic domains.

Increasingly, it is apparent that many ion channels possess regulatory (or beta) subunits and are modulated by a variety of cytosolic agents that interact with either the pore-forming or regulatory subunit. The K_{ATP} channel appears to be subject to more extensive modulation than most channels, perhaps because it possesses two very different types of subunit. Precisely how channel regulators such as nucleotides, sulfonylureas and K_{ATP} channel openers achieve their functional effects is likely to take some time to sort out. Elucidation of the three-dimensional structure of this large and complex hetero-octameric channel (either the whole complex or its constituent parts), however, would be a major step towards this goal.

GLOSSARY

Diabetes mellitus A metabolic disorder characterized by elevation of the fasting blood glucose concentration. There are two major forms of the disease: type 1 and type 2. Type 1 diabetes results from the autoimmune destruction of the pancreatic β-cells. Type 2 diabetes is characterized by the inability of the pancreatic β-cells to secrete sufficient insulin and by a reduced efficacy of insulin action. Type 2 diabetes is far more common and can affect up to 40% of elderly adults in some populations. It is treated by diet, drugs (such as sulfonylureas) and, if these fail, insulin.

Glucose intolerance A pre-diabetic condition in which the blood glucose concentration remains elevated for an abnormally long time after a meal (or glucose challenge).

Hyperglycemia A higher than normal blood glucose concentration. A fasting blood glucose concentration of above 5.5 mM is usually regarded as hyperglycemic.

Hyperinsulinemia A higher than normal blood insulin concentration.

Hypoxia Low oxygen level.

Ischemia Interruption of the blood supply.

First and second phase insulin secretion Insulin secretion shows a biphasic response to a glucose challenge: a rapid, large and transient response (first phase) that is followed by a smaller but sustained response (second phase).

Seizure An abnormal discharge of electrical activity in the brain.

ACKNOWLEDGMENTS

This work was supported by grants from the Royal Society, the Wellcome Trust, the Japan Ministry of Education, Science, Sports and Culture. FMA is the Royal Society GlaxoSmithKline Research Professor.

REFERENCES

Aguilar-Bryan, L. and Bryan, J. (1999) Molecular biology of adenosine triphosphate-sensitive potassium channels. Endocr. Rev. 20, 101–135.

Aguilar-Bryan, L., Nichols, C.G., Wechsler, S.W., Clement IV, J.P., Boyd III, A.E., González, G., Herrera-Sosa, H., Nguy, K., Bryan, J. and Nelson, D.A. (1995) Cloning of the β-cell high-affinity sulfonylurea receptor: A regulator of insulin secretion. Science 268, 423–426.

Alekseev, A.E., Kennedy, M.E., Navarro, B. and Terzic, A. (1997) Burst kinetics of co-expressed Kir6.2/SUR1 clones: Comparison of recombinant with native ATP-sensitive K^+ channel behavior. J. Membr. Biol. 159, 161–168.

Al-Shawi, M.K., Urbatsch, I.L. and Senior A.E. (1994) Covalent inhibitors of P-glycoprotein ATPase activity. J. Biol. Chem. 269, 8986–8992.

Ashcroft, F.M. and Gribble, F.M. (1998) Correlating structure and function in

ATP-sensitive K^+ channels. Trends Neurosci. 21, 288–294.

Ashcroft, F.M. and Gribble, F.M. (1999) ATP-sensitive K^+ channels and insulin secretion: their role in health and disease. Diabetologia 42, 903–919.

Ashcroft, F.M. and Gribble, F.M. (2000) New windows on the mechanism of action of K_{ATP} channel openers. Trends Pharmacol. Sci. 21, 439–445.

Ashcroft, F.M., Harrison, D.E. and Ashcroft, S.J.H. (1984) Glucose induces closure of single potassium channels in isolated rat pancreatic β-cells. Nature 312, 446–448.

Ashfield, R., Gribble, F.M., Ashcroft, S.J.H. and Ashcroft, F.M. (1999) Identification of the high-affinity tolbutamide site on the SUR1 subunit of the K_{ATP} channel. Diabetes 48, 1341–1347.

Askenasy, N. and Navon, G. (1997) Intermittent ischemia: energy metabolism, cellular volume regulation, adenosine and insights into preconditioning. J. Mol. Cell. Cardiol. 29, 1715–1730.

Azzaria, M., Schurr, E. and Gros, P. (1989) Discrete mutations introduced in the predicted nucleotide-binding sites of the *mdr1* gene abolish its ability to confer multidrug resistance. Mol. Cell Biol. 9, 5289–5297.

Babenko, A.P., Gonzalez, G., Aguilar-Bryan, L. and Bryan, J. (1998) Reconstituted human cardiac K_{ATP} channels. Circ. Res. 83, 1132–1143.

Babenko, A.P., Gonzalez, G. and Bryan, J. (1999) The tolbutamide site of SUR1 and a mechanism for its functional coupling to K_{ATP} channel closure. FEBS Lett. 459, 367–376.

Babenko, A.P., Gonzalez, G. and Bryan, J. (2000) Pharmaco-topology of sulfonylurea receptors. J. Biol. Chem. 275, 717–720.

Baukrowitz, T. and Fakler, B. (2000) K_{ATP} channels gated by intracellular nucleotides and phospholipids. Eur. J. Biochem. 267, 5842–5848.

Baukrowitz, T., Hwang, T.-C., Nairn, A.C. and Gadsby, D.C. (1994) Coupling of CFTR Cl^- channel gating to an ATP hydrolysis cycle. Neuron 12, 473–482.

Baukrowitz, T., Schulte, U., Oliver, D., Herlitze, S., Krauter, T., Tucker, S.J., Ruppersberg, J.P. and Fakler, B. (1998) PIP_2 and PIP as determinants for ATP inhibition of K_{ATP} channels. Science 282, 1141–1144.

Bear, C.E., Li, C., Galley, K., Wang, Y., Garami, E. and Ramjiisingh, M. (1997) Coupling of ATP hydrolysis with channel gating by purified, reconstituted CFTR. J. Bioenerg. Biomembr. 29, 465–473.

Béguin, P., Nagashima, K., Nishimura, M., Gonoi, T. and Seino, S. (1999) PKA-mediated phosphorylation of the human K_{ATP} channel: separate roles of Kir6.2 and SUR1 subunit phosphorylation. EMBO J. 18, 4722–4732.

Bell, G. and Polonsky, K.S. (2001) Diabetes mellitus and genetically programmed defects in beta-cell function. Nature 414, 788–791.

Bienengraeber, M., Alekseev, A.E., Abraham, M.R., Carrasco, A.J., Moreau, C., Vivaudou, M., Dzeja, P.P. and Terzic, A. (2000) ATPase activity of the sulfonylurea receptor: a catalytic function for the K_{ATP} channel complex. FASEB J. 14, 1943–1952.

Bränström, R., Corkey, B.E., Berggren, P.-O. and Larsson, O. (1997) Evidence for a unique long chain acyl-CoA ester binding site on the ATP-regulated potassium channel in mouse pancreatic beta cells. J. Biol. Chem. 272, 17390–17394.

Bränström, R., Leibiger, I.B., Leibiger, B., Corkey, B.E., Berggren, P.-O. and Larsson, O. (1998) Long chain coenzyme A esters activate the pore-forming subunit (Kir6.2) of the ATP-regulated potassium channel. J. Biol. Chem. 273, 31395–31400.

Carrasco, A.J., Dzeja, P.P., Alekseev, A.E., Pucar, D., Zingman, L.V., Abraham, M.R., *et al.* (2001) Adenylate kinase phosphotransfer communicates cellular energetic signals to ATP-sensitive potassium channels. Proc. Natl Acad. Sci. USA 98, 7623–7628.

Cartier, E.A., Conti, L.R., Vandenberg, C.A. and Shyng, S.L. (2001) Defective trafficking and function of K_{ATP} channels caused by a sulfonylurea receptor 1 mutation associated with persistent hyperinsulinemic hypoglycemia of infancy. Proc. Natl Acad. Sci. USA 98, 2882–2887.

Chan, C.B., De Leo, D., Joseph, J.W., McQuaid, T.S., Ha, X.F., Xu, F., Tsushima, R.G., Pennefather, P.S., Salapatek, A.M. and Wheeler, M.B. (2001) Increased uncoupling protein-2 levels in beta-cells are associated with impaired glucose-stimulated insulin secretion: mechanism of action. Diabetes 50, 1302–1310.

Chutkow, W.A., Simon, M.C., Beau, M.M.L. and Burant, C.F. (1996) Cloning, tissue expression, and chromosomal localization of SUR2, the putative drug-binding subunit of cardiac, skeletal muscle, and vascular K_{ATP} channels. Diabetes 45, 1439–1445.

Chutkow, W.A., Makielski, J.C., Nelson, D.J., Burant, C.F. and Fan, Z. (1999) Alternative splicing of *sur2* exon 17 regulates nucleotide sensitivity of the ATP-sensitive potassium channel. J. Biol. Chem. 274, 13656–13665.

Clement IV, J.P., Kunjilwar, K., Gonzalez, G., Schwanstecher, M., Panten, U., Aguliar-Bryan, L. and Bryan, J. (1997) Association and stoichiometry of K_{ATP} channel subunits. Neuron 18, 827–838.

Daut, J., Klieber, H.G., Cyrys, S. and Noack, T. (1994) K_{ATP} channels and basal coronary vascular tone. Cardiovasc. Res. 28, 811–817.

Dean, M., Rzhetsky, A. and Allikmets, R. (2001) The human ATP-binding cassette (ABC) transporter superfamily. Genome Res. 11, 1156–1166.

D'hahan, N., Jacquet, H., Moreau, C., Catty, P. and Vivadou, M. (1999a) A transmembrane domain of the sulfonylurea receptor mediates activation of ATP-sensitive K^+ channels by K^+ channel openers. Mol. Pharmacol. 56, 308–315.

D'hahan, N., Moreau, C., Prost, A.-L., Jacquet, H., Alekseev, A.E., Terzic, A. and Vivaudou, M. (1999b) Pharmacological plasticity of cardiac ATP-sensitive potassium channels toward diazoxide revealed by ADP. Proc. Natl Acad. Sci. USA 96, 12162–12167.

Drain, P., Li, L. and Wang, J. (1998) K_{ATP} channel inhibition by ATP requires distinct functional domains of the cytoplasmic C terminus of the pore-forming subunit. Proc. Natl Acad. Sci. USA 95, 13953–13958.

Dukes, I.D., Sreenan, S., Roe, M.W., Levisetti, M., Zhou, Y.-P., Ostrega, D., *et al.* (1998) Defective pancreatic beta-cell glycolytic signaling in hepatocyte nuclear factor-1alpha-deficient mice. J. Biol. Chem. 273, 24457–24464.

Dunne, M.J., Kane, C., Shepherd, R.M., Sanchez, J.A., James, R.F.L., Johnson, P.R.V., *et al.* (1997) Familial persistent hyperinsulinemic hypoglycemia of infancy and mutations in the sulfonylurea receptor. N. Engl. J. Med. 336, 703–706.

Efanova, I.B., Zaitsev, S.V., Zhivotovsky, B., Kohler, M., Orrenius, S. and Berggren, P.-O. (1998) Glucose and tolbutamide induce apoptosis in pancreatic beta-cells. A process dependent on intracellular Ca^{2+} concentration. J. Biol. Chem. 273, 33501–33507.

Fan, Z. and Makielski, J.C. (1997) Anionic phospholipids activate ATP-sensitive potassium channels. J. Biol. Chem. 272, 5388–5395.

Fan, Z. and Makielski, J.C. (1999) Phosphoinositides decrease ATP sensitivity of the cardiac ATP-sensitive K^+ channel: A molecular probe for the mechanism of ATP-sensitive inhibition. J. Gen. Physiol. 114, 251–270.

Findlay, I. and Dunne, M.J. (1986) ATP maintains ATP-inhibited K^+ channels in an operational state. Pflügers Arch. 407, 238–240.

Frayling, T.M., Bulamn, M.P., Ellard, S., Appleton, M., Dronsfield, M.J., Mackie, A.D., *et al.* (1997) Mutations in the hepatocyte nuclear factor-1alpha gene are a common cause of maturity-onset diabetes of the young in the UK. Diabetes 46, 720–725.

Froguel, P., Zouali, H., Vionnet, N., Velho, G., Vaxillaire, M., Sun, F., *et al.* (1993) Familial hyperglycemia due to mutations in glucokinase. Definition of a subtype of diabetes mellitus. N. Engl. J. Med. 328, 697–702.

Gadsby, D.C. and Nairn, A.C. (1999) Control of CFTR channel gating by phosphorylation and nucleotide hydrolysis. Physiol. Rev. 79, 77–107.

Gao, M., Cui, H.R., Loe, D.W., Grant, C.E., Almquist, K.C., Cole, S.P.C. and Deeley, R.G. (2000) Comparison of the functional characteristics of the nucleotide binding domains of multidrug resistance protein 1. J. Biol. Chem. 275, 13098–13108.

Ghosh, A., Ronner, P., Cheong, E., Khalid, P. and Matschinsky, F.M. (1991) The role of ATP and free ADP in metabolic coupling during fuel-stimulated insulin release from islet beta-cells in the isolated perfused rat pancreas. J. Biol. Chem. 266, 22887–22892.

Giblin, J.P., Leaney, J.L. and Tinker, A. (1999) The molecular assembly of ATP-sensitive potassium channels. J. Biol. Chem. 274, 22652–22659.

Gillis, K.D., Gee, W.M., Hammoud, A., Mcdaniel, M.L., Falke, L.C. and Misler, S. (1989) Effects of sulfonamides on a metabolite-regulated ATP-sensitive K^+ channel in rat pancreatic β-cells. Am. J. Physiol. 257, C1119–C1127.

Glaser, B. (2000) Hyperinsulinism of the newborn. Semin. Perinatol. 24, 150–163.

Glaser, B., Kesavan, P., Heyman, M., Davis, E., Cuesta, A., Buchs, A., *et al.* (1998) Familial hyperinsulinism caused by an activating glucokinase mutation. N. Engl. J. Med. 338, 226–230.

Gopel, S.O., Kanno, T., Barg, S. and Rorsman, P. (2000a) Patch-clamp characterisation of somatostatin-secreting delta-cells in intact mouse pancreatic islets. J. Physiol. (Lond.) 528, 497–507.

Gopel, S.O., Kanno, T., Barg, S., Weng, X.-G., Gromada, J. and Rorsman, P. (2000b) Regulation of glucagon release in mouse alpha-cells by K_{ATP} channels and inactivation of TTX-sensitive Na^+ channels. J. Physiol. (Lond.) 528, 509–520.

Gribble, F.M. and Ashcroft, F.M. (1999) Differential sensitivity of beta-cell and extrapancreatic K_{ATP} channels to gliclazide. Diabetologia 42, 845–848.

Gribble, F.M., Tucker, S.J. and Ashcroft, F.M. (1997a) The essential role of the Walker A motifs of SUR1 in K-ATP channel activation by Mg-ADP and diazoxide. EMBO J. 16, 1145–1152.

Gribble, F.M., Tucker, S.J. and Ashcroft, F.M. (1997b) The interaction of nucleotide with the tolbutamide block of cloned ATP-sensitive K^+ channel currents expressed in *Xenopus* oocytes: a reinterpretation. J. Physiol. 504, 35–45.

Gribble, F.M., Proks, P., Corkey, B.E. and Ashcroft, F.M. (1998a) Mechanism of cloned ATP-sensitive potassium channel activation by oleoyl-CoA. J. Biol. Chem. 273, 26383–26387.

Gribble, F.M., Tucker, S.J., Haug, T. and Ashcroft, F.M. (1998b) MgATP activates the β cell K_{ATP} channel by interaction with its SUR1 subunit. Proc. Natl Acad. Sci. USA 95, 7185–7190.

Gribble, F.M., Tucker, S.J., Seino, S. and Ashcroft, F.M. (1998c) Tissue specificity of sulfonylureas. Diabetes 47, 1412–1418.

Gribble, F.M., Loussouarn, G., Tucker, S.J., Zhao, C., Nichols, C.G. and Ashcroft, F.M. (2000) A novel method for measurement of submembrane ATP concentration. J. Biol. Chem. 275, 30046–30049.

Grover, G.J. and Garlid, K.D. (2000) ATP-sensitive potassium channels: a review of their cardioprotective pharmacology. J. Mol. Cell. Cardiol. 32, 677–695.

Hafkemeyer, P., Dey, S., Ambudkar, S.V., Hrycyna, C.A., Pastan, I. and Gottesman, M.M. (1998) Contribution to substrate specificity and transport of nonconserved residues in transmembrane domain 12 of human P-glycoprotein. Biochemistry 37, 16400–16409.

Hambrock, A., Loffler-Walz, C., Kurachi, Y. and Quast, U. (1998) Mg^{2+} and ATP dependence of K_{ATP} channel modulator binding to the recombinant sulphonylurea receptor, SUR2B. Br. J. Pharmacol. 125, 577–583.

Hambrock, A., Löffler-Walz, C., Kloor, D., Delabar, U., Horio, Y., Kurachi, Y. and Quast, U. (1999) ATP-sensitive K^+ channel modulator binding to sulfonylurea receptors SUR2A and SUR2B: Opposite effects of MgADP. Mol. Pharmacol. 55, 832–840.

Hani, E.H., Boutin, P., Durand, E., Inoue, H., Permutt, M.A., Velho, G. and Froguel, P. (1998) Missense mutations in the pancreatic islet beta cell inwardly rectifying K^+ channel gene (Kir6.2/BIR): a meta-analysis suggests a role in the polygenic basis of Type II diabetes mellitus in Caucasians. Diabetologia 41, 1511–1515.

Hart, L.M., Dekker, J.M., van Haeften, T.W., Ruige, J.B., Stehouwer, C.D., Erkelens, D.W., Heine, R.J. and Maassen, J.A. (2000) Reduced second phase insulin secretion in carriers of a sulphonylurea receptor gene variant associating with Type II diabetes mellitus. Diabetologia 43, 515–519.

Hayabuchi, Y., Davies, N.W. and Standen, N.B. (2001) Angiotensin II inhibits rat arterial K_{ATP} channels by inhibiting steady-state protein kinase A activity and activating protein kinase Ce. J. Physiol. (Lond.) 530, 193–205.

Hernandez-Sanchez, C., Basile, A.S., Fedorova, I., Arima, H., Stannard, B., Fernandez, A.M., Ito, Y. and LeRoith, D. (2001) Mice transgenically overexpressing sulfonylurea receptor 1 in forebrain resist seizure induction and excitotoxic neuron death. Proc. Natl Acad. Sci. USA 98, 3549–3554.

Heron, L., Virsolvy, A., Peyrollier, K., Gribble, F.M., Cam, A.L., Ashcroft, F.M. and Bataille, D. (1998) Human alpha-endosulfine, a possible regulator of sulfonylurea-sensitive K_{ATP} channel: Molecular cloning, expression and biological properties. Proc. Natl Acad. Sci. USA 95, 8387–8391.

Higgins, C.F. (1992) From microorganisms to man. Annu. Rev. Cell Biol. 8, 67–113.

Hilgemann, D.W. and Ball, R. (1996) Regulation of cardiac Na^+, Ca^+ exchange and K_{ATP} potassium channels by PIP2. Science 273, 956–959.

Himmelreich, U. and Dobson, G.P. (2000) Detection and quantification of free cytosolic inorganic phosphate and other phosphorus metabolites in the beating mouse heart muscle *in situ*. NMR Biomed. 13, 467–473.

Holland, I.B. and Blight, M.A. (1999) ABC-ATPases, adaptable energy generators fuelling transmembrane movement of a variety of molecules in organisms from bacteria to humans. J. Mol. Biol. 293, 381–399.

Hou, Y.-X., Cui, L., Riordan, J.R. and Chang, X.-B. (2000) Allosteric interactions between the two non-equivalent nucleotide binding domains of multidrug resistance protein MRP1. J. Biol. Chem. 275, 20280–20287.

Hrycyna, C.A., Ramachandra, M., Germann, U.A., Cheng, P.W., Pastan, I. and Gottesman, M.M. (1999) Both ATP sites of human P-glycoprotein are essential but not symmetric. Biochemistry 38, 13887–13899.

Huang, C.-L., Feng, S. and Hilgemann, D.W. (1998) Direct activation of inward rectifier potassium channels by PIP_2 and its stabilization by $G_{\beta\gamma}$. Nature 391, 803–806.

Huopio, H., Reimann, F., Ashfield, R., Komulainen, J., Lenko, H.L., Rahier, J., *et al.* (2000) Dominantly inherited hyperinsulinism caused by a mutation in the sulfonylurea receptor type 1. J. Clin. Invest. 106, 897–906.

Hwang, T., Nagel, G., Nairn, A. and Gadsby, D. (1994) Regulation of the gating of cystic fibrosis transmembrane conductance regulator Cl channels by phosphorylation and ATP Hydrolysis. Proc. Natl Acad. Sci. USA 91, 4698–4702.

Inagaki, N., Gonoi, T., Clement IV, J.P., Namba, N., Inazawa, J., Gonzalez, G., Aguilar-Bryan, L., Seino, S. and Bryan, J. (1995a) Reconstitution of I_{KATP}: An inward rectifier subunit plus the sulfonylurea receptor. Science 270, 1166–1169.

Inagaki, N., Tsuura, Y., Namba, N., Masuda, K., Gonoi, T., Horie, M., Seino, Y., Mizuta, M. and Seino, S. (1995b) Cloning and functional characterization of a novel ATP-sensitive potassium channel ubiquitously expressed in rat tissues, including pancreatic islets, pituitary, skeletal muscle, and heart. J. Biol. Chem. 270, 5691–5694.

Inagaki, N., Gonoi, T., Clement IV, J.P., Wang, C.-Z., Aguilar-Bryan, L., Bryan, J. and Seino, S. (1996) A family of sulfonylurea receptors determines the pharmacological properties of ATP-sensitive K^+ channels. Neuron 16, 1011–1017.

Inagaki, N., Gonoi, T. and Seino, S. (1997) Subunit stoichiometry of the pancreatic β-cell ATP-sensitive K^+ channel. FEBS Lett. 409, 232–236.

Inoue, H., Ferrer, J., Welling, C.M., Elbein, S.C., Hoffman, M., Mayorga, R., *et al.* (1996) Sequence variants in the sulfonylurea receptor (SUR) gene are associated with NIDDM in caucasians. Diabetes 45, 825–831.

Isomoto, S., Kondo, C., Yamada, M., Matsumoto, S., Higashiguchi, O., Horio, Y., Matsuzawa, Y. and Kurachi, Y. (1996) A novel sulfonylurea receptor forms with BIR (Kir6.2) a smooth muscle type ATP-sensitive K^+ channel. J. Biol. Chem. 271, 24321–24324.

Ito, K., Olsen, S.L., Qiu, W., Deeley, R.G. and Cole, S.P. (2001a) Mutation of a single conserved tryptophan in multidrug resistance protein 1 (mrp1/abcc1) results in loss of drug resistance and selective loss of organic anion transport. J. Biol. Chem. 276, 15616–15624.

Ito, K., Suzuki, H. and Sugiyama, Y. (2001b) Charged amino acids in the transmembrane domains are involved in the determination of the substrate specificity of rat mrp2. Mol. Pharmacol. 59, 1077–1085.

Iwasaki, N., Kawamura, M., Yamagata, K., Cox, N.J., Karibe, S., Ohgawara, H., Inagaki, N., Seino, S., Bell, G.I. and Omori, Y. (1996) Identification of microsatellite markers near the human genes encoding the β-cell ATP-sensitive K^+ channel and linkage studies with NIDDM in Japanese. Diabetes 45, 267–269.

Kane, C., Shepherd, R.M., Squires, P.E., Johnson, P.R.V., James, R.F.L., Millia, P.J., Aynsley-Green, A., Lindley, K.J. and Dunne, M.J. (1996) Loss of functional K_{ATP} channels in pancreatic β-cells causes persistent hyperinsulinemic hypoglycemia of infancy. Nat. Med. 2, 1344–1347.

Karschin, C., Ecke, C., Ashcroft, F.M. and Karschin, A. (1997) Overlapping distribution of K_{ATP} channel-forming Kir6.2 subunit and the sulfonylurea receptor SUR1 in rodent brain. FEBS Lett. 401, 59–64.

Kassem, S.A., Ariel, I., Thornton, P.S., Scheimberg, I. and Glaser, B. (2000) Beta-cell proliferation and apoptosis in the developing normal human pancreas and in hyperinsulinism of infancy. Diabetes 49, 1325–1333.

Ko, Y.H. and Pedersen, P.L. (1995) The first nucleotide binding fold of the cystic fibrosis transmembrane conductance regulator can function as an active ATPase. J. Biol. Chem. 270, 22093–22096.

Koster, J.C., Marshall, B.A., Ensor, N., Corbett, J.A. and Nichols, C.G. (2000) Targeted overactivity of beta cell K_{ATP} channels induces profound neonatal diabetes. Cell 100, 645–654.

Langheinrich, U. and Daut, J. (1997) Hyperpolarization of isolated capillaries from guinea-pig heart induced by K^+ channel openers and glucose deprivation. J. Physiol. 502, 397–408.

Lee, K., Ozanne, S.E., Hales, C.N. and Ashford, M.L.J. (1994) Effects of chemical modification of amino and sulfhydryl groups on K_{ATP} channel function and sulfonylurea binding in CRI-G1 insulin-secreting cells. J. Membrane Biol. 139, 167–181.

Li, C., Ramjeesingh, M., Wang, W., Garami, E., Hewryk, M., Lee, D., Rommens, J.M., Galley, K. and Bear, C.E. (1996) ATPase activity of the cystic fibrosis transmembrane conductance regulator. J. Biol. Chem. 271, 28463–28468.

Light, P. (1996) Regulation of ATP-sensitive potassium channels by phosphorylation. Biochim. Biophys. Acta 1286, 65–73.

Light, P.E., Bladen, C., Winkfein, R.J., Walsh, M.P. and French, R.J. (2000) Molecular basis of protein kinase C-induced activation of ATP-sensitive potassium channels. Proc. Natl Acad. Sci. USA 97, 9058–9063.

Lin, Y.F., Jan, Y.N. and Jan, L.Y. (2000) Regulation of ATP-sensitive potassium channel function by protein kinase A-mediated phosphorylation in transfected HEK293 cells. EMBO J. 19, 942–955.

Lindner, T., Gragnoli, C., Schulze, J., Rietzsch, H., Petzold, C., Schröder, H.-E., Cox, N.J. and Bell, G.I. (1997) The 31-cM region of chromosome 11 including the obesity gene tubby and ATP-sensitive potassium channel genes, *SUR1* and *Kir6.2*, does not contain a major susceptibility locus for NIDDM in 127 non-hispanic white affected sibships. Diabetes 46, 1227–1229.

Liss, B., Bruns, R. and Roeper, J. (1999) Alternative sulfonylurea receptor expression defines metabolic sensitivity of K-ATP channels in dopaminergic midbrain neurons. EMBO J. 18, 833–846.

Liu, R. and Sharom, F.J. (1996) Site-directed fluorescence labeling of P-glycoprotein on cysteine residues in the nucleotide binding domains. Biochemistry 35, 11865–11873.

Loo, T.W. and Clarke, D.M. (1995a) Rapid purification of human P-glycoprotein mutants expressed transiently in HEK293 cells by nickel-chelate chromatography and characterization of their drug-stimulated ATPase activities. J. Biol. Chem. 270, 21449–21452.

Loo, T.W. and Clarke, D.M. (1995b) Covalent modification of human P-glycoprotein mutants containing a single cysteine in either nucleotide-binding folds abolishes drug-stimulated ATPase activity. J. Biol. Chem. 270, 22957–22961.

Loo, T.W. and Clarke, D.M. (2001) Defining the drug-binding site in the human multidrug resistance P-glycoprotein using a methanethiosulfonate analog of verapamil, MTS-verapamil. J. Biol. Chem. 276, 14972–14979.

Lorenz, E. and Terzic, A. (1999) Physical association between recombinant cardiac ATP-sensitive K^+ channel subunits Kir6.2 and SUR2A. J. Mol. Cell Cardiol. 31, 425–434.

Lorenz, E., Alekseev, A.E., Krapivinsky, G.B., Carrasco, A.J., Clapham, D.E. and Terzic, A. (1998) Evidence for direct physical association between a K^+ channel (Kir6.2) and an ATP-binding cassette protein (SUR1) which affects cellular distribution and kinetic behavior of an ATP-sensitive K^+ channel. Mol. Cell. Biol. 18, 1652–1659.

Maassen, J.A. and Kadowaki, T. (1996) Maternally inherited diabetes and deafness: a new diabetes subtype. Diabetologia 39, 375–382.

Markworth, E., Schwanstecher, C. and Schwanstecher, M. (2000) ATP^{4-} mediates closure of pancreatic beta-cell ATP-sensitive potassium channels by interaction with 1 of 4 identical sites. Diabetes 49, 1413–1418.

Matsuo, M., Kioka, N., Amachi, T. and Ueda, K. (1999) ATP binding properties of the nucleotide binding folds of SUR1. J. Biol. Chem. 274, 37479–37482.

Matsuo, M., Tanabe, K., Kioka, N., Amachi, T. and Ueda, K. (2000a) Different binding properties and affinities for ATP and ADP among sulfonylurea receptor subtypes, SUR1, SUR2A, and SUR2B. J. Biol. Chem. 275, 28757–28763.

Matsuo, M., Trapp, S., Tanizawa, Y., Kioka, N., Amachi, T., Oka, Y., Ashcroft, F.M. and Ueda, K. (2000b) Functional analysis of a mutant sulfonylurea receptor, SUR1-R1420C, that is responsible for persistent hyperinsulinemic hypoglycemia of infancy. J. Biol. Chem. 275, 41184–41191.

Matsuoka, T., Matsushita, K., Katayama, Y., Fujita, A., Inageda, K., Tanemoto, M., Inanobe, A., Yamashita, S., Matsuzawa, Y. and Kurachi, Y. (2000) C-Terminal tails of sulfonylurea receptors control ADP-induced activation and diazoxide modulation of

ATP-sensitive K^+ channels. Circ. Res. 87, 873–880.

Mikhailov, M.V., Mikhailova, E.A. and Ashcroft, S.J. (2000) Investigation of the molecular assembly of beta-cell K_{ATP} channels. FEBS Lett. 482, 59–64.

Mikhailov, M.V., Mikhailova, E.A. and Ashcroft, S.J. (2001) Molecular structure of the glibenclamide binding site of the beta-cell K_{ATP} channel. FEBS Lett. 499, 154–160.

Miki, T., Tashiro, F., Iwanaga, T., Nagashima, K., Yoshitomi, H., Aihara, H., *et al.* (1997) Abnormalities of pancreatic islets by targeted expression of a dominant-negative K_{ATP} channel. Proc. Natl Acad. Sci. USA 94, 11969–11973.

Miki, T., Nagashima, K., Tashiro, F., Kotake, K., Yoshitomi, H., Tamamoto, A., Gonoi, T., Iwanaga, T., Miyazaki, J.-I. and Seino, S. (1998) Defective insulin secretion and enhanced insulin action in K_{ATP} channel-deficient mice. Proc. Natl Acad. Sci. USA 95, 10402–10406.

Miki, T., Iwanaga, T., Nagashima, K., Ihara, Y. and Seino, S. (2001a) Roles of ATP-sensitive K^+ channels in cell survival and differentiation in the endocrine pancreas. Diabetes 50, S48–51.

Miki, T., Liss, B., Minami, K., Shiuchi, T., Saraya, A., Kashima, Y., *et al.* (2001b) ATP-sensitive K^+ channels in the hypothalamus are essential for the maintenance of glucose homeostasis. Nat. Neurosci. 4, 507–512.

Moreau, C., Jacquet, H., Prost, A.-L., D'hahan, N. and Vivaudou, M. (2000) The molecular basis of the specificity of action of K_{ATP} channel openers. EMBO J. 19, 6644–6651.

Mukai, E., Ishida, H., Horie, M., Noma, A., Seino, Y. and Takano, M. (1998) The antiarrhythmic agent cibenzoline inhibits K_{ATP} channels by binding to Kir6.2. Biochem. Biophys. Res. Com. 251, 477–481.

Muller, M., Bakos, É., Welker, E., Váradi, A., Germann U.A., Gottesman, M.M., Morse, B.S., Roninson, I.B. and Sarkadi, B. (1996) Altered drug-stimulated ATPase activity in mutants of the human multidrug resistance protein. J. Biol. Chem. 271, 1877–1883.

Nagata, K., Nishitani, M., Matsuo, M., Kioka, N., Amachi, T. and Ueda, K. (2000) Non-equivalent nucleotide trapping in the two nucleotide binding folds of the human multidrug resistance protein MRP1. J. Biol. Chem. 275, 17626–17630.

Nestorowicz, A., Wilson, B.A., Schoor, K.P., Inoue, H., Glaser, B., Landau, H., Stanley, C.A., Thornton, P.S., Clement IV, J.P., Bryan, J., Aguilar-Bryan, L. and Permutt, M.A. (1996) Mutations in the sulfonylurea receptor gene are associated with familial hyperinsulinism in Ashkenazi Jews. Hum. Mol. Genet. 5, 1813–1822.

Nestorowicz, A., Inagaki, N., Gonoi, T., Schoor, K.P., Wilson, B.A., Glaser, B., Landau, H., Stanley, C.A., Thornton, P.S., Seino, S. and Permutt, M.A. (1997) A non-sense mutation in the inward rectifier potassium channel gene, *Kir6.2*, is associated with familial hyperinsulinism. Diabetes 46, 1743–1748.

Nichols, C.G. and Lederer, W.J. (1990) The role of ATP in energy-deprivation contractures in unloaded rat ventricular myocytes. Can. J. Physiol. Pharmacol. 68, 183–194.

Nichols, C.G. and Lederer, W.J. (1991) Adenosine triphosphate-sensitive potassium channels in the cardiovascular system. Am. J. Physiol. 261, H1675–1686.

Nichols, C.G., Shyng, S.-L., Nestorowicz, A., Glaser, B., Clement IV, J.P., Gonzalez, G., Aguilar-Bryan, L., Permutt, M.A. and Bryan, J. (1996) Adenosine diphosphate as an intracellular regulator of insulin secretion. Science 272, 1785–1787.

Ohno-Shosaku, T., Zunkler, B.J. and Trube, G. (1987) Dual effects of ATP on K^+ currents of mouse pancreatic beta-cells. Pflügers Arch. 408, 133–138.

Otonkoski, T., Ammala, C., Huopio, H., Cote, G.J., Chapman, J., Cosgrove, K., *et al.* (1999) A point mutation inactivating the sulfonylurea receptor causes the severe form of persistent hyperinsulinemic hypoglycemia of infancy in Finland. Diabetes 48, 408–415.

Prentki, M. and Corkey, B.E. (1996) Are the beta-cell signaling molecules malonyl-CoA and cystolic long-chain acyl-CoA implicated in multiple tissue defects of obesity and NIDDM? Diabetes 45, 273–283.

Prentki, M., Tornheim, K. and Corkey, B.E. (1997) Signal transduction mechanisms in nutrient-induced insulin secretion. Diabetologia 40, S32–41.

Proks, P. and Ashcroft, F.M. (1997) Phentolamine block of K_{ATP} channels is mediated by Kir6.2. Proc. Natl Acad. Sci. USA 94, 11716–11720.

Quayle, J.M., Nelson, M.T. and Standen, N.B. (1997) ATP-sensitive and inwardly rectifying potassium channels in smooth muscle. Physiol. Rev. 77, 1165–1232.

Raab-Graham, K.F., Cirilo, L.J., Boettcher, A.A., Radeke, C.M. and Vandenbergs, C.A. (1999) Membrane topology of the amino-terminal region of the sulfonylurea receptor. J. Biol. Chem. 274, 29122–29129.

Randle, P.J. (1993) Glucokinase and candidate genes for type 2 (non-insulin-dependent) diabetes mellitus. Diabetologia 36, 269–275.

Reimann, F., Gribble, F.M. and Ashcroft, F.M. (2000) Differential response of K_{ATP} channels containing SUR2A or SUR2B subunits to nucleotides and pinacidil. Mol. Pharmacol. 58, 1318–1325.

Rissanen, J., Markkanen, A., Karkkainen, P., Pihlajamaki, J., Kekalainen, P., Mykkanen, L., Kuusisto, J., Karhapaa, P., Niskanen, L. and Laakso, M. (2000) Sulfonylurea receptor 1 gene variants are associated with gestational diabetes and type 2 diabetes but not with altered secretion of insulin. Diabetes Care 23, 70–73.

Sakura, H., Ämmälä, C., Smith, P.A., Gribble, F.M. and Ashcroft, F.M. (1995) Cloning and functional expression of the cDNA encoding a novel ATP-sensitive potassium channel subunit expressed in pancreatic β-cells, brain, heart and skeletal muscle. FEBS Lett. 377, 338–344.

Sakura, H., Wat, N., Horton, V., Millns, H., Turner, R.C. and Ashcroft, F.M. (1996) Sequence variations in the human Kir6.2 gene, a subunit of the beta-cell ATP-sensitive K-channel: no association with NIDDM in white Caucasian subjects or evidence of abnormal function when expressed *in vitro*. Diabetologia 39, 1233–1236.

Sakura, H., Trapp, S., Liss, B. and Ashcroft, F.M. (1999) Altered functional properties of K_{ATP} channel conferred by a novel splice variant of SUR1. J. Physiol. 521, 337–350.

Sánchez, J.A., Gonoi, T., Inagaki, N., Katada, T. and Seino, S. (1998) Modulation of reconstituted ATP-sensitive K^+ channels by GTP-binding proteins in a mammalian cell line. J. Physiol. 507, 315–324.

Schwanstecher, M., Schwanstecher, C., Dickel, C., Chudziak, F., Moshiri, A. and Panten, U. (1994) Location of the sulphonylurea receptor at the cytoplasmic face of the β-cell membrane. Br. J. Pharmacol. 113, 903–911.

Schwappach, B., Zerangue, N., Jan, Y.N. and Jan, L.Y. (2000) Molecular basis for K_{ATP} assembly: transmembrane interactions mediate association of a K^+ channel with an ABC transporter. Neuron 26, 155–167.

Seghers, V., Nakazaki, M., DeMayo, F., Aguilar-Bryan, L. and Bryan, J. (2000) *Sur1* Knockout Mice. A model for K_{ATP} channel-independent regulation of insulin secretion. J. Biol. Chem. 275, 9270–9277.

Seino, S., Iwanaga, T., Nagashima, K. and Miki, T. (2000) Diverse roles of K_{ATP} channels learned from Kir6.2 genetically engineered mice. Diabetes 49, 311–318.

Senior, A.E. and Gadsby, D.C. (1997) ATP hydrolysis cycles and mechanism in P-glycoprotein and CFTR. Semin. Cancer Biol. 8, 143–150.

Senior, A.E., Al-Shawi, M.K. and Urbatsch, I.L. (1995) The catalytic cycle of P-glycoprotein. FEBS Lett. 377, 285–289.

Sharma, N., Crane, A., Clement IV, J.P., Gonzalez, G., Babenko, A.P., Bryan, J. and Aguilar-Bryan, L. (1999) The C terminus of SUR1 is required for trafficking of K_{ATP} channels. J. Biol. Chem. 274, 20628–20632.

Sharma, N., Crane, A., Gonzalez, G., Bryan, J. and Aguilar-Bryan, L. (2000) Familial hyperinsulinism and pancreatic beta-cell ATP-sensitive potassium channels. Kidney Int. 57, 803–808.

Shyng, S.-L. and Nichols, C.G. (1997) Octameric stoichiometry of the K_{ATP} channel complex. J. Gen. Physiol. 110, 655–664.

Shyng, S.-L. and Nichols, C.G. (1998) Membrane phospholipid control of nucleotide sensitivity of K_{ATP} channels. Science 282, 1138–1141.

Shyng, S.-L., Ferrigni, T. and Nichols, C.G. (1997) Regulation of K_{ATP} channel activity by diazoxide and MgADP. J. Gen. Physiol. 110, 643–654.

Shyng, S.-L., Ferrigni, T., Shepard, J.B., Nestorowicz, A., Glaser, B., Permutt, M.A. and Nichols, C.G. (1998) Functional analyses of novel mutations in the sulfonylurea receptor 1 associated with persistent hyperinsulinemic hypoglycemia of infancy. Diabetes 47, 1145–1151.

Shyng, S.-L., Barbieri, A., Gumusboga, A., Cukras, C., Pike, L., Davis, J.N., Stahl, P.D. and Nichols, C.G. (2000) Modulation of nucleotide sensitivity of ATP-sensitive potassium channels by phosphatidylinositol-4-phosphate 5-kinase. Proc. Natl Acad. Sci. USA 97, 937–941.

Song, D.K. and Ashcroft, F.M. (2001) ATP modulation of ATP-sensitive potassium channel ATP sensitivity varies with the type of SUR subunit. J. Biol. Chem. 276, 7143–7149.

Stanley, C.A., Lieu, Y.K., Hsu, B.Y., Burlina, A.B., Greenberg, C.R., Hopwood, N.J., Perlman, K., Rich, B.H., Zammarchi, E. and Poncz, M. (1998) Hyperinsulinism and hyperammonemia in infants with regulatory mutations of the glutamate dehydrogenase gene. N. Engl. J. Med. 338, 1352–1357.

Stirling, B., Cox, N.J., Bell, G.I., Hanis, C.L., Spielman, R.S. and Concannon, P. (1995) Linkage studies in NIDDM with markers near the sulphonylurea receptor gene. Diabetologia 38, 1479–1481.

Suzuki, M., Li, R.A., Miki, T., Uemura, H., Sakamoto, N., Ohmoto-Sekine, Y., *et al.* (2001) Functional roles of cardiac and vascular ATP-sensitive potassium channels clarified by Kir6.2-knockout mice. Circ. Res. 88, 570–577.

Szabo, K., Szakacs, G., Hegedus, T. and Sarkadi, B. (1999) Nucleotide occlusion in the human cystic fibrosis transmembrane conductance regulator. J. Biol. Chem. 274, 12209–12212.

Takada, Y., Yamada, K., Taguchi, Y., Kino, K., Matsuo, M., Tucker, S.J., Komano, T., Amachi, T. and Ueda, K. (1998) Non-equivalent cooperation between the two nucleotide-binding folds of P-glycoprotein. Biochim. Biophys. Acta 1373, 131–136.

Tanabe, K., Tucker, S.J., Matsuo, M., Proks, P., Ashcroft, F.M., Seino, S., Amachi, T. and Ueda, K. (1999) Direct photoaffinity labeling of the Kir6.2 subunit of the ATP-sensitive K^+ channel by 8-azido-ATP. J. Biol. Chem. 274, 3931–3933.

Tanabe, K., Tucker, S.J., Ashcroft, F.M., Proks, P., Kioka, N., Amachi, T. and Ueda, K. (2000) Direct photoaffinity labeling of Kir6.2 by [γ-^{32}P]ATP-[γ]4-azidoanilide. Biochem. Biophys. Res. Com. 272, 316–319.

Tanizawa, Y., Matsuda, K., Matsuo, M., Ohta, Y., Ochi, N., Adachi, M., *et al.* (2000) Genetic analysis of Japanese patients with persistent hyperinsulinemic hypoglycemia of infancy: nucleotide-binding fold-2 mutation impairs cooperative binding of adenine nucleotides to sulfonylurea receptor 1. Diabetes 49, 114–120.

Terzic, A., Jahangir, A. and Kurachi, Y. (1995) Cardiac ATP-sensitive K^+ channels: regulation by intracellular nucleotides and K^+ channel-opening drugs. Am. J. Physiol. 269, 525–545.

Thomas, P.M., Cote, G.J., Wohllk, N., Haddad, B., Mathew, P.M., Rabl, W., Aguilar-Bryan, L., Gagel, R.F. and Bryan, J. (1995) Mutations in the sulfonylurea receptor gene in familial persistent hyperinsulinemic hypoglycemia of infancy. Science 268, 426–429.

Thomas, P.M., Wohllk, N., Huang, E., Kuhnle, U., Rabl, W., Gagel, R.F. and Cote, G.J. (1996a) Inactivation of the first nucleotide-binding fold of the sulfonylurea receptor, and familial persistent hyperinsulinemic hypoglycemia of infancy. Am. J. Hum. Genet. 59, 510–518.

Thomas, P., Ye, Y. and Lightner, E. (1996b) Mutation of the pancreatic islet inward rectifier Kir6.2 also leads to familial persistent hyperinsulinemic hypoglycemia of infancy. Hum. Mol. Genet. 5, 1809–1812.

Tucker, S.J., Gribble, F.M., Zhao, C., Trapp, S. and Ashcroft, F.M. (1997) Truncation of Kir6.2 produces ATP-sensitive K^+ channels in the absense of the sulphonylurea receptor. Nature 387, 179–183.

Tucker, S.J., Gribble, F.M., Proks, P., Trapp, S., Ryder, T.J., Haug, T., Reimann, F. and Ashcroft, F.M. (1998) Molecular determinants of K_{ATP} channel inhibition by ATP. EMBO J. 17, 3290–3296.

Tusnády, G.E., Bakos, É., Váradi, A. and Sarkadi, B. (1997) Membrane topology distinguishes a subfamily of the ATP-binding cassette (ABC) transporters. FEBS Lett. 402, 1–3.

Ueda, K., Inagaki, N. and Seino, S. (1997) MgADP antagonism to Mg^{2+}-independent ATP binding of the sulfonylurea receptor SUR1. J. Biol. Chem. 272, 22983–22986.

Ueda, K., Komine, J., Matsuo, M., Seino, S. and Amachi, T. (1999) Cooperative binding of ATP and MgADP in the sulfonylurea receptor is modulated by glibenclamide. Proc. Natl Acad. Sci. USA 96, 1268–1272.

Uhde, I., Toman, A., Gross, I., Schwanstecher, C. and Schwanstecher, M. (1999) Identification of the potassium channel opener site on sulfonylurea receptors. J. Biol. Chem. 274, 28079–28082.

Urbatsch, I.L., Sankaran, B., Weber, J. and Senior, A.E. (1995) P-glycoprotein is stably inhibited by vanadate-induced trapping of nucleotide at a single catalytic site. J. Biol. Chem. 270, 19383–19390.

Verkarre, V., Fournet, J.C., de Lonlay, P., Gross-Morand, M.S., Devillers, M., Rahier, J., *et al.* (1998) Paternal mutation of the sulfonylurea receptor (SUR1) gene and maternal loss of 11p15 imprinted genes lead to persistent hyperinsulinism in focal adenomatous hyperplasia. J. Clin. Invest. 102, 1286–1291.

Virsolvy-Vergine, A., Bruck, M., Dufour, M., Cauvin, A., Lupo, B. and Bataille, D. (1988) An endogenous ligand for the central sulfonylurea receptor. FEBS Lett. 242, 65–69.

Wada, Y., Yamashita, T., Imai, K., Miura, R., Takao, K., Nishi, M., *et al.* (2000) A region of the sulfonylurea receptor critical for a modulation of ATP-sensitive K^+ channels by G-protein βγ-subunits. EMBO J. 19, 4915–4925.

Weinreich, F., Riordan, J.R. and Nagel, G. (1999) Dual effects of ADP and adenylylimidodiphosphate on CFTR channel kinetics show binding to two different nucleotide binding sites. J. Gen. Physiol. 114, 55–70.

Weiss, J.N. and Venkatesh, N. (1993) Metabolic regulation of cardiac ATP-sensitive K^+ channels. Cardiovasc. Drugs Ther. 7, 499–505.

Xie, L.-H., Horie, M. and Takano, M. (1999) Phospholipase C-linked receptors regulate the ATP-sensitive potassium channel by means of phosphatidylinositol 4,5-bisphosphate metabolism. Proc. Natl Acad. Sci. USA 96, 15292–15297.

Yamada, M., Isomoto, S., Matsumoto, S., Kondo, C., Shindo, T., Horio, Y. and Kurachi, Y. (1997) Sulfonylurea receptor 2B and Kir6.1 form a sulfonylurea-sensitive but ATP-insensitive K^+ channel. J. Physiol. 499, 715–720.

Yamada, K., Ji, J.J., Yuan, H., Miki, T., Sato, S., Horimoto, N., Shimizu, T., Seino, S. and Inagaki, N. (2001) Protective role of ATP-sensitive potassium channels in hypoxia-induced generalized seizure. Science 292, 1543–1546.

Yokoshiki, H., Sunagawa, M., Seki, T. and Sperelakis, N. (1998) ATP-sensitive K^+ channels in pancreatic, cardiac and vascular smooth muscle cells. Am. J. Physiol. 274, 25–37.

Zerangue, N., Schwappach, B., Jan, Y.N. and Jan, L.Y. (1999) A new ER trafficking signal regulates the subunit stoichiometry of plasma membrane K_{ATP} channels. Neuron 22, 537–548.

Zhang, C.Y., Baffy, G., Perret, P., Krauss, S., Peroni, O., Grujic, D., *et al.* (2001) Uncoupling protein-2 negatively regulates insulin secretion and is a major link between obesity, beta cell dysfunction, and type 2 diabetes. Cell 105, 745–755.

Zingman, L.V., Alekseev, A.E., Bienengraeber, M., Hodgson, D., Karger, A.B., Dzeja, P.P. and Terzic, A. (2001) Signaling in channel/enzyme multimers: ATPase transitions in SUR module gate ATP-sensitive K^+ conductance. Neuron 31, 233–245.

Zunkler, B.J., Lins, S., Ohno-Shosaku, T., Trube, G. and Panten, U. (1988) Cytosolic ADP enhances the sensitivity to tolbutamide of ATP-dependent K^+ channels from pancreatic β-cells. FEBS Lett. 239, 241–244.

CHAPTER 28

ABC Transporters and Human Eye Disease

Rando Allikmets

Introduction

Human ATP-binding cassette (ABC) transporter genes have emerged as increasingly important players in inherited diseases. Out of approximately 50 known human genes (see Chapter 3), at least 15 have been associated with a disease phenotype (Dean *et al.*, 2001). The widespread impact of ABC transporters on human health was anticipated due to the vital function of these proteins in all cell types. This chapter will focus on two ABC genes, *ABCA4* and *ABCC6*, which are both involved in diseases of the eye.

Diseases of the retina include a wide spectrum of photoreceptor-affecting phenotypes, which have been mapped to over 120 loci on the human genome (RetNet™ Retinal Information Network; http://www.sph.uth.tmc.edu/Retnet/home.htm). Currently, less than half of the causal genes have been identified, although substantial progress has been made in determining the genetic basis of monogenic eye disorders. Mutations in new genes responsible for some form of retinal degeneration are identified on a regular basis. However, the vast majority of these genes are involved in rare phenotypes in a limited number of patients.

When the ABC transporter gene *ABCA4* (formerly known as *ABCR)* was cloned and characterized in 1997 as the causal gene for autosomal recessive Stargardt disease (Allikmets *et al.*, 1997a), it seemed as if just another missing link was added to the extensive table of genetic determinants of rare monogenic retinal dystrophies. Now, more than three years later, mutations in the *ABCA4* gene continue to emerge as one of the predominant determinants of a wide variety of retinal degeneration phenotypes. The discovery of the association between mutations in the *ABCC6* gene and an eye phenotype (Bergen *et al.*, 2000; Le Saux *et al.*, 2000; Ringpfeil *et al.*, 2000; Struk *et al.*, 2000) added a second gene to the list of ABC transporters that are involved in retinal disorders.

ABCA4 in Retinal Dystrophies

Several laboratories independently described *ABCA4* in 1997 as the causal gene for Stargardt disease (STGD1 (MIM 248200)) (Allikmets *et al.*, 1997a; Azarian and Travis, 1997; Illing *et al.*, 1997). Autosomal recessive STGD (arSTGD) is a juvenile-onset macular dystrophy associated with rapid central visual impairment, progressive bilateral atrophy of the foveal retinal pigment epithelium, and characteristic frequent appearance of orange-yellow flecks around the macula and/or the midretinal periphery **(Figure 28.1)**. There is no definitive evidence of genetic heterogeneity of arSTGD; all families segregating the disorder have been linked to the *ABCA4* locus on human chromosome 1p13–p22 (Anderson *et al.*, 1995; Kaplan *et al.*, 1993). Consequently, the role of the

ABC Proteins: From Bacteria to Man
ISBN 0-12-352551-9

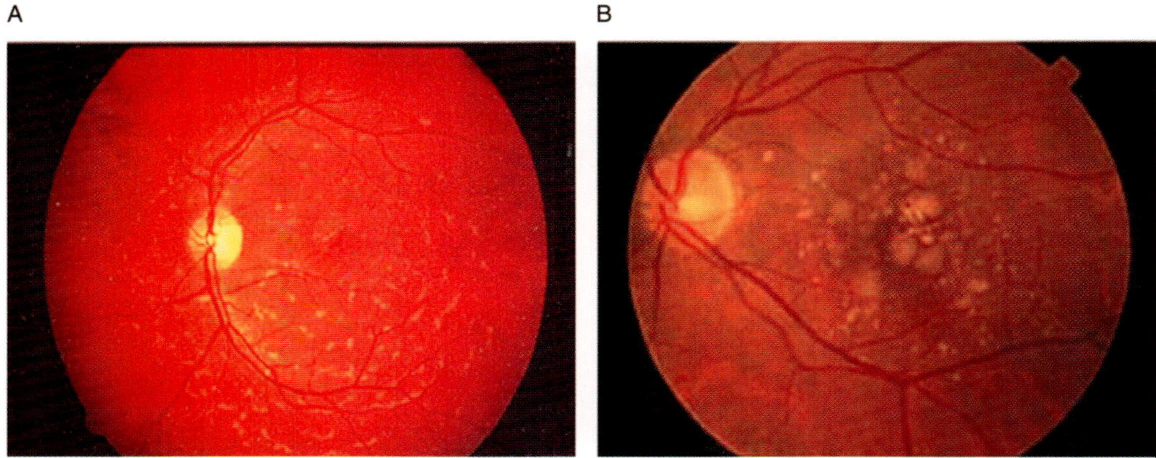

Figure 28.1. Fundus photographs of patients with Stargardt disease (A) and age-related macular degeneration (AMD) (B). Note macular dystrophy and characteristic orange-yellow flecks around the macula and the midretinal periphery in the case of Stargardt macular dystrophy, and degeneration of the macula and drusen (yellowish deposits around the macula) in the case of AMD.

ABCA4 gene in arSTGD has not been disputed, even despite a relatively low (usually ~60%) mutation detection rate of *ABCA4* in STGD patients (Lewis *et al.*, 1999; Maugeri *et al.*, 1999; Rivera *et al.*, 2000; Simonelli *et al.*, 2000).

Subsequently, several cases were reported where *ABCA4* mutations segregated with retinal dystrophies of a substantially different phenotype, such as autosomal recessive cone–rod dystrophy (arCRD) (Cremers *et al.*, 1998; Rozet *et al.*, 1998) and autosomal recessive retinitis pigmentosa (arRP) (Cremers *et al.*, 1998; Martinez-Mir *et al.*, 1998; Rozet *et al.*, 1999). arCRD and arRP have been characterized as groups of genetically heterogeneous diseases where several loci have been implicated by linkage (RetNet™). Clinical heterogeneity of these disorders further complicates the assessment of genetic determinants for each disease entity. Cone–rod dystrophy is characterized by more prominent cone degeneration, in comparison with rod degeneration, which is distinguished by more distinctive reduction of the photopic cone b-wave amplitude than the scotopic (rod b-wave) amplitude in the electroretinogram (ERG). Conversely, retinitis pigmentosa affects predominantly rod photoreceptors; the scotopic ERG is more severely reduced than the photopic ERG, and patients present with night blindness and loss of peripheral vision.

In all studies, disease-associated *ABCA4* alleles have revealed an extraordinary heterogeneity (Allikmets *et al.*, 1997a; Fishman *et al.*, 1999; Lewis *et al.*, 1999; Maugeri *et al.*, 1999; Rozet *et al.*, 1998; Simonelli *et al.*, 2000) **(Figure 28.2)**. The current tally of all *ABCA4* alleles suggests over 400 disease-associated *ABCA4* variants (R. Allikmets, unpublished data), allowing comparison of this gene to one of the best-known members of the ABC superfamily, the cystic fibrosis transmembrane conductance regulator (CFTR) (Riordan *et al.*, 1989) (see Chapter 29). What makes *ABCA4* an even more difficult diagnostic target than *CFTR* is that the most frequent disease-associated *ABCA4* alleles (e.g. G1961E, G863A/delG863, and A1038V) have been described in ~10% of STGD patients across all populations studied, whereas the delF508 allele of *CFTR* accounts for close to 70% of all cystic fibrosis alleles (Zielenski and Tsui, 1995).

Based on these findings, several investigators have proposed a model that suggests a direct correlation between the continuum of disease phenotypes and residual ABCA4 activity/function (Allikmets, 1999; Lewis *et al.*, 1999; Maugeri *et al.*, 1999; Shroyer *et al.*, 1999; van Driel *et al.*, 1998) **(Figure 28.3)**. According to the predicted effect on the ABCA4 transport function, Maugeri *et al.* (1999) classified *ABCA4* mutant alleles as 'mild', 'moderate' and 'severe'. Different combinations of these were predicted to result in distinct phenotypes in a continuum of disease manifestations, the severity of disease manifestation being inversely proportional to the residual ABCA4 activity **(Figure 28.3)**.

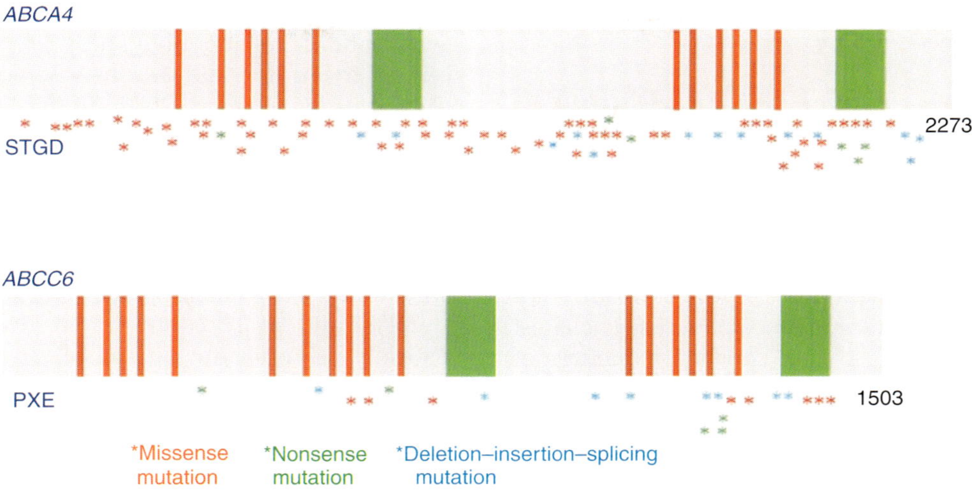

Figure 28.2. Mutations in **ABCA4** *and* **ABCC6** *genes. Schematic representation of mutation spectrum is shown for* **ABCA4** *in Stargardt disease (STGD) and for* **ABCC6** *in pseudoxanthoma elasticum (PXE). Note the high prevalence of evenly distributed missense alleles in* **ABCA4** *and C-terminal distribution of mainly deleterious mutations in* **ABCC6.** *The positions of the predicted transmembrane segments and the two NBDs in each gene are also indicated.*

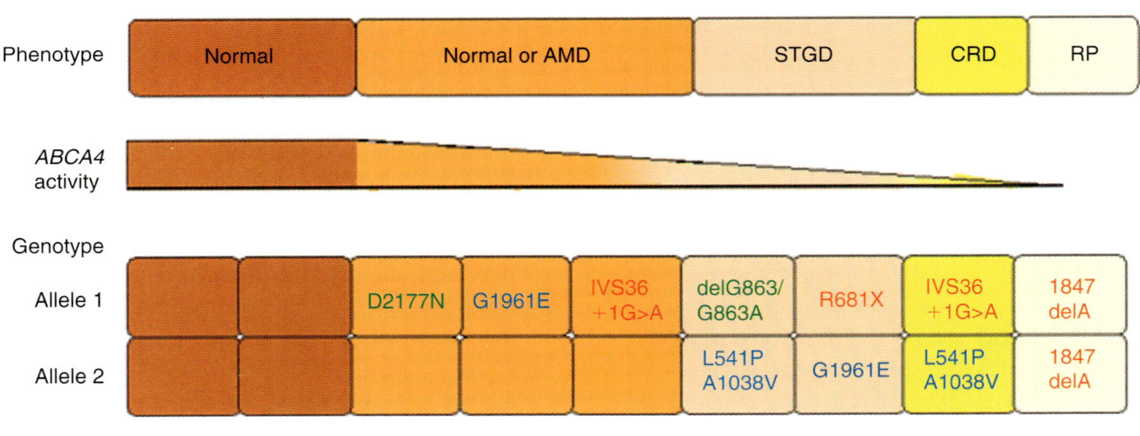

Figure 28.3. Genotype/phenotype model for **ABCA4.** *Modified from van Driel* **et al.** *(1998), Maugeri* **et al.** *(1999), and Shroyer* **et al.** *(1999).*

In addition, several studies have identified frequent complex alleles in both STGD and CRD patients (Lewis *et al.*, 1999; Maugeri *et al.*, 1999; Rivera *et al.*, 2000). The most prominent of these are L541P/A1038V and R943Q/G863A/delG863.

Recently, in an extension of their earlier study, the laboratory of Frans Cremers has determined the major role of mutant *ABCA4* alleles in arCRD (Maugeri *et al.*, 2000). This groundbreaking discovery of the major genetic component in a prominent fraction of retinal disease distinguishes autosomal recessive CRD as a disorder caused predominantly by genetic defects in one gene. This finding argues against the former assumption that arCRDs represent a genetically heterogeneous entity similar to arRP (RetNet™). The same study suggests that we revisit our current knowledge on the molecular genetics of arRP. The prediction that *ABCA4* alleles are responsible for ~8% of arRP (Maugeri *et al.*, 2000), making it the most prominent cause of the autosomal recessive form of retinitis pigmentosa, seems reasonable and is currently under further investigation.

ABCA4 IN AGE-RELATED MACULAR DEGENERATION (AMD)

The summarized data presented in the previous sections establish allelic variation in *ABCA4* as the most prominent cause of retinal dystrophies with Mendelian inheritance patterns. The latest estimates suggest the carrier frequency of *ABCA4* alleles in the general population is ~5% (Maugeri *et al.*, 1999; Yatsenko *et al.*, 2001; R. Allikmets, unpublished observation). This brings us to the hottest topic of ophthalmic genetics – the role of heterozygous *ABCA4* alleles in a complex trait, age-related macular degeneration (AMD, also designated ARMD2 (MIM 153800)). AMD, as a typical late-onset complex disorder, is caused by a combination of genetic and environmental factors **(Figure 28.1B)**. Its prevalence increases with age; among persons 75 years and older, mild or early forms occur in nearly 30% and advanced forms in about 7% of the population (Klein *et al.*, 1992; Vingerling *et al.*, 1995). Consequently, various forms of AMD affect over 10 million individuals in the United States alone.

In 1997, results of a joint study of four laboratories suggested an association of heterozygous *ABCA4* alleles with the AMD phenotype (Allikmets *et al.*, 1997b). This 'classical' case-control study of 167 AMD patients and 220 controls found *ABCA4* alterations in 16% of patients that were interpreted as associated with the disease phenotype because they were found in less than 1% of controls. Most alterations resulted in rare missense mutations, some of which had also been found in STGD1 patients (Allikmets *et al.*, 1997b). Subsequently, several reports disputed the conclusions of this study, stating that they were unable to replicate these findings and, therefore, to confirm the association (De La Paz *et al.*, 1999; Guymer *et al.*, 2001; Stone *et al.*, 1998). Problems with replication of an association study of a complex disease are not unexpected and discussion of the topic is beyond the scope of this review (see, for example, Long and Langley, 1999; O'Donovan and Owen, 1999). In short, difficulties arise mainly due to small sample size in studies of rare variants with modest effect on a complex trait.

Our hypothesis-generating finding that heterozygous *ABCA4* mutations may increase susceptibility to AMD was recently tested by an expanded collaborative study including 15 centers in Europe and North America (ABCR Consortium; Allikmets, 2000). In this study, the two most common AMD-associated variants, G1961E and D2177N, were genotyped in 1218 unrelated AMD patients and 1258 reportedly unaffected, matched controls. Together, these two non-conservative amino acid changes were found in one allele of *ABCA4* in 40 patients (~3.4%) and in 12 controls (~0.95%), a statistically significant difference ($p < 0.0001$) (Allikmets, 2000). The risk of AMD was estimated to be increased about threefold in carriers of D2177N and about fivefold in carriers of G1961E. In the context of common complex disorders, this represents an important contribution to the disease load. Since AMD affects millions of people worldwide and the described mutations represent only two out of thirteen reported earlier (Allikmets *et al.*, 1997b), the number of people at increased risk of developing age-related maculopathy as carriers for variant *ABCA4* alleles is substantial.

Finally, the following comments are offered on the meta-analysis of published data on the two most frequent *ABCA4* variants **(Table 28.1)**. It is apparent that the main reason for the controversial interpretation of the data is the *small sample size* in individual studies. If analyzed separately, none of the smaller studies, with the exception of Allikmets *et al.* (1997b), yields statistically significant results. A substantial increase in the sample size, as in the Consortium study, or in the proposed meta-analysis, results in a substantial increase of power of statistical analysis. Resulting *p* values, as well as relative risk estimates, leave no doubt that the association is statistically significant. It is noteworthy that the relative risk estimates calculated from the meta-analysis are slightly increased compared to the Consortium study (Allikmets, 2000) and are estimated at over 3 for the D2177N mutation and at approximately 5 for the G1961E variant.

These analyses clearly demonstrate the critical need for large cohorts of cases and matched controls for association studies of rare alleles. Considering all available data, heterozygous *ABCA4* alleles are estimated to increase susceptibility to AMD in about 8–10% of all cases. However, this estimate has to be viewed with caution, since the analysis of *ABCA4* variation in AMD is far from complete. It should be remembered that even in Stargardt disease approximately 30–40% of disease-associated

Table 28.1. Meta-analysis of published data on two *ABCA4* alleles

Study	D2177N		*p*	G1961E		*p*
	Case	Control		Case	Control	
Allikmets *et al.* (1997b)	7/167	1/220	0.012	6/167	0/220	0.006
ABCR Consortium (Allikmets, 2000)	21/1189	8/1258	0.005	19/1218	4/1258	0.0008
Guymer *et al.* (2001)	7/544	4/689	0.1	5/544	3/689	0.16
De La Paz *et al.* (1999)	2/164	0/56	0.55	N/A	N/A	N/A
Total	37/2064	13/2223	0.0002	30/1900	7/2167	<0.0001
Odds ratio (95% CI)			3.1 (1.6–5.9)			5.0 (2.2–11.3)

N/A, not applicable; *p* values were calculated from the one-sided Fisher's exact test, and odds ratios were calculated from the exact conditional hypergeometric distribution.

ABCA4 alleles go undetected (Allikmets, 1999). In addition, as emphasized above, founder alleles in some ethnic groups can seriously affect the analysis, suggesting large, multicenter-based studies of matched cases and controls as the only alternative method to achieve statistical significance. Consorted study design also helps to minimize the confounding effect of population stratification, the most serious reason for spurious associations (Allikmets, 2000).

Functional Studies of ABCA4

The ABCA4 protein was first described in the 1970s as an abundant component of photoreceptor outer segment disk rims (Papermaster *et al.*, 1976, 1978). Hence, it was called a Rim protein (RimP) for the following 20 years. Only in 1997 was the gene encoding RimP cloned and characterized as a member of the ABC transporter superfamily, suggesting a transport function of some substrate in photoreceptor outer segments (Allikmets *et al.*, 1997a; Illing *et al.*, 1997). All-*trans*-retinal, the isoform of rhodopsin chromophore, was identified as a potential substrate of ABCA4 by its ability to stimulate ATP hydrolysis by the purified reconstituted ABCA4 protein *in vitro*, suggesting that retinal could also be the physiological substrate for ABCA4 (Sun *et al.*, 1999). Studies of *Abca4* knockout mice fully support this hypothesis, and it has been proposed that ABCA4 is a 'flippase' for the protonated complex of all-*trans*-retinal and phosphatidylethanolamine (*N*-retinylidene-PE) (Weng *et al.*, 1999). Mice lacking the functional *Abca4* gene demonstrated delayed dark adaptation, increased all-*trans*-retinal following light exposure, elevated phosphatidylethanolamine (PE) in rod outer segments, accumulation of the protonated Schiff base complex of *N*-retinylidene-PE, and striking deposition of a major lipofuscin fluorophore in retinal pigment epithelium (RPE). Based on these findings, it was suggested that the *ABCA4*-mediated retinal degeneration may result from 'poisoning' of the RPE caused by A2-E accumulation, with secondary photoreceptor degeneration due to loss of the RPE support role (Weng *et al.*, 1999). A2-E, a pyridinium bis-retinoid, is derived from two molecules of vitamin A aldehyde and one molecule of ethanolamine, and has been characterized as one of the major components of retinal pigment epithelial lipofuscin. Accumulation of lipofuscin in the macular region of RPE is characteristic of aging eyes and is the hallmark of both STGD1 and AMD.

Together, these findings define ABCA4 as the 'rate-keeper' of retinal transport in the visual cycle, as illustrated in the proposed model shown in **Figure 28.4A**. ABCA4 is apparently not absolutely essential for this process, since individuals completely lacking the functional protein (e.g. some arRP patients) maintain some eyesight for several years. Over time, however, even mild dysfunction of ABCA4 affects the vision irreparably **(Figure 28.4B)**. Most recently, intriguing data that fully support *ABCA4* involvement in AMD were obtained from studies of *Abca4(+/−)* heterozygous mice (Mata *et al.*, 2001). A phenotype similar to that seen in *Abca4* knockouts (A2E accumulation in the RPE, etc.)

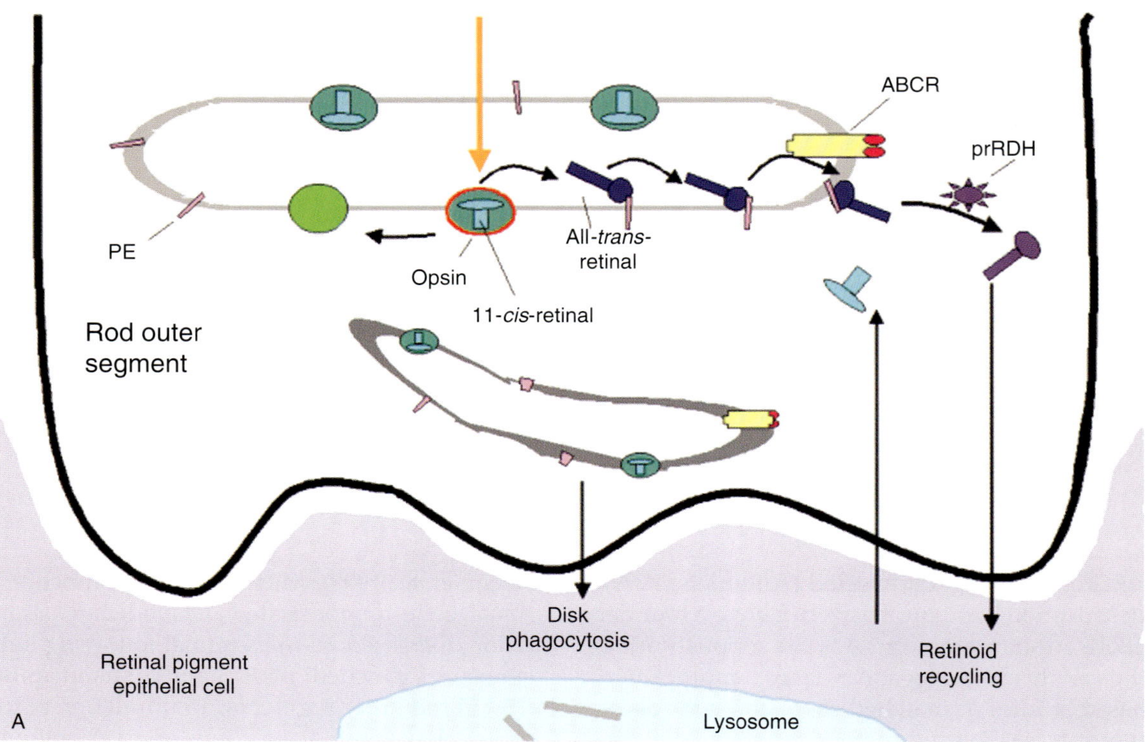

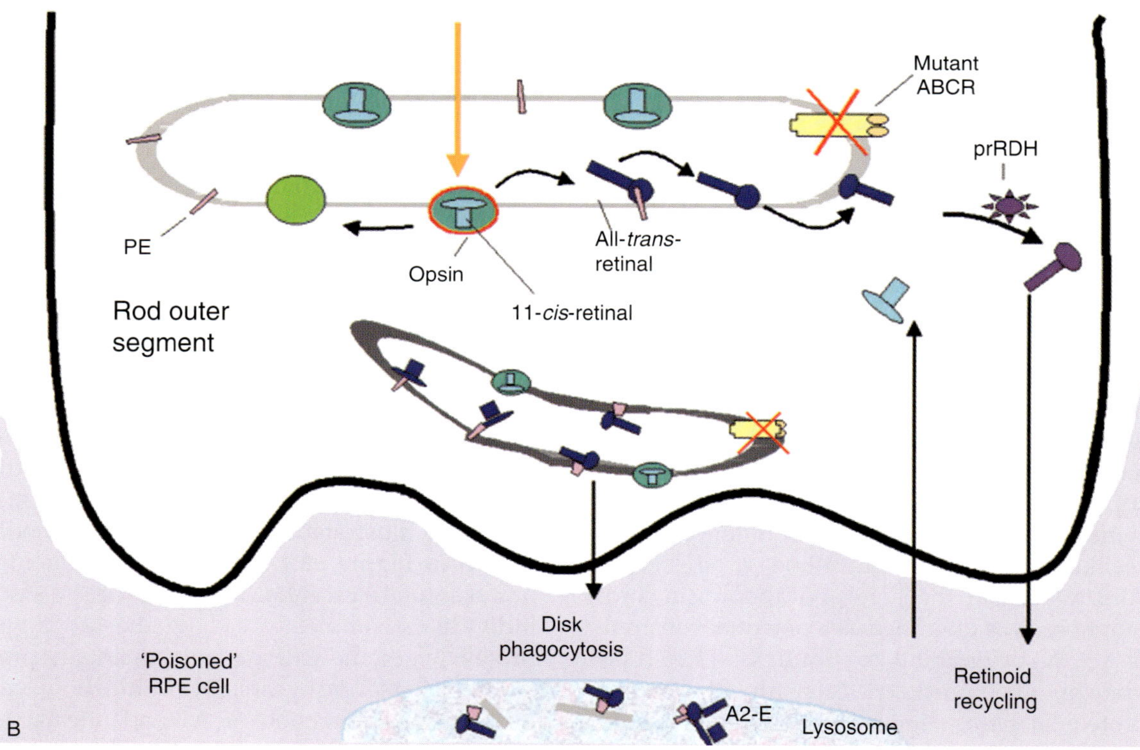

*Figure 28.4. Model for ABCA4 function in the visual cycle. A, Normal visual cycle in the case of functional ABCA4. Photoactivation of rhodopsin (orange arrow) results in the hydrolysis and release of all-***trans***-retinal into the photoreceptor outer segment disk membrane. ABCA4 either transports and/or presents the all-***trans***-retinal or its complex with phosphatidylethanolamine (RAL-PE) to retinol*

(continued)

was described in heterozygous mice, but its manifestation occurred at a slower, age-related, rate. The distinct, AMD-resembling phenotype in the *Abca4(+/−)* mouse model suggests that humans heterozygous for *ABCA4* mutations may be predisposed to A2E accumulation and concomitant retinal or macular disease (Mata *et al.*, 2001).

Remarkable allelic heterogeneity of the *ABCA4* gene has substantially complicated genetic analysis of its involvement in retinal disease, especially in the AMD complex trait. In a situation where a modest effect of a mutation can only be estimated by association analysis, the crucial question of the functional significance of a particular sequence variant often remains unanswered. Recent data from photoaffinity labeling and ATPase activity experiments from Jeremy Nathans' laboratory has dramatically advanced our knowledge in this field by determining the effect of close to 40 *ABCA4* mutations (Sun *et al.*, 2000). Thus, they demonstrated that both *ABCA4* variants analyzed in the Consortium study (Allikmets, 2000), G1961E and D2177N, affect the protein's ATPase activity *in vitro* **(Figure 28.5)**. The mutant G1961E protein, produced following the transfection of human embryonic kidney (293) cells with cloned cDNA, exhibits several-fold lower binding of 8-azido-ATP and dramatic inhibition of ABCA4 ATPase activity by retinal as compared to the wild-type protein. The D2177N variant had no effect on 8-azido-ATP binding, but exhibited a reproducible elevation in ATPase activity relative to the wild-type protein (Sun *et al.*, 2000). Consequently, the *ABCA4* variants considered to be associated with the AMD phenotype are not anonymous single nucleotide polymorphisms (SNPs), but rather mutations affecting ABCA4 function. These results will also challenge several suggestions that G1961E, the mutation most frequently found in STGD and AMD patients, is indeed a benign variant in linkage disequilibrium with another disease-causing mutation (Fishman *et al.*, 1999; Guymer *et al.*, 2001).

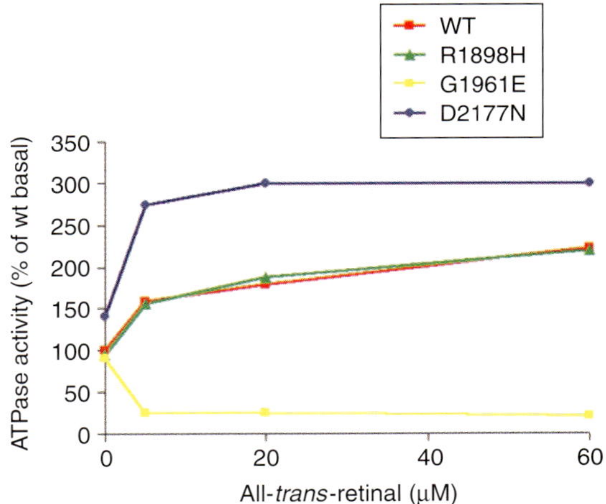

Figure 28.5. ***Effect of retinal on ATP hydrolysis by AMD-associated ABCA4 mutations. Modified from Sun* et al. *(2000). Note drastic inhibition of ATPase activity by the G1961E variant and elevation of the activity by the D2177N mutation, as compared to the wild type.***

Another issue that has been clarified is that of the functional significance of the G863A/delG863 variant. This variant is the most common single allele among STGD patients in northern Europe, and is also present in approximately 3% of the general population (Maugeri *et al.*, 1999). Although Maugeri *et al.* (1999) classified this variant as a 'mild' mutation, its role in retinal pathology has been disputed because of its high (>1%) frequency in the general population. The studies of Sun and colleagues (2000) clearly demonstrate a profound biochemical defect caused by either version of this mutation. Finally, both mutations found in the 'German' complex allele, L541P and A1038V, analyzed independently and in combination, render the ABCA4 protein defective (Sun *et al.*, 2000). In summary, functional studies fully support the proposed genotype/phenotype model of *ABCA4*, and offer several tools to advance our knowledge about the role of *ABCA4* in chorioretinal disease.

Figure 28.4. ***(continued)***
dehydrogenase (prRDH) on the cytosolic face of the disk. After reduction to all-*trans*-retinol the retinoid continues the visual cycle. The processed, RAL-PE free, disks are phagocytosed and digested by the retinal pigment epithelial cell. B, Altered cycle in the case of mutant* ABCA4*. Note accumulation of* N*-retinylidene-PE in rod outer segment disks and deposition of A2E in the retinal pigment epithelium (RPE). The accumulation of retinoids in phagolysosomes of the RPE leads to permanent A2E deposits followed by the RPE cell death and degeneration of photoreceptors.

ABCC6 AND PSEUDOXANTHOMA ELASTICUM

Pseudoxanthoma elasticum (PXE; MIM 264800) is a rare autosomal recessive (or dominant) disorder affecting the skin, eyes and cardiovascular system, with considerable morbidity and mortality. The disease affects the elastic fibers of affected organs, which become progressively calcified. The eyes are involved, displaying the characteristic appearance of angioid streaks, which result from fractures in Bruch's membrane, an elastin-rich sheath beneath the retina. As a result of fragmentation of this membrane, the blood vessels in the back of the eye break, resulting in bleeding and neovascularization. Consequently, the affected individuals experience progressive loss of visual acuity, which can be severe, although entire loss of vision is extremely rare. Thus, PXE has been considered as a prototypic heritable connective tissue disorder affecting the elastic fiber system.

Recently, PXE was linked to mutations in the *ABCC6* gene by four independent groups (Bergen *et al.*, 2000; Le Saux *et al.*, 2000; Ringpfeil *et al.*, 2000; Struk *et al.*, 2000). Genetic linkage analyses in various multiplex families have failed to suggest locus heterogeneity and therefore *ABCC6* seems to be the only gene underlying the PXE phenotype. The *ABCC6* gene consists of a total of 31 exons dispersed within ~73 kb of DNA on chromosome 16p13.1; the corresponding mRNA, ~6 kb, encodes a polypeptide of 1503 amino acids (Belinsky and Kruh, 1999; Kool *et al.*, 1999) (see also Chapter 21). ABCC6 is predicted to consist of three transmembrane regions comprising five, six and six transmembrane-spanning segments, respectively **(Figure 28.2)**. The majority of identified mutations reside in the COOH-terminal half of the protein, affecting primarily the intracellular domains. In contrast to the *ABCA4* gene, the majority of defects are deleterious mutations resulting in premature termination of translation, or mutations affecting the consensus splice sites, which are predicted to result in out-of-frame deletions in the mRNAs **(Figure 28.2)**. A particularly common allele carries a nonsense mutation R1141X, which has been independently described in families of various ethnic backgrounds (Bergen *et al.*, 2000; Le Saux *et al.*, 2000; Ringpfeil *et al.*, 2000; Struk *et al.*, 2000).

The endogenous function of ABCC6 is currently unknown. Initially, ABCC6 (also referred to as MRP6) was classified as a member of the multiple drug resistance-associated protein subgroup because of its homology to MRP1 (ABCC1), which has been well characterized as a transmembrane efflux pump primarily transporting amphipathic anticancer drugs, as well as glutathione, glucuronide and sulfate conjugated compounds (Borst *et al.*, 1999; Leslie *et al.*, 2001) (see Chapter 19). It was suggested, therefore, that the function of ABCC6 could also relate to cellular detoxification (Belinsky and Kruh, 1999). More recently, however, the substrate specificity of ABCC6 has been shown to be quite different from ABCC1 and other MRP-like proteins, and the only substrate demonstrated so far is BQ123, a small anionic peptide (Madon *et al.*, 2000). ABCC6 appears different from all other proteins of this subgroup also by its reported localization on both lateral and canalicular membranes of hepatocytes (Madon *et al.*, 2000) although this finding requires confirmation (see Chapter 21).

The expression of ABCC6 predominantly in the liver and kidney – organs not affected in PXE – raises the question of the relationship between the *ABCC6* mutations and the manifestations in PXE affecting the elastic fibers. As a hypothesis, one could propose that the absence of functional ABCC6 results in accumulation of certain metabolic compounds, resulting in progressive calcification of elastic fibers. This information, together with clinical observations suggesting environmental, hormonal and/or dietary modulation of the disease, raises the intriguing possibility that PXE is a primary metabolic disorder at the environment–genome interface (Uitto *et al.*, 2001).

PERSPECTIVES

The scientific progress in determining the role of the *ABCA4* gene in retinal pathology has been remarkable. We have significantly expanded our knowledge of the extensive range of phenotypes caused by various combinations of *ABCA4* mutations. *ABCA4* research has led to the formation of multicenter studies, encompassing large cohorts of ethnically diverse samples. Currently, ABCA4 is described as the transporter of *N*-retinylidene-PE, and there is an *in vitro* system(s) to study functional implications of all mutations. Finally, there is a mouse model that accurately reproduces many features of the human disorders. Most recent advances in the ABCA4 research include the generation of ABCR350 microarrays (Allikmets *et al.*, 2001), which, by containing all genetic variations of the *ABCA4* gene, can be used for systematic screening of patients with any and all *ABCA4*-associated pathology. Nevertheless, much more is yet to be accomplished in ABCC6 research. The generation and characterization of *Abcc6* knockout mice should provide important clues as to the endogenous cellular function of this MRP-related transporter.

With ABCA4, however, our efforts should now move to the next stage of research, directed towards finding therapeutic solutions for *ABCA4*-mediated chorioretinal disease by either improving the transport function of ABCA4 or by preventing accumulation of toxic products resulting from ABCA4 malfunction. Immediate areas of research may include gene therapy and determining synergistic activators for ABCA4. It is highly likely that even a slight improvement of ABCA4 function could delay the onset of related pathology and improve the quality of life of those individuals affected.

Acknowledgments

The author sincerely appreciates the work of all collaborators and colleagues involved in the research of the *ABCA4* gene, and excellent technical assistance by J. Tammur. Support by the Ruth and Milton Steinbach Fund, Research to Prevent Blindness Career Development Award, and NIH Grant EY-13435 is gratefully acknowledged.

References

Allikmets, R. (1999) Molecular genetics of age-related macular degeneration: current status. Eur. J. Ophthalmol. 9, 255–265.

Allikmets, R. (and the International ABCR Screening Consortium) (2000) Further evidence for an association of ABCR alleles with age-related macular degeneration. J. Hum. Genet. 67, 487–491.

Allikmets, R., Singh, N., Sun, H., Shroyer, N.F., Hutchinson, A., Chidambaram, A., *et al.* (1997a) A photoreceptor cell-specific ATP-binding transporter gene (ABCR) is mutated in recessive Stargardt macular dystrophy. Nat. Genet. 15, 236–246.

Allikmets, R., Shroyer, N.F., Singh, N., Seddon, J.M., Lewis, R.A., Bernstein, P.S., *et al.* (1997b) Mutation of the Stargardt disease gene (ABCR) in age-related macular degeneration. Science 277, 1805–1807.

Allikmets, R., Hutchinson, A., Jaakson, K., Kulm, M. and Pavel, H. (2001) Genotyping microarray (gene chip) for the ABCR (ABCA) gene. Invest. Ophthalmol. Vis. Sci. 42, S714.

Anderson, K.L., Baird, L., Lewis, R.A., Chinault, A.C., Otterud, B., Leppert, M. and Lupski, J.R. (1995) A YAC contig encompassing the recessive Stargardt disease gene (STGD) on chromosome 1p. Am. J. Hum. Genet. 57, 1351–1363.

Azarian, S.M. and Travis, G.H. (1997) The photoreceptor rim protein is an ABC transporter encoded by the gene for recessive Stargardt's disease (ABCR). FEBS Lett. 409, 247–252.

Belinsky, M.G. and Kruh, G.D. (1999) MOAT-E (ARA) is a full-length MRP/cMOAT subfamily transporter expressed in kidney and liver. Br. J. Cancer 80, 1342–1349.

Bergen, A.A., Plomp, A.S., Schuurman, E.J., Terry, S., Breuning, M., Dauwerse, H., *et al.* (2000) Mutations in ABCC6 cause pseudoxanthoma elasticum. Nat. Genet. 25, 228–231.

Borst, P., Evers, R., Kool, M. and Wijnholds, J. (1999) The multidrug resistance protein family. Biochim. Biophys. Acta 1461, 347–357.

Cremers, F.P., van de Pol, D.J., van Driel, M., den Hollander, A.I., van Haren, F.J., Knoers, N.V., *et al.* (1998) Autosomal recessive retinitis pigmentosa and cone-rod dystrophy caused by splice site mutations in the Stargardt's disease gene ABCR. Hum. Mol. Genet. 7, 355–362.

Dean, M., Rzhetsky, A. and Allikmets, R. (2001) The human ATP-binding cassette (ABC) transporter superfamily. Genome. Res. 11, 1156–1166.

De La Paz, M.A., Guy, VK., Abou-Donia, S., Heinis, R., Bracken, B., Vance, J.M., *et al.* (1999) Analysis of the Stargardt disease gene (ABCR) in age-related macular degeneration. Ophthalmology 106, 1531–1536.

Fishman, G.A., Stone, E.M., Grover, S., Derlacki, D.J., Haines, H.L. and Hockey, R.R. (1999) Variation of clinical expression in patients with Stargardt dystrophy and sequence variations in the ABCR gene. Arch. Ophthalmol. 117, 504–510.

Guymer, R.H., Heon, E., Lotery, A.J., Munier, F.L., Schorderet, D.F., Baird, P.N., *et al.* (2001) Variation of codons 1961 and 2177 of the Stargardt disease gene is not associated with age-related macular degeneration. Arch. Ophthalmol. 119, 745–751.

Illing, M., Molday, L.L. and Molday, R.S. (1997) The 220-kDa rim protein of retinal rod outer segments is a member of the ABC transporter superfamily. J. Biol. Chem. 272, 10303–10310.

Kaplan, J., Gerber, S., Larget-Piet, D., Rozet, J.M., Dollfus, H., Dufier, J.L., *et al.* (1993) A gene for Stargardt's disease (fundus flavimaculatus) maps to the short arm of chromosome 1. Nat. Genet. 5, 308–311.

Klein, R., Klein, B.E. and Linton, K.L. (1992) Prevalence of age-related maculopathy. The Beaver Dam Eye Study. Ophthalmology 99, 933–943.

Kool, M., van der Linden, M., de Haas, M., Baas, F. and Borst, P. (1999) Expression of human MRP6, a homologue of the multidrug resistance protein gene MRP1, in tissues and cancer cells. Cancer Res. 59, 175–182.

Le Saux, O., Urban, Z., Tschuch, C., Csiszar, K., Bacchelli, B., Quaglino, D., *et al.* (2000) Mutations in a gene encoding an ABC transporter cause pseudoxanthoma elasticum. Nat. Genet. 25, 223–227.

Leslie, E.M., Deeley, R.G. and Cole, S.P.C. (2001) Toxicological relevance of the multidrug resistance protein 1, MRP1 (ABCC1) and related transporters. Toxicology 167, 3–23.

Lewis, R.A., Shroyer, N.F., Singh, N., Allikmets, R., Hutchinson, A., Li, Y., *et al.* (1999) Genotype/phenotype analysis of a photoreceptor-specific ATP-binding cassette transporter gene, ABCR, in Stargardt disease. Am. J. Hum. Genet. 64, 422–434.

Long, A.D. and Langley, C.H. (1999) The power of association studies to detect the contribution of candidate genetic loci to variation in complex traits. Genome Res. 9, 720–731.

Madon, J., Hagenbuch, B., Landmann, L., Meier, P.J. and Stieger, B. (2000) Transport function and hepatocellular localization of mrp6 in rat liver. Mol. Pharmacol. 57, 634–641.

Martinez-Mir, A., Paloma, E., Allikmets, R., Ayuso, C., del Rio, T., Dean, M., *et al.* (1998) Retinitis pigmentosa caused by a homozygous mutation in the Stargardt disease gene ABCR. Nat. Genet. 18, 11–12.

Mata, N.L., Tzekov, R.T., Liu, X., Weng, J., Birch, D.G. and Travis, G.H. (2001) Delayed dark-adaptation and lipofuscin accumulation in abcr ± mice: implications for involvement of ABCR in age-related macular degeneration. Invest. Ophthalmol. Vis. Sci. 42, 1685–1690.

Maugeri, A., van Driel, M.A., van de Pol, D.J., Klevering, B.J., van Haren, F.J., Tijmes, N., *et al.* (1999) The 2588G→C mutation in the ABCR gene is a mild frequent founder mutation in the western European population and allows the classification of ABCR mutations in patients with Stargardt disease. Am. J. Hum. Genet. 64, 1024–1035.

Maugeri, A., Klevering, B.J., Rohrschneider, K., Blankenagel, A., Brunner, H.G., Deutman, A.F., *et al.* (2000) Mutations in the ABCA4 (ABCR) gene are the major cause of autosomal recessive cone-rod dystrophy. Am. J. Hum. Genet. 67, 960–966.

O'Donovan, M.C. and Owen, M.J. (1999) Candidate-gene association studies of schizophrenia. Am. J. Hum. Genet. 65, 587–592.

Papermaster, D.S., Converse, C.A. and Zorn, M. (1976) Biosynthetic and immunochemical characterization of large protein in frog and cattle rod outer segment membranes. Exp. Eye Res. 23, 105–115.

Papermaster, D.S., Schneider, B.G., Zorn, M.A. and Kraehenbuhl, J.P. (1978) Immunocytochemical localization of a large intrinsic membrane protein to the incisures and margins of frog rod outer segment disks. J. Cell Biol. 78, 415–425.

Ringpfeil, F., Lebwohl, M.G., Christiano, A.M. and Uitto, J. (2000) Pseudoxanthoma elasticum: mutations in the MRP6 gene encoding a transmembrane ATP-binding cassette (ABC) transporter. Proc. Natl Acad. Sci. USA 97, 6001–6006.

Riordan, J.R., Rommens, J.M., Kerem, B., Alon, N., Rozmahel, R., Grzelczak, Z., *et al.* (1989) Identification of the cystic fibrosis gene: cloning and characterization of complementary DNA. Science 245, 1066–1073.

Rivera, A., White, K., Stohr, H., Steiner, K., Hemmrich, N., Grimm, T., *et al.* (2000) A comprehensive survey of sequence variation in the ABCA4 (ABCR) gene in Stargardt disease and age-related macular degeneration. Am. J. Hum. Genet. 67, 800–813.

Rozet, J.M., Gerber, S., Souied, E., Perrault, I., Chatelin, S., Ghazi, I., *et al.* (1998) Spectrum of ABCR gene mutations in autosomal recessive macular dystrophies. Eur. J. Hum. Genet. 6, 291–295.

Rozet, J.M., Gerber, S., Ghazi, I., Perrault, I., Ducroq, D., Souied, E., *et al.* (1999) Mutations of the retinal specific ATP binding transporter gene (ABCR) in a single family segregating both autosomal recessive retinitis pigmentosa RP19 and Stargardt disease: evidence of clinical heterogeneity at this locus. J. Med. Genet. 36, 447–451.

Shroyer, N.F., Lewis, R.A., Allikmets, R., Singh, N., Dean, M., Leppert, M. and Lupski, J.R. (1999) The rod photoreceptor ATP-binding cassette transporter gene, ABCR, and retinal disease: from monogenic to multifactorial. Vision Res. 39, 2537–2544.

Simonelli, F., Testa, F., de Crecchio, G., Rinaldi, E., Hutchinson, A., Atkinson, A., *et al.* (2000) New ABCR mutations and clinical phenotype in Italian patients with Stargardt disease. Invest. Ophthalmol. Vis. Sci. 41, 892–897.

Stone, E.M., Webster, A.R., Vandenburgh, K., Streb, L.M., Hockey, R.R., Lotery, A.J. and Sheffield, V.C. (1998) Allelic variation in ABCR associated with Stargardt disease but not age-related macular degeneration. Nat. Genet. 20, 328–329.

Struk, B., Cai, L., Zach, S., Ji, W., Chung, J., Lumsden, A., *et al.* (2000) Mutations of the gene encoding the transmembrane transporter protein ABC-C6 cause pseudoxanthoma elasticum. J. Mol. Med. 78, 282–286.

Sun, H., Molday, R.S. and Nathans, J. (1999) Retinal stimulates ATP hydrolysis by purified and reconstituted ABCR, the photoreceptor-specific ATP-binding cassette transporter responsible for Stargardt disease. J. Biol. Chem. 274, 8269–8281.

Sun, H., Smallwood, P.M. and Nathans, J. (2000) Biochemical defects in ABCR protein variants associated with human retinopathies. Nat. Genet. 26, 242–246.

Uitto, J., Pulkkinen, L. and Ringpfeil, F. (2001) Molecular genetics of pseudoxanthoma elasticum: a metabolic disorder at the environment-genome interface? Trends Mol. Med. 7, 13–17.

van Driel, M.A., Maugeri, A., Klevering, B.J., Hoyng, C.B. and Cremers, F.P. (1998) ABCR unites what ophthalmologists divide(s). Ophthalmic Genet. 19, 117–122.

Vingerling, J.R., Dielemans, I., Hofman, A., Grobbee, D.E., Hijmering, M., Kramer, C.F. and de Jong, P.T. (1995) The prevalence of age-related maculopathy in the Rotterdam Study. Ophthalmology 102, 205–210.

Weng, J., Mata, N.L., Azarian, S.M., Tzekov, R.T., Birch, D.G. and Travis, G.H. (1999) Insights into the function of Rim protein in photoreceptors and etiology of Stargardt's disease from the phenotype in abcr knockout mice. Cell 98, 13–23.

Yatsenko, A.N., Shroyer, N.F., Lewis, R.A. and Lupski, J.R. (2001) Late-onset Stargardt disease is associated with missense mutations that map outside known functional regions of ABCR (ABCA4). Hum. Genet. 108, 346–355.

Zielenski, J. and Tsui, L.C. (1995) Cystic fibrosis: genotypic and phenotypic variations. Annu. Rev. Genet. 29, 777–807.

The Cystic Fibrosis Transmembrane Conductance Regulator (ABCC7)

John W. Hanrahan, Martina Gentzsch and John R. Riordan

CHAPTER 29

CFTR Mutation Causes Cystic Fibrosis

The cystic fibrosis transmembrane conductance regulator (CFTR) is the product of the gene mutated in patients with cystic fibrosis (CF), an autosomal recessive disease of relatively high frequency in the Caucasian population. The large number of families with CF enabled linkage analysis, and ultimately identification of the gene by positional cloning. Although ΔF508, the first disease-associated mutation discovered, is present on at least one allele in approximately 90% of patients, more than a thousand different mutations in the CFTR gene have been detected worldwide (http://www.genet.sickkids.on.ca/cftr/). The absence or dysfunction of CFTR results in aberrant ion and liquid homeostasis at epithelial surfaces of the respiratory, intestinal and reproductive tracts as well as other secretory and reabsorptive epithelia. CFTR is a channel that allows anions to diffuse through the membrane in either direction (absorptive or secretory) depending on the electrochemical gradients. Since chloride is the predominant inorganic anion *in vivo*, the CFTR pore conducts mainly Cl^- ions under physiological conditions, although significant bicarbonate transport may also occur in some tissues (see below).

The principal example of Cl^- absorption occurs in the sweat gland, where failure to reabsorb NaCl results in the 'salty sweat' that is diagnostic of the disease. More serious consequences, however, ensue in the gastrointestinal tract, where exocrine pancreatic function is lost owing to defective anion secretion by the small ducts, and reduced cAMP-stimulated chloride and H_2O secretion by intestinal crypts leads to mucus accumulation and severe malabsorption. Thus Cl^- ions move out of the cell through apical CFTR channels, which are the rate-limiting step during secretion **(Figure 29.1)**. Indeed, the crucial role that the CFTR chloride channel plays in fluid secretion in the intestine may be at the basis of the so-called 'heterozygote advantage', which could account for the relatively high mutant gene frequency despite the fact that the mutations are generally not propagated by homozygotes. Experiments with CFTR knockout mice provide some support for this theory (Cuthbert *et al.*, 1995; Gabriel *et al.*, 1994). According to this hypothesis heterozygotes may suffer less potentially fatal intestinal fluid loss and have better survival during toxic bacterial infections. Of greater pathological significance to homozygous patients is the lack of adequate hydration of the airways of the lung, where macromolecular secretions also accumulate owing to inefficient muco-ciliary clearance. Recurrent and persistent colonization by opportunistic microorganisms exacerbates inflammatory responses, leading to fibrosis and severely impaired airway function.

ABC Proteins: From Bacteria to Man
ISBN 0-12-352551-9

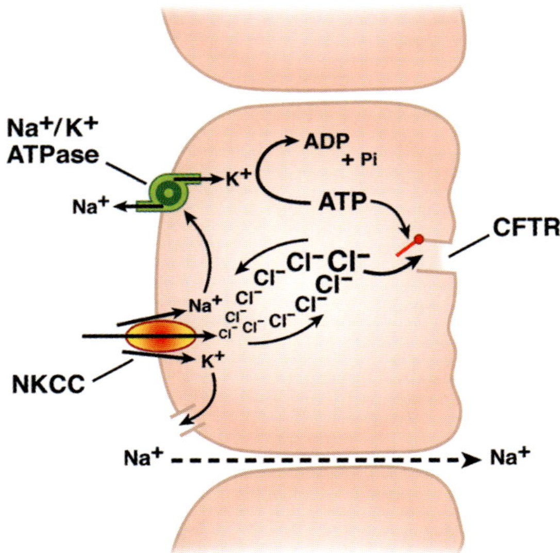

Figure 29.1. ***Role of CFTR in transepithelial chloride secretion.*** *Cross-section of the epithelium, with basolateral side shown on the left and apical side on the right. Monovalent ions Na^+, K^+ and $2Cl^-$ enter the cells from the blood side through the sodium, potassium, 2 chloride cotransporter (NKCC). Na^+ is actively pumped out by Na^+/K^+ ATPase at the basolateral membrane, and K^+ also leaves via a basolateral potassium channel. Cl^- is raised above electrochemical equilibrium and flows passively to the lumenal side through the CFTR chloride channel. Sodium ions follow paracellularly to maintain charge balance.*

CFTR IS AN ION CHANNEL

In addition to being at the basis of a common monogenic disease the other prominent feature of CFTR as an ABC protein is its function as an ion channel. On the basis of present knowledge, this property makes it unique among ABC proteins and initially caused some consternation because it apparently did not fit perfectly into the ABC 'transporter' mold (Higgins, 1992). However, since these initial discussions, compelling evidence of its anion channel activity has accumulated (see the section on CFTR as an epithelial chloride channel, below). Although neither the three-dimensional (3-D) structure nor the mechanisms of permeation (ion flow through the open channel pore) and gating (spontaneous transitions of the channel pore between open and closed states) are known as yet, some of the primary structural features that contribute to its regulated channel activity and distinguish it from related members of the ABCC subfamily are apparent. These features are very highly conserved in other mammals, amphibians, teleosts and cartilaginous fish, which are the most primitive organisms where CFTR homologues have been detected and where CFTR plays a central role in salt secretion **(Figure 29.1)**.

From this perspective, adaptation of the ABC structural architecture to the role of a nucleotide-regulated ion channel appears relatively recent when compared with other ABC proteins in much more ancient organisms (Saurin *et al.*, 1999). This adaptation to achieve a highly specific function may have been quite unique as there is a single CFTR gene in all organisms where it has been examined, except in salmon, where it may have duplicated as part of a whole genome duplication (Chen *et al.*, 2001). Moreover, there is a relative paucity of isoforms generated by alternate gene splicing or other mechanisms (Morales *et al.*, 1996). This contrasts with the other ABCC family member involved in ion conduction, the sulfonylurea receptor component of K_{ATP} channels (see Chapter 27), for which there are at least two separate genes (Inagaki *et al.*, 1996).

It is important to place the findings from directly studying CFTR, once the gene had been cloned and expressed, into the context of the electrophysiological changes described much earlier in CF epithelia (Knowles *et al.*, 1983; Quinton, 1983). Although there was increased sodium absorption in the airway epithelium of CF patients compared to normals (Knowles *et al.*, 1983), Quinton found reduced chloride conductance in CF sweat ducts relative to the normal, and proposed that it was the primary electrophysiological alteration in CF (Quinton, 1983). This seminal finding was pursued, in the ensuing decade prior to the discovery of the CFTR gene, in assays of chloride channel activity in normal and CF epithelial cells, primarily employing the patch clamp technique (see below). A chloride channel activated by cyclic AMP was found to be absent from CF epithelia (Li *et al.*, 1988; Schoumacher *et al.*, 1987). This channel had properties different from those eventually identified as those of the CFTR chloride channel (Berger *et al.*, 1991; Kartner *et al.*, 1991) and has still not been identified at the molecular level. However, this channel apparently requires a functional CFTR for its activation (Gabriel *et al.*, 1993). In addition to CFTR's primary, inherent chloride channel function, it has been reported to regulate the activity of other channels and transporters. This fascinating aspect of CFTR research has been reviewed elsewhere (Schwiebert *et al.*, 1999) and will be dealt with only briefly in this chapter together with a consideration of other proteins interacting with CFTR.

Whether CFTR might also conduct other, as yet unidentified substrates will not be answered here as the experimental evidence is not sufficient to confirm or exclude this possibility. Other important issues concerning the role of CFTR in human health and disease that will not be discussed include the contributions of mutations in the CFTR gene to clinical conditions other than cystic fibrosis (Noone and Knowles, 2001). These so-called 'CFTR-opathies' involve compromised functions of the major organs in which CFTR function is important, e.g. pancreatitis (Sharer *et al.*, 1998), disseminated bronchiectasis (Pignatti *et al.*, 1996), and congenital bilateral absence of the vas deferens (Osborne *et al.*, 1993).

This chapter is not intended to provide a comprehensive review and evaluation of all knowledge gained about CFTR. This has been done relatively recently (Supplement to *Physiological Reviews* 79 (1), 1999). Rather we consider it from the perspective of the other ABC proteins that are the subject of this book, emphasizing its common and distinguishing features. To do this we focus on what is known of its structure, its properties as a chloride channel, its phosphorylation and dephosphorylation, which control channel activity, its binding and hydrolysis of ATP and the consequences of these events for channel gating. Finally, because the wild-type CFTR polypeptide itself matures inefficiently during its biosynthesis and because the most common disease-causing mutant form succumbs to degradation by the quality-control system of the endoplasmic reticulum, we discuss what is known of the features of CTFR that may contribute to this behavior.

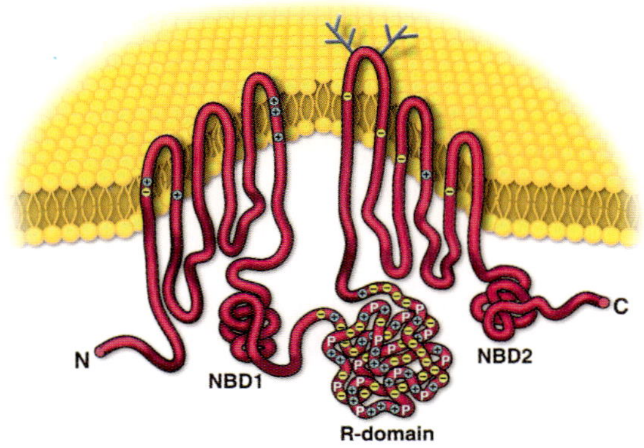

Figure 29.2. Cartoon of CFTR glycoprotein indicating its major distinguishing features: (1) the presence of the central highly charged R-domain, phosphorylation of which is necessary for nucleotide-regulated chloride channel activity, (2) the presence of charged amino acids within predicted membrane-spanning segments, several of which influence ion permeation.

CFTR STRUCTURE

PRIMARY STRUCTURE

The principal features of the CFTR protein sequence that distinguish it from other ABCC family members were recognized from the cDNA sequence (Riordan *et al.*, 1989). The most striking difference from the known eukaryotic ABC transporters such as P-glycoprotein (Pgp) is the presence of an extended hydrophilic region following nucleotide-binding domain 1 (NBD1). This contains many charged residues and an extraordinarily large number of precise dibasic and monobasic consensus sequences for phosphorylation by protein kinase A (PKA) (**Figure 29.2**; Riordan *et al.*, 1989). Phosphorylation and dephosphorylation of this region, named the R-domain, exerts major control over chloride channel activity as is discussed in detail later in the chapter. The second apparent distinguishing feature at the time of CFTR discovery was the presence of a significantly higher number of amino acids with charged side-chains within putative transmembrane spanning (TMS) sequences, compared with other ABC proteins known at that time. This feature is clearly not necessarily diagnostic of ion channels and there are now other ABC proteins known that have charged residues within the membrane spans (see Chapter 7). Nevertheless, several of these residues within TMS sequences of CFTR have been found to influence the properties of the ion pore, as described in the section on the CFTR channel pore, below.

It was of course the homology of the two NBD sequences with those of other known ABC proteins that placed CFTR in this family. Each of the features of the NBDs that are described in length elsewhere in this book are present in CFTR, with most similarity to other members of the ABCC subfamily. As was first pointed out for MRP1, the NBD1 of this protein, as well as that of CFTR, has a deletion of approximately 13 residues between the Walker A and B motifs compared with NBD2 and the NBDs of many other ABC proteins (Deeley and Cole, 1997). Another notable feature of the CFTR NBD1 sequence is the substitution of the glutamic acid residue that follows the Walker B aspartic acid by a serine **(Figure 29.3)**. This may be significant since this glutamate serves as a catalytic base in the hydrolysis of ATP by other NBDs, where it is present, and it is found in NBD2 of CFTR. The LSGGQ signature motif is strictly conserved in NBD1 but is quite degenerate in NBD2. The significance of other short sequence motifs within the NBDs is being increasingly appreciated from alignments with those NBDs of other ABC proteins for which 3-D structures have been determined. Some of these motifs are structural elements that contribute to the stability of the domains and of the protein as a whole. Four-residue hydrophobic 'patches' near the C-termini of NBDs of CFTR are essential for the maturation and stability of the protein (Gentzsch and Riordan, 2001). Other predictive suggestions have been made about conformation changes that may distinguish ATP- and ADP-bound forms of CFTR NBDs by alignment with ABC proteins where 3-D structures have been determined with either nucleotide bound (Berger *et al.*, 2002).

In addition to the R-domain and NBDs, which make up more than half of the protein mass, the N- and C-terminal tails and four cytoplasmic loops between predicted membrane spans also present on the cytoplasmic side of the plasma membrane. Sequences within the N-terminal tail have been found to influence channel gating and interactions with other proteins (Naren *et al.*, 1998). An amphipathic helix with acidic amino acids on one face between residues 46 and 60 influences channel gating (Naren *et al.*, 1999), whereas another short sequence nearer the N-terminus apparently mediates interaction with an annexin (Nelson *et al.*, 2001). The acidic helix also binds SNARE proteins and hence is suggestive of a role which provides a relationship between vesicular trafficking and CFTR channel gating (Naren *et al.*, 1997). The C-terminal extension of CFTR beyond the end of NBD2, beginning at approximately residue 1424, is

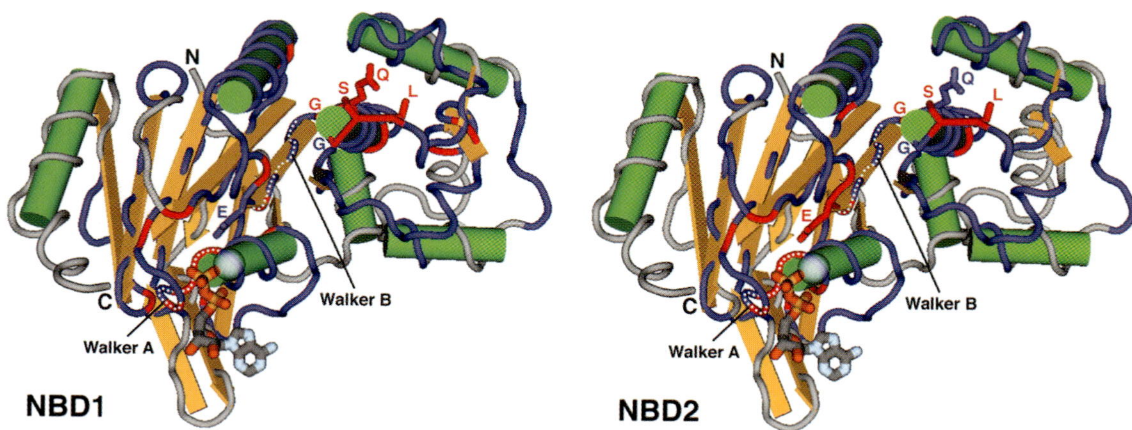

Figure 29.3. Patterns of similarity between NBD1 and NBD2 of CFTR and other ABC-ATPases. Gapped-BLAST alignment of NBD1 (left) and NBD2 (right) with HisP, MalK, LolD (MJ0796) and LivG (MJ1267) using the program Cn3D 3.0 (National Center for Biotechnology). Similarities are shown on the crystal structure of LivG (MMDB 16953), catalytic domain to the left, helical domain to the right of each NBD. Red represents identical residues, blue represents residues that are aligned and gray corresponds to sequences that do not align. The Mg^{2+} ion is shown as a gray sphere. In the ATP-binding protein LivG, residue E179 following Walker B appears to function as a catalytic base activating the hydrolytic water for the attack on the γ-phosphate of ATP (Karpowich et al.*, 2001). This glutamate is conserved in NBD2, HisP, MalK, LolD and LivG but is replaced by a serine in NBD1 of CFTR (red in the alignment with NBD2, but blue in the alignment with NBD1). The transport signature motif LSGGQ (residues 154–158 in LivG) is identical in all NBDs but not conserved in NBD2 of CFTR. In dimers derived from the ABC ATPase Rad50, the LSGGQ sequence contacts the nucleotide bound to the other subunit. If NBD1 and NBD2 of CFTR form a Rad50-like dimer, ATP bound to NBD1 would be contacted by an imprecise signature sequence of NBD2.*

apparently not essential functionally but contains several motifs involved in the trafficking and localization of the molecule within the cell. The best-characterized motif among these is the class I PDZ-domain binding sequence comprising the final three residues of the protein (Short *et al.*, 1998). Interactions with several different PDZ-domain-containing proteins play roles in the apical localization in epithelial cells (Milewski *et al.*, 2001), trafficking through the Golgi apparatus (Cheng *et al.*, 2002) and channel regulation (Raghuram *et al.*, 2001; Wang *et al.*, 2000), possibly by localization in a regulatory complex in proximity to protein kinase A (Huang *et al.*, 2001). The PDZ-binding domain is also involved in tethering together two CFTR polypeptides to modulate channel activity (Raghuram *et al.*, 2001; Wang *et al.*, 2000). Interestingly MRP2 (ABCC2), which also has a PDZ-binding C-terminus, also localizes to the apical membrane of polar cells whereas MRP1, which does not, resides in the basolateral membrane. At the opposite N-terminal end of the C-terminal extension is tyrosine 1424, which is involved in endocytosis of CFTR through interaction with a subunit of the clathrin adaptor protein complex, AP-2 (Weixel and Bradbury, 2001). A dileucine motif at residues 1430 and 1431 may also be involved in internalization. The C-terminal extension has additional motifs that are less well defined, including a site for phosphorylation by AMP activated protein kinase (Hallows *et al.*, 2000) and an acidic cluster just upstream of the PDZ-binding terminus. Mutagenesis of residues in the four cytoplasmic loops (CLs) between membrane-spanning helices generally has a greater effect on protein maturation than channel function (Seibert *et al.*, 1996a, 1996b, 1997) although they are believed to mediate the influence of conformation changes in the NBDs affecting the activity of the ion pore.

Another set of short sequence motifs is composed of RXR tripeptides in the N-tail, NBD1 and the R-domain (Chang *et al.*, 1999). These influence the fate of ΔF508 CFTR, perhaps by directly contributing to, or signaling, misfolding and mislocalization. The six extracytoplasmic loops (ELs) are much shorter than those on the cytoplasmic side, indicating that very little of the protein is exposed at the cell surface. Disease-associated single residue substitutions in the ELs result primarily in decreases in the channel open state (Hämmerle *et al.*, 2001). The longest of the ELs are the first in each half of the molecule, with that in the second half containing the two asparagine residues (894 and 900) that are the sites of N-glycosidic linkage of the two oligosaccharide chains on CFTR (Riordan *et al.*, 1989). These carbohydrate chains are not essential to the function of the molecule.

Secondary structure

There has been very little experimental determination of secondary structure elements in CFTR, none on the whole purified protein. The 12 membrane-spanning sequences are assumed to be α-helical but this is not confirmed in the intact molecule. Reasonable predictions can be made about the segments of the NBDs that align well with those of other NBDs whose 3-D structures have been determined, and some of these have been discussed (Berger and Welsh, 2000; Thomas and Hunt, 2001). As one example, the four-residue hydrophobic 'patch' near the C-terminal end of NBDs of CFTR and other ABC proteins appear to be within the penultimate β-strand of the domains (Gentzsch *et al.*, 2002). The α-helical structure of the acidic segment of the N-tail that modulates channel gating has been confirmed by nuclear magnetic resonance (NMR) spectroscopy of the corresponding synthetic peptide. As a means of testing the hypothesis that phosphorylation by PKA of the R-domain may cause a conformational change, circular dichroism (CD) spectra of an isolated recombinant R-domain were recorded before and after phosphorylation (Dulhanty and Riordan, 1994). Calculation of the proportion of different secondary structure elements by deconvolution methods indicated that the overall domain is relatively unstructured but that phosphorylation causes a decrease in the relatively small α-helical content. More recent considerations emphasized the lack of R-domain secondary structure (Ostedgaard *et al.*, 2001). The content of α and β secondary structure in an NBD1–R-domain fusion (Ko *et al.*, 1993) was considerably less than that of NBD1 alone (Neville *et al.*, 1998). The relatively unordered structure of the R-domain is consistent with the fact that despite its essential functional role, the sequence of the R-domain is less conserved among different species than any other part of the molecule. Only the consensus phosphorylation sites are conserved. Whether the functionally relevant consequences of phosphorylation involve changes in secondary or tertiary structure is unknown.

TERTIARY STRUCTURE

As we have no 3-D structure information on CFTR, the only clues to its tertiary structure are provided by membrane topology, for which there is some experimental evidence (Chang *et al.*, 1994). The 2-D segmentation discussed in the above section is probably correct; however, spatial relationships in the third dimension are all unknown. The relationship between the 3-D structure and our current 2-D sketches are likely to be as dissimilar as in the case of the ClC chloride channel, for which a structure at atomic resolution was recently obtained (Dutzler *et al.*, 2002). A 3-D image of the protein even at low resolution will be essential before great insight into the mechanism of action can be gained.

QUATERNARY STRUCTURE

While the CFTR polypeptide alone can generate a PKA- and ATP-regulated chloride channel (Bear *et al.*, 1992), it is not known with certainty how many of these polypeptides form the channel. This uncertainty applies to most ABC proteins with the exception of SUR1, a monomer of which associates with each of the four subunits of the tetrameric Kir6.2 potassium channel to produce an octameric K_{ATP} (Aguilar-Bryan *et al.*, 1998; Chapter 22). A second ABCC protein, MRP1, appears as a dimer in 2-D crystalline arrays (Rosenberg *et al.*, 2001), but there is no independent evidence that the conjugated organic anion exporter functions as a dimer. In contrast, Pgp, which has been perhaps the most extensively studied ABC protein from the point of view of its oligomeric structure, appeared monomeric in the same type of 2-D crystal analysis by electron diffraction (Rosenberg *et al.*, 1997). However, the Pgp particle diameters of 10–12 nm (Rosenberg *et al.*, 1997), similar to estimates made earlier by freeze-fracture electron microscopy of membranes containing Pgp (Arsenault *et al.*, 1988), are also quite similar to values obtained more recently (Eskandari *et al.*, 1998) for CFTR in *Xenopus* oocyte membranes (~9 nm). Despite these comparable dimensions of a monomeric Pgp and the CFTR particles, the authors of the latter study concluded that CFTR is dimeric. This interpretation agreed with that from a study of concatemerized constructs containing one wild-type CFTR sequence and a variant with aberrant regulatory properties (Zerhusen *et al.*, 1999). Detection of channels with intermediate regulatory behavior was believed to reflect the contribution of two CFTR polypeptides to a single channel. Using a similar approach, in which variants with very different ionic conductances were coexpressed in equal amounts, no channels of intermediate conductance were formed (Chen *et al.*, 2002). Although a portion of purified CFTR molecules chromatographed with a size corresponding to dimers, those in the size range of monomers yielded CFTR chloride channel activity (Ramjeesingh *et al.*, 2001). Differentially epitope-tagged CFTR species when coexpressed failed to be co-immunoprecipitated (Chen *et al.*, 2002) or appreciably crosslinked to form dimers by chemical crosslinkers (Chen and Riordan, 2002). CFTR solubilized with different detergents from membranes of cells either heterologously or endogenously expressing the protein migrated as a monomer on velocity gradient centrifugation in sucrose gradients (Chen *et al.*, 2002). Thus, at this stage definitive evidence of the stable quaternary structure of CFTR is not available.

However, the influence of bivalent PDZ-domain proteins on CFTR channel activity has raised the possibility that two CFTR molecules might be tethered together at least transiently to yield channels of increased open probability compared to untethered individual molecules (Raghuram *et al.*, 2001; Wang *et al.*, 2000). It is not clear if this tethering promotes a more extensive interaction interface between two polypeptides and if so how this might relate to the formation or regulation of the ion pore. This raises the issue of CFTR quaternary structure from the perspective of other ion channels, nearly all of which are oligomeric or assemblies of repeated units within a single polypeptide (Catterall, 2000). Cation channel pores are formed at interfaces between subunits. It is unknown if the CFTR pore is formed employing this architectural strategy or if it is assembled by elements within individual subunits. ClC chloride channels are dimeric but each monomer contains a functional pore that is removed from the dimer interface (Dutzler *et al.*, 2002). One report has suggested that there may even be separate pores in each half of a single CFTR polypeptide (Yue *et al.*, 2000).

Resolution of the quaternary structure of CFTR, which is unique among both ABC proteins and ion channels, will be essential to achieve an understanding of its structure and function.

CFTR IS AN EPITHELIAL CHLORIDE CHANNEL

CHLORIDE CONDUCTANCE DEFECT IN CYSTIC FIBROSIS

Electrophysiological experiments on isolated sweat ducts first revealed a chloride conductance defect in cystic fibrosis (Quinton, 1983). In normal ducts, transepithelial Na^+ reabsorption is active and involves diffusion through apical channels and extrusion by basolateral Na^+/K^+ ATPase pumps. Transepithelial Cl^- reabsorption maintains electroneutrality and is passive, mediated by a parallel conductance. The electrical effects of imposing transepithelial salt gradients across the duct suggested that Cl^- conductance is abnormally low in CF. This was confirmed when normal ducts were found to behave like those from CF individuals when Cl^- on both sides of the epithelium was replaced with the less permeant anion sulfate (Quinton, 1983). When Cl^- is unable to follow Na^+, the lumen-negative voltage hyperpolarizes by more than 30 mV and NaCl reabsorption is greatly reduced (Quinton, 1983; Quinton and Bijman, 1983). This low Cl^- conductance nicely explains why the sweat of CF patients is salty, a diagnostic feature of the disease. In the airways ion transport is more complex; salt and fluid are absorbed under resting conditions and secreted when stimulated by epinephrine, acetylcholine, tachykinins and purine nucleotides. The transepithelial potential across human nasal epithelium is −20 to −40 mV (lumen-negative) and, like the sweat duct, becomes hyperpolarized in CF patients. The transepithelial potential in CF is more sensitive to block by amiloride, an antagonist of the epithelial Na^+ channel (Garty and Palmer, 1997). This would be consistent with both reduced Cl^- conductance and elevated Na absorption. However, subsequent studies of cultured cells suggested a two- to fourfold increase in apical membrane Na conductance in CF (Boucher, 1994; Willumsen and Boucher, 1989). Although observed at the single channel level (Stutts *et al.*, 1995), the mechanism by which CFTR normally downregulates the Na^+ channel is not understood and is not universal since CFTR is required for normal Na^+ conductance in the sweat duct (Reddy *et al.*, 1999).

CFTR IS A LOW-CONDUCTANCE, NON-RECTIFYING CHLORIDE CHANNEL IN EPITHELIAL CELLS

The precise function of CFTR was not obvious when the gene was cloned. However, a cAMP-stimulated Cl^- channel with a nearly linear (i.e. non-rectifying, **Figure 29.4**) current–voltage relationship and conductance of 7–10 pS appeared when CFTR was expressed in Sf9 insect cells (Kartner *et al.*, 1991), fibroblasts (Berger *et al.*, 1991), Chinese hamster ovary (CHO) cells (Tabcharani *et al.*, 1991), and *Xenopus* oocytes

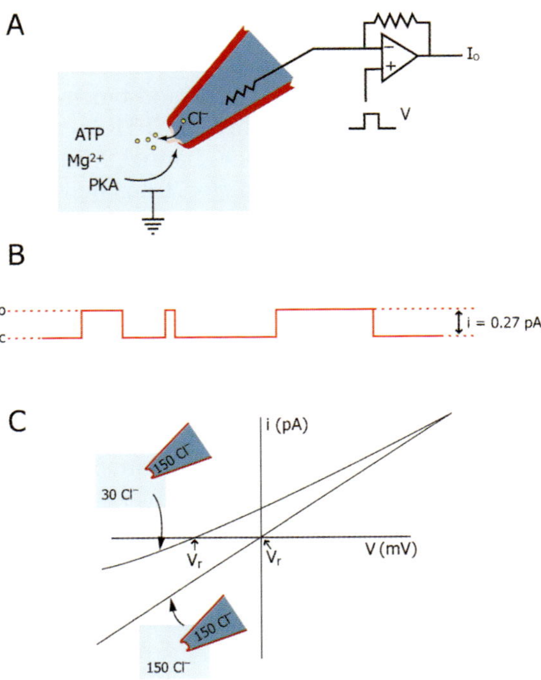

Figure 29.4. Patch clamp recording of CFTR channel current. A, Inside-out configuration showing pipette tip with excised patch that contains a single CFTR chloride channel. B, Hypothetical trace illustrating unitary current amplitude and slow gating. Openings are indicated by upward transitions, and result from nucleotide binding, probably at NBD2. Downward transitions are the closing events caused by dissociation of the nucleotide or its hydrolysis products. C, Current–voltage relationship calculated for a single channel. The reversal potential (V_r, zero-current potential) is 0 mV in symmetrical, high-Cl^- solutions. V_r shifts to a negative potential as predicted by the Nernst equation when bath Cl^- concentration is reduced.

(Bear *et al.*, 1991). These properties are identical to those of a channel reported in pancreatic ducts (Gray *et al.*, 1988, 1989), rat thyroid epithelium (Champigny *et al.*, 1990), and in T84 cells, the most widely used model cell line for epithelial Cl^- secretion (Tabcharani *et al.*, 1990). The similarities in unitary conductance, regulation, gating and pharmacology led to the proposal that CFTR is itself the low-conductance Cl^- channel (Kartner *et al.*, 1991). A different Cl^- channel with higher conductance that preferentially conducts anions into the cell (i.e. exhibits 'outward rectification' of current flow) had been previously identified as defective in CF (Frizzell *et al.*, 1986; Hwang *et al.*, 1989; Li *et al.*, 1988). Confusion was exacerbated when the macroscopic conductance first generated by heterologous CFTR expression was attributed to rectifying anion channels (Anderson *et al.*, 1991c). However, mutagenesis and biochemical reconstitution soon established CFTR as a non-rectifying, low-conductance channel. Mutations in the first and sixth predicted transmembrane segments of CFTR altered the selectivity of whole cell anion currents (Anderson *et al.*, 1991b). When CFTR protein was purified to homogeneity from Sf9 insect cells and reconstituted into planar lipid bilayers, it generated channels having low conductance like those recorded previously on epithelial cells (Bear *et al.*, 1992). What happened to the outwardly rectifying anion channel? Most attention has focused on other Cl^- channels that are consistently activated while still on the cell (as opposed to the less physiological conditions existing in isolated membrane patches). The outwardly rectifying anion channel is probably one of many membrane proteins that are influenced by the presence of CFTR, but there is no compelling evidence for a role in Cl^- secretion. It has been difficult to establish the physiological roles of particular Cl^- channels because they cannot be pharmacologically dissected using available inhibitors. For example, a cAMP-stimulated, DIDS (4,4′-diisothiocyanatostilbene-2,2′-disulfonate)-sensitive whole cell Cl^- current has been attributed to the outward rectifier, based on its sensitivity to external DIDS, which had little effect on CFTR in early studies of pancreatic ducts and T84 monolayers (Gray *et al.*, 1990; Tabcharani *et al.*, 1990). However, DIDS sensitivity depends on experimental conditions, and external DIDS does partially inhibit CFTR (Kartner *et al.*, 1991). Since the action of DIDS on CFTR is enhanced by large voltage ramps such as those applied to whole cell patches, inhibition of CFTR probably explains the DIDS-sensitive currents during cAMP stimulation that have been ascribed to outwardly rectifying channels.

Role of CFTR in pathobiology of cystic fibrosis

It is now widely accepted that CFTR is a non-rectifying Cl^- channel with low (i.e. 7–10 pS) conductance, but its physiological role in the airways remains hotly debated. In the airways, CFTR expression is highest in distal regions of the submucosal glands, where fluid secretion hydrates the mucus and helps to expel it onto the airway surface, and this may explain the early mucus impaction of glands seen in CF infants (Sturgess and Imrie, 1982). In one absorptive model, the ionic strength of airway surface liquid is increased in CF due to diminished salt absorption, which ultimately reduces the killing activity of antibacterial substances on the airway surface (Smith *et al.*, 1996). In another, it has been proposed that pathology in the airways results from Na^+ and fluid hyperabsorption caused by dysregulated Na^+ absorption (Matsui *et al.*, 1998). According to this scheme, CF symptoms are caused by the loss of CFTR protein and its effects on other proteins, particularly the epithelial Na^+ channel, rather than abnormally low chloride conductance. The mechanisms that underlie these regulatory effects of CFTR are not understood, and whether they even occur when the channels are expressed in *Xenopus* oocytes remains highly controversial (Konig *et al.*, 2001; Nagel *et al.*, 2001).

The CFTR Channel Pore

Structure and function of the pore

Since a high-resolution structure of CFTR is not available, the identity of pore-lining amino acids has been inferred by comparing electrophysiological properties of wild-type and mutant channels. A strong case can be made that the sixth transmembrane segment (TMS6) lines the pore, since mutations there affect anion selectivity (Anderson *et al.*, 1991b; Linsdell *et al.*, 1997a), conductance (Sheppard *et al.*, 1993;

Tabcharani *et al.*, 1993), multi-ion pore behavior (Tabcharani *et al.*, 1993), and blocker sensitivities (Linsdell and Hanrahan, 1996; McDonough *et al.*, 1994). Scanning cysteine accessibility mutagenesis (SCAM), in which TMS residues are systematically replaced by single cysteines and the effects of hydrophilic sulfhydryl reagents on permeation are assayed, suggested that TMS 1, 3 and 6 are mostly α-helical and pore-lining (Akabas, 2000). α-Helical structure was also demonstrated in TMS 3 and 4 when they were expressed as a helix-loop-helix construct and studied by circular dichroism, even when solubilized in SDS (Therien *et al.*, 2001). A model in which the amino acids distal to TMS 6 fold back into the membrane was proposed based on the accessibility of cysteines engineered in this region to externally applied hydrophilic methanesulfonate (MTS) reagents (Cheung and Akabas, 1997). Mutations that alter the charge at R352 in the proposed re-entrant loop dramatically alter anion:cation selectivity, consistent with the model (Guinamard and Akabas, 1999). However, further studies are needed as the interpretation of scanning cysteine accessibility results has been questioned, and the assumption that only pore-lining residues are accessible to such reagents has not been verified in other types of channel (Holmgren *et al.*, 1996; Mansoura *et al.*, 1998; Yang *et al.*, 1996). Mutating G314 in TMS 5 and S1118 in TMS 11 affects both anion permeation and channel gating, suggesting that they also contribute to the pore (McCarty, 2000; Zhang *et al.*, 2000).

A model of the CFTR pore has been proposed based on electrophysiological studies (McCarty, 2000), but there has not yet been direct biochemical confirmation for any pore-lining residue and the functional data obtained from some mutants are difficult to reconcile. Deleting the amino terminus and TMSs 1–4 apparently has little effect on channel function when expressed in *Xenopus* oocytes (Carroll *et al.*, 1995), yet there are several disease-associated mutations in this region that alter channel properties when expressed in mammalian cells or oocytes (Akabas *et al.*, 1994; Anderson *et al.*, 1991b; Mansoura *et al.*, 1998). Mutations in TMS 12 affect sensitivity to the open channel blocker diphenylamine-2-carboxylic acid (DPC), suggesting that it lines the pore (McDonough *et al.*, 1994), yet normal-looking channels are recorded when a mutant lacking the C-terminal half (i.e. lacking TMS 7–12 and NBD2) is expressed (Sheppard *et al.*, 1994). It was suggested that the N-terminal half of CFTR can form functional channels by dimerizing. There is no experimental evidence for such homodimers, which would lack a hypothetical salt bridge between R347 in TMS 6 and D924 in TMS 8, which was suggested to be essential for normal conductance (Cotten and Welsh, 1999). TMS 12 probably lines the pore, but its contribution must differ substantially from that of TMS 6 since alanine substitutions at T1134, M1137, N1138, S1141 and T1142 (amino acids that correspond to important residues in TMS 6) have little, or no effect on permeation (Gupta *et al.*, 2001). Flickery block of the CFTR channel by the pH buffer MOPS (3-morpholinopropane-sulfonic acid) suggests that the CFTR pore may have two conformationally distinct open states with different blocker sensitivities (Ishihara and Welsh, 1997). This possibility awaits confirmation using other methods, but it is intriguing and could potentially be used to dissect steps in channel-gating models.

PERMEATION IN THE PORE

CFTR is highly selective for monovalent anions over cations ($P_{Cl}/P_{cation} > 8$–14) (Bear and Reyes, 1992; Tabcharani *et al.*, 1990). Permeability ratios (P_X/P_{Cl}) generally follow the (inverse) lyotropic series, with large, weakly hydrated ions being most permeant. Lyotropic anions such as iodide also bind tightly under some conditions and have given inconsistent results even within laboratories. For example, reported P_I/P_{Cl} values range from near unity (Gray *et al.*, 1990; Kartner *et al.*, 1991), to 0.8 (Linsdell *et al.*, 2000), 0.4 (3; 10; 10; 22; 84; 92) and 0 (Champigny *et al.*, 1990). Inhibition of Cl^- current by I^- has been observed in mixtures of both ions (Tabcharani *et al.*, 1992). The properties of chimeras between human and *Xenopus* CFTR (which has inherently higher P_I/P_{Cl}) suggest that determinants of P_I/P_{Cl} may be situated in TMS 1–6 (Price *et al.*, 1996). When analyzed from the extracellular side using mixed solutions, the permeability ratios for polyatomic anions follow the sequence $NO_3^-(1.73) > Cl^-(1.0) > HCO_3^-(0.25) >>$ gluconate$^-$(0.03) (Gray *et al.*, 1990). A similar sequence $NO_3^- > Cl^- > HCO_3^- >$ formate$^- >$ acetate$^-$ is obtained under bi-ionic conditions, although permeability to large kosmotropic anions was higher from the cytoplasmic than extracellular side (Linsdell *et al.*, 1997a). Indeed, external pyruvate, propanoate, methane sulfonate, ethane sulfonate and gluconate are not measurably permeant (i.e. $P_X/P_{Cl} < 0.06$) when macroscopic currents are measured using excised patches exposed to PKA and ATP. Yet, currents carried by these anions are detectable from the cytoplasmic side (Linsdell and Hanrahan, 1998). The relationship between macroscopic permeability ratio and ion diameter suggests a minimum *functional* diameter of 5.3 Å from the outside (Linsdell *et al.*, 1997a), and ~12 Å from the inside (Linsdell and Hanrahan, 1998; Linsdell *et al.*, 1997a). HCO_3^- is the major ion transported in the pancreatic duct and may also be important in airway and colonic fluid secretion (Quinton, 2001), accounting for about half the anion secretion by airway submucosal glands. HCO_3^- permeates through the CFTR channel pore, though less well than Cl^- (Gray *et al.*, 1990; Hanrahan *et al.*, 1993; Linsdell *et al.*, 1997a; Poulsen *et al.*, 1994). The physiological significance of this route for bicarbonate flux in pancreatic ducts has been questioned (Lee *et al.*, 1999). Interestingly, external HCO_3^- inhibits CFTR-mediated Cl^- current with an apparent K_i of 7 mM (O'Reilly *et al.*, 2000). Such inhibition may be important during pancreatic bicarbonate secretion since it would reduce dissipation of the HCO_3^- gradient when luminal HCO_3^- concentration is elevated to ~150 mM, which is sixfold higher than plasma and >10-fold higher than cytoplasm.

Several lines of evidence suggest that the CFTR pore holds more than one anion simultaneously. The bi-ionic permeability ratio P_I/P_{Cl} is concentration dependent (Tabcharani *et al.*, 1992), a common feature of multi-ion pores (Eisenman and Horn, 1983; Hille, 1992). At least two different electrical distances have been calculated for voltage-dependent block by internal thiocyanate and gluconate respectively (0.2 and 0.4), and block by intracellular gluconate is relieved by raising the external Cl^- concentration, consistent with yet another external anion binding site. When wild-type channels are bathed with Cl^-/SCN^- mixtures, the conductance decreases from 7 to 2 pS as the SCN^- mole fraction is elevated from 0 to 7%, and then increases again as the SCN^- mole fraction is elevated further to 97% (Tabcharani *et al.*, 1993). This anomalous mole fraction effect (AMFE), and voltage-dependent block by cytoplasmic SCN^- are abolished when R347 is mutated to aspartate, and become pH dependent in the R347H mutant. The positive charge on R347 probably does not interact directly with anions since tight binding would slow permeation (e.g. Dutzler *et al.*, 2002). However, it may stabilize a binding site for anions. R347 may also have an impact on NBD function, as the ATPase activity of R347D is significantly reduced compared to wild-type CFTR (Kogan *et al.*, 2001).

The permeability and conductance ratios measured for SCN^-, I^-, and Br^- are very different (Linsdell *et al.*, 1997a). Dawson and colleagues have emphasized the importance of anion binding in CFTR (e.g. Dawson *et al.*, 1999). Conductance ratios are more strongly affected by mutations in TMS 1, 5 and 6 than permeability ratios, and therefore may be more sensitive to pore structure (Mansoura *et al.*, 1998). Mutating K335 near the extracellular end of TMS 6 to a negatively charged glutamate reduces conductance by about 50% (Anderson *et al.*, 1991b; Tabcharani *et al.*, 1993). Covalent modification of cysteines substituted at R334 or K335 with the negatively charged methanethiosulfonate reagent MTSES caused a similar decrease, whereas modification by positively charged reagents had the opposite effect (Smith *et al.*, 2001). Thus, the positive charge associated with R334 and K335 influences conductance, probably by elevating Cl^- concentration near the external mouth of the pore. Permeability ratios depend on barrier heights and the relative ease with which anions can enter the pore from the bulk solution; therefore they are dominated by anion–water interactions (i.e. hydration energy), whereas conductance ratios mainly reflect ion concentrations and ion binding within the pore. Ion permeation is often interpreted by assuming that ions hop between sites, or wells, that are essentially fixed within the pore and are separated by energetically unfavorable regions, or barriers. A four-barrier, three-site (4B3S) model can reproduce experimentally measured current–voltage relationships under a wide range of conditions, including variations in single channel conductance, reversal potentials, block by intracellular gluconate, and AMFEs in mixtures of SCN^- and Cl^- (Linsdell *et al.*, 1997b). Moreover, loss of the AMFE (as seen in the R347D mutant) can be simulated using this 4B3S model by adjusting the well that corresponds to an intracellular SCN^- block and the adjacent barrier. Admittedly such models are highly speculative and require many assumptions. However, it is heartening that most inferences regarding selectivity, multi-ion pore behavior and other biophysical properties of K^+ and Cl^- channels are strongly supported by the recent corresponding crystal structures (Doyle *et al.*, 1998; Dutzler *et al.*, 2002).

Ser/Thr Phosphorylation Activates the Channel

CFTR channels have negligible activity until they are exposed to MgATP and phosphorylated at multiple sites by PKA. This increases their open probability to about 0.4 (Berger *et al.*, 1991; Chang *et al.*, 1993; Cheng *et al.*, 1991; Gadsby and Nairn, 1999b; Neville *et al.*, 1997; Picciotto *et al.*, 1992; Tabcharani *et al.*, 1991). Nine of the 10 predicted dibasic PKA consensus sequences on CFTR (i.e. R, R/K, X, S/T) are in the R-domain between aa ~634 and 830, the main site of phosphorylation control **(Figure 29.5)**. Mutagenesis of the dibasic and most monobasic PKA sites strongly inhibits, but does not abolish, channel activity (Chang *et al.*, 1993; Cheng *et al.*, 1991; Seibert *et al.*, 1995). Thus, PKA regulation is redundant; partial activation persists when most sites are removed in various combinations and no one PKA site seems to be essential for channel activation (Chang *et al.*, 1993; Ostedgaard *et al.*, 2001). Preliminary mass spectroscopy of the R-domain suggests that S768 is phosphorylated first, followed by 700 and 795, then 712 and 737, and finally 660, 670,753 and 813 (Nairn *et al.*, 1996). The significance of this highly ordered phosphorylation is not known, and it is not known whether the same hierarchy occurs under different assay conditions and, most importantly, *in vivo*. Phosphorylation by PKA can alter the secondary structure of recombinant R-domain as evidenced by circular dichroism and mobility in PAGE gels (Dulhanty and Riordan, 1994). The conformational change may be due in large part to phosphorylation of S737, since substituting alanine at this site reduces PKA's effect on electrophoretic mobility (Borchardt *et al.*, 1996). Interestingly, replacing S737 or S768 with alanines *increases* channel activity, suggesting that phosphorylation of these two sites inhibits channel activation (Wilkinson *et al.*, 1997).

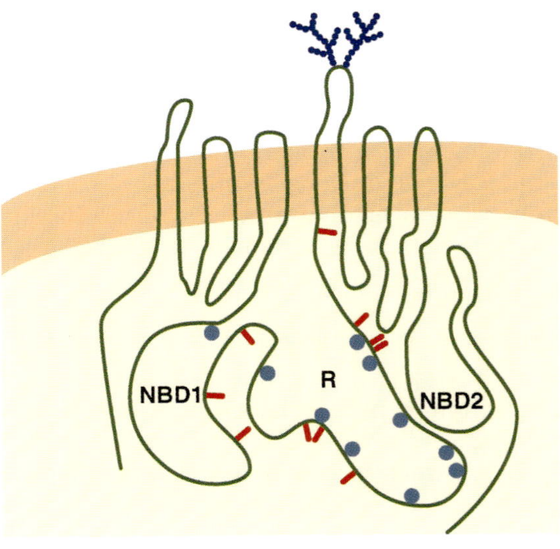

Figure 29.5. ***Phosphorylation control of CFTR. Distribution of consensus sequences for phosphorylation by protein kinases A and C within or neighboring the R-domain. Note the large number of potential sites, which are highly conserved among vertebrates. One site (S686) can be phosphorylated by both kinases. From the N- to C-terminus, the following predicted sites of phosphorylation are indicated: PKC – T582, T604, S641, T682, S686, S707, S790, T791, and S809; PKA – S660, S686, S700, S737, S753, S768, T788, S795, and S813. Only the most preferred PKA and PKC consensus sequences are shown; there are many other potential sites.***

CFTR responses to PKA stimulation are modulated in some complex way by PKC *in vivo* and *in situ*. Acute exposure to PKC or PKC activators enhances the rate and magnitude of subsequent activation by PKA (Button *et al.*, 2001; Dechecchi *et al.*, 1992; Jia *et al.*, 1997; Middleton and Harvey, 1998; Tabcharani *et al.*, 1991; Yamazaki *et al.*, 1999). PKC phosphorylates CFTR, albeit weakly, at nine potential sites distinct from those of PKA **(Figure 29.5)** (Picciotto *et al.*, 1992), but does not alter the secondary structure of recombinant R-domain significantly (Dulhanty and Riordan, 1994). Interestingly, mutating the potential PKC sites S686 and S790 in *Xenopus* CFTR does not alter its modulation by PKC even though they are the main sites phosphorylated by PKC in human CFTR (Button *et al.*, 2001). Stimulating PKC in HT29 intestinal cells does not increase the open probability of CFTR channels but increases their incidence in patches of plasma membrane 16-fold, suggesting that PKC activation may elevate channel density (Bajnath *et al.*, 1993). Recent studies of a mutant lacking all nine PKC sites on the R-domain demonstrate that PKC modulation of channel gating requires direct phosphorylation of CFTR itself rather than an ancillary protein, and at least one of the nine PKC sites also mediates the partial activation induced by PKC alone (Chappe *et al.*, 2002). Removing the PKA and PKC sites may inhibit ATPase activity since phosphorylation and dephosphorylation influences the ATPase activity of CFTR purified from Sf9 insect cells (Li *et al*, 1996). The nucleotide-binding and ATPase activities of phosphorylation site mutants have not yet been assayed biochemically.

CFTR channels deactivate quickly when they are excised from CHO cells (Tabcharani *et al.*, 1991) or airway cells (Becq *et al.*, 1994). This rapid rundown is probably due to dephosphorylation of PKA sites, because it does not occur when the PKA catalytic subunit is present in the bath solution, and channels can be fully restimulated by PKA catalytic subunit if it is added during the first few minutes (Berger *et al.*, 1991; Tabcharani *et al.*, 1991). The rundown has been used to assay the membrane-delimited phosphatases that regulate CFTR. Insensitivity of rundown to okadaic acid and calyculin A suggests that the phosphatase is not a protein phosphatase type 1 or 2A (PP1, PP2A). Rundown does not require exogenous calmodulin or calcium, which excludes PP2B, but is slowed by reducing magnesium to submillimolar concentrations, consistent with a role for a PP2C-like protein phosphatase. Importantly, the deactivation of CFTR-mediated short-circuit current across T84 epithelial cell monolayers (i.e. active Cl^- secretion) after wash-out of agonists is insensitive to calyculin A or okadaic acid at high concentrations (Luo *et al.*, 1998; Travis *et al.*, 1997). Thus, the pharmacological properties of rundown in intact cells are also consistent with a PP2C-like phosphatase and it is likely that the membrane-associated phosphatase is the main one regulating CFTR in intact cells.

CFTR and the PP2C-like phosphatase apparently exist in a complex that facilitates down-regulation of channel activity. A monoclonal anti-CFTR antibody co-precipitates PP2C from baby hamster kidney (BHK) cells stably expressing CFTR without co-precipitating PP1, PP2A or PP2B (Zhu *et al.*, 1999). Similarly, a polyclonal anti-PP2C antibody co-precipitates CFTR from BHK membrane extracts. A specific association between CFTR and a PP2C was confirmed by chemical crosslinking with dithiobis sulfosuccinimidyl propionate (DTSSP), a hydrophilic, cleavable, bifunctional reagent. Exposing CFTR-containing BHK cell lysates to DTSSP cross-linked histidine-tagged CFTR with a PP2C-like phosphatase, with both proteins appearing in the same elution fraction during Ni^{2+}-NTA agarose chelate chromatography (Dahan *et al.*, 1999; Zhu *et al.*, 1999). Crosslinking was specific for PP2C over PP1, PP2A and PP2B, which do not co-purify with CFTR. The CFTR–PP2C association has also been observed recently in pull-down assays in the absence of crosslinker (Zhu *et al.*, 1999). The association has only been studied in unstimulated cells, though it is conceivable that the complex between CFTR and PP2C is itself regulated by PKA or other factors.

CFTR BINDS AND HYDROLYZES ATP

While the first level of control of the CFTR channel is by phosphorylation, the second level is by ATP binding and hydrolysis at the NBDs although not necessarily in the identical manner as occurs in other ABC proteins. In fact, the extent to which nucleotide binding and hydrolysis have been characterized is still rather limited due to difficulties in obtaining active purified protein. Even isolated NBDs expressed heterologously in bacteria have been assayed primarily as fusions with other bacterial proteins to overcome problems of insolubility (Hartman *et al.*, 1992; Howell *et al.*, 2000; Ko *et al.*, 1993; Randak *et al.*, 1997).

THE ISOLATED NBDs

Data from experiments assessing nucleotide binding and hydrolysis by these individual NBD constructs has not yet provided a complete picture of the events occurring at each domain. Most information is available for NBD1, which when obtained and assayed as a fusion with glutathione *S*-transferase (GST) (Howell *et al.*, 2000) or maltose-binding protein (MBP) (Ko *et al.*, 1993) exhibits photolabeling by [α-^{32}P]-8-N_3ATP that can be displaced by ATP in the mM range, binding to ATP-agarose (Hartman *et al.*, 1992), binding of TNP-ATP with a K_d in the low μM range, and varying levels of ATPase activity (Duffieux *et al.*, 2000). Values reported for the K_m of hydrolysis range from 60 μM (Howell *et al.*, 2000) to 250 μM (Duffieux *et al.*, 2000; Ko and Pedersen, 1995) and for V_{max} from 30 (Ko and Pedersen, 1995) to 500 (Duffieux *et al.*, 2000) nmol mg^{-1} min^{-1}. These large ranges are presumably due to differences in construct domain boundaries, methods of expression, purification, solubility and assay.

One of the most recent preparations, which was not produced as a fusion with another protein, was obtained in soluble form without denaturation and renaturation and with improved domain boundaries, and hydrolyzed ATP at the highest rate (Duffieux *et al.*, 2000). However, the authors acknowledge that the preparation was approximately 90% pure and hence the possible contribution of contaminating proteins is not excluded. It is unlikely that the purity of any of the other preparations was higher. The best evidence that hydrolysis was catalyzed by isolated NBD1 is provided by the fact that parallel preparations with substitutions of the Walker A lysine residue had greatly reduced activity (Ko and Pedersen, 1995). Notably, however, this wild-type preparation had the lowest specific activity reported (30 nmol mg^{-1} min^{-1}).

In all of these investigations, authors have emphasized the difficulty of obtaining the NBD1 polypeptide in soluble, monodisperse form. It remains unclear if this is due primarily to the choice of inappropriate domain boundaries, a requirement for interaction with other domains in CFTR or other proteins, or other factors not adequately provided by bacterial hosts. However, this difficulty in generating a soluble native structure in isolation may not be surprising in view of the extreme susceptibility of this domain to even relatively minor mutagenic modifications (see below).

There have been far fewer studies of the isolated second NBD (Randak *et al.*, 1995, 1996, 1997). Assayed as an uncleaved GST fusion, NBD2 bound TNP-ATP with a somewhat lower affinity than NBD1; TNP-GTP was bound more strongly. As the amount of ADP produced in an ATPase assay was increased by added AMP, it was concluded that this domain had adenylate kinase as well as ATPase activity. However, there have been no other reports of either the isolated NBDs or the whole CFTR protein exhibiting nucleotide metabolizing activities other than ATPase.

THE INTACT CFTR PROTEIN

Turning to the intact CFTR protein, the practical problems of obtaining pure soluble native NBDs are replaced by another set of limitations stemming from the difficulty in obtaining high level expression of a form that can be purified in a functionally competent state. For this reason, there have been few studies of nucleotide binding and hydrolysis by the whole purified protein. To avoid the need for purified and reconstituted protein, photoaffinity labeling experiments using azido-nucleotides have been performed with membranes isolated from cells expressing CFTR. Importantly, 8-azido ATP supports CFTR channel gating at least as effectively as unmodified ATP. [α-^{32}P]-8-N_3ATP labeled CFTR in membranes of insect cells in which it had been expressed (Travis *et al.*, 1993), with more labeling at NBD1 than NBD2 (Szabo *et al.*, 1999). This asymmetry of the domains with respect to nucleotide reactivity was emphasized by the very much higher affinity of the nonhydrolyzable AMP-PNP for NBD1 than NBD2 (Aleksandrov *et al.*, 2001). Significantly, labeling of NBD1 does not require vanadate, which is normally responsible for the trapping of N_3ADP after N_3ATP hydrolysis by some other ATPases (Urbatsch *et al.*, 1995). Furthermore, labeling of NBD1 is not completely dependent on a divalent cation such as Mg^{2+}, whereas NBD2 labeling is (Aleksandrov *et al.*, 2002). However, experiments with [α-^{32}P]-8-N_3ATP may indicate either its own photo-binding or that of the hydrolysis product [α-^{32}P]-8-N_3ADP. In contrast, [γ-^{32}P]-8-N_3ATP-labeling reflects only binding of the nucleoside triphosphate. In this case NBD2 is labeled only if hydrolysis is prevented (e.g. at 0°C); under hydrolysis conditions it is not labeled. N_3ADP binds both domains but dissociates rapidly from NBD2 whereas it is retained at NBD1. Mutagenesis of Walker A lysines that interact with the β-phosphates of the nucleotides prevents labeling of the domain in which the substitution is made, but not in the other. Overall, these studies indicate that NBD1 is primarily a site of stable nucleotide binding whereas NBD2 is a site of rapid hydrolysis and product dissociation. While these photolabeling experiments provide interesting insights into nucleotide interactions at CFTR NBDs, they do not provide quantitative binding constants or kinetic parameters characterizing hydrolysis.

Direct assays of ATP hydrolysis using CFTR purified from insect cells, revealed K_m and V_{max} values of approximately 1 mM and 50 nmol mg^{-1} min^{-1}, respectively (Li *et al.*, 1996). The K_m was reduced about threefold by phosphorylation of the protein by PKA, and a disease-associated mutation in the NBD signature motif, G551D, greatly reduced hydrolysis. As another means of addressing the relative contribution of the two NBDs to the overall hydrolysis, the Walker A lysine mutants K464A and K1250A were purified and assayed

(Ramjeesingh *et al.*, 1999). Either mutation reduced the hydrolytic specific activities by at least 20-fold, with the K1250A variant appearing essentially inactive. This would seem to indicate that both domains are required for catalysis. When purified by different methods from mammalian cell membranes, CFTR was found to hydrolyze at a similar low rate of approximately 40 nmol $mg^{-1} min^{-1}$ (Aleksandrov *et al.*, 2002).

It is apparent that integration of the results from studies with isolated NBDs and the intact CFTR protein neither reveal an entirely coherent picture nor provide a clear understanding of the steps involved in nucleotide binding and hydrolysis. It is of course essential that this picture be clarified before the adaptation of the ABC structural architecture to the role of nucleotide-regulated ion channel can be better understood.

REGULATION OF CHANNEL GATING BY PURINE NUCLEOTIDES

CFTR chloride channel activity was amenable to electrophysiological methods for many years before biochemical assays of the interactions of nucleotides were performed. Consequently most of the evidence supporting models for nucleotide gating comes from patch clamp and bilayer studies that have been comprehensively reviewed (Gadsby and Nairn, 1999a; Nagel, 1999; Sheppard and Welsh, 1999; Zou and Hwang, 2001).

ATP HYDROLYSIS MAY NOT BE ESSENTIAL FOR CHANNEL ACTIVATION

In fact, the ability to precisely monitor the gating transitions of a single CFTR channel is powerful as it enables a more direct quantitative readout of protein function than do assays of transport by other ABC transporters. Initially it was found that MgATP or other purine nucleoside triphosphates were required to activate PKA phosphorylated channels (Anderson *et al.*, 1991a). This was readily reconcilable with the relatively low substrate specificity of other ABC ATPases and hence was consistent with the idea that hydrolysis was related to gating. Non-hydrolyzable ATP analogues such as AMP-PNP were found to be ineffective in initiating channel gating, although this finding was controversial (Carson and Welsh, 1993; Quinton and Reddy, 1992) and more recently was in fact shown to be effective under somewhat different conditions than those pertaining to ATP (Aleksandrov *et al.*, 2000). Nevertheless, the idea that hydrolysis was essential for channel opening became widely accepted and it was postulated that the gating transitions were not in thermal equilibrium, but rather dependent on the input of energy from ATP hydrolysis (Hwang *et al.*, 1994). This implied a mechanism unique among known ion channels, in which initiation of gating results from a structural perturbation caused by either ligand binding or an electrical field. This concept of energetic coupling between ATP hydrolysis and channel gating, either explicitly or implicitly espoused by many authors (Baukrowitz *et al.*, 1994; Carson and Welsh, 1993; Carson *et al.*, 1995; Gunderson and Kopito, 1994), was extended to suggest that hydrolysis at NBD1 and NBD2 drove channel opening and closing, respectively (Gadsby and Nairn, 1999a; Welsh *et al.*, 1998). This seemed a logical rationalization of ABC domain structure and ion channel gating transitions. The evidence supporting such a model, however, was less than compelling and did not include any direct measurements of ATP binding or hydrolysis as outlined in the previous section.

We briefly consider from the present perspective each piece of evidence favoring the interpretation (Gadsby and Nairn, 1999a). First, it was mentioned that since most disease-associated mutations are in NBD1, it might be the domain involved in channel opening. Although not a strong argument in itself, the predominance of NDB1 mutations reflected early observations. Mutations are now, of course, known to be broadly distributed across the entire sequence. Second, the notion that NBD1 mutations favor the closed state and NBD2 mutations the open state is not generally correct. The fact that the mutation of K1250, the Walker A lysine in NBD2, can still open was taken as evidence that NBD1 rather than NBD2 is responsible for opening. However, the fact that the corresponding NBD1 mutation, K464A, had remarkably little effect on opening did not seem to dissuade proponents of this general model. Third, a major tenet of the NBD1-open, NBD2-closed model was the ability of non-hydrolyzable AMP-PNP to prolong the channel open state when added in the presence of ATP (Carson and Welsh, 1993). This was postulated to occur by high-affinity binding to NBD2 and inhibition of hydrolysis there, which in one version of the model is required for the dissociation of the product of hydrolysis products from NBD1, which is necessary for channel closing (Baukrowitz *et al.*, 1994; Hwang *et al.*, 1994).

Variations on the above scheme have been based on the effects of nucleotides on the response of either macroscopic or single channel currents and implicitly incorporate the idea of alternating site catalysis of the NBDs analogous to the behavior of Pgp (Senior and Gadsby, 1997). As outlined in the previous section, there is now considerable evidence that this analogy is not appropriate. Indeed, some early observations were inconsistent with a requirement for hydrolysis for channel opening. These included channel gating in the absence of a divalent cation such as Mg^{2+} (Aleksandrov *et al.*, 2000; Ikuma and Welsh, 2000; Schultz *et al.*, 1996), which is absolutely essential for hydrolysis by all known ATPases including CFTR (Aleksandrov *et al.*, 2002), and the ability of high concentrations of AMP-PNP to support channel gating (Aleksandrov *et al.*, 2000; Quinton and Reddy, 1992). Although this latter effect was attributed entirely to the presence of contaminating ATP in the AMP-PNP preparations (Carson and Welsh, 1993), this has since been shown not to be the case (Aleksandrov *et al.*, 2000). Confirmation that either AMP-PNP or ATP can support gating in the absence of Mg^{2+} (Aleksandrov *et al.*, 2000) precludes an energetic coupling between hydrolysis and gating.

Thermodynamic analysis of the gating transitions revealed a high energy of activation of opening but a low value similar to that of diffusion in water for closing (Aleksandrov and Riordan, 1998; Mathews *et al.*, 1998). The energy input, however, reflects the entropy change on the binding of nucleotide or formation of the transition state for hydrolysis, and is not related to the energy of the γ-phosphate bond of ATP. Hence there is structural coupling between the binding event and the initiation of gating that is quite analogous to the conformational triggering of other ligand-gated channels. The ensuing opening and closing transitions are at thermal equilibrium. As is outlined in more detail below, the rate-limiting step in termination of gating is likely to be the dissociation of nucleotide, which could occur before hydrolysis, but is presumed to be much more efficient after cleavage of the γ-phosphate. This interpretation does not imply that ATP hydrolysis is unimportant to CFTR channel function, but rather views its primary role as providing efficient reversibility of the gating cycle.

THE REGULATION OF GATING IS PRIMARILY CONTROLLED BY NBD2

The contribution of each NBD in these events is not yet entirely clear but the studies of nucleotide binding and hydrolysis summarized in the previous section enable a more informed interpretation of their roles than was possible previously. Although one study of CFTR purified from insect cells found that substitution of Walker A lysines in either domain greatly diminished ATP hydrolysis by the whole molecule (Ramjeesingh *et al.*, 1999), recent experiments with CFTR in mammalian cell membranes indicated much more rapid hydrolysis at NBD2 than NBD1 (Aleksandrov *et al.*, 2002). Requirements for hydrolysis at NBD2 paralleled those for hydrolysis by the whole purified protein. Moreover, there is good correspondence between the influence of NBD2 mutations on ATP binding and hydrolysis there (Aleksandrov *et al.*, 2002) and on channel gating. That is, K1250A prevents both binding and hydrolysis at NBD2 and greatly slows both opening and closing. In contrast, K464A, which also abolishes binding at NBD1, only slightly reduces the channel opening rate. At this point it is important to mention that although 3-D structures of NBDs of some ABC transporters and related proteins such as RAD-50 (Hopfner *et al.*, 2000) have suggested interactions between NBDs (see also Chapter 6), this has not been formally demonstrated in CFTR. Allosteric coupling between NBDs has been demonstrated functionally, with at least one other ABCC protein, MRP1 (Gao *et al.*, 2000; Hou *et al.*, 2000; Nagata *et al.*, 2000), though not yet with CFTR. While NBD2 appears to be most directly linked to channel gating, it is reasonable to assume NBD1 must play some role. The strongest indication that it does is provided by the effect of mutations such as G551D that almost completely prevent gating. In fact, this mutation may impact the nucleotide interaction at NBD2 more than at NBD1 and hence its effect on gating would be via NBD2. Such an interpretation would be reasonable if CFTR NBDs formed a nucleotide sandwich as in RAD-50 (Hopfner *et al.*, 2000). In this model the LSGGQ motif of one NBD plays some role in the nucleotide interaction at the other domain. Since the exact consensus of this motif is fulfilled in NBD1 it may be well suited to participate in the nucleotide complex at NBD2, which appears to have most direct influence on the channel.

While photolabeling experiments have revealed a much stronger interaction of AMP-PNP with NBD1 than NBD2, it is still not known which of the following three effects of AMP-PNP are due to this high-affinity interaction: (1) reduction of the opening rate, (2) reduction of the closing rate, and (3) ability to open the channel in the absence of ATP. In fact, all of these effects apparently require concentrations higher than those necessary to saturate NBD1 (Aleksandrov *et al.*, 2002).

Although we are still far from completely understanding nucleotide regulation of the channel, some aspects have been clarified. It is very clear that the two NBDs are non-equivalent in many respects. The principal nucleotide interaction at NBD1 involves stable binding of nucleoside triphosphate that is independent of vanadate. This binding does not appear to be crucial for channel gating because it is prevented by the K464A mutation, which only slightly slows channel opening. MgATP is bound, hydrolyzed and the products rapidly released at NBD2 of wild-type but not K1250A, which severely slows channel opening and closing, implying a direct role of this domain in gating. Some role for ATP at NBD1 is anticipated but is not yet clear. Unlike other studies (Zou *et al.*, 2001) recent experiments indicate that binding at NBD1 and hydrolysis at NBD2 are not strictly interdependent, K464A abrogates the former without influencing the latter, and K1250A has just the opposite effect (Aleksandrov *et al.*, 2002). This does not exclude the possibility of more elaborate allosteric interactions between the domains, which would require other kinds of experiments to detect. In our current working hypothesis, CFTR is a ligand-gated channel in which reversibility of the gating process is made highly efficient by hydrolysis of the substrate. Channel opening occurs due to a change in the configuration of the closed state as occurs on ligand binding by other ligand-gated channels. It is not yet known if under physiological conditions, the conformational perturbation that is rate limiting for initiation of gating is due to the initial binding or formation of the transition state for hydrolysis. In this model, hydrolysis and product dissociation are necessary for termination of gating and the latter step appears to be rate limiting. Current evidence indicates that NBD2 is the more active player in this entire cycle. The role of NBD1 remains to be more clearly elucidated but ATP binding there may influence the frequency of gating cycles.

CFTR MATURATION AND TRAFFICKING IN THE SECRETORY PATHWAY

Attention was drawn to this topic because most CF patients have at least one copy of the ΔF508 mutation, located in the helical domain of the NBD, which causes CFTR to fold inefficiently and be retained in the ER. Thus, although there are many other mutations that compromise aspects of CFTR function as discussed above (gating, permeation, etc.), most CF is due to the failure of the protein to be transported to the apical membrane, where it is required for chloride conductance **(Figure 29.6)**. Like other membrane and secreted glycoproteins **(Figure 29.7)**, CFTR is core-glycosylated co-translationally

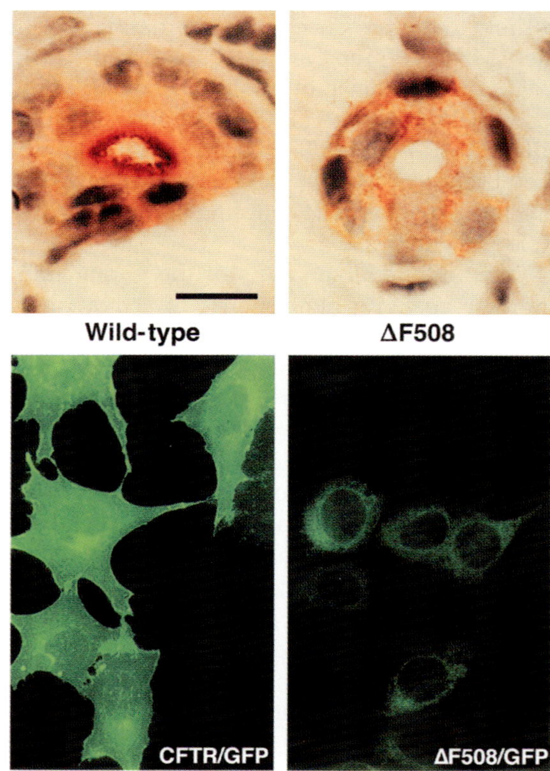

*Figure 29.6. Immunostaining of CFTR in sweat ducts in cryosections of skin biopsies from a non-CF individual and a homozygous **ΔF508** patient (upper panels). Fluorescence microscopy of wild-type and **ΔF508** green fluorescent protein fusions heterologously expressed in BHK-21 cells (lower panels).*

in the ER, from which it is exported in COP II-coated vesicles to vesicular-tubular clusters. From there the protein is transferred to the Golgi apparatus, where oligosaccharide chain trimming and extension occurs to form complex structures (Kopito, 1999; Riordan, 1999). When first expressed heterologously in mammalian cells, ΔF508 CFTR was observed to be synthesized as a full-length core-glycosylated molecule, which did not exit the ER or acquire complex oligosaccharide chains (Cheng *et al.*, 1990). Initially it was not known if the lack of carbohydrate processing was the cause, or the result, of mislocalization. However, there is

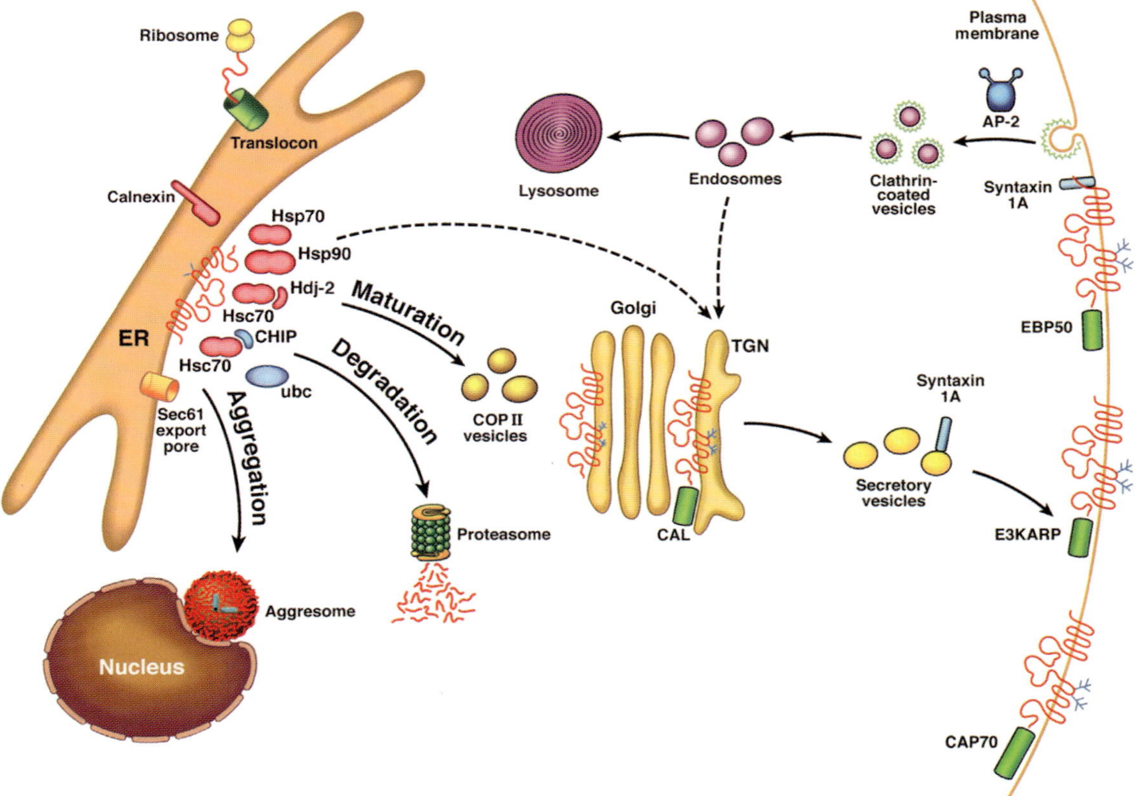

Figure 29.7. Schematic representation of CFTR processing and trafficking. CFTR molecules are synthesized on ER-associated ribosomes. Core oligosaccharide chains are attached while CFTR is incorporated into the ER membrane. At the ER the lumenal chaperone calnexin and the cytosolic chaperones Hsp90, Hsp70 and the Hsc70/Hdj-2 or Hsc70/CHIP complex interact transiently with unfolded or partially folded intermediates. Mature CFTR molecules are exported from the ER and are transported to the Golgi apparatus by COP II vesicles. An alternative pathway for mature CFTR molecules bypassing COP II vesicles and early Golgi compartments has been described recently (Yoo et al.*, 2002). The maturation of CFTR polypeptides is reflected by the attachment of complex oligosaccharide chains in the Golgi. Secretory vesicles deliver CFTR from the* trans*-Golgi network (TGN) to the plasma membrane. The PDZ-domain-containing protein CAL binds to the C-terminus of CFTR at the Golgi apparatus and may favor retention of CFTR. Other PDZ-domain proteins (EBP50, E3KARP, CAP70) bind to CFTR to anchor the protein to the plasma membrane or tether it in regulatory complexes. Syntaxin 1A, a component of the membrane trafficking machinery, binds CFTR and inhibits channel activity. Binding to the clathrin adaptor complex AP-2 mediates endocytosis in cathrin-coated vesicles. Endocytosed CFTR can be recycled or degraded in the lysosome. Molecules that fail to fold correctly at the ER are ubiquitinated and retrotranslocated through the Sec61 export pore. Ubiquitination occurs with the help of ubiquitin conjugating enzymes (ubc) and ubiquitin ligases, such as CHIP. In the cytosol, the ubiquitinated molecules are subjected to degradation by the proteasome. This pathway eliminates not only misfolded mutant CFTR, but also about 70% of wild-type proteins that do not attain a native conformation. When degradation by the proteasome is inhibited or saturated, aggregation of export-incompetent CFTR molecules can occur. The aggregated molecules are transported to structures named aggresomes near the centriole.*

now strong evidence in support of the latter possibility and the primary consequence of the absence of phe508 is defective folding of the polypeptide (Qu and Thomas, 1996). This behavior of the ΔF508 molecule, and indeed many other variants with single amino acid substitutions or short insertions or deletions in cytoplasmic domains, has been confirmed in all mammalian cell expression systems tested (Kopito, 1999; Riordan, 1999). In amphibian (Drumm *et al.*, 1991) and insect (Li *et al.*, 1993) cell systems, however, both maintained at lower temperatures than mammalian cells, there is partial maturation and transport of the ΔF508 protein to the cell surface. Thus, the maturation is at least partially temperature sensitive for folding (Denning *et al.*, 1992).

Other means of promoting maturation are being intensively sought since development of a small molecule drug with this effect would provide a potential therapeutic strategy. Some compounds at high concentrations, including osmolytes such as glycerol, can slightly improve maturation but are impractical for use *in vivo* (Brown *et al.*, 1996). Preliminary clinical trials with phenyl butyrate, which is already used in the treatment of thalassemia (Zeitlin, 2000), have been undertaken with CF patients on the basis of reports that it improves maturation in cell culture systems (Rubenstein and Zeitlin, 2000). In several other instances, where misfolding of mutant proteins is at the basis of other genetic diseases, the binding of high-affinity ligands has been discovered to improve folding and at least partially rescue the mutant phenotype (Fan *et al.*, 1999; Foster *et al.*, 1999; Klabunde *et al.*, 2000; Morello *et al.*, 2000). Unfortunately, with one possible exception (Dormer *et al.*, 2001), this approach has not yet been feasible because of a paucity of compounds known to specifically bind CFTR with high affinity (Schultz *et al.*, 1999), although high-throughput screens for such reagents (Galietta *et al.*, 2001) may still identify such drugs. It is ironic that misfolding of other ABCC proteins such as Pgp caused by *in vitro* mutagenesis can be somewhat aleviated by drug binding (Loo and Clarke, 2000); unfortunately it is not the multidrug resistance phenotype that needs to be rescued.

Any such pharmacological approach to the rescue of ΔF508 and other processing mutants in patients rests on the assumption that the abortive biogenesis observed in cultured cells also occurs in the epithelial cells of the affected tissues in patients. The failure of the ΔF508 protein to reach the apical membrane of sweat duct cells was clearly demonstrated (Kartner *et al.*, 1992), but conflicting results have been reported in intestinal and airway cells (Bronsveld *et al.*, 2001; Kalin *et al.*, 1999). The low abundance of CFTR in the latter makes the resolution of this conflict very demanding. However, the recent application of especially specific, high-affinity antibodies, which are capable of detecting mature and immature CFTR in biopsy and transplant specimens, detected both forms in non-CF tissues, but only the immature form in ΔF508 homozygotes (Kreda *et al.*, 2001).

Mechanistically the very stringent localization quality control applied to CFTR is not well understood at the molecular or the cellular level. The key issue to be understood is the recognition mechanism that distinguishes the mutant and wild-type nascent chains. Unraveling the steps in this identification process is complicated by the fact that even wild-type CFTR matures inefficiently; as little at 25% of the nascent chain synthesized is converted to the mature form with the remainder degraded at the ER by the 26S proteasome (Gelman *et al.*, 2002; Jensen *et al.*, 1995; Lukacs *et al.*, 1994; Ward *et al.*, 1995). With mutants such as ΔF508 this percent maturation is reduced to zero. Ubiquitination occurs even before synthesis of the entire polypeptide is complete (Sato *et al.*, 1998). Multiple molecular chaperones on both sides of the ER membrane interact with the nascent chain and there is evidence that they facilitate folding and also direct CFTR to the ubiquitin-proteasomal pathway (Loo *et al.*, 1998; Meacham *et al.*, 1999, 2001; Yang *et al.*, 1993). The calnexin/UDP glucose-glycosyl transference conformation-sensing mechanism (Ritter and Helenius, 2000) does not appear to play an important role, since preventing the interaction of calnexin with CFTR neither prevents wild-type CFTR from maturing nor promotes maturation of ΔF508 (Pind *et al.*, 1994). Several cytoplasmic chaperones do have an impact, especially Hsp70 and its co-chaperones (Meacham *et al.*, 1999, 2001), and Hsp90 (Loo *et al.*, 1998).

That the principal recognition events occur on the cytoplasmic rather than the lumenal side of the ER membrane is supported by the finding that the majority of disease-associated mutations in cytoplasmic domains cause misprocessing while none of those analyzed in extracytoplasmic loops have this effect (Hämmerle *et al.*, 2001; Seibert *et al.*, 1996a, 1996b, 1997). The related fact that different missense mutations across the entire cytoplasmic face of the protein prevent maturation indicates that achievement of a global native structure is required for ER export. However, since it is unlikely that tertiary structure can be detected directly by the cellular machinery it may be that discrete short sequence motifs that are either exposed or buried are the actual recognition elements. Positive export signals consisting of a short acidic consensus, DXE, play such a role in some secretory proteins (Nishimura *et al.*, 1999) and could do so in CFTR. Negative retention/retrieval signals, arginine-framed tripeptides discovered in the channel and ABC protein components of K_{ATP} (Zerangue *et al.*, 1999), apparently do contribute to the ER retention of ΔF508 (Chang *et al.*, 1999). The receptors for these positive and negative traffic signals have not yet been identified.

The vesicular trafficking events that move CFTR through proximal steps in the secretory pathway are apparently conventional whereas late steps may be novel at more distal stages (Bannykh *et al.*, 2000; Yoo *et al.*, 2002). Thus, COP II vesicles are responsible for the ER export of CFTR as with other secretory proteins. However, movement to the Golgi may take a less-well-characterized route that is not blocked by inhibitors of several of the small GTPases and the syntaxin 5 SNARE protein, which are essential for the conventional pathway to the *cis*-Golgi (Yoo *et al.*, 2002). Rather, the apparent involvement of the endosomal t-SNARE syntaxin 13 suggests that CFTR maturation may involve movement through the *trans*-Golgi-endosomal pathway (Yoo *et al.*, 2002). It is not yet known if this routing contributes to the fragility of wild-type CFTR in the secretory pathway and the complete inability of many mutants to successfully traverse it.

Disease-associated mutations in other human ABC proteins also cause misfolding and failure to be transported out of the ER. Examples that have been documented are listed in **Table 29.1**. However, many more will probably emerge as the mutations detected in patients are studied in heterologous expression systems. The extensive research on this aspect of cystic fibrosis, while not yet yielding an effective treatment, provides a helpful framework for progress towards this end in other ABC protein diseases.

TABLE 29.1. OTHER MISFOLDED MUTANT ABC PROTEINS IN HUMAN DISEASE

Protein	ABC family member	Disease
ABC1	ABCA1	Familial high-density lipoprotein deficiency (FHD)[a]
ABCR	ABCA4	Stargardt disease[b]
MDR3	ABCB4	Progressive familial intrahepatic cholestasis type 3 (PFIC) Intrahepatic cholestasis of pregnancy (ICP)[c]
MRP2	ABCC2	Dubin–Johnson Syndrome[d]
SUR1	ABCC8	Persistent hyperinsulinemic hypoglycemia of Infancy (PHHI)[e]
ALD	ABCD1	X-linked adrenoleukodystrophy[f]

[a]Brooks-Wilson *et al.*, 1999.
[b]Lewis *et al.*, 1999.
[c]Dixon *et al.*, 2000.
[d]Keitel *et al.*, 2000.
[e]Cartier *et al.*, 2001.
[f]Smith *et al.*, 1999.

REFERENCES

Aguilar-Bryan, L., Clement, J.P.t., Gonzalez, G., Kunjilwar, K., Babenko, A. and Bryan, J. (1998) Toward understanding the assembly and structure of KATP channels. Physiol. Rev. 78(1), 227–245.

Akabas, M.H. (2000) Cystic fibrosis transmembrane conductance regulator. Structure and function of an epithelial chloride channel. J. Biol. Chem. 275(6), 3729–3732.

Akabas, M.H., Kaufmann, C., Cook, T.A. and Archdeacon, P. (1994) Amino acid residues lining the chloride channel of the cystic fibrosis transmembrane conductance regulator. J. Biol. Chem. 269(21), 14865–14868.

Aleksandrov, A.A. and Riordan, J.R. (1998) Regulation of CFTR ion channel gating by MgATP. FEBS Lett. 431(1), 97–101.

Aleksandrov, A.A., Chang, X., Aleksandrov, L. and Riordan, J.R. (2000) The non-hydrolytic pathway of cystic fibrosis transmembrane conductance regulator ion channel gating. J. Physiol. 528 (Pt 2), 259–265.

Aleksandrov, L., Mengos, A., Chang, X., Aleksandrov, A. and Riordan, J.R. (2001) Differential interactions of nucleotides at the two nucleotide binding domains of the cystic fibrosis transmembrane conductance regulator. J. Biol. Chem. 276(16), 12918–12923.

Aleksandrov, L., Aleksandrov, A.A., Chang, X.-B. and Riordan, J.R. (2002) The first nucleotide binding domain of CFTR is a site of stable nucleotide interaction whereas the second is a site of rapid turnover. J. Biol. Chem. 3, 15419–15425.

Anderson, M.P., Berger, H.A., Rich, D.P., Gregory, R.J., Smith, A.E. and Welsh, M.J. (1991a) Nucleoside triphosphates are required to open the CFTR chloride channel. Cell 67, 775–784.

Anderson, M.P., Gregory, R.J., Thompson, S., Souza, D.W., Paul, S., Mulligan, R.C., Smith, A.E. and Welsh, M.J. (1991b) Demonstration that CFTR is a chloride channel by alteration of its anion selectivity. Science 253, 202–205.

Anderson, M.P., Rich, D.P., Gregory, R.J., Smith, A.E. and Welsh, M.J. (1991c) Generation of cAMP-activated chloride currents by expression of CFTR. Science 251, 679–682.

Arsenault, A.L., Ling, V. and Kartner, N. (1988) Altered plasma membrane ultrastructure in multidrug-resistant cells. Biochim. Biophys. Acta. 938(2), 315–321.

Bajnath, R.B., Groot, J.A., De Jonge, H.R., Kansen, M. and Bijman, J. (1993) Synergistic activation of non-rectifying small-conductance chloride channels by forskolin and phorbol esters in cell-attached patches of the human colon carcinoma cell line HT-29cl.19A. Pflugers Arch. 425(1–2), 100–108.

Bannykh, S.I., Bannykh, G.I., Fish, K.N., Moyer, B.D., Riordan, J.R. and Balch, W.E. (2000) Traffic pattern of cystic fibrosis transmembrane regulator through the early exocytic pathway. Traffic 1(11), 852–870.

Baukrowitz, T., Hwang, T.C., Nairn, A.C. and Gadsby, D.C. (1994) Coupling of CFTR Cl^- channel gating to an ATP hydrolysis cycle. Neuron 12(3), 473–482.

Bear, C.E. and Reyes, E.F. (1992) cAMP-activated chloride conductance in the colonic cell line, Caco-2. Am. J. Physiol. 262(1 Pt 1), C251–C256.

Bear, C.E., Duguay, F., Naismith, A.L., Kartner, N., Hanrahan, J.W. and Riordan, J.R. (1991) Cl^- channel activity in *Xenopus* oocytes expressing the cystic fibrosis gene. J. Biol. Chem. 266(29), 19142–19145.

Bear, C.E., Li, C., Kartner, N., Bridges, R.J., Jensen, T.J., Ramjeesingh, M. and Riordan, J.R. (1992) Purification and functional reconstitution of the cystic fibrosis transmembrane conductance regulator (CFTR). Cell 68, 809–818.

Becq, F., Jensen, T.J., Chang, X.B., Savoia, A., Rommens, J.M., Tsui, L.C., Buchwald, M., Riordan, J.R. and Hanrahan, J.W. (1994) Phosphatase inhibitors activate normal and defective CFTR chloride channels. Proc. Natl Acad. Sci. USA 91(19), 9160–9164.

Berger, A.L. and Welsh, M.J. (2000) Differences between cystic fibrosis transmembrane conductance regulator and HisP in the interaction with the adenine ring of ATP. J. Biol. Chem. 275(38), 29407–29412.

Berger, A.L., Ikuma, M., Hunt, J.F., Thomas, P.J. and Welsh, M.J. (2002) Mutations that change the position of the putative gamma-phosphate linker in the nucleotide binding domains of CFTR alter channel gating. J. Biol. Chem. 277(3), 2125–2131.

Berger, H.A., Anderson, M.P., Gregory, R.J., Thompson, S., Howard, P.W., Maurer, R.R., Mulligan, R., Smith, A.E. and Welsh, M.J. (1991) Identification and regulation of the CFTR-generated chloride channel. J. Clin. Invest. 88, 1422–1431.

Borchardt, J., Kole, J. and Cohn, J. (1996) Phosphorylation of CFTR Ser-737 by protein kinase A. Pediatr. Pulm. (Suppl. 13), 212 (Abstract).

Boucher, R.C. (1994) Human airway ion transport. Part two. Am. J. Respir. Crit. Care Med. 150(2), 581–593.

Bronsveld, I., Mekus, F., Bijman, J., Ballmann, M., de Jonge, H.R., Laabs, U., *et al.* (2001) Chloride conductance and genetic background modulate the cystic fibrosis phenotype of Delta F508 homozygous twins and siblings. J. Clin. Invest. 108(11), 1705–1715.

Brooks-Wilson, A., Marcil, M., Clee, S.M., Zhang, L.H., Roomp, K., van Dam, M., *et al.* (1999) Mutations in ABC1 in Tangier disease and familial high-density lipoprotein deficiency. Nat. Genet. 22(4), 336–345.

Brown, C.R., Hong-Brown, L.Q., Biwersi, J., Verkman, A.S. and Welch, W.J. (1996) Chemical chaperones correct the mutant phenotype of the delta F508 cystic fibrosis transmembrane conductance regulator protein. Cell Stress Chaperon. 1(2), 117–125.

Button, B., Reuss, L. and Altenberg, G.A. (2001) PKC-mediated stimulation of amphibian CFTR depends on a single phosphorylation consensus site: insertion of this site confers PKC sensitivity to human CFTR. J. Gen. Physiol. 117(5), 457–468.

Carroll, T.P., Morales, M.M., Fulmer, S.B., Allen, S.S., Flotte, T.R., Cutting, G.R. and Guggino, W.B. (1995) Alternate translation initiation codons can create functional forms of cystic fibrosis transmembrane conductance regulator. J. Biol. Chem. 270(20), 11941–11946.

Carson, M.R. and Welsh, M.J. (1993) 5′-adenylylimidodiphosphate does not activate CFTR chloride channels in cell-free patches of membrane. Am. J. Physiol. 265(1 Pt 1), L27–L32.

Carson, M.R., Travis, S.M. and Welsh, M.J. (1995) The two nucleotide-binding domains of cystic fibrosis transmembrane conductance regulator (CFTR) have distinct functions in controlling channel activity. J. Biol. Chem. 270(4), 1711–1717.

Cartier, E.A., Conti, L.R., Vandenberg, C.A. and Shyng, S.L. (2001) Defective trafficking and function of KATP channels caused by a sulfonylurea receptor 1 mutation associated with persistent hyperinsulinemic hypoglycemia of infancy. Proc. Natl Acad. Sci. USA 98(5), 2882–2887.

Catterall, W.A. (2000) From ionic currents to molecular mechanisms: the structure and function of voltage-gated sodium channels. Neuron 26(1), 13–25.

Champigny, G., Verrier, B., Gerard, C., Mauchamp, J. and Lazdunski, M. (1990) Small conductance chloride channels in the apical membrane of thyroid cells. FEBS Lett. 259(2), 263–268.

Chang, X.-B., Tabcharani, J.A., Hou, Y.-X., Jensen, T.J., Kartner, N., Alon, N., Hanrahan, J.W. and Riordan, J.R. (1993) Protein kinase A (PKA) still activates CFTR chloride channel after mutagenesis of all ten PKA consensus phosphorylation sites. J. Biol. Chem. 268(15), 11304–11311.

Chang, X.-B., Hou, Y.-X., Jensen, T. and Riordan, J.R. (1994) Mapping of the cystic fibrosis transmembrane conductance regulator membrane topology by glycosylation site insertion. J. Biol. Chem. 269(28), 18572–18575.

Chang, X.-B., Cui, L., Hou, Y., Jensen, T.J., Aleksandrov, A.A., Mengos, A. and Riordan J.R. (1999) Removal of multiple arginine-framed trafficking signals overcomes misprocessing of delta F508 CFTR present in most patients with cystic fibrosis. Mol. Cell 4(1), 137–142.

Chappe, V., Hinkson, D.A.R., Chang, X.-B., Riordan, J.R. and Hanrahan, J.W. (2002) Phosphorylation of protein kinase C consensus sites in NBD1 and the R domain control CFTR channel activation by PKA. J. Physiol. (in press).

Chen, J.-H. and Riordan, J.R. (2002) Chemical cross-linking reveals different CFTR associations when the protein is heterologously and endogenously expressed. Arch. Biochem. Biophys. (submitted).

Chen, J.-H., Chang, X.-B., Aleksandrov, A.A. and Riordan, J.R. (2002) CFTR is a monomer: biochemical and functional evidence. J. Membrane Biol. (on line DOI: 10.1007/500232-001-0174-2 (07/10/02).

Chen, J.M., Cutler, C., Jacques, C., Boeuf, G., Denamur, E., Lecointre, G., Mercier, B., Cramb, G. and Ferec, C. (2001) A combined analysis of the cystic fibrosis transmembrane conductance regulator: implications for structure and disease models. Mol. Biol. Evol. 18(9), 1771–1788.

Cheng, J., Moyer, B.D., Milewski, M., Loffing, J., Ikeda, M., Mickle, J.E., Cutting, G.R., Li, M., Stanton, B.A. and Guggino, W.B. (2002) A Golgi-associated PDZ domain protein modulates cystic fibrosis transmembrane regulator plasma membrane expression. J. Biol. Chem. 277(5), 3520–3529.

Cheng, S.H., Gregory, R.J., Marshall, J., Paul, S., Souza, D.W., White, G.A., O'Riordan, C.R. and Smith, A.E. (1990) Defective intracellular transport and processing of CFTR is the molecular basis of most cystic fibrosis. Cell 63, 827–834.

Cheng, S.H., Rich, D.P., Marshall, J., Gregory, R.J., Welsh, M.J. and Smith, A.E. (1991) Phosphorylation of the R domain by cAMP-dependent protein kinase regulates the CFTR chloride channel. Cell 66, 1027–1036.

Cheung, M. and Akabas, M.H. (1997) Locating the anion-selectivity filter of the cystic fibrosis transmembrane conductance regulator (CFTR) chloride channel. J. Gen. Physiol. 109(3), 289–299.

Cotten, J.F. and Welsh, M.J. (1999) Cystic fibrosis-associated mutations at arginine 347 alter the pore architecture of CFTR. Evidence for disruption of a salt bridge. J. Biol. Chem. 274(9), 5429–5435.

Cuthbert, A.W., Halstead, J., Ratcliff, R., Colledge, W.H. and Evans, M.J. (1995) The genetic advantage hypothesis in cystic fibrosis heterozygotes: a murine study. J. Physiol. 482(Pt 2), 449–454.

Dahan, D., Zhu, T., Evaglelidis, A. and Hanrahan, J.W. (1999) CFTR-protein phosphatase 2C (PP2C) association is independent of the carboxyl terminus PDZ binding motif. FASEB J. 13, A71(Abstract).

Dawson, D.C., Smith, S.S. and Mansoura, M.K. (1999) CFTR: mechanism of anion conduction. Physiol. Rev. 79(1 Suppl.): S47–S75.

Dechecchi, M.C., Rolfini, R., Tamanini, A., Gamberi, C., Berton, G. and Cabrini, G. (1992) Effect of modulation of protein kinase C on the cAMP-dependent chloride conductance in T84 cells. FEBS Lett. 311(1), 25–28.

Deeley, R.G. and Cole S.P. (1997) Function, evolution and structure of multidrug resistance protein (MRP). Semin. Cancer Biol. 8(3), 193–204.

Denning, G.M., Anderson, M.P., Amara, J.F., Marshall, J., Smith, A.E. and Welsh, M.J. (1992) Processing of mutant cystic fibrosis transmembrane conductance regulator is temperature-sensitive. Nature 358(6389), 761–764.

Dixon, P.H., Weerasekera, N., Linton, K.J., Donaldson, O., Chambers, J., Egginton, E., *et al.* (2000) Heterozygous MDR3 missense mutation associated with intrahepatic cholestasis of pregnancy: evidence for a defect in protein trafficking. Hum. Mol. Genet. 9(8), 1209–1217.

Dormer, R.L., Derand, R., McNeilly, C.M., Mettey, Y., Bulteau-Pignoux, L., Metaye, T., *et al.* (2001) Correction of delF508-CFTR activity with benzo(c)quinolizinium compounds through facilitation of its processing in cystic fibrosis airway cells. J. Cell Sci. 114(Pt 22), 4073–4081.

Doyle, D.A., Morais Cabral, J., Pfuetzner, R.A., Kuo, A., Gulbis, J.M., Cohen, S.L., Chait, B.T. and MacKinnon, R. (1998) The structure of the potassium channel: molecular basis of K^+ conduction and selectivity. Science 280(5360), 69–77.

Drumm, M.L., Wilkinson, D.J., Smit, L.S., Worrell, R.T., Strong, T.V., Frizzell, R.A., Dawson, D.C. and Collins, F.S. (1991) Chloride conductance expressed by ÐF508 and other mutant CFTRs in *Xenopus* oocytes. Science 254, 1797–1799.

Duffieux, F., Annereau, J.P., Boucher, J., Miclet, E., Pamlard, O., Schneider, M., Stoven, V. and Lallemand, J.Y. (2000) Nucleotide-binding domain 1 of cystic fibrosis transmembrane conductance regulator production of a suitable protein for structural studies. Eur. J. Biochem. 267(17), 5306–5312.

Dulhanty, A.M. and Riordan J.R. (1994) Phosphorylation by cAMP-dependent protein kinase causes a conformational change in the R domain of the cystic fibrosis transmembrane conductance regulator. Biochemistry 33(13), 4072–4079.

Dutzler, R., Campbell, E.B., Cadene, M., Chait, B.T. and MacKinnon, R. (2002) X-ray structure of a ClC chloride channel at 3.0 A reveals the molecular basis of anion selectivity. Nature 415(6869), 287–294.

Eisenman, G. and Horn, R. (1983) Ionic selectivity revisited: the role of kinetic and equilibrium processes in ion permeation through channels. J. Membr. Biol. 76(3), 197–225.

Eskandari, S., Wright, E.M., Kreman, M., Starace, D.M. and Zampighi, G.A. (1998) Structural analysis of cloned plasma membrane proteins by freeze-fracture electron microscopy. Proc. Natl Acad. Sci. USA 95(19), 11235–11240.

Fan, J.Q., Ishii, S., Asano, N. and Suzuki, Y. (1999) Accelerated transport and maturation of lysosomal alpha-galactosidase A in Fabry lymphoblasts by an enzyme inhibitor. Nat. Med. 5(1), 112–115.

Foster, B.A., Coffey, H.A., Morin, M.J. and Rastinejad, F. (1999) Pharmacological rescue of mutant p53 conformation and function. Science 286(5449), 2507–2510.

Frizzell, R.A., Rechkemmer, G. and Shoemaker, R.L. (1986) Altered regulation of airway epithelial cell chloride channels in cystic fibrosis. Science 233, 558–560.

Gabriel, S.E., Clarke, L.L., Boucher, R.C. and Stutts, M.J. (1993) CFTR and outward rectifying chloride channels are distinct proteins with a regulatory relationship. Nature 363, 263–268.

Gabriel, S.E., Brigman, K.N., Koller, B.H., Boucher, R.C. and Stutts, M.J. (1994) Cystic fibrosis heterozygote resistance to cholera toxin in the cystic fibrosis mouse model. Science 266(5182), 107–109.

Gadsby, D.C. and Nairn A.C. (1999a) Control of CFTR channel gating by phosphorylation and nucleotide hydrolysis. Physiol. Rev. 79(1 Suppl.): S77–S107.

Gadsby, D.C. and Nairn A.C. (1999b) Regulation of CFTR Cl^- ion channels by phosphorylation and dephosphorylation. Adv. Sec. Mess. Phosph. 33, 79–106.

Galietta, L.J., Springsteel, M.F., Eda, M., Niedzinski, E.J., By, K., Haddadin, M.J.,

Kurth, M.J., Nantz, M.H. and Verkman, A.S. (2001) Novel CFTR chloride channel activators identified by screening of combinatorial libraries based on flavone and benzoquinolizinium lead compounds. J. Biol. Chem. 276(23), 19723–19728.

Gao, M., Cui, H.R., Loe, D.W., Grant, C.E., Almquist, K.C., Cole, S.P. and Deeley, R.G. (2000) Comparison of the functional characteristics of the nucleotide binding domains of multidrug resistance protein 1. J. Biol. Chem. 275(17), 13098–13108.

Garty, H. and Palmer, L.G. (1997) Epithelial sodium channels: function, structure, and regulation. Physiol. Rev. 77(2), 359–396.

Gelman, M.S., Kannegaard, E.S. and Kopito, R.R. (2002) A principal role for the proteasome in ER-associated degradation of misfolded intracellular CFTR. J. Biol. Chem. 277, 11709–11714.

Gentzsch, M. and Riordan J.R. (2001) Localization of sequences within the C-terminal domain of the cystic fibrosis transmembrane conductance regulator which impact maturation and stability. J. Biol. Chem. 276(2), 1291–1298.

Gentzsch, M., Aleksandrov, A., Aleksandrov, L. and Riordan, J.R. (2002) Functional analysis of the C-terminal boundary of the second nucleotide binding domain of the cystic fibrosis transmembrane conductance regulator and structural implications. Biochem. J. 05/20/02 DOI: 10.1042/BJ20020511.

Gray, M.A., Greenwell, J.R. and Argent, B.E. (1988) Secretin-regulated chloride channel on the apical plasma membrane of pancreatic duct cells. J. Mem. Biol. 105(2), 131–142.

Gray, M.A., Harris, A., Coleman, L., Greenwell, J.R. and Argent, B.E. (1989) Two types of chloride channel on duct cells cultured from human fetal pancreas. Am. J. Physiol. 257, C240–C251.

Gray, M.A., Pollard, C.E., Harris, A., Coleman, L., Greenwell, J.R. and Argent, B.E. (1990) Anion selectivity and block of the small-conductance chloride channel on pancreatic duct cells. Am. J. Physiol. 259(5 Pt 1), C752–C761.

Guinamard, R. and Akabas, M.H. (1999) Arg352 is a major determinant of charge selectivity in the cystic fibrosis transmembrane conductance regulator chloride channel. Biochemistry 38(17), 5528–5537.

Gunderson, K.L. and Kopito, R.R. (1994) Effects of pyrophosphate and nucleotide analogs suggest a role for ATP hydrolysis in cystic fibrosis transmembrane regulator channel gating. J. Biol. Chem. 269(30), 19349–19353.

Gupta, J., Evagelidis, A., Hanrahan, J.W. and Linsdell, P. (2001) Asymmetric structure of the cystic fibrosis transmembrane conductance regulator chloride channel pore suggested by mutagenesis of the twelfth transmembrane region. Biochemistry 40(22), 6620–6627.

Hallows, K.R., Raghuram, V., Kemp, B.E., Witters, L.A. and Foskett, J.K. (2000) Inhibition of cystic fibrosis transmembrane conductance regulator by novel interaction with the metabolic sensor AMP-activated protein kinase. J. Clin. Invest. 105(12), 1711–1721.

Hämmerle, M.M., Aleksandrov, A.A. and Riordan, J.R. (2001) Disease-associated mutations in the extracytoplasmic loops of cystic fibrosis transmembrane conductance regulator do not impede biosynthetic processing but impair chloride channel stability. J. Biol. Chem. 276(18), 14848–14854.

Hanrahan, J.W., Tabcharani, J.A. and Grygorczyk, R. (1993) Patch clamp studies of apical membrane chloride channels. In: Cystic Fibrosis – Current Topics (ed. J.A. Dodge, D.J.H. Brock and J.H. Widdicombe), pp. 93–137. London: John Wiley and Sons.

Hartman, J., Huong, Z., Rado, T.A., Peng, S., Jilling, T., Muccio, D.D. and Sorscher, E.J. (1992) Recombinant synthesis purification and nucleotide binding characteristics of the first nucleotide binding domains of the cystic fibrosis gene product. J. Biol. Chem. 267, 6455–6458.

Higgins, C.F. (1992) ABC transporters: from microorganisms to man. Annu. Rev. Cell Biol. 8, 67–113.

Hille, B. (1992) Ionic Channels of Excitable Membranes. Sunderland, MA: Sinauer Associates.

Holmgren, M., Liu, Y., Xu, Y. and Yellen, G. (1996) On the use of thiol-modifying agents to determine channel topology. Neuropharmacology 35(7), 797–804.

Hopfner, K.P., Karcher, A., Shin, D.S., Craig, L., Arthur, L.M., Carney, J.P. and Tainer, J.A. (2000) Structural biology of Rad50 ATPase: ATP-driven conformational control in DNA double-strand break repair and the ABC-ATPase superfamily. Cell 101(7), 789–800.

Hou, Y.-X., Cui, L., Riordan, J.R. and Chang, X.-B. (2000) Allosteric interactions between the two non-equivalent nucleotide binding domains of multidrug resistance protein MRP1. J. Biol. Chem. 275, 20280–20287.

Howell, L.D., Borchardt, R. and Cohn, J.A. (2000) ATP hydrolysis by a CFTR domain: pharmacology and effects of G551D mutation. Biochem. Biophys. Res. Commun. 271(2), 518–525.

Huang, P., Lazarowski, E.R., Tarran, R., Milgram, S.L., Boucher, R.C. and Stutts, M.J. (2001) Compartmentalized autocrine signaling to cystic fibrosis transmembrane conductance regulator at the apical membrane of airway epithelial cells. Proc. Natl Acad. Sci. USA 98(24), 14120–14125.

Hwang, T., Lu, L., Zeitlin, P.L., Gruenert, D.C., Huganir, R. and Guggino, W.B. (1989) Cl^- channels in CF: Lack of activation by protein kinase C and cAMP-dependent protein kinase. Science 244, 1351–1353.

Hwang, T.-C., Baukrowtiz, G., Nagel, G., Horie, A.C. and Gadsby, D.C. (1994) Regulation of gating of cystic fibrosis transmembrane conductance regulator C1 channels by phosphorylation and ATP hydrolysis. Proc. Natl Acad. Sci. USA 91, 4698–4702.

Ikuma, M. and Welsh, M.J. (2000) Regulation of CFTR Cl^- channel gating by ATP binding and hydrolysis. Proc. Natl Acad. Sci. USA 97(15), 8675–8680.

Inagaki, N., Gonoi, T., Clement, J.P., Wang, C.Z., Aguilar-Bryan, L., Bryan, J. and Seino, S. (1996) A family of sulfonylurea receptors determines the pharmacological properties of ATP-sensitive K^+ channels. Neuron 16(5), 1011–1017.

Ishihara, H. and Welsh, M.J. (1997) Block by MOPS reveals a conformation change in the CFTR pore produced by ATP hydrolysis. Am. J. Physiol. 273(4 Pt 1), C1278–C1289.

Jensen, T.J., Loo, M.A., Pind, S., Williams, D.B., Goldberg, A.L. and Riordan, J.R. (1995) Multiple proteolytic systems, including the proteasome, contribute to CFTR processing. Cell 83, 129–135.

Jia, Y., Mathews, C.J. and Hanrahan, J.W. (1997) Phosphorylation by protein kinase C is required for acute activation of cystic fibrosis transmembrane conductance regulator by protein kinase A. J. Biol. Chem. 272(8), 4978–4984.

Kalin, N., Claass, A., Sommer, M., Puchelle, E. and Tummler, B. (1999) DeltaF508 CFTR protein expression in tissues from patients with cystic fibrosis [see comments]. J. Clin. Invest. 103(10), 1379–1389.

Karpowich, N., Martsinkevich, O., Millen, L., Yuan, Y.R., Dai, P.L., MacVey, K., Thomas, P.J. and Hunt, J.F. (2001) Crystal structures of the MJ1267 ATP binding cassette reveal an induced-fit effect at the ATPase active site of an ABC transporter. Structure(Camb) 9(7), 571–586.

Kartner, N., Hanrahan, J.W., Jensen, T.J., Naismith, A.L., Sun, S., Ackerley, C.A., *et al.* (1991) Expression of the cystic fibrosis gene in non-epithelial invertebrate cells produces a regulated anion conductance. Cell 64(4), 681–691.

Kartner, N., Augustinas, O., Jensen, T.J., Naismith, A.L. and Riordan, J.R. (1992) Mislocalization of delta F508 CFTR in cystic fibrosis sweat gland. Nat. Genet. 1(5), 321–327.

Keitel, V., Kartenbeck, J., Nies, A.T., Spring, H., Brom, M. and Keppler, D. (2000) Impaired protein maturation of the conjugate export pump multidrug resistance protein 2 as a consequence of a deletion mutation in Dubin–Johnson syndrome. Hepatology 32(6), 1317–1328.

Klabunde, T., Petrassi, H.M., Oza, V.B., Raman, P., Kelly, J.W. and Sacchettini, J.C. (2000) Rational design of potent human transthyretin amyloid disease inhibitors. Nat. Struct. Biol. 7(4), 312–321.

Knowles, M.R., Stutts, M.J., Spock, A., Fischer, N., Gatzy, J.T. and Boucher, R.C. (1983) Abnormal ion permeation through cystic fibrosis respiratory epithelium. Science 221, 1067–1070.

Ko, Y.H. and Pedersen P.L. (1995) The first nucleotide binding fold of the cystic fibrosis transmembrane conductance regulator can function as an active ATPase. J. Biol. Chem. 270(38), 22093–22096.

Ko, Y.H., Thomas, P.J., Delannoy, M.R. and Pedersen, P.L. (1993) The cystic fibrosis transmembrane conductance regulator. Overexpression, purification, and characterization of wild type and delta F508 mutant forms of the first nucleotide binding fold in fusion with the maltose-binding protein. J. Biol. Chem. 268(32), 24330–24338.

Kogan, I., Ramjeesingh, M., Huan, L.J., Wang, Y. and Bear, C.E. (2001) Perturbation of the pore of the cystic fibrosis transmembrane conductance regulator (CFTR) inhibits its ATPase activity. J. Biol. Chem. 276(15), 11575–11581.

Konig, J., Schreiber, R., Voelcker, T., Mall, M. and Kunzelmann, K. (2001) The cystic fibrosis transmembrane conductance regulator (CFTR) inhibits ENaC through an increase in the intracellular Cl^- concentration. EMBO Rep. 2(11), 1047–1051.

Kopito, R.R. (1999) Biosynthesis and degradation of CFTR. Physiol. Rev. 79(1 Suppl.): S167–S173.

Kreda, J.M., Mall, M., Seydewitz, H., Kuehr, J., Mengos, A., Jensen, T.J., Riordan, J.R. and Boucher, R. (2001) CFTR expression in epithelia from cystic fibrosis patients. Mol. Biol. Cell (Suppl. 12)(Abstract 2137).

Lee, M.G., Choi, J.Y., Luo, X., Strickland, E., Thomas, P.J. and Muallem, S. (1999) Cystic fibrosis transmembrane conductance regulator regulates luminal Cl^-/HCO_3^- exchange in mouse submandibular and pancreatic ducts. J. Biol. Chem. 274(21), 14670–14677.

Lewis, R.A., Shroyer, N.F., Singh, N., Allikmets, R., Hutchinson, A., Li, Y., Lupski, J.R., Leppert, M. and Dean, M. (1999) Genotype/Phenotype analysis of a photoreceptor-specific ATP-binding cassette transporter gene, ABCR, in Stargardt disease. Am. J. Hum. Genet. 64(2), 422–434.

Li, C., Ramjeesingh, M., Reyes, E., Jensen, T., Chang, X., Rommens, J.M. and Bear, C.E. (1993) The cystic fibrosis mutation (delta F508) does not influence the chloride channel activity of CFTR. Nat. Genet. 3(4), 311–316.

Li, C., Ramjeesingh, M., Wang, W., Garami, E., Hewryk, M., Lee, D., Rommens, J.M., Galley, K. and Bear, C.E. (1996) ATPase activity of the cystic fibrosis transmembrane conductance regulator. J. Biol. Chem. 271(45), 28463–28468.

Li, M., McCann, J.D., Liedtke, C.M., Nairn, A.C., Greengard, P. and Welsh, M.J. (1988) Cyclic AMP-dependent protein kinase opens chloride channels in normal but not cystic fibrosis airway epithelium. Nature 331, 358–360.

Linsdell, P. and Hanrahan, J.W. (1996) Disulphonic stilbene block of cystic fibrosis transmembrane conductance regulator Cl^- channels expressed in a mammalian cell line and its regulation by a critical pore residue. J. Physiol. 496(Pt 3), 687–693.

Linsdell, P. and Hanrahan, J.W. (1998) Adenosine triphosphate-dependent asymmetry of anion permeation in the cystic fibrosis transmembrane conductance regulator chloride channel. J. Gen. Physiol. 111(4), 601–614.

Linsdell, P., Tabcharani, J.A., Rommens, J.M., Hou, Y.X., Chang, X.B., Tsui, L.C., Riordan, J.R. and Hanrahan, J.W. (1997a) Permeability of wild-type and mutant cystic fibrosis transmembrane conductance regulator chloride channels to polyatomic anions. J. Gen. Physiol. 110(4), 355–364.

Linsdell, P., Tabcharani, J.A. and Hanrahan, J.W. (1997b) Multi-ion mechanism for ion permeation and block in the cystic fibrosis transmembrane conductance regulator chloride channel. J. Gen. Physiol. 110(4), 365–377.

Linsdell, P., Evagelidis, A. and Hanrahan, J.W. (2000) Molecular determinants of anion selectivity in the cystic fibrosis transmembrane conductance regulator chloride channel pore. Biophys. J. 78(6), 2973–2982.

Loo, M.A., Jensen, T.J., Cui, L., Hou, Y.-X., Chang, X.-B. and Riordan, J.R. (1998) Perturbation of Hsp90 interaction with nascent CFTR prevents its maturation and accelerates its degradation by the proteasome. EMBO J. 17, 6879–6887.

Loo, T.W. and Clarke, D.M. (2000) Blockage of drug resistance *in vitro* by disulfiram, a drug used to treat alcoholism. J. Natl Cancer Inst. 92(11), 898–902.

Lukacs, G.L., Mohamed, A., Kartner, N., Chang, X.-B., Riordan, J.R. and Grinstein, S. (1994) Confirmational maturation of CFTR but not its mutant counterpart (ÐF508) occurs in the endoplasmic reticulum and requires ATP. EMBO J. 13, 6076–6086.

Luo, J., Pato, M.D., Riordan, J.R. and Hanrahan, J.W. (1998) Differential regulation of single CFTR channels by PP2C, PP2A, and other phosphatases. Am. J. Physiol. 274(5 Pt 1), C1397–C1410.

Mansoura, M.K., Smith, S.S., Choi, A.D., Richards, N.W., Strong, T.V., Drumm, M.L., Collins, F.S. and Dawson, D.C. (1998) Cystic fibrosis transmembrane conductance regulator (CFTR) anion binding as a probe of the pore. Biophys. J. 74(3), 1320–1332.

Mathews, C.J., Tabcharani, J.A. and Hanrahan, J.W. (1998) The CFTR chloride channel: nucleotide interactions and temperature-dependent gating. J. Membr. Biol. 163(1), 55–66.

Matsui, H., Grubb, B.R., Tarran, R., Randell, S.H., Gatzy, J.T., Davis, C.W. and Boucher, R.C. (1998) Evidence for periciliary liquid layer depletion, not abnormal ion composition, in the pathogenesis of cystic fibrosis airways disease. Cell 95(7), 1005–1015.

McCarty, N.A. (2000) Permeation through the CFTR chloride channel. J. Exp. Biol. 203 (Pt 13), 1947–1962.

McDonough, S., Davidson, N., Lester, H.A. and McCarty, N.A. (1994) Novel pore-lining residues in CFTR that govern permeation and open-channel block. Neuron 13(3), 623–634.

Meacham, G.C., Lu, Z., King, S., Sorscher, E., Tousson, A. and Cyr, D.M. (1999) The Hdj-2/Hsc 70 chaperone pair facilitates early steps in CFTR biogenesis. EMBO J. 18, 1492–1505.

Meacham, G.C., Patterson, C., Zhang, W., Younger, J.M. and Cyr, D.M. (2001) The Hsc70 co-chaperone CHIP targets immature CFTR for proteasomal degradation. Nat. Cell Biol. 3(1), 100–105.

Middleton, L.M. and Harvey, R.D. (1998) PKC regulation of cardiac CFTR Cl^- channel function in guinea pig ventricular myocytes. Am. J. Physiol. 275(1 Pt 1), C293–C302.

Milewski, M.I., Mickle, J.E., Forrest, J.K., Stafford, D.M., Moyer, B.D., Cheng, J., Guggino, W.B., Stanton, B.A. and Cutting, G.R. (2001) A PDZ-binding motif is essential but not sufficient to localize the C terminus of CFTR to the apical membrane. J. Cell Sci. 114(Pt 4), 719–726.

Morales, M.M., Carroll, T.P., Morita, T., Schwiebert, E.M., Devuyst, O., Wilson, P.D., *et al.* (1996) Both the wild type and a functional isoform of CFTR are expressed in kidney. Am. J. Physiol. 270(6 Pt 2), F1038–F1048.

Morello, J.P., Salahpour, A., Laperriere, A., Bernier, V., Arthus, M.F., Lonergan, M., *et al.* (2000) Pharmacological chaperones rescue cell-surface expression and function of misfolded V2 vasopressin receptor mutants. J. Clin. Invest. 105(7), 887–895.

Nagata, K., Nishitani, M., Matsuo, M., Kioka, N., Amachi, T. and Ueda, K. (2000) Nonequivalent nucleotide trapping in the two nucleotide binding folds of the human multidrug resistance protein MRP1. J. Biol. Chem. 275(23), 17626–17630.

Nagel, G. (1999) Differential function of the two nucleotide binding domains on cystic fibrosis transmembrane conductance regulator. Biochim. Biophys. Acta 1461(2), 263–274.

Nagel, G., Szellas, T., Riordan, J.R., Friedrich, T. and Hartung, K. (2001) Non-specific activation of the epithelial sodium channel by the CFTR chloride channel. EMBO Rep. 2(3), 249–254.

Nairn, A.C., Qin, J., Chait, B.T. and Gadsby, D.C. (1996) Identification of sites in the R-domain of CFTR phosphorylated by cAMP-dependent protein kinase and dephosphorylated by protein phosphatases 2A and 2C. Pediatr. Pulm. (Suppl. 13), 211 (Abstract).

Naren, A.P., Nelson, D.J., Xie, W., Jovov, B., Pevsner, J., Bennett, M.K., Benos, D.J., Quick, M.W. and Kirk, K.L. (1997) Regulation of CFTR chloride channels by syntaxin and Munc18 isoforms. Nature 390(6657), 302–305.

Naren, A.P., Quick, M.W., Collawn, J.F., Nelson, D.J. and Kirk, K.L. (1998) Syntaxin 1A inhibits CFTR chloride channels by means of domain-specific protein–protein interactions. Proc. Natl Acad. Sci. USA 95(18), 10972–10977.

Naren, A.P., Cormet-Boyaka, E., Fu, J., Villain, M., Blalock, J.E., Quick, M.W. and Kirk, K.L. (1999) CFTR chloride channel regulation by an interdomain interaction. Science 286(5439), 544–548.

Nelson, D.J., *et al.* (2001) CFTR associates with the P11 subunit of the annexin II heterotetramer. Pediatr. Pulm. (Suppl. 18), 178 (abstract).

Neville, D.C., Rozanas, C.R., Price, E.M., Gruis, D.B., Verkman, A.S. and Townsend, R.R. (1997) Evidence for phosphorylation of serine 753 in CFTR using a novel metal-ion affinity resin and matrix-assisted laser desorption mass spectrometry. Protein Sci. 6(11), 2436–2445.

Neville, D.C., Rozanas, C.R., Tulk, B.M., Townsend, R.R. and Verkman, A.S. (1998) Expression and characterization of the NBD1-R domain region of CFTR: evidence for subunit–subunit interactions. Biochemistry 37(8), 2401–2409.

Nishimura, N., Bannykh, S., Slabough, S., Matteson, J., Altschuler, Y., Hahn, K. and Balch, W.E. (1999) A di-acidic (DXE) code directs concentration of cargo during export from the endoplasmic reticulum. J. Biol. Chem. 274(22), 15937–15946.

Noone, P.G. and Knowles, M.R. (2001) 'CFTR-opathies': disease phenotypes associated with cystic fibrosis transmembrane regulator gene mutations. Respir. Res. 2(6), 328–332.

O'Reilly, C.M., Winpenny, J.P., Argent, B.E. and Gray, M.A. (2000) Cystic fibrosis transmembrane conductance regulator currents in guinea pig pancreatic duct cells: inhibition by bicarbonate ions. Gastroenterology 118(6), 1187–1196.

Osborne, L.R., Lynch, M., Middleton, P.G., Alton, E.W., Geddes, D.M., Pryor, J.P., Hodson, M.E. and Santis, G.K. (1993) Nasal epithelial ion transport and genetic analysis of infertile men with congenital bilateral absence of the vas deferens. Hum. Mol. Genet. 2(10), 1605–1609.

Ostedgaard, L.S., Baldursson, O. and Welsh, M.J. (2001) Regulation of the cystic fibrosis transmembrane conductance regulator Cl^- channel by its R domain. J. Biol. Chem. 276(11), 7689–7692.

Picciotto, M.R., Cohn, J.A., Bertuzzi, G., Greengard, P. and Nairn, A.C. (1992) Phosphorylation of the cystic fibrosis transmembrane conductance regulator. J. Biol. Chem. 267, 12742–12752.

Pignatti, P.F., Bombieri, C., Benetazzo, M., Casartelli, A., Trabetti, E., Gile, L.S., Martinati, L.C., Boner, A.L. and Luisetti, M. (1996) CFTR gene variant IVS8–5T in disseminated bronchiectasis. Am. J. Hum. Genet. 58(4), 889–892.

Pind, S., Riordan, J.R. and Williams, D.B. (1994) Participation of the endoplasmic reticulum chaperone calnexin (p88, IP90) in the biogenesis of the cystic fibrosis transmembrane conductance regulator. J. Biol. Chem. 269(12), 12784–12788.

Poulsen, J.H., Fischer, H., Illek, B. and Machen, T.E. (1994) Bicarbonate conductance and pH regulatory capability of cystic fibrosis transmembrane conductance regulator. Proc. Natl Acad. Sci. USA 91(12), 5340–5344.

Price, M.P., Ishihara, H., Sheppard, D.N. and Welsh, M.J. (1996) Function of *Xenopus* cystic fibrosis transmembrane conductance regulator (CFTR) Cl channels and use of human–*Xenopus* chimeras to investigate the pore properties of CFTR. J. Biol. Chem. 271(41), 25184–25191.

Qu, B.H. and Thomas, P.J (1996) Alteration of the cystic fibrosis transmembrane conductance regulator folding pathway. J. Biol. Chem. 271(13), 7261–7264.

Quinton, P.M. (1983) Chloride impermeability in cystic fibrosis. Nature 301, 421–422.

Quinton, P.M. (2001) The neglected ion: HCO_3. Nat. Med. 7(3), 292–293.

Quinton, P.M. and Bijman, J. (1983) Higher bioelectric potentials due to decreased chloride absorption in the sweat glands of patients with cystic fibrosis. N. Engl. J. Med. 308, 1185–1189.

Quinton, P.M. and Reddy, M.M. (1992) Control of CFTR chloride conductance by ATP levels through non- hydrolytic binding. Nature 360(6399), 79–81.

Raghuram, V., Mak, D.D. and Foskett, J.K. (2001) Regulation of cystic fibrosis transmembrane conductance regulator single-channel gating by bivalent PDZ-domain-mediated interaction. Proc. Natl Acad. Sci. USA 98(3), 1300–1305.

Ramjeesingh, M., Li, C., Garami, E., Huan, L.J., Galley, K., Wang, Y. and Bear, C.E. (1999) Walker mutations reveal loose relationship between catalytic and channel-gating activities of purified CFTR (cystic fibrosis transmembrane conductance regulator). Biochemistry 38(5), 1463–1468.

Ramjeesingh, M., Li, C., Kogan, I., Wang, Y., Huan, L.J. and Bear, C.E. (2001) A monomer is the minimum functional unit required for channel and ATPase activity of the cystic fibrosis transmembrane conductance regulator. Biochemistry 40(35), 10700–10706.

Randak, C., Roscher, A.A., Hadorn, H.B., Assfalg-Machleidt, I., Auerswald, E.A. and Machleidt, W. (1995) Expression and functional properties of the second predicted nucleotide binding fold of CFTR fused to GST. FEBS Lett. 363, 189–194.

Randak, C., Neth, P., Auerswald, E.A., Assfalg-Machleidt, I., Roscher, A.A., Hadorn, H.B. and Machleidt, W. (1996) A recombinant polypeptide model of the second predicted nucleotide binding fold of the cystic fibrosis transmembrane conductance regulator is a GTP-binding protein. FEBS Lett. 398(1), 97–100.

Randak, C., Neth, P., Auerswald, E.A., Eckerskorn, C., Assfalg-Machleidt, I. and Machleidt, W. (1997) A recombinant polypeptide model of the second nucleotide-binding fold of the cystic fibrosis transmembrane conductance regulator functions as an active ATPase, GTPase and adenylate kinase. FEBS Lett. 410(2–3), 180–186.

Reddy, M.M., Light, M.J. and Quinton, P.M. (1999) Activation of the epithelial Na^+ channel (ENaC) requires CFTR Cl^- channel function. Nature 402(6759), 301–304.

Riordan, J.R. (1999) Cystic fibrosis as a disease of misprocessing of the cystic fibrosis transmembrane conductance regulator glycoprotein. Am. J. Hum. Genet. 64(6), 1499–1504.

Riordan, J.R., Rommens, J.M., Kerem, B., Alon, N., Rozmahel, R., Grzelczak, Z., *et al.* (1989) Identification of the cystic fibrosis gene: cloning and characterization of complementary DNA. Science 245(4922), 1066–1073.

Ritter, C. and Helenius, A. (2000) Recognition of local glycoprotein misfolding by the ER folding sensor UDP-glucose:glycoprotein glucosyltransferase. Nat. Struct. Biol. 7(4), 278–280.

Rosenberg, M.F., Callaghan, R., Ford, R.C. and Higgins, C.F. (1997) Structure of the multidrug resistance P-glycoprotein to 2.5 nm resolution determined by electron microscopy and image analysis. J. Biol. Chem. 272(16), 10685–10694.

Rosenberg, M.F., Mao, Q., Holzenburg, A., Ford, R.C., Deeley, R.G. and Cole, S.P. (2001) The structure of the multidrug resistance protein 1 (MRP1/ABCC1) crystallization and single-particle analysis. J. Biol. Chem. 276(19), 16076–16082.

Rubenstein, R.C. and Zeitlin, P.L. (2000) Sodium 4-phenylbutyrate downregulates Hsc70: implications for intracellular trafficking of DeltaF508-CFTR. Am. J. Physiol. Cell Physiol. 278(2), C259–C267.

Sato, S., Ward, C.L. and Kopito, R.R. (1998) Cotranslational ubiquitination of cystic fibrosis transmembrane conductance regulator *in vitro*. J. Biol. Chem. 273(13), 7189–7192.

Saurin, W., Hofnung, M. and Dassa, E. (1999) Getting in or out: early segregation between importers and exporters in the evolution of ATP-binding cassette (ABC) transporters. J. Mol. Evol. 48(1), 22–41.

Schoumacher, R.A., Shoemaker, R.L., Halm, D.R., Tallant, E.A., Wallace, R.W. and Frizzell, R.A. (1987) Phosphorylation fails to activate chloride channels from cystic fibrosis airway cells. Nature 330, 752–754.

Schultz, B.D., Bridges, R.J. and Frizzell, R.A. (1996) Lack of conventional ATPase properties in CFTR chloride channel gating. J. Membr. Biol. 151(1), 63–75.

Schultz, B.D., Singh, A.K., Devor, D.C. and Bridges, R.J. (1999) Pharmacology of CFTR chloride channel activity. Physiol. Rev. 79(1 Suppl.): S109–S144.

Schwiebert, E.M., Benos, D.J., Egan, M.E., Stutts, M.J. and Guggino, W.B. (1999) CFTR is a conductance regulator as well as a chloride channel. Physiol. Rev. 79, 145.

Seibert, F.S., Tabcharani, J.A., Chang, X.-B., Dulhanty, A.M., Mathews, C., Hanrahan, J.W. and Riordan, J.R. (1995) cAMP-dependent protein kinase-mediated phosphorylation of cystic fibrosis transmembrane conductance regulator residue ser-753 and its role in channel activation. J. Biol. Chem. 270(5), 2158–2162.

Seibert, F.S., Linsdell, P., Loo, T.W., Hanrahan, J.W., Clarke, D.M. and Riordan, J.R. (1996a) Disease-associated mutations in the fourth cytoplasmic loop of CFTR compromise biosynthetic processing and chloride channel activity. J. Biol. Chem. 271, 15139–15143.

Seibert, F.S., Linsdell, P., Loo, T.W., Hanrahan, J.W., Riordan, J.R. and Clarke, D.M. (1996b) Cytoplasmic loop three of cystic fibrosis transmembrane conductance regulator contributes to regulation of chloride channel activity. J. Biol. Chem. 271(44), 27493–27499.

Seibert, F.S., Jia, Y., Mathews, C.J., Hanrahan, J.W., Riordan, J.R., Loo, T.W. and Clarke, D.M. (1997) Disease-associated mutations in cytoplasmic loops 1 and 2 of cystic fibrosis transmembrane conductance regulator impede processing or opening of the channel. Biochemistry 36(39), 11966–11974.

Senior, A.E. and Gadsby, D.C. (1997) ATP hydrolysis cycles and mechanism in P-glycoprotein and CFTR. Semin. Cancer Biol. 8(3), 143–150.

Sharer, N., Schwarz, M., Malone, G., Howarth, A., Painter, J., Super, M. and Braganza, J. (1998) Mutations of the cystic fibrosis gene in patients with chronic pancreatitis. N. Engl. J. Med. 339(10), 645–652.

Sheppard, D.N. and Welsh, M.J. (1999) Structure and function of the CFTR chloride channel. Physiol. Rev. 79(1 Suppl.): S23–S45.

Sheppard, D.N., Rich, D.P., Ostedgaard, L.S., Gregory, R.J., Smith, A.E. and Welsh, M.J. (1993) Mutations in CFTR associated with mild-disease-form Cl^- channels with altered pore properties. Nature 362(6416), 160–164.

Sheppard, D.N., Ostedgaard, L.S., Rich, D.P. and Welsh, M.J. (1994) The amino-terminal portion of CFTR forms a regulated Cl^- channel. Cell 76(6), 1091–1098.

Short, D.B., Trotter, K.W., Reczek, D., Kreda, S.M., Bretscher, A., Boucher, R.C., Stutts, M.J. and Milgram, S.L. (1998) An apical PDZ protein anchors the cystic fibrosis transmembrane conductance regulator to the cytoskeleton. J. Biol. Chem. 273(31), 19797–19801.

Smith, J.J., Travis, S.M., Greenberg, E.P. and Welsh, M.J. (1996) Cystic fibrosis airway epithelia fail to kill bacteria because of abnormal airway surface fluid [published erratum appears in Cell 1996 Oct 18; 87(2), following 355]. Cell 85(2), 229–236.

Smith, K.D., Kemp, S., Braiterman, L.T., Lu, J.F., Wei, H.M., Geraghty, M., Stetten, G., Bergin, J.S., Pevsner, J. and Watkins, P.A. (1999) X-linked adrenoleukodystrophy: genes, mutations, and phenotypes. Neurochem. Res. 24(4), 521–535.

Smith, S.S., Liu, X., Zhang, Z.R., Sun, F., Kriewall, T.E., McCarty, N.A. and Dawson, D.C. (2001) CFTR: covalent and noncovalent modification suggests a role for fixed charges in anion conduction. J. Gen. Physiol. 118(4), 407–431.

Sturgess, J. and Imrie, J. (1982) Quantitative evaluation of the development of tracheal submucosal glands in infants with cystic fibrosis and control infants. Am. J. Pathol. 106(3), 303–311.

Stutts, M.J., Canessa, C.M., Olsen, J.C., Hamrick, M., Cohn, J.A., Rossier, B.C. and Boucher, R.C. (1995) CFTR as a cAMP-dependent regulator of sodium channels. Science 269, 847–850.

Szabo, K., Szakacs, G., Hegeds, T. and Sarkadi, B. (1999) Nucleotide occlusion in the human cystic fibrosis transmembrane conductance regulator. Different patterns in the two nucleotide binding domains. J. Biol. Chem. 274(18), 12209–12212.

Tabcharani, J.A., Low, D., Elie, D. and Hanrahan, J.W. (1990) Low conductance chloride channel activated by cAMP in the epithelial cell line T84. FEBS Lett. 270, 157–164.

Tabcharani, J.A., Chang, X.-B., Riordan, J.R. and Hanrahan, J.W. (1991) Phosphorylation-regulated Cl^- channel in CHO cells stably expressing the cystic fibrosis gene. Nature 352, 628–631.

Tabcharani, J.A., Chang, X.-B., Riordan, J.R. and Hanrahan, J.W. (1992) The cystic fibrosis transmembrane conductance regulator chloride channel: Iodide block and permeation. Biophys. J. 62, 1–4.

Tabcharani, J.A., Rommens, J.M., Hou, Y.X., Chang, X.B., Tsui, L.C., Riordan, J.R. and Hanrahan, J.W. (1993) Multi-ion pore behaviour in the CFTR chloride channel. Nature 366(6450), 79–82.

Therien, A.G., Grant, F.E. and Deber, C.M. (2001) Interhelical hydrogen bonds in the CFTR membrane domain. Nat. Struct. Biol. 8(7), 597–601.

Thomas, P.J. and Hunt, J.F. (2001) A snapshot of Nature's favorite pump. Nat. Struct. Biol. 8(11), 920–923.

Travis, S.M., Carson, M.R., Ries, D.R. and Welsh, M.J. (1993) Interaction of nucleotides with membrane-associated cystic fibrosis transmembrane conductance regulator. J. Biol. Chem. 268(21), 15336–15339.

Travis, S.M., Berger, H.A. and Welsh, M.J. (1997) Protein phosphatase 2C dephosphorylates and inactivates cystic fibrosis transmembrane conductance regulator. Proc. Natl Acad. Sci. USA 94(20), 11055–11060.

Urbatsch, I.L., Sankaran, B., Weber, J. and Senior, A.E. (1995) P-glycoprotein is stably inhibited by vanadate-induced trapping of nucleotide at a single catalytic site. J. Biol. Chem. 270(33), 19383–19390.

Wang, S., Yue, H., Derin, R.B., Guggino, W.B. and Li, M. (2000) Accessory protein facilitated CFTR–CFTR interaction, a molecular mechanism to potentiate the chloride channel activity. Cell 103(1), 169–179.

Ward, C.L., Omura, S. and Kopito, R.R. (1995) Degradation of CFTR by the ubiquitin-proteasome pathway. Cell 83, 121–127.

Weixel, K.M. and Bradbury, N.A. (2001) Mu 2 binding directs the cystic fibrosis transmembrane conductance regulator to the clathrin-mediated endocytic pathway. J. Biol. Chem. 276(49), 46251–46259.

Welsh, M.J., Robertson, A.D. and Ostegaard L.S. (1998) Structural biology. The ABC of a versatile engine. Nature 396(6712), 623–624.

Wilkinson, D.J., Strong, T.V., Mansoura, M.K., Wood, D.L., Smith, S.S., Collins, F.S. and Dawson, D.C. (1997) CFTR activation: additive effects of stimulatory and inhibitory phosphorylation sites in the R domain. Am. J. Physiol. 273(1 Pt 1), L127–L133.

Willumsen, N.J. and Boucher, R.C. (1989) Shunt resistance and ion permeabilities in normal and cystic fibrosis airway epithelia. Am. J. Physiol. 256, C1054–C1063.

Yamazaki, J., Britton, F., Collier, M.L., Horowitz, B. and Hume, J.R. (1999) Regulation of recombinant cardiac cystic fibrosis transmembrane conductance regulator chloride channels by protein kinase C. Biophys. J. 76(4), 1972–1987.

Yang, N., George, A.L., Jr and Horn, R. (1996) Molecular basis of charge movement in voltage-gated sodium channels. Neuron 16(1), 113–122.

Yang, Y., Janich, S., Cohn, J.A. and Wilson, J.M. (1993) The common variant of cystic fibrosis transmembrane conductance regulator is recognized by hsp70 and degraded in a pre-Golgi nonlysosomal compartment. Proc. Natl Acad. Sci. USA 90, 9480–9484.

Yoo, J.-S., Moyer, B.D., Bannykh, S.I., Yoo, H.-M., Riordan, J.R. and Balch, W.E. (2002) Non-conventional trafficking of the cystic fibrosis transmembrane conductance regulator (CFTR) through the early secretory pathway. J. Biol. Chem. 277, 1140–1149.

Yue, H., Devidas, S. and Guggino, W.B. (2000) The two halves of CFTR form a dual-pore ion channel. J. Biol. Chem. 275(14), 10030–10034.

Zeitlin, P.L. (2000) Pharmacologic restoration of delta F508 CFTR-mediated chloride current. Kidney Int. 57(3), 832–837.

Zerangue, N., Schwappach, B., Jan, Y.N. and Jan, L.Y. (1999) A new ER trafficking signal regulates the subunit stoichiometry of plasma membrane K_{ATP} channels. Neuron 22, 537–548.

Zerhusen, B., Zhao, J., Xie, J., Davis, P.B. and Ma, J. (1999) A single conductance pore for chloride ions formed by two cystic fibrosis transmembrane conductance regulator molecules. J. Biol. Chem. 274(12), 7627–7630.

Zhang, Z.R., McDonough, S.I. and McCarty, N.A. (2000) Interaction between permeation and gating in a putative pore domain mutant in the cystic fibrosis transmembrane conductance regulator. Biophys. J. 79(1), 298–313.

Zhu, T., Dahan, D., Evagelidis, A., Zheng, S., Luo, J. and Hanrahan, J.W. (1999) Association of cystic fibrosis transmembrane conductance regulator and protein phosphatase 2C. J. Biol. Chem. 274(41), 29102–29107.

Zou, X. and Hwang, T.C. (2001) ATP hydrolysis-coupled gating of CFTR chloride channels: structure and function. Biochemistry 40(19), 5579–5586.

CHAPTER 30

ABC Proteins, the Fascination, the Politics, the Potential for Applications for Improving Human Health

I. Barry Holland

This chapter is dedicated to the memory of a remarkable young scientist, Julian Boucher, a truly inspiring colleague, who died in 1999.

Studies of ABC proteins, in the form of HisP and MalK, were already well under way in the early 1980s and these are described in the excellent introductory overview and in other chapters of this volume. The select band of devotees involved in these initial studies were fascinated by the mechanism of histidine and maltose uptake in Gram-negative bacteria, certainly an esoteric subject. Everything changed dramatically in the mid-1980s with the realization that P-glycoprotein (Pgp), responsible for multidrug resistance and a serious obstacle to effective antitumor chemotherapy, was also an ABC transporter. This was followed quickly by the identification of the CFTR protein as a novel ABC transporter, and the subject has never looked back, with now thousands of ABC genes in the database and the avalanche continues as new genome sequences accumulate.

The Fascination

The reason for the fascination of ABC proteins and their associated partners, however, does not stop at the sheer size of this superfamily but is compelling, as this volume so demonstrably illustrates, because of the enormous breadth of biological processes that they embrace. More dramatically and still quite extraordinarily, despite more than 15 years now of close aquaintance with these proteins, remains the fact that these processes are driven by essentially the same ubiquitous ATPase. This is a molecular machine still easily recognizable by virtue of sequence motifs, sufficently unchanged as to be detectable by 'eye' despite more than three billion years of evolution and wide dissemination throughout all living organisms. If (when) finally we do get our hands on life forms from Mars and beyond we shall be very surprised if ABC proteins and ABC transporters are not represented.

The attraction and indeed beauty of ABC transporters is that their study unites on the one hand many varied disciplines, and on the other, more importantly, brings together scientists with interests in quite disparate organisms found in all conceivable niches on the planet. All of us are engaged in the hunt for the common principles that govern the mechanism whereby so many different molecules or ions trigger the different ATPase machines into action. In addition, we are extremely curious to understand how the resulting release of energy is used to facilitate the action or 'opening' of the corresponding transport pathway and, finally, how actual movement of molecules through the membrane is accomplished.

Recently the ABC picture has been even further enriched, if that were possible, by the realization that there also exist more distant cousins of the ABC membrane transporters. These use

ABC Proteins: From Bacteria to Man
ISBN 0-12-352551-9

ATP to effect some critical steps in polypeptide synthesis, DNA repair or recombination. It is not yet clear to the transporter aficionados how to reconcile the common principles of action of such ABCs with those involved in import or export. Nevertheless, these principles are surely there, involving perhaps the most fascinating secrets of these proteins: the mechanism of intra-molecular signaling between the ABC and the membrane domain (or DNA), and the nature of the crosstalk between ABC monomers which is required to activate and then to utilize the energy released in an ordered way.

THE POLITICS

The stimulation and attraction of working with ABC proteins, aside from the intellectual challenge of simply knowing how they work, is undoubtedly for many of us that some of these proteins are 'useful' in relation to the human condition. In some cases when the human ABC machine malfunctions, sadly this can bring morbidity and premature death. This surely provides the extra incentive and motivation for the scientist to figure out how such proteins function in the 'hope' of effecting 'cures'. At the same time it is in the nature of the broad canvas of academic research that scientists instinctively study all manner of topics, relevant or completely irrelevant (for the moment), simply because there are always new truths to be discovered everywhere. Certainly, however, studying an ABC protein with the most trivial of roles in the most obscure of organisms can be perceived as justified because it belongs to the superfamily that contains CFTR, Pgp and MRP, and hence the opportunity to contribute to curing cystic fibrosis, or the alleviation of problems of multidrug resistance in cancer chemotherapy, respectively. This is a fine, highly motivated sentiment and clearly in this case containing an element of truth. However, such sentiments are easily colored by unreal expectations, and an understandable degree of self-delusion, shared by scientist and public alike, in relation to what practical dividends may actually stem from basic research.

Increasingly, therefore, we are asked to justify our research in terms of the resulting benefits to society, leading us to succumb too frequently, although with the highest of intentions, to the employment of certain artifices to meet the demands of funding agencies. Unfortunately, this in turn leads to some unwelcome repercussions, with the perception of science suffering when we fail to deliver new products and therapies rapidly from the laboratory bench into the hospitals and pharmacies. In reality, in the real laboratory world of research directors, students and postdocs, fundamental research at the frontier is slow and painstaking, progress incremental, requiring infinite patience and ingenuity to test and discard many hypotheses before making real groundbreaking discoveries. Research is also about training oneself to think constructively and creatively and, above all for the experienced scientist, to inspire and guide the next generations to think creatively, to critically weigh evidence, and to formulate conclusions based on informed judgments. Happily, ABC protein research is a rich and fertile field in which to express and learn such skills.

Before moving on to the topic of the exploitation of basic knowledge of ABC proteins, a final comment on the realities facing current scientists. Academics, like our corporate colleagues, are increasingly subject to the same pressures to 'perform productively', to publish to fill quotas rather than to prove theories. Not surprisingly, this increases the tendency towards research without risk, publishable but non-contentious research that skims the initial descriptive cream of a new phenomenon or an old phenomenon in a new organism, before moving on to repeat the same formula. Digging deep into the fundamentals of a subject, where the going becomes slow, tough and above all risky, is not at all attractive. All these comments apply in the ABC field as to any other, and in surveying the mass of recent publications, for example, on prokaryote ABC proteins, it is clear that the overwhelming majority are simply describing new examples; we encourage more to wrestle with the basic principles, despite the obstacles.

DIFFICULTIES, REALITIES AND PROSPECTS FOR APPLICATIONS

Successful application or exploitation of knowledge gained from academic studies is not a simple matter, and like basic research also takes time, patience and flair, perhaps also an element of luck and certainly should also include exhaustive attention to detail. Bearing this in mind a number of such potential applications in relation to the ABC field are already in view. For example, we can certainly anticipate for the near future that many of us could be diagnosed as having an ABC protein not quite optimum for a long life of perfect control of cholesterol levels, a situation which could respond perhaps to some future molecular tweaking to relieve the pressure on our arteries. Whilst in principle we can already envisage, with regard to tackling such problems of human health at the genetic level, the exploitation of fundamental knowledge to achieve the necessary genetic engineering and gene therapy, actually carrying this into practice is far from trivial. All higher organisms are incredibly complex interconnected masses of metabolic and intercellular circuits, with gene expression differentially regulated in different tissues and phases of life, operating at the optimum balance evolved over many millions of years. Modifying or delivering replacements for defective ABC genes which function perfectly *in situ*, or designing medicines which precisely counter the expulsion of antitumor drugs, without disrupting other physiological functions of Pgp, MRPs or the other ABC proteins, is a tall order. Nevertheless, these are feasible and laudable objectives, which will require comprehensive, dedicated research in model microorganisms, in animals and finally in rigorous clinical trials in humans in order to fulfill them. Even then we cannot evade the reality that success cannot be guaranteed no matter how smart we are. We, our peer reviewers and our support providers, have to learn (or relearn) to accept therefore the concept of sometimes failing in such endeavors; equally importantly to accept the concept of starting over with a new strategy when needed, no matter how expensive or inglorious.

Notwithstanding the difficulties, exploitation of knowledge from fundamental and applied studies of ABC proteins should ultimately bring some long-term returns. Important benefits in diagnostics have already accrued in the screening for CFTR alleles in the population over the last decade. In fact the application of fundamental knowledge in the area of susceptibility testing and diagnostics will probably continue to lead the way in the discovery of new treatments for disease for some time to come. We certainly may anticipate greatly increased use of gene diagnostic probes for screening for potentially disadvantageous alleles of several ABC proteins, including the ABCA1 protein involved in cholesterol trafficking.

Treatments of human diseases arising directly from the results of academic or fundamental research and concomitant advances in technology are still very much in their infancy. Thus, they lag far behind treatments arising primarily from purely empirical discovery of drugs and procedures. A good example of this is the paradoxical fact that drugs like glibenclamide, used each day by millions of patients suffering from type 2 diabetes in order to stimulate insulin secretion, were identified and developed through empirical techniques many years before its target protein, SUR, was discovered and characterized. Nevertheless, we might anticipate for the future, still some way off, that from high-resolution structures of SUR, combined with better understanding of its molecular functioning and its precise contribution in the physiological context of insulin regulation, it may be possible to *design* drugs which slot precisely into a specific pocket of the target structure with minimal side effects. Interestingly, in the case of glibenclamide, the site of action has been traced to the membrane domain of SUR, a region of the molecule likely to constitute a much more specific target compared with the highly conserved ABC domain. Here perhaps is a constructive lesson for elaborating designer drugs effective against the highly specific transport domain of other ABC proteins, such as the multidrug transporters in humans, pathogenic microorganisms and parasites, rather than the ABC ATPase. There is clearly great interest now in developing drugs against such transporters, which on the one hand can limit the effectiveness of cancer chemotherapy, or, increasingly, pose a serious threat in the form of multidrug resistant microorganisms or other pathogens, on the other. At the present time we are limited to screening for such drugs by empirical procedures in the absence of the atomic level structures of the transporter. Such structures are an absolute requirement for future rational drug design.

For the moment no effective drugs against such ABC proteins have yet made it into clinical practice. In contrast, increasingly, human multidrug ABC transporters are being put to good use either as dominant selective markers for concomitant transfer of a 'corrective' gene in relation to gene therapy, or through transient expression in transfected bone marrow cells in order to provide protection against cytotoxic anticancer drugs during chemotherapy. Moreover, in recognition that Pgps and the MRP-type drug transporters are likely to play significant roles in the absorption, tissue distribution and elimination of many new potential drugs (see Chapter 18), pharmaceutical companies now include ABC transporter assays in early screens in drug development programs, in order to eliminate drugs that are transported by these drug pumps.

PREVENTION AND TREATMENTS OF DISEASE: THE MOLECULAR FUTURE

A major objective for the next twenty years in regard to human health care is of course to shift the balance decisively away from empirically based treatments and drug discovery, towards informed procedures for prevention and treatments. This will be based on fundamental knowledge of how the cells, tissues and organs of the human body actually work at the molecular level. This is manifestfully not because the empirically based procedures are not effective but that armed with informed insight we can hope ultimately to do far better. Conventional therapies have resulted in substantial increases in life expectancies for cystic fibrosis sufferers but still the disease takes away from us young persons with lives unfulfilled. No doubt more developments in conventional methods will give more progress yet, particularly perhaps in countering bacterial infections of the airways in CF patients. However, much hope is pinned on gene therapy or treatments to specifically rescue the function of the major mutant protein in the Caucasian population, the deletion F^{508}. In the latter case, the approach to novel treatments stems directly from basic studies of the CFTR protein and its gene in many academic laboratories, showing that this mutant protein folds incorrectly. Much to our frustration, however, our inability to understand *why* this mutant misfolds severely hampers our attempts to *design* a cure. We shall understand such riddles in the medium future but for the moment we must rely on less precise procedures, by administering empirically derived compounds which may bind the mutant CFTR *and* suppress the folding defect. Unfortunately, few ligands with high affinity for CFTR are so far available. Much effort has been put into the even more ambitious quest for a gene therapy for CF patients over the last decade and is still ongoing. Various approaches have been tried, including gene delivery into the airways by disarmed viral vectors or transfection by either naked DNA or DNA packaged with cationic liposomes. The feasibility of at least transient expression of the CFTR protein in respiratory epithelial cells from DNA administered by relatively simple procedures has been demonstrated and at least 20 clinical trials worldwide had already been reported by 1997. From these and subsequent trials it appears that these procedures are safe but so far not effective clinically.

PERSPECTIVES

Thus, whilst an effective treatment for cystic fibrosis is yet some way off, we must not forget that determination and optimism are the essential characteristics for most successful scientists. Similarly, as editors of this, we hope, landmark volume, celebrating the joys and excitement of prizing from nature the secrets of ABC systems, we take pride and hope in looking to the future for further major advances in fundamental knowledge. Encouragingly, the study of this ABC superfamily, albeit in most cases including the most refractory of macromolecules, membrane proteins, is now embracing and benefiting from the new frontier of biology, the exploitation of biophysical and sophisticated spectroscopic techniques, to yield vital high-resolution structural information. This is already showing dividends with crystal structures of several ABC domains very recently solved and now, the first of many, we hope, the structure of an entire molecule, MsbA. This we equally hope will be the prelude to the application of even more novel techniques which would reveal the dynamic properties of ABC proteins as they move their transport substrates through membranes and along polypeptide chains or DNA molecules.

Notwithstanding the need for patience and rigor in research (tempered by more realistic expectations) any successful applications designed to alleviate suffering and to enhance the human condition in relation to ABC proteins will require not only global understanding of their physiological role but also the molecular and atomic level detail required to understand the dynamics of how these proteins and their associates maneuver and change conformation as they function. Much more research is required to meet these demands, ideally with enlightened funding regimes. These should provide for and inspire 'riskier' creative thinking in basic research in the public sector, at least in wealthy states, unfettered by pressures to do *relevant* research. Support for such 'blue skies' research should, however, also include provision for better opportunities, *when and where appropriate*, for academics to collaborate in a whole variety of ways with the corporate sector in advancing the exploitation of their basic research. This is common practice in the United States, but is woefully underdeveloped in Europe. Finally, how we as a global society identify urgent problems of human health for priority attention and then how to mobilize our resources worldwide in the best way to meet the challenge are also in need of radical review, but that would be outside the scope of this text.

It only remains now at the end of this concluding chapter for myself, on behalf of all the editors, to thank most warmly all the participants involved in the preparation of this volume; equally we acknowledge and applaud the efforts of the many others, past and present in the laboratories of the book's major contributors, who in the end make all our achievements both possible and enjoyable; and of course we are heavily indebted to the ever larger community of scientists worldwide working on ABC proteins, who, whilst not having contributed directly to this volume, have provided a vast store of published work that we have plundered in the hope of producing a balanced and inspiring account of many if not all of these fascinating and important ABC proteins.

Index

Page references in *italics* indicate tables, figures or boxes; those in **bold** indicate main discussions.